Methods in Enzymology

Volume 194
GUIDE TO YEAST GENETICS AND
MOLECULAR BIOLOGY

METHODS IN ENZYMOLOGY

EDITORS-IN-CHIEF

John N. Abelson   Melvin I. Simon

DIVISION OF BIOLOGY
CALIFORNIA INSTITUTE OF TECHNOLOGY
PASADENA, CALIFORNIA

FOUNDING EDITORS

Sidney P. Colowick and Nathan O. Kaplan

Methods in Enzymology
Volume 194

Guide to Yeast Genetics and Molecular Biology

EDITED BY

Christine Guthrie

DEPARTMENT OF BIOCHEMISTRY AND BIOPHYSICS
UNIVERSITY OF CALIFORNIA, SAN FRANCISCO
SAN FRANCISCO, CALIFORNIA

Gerald R. Fink

WHITEHEAD INSTITUTE FOR BIOMEDICAL RESEARCH AND
DEPARTMENT OF BIOLOGY
MASSACHUSETTS INSTITUTE OF TECHNOLOGY
CAMBRIDGE, MASSACHUSETTS

ACADEMIC PRESS, INC.
Harcourt Brace Jovanovich, Publishers
San Diego New York Boston
London Sydney Tokyo Toronto

QP
601
.M49
v.194

This book is printed on acid-free paper. ∞

Copyright © 1991 by Academic Press, Inc.
All Rights Reserved.
No part of this publication may be reproduced or transmitted in any form or by any means, electronic or mechanical, including photocopy, recording, or any information storage and retrieval system, without permission in writing from the publisher.

Academic Press, Inc.
San Diego, California 92101

United Kingdom Edition published by
Academic Press Limited
24-28 Oval Road, London NW1 7DX

Library of Congress Catalog Card Number: 54-9110

ISBN 0-12-182095-5 (Hardcover) (alk. paper)
ISBN 0-12-310670-2 (Paperback) (alk. paper)

Printed in the United States of America
90 91 92 93   9 8 7 6 5 4 3 2 1

Table of Contents

Contributors to Volume 194	xi
Preface	xvii
Volumes in Series	xix
Process Guide	xxxv

Section I. Basic Methods of Yeast Genetics

1. Getting Started with Yeast	Fred Sherman	3
2. Micromanipulation and Dissection of Asci	Fred Sherman and James Hicks	21
3. Mapping Yeast Genes	Fred Sherman and Paul Wakem	38
4. Positional Mapping of Genes by Chromosome Blotting and Chromosome Fragmentation	Sandra L. Gerring, Carla Connelly, and Philip Hieter	57
5. Assay of Yeast Mating Reaction	George F. Sprague, Jr.	77
6. Monitoring Meiosis and Sporulation in *Saccharomyces cerevisiae*	Yona Kassir and Giora Simchen	94
7. Identifying Sporulation Genes, Visualizing Synaptonemal Complexes, and Large-Scale Spore and Spore Wall Purification	Rochelle Easton Esposito, Michael Dresser, and Michael Breitenbach	110
8. Putting the *HO* Gene to Work: Practical Uses for Mating-Type Switching	Ira Herskowitz and Robert E. Jensen	132
9. Spore Enrichment	Beth Rockmill, Eric J. Lambie, and G. Shirleen Roeder	146
10. Analysis and Manipulation of Yeast Mitochondrial Genes	Thomas D. Fox, Linda S. Folley, Julio J. Mulero, Thomas W. McMullin, Peter E. Thorsness, Lars O. Hedin, and Maria C. Costanzo	149

Section II. Cloning and Recombinant DNA

11. DNA of *Saccharomyces cerevisiae*	PETER PHILIPPSEN, AGATHE STOTZ, AND CHRISTINE SCHERF	169
12. High-Efficiency Transformation of Yeast by Electroporation	DANIEL M. BECKER AND LEONARD GUARENTE	182
13. Transmission of Plasmid DNA to Yeast by Conjugation with Bacteria	JACK A. HEINEMANN AND GEORGE F. SPRAGUE, JR.	187
14. Cloning Genes by Complementation in Yeast	MARK D. ROSE AND JAMES R. BROACH	195
15. Gene Isolation with λgt11 System	RICHARD A. YOUNG AND RONALD W. DAVIS	230
16. Gene Overexpression in Studies of *Saccharomyces cerevisiae*	JASPER RINE	239
17. Preparation of Clone Libraries in Yeast Artificial-Chromosome Vectors	DAVID T. BURKE AND MAYNARD V. OLSON	251

Section III. Making Mutants

18. Classical Mutagenesis Techniques	CHRISTOPHER W. LAWRENCE	273
19. Targeting, Disruption, Replacement, and Allele Rescue: Integrative DNA Transformation in Yeast	RODNEY ROTHSTEIN	281
20. *In Vitro* Mutagenesis and Plasmid Shuffling: From Cloned Gene to Mutant Yeast	ROBERT S. SIKORSKI AND JEF D. BOEKE	302
21. Recovery of Plasmids from Yeast into *Escherichia coli*: Shuttle Vectors	JEFFREY N. STRATHERN AND DAVID R. HIGGINS	319
22. Shuttle Mutagenesis: Bacterial Transposons for Genetic Manipulations in Yeast	MERL F. HOEKSTRA, H. STEVEN SEIFERT, JAC NICKOLOFF, AND FRED HEFFRON	329
23. Ty Mutagenesis in *Saccharomyces cerevisiae*	DAVID J. GARFINKEL AND JEFFREY N. STRATHERN	342
24. Transformation of Yeast Directly with Synthetic Oligonucleotides	RICHARD P. MOERSCHELL, GOUTAM DAS, AND FRED SHERMAN	362

Section IV. Biochemistry of Gene Expression

25. Vectors for Expression of Cloned Genes in Yeast: Regulation, Overproduction, and Underproduction	Jane C. Schneider and Leonard Guarente	373
26. Vectors for Constitutive and Inducible Gene Expression in Yeast	Mark Schena, Didier Picard, and Keith R. Yamamoto	389
27. Preparation of High Molecular Weight RNA	Karl Köhrer and Horst Domdey	398
28. Preparation and Analysis of Low Molecular Weight RNAs and Small Ribonucleoproteins	Jo Ann Wise	405
29. Measurement of mRNA Decay Rates in *Saccharomyces cerevisiae*	Roy Parker, David Herrick, Stuart W. Peltz, and Allan Jacobson	415
30. Labeling of RNA and Phosphoproteins in *Saccharomyces cerevisiae*	Jonathan R. Warner	423
31. Tackling the Protease Problem in *Saccharomyces cerevisiae*	Elizabeth W. Jones	428
32. Structural and Functional Analysis of Yeast Ribosomal Proteins	H. A. Raué, W. H. Mager, and R. J. Planta	453
33. High-Expression Vectors with Multiple Cloning Sites for Construction of *trpE* Fusion Genes: pATH Vectors	T. J. Koerner, John E. Hill, Alan M. Myers, and Alexander Tzagoloff	477
34. Production of Proteins by Secretion from Yeast	Donald T. Moir and Lance S. Davidow	491
35. Epitope Tagging and Protein Surveillance	Peter A. Kolodziej and Richard A. Young	508
36. Reverse Biochemistry: Methods and Applications for Synthesizing Yeast Proteins *in Vitro*	Kevin Struhl	520
37. *In Vitro* Protein Synthesis	Michael J. Leibowitz, Francis P. Barbone, and Denise E. Georgopoulos	536
38. RNA Polymerase II Transcription *in Vitro*	Neal F. Lue, Peter M. Flanagan, Raymond J. Kelleher III, Aled M. Edwards, and Roger D. Kornberg	545

39. Direct Sequence and Footprint Analysis of Yeast DNA by Primer Extension	JON M. HUIBREGTSE AND DAVID R. ENGELKE	550

Section V. Cell Biology

40. Immunofluorescence Methods for Yeast	JOHN R. PRINGLE, ALISON E. M. ADAMS, DAVID G. DRUBIN, AND BRIAN K. HAARER	565
41. Preparation of Yeast Cells for Thin-Section Electron Microscopy	BRECK BYERS AND LORETTA GOETSCH	602
42. Immunogold Labeling of Yeast Ultrathin Sections	MICHAEL W. CLARK	608
43. Analysis of Mitochondrial Function and Assembly	MICHAEL P. YAFFE	627
44. Methods for Studying the Yeast Vacuole	CHRISTOPHER J. ROBERTS, CHRISTOPHER K. RAYMOND, CARL T. YAMASHIRO, AND TOM H. STEVENS	644
45. Analysis of Polypeptide Transit through Yeast Secretory Pathway	ALEX FRANZUSOFF, JONATHAN ROTHBLATT, AND RANDY SCHEKMAN	662
46. *In Vitro* Protein Translocation across Microsomal Membranes of *Saccharomyces cerevisiae*	PABLO D. GARCIA, WILLIAM HANSEN, AND PETER WALTER	675
47. Analysis of Glycoproteins from *Saccharomyces cerevisiae*	P. ORLEAN, M. J. KURANDA, AND C. F. ALBRIGHT	682
48. Yeast Endocytosis Assays	VJEKOSLAV DULIC, MARK EGERTON, IBRAHIM ELGUINDI, SUSAN RATHS, BIRGIT SINGER, AND HOWARD RIEZMAN	697
49. Inducing and Assaying Heat-Shock Response in *Saccharomyces cerevisiae*	CHARLES M. NICOLET AND ELIZABETH A. CRAIG	710
50. Nucleolar-Specific Positive Stains for Optical and Electron Microscopy	MICHAEL W. CLARK	717
51. Staining of Actin with Fluorochrome-Conjugated Phalloidin	ALISON E. M. ADAMS AND JOHN R. PRINGLE	729
52. Staining of Bud Scars and Other Cell Wall Chitin with Calcofluor	JOHN R. PRINGLE	732

53. Isolation of Yeast Nuclei	JOHN P. ARIS AND GÜNTER BLOBEL	735
54. Analysis of Chromosome Segregation in *Saccharomyces cerevisiae*	JAMES H. SHERO, MICHAEL KOVAL, FORREST SPENCER, ROBERT E. PALMER, PHILIP HIETER, AND DOUGLAS KOSHLAND	749
55. Genetic Screens and Selections for Cell and Nuclear Fusion Mutants	VIVIAN BERLIN, JULIE A. BRILL, JOSHUA TRUEHEART, JEF D. BOEKE, AND GERALD R. FINK	774

Section VI. Fission Yeast

56. Molecular Genetic Analysis of Fission Yeast *Schizosaccharomyces pombe*	SERGIO MORENO, AMAR KLAR, AND PAUL NURSE	795

Section VII. Appendix

57. Genetic and Physical Maps of *Saccharomyces cerevisia*	ROBERT K. MORTIMER, DAVID SCHILD, C. REBECCA CONTOPOULOU, AND JONATHAN A. KANS	827

AUTHOR INDEX 865

SUBJECT INDEX 905

Contributors to Volume 194

Article numbers are in parentheses following the names of contributors.
Affiliations listed are current.

ALISON E. M. ADAMS (40, 51), *Department of Molecular and Cellular Biology, University of Arizona, Tucson, Arizona 85721*

C. F. ALBRIGHT (47), *Whitehead Institute for Biomedical Research, Cambridge, Massachusetts 02142 and Massachusetts Institute of Technology, Cambridge, Massachusetts 02139*

JOHN P. ARIS (53), *Laboratory of Cell Biology, Howard Hughes Medical Institute, The Rockefeller University, New York, New York 10021*

FRANCIS P. BARBONE (37), *Department of Molecular Genetics and Microbiology, UMDNJ-Robert Wood Johnson Medical School, Piscataway, New Jersey 08854*

DANIEL M. BECKER (12), *Department of Biology, Massachusetts Institute of Technology, Cambridge, Massachusetts 02139*

VIVIAN BERLIN (55), *Vertex Pharmaceuticals, Cambridge, Massachusetts 02139*

GÜNTER BLOBEL (53), *Laboratory of Cell Biology, Howard Hughes Medical Institute, The Rockefeller University, New York, New York 10021*

JEF D. BOEKE (20, 55), *Department of Molecular Biology and Genetics, The Johns Hopkins University School of Medicine, Baltimore, Maryland 21205*

MICHAEL BREITENBACH (7), *Institute of Microbiology and Genetics, University of Vienna, A-1090, Vienna, Austria*

JULIE A. BRILL (55), *Whitehead Institute for Biomedical Research, Cambridge, Massachusetts 02142 and Department of Biology, Massachusetts Institute of Technology, Cambridge, Massachusetts 02139*

JAMES R. BROACH (14), *Department of Molecular Biology, Lewis Thomas Laboratory, Princeton University, Princeton, New Jersey 08544*

DAVID T. BURKE (17), *Department of Molecular Biology, Princeton University, Princeton, New Jersey 08540*

BRECK BYERS (41), *Department of Genetics, University of Washington, Seattle, Washington 98195*

MICHAEL W. CLARK (42, 50), *Department of Biology, McGill University, Montreal, Quebec H3A 1B1, Canada*

CARLA CONNELLY (4), *Department of Molecular Biology and Genetics, The Johns Hopkins University School of Medicine, Baltimore, Maryland 21205*

C. REBECCA CONTOPOULOU (57), *Department of Molecular and Cell Biology, Division of Biophysics and Cell Biology, University of California, Berkeley, Berkeley, California 94720*

MARIA C. COSTANZO (10), *Section of Genetics and Development, Cornell University, Ithaca, New York 14853*

ELIZABETH A. CRAIG (49), *Department of Physiological Chemistry, University of Wisconsin-Madison, Madison, Wisconsin 53706*

GOUTAM DAS (24), *Astra Research Center, Malleswarm, Bangalore 56003, India*

LANCE S. DAVIDOW (34), *Molecular Genetics Department, Collaborative Research, Inc., Bedford, Massachusetts 01730*

RONALD W. DAVIS (15), *Department of Biochemistry, School of Medicine, Stanford University, Stanford, California 94305*

HORST DOMDEY (27), *Laboratorium für Molekulare Biologie, Genzentrum der Ludwig-Maximilians-Universität München, D-8033 Martinsried, Federal Republic of Germany*

CONTRIBUTORS TO VOLUME 194

MICHAEL DRESSER (7), *Program in Molecular and Cellular Biology, Oklahoma Medical Research Foundation, Oklahoma City, Oklahoma 73104*

DAVID G. DRUBIN (40), *Department of Molecular and Cell Biology, University of California, Berkeley, Berkeley, California 94720*

VJEKOSLAV DULIC (48), *Department of Molecular Biology, Research Institute of Scripps Clinic, La Jolla, California 92037*

ALED M. EDWARDS (38), *Department of Cell Biology, Stanford University School of Medicine, Stanford, California 94305*

MARK EGERTON (48), *Biochemistry Department, Biocenter of University of Basel, CH-4056 Basel, Switzerland*

IBRAHIM ELGUINDI (48), *Department of Biochemistry, Southwestern Medical Center, University of Texas, Dallas, Texas 75235*

DAVID R. ENGELKE (39), *Department of Biological Chemistry, University of Michigan, Ann Arbor, Michigan 48109*

ROCHELLE EASTON ESPOSITO (7), *Department of Molecular Genetics and Cell Biology, The University of Chicago, Chicago, Illinois 60637*

GERALD R. FINK (55), *Whitehead Institute for Biomedical Research, Cambridge, Massachusetts 02142 and Department of Biology, Massachusetts Institute of Technology, Cambridge, Massachusetts 02139*

PETER M. FLANAGAN (38), *Department of Cell Biology, Stanford University School of Medicine, Stanford, California 94305*

LINDA S. FOLLEY (10), *Section of Genetics and Development, Cornell University, Ithaca, New York 14853*

THOMAS D. FOX (10), *Section of Genetics and Development, Cornell University, Ithaca, New York 14853*

ALEX FRANZUSOFF (45), *Department of Cellular and Structural Biology, University of Colorado Medical School, Denver, Colorado 80262*

PABLO D. GARCIA (46), *Department of Physiology, University of California, San Francisco, San Francisco, California 94143*

DAVID J. GARFINKEL (23), *Laboratory of Eukaryotic Gene Expression, ABL-Basic Research Program, National Cancer Institute–Frederick Cancer Research and Development Facility, Frederick, Maryland 21701*

DENISE E. GEORGOPOULOS (37), *Department of Molecular Biology and Genetics, The Johns Hopkins University School of Medicine, Baltimore, Maryland 21205*

SANDRA L. GERRING (4), *Department of Molecular Biology and Genetics, The Johns Hopkins University School of Medicine, Baltimore, Maryland 21205*

LORETTA GOETSCH (41), *Department of Genetics, University of Washington, Seattle, Washington 98195*

LEONARD GUARENTE (12, 25), *Department of Biology, Massachusetts Institute of Technology, Cambridge, Massachusetts 02139*

BRIAN K. HAARER (40), *Department of Anatomy and Cell Biology, University of Michigan, Ann Arbor, Michigan 48109*

WILLIAM HANSEN (46), *Department of Physiology, University of California, San Francisco, San Francisco, California 94143*

LARS O. HEDIN (10), *Institute of Ecosystem Studies, The New York Botanical Garden, Millbrook, New York 12545*

FRED HEFFRON (22), *Department of Molecular Biology, Research Institute of Scripps Clinic, La Jolla, California 92037*

JACK A. HEINEMANN (13), *National Institutes of Health, Institute of Allergy and Infectious Diseases, Rocky Mountain Laboratory, Hamilton, Montana 59840*

DAVID HERRICK (29), *Department of Molecular Genetics and Microbiology, University of Massachusetts Medical School, Boston, Massachusetts 02125*

IRA HERSKOWITZ (8), *Department of Biochemistry and Biophysics, University of California, San Francisco School of Medicine, San Francisco, California 94143*

JAMES B. HICKS (2), *Scripps Clinic and Research Foundation, La Jolla, California 92037*

PHILIP HIETER (4, 54), *Department of Molecular Biology and Genetics, The Johns Hopkins University School of Medicine, Baltimore, Maryland 21205*

DAVID R. HIGGINS (21), *Laboratory of Eukaryotic Gene Expression, ABL-Basic Research Program, National Cancer Institute-Frederick Cancer Research and Development Facility, Frederick, Maryland 21701*

JOHN E. HILL (33), *Department of Cell Biology, New York University Medical Center, New York, New York 10016*

MERL F. HOEKSTRA (22), *Molecular Biology and Virology Laboratories, The Salk Institute for Biological Studies, La Jolla, California 91037*

JON M. HUIBREGTSE (39), *Laboratory of Tumor Virus Biology, National Cancer Institute, National Institutes of Health, Bethesda, Maryland 20892*

ALLAN JACOBSON (29), *Department of Molecular Genetics and Microbiology, University of Massachusetts Medical School, Boston, Massachusetts 02125*

ROBERT E. JENSEN (8), *Department of Cell Biology and Anatomy, The Johns Hopkins University School of Medicine, Baltimore, Maryland 21205*

ELIZABETH W. JONES (31), *Department of Biological Sciences, Carnegie Mellon University, Pittsburgh, Pennsylvania 15213*

JONATHAN A. KANS (57), *The National Center for Biotechnology Information, National Library of Medicine, National Institutes of Health, Bethesda, Maryland 20894*

YONA KASSIR (6), *Department of Biology, Technion-Israel Institute of Technology, Haifa 32000, Israel*

RAYMOND J. KELLEHER III (38), *Department of Cell Biology, Stanford University School of Medicine, Stanford, California 94305*

AMAR KLAR (56), *Laboratory of Eukaryotic Gene Expression, ABL-Basic Research Program, National Cancer Institute-Frederick Cancer Research and Development Facility, Frederick, Maryland 21701*

T. J. KOERNER (33), *Research Department, American Cancer Society, Atlanta, Georgia 30329*

KARL KÖHRER (27), *Laboratorium für Molekulare Biologie, Genzentrum der Ludwig-Maximilians-Universität München, D-8033 Martinsried, Federal Republic of Germany*

PETER A. KOLODZIEJ (35), *Whitehead Institute for Biomedical Research, Cambridge, Massachusetts 02142 and Department of Biology, Massachusetts Institute of Technology, Cambridge, Massachusetts 02139*

ROGER D. KORNBERG (38), *Department of Cell Biology, Stanford University School of Medicine, Stanford, California 94305*

DOUGLAS KOSHLAND (54), *Department of Embryology, Carnegie Institution of Washington, Baltimore, Maryland 21210*

MICHAEL KOVAL (54), *Department of Embryology, Carnegie Institution of Washington, Baltimore, Maryland 21210*

M. J. KURANDA (47), *Repligen Corporation, Cambridge, Massachusetts 02139*

ERIC J. LAMBIE (9), *Laboratory of Molecular Biology, University of Wisconsin, Madison, Wisconsin 53706*

CHRISTOPHER W. LAWRENCE (18), *Department of Biophysics, University of Rochester School of Medicine and Dentistry, Rochester, New York 14642*

MICHAEL J. LEIBOWITZ (37), *Department of Molecular Genetics and Microbiology, UMDNJ-Robert Wood Johnson Medical School, Piscataway, New Jersey 08854*

NEAL F. LUE (38), *Department of Cell Biology, Stanford University School of Medicine, Stanford, California 94305*

W. H. MAGER (32), *Department of Biochemistry, Vrije Universiteit de Boelelaan 1083, 1081 HV Amsterdam, The Netherlands*

THOMAS W. MCMULLIN (10), *Section of Genetics and Development, Cornell University, Ithaca, New York 14853*

RICHARD P. MOERSCHELL (24), *Department of Biochemistry, University of Rochester School of Medicine and Dentistry, Rochester, New York 14642*

DONALD T. MOIR (34), *Molecular Genetics Department, Collaborative Research, Inc., Bedford, Massachusetts 01730*

SERGIO MORENO (56), *Imperial Cancer Research Fund, Microbiology Unit, Department of Biochemistry, Oxford University, Oxford OX1 3QU, England*

ROBERT K. MORTIMER (57), *Department of Molecular and Cell Biology, Division of Genetics, University of California, Berkeley, Berkeley, California 94720*

JULIO J. MULERO (10), *Section of Genetics and Development, Cornell University, Ithaca, New York 14853*

ALAN M. MYERS (33), *Department of Biochemistry and Biophysics, Iowa State University, Ames, Iowa 50011*

JAC NICKOLOFF (22), *Laboratory of Radiobiology, Harvard School of Public Health, Boston, Massachusetts 02115*

CHARLES M. NICOLET (49), *Department of Physiological Chemistry, University of Wisconsin-Madison, Madison, Wisconsin 53706*

PAUL NURSE (56), *Imperial Cancer Research Fund, Microbiology Unit, Department of Biochemistry, Oxford University, Oxford OX1 3QU, England*

MAYNARD V. OLSON (17), *Department of Genetics, Washington University School of Medicine, St. Louis, Missouri 63110*

P. ORLEAN (47), *Department of Biochemisty, University of Illinois Urbana-Champaign, Urbana, Illinois 61801*

ROBERT E. PALMER (54), *Department of Embryology, Carnegie Institution of Washington, Baltimore, Maryland 21210*

ROY PARKER (29), *Department of Molecular and Cellular Biology, University of Arizona, Tucson, Arizona 85721*

STUART W. PELTZ (29), *Department of Molecular Genetics and Microbiology, University of Massachusetts Medical School, Boston, Massachusetts 02125*

PETER PHILIPPSEN (11), *Institute for Microbiology and Molecular Biology, University of Giessen, D-6300 Giessen, Federal Republic of Germany*

DIDIER PICARD (26), *Department of Cellular Biology, University of Geneva, 1211 Geneva 4, Switzerland*

R. J. PLANTA (32), *Department of Biochemistry, Vrije Universiteit de Boelelaan 1083, 1081 HV Amsterdam, The Netherlands*

JOHN R. PRINGLE (40, 51, 52), *Department of Biology, University of Michigan, Ann Arbor, Michigan 48109*

SUSAN RATHS (48), *Biochemistry Department, Biocenter of University of Basel, CH-4056 Basel, Switzerland*

H. A. RAUÉ (32), *Department of Biochemistry, Vrije Universiteit de Boelelaan 1083, 1081 HV Amsterdam, The Netherlands*

CHRISTOPHER K. RAYMOND (44), *Institute of Molecular Biology, University of Oregon, Eugene, Oregon 97403*

HOWARD RIEZMAN (48), *Biochemistry Department, Biocenter of University of Basel, CH-4056 Basel, Switzerland*

JASPER RINE (16), *Department of Molecular and Cellular Biology, Division of Genetics, University of California, Berkeley, Berkeley, California 94720*

CHRISTOPHER J. ROBERTS (44), *Institute of Molecular Biology, University of Oregon, Eugene, Oregon 97403*

BETH ROCKMILL (9), *Department of Biology, Yale University, New Haven, Connecticut 06511*

G. SHIRLEEN ROEDER (9), *Department of Biology, Yale University, New Haven, Connecticut 06511*

MARK D. ROSE (14), *Department of Molecular Biology, Lewis Thomas Laboratory, Princeton University, Princeton, New Jersey 08544*

JONATHAN ROTHBLATT (45), *Department of Biological Sciences, Dartmouth College, Hanover, New Hampshire 03755*

RODNEY ROTHSTEIN (19), *Department of Genetics and Development, Columbia College of Physicians and Surgeons, Columbia University, New York, New York 10032*

RANDY SCHEKMAN (45), *Division of Biochemistry and Molecular Biology, University of California, Berkeley, Berkeley, California 94720*

MARK SCHENA (26), *Department of Biochemistry, Beckman Center, Stanford University, Stanford, California 94305*

CHRISTINE SCHERF (11), *Institute for Microbiology and Molecular Biology, University of Giessen, D-6300 Giessen, Federal Republic of Germany*

DAVID SCHILD (57), *Cell and Molecular Biology, Division of Lawrence Berkeley Laboratory, Berkeley, California 94720*

JANE C. SCHNEIDER (25), *MSU-DOE Plant Research Laboratory, Michigan State University, East Lansing, Michigan 48824*

H. STEVEN SEIFERT (22), *Department of Microbiology and Immunology, Northwestern University School of Medicine, Chicago, Illinois 60611*

FRED SHERMAN (1, 2, 3, 24), *Department of Biochemistry, University of Rochester School of Medicine and Dentistry, Rochester, New York 14642*

JAMES H. SHERO (54), *Department of Molecular Biology and Genetics, The Johns Hopkins University School of Medicine, Baltimore, Maryland 21205*

ROBERT S. SIKORSKI (20), *Department of Molecular Biology and Genetics, The Johns Hopkins University School of Medicine, Baltimore, Maryland 21205*

GIORA SIMCHEN (6), *Department of Genetics, The Hebrew University of Jerusalem, Jerusalem 91904, Israel*

BIRGIT SINGER (48), *Biochemistry Department, Biocenter of University of Basel, CH-4056 Basel, Switzerland*

FORREST SPENCER (54), *Department of Molecular Biology and Genetics, The Johns Hopkins University School of Medicine, Baltimore, Maryland 21205*

GEORGE F. SPRAGUE, JR. (5, 13), *Institute of Molecular Biology, University of Oregon, Eugene, Oregon 97403*

TOM H. STEVENS (44), *Institute of Molecular Biology, University of Oregon, Eugene, Oregon 97403*

AGATHE STOTZ (11), *Department of Microbiology, Biocenter of University of Basel, CH-4056 Basel, Switzerland*

JEFFREY N. STRATHERN (21, 23), *Laboratory of Eukaryotic Gene Expression, ABL-Basic Research Program, National Cancer Institute–Frederick Cancer Research and Development Facility, Frederick, Maryland 21701*

KEVIN STRUHL (36), *Department of Biological Chemistry and Molecular Pharmacology, Harvard Medical School, Boston, Massachusetts 02115*

PETER E. THORSNESS (10), *Section of Genetics and Development, Cornell University, Ithaca, New York 14853*

JOSHUA TRUEHEART (55), *Department of Biochemistry, University of California, Berkeley, Berkeley, California 94720*

ALEXANDER TZAGOLOFF (33), *Department of Biological Sciences, Columbia University, New York, New York 10027*

PAUL WAKEM (3), *Department of Biochemistry, University of Rochester School of Medicine and Dentistry, Rochester, New York 14642*

PETER WALTER (46), *Department of Biochemistry and Biophysics, University of California, San Francisco, San Francisco, California 94143*

JONATHAN R. WARNER (30), *Department of Cell Biology, Albert Einstein College of Medicine, Bronx, New York 10461*

JO ANN WISE (28), *Department of Biochemistry, University of Illinois, Urbana-Champaign, Urbana, Illinois 61801*

MICHAEL P. YAFFE (43), *Department of Biology, University of California, San Diego, La Jolla, Caifornia 92093*

KEITH R. YAMAMOTO (26), *Department of Biochemistry and Biophysics, University of California, San Francisco, San Francisco, California 94143*

CARL T. YAMASHIRO (44), *Department of Biological Sciences, Stanford University, Stanford, California 94305*

RICHARD A. YOUNG (15, 35), *Whitehead Institute for Biomedical Research, Cambridge, Massachusetts 02142 and Department of Biology, Massachusetts Institute of Technology, Cambridge, Massachusetts 02139*

Preface

This volume presents a wide variety of techniques currently used to analyze yeast. The biology of yeast, once considered a lazy backwater, is now a burgeoning enterprise investigated by thousands of scientists worldwide. The realization that the elements of cell structure and function are basic to all eukaryotic cells has propelled yeast to center stage because it is so easily manipulated. Now problems common to all cells can be addressed thanks to hundreds of techniques that have enabled workers to surmount difficult experimental hurdles. Some of these procedures had to be developed specifically to deal with problems unique to yeast biology. For example, it was not long ago that dissection of yeast spores and isolaton of yeast DNA were considered arcane skills restricted to a few artisans. Progress meant altering and streamlining these procedures so that they could be performed easily, routinely, and reproducibly even by beginners.

This book will be a valuable resource both for beginners and for current practitioners. It should enable newcomers to set up a yeast laboratory and master basic manipulations. For current practitioners, the compilation of recipes in a single book, instead of dispersed in the notebooks of former postdoctorals, should simplify and codify current laboratory practices. Yeast molecular biology is progressing so rapidly that some of these procedures will, undoubtedly, have been altered by the time the book is published. Even in these cases, the thoughtful comments of the authors should help experimentalists avoid common pitfalls.

We thank the authors for their care in preparing the articles and for providing their unique perspectives.

CHRISTINE GUTHRIE
GERALD R. FINK

METHODS IN ENZYMOLOGY

VOLUME I. Preparation and Assay of Enzymes
Edited by SIDNEY P. COLOWICK AND NATHAN O. KAPLAN

VOLUME II. Preparation and Assay of Enzymes
Edited by SIDNEY P. COLOWICK AND NATHAN O. KAPLAN

VOLUME III. Preparation and Assay of Substrates
Edited by SIDNEY P. COLOWICK AND NATHAN O. KAPLAN

VOLUME IV. Special Techniques for the Enzymologist
Edited by SIDNEY P. COLOWICK AND NATHAN O. KAPLAN

VOLUME V. Preparation and Assay of Enzymes
Edited by SIDNEY P. COLOWICK AND NATHAN O. KAPLAN

VOLUME VI. Preparation and Assay of Enzymes (*Continued*)
Preparation and Assay of Substrates
Special Techniques
Edited by SIDNEY P. COLOWICK AND NATHAN O. KAPLAN

VOLUME VII. Cumulative Subject Index
Edited by SIDNEY P. COLOWICK AND NATHAN O. KAPLAN

VOLUME VIII. Complex Carbohydrates
Edited by ELIZABETH F. NEUFELD AND VICTOR GINSBURG

VOLUME IX. Carbohydrate Metabolism
Edited by WILLIS A. WOOD

VOLUME X. Oxidation and Phosphorylation
Edited by RONALD W. ESTABROOK AND MAYNARD E. PULLMAN

VOLUME XI. Enzyme Structure
Edited by C. H. W. HIRS

VOLUME XII. Nucleic Acids (Parts A and B)
Edited by LAWRENCE GROSSMAN AND KIVIE MOLDAVE

VOLUME XIII. Citric Acid Cycle
Edited by J. M. LOWENSTEIN

VOLUME XIV. Lipids
Edited by J. M. LOWENSTEIN

VOLUME XV. Steroids and Terpenoids
Edited by RAYMOND B. CLAYTON

VOLUME XVI. Fast Reactions
Edited by KENNETH KUSTIN

VOLUME XVII. Metabolism of Amino Acids and Amines (Parts A and B)
Edited by HERBERT TABOR AND CELIA WHITE TABOR

VOLUME XVIII. Vitamins and Coenzymes (Parts A, B, and C)
Edited by DONALD B. MCCORMICK AND LEMUEL D. WRIGHT

VOLUME XIX. Proteolytic Enzymes
Edited by GERTRUDE E. PERLMANN AND LASZLO LORAND

VOLUME XX. Nucleic Acids and Protein Synthesis (Part C)
Edited by KIVIE MOLDAVE AND LAWRENCE GROSSMAN

VOLUME XXI. Nucleic Acids (Part D)
Edited by LAWRENCE GROSSMAN AND KIVIE MOLDAVE

VOLUME XXII. Enzyme Purification and Related Techniques
Edited by WILLIAM B. JAKOBY

VOLUME XXIII. Photosynthesis (Part A)
Edited by ANTHONY SAN PIETRO

VOLUME XXIV. Photosynthesis and Nitrogen Fixation (Part B)
Edited by ANTHONY SAN PIETRO

VOLUME XXV. Enzyme Structure (Part B)
Edited by C. H. W. HIRS AND SERGE N. TIMASHEFF

VOLUME XXVI. Enzyme Structure (Part C)
Edited by C. H. W. HIRS AND SERGE N. TIMASHEFF

Volume XXVII. Enzyme Structure (Part D)
Edited by C. H. W. Hirs and Serge N. Timasheff

Volume XXVIII. Complex Carbohydrates (Part B)
Edited by Victor Ginsburg

Volume XXIX. Nucleic Acids and Protein Synthesis (Part E)
Edited by Lawrence Grossman and Kivie Moldave

Volume XXX. Nucleic Acids and Protein Synthesis (Part F)
Edited by Kivie Moldave and Lawrence Grossman

Volume XXXI. Biomembranes (Part A)
Edited by Sidney Fleischer and Lester Packer

Volume XXXII. Biomembranes (Part B)
Edited by Sidney Fleischer and Lester Packer

Volume XXXIII. Cumulative Subject Index Volumes I–XXX
Edited by Martha G. Dennis and Edward A. Dennis

Volume XXXIV. Affinity Techniques (Enzyme Purification: Part B)
Edited by William B. Jakoby and Meir Wilchek

Volume XXXV. Lipids (Part B)
Edited by John M. Lowenstein

Volume XXXVI. Hormone Action (Part A: Steroid Hormones)
Edited by Bert W. O'Malley and Joel G. Hardman

Volume XXXVII. Hormone Action (Part B: Peptide Hormones)
Edited by Bert W. O'Malley and Joel G. Hardman

Volume XXXVIII. Hormone Action (Part C: Cyclic Nucleotides)
Edited by Joel G. Hardman and Bert W. O'Malley

Volume XXXIX. Hormone Action (Part D: Isolated Cells, Tissues, and Organ Systems)
Edited by Joel G. Hardman and Bert W. O'Malley

Volume XL. Hormone Action (Part E: Nuclear Structure and Function)
Edited by Bert W. O'Malley and Joel G. Hardman

VOLUME XLI. Carbohydrate Metabolism (Part B)
Edited by W. A. WOOD

VOLUME XLII. Carbohydrate Metabolism (Part C)
Edited by W. A. WOOD

VOLUME XLIII. Antibiotics
Edited by JOHN H. HASH

VOLUME XLIV. Immobilized Enzymes
Edited by KLAUS MOSBACH

VOLUME XLV. Proteolytic Enzymes (Part B)
Edited by LASZLO LORAND

VOLUME XLVI. Affinity Labeling
Edited by WILLIAM B. JAKOBY AND MEIR WILCHEK

VOLUME XLVII. Enzyme Structure (Part E)
Edited by C. H. W. HIRS AND SERGE N. TIMASHEFF

VOLUME XLVIII. Enzyme Structure (Part F)
Edited by C. H. W. HIRS AND SERGE N. TIMASHEFF

VOLUME XLIX. Enzyme Structure (Part G)
Edited by C. H. W. HIRS AND SERGE N. TIMASHEFF

VOLUME L. Complex Carbohydrates (Part C)
Edited by VICTOR GINSBURG

VOLUME LI. Purine and Pyrimidine Nucleotide Metabolism
Edited by PATRICIA A. HOFFEE AND MARY ELLEN JONES

VOLUME LII. Biomembranes (Part C: Biological Oxidations)
Edited by SIDNEY FLEISCHER AND LESTER PACKER

VOLUME LIII. Biomembranes (Part D: Biological Oxidations)
Edited by SIDNEY FLEISCHER AND LESTER PACKER

VOLUME LIV. Biomembranes (Part E: Biological Oxidations)
Edited by SIDNEY FLEISCHER AND LESTER PACKER

VOLUME LV. Biomembranes (Part F: Bioenergetics)
Edited by SIDNEY FLEISCHER AND LESTER PACKER

VOLUME LVI. Biomembranes (Part G: Bioenergetics)
Edited by SIDNEY FLEISCHER AND LESTER PACKER

VOLUME LVII. Bioluminescence and Chemiluminescence
Edited by MARLENE A. DELUCA

VOLUME LVIII. Cell Culture
Edited by WILLIAM B. JAKOBY AND IRA PASTAN

VOLUME LIX. Nucleic Acids and Protein Synthesis (Part G)
Edited by KIVIE MOLDAVE AND LAWRENCE GROSSMAN

VOLUME LX. Nucleic Acids and Protein Synthesis (Part H)
Edited by KIVIE MOLDAVE AND LAWRENCE GROSSMAN

VOLUME 61. Enzyme Structure (Part H)
Edited by C. H. W. HIRS AND SERGE N. TIMASHEFF

VOLUME 62. Vitamins and Coenzymes (Part D)
Edited by DONALD B. MCCORMICK AND LEMUEL D. WRIGHT

VOLUME 63. Enzyme Kinetics and Mechanism (Part A: Initial Rate and Inhibitor Methods)
Edited by DANIEL L. PURICH

VOLUME 64. Enzyme Kinetics and Mechanism (Part B: Isotopic Probes and Complex Enzyme Systems)
Edited by DANIEL L. PURICH

VOLUME 65. Nucleic Acids (Part I)
Edited by LAWRENCE GROSSMAN AND KIVIE MOLDAVE

VOLUME 66. Vitamins and Coenzymes (Part E)
Edited by DONALD B. MCCORMICK AND LEMUEL D. WRIGHT

VOLUME 67. Vitamins and Coenzymes (Part F)
Edited by DONALD B. MCCORMICK AND LEMUEL D. WRIGHT

VOLUME 68. Recombinant DNA
Edited by RAY WU

VOLUME 69. Photosynthesis and Nitrogen Fixation (Part C)
Edited by ANTHONY SAN PIETRO

VOLUME 70. Immunochemical Techniques (Part A)
Edited by HELEN VAN VUNAKIS AND JOHN J. LANGONE

VOLUME 71. Lipids (Part C)
Edited by JOHN M. LOWENSTEIN

VOLUME 72. Lipids (Part D)
Edited by JOHN M. LOWENSTEIN

VOLUME 73. Immunochemical Techniques (Part B)
Edited by JOHN J. LANGONE AND HELEN VAN VUNAKIS

VOLUME 74. Immunochemical Techniques (Part C)
Edited by JOHN J. LANGONE AND HELEN VAN VUNAKIS

VOLUME 75. Cumulative Subject Index Volumes XXXI, XXXII, and XXXIV–LX
Edited by EDWARD A. DENNIS AND MARTHA G. DENNIS

VOLUME 76. Hemoglobins
Edited by ERALDO ANTONINI, LUIGI ROSSI-BERNARDI, AND EMILIA CHIANCONE

VOLUME 77. Detoxication and Drug Metabolism
Edited by WILLIAM B. JAKOBY

VOLUME 78. Interferons (Part A)
Edited by SIDNEY PESTKA

VOLUME 79. Interferons (Part B)
Edited by SIDNEY PESTKA

VOLUME 80. Proteolytic Enzymes (Part C)
Edited by LASZLO LORAND

VOLUME 81. Biomembranes (Part H: Visual Pigments and Purple Membranes, I)
Edited by LESTER PACKER

VOLUME 82. Structural and Contractile Proteins (Part A: Extracellular Matrix)
Edited by LEON W. CUNNINGHAM AND DIXIE W. FREDERIKSEN

VOLUME 83. Complex Carbohydrates (Part D)
Edited by VICTOR GINSBURG

VOLUME 84. Immunochemical Techniques (Part D: Selected Immunoassays)
Edited by JOHN J. LANGONE AND HELEN VAN VUNAKIS

VOLUME 85. Structural and Contractile Proteins (Part B: The Contractile Apparatus and the Cytoskeleton)
Edited by DIXIE W. FREDERIKSEN AND LEON W. CUNNINGHAM

VOLUME 86. Prostaglandins and Arachidonate Metabolites
Edited by WILLIAM E. M. LANDS AND WILLIAM L. SMITH

VOLUME 87. Enzyme Kinetics and Mechanism (Part C: Intermediates, Stereochemistry, and Rate Studies)
Edited by DANIEL L. PURICH

VOLUME 88. Biomembranes (Part I: Visual Pigments and Purple Membranes, II)
Edited by LESTER PACKER

VOLUME 89. Carbohydrate Metabolism (Part D)
Edited by WILLIS A. WOOD

VOLUME 90. Carbohydrate Metabolism (Part E)
Edited by WILLIS A. WOOD

VOLUME 91. Enzyme Structure (Part I)
Edited by C. H. W. HIRS AND SERGE N. TIMASHEFF

VOLUME 92. Immunochemical Techniques (Part E: Monoclonal Antibodies and General Immunoassay Methods)
Edited by JOHN J. LANGONE AND HELEN VAN VUNAKIS

VOLUME 93. Immunochemical Techniques (Part F: Conventional Antibodies, Fc Receptors, and Cytotoxicity)
Edited by JOHN J. LANGONE AND HELEN VAN VUNAKIS

VOLUME 94. Polyamines
Edited by HERBERT TABOR AND CELIA WHITE TABOR

VOLUME 95. Cumulative Subject Index Volumes 61–74, 76–80
Edited by EDWARD A. DENNIS AND MARTHA G. DENNIS

VOLUME 96. Biomembranes [Part J: Membrane Biogenesis: Assembly and Targeting (General Methods; Eukaryotes)]
Edited by SIDNEY FLEISCHER AND BECCA FLEISCHER

VOLUME 97. Biomembranes [Part K: Membrane Biogenesis: Assembly and Targeting (Prokaryotes, Mitochondria, and Chloroplasts)]
Edited by SIDNEY FLEISCHER AND BECCA FLEISCHER

VOLUME 98. Biomembranes (Part L: Membrane Biogenesis: Processing and Recycling)
Edited by SIDNEY FLEISCHER AND BECCA FLEISCHER

VOLUME 99. Hormone Action (Part F: Protein Kinases)
Edited by JACKIE D. CORBIN AND JOEL G. HARDMAN

VOLUME 100. Recombinant DNA (Part B)
Edited by RAY WU, LAWRENCE GROSSMAN, AND KIVIE MOLDAVE

VOLUME 101. Recombinant DNA (Part C)
Edited by RAY WU, LAWRENCE GROSSMAN, AND KIVIE MOLDAVE

VOLUME 102. Hormone Action (Part G: Calmodulin and Calcium-Binding Proteins)
Edited by ANTHONY R. MEANS AND BERT W. O'MALLEY

VOLUME 103. Hormone Action (Part H: Neuroendocrine Peptides)
Edited by P. MICHAEL CONN

VOLUME 104. Enzyme Purification and Related Techniques (Part C)
Edited by WILLIAM B. JAKOBY

VOLUME 105. Oxygen Radicals in Biological Systems
Edited by LESTER PACKER

VOLUME 106. Posttranslational Modifications (Part A)
Edited by FINN WOLD AND KIVIE MOLDAVE

VOLUME 107. Posttranslational Modifications (Part B)
Edited by FINN WOLD AND KIVIE MOLDAVE

VOLUME 108. Immunochemical Techniques (Part G: Separation and Characterization of Lymphoid Cells)
Edited by GIOVANNI DI SABATO, JOHN J. LANGONE, AND HELEN VAN VUNAKIS

VOLUME 109. Hormone Action (Part I: Peptide Hormones)
Edited by LUTZ BIRNBAUMER AND BERT W. O'MALLEY

VOLUME 110. Steroids and Isoprenoids (Part A)
Edited by JOHN H. LAW AND HANS C. RILLING

VOLUME 111. Steroids and Isoprenoids (Part B)
Edited by JOHN H. LAW AND HANS C. RILLING

VOLUME 112. Drug and Enzyme Targeting (Part A)
Edited by KENNETH J. WIDDER AND RALPH GREEN

VOLUME 113. Glutamate, Glutamine, Glutathione, and Related Compounds
Edited by ALTON MEISTER

VOLUME 114. Diffraction Methods for Biological Macromolecules (Part A)
Edited by HAROLD W. WYCKOFF, C. H. W. HIRS, AND SERGE N. TIMASHEFF

VOLUME 115. Diffraction Methods for Biological Macromolecules (Part B)
Edited by HAROLD W. WYCKOFF, C. H. W. HIRS, AND SERGE N. TIMASHEFF

VOLUME 116. Immunochemical Techniques (Part H: Effectors and Mediators of Lymphoid Cell Functions)
Edited by GIOVANNI DI SABATO, JOHN J. LANGONE, AND HELEN VAN VUNAKIS

VOLUME 117. Enzyme Structure (Part J)
Edited by C. H. W. HIRS AND SERGE N. TIMASHEFF

VOLUME 118. Plant Molecular Biology
Edited by ARTHUR WEISSBACH AND HERBERT WEISSBACH

VOLUME 119. Interferons (Part C)
Edited by SIDNEY PESTKA

VOLUME 120. Cumulative Subject Index Volumes 81–94, 96–101

VOLUME 121. Immunochemical Techniques (Part I: Hybridoma Technology and Monoclonal Antibodies)
Edited by JOHN J. LANGONE AND HELEN VAN VUNAKIS

VOLUME 122. Vitamins and Coenzymes (Part G)
Edited by FRANK CHYTIL AND DONALD B. MCCORMICK

VOLUME 123. Vitamins and Coenzymes (Part H)
Edited by FRANK CHYTIL AND DONALD B. MCCORMICK

VOLUME 124. Hormone Action (Part J: Neuroendocrine Peptides)
Edited by P. MICHAEL CONN

VOLUME 125. Biomembranes (Part M: Transport in Bacteria, Mitochondria, and Chloroplasts: General Approaches and Transport Systems)
Edited by SIDNEY FLEISCHER AND BECCA FLEISCHER

VOLUME 126. Biomembranes (Part N: Transport in Bacteria, Mitochondria, and Chloroplasts: Protonmotive Force)
Edited by SIDNEY FLEISCHER AND BECCA FLEISCHER

VOLUME 127. Biomembranes (Part O: Protons and Water: Structure and Translocation)
Edited by LESTER PACKER

VOLUME 128. Plasma Lipoproteins (Part A: Preparation, Structure, and Molecular Biology)
Edited by JERE P. SEGREST AND JOHN J. ALBERS

VOLUME 129. Plasma Lipoproteins (Part B: Characterization, Cell Biology, and Metabolism)
Edited by JOHN J. ALBERS AND JERE P. SEGREST

VOLUME 130. Enzyme Structure (Part K)
Edited by C. H. W. HIRS AND SERGE N. TIMASHEFF

VOLUME 131. Enzyme Structure (Part L)
Edited by C. H. W. HIRS AND SERGE N. TIMASHEFF

VOLUME 132. Immunochemical Techniques (Part J: Phagocytosis and Cell-Mediated Cytotoxicity)
Edited by GIOVANNI DI SABATO AND JOHANNES EVERSE

VOLUME 133. Bioluminescence and Chemiluminescence (Part B)
Edited by MARLENE DELUCA AND WILLIAM D. MCELROY

VOLUME 134. Structural and Contractile Proteins (Part C: The Contractile Apparatus and the Cytoskeleton)
Edited by RICHARD B. VALLEE

VOLUME 135. Immobilized Enzymes and Cells (Part B)
Edited by KLAUS MOSBACH

VOLUME 136. Immobilized Enzymes and Cells (Part C)
Edited by KLAUS MOSBACH

VOLUME 137. Immobilized Enzymes and Cells (Part D)
Edited by KLAUS MOSBACH

VOLUME 138. Complex Carbohydrates (Part E)
Edited by VICTOR GINSBURG

VOLUME 139. Cellular Regulators (Part A: Calcium- and Calmodulin-Binding Proteins)
Edited by ANTHONY R. MEANS AND P. MICHAEL CONN

VOLUME 140. Cumulative Subject Index Volumes 102–119, 121–134

VOLUME 141. Cellular Regulators (Part B: Calcium and Lipids)
Edited by P. MICHAEL CONN AND ANTHONY R. MEANS

VOLUME 142. Metabolism of Aromatic Amino Acids and Amines
Edited by SEYMOUR KAUFMAN

VOLUME 143. Sulfur and Sulfur Amino Acids
Edited by WILLIAM B. JAKOBY AND OWEN W. GRIFFITH

VOLUME 144. Structural and Contractile Proteins (Part D: Extracellular Matrix)
Edited by LEON W. CUNNINGHAM

VOLUME 145. Structural and Contractile Proteins (Part E: Extracellular Matrix)
Edited by LEON W. CUNNINGHAM

VOLUME 146. Peptide Growth Factors (Part A)
Edited by DAVID BARNES AND DAVID A. SIRBASKU

VOLUME 147. Peptide Growth Factors (Part B)
Edited by DAVID BARNES AND DAVID A. SIRBASKU

VOLUME 148. Plant Cell Membranes
Edited by LESTER PACKER AND ROLAND DOUCE

VOLUME 149. Drug and Enzyme Targeting (Part B)
Edited by RALPH GREEN AND KENNETH J. WIDDER

VOLUME 150. Immunochemical Techniques (Part K: *In Vitro* Models of B and T Cell Functions and Lymphoid Cell Receptors)
Edited by GIOVANNI DI SABATO

VOLUME 151. Molecular Genetics of Mammalian Cells
Edited by MICHAEL M. GOTTESMAN

VOLUME 152. Guide to Molecular Cloning Techniques
Edited by SHELBY L. BERGER AND ALAN R. KIMMEL

VOLUME 153. Recombinant DNA (Part D)
Edited by RAY WU AND LAWRENCE GROSSMAN

VOLUME 154. Recombinant DNA (Part E)
Edited by RAY WU AND LAWRENCE GROSSMAN

VOLUME 155. Recombinant DNA (Part F)
Edited by RAY WU

VOLUME 156. Biomembranes (Part P: ATP-Driven Pumps and Related Transport: The Na,K-Pump)
Edited by SIDNEY FLEISCHER AND BECCA FLEISCHER

VOLUME 157. Biomembranes (Part Q: ATP-Driven Pumps and Related Transport: Calcium, Proton, and Potassium Pumps)
Edited by SIDNEY FLEISCHER AND BECCA FLEISCHER

VOLUME 158. Metalloproteins (Part A)
Edited by JAMES F. RIORDAN AND BERT L. VALLEE

VOLUME 159. Initiation and Termination of Cyclic Nucleotide Action
Edited by JACKIE D. CORBIN AND ROGER A. JOHNSON

VOLUME 160. Biomass (Part A: Cellulose and Hemicellulose)
Edited by WILLIS A. WOOD AND SCOTT T. KELLOGG

VOLUME 161. Biomass (Part B: Lignin, Pectin, and Chitin)
Edited by WILLIS A. WOOD AND SCOTT T. KELLOGG

VOLUME 162. Immunochemical Techniques (Part L: Chemotaxis and Inflammation)
Edited by GIOVANNI DI SABATO

VOLUME 163. Immunochemical Techniques (Part M: Chemotaxis and Inflammation)
Edited by GIOVANNI DI SABATO

VOLUME 164. Ribosomes
Edited by HARRY F. NOLLER, JR. AND KIVIE MOLDAVE

VOLUME 165. Microbial Toxins: Tools for Enzymology
Edited by SIDNEY HARSHMAN

VOLUME 166. Branched-Chain Amino Acids
Edited by ROBERT HARRIS AND JOHN R. SOKATCH

VOLUME 167. Cyanobacteria
Edited by LESTER PACKER AND ALEXANDER N. GLAZER

VOLUME 168. Hormone Action (Part K: Neuroendocrine Peptides)
Edited by P. MICHAEL CONN

VOLUME 169. Platelets: Receptors, Adhesion, Secretion (Part A)
Edited by JACEK HAWIGER

VOLUME 170. Nucleosomes
Edited by PAUL M. WASSARMAN AND ROGER D. KORNBERG

VOLUME 171. Biomembranes (Part R: Transport Theory: Cells and Model Membranes)
Edited by SIDNEY FLEISCHER AND BECCA FLEISCHER

VOLUME 172. Biomembranes (Part S: Transport Membrane Isolation and Characterization)
Edited by SIDNEY FLEISCHER AND BECCA FLEISCHER

VOLUME 173. Biomembranes [Part T: Cellular and Subcellular Transport: Eukaryotic (Nonepithelial) Cells]
Edited by SIDNEY FLEISCHER AND BECCA FLEISCHER

VOLUME 174. Biomembranes [Part U: Cellular and Subcellular Transport: Eukaryotic (Nonepithelial) Cells]
Edited by SIDNEY FLEISCHER AND BECCA FLEISCHER

VOLUME 175. Cumulative Subject Index Volumes 135–139, 141–167

VOLUME 176. Nuclear Magnetic Resonance (Part A: Spectral Techniques and Dynamics)
Edited by NORMAN J. OPPENHEIMER AND THOMAS L. JAMES

VOLUME 177. Nuclear Magnetic Resonance (Part B: Structure and Mechanism)
Edited by NORMAN J. OPPENHEIMER AND THOMAS L. JAMES

VOLUME 178. Antibodies, Antigens, and Molecular Mimicry
Edited by JOHN J. LANGONE

VOLUME 179. Complex Carbohydrates (Part F)
Edited by VICTOR GINSBURG

VOLUME 180. RNA Processing (Part A: General Methods)
Edited by JAMES E. DAHLBERG AND JOHN N. ABELSON

VOLUME 181. RNA Processing (Part B: Specific Methods)
Edited by JAMES E. DAHLBERG AND JOHN N. ABELSON

VOLUME 182. Guide to Protein Purification
Edited by MURRAY P. DEUTSCHER

VOLUME 183. Molecular Evolution: Computer Analysis of Protein and Nucleic Acid Sequences
Edited by RUSSELL F. DOOLITTLE

VOLUME 184. Avidin-Biotin Technology
Edited by MEIR WILCHEK AND EDWARD A. BAYER

VOLUME 185. Gene Expression Technology
Edited by DAVID V. GOEDDEL

VOLUME 186. Oxygen Radicals in Biological Systems (Part B: Oxygen Radicals and Antioxidants)
Edited by LESTER PACKER AND ALEXANDER N. GLAZER

VOLUME 187. Arachidonate Related Lipid Mediators
Edited by ROBERT C. MURPHY AND FRANK A. FITZPATRICK

VOLUME 188. Hydrocarbons and Methylotrophy
Edited by MARY E. LIDSTROM

VOLUME 189. Retinoids (Part A: Molecular and Metabolic Aspects)
Edited by LESTER PACKER

VOLUME 190. Retinoids (Part B: Cell Differentiation and Clinical Applications)
Edited by LESTER PACKER

VOLUME 191. Biomembranes (Part V: Cellular and Subcellular Transport: Epithelial Cells)
Edited by SIDNEY FLEISCHER AND BECCA FLEISCHER

VOLUME 192. Biomembranes (Part W: Cellular and Subcellular Transport: Epithelial Cells)
Edited by SIDNEY FLEISCHER AND BECCA FLEISCHER

VOLUME 193. Mass Spectrometry
Edited by JAMES A. MCCLOSKEY

VOLUME 194. Guide to Yeast Genetics and Molecular Biology
Edited by CHRISTINE GUTHRIE AND GERALD R. FINK

VOLUME 195. Adenylyl Cyclase, G Proteins, and Guanylyl Cyclase (in preparation)
Edited by ROGER A. JOHNSON AND JACKIE D. CORBIN

VOLUME 196. Molecular Motors and the Cytoskeleton (in preparation)
Edited by RICHARD B. VALLEE

VOLUME 197. Phospholipases (in preparation)
Edited by EDWARD A. DENNIS

VOLUME 198. Peptide Growth Factors (Part C) (in preparation)
Edited by DAVID BARNES, J. P. MATHER, AND GORDON H. SATO

VOLUME 170. Nucleosomes
Edited by PAUL M. WASSARMAN AND ROGER D. KORNBERG

VOLUME 171. Biomembranes (Part R: Transport Theory: Cells and Model Membranes)
Edited by SIDNEY FLEISCHER AND BECCA FLEISCHER

VOLUME 172. Biomembranes (Part S: Transport Membrane Isolation and Characterization)
Edited by SIDNEY FLEISCHER AND BECCA FLEISCHER

VOLUME 173. Biomembranes [Part T: Cellular and Subcellular Transport: Eukaryotic (Nonepithelial) Cells]
Edited by SIDNEY FLEISCHER AND BECCA FLEISCHER

VOLUME 174. Biomembranes [Part U: Cellular and Subcellular Transport: Eukaryotic (Nonepithelial) Cells]
Edited by SIDNEY FLEISCHER AND BECCA FLEISCHER

VOLUME 175. Cumulative Subject Index Volumes 135–139, 141–167

VOLUME 176. Nuclear Magnetic Resonance (Part A: Spectral Techniques and Dynamics)
Edited by NORMAN J. OPPENHEIMER AND THOMAS L. JAMES

VOLUME 177. Nuclear Magnetic Resonance (Part B: Structure and Mechanism)
Edited by NORMAN J. OPPENHEIMER AND THOMAS L. JAMES

VOLUME 178. Antibodies, Antigens, and Molecular Mimicry
Edited by JOHN J. LANGONE

VOLUME 179. Complex Carbohydrates (Part F)
Edited by VICTOR GINSBURG

VOLUME 180. RNA Processing (Part A: General Methods)
Edited by JAMES E. DAHLBERG AND JOHN N. ABELSON

VOLUME 181. RNA Processing (Part B: Specific Methods)
Edited by JAMES E. DAHLBERG AND JOHN N. ABELSON

VOLUME 157. Biomembranes (Part Q: ATP-Driven Pumps and Related Transport: Calcium, Proton, and Potassium Pumps)
Edited by SIDNEY FLEISCHER AND BECCA FLEISCHER

VOLUME 158. Metalloproteins (Part A)
Edited by JAMES F. RIORDAN AND BERT L. VALLEE

VOLUME 159. Initiation and Termination of Cyclic Nucleotide Action
Edited by JACKIE D. CORBIN AND ROGER A. JOHNSON

VOLUME 160. Biomass (Part A: Cellulose and Hemicellulose)
Edited by WILLIS A. WOOD AND SCOTT T. KELLOGG

VOLUME 161. Biomass (Part B: Lignin, Pectin, and Chitin)
Edited by WILLIS A. WOOD AND SCOTT T. KELLOGG

VOLUME 162. Immunochemical Techniques (Part L: Chemotaxis and Inflammation)
Edited by GIOVANNI DI SABATO

VOLUME 163. Immunochemical Techniques (Part M: Chemotaxis and Inflammation)
Edited by GIOVANNI DI SABATO

VOLUME 164. Ribosomes
Edited by HARRY F. NOLLER, JR. AND KIVIE MOLDAVE

VOLUME 165. Microbial Toxins: Tools for Enzymology
Edited by SIDNEY HARSHMAN

VOLUME 166. Branched-Chain Amino Acids
Edited by ROBERT HARRIS AND JOHN R. SOKATCH

VOLUME 167. Cyanobacteria
Edited by LESTER PACKER AND ALEXANDER N. GLAZER

VOLUME 168. Hormone Action (Part K: Neuroendocrine Peptides)
Edited by P. MICHAEL CONN

VOLUME 169. Platelets: Receptors, Adhesion, Secretion (Part A)
Edited by JACEK HAWIGER

Process Guide

This Guide is intended as an integrative supplement to the more extensive Subject Index. Specific topics for which detailed protocols and/or substantive discussions are provided are listed by chapter number.

A

Antibodies
  antigen production
    lacZ, 25
    λgt11, 15
    trpE, 33
  epitope tagging, 22, 35
  immunoblotting, colony, 44
  immunoprecipitation, 35, 45
  immunopurification, 40

B

Banks (Libraries)
  by mail, 14
  de novo construction
    human (YAC), 17
    yeast genomic, 14

C

Cloning, *see also* Vectors
  by complementation, 14
  by high-copy, cross-suppression, 14, 16
  by λgt11, 15
  mitochondrial genes, 10
  troubleshooting guide, 14

D

DNA
  assays
    diphenylamine, 6
    binding protein, 36, 39
  isolation
    human, 17
    yeast
      genomic, 11, 14, 21
      plasmid, 21
  labeling, *see* Labeling
  restriction patterns, 11
  sequencing, 10, 39
  size fractionation, 14, 17
  synthesis, 6
  types, 1, 11

E

Endocytosis, 48

G

Genetics, *see also* Mapping, Mutagenesis
  ascus dissection, 2
  chromosome segregation, 54
  complementation, 1, 55
  cytoduction, 10, 55
  mating, 1, 5
  mating-type switching, 8, 55
  micromanipulators, 2
  nomenclature, 1
  recombination, 2
  sporulation, 2, 6
  strain storage, 1
Golgi membranes, *see* Membranes
Gene expression, control of
  high-copy vectors, 14, 16, 25
  promoter fusions
    constitutive
      ADH, PGK, 25
      GPD, 26
    regulatable
      glucocorticoid, 26
      heat-shock, 49
      GAL, 22, 25
      PHO5, 25
  uses of overexpression, 16

H

Heat shock, 49

L

Labeling
  DNA, meiotic, 6
  protein
    [^{35}S]methionine, 35, 43, 45, 49
    mitochondrial, 10
    ortho[^{32}P]phosphate, 30, 35
    [^{35}S]sulfate, 45, 48
  RNA
    [methyl-^{3}H]methionine, 30
    ortho[^{32}P]phosphate, 28, 30
    run-on transcription, 30
    [^{14}C]- or [^{3}H]uracil, 30
Libraries, *see* Banks

M

Mapping
  genetic, 3
  2 μm (micron), 3
  physical
    chromosome fragmentation, 3, 4, 8
    chromosome separation and blotting, 3, 4
    targeted integration, 18
Media
  bacterial, 13, 15, 22
  complex
    fermentable (YPD, YEPD), 1, 9, 10
    nonfermentable (YPG, YPDG, YPEG, YPDGE), 1, 10, 43
  counterselection
    against *URA3,LYS2,CAN1,CYH2,* 20
    chlorolactate, 24
    inositol, 18
  galactose indicator, 1, 23
  low phosphate, 30, 35
  low sulfate, 10, 35, 43, 45, 48
  maltose indicator, 1
  minimal (SD), 1, 10, 48
  slants (YPAD), 1
  sporulation, 1, 6, 9
  synthetic complete (SC), 1, 23, 30
Membranes
  fractionation, 45
  Golgi, 44
  markers, 44, 45
Microscopy
  digital imaging (DIM), 54
  electron (EM), thin section, 41, 42
  immunofluorescence, 40
  immunogold, 42
  light, 2
Mitochondria
  assays, functional, 43
  DNA isolation, 10
  isolation, 43
  labeling of protein, 10, 43
  mutants, 10
  protein import *in vitro,* 43
  RNA isolation, 10
  transformation, 10
  translocation, 43
Mutagenesis
  ethidium bromide, 10
  gene disruption, 19
  gene replacement,
    integrative, 19
    plasmid shuffle, 20, 25
  oligonucleotide, 24
  recovery of mutants
    counterselection, 20, 23, 24
    gap repair, 19
    inositol enrichment, 18
    plasmid isolation, 21
    secretion mutants, 34
  transposon
    bacterial, 19, 22
    yeast Ty, 23
  in vivo (EMS, NG, UV), 18
  in vitro, hydroxylamine, 20

N

Nuclei
  isolation, 28, 38, 53
  staining
    DAPI, 54, 55
    Giemsa, 6

P

Pheromones
  assays, 5, 48
  α factor, ^{35}S-labeled, 48
Proteases
  inhibitors, 31, 38, 53
  mutants, 31
  types, 31
Proteins, *see also* Translation, Secretion
  extraction, 43, 47, 53

fusion, 33, 34, 40
immunoprecipitation, *see* Antibodies
labeling, *see* Labeling
ribosomal, 32

R

RNA
  analysis
    Northern blotting, 28
    size fractionation, 28
  isolation
    guanidine thiocyanate, 28
    nuclear, 28
    phenol/SDS, 27
    poly(A)$^+$, 27
    spore, 27
  labeling, *see* Labeling
  turnover, 29

S

Secretion
  fusion proteins, 34
  glycosylation, 47
  mutants, 45
  secretory, vesicle, isolation, 45
Spheroplasts, generation, 40, 43, 44, 45, 46, 53
Spores
  generation, 2, 6, 7, 9
  genetic mapping, 2
  germination, 2
  staining, 7

T

Transcription
  run-on, 30
  in vitro
    yeast pol II, 38
    SP6, T7, 36, 46
  in vivo, inhibition, 29
Transformation
  bacteria, 14, 21
  yeast
    bacterial conjugation, 13
    bombardment, 10
    electroporation, 12
    lithium acetate, 12
    spheroplasting, 12, 17
Translation
  yeast *in vitro* system, 37
  heterologous systems, 36
Translocation, *see also* Secretion, Membranes
  microsomal, 46
  mitochondrial, 43
Transposition, *see* Mutagenesis, transposon

V

Vacuoles, isolation, 44
Vectors, *see also* Gene expression
  pATH-trpE, 33
  pUC, 19
  secretion, 34
  shuffle, 20
  shuttle mutagenesis, 22
  specialized, 14, 16
  Ty, 23
  YAC, 17, 54
  YCp, 14,25
  YEp, 14, 16, 25
  YIp, 19

Z

Zygotes, isolation, 1, 5

Section I
Basic Methods of Yeast Genetics

[1] Getting Started with Yeast

By FRED SHERMAN

Virtues of Yeast

The yeast *Saccharomyces cerevisiae* is recognized as an ideal eukaryotic microorganism for biological studies. Although yeasts have greater genetic complexity than bacteria, they share many of the technical advantages that permitted rapid progress in the molecular genetics of prokaryotes and their viruses. Some of the properties that make yeast particularly suitable for biological studies include rapid growth, a budding pattern resulting in dispersed cells, the ease of replica plating and mutant isolation, a well-defined genetic system, and, most importantly, a highly versatile DNA transformation system. Being nonpathogenic, yeast can be handled with few precautions. Large quantities of normal bakers' yeast are commercially available and can provide an inexpensive source for biochemical studies.

Strains of *Saccharomyces cerevisiae*, unlike most other microorganisms, have both a stable haploid and diploid state, and they are viable with a great many markers. The development of DNA transformation has made yeast particularly accessible to gene cloning and genetic engineering techniques. Structural genes corresponding to virtually any genetic trait can be identified by complementation from plasmid libraries. Plasmids can be introduced into yeast cells either as replicating molecules or by integration into the genome. Integrative recombination of transforming DNA in yeast proceeds exclusively via homologous recombinations, in contrast to most other organisms. Cloned yeast sequences, accompanied with foreign sequences on plasmids, can therefore be directed at will to specific locations in the genome.

In addition, homologous recombination coupled with the high levels of gene conversion in yeasts has led to the development of techniques for the direct replacement of genetically engineered DNA sequences into normal chromosome locations. Thus, normal wild-type genes, even those having no previously known mutations, can be conveniently replaced with altered and disrupted alleles. The phenotypes arising after disruption of yeast genes have contributed significantly toward understanding of the function of certain proteins *in vivo*. Many investigators have been shocked to find viable mutants with few or no detrimental phenotypes after disrupting "essential" genes. Genes can be directly replaced at high efficiencies in yeasts and other fungi, but not in any other eukaryotic organisms. Also unique to yeast, transformation can be carried out directly with synthetic oligonucleotides, permitting the convenient production of numerous altered forms of proteins. These techniques have been extensively exploited in the analysis of gene regulation, structure–function relationships of proteins, chromosome structure, and other general questions in cell biology.

The overriding virtues of yeast are illustrated by the fact that mammalian genes are routinely being introduced into yeast for systematic analyses of the functions of the corresponding gene products. Furthermore, the replication of artificial circular and linear chromosomes has allowed detailed studies of telomeres, centromeres, length dependencies, and origins of replication. Mitochondrial DNA can now be altered in defined ways by transformation, adding to the already impressive array of genetic and biochemical techniques that have allowed detailed analysis of this organelle. The ease with which the genome of yeast can be manipulated is truly unprecedented for any other eukaryote. Many of these techniques are reviewed in this volume.

Information on Yeast

A general introduction to a few selected topics on yeast can be found in other sources (see, e.g., Ref. 1). An article by Rine and Carlson[2] outlines methods of yeast genetics with an emphasis on the use of recombinant DNA procedures. The extended survey of metabolism, cellular regulation, and cell division in *S. cerevisiae* by Hanes et al.[3] contains 450 citations. Bennett et al.[4] recently reviewed the cell cycle, chromosome structure and function, and related topics concerning *S. cerevisiae*. Comprehensive reviews of the genetics and molecular biology of *S. cerevisiae* are contained in two volumes entitled "Molecular Biology of the Yeast *Saccharomyces*,"[5] which are being revised and expanded.[6] Overviews of numerous subjects

[1] J. D. Watson, N. H. Hopkins, J. W. Roberts, J. A. Steitz, and A. M. Weiner, "Molecular Biology of the Gene," Chaps. 18 and 19. Benjamin/Cummings, Menlo Park, California, 1987.
[2] J. Rine and M. Carlson, *in* "Gene Manipulation in Fungi" (J. W. Bennett and L. L. Lasure, eds.), p. 125. Academic Press, New York, 1985.
[3] S. D. Hanes, R. Koren, and K. A. Bostian, *Crit. Rev. Biochem.* **21**, 153 (1986).
[4] C. Bennett, E. Perkins, and M. A. Resnick, *in* "Microbial Genetics" (U. Streipes and R. Yasbin, eds.). Alan R. Liss, New York, 1990.
[5] J. N. Strathern, E. W. Jones, and J. R. Broach (eds.), "Molecular Biology of the Yeast *Saccharomyces*," I: Life Cycle and Inheritance; II: Metabolism and Gene Expression. Cold Spring Harbor Laboratory, Cold Spring Harbor, New York, 1981.
[6] J. R. Broach, E. W. Jones, and J. R. Pringle (eds.), "The Molecular Biology of the Yeast *Saccharomyces*," Vols. 1 and 2. Cold Spring Harbor Laboratory, Cold Spring Harbor, New York, 1990.

are also covered in the volumes edited by Rose and Harrison,[7,8] Prescott,[9] Spencer et al.,[10] Hicks,[11] and Barr et al.,[12] as well as recent review articles.[13-18] The text "Fungal Genetics"[19] covered general principles, especially regarding classical genetics of fungi, and contains a useful glossary. Many techniques useful in yeast studies have been covered in this series and elsewhere.[20,21]

Strains of Saccharomyces cerevisiae

Although genetic analyses have been undertaken with a number of taxonomically distinct varieties of yeast, extensive studies have been restricted primarily to the many freely interbreeding species of the budding yeast *Saccharomyces* and to the fission yeast *Schizosaccharomyces pombe*. Although the Latin binomial *Saccharomyces cerevisiae* is commonly used to designate many of the laboratory stocks of *Saccharomyces* used throughout the world, it should be pointed out that most of these strains originated from the interbred stocks of Winge, Lindegren, and others who employed fermentation markers not only from *S. cerevisiae* but also from *S. bayanus, S. carlsbergensis, S. chevalieri, S. chodati, S. diastaticus,*

---

[7] A. H. Rose and J. S. Harrison (eds.), "The Yeasts," Volume I: Biology of the Yeasts, 1969; Volume 2: Physiology and Biochemistry of Yeasts, 1971; Volume 3: Yeast Technology, 1970. Academic Press, New York.

[8] A. H. Rose and J. S. Harrison (eds.), "The Yeasts," 2nd Ed., Volume 1: Biology of Yeasts, 1987; Volume 2: Yeast and the Environment, 1988. Academic Press, New York.

[9] D. Prescott (ed.), "Methods in Cell Biology," Vol. 11, 1975; Vol. 12, 1976. Academic Press, New York.

[10] J. F. T. Spencer, D. M. Spencer, and A. R. W. Smith (eds.), "Yeast Genetics, Fundamental and Applied Aspects." Springer-Verlag, New York, 1983.

[11] J. Hicks (ed.), "Yeast Cell Biology." Alan R. Liss, New York, 1986.

[12] P. J. Barr, A. J. Brake, and P. Valenzuela (eds.), "Yeast Genetics Engineering." Butterworth, Boston, 1989.

[13] L. Guarente, *Annu. Rev. Genet.* **21,** 425 (1987).

[14] M. Johnston, *Microbiol. Rev.* **51,** 458 (1987).

[15] I. Herskowitz, *Microbiol. Rev.* **52,** 536 (1988).

[16] E. C. Friedberg, *Microbiol. Rev.* **52,** 70 (1988).

[17] F. Cross, L. H. Hartwell, C. Jackson, and J. B. Konopka, *Annu. Rev. Cell. Biol.* **4,** 429 (1988).

[18] K. Struhl, *Annu. Rev. Biochem.* **58,** 1051 (1989).

[19] J. R. S. Fincham, P. R. Day, and A. Radford, "Fungal Genetics." Univ. of California Press, Berkeley and Los Angeles, 1979.

[20] R. Wu, L. Grossman, and K. Moldave (eds.), this series, Vol. 101.

[21] I. Campbell and J. H. Duffus (eds.), "Yeast, a Practical Approach." IRL Press, Oxford, 1988.

etc.[22,23] Nevertheless, it is still recommended that the interbreeding laboratory stocks of *Saccharomyces* be denoted as *S. cerevisiae,* in order to distinguish them from the more distantly related species of *Saccharomyces.*

Care should be taken in choosing strains for genetic and biochemical studies. Unfortunately there are no truly wild-type *Saccharomyces* strains that are commonly employed in genetic studies. Also, most domesticated strains of brewers' yeast, probably many strains of bakers' yeast, and true wild-type strains of *S. cerevisiae* are not genetically compatible with laboratory stocks. It is usually not appreciated that many "normal" laboratory strains contain mutant characters, a fact not too surprising since they were derived from pedigrees involving mutagenized strains. The haploid strain S288C is often used as a normal standard because it gives rise to well-dispersed cells and because many isogenic mutant derivatives are available.[23] However, S288C contains an abnormally low amount of cytochrome c and assimilates several carbon sources at reduced rates. In contrast to true wild-type and domesticated bakers' yeast, which give rise to less than 2% ρ^- colonies (see below), many laboratory strains produce high frequencies of ρ^- mutants. Another strain, D273-10B, has been extensively used as a typical normal yeast, especially for mitochondrial studies. One should examine the specific characters of interest before initiating a study with any strain.

Many strains containing characterized auxotrophic, temperature-sensitive, and other markers can be obtained from the Yeast Genetics Stock Center [Donner Laboratory, University of California, Berkeley, CA 94720; (415)-642-0815]. Other sources of yeast strains include the American Type Culture Collection (12301 Parklawn Drive, Rockville, MD 20852), the National Collection of Yeast Cultures (Food Research Institute, Colney Lane, Norwich NR4 7UA, UK), the Centraalbureau voor Schimmelcultures (Yeast Division, Julianalaan 67a, 2628 BC Delft, Netherlands), and the Czechoslovak Collection of Yeasts (Institute of Chemistry, Slovak Academy of Sciences, Dubravaska cesta, 809 33 Bratislava, Czechoslovakia). Before using strains obtained from these sources or from any investigator, it is advisable to test the strains and verify their genotypes.

Genome of *Saccharomyces cerevisiae*

Saccharomyces cerevisiae contains a haploid set of 16 chromosomes that have been well characterized genetically[24] and physically.[25–27] A single

[22] C. C. Lindegren, "The Yeast Cell: Its Genetics and Cytology." Educational Publ., St. Louis, Missouri, 1949.
[23] R. K. Mortimer and J. R. Johnson, *Genetics* **113,** 35 (1986).
[24] R. K. Mortimer, D. Schild, C. R. Contopoloulou, and J. A. Kans, *Yeast* **5,** 321 (1989).

marker, *KRB1*, has been assigned to chromosome XVII,[28] although there is no physical evidence for this chromosome. The 16 chromosomes range in size from 200 to 2200 kilobases (kb), with a total length of 14,000 kb. A total of 769 genes have been mapped,[24] constituting approximately one-half of the reported mutations. The chromosomal genome is densely packed with an estimated 6500 genes having an average size of 2 kb and few introns. Ribosomal RNA is coded by approximately 120 copies of a single tandem array on chromosome XII. In addition, chromosomes contain movable DNA elements, retrotransposons, that vary in number and position in different strains of *S. cerevisiae*, with most laboratory strains having approximately 30.

Other nucleic acid entities, listed in Table I,[29-32] can also be considered part of the yeast genome. Mitochondrial DNA encodes components of the mitochondrial translational machinery and approximately 15% of the mitochondrial proteins. ρ^0 mutants completely lack mitochondrial DNA and are deficient in the respiratory polypeptides synthesized on mitochondrial ribosomes, namely, cytochrome *b* and subunits of cytochrome oxidase and ATPase complexes. Even though ρ^0 mutants are respiratory deficient, they are viable and still retain mitochondria, although morphologically abnormal ones.

The 2-μm circle plasmids, present in most strains of *S. cerevisiae*, apparently function solely for their own replication. Generally *cir*^{0} strains, which lack 2-μm DNA, have no observable phenotype; however, a certain chromosomal mutation, *nib1*, causes a reduction in division potential of *cir*$^+$ strains.[33,34]

Similarly, almost all *S. cerevisiae* strains contain double-stranded (ds) RNA viruses that are not usually extracellularly infective but are transmitted by mating. This dsRNA, constituting approximately 0.1% of total nucleic acid, determines a toxin and components required for the viral transcription and replication. *KIL-o* mutants, lacking class M dsRNA

---

[25] S. L. Gerring, C. Connelly, and P. Hieter, this volume [4].
[26] F. Sherman and L. P. Wakem, this volume [3].
[27] P. Philippsen, A. Stotz, and C. Scherf, this volume [11].
[28] R. B. Wickner, F. Boutelet, and F. Hilger, *Mol. Cell. Biol.* **3**, 415 (1983).
[29] B. Dujon, *in* "Molecular Biology of the Yeast *Saccharomyces* Life Cycle and Inheritance" (J. N. Strathern, E. W. Jones, and J. R. Broach, eds.), p. 505. Cold Spring Harbor Laboratory, Cold Spring Harbor, New York, 1981.
[30] J. R. Broach, this series, Vol. 101, p. 307.
[31] G. F. Carle and M. Olson, *Proc. Natl. Acad. Sci. U.S.A.* **82**, 3756 (1985).
[32] R. B. Wickner, *FASEB J.* **3**, 2257 (1989).
[33] C. Holm, *Cell (Cambridge, Mass.)* **29**, 585 (1982).
[34] R. Sweeny and V. A. Zakian, *Genetics* **122**, 749 (1989).

(Table I) and consequently the killer toxin, are readily induced by chemical and physical agents.

Only mutations of chromosomal genes exhibited Mendelian 2:2 segregation in tetrads after sporulation of heterozygous diploids[19]; this property is dependent on the disjunction of chromosomal centromeres. In contrast, non-Mendelian inheritance is observed for the phenotypes associated with the absence or alteration of other nucleic acids described in Table I.

Genetic Nomenclature

Chromosomal Genes

The following genetic nomenclature for the yeast *S. cerevisiae* is now universally accepted. Gene symbols are consistent with the proposals of Demerec *et al.*,[35] whenever possible, and are designated by three italicized letters (e.g., *arg*). Contrary to the proposals of Demerec *et al.*[35] for *Escherichia coli,* yeast genetic loci are identified by numbers, not letters, following the gene symbols (e.g., *arg2*). Dominant alleles are denoted by using uppercase italics for all letters of the gene symbol (e.g., *ARG2*). Lowercase letters denote the recessive allele (e.g., the auxotroph *arg2*). Wild-type genes are designated with the superscript plus symbol ($sup6^+$ or $ARG2^+$). Alleles are designated by a number separated from the locus number by a hyphen (e.g., *arg2-14*). Locus numbers should be consistent with the original assignments; however, allele numbers may be specific to a particular laboratory. For example, two different isolates from two different laboratories could be denoted *can1-1,* where both are mutations of the *CAN1* locus. The symbol Δ can denote complete or partial deletions (e.g., *his3-Δ1*). Insertion of genes follows the bacterial nomenclature by using the symbol ::. For example, *cyc1::URA3* denotes the insertion of the *URA3* gene at the *CYC1* locus, in which *URA3* is dominant (and functional) and *cyc1* is recessive (and defective).

Phenotypes are sometimes denoted by cognate symbols in roman type and by the superscripts + and −. For example, the independence and requirement for arginine can be denoted by Arg^+ and Arg^-, respectively.

The following examples illustrate the conventions used in the genetic nomenclature for *S. cerevisiae:*

ARG2	Locus or dominant allele
arg2	Locus or recessive allele that confers a requirement for arginine
ARG2$^+$	Wild-type allele

[35] M. Demerec, E. A. Adelberg, A. J. Clark, and P. E. Hartman, *Genetics* **54**, 61 (1966).

TABLE I
GENOME OF DIPLOID *Saccharomyces cerevisiae* CELL[a]

Characteristic	Mendelian		Non-Mendelian					
	Nuclear Double-stranded DNA			Double-stranded RNA RNA molecules				
	Chromosomes	2-μm plasmid	Mitochondrial	L-A	M	L-BC	T	W
Relative amount (%)	85	5	10	80	10	9	0.5	0.5
Number of copies	2 sets of 16	60–100	~50 (8–130)	1000	170	150	10	10
Size (kb)	14,000 (200–2200)	6.318	70–76	4.576	1.8	4.6	2.7	2.25
Deficiencies in mutants	All kinds	None	Cytochromes $a \cdot a_3, b$	Killer toxin			None	

[a] Adapted from Refs. 29–32.

arg2-9   Specific allele or mutation
Arg$^+$ Strain not requiring arginine
Arg$^-$ Strain requiring arginine

Most alleles can be unambiguously assigned as dominant or recessive by examining the phenotype of the heterozygous diploid crosses. However, dominant and recessive traits are defined only with pairs, and a single allele can be both dominant and recessive. For example, because the alleles *CYC1*$^+$, *cyc1-717,* and *cyc1-Δ1* produce, respectively, 100, 5, and 0% of the gene product, the *cyc1-717* allele can be considered recessive in the *cyc1-717/CYC1*$^+$ cross and dominant in the *CYC1-717/cyc1-Δ1* cross. Thus, sometimes it is less confusing to denote all mutant alleles in lowercase letters, especially when considering a series of mutations having a range of activities.

There are a number of exceptions to these general rules. Gene clusters, complemention groups within a gene, or domains within a gene having a different characteristics can be designated by capital letters following the locus number; for example, *HIS4A, HIS4B, and HIS4C* denote three regions of the *HIS4* locus that correspond to three domains of the single polypeptide chain, which in turn determines three different enzymatic steps.

Although superscript letters should be avoided, it is sometimes expedient to distinguish genes conferring resistance and sensitivity by the superscripts R and S, respectively. For example, the genes controlling resistance to canavanine sulfate *(can1)* and copper sulfate *(CUP1)* and their sensitive alleles could be denoted, respectively, as *can*^{R}*1*, *CUP*^{R}*1*, CAN^S*1*, and *cup*^{S}*1*.

Wild-type and mutant alleles of the mating-type locus and related loci do not follow the standard rules. The two wild-type alleles of the mating-type locus are designated *MAT*a and *MAT*α. The two complementation groups of the *MAT*α locus are denoted *MAT*α*1* and *MAT*α*2*. Mutations of the *MAT* genes are denoted, for example, *mat*a*-1* and *mat*α*1-1*. The wild-type homothallic alleles at the *HMR* and *HML* loci are denoted *HMR*a, *HMR*α, *HML*a, and *HML*α.[36] Mutations at these loci are denoted, for example, *hmr*a*-1*, *hml*α*-1*. The mating phenotypes of *MAT*a and *MAT*α cells are denoted simply a and α, respectively.

Dominant and recessive suppressors should be denoted, respectively, by three uppercase or three lowercase letters, followed by a locus designation (e.g., *SUP4, SUF1, sup35,* and *suf11*). In some instances UAA ochre suppressors and UAG amber suppressors are further designated, respectively, o and a following the locus. For example, *SUP4*-o refers to suppres-

---

[36] I. Herskowitz and R. E. Jensen, this volume [8].

sors of the *SUP4* locus that insert tyrosine residues at UAA sites; *SUP4*-**a** refers to suppressors of the same *SUP4* locus that insert tyrosine residues at UAG sites. The corresponding wild-type locus coding for the normal tyrosine tRNA and lacking suppressor activity can be referred to as $sup4^+$. Intragenic mutations that inactivate suppressor can denoted, for example, $sup4^-$ or *sup4*-o-*1*. The nomenclature describing suppressor and wild-type alleles in yeast is unrelated to the bacterial nomenclature. For example, an ochre *E. coli* suppressor that inserts tyrosine residues at both UAA and UAG sites is denoted as su_4^+, and the wild-type locus coding for the normal tyrosine tRNA and lacking suppressor activity can be referred to as Su_4, su_4^-, or *supC*. Frameshift suppressors are denoted as *suf* (or *SUF*), whereas metabolic suppressors are denoted with a variety of specialized symbols, such as *ssn* (suppressor of *snf1*), *srn* (suppressor of *rna1-1*), and *suh* (suppressor of *his2-1*).

Because the sites of mutations are usually used for genetic mapping, published chromosome maps usually contain the mutant allele. For example, chromosome III contains *his4* and *leu2*, whereas chromosome IX contains *SUP22* and *FLD1*. However, multiple dominant wild-type genes that control the same character (e.g., *SUC1* and *SUC2*), are used in genetic mapping, and such chromosomal loci are denoted in capital letters on genetic maps. Capital letters are also used to designate certain DNA segments whose locations have been determined by a combination of recombinant DNA techniques and classic mapping procedures, for example, *RDN1*, the segment encoding ribosomal RNA. *Escherichia coli* genes inserted into yeast are usually denoted by the prokaryotic nomenclature (e.g., *lacZ*). In order to prevent duplications, new gene symbols should be approved by the committee for genetic nomenclature, headed by Dr. R. K. Mortimer [Department of Molecular and Cellular Biology, Division of Biophysics and Cell Physiology, University of California, Berkeley, CA 94720; (414)-643-8877].

Non-Mendelian Determinants

Where necessary, non-Mendelian genotypes can be distinguished from chromosomal genotypes by enclosure in square brackets (e.g., *[KIL-o] MATα trp1-1*). Although it is advisable to employ the above rules for designating non-Mendelian genes and to avoid using Greek letters, the use of well-known and generally accepted Greek symbols should be continued; thus, the original symbols ρ^+, ρ^-, ψ^+, and ψ^- or their transliteration, rho^+, rho^-, psi^+, and psi^-, respectively, should be retained. Detailed designations for mitochondrial mutants[29,37] and killer strains[38] have been presented.

TABLE II
SOME NON-MENDELIAN DETERMINANTS OF YEAST[a]

Wild type	Mutant or polymorphic variant	Genetic element	Mutant phenotype
ρ^+	ρ^-	Mitochondrial DNA	Deficiency of cytochromes $a \cdot a_3$, b, and respiration
KIL-k_1	KIL-o	RNA plasmid	Sensitive to killer toxin
cir^+	cir^o	2-μm circle plasmid	None
ψ^+	ψ^-	Unknown	Decreased efficiency of certain suppression
URE3	$ure3^-$	Unknown	Deficiency in ureidosuccinate utilization

[a] Adapted from Refs. 29, 30, 32, 39, and 40. Other non-Mendelian determinants have been reported.[34,41]

Some of the known non-Mendelian determinants in yeast are listed in Table II.[29,30,32,39,40]

Growth of Yeast

For experimental purposes, yeast are usually grown at 30° on the complete medium, YPD (Table III), or on synthetic media, minimal (SD) or complete (SC) (Tables IV[42] and V). For industrial or certain special purposes when large amounts of high titers are desirable, yeast can be grown in less expensive media with high aeration and pH control.[43] The ingredients of standard laboratory media are presented in Tables III–VI. Synthetic media[42] are conveniently prepared with Bacto-yeast nitrogen base without amino acids (Difco Laboratories, Detroit, MI), containing the constituents presented in Table IV. Nutritional requirements of mutants are supplied with the nutrients listed in Table V. Growth on nonfermentable carbon sources can be tested on YPG medium (Table III), and

---

[37] L. A. Grivell, in "Genetic Maps" (S. J. O'Brien, ed.), p. 234. Cold Spring Harbor Laboratory, Cold Spring Harbor, New York, 1984.
[38] R. B. Wickner, in "Molecular Biology of the Yeast *Saccharomyces* Life Cycle and Inheritance" (J. N. Strathern, E. W. Jones, and J. R. Broach, eds.), p. 415. Cold Spring Harbor Laboratory, Cold Spring Harbor, New York, 1981.
[39] M. Aigle and F. Lacroute, *Mol. Gen. Genet.* **136**, 327 (1975).
[40] M. F. Tuite, P. M. Lund, A. B. Futcher, M. J. Dobson, B. S. Cox, and C. S. McLaughlin, *Plasmid* **8**, 103 (1982).
[41] S. W. Liebman and J. A. All-Robyn, *Curr. Genet.* **8**, 567 (1984).
[42] L. J. Wickersham, *U.S. Dept. Agric. Tech. Bull.*, No. 1029 (1951).
[43] J. White, "Yeast Technology." Wiley, New York, 1954.

TABLE III
COMPLEX MEDIA

Medium	Component	Composition
YPD (for routine growth)	1% Bacto-yeast extract	10 g
	2% Bacto-peptone	20 g
	2% Dextrose	20 g
	2% Bacto-agar	20 g
	Distilled water	1000 ml
YPG [containing nonfermentable carbon source (glycerol) that does not support growth of ρ^- or *pet* mutants]	1% Bacto-yeast extract	10 g
	2% Bacto-peptone	20 g
	3% (v/v) Glycerol	30 ml
	2% Bacto-agar	20 g
	Distilled water	970 ml
YPDG (used to determine proportion of ρ^- cells; ρ^+ and ρ^- colonies appear, respectively, large and small on this medium)	1% Bacto-yeast extract	10 g
	2% Bacto-peptone	20 g
	3% (v/v) Glycerol	30 ml
	0.1% Dextrose	1 g
	2% Bacto-agar	20 g
	Distilled water	970 ml
YPAD (for preparation of slants; adenine is added to inhibit reversion of *ade1* and *ade2* mutations)[a]	1% Bacto-yeast extract	10 g
	2% Bacto-peptone	20 g
	2% Dextrose	20 g
	0.003% Adenine sulfate	40 mg
	Distilled water	1000 ml
	2% Bacto-agar	20 g

[a] Medium is dissolved in a boiling water bath, and 1.5-ml portions are dispensed with an automatic pipettor into 1-dram (3-ml) vials. The caps are screwed on loosely, and the vials are autoclaved. After autoclaving, the rack is inclined so that the agar is just below the neck of the vial. The caps are tightened after 1 or 2 days.

fermentation markers can be determined with indicator media (Table VI) on which acid production induces color changes.

Strains of *S. cerevisiae* can be sporulated at 30° on the media listed in Table VII; most strains will readily sporulate on the surface of sporulation medium after replica plating fresh cultures from a YPD plate.[44] (Also see Refs. 45 and 46.)

Media for petri plates are prepared in 2-liter flasks, with each flask containing no more than 1 liter of medium, which is sufficient for approximately 40 standard plates. Unless stated otherwise, all components are

[44] R. R. Fowell, *Nature (London)* **170**, 578 (1952).
[45] B. Rockmill, E. J. Lambie, and G. S. Roeder, this volume [9].
[46] R. E. Esposito, M. Dresser, and M. Breitenbach, this volume [7].

TABLE IV
SYNTHETIC MINIMAL (SD) MEDIA[a]

Component	Composition
0.67% Bacto-yeast nitrogen base (without amino acids)	6.7 g
2% Dextrose	20 g
2% Bacto-agar	20 g
Distilled water	1000 ml
	Amount per liter
Carbon source	
Dextrose	20 g
Nitrogen source	
Ammonium sulfate	5 g
Vitamins	
Biotin	20 μg
Calcium pantothenate	2 mg
Folic acid	2 μg
Inositol	10 mg
Niacin	400 μg
p-Aminobenzoic acid	200 μg
Pyridoxine hydrochloride	400 μg
Riboflavin	200 μg
Thiamin hydrochloride	400 μg
Compounds supplying trace elements	
Boric acid	500 μg
Copper sulfate	40 μg
Potassium iodide	100 μg
Ferric chloride	200 μg
Manganese sulfate	400 μg
Sodium molybdate	200 μg
Zinc sulfate	400 μg
Salts	
Potassium phosphate monobasic	850 mg
Potassium phosphate dibasic	150 mg
Magnesium sulfate	500 mg
Sodium chloride	100 mg
Calcium chloride	100 mg

[a] This synthetic medium is based on media described by Wickersham[42] and is marketed, without dextrose, by Difco Laboratories (Detroit, MI) as "Yeast nitrogen base without amino acids."

autoclaved together for 15 min at 250°F (i.e., 120°C) and 15 pounds (i.e., 1 atm) pressure. The plates should be allowed to dry at room temperature for 2–3 days after pouring. The plates can be stored in sealed plastic bags

TABLE V
SYNTHETIC COMPLETE (SC) MEDIA[a]

Constituent	Final concentration (mg/liter)	Stock per 1000 ml	Amount of stock (ml) for 1 liter
Adenine sulfate	20	200 mg[b]	10
Uracil	20	200 mg[b]	10
L-Tryptophan	20	1 g	2
L-Histidine-HCl	20	1 g	2
L-Arginine-HCl	20	1 g	2
L-Methionine	20	1 g	2
L-Tyrosine	30	200 mg	15
L-Leucine	30	1 g	3
L-Isoleucine	30	1 g	3
L-Lysine-HCl	30	1 g	3
L-Phenylalanine	50	1 g[b]	5
L-Glutamic acid	100	1 g[b]	10
L-Aspartic acid	100	1 g[b,c]	10
L-Valine	150	3 g	5
L-Threonine	200	4 g[b,c]	5
L-Serine	400	8 g	5

[a] Synthetic complete medium contains synthetic minimal medium (SD) with various additions. It is convenient to prepare sterile stock solutions which can be stored for extensive periods. All stock solutions can be autoclaved for 15 min at 250°F. The appropriate volume of stock solutions is added to the ingredients of SD medium, and sufficient distilled water is added so that the total volume is 1 liter. The threonine and aspartic acid solutions should be added separately after autoclaving. Concentrations of the stock solutions (amount per 100 ml) are given. Some stock solutions should be stored at room temperature in order to prevent precipitation, whereas the other solutions may be refrigerated. It is best to use HCl salts of amino acids.
[b] Store at room temperature.
[c] Add after autoclaving the media.

for over 3 months at room temperature. The agar is omitted for liquid media.

Different types of synthetic media, especially omission media, can be prepared by mixing and grinding dry components in a ball mill. Small batches of liquid cultures can be grown in shake flasks using standard bacteriological techniques with high aeration. High aeration can be achieved by vigorously shaking cultures having liquid volumes less than 20% of the flask volume.

"Normal" laboratory haploid strains have a doubling time of 90 min in YPD medium and approximately 140 min in synthetic media during the

TABLE VI
INDICATOR MEDIA

Indicator medium	Component	Composition
MAL[a]	1% Bacto-yeast extract	10 g
	2% Bacto-peptone	20 g
	2% Maltose	20 g
	Bromcresol purple solution (0.4% stock)	9 ml
	2% Bacto-agar	20 g
	Distilled water	1000 ml
GAL[b]	1% Yeast extract	10 g
	2% Peptone	20 g
	2% Agar	20 g
	Bromthymol blue (4 mg/ml stock)	20 ml
	Distilled water	880 ml
	2% Galactose (20% stock)[b]	100 ml

[a] Fermentation indicator medium used to distinguish strains which ferment or do not ferment maltose. Owing to the pH change, the maltose-fermenting strains will produce a yellow halo on a purple background. A 0.4% bromcresol purple solution is prepared by dissolving 20 mg of the indicator in 50 ml of ethanol.
[b] Galactose-fermenting strains will produce a yellow halo on a blue background. After autoclaving, add 100 ml of a filter-sterilized 20% galactose solution.

exponential phase of growth. However, strains with greatly reduced growth rates in synthetic media are often encountered. Usually strains reach a maximum density of 2×10^8 cells/ml in YPD medium. Titers 10 times this value can be achieved with special conditions, such as pH control, continuous additions of balanced nutrients, filter-sterilization of media, and extreme aeration that can be delivered in fermentors.

The sizes of haploid and diploid cells vary with the phase of growth[47] and from strain to strain. Typically, diploid cells are 5×6 μm ellipsoids and haploid cells are 4 μm diameter spheroids.[48] The volumes and gross composition of yeast cells are listed in Table VIII. During exponential growth, haploid cultures tend to have higher numbers of cells per cluster compared to diploid cultures. Also, haploid cells have buds that appear adjacent to the previous one, whereas diploid cells have buds that appear at the opposite pole.[49]

[47] A. E. Wheals, in "The Yeast" (A. H. Rose and J. S. Harrison, eds.), 2nd Ed., Vol. 1, p. 283. Academic Press, New York, 1987.
[48] R. K. Mortimer, *Radiat. Res.* **9**, 312 (1958).
[49] D. Freidfelder, *J. Bacteriol.* **80**, 567 (1960).

TABLE VII
SPORULATION MEDIA

Sporulation medium	Component	Composition
Presporulation[a]	0.8% Bacto-yeast extract	0.8 g
	0.3% Bacto-peptone	0.3 g
	10% Dextrose	10 g
	2% Bacto-agar	2 g
	Distilled water	100 ml
Sporulation[b]	1% Potassium acetate	10 g
	0.1% Bacto-yeast extract	1 g
	0.05% Dextrose	0.5 g
	2% Bacto-agar	20 g
	Distilled water	1000 ml
Minimal sporulation[c]	1% Potassium acetate	10 g
	2% Bacto-agar	20 g
	Distilled water	1000 ml

[a] Strains are grown 1 or 2 days on this medium before transferring to sporulation medium. This is only necessary for strains that do not sporulate well when incubated on sporulation medium directly.

[b] Strains will undergo several divisions on this medium and then sporulate after 3 to 5 days of incubation. Sporulation of auxotrophic diploids is usually increased by adding the nutritional requirements to the sporulation medium at 25% of the levels given above for SD complete medium.

[c] Nutritional requirements are added as needed for auxotrophic diploids as for sporulation medium above (25% of level for SD complete medium).

TABLE VIII
SIZE AND COMPOSITION OF YEAST CELLS

Characteristic	Haploid cell	Diploid cell
Volume (μm^3)	70	120
Composition (pg/cell)[a]		
Wet weight	60	80
Dry weight	15	20
DNA	0.017	0.034
RNA	1.2	1.9
Protein	6	8

[a] One picogram equals 10^{-12} g.

Strain Preservation

Yeast strains can be stored for short periods of time at 4° on YPD medium in petri dishes or in closed vials (slants). Although most strains remain viable at 4° for at least 1 year, unusually sensitive mutants die after several months. Yeast strains can be stored indefinitely in 15% (v/v) glycerol at −60° or lower temperature. (Yeast tend to die after several years if stored at temperatures above −55°.[50])

Many workers use 2-ml vials (35 × 12 mm) containing 1 ml of sterile 15% (v/v) glycerol. The strains are first grown on the surfaces of YPD plates; the yeast is then scraped up with sterile applicator sticks and suspended in the glycerol solution. The caps are tightened and the vials shaken before freezing. The yeast can be revived by transferring a small portion of the frozen sample to a YPD plate.

Other Techniques for Genetic Analysis

Major techniques used for genetic analysis are covered in this volume, including dissection of asci,[51] gene mapping,[25,26] and mutagenesis.[52-56] Other simple procedures used in yeast studies are described below.

Replica Plating

Testing of strains on numerous media can be carried out by the standard procedure of replica plating with velveteen.[57] However, subtle differences in growth are better revealed by transferring diluted suspensions of cells with specially constructed spotting apparatuses. An array of inoculating rods, fastened on a metal plate, is dipped into microtiter or other compartmentalized dishes, containing yeast suspensions. Small and uniform aliquots can be repetitively transferred to different types of media.

Mating and Complementation

A few crosses can be simply carried out by mixing equal amounts of the *MAT*a and *MAT*α strains on a YPD plate and incubating at 30° for at least 6 hr and preferably overnight. Prototrophic diploid colonies can then be

---

[50] A. M. Well and G. G. Stewart, *Appl. Microbiol.* **26**, 577 (1973).
[51] F. Sherman and J. B. Hicks, this volume [2].
[52] C. Lawrence, this volume [18].
[53] R. S. Sikorski and J. D. Boeke, this volume [20].
[54] M. F. Hoekstra, H. S. Seifert, J. Nickoloff, and F. Heffron, this volume [22].
[55] D. J. Garfinkel and J. N. Strathern, this volume [23].
[56] R. P. Moerschell, G. Das, and F. Sherman, this volume [24].
[57] J. Lederberg and E. M. Lederberg, *J. Bacteriol.* **63**, 399 (1952).

selected on appropriate synthetic media if the haploid strains contain complementing auxotrophic markers. Similarly, testing of mating types or other markers of meiotic progenies, which requires the selection of numerous diploid hybrids, can be carried out by replica plating, using any one of a number techniques such as cross-streaking and spotters. Prototrophic diploids also can be selected by overlaying a mixture of the two haploid strains directly on minimal plates, although the frequencies of matings may be slightly reduced. If the diploid strain cannot be selected, zygotes can be isolated from the mating mixture with a micromanipulator. Zygotes, which can be identified by a characteristic thick zygotic neck, are best isolated 4–6 hr after mixing, when the mating process has just been completed[58]; diploids isolated by micromanipulation should be verified by sporulation and the lack of mating.

Formation of prototrophic diploids, indicative of complementation, is used to test *MATa* and *MATα* mating types and to determine unknown markers in new mutants and meiotic segregants. Mating-type tests are best carried out with *MATa* and *MATα* tester strains, each containing markers not in the strains to be tested.

Complementation analysis consists of testing diploid strains that were constructed from two haploid mutants which have the same mutant phenotype, such as a specific amino acid requirement, sensitivity to UV light, or other mutant characters. If the mutant character is found in the diploid and if the two mutant genes are recessive, it can be concluded that the two mutant genes are allelic, that is, the lesions are in genes controlling the same function or, in most cases, the same polypeptide chain. In rare and special instances, a double heterozygous diploid strain may exhibit the phenotype of the recessive marker, confusing this test of complementation.

However, because of allelic (or intragenic) complementation, the growth of double heterozygous diploids does not always indicate that the two mutations are in different genes. Some cases of allelic complementation occur when the normal enzyme is composed of two or more identical subunits. The enzymes formed by allelic complementation are mutant proteins containing two different altered polypeptides in which each of the mutant polypeptides compensates for the defects of the other to produce a catalytically active protein. Allelic complementation can be pronounced when the enzyme contains separate domains carrying out different catalytic functions, such as the *HIS4A, HIS4B,* and *HIS4C* regions. Allelic complementation is frequent in yeast. For example, mutations in 5 out of 10 genes controlling histidine biosynthesis show extensive allelic complementation. Because of allelic complementation, frequencies of meiotic

[58] E. P. Sena, D. N. Radin, and S. Fogel, *Proc. Natl. Acad. Sci. U.S.A.* **70,** 1373 (1973).

recombination are required to determine if two complementing mutants are alleles of the same gene. The frequencies of recombination are extremely low if the mutations are in the same gene, whereas the frequencies of normal meiotic segregants can be as high as 25% if the mutations are in different genes.

Complementation tests are required for scoring meiotic progeny from hybrids heterozygous for two or more markers controlling the same character. For example, a $HIS3^+$ $his4^-$ × $his3^-$ $HIS4^+$ diploid will produce tetratype tetrads having the following genotypes:

$HIS3^+$ $HIS4^+$
$HIS3^+$ $his4^-$
$his3^-$ $HIS4^+$
$his3^-$ $his4^-$

Intragenic complementation tests are required to determine the segregation of the *his* alleles in the $HIS3^+$ $his4^-$, $his3^-$ $HIS4^+$, and $his3^-$ $his4^-$ segregants. These tests are carried out with $MAT\mathbf{a}$ and $MAT\alpha$ tester strains having either $HIS3^+$ $his4^-$ or $his3^-$ $HIS4^+$ markers. Diploids homozygous for either $his3^-$ or $his4^-$ will not grow on histidine-deficient medium as indicated in the tabulation below:

	×$HIS3^+$ $his4^-$	×$his3^-$ $HIS4^+$
$HIS3^+$ $HIS4^+$ +	+	+
$HIS3^+$ $his4^-$ −	−	+
$his3^-$ $HIS4^+$ −	+	−
$his3^-$ $his4^-$ −	−	−

Random Spores

Although dissection of asci with recovery of all four ascospores is the preferred procedure for obtaining meiotic progeny,[51,52] random spores can be used when isolating rare recombinants, or when analyzing a large number of crosses. Several techniques have been devised for eliminating or reducing unsporulated diploid cells from the culture. The proportion of random spores can be increased by sporulating the diploid strain on a medium containing a high concentration of potassium acetate (2%, w/v), which kills vegetative cells of many sporulated strains.[59] The sporulated culture is treated with Glusulase (snail juice) (NEN Research Products, Wilmington, Delaware), the spores are separated by a minimal level of sonication, and various dilutions are plated for single colonies. Further-

---

[59] R. J. Rothstein, R. E. Esposito, and M. S. Esposito, *Genetics* **85**, 35 (1977).

more, vegetative cells can be preferentially killed by treating a sporulated culture with an equal volume of diethyl ether.[45] Another convenient method for producing random spores relies on the selection against vegetative diploid cells that are heterozygous for can^R1, the recessive marker confirming resistance to canavanine sulfate.[60] If the $can^R1/+$ diploid is sporulated and the spores are separated and plated on canavanine medium, one-half of the haploid spores will germinate and grow, whereas the remaining haploids and all of the diploids will not.

Acknowledgments

The writing of this chapter was supported by Public Health Research Grant GM12702 from the National Institutes of Health.

[60] F. Sherman and H. Roman, *Genetics* **48**, 255 (1963).

[2] Micromanipulation and Dissection of Asci

By FRED SHERMAN and JAMES HICKS

Separation of the four ascospores from individual asci by micromanipulation is required for meiotic genetic analyses and for the construction of strains with specific markers. In addition, micromanipulation is used to separate zygotes from mass-mating mixtures and, less routinely, for positioning of vegetative cells and spores for mating purposes and for single-cell analyses. The relocation and transfer of ascospores, zygotes, and vegetative cells are almost exclusively carried out on agar surfaces with a fine glass microneedle mounted in the path of a microscope objective and controlled by a micromanipulator. Although specialized equipment and some experience are required to carry out these procedures, most workers can acquire proficiency with a few days of practice.

Micromanipulators

Micromanipulators used for yeast studies operate with control levers or joysticks that can translate hand movements into synchronously reduced movements and microtools.[1,2] Most instruments were designed so that

[1] F. Sherman, *Methods Cell Biol.* **11**, 189 (1975).
[2] H. M. El-Badry, "Micromanipulators and Micromanipulation." Academic Press, New York, 1963.

movement of the tool in the horizontal (x and y) plane is directly related to the movement of the control handle, whereas the vertical (z direction) tool movement is controlled by rotating a knob, located either on or near the horizontal control handle. Other designs have various combinations in which the joystick controls the x and z planes, a screw controls the z movement, the mechanical stage controls the x and y planes, etc. Many commercially available micromanipulators having single control levers are listed in Table I, along with distributors and prices in the United States. The de Fonbrune types and the ones distributed by Lawrence and Rainin companies are commonly used for genetic studies in the United States. The Singer MKIII micromanipulator is popular in the United Kingdom. The others are used by a limited number of investigators.

Transmission of hand motions to the tool is based on pneumatic principles (the de Fonbrune and Stoelting units) or on several ingenious mechanical principles involving direct coupling to sliding components (the units distributed by the Lawrence, Rainin, Singer, Zeiss, Leitz, Narishige,

TABLE I
COMMERCIALLY AVAILABLE MICROMANIPULATORS WITH SINGLE-LEVER CONTROLS

Distributor	Micromanipulator	Catalog no.	Price[a] (approximate)
Technical Products International (St. Louis, MO)	de Fonbrune-type micromanipulator	MM1-1	$4750
Lawrence Precision Machine (Hayward, CA)	Mark I micromanipulator		$1000
	Mechanical stage I		$2400
Rainin Instrument Co. (Woburn, MA)	Micromanipulator		$380
	Micromanipulator and microscope	RIC-TDM	$5333
Singer Instrument Co. Ltd. (Watchet, Somerset, UK)	Singer MKIII micromanipulator		$6750
	Singer MSM System		$20,800
C. H. Stoelting Co. (Chicago, IL)	Joystick micromanipulator	51500	$2750
		51502	$2600
Carl Zeiss Instruments (Thornwood, NY)	Zeiss micromanipulator MR	471843	$5000
Wild Leitz (Rockleigh, NJ)	Leitz manual micromanipulator	520137	$6269
Narishige USA Inc. (Greenvale, NY)	Joy stick manipulator	MN-100	$1070
Research Instruments Limited (Cornwall, UK)	Single control lever micromanipulator	TL0500	$4780

[a] Approximate prices in the United States (September, 1989).

FIG. 1. The de Fonbrune-type micromanipulator. The control unit (right) and the receiver (left) are interconnected by flexible tubing on opposite sides of a Leitz Laborlux II microscope. The complete assembly is on a vibration eliminator (Vibration Damping Mount, Cat. No. 9705/601, The Laboratory Apparatus Co., Cleveland, OH).

and Research Instruments companies). The Jena sliding micromanipulator, which relies on two metal plates separated by a layer of lubricant, has been used in the United States and Europe for yeast studies. (However, this Jena micromanipulator is no longer commercially available.)

The de Fonbrune micromanipulator, shown in Fig. 1, pneumatically transmits a fine degree of motion from a single joystick.[3] The micromanipulator consists of two free-standing units: (1) a joystick controlling three piston pumps that is connected by tubing to (2) three diaphragms or aneroids that actuate a lever holding the microtool. Lateral movement of the joystick controls the x and y horizontal movements, whereas rotation of the joystick controls vertical z movement. The ratio of distances of the

[3] P. de Fonbrune, "Technique de Micromanipulation." Masson, Paris, 1949.

joystick and microtool movement is adjustable from approximately 1:40 to 1:2000. The Stoelting micromanipulator design is similar to that of the de Fonbrune, but it has a fixed ratio of approximately 1:50.

The unit manufactured by Lawrence Precision Machine consists of two components (Fig. 2), a needle holder or micromanipulator for vertical (z) movement, and a very fine mechanical stage or "two axes" stage used to move the petri dish or chamber horizontally (x and y) during the dissection. Although the micromanipulator can be used with other microscope stages, the Lawrence stage has especially fine control. One complete turn of the micrometer wheels moves the stage 1 or 2 mm, depending on which screw pitch is installed. The 2 mm per turn screw pitch is recommended for dissection on petri plates. The cells are positioned in the horizontal plane by the two controls on the stage, while the microneedle vertical position is controlled by a screw on the micromanipulator. Vibration is at a minimum, since the micromanipulator is clamped directly to the stage.

The micromanipulator and microscope unit "Tetrad Dissection System," distributed by Rainin Instrument Co. (Fig. 3), is based on the design of Sherman[4] and is primarily intended for dissection of asci. The joystick controls the fine y and z motion, and the microscope stage is used for controlling the x and y positions. The direct coupling of the manipulator to the microscope minimizes vibration. The unit is supplied with a Zeiss 16 microscope optically aligned for dissection on petri plates.

The Singer MKIII micromanipulator is a robust mechanical micromanipulator used for yeast genetic studies in several laboratories in the United Kingdom. In the Singer MKIII micromanipulator, as in the de Fonbrune micromanipulator, lateral hand movements control the horizontal (x and y) movements of the microtool, while rotation of the same lever controls vertical (z) movements. As in the de Fonbrune micromanipulator, the ratio of movement is variable.

Recently, the Singer MSM System was specifically developed for the dissection of asci. This work station (Fig. 4) comes complete with everything required for dissecting asci, including a specially built microscope, a computer-controlled motorized stage, and microneedles. A television camera can be conveniently attached to the unit.

The other micromanipulators listed in Table I have been used extensively for microsurgical studies on living cells and on occasion for yeast studies. An inexpensive micromanipulator, described by Sherman,[4] was designed for yeast genetic studies and has been used extensively in the United States. The components can be purchased for less than $600 [MPM-0-100 micromanipulator ($320), Allen Benjamin, Inc., 1950 East

---

[4] F. Sherman, *Appl. Microbiol.* **26**, 829 (1973).

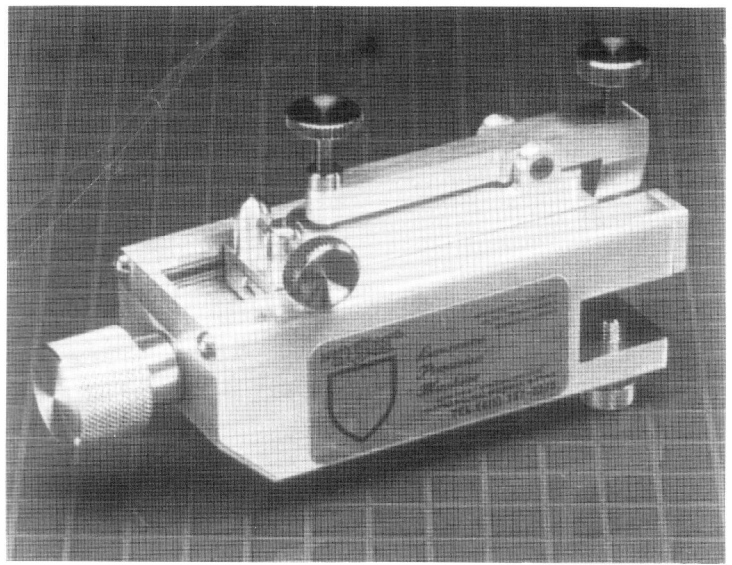

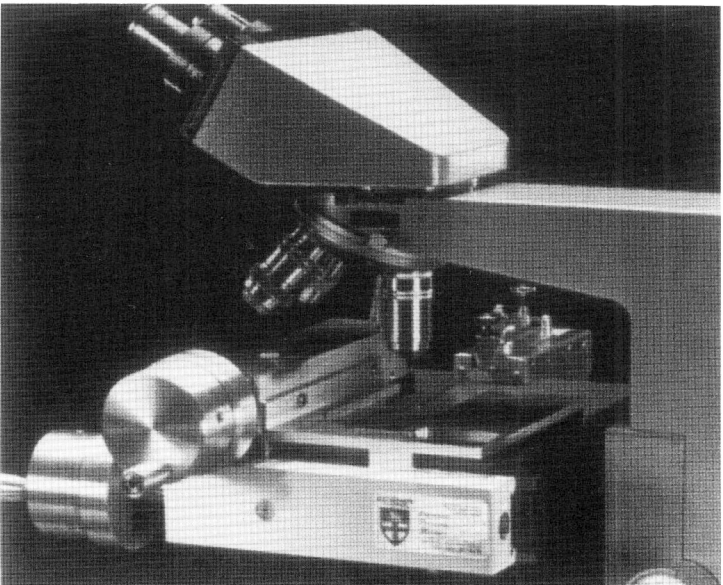

FIG. 2. Lawrence Precision Machine micromanipulator. The needle holder or micromanipulator (top) is mounted on a fine mechanical stage (bottom) used to move the petri dish or chamber during the dissection.

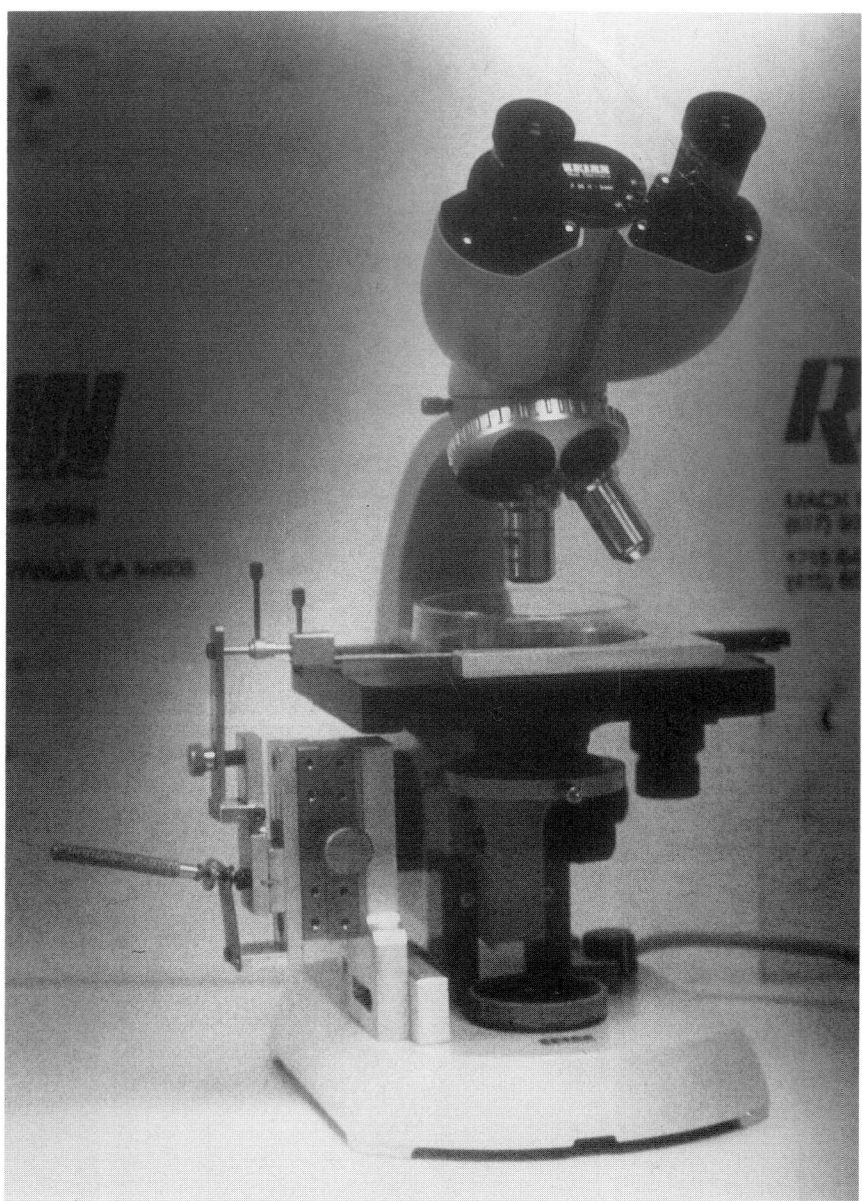

FIG. 3. The Tetrad Dissection System, distributed by Rainin Instrument Co., showing the micromanipulator mounted on the microscope.

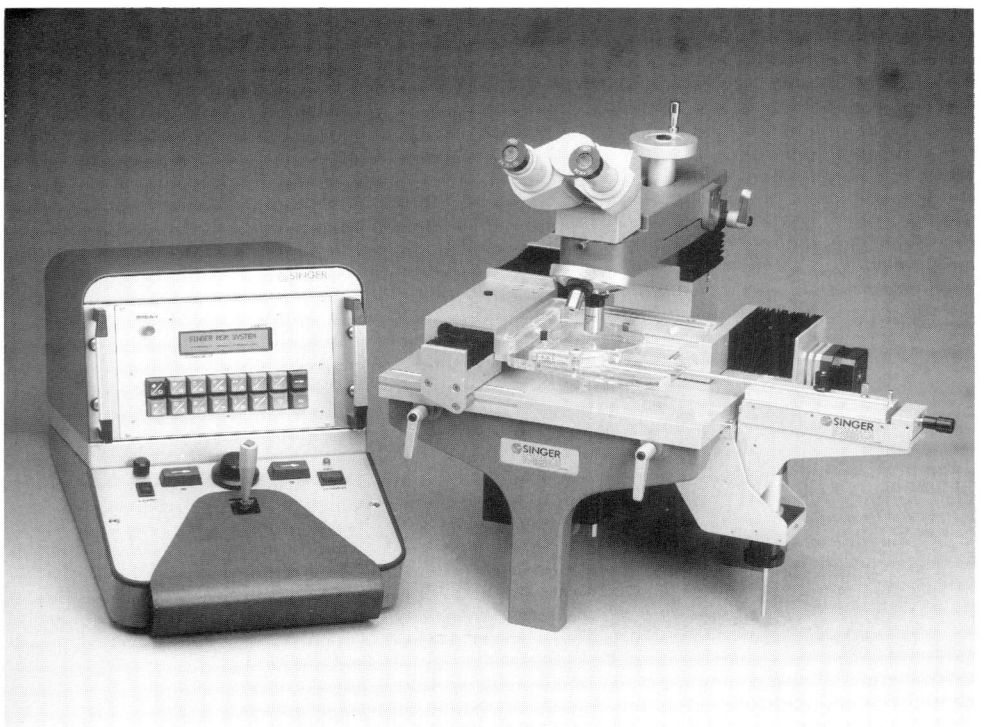

FIG. 4. The Singer MSM System. The specially constructed microscope comes with 16× eyepieces and 20× and 3.2× objectives and has a hinged overarm that allows convenient changing of plates. The stage of the microscope is driven by electric motors that are controlled by a microprocessor (shown at left), allowing repeated positioning with an accuracy of 4 μm and the automatic recall of any matrix point. The micromanipulation movements are controlled with a pendant joystick and coaxial ring shown at lower right.

First Avenue, Tempe, AZ 85281; 55736 dovetail slide ($175), C. H. Stoelting Co., 620 Wheat Lane, Wood Dale, IL 60191; MM1-6 microclamp ($77), Technical Products International, Inc., 13795 Rider Trail, Suite 104, St. Louis, MO 63045]. The assembly of the unit requires some custom machining.

It is difficult to recommend objectively one specific micromanipulator, as most workers strongly prefer the model with which they have had the most experience. It has been our experience that the skill of ascus dissection can be taught more quickly with the de Fonbrune-type micromanipulator. The de Fonbrune-type, the Singer MKIII, and other micromanipulators not attached directly to the microscope stage require more space and, in some instances, heavy base plates or vibration eliminators (see Fig. 1). The micromanipulators distributed by Lawrence Precision Machine and Rainin are compact units that are directly mounted to the microscope and exhibit little vibration. The Rainin unit includes a microscope already assembled for micromanipulation and can be used directly without additional modifications. The Singer MSM System, although expensive, is the ultimate apparatus for dissection of asci.

Microscopes

It should be stressed that microscope requirements depend on the type of micromanipulator. Micromanipulators not directly attached to the stage should be used in conjunction with microscopes having fixed stages and tube focusing, such as the Leitz Laborlux II microscope. Fine mechanical stages with graduations are essential with all micromanipulators. It is convenient to have long working distance objectives for magnifications in the range of 150–300×. Long working distances can be achieved with 10× and 15× objectives, and the appropriate magnifications with 20× or 25× eye pieces.

Microneedles

Separation of ascospores, zygotes, and vegetative yeast cells can be carried out with simple glass microneedles attached to any one of the micromanipulators described above. Microneedles can be made either individually[5] or from a stock of glass fibers.[6]

Individually prepared glass microneedles can be easily made with the small flame from the pilot light of an ordinary bunsen burner. A 2-mm-diameter glass rod is drawn out to a fine tip with the flame of the bunsen burner; by using the pilot flame, an even finer tip is drawn out at a right angle with an auxiliary piece of glass rod as illustrated in Fig. 5. The end is broken off so that the tip has a diameter of 10 to 100 μm and a length of a few millimeters. The drawn-out tip can be cut with a razor blade or broken between the surface and edge of two glass slides. It is critical that the microneedles have a flat end, which sometimes requires several attempts. The exact diameter is not critical, and various investigators have different preferences. Spores are more readily picked up and transferred with micro-

[5] C. C. Lindegren, "The Yeast Cell: Its Genetics and Cytology." Educational Publ., St. Louis, Missouri, 1949.
[6] K. E. Scott and R. Snow, *J. Gen. Appl. Microbiol.* **24,** 295 (1978).

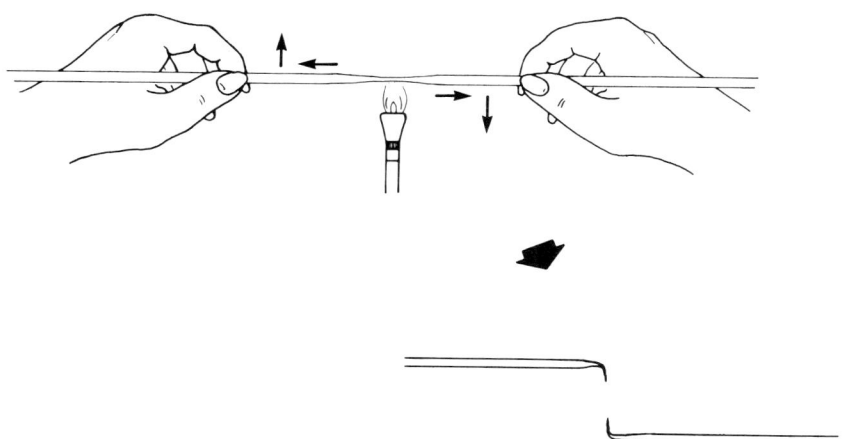

FIG. 5. Construction of microneedles. Microneedles required for the separation of ascospores can be made by first drawing out a 2-mm glass rod to a fine tip and then drawing out the end to an even finer tip at a right angle.

needles having tips of larger diameters, whereas manipulations in crowded areas having high densities of cells are more manageable with microneedles having smaller diameters. An approximately 40-μm-diameter microneedle is an acceptable compromise. The length of the perpendicular end should be compatible with the height of the petri dish or chamber; too short an end may result in optical distortions from the main shank of the microneedle. Longer microneedles are required for manipulations on the surfaces of petri dishes. The microneedle micromanipulator is attached to and positioned under the microscope objective as shown in Fig. 6.

Construction of microneedles from glass fibers involves two steps: (1) preparation of a stock of glass fibers and (2) the gluing of a short segment of glass fiber perpendicular to a glass mounting rod. The glass mounting rod is made by first heating a 2-mm rod in a burner and pulling slowly to form a taper. When the rod has sufficient taper, the end is pulled quickly at right angles, similar to the procedure shown in Fig. 5; the end is broken so that the right angle projection is approximately 2 mm. The mounting rod should be cut with a file to approximately the size required to fit on the microscope stage, taking into account the distance from the manipulator to the center of the microscope field of view.

Glass fibers are made by drawing thin filaments from a 2-mm glass rod. The 2-mm glass rod is heated until it is white hot. As the rod is removed from the flame, it is rapidly pulled apart to produce an extremely thin thread of 20 to 40 μm in diameter. (A human hair, a convenient size

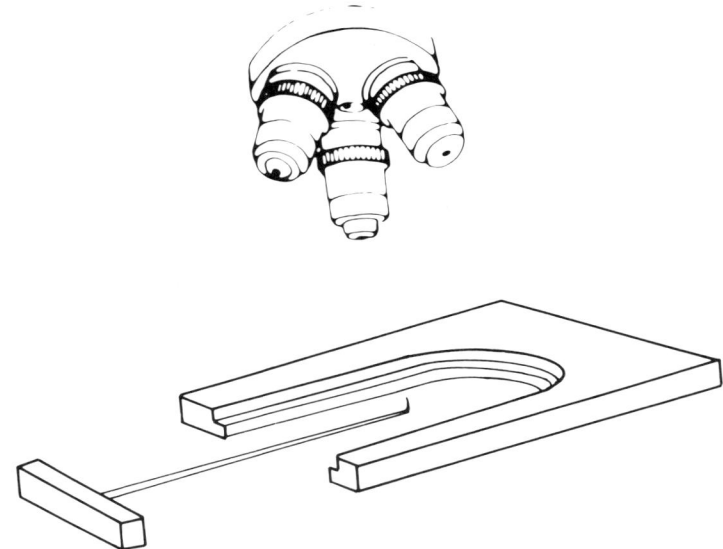

FIG. 6. Relative positions of the microneedle, a rig for holding a petri dish, and the microscope objectives.

standard, is approximately 20 μm in diameter.) The thread of glass is most easily produced if the flame is narrow and extremely hot. Many burners contain a pilot flame that is suitable for production of glass fibers. The thin thread can be broken with the fingers or cut with scissors or coverslips. The segments are placed on a microscope slide for examination under a dissecting microscope (Fig. 7). Segments of about 1 cm are usually desired for dissection on a petri dish with a standard micromanipulator. The exact length is not too important at this point, because the microneedle eventually can be cut to size with a coverslip. Optical glass fibers, which are commercially available [0.002 inch diameter, Cat. No. F31.735, $35.95, Edmund Scientific (101 East Gloucester Pike, Barrington, NJ 08007)], and which have a uniform size, have been used to prepare microneedles[7]; however, some workers believe that ascospores exhibit less adherence to the optical glass fibers.

The segments of glass are examined under a low-power dissecting microscope to determine which will make a good needle, that is, which have tips with a flat surface perpendicular to the long axis of the needle and no burrs or cracks in the tip (Fig. 7). However, a needle with minor

[7] D. J. Eichinger and J. D. Boeke, *Yeast*, in press (1990).

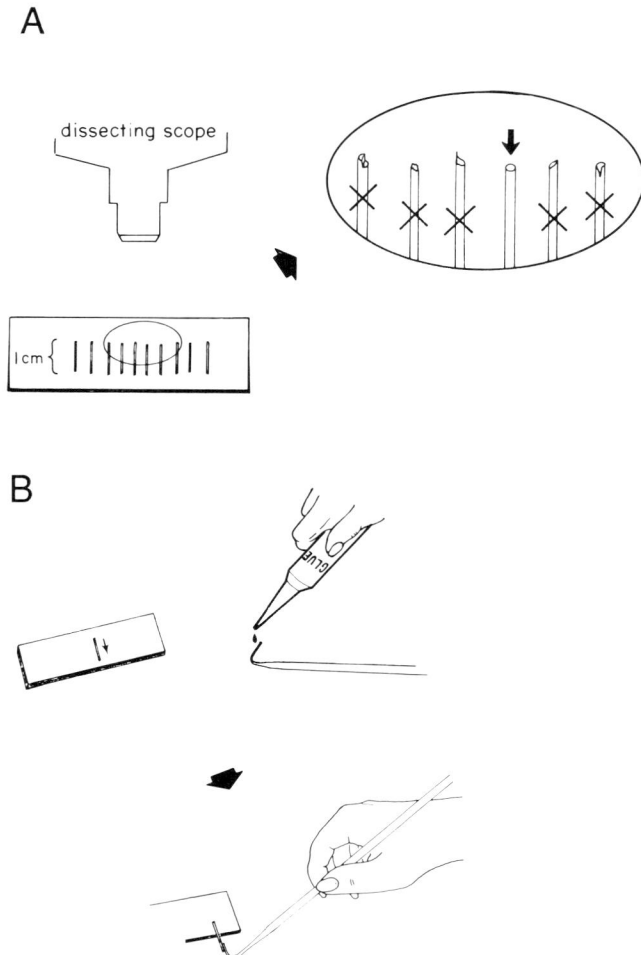

FIG. 7. Use of glass fibers for constructing microneedles. (A) A glass fiber approximately 40 μm in diameter is broken into segments approximately 1 cm long with a razor blade and examined with a microscope. (B) A segment containing a flat end is attached at a right angle to a mounting rod with cyanoacrylic glue (Duro SUPER GLUE).

imperfections (e.g., a half-circle) sometimes will work if it has a flat working surface.

The glass-fiber segment with the best tip is moved down on the slide so that the good end is on the slide and the end to be glued is hanging off the edge. A small drop of Duro SUPER GLUE (cyanoacrylic glue) is applied to the mounting rod, and the glass fiber is glued to the end as shown in Fig. 7. The easiest way to apply the glue is to place a drop on a microscope slide and then to dip the whisker of the mounting rod into the drop. After contact, the glass fiber will usually come off the slide and stick to the mounting rod without coaxing. If the glass fiber is not perpendicular to the stock, one may quickly adjust the angle before the glue sets.

The needle is mounted into the micromanipulator and centered in the field. The adjustment of the needle is made most easily first at low magnification and then at higher magnification. As recommended above, asci are usually dissected at 150× or greater magnification.

Dissection of Asci

Digestion with Snail Juice

Sporulated cultures usually consist of unsporulated vegetative cells, four-spored asci, three-spored asci, etc. Dissection of asci requires the identification of four-spored asci and the relocation of each of the four ascopores to separate positions where they will form isolated spore colonies. The procedure requires digestion of the ascus wall with snail juice or other commercial enzyme without dissociating the four spores from the ascus.[8] (With very unusual strains that are particularly sensitive to enzyme treatments, the separation of ascospores can be carried out by rupturing the ascus wall with a microneedle.)[5,9]

Snail juice can be obtained commercially as Glusulase (NEN Research Products, Wilmington, Delaware; Cat. No. NEE-154) and *Suc d'Helix pomatia* (L'Industrie Biologique Francaise, Genevilliers, France). Several other commercial enzymes have been successfully used, including Zymolyase (ICN, Costa Mesa, CA). Approximately 0.2 ml of an aqueous suspension of a sporulated culture is usually treated for 10–30 min, depending on the particular yeast strain and on the strength of the Glusulase or other digestive agent. Glusulase should be diluted in water from 1:4 to 1:100, depending on the particular batch of enzyme and sensitivity of the yeast strain.

The progress of the digestion can be followed by removing a loopful of the digest to a glass slide and examining it under phase contrast at 400× magnification. The sample is ready for digestion when the spores in most of the asci are visible as discrete spheres, arranged in a diamond shape. Typically digested asci are seen in Fig. 8A. If a majority of the asci are still arranged in tightly packed tetrahedrons or diamond shapes in which the spores are not easily resolved, digestion is incomplete and the spores will

---

[8] J. R. Johnston and R. K. Mortimer, *J. Bacteriol.* **78**, 292 (1959).
[9] Ö. Winge and O. Lausten, *C. R. Trav. Lab. Carlsberg, Ser. Physiol.* **22**, 99 (1937).

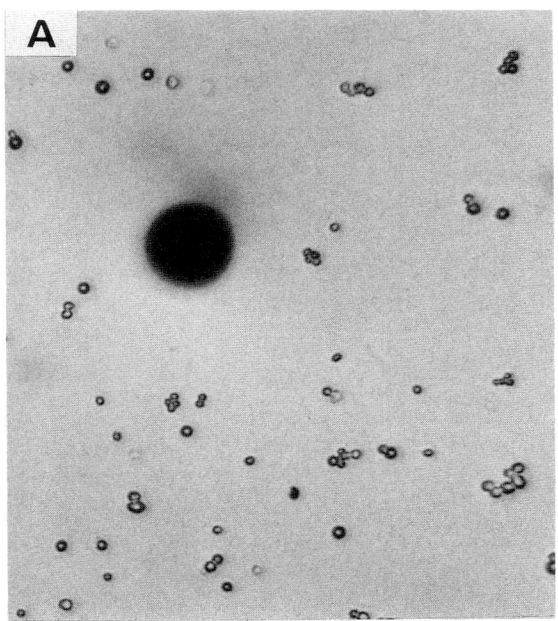
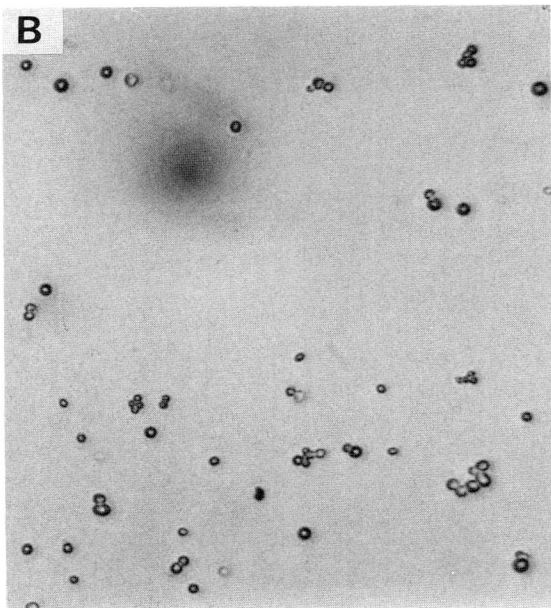

FIG. 8. A field of sporulated culture. (A) A four-spored cluster is seen to the right of the microneedle tip. (B) The cluster was picked up on the microneedle, which was lowered beneath the focal plane. The ascospores and the tip of the microneedle are, respectively, approximately 5 and 50 μm in diameter.

not be easily separated by microdissection. It is convenient to use a Glusulase concentration that will digest the ascus wall in approximately 15 min. Extensive treatment sometimes can decrease the viability and dissociate the four-spore clusters. The culture is suspended by gently rotating the tube; an aliquot is transferred with a wire loop to the surface of a petri plate or agar slab. It is important not to agitate the spores once they have been treated. If the treated spores are vortexed or shaken, the integrity of the ascus cannot be assured as the contents of one ascus may disperse and reassemble with the contents of another.

Separation of Ascospores

Micromanipulation can be implemented directly on the surfaces of ordinary petri dishes filled with nutrient medium or in special chambers on thin agar slabs. The petri dish (or chamber) is positioned so that the inoculum is in the microscope field over the microneedle. Examination of the streak should reveal the presence of the desired four-spored clusters as well as smaller clusters and vegetative cells. A typical preparation is shown in Fig. 8. A cluster of four spores is picked up on the microneedle by positioning microneedle tip next to the four-spored cluster on the surface of the agar. The microneedle is moved in a sweeping action, first touching the agar surface and then lowering in a single motion. The absence of the four spores from the agar surface indicates that they have been transferred to the microneedle. Several attempts may be required to pick up all four ascospores.

The microneedle can be considered a platform to which the spores are transferred. It is obvious from the relative sizes of the microneedle and spores (Fig. 8) that the microneedle does not "poke" the tetrad of spores to pick them up. The flat surface of the microneedle does not interact with the spores themselves, but rather with the water layer on the surface of the agar. When the microneedle approaches the surface of the agar, a miniscus forms and a halo of refracted light can often be seen around the shadow of the microneedle. At this time a column of water connects the microneedle and the agar (Fig. 9). The spores disappear from view into the miniscus. The combined sideways and downward sweeping motion is an attempt to coax the spores into the half of the miniscus that remains on the microneedle surface as it breaks away. Success in this endeavor is assayed by the disappearance of the spores from the visual field in the microscope. At the new position the process is repeated, this time with the hope that spores go from the microneedle miniscus to the surface of the agar.

Once the four spores have been transferred to the first position, it is necessary to separate at least one spore from the rest so that it can be left

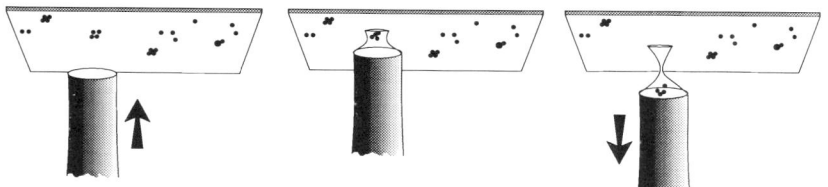

FIG. 9. Transfer of four spores from an agar surface to the platform of a microneedle, by way of a water miniscus.

behind. A simple technique for achieving this goal is to move the microneedle onto the surface of the agar, forming the crisp image and halo, directly next to the cluster of spores and to vibrate the microneedle by tapping more or less gently on the table near the microscope. The spores will often be separated by several microneedle diameters by this action. Three spores can be collected by sweeping the surface of the agar with the needle tip, and the process is repeated at the next three stops.

Note the position on the mechanical stage and place the four spores on the surface of the agar at least 5 mm from the streak. Pick up three spores and move the dish away from the streak another 5 mm (Fig. 10). Deposit the three spores and pick up two spores. Move the chamber an additional 5 mm; deposit the two spores and pick up one spore. Move the chamber 5 mm more and plant the remaining spore. Move the chamber 5 mm from

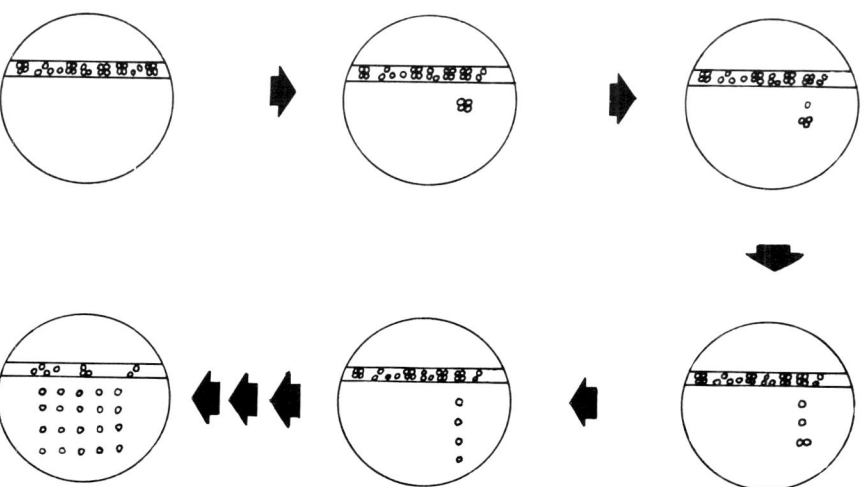

FIG. 10. The steps for sequentially separating the cluster of four ascospores approximately 5 mm apart on petri dishes.

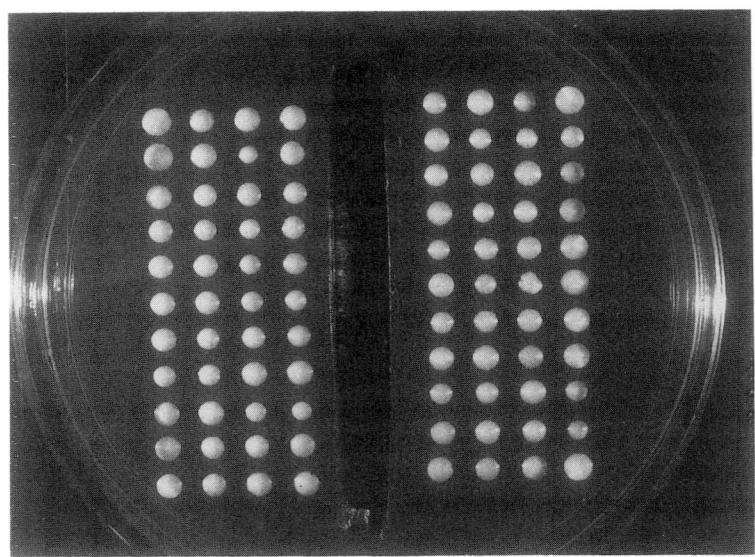
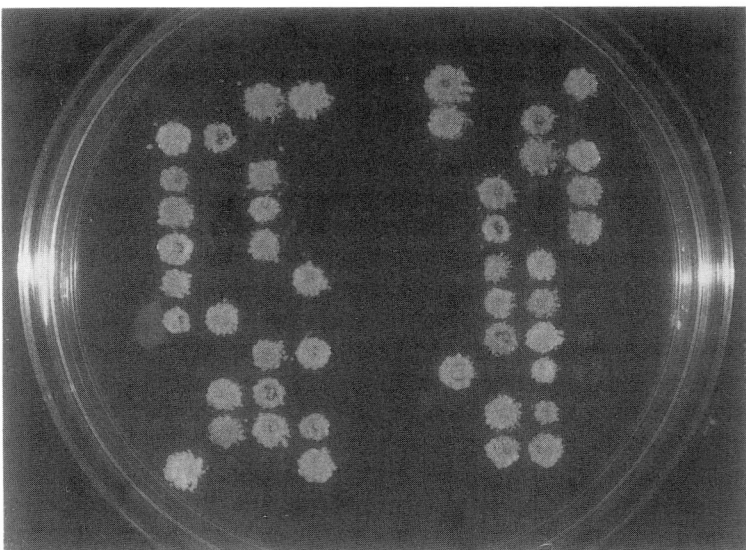

FIG. 11. Spore colonies derived from asci separated on the surface of a petri dish (top). The central area of the dish, containing a streak of the sporulated culture, was cut out and removed after dissection. The spore colonies were replica plated to a synthetic medium lacking a nutrient (bottom). The 2:2 segregation of a heterozygous marker is revealed by the growth pattern on selective medium. The complete viability and uniform colony size shown in this figure are not typical of the meiotic progeny from most diploids; however, these properties can be chosen during the course of strain construction.

the line of the four spores and select another four-spore cluster. Separate the spores as before at 5 mm intervals. Continue until a sufficient number of asci are dissected or until the entire dish is covered.

After picking up the four spaces from an ascus, it is often convenient to set the stage micrometer so that each group of four spore colonies falls on cardinal points such as 15, 20, and 25. This makes it easier to keep track of progress and prevents the spore colonies from growing too close together. Likewise, positions on the y axis can be marked on the stage micrometer so that the four spore colonies from each ascus are evenly spaced. Take care not to break the microneedle when removing the dish or chamber from the stage. The thin agar slab is transferred from the chamber to the surface of a nutrient plate, which is then incubated for 3 days until the spore colonies are formed. Petri dishes containing separated spores are similarly incubated. As shown in Fig. 11, colonies derived from ascospores that were separated on a petri dish can be replica plated directly to media for testing nutritional requirements.

Although considerable patience is required to master ascus dissection, most workers are able to carry out this procedure after a few days of practice.

Isolation of Cells

In addition to ascus dissection, micromanipulation is occasionally required for separating zygotes from mating mixtures, for pairing vegetative cells and spores for mating, and for separating mother and daughter cells during vegetative growth. Zygotes usually can be picked up on microneedles, although vegetative cells usually cannot. However, vegetative cells can be separated simply by dragging them across the agar surface with microneedles. The use of microneedles is rather effective, since the cells usually follow closely in the wake of the microneedle as it is moved along the liquid surface film of the agar.

Acknowledgments

The writing of this chapter was supported by Public Health Research Grant GM12702 from the National Institutes of Health. We wish to thank Dr. G. R. Fink and Ms. Julia Khorana (Whitehead Institute for Biomedical Research, MIT) for the drawings used in Figs. 5–7 and 10, and Lawrence Precision Machine, Rainin Instrument Co., Singer Instrument Co., and Dr. D. Campbell (University of Rochester Medical School), for the photographs used in Figs. 2, 3, 4, and 11, respectively. We also wish to thank Dr. D. Campbell for useful suggestions concerning the writing of this chapter.

[3] Mapping Yeast Genes

By FRED SHERMAN and PAUL WAKEM

The complete characterization of a gene should include the determination of its chromosomal position in the yeast genome. Such information could be useful for a number of reasons, such as establishing the identity between independently derived mutations and cloned genes. For example, the *cyc9, tup1, umr7,* and *flk1* mutations all were derived by selecting for different phenotypes. These presumably diverse types of mutations were discovered to be alleles of the same locus only after finding that they mapped to the same chromosomal position.[1,2] Thus, mapping of a new gene could reveal that it corresponds to a previously well-investigated gene. Also, close linkage to other characterized genes allows for the convenient retrieval of the DNA segment encompassing the gene in question. For example, because a *CYC3* mutant was shown to be approximately 2 centimorgans (cM) from *PYK1*, and because *PYK1* and adjacent genomic regions were available on plasmids, it was a simple matter to identify the DNA segment corresponding to the *CYC3* locus.[3] Mapping a gene is desirable, not only for preventing the unnecessary duplication of effort, but also for contributing to background information useful for future studies.

Mapping Strategies

As summarized in Fig. 1, locating the chromosomal position of genes can involve different approaches, depending primarily on whether the gene has been cloned. We wish to emphasize that recombinant DNA procedures have largely replaced and certainly supplemented traditional genetic methods for determining the chromosomal positions of genes. After it is established that a mutant phenotype is controlled by a single gene (Procedure 1), every effort should be made to clone the gene. The cloned gene can be used to prepare hybridization probes for determining its position on a chromosome (Procedure 3) and a fragment (Procedure 4). If necessary, the cloned gene can be used for chromosomal fragmentation (Procedure 5) and for determining its physical position on the chromosome. Furthermore, cloned genes are obviously required for other standard characterizations, such as transcriptional analysis and DNA sequencing.

[1] R. J. Rothstein and F. Sherman, *Genetics* **94**, 871 (1980).
[2] H. C. Stark, D. Fugit, and D. B. Mowshowitz, *Genetics* **94**, 921 (1980).
[3] M. E. Dumont, J. F. Ernst, and F. Sherman, *EMBO J.* **6**, 235 (1987).

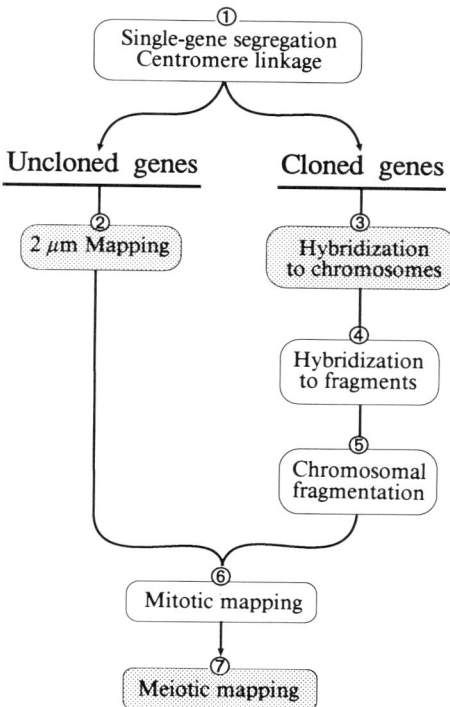

FIG. 1. Strategies for mapping genes. A meiotic analysis (Procedure 1) should be carried out with uncharacterized mutations, especially those derived from heavily mutagenized cells. The analysis will reveal if the mutant phenotype is controlled by a single gene and, at the same time, if it is centromere linked. Attempts should be made to clone the gene before proceeding with detailed mapping studies. If the gene cannot be cloned, the chromosomal assignment can be conveniently determined by the 2-μm mapping method (Procedure 2). Often a cloned gene can be simply mapped by first determining the chromosome on which it resides by hybridization studies (Procedure 3 or 4) and subsequently employing a meiotic analysis (Procedure 7) to determine its chromosomal location relative to known markers. (Procedure 4 should be available by the end of 1990.) The physical location of the cloned segment can also be determined by chromosomal fragmentation (Procedure 5), especially if other methods are unsuccessful. Mitotic mapping (Procedure 6) reveals on which chromosome arm the gene is located and is useful if the chromosome lacks sufficient markers for easily revealing meiotic linkage.

Although the ease of using cloned segments for chromosomal assignment is well established, in several unusual cases genes cannot be cloned by simply transforming and complementing a mutant strain. In fact, certain mutant alleles were first mapped by genetic procedures and then cloned by examining segments adjacent to linked markers.[4] Chromosomal assignment of uncloned genes is best determined by the 2-μm mapping method (Procedure 2), although a number of other methods have been suggested.[5-8]

Strategy: Current

Currently, the most expedient strategy to map a gene is first to clone it, determine on which chromosomal fragment the gene resides by hybridization to separated chromosomes (Procedure 3), and then carry out a meiotic analysis with markers on this chromosome (Procedure 7). Additional tests are possible, and optimal strategies depend on other available information, including the outcome of initial tests, the density of available markers on the particular chromosome, and the state of the development of new procedures. For example, when a chromosomal assignment is not clearly determined by hybridization (Procedure 3), the correct chromosome may be identified with the 2-μm mapping method (Procedure 2). Workers should be at least acquainted with the wide range of procedures that make yeast unusually amenable to genomic analysis.

Strategy: Near Future

Approximately 90% of the nuclear genome has been represented by approximately 10^3 yeast inserts in bacteriophage λ. The location of a gene within a chromosomal region can be determined by hybridizing a cloned fragment to this set of phage λ (Procedure 4). This phage λ library, which will be available before the end of 1990, would be used instead of separated chromosomes (Procedure 3).

Strategy: Far Future

We have, of course, recommended these strategies on the basis of present-day technologies and information. The DNA sequence of the entire yeast genome will undoubtedly be completed before the year 2000. Therefore, locating cloned segments to positions on chromosomes would be efficiently carried out in the future by simply determining the DNA

---

[4] R. Fleer, C. M. Nicolet, G. A. Pure, and E. C. Friedberg, *Mol. Cell. Biol.* **7**, 1180 (1987); R. K. Mortimer, D. Schild, C. R. Contopoulou, and J. A. Kans, *Yeast* **5**, 321 (1989).
[5] R. Wickner, *Genetics* **92**, 803 (1979).
[6] R. K. Mortimer, R. Contopoulou, and D. Schild, *Proc. Natl. Acad. Sci. U.S.A.* **78**, 5778 (1981).
[7] J. S. Wood, *Mol. Cell. Biol.* **2**, 1080 (1981).
[8] S. Klapholz and R. Esposito, *Genetics* **100**, 387 (1982).

sequence of a small portion of the segment and by a computer search for the corresponding sequence within the database.

Procedures

Procedure 1: Single-Gene Segregation and Centromere Linkage

A new recessive mutation should first be tested by genetic complementation, which involves crossing the unknown mutant to known, characterized mutants and examining the phenotype of the diploid. New mutants should be crossed to a series of known mutants having the same or similar phenotypes. Lack of complementation of two recessive mutations is almost always indicative of allelism. Meiotic analysis (Procedure 7) of the presumed homozygous diploid should reveal complete linkage and therefore identity. However, complementation of recessive mutations does not establish that they correspond to different genes. A meiotic analysis could be carried out when allelic complementation is suspected, especially when the diploid appears to have a partial mutant phenotype.

The second step in characterizing an unknown mutation should involve a meiotic analysis to determine if the mutant phenotype is controlled by a single gene. This is particularly critical when the mutant is derived from heavily mutagenized cells, such as those commonly used to obtain temperature-sensitive mutations and related defects. There have been numerous examples where temperature-sensitive growth and a particular enzyme deficiency segregated independently from each other, indicating mutations of two separate genes.

The mutant haploid strain should be crossed to a strain carrying at least one centromere-linked marker, such as *trp1*. Thus, a meiotic analysis would reveal both single-gene segregation and centromere linkage. The diploid should then be sporulated, the asci dissected, and the haploid segregants tested according to the methods outlined below (Procedure 7, meiotic mapping) and elsewhere in this volume.[9] A 2:2 segregation of the mutant phenotype is indicative of a single-gene mutation. Less than two-thirds second division segregation (less than two-thirds tetratype asci relative to *trp1*; see Procedure 7, meiotic mapping) is indicative of centromere linkage of the unmapped gene. If centromere linkage is suspected, the mutant should be crossed to a set of centromere-tester strains that have markers near the centromeres of each of the 16 chromosomes (Table

---

[9] F. Sherman and J. Hicks, this volume [2].

I).[10-12] The unmapped gene should exhibit linkage to one of the centromere-linked markers and should be further analyzed with additional markers on the assigned chromosome as outlined in Procedure 7, meiotic mapping.

If the mutant gene is not centromere linked, it is advisable to next determine on which chromosome it resides. Chromosomal assignments of cloned genes are conveniently determined by hybridization to separated chromosomes as outlined in Procedure 3. If the gene cannot be cloned, or if there is no intention of cloning the gene, the chromosomal assignment can be determined with the 2-μm mapping procedure.

Procedure 2: 2-μm Mapping

The 2-μm mapping procedure for the convenient assignment of any recessive mutation to its chromosome does not require a cloned fragment but relies instead on a set of *cir°* tester strains, each containing 2-μm plasmid DNA integrated at or near the centromere of a different chromosome.[13] (Strains containing endogenous 2-μm plasmids are denoted *cir⁺* strains, whereas strains lacking 2-μm plasmids are denoted *cir°*.) Plasmids containing the 2-μm inverted repeat sequence and segments of yeast DNA are integrated into the genome at the region of homology of *cir°* tester strains. The 2-μm plasmid DNA is stably maintained as an integrant because the plasmid DNA lacks the *FLP* (flip) gene required for 2-μm site-specific recombination and the *cir°* cells contain no resident 2-μm plasmids to provide FLP function. Specific mitotic chromosome loss can be induced by constructing a *cir°* × *cir⁺* cross. The FLP recombination function provided by the 2-μm circles of the *cir⁺* parent can recognize a site in the 2-μm repeat sequence of the integrant and catalyze a site-specific recombination event. This results in the loss of the integrant as well as chromosomal DNA distal to the site of integration. If the 2-μm plasmid is integrated at or near the centromere, the entire chromosome is lost at a high frequency. These properties of chromosomes with integrated 2-μm plasmids have been described in detail.[14-16]

A recessive mutation in a *cir⁺* strain can be assigned to its chromosome by crossing to the *cir°* tester strains. Subclones of the *cir°*/*cir⁺* diploids will

[10] R. K. Mortimer, D. Schild, C. R. Contopoulou, and J. A. Kans, this volume [57].
[11] D. Campbell, J. S. Doctor, J. H. Feuersanger, and M. M. Doolittle, *Genetics* **98**, 239 (1981).
[12] R. F. Gaber, L. Mathison, I. Edelman, and M. R. Culbertson, *Genetics* **103**, 389 (1983).
[13] L. P. Wakem and F. Sherman, *Genetics* **125**, 333 (1990).
[14] S. C. Falco, Y. Li, J. R. Broach, and D. Botstein, *Cell (Cambridge, Mass.)* **29**, 573 (1982).
[15] S. C. Falco, M. Rose, and D. Botstein, *Genetics* **105**, 843 (1983).
[16] S. C. Falco and D. Botstein, *Genetics* **105**, 857 (1983).

TABLE I
STRAINS OF *Saccharomyces cerevisiae* USEFUL FOR
DETERMINING CENTROMERE LINKAGE AND FOR MEIOTIC
ANALYSIS[a,b]

Multiply marked mapping strains[10]	
X4119-19C	*MA*T**a** *his7* (II) *tyr1* (II) *cdc9* (IV) *trp4* (IV) *aro1B* (IV) *hom2* (IV) *rad2 (VII) thr1* (VIII) *lys11* (IX) *gal2* (XII) *ade2* (XV)
X4119-15D	*MAT*α *cdc9* (IV) *trp4* (IV) *aro1B* (IV) *hom2* (IV) *ade8* (IV) *his1* (V) *lys11* (IX) *gal2* (XII)
STX145-13D	*MA*T**a** *cdc19* (I) *tyr1* (II) *gal1* (II) *trp1* (IV) *rad4* (V) *met14* (XI) *ura1* (XI) *lys9* (XIV) *pet8* (XIV) *ade2* (XV) *gal5*
STX145-15D	*MAT*α *cdc19* (I) *tyr* (II) *lys2* (II) *gal1* (II) *trp1* (IV) *rad4* (V) *met14* (XI) *ura1* (XI) *gal2* (XII) *lys9* (XIV) *pet8* (XIV) *ade2* (XV) *gal5*
STX66-4A	*MA*T**a** *rad18* (III) *lys4* (IV) *trp1* (IV) *prt3 CUP1* (VIII) *gal2* (XII) *ade2* (XV) *met2* (XIV) *pha2* (XIV)
STX82-3A	*MAT*α *rad18* (III) *lys4* (IV) *prt3 CUP1* (VIII) *gal2* (XII) *ade2* (XV) *met2* (XIV) *pha2* (XIV) *pet2* (XIV)
STX146-19A	*MA*T**a** *leu2* (III) *rad55* (IV) *thr1* (VIII) *cdc11* (X) *trp3* (XI) *met1* (XI) *his3* (XV) *pet17* (XV) *arg1* (XV) *ade2* (XV)
STX153-10C	*MAT*α *ade1* (I) *his7* (II) *tyr1* (II) *rad55* (IV) *cdc11* (X) *ura1* (XI) *asp5* (XII) *met6* (V) *his3* (XV) *pet17* (XV) *gal2* (XII)
STX84-5A	*MA*T**a** *ade1* (I) *rad57* (IV) *cdc4* (VI) *ura3* (V) *arg4* (VIII) *gal2* (XII) *pet8* (XIV) *aro7* (XVI)
STX75-3C	*MAT*α *ade1* (I) *his4* (III) *rad57* (IV) *cdc4* (VI) *ura3* (V) *leu1* (VII) *arg4* (VIII) *his5* (IX) *pet8* (XIV) *aro7* (XVI)
STX147-9B	*MA*T**a** *lys2* (II) *his7* (II) *tyr1* (II) *gal1* (II) *trp1* (IV) *cly8* (VII) *ade5* (VII) *aro2* (VII) *met13* (VII) *lys5* (VII) *trp5* (VII) *cyh2* (VII) *arg4* (VIII) *lys1* (IX) *ura4* (XII) *gal2* (XII) *rad56* (XVI) [ρ⁻]
STX147-4C	*MAT*α *ade1* (I) *his7* (II) *tyr1* (II) *gal1* (II) *cly8* (VII) *ade5* (VII) *aro2* (VII) *met13* (VII) *lys5* (VII) *cyh2* (VII) *arg4* (VIII) *lys1* (IX) *ura4* (XII) *gal2* (XII) *ade2* (XV) *rad56* (XV)
STX83-17D	*MA*T**a** *leu2* (III) *pet14* (IV) *rad50-1* (XIV) *rna3* (IV) *trp3* (XI) *gal2* (XII) *arg1* (XV) *ade2* (XV)

(*continued*)

TABLE I (continued)

X4120-19D	MATα lys2 (II) leu2 (III) pet14 (IV) rna3 (IV) ade8 (IV) aro1D (IV) met10 (VI) ade5 (VII) leu1 (VII) CUP1 (VIII) trp3 (XI) gal2 (XII) arg1 (XV) Rad⁻
X4126-6D	MATa gal1 (II) his4 (III) leu2 (III) CUP1 (VIII) his5 (IX) ilv3 (X) ura1 (XI) ade3 (VII) gal2 (XII) rad52 (XIII) rna1 (XIII)
STX77-6C	MATα gal1 (II) his4 (III) trp1 (IV) hom3 (V) ura3 (V) CUP1 (VIII) ilv3 (X) ade3 (VII) rad52 (XIII) rna1 (XIII)
STX155-3C	MATa ade1 (I) gal1 (II) lys5 (VII) met13 (VII) aro2 (VII) his6 (IX) ura2 (X) gal2 (XII) lys7 (XIII) prt1 (XV) rad1 (XVI) met4 (XIV)
STX155-9B	MATα gal1 (II) lys5 (VII) his6 (IX) ura2 (XII) gal2 (XII) lys7 (XIII) prt1 (XV) ade2 (XV) rad1 (XVI) met4 (XIV)
X4037-14C	MATa gal1 (II) leu2 (III) arg9 (V) ilv3 (X) met14 (XI) lys7 (XIII) pet17 (XV) trp1 (IV) (temperature sensitive)
X4036-10D	MATa ade1 (I) ade2 (XV) ura3 (V) cdc14 (VI) arg4 (VIII) his6 (IX) his5 (IX) pet8 (XIV) aro7 (XII) gal2 (XII)
X4034-22C	MATa ade1 (I) gal1 (II) leu2 (III) trp1 (IV) ura3 (V) cdc14 (VI) leu1 (VII) his2 (VI) arg4 (VIII) his6 (IX) lys7 (XIII) met14 (XI) asp5 (XII) pet17 (XV)
X3127-27D	MATα ade1 (I) leu2 (III) trp1 (IV) leu1 (VII) thr1 (VIII) arg4 (VIII) lys7 (XIII) pet17 (XV)
X4031-15A	MATα ade1 (I) gal1 (II) trp1 (IV) ura3 (V) his2 (VI) leu1 (VII) arg4 (VIII) his6 (IX) met14 (XI) asp5 (XII) aro7 (XVI)
X3144-11A	MATα leu2 (III) trp1 (IV) arg9 (V) his6 (IX) ilv3 (X) met14 (XI) pet8 (XIV) pet19
Centromere-marked mapping strains[11]	
GT153-6A	MATα ade1 (I) gal7 (II) leu2 (III) trp1 (IV) ura3 (V) his2 (VI) leu1 (VII) arg4 (VIII) his6 (IX) ilv3 (X) met14 (XI) asp5 (XII) lys7 (XIII) lys9 (XIV) ade2 (XV) aro7 (XVI) met2
GT153-63B	MATa ade1 (I) gal7 (II) leu2 (III) trp1 (IV) ura3 (V) his2 (VI) leu1 (VII) arg4 (VIII) his6 (IX) ilv3 (X) met14 (XI) asp5 (XII) lys7 (XIII) lys9 (XIV) ade2 (XV) aro7 (XVI) met2

(continued)

TABLE I *(continued)*

Mapping strains[12]

Strain	Genotype
A141-1D	*MA*T**a** *leu2-3* (III) *met1* (XI) *ade6* (IX) *cdc11* (X) *pet17* (XV) *lys11* (IX) *his1* (V)
A141-37C	*MA*Tα *leu2-3* (III) *met1* (XI) *ade6* (IX) *cdc11* (X) *pet17* (XV) *lys11* (IX) *his1* (V)
A298-65C	*MA*T**a** *leu2-3* (III) *pet2* (XIV) *arg4* (VIII) *ade8* (IV) *aro1C* (IV) *trp4* (IV) *rna3* (IV) Met⁻ Thr⁻
A298-61D	*MA*Tα *leu2-3* (III) *pet2* (XIV) *arg4* (VIII) *ade8* (IV) *aro1C* (IV) *trp4* (IV) *rna3* (IV) Ura⁻ Thr⁻
A343-1A	*MA*T**a** *leu2-3* (III) *pet14* (IV) *arg1* (XV) *lys7* (XIII) *ura1* (XI) *ade3* (VII) *met6* (V)
A343-6A	*MA*Tα *leu2-3* (III) *pet14* (IV) *arg1* (XV) *lys7* (XIII) *ura1* (XI) *ade3* (VII) *met6* (V)
A236-57B	*MA*T**a** *leu2-3* (III) *trp1* (IV) *met4* (XIV) *aro7* (XVI) *his3* (XV) *lys11* (IX) *SUC2* (IX) *MAL3* (II) *can1* (V)
A236-24C	*MA*Tα *leu2-3* (III) *trp1* (IV) *met4* (XIV) *aro7* (XVI) *his3* (XV) *lys11* (IX) *SUC2* (IX) *MAL3* (II) *can1* (V)
A331-4D	*MA*T**a** *leu2-3* (III) *trp2* (V) *his6* (IX) *ura1* (XI) *ino1* (X) *MAL1* (VII) *tsl1* (I)
A331-4B	*MA*Tα *leu2-3* (III) *trp2* (V) *his6* (IX) *ura1* (XI) *ino1* (X) *MAL1* (VII) *tsl1* (I)
A250-19B	*MA*T**a** *leu2-3* (III) *asp5* (XII) *ilv3* (X) *MAL4* (XI) *SUC*
A256-99A	*MA*T**a** *leu2-3* (III) *trp5* (VII) *lys9* (XIV) *met10* (VI) *ade1* (I) *pet9* (II)
A333-1B	*MA*Tα *leu2-3* (III) *trp5* (VII) *his5* (IX) *lys9* (XIV) *met10* (VI) *ade1* (I) *thr4* (III) *cdc9* (IV)
A193-16C	*MA*T**a** *leu2-3* (III) *met13* (VII) *ade2* (XV) *cdc4* (VI) *pet3* (VIII) *ura4* (XII) *his4-15* (III) *lys2* (II)
A193-23A	*MA*Tα *leu2-3* (III) *met13* (VII) *ade2* (XV) *cdc4* (VI) *pet3* (VIII) *ura4* (XII) *his4-15* (III) *lys2* (II)
A121-3A	*MA*T**a** *leu2-3* (III) *met14* (XI) *ade5* (VII) *pet8* (XIV) *ura3* (V) *his7* (II) *lys1* (IX)
A121-3D	*MA*Tα *leu2-3* (III) *met14* (XI) *ade5* (VII) *pet8* (XIV) *ura3* (V) *his7* (II) *lys1* (IX)
A334-27B	*MA*T**a** *leu2-3* (III) *pha2* (XIV) *petx* (XIV) *prt1* (XV) *arg8* (XV)
A334-27B	*MA*Tα *leu2-3* (III) *pha2* (XIV) *petx* (XIV) *prt1* (XV) *arg8* (XV) His⁻ Ura⁻

(continued)

TABLE I (continued)

Mapping strains[a]	
K382-23A	*MAT*a *spo11 ura3* (V) *can1* (V) *cyh2* (VII) *ade2* (XV) *his7* (II) *hom3* (V)
K382-19D	*MAT*α *spo11 ura3* (V) *can1* (V) *cyh2* (VII) *ade2* (XV) *his7* (II) *hom3* (V) *tyr1* (II)
K398-4D	*MAT*a *spo11 ura3* (V) *ade6* (VII) *arg4* (VIII) *aro7* (XVI) *asp5* (XII) *met14* (XI) *lys2* (II) *pet17* (XV) *trp1* (IV)
K381-9D	*MAT*α *spo11 ura3* (V) *ade6* (VII) *arg4* (VIII) *aro7* (XVI) *asp5* (XII) *met14* (XI) *lys2* (II) *pet17* (XV) *trp1* (IV)
K399-7D	*MAT*a *spo11 ura3* (V) *his2* (VI) *leu1* (VII) *lys1* (IX) *met4* (XIV) *pet8* (XIV)
K393-35C	*MAT*α *spo11 ura3* (V) *his2* (VI) *leu1* (VII) *lys1* (IX) *met4* (XIV) *pet8* (XIV)
K396-11A	*MAT*a *spo11 ura3* (V) *ade1* (I) *his1* (V) *leu2* (III) *lys7* (XIII) *met3* (X) *trp5* (VII)
K396-22B	*MAT*α *spo11 ura3* (V) *ade1* (I) *his1* (V) *leu2* (III) *lys7* (XIII) *met3* (X) *trp5* (VII)
K381-15C	*MAT*a *ura3* (V) *ade6* (VII) *arg4* (VIII) *aro7* (XVI) *asp5* (XII) *met14* (XI) *lys2* (II) *pet17* (XV) *trp1* (IV)
K381-10A	*MAT*α *ura3* (V) *ade6* (VII) *arg4* (VIII) *aro7* (XVI) *asp5* (XII) *met14* (XI) *lys2* (II) *pet17* (XV) *trp1* (IV)
K393-27C	*MAT*a *ura3* (V) *his2* (VI) *leu1* (VII) *lys1* (IX) *met4* (XIV) *pet8* (XIV)
K393-2D	*MAT*α *ura3* (V) *his2* (VI) *leu1* (VII) *lys1* (IX) *met4* (XIV) *pet8* (XIV)
K396-27B	*MAT*a *ura3* (V) *ade1* (I) *his1* (V) *leu2* (III) *lys7* (XIII) *met3* (X) *trp5* (VII)
K396-11B	*MAT*α *ura3* (V) *ade1* (I) *his1* (V) *leu2* (III) *lys7* (XIII) *met3* (X) *trp5* (VII)

[a] Chromosomal assignments of some markers are indicated in parentheses.
[b] Strains are available from the Yeast Genetics Stock Center (Berkeley, CA).

lose at high frequencies the integrated 2-μm plasmid DNA plus the chromosome into which integration occurred. The recessive mutation will be revealed only in the diploid formed with the *cir*° tester strain that contains an integrant at the centromeric region of the same chromosome on which the mutation is located. The mutation will remain heterozygous in all of the other diploid strains.

The tester strains were constructed with derivatives of the plasmid

YEp24 which contained the $URA3^+$ gene, a 2-μm segment, pBR322, and different yeast segments corresponding to regions at or near the centromere of each of the 16 chromosomes. The YEp24 derivatives were integrated into a yeast strain, and the [cir^o] MATa and [cir^o] MATα strains listed in Table II were prepared.

A recessive mutation, m^-, in a cir^+ strain is crossed to each of the 16 cir^o tester strains of opposite mating type (Table II). A cir^o/cir^+ diploid strain for each of the 16 crosses is isolated on selective medium, then subcloned onto a nonselective medium. An isolated subclone is diluted and plated on a nonselective medium at a concentration which results in about 50 to 100 isolated colonies per plate. Several hundred colonies of each diploid isolate are replica plated onto a medium which detects the unmapped mutation, m^-. The phenotype conferred by the recessive mutation m^- will be manifested only in the diploid strain derived from the cir^o tester strain with an integrant in the same chromosome that the m^- mutation resides. A recessive mutation should be uncovered at a frequency between 1 and 50%, depending on both the particular chromosomal tester strain used and the position of the mutation on the chromosome. In addition, mutations on chromosome IX can be assigned to either the left or right chromosome arm by using strains B-7175 and B-7176 (Table II). In diploids containing B-7175, mutations on the left arm of chromosome IX are expressed at frequencies greater than 10%, whereas mutations on the right arm are expressed at frequencies of less than 10%. In contrast, in diploids containing B-7176, mutations on the left and right arm of chromosome IX are expressed at frequencies of less than and greater than 10%, respectively.

Procedure 3: Hybridization to Chromosomes

Pulsed-field gel electrophoresis and special methods for preparing high molecular weight DNA from yeast cells have allowed the separation on agarose gels of full-length chromosome-sized DNA from *Saccharomyces cerevisiae*. The following procedures that generate electric fields in alternating orientations have been introduced: pulsed-field gradient gel electrophoresis (PFGGE);[17] orthogonal-field alternation gel electrophoresis (OFAGE);[18] field-inversion gel electrophoresis (FIGE);[19] contour-clamped homogeneous electric-field gel electrophoresis (CHEF);[20] transverse alter-

---

[17] D. C. Schwartz and C. R. Cantor, *Cell (Cambridge, Mass.)* **37**, 67 (1984).
[18] G. F. Carle and M. V. Olson, *Nucleic Acids Res.* **12**, 5647 (1984).
[19] G. F. Carle, M. Frank, and M. V. Olson, *Science* **232**, 65 (1986).
[20] G. Chu, D. Vollrath, and R. Davis, *Science* **234**, 1582 (1986).

TABLE II
Mapping Strains of *Saccharomyces cervisiae*[a]

Strain	Tester	Genotype
B-7588	CHR1::$URA3^+$	[cir°] MATa ura3-52 leu2-3,112 trp1-289 met2
B-7170	CHR2::$URA3^+$	[cir°] MATa ura3-52 leu2-3,112 trp1-289 his3-$\Delta 1$ met2 Cyh^R
B-7171	CHR3::$URA3^+$ $LEU2^+$	[cir°] MATa ura3-52 leu2-3,112 trp1-289 his3-$\Delta 1$ met2 Cyh^R
B-7589	CHR4::$URA3^+$	[cir°] MATa ura3-52 leu2-3,112 trp1-289 his3-$\Delta 1$ met2 Cyh^R
B-7590	CHR5::$URA3^+$	[cir°] MATa ura3-52 leu2-3,112 trp1-289 his3-$\Delta 1$ met2 Cyh^R
B-7591	CHR6::$URA3^+$	[cir°] MATa ura3-52 leu2-3,112 trp1-289 his3-$\Delta 1$ met2
B-7173	CHR7::$URA3^+$	[cir°] MATa ura3-52 leu2-3,112 trp1-289 his3-$\Delta 1$ met2 Cyh^R
B-7174	CHR8::$URA3^+$	[cir°] MATa ura3-52 leu2-3,112 trp1-289 his3-$\Delta 1$ met2 Cyh^R
B-7175	CHR9::$URA3^+$	[cir°] MATa ura3-52 leu2-3,112 trp1-289 his3-$\Delta 1$ met2 Cyh^R
B-7176	CHR9::$URA3^+$	[cir°] MATa ura3-52 leu2-3,112 trp1-289 his3-$\Delta 1$ met2 Cyh^R
B-7593	CHR10::$URA3^+$	[cir°] MATa ura3-52 leu2-3,112 trp1-289 his3-$\Delta 1$ met2
B-7178	CHR11::$URA3^+$	[cir°] MATa ura3-52 leu2-3,112 trp1-289 his3-$\Delta 1$ met2 Cyh^R
B-7595	CHR12::$URA3^+$	[cir°] MATa ura3-52 leu2-3,112 trp1-289 met2 Cyh^R
B-7255	CHR13::$URA3^+$	[cir°] MATa ura3-52 leu2-3,112 trp1-289 his3-$\Delta 1$ met2 Cyh^R
B-7596	CHR14::$URA3^+$	[cir°] MATa ura3-52 leu2-3,112 trp1-289 his3-$\Delta 1$ met2 Cyh^R
B-7180	CHR15::$URA3^+$	[cir°] MATa ura3-52 leu2-3,112 trp1-289 his3-$\Delta 1$ met2 Cyh^R
B-7598	CHR16::$URA3^+$	[cir°] MATa ura3-52 leu2-3,112 trp1-289 met2 Cyh^R
B-7599	CHR1::$URA3^+$	[cir°] MATα ura3-52 leu2-3,112 trp1-289 met2
B-7600	CHR2::$URA3^+$	[cir°] MATα ura3-52 leu2-3,112 trp1-289 his3-$\Delta 1$ met2 Cyh^R
B-7601	CHR3::$URA3^+$ $LEU2^+$	[cir°] MATα ura3-52 leu2-3,112 trp1-289 his3-$\Delta 1$ met2 Cyh^R
B-7602	CHR4::$URA3^+$	[cir°] MATα ura3-52 leu2-3,112 trp1-289 his3-$\Delta 1$ met2 Cyh^R
B-7603	CHR5::$URA3^+$	[cir°] MATα ura3-52 leu2-3,112 trp1-289 his3-$\Delta 1$ met2 Cyh^R
B-7604	CHR6::$URA3^+$	[cir°] MATα ura3-52 leu2-3,112 trp1-289 his3-$\Delta 1$ met2 Cyh^R
B-7605	CHR7::$URA3^+$	[cir°] MATα ura3-52 leu2-3,112 trp1-289 his3-$\Delta 1$ met2 Cyh^R
B-7606	CHR8::$URA3^+$	[cir°] MATα ura3-52 leu2-3,112 trp1-289 his3-$\Delta 1$ met2 Cyh^R
B-7607	CHR9::$URA3^+$	[cir°] MATα ura3-52 leu2-3,112 trp1-289 his3-$\Delta 1$ met2 Cyh^R
B-7608	CHR10::$URA3^+$	[cir°] MATα ura3-52 leu2-3,112 trp1-289 his3-$\Delta 1$ met2 Cyh^R
B-7609	CHR11::$URA3^+$	[cir°] MATα ura3-52 leu2-3,112 trp1-289 his3-$\Delta 1$ met2 Cyh^R
B-7610	CHR12::$URA3^+$	[cir°] MATα ura3-52 leu2-3,112 trp1-289 met2 Cyh^R
B-7611	CHR13::$URA3^+$	[cir°] MATα ura3-52 leu2-3,112 trp1-289 his3-$\Delta 1$ met2 Cyh^R
B-7612	CHR14::$URA3^+$	[cir°] MATα ura3-52 leu2-3,112 trp1-289 met2
B-7613	CHR15::$URA3^+$	[cir°] MATα ura3-52 leu2-3,112 trp1-289 his3-$\Delta 1$ met2 Cyh^R
B-7614	CHR16::$URA3^+$	[cir°] MATα ura3-52 leu2-3,112 trp1-289 met2 Cyh^R

[a] Strains are available from the Yeast Genetics Stock Center (Berkeley, CA). From Ref. 13.

nating-field electrophoresis (TAFE);[21] and rotary gel electrophoresis (RGE).[22] FIGE uses a conventional gel electrophoresis apparatus and alternates the direction of the fields by 180°, with the forward pulse time typically 3 times that of the backward pulse time. In addition, FIGE

[21] K. Gardiner, W. Laas, and D. Patterson, *Somatic Cell Mol. Genet.* **12**, 185 (1986).
[22] P. Serwer, *Electrophoresis* **8**, 301 (1987).

requires variation of the switch-time intervals during the course of the run, using a computer-assisted timer. PFGGE, OFAGE, CHEF, TAFE, and RGE require special electrophoresis chambers, but they utilize a simple interval timer. The PFGGE, CHEF, TAFE, and RGE configurations are commercially available. PFGGE and OFAGE produce DNA bands that are sharper, but the trajectory of the DNA is not straight and varies from lane to lane. CHEF, TAFE, FIGE, and RGE produces straight lines, but the DNA bandwidths can be wider.

Chromosomal-sized DNA from *S. cerevisiae* is easily resolved with all of these methods (PFGGE, OFAGE, FIGE, CHEF, and RGE). In fact, *S. cerevisiae* DNA is well-studied and is often used for molecular size markers when investigating chromosomal-sized DNA or large fragments from other sources. The commercially available CHEF system, Hex · a · field Apparatus (BRL, Gaithersburg, MD), is a convenient, compact system that uses gels with 10 lanes. The CHEF-DR II Megabase DNA Electrophoresis System (Bio-Rad Laboratories, Richmond, CA) allows the separation of over 12-megabase DNA segments as straight lanes and can accommodate up to 30 samples.

In addition, preparations of separated chromosomal DNA are commercially available as agarose gel wafers, consisting of DNA entrapped in dried agarose gels (*Saccharomyces cerevisiae* CHROMO-DI-HYBRIDIZER, Cat. No. 7002-1, Clontech Laboratories, Inc., Palo Alto, CA). Although expensive, this product is advisable for occasional users who do not have pulsed-field electrophoresis equipment.

The assignment of a cloned fragment is easily accomplished by hybridization of a labeled probe to a DNA blot of separated chromosome-sized DNA molecules, as described elsewhere in this volume.[23]

Procedure 4: Hybridization to Fragments

A physical map of the entire yeast genome is currently being constructed by M. V. Olson and co-workers.[24] Two different approaches are being used; one consists of the construction of a low-resolution *Not*I and *Sfi*I restriction map, and the other entails the assignment and ordering of yeast inserts in bacteriophage λ clones to these fragments. Construction of the restriction map with the *Sfi*I and *Not*I enzymes, which have 8-base pair (bp) target sites, is essentially complete (Fig. 2).[25] The yeast nuclear genome contains 61 *Sfi*I sites (considering only the two flanking sites in the

---

[23] S. L. Gerring, C. Connelly, and P. Hieter, this volume [4].
[24] M. V. Olson, unpublished.
[25] A. J. Link and M. V. Olson, unpublished.

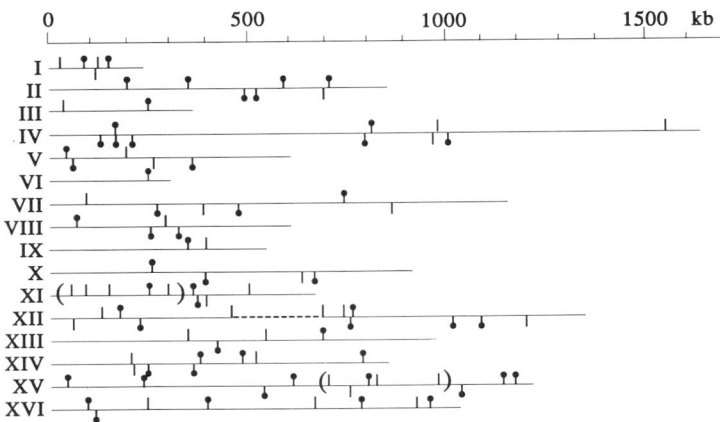

FIG. 2. Physical map of the entire yeast genome, showing *Sfi*I and *Not*I sites on each of the 16 chromosomes. Lines above and below the chromosomes denote, respectively, *Sfi*I and *Not*I sites in DNA from *S. cerevisiae* strain AB972. Filled circles indicate that a phage λ clone encompassing the site has been identified. The order of fragments enclosed in brackets is unknown. The ribosomal region on chromosome XII is denoted by dashes. Because the map is based on digests with solely *Sfi*I and solely *Not*I, the relative positions of nearby restriction sites are ambiguous and are arbitrarily presented.[25]

rDNA cluster) and 38 *Not*I sites, which have been ordered to create a complete physical map of the 16 yeast chromosomes. Ambiguities in the order of fragments remain on chromosomes XI (involving 4 *Sfi*I fragments) and XV (involving 3 *Sfi*I fragments). The 61 *Sfi*I sites define 77 fragments, all of which have been sized and most of which have been identified by specific hybridization probes. The 54 fragments produced by the 38 *Not*I sites have also been analyzed in a similar fashion.

The low-resolution map is used to assign contiguous fragments from a λ bacteriophage yeast gene bank, which consists of approximately 5000 inserts with an average size of 15 kb.[26] Most of the *Sfi*I and *Not*I junctions have been identified within the phage λ clones (Fig. 2). So far, approximately 90% of the nuclear genome is represented among approximately 1000 chosen phage λ clones, which have been assigned to the *Sfi*I–*Not*I restriction map. [These 1000 phage λ clones have been deposited with the American Type Culture Collection (ATCC) for distribution.] These 1000 yeast inserts in phage λ are distributed to three filters for convenient hybridization tests. Thus, the location of a gene within a chromosomal region can be simply determined by hybridizing a labeled, cloned fragment

[26] M. Olson, J. Dutchik, M. Graham, G. Brodeur, C. Helms, M. Frank, M. MacCollin, R. Scheinman, and T. Frank, *Proc. Natl. Acad. Sci. U.S.A.* **83**, 7826 (1986).

to a set of phage λ clones that cover the entire genome or a particular chromosome. So far, over 30 yeast genes have been mapped by this procedure.[24] Details of this method will be available before the end of 1990. Furthermore, a high-resolution map, consisting of *Hin*dIII and *Eco*RI sites of the phage λ clones, is being constructed. After completion of the physical maps, a cloned segment may be accurately related to the yeast genome from its *Hin*dIII and *Eco*RI restriction sites and from its pattern of hybridization to yeast inserts in phage λ.

Procedure 5: Chromosomal Fragmentation

Vollrath *et al.*[27] have described an elegant method for physically mapping any cloned DNA segment by fragmenting a yeast chromosome into proximal and distal pieces by integrative transformation and sizing the resulting chromosomal fragments on pulsed-field gels. This procedure is described in detail elsewhere in this volume.[23] Because of the availability of the more convenient Procedure 4 (hybridization to fragments), the fragmentation of chromosomes may be useful only in special cases when other procedures fail.

Procedure 6: Mitotic Mapping

Mitotic crossing-over in diploid strains results in the homozygosity of all markers located in the same chromosome arm distal to the point of exchange. Mitotic mapping is most useful for determining the chromosomal arm on which a gene resides, especially if the chromosome has already been identified. If a gene shows no centromere linkage, it is sometimes more efficient to first identify on which chromosome arm the gene is located before undertaking a detailed meiotic analysis. Mitotic mapping also reveals the linkage of genes that are far apart on the same arm, whereas the linkage of widely separated genes is less evident with meiotic mapping.

Since the spontaneous rate of mitotic recombinations is very low, it is necessary to use either a selective system or to increase the rate of mitotic crossing-over by treating cells with a recombinogenic agent or mutagen, as described below. If the treatment is not too severe, multiple exchanges are rare, and most recombinations are due to a single mitotic crossing-over.

Spontaneous or induced homozygosity of heterozygous markers in diploid strains is primarily due either to mitotic crossing-over or to mitotic gene conversion (see Fig. 3). Mitotic crossing-over results in the reciprocal

[27] D. Vollrath, R. Davis, C. Connelly, and P. Hieter, *Proc. Natl. Acad. Sci. U.S.A.* **85,** 6027 (1988).

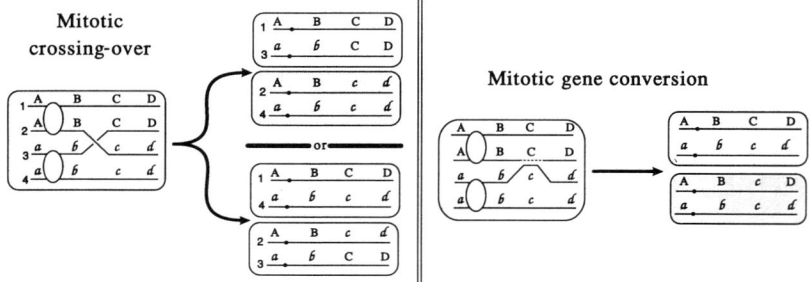

FIG. 3. Segregation of heterozygous markers *ABCD/abcd* in a diploid cell after mitotic crossing-over and mitotic gene conversion. **Mitotic crossing-over** results in equal frequencies of two types of sectored colonies, depending on the assortment of the centromeres: **(top)** one sector is homozygous for *c* and *d*; **(bottom)** both sectors are heterozygous for all markers, but there is a reversal of the gene pairs *c/C* and *d/D* in one sector. **Mitotic gene conversion** results in homozygosity of the single marker *c*. Sectors homozygous for recessive markers are stippled.

exchanges of genes located distal to the point of the event on the chromosome. After cell division, there is a 50% chance that the cell will give rise to a sectored colony which is homozygous for the allelic pairs of distal markers. This is revealed by the requirement or phenotype of the recessive genes in a sector of the colony. There is also an equal chance that the cell will give rise to a colony that is entirely heterozygous, which will be phenotypically indistinguishable from the parental strain. In this case, however, a sector of the colony will maintain its parental configuration while the other sector will have a reversal of the linked genes which were distal to the point of exchange. In contrast, mitotic gene conversion results in a nonreciprocal exchange of a single gene, and it is believed to be analogous to irregular segregations which are observed at low frequencies after meiosis (meiotic gene conversion). If a sector is homozygous for only a single gene, one would have to perform genetic analysis on the opposite sector in order to determine if the event was due to crossing-over or gene conversion. However, if two or more adjacent genes sector together, the event in all likelihood is due to mitotic crossing-over.

There is an approximately linear relationship between the frequency of induced homozygosity of a gene and its centromere distance, a result consistent with mitotic crossing-over. However, the above relationship breaks down and multiple events obscure the linkage of adjacent genes if mutagenic treatments are too severe. Although mitotic mapping is best carried out with X-irradiation because of linear responses and the low proportion of mitotic gene conversion,[28] UV-irradiation is safer and adequate for most purposes, and UV sources are more readily available.

Once the gene has been assigned to a chromosome, haploid strains containing the mutation and markers situated on both arms of the chromosome should be constructed; at least some of the markers should be centromere linked. The construction of such strains will also result in at least a limited amount of information on meiotic linkage. The haploid strain having the markers is crossed to any other strain to produce a heterozygous diploid strain. Approximately 100–200 cells plated on YPD medium[29] are irradiated with 30 and 60 J m^{-2}. After incubation, the cultures are replica plated on appropriate media, and the recessive markers are scored. Ideally, homozygosity will be observed for the mutation and for markers on one arm but not the other.

Mitotic mapping can also be carried out with strains having the markers in trans, in which case homozygous sectors will occur on opposite sides of the same colony. Such sectors can be more easily detected by replica plating flat, large colonies that arise on drier plates and YPG medium.[29] Easily detected markers which reside on the same chromosome as the unmapped gene should be used to simplify mitotic mapping. For example, the *ade2* and *ade1* markers lead to the accumulation of a red pigment, whereas other *ade* markers prevent the accumulation. Therefore, the homozygosity of these markers can be used to reveal sectors by visual inspection.[30,31]

Procedure 7: Meiotic Mapping

Meiotic analysis is the traditional method for genetically determining the order and distances between genes of organisms having well-defined genetic systems. Yeast is especially suited for meiotic mapping because the four spores in an ascus are the products of single meiotic event, and the genetic analysis of these tetrads provides a sensitive means for determining linkage relationships of genes presented in the heterozygous condition. It is also possible to map a gene relative to its centromere if known centromere-linked genes are present in the cross. Although the isolation of the four spores from an ascus is one of the more difficult techniques in yeast genetics, requiring a micromanipulator and considerable practice,[9] tetrad analysis is not only useful for linkage studies but also for constructing strains necessary in genetic and biochemical experiments.

[28] S. Nakai and R. K. Mortimer, *Mol. Gen. Genet.* **103,** 329 (1969).
[29] F. Sherman, this volume [1].
[30] H. Roman, *Cold Spring Harbor Symp. Quant. Biol.* **21,** 175 (1956).
[31] F. K. Zimmermann, *Mutat. Res.* **21,** 263 (1973).

Tetrad type	Genes on homologous chromosomes	Genes on nonhomologous chromosomes
Parental ditype (PD)	No crossover	50% where both segregate at 1st division 25% where both segregate at 2nd division
A B A B a b a b		
Nonparental ditype (NPD)	4-strand double crossover 25% of all double crossovers	50% where both segregate at 1st division 25% where both segregate at 2nd division
A b A b a B a B		
Tetratype (T)	Single crossover	100% where one segregates at 1st division & one at 2nd division 50% where both segregate at 2nd division
A B A b a B a b		

There are three classes of tetrads from a hybrid which is heterozygous for two markers, $AB \times ab$: PD (parental ditype), NPD (nonparental ditype), and T (tetratype). The following ratios of these tetrads can be used to deduce gene and centromere linkage:

	PD	NPD	T
	AB	aB	AB
	AB	aB	Ab
	ab	Ab	ab
	ab	Ab	aB
Random assortment	1 :	1 :	4
Linkage	>1 :	<1	
Centromere linkage	1 :	1 :	<4

There is an excess of PD to NPD asci if two genes are linked. If two genes are on different chromosomes and are linked to their respective centromeres, there is a reduction of the proportion of T asci. If two genes are on different chromosomes and at least one gene is not centromere-linked, or if two genes are widely separated on the same chromosome, there is independent assortment and the PD:NPD:T ratio is 1:1:4. The origin of different tetrad types is illustrated in Fig. 4.

The number of tetrads required for determining linkage is dependent on the nearness of the markers. Useful parameters for deducing statistically significant deviations from 1:1 for various PD/NPD ratios are presented in Table III.[32]

The frequencies of PD, NPD, and T tetrads can be used to determine the map distance in centimorgans (cM) between two genes if there are two

[32] D. D. Perkins, *Genetics* **38,** 187 (1952).

FIG. 4. Origin of different tetrad types. Different tetrad types are produced with genes on (1) homologous or (2) nonhomologous chromosomes from the cross $AB \times ab$. When PD > NPD, then the genes are on homologous chromosomes because of the rarity of NPD, which arise from four-strand double crossovers. The tetratype (T) tetrads arise from single crossovers. See text for the method of converting the %T and %NPD tetrads to map distances when genes are on homologous chromosomes. If the genes are on nonhomologous chromosomes, then PD = NPD because of independent assortment. Tetratype tetrads of genes on nonhomologous chromosomes arise by crossovers between either of the genes and their centromere, as shown at lower right. The %T can be used to determine centromere distances if the centromere distance is known for one of the genes (see text).

TABLE III
SMALLEST PD/NPD RATIOS SHOWING SIGNIFICANT DEVIATIONS IN ONE DIRECTION FROM 1:1[a]

Total no.	Ratios at various levels of significance			Total no.	Ratios at various levels of significance		
	5%	2.5%	1%		5%	2.5%	1%
5	5:0	—	—	28	19:9	20:8	21:7
6	6:0	6:0	—	29	20:9	21:8	22:7
7	7:0	7:0	7:0	30	20:10	21:9	22:8
8	7:1	8:0	8:0	31	21:10	22:9	23:8
9	8:1	8:1	9:0	32	22:10	22:10	23:9
10	9:1	9:1	10:0	33	22:11	23:10	24:9
11	9:2	10:1	10:1	34	23:11	24:10	25:9
12	10:2	10:2	11:1	35	23:12	24:11	25:10
13	10:3	11:2	12:1	36	24:12	25:11	26:10
14	11:3	12:2	12:2	37	24:13	25:12	26:11
15	12:3	12:3	13:2	38	25:13	26:12	27:11
16	12:4	13:3	14:2	39	26:13	27:12	28:11
17	13:4	13:4	14:3	40	26:14	27:13	28:12
18	13:5	14:4	15:3	41	27:14	28:13	29:12
19	14:5	15:4	15:4	42	27:15	28:14	29:13
20	15:5	15:5	16:4	43	28:15	29:14	30:13
21	15:6	16:5	17:4	44	28:16	29:15	31:13
22	16:6	17:5	17:5	45	29:16	30:15	31:14
23	16:7	17:6	18:5	46	30:16	31:15	32:14
24	17:7	18:6	19:5	47	30:17	31:16	32:15
25	18:7	18:7	19:6	48	31:17	32:16	33:15
26	18:8	19:7	20:6	49	31:18	32:17	34:15
27	19:8	20:7	20:7	50	32:18	33:17	34:16

[a] The table can be used to conveniently determine statistically significant deviations of PD = NPD, that is, to deduce if PD > NPD and therefore if the genes are linked. For example, when a total of 7 tetrads are analyzed, all must be PD to conclude that linkage exists at the 1% level of significance. On the other hand, only 6 out of 25 tetrads must be NPD to deduce linkage at the 1% level of significance. From Ref. 32.

or fewer exchanges within the interval:[33]

$$cM = \frac{100}{2}\left[\frac{T + 6NPD}{PD + NPD + T}\right]$$

The equation for deducing map distances in centimorgans is accurate for distances up to approximately 35 cM. For larger distances up to approxi-

---

[33] D. D. Perkins, *Genetics* **34**, 607 (1949).

mately 75 cM, the value can be corrected by the following empirically derived equation[34]:

$$\text{cM (corrected)} = \frac{(80.7)(\text{cM}) - (0.883)(\text{cM})^2}{83.3 - \text{cM}}$$

Similarly, the distance between a marker and its centromere can be approximated from the percentage of T tetrads with a tightly linked centromere marker, such as *trp1*:

$$\text{cM}' = \frac{100}{2}\left[\frac{\text{T}}{\text{PD} + \text{NPD} + \text{T}}\right]$$

In practice, the meiotic mapping of genes is best carried out with as many known markers as can be introduced in the fewest crosses. Several workers have systematically constructed sets of strains containing markers distributed over the entire genome and on specific chromosomes. Meiotic mapping and detection of linkage have been greatly aided by these strains, which are listed in Table I.

Acknowledgments

We wish to thank Drs. M. V. Olson (Washington University) and P. Hieter (The Johns Hopkins University) for unpublished information and Dr. D. Campbell (University of Rochester Medical School) for useful suggestions concerning the writing of this chapter. The writing of this chapter was supported by U.S. Public Health Service Research Grant R01 GM12702.

[34] C. Ma and R. K. Mortimer, *Mol. Cell. Biol.* **3**, 1886 (1983).

[4] Positional Mapping of Genes by Chromosome Blotting and Chromosome Fragmentation

By SANDRA L. GERRING, CARLA CONNELLY, and PHILIP HIETER

Introduction

Mapping genes defined by newly identified mutations is of primary importance in the genetic and molecular analysis of *Saccharomyces cerevisiae*. As many genes in this organism are presently under study, mapping a mutation to a previously identified locus can provide a wealth of information concerning the function of the corresponding gene product. This is especially true if the gene was identified in two quite different screens.

Thus, gene mapping can create the opportunity to exchange mutant alleles and reagents between the respective laboratories and to prevent unnecessary duplication of effort. Mapping a new gene under study should therefore be an early and essential priority.

The majority of genes identified in *S. cerevisiae* have been mapped using tetrad analysis to show genetic linkage to previously identified loci.[1] Although genetic methods are available which can simplify this approach by first narrowing a genetic locus down to a single chromosome,[1] tetrad analysis is relatively time consuming. Since the cloning of genes corresponding to unmapped mutations is often easy, new mapping methods have been developed which utilize these cloned DNA segments.

Pulsed-field gel electrophoresis of chromosome-sized DNA molecules[2-4] and chromosome fragmentation *in vivo*[5] are two recently developed techniques that facilitate the rapid mapping of cloned genes in *Saccharomyces cerevisiae*. The former technique, in conjunction with Southern hybridization, allows the rapid assignment of a cloned gene to a chromosome. Therefore, identity to genes on all other chromosomes is immediately excluded. The latter technique allows the assignment of a gene to an arm of a chromosome and gives the physical distance of the gene from each telomere on the chromosome as well. By assuming that the physical and genetic maps are colinear and proportional for most regions of the genome, the physical position can be used to identify a small subset of previously mapped mutations as potential alleles of the gene being mapped. Identity or nonidentity can then be checked rapidly by a variety of means including complementation tests by mating or DNA-mediated transformation and allelism tests by tetrad analysis. In addition, knowledge of the physical position greatly facilitates the choice of linked genetic markers for meiotic mapping (if required).

In this chapter we introduce the techniques of chromosome blotting and chromosome fragmentation, provide protocols required to implement them, and discuss their advantages, disadvantages, and potential applications. The two techniques rely on several recent advances including pulsed-field gel electrophoresis, the establishment of an electrophoretic karyotype of *Saccharomyces cerevisiae,* and a vector system that allows the construction of chromosome fragments proximal or distal to cloned DNA segments.

[1] R. K. Mortimer and D. Schild, *Microbiol. Rev.* **49**, 181 (1985).
[2] D. Schwartz and C. Cantor, *Cell (Cambridge, Mass.)* **37**, 67 (1984).
[3] G. Carle and M. Olson, *Nucleic Acids Res.* **12**, 5647 (1984).
[4] G. Chu, D. Vollrath, and R. W. Davis, *Science* **234**, 1582 (1986).
[5] D. Vollrath, R. Davis, C. Connelly, and P. Hieter, *Proc. Natl. Acad. Sci. U.S.A.* **85**, 6027 (1988).

Chromosome Blotting Method

Principle

Genetic mapping of a new mutation can be simplified by first assigning the gene to one of the 16 yeast chromosomes. The cloning of the wild-type version of a gene corresponding to a new mutation is often straightforward (see [14] in this volume).[6] Chromosome assignment is then easily accomplished by hybridization of a labeled probe to a Southern blot of separated chromosome-sized DNA molecules.

Pulsed-field gel electrophoresis separates large DNA molecules (which are not resolved using standard gel electrophoresis) by placing them in periodically alternating electrical fields.[2] Pulsed-field gel techniques depend on the size-dependent retardation of large DNA molecules [> 50 kilobases (kb)]. The method is thought to work by taking advantage of differences in relaxation time exhibited by different sized DNA molecules as they reorient in an alternating electric field. A simple way to think about the phenomenon is that smaller molecules can reorient or "turn corners" in the gel matrix more rapidly than larger molecules and that the rate of reorientation is roughly proportional to size. As a result, electrophoretic mobilities in the range from zero to several thousand kilobases become a function of size.[2-4]

Pulsed-field gel electrophoresis has been used in conjunction with a method of preparing intact chromosome-sized DNA molecules from yeast spheroplasts[2] to develop a *Saccharomyces cerevisiae* electrophoretic karyotype.[7] The original strain used for the development of this karyotype (strain AB972[7]) contained multiple chromosome-sized DNA molecules which comigrated. Other strains that exhibited chromosome-length polymorphisms (CLPs) were found to resolve some of these chromosomes (see Fig. 1). Thus, use of a set of three *Saccharomyces cerevisiae* strains was necessary to resolve all chromosomes by OFAGE (orthogonal-field alternation gel electrophoresis) except the comigrating doublet of chromosomes VII and XV, for which no CLP was found.[7] In addition, chromosome XII was observed not to enter the gel, though this problem was solved with the development of CHEF (contour-clamped homogeneous electric field) gel electrophoresis.[4] Recently, an *S. cerevisiae* strain was developed by crossing three chromosome-length polymorphisms into a common background and then fragmenting chromosome VII into two resolved DNA bands.[8] In this strain (YPH149) (Table I) all 16 chromosomes are resolved (as 17

---

[6] M. Rose, this series, Vol. 152, p. 481.
[7] G. Carle and M. Olson, *Proc. Natl. Acad. Sci. U.S.A.* **82,** 3756 (1985).
[8] C. Connelly and P. Hieter, unpublished, 1990.

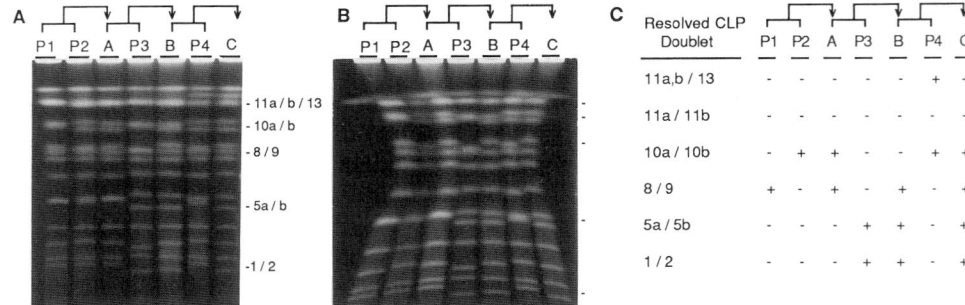

FIG. 1. Construction of YPH80 and its electrophoretic karyotype. (A) Electrophoretic karyotypes (by CHEF gel electrophoresis) of strains used to construct YPH80. Three crosses were performed as indicated and haploid segregants screened on pulsed-field gels for appropriate karyotypes. The parental strains (see Ref. 7) are as follows: P1, AB972; P2, YNN281 (YPH1); P3, A364a; P4, YPH45. The haploid segregants are as follows: A, YPH78; B, YPH79; C, YPH80. Chromosome-sized DNA was stained with ethidium bromide. (B) Electrophoretic karyotypes of all strains (by OFAGE). Note that chromosome XII (band 13) does not enter the gel. (C) Summary of electrophoretic karyotype with respect to resolved chromosome doublets arising from chromosome-length polymorphisms (CLPs) in the strains shown in A and B. YPH80 resolves all chromosomes as distinct bands except VII and XV (band 11).

linear DNA bands) by CHEF electrophoresis, allowing unambiguous chromosomal assignment of a DNA probe in a single experiment (see Fig. 2).

Preparation of Chromosome-Sized DNA in Agarose

The method below for preparation of chromosome-sized DNA in agarose gels is a minor modification of the originally published protocol.[2] Stationary-phase yeast cultures are immobilized in low-melt agarose and further processed by diffusing various reagents into the agarose plugs. Besides the ease of the protocol, the main advantage is that the fragile high molecular weight DNA is supported, and thus protected from breakage. A

TABLE I
GENOTYPES OF *Saccharomyces cerevisiae* STRAINS

Strain	Genotype
YPH49	**a**/α *ura3-52/ura3-52 lys2-801/lys2-801 ade2-101/ade2-101 trp1-Δ1/trp1-Δ1*
YPH80	α *ura3-52 lys2-801 ade2-101 his7 trp1-Δ1*
YPH149	α *ura3-52 lys2-801 ade2-101 his7 trp1Δ1* CFVII(*RAD2*.p.YPH149) [(CFVII(*RAD2*.d.YPH146.*TRP1*)] ρ^-

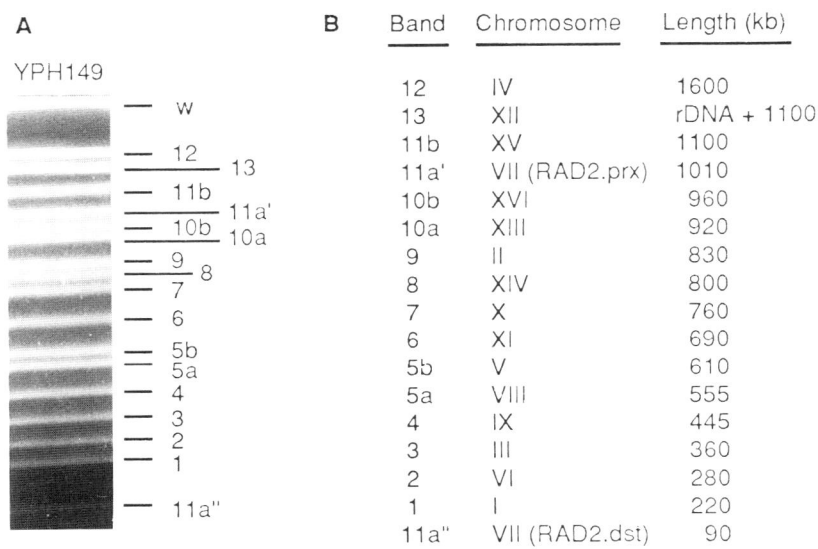

FIG. 2. Electrophoretic karyotype of YPH149 by CHEF gel electrophoresis. (A) The electrophoretic karyotype of the mapping strain YPH149 along with band number assignments is shown. In this strain band 11 (XV and VII in YPH80, Fig. 1) has been resolved into bands 11a' [CFVII (*RAD2*.p.YPH149)], 11a" [CFVII(*RAD2*.d.YPH146.*TRP1*)], and 11b (chromosome XV) by fragmenting chromosome VII at *RAD2*. Chromosome-sized DNA was stained with ethidium bromide. (B) Band numbers, chromosome assignments, and chromosome sizes are listed.

high concentration of EDTA also protects the DNA from the action of nucleases.

Reagents

LET: 0.5 M EDTA, 10 mM Tris, pH 7.5

NDS: 0.5 M EDTA, 10 mM Tris, pH 7.5, 1% N-lauroylsarcosine, pH 9.5, 2 mg/ml proteinase K

EDTA/Tris solution: 50 mM EDTA, 10 mM Tris, pH 7.5

Zymolyase stock: 20 mg/ml (20T) in 10 mM sodium phosphate, pH 7.5

LMP agarose: 1% low-melting point agarose in 125 mM EDTA, pH 7.5

1. Grow 5 ml yeast culture in rich (YPD) broth to stationary phase (OD$_{600}$ 10–14).

2. Transfer 1.0 ml to a microcentrifuge tube (PGC Scientific 2.2-ml tube, #509-220).

3. Pellet cells (10 sec) in microfuge and resuspend in 1 ml EDTA/Tris solution.
4. Wash 2 times with 1 ml EDTA/Tris solution.
5. Resuspend in 0.15 ml of EDTA/Tris solution plus 1 μl Zymolyase (for multiple tubes make a cocktail).
6. Place in 42° water bath/heatblock for 30 sec.
7. Add 0.25 ml of 42° LMP agarose.
8. Gently mix (avoid bubbles) by pipetting in a large-bore pipette tip and place immediately on ice.
9. After the agarose plugs have hardened, add 0.4 ml LET as an overlay.
10. Incubate 8–10 hr or overnight, 37°.
11. Transfer plug to a 12 × 75 mm Falcon tube containing 0.4 ml NDS. This is accomplished by using a small spatula to release the plug from the casting tube onto a clean glass plate.
12. Incubate overnight at 50°.
13. Remove NDS and dialyze the agar plug 4 times (1 hr each wash) by soaking in 2 ml EDTA/Tris solution. Remove each wash with a Pasteur pipette.
14. Store at 4° in EDTA/Tris solution. Plugs stored in this fashion are stable for years.

Running Orthogonal-Field Gels

Reagents

10× TBE: 108 g/liter Tris base, 55 g/liter boric acid, 9.3 g/liter disodium EDTA, pH adjusted to 8.3

1. Set refrigeration on external cooling bath to 4° (temperature in gel chamber should be 10° when voltage is on during the run).
2. Replace silicone tubing in pump if a persistaltic pump is used. If an impeller pump is used this is not necessary.
3. Add 2 liters precooled (4°) 0.5× TBE buffer to gel chamber.
4. Turn on pump and set temporarily at a high flow rate to drive all bubbles out of the system.
5. Boil 50 ml of 1% agarose in 0.5× TBE.
6. Cool agarose to 60°.
7. Pour gel by pipetting 30 ml directly onto 4 inch × 4 inch sandblasted glass plate on a level surface. The comb should be placed directly on the surface of the glass plate.
8. On a glass plate cut off about 50 μl of chromosomal DNA–agarose plug. Blot dry with a Kimwipe.

9. Gently place plug in the bottom of an Eppendorf tube in a 65° heat block.
10. Incubate 5 min or until plug is melted (if the plug does not melt, additional washing in EDTA/Tris solution may be required).
11. Load the melted plug directly into a dry well with a glass capillary pipette, avoiding air bubbles.
12. The loaded gel should be submerged soon after loading or the wells will deform.
13. Taking HEED OF ALL SAFETY PRECAUTIONS, place gel into OFAGE apparatus.
14. Allow gel to equilibrate to buffer temperature (5 min).
15. Perform electrophoresis at 275 V, 10°, for 14 hr, with 40-sec pulses.
16. Stain gel for 30 min in 200 ml of 0.5 μg/ml ethidium bromide dissolved in water.
17. Destain gel in water for 30 min and photograph.

Running Contour-Clamped Homogeneous Electric Field Gels

1–6. Follow instructions 1–6 for running orthogonal-field gels. The running conditions are set for a buffer volume which just immerses the gel (our CHEF apparatus takes 1.4 liters).
7. Pour gel in CHEF apparatus to specifications.
8–14. Follow instruction for OFAGE above.
15. Perform electrophoresis at 200 V, 10° (temperature of buffer in gel chamber when voltage is on), for 14 hr, with 60-sec pulse frequency, followed by 200 V, 10°, for 10 hr, with 90-sec pulse frequency.
16–17. Follow instructions for OFAGE.

Gel Preparation for Southern Transfer and Hybridization

Reagents

Acid solution: 50 mM HCl
Denaturation solution: 1 N NaOH, 1.5 M NaCl
Neutralization solution: 1 M Tris, pH 7.5, 1.5 M NaCl
Hybridization buffer: 6×SSC, 0.1% (w/v) SDS, 4× Denhardt's, 30 mM Tris, pH 7.5, 0.1 mg/ml denatured herring sperm DNA
Wash buffer: 0.1× SSC, 0.05% (w/v) SDS
100× Denhardt's: 2% (w/v) bovine serum albumin, 2% (w/v) Ficoll, 2% (w/v) poly(vinylpyrrolidone)
20× SSC: 3 M NaCl, 0.3 M sodium citrate, pH 7.0

1. Nick the chromosomal DNA by gently shaking the gel in 200 ml acid solution for 15 min at room temperature.
2. Denature in 200 ml denaturation solution for 30 min at room temperature.
3. Neutralize in 200 ml neutralization solution for 30 min at room temperature.
4. Transfer DNA to nitrocellulose by standard Southern transfer method and bake filter at 80° for 2 hr. Alternatively, DNA can be transferred to nylon filters using standard procedures.
5. Prehybridize by presoaking filters in 4× Denhardt's solution, 3× SSC in a baking dish at 70° for 2 hr.
6. Hybridize for 36 hr in hybridization buffer at 68°.
7. Wash filters in wash buffer (500 ml/wash) as follows: 1 time at room temperature, 5 min, 3 times at 55° for 30 min each.
8. Allow filters to air dry and expose to X-ray film.

Application of the Chromosome Blotting Method and Comments

The electrophoretic karyotype patterns obtained using a CHEF apparatus and an OFAGE apparatus do not differ significantly (Fig. 1). One difference is that, using the conditions described above, chromosome XII (band 13) enters the gel when the plugs are run in contour-clamped homogeneous electric fields (CHEF) but not in orthogonally alternating fields (OFAGE). In addition, the trajectory of the DNA using an OFAGE apparatus is not straight.

The electrophoretic karyotypes of the two strains used for mapping are shown in Fig. 1, lane C (YPH80), and Fig. 2 (YPH149). YPH80 resolves all chromosomes except the chromosome XV/VII comigrating doublet (Fig. 1, band 11a/11b). This strain carries no homology to pBR322 sequences, so purification of probe sequences from vector sequences is not necessary. However, a disadvantage of YPH80 is that genes mapping to chromosome XV or VII (15% of genome) cannot be assigned unambiguously. YPH149 involves all chromosomes, but carries homology to pBR322 on two "chromosome fragments" (11a', 11a"; Fig. 2). Thus, a fully resolved electrophoretic karyotype is available in which the 16 chromosomes are represented on 17 resolved DNA molecules. When using strain YPH149, unambiguous assignment can be made in a single experiment, but probe sequences must be purified from vector sequences prior to hybridization.

The chromosome blotting procedure was used to assign the cloned *CTF1* gene[9] to a chromosome. As seen in Fig. 3, the clone hybridizes to band 10b in YPH149, which corresponds to chromosome XVI. Identity of

[9] S. L. Gerring, F. Spencer, and P. Hieter, submitted.

[4] POSITIONAL MAPPING OF CLONED GENES 65

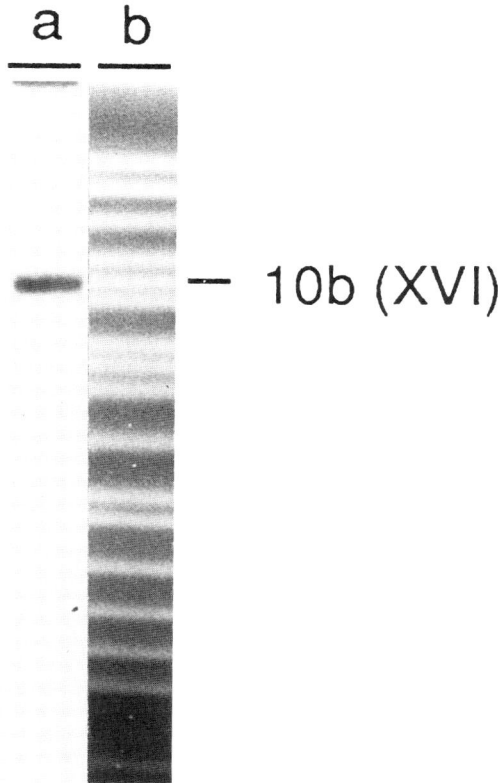

FIG. 3. Mapping of *CTF1* to chromosome XVI by hybridization to a CHEF gel blot of YPH149. An ethidium-stained CHEF gel of YPH149 is shown in (b). All 16 chromosomes are resolved in 17 bands. A blot of the gel was probed with gel-purified *CTF1* DNA (a). The blot and ethidium-stained gel are aligned to reveal that *CTF1* is on band 10b, which corresponds to chromosome XVI (Fig. 2).

CTF1 to genes on all other chromosomes was thus immediately excluded. Examination of the genes present on chromosome XVI provided a list of previously identified loci that the *CTF1* gene might be identical to. One of these, *chl1* (*ch*romosome *l*oss 1)[10] was phenotypically similar to *ctf1* mutants and was an obvious candidate.

Assignment of the hybridizing chromosome band by measurement from the gel origin is sometimes difficult. In such cases it is useful to use a second probe to allow unambiguous assignment. A blot probed with the DNA to be mapped is reprobed with DNA from a known chromosome,

[10] P. Liras, J. McCusker, S. Mascioli, and J. Haber, *Genetics* **88**, 651 (1978).

and the extra signal is used to line up the bands in the gel with reference to the known chromosome band. Any probe with a known chromosomal assignment can be used. We usually use pBR322 DNA which will hybridize to chromosome fragment bands 11a' and 11a" of YPH149 (Fig. 2), providing two independent measurements for determining the position of the hybridizing chromosome band.

Chromosome Fragmentation Method

Principle

Genetic mapping techniques describe chromosomal position in terms of distance (in recombinational map units) between a gene and its centromere, or between a gene and previously mapped loci. We have recently described a method which characterizes chromosomal position in terms of physical distance (in kilobases) between a gene and each of the telomeric ends of a linear chromosomal DNA molecule.[5] The method involves breaking the chromosome at the site of the gene and measuring the lengths of two chromosome fragments (proximal and distal to the gene) on pulsed-field gels.

Chromosome fragmentation relies on the fact that free DNA ends are highly recombinogenic in *Saccharomyces cerevisiae* and interact directly with their homologous sequences.[11] Dunn *et al.* showed that free ends containing the telomere-adjacent Y' sequences can efficiently "heal" (i.e., obtain telomere sequences by recombination with endogenous chromosomes) to a telomere.[12] We took advantage of this knowledge to design two vectors, YCF3 and YCF4,[5] which are useful in positionally mapping genes by chromosome fragmentation (see Ref. 5 and Fig. 4A). A DNA segment is inserted into the polylinker of these vectors in both orientations (Fig. 4A), and the four plasmid constructs are linearized at a unique site between the mapping segment and the Y' sequence. On transformation into yeast, the free ends interact with homologous sequences in the genome, and, depending on the insert orientation and vector, one stable "chromosome fragment" (CF) is generated which contains all chromosomal sequences proximal to the mapping segment and one stable CF is made with all chromosome sequences distal to the mapping segment (Fig. 4B). The lengths of these two stable fragments can be measured on pulsed-field gels, and this indicates the distance (in kilobases) of the cloned gene from each telomere on the chromosome.

[11] T. Orr-Weaver, J. Szostak, and R. Rothstein, *Proc. Natl. Acad. Sci. U.S.A.* **78,** 6354 (1981).
[12] B. Dunn, P. Szauter, M. Pardue, and J. Szostak, *Cell (Cambridge, Mass.)* **39,** 191 (1984).

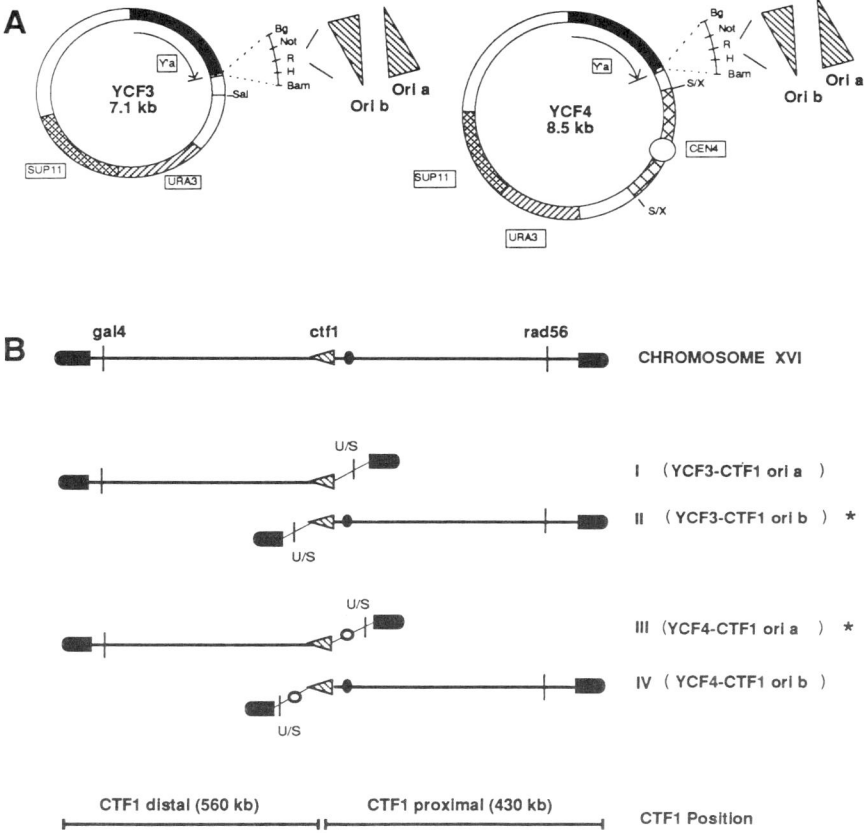

FIG. 4. (A) Yeast chromosome fragmentation (YCF) vectors. YCF3 and YCF4 are derived from pBR322 (unshaded areas) and contain yeast telomere adjacent sequences (Y' sequences, in black) in addition to the *URA3* and *SUP11* genes. *SUP11* is an ochre-suppressing tRNA which allows detection of the presence (*ade2-101* red phenotype is suppressed to pink in diploid strains) or absence (red phenotype) of YCF3- or YCF4-derived chromosome fragments. YCF3 is acentric whereas YCF4 is centric. Orientation (ori a or ori b) of the *CTF1* EcoRI mapping fragment in YCF3 or YCF4 is indicated by the hatched triangle. (B) Chromosome fragmentation with YCF3- and YCF4-*CTF1* mapping constructs. *CTF1* was cloned into both YCF3 and YCF4 in two orientations, a and b (denoted by hatched triangle). On linearization of the plasmids between Y' and *CTF1* and transformation into yeast, the Y' sequences heal to a telomere. Depending on the orientation of *CTF1* in the vectors, chromosome fragments proximal or distal to *CTF1* are obtained. The location of the *URA3* and *SUP11* genes (designated U/S) which are embedded in vector sequences (thin lines) are shown. Constructs YCF3-*CTF1* ori b and YCF4-*CTF1* ori a yield stable monocentric fragments (designated by asterisks).

Procedure

A DNA segment (usually 1–2 kb) to be mapped is cloned in two orientations into the polylinker of each of the two vectors YCF3 and YCF4 (Fig. 4A), yielding four constructs. The *Not*I site (an extremely infrequent restriction site) is usually used to linearize each of the constructs between the mapping segment and the Y' sequence, and each is transformed independently into YPH49. On transformation, the Y' sequence recombines with genomic Y'-telomeric sequences and heals to a telomere. The mapping segment, at the other end of the transformed linear DNA, recombines with its unique homologous genomic sequence, yielding a chromosome fragment. As depicted in Fig. 4B, the orientation of the mapping segment in the two vectors will determine whether the fragment extends in a proximal or distal direction from the mapping segments.

All four linearized constructs will usually yield Ura$^+$ transformants. However, only *one* of the two orientations of mapping segment in *each* vector will give rise to transformants containing stable, monocentric chromosome fragments. The other orientations give rise to transformants containing acentric (Fig. 4B, fragment I) or dicentric fragments (Fig. 4B, fragment IV) which are highly unstable. Strains containing the stable, monocentric fragments (fragments II and III) can be analyzed to determine the physical location of the gene being mapped.

The colony color sectoring assay[13] allows easy distinction between transformants containing mitotically stable monocentric fragments and those containing mitotically unstable acentric and dicentric fragments. All the mapping strains used contain the *ade2-101* ochre mutation which causes an accumulation of red pigment in cells, leading to the formation of red colonies on the appropriate growth medium. This mutation is partially suppressed in a diploid by one copy of an ochre-suppressing tRNA gene, *SUP11,* leading to the generation of homogeneously pink colonies. Two copies of the *SUP11* gene in a diploid fully suppresses the mutation, and the colonies formed will be white.

Only two of the constructs (II and III in the example given, Fig. 4B) give rise to stable monocentric fragments, and these will yield transformants that form homogeneously pink colonies when plated on nonselective medium (Step 18 below). The fragments produced with the other two constructs will give either acentric (I, Fig. 4B) or dicentric fragments (IV, Fig. 4B) which will be unstable and will give rise to red/pink/white sectored colonies on nonselective medium. It is important to realize that up to 50%

---

[13] P. Hieter, C. Mann, M. Snyder, and R. Davis, *Cell (Cambridge, Mass.)* **40**, 381 (1985).

of *all* transformants will derive from recircularized plasmid. Since these are small and contain weak *ARS* function (in contrast to the large linear chromosome fragments), they will be highly unstable[13] when tested for mitotic stability of the *SUP11* marker in sectored colonies. Thus, it is the *presence* of stable pink colonies (among transformants derived from two of the four YCF3 and YCF4 constructs) in the mitotic stability test (Step 18 below) that distinguishes events II and III (stable monocentric fragments) from I and IV (unstable acentric or dicentric fragments).

Chromosome-sized DNA is prepared from homogeneously pink colonies containing the mitotically stable YCF3- and YCF4-derived chromosome fragments, and the sizes of the proximal and distal fragments are determined from their mobility in a pulsed-field gel. Because gel mobility is essentially a linear function of length for the chromosome-sized DNA, size determination is accomplished by measuring the fractional mobility between two bands and computing the CF length from the known sizes of the YPH49 chromosomes (Fig. 5). Finally, to distinguish between the left and right arm of the chromosome, hybridization to the chromosome fragments is performed with a known telomere-adjacent probe from either arm of the chromosome.

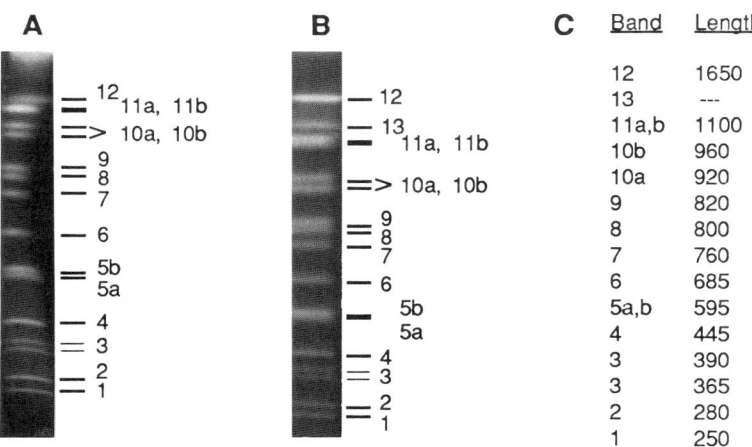

FIG. 5. Electrophoretic karyotype and chromosome lengths of fragmentation mapping strain YPH49. Chromosome-sized DNA was prepared and pulsed-field gels run and photographed as described in the text. (A) OFAGE. (B) CHEF gel electrophoresis. (C) Chromosome band numbers and corresponding DNA lengths, which can be used to determine the lengths of chromosome fragments generated in this strain, are shown.

Reagents

Uracil-minus defined medium: 0.67% (w/v) Bacto-yeast nitrogen base without amino acids, 2% (w/v) dextrose, 2% (w/v) Bacto-agar, 40 μg/mg lysine, 30 μg/ml tryptophan, 6 μg/ml adenine (the plates are limiting for adenine and speed up development of colony color)

Uracil-containing defined medium: same as above except 20 μg/ml uracil is added

Lithium acetate: 0.1 M lithium acetate dissolved in TE

TE: 10 mM Tris base, pH 7.5, 1 mM EDTA

1. Clone a fragment of the gene to be mapped into each of the YCF3 and YCF4 vectors (Fig. 4A) in two orientations. Be sure to leave available a unique restriction site in the polylinker so that the final construct can be linearized between the Y' sequence and the cloned gene. (This is usually the *Not*I site.)

2. From a fresh overnight culture of YPH49, inoculate 50 ml of YPD liquid medium to an OD$_{600}$ of 0.3. Grow to an OD$_{600}$ of approximately 1.0 (about 5 hr at 30°).

3. Collect cells (3500 rpm, 5 min, room temperature).

4. Resuspend cells in 10 ml lithium acetate, collect as in Step 3.

5. Resuspend cells in 10 ml lithium acetate, incubate 1 hr at 30°.

6. Collect and resuspend in 0.5 ml lithium acetate. (These competent cells can be stored at 4° for up to 1 week or frozen in 20% glycerol for months.)

7. Linearize 3 μg of each of the four constructs with an enzyme that cuts between the Y' sequence and the mapping segment (usually *Not*I).

8. Aliquot 50 μl of competent cells into Eppendorf tubes and add 3 μg (in up to 10 μl) linearized DNA to each. Carrier DNA is not needed in the transformation mix.

9. Incubate 10 min at 30°.

10. Add 0.5 ml of 40% (w/v) polyethylene glycol 4000, 10 mM Tris, pH 7.5. Vortex to resuspend cells completely.

11. Incubate 1 hr at 30°.

12. Heat to 42° for 5 min.

13. Add 1 ml sterile water and mix.

14. Collect by spinning for 5 sec in a microcentrifuge at room temperature.

15. Resuspend in 1 ml sterile water.

16. Collect (Step 14) and resuspend in 100 μl sterile water.

17. Plate out on uracil-minus defined medium. Incubate at 30° for 3–4 days.

18. Streak out for single colonies 24 independent transformants for each construct onto nonselective plates (YPD plates or, preferably, uracil-containing defined medium). Incubate at 30° for 3–4 days.

19. Notice which transformations give stably maintained fragments as determined by the colony sectoring assay (homogeneously pink colonies). One orientation of the mapping segment in YCF3 and the opposite orientation of the mapping segment in YCF4 should yield stable uniformly pink colonies.

20. Prepare chromosome-sized DNA in plugs (see protocol above) from four *stable* transformants (Step 19 above) for each vector and run on a pulsed-gel apparatus (see above). Use YPH49 as a parental standard.

21. Determine the sizes of the fragments by comparison with a ladder or with the sizes of the chromosomes of the mapping strain (Fig. 5).

22. Transfer the gel to nitrocellulose using the chromosome blotting method (protocol above) and hybridize with a telomere-adjacent probe that is present on the same chromosome. In essence any probe on that chromosome is applicable, but the telomere-adjacent ones will give unambiguous assignments.

23. Hybridization to the YCF3-generated proximal fragment indicates that the mapping segment is on the chromosome arm opposite the probe. Hybridization to the YCF4-generated distal fragment indicates it is on the same arm as the probe (e.g., see Fig. 6).

Application of Chromosome Fragmentation Method and Comments

Chromosome fragmentation was used to positionally map *CTF1* on chromosome XVI (Fig. 6 and Ref. 9). Approximately the same number of transformants were obtained with 3 µg of linearized centric and acentric mapping construct (Table II). However only YCF3–*CTF1*ori b and YCF4–*CTF1*ori a gave a significant proportion of stable pink transformants on further analysis. Plugs made from two independent transformants containing each of these were subjected to OFAGE (Fig. 6) to determine the CF sizes. YCF3–*CTF1* ori b yielded a fragment proximal to *CTF1* (III, Fig. 4B), and YCF4–CTF1 ori a yielded a fragment distal to CTF1 (III, Fig. 4B). By comparison with other chromosome-sized DNA molecules in YPH49 (Fig. 5), the distal and proximal fragments were estimated to be 560 and 430 kb, respectively. These results mapped *CTF1* 560 kb from the telomere on its own arm and 430 kb from the telomere on its opposite arm. Probing with *GAL4* indicated that the *distal* fragment was derived from the left arm of chromosome XVI. Therefore, *CTF1* was on the left arm of chromosome XVI, 430 kb from the right arm telomere and 560 kb from the left arm telomere. *CHL1*,[10] a candidate for identity to *CTF1* after it was assigned to chromosome XVI, is located in the same

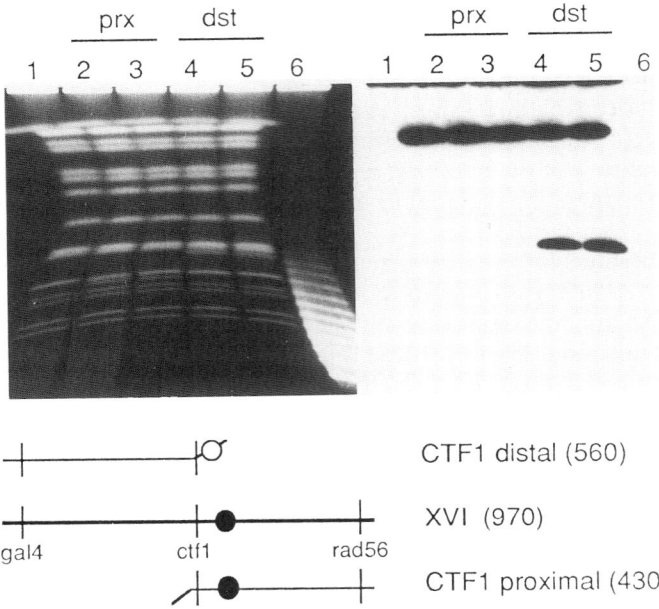

FIG. 6. Physical mapping of *CTF1* by chromosome fragmentation. Plugs containing chromosome-sized DNA were made from stable pink colonies derived from fragmentation events using YCF3–*CTF1* ori b and YCF4–*CTF1* ori a (Fig. 4B). An ethidium-stained gel produced by OFAGE is shown at left. Lane 1 contains YPH49 alone. Lanes 2 and 3 contain DNA from two independent transformants using YCF3–*CTF1* ori b. A new *CTF1* proximal CF is visible as a 430-kb band. Lanes 4 and 5 contain DNA from two independent transformants using YCF4–*CTF1* ori a. A 560-kb *CTF1* distal CF appears as a new band. At right is shown the same gel probed with *GAL4* DNA. Hybridization to the distal fragment indicates that *CTF1* is on the same arm as *GAL4*.

region (by extrapolation of the physical distances to the genetic map). The two genes were subsequently shown to be identical by complementation and genetic linkage.[9]

The main advantage of this method for mapping genes is its generality. There is no need to construct multiple heterozygous strains in order to determine linkage to previously mapped genes. A disadvantage is the prerequisite of first cloning the gene corresponding to a mutation. In general, the physical map distance corresponds quite well with distances determined meiotically by the relationship 1 (cM) centimorgan equals 3.6 kb.[1] After the approximate location on the genetic map has been determined by extrapolating the physical distances to genetic distances, more directed meiotic analysis can be performed, if necessary. Alternatively, other cloned genes can be easily placed proximally or distally to the

TABLE II
ANALYSIS OF TRANSFORMANTS OBTAINED FROM
CONSTRUCTS[a]

Construct	Number of transformants	Stable pink transformants
YCF3-CTF1 ori a	520	2/24
YCF3-CTF1 ori b	447	13/24
YCF4-CTF1 ori a	212	11/24
YCF4-CTF1 ori b	181	1/24

[a] Three micrograms of each of the four plasmids linearized with NotI was transformed into YPH 49. The total number of transformants and number of stable pink colonies was determined for each. Note that the four constructs correspond to those in Fig. 4B and that only the constructs which formed monocentric fragments gave a significant number of stable pink transformants.

mapped gene by hybridization to Southern blots of the proximal or distal chromosome fragments.

Chromosome fragmentation causes aneuploidy of sequences distal or proximal to the site of recombination. The segmental aneuploidy produced depends on the fate of the chromosome with which recombination takes place. We have observed[8] that fragmentation by the centromeric vector (YCF4) usually (>90% of independent transformants) results in partial trisomy of the distal segment ($2N$ + CF), whereas fragmentation by the acentric vector (YCF3) usually (>90% of independent transformants) results in partial monosomy of the distal segment ($2n - 1$ + CF). Less frequently, the centromeric vector results in partial monosomy of the proximal segment (target chromosome lost), and the acentric vector results in partial trisomy of the proximal segment (target chromosome not lost). Although yeast is highly tolerant of whole chromosomal aneuploidy ($2n$ + 1 and $2n - 1$),[14,15] specific phenotypes may be observed for segmental aneuploids generated during chromosome fragmentation. The specific example of chromosome instability associated with a 435-kb partial monosomy for sequences distal to LYS2 has been reported.[5] Because chromosome fragments can be generated for defined segments of the genome, this technology may be used in studying the phenotypic consequences of specific aneuploidies in a systematic way.

[14] B. Schaffer, I. Brearly, R. Littlewood, and G. Fink, *Genetics* **67**, 483 (1971).
[15] E. Parry and B. Cox, *Genet. Res.* **16**, 333 (1970).

TABLE III
Fragmentation Events[a]

Locus	CHR	Arm	Vector	PRX/DIS	Size (kb)	Source or Ref.
PHO11	I	Right	YCF4	Distal	28	H. Steensma, P. deJonge, A. Kaptein, and D. Kaback, *Curr. Genet.* **16**, 131 (1989)
GAL1	II	Right	YCF3	Proximal	320	D. Vollrath and P. Hieter, *Proc. Natl. Acad. Sci. U.S.A.* **85**, 6027 (1988)
GAL1	II	Right	YCF4	Distal	560	D. Vollrath and P. Hieter, *Proc. Natl. Acad. Sci. U.S.A.* **85**, 6027 (1988)
LYS2	II	Right	YCF1	Proximal	365[b]	D. Vollrath and P. Hieter, *Proc. Natl. Acad. Sci. U.S.A.* **85**, 6027 (1988)
LYS2	II	Right	YCF2	Distal	475	D. Vollrath and P. Hieter, *Proc. Natl. Acad. Sci. U.S.A.* **85**, 6027 (1988)
MEL1	II	Left	YCF3	Proximal	810	D. Vollrath and P. Hieter, *Proc. Natl. Acad. Sci. U.S.A.* **85**, 6027 (1988)
MEL1	II	Left	YCF4	Distal	25	D. Vollrath and P. Hieter, *Proc. Natl. Acad. Sci. U.S.A.* **85**, 6027 (1988)
CEN3	III	—	YCF1	Left	140	C. Connelly and P. Hieter[c]
CEN3	III	—	YCF3	Left	140	D. Koshland, C. Connelly, and P. Hieter[c]
CEN3	III	—	YCF3	Right	280[d]	D. Koshland, C. Connelly, and P. Hieter[c]
D8B	III	Left	YCF1	Proximal	290[e]	C. Connelly and P. Hieter[c]
D8B	III	Left	YCF2	Distal	130	C. Connelly and P. Hieter[c]
H9G1-1	III	Right	YCF1	Proximal	260[f]	C. Connelly and P. Hieter[c]
H9G1-1	III	Right	YCF2	Distal	160	C. Connelly and P. Hieter[c]
HML	III	Left	YCF3	Proximal	?	A. Rose[c]
HMR	III	Right	YCF3	Proximal	?	A. Rose[c]
CEN4	IV	—	YCF4	Left	450	C. Connelly and P. Hieter[c]
CTF12	IV	Right	YCF3	Proximal	960	F. Spencer, C. Connelly, and P. Hieter[c]
CTF12	IV	Right	YCF4	Distal	710	F. Spencer, C. Connelly and P. Hieter[c]
MFA1	IV	—	YCF3	Proximal	1400	S. Sapperstein and S. Michaelis[c]
MFA1	IV	—	YCF4	Distal	200	S. Sapperstein and S. Michaelis[c]
RPC53	IV	Left	YCF3	Proximal	210	C. Mann and I. Treich[c]
RPC53	IV	Left	YCF4	Distal	±1100	C. Mann and I. Treich[c]
RAD3	V	Right	YCF3	Proximal	540	B. Ferguson and W. Fangman[c]
RAD3	V	Right	YCF4	Distal	55[g]	B. Ferguson and W. Fangman[c]
CEN6	VI	—	YCF3	Right	155	J. Shero, and P. Hieter[c]
CEN7	VII	—	YCF1	Left	525	C. Connelly and P. Hieter[c]
CEN7	VII	—	YCF1	Right	600	C. Connelly and P. Hieter[c]
CYH2	VII	Left	YCF3	Proximal	920[h]	B. Garvik and L. Hartwell[c]
GCD12	VII	Left	YCF3	Proximal	640	C. Paddon and A. Hinnebusch, *Genetics* **122**, 543 (1989)
GCD12	VII	Left	YCF4	Distal	480	C. Paddon and A. Hinnebusch, *Genetics* **122**, 543 (1989)
RAD2	VII	Right	YCF1	Proximal	1030	C. Connelly and P. Hieter[c]
RAD2	VII	Right	YCF2	Distal	90	C. Connelly and P. Hieter[c]

(continued)

TABLE III *(continued)*

Locus	CHR	Arm	Vector	PRX/DIS	Size (kb)	Source or Ref.
CDC23	VIII	Right	YCF3	Proximal	455	R. Sikorski, C. Connelly, and P. Hieter[c]
CDC23	VIII	Right	YCF4	Distal	150	R. Sikorski, C. Connelly, and P. Hieter[c]
PHO11	VIII	—	YCF4	Distal	44[i]	H. Steensma, P. deJonge, A. Kaptein, and D. Kaback, *Curr. Genet.* **16**, 131 (1989)
GAL2	XII	Right	YCF3	Proximal	300	D. Jaeger, J. Hegemann, and P. Philippsen[c]
CTF13	XIII	Right	YCF3	Distal	475	K. Floy, F. Spencer, and P. Hieter[c]
CTF13	XIII	Right	YCF4	Proximal	445	K. Floy, F. Spencer, and P. Hieter[c]
RNase H	XIII	Right	YCF1	Proximal	750	R. Crouch[c]
RNase H	XIII	Right	YCF2	Distal	210	R. Crouch[c]
SPT21	XIII	Right	YCF3	Proximal	640	G. Natsoulis and J. Boeke[c]
SPT21	XIII	Right	YCF4	Distal	310	G. Natsoulis and J. Boeke[c]
CEN14	XIV	—	YCF3	Left	650	C. Connelly and P. Hieter[c]
CEN14	XIV	—	YCF3	Right	170	C. Connelly and P. Hieter[c]
CMS1	XIV	Left	YCF3	Proximal	730	J. Shero, C. Connelly, and P. Hieter[c]
CMS1	XIV	Left	YCF4	Distal	80	J. Shero, C. Connelly, and P. Hieter[c]
CEN15	XV	—	YCF1	Right	790	C. Connelly and P. Hieter[c]
CEN15	XV	—	YCF1	Left	325	C. Connelly and P. Hieter[c]
HIS3	XV	Right	YCF1	Proximal	750	C. Connelly and P. Hieter[c]
HIS3	XV	Right	YCF2	Distal	350	C. Connelly and P. Hieter[c]
CTF1	XVI	Left	YCF3	Proximal	430	S. L. Gerring and P. Hieter[c]
CTF1	XVI	Left	YCF4	Distal	560	S. L. Gerring and P. Hieter[c]
RPC40	XVI	Right	YCF3	Proximal	±250	C. Mann and I. Treich[c]
RPC40	XVI	Right	YCF4	Distal	±900	C. Mann and I. Treich[c]

[a] CHR, chromosome; PRX, proximal; DIS, distal.
[b] Increased chromosome loss.
[c] Unpublished.
[d] Also 255 depending on CLP.
[e] Also 265 depending on CLP.
[f] Also 235 depending on CLP.
[g] Lost 0.1%/generation.
[h] The partial monosome grew extremely poorly. The fragmented chromosome was extremely unstable and gave rise to rearrangements.
[i] New locus.

Table III lists all known fragmentation events and should serve as a guide to indicate which chromosomal fragmentation events may be associated with aneuploid phenotypes. In addition, exchange of these strains carrying various fragments should be of assistance to localizing cloned genes to subregions of chromosomes by direct probing. Data from fragmentation experiments at different loci on the same chromosome can be used to determine the physical distance between two loci by subtraction and to construct physical maps of chromosomes. For example, the distance of *HIS3* from the centromere of chromosome XV can be calculated at 440 kb by subtracting the length of the *HIS3* distal fragment (350 kb) from the length of the CEN15 R fragment (790 kb). Similarly, the distance between H9G and D8B on chromosome III can be calculated at 130 kb (H9G proximal 260 kb minus D8B distal 130 kb). Chromosome fragments may also be useful for localizing unmapped mutations for which the cloned gene is unavailable by testing for complementation of mutant phenotypes by the genomic subsegments present on particular fragments.

Fragment Nomenclature

The following convention is used to describe a chromosome fragment: CF# (DNA.dir.STRAIN). CF indicates that a chromosome fragment is being described, # refers to the chromosome from which the fragment is derived, and DNA refers to the clone used to make the fragment. Dir is used to describe which part of the chromosome is on the fragment and can be one of four possibilities (d, p, R, or L). d refers to fragments derived from YCF4 fragmentation events containing chromosomal sequences distal to the mapping segment. p refers to fragments derived from YCF3 fragmentation events containing chromosomal sequences proximal to the mapping segment. L and R refer to fragments containing all chromosomal sequences on the left or right arms (i.e., those derived from fragmentation events in which a centromere DNA-containing restriction fragment was inserted into YCF3 in either of two orientations). Finally, STRAIN is the name given to the strain in which the fragment was first constructed. For example, fragments derived from fragmentation at *CTF1* on chromosome XVI would be designated CFXVI(*CTF1*.d.YPH582) and CFXVI(*CTF1*.p.YPH583) for the distal and proximal fragments, respectively. Fragments derived from fragmentation at the centromere of chromosome VII would be designated CFVII(CEN7.R.YPH120) and CFVII(CEN7.L.YPH126) for the right and left arm fragments, respectively.

When YCF3 and YCF4 are used, the fragments are marked with *URA3 SUP11,* and this is implicit in the nomenclature. If, however (as is the case for YPH149), the marker has been changed to another (*TRP1* in this case), this can be noted after the STRAIN designation. Thus, in the case of YPH149, the *RAD2* distal CF is called CFVII(*RAD2*.d.YPH146.*TRP1*).

Availability of Chromosome Fragment Strains

Dr. Robert Mortimer has designated a separate collection of chromosome fragment-containing YPH49-derived strains within the Yeast Genetics Stock Culture Center. It is requested that all strains in which chromosome fragments are generated in the future be sent to the Stock Center with descriptive information as indicated in Table III. Updated versions of Table III, as well as CF-containing yeast strains, can be obtained directly by writing or calling the Yeast Genetics Stock Center [Donner Laboratory, University of California, Berkeley, CA 94720; (415)-642-0815].

[5] Assay of Yeast Mating Reaction

By GEORGE F. SPRAGUE, JR.

Background and Principle of Assays

Haploid cells of the yeast *Saccharomyces cerevisiae* exhibit either of two cellular phenotypes, the mating types **a** or α. These cells can reproduce vegetatively by a mitotic cell cycle. However, when cells of opposite mating type are cocultured, they exit the cell cycle and participate in a mating process that results in cell and nuclear fusion to create an **a**/α diploid zygote (reviewed in Refs. 1–3). Like the haploid cells, **a**/α cells can reproduce by mitosis, but, unlike **a** or α cells, **a**/α cells cannot mate. Instead they have a new property, the ability to undergo meiosis and sporulation when nutrients are limiting, thereby regenerating the two haploid cell types.

The overall mating process can be assayed easily by several procedures that detect zygote formation (see below). In addition, particular facets of the mating reaction can be assayed without requiring that mating actually take place. Assays of these partial mating reactions are especially useful if mutations affecting the mating process are being studied. To provide a framework for understanding these assays I first briefly review the mating process, highlighting the biochemical and physiological events that can be monitored easily.

The mating process requires that **a** and α cells first communicate with each other and then interact physically so that cell and nuclear fusion can occur. Communication is achieved via the cell-type-specific production of secreted mating factors (pheromones) and receptors for those pheromones.

---

[1] G. F. Sprague, Jr., L. C. Blair, and J. Thorner, *Annu. Rev. Microbiol.* **37**, 623 (1983).
[2] F. Cross, L. H. Hartwell, C. Jackson, and J. B. Konopka, *Annu. Rev. Cell Biol.* **4**, 611 (1988).
[3] I. Herskowitz, *Microbiol. Rev.* **52**, 536 (1988).

In particular, only α cells secrete the 13-residue polypeptide α-factor pheromone,[4] which binds to a receptor present only on the surface of a cells.[5] Likewise, only a cells secrete the polypeptide a-factor pheromone,[6] which interacts with a receptor present on the surface of α cells.[7-9] The crucial role of the pheromones and receptors in mating is revealed by the finding that mutations in their structural genes lead to a nonmating phenotype. For example, α cells harboring mutations in the a-factor receptor structural gene, *STE3*, have a mating efficiency of 10^{-6} compared with wild-type α cells.[7,9] As expected, mutation of *STE3* confers an α-specific mating defect; a *ste3* mutants are mating proficient because a cells do not require this receptor to mate.

Synthesis of two other types of proteins is also limited to α cells or to a cells. First, each cell type produces a unique agglutinin, a cell surface glycoprotein that enables the cell to adhere tightly to cells of the opposite mating type.[10,11] The agglutinins apparently have only a subtle role in mating as mutation of the α-agglutinin structural gene causes only a modest reduction in mating efficiency, at least when cells are mated on a solid surface.[12,13] Second, a cells secrete a protease, termed barrier because of the nature of the assay that first revealed its presence, that specifically degrades α-factor.[14,15] An α cell activity that specifically inactivates a-factor has been described recently.[16] The key point is that the features that distinguish a and α cells are conferred by proteins present at the cell surface, an appropriate setting given that this is where the initial interaction between the mating pair will occur.

Subsequent steps in mating utilize intracellular proteins that are present in both a and α cells. In particular, binding of pheromone to receptor activates a signal transduction pathway that is shared by the two cell types.[17,18] Propagation of a signal along this pathway elicits physiological changes in the responding cell that enable it to mate efficiently:

[4] W. Duntze, V. L. MacKay, and T. R. Manney, *Science* **168**, 1472 (1970).
[5] D. D. Jenness, A. C. Burkholder, and L. H. Hartwell, *Cell (Cambridge, Mass.)* **35**, 521 (1983).
[6] L. E. Wilkinson and J. R. Pringle, *Exp. Cell Res.* **89**, 175 (1974).
[7] V. L. MacKay and T. R. Manney, *Genetics* **76**, 273 (1974).
[8] N. Nakayama, A. Miyajima, and K. Arai, *EMBO J.* **4**, 2643 (1985).
[9] D. C. Hagen, G. McCaffrey, and G. F. Sprague, Jr., *Proc. Natl. Acad. Sci. U.S.A.* **83**, 1418 (1986).
[10] R. Betz, W. Duntze, and T. R. Manney, *FEMS Microbiol. Lett.* **4**, 107 (1978).
[11] G. Fehrenbacher, K. Perry, and J. Thorner, *J. Bacteriol.* **134**, 893 (1978).
[12] K. Suzuki and N. Yanagishima, *Curr. Genet.* **9**, 185 (1985).
[13] P. N. Lipke, D. Wojciechowicz, and J. Kurjan, *Mol. Cell. Biol.* **9**, 3155 (1989).
[14] J. B. Hicks and I. Herskowitz, *Nature (London)* **260**, 246 (1976).
[15] V. L. Mackay, S. K. Welch, M. Y. Insley, T. R. Manney, J. Holly, G. C. Saari, and M. L. Parker, *Proc. Natl. Acad. Sci. U.S.A.* **85**, 55 (1988).
[16] M. Steden, R. Betz, and W. Duntze, *Mol. Gen. Genet.* **219**, 219 (1989).

1. Transcription of a number of genes whose products are involved in mating is increased. Of particular note is the *FUS1* gene whose transcription increases 50 times or more when cells are exposed to pheromone.[19,20] The FUS1 product appears to be a cell surface protein that catalyzes a late step in cell fusion.

2. In addition to the increase in FUS1 product, other changes occur at the cell surface. For example, increased amounts of agglutinin are deposited[10,11]; see also Point 5 below.

3. In ways that are as yet uncharacterized biochemically, the nucleus is made ready for fusion with the nucleus donated by the mating partner.[21]

4. The cell division cycle arrests in the G_1 phase.[6,22,23]

5. Each member of the mating pair elongates toward its partner.[24-26] When cells are exposed to pheromone in the absence of a nearby mating partner, this projection formation can become exaggerated, leading to the formation of morphologically aberrant, pear-shaped cells (sometimes called shmoos).

As a consequence of these events, the cell cycles of the mating pair are synchronized, and proteins that catalyze and nuclear fusion are poised to act; mating to form an **a**/α zygote can proceed efficiently.

As already noted, **a** and α cells are distinctive because each cell type produces a set of unique proteins. In addition, both haploid cell types are further distinguished from diploid **a**/α cells because the haploid cells synthesize components of the common signal transduction pathway that **a**/α cells fail to synthesize. For example, an early step in the signal transduction pathway involves a heterotrimeric GTP-binding protein (G protein) composed of Gα, Gβ, and Gγ subunits encoded by the *GPA1* (*SCG1*), *STE4*, and *STE18* genes.[27-30] Transcription of these genes occurs in **a** and α cells

---

[17] A. Bender and G. F. Sprague, Jr., *Cell (Cambridge, Mass.)* **47**, 929 (1986).
[18] N. Nakayama, A. Miyajima, and K. Arai, *EMBO J.* **6**, 249 (1987).
[19] J. Trueheart, J. D. Boeke, and G. R. Fink, *Mol. Cell. Biol.* **7**, 2316 (1987).
[20] G. McCaffrey, F. J. Clay, K. Kelsay, and G. F. Sprague, Jr., *Mol. Cell. Biol.* **7**, 2680 (1987).
[21] M. D. Rose, B. R. Price, and G. R. Fink, *Mol. Cell. Biol.* **6**, 3490 (1986).
[22] E. Bucking-Throm, W. Duntze, L. H. Hartwell, and T. R. Manney, *Exp. Cell Res.* **76**, 99 (1973).
[23] L. M. Hereford and L. H. Hartwell, *J. Mol. Biol.* **84**, 445 (1974).
[24] L. H. Hartwell, *Exp. Cell Res.* **76**, 111 (1973).
[25] P. N. Lipke, A. Taylor, and C. E. Ballou, *J. Bacteriol.* **127**, 610 (1976).
[26] J. D. Rine, Ph.D. Dissertation, University of Oregon (1979).
[27] I. Miyajima, M. Nakafuku, N. Nakayama, C. Brenner, A. Miyajima, K. Kaibuchi, K.-I. Arai, Y. Kaziro, and K. Matsumoto, *Cell (Cambridge, Mass.)* **50**, 1011 (1987).
[28] C. Dietzel and J. Kurjan, *Cell (Cambridge, Mass.)* **50**, 1001 (1987).
[29] K.-Y. Jahng, J. Ferguson, and S. I. Reed, *Mol. Cell. Biol.* **8**, 2484 (1988).
[30] M. Whiteway, L. Hougan, D. Dignard, D. Y. Thomas, L. Bell, G. C. Saari, F. J. Grant, P. O. O'Hara, and V. L. MacKay, *Cell (Cambridge, Mass.)* **56**, 467 (1989).

but not in **a**/α cells. Thus, the generation of mating-competent cells requires the differential expression of three sets of genes: α-specific genes, **a**-specific genes, and haploid-specific genes. The appropriate pattern of gene expression is achieved through the action of regulatory proteins encoded by the mating-type locus (*MAT*), the only locus whose genetic content differs among the cell types. Although the details are beyond the scope of this chapter, the two *MAT* alleles, *MAT***a** and *MAT*α, encode three regulatory activities, α1, α2, and **a**1–α2, that control expression of the three gene sets.[1,31–33]

Mating Assays

Mating assays fall into two categories, qualitative or quantitative. In either case, the most convenient method to assess diploid formation relies on complementation of the nutritional deficiencies of the cells being mated. For example, the mating type of a strain that requires compound X for growth can be established by determining whether it can form prototrophs when mixed with tester **a** and α strains that require compound Y for growth. In the absence of useful nutritional requirements, other methods can be applied to assess diploid formation, detailed below. For most purposes, qualitative mating tests provide all the needed information. They can be used to establish the mating type of a strain, to follow the segregation of mating type among the meiotic products of a diploid, or to construct a new diploid so that the genetic properties of a mutation can be studied. However, in some instances, it is important to use a quantitative assay to measure the efficiency with which a strain mates, particularly if qualitative tests reveal a mating deficiency.

Qualitative Tests

Complementation of Nutritional Requirements
Procedure

1. Prepare a master plate of the strains to be tested by applying small patches of the strains to the surface of a YEPD plate [1% (w/v) Bacto-yeast extract, 2% (w/v) Bacto-peptone, 2% (w/v) dextrose, 2% (w/v) Bacto-agar; same as YPD of Ref. 34]. For the purposes of this illustration, imagine that

[31] K. Nasmyth and D. Shore, *Science* **237**, 1162 (1988).
[32] G. F. Sprague, Jr., *Adv. Genet.* **27**, 33 (1990).
[33] I. Herskowitz and R. E. Jensen, this volume [8].
[34] F. Sherman, this volume [1].

the strains are spore clones derived from a diploid homozygous for an *ade6* mutation. Incubate overnight at 30°.

2. Spread about 10^6 cells of mating type α on the surface of a YEPD plate and spread 10^6 mating type **a** cells on the surface of a second YEPD plate. It is useful to establish standard mating type tester strains that have nutritional defects not commonly found in other strains that are used routinely. For example, our laboratory uses **a** *lys1* and α *hom3* strains as standard testers.

3. Replica plate the master plate onto each of the YEPD plates spread with mating type tester strains. Use a separate velvet for each tester. In addition, prepare a fresh version of the master plate by replica plating to a YEPD plate not spread with tester cells. The latter plate will provide a control for evaluating the mating of the patches with the tester strains in Step 5. Incubate overnight at 30°.

4. Replica plate the three YEPD plates to synthetic minimal medium plates [SD; 0.67% (w/v) Bacto-yeast nitrogen base without amino acids, 2% (w/v) dextrose, 2% (w/v) Bacto-agar; Ref. 34]. Incubate overnight at 30°.

5. Score for the presence of prototrophs. Patches derived from the new master plate should not show appreciable growth on the SD plate. A positive mating reaction with a tester strain should lead to confluent growth at the position of a patch.

Alternative procedure. Steps 2, 3, and 4 can be combined into one step, as follows (see Fig. 1). Prepare a suspension of 10^8 or more cells of each mating type tester in fresh YEPD broth. The testers can be stored on keeper plates in the refrigerator for 2 to 3 weeks. Suspensions are made as needed by transferring cells to YEPD broth. Spread 5×10^7 (or more) cells of each tester on the surface of separate SD plates and replica plate the master plate to both testers. Incubate for 1 or 2 days and score as described above.

Zygote Isolation. It is not always possible to isolate diploids or to test mating phenotype by relying on complementation of nutritional markers; for example, a strain of interest may be a prototroph. In such cases, diploid formation can be assessed by direct microscopic examination. Cells to be mated are positioned next to each other in pairs by micromanipulation on YEPD plates or on YEPD agar slabs held in dissection chambers.[35] After 2–3 hr of incubation, the pairs are examined and formation of dumbbell-shaped zygotes scored. Alternatively, cells to be mated are mixed in a small patch on a YEPD plate and incubated 2–3 hr. A toothpick is then used to spread a small amount of the mixture onto a YEPD plate or agar slab. Zygotes are visualized by microscopy. If the purpose of the mating is to

---

[35] F. Sherman and J. Hicks, this volume [2].

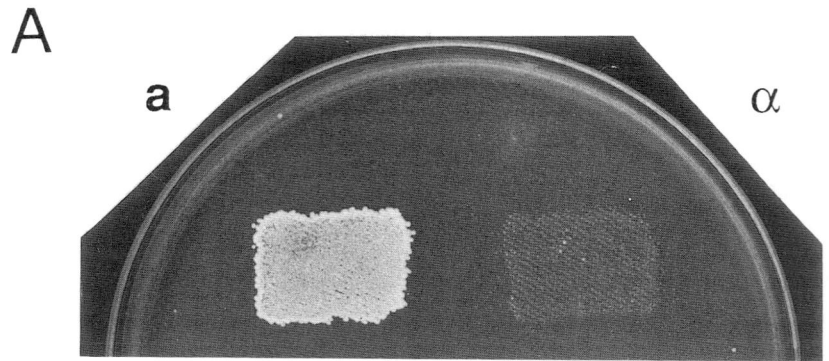

mated with an α strain

mated with an a strain

FIG. 1. Qualitative mating test. The alternative procedure described in the text was used to determine the mating properties of two patches of cells. (A) The patches were replica plated to a lawn of α cells. (B) The patches were replica plated to a lawn of a cells. The patch at left, labeled a, formed prototrophs with the α lawn but not with the a lawn, so this patch exhibits the a phenotype. The patch at right, labeled α, formed prototrophs with the a lawn but not with the α lawn, so this patch exhibits the α phenotype.

form a diploid to be used in subsequent studies, rather than simply to test the mating type of a strain, zygotes can be separated from the mix of cells by micromanipulation and allowed to form a colony.

Isolation of Diploids by Pheromone Tests. A final method to isolate $\mathbf{a}/\alpha$ diploids when complementation of nutritional defects is not feasible relies on the fact that $\mathbf{a}/\alpha$ diploids do not secrete either α-factor or $\mathbf{a}$-factor. Roughly equal numbers of the two strains are mixed on a YEPD plate and incubated for 3–4 hr at 30°. The mix is then suspended in YEPD broth, diluted, and plated on YEPD plates at a density that will yield 50–300 colonies per plate. The colonies are then tested for pheromone production as described below. Colonies that do not produce either pheromone are presumably $\mathbf{a}/\alpha$ diploids. This assignment can be confirmed by determining whether they can sporulate.[36]

Quantitative Test

Procedure[7,37,38]

1. Grow a culture of the strain whose mating efficiency is to be determined to a density of about 10^7 cells/ml in YEPD broth. For the purposes of this illustration, consider an **a** *his4 leu2* strain. Prepare a similar culture of an appropriate mating type tester strain, for example, α *lys2 trp1*.

2. Mix 2×10^6 cells of the **a** strain with 10^7 cells of the α tester and collect the cells on a 0.45-μm pore, 25-mm diameter nitrocellulose filter disk (type HA, Millipore, Bedford, MA). In addition, collect cells of each strain on separate filters; these filters will serve as controls for reversion or contamination of the two cultures.

3. Place the filters on the surface of a YEPD plate and incubate at 30° for 5 hr.

4. Resuspend the cells on each filter in 1 ml of SD broth, sonicate for 5–10 sec to disperse clumps, and dilute in SD broth. Plate the cells from the mating mix filter on SD medium to titer $\mathbf{a}/\alpha$ diploids. Likewise, plate cells from the single strain filters on SD medium to check for reversion or contamination. The mating efficiency can be calculated in two ways. (1) Determine the titer of $\mathbf{a}/\alpha$ cells plus **a** cells whose mating efficiency is being determined by plating the cells from the mating mix filter on SD medium supplemented with histidine and leucine. Mating efficiency is expressed as the titer of $\mathbf{a}/\alpha$ cells divided by the titer of $\mathbf{a}/\alpha$ cells plus **a** cells. (2)

---

[36] Y. Kassir and G. Simchen, this volume [6].
[37] L. H. Hartwell, *J. Cell Biol.* **85**, 811 (1980).
[38] K. L. Clark and G. F. Sprague, Jr., *Mol. Cell. Biol.* **9**, 2682 (1989).

Determine the titer of the **a** strain by plating the cells resuspended from the single strain filter on SD medium supplemented with histidine and leucine. Mating efficiency is expressed as the titer of **a**/α cells divided by the titer of **a** cells.

Assay for Pheromone Production

A simple petri plate assay can be used to test for pheromone production. The assay is based on a halo test, first developed to detect killer activity.[39] If patches (or colonies) secrete pheromone, they inhibit the growth of a tester strain, resulting in a clear zone (halo) surrounding the patch.[40] This plate assay uses special pheromone tester strains that are hypersensitive to pheromone. The α-factor testers are **a** *sst1* (*bar1*)[40-42] strains, which lack the barrier protease that degrades α-factor. The **a**-factor testers are α *sst2* strains; these strains are defective in adaptation to the pheromone-generated signal and therefore remain arrested in G_1 for many hours. **a** *sst2* strains could be used as α-factor testers, but it is not necessary to do so. *sst2* mutants accumulate *ste* mutations on subculture. Therefore, it is prudent to test the mating competence of colonies of the α *sst2* tester and use only these colonies that give a strong mating response in the qualitative test described above.

Procedure

1. Prepare a master plate by making small patches of the strains to be tested on the surface of a YEPD plate. Incubate overnight at 30°. Colonies also produce sufficient amounts of pheromone to be detected by the plate assay.

2. Spread 10^5 to 5×10^5 cells of a pheromone tester strain on the surface of a BBMB plate[43] [YEPD that also contains 0.1 M citrate, pH 4.5, and 0.03% (w/v) methylene blue; add the citrate and methylene blue from concentrated solutions after autoclaving the YEPD]. The dye accents the halos, and the low pH may increase the sensitivity of the assay.

3. Replica plate the master plate to the tester lawns and incubate at room temperature for 1 or 2 days. Score for the presence of halos (Fig. 2). Incubation at room temperature gives larger halos than incubation at 30°.

[39] G. R. Fink and C. A. Styles, *Proc. Natl. Acad. Sci. U.S.A.* **69**, 2846 (1972).
[40] G. F. Sprague, Jr., and I. Herskowitz, *J. Mol. Biol.* **153**, 305 (1981).
[41] R. K. Chan and C. A. Otte, *Mol. Cell. Biol.* **2**, 11 (1982).
[42] R. K. Chan and C. A. Otte, *Mol. Cell. Biol.* **2**, 21 (1982).
[43] L. C. Blair, Ph.D. Dissertation, University of Oregon (1979).

a-factor tester lawn

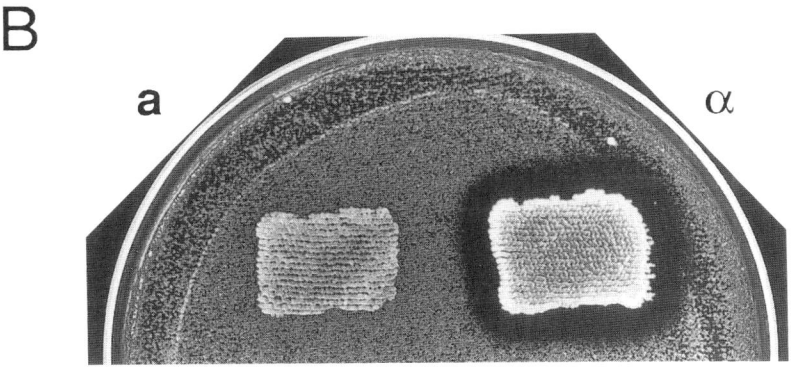

α-factor tester lawn

FIG. 2. Halo assay for pheromone production. The same patches used in Fig. 1 were replica plated to lawns of **a**-factor tester cells (A) and to lawns of α-factor tester cells (B). The patch labeled **a** produced **a**-factor, as revealed by the lack of growth of the **a**-factor tester lawn in the region surrounding the patch. The patch labeled α produced α-factor.

This added sensitivity is not always critical; incubation at 30° should be adequate for many purposes.

As mentioned above, the halo assay is also capable of detecting killer activity. Pheromone tester strains should therefore be killer resistant. Even if this precaution is taken, the finding that a mutant strain has acquired the ability to produce an active pheromone that could not be produced by its parent should be confirmed by an independent assay for pheromone production. In brief, the mutant should serve as a source of pheromone that is used to treat wild-type tester cells in any of the pheromone response assays described below.

Assay for Pheromone Response

As noted in the background section, treatment of **a** or α cells with the appropriate pheromone elicits a number of physiological changes. A couple of these changes are quite easy to assay, namely, *FUS1* induction and cell cycle arrest in G_1, along with attendant shmoo formation. At least in the case of response to α-factor, the two changes occur with different K_m values for pheromone,[44] so they may be useful in different situations.

FUS1 Induction

FUS1 induction can be examined either by Northern analysis to measure transcript levels before and after pheromone treatment or by using a strain harboring a plasmid-borne *FUS1-lacZ* gene fusion to measure β-galactosidase levels before and after treatment.[19,20]

Procedure (transcript analysis)

1. An exponential culture (5×10^6 to 10^7 cells/ml) growing in YEPD is divided into two parts. One part is not treated with pheromone, the other is treated. α-Factor is available commercially (Sigma, St. Louis, MO) and is added at a concentration of 0.5 to 5.0 μM. **a**-Factor is not available commercially, so a cell-free filtrate of a saturated culture of **a** cells is the routine pheromone source.[45] In this case, equal volumes of the cell culture and the pheromone filtrate are mixed. If a filtrate is the source of pheromone, then the part of the cell culture that is not treated with pheromone should also be mixed with an equal volume of either fresh YEPD or the filtrate of a saturated culture of a strain that does not secrete pheromone (e.g., a *mat*α*1* mutant[46] or a strain carrying mutations in the pheromone structural genes[47,48]). If necessary, the concentration of pheromone present in different preparations can be measured by performing 2-fold serial dilutions and applying 5 or 10 μl to a BBMB plate previously spread with a known number of pheromone tester cells. The lowest concentration that gives a detectable halo is defined as 1 unit of pheromone/ml.[22]

---

[44] S. A. Moore, *J. Biol. Chem.* **258,** 13849 (1983).
[45] D. C. Hagen and G. F. Sprague, Jr., *J. Mol. Biol.* **178,** 835 (1984).
[46] J. N. Strathern, J. B. Hicks, and I. Herskowitz, *J. Mol. Biol.* **147,** 357 (1981).
[47] J. Kurjan, *Mol. Cell. Biol.* **5,** 787 (1985).
[48] S. Michaelis and I. Herskowitz, *Mol. Cell. Biol.* **8,** 1309 (1988).

2. After 60 min of incubation with shaking at 30°, the cells are harvested, RNA isolated, and a Northern blot prepared. It is most convenient to use a single-strand SP6 or T7 polymerase-generated radioactive probe to visualize the *FUS1* transcript. Compare *FUS1* transcript levels to the levels of transcript from a gene such as *URA3* whose expression is not influenced by pheromone.

Procedure (β-galactosidase induction)

1. Grow the plasmid-bearing strain overnight to saturation in a medium that selects for maintenance of the plasmid (e.g., SD supplemented with all required nutrients except the one being used to maintain the plasmid).
2. Harvest the cells by centrifugation and suspend them in YEPD at a density of about 10^7 cells/ml. Grow at 30° for one doubling.
3. Divide the culture in half and add pheromone as described above for transcript analysis. Incubate at 30° for 2.5 hr.
4. Assay for β-galactosidase.

Cell Cycle Arrest

Procedure

1. Add pheromone as described above for transcript analysis to an exponential culture growing in YEPD broth.
2. Remove samples at 0, 1, 2, and 3 hr and fix by mixing with an equal volume of 0.15 M NaCl, 7.4% formaldehyde.
3. Sonicate each sample to disrupt clumps and examine by phase-contrast microscopy. Count the number of shmoos and unbudded cells and the number of budded cells (Fig. 3). Unbudded cells and shmoos are considered to be in the G_1 phase of the cell cycle.[24] Exponential cultures typically contain 40–50% unbudded cells. After pheromone treatment, 90% (or more) of the cells may be unbudded or shmooed. At least 200 cells should be examined from each sample. Samples can be concentrated by centrifugation if the original sample is too dilute for easy counting.

Confrontation Assay for Response to α-Factor[7]

Procedure

1. Place a heavy streak of α cells on the surface of a YEPD agar slab held on a glass slide in a tetrad dissection chamber[35] or on the surface of a YEPD plate. Incubate at 30° for 2 hr to precondition the agar with α-factor.
2. Place about 20 test **a** cells at a distance of two to four cell diameters from the α-factor source by micromanipulation. Use unbudded cells, if

A

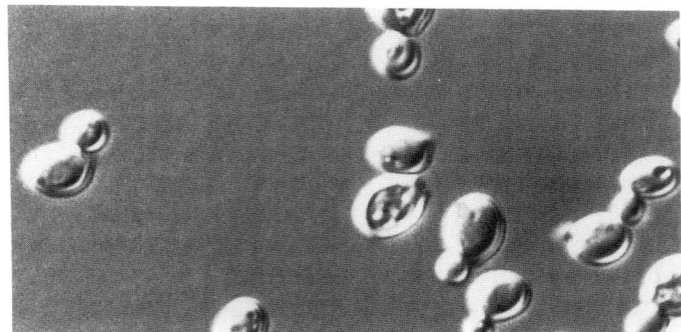

B

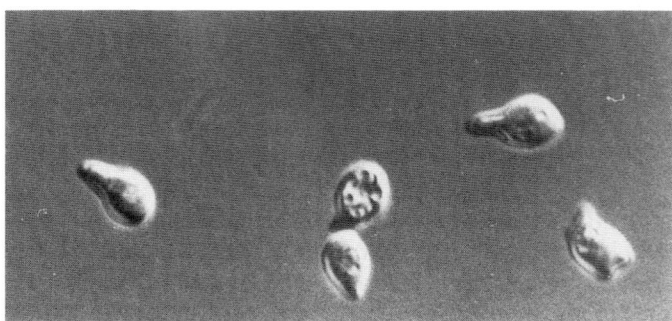

FIG. 3. Cellular morphology of α-factor-treated cells. An exponential culture of a *bar1* (*sst1*) cells (A) was exposed to α-factor (0.5 μM) for 4 hr (B). Note that the pheromone-treated cells are arrested as unbudded cells and have taken on the pear-shaped morphology.

possible. Place the same number of test **a** cells elsewhere on the agar, well away from the α-factor source. The **a** cells can be wild type at the *SST1* and *SST2* genes. Incubate at 30° and examine the cells every hour for 2–3 hr. The cells near the α-factor source will complete any cell cycle that they have begun, but they will not initiate new cell cycles. Thus, they will arrest as unbudded cells and with time will take on the shmoo morphology. The test cells not placed near the α-factor source will continue to bud (see Fig. 5).

Note: The confrontation assay does not work well for **a**-factor, presumably because it is very hydrophobic[49] and does not diffuse well in agar.

Although in principle it might be possible to use α *sst2* cells as testers, in practice this does not work satisfactorily. Cultures of α *sst2* cells contain many morphologically aberrant cells,[42] presumably because they are responding to endogenous **a**-factor produced by cells that have undergone a rare change to the **a** mating type. This may occur by mating-type interconversion[33] or simply as a result of failure to express that *MAT*α locus, since it is known that the absence of information at the mating-type locus leads to a cell that mates as **a**.[46] Perhaps an α *sst2* strain that also carries a mutation that would prevent **a**-factor production (e.g., *ste6* or *mfa1* and *mfa2*) would be a suitable α test cell for an **a**-factor confrontation assay, but such a strain has not been constructed.

Pheromone Response Visualized on Petri Plates

Procedure

1. A final method to assess the ability of a strain to respond to pheromone is a reversal of the halo assay for pheromone production described above. In this case a strain to be tested is spread on the surface of a BBMB plate at a known cell density (10^5 to 5×10^5 cells). In addition, control plates are prepared, one spread with the same number of wild-type cells and another spread with *sst1* or *sst2* mutants.

2. Apply known quantities of pheromone in separate spots to the surface of the plates. For wild-type cells, 10, 25, 50, and 100 ng of α-factor will give a range of halo sizes.

3. Incubate and score as described for the pheromone production assay.

Assay for Barrier

Barrier activity can be assayed on petri plates or by a modification of the confrontation assay.

Barrier Activity Visualized on Petri Plates

Procedure[40]

1. Spread the surface of a BBMB plate with 10^5 to 5×10^5 **a** *sst1* cells.
2. With a sharp, round toothpick, apply a thin strip (0.5–1.0 by 10–15 mm; the thinner, the better) of a test strain whose barrier phenotype is

---

[49] R. J. Anderegg, R. Betz, S. A. Carr, J. W. Crabb, and W. Duntze, *J. Biol. Chem.* **263**, 18236 (1988).

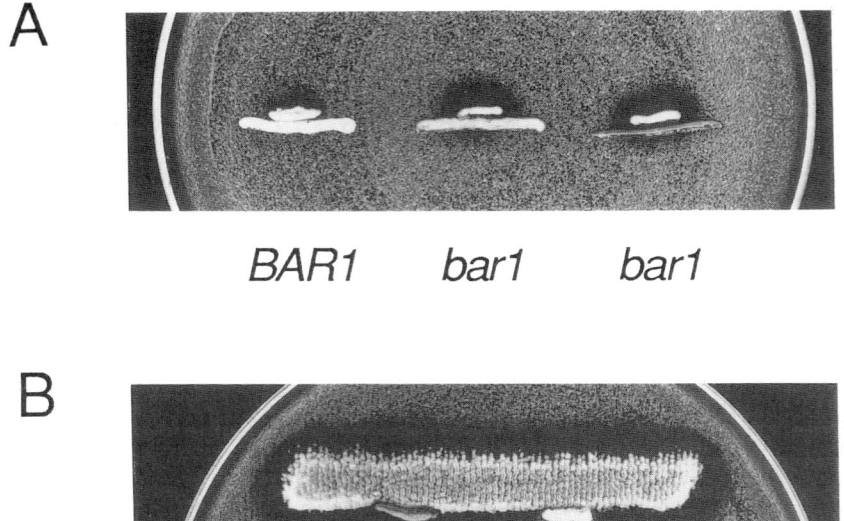

FIG. 4. Plate barrier assays. (A) The first procedure described in the text is pictured. The long strips are test **a** strains either wild type (*BAR1*) or mutant (*bar1*) at the gene that confers the barrier phenotype. (Note: *BAR1* and *SST1* are the same gene.[40,42]) The short strips are α-factor-producing cells. A large α-factor halo that extends below the test strip of **a** cells indicates a Bar⁻ phenotype. (B) The alternative procedure described in the text is pictured. The large rectangular patch of cells are α cells producing α-factor. Extension of the α-factor halo below the short strips of test **a** cells indicates a Bar⁻ phenotype.

being examined. Adjacent to this strip, apply a thin strip of α-factor-producing cells (0.5–1.0 by 5 mm). Incubate at room temperature 1 or 2 days. If the test strain is Bar⁻, the halo surrounding the α cell strip will be large and will usually traverse the test strip. If the test strain is Bar⁺, the halo surrounding the α cell strip will be small and will not traverse the test strip (Fig. 4A).

Alternative Procedure[50]

1. Apply a band of α cells (~8 mm wide) to a YEPD plate. Incubate overnight at 30°.

[50] J. Schultz and M. Carlson, *Mol. Cell. Biol.* **7**, 3637 (1987).

2. Spread 10^5 to 5×10^5 **a** *sst1* cells on the surface of a BBMB plate and replica plate the α cell band to the BBMB plate.

3. Adjacent to the α cell band apply narrow strips (0.5–1.0 mm wide) of test **a** cells. Incubate for 1 or 2 days at room temperature. If the test strain is Bar$^-$, the halo surrounding the α band will traverse the test strip. If the test strain is Bar$^+$, the halo surrounding the α band will not traverse the test strip (Fig. 4B).

Barrier Confrontation Assay[14]

The barrier confrontation assay is a modification of the α-factor confrontation assay. A strip of barrier test cells is placed between the α-factor source and the individual **a** cells that are used to assay for the presence of α-factor.

Procedure

1. Apply a heavy streak of α-factor-producing cells to the surface of an agar slab as described above for the α-factor confrontation assay. Apply a parallel strip of cells of unknown barrier phenotype. The two strips overlap in the center of the slab but extend in opposite directions beyond the overlap (see Fig. 5 for geometry). Incubate at 30° for 4 hr to precondition the agar with α-factor and barrier.

2. Apply by micromanipulation individual **a** cells at three locations along the slab: (i) Two or three cell diameters from the barrier test strip, on the face opposite the α-factor-producing strip; (ii) at the same distance from the α-factor-producing strip but at a position where the barrier test strip does not intervene; and (iii) two or three cell diameters from the barrier test strip at a position where the α-factor-producing strip does not overlap the barrier test strip.

3. After 3–4 hr incubation, determine whether the individual **a** cells at each of the three positions have arrested as unbudded cells (and perhaps formed shmoos) or have initiated new budding cycles. Position (i) tests the barrier phenotype. If the test strip is Bar$^-$, the individual **a** cells will arrest as unbudded cells. Position (ii) ensures that there is sufficient α-factor at that distance from the source to cause cell division arrest. Position (iii) ensures that the test barrier strip was not secreting α-factor (Fig. 5).

Mating Nonmating Mutants

In a number of instances, mutants originally isolated to study processes unrelated to mating have been found to be defective in mating as well (e.g.,

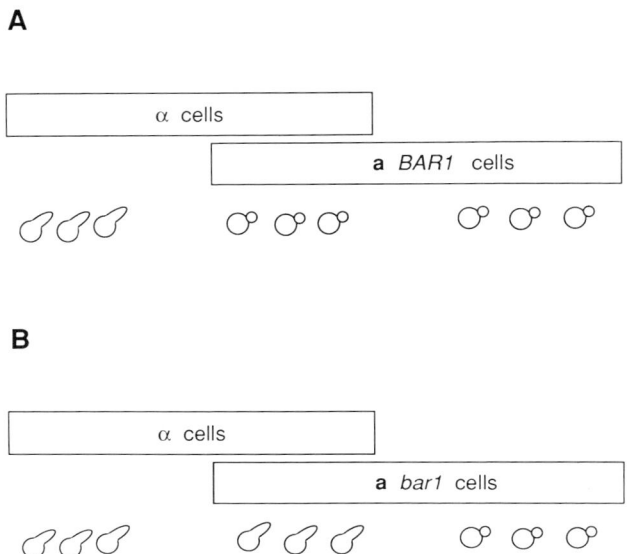

FIG. 5. Diagram of barrier confrontation assay. Strips of α cells and test **a** cells are shown as rectangles. Individual **a** cells have been positioned in three environments by micromanipulation. The diagram is not to scale; in practice the strips are 1–3 mm wide, whereas the cells are only 5–10 μm in diameter. (A) The individual **a** cells located in the middle position are drawn as budded cells, indicating that the test **a** strip is Bar⁺. (B) The individual **a** cells located in the middle position are drawn as shmoos, indicating that the test **a** strip is Bar⁻.

SSN6,[50] *TUP1*,[51] *KEX2*,[52] and *RAM1*[53]). Fortunately, mating-defective mutants generally are able to mate, albeit at a much reduced frequency, so rare diploids can be isolated and used for genetic analysis of the mutant phenotype.[7] The isolation of these rare diploids again relies on complementation of nutrition defects of the mutant strain and its mating partner. Because diploids may arise at a frequency of 10^{-6} or less, the mutant strain should ideally carry two or more nutritional defects so that revertants are not more prevalent than diploids.

The need to perform rare mating disappears if the wild-type gene that complements a mating defect has been cloned on a plasmid. In this case the plasmid can be introduced into a strain of interest[54] and diploids

[51] R. B. Wickner, *J. Bacteriol.* **117**, 252 (1974).
[52] M. J. Leibowitz and R. B. Wickner, *Proc. Natl. Acad. Sci. U.S.A.* **73**, 2061 (1976).
[53] S. Powers, S. Michaelis, D. Broek, S. Santa Anna-A., J. Field, I. Herskowitz, and M. Wigler, *Cell (Cambridge, Mass.)* **47**, 413 (1986).
[54] D. M. Becker and L. Guarente, this volume [12].

isolated by the qualitative mating procedure. Once diploids have been identified, derivatives that have lost the plasmid can be isolated and used for subsequent analyses.

Rare mating procedure

1. Apply about 10^6 cells of both the mutant strain and the mating partner to the surface of a YEPD plate. Incubate for 2 or 3 days at room temperature. For a given mutant strain, repeat the mating with cells derived from three or four independent colonies. A particular colony may contain revertants or extragenic suppressors of the mating defect, and matings involving these cells would likely be more prevalent than rare matings between the mutant and the partner strain. Thus, if the first diploid chosen for dissection proves to be reverted for the original mutation, other diploids will already be at hand to be used in tetrad analysis. At least one of the independent diploids should exhibit a segregation pattern that indicates that the diploid resulted from the union of a mutant and partner cell without the occurrence of a reversion or suppressor mutation in the mutant cell.

2. Replica plate the mating plates to selective medium on which only diploids will grow (e.g., SD). Incubate at 30° for 2 days.

3. Pick prototrophic papillae and purify by streaking on the same medium.

4. Sporulate and dissect tetrads.[36]

Alternative procedure. If it is not possible to isolate rare diploids by the mating procedure, cell fusion using spheroplasts[55] can be used to obtain diploids. Again, diploids are selected by complementation of the nutritional defects of the two strains.

1. Prepare spheroplasts as for DNA transformation.[54]

2. Mix approximately 10^7 to 5×10^7 spheroplasts of each strain, centrifuge gently to pellet the cells, and suspend in 0.5 ml of 40% polyethylene glycol 4000, 5% ethylene glycol, 10 mM Tris-HCl (pH 7.6), and 10 mM CaCl$_2$.

3. Incubate at room temperature for 20 min. Add the mixture to regeneration agar and plate on SD medium containing 1 M sorbitol to select prototrophic fusants.

---

[55] R. Schnell, L. D'Ari, M. Foss, D. Goodman, and J. Rine, *Genetics* **122**, 29 (1989).

[6] Monitoring Meiosis and Sporulation in *Saccharomyces cerevisiae*

By YONA KASSIR and GIORA SIMCHEN

Introduction

Meiosis occurs in all eukaryotic organisms and is an attractive experimental system for the study of cellular differentiation. The events leading to meiosis and sporulation in yeast may be regarded as a developmental pathway, which is regulated by a cascade of regulatory genes. The products of a single meiosis are four haploid spores, the tetrad, held together in a single ascus. During meiosis, new combinations of chromosomes and genes are generated, through independent segregation of nonhomologous chromosomes and intrachromosomal recombination. Sporulation and meiosis are therefore employed to construct new strains and to study allelism as well as linkage relationships of genes.

In this chapter we provide an account of methods used for studying meiosis and sporulation as well as information sought by researchers who wish to use meiotic analysis for studying other aspects of yeast biology. Some of the sections should help researchers overcome problems of poor sporulation or no sporulation at all; others provide information that will permit one to monitor the events of meiosis as well as to identify the meiotic stage at which certain mutant strains arrest.

Media and Procedures for Sporulation

Media. Growth media such as YEPD, SD, SC, SC − ADE, SC + CAN, and SC + CYH are described elsewhere in [1] in this volume.

YEPA: 20 g potassium acetate, 20 g peptone, 10 g yeast extract in 1 liter distilled water

PSP2: 6.7 g yeast nitrogen base without amino acids, 1 g yeast extract, 10 g potassium acetate in 1 liter of 50 mM potassium phthalate buffer (pH 5.0)

SPM: 3.0 g potassium acetate, 0.2 g raffinose in 1 liter distilled water

SPO: 2.5 g yeast extract, 15 g potassium acetate in 1 liter distilled water; after autoclaving add sterile solutions containing the following: 0.5 g glucose, 40 mg each of adenine, uracil, and tyrosine, 20 mg each of histidine, leucine, lysine, tryptophan, methionine, and arginine, 100 mg phenylalanine, and 350 mg threonine

SPO X-Gal: to 1 liter of SPO medium (after autoclaving) add 40 mg of X-Gal (5-bromo 4-chloro 3-indolyl β-D-galactoside) in 2 ml N,N-diethylformamide

Sporulation on Plates

Cells are grown on YEPD or SC plates for 2–4 days and thereafter plated or spread on SPO plates. Spore formation is observed after 2–7 days. Strains differ in the time taken to accomplish meiosis and sporulation. The fastest known strain, SK1, finishes the process in 12 hr,[1] whereas in slow-sporulating strains sporulation can take up to 2 weeks. Differences also exist between strains with regard to sporulation frequencies, which in some, but few, strains reach 90%. These differences are mostly genetic but have not been studied systematically. Backcrosses to a "good" sporulating strain clearly shorten sporulation time of a slower strain and improve its sporulation frequency.

Synchronous Sporulation in Liquid Media

Cells are grown in liquid PSP2 (supplemented with the required amino acids) for at least five generations, to a titer of $0.8-1.2 \times 10^7$ cells/ml. Cells are washed once in sterile distilled water and resuspended in SPM. In order to achieve satisfactory and synchronous sporulation it is important to have good aeration; thus, cells should be grown with shaking in a flask that is 10 times larger than the volume of the culture. Some strains do not grow well in PSP2 and should therefore be grown in YEPA medium. With good standard strains asci start to appear at 12 hr and reach the maximal frequency of 70–90% asci at 24–48 hr.[2]

Another method for sporulation in liquid medium which gives somewhat poorer synchrony is the following. Grow cells in liquid YEPD to stationary phase, or take enough cells from a YEPD plate and transfer to 2 ml of SPM at a cell density of $1-2 \times 10^7$ cells/ml. Incubate on a roller for 2–7 days.

The preferred temperature for sporulation is 30°, but good levels of sporulation can be achieved between 23° and 34° (these temperatures are used for *ts* mutants, for instance[3]). At the lower temperatures, sporulation is slower. Most strains may not complete sporulation at a temperature above 34° and arrest at the pachytene stage.[4]

[1] D. H. Williamson, L. H. Johnston, D. F. Fennel, and G. Simchen, *Exp. Cell Res.* **145**, 209 (1983).
[2] G. Simchen, R. Pinon, and Y. Salts, *Exp. Cell Res.* **75**, 207 (1972).
[3] Y. Kassir and G. Simchen, *Genetics* **90**, 49 (1978).
[4] B. Byers and L. Goetsch, *Mol. Gen. Genet.* **187**, 47 (1982).

Monitoring Meiotic Events

The process of meiosis consists of several stages during which different events take place. Certain mutants may affect some events of meiosis and not others, or they may arrest at a defined stage and not continue with subsequent events. The dependence between various events of meiosis is not yet fully understood and needs to be further clarified in wild-type and in mutant strains. Several experimental methodologies may be used to monitor meiosis.

Readiness and Commitment to Meiosis[2]

Meiosis is normally induced by shifting cells to SPM, a medium without a source of nitrogen and with acetate as the sole carbon source. Meiosis is not induced in water. Readiness is defined as the stage at which cells which have been transferred to sporulation medium, when shifted to water, proceed with the meiotic division. Commitment is the stage at which cells which have been transferred to sporulation medium are committed to proceed with the meiotic division even when shifted back to vegetative medium.

The experiment is done in the following manner. At time 0, cells are shifted from PSP2 to SPM. At various times thereafter cells are washed in water and resuspended either in water or in vegetative medium such as YEPA or PSP2. After 24–48 hr the percentage of asci in the various subcultures is determined. Under these conditions, the asci formed remain intact, and the spores do not germinate (spores do not germinate in acetate-based media).

Recombination and Haploidization

Following meiosis one can dissect tetrads or break the asci by sonication and plate random spores (see [2] in this volume). Recombination is monitored in these spores by replica plating to examine the genotype of each spore.[5] *MATa/MATα* diploids are nonmaters, whereas the haploid progeny are either **a** or α maters.[6] The colonies generated by the spores are therefore examined for their mating ability with *MATa* or *MATα* tester strains (see [5] in this volume).

---

[5] A detailed review on genetic mapping using meiotic recombination is given by R. K. Mortimer and D. Schild, *in* "The Molecular Biology of the Yeast *Saccharomyces*" (J. N. Strathern, E. W. Jones, and J. R. Broach, eds.), Vol. 1, p. 11. Cold Spring Harbor Laboratory, Cold Spring Harbor, New York, 1981.

[6] H. L. Roman and S. M. Sands, *Proc. Natl. Acad. Sci. U.S.A.* **39**, 171 (1954).

Commitment to Recombination and Haploidization

Intragenic recombination between two auxotrophic heteroallelic mutations in any gene, for example, *ade2-1/ade2R-8,* results in a prototrophic cell, in this case a cell that grows on medium lacking adenine. Haploidization is monitored by the uncovering of recessive resistance markers. Diploid cells which are heterozygous for a recessive drug resistance marker such as *can1* or *cyh2* are unable to grow on plates that contain canavanine or cyclohexamide, respectively, but one-half of the haploid progeny are Can^R or Cyh^R and will grow on such plates. (*CAN1* codes for the arginine permease, and cells which contain the *can1* allele are able to grow on media containing canavanine, an analog of arginine. *CYH2* codes for one of the ribosomal proteins, and *cyh2* mutants are insensitive to the presence of cyclohexamide in the medium.) Commitment to recombination and haploidization is monitored in the following manner. At various times after transfer of cells to SPM, samples are taken and plated at appropriate dilutions on YEPD, SC − ADE, SC + CAN, or SC + CYH plates. After incubation for 2–4 days, the colonies on each plate are counted. Cell viability is calculated from the number of colonies on the YEPD plates (compared to the time 0 plating); intragenic recombination is calculated from the number of colonies on the SC − ADE plates, divided by the number on YEPD; and haploidization is calculated from the number of colonies on the drug-containing plates. For example, recombination between *ade2-1* and *ade2-R8* increases from 1×10^{-6} in mitotic cultures and at time 0 to $2-4 \times 10^{-4}$ in meiotic cultures.

The results from these plating experiments do not necessarily reflect the times at which recombination and haploidization took place, but rather the times in meiosis at which the cells became *committed* to these two events.[7] The actual recombination and haploidization could have occurred later, after the cells were plated onto the selective media.[3,7]

Premeiotic DNA Synthesis

Radioactive Labeling of DNA.[2] Sporulating cells of *Saccharomyces cerevisiae* do not take up radioactive label efficiently from the medium. It is preferable therefore, to prelabel the cells while growing in the presporulation medium. Premeiotic DNA synthesis uses nucleotides from breakdown of RNA. Therefore, labeled uracil which is incorporated into RNA, and hence enters internal pools of nucleotides, is used. Cells are grown in PSP2 medium supplemented with [2-^{14}C]uracil (0.8 μCi/ml, specific activ-

---

[7] R. E. Esposito and M. S. Esposito, *Proc. Natl. Acad. Sci. U.S.A.* **71**, 3172 (1974).

ity 60 mCi/mmol). Amino acids required by the strain should be added. If the strain is auxotrophic for uracil one should add only 3 μg/ml cold (nonradioactive) uracil (excessive supplementation reduces the uptake of radioactive uracil). There is no need to add cold uracil to URA$^+$ cells. Cells are grown for at least five generations, washed once with water, and resuspended in SPM. No label is added to the sporulation medium. Samples of 0.5 ml are taken (in triplicate) at intervals of 1–2 hr. An equal volume of 1 N NaOH is added to each sample (to degrade the RNA), and the mixture is incubated overnight (18–24 hr) at 36°. Degradation of RNA can also be achieved by 30 min of incubation at 65°. An equal volume of cold 20% trichloroacetic acid (TCA) is added, and the mixture is incubated on ice for 20 min. The precipitated DNA is collected on GF/C filters, washed once with ice-cold 5% TCA containing 0.1 g/liter "cold" uracil, and subsequently with cold 95% ethanol containing 0.1 g/liter "cold" uracil. The filters are dried, immersed in scintillation fluid, and counted [counts per minute (cpm)]. Cells undergoing normal meiosis show a basal level of radioactive label in DNA (TCA-precipitable counts) at the time of transfer to SPM (accumulated during growth in PSP2 with [^{14}C]uracil) and close to twice that level of TCA-precipitable counts in early stages of meiosis, reflecting a full round of premeiotic DNA replication.

Diaminobenzoic Acid: Colorimetric Assay of DNA.[2] Sporulate cells in liquid SPM by the normal procedure, and at various times withdraw 1-ml samples (it is advisable to have 3–5 repeats for each sampling time). Add an equal volume of 2 N NaOH to each sample and incubate the mixture for 24 hr at room temperature (samples can be kept at 4° for long periods). Add an equal volume of 50% cold TCA, incubate the mixture for 30 min on ice, and pellet the DNA by centrifugation at 10,000 g for 10 min at 4°. Wash the pellet once with 5% TCA, once with 0.1 M potassium acetate in 95% ethanol, once with 100% ethanol, and then dry the pellet overnight at 70°. Dissolve 0.625 g diaminobenzoic acid (DABA) in 1 ml warm water. In order to clear the DABA solution, add activated charcoal (Norit) and spin down. Resuspend the DNA pellet in 0.1 ml of the DABA solution, cap the tubes well, and incubate for 30 min at 60°. Add 2 ml of 0.6 N perchloric acid (PCA). Measure the optical density in a fluorospectrophotometer at excitation wavelength of 408 nm and emission wavelength of 508 nm.

Diphenylamine Assay[8] *of DNA.* Take a sample of 10^8 cells, pellet the cells, and wash once in cold sterile water. The cell pellet can be stored at −20°. Resuspend the pellet in 4 ml of PCA wash solution (0.2 N perchloric acid in 50% ethanol), and incubate for 40 min at room temperature. Pellet the cells by centrifugation at 4000 g for 15 min at 4°. Aspirate

---

[8] K. Burton, this series, Vol. 12B, p. 163.

off the supernatant and wash the pellet with PCA wash solution. Resuspend the pellet in 4 ml of freshly prepared lipid extraction solution (75% ethanol, 25% anhydrous ether) and incubate at 60° for 15 min. Centrifuge at 4° and extract the pellet again. Dry the pellet at 50° overnight. Resuspend the pellet in 1.25 ml of 1.5 N PCA made up in water, incubate at 70° for 30 min, and cool to room temperature. Add 0.75 ml DPA reagent [0.01% paraldehyde, 4% diphenylamine (DPA)], mix, and incubate at 30° for at least 18 hr. Add 0.5 ml amyl acetate, mix, and spin to pellet debris. Read a sample from the upper amyl acetate phase in a spectrophotometer at 595 nm.

Nuclear Staining

The two principal methods for staining of nuclei in meiosis use either DAPI (4′,6-diamidino-2-phenylindole) or Giemsa stain. The former is described elsewhere.[1] As the DAPI staining requires a fluorescence microscope, which is not always available, we present here the Giemsa staining method.[9]

Mix 0.5 ml cells (6×10^6 cells) with 4.5 ml of 4% formaldehyde in 0.15 M NaCl. Spin down the cells and place a drop of cells from the pellet on a glass slide. When the cells have almost dried, fix them by mixing a toothpickful of egg white with the cells while spreading them on the slide. Incubate the slides in 70% ethanol for from 2 min to 48 hr. Transfer the slides to 1% NaCl and incubate at 60° for 1 hr. Transfer the slides to 100 μg/ml RNase in 50 mM Tris, pH 7.4, 5 mM EDTA, and incubate at 60° for 2 hr. Transfer to a Giemsa solution that contains 0.35 ml Giemsa stain (Gurr) and 10 ml Gurr buffer, pH 6.9 (Hopkins and Williams Ltd.), and incubate at room temperature for 90 min. Wash the slides with the Gurr buffer and dry them. Observe cells and stained nuclei with a light microscope.

Utilization of Mutations in Study of Meiosis

In order to understand the relationships between the different events of meiosis, and to gain more insight into these special events, one may use mutations that block meiosis at different stages. Three groups of genes have been shown to be required for meiosis:[10] *(a)* Meiosis-specific genes: a mutation in such a gene has no mitotic phenotype but affects the meiotic

---

[9] C. F. Robinow and J. J. Marak, *J. Cell Biol.* **29**, 129 (1966).
[10] R. E. Esposito and S. Klapholz, *in* "The Molecular Biology of the Yeast *Saccharomyces*" (J. N. Strathern, E. W. Jones, and J. R. Broach, eds.), Vol. 1, p. 211. Cold Spring Harbor Laboratory, Cold Spring Harbor, New York, 1981.

pathway; *(b) CDC* genes: a mutation in such a gene affects the mitotic cell cycle, and many of the *CDC* genes that are required for the nuclear cycle[11] also affect the meiotic cell cycle;[12] and *(c) RAD* genes: a mutation in such a gene causes radiation sensitivity to the cell, and some of the *rad* mutants have also been found to arrest cells in the meiotic pathway, for instance, *rad6*,[13] *rad50*,[14] and *rad52*.[14] There are other genes which code for proteins that are important for meiotic cells, as well as for vegetative cells, for example, *TOP2*,[15] *TUB2*,[16] *YPT1*,[17] and *BCY1*[18]. These may also be regarded as mitotic cell cycle genes.

In order to study the effect of a mutation on meiosis, one must first determine the terminal phenotype at which the mutant arrests. When the mutation is conditional, one applies the restrictive conditions at the time of transfer to SPM, or after 1–2 hr in SPM, in order to allow the completion of the mitotic division. Staining of nuclei (as described above) reveals the number of nuclei per cell, and thus may identify several stages of arrest: cells that arrested prior to the first meiotic division contain a single nucleus; those that completed only the first meiotic division contain two nuclei; and cells in which the nuclei underwent both meiotic divisions, but which were arrested prior to ascus formation, contain four nuclei. Asci may be of two kinds: two-spored asci or four-spored asci.

Further important parameters to examine are whether the mutant cells are able to complete the premeiotic DNA replication, commitment to recombination, recombination, and commitment to haploidization (as described above). Electron microscopy can reveal ultrastructures such as the spindle plaque body and the presence or absence of the synaptonemal complex (see [7] in this volume).

Analysis of *ts* mutations gives more information. Temperature shift-up experiments reveal whether the gene product is required more than once in meiosis as well as the time at which it is required. Shift-down experiments may give some clues as to the functions that do take place at the restrictive temperature and the reversibility of the specific meiotic defect.[3,10]

Studies of this nature reveal the independence of the various processes

---

[11] J. R. Pringle and L. H. Hartwell, *in* "The Molecular Biology of the Yeast *Saccharomyces*" (J. N. Strathern, E. W. Jones, and J. R. Broach, eds.), Vol. 1, p. 97. Cold Spring Harbor Laboratory, Cold Spring Harbor, New York, 1981.

[12] G. Simchen, *Genetics* **76**, 745 (1974).

[13] J. C. Game, T. J. Zamb, R. J. Braun, M. Resnick, and R. M. Roth, *Genetics* **94**, 51 (1980).

[14] J. C. Game and R. K. Mortimer, *Mutat. Res.* **24**, 281 (1974).

[15] C. Holm, T. Goto, J. Wang, and D. Botstein, *Cell (Cambridge, Mass.)* **41**, 553 (1985).

[16] N. F. Neff, J. H. Thomas, P. Grisafi, and D. Botstein, *Cell (Cambridge, Mass.)* **33**, 211 (1983).

[17] N. Segev and D. Botstein, *Mol. Cell. Biol.* **7**, 2367 (1987).

[18] K. Matsumoto, I. Uno, and T. Ishikawa, *Cell (Cambridge, Mass.)* **32**, 417 (1983).

of meiosis. For example, commitment to meiotic recombination occurs concomitantly with DNA replication, but one can separate the two events, and cells can undergo premeiotic DNA replication without commitment to meiotic recombination.[3,10]

Sporulation of a/a and α/α Diploids

In many genetic experiments *MAT* homoallelic diploids are obtained either incidently or by design. Normally, *MAT*a/*MAT*a or *MAT*α/*MAT*α diploids are sporulation deficient[6] and therefore cannot be analyzed genetically. Several methods are now available to overcome this difficulty. In this section we first describe the situations in which *MAT*a/*MAT*a and *MAT*α/*MAT*α diploids are obtained and then describe the methods which may be used to sporulate these strains.

Situations Leading to Formation of **a/a** *and* α/α *Diploids*

1. Following meiosis of *MAT*a/*MAT*a/*MAT*α/*MAT*α tetraploid cells one obtains three types of progeny: *MAT*a/*MAT*a, *MAT*α/*MAT*α, and *MAT*a/*MAT*α.[19]

2. Rare matings between *MAT*a and *MAT*a cells or *MAT*α and *MAT*α cells.[20,21] Such matings, however, can also result in the formation of *MAT*a/*MAT*α diploids (owing to rare *MAT* interconversion) as well as aberrations in chromosome III on which the mating-type gene is located.[22] It is advisable therefore to sporulate the resulting diploids and characterize the *MAT* genotype of the progeny.

3. Spontaneous, or UV-induced mitotic recombination in *MAT*a/*MAT*α diploids. Isolation of the homoallelic derivatives is done either by screening for mating-capable colonies[23] or by selection for homozygosis of the recessive drug resistance mutation $cry1^R$ which is linked to *MAT*.[24] Cells which contain the *cry1* allele are able to grow on plates containing cryptopleurine. The original diploid colony should be heterozygous for $cry1^R$, and mitotic recombination between *CRY1* and the centromere results in homozygosis for both $cry1^R$ and the linked *MAT* allele.

4. Replacement of the *MAT* information in a *MAT*a/*MAT*α diploid by transformation with a DNA fragment that contains the *MAT* gene. A

[19] H. Roman, M. N. Phillips, and S. M. Sands, *Genetics* **40**, 546 (1955).
[20] V. I. Mackay and T. R. Manney, *Genetics* **76**, 272 (1974).
[21] J. N. Strathern, J. Hicks, and I. Herskowitz, *J. Mol. Biol.* **147**, 35 (1981).
[22] Y. Kassir and G. Simchen, *Genetics* **109**, 481 (1985).
[23] G. Simchen, Y. Salts, and R. Pinon, *Genetics* **73**, 531 (1973).
[24] P. G. Grant, L. Sanchez, and A. Jimerez, *J. Bacteriol.* **12**, 1308 (1974).

replacement with the *MAT*a allele will result in a *MAT*a/*MAT*a diploid, whereas a replacement with the *MAT*α will result in a *MAT*α/*MAT*α diploid.

Methods for Sporulating a/a and α/α Diploids

1. *MAT*a/*MAT*a and *MAT*α/*MAT*α become sporulation proficient if they are homozygous for *rme1*.[25] (*RME1* is a negative regulator of meiosis which is repressed by a complex of the *MAT*a1 and *MAT*α2 gene products.[25,26]) Some laboratory strains contain the *RME1* allele, whereas others contain *rme1*. One can introduce the *rme1* allele by a cross or by transformation with a DNA fragment that contains a disrupted allele of *RME1*.[26]

2. Transformation of *MAT*a/*MAT*a or *MAT*α/*MAT*α diploids with a plasmid that contains the *MAT*α or *MAT*a gene, respectively, will enable the transformants to sporulate. Moreover, transformation with the *HML*α or *HMR*a gene on a multicopy plasmid will also result in sporulation proficiency. Normally, the *HMR* and *HML* information is repressed by the four *SIR* genes,[27] but the presence of one of the *HM* genes on a multicopy plasmid may titrate out the *"SIR"* repression so that the information from the *HM* gene on the plasmid or from the genome is expressed and meiosis occurs.[22]

3. The *HO* allele on a plasmid induces the homothallic interconversion process, which results in the production of a *MAT*a/*MAT*α cell from a *MAT*a/*MAT*a or *MAT*α/*MAT*α cell.[28] *HO* can be introduced into the cell by transformation (see [8] in this volume), and following interconversion one can select the colonies that have lost the plasmid and then induce meiosis (see [8] in this volume). The disadvantage of this method is the loss of the original genotype (with respect to *MAT*).

4. *MAT*a/*MAT*a and *MAT*α/*MAT*α cells that contain the gene *IME1* on a multicopy plasmid can sporulate. The *IME1* gene is a positive regulator of meiosis which is negatively regulated by *RME1*. Thus, on a multicopy plasmid it titrates out *RME1*, and meiosis is induced.[29]

5. The gene products of *MAT*a1 and *MAT*α2 can be transmitted from one nucleus to another in a zygote in which nuclear fusion has not yet occurred. If one nucleus is *MAT*a/*MAT*a and the other *MAT*α/*MAT*α, then the two nuclei may enter meiosis separately and produce eight-spored

[25] Y. Kassir and G. Simchen, *Genetics* **82**, 187 (1976).
[26] A. P. Mitchell and I. Herskowitz, *Nature (London)* **319**, 738 (1986).
[27] J. D. Rine and I. Herskowitz, *Genetics* **116**, 9 (1987).
[28] R. E. Jensen, G. F. Sprague, Jr., and I. Herskowitz, *Proc. Natl. Acad. Sci. U.S.A.* **80**, 3035 (1983).
[29] Y. Kassir, D. Granot, and G. Simchen, *Cell (Cambridge, Mass.)* **52**, 853 (1988).

asci.[30] Two methods exist for obtaining zygotes without nuclear fusion. First, if one of the strains is also *kar1/kar1*, then in most cases the zygotes will not contain fused nuclei.[31] Second, one can mate fresh logarithmic cells for 3-4 hr on YEPD plates and then immediately spread them onto SPM plates. This "quick mating" procedure catches some of the zygotes in a stage with two distinct nuclei, and the transfer to SPM induces meiosis in the individual nuclei of these heterokaryons.[32]

Identification of Alleles and Mutations at *MAT*, *HMR*, and *HML*

Three copies of mating-type information, *MAT*, *HMR*, and *HML*, reside at three sites on chromosome III; however, only the information at the *MAT* locus is normally expressed.[33-35] One can discriminate between the presence of the **a** or α allele at any of these sites by Southern analysis.[34,35] In many instances mutations in *MAT*, *HMR*, or *HML* are the source of alteration in meiotic behavior. These mutations, however, cannot be distinguished by Southern analysis, and a genetic approach is required for their identification.

The *mata1* mutation has little or no effect in haploids, but *mata1/MATα* diploids are α maters and sporulation deficient.[25] Sporulation of such a diploid by one of the methods presented above will result in the production of tetrads which on dissection segregate as 2 **a** maters:2 α maters.

Haploids carrying the *matα1* mutation are nonmaters (sterile). Forced mating of a *matα1* cell to a *MAT***a** *RME1* cell will result in a nonmater diploid which is capable of undergoing sporulation.[20,21] Sporulation of *MAT***a**/*matα1* diploids gives rise to segregation of 2 sterile spores:2 mater spores in each tetrad.

Haploids with the *matα2* mutation are sterile (nonmaters), and *MAT***a**/*matα2* diploids are nonmater sporulation-deficient cells.[20,21] The genotype of the diploid should also be verified by sporulation and tetrad analysis using one of the techniques described in the section on Sporulation of **a**/**a** and α/α Diploids.

The *matα1α2* double mutant (alf) is an **a** mater and, like the *mata1* mutant, on mating with a *MATα* tester strain gives rise to an α mater

---

[30] A. J. S. Klar, *Genetics* **94**, 597 (1980).
[31] J. Conde and G. Fink, *Proc. Natl. Acad. Sci. U.S.A.* **73**, 3651 (1976).
[32] J. Margolskee, personal communication (1986).
[33] J. B. Hicks, N. J. Strathern, and A. J. S. Klar, *Nature (London)* **282**, 478 (1979).
[34] K. A. Nasmyth and K. Tatchell, *Cell (Cambridge, Mass.)* **19**, 753 (1980).
[35] J. N. Strathern, E. Spatola, C. McGill, and J. B. Hicks, *Proc. Natl. Acad. Sci. U.S.A.* **77**, 2839 (1980).

sporulation-deficient diploid cell.[21] How does one distinguish between these two mutations? Sporulation of both *mata1/MATα* and *matα1α2/MATα* diploids results in segregation of 2 **a** mater:2 α mater progeny, but the latter strain also gives rise to *matα1* and *matα2* recombinant progeny[21] (1–2% recombination between α1 and α2), which are nonmaters (sterile).

The same types of mutations can be obtained in the **a** and α alleles at *HMR* and *HML*, respectively.[36] In order to identify these mutations, namely, *hmr***a***1*, *hmlα1*, *hmlα2*, and *hmlα1α2*, one should transfer the information from the *HM* loci to *MAT* by the homothallic interconversion process and then study the new allele at *MAT*. For example, in order to identify whether the mutant strain contains the *HMR***a** or *hmr***a***1* allele, a putative *HMLα MATα hmr***a***1* haploid cell is transformed with the *HO* plasmid. As a result of homothallic interconversions followed by mating, the transformant will become a *MAT***a**/*MATα* diploid. Tetrad analysis of this diploid that contains the *HO* plasmid will indicate the presence of the *HMR***a** or *hmr***a***1* allele. The former will give rise to 4 nonmater Spo⁺ colonies, whereas the later will give rise to 2 nonmater Spo⁺ and 2 α mater Spo⁻ colonies. In both cases the spores themselves are 2 *MAT***a**:2 *MATα*, but in the presence of *HMR***a**, *HMLα*, and the plasmid carrying the *HO* gene, they become *MAT***a**/*MATα* diploids (as a result of homothallic interconversion, followed by mating). However, if the strain contains the mutant *hmr***a***1* strain, the *MATα* spores will show conversion to *mat***a***1* and will mate with a sib *MATα* cell to form a *mat***a***1/MATα* diploid which is α mater and Spo⁻. The *MAT***a** spores will show conversion to *MATα*, and a normal *MAT***a**/*MATα* (nonmater Spo⁺) diploid will be produced.

Identifications of Mutations and Genes Regulating Entry into Meiosis

Meiosis is normally induced in *MAT***a**/*MATα* diploids on transfer to an acetate medium, free of nitrogen and glucose (SPM). Several genes which participate in the decision a cell makes to enter meiosis have already been identified. We first describe the pathway by which *MAT* transmits its signal for initiation of meiosis and then discuss the starvation signal.

MAT Signal Transduction Pathway

The simplest model for initiation of meiosis is the following.[37] The gene products of *MAT***a***1* and *MATα2* are required to turn off the transcription

---

[36] L. C. Blair, P. J. Kushner, and I. Herskowitz, in "Eucaryotic Gene Regulation" (R. Axel, T. Maniatis, and C. F. Fox, eds.), p. 13. Academic Press, New York, 1979.

[37] Y. Kassir and G. Simchen, *Curr. Genet.* **15,** 167 (1989).

of *RME1*. *RME1* is a repressor of meiosis which blocks the transcription of *IME1*. The transcription of *IME1* is induced by nitrogen starvation (the second signal of meiosis) in the presence of acetate as the sole carbon source. *IME1* is a positive regulator of meiosis which is required for the transcription of *IME2 (SME1)* and a second, unidentified gene.[38,39]

In order to isolate the different genetic components which determine initiation of meiosis, one may use several methodologies as follows. One approach is to isolate and analyze sporulation-proficient derivatives or mutants from *mata1/MATα MATa/MATa,* or *MATα/MATα* strains. A new mutation may affect a gene that is specifically involved in initiation of meiosis, for instance, *RME1*,[25] or may have a nonspecific effect, for example, causing the expression of the cryptic mating-type information at *HMR* and *HML*. The latter mutations are therefore dependent on the information in the silent HM cassettes.

We first briefly list the types of events that cause the expression of the *HM* loci:

1. A heterothallic interconversion event will result in the generation of a *MATa/MATα* diploid from a *mata1/MATα* cell.[22]

2. Expression of the information at one of the *HM* loci may result from the following events: a deletion in chromosome III that joins part of *HMRa* to *MAT* results in a nonmater sporulation-proficient strain;[40] reciprocal recombination between *MATa* and *HMLα* results in a ring chromosome III that lacks the right arm distal to *MAT* and produces a nonmater sporulation-proficient strain;[41] a mutation at the E or I site of *HMR* (these sites were shown to be essential and important, respectively, for the "silencing" of *HMR* and *HML*[42]) is cis dominant, and a *mata1/MATα* cell harboring such a mutation is nonmater Spo$^+$ or α mater SPO$^+$;[22,42] a duplication of *HMRa* that retains the E site but instead of the I site contains sequences that are normally distal to *MAT* can also result in a sporulation-proficient derivative (the *SAD* duplication[43-46]).

---

[38] H. E. Smith and A. P. Mitchell, *Mol. Cell. Biol.* **9**, 2142 (1989).
[39] M. Yoshida, H. Kawaguchi, Y. Sakata, K. Kominami, M. Hirano, H. Shima, R. Akada, and I. Yamashita, *Mol. Gen. Genet.* **221**, 176 (1990).
[40] C. Hawthorne, *Genetics* **48**, 1727 (1963).
[41] J. N. Strathern, C. S. Newlon, I. Herskowitz, and J. B. Hicks, *Cell (Cambridge, Mass.)* **18**, 309 (1979).
[42] J. Abraham, J. Feldman, K. H. Nasmyth, J. N. Strathern, A. J. S. Klar, J. A. Broach, and J. B. Hicks, *Cold Spring Harbor Symp. Quant. Biol.* **47**, 989 (1983).
[43] A. K. Hopper and V. L. MacKay, *Mol. Gen. Genet.* **180**, 301 (1980).
[44] Y. Kassir and I. Herskowitz, *Mol. Gen. Genet.* **180**, 315 (1980).
[45] Y. Kassir, J. B. Hicks, and I. Herskowitz, *Mol. Cell. Biol.* **3**, 871 (1983).
[46] J. Hicks, J. Strathern, A. Klar, S. Ismail, and J. Broach, *Mol. Cell. Biol.* **4**, 1278 (1984).

3. There are four genes designated *SIR* (silent information repressor), and a recessive mutation in any one of them results in the expression of the information at *HML* and *HMR*.[27] A sporulation-proficient derivative of a *mata1/MATα* diploid may therefore result from a recessive mutation in a *SIR* gene.[22,47,48]

The methodologies by which one may identify such mutations are beyond the scope of this chapter and are described elsewhere.[22] A recessive mutation that allows sporulation of both *MATa/MATa* and *MATα/MATα* diploids, and which is not dependent on the information at *HMR* and *HML*, might be a mutation in *RME1*[25,49] or in another unidentified gene. Allelism to a known *rme1* strain is checked by crossing *MATα* segregants with a *mata 1 rme1* tester strain and by crossing *MATa* segregants with a *mata1 rme1sir1-1* tester strain (the latter is a bimater and can mate with the *MATa* strain).[22] If the resulting *mata1/MATα* and *mata1/MATa* diploids are sporulation proficient, one may conclude that the new mutation is in the *RME1* gene. One must verify this by dissecting tetrads and showing that all the progeny are indeed *rme1*. This is done by mating the haploid segregants to the appropriate tester strains and checking the sporulation capabilities of the resulting diploids.

A dominant mutation which is not *SAD* and is not linked to *HMR* and *HML* might be a mutation in the promoter of *IME1*,[29] which confers insensitivity to repression by *RME1*, and thus allows constitutive expression. Such a mutation may also identify a new gene which is required for the initiation of meiosis. In order to test allelism with *ime1*$^-$, the haploid mutant strain (putatively designated X^-) should be mated to an *ime1-0* tester strain, the resulting diploid sporulated, and tetrads dissected and analyzed. If the mutation is indeed in the *IME1* gene we expect to get 2 *ime1-0*:2 X^- spores. The segregants will be mated with an *ime1-0* tester strain. *ime1-0* segregants are expected to give rise to *MATa/MATα* sporulation-deficient diploids, whereas the X^- segregants should give rise to sporulation-proficient *MATa/MATα* diploids. The presence of the X^- mutation is examined by crossing the segregants with *mata1* or *mata1 sir1-1* strains (the latter is a bimater and can mate with *MATa* segregants). The *mata1/MATα* and *mata1/MATa* diploids will be Spo$^+$ if they contain X^- Inability to obtain wild-type recombinants that are neither *ime1-0* nor X^- indicates that the X^- mutation is in the gene *IME1*, or that it is very closely linked to it. In order to prove that the new mutation is indeed in the *IME1*

---

[47] J. E. Haber and J. P. George, *Genetics* **93**, 13 (1979).
[48] J. P. Margolskee, *Mol. Gen. Genet.* **211**, 430 (1988).
[49] J. D. Rine, G. F. Sprague, Jr., and I. Herskowitz, *Mol. Cell. Biol.* **1**, 958 (1981).

gene, one might clone it by gap repair of a CEN plasmid that carries the *IME1* region but is gapped to omit the *IME1* gene with its promoter. When the repaired plasmid is introduced into a *mata1/MATα* strain, it should enable the cell to sporulate. (A single-copy CEN plasmid with the normal gene *IME1* does not promote sporulation ability in a *mata1/MATα* strain.[29])

Another method that proved successful in the isolation of genes required for initiation of meiosis is to look for a gene on a multicopy plasmid that bypasses the requirement for a specific control. For example, the *IME1* gene was identified by transforming *mata1/MATα* diploid with a yeast DNA library on a multicopy vector and screening for sporulation-proficient transformants.[29] The *IME2* gene was similarly identified by transforming a *MATa/MATα* diploid which carried the gene *RME1* on a multicopy vector, and thus was sporulation deficient, with a yeast DNA library on another multicopy vector and screening for transformants which induced meiotic recombination.[38]

A third approach for the isolation of mutations in genes that are required for initiation of meiosis is to look on SPM medium for sporulation-deficient mutants and, among them, to screen for the ones that arrest early in meiosis, namely, prior to the premeiotic DNA replication.

Starvation Signal Transduction Pathway

We do not know yet how the starvation signals induce meiosis. In order to facilitate the study of this pathway, one would like to identify the genes or gene products that take part in this signal transduction. Genes whose products negatively control the induction of meiosis are expected to give rise to recessive mutations that initiate meiosis without the requirement for starvation. One can therefore isolate mutants that sporulate on YEPD, YEPA, SPM + glucose, or SPM + NH_4Cl plates.[50] Using other methodologies that are not directly concerned with meiosis, several genes have been identified in which mutations were found to have such an effect (*CDC25, CDC35,* and *RAS2*); these genes belong to the pathway leading to activation of the cAMP-dependent protein kinase. A recessive mutation in one of these genes results in a low level of cAMP, G_1 arrest, and induction of meiosis without starvation.[18,51-53] The *spd1* mutation also causes meiosis

---

[50] I. W. Dawes, *Nature (London)* **255,** 707 (1975).
[51] V. Shilo, M.Sc. Thesis, The Hebrew University, Jerusalem, Israel (1976).
[52] V. Shilo, G. Simchen, and B. Shilo, *Exp. Cell Res.* **112,** 241 (1978).
[53] K. Tatchell, L. C. Robinson, and M. Breitenbach, *Proc. Natl. Acad. Sci. U.S.A.* **82,** 3785 (1985).

in the absence of the starvation signal, but the nature of this mutation is not understood.[50,54] Following isolation of mutants, one should first examine the new mutation for allelism and epistasis to the above-mentioned genes.

Mutations that result in failure to respond to the starvation signal and failure to arrest at G_1 are also known, and they result in sporulation deficiency as well; to this class belong the mutations *bcy1* and *RAS2-val19*.[18,55] The initial effect of these mutants is high activity of the cAMP-dependent protein kinase.[18,55] Thus, some of the Spo⁻ mutants, which are obtained by other screens might affect this class of genes. They may be recognized by their inability to accumulate as unbudded cells after transfer to SPM medium.

Genes whose functions are positively regulated by starvation and which by themselves are positive regulators of meiosis are also expected to give rise to Spo⁻ mutants. Such a gene can be identified by its ability to induce meiosis in vegetative conditions (YEPA and/or YEPD and/or SPM + glucose and/or SPM + NH_4Cl media) when present in the cell on a multicopy vector. The *SME1 (IME2)* gene was isolated in this manner,[39] and the gene *IME1* discussed above also belongs to this class.[29] The transcription of *IME1* and *SME1 (IME2)* is repressed by the presence of nitrogen and glucose in the medium. Another method is therefore to screen for mutations or plasmids that specifically affect the transcription of *IME1* or *IME2*. *IME* expression would be monitored by fusing it to *lacZ* (see below and [7] in this volume).

Screening and Isolation of Sporulation-Specific Mutants or Genes

In this section we wish to discuss the methodologies developed for the screening of sporulation-proficient or -deficient derivatives.

Mutagenesis of Haploid Spores of a Strain That Contains the HO Allele.[56] Following growth, the spores give rise to nonmater homozygous diploid colonies that have resulted from homothallic interconversion. Thus, a recessive mutation that has been induced in the haploid spore will become homozygous, and the mutant phenotype will be expressed in the diploid colony. The colonies are replicated onto SPO plates, and sporulation capability is examined either by screening for the presence or absence of asci or by using a late meiotic gene fused to *lacZ* and screening for

[54] F. Vezinhet, J. H. Kinnaird, and I. W. Dawes, *J. Gen. Microbiol.* **115**, 391 (1979).
[55] T. Kataoka, S. Powers, C. McGill, O. Fasano, J. Strathern, J. Broach, and M. Wigler, *Cell (Cambridge, Mass.)* **37**, 437 (1984).
[56] M. S. Esposito and R. E. Esposito, *Genetics* **61**, 79 (1969).

β-galactosidase activity. On SPO X-Gal plates Spo⁻ mutants will remain white, whereas Spo⁺ colonies will become blue.

UV-Induced Mutagenesis. UV-induced mutagenesis has been shown to be recombinogenic; thus following UV irradiation of diploid cells one can isolate both dominant and recessive mutations.[3] This permits the use of heterothallic diploid strains with heterozygous recessive selectable markers, for instance, *MATa/MATα can1/CAN1 cyh2/CYH2*. After 4 days of incubation on SPO plates, the colonies are replica plated onto plates that contain both canavanine and cyclohexamide. Sporulation-proficient cells give rise to Can^R Cyh^R *(can1 cyh2)* haploid segregants that can grow on such media, whereas sporulation-deficient colonies will not grow on SC + CAN + CYH plates. Following UV-induced mutagenesis of diploid strains one can also screen for sporulation by screening for the expression of *lacZ* fused to a meiotic gene.

If one wishes to isolate a mutation that affects the cells at a specific meiotic stage or earlier, for example, a mutation affecting the recombination process, one may screen for absence of recombinants. The strain should be heteroallelic for two noncomplementing auxotrophic mutations in a gene such as *HIS4* or *ADE2;* the colonies are replicated from SPO plates onto plates that lack histidine or adenine. Recombination-proficient colonies will give rise to papillae, whereas recombination-deficient mutants will not show papillae on these media. The same methodology can be used in order to screen for multicopy plasmids that alter the regulation of initiation of meiosis.

Interactions between Sporulation-Specific Genes

As has been mentioned above, sporulation-specific genes are isolated in order to understand the pathways which lead to meiosis. Further understanding might come from the interactions between the various genes that affect sporulation. Epistasis relationships may be revealed by constructing double mutants and examining the sporulation capabilities of the double mutants at various conditions. Studies of meiosis in double mutants are meaningful in cases where each single mutant shows a different phenotype. Strains which contain two sporulation-specific genes on a multicopy plasmid may also tell us about the interaction between the two genes, whether a particular phenotype becomes more extreme, or whether the overexpression of one gene can counteract the effects of overexpression of the second gene. Interactions are also indicated when overexpression of a particular gene suppresses a mutation in another gene. Another approach is to study the expression of one gene at the level of transcription or as the β-galactosidase activity of *lacZ* fusion of that gene in strains carrying mutations or multicopy plasmids of other meiosis-specific genes. The combined approaches, which are also used to elucidate other processes in yeast biology, have already revealed some elements in the regulatory pathway of yeast meiosis.[26,29,38,39]

Acknowledgments

We thank Dr. Shoshana Klein for helpful comments on the manuscript. Work done in our laboratory has been supported by grants from the United States–Israel Binational Science Foundation (BSF).

[7] Identifying Sporulation Genes, Visualizing Synaptonemal Complexes, and Large-Scale Spore and Spore Wall Purification

By ROCHELLE EASTON ESPOSITO, MICHAEL DRESSER, and MICHAEL BREITENBACH

Introduction

Yeast sporulation is an important model system in eukaryotic cell differentiation. Despite its unicellular nature, the sporulating yeast cell undergoes a genuine morphogenic process that is controlled by both external and internal factors, e.g., nutritional state and cell type. During sporulation, the process of meiosis is tightly coupled to spore formation (reviewed in Refs. 1–8). Meiosis itself is a central feature in the sexual reproduction of eukaryotes, and most of the characteristics of meiosis are

---

[1] R. E. Esposito and S. Klapholz, in "Molecular Biology of the Yeast *Saccharomyces*" (J. N. Strathern, E. W. Jones, and J. R. Broach, eds.), Vol. 1, p. 211. Cold Spring Harbor Laboratory, Cold Spring Harbor, New York, 1981.
[2] M. Breitenbach and E. Lachkovics, in "Secondary Metabolism and Differentiation in Fungi" (J. W. Bennett and A. Ciegler, eds.), p. 307. Dekker, New York, 1983.
[3] W. Dawes, in "Yeast Genetics: Fundamental and Applied Aspects" (J. F. T. Spencer, D. M. Spencer and A. R. W. Smith, eds.), p. 29 Springer-Verlag, New York, 1983.
[4] P. T. Magee, in "Meiosis" (P. B. Moens, ed.), p. 355. Academic Press, New York, 1987.
[5] Z. Olempska-Beer, *Anal. Biochem.* **164**, 278 (1987).
[6] M. A. Resnick, in "Meiosis" (P. B. Moens, ed.) p. 157. Academic Press, New York, 1987.
[7] J. R. Dickinson, *Microbiol. Sci.* **5**, 121 (1988).
[8] J. J. Miller, in "The Yeasts" (A. H. Rose and J. S. Harrison, eds.), 2nd Ed., Vol. 3, p. 489. Alden Press, London, 1989.

conserved in all eukaryotes. In yeast, the dormant haploid spores that result after meiosis are also resistant to harsh environmental conditions such as heat, digestive enzymes of animals, and starvation.[9]

In the past, yeast has been considered a favorable system for the study of meiosis and spore development because of its convenient genetics, ease of isolating mutants, and biochemistry. In this chapter we describe recent techniques that provide yeast with other unique advantages for the study of meiotic recombination, chromosome segregation, and the subsequent packaging of meiotic products into resistant spores. In our view, a key element to progress in this field is the invention of new and powerful mutant screens and/or selective procedures that allow identification of mutants defective in specific discrete steps of meiosis and spore formation. Here, we focus on a novel approach to detecting sporulation mutants based on *spo13* meiosis, cloning procedures for sporulation genes, and two recent methods that are useful for characterizing mutants, namely, visualization of synaptons and purification of spores and spore walls on density gradients. Conditions optimizing sporulation efficiency and synchrony of sporulation as well as other methods used in the analysis of landmark events are given elsewhere.[5,8,10-17]

Novel Mutant Recovery Systems

Rationales of Previous Approaches

In earlier systematic mutant hunts, defects at various stages of the sporulation process were sought by assaying whether the final meiotic products were affected. The following criteria were used: (1) reduction in the number of spores per ascus or total asci, (2) reduction in spore viability, and (3) derepression of spore formation under nonstarvation conditions. Later it was recognized that the mutant phenotypes in each class were pleiotropic and not clearly distinct from one another. Thus, at the Tenth International Conference on Yeast Genetics and Molecular Biology, the

---

[9] K. H. Ho and J. J. Miller, *Can. J. Microbiol.* **24**, 312 (1978).
[10] I. W. Dawes and I. D. Hardie, *Mol. Gen. Genet.* **131**, 281 (1974).
[11] J. Golombek and E. Wintersberger, *Exp. Cell Res.* **86**, 199 (1974).
[12] M. S. Esposito and R. E. Esposito, in "Methods in Cell Biology" (D. M. Prescott, ed.), Vol. 11, p. 303. Academic Press, New York, 1975.
[13] J. E. Haber and H. O. Halvorson, in "Methods in Cell Biology" (D. M. Prescott, ed.), Vol. 11, p. 46. Academic Press, New York, 1975.
[14] A. Hartig and M. Breitenbach, *FEMS Lett.* **1**, 79 (1977).
[15] K. Ueki, M. Abe, K. Tada, and N. Sando, *J. Gen. Microbiol.* **129**, 3619 (1983).
[16] Y. Kassir and G. Simchen, this volume [6].
[17] B. Rockmill, E. J. Lambie, and G. S. Roeder, this volume [9].

spo gene symbol was defined as sporulation-abnormal and assigned to all of the variants.[18] Two other criteria were subsequently applied: (4) failure to exhibit meiotic recombinants in return-to-growth studies and (5) altered control or regulation of meiosis by cell type (reviewed in Refs. 1 and 3; see also [6] and [9] in this volume).

The classic approach to isolating Spo⁻ mutants has involved mutagenizing haploid spores of a homothallic strain.[19] Spore outgrowth, mating-type switching, and mating between cells in a spore colony results in homozygosis of all genes except for *MAT*. The surviving diploids are then directly tested for their ability to sporulate and both recessive and dominant mutations readily detected. Temperature-sensitive mutants that affect sporulation only at a restrictive temperature have been particularly useful to facilitate outcrossing and genetic analysis of the primary mutants. For example, spores obtained from the mutants at the permissive temperature can be easily mated to heterothallic strains to obtain the mutations in standard haploid backgrounds. It is now possible to simplify this procedure using strains that carry the cloned *HO* gene on a plasmid. Once spores are mutagenized and diploid cells formed after *MAT* switching and mating, the *HO* gene can be lost by transfer to a medium nonselective for the plasmid. Isogenic heterothallic *MAT*a and *MAT*α derivatives can be obtained directly by sporulating these diploid derivatives at a permissive temperature without outcrossing to another strain.

Early on, it was realized that the utility of the homothallic system for recovering recessive mutants in homozygous diploids could not be conveniently applied to mutant hunts specifically designed to detect defects in meiotic recombination. This is due to the fact that mating-type switching and mating of cells derived from a single spore result in diploids that lack heterozygous markers for detecting genetic exchange. Current technology can also now be applied to circumvent this difficulty. For example, artificially constructed gene duplications with different mutations in each gene, described in a later section, may be introduced into a single chromosome in the parental strain and exchange assayed in the diploidized spore progeny during further rounds of meiosis. In the early 1970s the approach devised to solve this problem utilized a haploid strain containing two copies of chromosome III with appropriate heteroallelic and heterozygous genetic markers to monitor recombination between them.[20] Such $n + 1$ disomic strains, heterozygous for *MAT*a/*MAT*α, are capable of premeiotic

[18] R. E. Esposito, *in* "Workshop Reports of the Tenth International Conference on Yeast Genetics and Molecular Biology" (A. Goffeau and J. Wiame eds.), p. 19. Tenth International Conference, Laboratoire d'Enzymologie, Louvain-la-Neuve, Belgium, 1980.
[19] M. S. Esposito and R. E. Esposito, *Genetics* **61**, 79 (1969).
[20] R. Roth and S. Fogel, *Mol. Gen. Genet.* **112**, 259 (1971).

DNA synthesis and meiotic recombination, although not mature viable spore formation. Putative recessive meiotic Rec⁻ mutations are detected when these $n + 1$ cells are allowed to initiate meiosis on sporulation medium and transferred to selective growth media diagnostic for recombinant genotypes. Finally, previous and current mutant selection schemes[21-23] for detecting alleles that bypass mating-type control are reviewed elsewhere in this volume[16] and are not addressed here.

spo13 System

The discovery of the *spo13* mutant[24] provided a significant advance in developing new methods to identify mutants of meiosis and spore development. Sporulating diploid yeast with a recessive defect in the gene *SPO13* undergoes an atypical meiosis consisting of one rather than two meiotic divisions. Premeiotic DNA synthesis, chromosome paring, and recombination during prophase of meiosis I occur as in wild-type cells. In most diploid *spo13* strains, the vast majority of cells (~95%), then fail to separate homologs at mitosis I (i.e., reductional division of chromosomes). Instead, the replicated chromatids or each chromosome separate as in mitosis (i.e., equational division of chromosomes), and the products become enclosed in two diploid ascospores as shown in Fig. 1.[25]

Apart from the intriguing nature of the *SPO13* function in meiosis, the phenotype of the mutant provides a useful tool for other analyses: (1) it suppresses the inability of haploid cells expressing both *MATa* and *MATα* to form viable spores and allows an unusual meiosis termed "haploid meiosis," [26] and (2) it suppresses the lethal effects of certain recombination-deficient (Rec⁻) and segregation-deficient (Seg⁻) mutants and allows them to form viable meiotic products for further study.[1,27-30] These uses of *spo13* are described in more detail in the next several sections. Here, we wish to emphasize its ability to allow appropriately marked haploid cells, otherwise incapable of successfully completing two meiotic divisions, to

---

[21] A. K. Hopper and B. D. Hall, *Genetics* **80**, 41 (1975).
[22] Y. Kassir, D. Granot, and G. Simchen, *Cell (Cambridge, Mass.)* **52**, 853 (1988).
[23] H. Smith and A. P. Mitchell, *Mol. Cell. Biol.* **9**, 2142 (1989).
[24] S. Klapholz and R. E. Esposito, *Genetics* **96**, 567 (1980).
[25] S. Klapholz and R. E. Esposito, *Genetics* **96**, 589 (1980).
[26] J. E. Wagstaff, S. Klapholz, and R. E. Esposito, *Proc. Natl. Acad. Sci. U.S.A.* **79**, 2986 (1982).
[27] R. E. Malone and R. E. Esposito, *Mol. Cell. Biol.* **1**, 891 (1981).
[28] S. Klapholz, C. S. Waddell, and R. E. Esposito, *Genetics* **110**, 187 (1985).
[29] B. Rockmill and G. S. Roeder, *Proc. Natl. Acad. Sci. U.S.A.* **85**, 6067 (1988).
[30] N. M. Hollingsworth and B. Byers, *Genetics* **121**, 445 (1989).

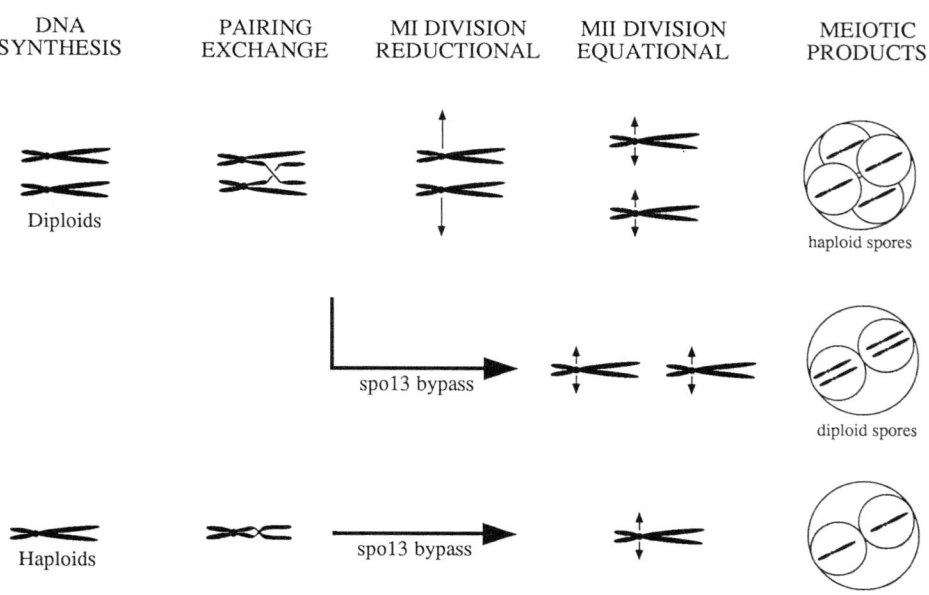

FIG. 1. Schematic diagram of *spo13* single division meiosis in diploid and haploid cells. Haploids require the expression of *MAT*a and *MAT*α mating types to undergo *spo13* haploid meiosis (see text for details).

undergo various landmarks of meiotic development during the single division and to produce viable haploid gametes.

Normally, *MAT*a or *MAT*α haploid cells are unable to initiate the earliest meiotic events, presumably because the a1–α2 repressor is required to turn off transcription of *RME1*, a key negative regulator of meiosis.[31] When haploid cells are able to express both mating type genes, they attempt meiosis but do not form mature spores. This occurs, for example, in (1) $n + 1$ *MAT*a/*MAT*α disomic strains as noted above,[26] (2) haploids containing a *sir* mutation allowing transcription of the silent mating-type cassettes,[32] and (3) haploids carrying a plasmid bearing the opposite *MAT* allele.[33] These strains are capable of premeiotic DNA synthesis, meiotic recombination, and the two meiotic divisions, and they form tri- and tetranucleate cells with immature spores. Their failure to produce mature viable asci is assumed to be due to extensive aneuploidy

[31] A. P. Mitchell and I. Herskowitz, *Nature (London)* **319**, 738 (1986).
[32] J. E. Wagstaff, S. Klapholz, C. S. Waddell, L. Jensen, and R. E. Esposito, *Mol. Cell. Biol.* **5**, 3532 (1985).
[33] S. Gottlieb and R. E. Esposito, *Cell (Cambridge, Mass.)* **56**, 771 (1989).

resulting from random segregation of unpaired chromosomes at meiosis I.[26]

When a *spo13* mutation is introduced into such strains, in a majority of cells meiosis I reductional segregation is bypassed, and the sister chromatids of each chromosome are partitioned equally to each daughter nucleus allowing the formation of dyad asci containing two similar haploid spores as shown in Fig. 1. The single division meiosis in *spo13* mutants, thus, provides a permissive condition for haploids expressing both mating-type alleles to produce viable spores. Haploid meiosis offers a simple solution to identify and analyze meiotic products of recessive Spo⁻ mutants. The advantages of this system are as follows: (1) haploid cells can be mutagenized, (2) procedures for obtaining homozygous diploid cells are not needed and, instead, the original haploids can be directly tested for all growth- and differentiation-related phenotypes, and (3) conditional as well as nonconditional mutants can be easily studied as the mutagenized haploids can be outcrossed to wild-type mating partners. Thus far, haploid meiosis has been used to isolate recessive mutants affecting meiotic recombination,[30] spore wall biogenesis,[34,35] and spore germination.[36]

Introduction of spo13 System into Laboratory Strains

The *spo13* single division meiosis system can be readily introduced into standard laboratory strains using the plasmid p*(spo13)*16 (Fig. 2).[37] This plasmid carries the *URA3* gene inserted into the coding region of *SPO13*, facilitating the creation of a genomic disruption[38] at the chromosomal location of *SPO13* on chromosome VIIIR. The gene disruption mutant yields exclusively two-spored asci.[37] Alternatively, the *spo13-1* ochre mutation may be crossed into the desired strain. In most of our laboratory stocks, *spo13-1* results in approximately 98% two-spored asci; in some genetic backgrounds, and in the presence of paromomycin, the percentage of four-spored asci may be greater, presumably owing to partial suppression of the ochre allele.[24,39] When a high yield of tetrads is required for other analyses, the wild-type *SPO13* gene on the CEN plasmid p*(SPO13)*76 may be used to complement either the disruption or the

---

[34] P. Briza, J. Segall, A. Ellinger, and M. Breitenbach, *Genes Develop.*, in press (1990).
[35] M. Breitenbach and P. Briza, unpublished (1990).
[36] M. Breitenbach, unpublished (1990).
[37] H. T. Wang, S. Frackman, J. Kowalisyn, R. E. Esposito, and R. Elder, *Mol. Cell Biol.* **7**, 1425 (1987).
[38] R. Rothstein, this series, Vol. 101, p. 202.
[39] J. E. Wagstaff, Ph.D. Dissertation, University of Chicago, p. 22 (1983).

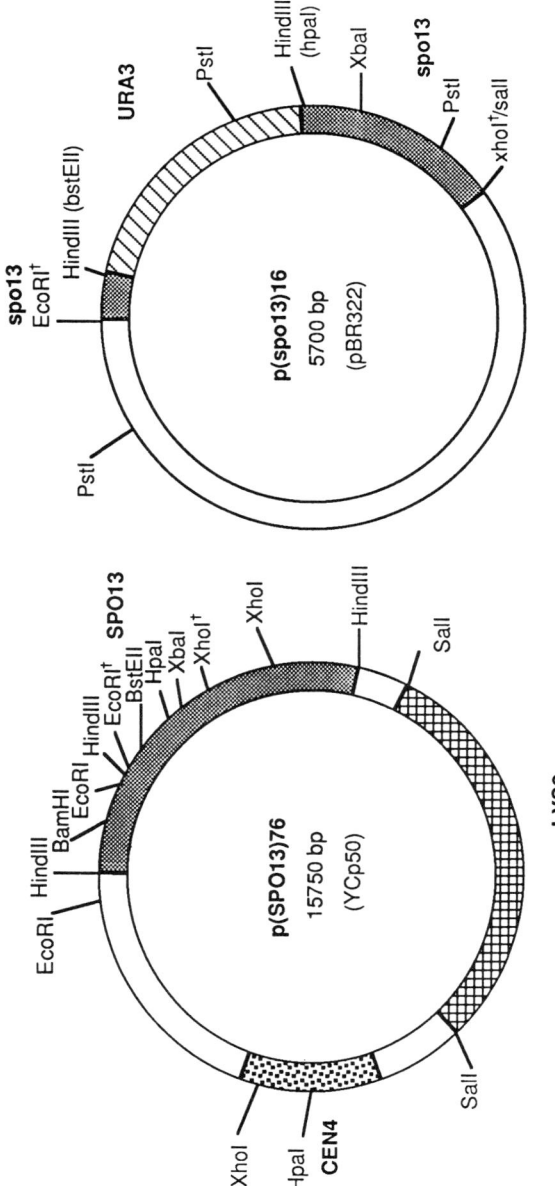

FIG. 2. Plasmid p(SPO13)76 carries the wild-type *SPO13* gene. The disruption plasmid, p(*spo13*)16, contains an *EcoRI–XhoI* subclone (†) of *SPO13* in which 570 bases of the *SPO13* coding region have been replaced by a 1.2-kb *URA3* fragment. Restriction enzyme sites written with an initial lowercase letter are no longer functional, and those in parentheses have been changed to *Hin*dIII sites.

spo13-1 mutation. This plasmid was constructed by S. Gottlieb in the laboratory of R. E. Esposito and is shown in Fig. 2.

As described above, haploid meiosis requires the stable expression of *MAT*a and *MAT*α. One simple way this can be achieved is by transforming haploid *spo13* mutant strains with a CEN plasmid containing the opposite *MAT* allele.[33] After meiotic products are produced, vegetative derivatives that have lost the plasmid can be recovered and the resulting haploids used in standard crosses. Another convenient method employs a haploid strain carrying both *spo13* and a temperature-sensitive *(ts) sir* mutation. One such *MAT*α strain, PB2-1C, contains *sir4^{ts}, his6, leu2,* and *trpl.*[34,35] PB2-1C sporulates at a reasonably high level at the restrictive temperature for *sir4* (55% at 25°), mates with near wild-type efficiency at the permissive temperature (22°), and can be transformed with multicopy yeast plasmids at moderate efficiency. Finally, a chromosome III disome can be made in *spo13* strains using the *kar* method[30] or by conventional crosses between a *spo13* haploid and a *SPO13* strain containing a disome. Disomic strains are particularly useful to study mutants affecting genetic recombination. Subsequent genetic analysis can be done with haploid derivatives that lose one chromosome of the disome. Such strains are easily recovered after haploid meiosis because a certain proportion of cells undergo aberrant segregation (one homolog segregates to one pole and the other separates its sister chromatids to opposite poles).[26]

Use of spo13 to Identify and Analyze Rec⁻ Mutations

How are genes specifically important for meiotic chromosome behavior identified? Analysis of genes controlling meiotic recombination, by the study of Rec⁻ mutations which reduce or abolish exchange, is hampered by the fact that proper meiosis I reductional chromosome segregation generally depends on normal exchange between homologs. In the absence of exchange, homologs segregate randomly, forming spores that contain abnormal chromosome numbers which are largely inviable and unavailable for further study. Two methods have been used to suppress Rec⁻-dependent meiotic lethality so that genes required for exchange can be identified (reviewed in Ref. 1). The key element in both approaches involves making recombination dispensable for the formation of viable cells during meiosis by eliminating meiosis I segregation. This has been done by (1) interrupting cells during meiotic development, and returning them to mitotic growth,[40,41] and (2) the use of *spo13* single division meiosis.[27,28] In both

---

[40] S. Prakash, L. Prakash, W. Burke, and B. A. Montelone, *Genetics* **94,** 31 (1980).
[41] M. A. Resnick, J. M. Kasimos, J. G. Game, R. J. Brown, and R. M. Roth, *Science* **212,** 543 (1981).

cases, prophase of meiosis I is followed by an equational mitotic or meiosis II-like division which does not require recombination to occur properly. In the first case, diploid cells are introduced into a medium that induces meiosis and returned to growth after various periods of time. In Rec$^+$ cells, given appropriate heterozygous and heteroallelic markers, a dramatic increase (10^2 to 10^4) in the recovery of recombinant progeny can be detected after a few hours of exposure to sporulation conditions. The failure to recover recombinants by this procedure has been used to define putative Rec$^-$ mutants. The limitation of this method is that it does not distinguish between mutants that are Rec$^-$ and those that are Rec$^+$ but unable to return to growth after commitment to recombination.

The *spo13* single division meiosis in diploid cells allows certain Rec$^-$ mutants to progress through meiosis II and form viable nonrecombinant gametes. We have proposed that those Rec$^-$ mutants which can be rescued into viable products by *spo13-1* are defective in early Rec$^-$ functions required to initiate exchange. Those that cannot be rescued are presumed defective in late Rec$^-$ functions in the repair and/or resolution of recombination intermediates producing chromosomes with breaks or with structures that cannot be disjoined. Double mutants defective in both early and late Rec functions are rescued. This general method of classification provides an approach to define Rec$^-$ genes and understand the genetic pathways utilized in meiotic recombination between homologs.[1,27,28,42]

Further insights are offered by haploid meiosis analysis. Haploid meiosis facilitates the identification of genes that control intrachromosomal recombination, since sister–chromatid exchange and intrachromatid exchange can be assayed on an individual chromosome in the absence of its homolog. Three common systems used to monitor intrachromosomal exchange between repeated genes during *spo13* meiosis are shown in Fig. 3 and described in detail elsewhere.[32,33,43] They include rDNA recombination assayed by the fate of unique gene insertions into rDNA[44,45] and exchange between duplicated *his4* genes[46] or 5′ and 3′ truncated *his3* genes.[47] Using these and other assays, the effects of known and newly identified Rec$^-$ genes can be compared in haploid and diploid *spo13* meiosis to determine whether the same Rec gene functions and pathways are utilized for recombination both within and between chromosomes. These analyses are complemented by cytological studies visualizing synap-

[42] R. E. Malone, *Mol. Gen. Genet.* **189**, 405 (1983).
[43] S. Gottlieb, J. Wagstaff, and R. E. Esposito, *Proc. Natl. Acad. Sci. U.S.A.* **86**, 7072 (1989).
[44] T. D. Petes, *Cell (Cambridge, Mass.)* **80**, 765 (1980).
[45] J. W. Szostak and R. Wu, *Nature (London)* **284**, 426 (1980).
[46] J. A. Jackson and G. R. Fink, *Nature (London)* **292**, 306 (1981).
[47] M. Fasullo and R. Davis, *Proc. Natl. Acad. Sci. U.S.A.* **84**, 6215 (1987).

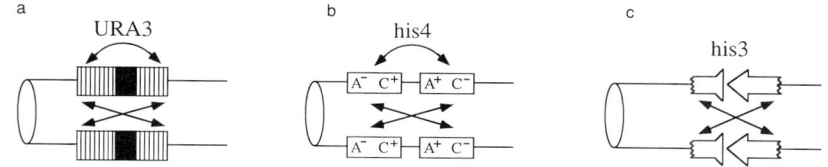

FIG. 3. Three systems commonly used to assay intrachromosomal recombination: (a) between rDNA repeats monitored by the fate of a unique gene insert (e.g., *URA3*) into rDNA,[44,45] (b) between duplicated *his4* genes,[46] and (c) between 5' and 3' truncated *his3* genes.[47] Arrows show the points of exchange. See Ref. 43 for details of the use of these systems to assay recombination during haploid meiosis.

tonemal complexes (described below) and molecular studies aimed at cloning the genes, isolating their products, and studying their effects on synapsis, breakage, and repair of DNA.

Isolation of New Rec Mutants

New recessive Rec mutants are also being isolated using the haploid meiosis system. In such schemes, a disomic strain is frequently employed that contains markers to assay both intrachromosomal recombination and/or exchange between homologs. The general procedure is to mutagenize the haploid, patch the survivors on YPD plates, and replica plate to sporulation medium. The sporulation plates are incubated for about 5 days, then replica plated to diagnostic media to assay for the presence or absence of recombinant genotypes among the viable spore products. Conditional mutants are detected by incubating the sporulation plates at different temperatures, generally from 20° to 34°.

Some genetic systems used to monitor exchange during haploid meiosis are designed to recover specific types of Rec mutants. For example, Hollingsworth and Byers[30] developed a method to isolate mutants defective in pairing and synaptonemal complex formation. Their initial rationale was based on the finding that intrachromosomal recombination on chromosome III is elevated during haploid *(n)* meiosis when there is no homologous pairing partner, compared to diploid (2*n*) meiosis when the chromosome is paired, suggesting that Rec events within and between chromosomes are competitive.[32,46] They reasoned that in a disomic (*n* + 1) *spo13* haploid, mutants defective in pairing between homologs might exhibit an increase in meiotic intrachromosomal exchange, similar to the situation in which the homolog is absent. In order to readily detect this phenotype they inserted the dominant cycloheximide-sensitive gene *(CYH2)* between repeated sequences located on one chromosome of the disome in a *cyh2*[R] *spo13* strain. The level of meiotic intrachromosomal recombination between the repeats leading to excision of *CYH2* and formation of a cycloheximide-resistant spore was scored by replica plating sporulation plates to cycloheximide-containing medium.

During *spo13* single division meiosis cycloheximide-resistant spores may also arise from reductional division of the disome or aberrant segregation in which one homolog segregates equationally and the other reductionally.[25] The frequency of such events varies with strain background. For example both types of segregation events are known to occur more often for a given pair of homologs in haploid ($n + 1$) strains than in diploid ($2n$) cells.[25,26] Their presence also depends on recombination proficiency since both Rec⁻ haploids and diploids generally exhibit substantially reduced levels of these segregations (e.g., > 95% equational segregation).[26-28,30] The background level of spores resulting from reductional or aberrant segregation in a Rec⁺ ($n + 1$) parental strain can be significantly reduced by utilizing heterozygous recessive markers in repulsion, flanking the *CYH* gene; selection for wild-type spores for these markers will enrich for meiotic products containing both members of the disome (+—/—+) owing to equational segregation.[30] This type of selection will also enrich for Rec⁻ mutants since they exhibit higher levels of equational segregation. In the particular mutant hunt scheme of Hollingsworth and Byers,[30] the detection of the *hop1* Rec⁻ mutant was primarily due to this shift in segregation pattern. In more general mutant hunts for Rec⁻ mutants, heteroallelic markers and selection for prototrophic recombinants after haploid meiosis are most convenient.

Use of spo13 to Identify and Analyze Segregation Genes

The *spo13* mutation provides a useful tool not only to define and order early and late Rec genes but also to identify and order segregation (Seg) genes that function in meiosis I and II. For example, Seg⁻ mutations whose meiotic lethality is suppressed by *spo13* are considered defective in late meiosis I functions, after the *spo13* block. One mutant in this class is *spo12*, which is lethal in haploid meiosis.[1,26] Its wild-type allele is clearly not required for the meiosis II division in *spo13* mutants. Seg⁻ mutants not suppressed by *spo13* are thought to be defective at an earlier stage in meiosis I, prior to the *spo13* block, or later in meiosis II and/or spore formation. These alternatives can be further distinguished cytologically; single mutants that remain mononucleate are classified as having an initial defect in meiosis I (e.g., *spo1*), whereas those that form binucleate or tri- and tetranucleate cells are considered aberrant in meiosis II and spore formation, respectively (e.g., *spo14*).[48]

[48] S. M. Honigberg, S. Klapholz, M. Townsend, and R. E. Esposito, *GSA Yeast Genet. Mol. Biol. Meeting Abstr.* p. 97 (1989).

Isolation of Spore Wall Mutants

The haploid meiosis system has also recently been used to isolate mutants defective in spore wall biogenesis.[34,35] The method described below was made possible by the finding that spore walls contain relatively large amounts of dityrosine, a natural fluorescence marker that can be easily scored.[49] Dityrosine plays a major structural role by cross-linking polypeptides in the spore wall outer layer. This contributes to the resistance of the spore to lytic enzymes such as glusulase and zymolyase and hydrophobic solvents like diethyl ether.[34,35,49] The procedure given here is useful not only to recover spore wall mutants but also mutations in genes required at other stages of sporulation that prevent the formation of mature spores.

Early log-phase cells of a strain suitable for haploid meiosis, such as PB2-1C, are mutagenized, plated for single colonies, and patched onto YPD master plates. The masters are incubated overnight, then replica plated to sporulation medium and to sterile nitrocellulose filters (BA85, Schleicher and Schuell, Dueren, West Germany) placed on the surface of YPD plates prewetted with approximately 200 µl of sterile water. After 24 hr the filters are transferred to sporulation plates, incubated for 3 days at 25°, and soaked in petri dishes containing a mixture of 70 µl glusulase, 200 µl water, and 15 µl mercaptoethanol for 5 hr at 30°. Finally, the filters are transferred to petri dishes containing 300 µl of concentrated aqueous ammonia and photographed with a Polaroid MP40 camera under UV light (302 nm, laboratory UV handlamp) using a UV filter and a Kodak Wratten filter (No. 98, 10 × 10 cm, Cat. No. 174-0893, transparent in the 390–450 nm range). Mutagenesis to approximately 50% survival results in about 10% nonfluorescent variants (Fig. 4). Approximately 0.5% of them form immature asci, as viewed by phase-contrast microscopy of wet mounts from the sporulation plates, and are presumptive spore wall mutants. The remaining nonfluorescent isolates are lacking spores because of respiratory deficiency or Spo⁻ lesions affecting earlier stages of sporulation.

The presumptive spore wall mutants are tested for their dityrosine content by hydrolyzing asci taken from the nitrocellulose filters in 6 N HCl at 110° for 12 hr in a closed vessel after flushing with nitrogen (i.e., standard conditions for amino acid analysis). After removing HCl under reduced pressure in a rotary evaporator, the lysate is taken up in 10 µl of water and amino acids identified on thin-layer plates (Kieselgel 60, Merck, Darmstadt, Germany) using the solvent system chloroform/methanol/17% aqueous ammonia (2:2:1, v/v/v).[50] A dityrosine standard prepared by

---

[49] P. Briza, G. Winkler, H. Kalchhauser, and M. Breitenbach, *J. Biol. Chem.* **261**, 4288 (1986).

[50] R. Amadò, R. Aeschbach, and H. Neukom, this series, Vol. 107, p. 377.

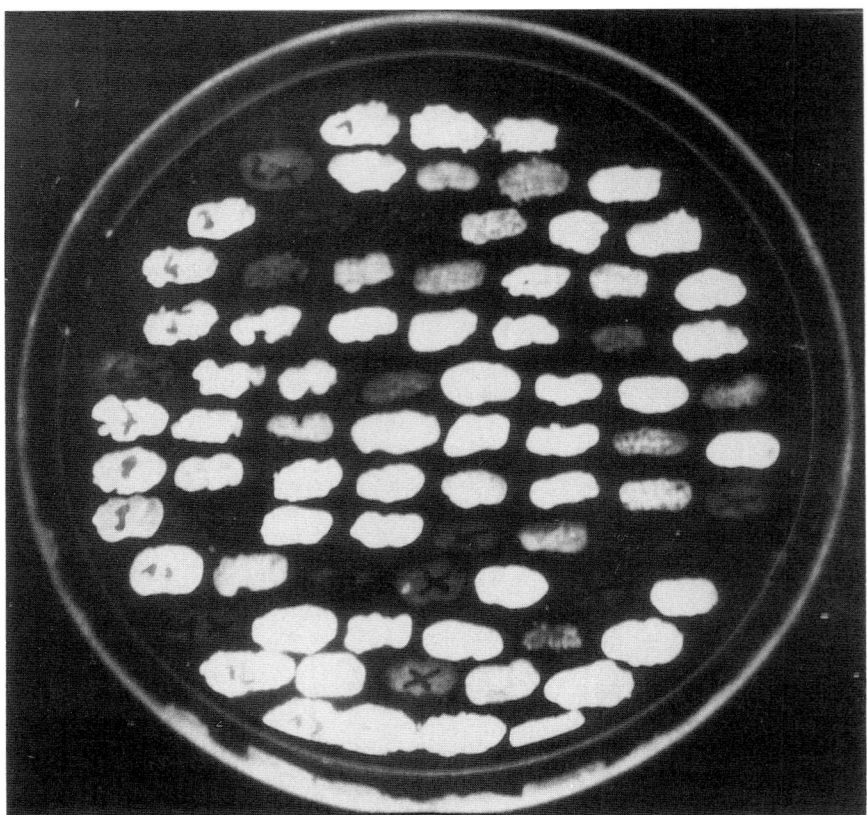

FIG. 4. Natural fluorescence of dityrosine in spore walls. The nitrocellulose filter shown contains patched sporulated survivors of mutagenesis that have gone through haploid meiosis. It was photographed under UV light as described in the text. Colonies appearing bright are wild type. Colonies appearing dark or showing very little fluorescence are dityrosine-lacking mutants, most of which are nonsporulators.

oxidizing L-tyrosine (Sigma, St. Louis, MO) with H_2O_2/horseradish peroxidase and purifying on DEAE and Sephadex G-10 columns[50] is used on the same plates to check if the mutant spores contain residual dityrosine. The thin-layer chromatography plates are illuminated with UV light in a dark room, as described earlier, to visualize the blue fluorescence of the dityrosine.

Mutants containing no or very little residual dityrosine are further analyzed by a number of techniques.[34] The most important of these is thin-section transmission electron microscopy of intact asci to reveal the layers of the spore wall.[34,49,51-53] The mutants isolated so far produce immature spores that lack either the outermost layer or both outer layers of the spore wall and that are sensitive to glusulase or zymolyase and to diethyl ether killing; however, the mutants are viable and still exhibit moderate heat resistance.[8,9] None of the mutants displays growth defects on various carbon sources (glucose, acetate, glycerol, ethanol) or at the temperatures tested (23°, 30°, and 37°). The known chemistry of the spore wall suggests that approximately 10 genes may be specifically involved in biosynthesis of the outer spore wall layers.[34] Thus far, three genes have been identified.[34,54] Two are needed for synthesis of a dityrosine-containing polymer of the outer spore wall layer. The remaining gene, which appears to act indirectly, is required for the synthesis of the chitosan layer of the spore wall and also regulates chitin synthesis in vegetative cells. Most of the known mutants defective in the dityrosine-containing layer are spore autonomous, that is, they are expressed in the spore even in sporulated heterozygous diploids. It is therefore difficult to dissect tetrads and isolate segregants carrying the mutant genes. The recessive nature of the genes, however, can be determined by classic tetraploid analysis or more simply by crossing the mutants to another *spo13* strain and examining the phenotypes of the diploid heterozygous spores. Direct cloning may also be used, as described below.

Cloning of Spo⁻ Genes

The cloning of genes required for meiotic recombination and segregation and spore wall biosynthesis utilizes standard procedures[55] with the added feature of specific enrichments to recover plasmids containing the desired genes. Only those features of the cloning procedures that are relevant to Spo genes are noted below. First, in all cases, transformed mutants must be sporulated to detect complementation of the Spo⁻ defect. If the libraries used are present in unstable vectors (lacking a CEN sequence) and pregrowth on selective medium is required to maintain plasmids, sporulation is likely to be reduced as ascus formation is generally maximum when pregrowth is on rich medium. In this instance, brief incubation in YPD (6–8 hr) prior to sporulation, rather than the typical 48 hr of YPD pregrowth, is helpful. Second, during selection of the wild-type gene for most Spo⁻ mutants, elimination of unsporulated cells that may have genotypes

---

[51] P. Briza, A. Ellinger, G. Winkler, and M. Breitenbach, *J. Biol. Chem.* **263**, 11569 (1988).
[52] N. J. W. Kreger-van-Rij, *Arch. Microbiol.* **117**, 73 (1978).
[53] B. Byers and L. Goetsch, this volume [41].
[54] C. Bulawa, M. Pammer, and M. Breitenbach, unpublished (1990).
[55] M. D. Rose and J. R. Broach, this volume [14].

similar to the spores is often useful. If this is desirable, steps can be incorporated after sporulation to lyse or kill the vegetative cells with glusulase and/or diethyl ether, respectively, and to purify asci or spores on gradients.[10,17,35,56] Finally, spore genotypes themselves can be selected. When *spo/spo* diploids are transformed, haploid spores can be selected by plating on medium selective for recessive drug resistance markers from heterozygous drug-sensitive cells (e.g., *can1*[R] or *cyh2*; see Ref. 21 and also reviews in Refs. 1 and 3). For haploid meiosis this option is not available, and eliminating the vegetative haploids and physically purifying asci or spores are more critical. In the case of Rec⁻ strains, direct selection for Rec⁺ meiotic products can be performed.[57]

The method used to clone spore wall genes is given below as an illustration of how these approaches are used. Standard procedures are employed to transform the mutant strains with yeast libraries.[55] When the spheroplast method is used, transformants are recovered from the top agar by squeezing through a 1-mm injection needle and replated to form a dense lawn on plates selective for the plasmid. As in the original assay for dityrosine fluorescence, the cells are replica plated to nitrocellulose filters on the surface of a medium selective for the plasmid or YPD for stable (e.g., CEN) plasmids. If selective medium is used, after 2 days of incubation the filter is transferred to YPD for 6–8 hr to provide optimal pregrowth plate conditions for sporulation. The filter is then transferred to sporulation medium for 4–5 days. To enrich for wild-type transformants, spore wall mutants and unsporulated cells are lysed by suspending approximately one-quarter of the cells from the sporulation plate in 5 ml of 0.1 M sodium acetate (pH 5.0), 20 μl mercaptoethanol, and 50 μl glusulase and incubated at 30° with shaking for 6–12 hr. The remaining spores and cells are centrifuged, washed, and plated, and the survivors are screened for spore dityrosine fluoresence as in the original mutant hunt. In this case, however, colonies do not have to be patched individually since those that regain fluorescence can be easily detected within a lawn of cells.

Spore and Spore Wall Purification

The physicochemical properties of yeast spores are sufficiently different from those of vegetatively growing yeast cells to allow efficient discrimination and/or physical separation. As previously mentioned, for genetic screens it is often desirable to selectively kill vegetative cells with near 100%

[56] I. W. Dawes, J. F. Wright, F. Vezinhet, and N. Ajam, *J. Gen. Microbiol.* **119**, 165 (1980).
[57] C. Atcheson, B. DiDomenico, S. Frackman, R. E. Esposito, and R. Elder, *Proc. Natl. Acad. Sci. U.S.A.* **84**, 8035 (1987).

survival of spores. A well-established method to achieve this goal is treatment with aqueous diethyl ether or vapors of diethyl ether.[10,17] The molecular basis for the ether resistance of yeast spores appears to be due to the dityrosine-containing outer layer of the yeast spore wall.[49,51]

Selective survival of spores, however, is not sufficient for biochemical investigations which require efficient physical separation of spores from nonsporulated cells. The method described below utilizes a density gradient medium, Percoll (Pharmacia, Uppsala, Sweden), that does not penetrate and therefore does not damage or kill yeast cells or spores. Routine preparations of 10 g of spores purified to homogeneity can be achieved on Percoll gradients.[58] An older method[59] of separating spores from vegetative cells by shaking with a vegetable oil phase, while simpler, is less efficient. Other procedures, such as those of Rousseau and Halvorson[60] and Savarese[61] are not generally applicable, as they employ strains that sporulate to nearly 100%. A small-scale purification of yeast ascospores, based on the affinity of spores for the walls of plastic centrifuge tubes, is given elsewhere in this volume.[17] Finally, it should be noted that intact asci can also be separated from vegetative cells using gradient centrifugation in Urografin[56] and Percoll (not described here).

Preparation of Pure and Viable Yeast Ascospores

A sporulated culture is centrifuged and washed in large (0.5 or 1 liter) centrifuge bottles at 4000 rpm in a Sorvall high-speed laboratory centrifuge (methods to achieve a high percentage of four-spored asci are described elsewhere[16]). The procedure is usually performed with large volumes of sporulation culture (up to 10 liters), although smaller volumes (0.5 liter) can also be used. The pellet, consisting of mature asci and nonsporulated cells, is weighed, and 5 ml of 0.1 M sodium phosphate buffer (pH 7.2), 2 μl mercaptoethanol, and 0.8 mg of zymolyase 20,000 (Kirin Brewery, Japan) is added per gram of pellet. The mixture is shaken for 2–3 hr at 30°, and the absence of ascus walls is checked in a phase-contrast microscope. Per gram of pellet, 5 ml of 0.5% Triton X-100 is then added, the mixture centrifuged at 4000 rpm for 10 min, and the pellet washed 3 times with 0.5% Triton X-100. The pellet is finally suspended in a small volume of 0.5% Triton X-100 and pipetted on top of a step gradient of Percoll consisting of 10 ml of each of four layers (given in Table I) in Sorvall SS34 centrifuge tubes.

[58] A. Hartig, R. Schroeder, E. Mucke, and M. Breitenbach, *Curr. Genet.* **4**, 29 (1981).
[59] C. C. Emeis and H. Gutz, *Z. Naturforsch. B: Chem. Sci.* **13**, 647 (1958).
[60] P. Rousseau and H. O. Halvorson, *J. Bacteriol.* **100**, 1426 (1969).
[61] J. J. Savarese, *Can. J. Microbiol.* **20**, 1517 (1974).

TABLE I
COMPOSITION OF PERCOLL STEP GRADIENT TO
ISOLATE YEAST ASCOSPORES

Layer	Composition of each layer (ml)		
	Percoll	Triton X-100 (0.5%)	Saccharose (2.5 M)
1 (bottom)	8	1	1
2	7	2	1
3	6	3	1
4 (top)	5	4	1

A maximum of about 5 ml of raw spore preparation can be separated in one Sorvall SS34 tube. The tubes are centrifuged for 45 min at 10,000 rpm in a Sorvall SS34 rotor at 4°. Vegetative cells and cellular debris form a diffuse band in layers 3 and 4. Spores form a dense pellet. The spore pellet is withdrawn and washed 3 times in 0.5% Triton X-100. Wild-type spores can be stored in 0.5% Triton X-100 at 4° at about 10^9 spores/ml for up to 1 year without loss of plating efficiency.

After a single gradient purification, the spore preparation contains about 1% vegetative cells. A convenient way of counting the few vegetative cells remaining in the spore preparation is staining with fluorescein isothiocyanate–concanavalin A (FITC–Con A),[51] counting the fluorescent vegetative cells in a fluorescence microscope, and comparing their number to the total number of spores and cells. Spores do not stain with FITC–Con A owing to their protective outer layer, which shields the inner glycan/mannan layers. Repeating the gradient purification described above results in only 0.01% remaining vegetative cells.

Purification of Walls of Spores or Vegetative Cells

Ten grams of purified spores (see above) is treated with 20 ml of glass beads (diameter 0.45 mm) and 10 ml of 0.5% Triton X-100 in a bead beater (Braun, Melsungen, West Germany) for 3 min at setting 1. The spore homogenate is taken up in a drawn-out Pasteur pipette, and the glass beads are washed twice in 5% Triton X-100. The combined aqueous phases are centrifuged for 15 min at 3000 rpm (Sorvall SS34 rotor). The pellet consists mainly of wall fragments and remaining intact spores. Cytoplasmic material and most of the membrane material of the spores remain in the supernatant. The pellet is resuspended in 60% Percoll/2% Triton X-100 in water and centrifuged in a Beckman SW 25.2 rotor at 23,000 rpm for

1 hr (self-generating Percoll gradient). The remaining intact spores form a pellet, whereas spore walls form a distinct turbid band in the gradient. The band corresponding to the wall fraction is collected, and Percoll and Triton X-100 are removed by repeated washing in water and centrifuging in a Sorvall SS34 rotor at 5,000 rpm. The spore wall preparation consists almost exclusively of wall fragments as demonstrated by standard transmission electron microscopy.[53] Walls of vegetative cells can be purified in essentially the same way. However, in the latter case, breaking of vegetative cells under the conditions described above takes only 1.5 min. A solution of 80% Percoll/2% Triton X-100 in water is used for the self-generating Percoll gradient. Remaining intact cells form a band in the gradient, and a fraction consisting of vegetative cell walls is pelleted.

Visualization of Synaptonemal Complexes by Nuclear Spreading

Chromosome pairing and synaptonemal complex formation are cytological correlates of the elevated levels of genetic recombination that take place during meiosis. One approach to understanding the molecular details of meiotic chromosome behavior is to analyze the roles of defined structural components by combining cytological and genetic methods. Procedures that allow efficient and detailed observation of large numbers of meiotic nuclei from large eukaryotes[62] have been adapted for use in yeast, where chromosome condensation and pairing and synaptonemal complex assembly and disassembly can be analyzed by light and electron microscopy.[63] Chromosomes, synaptonemal complexes, nucleoli, and spindle pole bodies each undergo characteristic changes in morphology during the course of the first meiotic prophase, and each is accessible to immunocytological analysis.[64-66] Thus, it is now possible to localize defined gene products efficiently to specific meiotic nuclear structures and to assess their roles cytologically as well as genetically.

Spread preparations of meiotic and vegetative nuclei are prepared from spheroplasts by bursting the cells in a hypotonic medium designed to preserve the nuclear structures of interest, then drying down the nuclei onto the surface of microscope slides coated with poly(L-lysine) for light microscopy or with polystyrene for electron microscopy.[63] Meticulous

[62] M. E. Dresser and M. J. Moses, *Chromosoma* **76,** 1 (1980).
[63] M. E. Dresser and C. N. Giroux, *J. Cell Biol.* **106,** 567 (1988).
[64] M. E. Dresser, D. Pisetsky, R. Warren, G. McCarty, and M. Moses, *J. Immunol. Methods* **104,** 111 (1987).
[65] C. N. Giroux, M. E. Dresser, and H. F. Tiano, *Genome* **31,** 88 (1989).
[66] M. E. Dresser and C. N. Giroux, unpublished observations (1988).

preparation of the slide surfaces and of the fixative are key elements in gaining optimal results.

Preparation of Microscope Slides

The glass microscope slides should have a minimum of surface scratches and pits and should be thoroughly washed in hot water with a nonabrasive detergent (e.g., Bon Ami cleanser), rinsed in distilled water, soaked and wiped with clean cheesecloth in 95% ethanol, then dried and polished with hard lens paper.

Coating with Poly(L-Lysine)

Cleaned slides are flooded with 1 mg/ml poly(L-lysine) (Sigma, type 1-B, molecular weight 70,000 or higher) in 0.1 M sodium bicarbonate buffer, pH 9.0, for 10 sec, drained (the solution can be reused), rinsed with distilled water, and air dried. The slides can be stored for months until use.[67]

Coating with Polystyrene

Small pieces of plastic broken from a Falcon Optilux petri dish (the only source that has proved successful repeatedly) provide the polystyrene. Cleaned slides are dipped in 0.5% (w/v) polystyrene in chloroform, withdrawn with a single smooth motion, drained by touching the end to the edge of the jar, then laid flat and allowed to dry, leaving an even coat of plastic up three-fourths of the length of the slide. Clear fingernail polish is painted along the edges of the slide and in a band across the top to prevent the film from lifting off the slide prematurely. The slides can be stored at room temperature for months until use.[68]

Preparation of Fixative

Fixative is prepared fresh on the day of use by adding 0.8 g paraformaldehyde (Sigma) and 60 μl of 1 N KOH to 20 ml distilled water and stirring over low heat (the solution should never exceed 55°) until clear. After cooling, the pH is brought to 7.0 by adding 1 N HCl. If more than 30 min is required for the solution to clear, a new lot of paraformaldehyde should be obtained.

[67] D. Mazia, G. Schatten, and W. J. Sale, *J. Cell Biol.* **66**, 198 (1975).
[68] B. Felluga and G. B. Martinucci, *J. Submicrosc. Cytol.* **8**, 347 (1976).

Preparation of Spread Meiotic Nuclei

To make spread preparations of nuclei containing synaptonemal complexes, cells in sporulation medium are harvested about the time that the first tetranucleate cells appear in the population (8–10 hr, depending on the strain). Samples of $2-4 \times 10^8$ cells are washed and then resuspended in 10 ml of 2% (w/v) potassium acetate/0.8 M sorbitol/pH 7.0. Dithiothreitol is added to a final concentration of 10 mM for 10 min at 30°, then zymolyase 100T (ICN Biomedicals, Inc. Costa Mesa, CA) is added to 25 μg/ml final concentration to begin spheroplasting the cells. Spheroplasting is monitored by diluting a sample of cells into the potassium acetate/sorbitol with 2% (w/v) sarkosyl (which lyses the spheroplasts) and comparing the number of cells in a given volume to the number of cells in medium without sarkosyl. When 70–90% of the cells are spheroplasted, the cells are washed with ice-cold 0.1 M 2-(N-morpholino)ethanesulfonic acid/1 M sorbitol/1 mM EDTA/0.5 mM MgCl$_2$, pH 6.4,[69] then kept on ice as a pellet with the supernatant removed.

To burst and fix the spheroplasts, they are resuspended in 25 parts (v/v) ice-cold 0.1 M 2-(N-morpholino)ethanesulfonic acid/1 mM MgCl$_2$, pH 6.4 (i.e., no sorbitol), and, after 5 sec, the suspension is mixed with fixative at a volume ratio of 1 part suspension to 7 parts fixative. Aliquots of 100 μl are placed on microscope slides, prepared as above, and covered with a 24 × 50 mm coverslip to ensure even coverage of the nuclei over the slide surface. After 10 min at room temperature, the coverslips are gently rinsed off with fresh fixative, and 0.35 ml more of fresh fixative is added to the slides for another 5 min. This final fixative is gently drained from the surface of the slide, then the preparations are rinsed with 5 ml of 0.4% Photoflo (Kodak, Rochester, NY) and allowed to air dry. Preparations for immunocytological analysis are most accessible to antibodies before drying although labeling can be preserved by briefly drying in a protective agent, for example, Ficoll 400,000.[64,66]

Light Microscopic Analysis

In adequate preparations, each of the stages of meiotic prophase are evident using phase microscopy (40× objective) on unstained material. Visualization of the chromatin is aided by staining with 4',6-diamidino-2-phenylindole (DAPI), simply by mounting the preparation in 90% glycerol/10% 1 M sodium bicarbonate, pH 9.0/0.1 mg/ml 1,4-phenylenedi-

---

[69] L. Goetsch and B. Byers, *Mol. Gen. Genet.* **187**, 54 (1982).

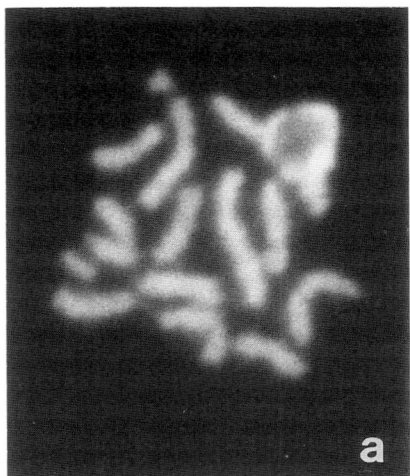
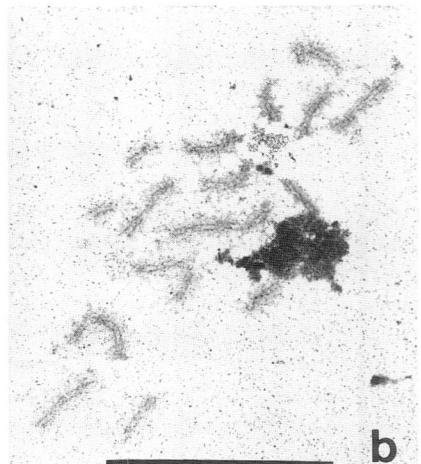

FIG. 5. (a) Fluorescence light micrograph of chromosome bivalents in a DAPI-stained yeast meiotic nucleus. Diffuse chromatin in the nucleolus organizer region in the middle of bivalent XII is evident (upper right). (b) Electron micrograph of synaptonemal complexes in a silver-stained yeast meiotic nucleus. The nucleolus is heavily stained (right), as are the spindle pole bodies (above and to the left of the nucleus). Bar: 5 µm.

amine/0.5 µg/ml DAPI, waiting for about 30 min, then using fluorescence microscopy[66] (Fig. 5). Fluorescence microscopy following staining with acridine orange reveals the nucleolus as an orange-red fluorescent patch among the yellow-green fluorescent chromatin,[63] but the conditions for proper staining are more difficult to control than for DAPI. Synaptonemal complexes are easily contrasted for light and electron microscopy by silver staining[62] using a simple one-step method.[70]

Electron Microscopic Analysis

Nuclei on polysytrene-coated slides are prepared for electron microscopic analysis by staining with any of a number of established methods, for example, with silver (see Fig. 5 and references cited above), phosphotungstic acid,[71] uranyl acetate,[69] or simply by depending on the mass of the unstained nuclear structures to provide sufficient contrast. For preparations with a low yield of nuclei, transfer to grids is best carried out after scanning the slide with a light microscope and using a felt-tip pen to make

[70] W. M. Howell and D. A. Black, *Experientia* **36,** 1014 (1980).
[71] M. J. Moses, *Chromosoma* **60,** 99 (1977).

small dots on the plastic film adjacent to the areas of interest. The plastic film is scored all around the edge with a razor blade, then the slide is lowered slowly, and at a low angle to the surface, into a dish containing distilled water. If the plastic fails to come off the slide and to float onto the surface of the water, a drop of hydrofluoric acid (use caution) touched to the water near the slide will usually loosen it. Once the film is free on the surface, electron microscope grids are placed shiny side down over the areas of interest. A piece of Parafilm (American Can Co.) is rolled onto the grids/film, forcing out air bubbles, and then lifted from one side to remove the adhering grids/film from the water (practice helps). This sandwich is then placed with the film side up on a flat surface to dry. Each grid is removed by using the tips of the grid forceps to inscribe a circle in the plastic around the grid, then placed film side up on a clean glass microscope slide and scanned with phase or bright-field light microscopy to confirm the presence and location of nuclei (grids with a unidirectional center mark are useful, e.g., Ted Pella GC-50 or GC-100). The plastic is quite stable in the electron beam and generally does not require carbon coating. Note that the nuclei are "underneath" the plastic film on the grid unless a different scheme is used to transfer the film to grids.

Concluding Remarks

In this chapter we have presented new methods for detecting and analyzing mutants affecting meiosis and spore formation. In our view, the ultimate goal of research in this area is to understand the genetic program(s) operating during meiotic cell differentiation and spore maturation. We hope that the new tools presented here for identifying and characterizing such mutants will help researchers come closer to realizing this goal.

Acknowledgments

We wish to thank Cathy Atcheson for helpful comments on the manuscript and Chris Fritze and Lela Buckingham for preparing the figures. The photograph of dityrosine fluorescence was prepared by P. Briza. The cytological methods employed in visualizing synaptonemal complexes were developed by M.D. in the laboratories of M. J. Moses and C. N. Giroux. This work was supported by National Institutes of Health Grants GM29182 and HD19252 (to R.E.E.), an NRC Associateship (awarded to M.D.), and the Austrian "Fonds zur Förderung der Wissenschaftlichen Forschung," Project S29/03 (to M.B.).

[8] Putting the HO Gene to Work: Practical Uses for Mating-Type Switching

By IRA HERSKOWITZ and ROBERT E. JENSEN

Background and Principles of the Methods

Yeast strains are of two sorts: cells either have a stable or unstable mating type (reviewed in Refs. 1 and 2). In most laboratory strains (which carry the *ho* allele) the mating type is stable: α cells (genetically *MAT*α) give rise to cells that are *MAT*α, and **a** cells (genetically *MAT***a**) give rise to cells that are *MAT***a**. In strains carrying the *HO* gene, the mating type of a haploid cell is unstable and can change from **a** to α or from α to **a** nearly every cell division.

Vegetative cells or spores that contain the *HO* gene form colonies that contain **a**/α diploid cells. This occurs in a two-step process: first, the initial cell gives rise to progeny that have the opposite mating type. Then, siblings of opposite mating type mate with each other to form **a**/α diploids[3] (Fig. 1). These cells remain stably diploid because expression of the *HO* gene is turned off in **a**/α cells.[4] Note that the **a**/α diploid is homozygous for every nucleotide in the genome except for the mating type locus! The efficiency of this process is high: a colony produced by a cell carrying *HO* contains at least 10–50% **a**/α cells.

The process by which cells change mating type is a remarkable one involving a stereotyped genetic rearrangement catalyzed by the *HO* gene. α cells give rise to **a** cells by changing the allele of the mating-type locus. This occurs by a recombinational event between the allele at the mating-type locus and silent versions of the mating-type locus alleles located elsewhere in the genome (Fig. 2). The *HML* locus usually contains a silent equivalent of *MAT*α, and the *HMR* locus usually contains a silent equivalent of *MAT***a**.[5,6] These blocks of information, termed "genetic cassettes," are

---

[1] I. Herskowitz and Y. Oshima, *in* "The Molecular Biology of the Yeast *Saccharomyces cerevisiae*" (J. N. Strathern, E. W. Jones, and J. R. Broach, eds.), Volume 1, p. 181, Cold Spring Harbor Laboratory Press, Cold Spring Harbor, New York, 1981.
[2] I. Herskowitz, *Microbiol. Rev.* **52**, 536 (1988).
[3] J. B. Hicks and I. Herskowitz, *Genetics* **83**, 245 (1976).
[4] R. Jensen, G. F. Sprague, Jr., and I. Herskowitz, *Proc. Natl. Acad. Sci. U.S.A.* **80**, 3035 (1983).
[5] J. N. Strathern, E. Spatola, C. McGill, and J. B. Hicks, *Proc. Natl. Acad. Sci. U.S.A.* **77**, 2839 (1980).
[6] K. A. Nasmyth and K. Tatchell, *Cell (Cambridge, Mass.)* **19**, 753 (1980).

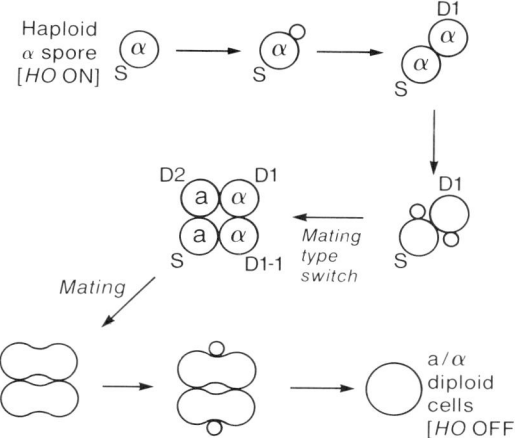

FIG. 1. Formation of stable a/α diploid cells from haploid *HO* cells by mating-type interconversion. The diagram shows the first two cell divisions of a spore (S) that is initially *MAT*α and that carries the *HO* gene [see Ref. 3 and J. Strathern and I. Herskowitz, *Cell (Cambridge, Mass.)* **17**, 371 (1979)]. Its first daughter cell is denoted D1; the first daughter of D1 is denoted D1-1. After the S cell undergoes two cell divisions, 60–70% of microcolonies contain two cells exhibiting the a mating type and two cells exhibiting the original mating type. (In the other 30–40% of microcolonies at the four-cell stage, all four cells exhibit α mating type. These cells can change mating type in subsequent cell divisions; see below.) The a and α siblings then mate to form two a/α diploid zygotes. These cells remain stably a/α because the *HO* gene is turned off.[4] Mating-type switching is restricted to mother cells (e.g., the S cell at the two-cell stage but not the S cell at the one-cell stage) and occurs in approximately 60–70% of cell divisions. Because of the stochastic nature of mating-type switching, colonies derived from haploid *HO* cells are composed of a mixture of a, α, and a/α cells.

expressed when situated at the playback locus (the mating-type locus) and silent when situated at *HML* and *HMR*.

The *HO* gene codes for a site-specific endonuclease, which recognizes a site at the mating-type locus and produces a double-strand break at that position.[7–9] It is clear that *HO* codes for the endonuclease, since *HO* protein produced in *E. coli* exhibits endonuclease activity.[9] The cleavage site for *HO* is located just to the right of the distinctive segments (*Y*a and *Y*α) found at *MAT*a and *MAT*α, respectively (see Fig. 2). Thus, cleavage

---

[7] J. N. Strathern, A. J. S. Klar, J. B. Hicks, J. A. Abraham, J. M. Ivy, and K. A. Nasmyth, *Cell (Cambridge, Mass.)* **31**, 183 (1982).
[8] R. Kostriken, J. N. Strathern, A. J. S. Klar, and F. Heffron, *Cell (Cambridge, Mass.)* **35**, 167 (1983).
[9] R. Kostriken and F. Heffron, *Cold Spring Harbor Symp. Quant. Biol.* **49**, 89 (1984).

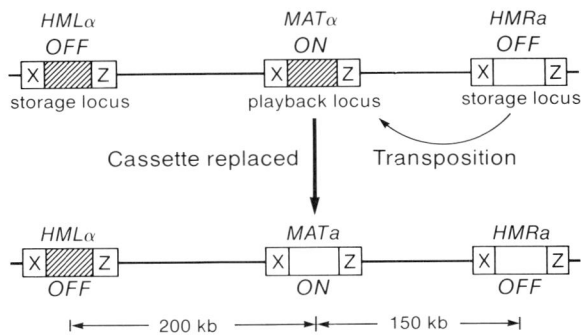

FIG. 2. Switching from α to a by changing genetic cassettes at the mating-type locus. The top line shows the arrangement of cassettes on chromosome III in an α cell. The cassette at *MAT* is expressed; those located at *HML* and *HMR* are repressed (see Refs. 1 and 2). Switching to a occurs by removing the α cassette from *MAT* and replacing it with information from *HMR*a. The central regions of the cassettes (shown as striped or open rectangles) represent distinct nucleotide sequences (Yα and Ya, respectively). The Yα region is 747 base pairs (bp); the Ya region is 642 bp [C. R. Astell, L. Ahlstrom-Jonasson, M. Smith, K. Tatchell, K. A. Nasmyth, and B. D. Hall, *Cell (Cambridge, Mass.)* **27**, 15 (1981)]. The regions adjacent to Y (the X and Z regions, which are approximately 700 and 200 bp, respectively) are involved in recombination between cassettes at *MAT* and *HML* or *HMR*. Mating-type interconversion is initiated by *HO* endonuclease, which produces a double-strand break at *MAT* (see Fig. 3). Subsequent repair of the double-strand break leads to a duplicative transposition of information from *HML* or *HMR* or *MAT*. The distances between *MAT* and the silent cassettes are indicated. Additional information on the structure of the *HML* loci is given in Ref. 5 and by C. R. Astell, L. Ahlstrom-Jonasson, M. Smith, K. Tatchell, K. A. Nasmyth, and B. D. Hall, *Cell (Cambridge, Mass.)* **27**, 15 (1981). (From Ref. 2, with permission.)

occurs in the Z region, near the Y–Z border. A detailed view of the cleavage site (termed the Y/Z site) is shown in Fig. 3. It is interesting to note that the *HML* and *HMR* loci also contain the cleavage site for *HO* endonuclease, but cleavage does not occur because the same products that repress transcription of these cassettes also prevent them from being cut.[10]

After *HO* endonuclease has produced a double-strand break at *MAT* (e.g., at *MAT*α), DNA repair ensues and usually results in replacement of the α cassette by an a cassette. The *HMR*a locus is the donor of the a information in this case. As shown in Fig. 2, the information at *HMR*a is not changed as a result of mating-type switching. Mating-type interconversion is thus a nonreciprocal transfer of information from one position (*HML* or *HMR*) to another *(MAT)*. This process is remarkably efficient: approximately 65% of all of the cells in a population that are competent to

[10] A. J. S. Klar, J. N. Strathern, and J. B. Hicks, *Cell (Cambridge, Mass.)* **25**, 517 (1980).

A
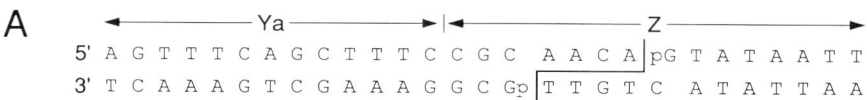

B
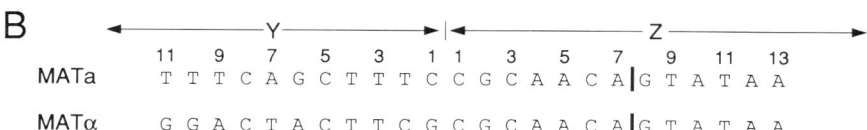

FIG. 3. *HO* endonuclease cleavage site. *HO* endonuclease cleaves in the Z region to produce a double-strand break with a 3' single-stranded region as shown in (A). The 24-bp site (Y11–Z13) from *MAT*a (B) is sufficient to act as a site for *HO* endonuclease *in vivo* [J. A. Nickoloff, E. Y. Chen, and F. Heffron, *Proc. Natl. Acad. Sci. U.S.A.* **83**, 7831 (1986)]. The corresponding region of *MAT*α is also shown.

switch mating types do so.[3,11] Sometimes (in ~20% of cases) the double-strand break at *MAT*α is repaired using the information from *HML*α, in which case no change in mating type is detected, even though there has been a genetic rearrangement.[12,13]

This chapter briefly presents some of the practical uses for the cloned *HO* gene, its target site (the Y/Z site), and the process of mating-type interconversion itself. All involve simple variations on the same general schemes. Some of the major uses are the following: (1) There are many cases in which it is useful to have both *MAT*a and *MAT*α versions of a given strain. This can be done simply by introducing the cloned *HO* gene on a plasmid, allowing cells to switch mating types, and then recovering cells of both mating types after the plasmid carrying *HO* has segregated away. This is a superb way of constructing an isogenic set of strains (a, α, and a/α) that differ only at the mating-type locus. (2) The *HO* endonuclease can be used to deliver double-strand breaks in a controlled manner for studies of recombination and repair. These studies exploit the pGAL-*HO* plasmid, in which the *HO* gene is inducible by galactose. (3) Because the cut site for *HO* endonuclease is rare, *HO* may be useful for certain manipulations of large DNA fragments, for example, for breaking YAC plasmids into smaller pieces and for producing segments with defined ends. (4) In some cases, one desires to inactivate the *HO* gene so that a strain no longer switches its mating type (i.e., to convert a homothallic to a heterothallic

[11] J. Strathern and I. Herskowitz, *Cell (Cambridge, Mass.)* **17**, 371 (1979).
[12] J. Rine, R. Jensen, D. Hagen, L. Blair, and I. Herskowitz, *Cold Spring Harbor Symp. Quant. Biol.* **45**, 951 (1981).
[13] A. J. S. Klar, J. B. Hicks, and J. N. Strathern, *Cell (Cambridge, Mass.)* **28**, 551 (1982).

strain). This can be done by gene replacement, using a mutated form of the cloned *HO* gene.

Switching Mating Types

Rationale

The methods described in this section involve using a cloned *HO* gene to stimulate mating-type switching. Most strains carry the *HMLα* and *HMR*a alleles and thus have the appropriate silent cassettes available for mating-type interconversion. A functional *HO* gene is introduced into a strain by transformation. The *HO* gene might be governed by its normal regulatory signals (expressed in **a** and α cells but not in **a**/α cells) or under inducible control by galactose (in the plasmid pGAL-*HO*). Cells are given the opportunity for mating-type interconversion to occur and for siblings of opposite mating type to mate with each other. From the mixed population of cells in a colony, **a**, α, and **a**/α cells are isolated.

Uses

Construction of Isogenic Strains. For studies of cell specialization, for example, analysis of differences between **a** and α cells or between haploid cells and **a**/α diploids, it is often important to have cells that are perfectly isogenic except for the mating-type locus. In such strains, any phenotypic difference must be due to the mating-type locus and not to a mutation elsewhere in the genome. There are two other reasons for creating **a**/α cells from haploids. First, it allows one to determine whether a mutation isolated in a haploid strain affects sporulation. Second, because **a**/α cells have so many differences from **a** and α haploid strains (see Ref. 2), we encourage investigators to examine their favorite phenotypes in **a**/α cells to see if there are any differences from haploid strains.

The haploid strain of interest is transformed with a plasmid carrying *HO*, for example, YCp50-*HO*, which is a centromere plasmid carrying the *URA3* marker.[14] The transformants (selected as Ura$^+$) will form colonies containing cells of the original mating type, cells that have switched, and diploids resulting from mating between siblings. Hence, each Ura$^+$ colony will contain **a**, α, and **a**/α diploids. Transformants should be restreaked on rich medium, to allow the plasmid to segregate away (and produce cells that are thus Ura$^-$), and then individual colonies tested for cell type (**a**, α, or **a**/α) by standard assays such as mating factor production (using the halo

---

[14] D. W. Russell, R. Jensen, M. J. Zoller, J. Burke, B. Errede, M. Smith, and I. Herskowitz, *Mol. Cell. Biol.* **6**, 4281 (1986).

assays described in [5], this volume). Depending on the strain, one can expect at least 10–20% diploids and a mixed population of colonies that mate as **a** or α.

In carrying out assays for mating factor production, it is important that the colonies or patches being tested be pure populations and not, for example, mixtures of **a** and α cells. A mixture of this type often will not exhibit α-factor activity because **a** cells produce *BAR* protease, which degrades α-factor. The safest strategy is to isolate several nonmating colonies (presumptively **a**/α diploids) and then sporulate them. If the strain was a true diploid (and not a triploid or tetraploid), then it should yield 2**a**:2α segregants, which are haploid. This procedure avoids the possibility that **a** or α cells isolated from a colony that has undergone mating-type interconversion are homozygous diploids. Another *HO*-containing plasmid that can be used for these manipulations is the *LEU2*$^+$ plasmid YEp13-*HO*.[4,14]

There are situations in which the presence of an *HO* gene is not sufficient to cause switching of the mating-type locus. For example, a strain that is *HML*a *MAT*a *HMR*a will not be able to switch to *MAT*α, nor will a strain that is *HML*α *MAT*α *HMR*α be able to switch to *MAT*a.[2,15] Similarly, there are some naturally occurring α strains (such as *S. diastaticus*)[2,16] in which *MAT*α contains a mutation in the *HO* endonuclease cleavage site.[17] If *HO* cannot be used to switch mating types, another strategy for switching mating types is to carry out one-step gene replacement (see [19], this volume) with the desired mating-type locus allele itself. The *MAT* alleles are carried on a *Hin*dIII–*Hin*dIII segment.[18] Hence an α cell can be switched to **a** by cotransformation, using the *MAT*a *Hin*dIII fragment (in 50-fold molar excess) and a separate plasmid which carries *URA3, LEU2, TRP1,* or some other selectable marker as a companion for cotransformation.

Facilitation of Complementation Tests. In carrying out complementation tests, it is often valuable to have **a** and α derivatives of the primary isolates of mutants. If a mutant is isolated in an **a** cell, an α mating-type locus can be introduced by genetic crosses, but this is cumbersome, and, more importantly, it may not be known with certainty which segregants have the original mutation. In contrast, α derivatives can be readily generated by switching mating types. Furthermore, the particular constellation

---

[15] S. Harashima, Y. Nogi, and Y. Oshima, *Genetics* **77**, 639 (1974).
[16] I. Takano, T. Kusumi, and Y. Oshima, *Mol. Gen. Genet.* **126**, 19 (1973).
[17] B. Weiffenbach, D. T. Rogers, J. E. Haber, M. Zoller, D. W. Russell, and M. Smith, *Proc. Natl. Acad. Sci. U.S.A.* **80**, 3401 (1983).
[18] C. R. Astell, L. Ahlstrom-Jonasson, M. Smith, K. Tatchell, K. A. Nasmyth, and B. D. Hall, *Cell (Cambridge, Mass.)* **27**, 15 (1981).

of markers in the original mutant strain (various auxotrophies, unknown mutations, etc.) is preserved. [An exception to this guarantee is when a mutation is at or near the mating-type locus itself. This proved to be the case for mutations of the *BUD5* gene, which is adjacent to the mating-type locus and which can be switched from *bud5*$^-$ to *BUD5*$^+$ by mating-type switching (John Chant, personal communication, 1990).] The method of producing **a** and α derivatives of mutants for complementation testing is as above.

Testing for Bilateral Mating Defect. Some mutants exhibit a severe mating defect only when they are mated to another strain that is defective in the same gene or related aspect of mating. This behavior is termed a bilateral mating defect because the genetic defects must be present in both partners. Examples of mutants with this behavior are strains defective in the *FUS1* and *FUS2* genes, which mate much better to wild-type strains than to partners with defects in the *FUS* genes.[19] Mating-type interconversion can be used to produce **a** and α versions of the mutants, which can then be tested for their ability to mate with each other. Such strains are created by transformation with an *HO* plasmid as described above. Because both strains have the same auxotrophic mutations, mating must be assessed microscopically for formation of dumbbell-shaped zygotes.

Construction of Polyploid Strains. Various cell biological techniques such as immunofluorescence and photomicroscopy are improved if large yeast cells are used. Diploid and tetraploid cells are larger than haploids[20] and therefore are appropriate for these applications; however, it is important that the diploid or tetraploid cells be genetically homogeneous, in other words, that they be homozygous. Most yeast strains used for industrial fermentations are polyploid. Use of mating-type interconversion to create strains of higher ploidy makes it possible to create polyploid strains of defined genotype (discussed further in Ref. 1).

Transform the desired strain with pGAL-*HO*, which is a derivative of YCp50 containing *URA3* (Fig. 4). Inoculate a minimal medium plate containing galactose as carbon source with transformants to produce a patch of cells. Over 1–2 days of incubation, the haploid will undergo mating-type interconversion and mate with neighbors to form an **a**/α diploid. Switching of mating type will continue in these cells because expression of *HO* from pGAL-*HO* is not turned off in **a**/α cells. Hence **a**/**a** and α/α cells are produced, and they can mate with haploids or with each other to form **a**/**a**/α, **a**/α/α, and **a**/**a**/α/α cells. The process is stopped by

[19] J. Trueheart, J. D. Boeke, and G. R. Fink, *Mol. Cell. Biol.* **7**, 2316 (1987).
[20] R. K. Mortimer, *Radiat. Res.* **9**, 312, (1958).

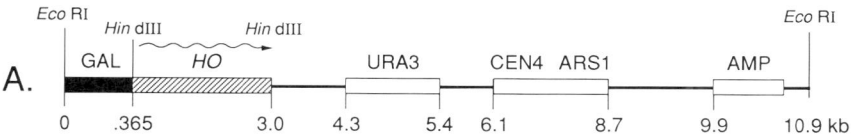

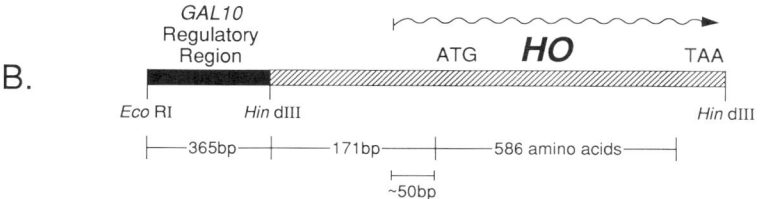

FIG. 4. Structure of pGAL-*HO*. pGAL-*HO* contains the yeast *HO* gene under control by the regulatory region of the *GAL10* gene [R. E. Jensen, Jr., Ph.D. Dissertation, University of Oregon (1983)]. It is a derivative of YCp50 [M. D. Rose, P. Novick, J. H. Thomas, D. Botstein, and G. R. Fink, *Gene* **60**, 237 (1987)] constructed in the following manner: the 29-bp *Eco*RI–*Hin*dIII segment of YCp50 was replaced with a 365-bp segment containing the *GAL10* regulatory region. This resulted in a plasmid (pGAL) containing a unique *Hin*dIII site, into which was placed the 2.592-kb *Hin*dIII–*Hin*dIII segment containing the entire *HO* coding sequence and 171 bp prior to the initiator ATG.[14] (A) Structure of the final plasmid, pGAL-*HO*. (B) Additional details of the regulatory region. Maps are not drawn to scale. The *HO* transcript starts at approximately −50 relative to the translation initiation codon.[14] pGAL-*HO* was constructed as follows: pRY20, a plasmid containing the *GAL10* regulatory region on a 365-bp *Ava*I–*Hin*dIII DNA fragment, was obtained from R. Yocum [L. Guarente, R. R. Yocum, and P. Gifford, *Proc. Natl. Acad. Sci. U.S.A.* **79**, 7410 (1982)]. It was digested with *Ava*I, and the resulting overhang was filled in using Klenow polymerase. Similarly, YCp50 (obtained from C. Mann) was digested with *Eco*RI and the DNA ends filled in with Klenow polymerase. Both plasmids were digested with *Hin*dIII, the fragments mixed and then ligated using T4 DNA ligase, and transformed into *Escherichia coli*. A plasmid pGAL, which contains a 365-bp *Eco*RI–*Hin*dIII *GAL10* fragment inserted into YCp50, was identified. (An *Eco*RI site is regenerated by the blunt-end ligation of a filled-in *Ava*I site to a filled-in *Eco*RI site.) pGAL contained a single *Hin*dIII site, into which a 2.5-kb DNA fragment carrying the *HO* gene was inserted to form pGAL-*HO*.

streaking the mixture on glucose plates (repressing conditions) and by subculturing the colonies to allow the plasmid to be lost. Cells in individual colonies should be examined for cell type (**a**, α, or **a**/α) and for cell size, which will provide an indication of ploidy. Large cells with **a** mating type will presumably be **a**/**a**, **a**/**a**/**a**, or **a**/**a**/**a**/**a**, whereas large nonmating cells (**a**/α cell type) might be **a**/**a**/α, **a**/α/α, **a**/**a**/α/α, or other genotypes. Probably the best way to determine ploidy of a cell, for example, a presumptive **a**/**a** diploid, is to cross it to a known α haploid and α/α diploid strain. The **a**/**a**/α triploid will yield meiotic segregants that germinate at low efficiency

(~15%),[21] whereas the **a**/**a**/α/α tetraploid will yield meiotic segregants that germinate at high efficiency.[22]

Use of pGAL-*HO* to Deliver Double-Strand Breaks

Rationale

The methods described in this section involve the use of *HO* endonuclease to deliver double-strand breaks to its target site in a controlled manner. *HO* endonuclease is expressed from the pGAL-*HO* plasmid, in which *HO* expression is induced by galactose and absent when cells are grown in glucose. The purpose of the double-strand breaks here is to examine the role of double-strand breaks in stimulating recombination and to study repair of these breaks. Cells carrying pGAL-*HO* are treated as follows: (1) they are grown first in glucose minimal medium (lacking uracil, in order to select for maintenance of the plasmid), then (2) the cultures are shifted to glycerol or lactate medium to remove glucose repression and (3) shifted to galactose medium to turn on or otherwise increase *HO* expression. [Galactose induction is much more rapid when strains are transferred from glycerol or lactate than when they are transferred directly from glucose medium (R. Jensen, unpublished observations, 1983).] At various times thereafter, cells are analyzed for recombination or repair, either by scoring recombinants or by physical methods.

Uses

Studies of Recombination. *HO* is used to produce a double-strand break in the genome at a known location by placing the *HO* cleavage site at positions of interest. Four different arrangements are shown in Fig. 5. The cut site can be either a 117-bp segment from *MAT***a**, which includes the cut site,[23] or a 24-bp segment of *MAT***a** (Fig. 3)[24-28] which has been shown to be sufficient for stimulating recombination *in vivo*.[23]

In the arrangement diagrammed in Fig. 5A, recombination between two homologous chromosomes is monitored. The cut site is present on one chromosome (e.g., in *HIS4* or at *MAT*) and not on its homolog.[24]

[21] E. M. Parry and B. S. Cox, *Genet. Res.* **16**, 333 (1970).
[22] H. Roman, M. M. Philips, and S. M. Sands, *Genetics* **40**, 546 (1955).
[23] J. A. Nickoloff, E. Y. Chen, and F. Heffron, *Proc. Natl. Acad. Sci. U.S.A.* **83**, 7831 (1986).
[24] A. L. Kolodkin, A. J. S. Klar, and F. W. Stahl, *Cell (Cambridge, Mass.)* **46**, 733 (1986).
[25] A. Ray, I. Siddiqi, A. L. Kolodkin, and F. W. Stahl, *J. Mol. Biol.* **201**, 247 (1988).
[26] J. A. Nickoloff, J. D. Singer, M. F. Hoekstra, and F. Heffron, *J. Mol. Biol.* **207**, 527 (1989).
[27] N. Rudin, E. Sugarman, and J. E. Haber, *Genetics* **122**, 519 (1989).
[28] A. Ray, N. Machin, and F. W. Stahl, *Proc. Natl. Acad. Sci. U.S.A.* **86**, 6225 (1989).

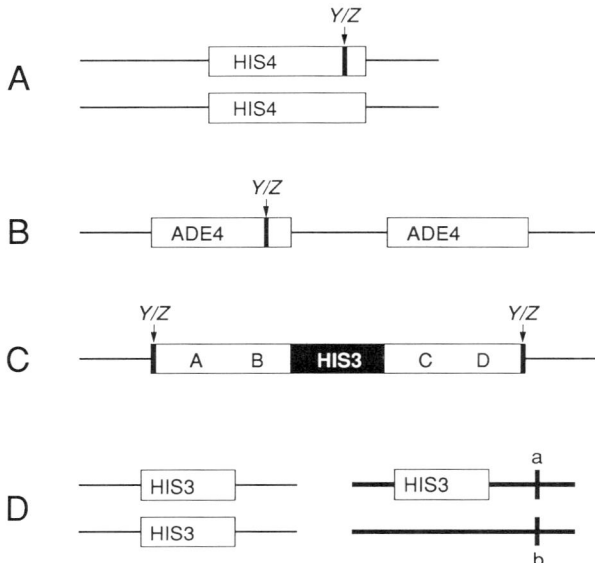

FIG. 5. Different arrangements of *HO* cut sites. *HO* cut sites (either a small segment containing the cut site or a synthetic 24-bp site; see Fig. 3) can be placed so that the effect of a double-strand break on recombination can be assessed in different situations. (A) The *HO* cut site is located in one allele and not in a second. (After Ref. 24.) (B) The *HO* cut site is located in one of two alleles that are present as a duplication on the same chromosome. (After Refs. 25–27.) (C) *ABCD* is a segment of, for example, human DNA, into which the yeast *HIS3* gene has been inserted as indicated. *HO* cut sites have been inserted at the ends of the human DNA. After induction of *HO* synthesis, the *ABCD* fragment is liberated and can then recombine with homologous chromosomes. If the liberated fragment carries a selectable gene, for example, *HIS3*, then recombination with the homologous chromosome can be scored as a *HIS3*+ recombinant. [After G. Duyk, J. Rine, and D. Cox, personal communication (1990).] (D) The *HO* cut site is placed either at position *a* or at position *b*, which is adjacent to a gene (*HIS3* in the diagram) that has a partner at a nonhomologous position in the genome with which it can potentially recombine. (After Ref. 28.)

In the arrangement of Fig. 5B, recombination between two homologous segments located on the same chromosome is monitored. Once again, the cut site is present in one allele and not the other. Studies have been performed using *ADE4*,[25] *URA3*,[26] and *lacZ*.[27]

In another arrangement (Fig. 5C), the segment that is to undergo recombination is flanked by cut sites. The uses of such an arrangement are discussed below.

Additional arrangements to test various hypotheses concerning recombination are, of course, also possible. A sophisticated use is shown in Fig.

5D, in which a double-strand break is produced adjacent to a segment that is present also on two homologous chromosomes.[28] (These studies show that the double-strand break stimulates triparental recombination among the homologous segments.) The effect of a double-strand break in stimulating recombination has been analyzed in both meiotic[24,26] and mitotic cells.[25-27]

The procedures used for mitotic recombination are as follows.[25,27,28] A single colony of the strain grown in glucose minimal medium (SD) lacking uracil is suspended in YEPL broth (YEP broth with 3 M lactate at pH 6.5) and incubated with shaking for 8 hr.[28] Aliquots are plated on YEP–galactose plates, and incubated for 18 hr. Cells are washed off plates, washed and diluted in water, and plated for viable counts (on SD complete medium) and for recombinants (on appropriate SD drop-out media). The frequency of retention of the pGAL-HO plasmid is typically about 70% of viable cells and is measured by colony count on SD–Ura medium. (A variation on this protocol is described in Ref. 25.)

In a second method,[27] a stationary-phase culture is grown in medium selective for retention of the pGAL-HO plasmid and diluted 1:100 into YEPL medium to approximately 5×10^5 cells/ml. Cultures are grown at 30° with vigorous aeration to $1-3 \times 10^7$ cells/ml, at which time galactose is added and the incubation continued. Cells are removed at various times for subsequent analysis.

The procedures used for meiotic recombination are as follows.[24,25] Cultures are grown on SD–Ura medium for 2 days, followed by overnight incubation on presporulation medium (see [1], this volume). Cells are then transferred to sporulation plates with or without galactose and incubated. Dissected spores are incubated at 26° and scored after 3–4 days. There is some concern that this procedure may induce double-strand breaks prior to meiosis (discussed in Ref. 24). If this proves to be the case, then it should be possible to restrict HO activity to meiosis by putting HO under control of a SPO gene that is expressed only during meiosis (Larry Gilbertson, personal communication, 1990).

Studies on DNA Repair. The observation that strains defective in the $RAD52$ gene are inviable in the presence of HO[29] demonstrates that the product of $RAD52$ is necessary for repair of the HO-induced cut. Whereas X-irradiation results in a complex spectrum of lesions (see Ref. 30 for discussion), a double-strand break induced by HO (or by EcoRI endonuclease)[30] is a chemically defined event and ought to be valuable for discriminating among different repair-deficient mutants. Merl Hoekstra and Fred

---

[29] R. E. Malone and R. E. Esposito, *Proc. Natl. Acad. Sci. U.S.A.* **77**, 503 (1980).
[30] G. Barnes and J. Rine, *Proc. Natl. Acad. Sci. U.S.A.* **82**, 1354 (1985).

Heffron (personal communication, 1990), have identified mutants that are deficient in repair of *HO*-induced double-strand breaks as galactose-sensitive mutants of strains that carry a pGAL-*HO* plasmid. The ability to induce the DNA damage by addition of galactose (to induce *HO* endonuclease synthesis) makes it possible to produce a homogeneous population of cells each initially with the same DNA lesion. This technique is being used to analyze the time course of repair of cleavage of the mating-type locus and for identification of possible intermediates in the repair process.[31]

The pGAL-*HO* plasmid is introduced into the *rad*⁻ mutant by transformation, selecting for transformants on glucose medium. Cultures can then be grown up in glucose minimal medium (to select for maintenance of the plasmid), transferred to glycerol or lactate medium, and then induced by galactose. Induction can be brought about in two different ways: exposure of cells to galactose for a defined period of time (by resuspending the culture in S-galactose medium) or plating cells on S-galactose medium.

Use of *HO* Endonuclease in Studying Chromosome Structure

Rationale

Because the recognition sequence for *HO* endonuclease is large (spanning a region of 24 bp)[32] the frequency of cut sites is relatively rare. Hence *HO* endonuclease should be useful for analysis of large DNA fragments and, in general, for determining the positions of cloned genes on physically defined chromosomes.

Uses

Cleaving Large Chromosomes. One use of *HO* endonuclease might be for analyzing DNA segments that are 100–200 kilobases (kb) in size, for example, fragments of the size carried by YAC vectors. Strains in which the DNA is to be analyzed would carry pGAL-*HO*. Cultures would be grown in glucose minimal medium, then transferred to galactose or maintained in glucose prior to DNA extraction. DNA would be analyzed by physical methods (see [54], this volume). A difference in bands observed between the galactose-grown and glucose-grown cultures would indicate bands that have been generated by the *HO* endonuclease.

Chromosomes that lack *HO* cut sites can acquire such sites by gene

---

[31] B. Connolly, C. I. White, and J. E. Haber, *Mol. Cell. Biol.* **8,** 2342 (1988).
[32] J. A. Nickoloff, J. D. Singer, and F. Heffron, *Mol. Cell. Biol.* **10,** 1174 (1990).

replacement (described in Ref. 33). For example, a cloned segment (e.g., of 5 kb) could be positioned on a much larger segment (e.g., a YAC of 250 kb) by adding the *HO* cut site to the cloned segment and then integrating it into the YAC chromosome by homologous recombination. The recombinant chromosome will thus acquire the *HO* cut site; hence, cleavage *in vitro* or *in vivo* by *HO* endonuclease can provide physical information on the position of the cloned segment. Additional procedural information and rationale for this type of physical mapping are given in Ref. 33.

Fragment Liberation. *HO* endonuclease can also be used in a scheme for identifying which of a collection of YAC plasmids is homologous to a cloned DNA segment (G. Duyk, J. Rine, and D. Cox, personal communication, 1990). This technique detects homology by recombination rather than by hybridization with radioactive tracers. Consider a cloned segment of human DNA (with structure *ABCD;* Fig. 5C) which will be used as the probe. The operational probe is constructed from this sequence by insertion of a selectable marker (e.g., *HIS3*) between *B* and *C* and addition of *HO* cut sites to the left of *A* and to the right of *D*. Cleavage by *HO* endonuclease will liberate a linear DNA segment that has two recombinogenic ends. The probe (carried on a high copy number plasmid) is introduced into *his3⁻* yeast strains each carrying a different YAC plasmid. *HO* is induced in the diploids using pGAL-*HO* to liberate the probe segment, which then "scans" the YAC chromosomes for homology and recombines with homologous DNA to introduce the *HIS3* gene stably into the YAC. This scheme requires that the plasmid carrying the probe (which is *HIS3⁺*) be eliminated after recombination has taken place; this can be done if the plasmid carries the *URA3* gene, so that it can be selected against with 5-fluoroorotic acid.[34]

Use of Cloned Inactive *HO* Gene to Convert Homothallic Strains to Heterothallism

Rationale

In some cases, one wishes to prevent strains from switching mating types. For example, a natural strain of polyploid strain used for industrial purposes may carry *HO*. Another possibility is that a mutant has been isolated in a laboratory strain carrying *HO*. For example, mutants defective in sporulation were isolated in strains that carry *HO*, so that a/α diploids

---

[33] D. Vollrath, R. W. Davis, C. Connelly, and P. Hieter, *Proc. Natl. Acad. Sci. U.S.A.* **85**, 6027 (1988).
[34] J. D. Boeke, F. Lacroute, and G. R. Fink, *Mol. Gen. Genet.* **197**, 345 (1984).

would be homozygous for the particular spo^- mutation.[35] It may be desirable to inactivate the *HO* gene in these various strains, so that **a** and α derivatives carrying a particular mutation can be studied. Of course, the *HO* gene can be removed by conventional genetic crosses. However, inactivating the *HO* gene by transformation methods may sometimes be useful because, as we have noted above, gene replacement does not disturb the other markers in a strain.

The strategy for inactivating *HO* is the standard one-step gene replacement method (see [19], this volume). The inactive forms of the *HO* gene carry a *LEU2* or *lacZ* segment within the *HO* coding sequence (see Refs. 4, 14, and R. Jensen, unpublished, 1983). In both cases the foreign DNA has been inserted at the *Pst*I site at position 245 of the *HO* gene. In the case of the *lacZ* insert, an *HO*–lacZ hybrid protein is created.[14] The methods are as follows.

In the first method, the source of the inactivated *HO* gene is plasmid YIp5-*HO*::*LEU2* (YIp5-B2-Leu2 in Ref. 36). Recipient *leu2*$^-$ *HO* strains are transformed with plasmid DNA that has been linearized with *Bam*HI. Leu$^+$ transformants are selected and then scored for the status of their *HO* gene. In most cases, the recipient is an **a**/α diploid; hence, this strain is sporulated and meiotic products examined. Strains carrying *HO* will grow into colonies that contain cells able to sporulate, whereas strains defective in *HO* will grow into colonies that do not contain cells able to sporulate. HO^- strains invariably grow into colonies that exhibit an **a** or α mating type. In contrast, *HO* strains grow into colonies that can be a mixed population: in addition to **a**/α cells, they may contain **a** and α cells and thus give a mating response with both **a** and α tester cells. Because of the stochastic nature of the switching process, sometimes the colony derived from an *HO* spore will contain only **a**/α cells (and thus will not give a mating response at all), or it may contain a mixture of **a**/α cells along with **a** and α cells. Additional information on scoring of *HO* can be found in Ref. 3.

In the second method, the source of the inactivated *HO* gene is pHO-c12-*lacZ,* which results in the fusion of the first 82 amino acids of *HO* to the amino terminus of *E. coli* β-talactosidase. Yeast cells containing this fusion form blue colonies on appropriate indicator plates (containing the dye X-Gal; Ref. 14). In this case, the HO^- mutation is introduced by the procedure of cotransformation. The procedure is as follows.[14] The recipient strain (*HO leu2*$^-$) is transformed with a mixture of circular (uncleaved) *LEU2*$^+$ plasmid YEp13 (to select for cells that have taken up

---

[35] M. S. Esposito and R. E. Esposito, *Genetics* **61,** 79 (1969).
[36] R. E. Jensen, Jr., Ph.D. Dissertation, University of Oregon (1983).

DNA) and a linear 5.5-kb HindIII fragment from pHO-c12-*lacZ*. The linear fragment is in an excess of 100- to 1000-fold over the circular plasmid. Leu$^+$ transformants are selected and screened for Ho$^-$ as described above. In addition, the Ho$^-$ transformants now produce β-galactosidase in a regulated manner (in **a** and α cells but not in **a**/α cells).[4,14] Approximately 3% of the Leu$^+$ transformants are expected to be Ho$^-$.

Strains that are *ho* can be converted to *HO* by a similar method. In this case, cotransformation is performed with the 2.6-kb HindIII fragment carrying *HO*.

Acknowledgments

We thank Janet Chenevert, Larry Gilbertson, and Fred Heffron for comments on the manuscript; Geoffrey Duyk, Jasper Rine, and David Cox for allowing us to cite unpublished work; Kerrie Andow and Aaron Neiman for preparation of figures; and Kathy Peterson for help with preparation of the manuscript.

[9] Spore Enrichment

By Beth Rockmill, Eric J. Lambie, and G. Shirleen Roeder

Few strains of *Saccharomyces cerevisiae* sporulate with greater than 90% efficiency. Yet it is often useful to monitor only those cells that have completed sporulation for parameters such as viability, recombination frequency, and biochemical properties. We describe two methods for distinguishing spores among incompletely sporulated populations of cells, each method with its own set of applications. The first, the ether test, is a qualitative method designed to permit screening of colonies for spore viability and/or sporulation. The second, the hydrophobic separation of spores, is a quantitative method in which spores are physically separated from unsporulated cells.

Ether Test

The ether test is a convenient plate assay for sporulation or spore viability based on the observation that vegetative (unsporulated) cells are more sensitive than spores to killing by diethyl ether vapors.[1] When a mixed population of vegetative cells and spores is exposed to ether vapors, the unsporulated cells die and only the spores survive. Thus, colonies that

[1] W. Dawes and I. D. Hardie, *Mol. Gen. Genet.* **131**, 281 (1974).

contain viable spores can be distinguished from those that do not. The ether test has been used successfully to identify sporulation-proficient, meiotic-lethal mutants (*red1* and *mer1*) and to clone the corresponding genes.[2,3] The ether test should also prove useful for identifying mutants defective in germination and sporulation.

The ether test is performed on sporulated cells replica plated to rich medium and exposed to ether vapors. Patches of cells sporulated on solid medium (2% potassium acetate, 0.1% dextrose, 0.1% yeast extract, 1.5% agar, and supplemented with amino acids, uracil, and adenine) are replica plated to YEPD medium (2% dextrose, 2% Bacto-peptone, 1% yeast extract, 1.5% agar, and optionally supplemented with 38 mg/liter adenine) just prior to the ether incubation. The YEPD medium is prepared in glass petri dishes because ether dissolves polystyrene. Squares of filter paper (Whatman #1; approximately 4 × 4 cm) are cut, and one is placed in the inverted lid of each dish. The dishes are set inverted (medium up) in an ether-resistant plastic box (e.g., Freezerette boxes, Republic Moulding Corp., Chicago, IL) which is placed in a fume hood. Approximately 0.75 ml of diethyl ether is pipetted onto each filter. Several milliliters of ether is also pipetted into a small beaker which is placed in the box to maintain the vapor pressure. The box is covered and incubated for 15 min at room temperature, after which the ether is reapplied to the filter for a second 15-min incubation. The filters are then removed and the plates are set with lids ajar in the hood until the ether odor is gone (30–45 min). The plates are incubated overnight at 30° and the patches are scored for growth the following day (see Fig. 1).

Over a 1000-fold enrichment for spores over unsporulated cells is usually achieved. Since variability may exist (even among plates tested and incubated together), positive and negative controls are recommended for each plate. Often, cells at the periphery of a plate have a growth advantage. Strains of different genetic backgrounds may require a slight increase or decrease in the time of exposure to ether.

Hydrophobic Spore Isolation

Relatively pure preparations of spores can be obtained for use in measuring meiotic recombination frequencies.[4] The hydrophobic spore isolation method has been especially valuable in measuring meiotic recombination in meiotic mutants[2,3] in which the rates of meiotic exchange are not

[2] B. Rockmill and G. S. Roeder, *Proc. Natl. Acad. Sci. U.S.A.* **85,** 6067 (1988).
[3] J. E. Engebrecht and G. S. Roeder, *Genetics* **121,** 237 (1989).
[4] E. J. Lambie and G. S. Roeder, *Cell (Cambridge, Mass.)* **52,** 863 (1988).

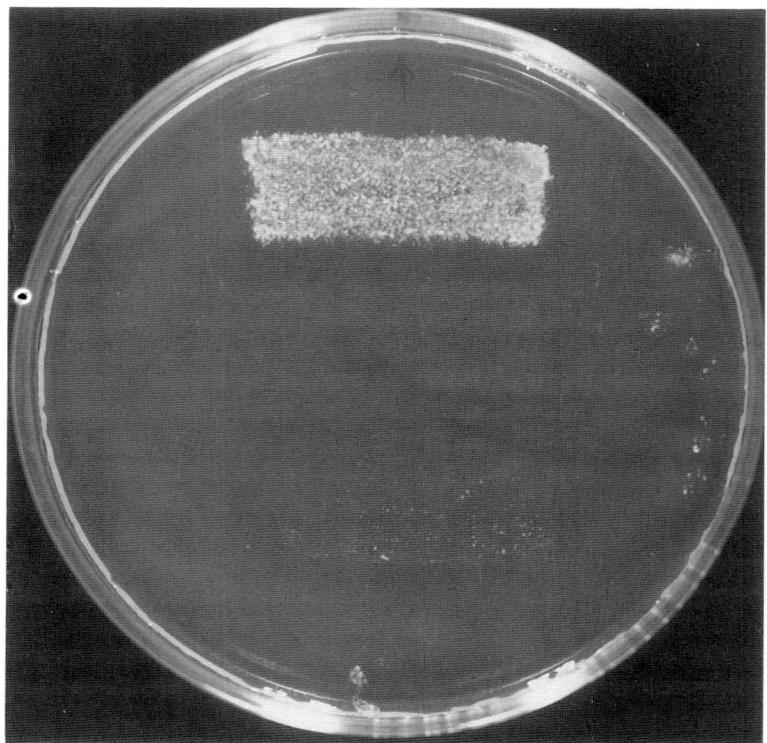

FIG. 1. Ether test of sporulated cells. (Top) Wild-type diploid (90% sporulation). (Middle) *red1*[2] homozygous diploid (meiotic lethal; 60% sporulation). (Bottom) *mer1*[3] homozygous diploid (meiotic lethal; 20% sporulation).

much higher than the mitotic background level. Pure spore suspensions are also useful in molecular and biochemical studies of spores and germinating cells.

The procedure to isolate spores is based on the relative hydrophobicity of spores compared to vegetative cells.[5] Cultures are sporulated in liquid (1–2% potassium acetate) at a concentration of approximately 2×10^7 cells/ml for 3 days. The cells are harvested by centrifugation in a microcentrifuge (10 sec), resuspended at a concentration of approximately 5×10^8 cells/ml in an aqueous solution of 100 µg/ml Zymolase 10,000 (Seikagaku Kogyo Co., Tokyo, Japan), and incubated at 30° until the ascus walls

[5] C. C. Emeis and H. Gutz, *Z. Naturforsch., B: Chem. Sci.* **13**, 647 (1958), *in* J. E. Haber and H. O. Halvorson, *Methods Cell Biol.* **11**, 45 (1975).

are digested (~20 min; time may vary depending on genetic background). Aliquots of 500 µl are centrifuged at 14,000 g for 30 sec in 1.5-ml polypropylene microcentrifuge tubes. The supernatant is discarded, and the cells are resuspended in 1.0 ml of water, centrifuged, and resuspended in 100 µl of water. Each tube is then agitated for approximately 2 min in an upright position, using a Vortex mixer at maximum speed (Fisher Scientific Co., Fairlawn, NJ). During this treatment, the spores clump together and stick to the walls of the tube, which are relatively hydrophobic. Vegetative cells do not accumulate on the walls of the tube and are removed by discarding the aqueous cell suspension and rinsing the tube several times with water. (The tube may be vortexed at low speed during rinsing.) The spores are then resuspended by adding 1.0 ml of 0.01% Nonidet P-40 (Sigma, St. Louis, MO) and sonicating on ice for 1–3 min, using a Braun-Sonic 2000 equipped with a microtip (power setting 40 W).

Spore preparations are typically greater than 99% pure. The purity of the spore suspension may be monitored by examination with a light microscope. The preparation can be stored at 4°; however, reclumping of spores and progressive decline of spore viability have been observed. This method can be used to isolate spores from strains that sporulate poorly, irrespective of genetic background.

[10] Analysis and Manipulation of Yeast Mitochondrial Genes

By THOMAS D. FOX, LINDA S. FOLLEY, JULIO J. MULERO, THOMAS W. MCMULLIN, PETER E. THORSNESS, LARS O. HEDIN, and MARIA C. COSTANZO

The mitochondrial genome of *Saccharomyces cerevisiae* has circular genetic and restriction maps. The overall DNA length is approximately 75 to 80 kilobases (kb), depending on the strain, although mitochondrial (mt) DNA isolated from yeast consists primarily of random linear fragments of subgenomic size. There are roughly 40 to 50 copies of mtDNA per cell.[1] The chromosome is known to encode seven major subunits of energy-transducing complexes, one ribosomal protein, several minor proteins

---

[1] B. Dujon, in "The Molecular Biology of the Yeast *Saccharomyces*" (J. N. Strathern, E. W. Jones, and J. R. Broach, eds.) Vol. 1, p. 505. Cold Spring Harbor Laboratory, Cold Spring Harbor, New York, 1981.

involved in intron splicing and gene conversion, large and small rRNAs, 24 tRNAs, and an RNA involved in tRNA processing.[1-3] The genome is easily amenable to genetic and molecular analysis, making yeast an organism of choice for understanding the role of the mitochondrial genetic system in eukaryotic cell biology. In addition to the reviews cited above, an excellent source for background and many detailed procedures is Volume 56 of this series.

Isolation of Mitochondrial Mutations

Most mutations affecting the mitochondrial genome render the cell unable to derive energy from respiration. Similar respiratory defects are also caused by mutations in numerous nuclear genes, usually termed *pet* mutations. This phenotype is most easily scored as an inability to grow on nonfermentable carbon sources: such mutants grow on YPD medium (1% yeast extract, 2% peptone, 2% dextrose) but fail to grow on YPEG [same as YPD but with 3% (v/v) ethanol plus 3% (v/v) glycerol replacing dextrose]. (Note: Many wild-type laboratory strains that can respire nevertheless fail to grow well on *minimal* medium containing ethanol and glycerol.) Mitochondrial mutations causing resistance to drugs that block respiratory growth (but not fermentative growth) have also been analyzed, as discussed in detail elsewhere.[1]

Mitochondrial mutations fall into basically two classes: limited lesions affecting single functions (termed mit^-), and large deletion mutations that block all mitochondrial protein synthesis (termed cytoplasmic petite or rho^-). Both wild-type and mit^- strains are termed rho^+, since they have functional mitochondrial genetic systems. The rho^- mutations occur at high frequency spontaneously (up to a few percent of cells in a growing culture), can be induced at very high frequency by many drug treatments, and are useful tools for the isolation and analysis of strains carrying more limited lesions. Of principal importance is the fact that when a rho^+ strain carrying a limited lesion (mit^-) is crossed to a rho^- strain that retains wild-type mtDNA sequences corresponding to that lesion, recombination between the mitochondrial genomes occurs and results in the generation of respiring diploid cells at high frequency (marker rescue).[4] (Note: in virtually all cases respiring progeny are generated by recombination, not complementation, for reasons discussed in detail below.) The limit case of

---

[2] L. A. Grivell, in "Genetic Maps" (S. J. O'Brien, ed.), p. 290. Cold Spring Harbor Laboratory, Cold Spring Harbor, New York, 1987.
[3] M. de Zamaroczy and G. Bernardi, *Gene* **47**, 155 (1986).
[4] P. P. Slonimski and A. Tzagoloff, *Eur. J. Biochem.* **61**, 27 (1976).

mtDNA deletions is the loss of the entire genome. These useful strains, entirely lacking mtDNA and termed rho°, obviously fail to yield respiring diploids when mated to mit^- mutants.

rho^- strains have the further curious property that the mtDNA sequences they retain are reiterated into concatemers whose length is about the same as wild-type mtDNA.[1] Since rho^- strains can retain and amplify very short segments of mtDNA (as small as several tens of base pairs), they provide a natural molecular cloning system for mtDNA sequences.

Treatment with Ethidium Bromide to Induce rho° and rho^- Strains

rho° derivatives are isolated as follows.[5] Grow the strain of interest from a small inoculum to saturation in minimal medium (0.67% Difco yeast nitrogen base plus any needed supplements) containing 2% glucose (SD) plus 25 μg/ml ethidium bromide (filter-sterilized). (Complete medium does not work as well, probably because oligonucleotides bind ethidium bromide and lower its effective concentration.) Inoculate a second culture from the first in the same medium and grow it to saturation. Streak for single colonies on YPD. Essentially every clone will be a rho°. This can be checked most easily by using highly purified mtDNA (see below) to probe Southern blots of total DNA from candidate strains.

rho^- strains can be induced by less extensive treatment with ethidium bromide as follows.[6] Cells growing logarithmically in YPD are washed and resuspended in 0.1 M potassium phosphate buffer (pH 6.5) at a concentration of approximately 10^6/ml. Ethidium bromide is added to concentrations between 1 and 10 μg/ml, and the cells are incubated at 30°, with agitation, for approximately 8 hr. Following incubation, the cells are diluted in water and plated on YPD for single colonies. Sensitivity to the treatment will vary from strain to strain, and concentrations and incubation times can be varied to obtain the desired results. The treatment can also be moderated by adding cycloheximide (10 μg/ml) to the resuspended cells before addition of ethidium bromide.[6] In general, more extensive treatment yields rho^- strains retaining shorter pieces of mtDNA and a greater proportion of rho° strains than does milder treatment.

Identification of Desired rho^- Clones

The simplest way to identify a rho^- clone retaining a particular region of mtDNA is to determine whether it can cause marker rescue after mating

[5] E. S. Goldring, L. I. Grossman, D. Krupnick, D. R. Cryer, and J. Marmur, *J. Mol. Biol.* **52**, 323 (1970).
[6] J. Deutsch, B. Dujon, P. Netter, E. Petrochilo, P. P. Slonimski, M. Bolotin-Fukuhara, and D. Coen, *Genetics* **76**, 195 (1974).

to a tester strain with a limited lesion in the region of interest. rho^- clones are replica plated to a freshly spread lawn of the tester strain on YPD (and to YPEG as a control), allowed to grow for 2 days at 30° and then printed to YPEG.

Freshly induced (primary) rho^- clones will generally contain many cells that fail to retain the marker of interest. Therefore, it is important to subclone at least twice by restreaking and mating to the tester.[7,8] Many rho^- clones that are positive in the first mating will prove to be highly unstable and cannot be recovered as single colonies on restreaking (i.e., subcloned). The mitotic stability of rho^- mitochondrial genomes is idiosyncratic, and the fraction of cells in a culture that retain a marker of interest can vary from less than 1 to 99%. Although relatively unstable rho^- strains are frequently very useful for purely genetic analysis, there are obvious difficulties in preparing purified mtDNA (see below) from them. One caution should be noted: rho^- mitochondrial genomes have frequently proved to be unstable during refrigerator storage. Therefore, it is important to store such strains promptly by freezing in 15% glycerol at $-70°$. Under these conditions they remain stable and viable for at least 12 years.

Isolation of Mutations Affecting Single Genes (mit⁻) in rho⁺ Strains

The chief difficulty in isolating simple mitochondrial mutants is that they occur far less frequently than rho^- mutants, which have a very similar nonrespiratory phenotype. Initially they were identified by virtue of the fact that, unlike rho^- strains, mit^- mutants frequently revert[9] and have active mitochondrial protein synthesis, which can be assayed biochemically.[10] It was subsequently found that the nuclear mutation $op1$ (which itself leads to a nonrespiratory phenotype and is also called $pet9$) causes lethality of rho^- but not mit^- mutants.[11] Thus, mit^- mutant clones can be identified in a population of $op1$ cells by mating to a Pet⁺ $rho°$ tester: the mit^- clones fail to yield respiring diploids.

At present, we believe the easiest and most general way to identify mit^- mutations in a given region, and to distinguish them from the high background of rho^- mutants, is by virtue of the fact that mit^- strains will recombine with a defined rho^- tester covering that region to give respiring diploids whereas the background rho^- strains will not. We have found that this scheme, coupled with the use of the inositol-starvation enrichment

---

[7] H. Fukuhara and M. Rabinowitz, this series, Vol. 56, p. 154.
[8] P. Borst, J. P. M. Sanders, and C. Heyting, this series, Vol. 56, p. 182.
[9] U. Flury, H. R. Mahler, and F. Feldman, *J. Biol. Chem.* **249,** 6130 (1974).
[10] A. Tzagoloff, A. Akai, and R. B. Needleman, *J. Bacteriol.* **122,** 826 (1975).
[11] Z. Kotylak and P. P. Slonimski, *in* "Mitochondria 1977" (W. Bandlow, R. J. Schweyen, K. Wolf, and F. Kaudewitz, eds.), p. 83. de Gruyter, Berlin, 1977.

procedure[12] to enrich a population for nonrespiring cells,[13] allows convenient isolation of *mit*⁻ mutants, even without mutagenesis. The starting strain must have inositol auxotrophic mutations in a genetic background that allows growth on minimal medium plus ethanol (3%, v/v) and glycerol (3%, v/v) (SEG). One strain that has worked well in this scheme, LH114AC (*MAT*α *ade2 ino1-13 ino4-8* [*rho*⁺]),[14] contains the mtDNA of the standard wild-type strain D273-10B.[2]

The following is a sample protocol for the isolation of spontaneous mutations in the gene encoding cytochrome oxidase subunit II, *oxi1* (all incubations at 30°):

1. Spread lawns of LH114AC on YPD plates and allow to grow.
2. Print the lawns to plates containing SEG medium supplemented with inositol.
3. Incubate 6 to 8 days, allowing nonrespiring cells to become dormant.
4. Print the SEG plates to minimal medium lacking inositol,[15] containing 3% ethanol plus 3% glycerol as a carbon source.
5. Incubate 7 to 8 days, allowing respiring cells to die from inositol starvation.
6. Print the inositol starvation plates to YPD and incubate to allow survivors to form colonies.
7. Print the colonies on YPD to the following plates: (a) YPEG, to detect respiring survivors; (b) YPD with a lawn of a *MAT*a *rho*⁻ strain that retains the *oxi1*⁺ gene, for example, strain 8-12;[16] (c) YPD with a lawn of a *MAT*a *rho*° strain.
8. Allow the colonies to mate with the lawns for 2 days.
9. Print the mated colonies to YPEG and incubate 2 days.

Most of the nonrespiring clones will be *rho*⁻, and will fail to yield respiring diploids after mating. However, the *oxi1*⁻ (*mit*⁻) mutants will yield respiring diploids when mated to the *rho*⁻ strain carrying *oxi1*⁺, but not the *rho*° strain. (Nuclear *pet* mutants will yield respiring diploids after mating with both testers.)

Mutagenesis of mtDNA can be carried out by growing cells in YPD medium containing $MnCl_2$.[17,18] In addition to inducing *mit*⁻ mutations, this treatment will also generate *rho*⁻ deletions, so crosses with defined

[12] S. A. Henry, T. F. Donahue, and M. R. Culbertson, *Mol. Gen. Genet.* **143,** 5 (1975).
[13] T. Mason, M. Breitbart, and J. Meyers, this series, Vol. 56, p. 131.
[14] M. C. Costanzo, E. C. Seaver, and T. D. Fox, *EMBO J.* **5,** 3637 (1986).
[15] C. W. Lawrence, this volume [18].
[16] T. D. Fox, *J. Mol. Biol.* **130,** 63 (1979).
[17] A. Putrament, H. Baranowski, and W. Prazmo, *Mol. Gen. Genet.* **126,** 357 (1973).
[18] G. Coruzzi, M. K. Trembath, and A. Tzagoloff, this series, Vol. 56, p. 95.

tester strains must still be used to identify mutants of interest. The sensitivity of different yeast strains to this salt varies and must be determined empirically for optimal results. We have used overnight growth in liquid YPD containing 2 to 7 mM MnCl$_2$, depending on the strain. Ethylmethane sulfonate is also an effective mutagen for mtDNA.[15,19]

Behavior of Mitochondrial DNA in Crosses and Construction of Strains

Although a full discussion of mitochondrial transmission genetics is beyond the scope of this chapter (and is reviewed elsewhere[1,20]), a brief discussion of simple genetic analysis and the construction of strains is necessary. The mating of two yeast cells carrying different mtDNAs generates a heteroplasmic zygote containing both mitochondrial genomes (about 40 to 50 copies each) in what can be considered, at least formally, the same compartment. In general, the heteroplasmic state is highly unstable. Mitotic growth of the zygote is accompanied by rapid segregation of homoplasmic cells containing either one of the parental mtDNAs or a product of recombination between them (although exceptions have been reported[21]). Thus, if wild-type and *mit*$^-$ strains are mated, and the resulting diploids allowed to grow for a few generations, both parental mitochondrial genotypes will be recovered as mitotic segregants among diploid clones with approximately equal frequency.

If two nonidentical *mit*$^-$ strains are mated, both nonrespiring parental types will be recovered, as well as respiring diploids produced by recombination (and nonrespiring recombinant double mutants). Recombination in yeast mitochondria is extraordinarily frequent. For example, mating of strains carrying mutations in *oxi1* separated by fewer than 750 bp produces up to 10% respiring diploid progeny in crosses analyzed quantitatively.[22] As mentioned above, a cross between a *rho*$^-$ strain carrying a wild-type gene and a *mit*$^-$ strain defective in that gene will similarly yield respiring recombinants at very high frequency, allowing almost all such crosses to be scored as + or − in a patch test.

The phenomena of frequent recombination and rapid mitotic segregation of heteroplasmic states make it impossible to maintain "diploid" or "merodiploid" mitochondria under most circumstances. Thus, routine trans-complementation analysis, scored at the level of growth, has never been reported. Some heteroplasmic combinations can be maintained selectively, despite their high instability, in exceptional cases where a *rho*$^+$ and

[19] U. Smolinska, *Mutat. Res.* **179**, 167 (1987).
[20] P. S. Perlman, C. W. Birky, Jr., and R. L. Strausberg, this series, Vol. 56, p. 139.
[21] A. S. Lewin, R. Morimoto, and M. Rabinowitz, *Plasmid* **2**, 474 (1979).
[22] B. Weiss-Brummer, R. Guba, A. Haid, and R. J. Schweyen, *Curr. Genet.* **1**, 75 (1979).

rho^- genome are both required for respiration but cannot form stable recombination products by homologous crossing-over.[23,24] However, the requirement for selective maintenance of respiring cells precludes the construction and analysis of nonrespiring (noncomplementing) heteroplasmic strains. These problems can be overcome to some extent by a kinetic "physiological" complementation assay, in which the rate of recovery of oxygen consumption in freshly formed heteroplasmic zygotes is measured.[25]

The construction of haploid strains carrying mitochondrial mutations in different nuclear genetic backgrounds is complicated by the fact that nonrespiring diploids fail to sporulate. Thus, one cannot simply mate a mit^- strain with a rho^o strain to generate a homoplasmic diploid and then isolate meiotic progeny. One way around this problem is to mate a mit^- strain with a respiring rho^+ strain and then immediately induce sporulation in the zygotes, which are still heteroplasmic.[26] Although most spores from such a cross will be wild type, approximately 30% of the tetrads will contain at least one (and frequently four) mit^- spores. Segregation of nuclear genes is normal. The procedure for this "forced" sporulation is as follows:

1. Grow both haploid strains to log phase in liquid YPD.
2. Mix equal volumes of the cultures.
3. Sediment the cells by brief centrifugation.
4. Spread the resulting slurry on the surface of a YPD plate and incubate at 30° until many zygotes have formed (usually 3–4 hr).
5. Scrape the cells from the surface of the YPD plate, spread them on a sporulation plate (1% potassium acetate, 0.1% yeast extract, 0.05% dextrose, 2% agar), and incubate at room temperature.

Another method of moving mitochondrial mutations to new nuclear genetic backgrounds is by using strains with the *kar1-1* mutation to block nuclear fusion after mating and allow cytoduction.[27] Typically, the strain carrying the mitochondrial mutation is mated to a rho^o derivative of a *kar1-1* strain, and the resulting zygotes are isolated by microdissection. Cells from the resulting colonies are cloned by streaking and scored for the nuclear markers of the *kar1-1* haploid and for the mitochondrial mutation

---

[23] P. P. Müller, M. K. Reif, S. Zonghou, C. Sengstag, T. L. Mason, and T. D. Fox, *J. Mol. Biol.* **175**, 431 (1984).
[24] T. D. Fox, J. C. Sanford, and T. W. McMullin, *Proc. Natl. Acad. Sci. U.S.A.* **85**, 7288 (1988).
[25] F. Foury and A. Tzagoloff, *J. Biol. Chem.* **253**, 3792 (1978).
[26] A. Tzagoloff, A. Akai, R. B. Needleman, and G. Zulch, *J. Biol. Chem.* **250**, 8236 (1975).
[27] J. Conde and G. R. Fink, *Proc. Natl. Acad. Sci. U.S.A.* **73**, 3651 (1976).

(usually by mating to an appropriate tester as described above). Once a *karl-1* strain carrying the mitochondrial mutation is isolated, it can be used to transfer that mtDNA to any rho^0 strain by subsequent rounds of cytoduction.

Analysis of Mitochondrial Gene Expression by *in Vivo* Labeling of Translation Products

In assessing the effects of mitochondrial (or nuclear) mutations on the expression of individual mitochondrial genes and the nature of their products, it is useful to examine mitochondrially encoded proteins directly. Mitochondrial translation products can be selectively ^{35}S-labeled in cycloheximide-treated cells since cytoplasmic ribosomes are sensitive to this drug but mitochondrial ribosomes are not.[28] A crude preparation of mitochondria can then be prepared, subjected to denaturing gel electrophoresis, and autoradiographed to reveal the handful of major proteins encoded on the mitochondrial genome. The following procedure for selective labeling and analysis of mitochondrial translation products is modified slightly from that described in a previous volume in this series[29] (all incubations at 30°):

1. Grow strains to saturation in 2 ml YPGal medium (same as YPD but with 2% galactose substituted for dextrose). YPEG can be used for respiring strains. YPD can be used for strains that fail to grow on galactose, but incorporation is lower owing to glucose repression.

2. Add 2 ml fresh medium (same as Step 1) and incubate with shaking for 2.5 hr.

3. Pellet cells, wash twice with sterile distilled water, and resuspend in 2 ml medium for labeling. If labeling with $^{35}SO_4$ use low sulfate medium (LSM)[30] containing 2% galactose (or other carbon source). If labeling with [^{35}S]methionine, either LSM or standard minimal medium (0.67% Difco yeast nitrogen base plus carbon source) may be used.

[28] G. Schatz and J. Saltzgaber, *Biochem. Biophys. Res. Commun.* **25**, 996 (1969).
[29] M. Douglas, D. Finkelstein, and R. A. Butow, this series, Vol. 56, p. 58.
[30] For LSM medium, the following stock solutions are prepared: (A) Rare elements (per liter): boric acid, 500 mg; $CuSO_4$, 40 mg; KI, 100 mg; $MnSO_4$, 500 mg; $Na_2MoO_4 \cdot 2H_2O$, 235 mg; and $ZnSO_4$, 400 mg. (B) Ferric chloride: $FeCl_3 \cdot 6H_2O$, 200 mg/liter. (C) Vitamins (per 200 ml): Calcium pantothenate, 80 mg; thiamin-HCl, 80 mg; pyridoxine, 80 mg; nicotinic acid, 20 mg; and biotin, 0.8 mg. (D) Inositol: *myo*-inositol, 20 g/liter. (E) Mineral salts (per 900 ml): $(NH_4)H_2PO_4$, 60 g; $MgCl_2$, 1.9 g; NH_4Cl, 16.2 g; KH_2PO_4, 10 g; NaCl, 1 g; $CaCl_2 \cdot 2H_2O$, 1.32 g. Autoclave solutions A–E separately. After solutions cool, mix together as follows for 10× low sulfate medium (10× LSM): (A) rare elements, 10 ml; (B) $FeCl_3$, 10 ml; (C) vitamins, 50 ml; (D) inositol, 10 ml; (E) mineral salts, 900 ml. For labeling, 10× LSM is diluted to 1× with water, and the appropriate carbon source is added.

4. Incubate with shaking for 30 min.
5. Add cycloheximide to a final concentration of 0.1 mg/ml (from a stock solution of 10 mg/ml in ethanol).
6. Incubate with shaking for 5 min.
7. Add 0.5 mCi $^{35}SO_4$ or 0.1 mCi of [^{35}S]methionine [excellent results are also obtained with crude hydrolyzates of ^{35}S-labeled bacteria, available from ICN Biomedicals, Inc. (Costa Mesa, CA) as Trans ^{35}S-Label].
8. Incubate with shaking for 1 hr.
9. Add 2 ml chase solution (1% casamino acids, 2 mg/ml Na_2SO_4).
10. Incubate with shaking for 10 min.
11. Pellet cells, wash once with chase solution, twice with MTE buffer (0.25 M mannitol, 20 mM Tris–SO_4, pH 7.4, 1 mM EDTA), and resuspend in 0.2 ml MTE on ice.
12. Add an equal volume of 0.45-mm glass beads and vortex at high speed for 5 min.
13. Remove the liquid and wash the glass beads 4 times with 0.2 ml MTE per wash.
14. Centrifuge the combined supernatant and washes at low speed (4000 rpm in a Sorvall SS34 rotor for 10 min, 4°).
15. Remove the supernatant and centrifuge at high speed (20,000 rpm in a Sorvall SS34 rotor for 10 min, 4°) to pellet crude mitochondria.
16. Discard the supernatant. Resuspend the pellet in 0.5 ml MTE and centrifuge again at high speed.
17. Resuspend pellet in 20 μl MTE.

Electrophoretic analysis is carried out in the standard sodium dodecyl sulfate (SDS)–discontinuous buffer system.[29,31] A sample of the resuspended mitochondria is dissolved in sample buffer containing SDS at room temperature (do not boil the samples as this can cause aggregation of several of the highly hydrophobic mitochondrial proteins). Gels containing 12 or 15% polyacrylamide can be used, although 9 to 15% linear polyacrylamide gradient gels give better results.

In addition to examining mitochondrial gene expression in stable haploid strains, this technique can also be applied to a population of newly formed zygotes to detect gene expression from a *rho⁻* genome.[32] On fusion of the mitochondria in zygotes formed by mating a *rho⁻* to a *rho⁺* strain, the information on the *rho⁻* genome is expressed by the transcription–translation machinery of the *rho⁺* mitochondrion. Thus, gene expression can be examined transiently in heteroplasmic cells that cannot be stably maintained owing to mitotic segregation.[33]

[31] U. K. Laemmli, *Nature (London)* **227**, 680 (1970).
[32] R. L. Strausberg and R. A. Butow, *Proc. Natl. Acad. Sci. U.S.A.* **74**, 2715 (1977).
[33] M. C. Costanzo and T. D. Fox, *Proc. Natl. Acad. Sci. U.S.A.* **85**, 2677 (1988).

In another chapter on mitochondria in this volume,[34] a protocol for pulse labeling (in the absence of cycloheximide) is presented that allows the analysis of products of both nuclear and mitochondrial genes. Although that procedure is more general, it is less convenient for the specific detection of mitochondrial translation products.

Isolation and Analysis of Mitochondrial DNA and RNA

Yeast mtDNA is predominantly A + T and can be easily separated from nuclear DNA by density gradient centrifugation in the presence of dyes, such as 4′,6-diamidino-2-phenylindole (DAPI) or bisbenzimide (Hoechst No. 33258), that preferentially bind (A + T)-rich DNA.[35]

Preparation of Pure Mitochondrial DNA

1. Resuspend 70 g of washed, early stationary-phase cells in 130 ml of 1 M sorbitol and add 0.5 ml mercaptoethanol and 0.2 g Zymolyase 5000.
2. Shake gently at 30° for 4 to 12 hr (check for spheroplasts).
3. Harvest spheroplasts by centrifuging in a large Sorvall rotor for 10 min at 7000 rpm, 4°.
4. Resuspend spheroplasts in 190 ml of 20 mM Tris, pH 8, 50 mM EDTA, and add 8 g SDS.
5. Stir at 65° for 30 min.
6. Add 5 ml of 5 M NaCl and incubate on ice for 90 min.
7. Centrifuge for 15 min at 15,000 rpm at 4° in a Sorvall SS34 rotor.
8. Pour the supernatant into a tared graduated cylinder and add 0.79 g of NaI per milliliter of supernatant (final density 1.49 g/ml).[36]
9. Centrifuge for 15 min at 15,000 rpm at 4° in a Sorvall SS34 rotor.
10. Remove the thick, white protein layer from the top.
11. Add 50 µl bisbenzimide solution (10 mg/ml) and stir for 30 min at room temperature.
12. Seal in 40-ml centrifuge tubes (Beckman Quick Seal) and centrifuge in a Beckman VTi50 rotor at 20°, 45,000 rpm, for 48 hr.
13. Illuminate tubes with long-wavelength UV light and harvest the diffuse upper band by side puncture.
14. Dialyze 2 times against 2 liters of 10 mM Tris-HCl, pH 8.0, 1 mM EDTA (TE80) and measure volume.
15. Per milliliter of dialyzed mtDNA add 0.17 µl bisbenzimide solution and 0.98 g CsCl.

[34] M. P. Yaffe, this volume [43].
[35] D. H. Williamson and D. J. Fennell, *Methods Cell Biol.* **12**, 335 (1975).
[36] J. P. M. Sanders, R. A. Flavell, and P. Borst, *Biochim. Biophys. Acta* **312**, 441 (1973).

16. Seal in centrifuge tubes and centrifuge in VTi50 rotor at 20°, 45,000 rpm, for 22 hr.

17. Illuminate tubes with long-wavelength UV light and harvest the upper band by side puncture. (To obtain more highly purified mtDNA simply recentrifuge this fraction in CsCl as above.)

18. Extract the solution with 2-propanol to remove bisbenzimide.

19. Dialyze 3 times against 2 liters TE80. The yield of mtDNA is approximately 0.5 mg.

Detection of Mitochondrial DNA and Small Scale Isolation[35]

1. Prepare spheroplasts from 3 g cells and wash with 1 M sorbitol.
2. Resuspend in 0.5 ml of 1 M sorbitol.
3. Add 6 ml of 50 mM Tris-HCl, 50 mM EDTA, pH 8.
4. Add 0.5 ml 20% sarkosyl and mix well.
5. Incubate 10 min at room temperature.
6. Centrifuge 10 min at 5000 rpm in a Sorvall SS34 rotor.
7. To 8 ml of supernatant add 0.2 ml 10 mg/ml bisbenzimide and mix.
8. Add 7.7 g CsCl.
9. Centrifuge at 40,000 rpm in a Beckman Ti50 rotor for 24–48 hr.
10. Visualize DNA under long-wavelength UV light. The upper band is mtDNA.

Mitochondrial DNA prepared by the protocol employing both NaI and CsCl gradients is of higher purity and is a more reliable substrate for biochemical reactions, such as restriction enzyme digestion, than mtDNA isolated by a single CsCl gradient. Of course, for simple Southern analysis of mtDNA, whole cell DNA preparations are fine.

Restriction Analysis and Cloning in Bacterial Vectors

Although mtDNA is just DNA, there are a few special problems in handling it, probably because of its extremely high A + T content and the fact that it contains large numbers of nicks, gaps, and breaks. First, ethanol precipitation of purified mtDNA before digestion with restriction enzymes should be avoided since such precipitation frequently generates aggregates that fail to enter electrophoresis gels.[8] Second, because mtDNA isolated from yeast consists of many random linear fragments of subgenomic length, long restriction fragments (larger than about 5 kb) will be found in substoichiometric amounts relative to shorter fragments, and electrophoretic gels will have a background smear between bands. Finally, recombinant clones containing some longer fragments of mtDNA have been found to be unstable and suffer deletions during propagation in bacteria.[37] Thus, it is important to check that the restriction map of cloned fragments matches that of the corresponding bona fide mtDNA.

DNA Sequence Analysis of Mitochondrial Mutations

Although short fragments of mtDNA can be easily cloned and sequenced by standard procedures, it is usually not necessary to clone the DNA. Restriction fragments of purified mtDNA can be end-labeled and sequenced directly by chemical degradation.[38,39] More recently we have found that mtDNA present in crude preparations of total yeast DNA[40] serves well as a template for dideoxynucleotide chain-termination reactions,[41] using 5'-end-labeled oligonucleotide primers instead of incorporating labeled nucleotides during polymerization. We have used the Sequenase kit (United States Biochemical Corp., Cleveland, OH), modified for the omission of labeled nucleotides as described below. This allows rapid analysis of mitochondrial mutations by the following protocol (routine nucleic acid manipulations as described[42]):

1. Harvest cells from a saturated 40 ml culture by centrifugation at 5000 rpm for 5 min.
2. Resuspend the cells in 3 ml of 0.9 M sorbitol, 0.1 M EDTA, pH 7.5.
3. Add 0.1 ml of 7.5 mg/ml Zymolyase 20,000 (in 1 M sorbitol, 0.1 M EDTA) and incubate at 37° for 70 min.
4. Pellet the cells by centrifugation as in Step 1 and resuspend them in 5 ml of 50 mM Tris-HCl, pH 7.4, 20 mM EDTA.
5. Add 0.5 ml of 10% SDS.
6. Incubate the mixture at 65° for 30 min.
7. Add 1.5 ml of 5 M potassium acetate and place on ice for 60 min.
8. Centrifuge in a Sorvall SS34 rotor at 10,000 rpm for 10 min.
9. Transfer the supernatant to a clean tube and add 2 volumes of ethanol at room temperature. Mix and centrifuge at 5000 rpm for 15 min at room temperature.

[37] P. E. Berg, A. Lewin, T. Christianson, and M. Rabinowitz, *Nucleic Acids Res.* **6**, 2133 (1979).
[38] A. M. Maxam and W. Gilbert, this series, Vol. 65, p. 499.
[39] T. D. Fox, *Proc. Natl. Acad. Sci. U.S.A.* **76**, 6534 (1979).
[40] F. Sherman, G. R. Fink, and J. B. Hicks, "Methods in Yeast Genetics," p. 125. Cold Spring Harbor Laboratory, Cold Spring Harbor, New York, 1986.
[41] F. Sanger, S. Nicklen, and A. R. Coulson, *Proc. Natl. Acad. Sci. U.S.A.* **74**, 5463 (1977).
[42] T. Maniatis, E. F. Fritsch, and J. Sambrook, "Molecular Cloning," Cold Spring Harbor Laboratory, Cold Spring Harbor, New York, 1982.

10. Dry the pellet and resuspend it in 3 ml of 10 mM Tris, 1 mM EDTA, pH 7.4 (TE).

11. Centrifuge at 10,000 rpm for 15 min at 4° and keep the supernatant.

12. Add 150 μl of a 1 mg/ml solution of pancreatic RNase and incubate at 37° for 30 min.

13. Add 1 volume of 2-propanol and mix gently. Lift out the fibrous DNA precipitate.

14. Resuspend the precipitate in 0.5 ml TE and extract 2 times with phenol/chloroform/isoamyl alcohol (25:24:1).

15. Extract 2 times with chloroform, precipitate the DNA with ethanol, and resuspend the DNA in 80 μl TE.

16. Digest the DNA to completion in a volume of 100 μl with a restriction enzyme that will not interfere with the desired sequence determination. We have used *Hin*dIII and *Pvu*II for this purpose. (One rationale for the digestion is to reduce the viscosity arising from the bulk of nuclear DNA, allowing the sample to be resuspended in small volumes.)

17. Precipitate the DNA at room temperature for 5 min by addition of 120 μl 2-propanol.

18. Spin briefly (10 sec) in a microcentrifuge to pellet the DNA, pour off the supernatant, add 0.3 ml 2-propanol, vortex, and spin again.

19. Dry the DNA pellet and resuspend in 7 μl TE.

20. Prepare 5'-^{32}P-labeled primer.[42]

21. Mix 7 μl of restricted total yeast DNA (from Step 19), 2 μl of 5× Sequenase buffer (200 mM Tris-HCl, pH 7.5, 100 mM MgCl$_2$, 250 mM NaCl), and 1 μl (1 pmol) of 5'-^{32}P-labeled primer.

22. Boil for 3 min and transfer immediately to a dry ice/ethanol bath for 1 min.

23. Thaw on ice (samples may be stored on ice for several hours).

24. Incubate at room temperature for 5–10 min. Then add 1 μl of 0.1 M dithiothreitol and 2 μl of a solution containing dATP, dGTP, dCTP, and dTTP, each at a concentration of 0.75 μM.

25. Continue following the standard Sequenase protocol.

Preparation of Mitochondrial RNA

Separation of mitochondrially coded RNAs from other cellular RNAs can be achieved by isolating them from highly purified organelles. The preparation of mitochondrial fractions is described elsewhere in this volume.[34] To extract RNA (and DNA),[43] the organelles can be suspended at a concentration of 10 mg protein/ml in a solution containing 50 mM Tris-HCl, pH 7.4, 10 mM EDTA, 1% SDS, and 100 μg/ml proteinase K. After

---

[43] P. Boerner, T. L. Mason, and T. D. Fox, *Nucleic Acids Res.* **9**, 6379 (1981).

incubation for 1 hr at room temperature, NaCl is added to a concentration of 150 mM, the solution is extracted several times with phenol/chloroform/isoamyl alcohol (25:24:1), and the nucleic acids are precipitated by addition of 2 volumes of ethanol. Of course, most routine analyses of mitochondrial RNAs involving hybridization can be performed on preparations of total cellular RNA.[44]

Although mitochondrial RNA is just RNA, anomalous hybridization results have been reported in some cases. For example, labeled DNA probes have failed to detect RNAs on Northern blots, despite the fact that the probes could be shown to be complementary by sequence analysis of cDNA.[45] Such spurious absence of hybridization is probably due to the extremely high A + U content of several RNA species.

Mitochondrial Genetic Transformation and Gene Replacement

Standard methods for transformation of yeast nuclear genes have never been shown to promote the entry of DNA into mitochondria. However, high-velocity bombardment[46] of yeast with tungsten microprojectiles carrying mtDNA sequences can result in delivery of the DNA to the mitochondrial matrix.[47] Furthermore, rho^0 strains can be converted to stable "synthetic rho^-" strains by transformation with bacterial plasmids carrying mitochondrial genes.[24] Following transformation into the mitochondria, the plasmids exhibit all the genetic properties of natural rho^- mtDNA outlined above, including the ability to recombine with rho^+ mtDNA. Thus, synthetic rho^- strains provide a convenient means by which to replace wild-type genes in a rho^+ strain with mutants generated *in vitro,* as described below.

The basic scheme for the isolation of a synthetic rho^- strain is to bombard a lawn of rho^0 cells on the surface of a petri plate with plasmids carrying both a selectable nuclear marker (e.g., *URA3*) and replication origin as well as the mitochondrial gene of interest (cotransformation). The nuclear and mitochondrial genes may either be on separate plasmids or together on the same plasmid. Colonies derived from cells that were hit and survived are initially selected as nuclear transformants (e.g., Ura^+). These colonies are then screened for the presence of the mitochondrial gene in

---

[44] K. Köhrer and H. Domdey, this volume [27].
[45] H. P. Zassenhaus, F. Farrelly, M. E. S. Hudspeth, L. I. Grossman, and R. A. Butow, *Mol. Cell. Biol.* **3**, 1615 (1983).
[46] T. M. Klein, E. D. Wolf, R. Wu, and J. C. Sanford, *Nature (London)* **327**, 70 (1987).
[47] S. A. Johnston, P. Q. Anziano, K. Shark, J. C. Sanford, and R. A. Butow, *Science* **240**, 1538 (1988).

mitochondria by scoring their ability to yield respiring diploids when mated to an appropriate mit^- tester strain.

Several synthetic rho^- strains have been generated by mitochondrial transformations with different plasmids, using a high-velocity microprojectile bombardment device (Du Pont, Inc., Wilmington, DE) powered by a gunpowder charge (other designs[48] are likely to work also). However, the efficiency of mitochondrial transformation varies widely from experiment to experiment and can be very low (from 1 mitochondrial transformant/ 500 nuclear transformants to < 1/10,000). The sources of this variation are unclear at present. Furthermore, the efficiency of nuclear transformation is highly dependent on the yeast strain used. The best strains we have used, related to DBY947,[47,49] yield several hundred nuclear transformants per bombarded plate, while the worst, related to D273-10B,[2] yield about ten. Thus, there appears to be much room for improvement in the transformation procedure described below.

Preparation of Cells to Be Bombarded

1. Grow the rho^o strain in complete medium (YP) plus 2% raffinose to late stationary phase.
2. Pellet the cells by centrifugation at 5000 rpm for 5 min.
3. Resuspend cells in water to one-twentieth the original culture volume.
4. Spread the cells, 0.1 ml/plate, onto petri plates containing minimal glucose medium (0.67% Difco yeast nitrogen base, 2% dextrose, 2% Difco Bacto-agar) plus 1 M sorbitol and supplemented to provide the appropriate prototrophic selection. These lawns may be stored refrigerated for several days before bombardment.

Preparation of Microprojectiles and Bombardment

1. Sterilize 0.5-μm tungsten particles (obtained from Johnson Matthey/AESAR Group, Seabrook, NH) by suspension in ethanol, pellet them by brief centrifugation, and resuspend at 50 mg/ml in sterile 50% (v/v) glycerol.
2. Add 25 μl of suspended tungsten particles to 3–5 μg of plasmid DNA (in a volume not greater than 3 μl). Use either an equimolar mixture of plasmids carrying nuclear and mitochondrial markers or a single plasmid carrying both markers.

---

[48] P. Christou, D. E. McCabe, and W. F. Swain, *Plant Physiol.* **87**, 671 (1988).
[49] N. F. Neff, J. H. Thomas, P. Grisafi, and D. Botstein, *Cell (Cambridge, Mass.)* **33**, 211 (1983).

3. Add 25 μl of 2.5 M CaCl$_2$ and 10 μl of 1 M spermidine free base (filter-sterilized).

4. Incubate at room temperature for 10 min, then sediment the particles by brief centrifugation.

5. Remove supernatant and resuspend the particles in 25 μl of 100% ethanol by strong agitation (mixing with a pipette tip) and brief sonication (hold the plastic tube directly against the sonicator probe). The resulting suspension is sufficient for three bombardments.

6. Distribute the particles evenly on three macroprojectiles, allowing the ethanol to evaporate.

7. Load the macroprojectile and charge into the barrel, place the stopping plate at the end of the barrel, and place a petri plate under the barrel.

8. Fire.

9. Incubate the bombarded plate at 30° until colonies appear.

An alternative method of coating particles with DNA is to simply suspend both in distilled water and dry off the water.[48] The particles are then suspended in ethanol and treated as described above. Using this procedure to coat tungsten particles, we have obtained nuclear yeast transformants in preliminary experiments.

Identification of Synthetic rho⁻ Clones

The desired synthetic *rho*⁻ clones are identified in exactly the manner described above for the identification of natural *rho*⁻ clones, namely, by crossing with *mit*⁻ tester strains whose lesions are known to be correctable by recombination with the plasmid-derived mitochondrial sequences.

1. Print the bombarded plates to a freshly spread lawn (on a YPD plate) of an appropriate *mit*⁻ tester strain.

2. Incubate 2 days at 30° to allow mating and recombination.

3. Print to YPEG medium to detect respiring diploids.

4. Pick colonies off the bombarded plate that correspond in position to respiring diploids, restreak them on YPD and retest by mating to the *mit*⁻ tester. The subcloning should be repeated. (Clones that have lost the plasmid with the nuclear marker but retain the mitochondrial marker can be obtained during subcloning.)

Replacement of Wild-Type by a Mutant Gene in rho⁺ Mitochondrial DNA

Synthetic *rho*⁻ strains carrying a plasmid with a defective mitochondrial gene can be identified as described above, as long as the plasmid carries wild-type genetic information corresponding to the mutant site in the *mit*⁻ tester strain to which it is crossed. Once such a synthetic *rho*⁻

strain has been obtained, the defective mitochondrial gene can be transferred to a rho^+ chromosome by double recombination, simply by mating the synthetic rho^- strain to a wild type. Among the diploid progeny of such a cross, a significant fraction of the clones will have the plasmid-derived defect in rho^+ mtDNA. (To obtain haploid mutant strains, the synthetic rho^- strain can be mated to a *kar1-1* rho^+ strain and the recombinant cytoductants selected.)

We have used this strategy to isolate a strain with an AUA codon in place of the normal translation initiation codon (AUG) of the gene encoding cytochrome oxidase subunit III (coxIII). A synthetic rho^- strain derived from a plasmid with the oligonucleotide-directed mutation was identified as described above, using a strain with a mutation in the downstream region of the coxIII gene as a *mit*$^-$ tester. This synthetic rho^- was mated to a wild-type rho^+ strain, and diploid clones were selected on minimal glucose medium by complementing nuclear auxotrophies. At least 20% of the diploids had a respiratory-deficient phenotype. Two independent clones were examined by DNA sequence analysis and shown to have the plasmid-derived AUG to AUA mutation. Because this strategy depends only on the well-established genetic behavior of rho^- mtDNA, it should be generally applicable to the manipulation of the yeast mitochondrial genome.

Acknowledgments

L.S.F. was a postdoctoral fellow of the Jane Coffin Childs Memorial Fund for Medical Research. J.J.M. was supported by a predoctoral fellowship from the National Science Foundation. T.W.M. was a Postdoctoral Associate of the Cornell Biotechnology Program. P.E.T. was a postdoctoral fellow of the American Cancer Society. Research in the authors' laboratory was supported by a grant (GM29362) from the National Institutes of Health.

Section II
Cloning and Recombinant DNA

[11] DNA of *Saccharomyces cerevisiae*

By PETER PHILIPPSEN, AGATHE STOTZ, and CHRISTINE SCHERF

Experimenters beginning to work with the DNA of *Saccharomyces* strains will find many of the current DNA isolation methods quite easy to follow. On the other hand, they may be confused by the pattern of bands seen following restriction endonuclease digestion and agarose gel electrophoresis of the DNA. The proper evaluation of these so-called restriction spectra is important for many experiments. Therefore, discussion of DNA isolation procedures is followed by a description of useful information derived from analysis of restriction spectra. The main focus is on DNA of *Saccharomyces cerevisiae* strains used in genetic studies. In addition, we describe our experience with domesticated strains or strains isolated from natural habitats and with species other than *S. cerevisiae*.

DNA Isolation Procedures

The nucleus of a haploid *S. cerevisiae* cell contains approximately 14,000 kilobases (kb) of DNA subdivided into 16 linear chromosomal DNAs with sizes ranging from 250 to 2000 kb.[1,2] In addition, many strains carry 50 to 100 copies of a 6.3-kb plasmid which exists in two sequence arrangements and which is usually referred to as 2-μm plasmid.[3] The 75 to 80 kb of mitochondrial DNA is circular, and each cell harbors 20 to 40 copies depending on the growth conditions.[4] For most experiments preparations of total DNA can be used. Several published procedures for isolation of total DNA from *Saccharomyces* have been used successfully. We describe one procedure in detail that is adopted from several methods,[5,6] and comment on the essential points. Volumes can be scaled up or down and, as with other rapid procedures, the DNA obtained is fragmented. However, the DNA is of sufficient length (50–250 kb) for most cloning and hybridization experiments.

---

[1] G. F. Carle and M. V. Olson, *Proc. Natl. Acad. Sci. U.S.A.* **83**, 3756 (1985).
[2] S. L. Gerring, C. Connelly, P. Hieter, this volume [4].
[3] J. R. Broach, in "The Molecular Biology of the Yeast *Saccharomyces*" (J. N. Strathern, E. W. Jones, and J. R. Broach, eds.), Vol. 1, p. 445. Cold Spring Harbor Laboratory, Cold Spring Harbor, New York, 1981.
[4] T. D. Fox, L. S. Folley, J. J. Mulero, T. W. McMullin, P. E. Thorsness, L. O. Hedin, and M. C. Costanzo, this volume [10].
[5] D. R. Cryer, R. Eccleshall, and J. Marmur, *Methods Cell. Biol.* **12**, 39 (1975).
[6] F. Winston, F. Chumley, and G. R. Fink, this series, Vol. 101, p. 211.

Basic Protocol

1. Grow cells in 30 ml YPD (2% yeast extract, 1% peptone, 2% dextrose) at 30° to a cell density of approximately 2×10^8 cells/ml (early stationary phase).

2. Spin cells at 5000 rpm for 5 min in a Sorvall SS34 or similar rotor. Use 30-ml polypropylene tubes with screw caps. Centrifugation steps are performed at 4°; all other steps are carried out at room temperature unless otherwise specified.

3. Resuspend cells in 10 ml water and spin at 5000 rpm for 5 min.

4. Resuspend cells in 3.0 ml of 0.9 M sorbitol, 0.1 M EDTA, 50 mM dithiothreitol, pH 7.5. Use the same tube until Step 9.

5. Add 0.5 mg DNase-free β-1,3-glucanase dissolved in 200 μl 0.9 M sorbitol (Zymolyase 20,000 from Seikagaku Kogyo, Ltd., Tokyo, Japan, or Miles Corp., Naperville, IL, or yeast lytic enzyme from ICN Biomedicals, Inc., Costa Mesa, CA) and incubate with occasional shaking at 37°. Conversion of the oval cells to round spheroplasts is followed with a light microscope or is quantitatively monitored by mixing 50-μl aliquots taken at 10-min intervals with 950 μl of water and reading the OD_{800} after 20 sec. The conversion is complete when no more oval cells are visible or when the OD_{800} has decreased by 80–90%, respectively. This takes 15–120 min and depends on the strain used, on the growth medium, and on the growth phase. Alternatively, complete formation of spheroplasts can be checked by adding 100 μl of 1% sodium dodecyl sulfate (SDS) to 100 μl of the turbid suspension. A clear solution should form immediately.

6. Spin spheroplasts at 5000 rpm for 5 min in a Sorvall SS34 rotor and carefully discard the supernatant.

7. Resuspend spheroplasts in 3.0 ml of 50 mM Tris-HCl, 50 mM EDTA, pH 8.0, by slowly and repeatedly drawing the spheroplasts into a pipette and releasing them into the tube. Then mix with 0.3 ml of 10% SDS and incubate at 65° for 30 min.

8. Add 1.0 ml of 5 M potassium acetate, mix, and let sit on ice for 60 min or longer. The white precipitate that forms consists mainly of insoluble potassium dodecyl sulfate and denatured proteins.

9. Spin at 15,000 rpm for 30 min and transfer the supernatant (~4 ml) to a 10- or 15-ml Corex tube or to a polypropylene tube.

10. Add 4.0 ml ice-cold absolute ethanol. On mixing, the nucleic acids (2% DNA and 98% RNA) and some residual proteins will immediately precipitate at room temperature.

11. Spin in Sorvall SM-24 or similar rotor at 10,000 rpm for 10 min and discard the supernatant. Allow the salt to diffuse out of the pellet by incubation with 4.0 ml of 50% ethanol for at least 5 min (vortex to break up the pellet). Spin at 10,000 rpm for 10 min.

12. Discard the supernatant, dry the pellet under reduced pressure, and redissolve the DNA and RNA in 3.0 ml of 10 mM Tris, 1 mM EDTA, pH 7.5. This may take more than 1 hr. A 10-min incubation at 42° can help. If a small pellet remains, spin at 10,000 rpm for 10 min and transfer supernatant to a new tube. At this point a 10-μl aliquot may be assayed by agarose gel electrophoresis as shown in Fig. 1B.

13. Add 150 μl of 1 mg/ml DNase-free pancreatic RNase (Boehringer Mannheim, FRG) and incubate at 37° for 30 min. The stock of RNase is dissolved in 10 mM sodium acetate, pH 7.0, and kept at −20°.

14. Add 3.0 ml of 2-propanol, mix, and remove the precipitated DNA with a small plastic forceps, a glass hook, or by centrifugation. Wash the DNA with 50% 2-propanol and dry under reduced pressure.

15. Redissolve the DNA in 0.5 ml of 10 mM Tris-HCl, 1 mM EDTA, pH 7.5, and check the DNA by agarose gel electrophoresis as shown in Fig. 1A. The yield should be 60–90 μg for a haploid strain. For quantitation, dilute 50 μl with 0.95 ml of 10 mM Tris-HCl, 10 mM NaCl, pH 8.0, and measure the optical density at 260, 280, and 300 nm. If the ratio of OD_{260}/OD_{280} is 1.9 to 2.0 and the OD_{300} close to zero, an OD_{260} of 0.2 corresponds to 10 μg DNA per milliliter. The average fragment size of the DNA is 100 kb. The DNA should be stored at +4° or −70°, not at −20°.

Comments

Step 1. The growth temperature for heat-sensitive strains is 20° to 23° and for cold-sensitive strains 30° to 35°. When cells have to be grown in minimal medium or selective medium the cell density does not go beyond 10^8 cells/ml in shaking flasks. Good yields are obtained when 50 mM sodium phosphate, pH 6.4, is added to the minimal medium. Cell densities can, in principle, be determined by measuring the OD_{600} of the culture. However, there is no universal correlation between cell density and OD_{600} since this correlation depends on the strain used and on the growth conditions.

Step 4. The addition of thiol reagents such as dithiothreitol or 2-mercaptoethanol is not essential. They supposedly break S–S bridges in the cell wall and slightly decrease the time for spheroplast formation.

Step 5. If cells start lysing (too much glucanase used) one may add one-tenth volume of 10% SDS, incubate 30 min at 65°, and proceed directly with Step 8. Insufficient conversion to spheroplasts will decrease the overall yield of DNA and sometimes leads to an overrepresentation of the 2-μm plasmid. The glucanase is active between pH 6 and 9 with a pH optimum of 7.5. It can be inactivated by a 5-min treatment at 60°. The glucanase preparation usually contains protease activity and traces of phosphatase activity but no DNase or RNase. The glucan cell wall of many

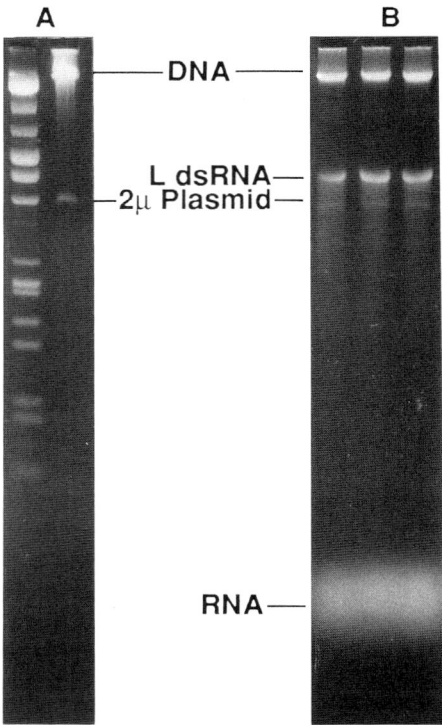

FIG. 1. Separation of *S. cerevisiae* nucleic acids by electrophoresis in 0.7% agarose gels. (A) aliquot of purified DNA (Step 15) run next to a mixture of λ *Hin*dIII and λ *Hin*dIII-*Eco*RI fragments. Sizes, from top to bottom are as follows: 23.7 and 21.2 kb (one band), 9.42 kb, 6.56 kb, 5.15 and 4.97 kb (one band), 4.36 and 4.27 kb (one band), 3.53 kb, 2.97 kb (faint partial band), 2.32 kb, 2.03 kb, 1.90 kb, 1.58 kb, 1.37 kb, 0.95 kb, 0.83 kb, and 0.56 kb. This *S. cerevisiae* DNA preparation shows traces of fragments in the 10–25 kb range. In most preparations the nuclear DNA does not contain such small fragments, and the DNA band resembles that of B. (B) DNA and RNA precipitated from the supernatant of Step 9 with half, equal, and double (left to right) the volume of ethanol. After a short spin the pellets were dissolved and aliquots were run on the same gels as the DNA samples of A. The RNA is often less degraded at this step and, therefore, stains more intensely with ethidium bromide than shown here. Sometimes intact 18 S rRNA (1.75 kb) and 25 S rRNA (3.4 kb) are visible as strong bands.

yeasts including *Ashbya, Candida, Hansenula, Kluyveromyces, Lipomyces, Pichia, Torulopsis, Saccharomycopsis,* and *Saccharomyces* can be degraded.

Step 7. It is essential to obtain a homogeneous and not too concentrated suspension of spheroplasts before the SDS is added.

Step 8. The slightly turbid solution contains in addition to the nucleic acids all soluble proteins, polysaccharides, lipids, and small molecular weight metabolites. Several procedures employ a treatment with 100 µg proteinase K at this step followed by repeated extractions with phenol, chloroform, and isoamyl alcohol (25:24:1). These extractions remove more material from the DNA solution than the precipitation with potassium acetate. However, the precipitation step is much more convenient, and the DNA obtained can be cleaved with all restriction endonucleases.

Step 10. Despite common belief it is not necessary to add twice the volume of ethanol. The concentration of nucleic acids is so high that even half the volume of ethanol would lead to a complete precipitation of DNA and RNA as shown in Fig. 1B. Unwanted material which may precipitate with 66% ethanol will stay in solution.

Step 11. Without removal of the salt it will take longer to dissolve the nucleic acid pellet and the double-stranded RNA may not be completely degraded by RNase (Step 13).

Step 12. The separation on an agarose gel before treatment with RNase will show whether the strain carries the very abundant 4.2-kb linear double-stranded RNA or not.[7] The intensity of this RNA band may exceed that of the chromosomal DNA in ethidium bromide-stained gels. In addition the 2-µm plasmid may be seen as a faint band (Fig. 1B). Depending on the electrophoresis conditions, the supercoiled 2-µm plasmid can comigrate with the 4.2-kb RNA.

Step 13. The RNase treatment does not remove the RNA as often assumed. It converts the RNA to mono- and oligonucleotides by cleavage at pyrimidines. This allows a satisfactory separation of DNA from the vast excess of RNA by precipitation with 2-propanol. Small RNA molecules can be also separated from DNA by passage of the nucleic acid solution through BioGel A15 (Bio-Rad, Richmond, CA).

Step 15. In standard 0.7% agarose gel electrophoresis the uncleaved DNA should run as one band slightly slower than the 23.7-kb marker fragment of λ DNA cleaved with HindIII (Fig. 1). Under these conditions DNA from 25 to over 500 kb migrates as one band. A faint fluorescence at the position of degraded RNA should not be underestimated with respect to RNA contamination since ethidium bromide intercalates only with the small fraction of RNA which is double stranded. Long-term storage at −20° is not recommended because the salt in the DNA freezes out, and the freezing point of a saturated salt solution is about that of the cycling

---

[7] R. B. Wickner, in "The Molecular Biology of the Yeast *Saccharomyces*" (J. N. Strathern, E. W. Jones, and J. R. Broach, eds.), Vol. 1, p. 415. Cold Spring Harbor Laboratory, Cold Spring Harbor, New York, 1981.

temperature of $-20°$ freezers. Frequent freezing and thawing should be avoided.[8] (Reference 8 also contains further valuable information for storing and processing DNA.)

Alternative Small-Scale Isolation Procedure

A very attractive modification for DNA isolations employs embedding of the washed cells in small blocks or beads of 1% low melting point agarose together with the glucanase.[2,9-11] The spheroplasts are lysed in agarose by addition of SDS or sarcosyl followed by a treatment with proteinase K. The embedding in agarose allows separation of the immobile chromosomal DNA from other cellular material simply by washing the agarose particles. The DNA can be cleaved with restriction nucleases and also ligated to vector DNA in the solid or remolten agarose. Agarose without inhibitors for DNA-modifying enzymes is offered by several manufacturers (e.g., FMC BioProducts, Rockland, ME).

Isolation and Fractionation of Yeast DNA by Cesium Chloride Gradients

Equilibrium density gradients in CsCl yield exceptionally clean DNA. Treat cells from 300-ml cultures as described in the basic protocol up to Step 8. (Multiply volumes and ingredients by 10.) Extract twice with 10 ml phenol/chloroform/isoamyl alcohol (25:24:1), adjust the aqueous phase to 0.5 M NaCl, and add the same volume of ethanol. Spin, wash the pellet in 50% ethanol, and dissolve the nucleic acids in 3 ml of 10 mM Tris, 1 mM EDTA, pH 7.5. Do not treat with RNase since only large RNA separates well from DNA in CsCl gradients.

For isolation of ribosomal DNA (γ-satellite) adjust the solution with CsCl to a density of 1.70 g/ml (total volume 10–15 ml) and run in a fixed-angle rotor at 200,000 g for 60–70 hr. The main nuclear DNA (40% G + C) and the 2-μm plasmid band at 1.699 g/ml, the ribosomal DNA (56% G + C) at 1.704 g/ml, and the mitochondrial DNA (10% G + C) at 1.683 g/ml.[5] Although the rDNA cluster is part of chromosome XII, this DNA bands separately because of its slightly higher G + C content and because the chromosomal DNA is fragmented. For better separation of mitochondrial DNA from nuclear DNA, add bisbenzimide to the CsCl gradients.[4] For isolation of 2-μm plasmid DNA add ethidium bromide (final concentration 0.5 mg/ml) and adjust the solution with CsCl to 1.55

[8] R. W. Davis, D. Botstein, and J. R. Roth, *in* "Advanced Bacterial Genetics," p. 211. Cold Spring Harbor Laboratory, Cold Spring Harbor, New York, 1980.
[9] D. C. Schwartz and C. R. Cantor, *Cell (Cambridge, Mass.)* **37**, 67 (1981).
[10] D. A. Jackson and P. R. Cook, *EMBO J.* **4**, 913 (1985).
[11] M. N. K. Linskens and J. A. Huberman, *Mol. Cell. Biol.* **8**, 4927 (1988).

g/ml. Run in swinging-bucket rotor at 150,000 g for 48-72 hr, isolate the plasmid (lower band) by side puncture with a syringe, and treat the DNA solution as in standard plasmid isolation procedures. It is always advisable to run the isolated DNA a second time in a CsCl gradient in order to achieve complete separation from RNA. Although RNA bands at higher density than DNA, traces of tRNA and small RNA fragments are still present in the DNA band after the first run.

Correlation between Bands in Restriction Spectra and Repeated DNA Sequences

Approximately 85% of the sequences in yeast chromosomes are unique. Some genes exist in pairs, for example, genes for histones, α-tubulin, translation elongation factor 1α, mating pheromones, several ribosomal proteins, and some glycolytic enzymes. Up to 15% of the nuclear DNA consists of repeated sequences, some of which can be seen as weak or strong bands after cleavage of the DNA with restriction endonucleases and separation of the fragments in agarose gels. These restriction spectra are specific not only for individual restriction endonucleases but also for the genomes of different *Saccharomyces* species. Interpretation of the band patterns can be quite useful. First, the extent of cleavage (partial or complete) can be deduced for many restriction endonucleases. Second, the average copy number of newly introduced plasmids can be estimated. Third, restriction site polymorphisms in repeated DNA can be detected.

Figure 2 summarizes the sequence organization and the location of restriction sites for the four classes of repeated DNA (rDNA, 2-μm plasmid, Ty elements, telomeric Y' sequences) which are the origin of most bands seen in restriction spectra. The ribosomal DNA consists of a cluster of usually 100 to 120 tandem copies of a 9.08-kb repeat unit.[12] Cell clones with higher and lower copy number have been observed.[13] The 2-μm plasmid is present in 50 to 100 copies in many but not all *S. cerevisiae* strains. We tested 50 domesticated and wild strains as well as laboratory strains and found 2-μm plasmids in 75% of the strains, occasionally with small deletions or loss of restriction sites. The two sequence arrangements shown for the 2-μm plasmid originate from recombination events between the two inverted repeated sequences (IR).[3] Another class of repeated DNA are the mobile elements Ty1, Ty2, Ty3, and Ty4.[14-16] Ty1 and Ty2 are

[12] J. Warner, *Microbiol. Rev.* **53,** 256 (1989).
[13] G. F. Carle and M. V. Olson, personal communication.
[14] J. D. Boeke, in "Mobile DNA" (D. E. Berg and M. M. Howe, eds.), p. 335. American Society for Microbiology, Washington, D.C., 1989.

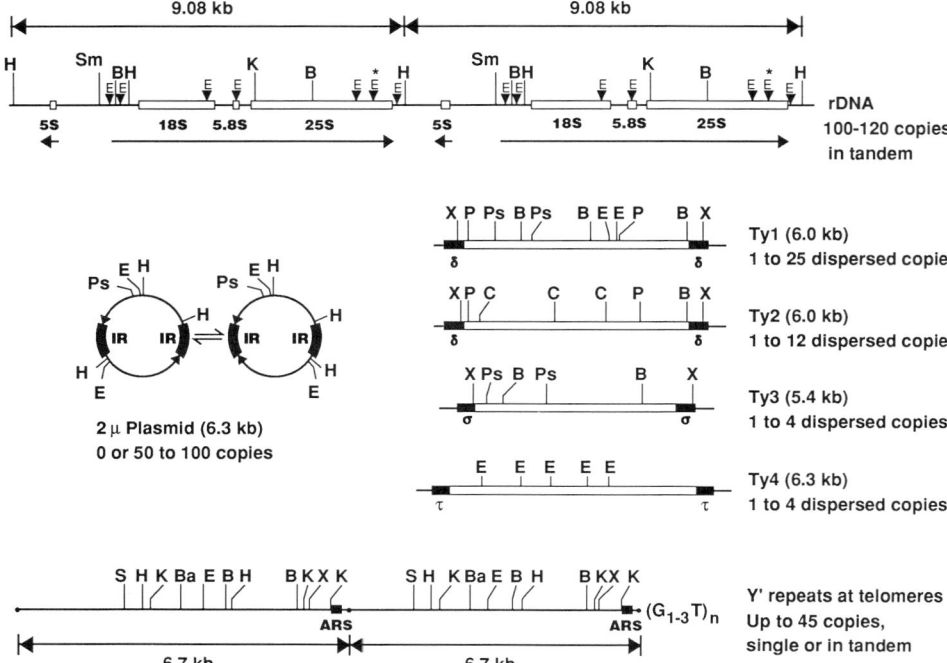

FIG. 2. Restriction maps of repeated nuclear DNA in *S. cerevisiae*. For the ribosomal DNA the arrangement of the rRNA coding sequences is shown. Arrows underneath the two repeat units represent the direction of transcription for the 5 S rRNA and the 35 S precursor rRNA. Arrows in the 2-μm plasmid indicate the relative orientation of the unique sequences (2.77 kb upper part and 2.36 kb lower part) between the 0.60-kb inverted repeats [J. L. Hartley and J. E. Donelson, *Nature (London)* **286**, 860 (1980)]. The terminal direct repeats of Ty elements are shown as black bars. The black bars in the two telomeric Y' repeats mark the position of autonomously replicating sequences. The end of the chromosomes is determined by the repeated $G_{1-3}T$ sequence motif [J. J. Sahampay, J. W. Szostak, and E. H. Blackburn, *Nature (London)* **310**, 154 (1984)]. Abbreviations for restriction sites are as follows: B for *Bgl*II, Ba for *Bam*HI, C for *Cla*I, E for *Eco*RI, H for *Hin*dIII, K for *Kpn*I, P for *Pvu*II, Ps for *Pst*I, and X for *Xho*I.

closely related. Laboratory strains used in genetic studies usually carry 30 to 35 copies dispersed in the genome. However, we have found natural isolates carrying just one or a few copies. Ty3 and Ty4 are present only in 1 to 4 copies.[15,16] Ty elements can be lost from the chromosome by homolo-

---

[15] D. J. Clark, V. W. Bilanchone, L. J. Haywood, S. L. Dildine, and S. B. Sandmeyer, *J. Biol. Chem.* **263**, 1413 (1988).
[16] R. Stucka, H. Lochmüller, and H. Feldmann, *Nucleic Acids Res.* **17**, 4993 (1989).

gous recombination between their 0.34-kb terminal direct repeats. Single copies of the terminal repeats remain at the previous location of the Ty element.[14] This explains the many copies of solo δ sequences (50–100), solo σ sequences (20–30), and solo τ sequences (15–25) found in all strains even in those with low Ty copy number. The fourth class of repeated DNA visible in restriction spectra consists of the 6.7-kb telomeric Y' sequences.[17] These are present as single copies or short clusters of 2 to 4 copies at the ends of almost all 16 chromosomes in *S. cerevisiae* laboratory strains, with about 45 copies per haploid genome. Some natural isolates of *S. cerevisiae* carry very few copies of the Y' sequence.[18]

Figure 3 shows examples of restriction spectra obtained with DNA of strain S288C, a strain commonly used in molecular and genetic studies. The DNA was isolated according to the procedure described above and was cleaved with five restriction endonucleases at two concentrations each. The *Eco*RI spectrum is dominated by ribosomal DNA fragments (R).[19] Most *S. cerevisiae* strains carry seven *Eco*RI sites in their rDNA. The site marked with an asterisk in the map of Fig. 2 is cleaved extremely slowly. Since the corresponding partial fragment of 0.94 kb (R*) is still visible in the spectrum, the cleavage was not quite complete. The second largest rDNA band (2.4 kb) varies slightly in size among *S. cerevisiae* strains owing to small deletions and insertions in the nontranscribed region adjacent to the 5 S rRNA gene.[20] In some strains this fragment is fused to the 0.59 kb fragment because the *Eco*RI site 3' of the 25 S rRNA coding sequence is missing (see examples in Fig. 4). The four *Eco*RI fragments originating from the two forms of 2-μm plasmids also are visible (P).[21] One of these comigrates with the 2.4-kb rDNA fragment.

The two strong bands in the *Hin*dIII spectrum at 6.4 and 2.7 kb originate from rDNA. The DNA was cleaved to completion since no partial band at 9.1 kb is seen. Five bands with 2-μm plasmid DNA also are visible.[21] Each of the two plasmid isomeres carries three *Hin*dIII sites. Six fragments are generated, two of which are identical, yielding the slightly more intense band at 1.3 kb. Y' sequences contain two *Hin*dIII sites 2.1 kb apart. The corresponding band runs close to the 2.2-kb plasmid band. The

---

[17] C. S. M. Chan and B. K. Tye, *Cell (Cambridge, Mass.)* **33,** 563 (1983).
[18] D. Jäger and P. Philippsen, *Mol. Cell. Biol.* **9,** 5754 (1989).
[19] Sizes of the rDNA *Eco*RI fragments are 2.79 (25 S rRNA), 2.46 (spacer and 5 S rRNA), 2.02 (18 S rRNA), 0.66 (18 S rRNA), 0.59 (25 S rRNA), 0.35 (25 S rRNA), and 0.22 kb (promoter region).
[20] R. Jemtland, E. Maehlum, O. S. Gabrielsen, and T. B. Oyen, *Nucleic Acids Res.* **14,** 5145 (1986).
[21] The *Eco*RI fragments of the 2-μm plasmid are 4.1, 3.9, 2.4, and 2.2 kb. The *Hin*dIII fragments of this plasmid are 4.1, 2.8, 2.2, 1.3, and 0.9 kb.

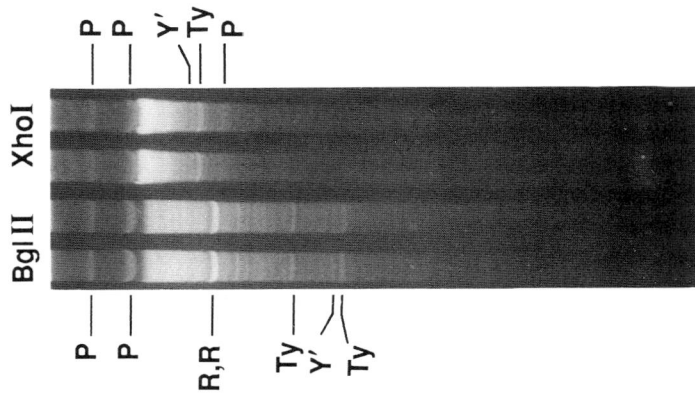
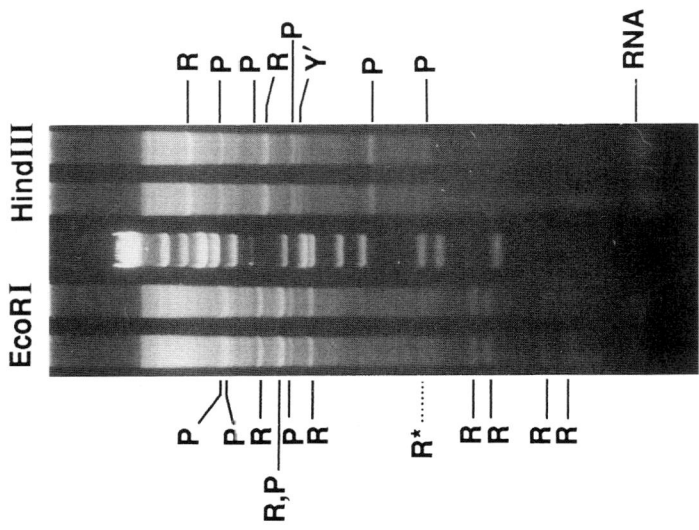

*Bgl*II spectrum shows one strong band at 4.55 kb which contains two rDNA fragments of equal size. The cleavage is complete since no partial band at 9.1 kb is seen. The 2-μm plasmid is not cleaved with *Bgl*II, and the open circles of plasmid monomers and plasmid dimers are visible as slowly migrating bands between the sample pocket and the largest linear DNA. Ty1 elements carry three *Bgl*II sites, 1.65 and 2.25 kb apart. The corresponding fragments are seen as weak bands. Another weak band originates from the 1.7-kb *Bgl*II fragment in Y' sequences.

The *Xho*I and *Bam*HI spectra do not contain strong bands since neither enzyme cleaves in the rDNA or in the 2-μm plasmid. Uncleaved supercoiled and open circle forms of 2-μm plasmids are seen. The most prominent band (5.6 kb) in the *Xho*I spectrum originates from cleavage in the δ sequences of Ty1 and Ty2 elements. The intensity corresponds to 25 or 30 copies, taking into account that some of the approximately 35 Ty elements in this strain lack *Xho*I sites. The very weak band above the Ty band most likely represents the number of tandemly repeated Y' sequences. *Xho*I cleaves once in this sequence, as does *Bam*HI. Indeed a band is present at the same position in the *Xho*I and the *Bam*HI spectrum. No Ty-specific repeated fragments are expected to be seen with *Bam*HI since this enzyme cleaves only a few Ty elements and in each case known only once.

A few clearly visible bands were left unassigned. Some of these may originate from mitochondrial DNA. It is generally believed that this DNA becomes much more fragmented during the isolation procedure than nuclear DNA, and if bands carry mitochondrial DNA fragments their intensities may therefore not reflect the number of mitochondrial genomes. Some unassigned bands could represent restriction site variants of known repeated sequences. There is ample of evidence for such variations in Ty elements and Y' sequences.[14,17] Other repeated DNA like tRNA genes (up to 12 dispersed copies for each of the 46 types of tRNAs) or the single δ, σ, and τ sequences are too short to account for bands in the restriction spectra of Fig. 3.

The restriction spectra of *Saccharomyces* species other than *S. cerevisiae* differ from those shown in Fig. 3. This is predominantly due to the

---

FIG. 3. Restriction spectra of DNA from *S. cerevisiae* strain S288C. About 1 μg of DNA was cleaved overnight with 1 and 5 units of the restriction nucleases indicated. The fragments were separated in a 0.7% agarose gel and stained with 1 μg/ml ethidium bromide. The gel photo was cut into three parts to allow assignment of bands originating from rDNA (R), 2-μm plasmids (P), *Ty* elements (Ty), or Y' sequences (Y'). The size marker lane between the *Eco*RI and *Hin*dIII spectra is the same as in Fig. 1. The DNA in the left lane of the *Bam*HI spectrum appears uncleaved; probably much less than 1 unit *Bam*HI was added. Further details are discussed in the text.

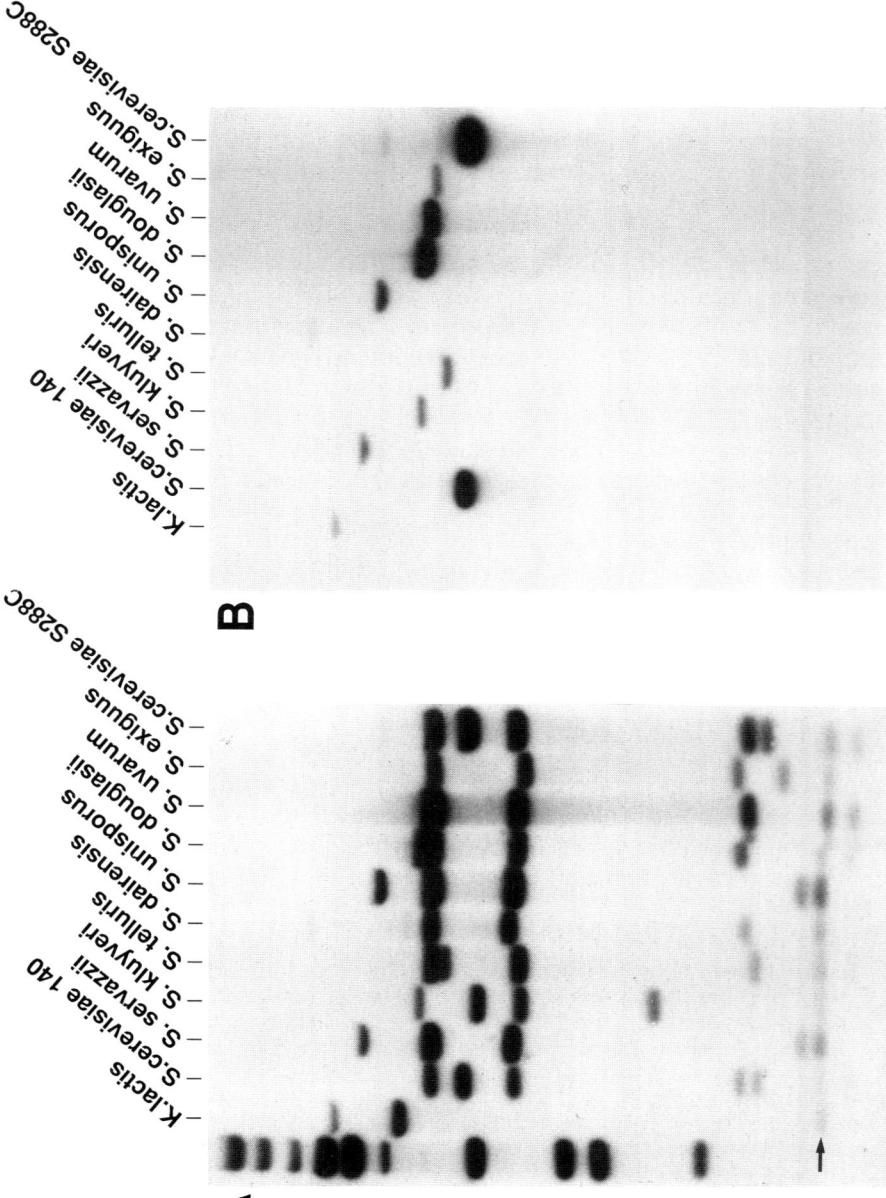

expected sequence divergence among the genomes, which also affects restriction sites in repeated DNA. In addition, repeated sequences not essential for the life cycle (2-μm plasmids, Ty elements, and Y' sequences) may be absent. The strongest bands in *Eco*RI restriction spectra represent almost exclusively rDNA, as concluded from hybridization experiments. Examples are shown in Fig. 4 for the proposed seven *Saccharomyces* species[22] and a few other strains. In order to identify regions of high sequence conservation or high sequence divergence in the rDNA, the *Eco*RI spectra were hybridized with the total rDNA repeat unit of *S. cerevisiae* (Fig. 4A) and a 0.6-kb fragment carrying the 5 S rRNA gene and part of the nontranscribed spacer (Fig. 4B). Despite the variations all strains, even *Kluyveromyces lactis,* carry the two *Eco*RI sites close to the 3'-end of the 25 S rRNA coding sequence which generate the 0.35-kb fragment marked by the arrow. This extremely conserved region most likely plays an important role in ribosome assembly or function. The sequence of the spacer region which is represented by the second largest *Eco*RI fragment in *S. cerevisiae* rDNA (see Fig. 2) is not at all conserved, as evident from variations in length and intensity of those bands which hybridize to the spacer probe (Fig. 4B). Still, the weakly hybridizing bands are strong bands in the restriction spectra. Similar comparisons of single-copy sequences in these *Saccharomyces* species often show no or only weak signals under stringent hybridization conditions, indicating that the average sequence divergence is 30% or more. This is in agreement with DNA renaturation studies performed before the tools of recombinant DNA techniques became available.[23] When conserved genes are used as

[22] J. A. Barnett, R. W. Payne, and D. Yarrow, "Yeast: Characteristics and Identification." Cambridge Univ. Press, Cambridge, 1983.
[23] J. N. Bicknell and H. C. Douglas, *J. Bacteriol.* **101,** 505 (1970).

FIG. 4. *Eco*RI fragments from ribosomal DNA of *Saccharomyces* and *Kluyveromyces* strains. About 1 μg total DNA was cleaved with *Eco*RI and separated on a 0.7% agarose gel. The fragments were denatured and blotted in both directions onto nitrocellulose. The left filter was hybridized with λgt21-Sc310 carrying one *S. cerevisiae* rDNA repeat [P. Philippsen, M. Thomas, R. A. Kramer, and R. W. Davis, *J. Mol. Biol.* **123,** 387 (1978)]. The right filter was hybridized with a subclone of λgt21-Sc310 which carries 0.6 kb of the 5 S rRNA gene and adjacent DNA. The size marker lane at left is the same as in Fig. 1; however, only fragments with homology to λgt21 are seen. The slowest migrating band of *S. telluris, S. douglasii, S. uvarum,* and *S. exiguus* rDNA represents two fragments, one from the 25 S rRNA coding region and the other from the spacer region. The strains were obtained from the following sources: S288C from the Yeast Genetics Stock Center, Berkeley, CA; *S. douglasii* from Don Hawthorne, Seattle, WA; all other strains from the Centraalbureau voor Schimmelcultures, Delft, Holland.

DNA probes (e.g., rDNA or the gene for the translation elongation factor $1\alpha^{24}$), hybridization signals are seen with DNA of all *Saccharomyces* species and even other yeast genera.

Acknowledgments

We thank Sue Klapholz, Jürg Gafner, and Hans Hegemann for useful comments and Valerie Lui for typing the manuscript. This work was supported by the Swiss National Science Foundation and by the Deutsche Forschungsgemeinschaft (SFB 272).

[24] F. Schirmaier and P. Philippsen, *EMBO J.* **3**, 3311 (1984).

[12] High-Efficiency Transformation of Yeast by Electroporation

By DANIEL M. BECKER and LEONARD GUARENTE

Introduction

A prerequisite for molecular biological manipulation of any organism is a reliable and efficient means for introducing exogenous DNA into the cell. Yet each of the techniques in general use for transforming yeast, namely, lithium acetate transformation[1] and spheroplast transformation,[2] suffers from significant limitations. Lithium acetate transformation, although relatively fast and simple, provides only a low efficiency of DNA transfer ($\sim 10^3$ colonies/μg of episomal plasmid). Spheroplast transformation, while more efficient ($\sim 1-5 \times 10^4$ colonies/μg), is complicated and time consuming. We present here a method for transforming yeast by electroporation that is extremely simple and an order of magnitude more efficient than spheroplast transformation.

A number of groups have attempted previously to introduce macromolecules into yeast by electroporation,[3-5] but efforts to introduce plasmids have failed to yield transformation efficiencies greater than approxi-

[1] H. Ito, Y. Fukada, K. Murata, and A. Kimura, *J. Bacteriol.* **153**, 163 (1983).
[2] A. Hinnen, J. B. Hicks, and G. R. Fink, *Proc. Natl. Acad. Sci. U.S.A.* **75**, 1929 (1978).
[3] H. Hashimoto, H. Morikawa, Y. Yamada, and A. Kimura, *Appl. Microbiol. Biotechnol.* **21**, 336 (1985).
[4] I. Karube, E. Tamiya, and H. Matsuoka, *FEBS Lett.* **182**, 90 (1985).
[5] I. Uno, K. Fukami, H. Kato, T. Takenawa, and T. Ishikawa, *Nature (London)* **333**, 188 (1988).

mately $1.5 \times 10^3/\mu g$ of DNA. The low efficiency is surprising, as 35–75% of yeast cells in a population can be demonstrated to take up macromolecules after an electric pulse.[6] This paradox suggested to us that the failure might be attributable, at least in part, to inadequate stabilization of the membrane of cells that had in fact been permeabilized and transformed. Further, recent publication of a protocol for high-efficiency electroporation of bacteria[7] suggested solutions to other problems inherent in the electroporation of organisms as small as yeast. We have, therefore, designed our protocol by adapting the principles of bacterial electroporation to transformation of *Saccharomyces cerevisiae,* taking care, in addition, to provide continuous osmotic support of the electrically compromised cells.

Preparation of Electrocompetent Cells

1. Inoculate 500 ml YPD in a 2-liter flask with an aliquot from an overnight culture. Grow with vigorous shaking at 30° to an OD_{600} of 1.3–1.5.

We usually start two to four cultures simultaneously with inocula of different volumes the night before the transformation, in order to guarantee one culture at the proper density at a reasonable time the next morning. With the strain that we use most commonly, BWG1-7a (*MATa leu2-3,112 ura3-52 his4-519 ade1-100*), an optical density of 1.3–1.5 at 600 nm corresponds to a density of approximately 1×10^8 cells/ml. The growth phase of the culture is extremely important: testing parallel cultures at optical densities varying over a 5-fold range (0.245 to 1.3), we observe a 60-fold increase in transformation efficiency ($6.5 \times 10^3/\mu g$ to $3.8 \times 10^5/\mu g$). The overnight culture need not be fresh, and the culture can also be started with an inoculum directly from a plate.

2. Divide the culture into two 250-ml centrifuge bottles and spin at 5000 rpm for 5 min at 4° in a Sorvall GSA rotor. Discard the supernatant.

This and the subsequent centrifugations achieve two purposes: they concentrate the cells 500- to 1000-fold and reduce the conductivity of the culture dramatically. The exact volumes, rotors, and centrifugation times are not critical as long as both of the goals are met. It is important, however, to keep the culture and all solutions cold.

3. Resuspend in a total of 500 ml ice-cold sterile water.

If the culture is divided in half, as suggested, resuspend each bottle in

---

[6] J. C. Weaver, G. I. Harrison, J. G. Bliss, J. R. Mourant, and K. T. Powell, *FEBS Lett.* **229**, 30 (1988).
[7] W. J. Dower, J. F. Miller, and C. W. Ragsdale, *Nucleic Acids Res.* **16**, 6127 (1988).

250 ml. Resuspension is most easily accomplished by vortexing with about 100 ml, then adding the remainder of the water and shaking the bottle.

4. Centrifuge at 5000 rpm for 5 min at 4° and discard the supernatant.
5. Resuspend in a total of 250 ml ice-cold sterile water.

At this step, we pool the two aliquots (125 ml each) of the original culture into a single bottle.

6. Centrifuge as above, and discard the supernatant.
7. Resuspend in 20 ml ice-cold 1 M sorbitol (182 g/liter in sterile, distilled water). Transfer to a chilled 30-ml centrifuge tube.

We vortex initially, then complete the resuspension by pipetting up and down in the sterile 25-ml pipette used to transfer the culture to the smaller centrifuge tube.

8. Spin at 5000 rpm for 5 min at 4° in a Sorvall SS34 rotor. Discard the supernatant.
9. Resuspend by adding 0.5 ml ice-cold 1 M sorbitol. Store on ice.

We resuspend by pipetting up and down in a 1-ml serologic pipette. The final volume varies from about 1 to 1.5 ml.

Electroporation

1. Aliquot 40 μl of yeast suspension per transformation to a sterile Eppendorf tube. Add $\leq$ 100 ng DNA in at most 5 μl to the tube, mix gently, and incubate on ice about 5 min.

Increasing the volume of yeast from 40 to 160 μl has no effect on the number of transformants recovered, suggesting that the number of transformation events under these conditions is not limited by the number of electrocompetent yeast in the reaction. Further, these data provide additional evidence that the increase in efficiency observed with increasing optical density is a function not of cell density per se, but of the growth phase of the culture. Note, too, that none of the above preparative steps disrupts the yeast cell wall; electroporation of spheroplasts prepared with Glusulase and plated in top agar provides efficiencies no better than those obtained with intact yeast with the procedure described here. The DNA should be in a low ionic strength buffer such as TE (10 mM Tris-HCl, 1 mM EDTA, pH 8.0) and must be in as small a volume as possible. Addition of carrier DNA reduces the transformation efficiency. Incubation time can be varied to convenience.

2. Transfer to a cold 0.2-cm sterile electroporation cuvette. Tap to the bottom.
3. Pulse at 1.5 kV, 25 μF, 200 Ω.

We use a commercial apparatus with commercially available cuvettes [Bio Rad Gene Pulser with Pulse Controller, with Bio Rad 0.2-cm cuvettes (Richmond, CA)]. This device discharges an exponential pulse through the cuvette. With electrodes separated by 0.2 cm and the initial voltage set to 1.5 kV, the initial field strength E_o is 7.5 kV/cm. With the resistance of the yeast/DNA mixture high, the time constant for decay is determined almost exclusively by the capacitance and resistance set in parallel, and it varies slightly around the theoretical value of 5 msec. We have not tested devices from other manufacturers, but we expect that any device which can reproduce these electric parameters should suffice.

4. Immediately add 1 ml cold 1 M sorbitol to the cuvette. Using a sterile Pasteur pipette, gently mix the contents of the cuvette and transfer to a culture tube.

We have found that including other components with the sorbitol, such as diluted YPD or $CaCl_2$, reduces the efficiency.

Plating

1. Plate aliquots of the transformation by spreading on selective plates containing 1 M sorbitol.

No incubation is required after resuspension and transfer of the reaction from the cuvette; we plate as soon as possible after electroporation. The manner in which the transformed yeast are plated is crucial to the transformation efficiency. The data in Table I are from an experiment in which two transformations (A and B) are performed in parallel using aliquots from the same preparation of electrocompetent cells. Each aliquot is transformed with 100 ng of plasmid, and portions of the transformed cells are plated in duplicate using a variety of techniques. Numbers represent averages of duplicate platings and are expressed as transformants per microgram.

TABLE I
REPLICA PLATING

Composition	Transformants
Top agar lacking sorbitol on a plate lacking sorbitol	A 1.5×10^2
	B 6.0×10^2
Top agar lacking sorbitol on a plate containing sorbitol	A 1.1×10^3
	B 2.9×10^3
Top agar containing sorbitol on a plate containing sorbitol	A 1.3×10^5
	B 4.5×10^4
Spread on a plate lacking sorbitol	A 2.0×10^4
	B 2.6×10^4
Spread on a plate containing sorbitol	A $> 3.0 \times 10^5$[a]
	B $> 3.0 \times 10^5$[a]

[a] The density of plating precludes more accurate quantitation.

Transformation of yeast by this procedure involves a minimum of manipulations. Preparation of cells in advance of electroporation involves four 5-min centrifugations. There is minimal preincubation with DNA and no carrier nucleic acid, and the pulse itself takes only a moment. Subsequent outgrowth is not required, and plating by spreading is sufficient to provide maximal efficiency. Using this procedure, we routinely obtain efficiencies of $2-5 \times 10^5$ colonies/μg using episomal plasmids of various sizes. This is a 10-fold higher efficiency than that which we observe, with the same plasmids, using spheroplast transformation. Although BWG1-7a and its derivatives are known to transform relatively well by all techniques, comparable results have been obtained using other strains, including petites.

We note, however, decreasing transformation efficiencies with increases of input DNA above 100 ng, limiting the utility of this protocol in certain circumstances, so we include below the protocols used in our laboratory for transformation of yeast by lithium acetate (based on Ref. 1) and by spheroplasting (based on Ref. 2).

Lithium Acetate Transformation

1. Inoculate 100 ml YPD and grow with vigorous shaking at 30° to an OD_{600} of 1.0-2.0.
2. Spin at 3000 rpm for 5 min. Resuspend in 5 ml TE (10 mM Tris-HCl, pH 7.5, 1 mM EDTA). Repeat the spin twice more, resuspending first in 5 ml of TE containing 100 mM lithium acetate, then in 1 ml TE/lithium acetate.
3. Shake at 30° for 60 min.
4. Add 400 μg carrier (calf thymus) DNA.
5. Aliquot 100 μl per transformation to sterile Eppendorf tube.
6. Add 1 μg transforming DNA.
7. Incubate for 30 min, 30°.
8. Add 700 μl PEG solution [35% (w/v) PEG 4000, 100 mM lithium acetate, TE, pH 7.5].
9. Incubate for 50 min at 30°.
10. Heat shock for 5 min at 42°.
11. Pellet 4 sec. Aspirate supernatant. Wash with 500 μl TE. Resuspend in 100 μl TE. Plate on selective media.

Spheroplast Transformation

1. Inoculate 100 ml YPD in a 2-liter flask with an aliquot from an overnight culture. Grow with vigorous shaking at 30° to an OD_{600} of 0.6-0.8.

2. Spin at 3000 rpm for 5 min at room temperature. Decant supernatant, and resuspend in 5 ml of 1.2 M sorbitol. Spin again, resuspending in 5 ml of 1.2 M sorbitol.

3. Add 50 μl Glusulase (Du Pont, Wilmington, DE).

4. Incubate at 30° for 1 hr, swirling gently every 10 min.

It is possible at this point to make a visual check of the effectiveness of spheroplasting by diluting a small volume of Glusulase-treated cells ($\sim$ 5 μl) into a drop of 5% SDS on a microscope slide and observing the formation of ghosts. We routinely proceed without this check.

5. Rinse spheroplasts 4 times with 5 ml of 1.2 M sorbitol, spinning at 3000 rpm, room temperature, each time.

6. Resuspend the final pellet in 0.6 ml STC (1 M sorbitol, 10 mM Tris-HCl, pH 7.5, 10 mM CaCl$_2$).

7. Aliquot 100 μl per transformation tube (Falcon 2059). Add 1 μg transforming DNA and 9 μg carrier (calf thymus) DNA.

8. Incubate for 5 min at room temperature.

9. Add 4 ml PEG solution [20% (w/v) PEG 4000, 10 mM CaCl$_2$, 10 mM Tris-HCl, pH 7.4]. Mix gently.

10. Incubate for 10 min at room temperature.

11. Pellet in a tabletop centrifuge for 5 min at room temperature. Resuspend in 150 μl SOS (1 M sorbitol, 6.5 mM CaCl$_2$, $\frac{1}{3}$ strength YPD, $\frac{1}{3}$ strength auxotrophy supplement).

12. Incubate at 30° for 20–60 min.

13. Plate in 5–6 ml regeneration top agar (1 × SD, 1.2 M sorbitol, 2% agar, auxotrophy supplements) on a plate containing 1.2 M sorbitol.

Acknowledgments

This work was supported by grants from the National Institutes of Health (to L.G.) and by a postdoctoral fellowship from the American Cancer Society (to D.M.B.).

[13] Transmission of Plasmid DNA to Yeast by Conjugation with Bacteria

By Jack A. Heinemann and George F. Sprague, Jr.

Introduction

It has been known since the 1940s that genetic information can be transmitted from one bacterial cell to another by a process termed conjugation.[1] The donor cell in a conjugation reaction harbors a conjugative plasmid that mediates the process (for review, see Refs. 2 and 3). This plasmid encodes functions that promote its transfer to the recipient and sometimes promote the transfer of other plasmids which happen to be present in the donor. Recently we have discovered that *Escherichia coli* cells harboring a conjugative plasmid have an unexpected property — they can conjugate with the yeast *Saccharomyces cerevisiae*.[4]

Conjugation between *E. coli* and *S. cerevisiae* can be exploited as a quick and easy method to introduce plasmid DNA into yeast. This method should prove useful and practical because most molecular cloning is first done in *E. coli* and because bacteria mate with yeast at high frequencies. In this chapter, we first describe the functions encoded by conjugative plasmids that are required to make a bacterial cell competent to transfer DNA and then outline two procedures that can be used to generate yeast transconjugants.

There are two types of conjugative plasmids, self-transmissible and mobilizable, which differ in the constellation of conjugation functions they encode. Self-transmissible plasmids encode a number of *tra* genes, the products of which promote DNA transmission from donor to recipient. The *tra* genes can be divided into two broad groups. One group specifies proteins involved in cell surface interactions and membrane pore formation between donor and recipient. The other group, referred to as *mob* genes, specifies proteins involved in DNA "mobilization." DNA transfer is initiated by a *mob*-encoded strand-specific nicking activity that acts at a specific site, *oriT*, to catalyze helix unwinding. As a result, a single DNA strand is displaced, which can be transferred to a recipient cell, and a strand remains in the donor to serve as a template for replication. Thus, the *mob* products prepare the DNA for transfer whereas the *tra* products promote cell – cell contact and provide the physical context for transfer.

Mobilizable plasmids encode only a subset of the activities outlined above. Typically they rely on the *tra* functions specified by a self-transmissible plasmid to achieve transfer. However, a mobilizable plasmid must contain an *oriT* site. Because the *mob/oriT* interaction is specific, the *oriT* site must be compatible with *mob* functions present in the donor if the plasmid is to be transferred. In fact, naturally occurring mobilizable plasmids such as ColE1 encode *mob* activities tailored to their *oriT*.

The conjugation system that we have developed allows the transfer of *E. coli/S. cerevisiae* shuttle plasmids from bacterial cells to yeast cells.

[1] J. Lederberg and E. L. Tatum, *Nature (London)* **158**, 558 (1946).
[2] N. Willets and B. Wilkins, *Microbiol. Rev.* **48**, 24 (1984).
[3] C. M. Thomas and C. A. Smith, *Annu. Rev. Microbiol.* **41**, 77 (1987).
[4] J. A. Heinemann and G. F. Sprague, Jr., *Nature (London)* **340**, 205 (1989).

TABLE I
PLASMIDS

Plasmid	Conjugation function[a]	Yeast replication determinant[b]	Selectable markers[c] In bacteria	In yeast
pDPT51	tra-R, mob-R, mob-C, oriT-R	None	Tp^r, Am^r	None
YEp13	oriT-C	2 μm	Tc^r, Am^r	LEU2
YEp24	oriT-C	2 μm	Tc^r, Am^r	URA3
YCp50	oriT-C	Centromeric	Tc^r, Am^r	URA3
YIp5	oriT-C	None; integrative plasmid	Tc^r, Am^r	URA3

[a] pDPT51 is a self-transmissible plasmid of the *incP* group, encoding all cis- and trans-acting functions specific to its transmission between cells (*tra*-R, *mob*-R, *oriT*-R, where the suffix R indicates that the origin of the conjugation functions is plasmid R751, the parent of pDPT51) and trans-acting functions specific for the mobilization of pBR322 derivatives (*mob*-C, where C indicates that the *mob* functions are derived from the ColE1 plasmid). The yeast shuttle vectors are each derivatives of pBR322 and possess the cis-acting function essential for mobilization (*oriT*-C).

[b] 2 μm indicates that the shuttle vector contains sequences from the endogenous yeast 2 μm plasmid that allow the shuttle vector to replicate independently of the host chromosomes by the same mechanism as 2 μm plasmids. Centromeric indicates that the shuttle vector contains an autonomous replication sequence (*ARS*) and a chromosomal centromere sequence (*CEN*), which together allow replication and segregation of plasmids to mother and daughter cells during mitosis. Integrative plasmid indicates that the shuttle vector contains no sequences that allow for establishment as an extrachromosomal element.

[c] Tp^r indicates that bacteria harboring the plasmid are resistant to trimethoprim, Am^r indicates resistance to ampicillin, and Tc^r indicates resistance to tetracycline. In addition, some plasmids contain the yeast *LEU2* or *URA3* genes. These plasmids can be selected in yeast because they complement nutritional auxotrophies conferred by mutations in the chromosomal *LEU2* or *URA3* genes.

Genetic experiments demonstrate that all three classes of conjugation functions, *tra*, *mob*, and *oriT*, are required for conjugation between bacteria and yeast.[4,5] Therefore, the donor bacterial strain must harbor a self-transmissible plasmid as well as the mobilizable shuttle plasmid of interest. The procedures described below were developed for shuttle plasmids that rely on the *mob*/*oriT* system from ColE1. The *Hha*I restriction endonuclease fragment of pBR322 demarcated by nucleotides 2212 and 2353[6] contains the *oriT* sequence of the plasmid pMB1, a relative of ColE1. Plasmids such as YEp13, YEp24, YCp50, and YIp5[7–10] (Table I), which retain this

[5] J. A. Heinemann and G. F. Sprague, Jr., unpublished (1989).
[6] J. Finnegan and D. Sherratt, *Mol. Gen. Genet.* **185,** 344 (1982).
[7] J. R. Broach, J. N. Strathern, and J. B. Hicks, *Gene* **8,** 121 (1979).
[8] D. Botstein, S. C. Falco, S. E. Stewart, M. Brennan, S. Scherer, D. T. Stinchcomb, K. Struhl, and R. W. Davis, *Gene* **8,** 17 (1979).

fragment, can therefore be mobilized by the *mob* functions of ColE1. (Note: pUC-based plasmids are deleted for a portion of this *Hha*I fragment and presumably are mobilized poorly, if at all, but we have not tested such plasmids in our system.) Alternatively, one could use a shuttle plasmid that relies on a different *mob/oriT* mobilization system.[4] The frequency of DNA transmission to yeast appears to be inversely proportional to the size of the molecule intended for transfer.[4] Therefore, the shuttle vector should consist of only relevant material: markers and replication sequences for selection in both yeast and bacteria, the gene of interest, and *oriT*.

Two different procedures have been used to generate yeast transconjugants. One procedure allows the direct isolation of transconjugants by performing conjugation on medium selective for their growth (Procedure 1). The second procedure allows conjugation to take place under conditions permissive for the growth of both donor and recipient (Procedure 2). Transconjugants are identified by subsequent plating on selective medium. Before describing the procedures in detail, we first summarize the behavior of representative shuttle plasmids in the conjugation protocols.

The transmission frequencies observed using both methods and using four distinct shuttle plasmids are shown in Table II. These plasmids differ in their mode of replication, in copy number, and in selectable markers. YEp13 is transmitted at the highest frequency, especially when Procedure 1 (direct selection) is used. YEp13 is a multicopy plasmid that contains *LEU2* as the selectable marker for yeast and is replicated by virtue of an origin of replication derived from the yeast 2-μm plasmid. YEp24 contains *URA3* and the 2-μm origin. It is transmitted less efficiently than YEp13 when Procedure 1 is used but is transmitted at essentially the same frequency when conjugation is allowed to occur on a permissive medium (Procedure 2). One possible explanation for this difference is to suppose that cells starved for uracil are less able to establish prototrophy by complementation than cells starved for leucine, a deficiency partially alleviated by preincubation on YEPD medium.

The third plasmid tested, YCp50, contains the *URA3* gene and an *ARS* sequence for replication. It is maintained as a single copy in yeast because it also harbors a centromere *(CEN)* sequence. This plasmid is transmitted poorly by both procedures. However, YCp50 is also transmitted less frequently than YEp24 and YEp13 in conjugation between bacteria, suggesting that the low frequency seen in a bacterium by yeast conjugation is a reflection of a defect in mobilization of the plasmid rather than a defect in establishing the plasmid in yeast. The *CEN ARS* sequences present in

---

[9] M. Johnston and R. W. Davis, *Mol. Cell. Biol.* **4**, 1440 (1984).
[10] K. Struhl, D. T. Stinchcomb, S. Scherer, and R. W. Davis, *Proc. Natl. Acad. Sci. U.S.A.* **76**, 1035 (1979).

TABLE II
FREQUENCY OF TRANSMISSION OF DIFFERENT SHUTTLE VECTORS BY pDPT51
IN BACTERIA × YEAST CROSSES

Plasmid	Yeast replication determinant[a]	Selectable marker[b]	Procedure 1: Transconjugant cells per donor[c] (per recipient)	Procedure 2: Transconjugant cells per recipient[d]
YEp13	2 μm	LEU2	1×10^{-3} (2×10^{-6})	1×10^{-4}
YEp24	2 μm	URA3	4×10^{-6} (8×10^{-7})	3×10^{-5}
YCp50	Centromeric	URA3	6×10^{-7} (3×10^{-7})	3×10^{-7}
YIp5	None; integrative plasmid	URA3	$<1 \times 10^{-8}$	$<6 \times 10^{-8}$

[a] 2 μm indicates that the shuttle vector contains sequences from the endogenous yeast 2-μm plasmid that allow the shuttle vector to replicate independently of the host chromosomes by the same mechanism as 2-μm plasmids. Centromeric indicates that the shuttle vector contains an autonomous replication sequence (ARS) and a chromosomal centromere sequence (CEN), which together allow replication and segregation of plasmids to mother and daughter cells during mitosis. Integrative plasmid indicates that the shuttle vector contains no sequences that allow for establishment as an extrachromosomal element.

[b] Plasmids contain the yeast LEU2 or URA3 genes. These plasmids can be selected in yeast because they complement nutritional auxotrophies confered by mutations in the chromosomal LEU2 or URA3 genes.

[c] Average frequency of transmission of plasmids from bacteria to yeast as observed by the direct selection protocol. Frequencies are reported on a per donor basis because the yeast recipient is plated in vast excess to ensure physical contact between conjugants. Frequencies on a per recipient basis are given in parentheses.

[d] Average frequency of transmission of plasmids from bacteria to yeast observed following preincubation of conjugants on permissive medium. Frequencies are reported on a per recipient basis because, over the period of preincubation, the donor doubling time bears no relation to the doubling time of yeast transconjugants. Bacteria reproduce faster than yeast, so frequencies per donor underrate the frequency at which bacteria mate with yeast.

YCp50 are inserted near *oriT* and may interefere with origin function. We have not tested the transmission frequency of other *CEN* plasmids. The final plasmid tested, YIp5, contains *URA3* but lacks a sequence to allow autonomous replication in yeast. YIp5 never yielded transconjugants. This is not unexpected, however, since in this case two requirements must be met to establish a transconjugant: the plasmid must be transferred and it must integrate into a host replicon.

The procedures that follow use donor bacteria that contain the conjugative plasmid pDPT51, a member of the *incP* family.[11] This plasmid is a derivative of R751, modified to encode the *mob* functions capable of

[11] D. P. Taylor, S. N. Cohen, W. G. Clark, and B. L. Marrs, *J. Bacteriol.* **154**, 580 (1983).

recognizing the *oriT* sequence from the plasmids ColE1 and pMB1 (see Ref. 11 for a restriction map of pDPT51). In addition, pDPT51 encodes a distinct *mob/oriT* system (Table I) and also, of course, encodes *tra* functions required for cell contact and DNA transfer.

To simplify the description of the procedures, we shall consider that the goal is to transfer the shuttle plasmid YEp13 from bacteria to yeast. YEp13 contains ampicillin and tetracycline resistance genes for selection in bacteria and *LEU2* for selection in yeast, and it replicates autonomously in both organisms. The procedures can be modified for other shuttle vectors. A donor strain carrying both pDPT51 and YEp13 can be created in either of two ways. First, a bacterial strain harboring the conjugative plasmid can be transformed with YEp13. Second, a bacterial strain harboring YEp13 can be conjugated with a donor bacterial strain harboring the appropriate conjugative plasmid. Use Procedure 2 as outlined below for bacteria by yeast conjugation, rather than Procedure 1, to conjugate two bacterial strains.

Procedure 1: Direct Isolation of Transconjugants on Selective Medium

1. Prepare saturated cultures of the recipient yeast strain [e.g., SY1229 (*MATa leu2 his3 gal2 can1 ura3*)] and the donor bacterial strain [e.g., the C600 derivative SB21 *(hsdR hsdM leuB6 thr)*, although RR1 and DH1 also work well]. Grow the yeast recipient in liquid YEPD medium (2% peptone, 1% yeast extract, 2% glucose).[12] Grow the bacterial donor harboring pDPT51 and YEp13 in liquid LBH medium (1% tryptone, 0.5% NaCl, 0.5% yeast extract, 1 mM NaOH[13]) supplemented with trimethoprim (200 μg/ml, selection for pDPT51) and tetracyline (20 μg/ml, selection for YEp13). Alternatively, the donor can be grown in supplemented minimal medium [OMBG: 10.5 g/liter K_2HPO_4, 4.5 g/liter KH_2PO_4, 50 mg/liter $MgSO_4$, 0.1 g/liter $(NH_4)_2SO_4$, 0.5 g/liter sodium citrate, 0.2% glucose, 1 mg/liter thiamin; same as minimal agar of Ref. 14, supplemented with 0.2 g/liter threonine]. This enables selection of YEp13 because *LEU2* complements *leuB* mutations.

The versatility of the conjugation procedure could be extended by the generation of derivatives of pDPT51 that confer on the host bacterium a different spectrum of antibiotic resistance. This would allow simultaneous

---

[12] F. Sherman, this volume [1].
[13] I. Herskowitz and E. Signer, *J. Mol. Biol.* **47**, 545 (1970).
[14] J. H. Miller, "Experiments in Molecular Genetics." Cold Spring Harbor Laboratory, Cold Spring Harbor, New York, 1972.

antibiotic selection of both the conjugative and shuttle plasmids in bacteria. To this end, an attempt is in progress to replace the gene encoding ampicillin resistance in pDPT51.

2. Prior to mating, dilute the bacterial culture 100-fold (to $\sim 1 \times 10^7$ cells/ml) and the yeast culture 25-fold (to $\sim 1 \times 10^7$ cells/ml) into fresh medium of the same composition as used to prepare the saturated cultures. Grow the diluted cultures to a density of about 5×10^8 cells/ml for bacteria and a density of about 5×10^7 cells/ml for yeast. Pellet the cells by centrifugation (3000 g for 5 min at 4°, SS34 rotor). Five milliliters of the bacterial culture and 20 ml of the yeast culture yield useful amounts of the organisms.

3. Suspend the cells in 5 ml of TNB (50 mM Tris, pH 7.6, 0.05% NaCl). Pellet the donor cells a second time to remove residual antibiotics and resuspend them in 5 ml of TNB. At this point, cells can either be used immediately or stored on ice for up to several hours. An empirical determination of the optimal donor to recipient ratio for any given shuttle vector must be made. Different titers of donor and recipient are made by dilution in TNB. The optimal frequency of transmission of YEp13 occurs at a donor to recipient ratio of 1:1000 (10^5 bacterial cells/ml and 10^8 yeast cells/ml). For other shuttle vectors, such as YEp24 and YCp50, a 1:1 ratio is standardly used because transconjugants are rare or absent at lower donor titers.

4. Mix equal volumes of donor and recipient cell suspensions and apply the mix to the surface of a solid medium that permits the growth of only transconjugant yeast [e.g., SD + His + Ura, which is SD (0.67% yeast nitrogen base, 2% glucose)[12] supplemented with 20 µg/ml histidine and 20 µg/ml uracil]. Mating only occurs on a solid substrate under conditions that ensure yeast/bacteria contact.[4] To achieve these conditions, 0.1 to 0.5 ml of the cell mix, which should contain approximately 5×10^7 yeast cells, is spread over the surface of a petri plate about 9 cm in diameter. Between 10 and 1000 transconjugants will appear (depending on the shuttle vector; see Table II). Alternatively, transconjugants arising from a number of separate matings can be selected on the same petri dish by applying 10 µl of each conjugation mix at separate positions on the plate. This volume yields about 10 transconjugant colonies per application when YEp13 is the shuttle vector. Colonies appear in approximately 3 days at 30°. This direct selection procedure ensures that all colonies arising from recipient exconjugant cells are unique clones.

Transconjugants can also be selected on medium that allows the growth of both donor bacteria and transconjugant yeast cells (SD − Leu, which is SD supplemented with 20 µg/ml adenine, 60 µg/ml lysine, 20 µg/ml arginine, 20 µg/ml histidine, 20 µg/ml tryptophan, 30 µg/ml tyrosine, 20 μg/ml threonine, 20 μg/ml methionine, 50 μg/ml phenylalanine, and 20 μg/ml uracil). Transconjugant yeast are readily identified by virtue of the unique cology morphology of yeast compared to bacteria. This environment produces YEp13 transconjugants at the same frequency as above but may boost the frequency of transfer under some circumstances, such as when transferring large plasmids.

Procedure 2: Conjugation on Permissive Media

As shown in Table II, shuttle vectors based on the plasmid YEp13 are transmitted at high frequency using the direct selection procedure. However, plasmids such as YEp24 and YCp50, which contain *URA3* as the selectable yeast marker, are not transmitted as frequently. In an effort to increase the transmission frequency of such vectors, we developed a protocol that allows conjugation to occur under conditions permissive for the growth of both donor and recipient.

1. Prepare conjugants as described in Steps 1, 2, and 3, above.
2. Mix equal volumes of donor and recipient cell suspensions and apply 20 μl of the mix to the surface of a nonselective solid medium such as YEPD.
3. After incubation for at least 8 hr, scrape the cells off the plate and resuspend in TNB. Apply the suspension to the surface of a solid medium on which only yeast transconjugants can grow. Colonies appear in approximately $2\frac{1}{2}$ days.

Comments

As the results summarized in Table II reveal, conjugation is a useful method for introduction of autonomously replicating plasmids into yeast. In fact, transfer of plasmid DNA, at least in the case of 2-μm-based plasmids, occurs with a frequency high enough to select transconjugants by replica plating. This could prove to be a simple method for introducing a clone bank maintained in *E. coli* into yeast. A replica plating procedure could also be used to introduce a plasmid into a large number of yeast colonies derived, for instance, from a mutagenized population. To introduce plasmids into yeast by means of replica plating, apply fresh colonies of bacterial donors (~1 mm in diameter) to a lawn of yeast recipients prepared as in Procedures 1 and 2 and spread on medium selective for the transconjugants. Alternatively, fresh yeast colonies [~24 hr old (Joachim Li, personal communication, 1989)] can be replicated to a lawn of freshly prepared bacterial donors to generate transconjugants.

A preliminary experiment also suggests that bacterial conjugation could be adapted to select integration or one-step gene replacement events as well. The experiment is motivated by the expectation that DNA is transferred to yeast as a linear single strand, which must become double stranded and circular to replicate autonomously. Because in yeast free ends of DNA recombine at high frequencies with homologous sequences, we asked whether transmission of YIp5 could be detected when the recipient yeast harbored YEp13. YEp13, of course, has sequences homologous to the presumed ends of the transferred YIp5, that is, the *oriT* site. In this experiment a single YIp5 transconjugant was obtained, and the YIp5 was integrated into YEp13. Although the frequency is very low ($\sim 1 \times 10^{-8}$), it does raise the possibility that plasmids could be made to integrate at their homologous chromosomal location if the plasmid *oriT* site is flanked by sequences homologous to a particular chromosome segment. It might even be possible to use conjugation to initiate one-step gene replacements by flanking with *oriT* sequences the plasmid-borne DNA segment that is to replace a homologous chromosome segment.

[14] Cloning Genes by Complementation in Yeast

By MARK D. ROSE and JAMES R. BROACH

Introduction

As an experimental organism, *Saccharomyces cerevisiae* provides a providential convergence of an extensively elaborated genetic system, a facile transformation method, and highly efficient *in vivo* homologous recombination. Among other benefits, this convergence makes the cloning of genes responsible for specific phenotypes in yeast exceptionally easy. In this chapter we provide a practical guide for obtaining and using genomic and cDNA banks to isolate specific genes by complementation in *Saccharomyces cerevisiae*.

This chapter is divided into several sections. In the first part, we describe the various approaches that can be used to identify specific clones using transformation of yeast with plasmid-borne genomic or cDNA fragments. In the second section, we survey the types of vectors available for use in yeast and highlight the current vectors of choice in each of the various categories. In the third section, we provide a step-by-step guide for acquiring genomic or cDNA banks, covering protocols for constructing genomic banks *de novo* as well as procedures for obtaining and recovering preexisting plasmid-borne banks. Finally, we present a trouble-shooting guide indicating potential pitfalls in using the banks and avenues to pursue if initial attempts at cloning come up short.

Protocols for many of the techniques required for cloning by complementation in yeast are covered in other sections of this volume and are not duplicated in this chapter. These include general procedures for preparation of DNA from yeast,[1] transformation of yeast,[2] recovery of plasmids from yeast,[3] and integrative transformation.[4] In addition, other volumes in this series contain chapters that explore various aspects of the procedures described here. These include a recent chapter on cloning by complementation in yeast,[5] a source from which we have borrowed liberally in compiling this chapter, as well as several chapters on yeast vectors.[6-8] From a pragmatic standpoint, though, we have written this chapter so that successful cloning can be accomplished solely using protocols provided in this volume.

Strategies for Cloning

Complementation of Recessive Alleles

The most straightforward approach to cloning genes from plasmid-borne banks is complementation of a recessive marker. A recipient strain is constructed that carries a recessive mutation in the gene of interest as well as a nonreverting null allele of the chromosomal cognate of the selectable marker carried on the plasmid vector. This strain is then transformed with pools of plasmids from a bank constructed from wild-type genomic DNA. Transformants are recovered by selecting for complementation by the vector-borne selectable marker. The rare clone carrying the gene of interest is then identified either by coselection for complementation of the mutation in the gene of interest or, more often, by replica plating to a medium that distinguishes mutant from wild-type phenotypes for the gene of interest. The colony yielding complementation is recovered, plasmid DNA

---

[1] P. Philippsen, A. Stotz, and C. Scherf, this volume [11].
[2] D. M. Becker and L. Guarente, this volume [12].
[3] J. N. Strathern and D. R. Higgins, this volume [21].
[4] R. Rothstein, this volume [19].
[5] M. D. Rose, this series, Vol. 152, p. 481.
[6] S. A. Parent, C. M. Fenimore, and K. A. Bostian, *Yeast* **1**, 83 (1985).
[7] A. B. Rose and J. R. Broach, this series, Vol. 185.
[8] K. A. Armstrong, T. Som, F. C. Volkert, and J. R. Broach, *in* "Yeast Genetic Engineering" (P. J. Barr, A. J. Brake, and P. Valenzuela, eds.). Butterworth, Stoneham, Massachusetts, 1988.

extracted, and identity between the complementing fragment and the gene of interest confirmed as described below.

This approach was first used to isolate the *CAN1* gene of yeast[9] and has since been used repeatedly to clone literally hundreds of different yeast genes. For most genes in which recessive, easily scored mutations are available, this method almost always works without difficulty. Problems have arisen in attempting to clone genes that subsequently have been shown to be lethal in yeast at multiple copies, to inhibit growth of *Escherichia coli,* or to be linked to yeast centromeres or telomeres. Solutions to such problems, should they arise, are discussed in the last section of this chapter.

Cloning Dominant Alleles

Cloning genes that are defined by dominant alleles is a straightforward extension of cloning by complementation of recessive alleles. The only difference is that the clone bank has to be constructed *de novo* from genomic or cDNA prepared from the strain carrying the dominant mutation. Accordingly, one cannot use any of the preexisting banks described below, all of which were prepared from DNA taken from essentially wild-type strains. Protocols for preparing clone banks *de novo* are provided below.

High-Copy Cross-Suppression

An extension of the complementation approach described above can occasionally permit isolation of genes involved in a particular biological process but for which no mutations of the genes exist. This follows from the observation that complementation of a particular mutant gene can sometimes be obtained by the presence of not only the wild-type copy of the gene but also multiple copies of some other gene (see [16] in this volume). This high-copy cross-suppression is often simply a distraction, complicating the identification of the actual gene sought. In many cases, cross-complementation is of little relevance to the biological process being addressed. This would be true, for instance, if cross-complementation arose simply from informational suppression, owing, for example, to increased missense or nonsense suppression from overexpression of a particular tRNA gene or ribosomal protein.[10-12] The fact that cross-complemen-

---

[9] J. R. Broach, J. N. Strathern, and J. B. Hicks, *Gene* **8,** 121 (1979).
[10] G. A. Pure, G. W. Robinson, L. Naumovski, and E. C. Friedberg, *J. Mol. Biol.* **183,** 31 (1985).
[11] J. M. Song, S. Picologlou, C. M. Grant, M. Firoozan, M. F. Tuite, and S. Liebman, *Mol. Cell. Biol.* **9,** 4571 (1989).
[12] E. Vallen and M. D. Rose, unpublished (1989).

tation occurs with some regularity is the reason that complementation per se is not *prima facie* evidence for the identity of a cloned gene. Additional steps to establish identity, as described below, are an essential component of cloning by complementation.

Isolating Regulated Promoters

In the absence of any direct information about the identity of a gene or its gene product, one recourse is to isolate a set of genes whose regulation fulfills some interesting set of criteria. One approach to achieving this end has been to clone random genomic fragments into a plasmid carrying an enhancerless promoter that drives expression of a readily scored gene, such as *lacZ*. Random transformants are then examined for conditional expression of *lacZ* in response to the desired signal. This procedure was pioneered by Ruby and Szostak, who used it to isolate sequences that mediated induction of expression after treatment of cells with DNA-damaging agents.[13]

Using Complementation in Yeast to Isolate Specific Genes from Other Organisms

Although the preceding discussion refers exclusively to isolation of yeast genes, several specialized yeast vectors extend the applicability of complementing yeast mutations to isolation of genes from a wide variety of species. In these systems, random cDNA molecules from any source are cloned directly into a vector between signals for expression in yeast.[14,15] In this manner, cDNA libraries from essentially any organism can be screened directly in yeast for activities that will complement specific yeast mutations.

Other Approaches to Cloning Specific Genes in Yeast

A variety of other approaches for identifying and cloning specific genes from yeast have been successfully used but are beyond the scope of this chapter. These include phage library screens, using either nucleic acid probes or specific antibodies,[16] differential colony or plaque hybridization, complementation in *E. coli,* and transposon tagging.[17] Several of these

[13] S. W. Ruby, and J. W. Szostak, *Mol. Cell. Biol.* **5,** 75 (1985).
[14] G. L. McKnight and B. L. McConaughy, *Proc. Natl. Acad. Sci. U.S.A.* **80,** 4412 (1983).
[15] J. Colicelli, C. Birchmeier, T. Michael, K. O'Neill, M. Riggs, and M. Wigler, *Proc. Natl. Acad. Sci. U.S.A.* **86,** 3599 (1989).
[16] R. A. Young and R. W. Davis, this volume [15].
[17] D. J. Garfinkel, M. F. Mastrangelo, N. J. Sanders, B. K. Shafer, and J. N. Strathern, *Genetics* **120,** 95 (1988).

Yeast Plasmid Vectors

Types of Yeast Vectors

Only two general types of vectors are routinely used to construct plasmid banks for transformation of yeast: YEp (yeast episomal plasmid) and YCp (yeast centromeric plasmid) vectors. All vectors of both classes carry a gene whose presence can be selected in yeast as well as a sequence that promotes autonomous replication in yeast (an *ARS* element). In addition, all vectors are constructed from bacterial plasmids and can be selected and propagated in *E. coli*. The yeast selectable marker is routinely any one of a number of cloned yeast genes for which strains containing nonrevertible null alleles of the gene are readily available. The most common markers are *LEU2, HIS3, URA3,* and *TRP1*. A number of YEp vectors are also currently available with a variety of dominant selectable markers, which allow transformation of essentially any yeast strain. *ARS* elements have been shown to be yeast origins of DNA replication, and the presence of these elements on a vector promotes high-frequency transformation of yeast, generally at a level of several thousand transformants per microgram of vector DNA.[18,19] For purposes of vector construction and use, little distinction exists between any of the available *ARS* elements.

The feature that distinguishes YEp and YCp vectors from *ARS* plasmids is the presence of an additional component that imparts enhanced mitotic and meiotic stability and fixed copy levels. YCp plasmids carry a chromosomal centromere, which allows the plasmid to engage the spindle apparatus of the cell. As a consequence, the copy number of the plasmid is reduced to one or two per cell, and the plasmid acquires much greater stability than *ARS* plasmids.[20] Typically, an overnight culture of a YCp-containing strain grown without selection will be composed of 90–99% plasmid-containing cells. This contrasts with the stability of *ARS* plasmids, which normally yield 1–5% plasmid-bearing cells under similar growth regimes. The stability of YCp plasmids depends to some degree on the size of the plasmid, with stability increasing with increasing size.

The singular feature of YEp vectors is their ability to engage the partitioning system of the endogenous yeast plasmid, the 2-μm circle, to achieve

---

[18] B. J. Brewer and W. L. Fangman, *Cell (Cambridge, Mass.)* **51**, 463 (1987).
[19] D. T. Stinchcomb, K. Struhl, and R. W. Davis, *Nature (London)* **282**, 39 (1979).
[20] L. Clarke and J. Carbon, *Nature (London)* **287**, 504 (1980).

TABLE I
SELECTED YEp AND YCp PLASMID VECTORS

Name	Yeast marker	E. coli genes	Parental plasmid	CEN or 2-μm genes	Other DNA	Size (kb)	Cloning sites[a]	Comments	Ref.
YEp vectors									
YEp13	LEU2	amp, tet	pBR322	REP3, ori	—	10.7	B, U, Y, H	Small EcoRI 2-μm B form fragment and 4.0-kb PstI LEU2 in pBR322; pPL200 is a derivative in which the LEU2 SalI site has been filled in, leaving a unique SalI site in tet	9, 28
YEp24	URA3	amp	pBR322	REP3, ori	—	7.8	B, L, C, A, J, U, S, M, F, X	Small EcoRI 2-μm B form fragment and 1.2-kb URA3 in pBR322; detailed restriction map available in Biolabs/BRL catalogs	29
pHV1	HIS3	amp	pUC	REP3, ori	—	5.4	E, Q, M, S, F	—	30
YEpL3	LEU2	amp, lacZα	pMB1	REP3, ori	f1 ori	6.8	M, B, S, P, H	Color screen for plasmids containing	24

pEMBLYe24	URA3	amp, lacZα	pMB1	REP3, ori	f1 ori	7.4	B, H, S	HindIII fragment from YEp24 containing URA3 and 2-μm sequences in pEMBL9; color screen for plasmids containing inserts; f1 origin for isolating single-stranded DNA	23
YEplac112	TRP1	amp, lacZα	pUC19	REP3, ori	—	4.9	H, F, P, S, X, B, M, K, Q, E	XbaI, HindIII, and PstI sites were removed from the TRP1 gene by site-directed mutagenesis; XbaI site in the 2-μm circle was removed by filling in; color screen for plasmids containing inserts	25

(Continued)

TABLE I (Continued)

Name	Yeast marker	E. coli genes	Parental plasmid	CEN or 2-μm genes	Other DNA	Size (kb)	Cloning sites[a]	Comments	Ref.
YEplac181	LEU2	amp, lacZα	pUC19	REP3, ori	—	5.7	H, F, P, S, X, B, M, K, Q, E	KpnI and EcoRI sites were removed from LEU2 by site-directed mutagenesis; 2-μm XbaI site was removed by filling in; color screen for plasmids with inserts	25
YEplac195	URA3	amp, lacZα	pUC19	REP3, ori	—	5.2	H, F, P, S, X, B, M, K, Q, E	PstI site was removed from URA3 by site-directed mutagenesis; 2-μm XbaI site was removed by filling in; color screen for plasmids with inserts	25
pWH5	LEU2	amp, λP_R-tet	pBR322	REP3, ori	λcI repressor	10.6	L, M, H	Inserts in the cloning sites destroy the cI repressor, allow-	26

Name	Marker	Backbone	CEN	ARS	Size	Sites	Description	Ref
YCp vectors								
YCp50	URA3	pBR322	CEN4	ARS1	8.0	E, H, B, F, S, R, C	ing expression of the *tet* gene, which is under control of the λpR promoter; Also transforms *Schizosaccharomyces pombe* Derived from YIp5 by insertion of *CEN4* and *ARS1* into the *Pvu*II site; available from ATCC under entry 37419	21
pSB32	LEU2	pBR322	CEN4	ARS1	8.3	H, B, F, S	Derived from YCp50	
pRS313	HIS3	pBLUESCRIPT	CEN6	ARSH4, f1 ori $P_{T3, T7}$	5.0	Y, S, C, E, B, Z, X, N, Q	Color screen for insert containing plasmids; T3 and T7 promoters to make RNA probes from inserted DNA; f1 *ori* to generate single-stranded DNA	22

(Continued)

TABLE I *(Continued)*

Name	Yeast marker	E. coli genes	Parental plasmid	CEN or 2-μm genes	Other DNA	Size (kb)	Cloning sites[a]	Comments	Ref.
pRS314	TRP1	amp, lacZα	pBLUESCRIPT	CEN6	ARSH4, f1 ori, $P_{T3, T7}$	4.8	Y, S, C, E, B, Z, K, N, Q, P	Color screen for insert containing plasmids; T3 and T7 promoters to make RNA probes from inserted DNA; f1 ori to generate single-stranded DNA	22
pRS315	LEU2	amp, lacZα	pBLUESCRIPT	CEN6	ARSH4, f1 ori, $P_{T3, T7}$	6.0	Y, S, H, P, B, Z, X, N, Q	Color screen for insert containing plasmids; T3 and T7 promoters to make RNA	22

Plasmid	Marker	Selection	Backbone	CEN	ARS/ori	Size (kb)	Sites	Description	Ref.
pRS316	URA3	amp, LacZα	pBLUESCRIPT	CEN6	ARSH4, f1 ori, P$_{T3, T7}$	4.9	Y, S, C, E, B, Z, X, N, Q, K, H	Color screen for insert containing plasmids; T3 and T7 promoters to make RNA probes from inserted DNA; f1 ori to generate single-stranded DNA	22
pMR366	URA3	amp, tet	pSC101	CEN4	ARS1	8.9	B, C, E, F, H, R, S	Fusion of pSC101 and YCp50; low copy number in E. coli	27

[a] A, HpaI; B, BamHI; C, ClaI; D, NdeI; E, EcoRI; F, SphI; G, BglII; H, HindIII; I, NarI; J, NheI; K, KpnI; L, BclI; M, SmaI; N, NotI; O, NcoI; P, PstI; Q, SacI; R, NruI; S, SalI; T, StuI; U, PvuII; V, EcoRV; W, ScaI; X, XbaI; Y, XhoI; Z, SpeI.

stable, high-copy propagation.[8] The sequence responsible for this property is a small region derived from the 2-μm circle, designated REP3 or STB. This sequence functions as a cis-acting partitioning element that facilitates equal distribution of plasmid molecules between mother and daughter cells at mitosis. This partitioning activity also requires two trans-acting products encoded by the 2-μm circle genes, REP1 and REP2. As most standard yeast strains harbor the 2-μm circle, these products are normally present in the recipient cell and need not be encoded by the YEp vector.

Most YEp vectors exhibit mitotic stability approaching that of YCp vectors, with nonselective overnight cultures yielding 80–95% plasmid-bearing cells. In addition, YEp vectors are maintained at relatively high copy levels, usually between 10 and 40 copies per cell. Discussion of the underlying mechanisms for maintenance of high plasmid copy levels is beyond the scope of this chapter and is addressed at length elsewhere.[8] However, it should be noted that vector copy levels are not rigidly fixed but rather exhibit considerable plasticity. Both the fragment size and particular sequences cloned into a YEp vector can influence stability and copy levels. As a general rule of thumb, the smaller the YEp plasmid, the higher the stability and copy level. In addition, insertion of a gene whose presence is deleterious to the host strain yields diminished stability and reduced average copy levels, even under growth conditions selective for retention of the plasmid.

A list of commonly used YCp and YEp vectors, along with their vital statistics, is provided in Table I.[21-30] Several of these vectors are diagrammed in Figs. 1 and 2.

Advanced Vector Technology

A number of additional features have been incorporated into various YEp and YCp vectors to make them easier to use or to provide facile solutions to specific cloning problems. Most of the vectors listed in Table I, for instance, have been designed to facilitate recovery of insert-containing

---

[21] M. D. Rose, P. Novick, J. H. Thomas, D. Botstein, and G. R. Fink, *Gene* **60,** 237 (1987).
[22] R. S. Sikorski and P. Hieter, *Genetics* **122,** 19 (1989).
[23] C. Baldari and G. Cesareni, *Gene* **35,** 27 (1985).
[24] G. Cesareni and J. A. H. Murray *in* "Genetic Engineering" (J. K. Setlow, ed.), Vol. 9, p. 135. Plenum, New York, 1987.
[25] R. D. Gietz and A. Sugino, *Gene* **74,** 527 (1988).
[26] A. Wright, K. Maundrell, W.-D. Heyer, D. Beach, and P. Nurse, *Plasmid* **15,** 156 (1986).
[27] M. D. Rose, L. M. Misra, and J. P. Vogel, *Cell (Cambridge, Mass.)* **57,** 1211 (1989).
[28] P. A. Lagosky, G. R. Taylor, and R. H. Haynes, *Nucleic Acids Res.* **15,** 10355 (1987).
[29] D. Botstein, S. C. Falco, S. E. Stewart, M. Brennan, S. Scherer, D. T. Stinchcomb, K. Struhl, and R. W. Davis, *Gene* **8,** 17 (1979).
[30] J. Nikawa and P. Sass, unpublished (1988).

plasmids. Insertions into plasmids YEp13 and YCp50 can be identified by sensitivity to tetracycline of the resultant bacterial transformants.[9,21] The YEplac, pEMBLY, and pRS series of vectors incorporate the *lacZ* α-complementing element and polylinker from the M13 vectors series so that insert-containing plasmids can be distinguished from vector alone by the standard blue/white colony color assay.[22-25] The YEp vector pWH5 goes one step further by incorporating a positive selection for plasmids with inserts.[26] In this plasmid, inserts in the cloning site inactivate the cI repressor, allowing expression of the *tet* gene, which is under control of the λpR promoter. The pEMBLY and pRS series also carry the f1 origin of replication to allow immediate recovery of single-stranded DNA from isolated cloned segments.[22-24] Finally, plasmid pMR366 is a YCp vector constructed from the stringent bacterial plasmid pSC101 (Ref. 27). As such, this vector is maintained at low copy levels in both *E. coli* and yeast, which

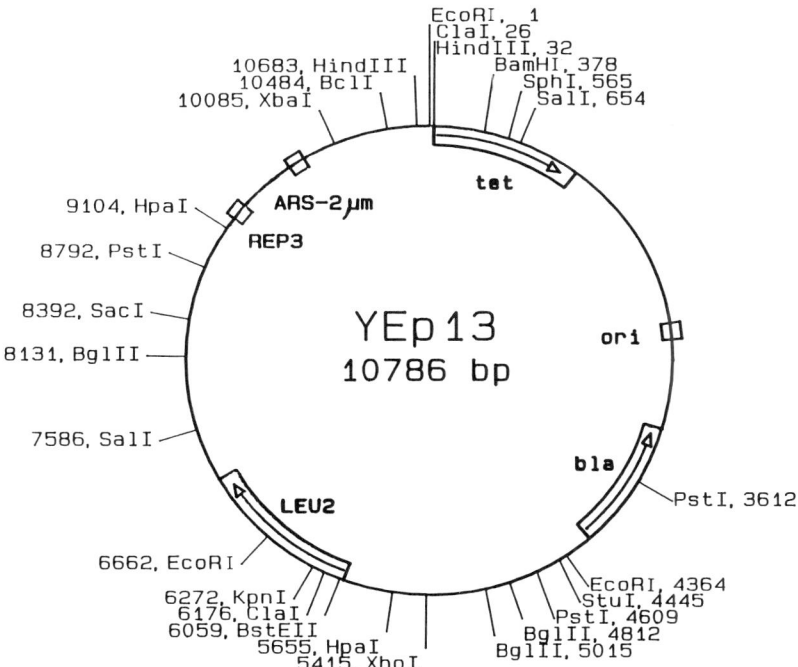

FIG. 1. Selected YEp vectors. Restriction map and genome organization for four 2-μm circle-based cloning vectors. Descriptions and references for these and other similar vectors are presented in Table I. Plasmids pHV1, pUV2, and pTV3 were constructed by J. Nikawa and P. Sass (unpublished, 1988). In the maps of these three vectors, all the sites for restriction enzymes shown are included.

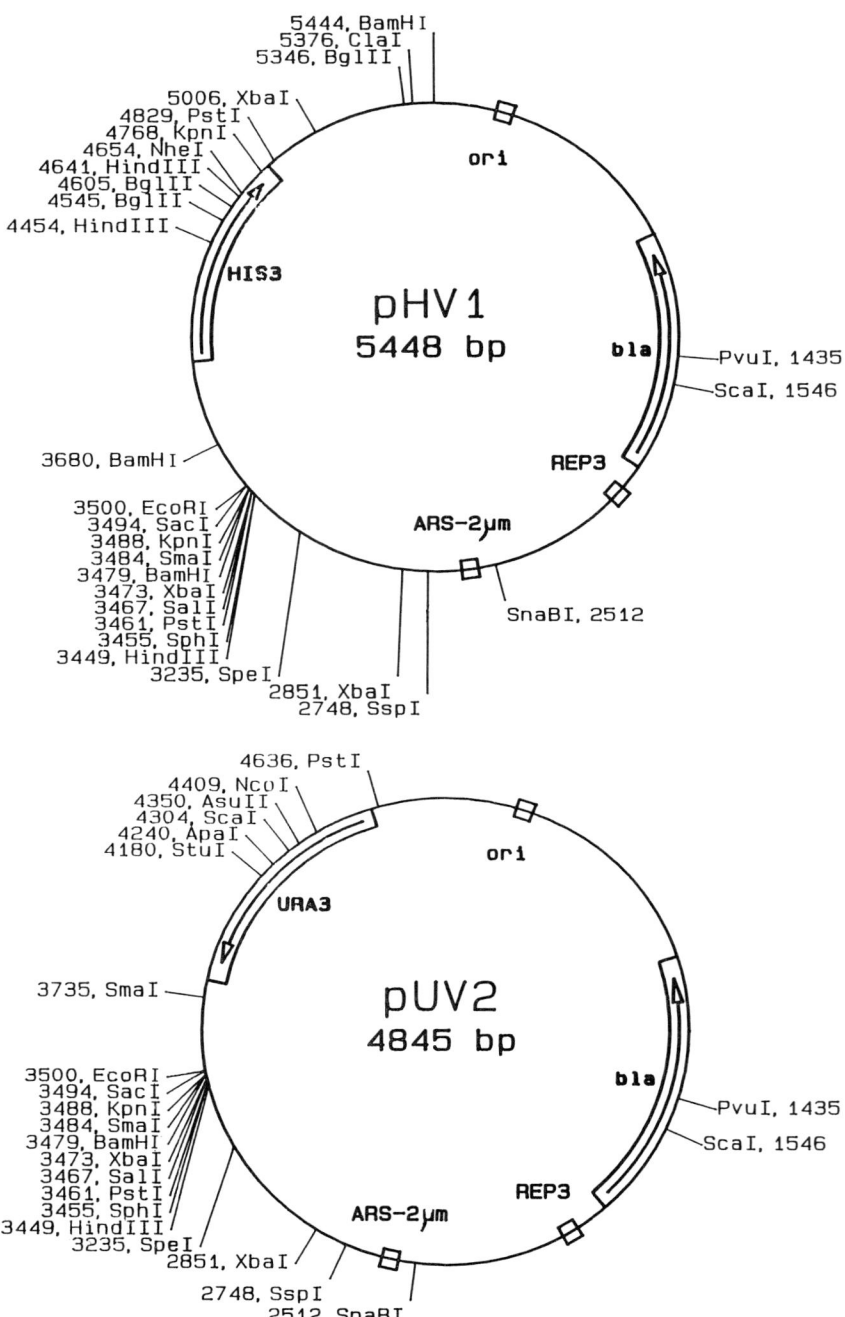

FIG. 1. See legend on p. 207.

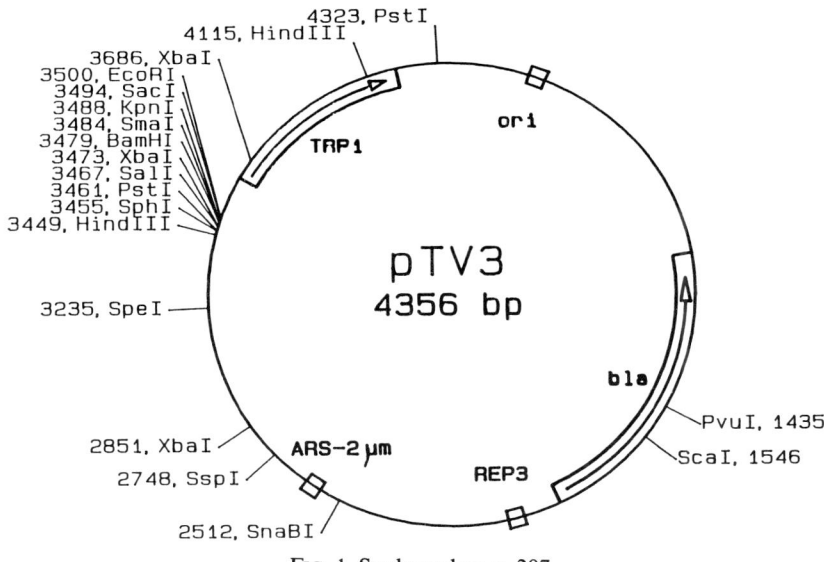

FIG. 1. See legend on p. 207.

facilitates recovery of fragments that are deleterious at high copy levels in either organism.

A variety of additional vectors, which can fulfill other cloning or expression requirements, are available, but are not listed in Table I. Extensive lists of these other vectors can be found in reviews by Rose and Broach[7] and by Parent et al.[6] Provided in these reviews are, for example, vectors that carry dominant selectable markers for yeast transformation, including markers conferring resistance to hygromycin, chloramphenicol, phleomycin, tunicamycin, polymyxin, methotrexate, or G418. As noted above, such vectors circumvent the need for introducing a recessive marker for plasmid selection into a strain of interest. In addition, vectors are available that propagate in yeast at exceptionally high copy levels. Finally, a variety of plasmids designed for expression of cloned genes in yeast have been constructed. For any of these specialized needs, one should refer to the above-mentioned reviews.

Selecting the Right Vector

All of the vectors listed in Table I are equally suitable for use in constructing yeast genomic banks. In determining which of the vectors to use, several factors should be considered. The first consideration is whether the vector should be a high-copy YEp plasmid or a low-copy YCp plasmid.

In general, if the goal is to isolate a gene by complementation of a recessive allele in yeast or to isolate a dominant allele of a gene from a particular mutant strain, then a YCp vector would be preferable. This would reduce the potential for isolating cross-suppressing clones, which would tend to complicate subsequent analysis. In addition, such a vector would diminish the possibility that the clone of interest was underrepresented, in the unlikely event that the gene of interest caused growth inhibition when propagated at high copy levels. On the other hand, if the goal is to identify cross-suppressing clones or to isolate genes on the basis of high-copy-induced novel phenotypes, then clearly a YEp vector would be the plasmid of choice.

The second variable in choosing the appropriate vector is the yeast selectable marker carried on the plasmid. As noted in Table I both YCp and YEp vectors are available with any one of four standard yeast selectable markers. Accordingly, the deciding factor in this selection is the ease of obtaining or constructing the recipient strain for transformation. Strains

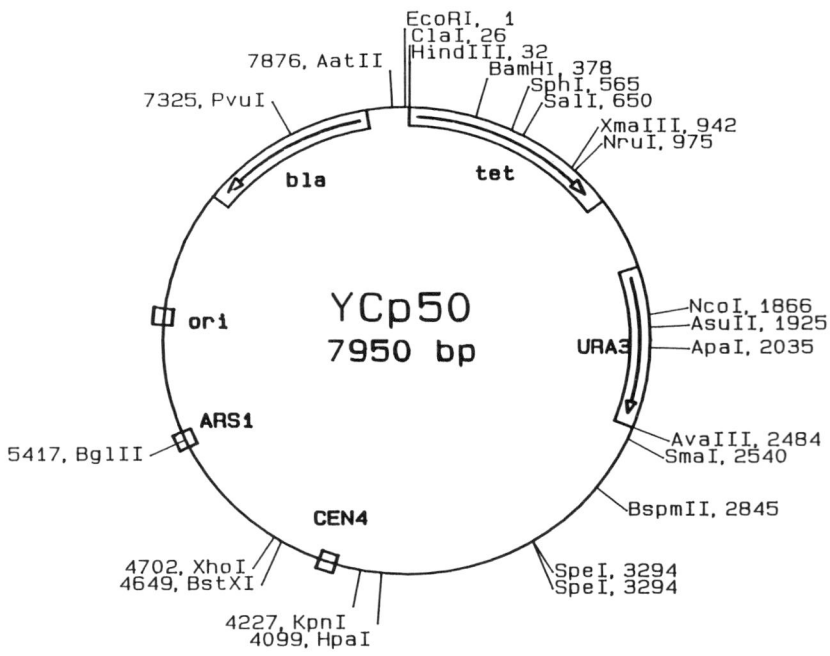

FIG. 2. Selected YCp vectors. Restriction map and genome organization of several centromere-based cloning vectors. Details and references are presented in Table I, as are descriptions and references for other similar vectors.

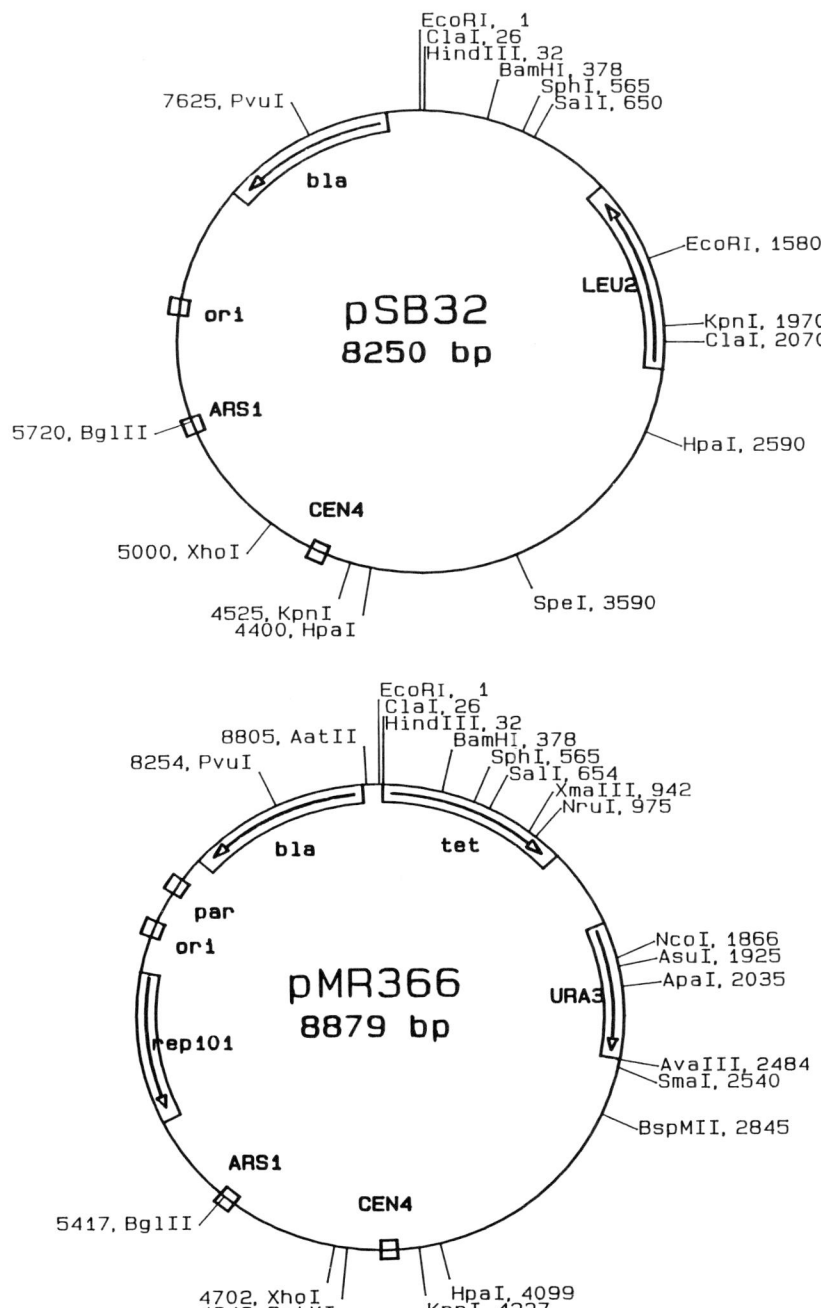

FIG. 2. *(Continued)*

with nonreverting alleles in any or all of these markers are readily available and can be crossed into a background containing a mutation in the gene of interest. Alternatively, *ura3* alleles can be introduced into the strain of interest, either by direct selection of a mutant clone or by transplacement of a nonreverting mutant allele; in both cases, *ura3* cells are selected by resistance to 5-fluoroorotic acid.

Additional considerations involve the ease and efficiency of constructing the library and analyzing the positive clones that emerge from the selection. In general, the greatest utility will be found with vectors that place inserts into a polylinker, require a simple method for scoring the frequency of inserts, and provide a simple method for preparation of single-stranded DNA for later analysis.

Constructing cDNA libraries to use for complementing specific mutations in yeast requires a yeast expression vector, in which a cloning site resides between signals for transcription initiation and termination in yeast. Three vectors that have been used for this purpose are shown in Fig. 3. In general, though, most of the expression vectors described by Rose and

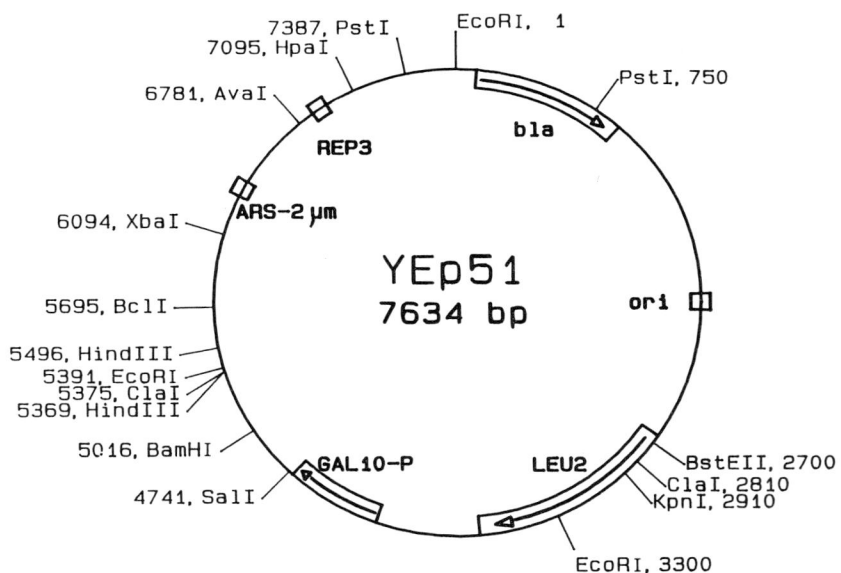

FIG. 3. Selected expression vectors suitable for use in constructing cDNA banks. Detailed descriptions of the organization and use of plasmids YEp51 and YEp52 are found in J. R. Broach, Y. Y. Li, L. C. C. Wu, and M. Jayarum, in "Experimental Manipulation of Gene Expression" (M. Inouye, ed.), Academic Press, New York, 1983; for plasmid pADNS, see Ref. 27.

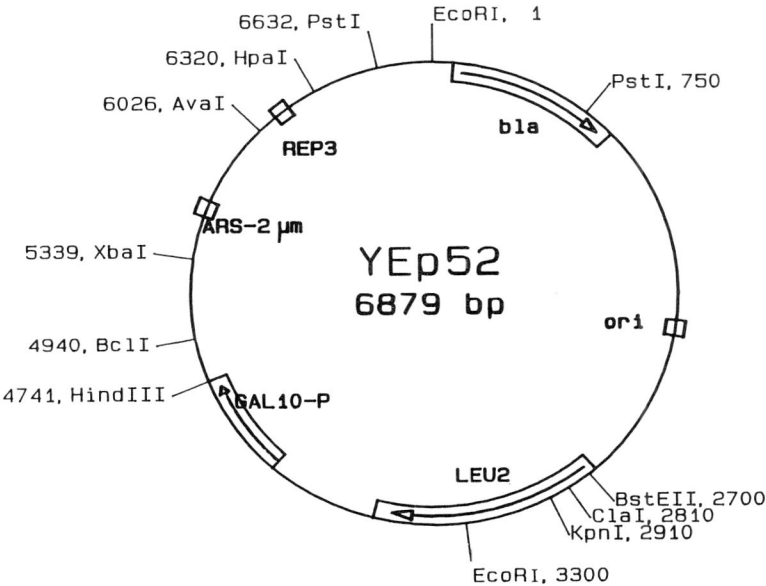
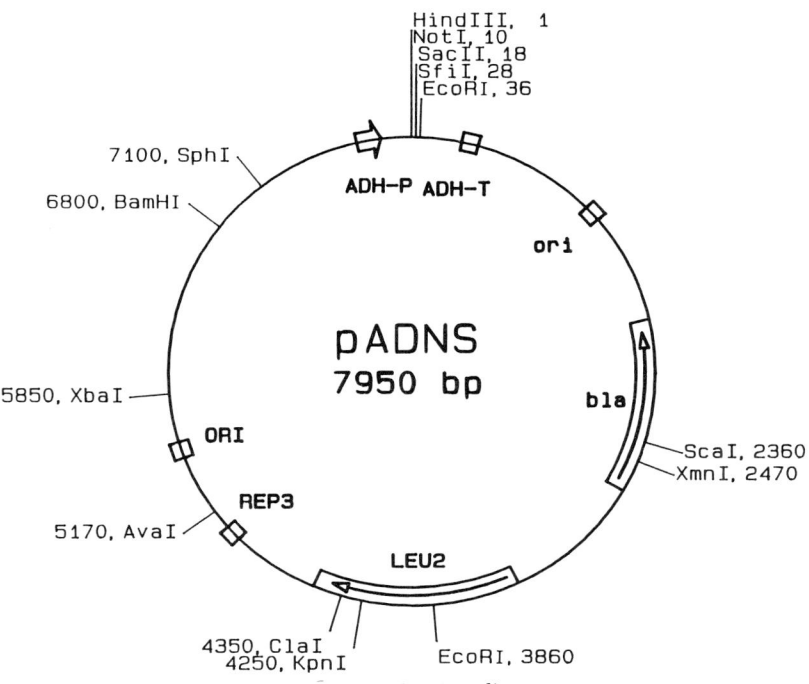

FIG. 3. *(Continued)*

Broach[7] should prove suitable for constructing cDNA banks by the approach described by Okayama and Berg as modified for yeast by McKnight and McConaughy.[14] A more recent procedure for constructing expressible cDNA banks, described by Colicelli et al.,[15] requires an expression vector, such as pADNS, in which the suitably positioned cloning site contains at least one rare restriction site.

Acquiring Plasmid-Borne Genomic Banks

Banking by Mail

The most facile means of obtaining a yeast genomic bank is to procure an established bank through the mail. Table II lists a number of previously constructed banks, many of which are currently available.[31-41]

Plasmid banks are sent in many different forms, from pure DNA to various stable cultures of bacteria. The overriding concern is maintaining the highest level of independence and complexity in the library. Once the informational content of the library has been degraded, its quality cannot be restored. Accordingly, one should attempt to obtain samples of the original transformants of the library from the original creators, without any intermediate amplification steps. Similarly, if the bank is sent as multiple small subbanks, the subbanks should not be pooled. This would tend to reduce the number of independent clones. Typically, for logistical reasons or because of diminishing stocks, clone banks are sent after having been amplified at least once. However, banks are often sent as stabs of the primary transformants in nutrient agar.

If the clone bank sample is DNA, it should be transformed into an appropriate *E. coli* strain and the transformants pooled as described below. Half of the sample of pooled transformants should be frozen for permanent storage. The representation of the library should be checked as described below for newly constructed banks. If the sample is some form of bacterial

---

[31] I. Sadler, K. Suda, G. Schatz, F. Kaudewitz, and A. Haid, *EMBO J.* **3**, 2137 (1984).
[32] H. Hashimoto, Y. Kikuchi, Y. Nogi, and T. Fukasawa, *Mol. Gen. Genet.* **191**, 31 (1983).
[33] A. Rosenbluh, M. Mevarech, Y. Koltin, and J. A. Gorman, *Mol. Gen. Genet.* **200**, 500 (1985).
[34] K. O'Mally, P. Pratt, J. Robertson, and M. Lilly, *J. Biol. Chem.* **257**, 2097 (1982).
[35] J. Nikawa and S. Yamashita, *Eur. J. Biochem.* **143**, 251 (1983).
[36] M. Carlson and D. Botstein, *Cell* **28**, 145 (1982).
[37] R. L. Burke, P. Tekamp-Olson, and R. Najarian, *J. Biol. Chem.* **258**, 2193 (1983).
[38] F. Spencer and P. Hieter, unpublished (1988).
[39] Y. Ohya, S. Yamamoto, Y. Ohsumi, and Y. Anraku, *J. Bacteriol.* **165**, (1986).
[40] M. J. Stark and J. S. Milner, *Yeast* **5**, 35 (1989).
[41] W. Kruger and I. Herskowitz, unpublished (1989).

TABLE II
PREVIOUSLY CONSTRUCTED CLONE BANKS ON YEAST SHUTTLE VECTORS

Vector	Selectable marker	Species	DNA source Strain	Genotype	Insert size (kb)	No. of clones	Genome equivalent	Ref.	Source
YEp13	LEU2	S. cerevisiae	AB320	HO ade2-1 lys2-1 trp5-2 leu2-1 can1-100 met4-1 ura(1,3)	7	5×10^4	21	28	K. Nasymth
YEp13	LEU2	S. cerevisiae	D273-10B	MATα	10–25	—	—	31	—
YEp13	LEU2	S. cerevisiae	MT13	a/α HO/HO ura3/+ his4/his4 lys1/lys1	—	3×10^4	—	32	—
YEp13	LEU2	Candida albicans	B792	—	10	1.6×10^4	5	33	—
YEp13	LEU2	S. cerevisiae	DC5	MATα leu2-3, 112 his3 can1-11	>4	—	—	34	M. Douglas
YEp13	LEU2	S. cerevisiae	X2180-1A	MATa SUC2 gal2	5–15	10^4	6	35	—
YEp13	LEU2	S. cerevisiae	1403-7A	MATa trp1 ura3 MAL4	5–20	1.2×10^4	7	17	—
YEp24	URA3	S. cerevisiae	DBY939	MATα suc2-215 gal2	>10	2×10^4	12	36	M. Carlson
YEp24	URA3	S. cerevisiae	S288C	MATα SUC2 gal2	—	—	—	37	R. Burke
pFL1	URA3	S. cerevisiae	—	—	—	—	—	—	F. Lacroute
pSB32	LEU2	S. cerevisiae	YNN214	MATα ura3-52 lys2-801 ade2-101	9–12	10^4	7	38	P. Hieter
YCpG11	TRP1	S. cerevisiae	A5-8-1A	MATα leu1	5–15	—	—	39	—
YCp50	URA3	Kluyveromyces lactis	—	—	5–8	—	4	40	M. Stark
YCp50	URA3	S. cerevisiae	GRF88	MATa his4-38 gal2	10–15	2.2×10^4	16	21	ATCC
YCp50	URA3	S. cerevisiae	GRF88	MATa his4-38 gal2	15–20	9×10^3	9	21	ATCC
YCp50	URA3	S. cerevisiae	GRF88	MATa his4-38 gal2	20–30	8×10^3	13	21	ATCC
pMR366	URA3	S. cerevisiae	XJJ10-8B	MATα ade2 his4-401 leu2 ura3-52 lys2 HOL1 rna16-1	2–25	1.4×10^4	9	41	I. Herskowitz

culture, either a frozen aliquot or a stab, the cells must be grown up to prepare DNA. Since different transformants will grow at different rates, extended growth in culture tends to diminish complexity. The following protocol is recommended for recovering banks from bacterial cultures.

1. Titer the culture to determine the viability. For a stab, a small (e.g., 1 ml) suspension should first be made in Luria broth (LB). For a frozen culture, remove a small chip of frozen cells with a sterile dowel stick and thaw on ice.

2. Once the viability has been determined, prepare multiple dilutions of the culture to obtain 100–300 colonies per plate. Maintain selection for the plasmids at all times. To ensure complete representation of the bank, plate out at least 3 times the number of colonies as originally went into making up the bank.

3. Wash the colonies off the plate, concentrate, and use for both frozen storage and as inoculum for large-scale plasmid preparations.

4. After preparing DNA from the large-scale culture, check the representation of the library by selecting for complementation of some auxotrophic marker in yeast.

Constructing Banks de Novo

Initial Considerations. Although the most efficient method of acquiring a genomic plasmid bank is via the mail, certain situations, such as the isolation of a gene defined by a dominant mutation, require construction of a genomic library *de novo*. This section addresses some of the parameters that must be considered during the construction of the library and includes a protocol for doing so.

Having selected the appropriate vector and DNA source, being sensitive to the considerations listed above, the next step is selecting a protocol for preparing the bank. Unless a great deal is already known about the structure of the gene of interest, the optimal plasmid bank must be constructed from randomly cleaved overlapping fragments of genomic DNA. One method of creating such fragments is by physically shearing the genomic DNA. Synthetic DNA linkers are then ligated onto the fragment ends to create sticky ends compatible with those of the vector DNA. Rather than physically shearing the DNA, an alternate method is to use partial digestion by a mixture of 4-base pair (bp) recognition site restriction enzymes that cleave to form blunt ends. In both methods, owing to the additional steps, the addition of synthetic linkers can be relatively inefficient.

To avoid the linker addition step, many libraries are currently constructed by partial digestion of the genomic DNA with a 4-bp recognition site restriction enzyme that leaves an overhang that is complementary to that left by cleavage in the vector. This approach has been very successful, but care must be taken to ensure the creation of a large, relatively unbiased library. First, the starting DNA to be fragmented must be of significantly higher molecular weight than the planned insert DNA. Each cleaved molecule with an end created by shearing is a competitive inhibitor of the formation of the desired plasmids. Second, restriction enzyme sites are fixed in location, and some sites will be cleaved preferentially at limiting enzyme concentration. This means that a large potential exists for a significant bias in the distribution of restriction fragments. To avoid this, several different conditions of restriction enzyme cleavage are used and the reaction products pooled together.

The partially digested DNA is then fractionated by size to select an optimal distribution of inserts. The desired size of insert DNA is influenced by the number of plasmids required to find the gene, by the ease of its selection, and by the contrasting ease of subsequent localization and analysis of the gene. Larger average inserts entail fewer plasmids to screen, which is of particular interest if the assay is difficult or if transformation of the selected recipient strain is inefficient. Smaller inserts facilitate subsequent steps of analysis. However, isolation of several different overlapping inserts that all complement a gene of interest often allows one to identify readily the relevant region even in banks constructed with large inserts. Given the small size of the yeast genome, a library with 15-kb inserts should contain roughly 0.1% positive clones.

Once the size-fractionated DNA has been obtained, it must be ligated to the vector. The parameters that govern this ligation event include the molar ratio of fragments to vector, the intramolecular end concentrations of the fragments and vector, and the absolute concentrations of each of these species. Concentrations of the reactants must be chosen to be high enough to initially favor the bimolecular ligation of vector to fragment but not so high as to allow appreciable formation of higher order products, such as unproductive vector dimers. Thus, in contrast to the case of constructing banks in cosmids or bacteriophage λ, in which the ligated products are subsequently packaged, the DNA concentration in ligations for plasmid banks must be held low. If the concentration is too low, however, unproductive unimolecular reactions predominate.

A key approach to simplifying the ligation reaction is to prevent the unimolecular reactions from forming products. One way to do this is by using dephosphorylated vector molecules. Since the vectors cannot self-ligate, the unproductive unimolecular and bimolecular reactions are inhibited. Therefore, the vector concentration can be chosen based solely on the parameters of the insert DNA. The concentration of insert DNA is chosen to be below its intramolecular end concentration to discourage insert DNA polymerization. The vector concentration is then chosen to be above the intramolecular end concentration of the insert DNA to favor the formation of the heterologous dimers. If possible, the molar ratio of vector to insert DNA is chosen to be near unity so as to avoid the formation of unproductive trimers.

An alternative approach involves partially filling in the overhanging ends, such that neither the insert nor vector fragments can self-ligate. For example, the enzyme *Sau*3A leaves 5'-GATC as the overhanging end. Addition of GA to the 3' end of the complementary strand leaves only the terminal 5'-GA to base pair. The enzyme *Sal*1 leaves 5'-TCGA as its overhang. Addition of the complementary TC leaves only 5'-TC for base pairing. Thus, molecules derived by the first treatment cannot pair with each other, nor can molecules derived by the second procedure. However, molecules derived by the first procedure can base pair to form dimers with molecules derived by the second procedure. Similar combinations can be arranged with other restriction enzymes as long as the two 5' bases of one 5' overhang are complementary to the two 5' bases of the other overhang.

Neither of the above techniques usually leads to regeneration of the 6-bp restriction site that was originally cleaved in the vector. To allow the efficient recovery of the ends of the insert, it is prudent to use a vector site that is immediately flanked by restriction sites as is found in the polylinker of several useful vectors.

Preparation of High Molecular Weight Yeast DNA. A procedure for the preparation of yeast DNA is described elsewhere in this volume.[1] Alternatively, the following modification of the method of Cryer *et al.*[42] has been used with good success.

1. Grow strain to 2×10^8 cells/ml in YEPD. Use at least 1 liter of cells.

2. Wash once in one-fifth volume of ice-cold 50 mM Na$_2$EDTA and resuspend in 50 mM Tris-HCl, pH 9.5, 2% (v/v) 2-mercaptoethanol (2-ME) for 15 min at room temperature.

3. Resuspend cells in 1/50th volume of 1 M sorbitol, 1 mM Na$_2$EDTA, pH 8.5, and 50 μg/ml Zymolyase 100T (ICN, Costa Mesa, CA). Incubate at 37° with gentle shaking until greater than 95% of the cells burst on dilution of a small sample into a solution of 1% sarkosyl (examined with a microscope).

4. Harvest spheroplasts by centrifugation and resuspend in 1/200th volume of lysis buffer (0.1 M Tris-HCl, pH 9.5, 0.1 M Na$_2$EDTA, 0.15 M NaCl, 2% 2-ME).

[42] D. R. Cryer, R. Eccleshall, and J. Marmur, *Methods Cell Biol.* **12**, 39 (1975).

5. Freeze the cell pellet in liquid nitrogen. Spheroplasts can be stored at −70°.

6. Thaw spheroplasts on ice and lyse by 3-fold dilution into lysis buffer made 4% (v/v) with sarkosyl and incubate at 45° for 20 min.

7. Add an equal volume of modified lysis buffer (lysis buffer prepared at pH 8.0 rather than pH 9.5 and containing 4% sarkosyl) and incubate at 70° for 15 min.

8. Add DNase-free RNase (prepared by boiling for 10 min in 50 mM potassium acetate, pH 5.5) to 0.1 mg/ml and incubate at 45° for 1 hr.

9. Add Pronase (Sigma, B grade) to a final concentration of 1.33 mg/ml in two aliquots at hourly intervals and incubate the mixture at 45° for a total of 2 hr.

10. Heat at 70° for 15 min.

11. Extract with an equal volume of chloroform/isoamyl alcohol (24:1, v/v) by gentle rocking at room temperature until a white emulsion forms. Separate phases by centrifugation at 20,000 g at room temperature. Incubate the aqueous supernatant at 45° to remove traces of chloroform.

12. To isolate the high molecular weight DNA from smaller sheared fragments, 2-μm DNA, and most of the mitochondrial DNA, the samples are centrifuged on preparative sucrose gradients (5–20% sucrose, 20 mM Tris-HCl, pH 8.0, 20 mM Na$_2$EDTA, 0.2 M NaCl, and 0.1% sarkosyl). Ten-milliliter samples are layered onto 24-ml gradients over a 3-ml cushion of Angio-Conray (Conray Sterile Solution, Picker, Highland Heights, OH), or other similar high-density cushion, made up to 20% sucrose.

13. Spin gradients in a Beckman SW 27 rotor at 13,500 rpm for 17 hr. Collect 1-ml samples from the top of the gradients using a wide-bore pipette (>1 mm orifice). Try to avoid the very viscous milky layer composed of cell wall material at the bottom of the gradient. The DNA should band a few fractions above this layer. Samples containing the DNA can be determined either by the viscosity of the sample or by mixing small aliquots with equal volumes of 1 μg/ml ethidium bromide and observing the fluorescence of droplets on irradiation at 354 nm. Pool DNA fractions and dialyze against 0.15 M NaCl, 10 mM Tris-HCl, pH 8.0, and 1 mM Na$_2$EDTA.

14. Determine the volume of the dialyzed sample and add CsCl to 10 g per 8 ml of solution (refractive index 1.400). Centrifuge at 50,000 rpm for 36 hr in a Type 50 Beckman rotor or equivalent.

15. Collect DNA by dripping through a 16-gauge needle. Determine DNA-containing fractions as above. Cesium chloride is removed by dialysis against 10 mM Tris-HCl, pH 8.0, 1 mM Na$_2$EDTA (TE). Determine the DNA concentration by absorbance at 254 nm. The yield should be 300 to 400 μg per 10 g wet cell pellet.

Size Fractionation of Partially Digested Genomic DNA. The procedure for preparation of partially digested, size-fractionated genomic DNA is modified from Maniatis *et al.*[43] The use of *Sau*3A is assumed, but other 4-bp recognition site enzymes that leave an overhang that is the same as a 6-bp enzyme can be used.

1. Prepare a series of 2-fold dilutions of restriction enzyme prepared in enzyme storage buffer. For *Sau*3A use 50 mM KCl, 10 mM Tris-HCl, pH 7.4, 0.1 mM Na$_2$EDTA, 1 mM dithiothreitol, 200 units (U)/ml bovine serum albumin (BSA), 50% glycerol.

2. Prepare several 1-μg aliquots of DNA in digestion buffer (100 mM NaCl, 10 mM Tris-HCl, pH 7.5, 10 mM MgCl$_2$, 100 μg/ml BSA). Starting with a total of 1 U of enzyme per microgram DNA, add aliquots of each dilution to each tube. Do not exceed 10% final concentration of glycerol. Incubate at 37° for 1 hr.

3. Stop reaction with DNA gel-loading buffer containing SDS and Na$_2$EDTA. Electrophorese samples on an agarose gel containing 0.5 μg/ml ethidium bromide along with appropriate DNA size standards.

4. Determine the enzyme concentration that yields a maximum of fluorescence within the desired size range.

5. Digest 100 μg of DNA in scaled-up reactions at each of three diffferent enzyme concentrations. Use the concentration determined in Step 4, 0.7×, and 0.5× for the same times as above. The concentration determined by fluorescence measures the mass distribution and not the number of fragments within a size range. To increase the number of fragments at the high end of the distribution, the lower concentrations are used. In addition, using several different concentrations helps avoid restriction site cleavage bias. Be careful to scale up all parameters of the digestion equally. Dilute the enzyme as above, before adding to the DNA.

6. Stop the reaction by addition of Na$_2$EDTA to 25 mM and NaCl to 0.2 M. Pool the DNA samples. Remove proteins by phenol extraction and ethanol precipitation.

7. Dissolve the air-dried pellet in TE, pH 8.0, at 0.5 to 1 mg/ml. Layer on top of one or two preformed sucrose gradients (10–40% sucrose, 1 M NaCl, 20 mM Tris-HCl, pH 8.0, 10 mM Na$_2$EDTA). Centrifuge at 26,000 rpm for 24 hr in a Beckman SW 27 rotor or equivalent.

8. Collect 10-drop fractions from the bottom by dripping through a 20-gauge needle. Add 10 μl of every third fraction to DNA gel loading buffer and fractionate on a 0.4% agarose gel along with appropriate size standards.

[43] T. Maniatis, E. F. Fritsch, and J. Sambrook, "Molecular Cloning: A Laboratory Manual." Cold Spring Harbor Laboratory, Cold Spring Harbor, New York, 1982.

9. Pool samples containing DNA of appropriate size (e.g., 10-20 kb). Dialyze 24 hr against 3 changes of TE, pH 8.0, to remove sucrose.

10. Concentrate the DNA by repeated extraction with equal volumes of *sec*-butanol. Once the volume is reduced to below 0.5 ml, add sodium acetate to 0.3 M and 2 volumes of cold ethanol to precipitate the DNA. Reprecipitate the DNA, rinse with 70% ethanol, and air dry the pellet. Resuspend in 50 μl TE, pH 8.0. Measure the DNA concentration by running a small fraction on an agarose gel along with standards of known concentration. Expect a yield of 5-10% of the input DNA.

Assembly of Library. The first part of this section describes construction of the library using dephosphorylated vector DNA. The second part describes construction using partially filled-in overhangs. The example given is for an approximately 8-kb vector, YCp50, using *Sau*3A-digested DNA. For many of the steps in which enzymes are titrated, two different assays are described, gel electrophoresis and transformation into *E. coli*. In general, the transformation assays are much more sensitive but give less information regarding the detailed molecular events. Therefore, the gel assays are included for trouble-shooting purposes, but the transformation assays are more useful for optimizing the reactions.

Dephosphorylated Vector

1. Cleave the vector DNA with *Bam*HI. Stop the reaction by addition of Na$_2$EDTA and phenol extraction. Cleavage must be thorough, since uncut molecules will be carried through the entire procedure and give rise to plasmids without inserts. Use a variety of enzyme concentrations and assay cleavage both by gel electrophoresis and by transformation into *E. coli*.

Some enzyme and DNA preparations contain residual levels of exonuclease activity. Measure the ability of the cut DNA to be ligated by diluting 0.2 μg into 100 μl ligation buffer (dilute conditions to favor intramolecular events); add ligase and incubate overnight at 15°. To one-half of the sample add carrier tRNA, ethanol precipitate, and run on an agarose gel with ethidium bromide to check for the appearance of reclosed circles and multimers. Use the second half of the sample to transform *E. coli* along with uncut and unligated samples as controls. If the ends are intact, at least 50% of the transformation activity will be recovered. Pick the lowest concentration of restriction enzyme that yields less than 0.1% transformation compared to uncut DNA, which can still be religated. Scale up the digestion conditions and prepare at least 20 μg of cleaved vector DNA. Purify the DNA by phenol extraction and ethanol precipitation. Rinse the pellet with 70% ethanol and air dry.

2. Remove the 5'-phosphates from the vector using calf intestinal phosphatase (CIP). Dissolve the DNA pellet in minimal volume of water. Dilute 2.5 μg into 500 μl of 50 mM Tris-HCl, pH 8.0. Split into 10 tubes. To each add 5 μl of a series of freshly prepared 2-fold serial dilutions CIP (Boehringer-Mannheim, Mannheim, FRG), in the same buffer. Begin with 0.1 U/μg DNA. Incubate at 37° for 1 hr. Terminate the reactions by addition of Na$_2$EDTA to 25 mM. Heat samples at 70° for 5 min. Add 5 μg carrier tRNA, NaCl to 0.2 M, and extract twice with phenol. Ethanol precipitate to remove traces of phenol. Assay the phosphatase reaction by a ligation assay; dilute DNA up to 250 μl ligation buffer and ligate overnight as above. For half of the sample measure the transformation efficiency into *E. coli* relative to uncut DNA, cut DNA, and cut DNA ligated without CIP treatment. Ethanol precipitate the remaining portion of the samples and run on an agarose gel with ethidium bromide. Determine the minimal concentration of CIP that reduces ligation by greater than 95%. Scale up the reaction to prepare 10 μg of protein-free, cleaved, dephosphorylated vector DNA.

3. Ligate insert and vector DNAs. Trial ligations should be performed to ensure the integrity of the vector ends, the efficiency of the transformation protocol, and the optimal ratio of vector to insert DNA. As mentioned above, the sizes of the vector and insert fragments and the DNA concentration have important effects on the efficiency of ligation to form the desired plasmids. We have used the following parameters with good results for an 8-kb plasmid. Mix the vector DNA with insert DNA in a 2:1 ratio by weight. For 10- to 15-kb inserts perform the ligation at 30 μg/ml total DNA concentration, at 22.5 μg/ml for 15- to 20-kb inserts, and at 15 μg/ml for 20- to 30-kb inserts. Heat to 37° for 5 min to separate sticky ends and ligate overnight at 15°. Transform *E. coli* by the most efficient protocol available. Compare the efficiency of transformation with and without ligase and with and without added insert DNA. If all is working well, the addition of the insert DNA should lead to a modest (5- to 10-fold) stimulation of the transformation frequency.

The absence of any stimulation of ligation of the vector in the presence of insert does not necessarily mean that the ligation did not work. The size of the fragments being ligated, the low molar ratio of insert relative to vector, and the dephosphorylation of the ends tend to make this step relatively inefficient. In addition, any trace of exonuclease activity (e.g., from the ligase) will allow removal of the dephosphorylated vector overhangs. Once the ends become blunt they can be ligated together by T4 DNA ligase. This can be demonstrated by showing that the restriction site that was cleaved is not regenerated in plasmids recovered from several transformants. Alternatively, if the site lies within a drug resistance gene, the gene will become inactivated in the transformants. Taking these effects into account, the best controls for the ligation of the dephosphorylated vector are the following: (1) add an equal mass of DNA which has been completely digested with *Sau*3A and (2) add an equal mass of DNA with incompatible ends (e.g., from shearing). The size-fractionated DNA should stimulate the ligation about 10-fold and the *Sau*3A-digested DNA about 100-fold relative to the sheared DNA.

The frequency of transformation may be quite low by standard $CaCl_2$ treatment methods. Accordingly, everything possible to boost transformation above 10 transformants/ng of vector DNA should be used. The cumulative result of various parameters, each of which exert only small effects, can be quite large. Before committing the library DNA, test the strain to be transformed for efficiency and keep overnight on ice (being sure that this does not harm transformation in your favorite strain). Plating out transformants at 4-fold lower density will give higher transformant yields. More than 50–100 ng of total DNA per transformation saturates the system. Pay careful attention to maintaining the cells at 0° before heat shock. All pipettes, tubes, solutions, and centrifuge rotors that contact the cells should be prechilled. In addition, some commercially prepared transformation systems boast very high efficiencies, and their use might be considered.

4. Pool the transformants into batches of several thousand colonies by washing colonies off of the transformation plate into a small amount of LB. Centrifuge cells to concentrate, resuspend in 15% glycerol, and freeze half of the cells in several 1-ml aliquots at −70°. Use the remainder as the inoculum for a large-scale preparation of plasmid DNA.

In the long run it is most efficient to create several independent pools, each containing about 3 times the number of plasmids that would be required to make up the yeast genome. Each pool should have a greater than 95% probability of containing the gene of interest; however, by taking only one positive transformant from each pool, independence is assured. This is important as the recovery of the plasmids from yeast and their characterization can be time consuming. Examination of several different overlapping clones facilitates the localization of the complementing gene.

5. Check the library first by determining the number of plasmids containing inserts. If a site within a drug resistance element was used to clone into, then determine the ratio of resistant to sensitive transformants. Plate out dilutions of the primary transformants on LB agar to achieve 100–300 colonies per plate. Replica plate to drug-containing medium. To ensure that the drug-sensitive transformants contain inserts, prepare plasmid DNA from 20 random transformants and digest with a restriction enzyme. Greater than 90% of the plasmids should be drug sensitive and contain inserts.

6. Check the representation of the library by measuring the frequency

of some expected gene or sequence. Use colony hybridization to check the frequency of ribosomal RNA sequences. As there are 100 to 200 repeats of rDNA, these sequences should be present in about 10% of the inserts. For a 15-kb average insert size, any unique gene should be present at about 0.1%.

7. To recover the library from storage, see the section above on banking by mail.

Partially Filled-In Overhangs

1. Prepare completely digested vector DNA as in Step (1) above. (For convenience, this protocol assumes cleavage at a *Sal*I or *Xho*I site. Other sites could be used, as noted above.)

2. Dissolve 20 μg vector DNA in 20 μl of 0.1 × TE, pH 8.0. Add 20 μl of 10× buffer [60 mM Tris-HCl, pH 7.5, 60 mM NaCl, 60 mM MgCl$_2$, 0.5% gelatin, and 15 mM dithiothreitol (DTT)]. Add 4 μl of 10 mM dCTP and 4 μl of 10 mM dTTP. Bring the volume to 200 μl with water and add 10 U of the Klenow fragment of DNA polymerase I. Incubate at room temperature for 15 min. To stop the reaction, add Na$_2$EDTA to 25 mM and heat to 70° for 5 min. Add NaCl to 0.2 M, phenol extract once, and ethanol-precipitate the DNA. Prepare the partially *Sau*3A-digested genomic DNA in a similar fashion but using dATP and dGTP. Test the efficacy of the fill-in reactions by the gel electrophoresis and transformation ligation assays described above.

3. As neither molecule can circularize, the molar ratio of vector to insert DNA can be adjusted to 1 (e.g., 5 μg of 8-kb vector plus 10 μg of 15- to 20-kb insert DNA in 1 ml ligation mix). Maintain the low DNA concentration to avoid formation of unproductive multimers. Test the ligation by demonstrating stimulation of ligation following addition of insert DNA. Use DNA that has been completely digested with *Sau*3A, and filled in as above, as a positive control. Use DNA that has not been filled in as a baseline control.

4–6. Transform and characterize as described above.

Using Yeast Plasmid Banks

Recipient Strains

The only requirement for the mutation in the gene of interest is that it yield a phenotype that can be readily distinguished from that of a strain carrying the wild-type allele of the gene. In almost all situations to date, the selectable marker carried on the plasmid is one of four commonly used cloned yeast genes, *LEU2, URA3, HIS3,* or *TRP1.* Accordingly, the recipient strain would carry a mutation in one of these genes and be auxotrophic for the corresponding amino acid or nucleotide.

As noted above, some cloning vectors are now available with *LYS2* or *ADE8,* although no banks constructed with these vectors are currently available. In addition, vectors carrying dominant selectable markers have been described. Using a bank constructed with such a vector could preclude the requirement for a strain carrying any marker other than the one in the gene of interest.

Selecting the Transformants

The library can now be used to transform the appropriate strain. The best approach is to select the transformants initially via the plasmid marker. Only after the transformants are obtained should they be screened for coacquisition of the desired trait. This method is preferred since the plasmid marker is usually selected via a nonreverting marker, whereas the gene of interest is either revertible or too subtle to select directly. Moreover, screening allows a measurement of the gene frequency in the library. Some genes may be underrepresented owing to toxicity in yeast or *E. coli,* and a low frequency will be the first indication of a problem.

For mutations that revert easily, authentic clones can be identified as having a uniform phenotype over the entire transformant colony. Revertant papillae are sparser and give rise to nonuniform or pebbly colonies. Such problematical screens can often be improved by reducing inoculum size so that revertant papillae are rare. This can be done by replica plating through a series of plates where the first replica immediately serves as the master for the second. In addition, examination of the plates at early times will make it easier to distinguish the papillae before they have grown together.

To prove that the candidate transformant is not a revertant, a cosegregation test is performed to show that the phenotype is actually conferred by the plasmid.

1. Inoculate a 10-ml YEPD overnight culture with a single, purified transformant colony.

2. After 18 to 24 hr at 30°, prepare serial dilutions of the culture in sterile water. Spread aliquots onto YEPD agar plates to produce 100–300 colonies. Incubate 2 days at 30°.

3. Once the colonies are large enough, replica plate to two selective media, one that will identify those colonies that have the plasmid (e.g., −URA) and one that will identify those that are complemented for the mutation of interest.

4. Score the colonies for whether the different recessive mutant phenotypes are expressed. Authentic transformants show cosegregation of both phenotypes. Transformants that segregate only Ura⁻ colonies that are still phenotypically suppressed for the gene of interest are revertants. Transformants that segregate Ura⁺ colonies (i.e., still contain a plasmid) but which now display the recessive unsuppressed phenotype may have been transformed by more than one plasmid. These are worthy of further examination.

5. Prepare DNA from individual, authentic transformants. This DNA can be used to transform *E. coli* for amplification and physical analysis.

6. Retransform the original yeast strain with plasmid DNA from *E. coli* to verify that the plasmid is solely responsible for the phenotype.

Recovering Plasmid from Transformants

Procedures for recovering plasmid DNA from yeast transformants is described elsewhere in this volume.[3]

Confirming Cloned Genes

Localizing Complementing Activity

Once a set of plasmid transformants in *E. coli* has been obtained, the next steps are to localize the gene and prove its identity. The first indications of the location of the functional genes will come from examination of the restriction map of the different complementing plasmids. If a single gene can complement the mutation, then all of the plasmids should contain overlapping DNA fragments, and the gene should be contained within the overlap region. The presence of more than one gene will be readily apparent as the different plasmids will assort into different groups according to the restriction fragments that they have in common.

There are several sophisticated methods for the further localization of the functional gene, including insertional inactivation by transposon mutagenesis[44,45] (see [22] in this volume) and linker insertion.[46] A simple method is to subclone portions of the complementing DNA fragment into a yeast plasmid. In all cases the yeast strain is then transformed with the

---

[44] H. S. Seifert, E. Y. Chen, M. So, and F. Heffron, *Proc. Natl. Acad. Sci. U.S.A.* **83**, 735 (1986).
[45] O. Hiusman, W. Raymond, K. U. Froehlich, N. Kleckner, D. Botstein, and M. A. Hoyt, *Genetics* **116**, 191 (1987).
[46] D. Shortle, *Gene* **22**, 181 (1983).

new constructs, and the transformants are examined to determine whether the plasmid still alters the phenotype of the cell. In the case of subcloning, a small fragment containing the complementing gene can usually be obtained. Restriction enzymes that produce no functional subclonable fragments must have at least one site within the gene. By this means restriction sites that lie within the gene can be determined.

Determining the Genetic Locus of the Complementing DNA

Proof that the complementing gene is the same as the gene which contains the mutation requires a genetic test of location. Homologous recombination causes a YIp plasmid to integrate at the location in the genome of the yeast DNA it contains.[4] This can be the site of the complementing DNA or the location of the selectable marker. Linearizing the plasmid by cleaving the yeast DNA will cause integration to occur preferentially at the site of cleavage. (Note: For the purpose of clarity, we assume in this discussion that the wild type is dominant. For cloning dominant mutations the logic is identical, but substitute the mutant strain in place of the wild type.)

1. Subclone the complementing fragment into an appropriate YIp vector.

2. Transform a mutant strain selecting for the vector-derived selectable marker.

3. As gene conversion events can transfer the mutation to the input DNA, check that the transformants have acquired the dominant phenotype associated with the complementing plasmid.

4. Cross the transformants to two strains: one carrying the wild-type alleles for both the gene of interest and the plasmid marker gene, the second carrying mutant alleles for both genes.

5. Sporulate the diploids and analyze the phenotypes of the tetrad spores. Different patterns will result depending on whether the normal genetic locus of the complementing gene is the same or different from the locaton of the mutation.

6. If the complementing gene and the mutation are at the same locus, then the diploid produced from the cross to the wild-type strain will be homozygous for the wild-type allele of the complementing gene. All tetrads from the cross will be 4:0 for the wild-type phenotype. If the plasmid integrates elsewhere, then the mutant phenotype will reappear in recombinant spores. For a complementing gene unlinked to the original marker, the frequency of 4:0, 3:1, and 2:2 tetrads will be 1:4:1 (see [1] in this volume). Likewise, examination of the plasmid vector marker should show

segregation of the recessive mutant phenotype depending on whether the plasmid integrates at the marker, at the complementing gene, or elsewhere (for the case in which the plasmid has integrated at the site of the original mutation, the plasmid vector marker will segregate 4:0, 3:1, and 2:2). If both markers show segregation of the recessive mutant phenotypes, then the plasmid has integrated at a location separable from both of them. This is prima facie evidence that the cloned DNA segment has homology to some region of the genome other than that of the gene of interest.

The cross to the double mutant strain is necessary to demonstrate that the two wild-type genes that come into the strain via the plasmid are actually linked to one another (all spores are parental ditypes) and that they segregate 2:2. This controls for the possibility of more than one integration event. This cross yields no information about the location of the cloned DNA. Any single integration event at any site in the genome will always yield 100% parental ditypes.

Failure of the plasmid to integrate at the genetic locus of the mutation may indicate that the cloned gene does not correspond to the mutated gene, that two or more homologous genes of interest are present in the genome, or that the cloned DNA fragment bears a portion of repeated DNA such as a Ty element. Therefore, a Southern blot hybridization analysis using the cloned gene as a radioactive probe should be performed soon after obtaining the gene of interest. If a repeated element is present on the fragment, then two recourses are available. A smaller functional subclone can be used that lacks the repetitive DNA. Alternately, the integration event may be directed to the nonrepetitive DNA sequences by cleaving the plasmid prior to transformation with a restriction enzyme that cuts the plasmid in the unique yeast DNA sequences.

Inactivation and Gene Replacement

In some cases, proof of the correctness of the cloned gene will come from sequence analysis and demonstration of sequence similarity to another previously cloned gene. In other cases, where no mutation is known and no homolog has been identified, judgment of the "correctness" of the cloned gene must rely on the phenotype produced by mutations made *in vitro* and used to replace the wild-type allele. The production and use of such mutations are covered thoroughly elsewhere in this volume.

Trouble-Shooting Guide

Although the cloning of yeast genes has been extremely successful, a number of circumstances can arise to make the isolation of particular

Incomplete Bank

No bank is perfect. Several restrictions on the creation of the bank result in significant biases in the representation of different DNA sequences. Some of these arise from phenotypic effects on the host strains, and these are considered below. Some of the restrictions arise from structural incompatibilities with the vector. For example, genes very close to centromeres will be underrepresented in the yeast transformants owing to the problems associated with the formation of a dicentric plasmid. Similarly, genes close to telomeres will be underrepresented owing to the lesser probability of a Sau3A site lying distal to the gene. To isolate these genes, alternate vectors should be considered. For the centromere genes, YRp plasmids would suffice and allow the cloning of the adjacent centromere. For telomere-linked genes, use of sheared DNA should remove the bias.

A more subtle problem is the underrepresentation of genes owing to a lack of surrounding sites. Although there are no examples of this to date, the problem can be alleviated either by the use of different restriction sites or by the creation of a library based on sheared DNA fragments. In many instances the underrepresentation can be dealt with simply by looking at enough transformants to isolate the rare plasmid of interest. To this end it is worth constructing banks that are much larger than what is anticipated to be required based on simplistic calculations of representation.

Lethal in Escherichia coli

A number of genes have been described that appear to confer considerable toxicity to *E. coli*. These range from completely lethal [*RAD4* (Ref. 47)] through highly toxic [*HOL1* (Ref. 48)] to mildly deleterious [*KAR2* (Ref. 49)]. Two approaches have been used to deal with this problem. The first utilizes plasmids that have low copy number in *E. coli*, such as plasmid pMR366 (cf. Table I). The reduced level of expression tends to alleviate the toxicity. A plasmid library has been constructed in this vector[41] (Table II). The second approach circumvents the problem by avoiding *E. coli* altogether. To this end ligation reactions are transformed directly into yeast, without *E. coli* as an intermediary. Owing to the difficulty in preparing large quantities of plasmid DNA from yeast, subsequent analysis

---

[47] R. Fleer, C. M. Nicolet, G. A. Pure, and E. C. Friedberg, *Mol. Cell. Biol.* **7**, 1180 (1987).
[48] R. Gaber, unpublished (1989).
[49] M. D. Rose and G. R. Fink, *Cell (Cambridge, Mass.)* **48**, 1047 (1987).

is necessarily indirect. In some cases subclones may be obtained that either inactivate the gene or separate it from flanking toxic genes.

High Copy Number Confusion in Yeast

The overexpression of genes on high-copy-number plasmids may result in novel phenotypes. This can be manifested either as the inability to isolate the gene of interest because it is lethal [e.g., *KAR1* (Ref. 49)] or as the isolation of other genes that suppress the mutation. To address both of these issues, libraries have been constructed in centromere-containing vectors. In some cases, however, even a single extra copy of a gene can cause suppression of mutations in other genes.[12]

[15] Gene Isolation with λgt11 System

By RICHARD A. YOUNG and RONALD W. DAVIS

The λgt11 vector–host system was developed to permit the efficient isolation of genes by using antibody probes of proteins expressed from foreign DNA.[1,2] It has also been used to isolate genes by screening for various protein activities.[3-5] The design of the λgt11 system, features of the recombinant DNA library and the probe that are important for successful screening, as well as methods used to isolate yeast recombinant DNA clones are reviewed here. Because of the utility of yeast for examining the functions of proteins that are shared with more complex eukaryotes, an increasing number of yeast genes will be isolated with antibodies that were originally raised against proteins from eukaryotes other than yeast, and we offer suggestions on how to minimize problems that may arise from the use of these and other antibodies.

Expression of Foreign DNA in λgt11 Recombinants

The ability to express foreign polypeptides from recombinant DNA in *Escherichia coli* provides a means to isolate specific recombinant DNA

[1] R. A. Young and R. W. Davis, *Proc. Natl. Acad. Sci. U.S.A.* **80**, 1194 (1983).
[2] R. A. Young and R. W. Davis, *Science* **222**, 778 (1983).
[3] J. Sikela and W. Hahn, *Proc. Natl. Acad. Sci. U.S.A.* **84**, 3038 (1987).
[4] D. Young, personal communication, 1989.
[5] H. Singh, J. H. LeBowitz, A. S. Baldwin, and P. A. Sharp, *Cell (Cambridge, Mass.)* **52**, 415 (1988).

sequences by virtue of the antigenicity or activity of the expressed gene product. The success of this approach requires that polypeptides specified by each of the individual recombinant DNA molecules in a library of recombinants be expressed at detectable levels. There are three major problems associated with obtaining expression of foreign DNA as a stable polypeptide. The first problem is that most foreign DNA does not contain the transcription control signals required for expression in *E. coli*. Thus, the foreign gene must be placed under the control of an *E. coli* promoter that is efficiently recognized by *E. coli* RNA polymerase.

The second problem with expressing foreign DNA is that unusual polypeptides are often rapidly degraded in *E. coli*.[6-9] The severity of this problem differs with each antigen; some foreign proteins are very stable, many are highly unstable. The instability of foreign proteins can be reduced in many cases by fusing the antigen to a stable host protein and by using host mutants deficient in proteolysis.[1, 10-12] Fusion of unstable foreign proteins to the carboxy terminus of the stable *E. coli* protein β-galactosidase has been shown to enhance the stability of some foreign proteins.[9,13] More importantly, the stability of the fusion product of β-galactosidase and foreign protein can be markedly increased (> 100-fold in some cases) in *lon* mutants of *E. coli*.[1] *lon* mutant strains are deficient in one of the ATP-dependent proteases which are responsible for the destruction of abnormal proteins.[14,15] This particular protease deficiency is especially useful since the presence of the mutation does not appear to alter the normal growth properties of the cell and because the *lon* protease appears to have some specificity for the class of abnormal polypeptides to which β-galactosidase fusions belong.

The third major problem with foreign synthesis in *E. coli* is that the

[6] A. L. Goldberg and A. C. St. John, *Annu. Rev. Biochem.* **45**, 747 (1976).

[7] P. Charnay, M. Gervais, A. Louise, F. Galibert, and P. Tiollais, *Nature (London)* **286**, 893 (1980).

[8] J. C. Edman, R. A. Hallewell, P. Valenzuela, H. M. Goodman, and W. J. Rutter, *Nature (London)* **291**, 504 (1981).

[9] H. Kupper, W. Keller, C. Jurtz, S. Forss, H. Schaller, R. Franze, K. Strommaier, O. Marquardt, V. G. Zaslavsky, and P. H. Hofschneider, *Nature (London)* **289**, 555 (1981).

[10] K. Itakura, T. Hirose, R. Crea, A. D. Riggs, H. L. Heyneker, F. Bolivar, and H. W. Boyer, *Science* **198**, 1056 (1977).

[11] D. V. Goeddel, D. G. Kleid, F. Bolivar, H. L. Heyneker, D. G. Yansura, R. Crea, T. Hirose, A. Kraszewski, K. Itakura, and A. D. Riggs, *Proc. Natl. Acad. Sci. U.S.A.* **76**, 106 (1979).

[12] A. R. Davis, D. P. Nayak, M. Ueda, A. L. Hiti, D. Dowbenko, and D. G. Kleid, *Proc. Natl. Acad. Sci. U.S.A.* **78**, 5376 (1981).

[13] K. Stanley, *Nucleic Acids Res.* **11**, 4077 (1983).

[14] A. Bukhari and D. Zipser, *Nature (London) New Biol.* **243**, 238 (1973).

[15] D. W. Mount, *Annu. Rev. Genet.* **14**, 279 (1980).

presence of these unusual proteins is often harmful or even lethal to the cell. Demanding high levels of gene expression can compound this problem, since constitutive high-level expression of even normal components of the cell can often be lethal.[16] A suitable solution to this problem has been to ensure that expression of the foreign protein is transient. Thus, the expression of the DNA encoding the foreign protein is repressed during early log-phase growth of the host cell culture. Near the end of this period, when the transcriptional and translational apparatus is still fully active, the expression of the foreign protein is induced, and satisfactory levels of even toxic proteins are produced before cells become inviable.

These concepts, designed to maximize the levels to which foreign proteins can accumulate in *E. coli*, have been incorporated into the λgt11 expression vector–host system.[1] The λ phage expression vector was constructed to permit insertion of foreign DNA into the β-galactosidase structural gene *lacZ* under the control of the *lac* operator. The recombinant λ phage are propagated lytically, and expression of the foreign DNA is repressed by the presence of the *lacI* gene product in specific *E. coli* host cells. Production of the foreign antigen fused to β-galactosidase can be rapidly induced by the addition of isopropyl-β-D-thiogalactoside (IPTG) to the culture medium. The presence of the *lon* mutation permits accumulation of otherwise unstable novel proteins to levels which facilitate detection by immunological or other methods (e.g., DNA binding).

The DNA Library

Successfully isolating genes with the λgt11 system depends on the quality of the recombinant DNA library (a "good" library is large enough to contain multiple copies of the DNA of interest) and the quality of the probe. Two types of libraries, genomic DNA and cDNA, have been used to isolate genes with antibody probes. For yeast, a genomic DNA library is much preferred over a cDNA library. A sheared genomic DNA library maximizes the probability that all coding sequences are equivalently represented. The small genome of yeast [12,000–15,000 kilobases (kb)] makes it highly likely that any one gene of interest can be found with a screen of 10^6 or fewer different recombinants. In practice, most yeast genes are found in a screen of only 10^5 recombinants. The relatively limited number of introns in yeast should not be a problem if the quality of the recombinant library permits an adequate number of insert break points in the DNA.

---

[16] A. Shatzman, Y.-S. Ho, and M. Rosenberg, *in* "Experimental Manipulation of Gene Expression," p. 1, Academic Press, New York, 1983.

The quality of the library and the probe are the major determinants of the frequency with which positive signals are obtained, but the size of the gene of interest can influence the frequency of positives. This is because larger genes provide a greater target size for insert DNA break points and, where polyclonal antibodies are used as probes, because the larger genes encode a greater number of potential antigenic determinants. Detailed protocols for the construction of cDNA and genomic DNA libraries in λgt11 are published elsewhere,[17,18] and λgt11 yeast genomic DNA libraries are commercially available (Clontech, Palo Alto, CA).

Antibody Probes

The ability to isolate a gene of interest will depend on the quality of the antibody probe. Independent of the type of antibody (polyclonal or monoclonal), or its origin (it may have been raised against a purified yeast protein or a protein from another organism), it is important to investigate the behavior of the antibody on a Western blot of a yeast crude extract. Even highly specific antibodies sometimes bind to irrelevant proteins, albeit with much lower avidity. The results of a Western blot will reveal how many different proteins are bound by the antibody and the signal intensity obtained with the protein of interest relative to other cross-reacting proteins. Note that the information can be misleading, as it depends on the relative amounts of the different proteins present in the crude extract, as well as the relative avidity of the antibody for the various proteins. However, a Western blot will provide some clues to the specificity of the antibody and allows one to determine the experimental condition that produces the maximum signal-to-noise ratio for the antibody.

Several factors influence the successful use of antibody probes of λgt11 recombinant DNA expression libraries. The amount of antigen that adheres to the nitrocellulose filter during a plaque lift varies with each fusion protein, apparently related to its level of expression and stability in the host cells. Successful screens with antibody probes have been accomplished with plaques containing as little as 30–60 pg of antigen, although some plaques contain as much as 200–800 pg of protein. The ability to detect limited amounts of antigen will depend on the titer and the binding characteristics of the antibody. The best signal-to-noise ratios are produced by high-titer, high-affinity antibodies used at low dilution (~ 1 : 1000).

[17] R. A. Young, B. Bloom, C. Grosskinsky, J. Ivanyi, D. Thomas, and R. W. Davis, *Proc. Natl. Acad. Sci. U.S.A.* **82**, 2583 (1985).

[18] T. V. Huynh, R. A. Young, and R. W. Davis. in "DNA Cloning Techniques: A Practical Approach" (D. Glover, ed.), Vol. 1, p. 49. IRL Press, Oxford, 1985.

There are two types of antigenic determinants bound by antibodies.[19,20] A segmental determinant occurs within a continuous segment of the polypeptide. An assembled topographic determinant consists of amino acid residues separated in the primary sequence but brought together in the surface topography of the native protein through folding. Antibodies that recognize proteins on Western blots are thought to recognize segmental determinants. In contrast, antibodies that bind to assembled topographic determinants may not be useful in isolating λgt11 clones because these determinants may not form the appropriate structure in a foreign environment. In practice, antibodies that produce good signals on Western blots generally produce good signals in the λgt11 screening procedure when used at similar dilutions.

Both polyclonal and monoclonal antibodies have been used successfully to isolate λgt11 clones. Because they often recognize multiple determinants, polyvalent antibodies may identify a larger fraction of the set of all clones that express a protein of interest. Both monoclonal and polyclonal antibodies can bind multiple proteins by virtue of cross-reactive epitopes, and isolation of the correct DNA must be confirmed by additional criteria. These additional criteria may include the length of the RNA of interest, as determined by Northern analysis, and the size of the protein(s) encoded by the clone, as determined by *in vitro* translation of yeast mRNA isolated by hybridization with the cloned DNA. Of course, sequence analysis of the DNA and microsequence analysis of a portion of the protein of interest will provide the most definitive confirmation. More detailed discussion of the use of antibodies in immunoscreening λgt11 recombinant DNA libraries can be found elsewhere.[21,22]

Protein Activity

Genes have been isolated from λgt11 libraries by screening for various protein activities. Sikela and Hahn[3] used calmodulin as a ligand to screen a λgt11 cDNA library for genes that encode calmodulin-binding proteins. Singh and co-workers[5] have isolated a gene encoding an enhancer binding protein by screening a λgt11 library with oligonucleotide probes containing the DNA-binding site for the protein. This approach has been used with a

---

[19] D. C. Benjamin, J. A. Berzofsky, I. J. East, F. R. N. Gurd, C. Hannum, S. J. Leach, E. Margoliash, J. G. Michael, A. Miller, E. M. Prager, M. Reichlin, E. E. Sercarz, S. J. Smith-Gill, P. E. Todd, and A. C. Wilson, *Annu. Rev. Immunol.* **2**, 67 (1985).

[20] J. A. Berzofsky, *Science* **229**, 932 (1985).

[21] M. Snyder, S. Ellege, D. Sweetser, R. A. Young, and R. W. Davis, this series, Vol. 154, p. 107.

[22] R. Mierendorf, C. Percy, and R. A. Young, this series, Vol. 152, p. 458.

λgt11 yeast genomic DNA library to isolate *Saccharomyces cerevisiae* DNA encoding *REB1*, a protein that binds to the transcriptional enhancer of the ribosomal RNA promoter.[23] Finally, genes have been cloned from λgt11 libraries that complement *E. coli* mutants.[4] In principle, a wide variety of protein activities could be used to detect the presence of a gene. In instances where it is possible to detect a protein by its activity, it may be possible to map the active polypeptide domain using techniques similar to those used to map epitopes.[24,25] It is important to remember that in *E. coli* many foreign proteins are rapidly degraded to small polypeptides that may lack enzyme activity but continue to be antigenic. Thus, the frequency with which signals are obtained when screening for activity may be lower than the frequency of positives observed in antibody screening.

Screening Procedure

The following procedure is preferred in our laboratory for screening with antibody probes. For screening with DNA probes, consult Ref. 5.

1. Grow plating cells. Streak *E. coli* Y1090 for single colonies on LB plates (pH 7.5) containing 50 g/ml ampicillin and incubate at 37°. Beginning with a single colony, grow Y1090 to saturation in LB (10 g/liter Bacto-tryptone, 5 g/liter Bacto-yeast extract, 10 g/liter NaCl, pH 7.5) plus 0.2% maltose at 37° with good aeration. Add $MgSO_4$ to 10 mM.

2. Infect cells with library. For each 90-mm plate (15 g Bacto-agar/liter), mix 0.2 ml of the Y1090 culture with 0.1 ml of λ diluent (10 mM Tris, pH 7.5, 10 mM MgCl) containing up to 3×10^4 plaque-forming units (pfu) (we favor 10^4 pfu) of the λgt11 library. For 150-mm plates, use 0.6 ml of the Y1090 culture with up to 10^5 pfu (we favor 3×10^4 pfu) in λ diluent. Adsorb the phage to the cells by incubation at 37° for 15 min.

3. Plate cells. Add 2.5 ml (for a 90-mm plate) or 7.5 ml (for a 150-mm plate) LB soft agar (8 g Bacto-agar/liter LB, pH 7.5) to the culture and plate. Incubate the plates at 42° for 3–4 hr. Plaques should be just visible at this point.

4. Overlay nitrocellulose filter. Move the plates to a 37° incubator. Quickly overlay each plate with a damp nitrocellulose filter disk which has been wetted in 10 mM IPTG. Schleicher and Schuell (Keene, NH) BA85 filters are particularly good for this protocol, although old filters can lose their binding properties. If necessary, blot the nitrocellulose to ensure that

---

[23] Q. Ju and J. Warner, personal communication, 1989.
[24] V. Mehra, D. Sweetser, and R. A. Young, *Proc. Natl. Acad. Sci. U.S.A.* **83**, 7013 (1986).
[25] M. Snyder, S. Elledge, and R. W. Davis, *Proc. Natl. Acad. Sci. U.S.A.* **83**, 730 (1986).

no pools of liquid remain before overlaying on the plate. Incubate for 3-4 hr longer at 37°.

5. Prepare antigen-bound filter for antibody screen. Move the plates to room temperature, mark the position of the filter on the plate with a needle, and remove the filters carefully. If the top agar tends to stick to the filter rather than the bottom agar, chill the plates at 4° for 10-20 min.

Do not allow the filters to dry out during any of the subsequent steps. Perform all of the following washing and incubation steps at room temperature with gentle shaking. Rinse the filters briefly in TBST (50 mM Tris-HCl, pH 8, 150 mM NaCl, and 0.05% Tween 20). Incubate the filters, antigen side up, in TBST plus 0.5% bovine serum albumin (BSA) for 5 min. Use 5 ml per 82-mm filter and 10 ml per 132-mm filter.

6. Antibody binding. Incubate the filters in TBST plus 0.1% BSA and antibody with gentle shaking for 1 hr at room temperature. Use 5 ml per 82-mm filter and 10 ml per 132-mm filter.

7. Wash. Wash the filters in TBST 3 times, 5 min each time.

8. Detection of bound antibody. Two different nonradioactive detection protocols are described for detection of the primary antibody. Protocols incorporating horseradish peroxidase-coupled or alkaline phosphatase-coupled secondary antibody produce signals directly on the nitrocellulose filter, thereby exactly reproducing the pattern of plaques on the plate (a faint background is produced by each plaque). This allows the precise location of the single plaque producing the signal, reducing the subsequent work involved in plaque purification. Moreover, the frequency of false positives is very low, and these differ in appearance from the donut-shaped, plaque-sized genuine positive signals. Signal-to-noise ratios and sensitivities obtained by using the two protocols do not differ substantially. The first of the two protocols is less time consuming.

Alkaline Phosphatase-Conjugated Second Antibody

1. Transfer the washed, antibody-bound filters to TBST containing affinity-purified, alkaline phosphatase-conjugated goat IgG (Promega Biotec, Madison, WI; 5 μl antibody/5 ml of TBST). Incubate for 30 min with gentle agitation.

2. Wash the filters in TBS (no Tween 20) 3 times, 5 min each time.

3. Incubate with the substrate solution according to the manufacturer's instructions.

4. After the color develops, wash with 2 changes of distilled water and allow to dry.

Biotinylated Antibody Followed by Avidin-Conjugated Horseradish Peroxidase

1. Transfer filters to TBST containing biotinylated second antibody, used according to the manufacturer's instructions.
2. Wash the filters in TBST 3 times, 5 min each time.
3. Transfer the filters to TBST containing Vectastain ABC reagent (Vector Laboratories, Burlingame, CA). Incubate for 30 min with gentle agitation.
4. Wash in TBS (no Tween 20) 3 times, 5 min each time.
5. Incubate the filters in the peroxidase substrate solution. After the color develops, wash with two changes of distilled water and allow to dry.

The construction of Y1090 is described in Ref. 2. λgt11 is described in Ref. 1. An $hsdR^-$ derivative of Y1090 is available from Promega Biotec.

Comments on Screening Procedure

1. Use fresh *E. coli* Y1090 (e.g., an overnight culture) to minimize the number of inviable cells and the lag time.
2. The plaque size can be adjusted by reducing or increasing the amount of cells plated.
3. Ideally, the plates should be fresh because phage diffusion will be greatest in moist plates, producing large plaques, and large plaque size is desirable. However, the excessive moisture in fresh plates will pool on the agar surface after placing plates in the incubator, and plaques will smear. The best solution is to plate infected cells on fresh (0- to 2-day-old) plates, place them in the 42° incubator, and then check for pools of liquid after 15–20 min. The tops of plates with pooled liquid can be removed or tilted ajar for a few minutes to permit evaporation. *Do not overdry.* Pools can remain on the surface of plates for up to 40 min without causing smearing of the plaques.
4. Plaques should just be visible before overlaying the nitrocellulose filter. If λgt11 plaques are not clearly visible within 4 hr after plating, there is a problem with the cells, phage, or plates.
5. If top agar sticks to the nitrocellulose filters, even when the filters are removed carefully and slowly, and if chilling the plates at 4° does not eliminate the problem, the plates still can be used. Although it will be more difficult to find a specific plaque on the plate if the top agar does not remain intact, large numbers of phage remain on the surface of the bottom agar after the top agar is removed, allowing the isolation of phage in the area of a plaque. During the washing and blocking step, almost any source

of protein can be substituted for BSA, if desired, as long as it is not bound by antibody.

6. Incubate the filter with the antigen side up. If necessary, the filter can be incubated with antibody overnight without significant changes in signal. However, remember that phage continue to diffuse in the agar during this time, and that this may affect the level of clonal purity, particularly in later screens. The antibody binding conditions can be changed if preliminary experiments with Western blots indicate that alternate conditions improve the signal-to-noise ratio.

7. Most investigators rock the filters back and forth during antibody binding and washing steps. This is not absolutely necessary but helps somewhat. Agitating the filters in a circular motion produces undesirable effects, as reagents tend to concentrate around the outer areas of the filter.

8. With nonradioactive protocols, signals should be obtained within 15 min and are sometimes seen within only 1–2 min after the addition of the substrate solution. ^{125}I-Labeled protein A protocols,[1,2] while sensitive, are prone to producing false-positive signals and are frequently abandoned by most investigators.

Additional Reading

Essential features of the λgt11 expression system and a generally successful screening method have been reviewed here. Additional details of the system and variations on the screening method can be found in Snyder et al.[21]

Phage and Bacterial Strains

λgt11 = *lac5 Δ(shin*dIII*λ2-3) sriλ3° c*I857 *srIλ4° nin5 srIλ5° Sam100*
Y1090 = *E. coli ΔlacU169 proA$^+$ Δlon araD139 strA*
   supF [*trpC22*::Tn*10*] (pMC9)
pMC9 = pBR322-*lacI^Q*

[16] Gene Overexpression in Studies of *Saccharomyces cerevisiae*

By JASPER RINE

The overproduction of proteins in yeast has been of practical importance to the biotechnology industry and of technical utility in basic research. In addition, the phenotypes conferred on cells by overproduction of certain proteins offer important insights into fundamental aspects of yeast biology; this chapter deals primarily with such uses of overproduction. The biological consequences of overproducing a protein are often not due to the increase in the level of that protein per se, but to the altered balance between the protein begin overproduced and other proteins within the cell. This chapter discusses the use of altered gene dosage as it applies to genetic studies of yeast, emphasizing a conceptual rather than technical perspective.

Methods of Achieving Overproduction

The tools of overproduction are of three types: (1) fusions of an open reading frame to a more powerful promoter such as those of *GAL1*, *GAL10*, and *ADH1*, as described in [25] and [26] in this volume, (2) use of plasmid vectors that replicate in multiple copies per cell, as described in [14] in this volume, and (3) use of Ty transposition vectors to amplify gene copy number by semirandom insertion of gene copies into the genome.[1] Of course, the latter two applications are limited to genes that are not capable of autogenous regulation. However, the use of segmental aneuploidy as a general mapping strategy in *Drosophila* genetics[2] indicates that the products encoded by genes are generally proportional in abundance to the copy number of the genes. Therefore, genes capable of regulating their own expression may also be the exception rather than the rule in yeast.

Multicopy plasmid vectors contain the origin of replication of the endogenous 2-μm plasmid and depend on proteins encoded by the endogenous plasmid for their high-copy maintenance. Since the vast majority of laboratory strains contain 2-μm plasmids, these vectors are widely applicable. In the rare strains lacking the endogenous plasmids, 2-μm-based vec-

---

[1] J. D. Boeke, H. Xu, and G. R. Fink, *Science* **239**, 280 (1988).
[2] D. L. Lindsley, L. Sandler, B. S. Baker, A. T. C. Carpenter, R. E. Denell, J. C. Hall, P. A. Jacobs, G. L. G. Miklos, B. K. Davis, R. C. Gethman, R. W. Hardy, A. Hessler, S. M. Miller, H. Nozawa, D. M. Parry, and M. Gould-Somero, *Genetics* **71**, 157 (1972).

tors that have only the *ori* region behave as conventional *ARS* plasmids and are too mitotically unstable for most purposes. Far and away the most commonly used multicopy plasmid and multicopy plasmid library is YEp24 and its associated library,[3] although more recent plasmids with polylinkers, such as pSEY8[4] or YEp351 and YEp352,[5] would provide technical advantages were libraries available in them. The library constructed in YEp13[6] is difficult to find in an unamplified condition and is rather unwieldy owing to a poor selection of restriction sites. In strains containing the endogenous 2-μm plasmid, proteins encoded by genes carried on YEp24 are commonly overproduced approximately 10-fold,[7] although greater levels of overproduction are occasionally reported.

A special class of 2-μm-based plasmids are those that are maintained at particularly high copy number, such as JDB207, which is replicated at approximately 200 copies per cell.[8] These plasmids contain a version of the yeast *LEU2* gene truncated at the 5' end that is likely to be responsible for maintaining these plasmids at high copy number to provide enough *LEU2* function to permit growth on selective medium. It is often desirable to use several micrograms of DNA in each transformation of spheroplasts using these plasmids in order to provide enough copies of the *LEU2* gene to enable the transformants to grow. Transformations with these plasmids using lithium acetate-treated cells as recipients may have technical limitations owing to the smaller number of DNA molecules taken up per cell using this procedure. A potential drawback of these plasmids is that in several cases clones that have no effect on viability at 10 copies per cell have a dramatic effect on viability at 200 copies per cell. There seems to be no way to predict reliably which clones will and which will not affect growth at very high copy number. Therefore it may be worth considering the use of plasmids whose high-copy maintainance can be regulated.

A method for controlling the copy number at which 2-μm-based plasmids are maintained in cells exploits the properties of heterologous thymidine kinase in yeast. Yeast lack thymidine kinase and hence are sensitive to the drugs amethopterin and sulfanilimide. Expression of herpes simplex virus thymidine kinase (HSV TK) confers resistance to these drugs.[9] 2-μm-

---

[3] M. Carlson and D. Botstein, *Cell (Cambridge, Mass.)* **28,** 145 (1982).
[4] S. D. Emr, A. Vassarotti, J. Garrett, B. L. Geller, M. Takeda, and M. G. Douglas, *J. Cell Biol.* **102,** 523 (1986).
[5] J. E. Hill, A. M. Myers, T. J. Koerner, and A. Tzagoloff, *Yeast* **2,** 162 (1986).
[6] K. A. Nasmyth and K. Tatchell, *Cell (Cambridge, Mass.)* **19,** 753 (1980).
[7] J. Rine, W. Hansen, E. Hardeman, and R. W. Davis, *Proc. Natl. Acad. Sci. U.S.A.* **80,** 6750 (1983).
[8] J. R. Broach, this series, Vol. 101, p. 307.
[9] J. B. McNeil and J. D. Friesen, *Mol. Gen. Genet.* **184,** 386 (1981).

based plasmids containing the HSV TK gene are maintained at 10 copies per cell during growth in the absence of the drugs and at 100 copies per cell during growth in the presence of the drugs.[10] Thus, the restriction fragment containing the HSV TK gene may be useful as a portable regulator of plasmid copy number.

There are two currently used alternatives to 2-μm-based multicopy maintenance of genes in *Saccharomyces cerevisiae*. A restriction fragment containing the *TRP1* gene and associated *ARS* element, when circularized and transformed into yeast, is maintained at approximatley 200 copies per cell.[11] Plasmids replicated by the *HMR*-E silencer are maintained at approximately 30 copies per cell.[12] Although these plasmids offer the potential for alternate means of achieving elevated copy number, in neither case is the molecular basis of the high-copy maintenance understood. Thus, it is too early to evaluate their general utility for overproduction studies.

A recent alternative to the use of multicopy plasmids for achieving gene amplification is the use of Ty transposition vectors to insert multiple copies of a gene, flanked by Ty element repeats, at nearly random locations in the yeast genome.[1] Although not yet widely used, this method offers the potential for more stable propagation of the additional gene copies than plasmids typically offer. However, each extra copy of the gene is also an insertion mutation. Therefore, this approach will probably prove useful in cases where stability is more important than isogenicity.

Uses of Overexpression

Selection of Recombinant Clones through Gene Dosage Effects

If a recessive mutation is available in the gene of interest, the best route for isolating the gene is by complementing the mutation with plasmids from a plasmid library. However, fewer than 20% of yeast genes have been identified by mutation,[13] and as many as 10% of yeast genes may be duplicated, precluding the easy identification of recessive mutations in these genes. The use of multicopy plasmid vectors offers an alternate route to gene isolation that does not require a mutation in the gene of interest. In one such set of experiments, transformants containing elevated copy numbers of the gene encoding an enzyme of interest were identified by their

---

[10] G. R. Zealy, A. R. Goodey, J. R. Piggott, M. E. Watson, R. C. Cafferkey, S. M. Doel, B. L. A. Carter, and A. E. Wheals, *Mol. Gen. Genet.* **211**, 155 (1988).
[11] V. A. Zakian and J. F. Scott, *Mol. Cell. Biol.* **2**, 221 (1982).
[12] W. J. Kimmerly and J. Rine, *Mol. Cell. Biol.* **7**, 4225 (1987).
[13] R. K. Mortimer, D. Schild, C. R. Contopoulou, and J. A. Kans, *Yeast* **5**, 321 (1989).

increased resistance to inhibitors of that enzyme.[7] Since multicopy plasmids of the YEp24 variety typically provide 10-fold overproduction of the gene products encoded by the plasmid, the increased inhibitor resistance is easily detected and can even be used to select the desired transformants from a population rather than screening for them. The transformants obtained in this way can be assayed for increased activity level of the enzyme that is the target of the inhibitor.

Increased enzyme activity need not be due to increased level of the protein. In principle, the clones identified by inhibitor resistance could arise from increased dosage of a positive regulator of the enzyme that acted at the transcriptional, translational, or posttranslational level. In practice, however, the structural genes for enzymes have been identified in a number of studies, whereas putative regulators have not yet been recovered. For example, the structural genes for hydroxymethylglutaryl-CoA reductase,[14] N-acetylglucosamine transferase,[7] and cytochrome P-450 lanosterol demethylase[15] have been isolated by this approach. In the case of HMG-CoA reductase, this approach to isolating the gene was particularly useful since the duplicate copies of the structural gene would have made identification of the genes by mutational analysis difficult. Preliminary indications suggest that this technique may be very useful in Candida,[16] which lacks the benefits of meiotic genetics.

For inhibitors that block the growth of yeast, the standard procedure is to identify the minimum concentration of the inhibitor that can block the growth of yeast cells. A population of transformants containing a multicopy plasmid library is then placed onto medium containing the inhibitor at 3-fold the minimum inhibitory concentration. Positive candidates are the only cells capable of forming a colony. Although many inhibitors do not completely arrest the growth of yeast, they are still useful in this method of gene isolation. Since approximately 25 generations separate a single cell from a typical colony, even a weak inhibitor that merely doubles the generation time of nonoverproducing transformants relative to overproducers would result, over 25 generations, in a considerable colony size difference between overproducers and wild type. In fact, a gene encoding HMG-CoA reductase was isolated with such a weak inhibitor by picking the large colonies on medium containing many tiny colonies.[7]

The chief difficulty encountered by this approach is determining whether a clone that causes increased enzyme activity carries the structural gene. (Presumably, the structural gene has not been previously isolated,

[14] M. E. Basson, M. Thorsness, and J. Rine, *Proc. Natl. Acad. Sci. U.S.A.* **83,** 5563 (1986).
[15] V. F. Kalb, J. C. Loper, C. R. Dey, C. W. Woods, and T. R. Sutter, *Gene* **45,** 237 (1986).
[16] D. R. Kirsch, M. H. Lai, and J. O'Sullivan, *Gene* **68,** 229 (1988).

precluding the opportunity to determine whether the clone directs integration into the genome at the locus of the structural gene.) If an antibody to the enzyme is available, hybrid selection of the complementary RNA followed by *in vitro* translation and immunoprecipitation should reveal whether the clone contains the structural gene for the enzyme (see [37] in this volume). Other approaches to establishing the identity of the clone include establishing sequence similarity to the enzyme or part of the enzyme from other species and determining whether the clone is functionally expressed in *Escherichia coli* (e.g., Ref. 17).

Multicopy plasmid libraries also offer an alternative to screening *ts* collections by direct biochemical assay. A number of genes have been identified by brute-force enzymatic screening of the *ts* mutant collection established by McLaughlin and Hartwell as described by Sternglanz and co-workers.[18] This approach is limited to those genes whose products are capable of mutating to a thermolabile form. (In practice, nonessential genes have been isolated that just happen to be defective in a colony also containing a *ts* lethal mutation owing to the heavy mutagenesis used to establish this collection.) Furthermore, mutants isolated from this collection often contain a number of different *ts* lesions and may thus require considerable work before a corresponding clone can be obtained. For activities with a relatively simple and quantitatively reliable assay, direct enzymatic screening of a collection of multicopy-plasmid transformants may be the most efficient route to isolating a particular gene. The enzymatic screen should be suitable for screening 7000 transformants with a library whose insert size averages 10 kilobases (kb) in order to achieve statistically 5-fold coverage of the genome.

Altering Gene Dosage to Understand Specific Macromolecular Interactions

A powerful example of how overproduction of gene products can reveal interesting dimensions of gene regulation is provided by studies of the galactose regulon in yeast. In this case, the GAL80 protein acts as a negative regulator by binding directly to the GAL4 protein, a transcriptional activator, and blocking the ability of GAL4 to function. Galactose, or more likely a metabolite of galactose, acts as an inducer by freeing GAL4 from GAL80 (reviewed in Ref. 19). This regulatory relationship is dependent on the stoichiometry of GAL80 and GAL4. Overproduction of GAL4 results in galactose-independent transcription of the galactose regu-

---

[17] J. W. Wallis, G. Chrebet, G. Brodsky, M. Rolfe, and R. Rothstein, *Cell (Cambridge, Mass.)* **58**, 409 (1989).
[18] S. DiNardo, K. Voelkil, and R. Sternglanz, *Proc. Natl. Acad. Sci. U.S.A.* **81**, 2616 (1984).
[19] M. Johnston, *Microbiol. Rev.* **51**, 458 (1987).

lon. Overproduction of both GAL4 and GAL80 reestablishes galactose-dependent control.[20] Thus, the role of the inducer is indirect, and GAL4 protein can activate transcription without direct benefit of a small molecule effector. The potential applications of this approach to perturbing specific protein-protein interactions are certainly not limited to studies of transcription.

Altered gene dosage is proving to be an important tool in studying the influence of the structure of chromatin on gene expression. In fact, the first proof that changes in chromatin structure could cause changes in gene expression came from the discovery that either increased or decreased dosage of histone genes could suppress the phenotype caused by solo δ elements in the promoter region of the HIS4 gene.[21] The sensitivity of the phenotype of regulatory mutations to changes in histone gene dosage may prove to be a useful early step in the characterization of regulatory mutants.

A potential application of overexpression in studies of transcriptional regulation would be the identification of the gene encoding a protein that binds to and functions at a particular site. This approach is likely to require a single base-pair mutation in the regulatory site that confers a readily distinguishable phenotype. Multicopy plasmid libraries would be screened in the site mutant for genes whose elevated dosage overproduces the binding protein that would restore regulation at the altered site. To my knowledge, this approach has not yet been applied successfully in yeast. This lack of success may be due, until the recent past, to relatively few single base-pair mutations in regulatory sites. Considering the difficulty of obtaining compensatory mutations in a DNA-binding protein that restore binding to a mutant site,[22] and the low frequency at which these mutations occur, it is sensible to screen 7000 transformants from a multicopy library for a suppressor clone prior to initiating a mutant screen for compensatory changes in the binding protein. This approach may also find utility in studies of protein-protein interactions as well.

A related opportunity for multicopy manipulations in yeast genetics is titration of regulatory proteins by increases in the copy number of the sites recognized by the proteins. To date, most genes exhibit normal regulation when present on plasmids maintained at roughly 10 copies per cell, and, hence, there must be sufficient quantities of regulatory proteins to accommodate the increased number of sites. It is too early to know whether

[20] Y. Nogi, H. Shimada, Y. Matsuzaki, H. Hashimoto, and T. Fukasawa, *Mol. Gen. Genet.* **195**, 29 (1984).
[21] A. C. Clark, D. Norris, M. A. Osley, J. S. Fassler, and F. Winston, *Genes Dev.* **2**, 150 (1988).
[22] P. Youderian, A. Vershon, S. Bouvier, R. T. Sauer, and M. M. Susskind, *Cell (Cambridge, Mass.)* **35**, 777 (1983).

regulatory sites carried on vectors maintained at 200 copies per cell would be capable of titrating regulatory proteins.

Altering Gene Dosage to Analyze Biological Pathways

Altered gene dosage can be used to deduce dependency relationships among genes in transcriptional cascades. For example, in the control of meiosis, *RME1* acts negatively to prevent meiosis and *IME1* and *IME2* act positively to promote meiosis. *IME1* is required for efficient *IME2* transcription, but might have been required for transcription of other genes in addition to *IME2* that could also be required for meiosis. However, the discovery that increased dosage of *IME2* alone allowed meiosis in the complete absence of *IME1* indicates that the major role of *IME1* is to activate expression of *IME2*.[23] This example illustrates an interesting situation that can be readily exploited to establish the order of functions in a dependent pathway. Consider a situation in which mutation of any one of three different genes confers the same loss-of-function phenotype. If one of these genes, when present in high copy number, has the opposite phenotype, for example, constitutive expression of the process under study, the overproduction phenotype can be used in the manner of double-mutant analysis to determine which phenotype is epistatic. For example, if overproduction of gene B in mutant A results in the constitutive phenotype whereas overproduction of gene B in mutant C results in the loss-of-function phenotype, then the most likely order of function is A then B then C. Further conceptual considerations of this type of analysis have been described.[24]

Altered gene dosage has proved to be particularly useful in studies of signal transduction pathways. A large number of the *STE* genes in yeast are involved in detecting mating pheromones in the vicinity of a yeast cell and signaling the presence of the pheromone to the inside of the cell to arrest division and prepare for mating. *STE2* and *STE3* encode the α- and a-factor receptors, respectively;[25] *SCG1/GPA1* encodes the α subunit[26,27] and *STE4* and *STE18* encode the β and γ subunits of a trimeric G protein involved in mating;[28] *STE7* and *STE11* are likely to encode kinases in-

---

[23] H. E. Smith and A. P. Mitchell, *Mol. Cell. Biol.* **9**, 2142 (1989).
[24] L. M. Hereford and L. H. Hartwell, *J. Mol. Biol.* **84**, 445 (1974).
[25] D. D. Jenness, A. C. Burkholder, and L. H. Hartwell, *Cell (Cambridge, Mass.)* **35**, 521 (1983).
[26] C. Dietzel and J. Kurjan, *Cell (Cambridge, Mass.)* **50**, 1001 (1987).
[27] I. Miyijama, M. Nakafuka, N. Nakafuka, C. Brenner, A. Miyajima, K. Kaibuchi, K. Arai, Y. Kaziro, and K. Matsumoto, *Cell (Cambridge, Mass.)* **50**, 1011 (1987).
[28] M. Whiteway, R. Freedman, A. S. Van, J. W. Szostak, and J. Thorner, *Mol. Cell. Biol.* **7**, 3713 (1987).

volved in signaling;[29] and *STE12* is a DNA-binding protein that is likely to regulate gene expression.[30] *STE5* is necessary for transmitting the signal from receptor to the nucleus and genetic epistasis tests have placed *STE5* downstream of the trimeric G protein subunits.[31] In the early efforts to clone the *STE* genes from multicopy plasmid libraries, both the *STE4* and *STE5* clones were found to be able to complement mutations in *STE4*, whereas only the *STE5* clone could complement mutations in *STE5*.[32]

This example is instructive owing to the substantial information available about the product of *STE4*. Commonly, the ability of increased levels of one gene to compensate for mutations in another gene have been interpreted as evidence that the products of the two genes serve a related function. This explanation clearly applies to the ability of elevated levels of *COX5b* to compensate for decreased levels of cytochrome oxidase subunit 5 in *cox5a* mutants[33] and other cases of this type. However the *STE4–STE5* case appears to be quite different, for although they function in the same pathway, STE4 and STE5 proteins can hardly be considered to perform a common function. Activation of STE4 (and STE18) protein on pheromone binding leads to activation of STE5 protein and subsequent mating competence. Based on the dosage studies, increased levels of STE5 at least partially bypass the need for STE4. To generalize, increased dosage of a downstream gene can decrease dependence on an upstream activator. This principle can be expected to apply to a variety of cases. However, recent studies of the RAS-regulated pathway have raised the general concern that multicopy plasmids can suppress mutations either by acting downstream of a mutant activator[34] or by bypassing the need for a function or pathway altogether.[35]

The relationship between the G protein subunits provides another example of the importance of relative dosage of proteins. Overproduction of β subunits (STE4 protein) causes constitutive mating factor signaling. In contrast, overproduction of either the γ subunit (STE18 protein) alone or α subunit (GPA1 protein) and β subunit together does not. This result is consistent with the view that the α subunit is a negative regulator of $\beta\gamma$. A simple, yet incorrect, interpretation of this result would be that the β subunit alone is causing the constitutive mating factor signal. However, the constitutive signal caused by increased β depends on a functional γ sub-

---

[29] M. A. Teague, D. Chaleff, and B. Errede, *Proc. Natl. Acad. Sci. U.S.A.* **83**, 7371 (1986).
[30] J. W. Dolan, C. Kirkman, and S. Fields, *Proc. Natl. Acad. Sci. U.S.A.* **86**, 5703 (1989).
[31] D. Blinder and D. D. Jenness, *Mol. Cell. Biol.* **9**, 3720 (1989).
[32] V. L. MacKay, this series, Vol. 101, p. 325.
[33] C. E. Trueblood and R. O. Poyton, *Mol. Cell. Biol.* **7**, 3520 (1987).
[34] T. Toda, S. Cameron, P. Sass, M. Zoller, and M. Wigler, *Cell (Cambridge, Mass.)* **50**, 277 (1987).
[35] T. Toda, S. Cameron, P. Sass, and M. Wigler, *Genes Dev.* **2**, 517 (1988).

unit. Thus, the increased dose of β is merely titrating the α subunit, freeing the previously inactive $\beta\gamma$ subunits. Apparently γ subunit by itself cannot titrate α subunits.[36a] The general principle from these studies is that phenotypes caused by increased dosage of a regulator may result from indirect effects.

Altered Gene Dosage and Macromolecular Fidelity

The information needed to guide the assembly of complex structures in biology is often contained within the components that make up the structure. Thus, a ribosome or a bacteriophage can be assembled *in vitro* from individual components. Nevertheless, assembly is sometimes very sensitive to the relative ratio of components, as first revealed by studies of bacteriophage T4 mutants.[36b] This principle has wide applicability in yeast genetics, as illustrated by the following example.

Yeast chromosomes are complex structures made of a variety of different components whose relative levels would be expected to influence the proper assembly of the chromosome. Hartwell and colleagues reasoned that an improperly assembled chromosome might be detected by a decrease in the fidelity of mitotic chromosome transmission. Using an appropriately marked disomic strain as a detector, they found that overproduction of the histone pair H2A and H2B or overproduction of the histone pair H3 and H4 both resulted in an elevated rate of chromosome loss, indicating a decreased fidelity of chromosome segregation. The decreased fidelity was due to an altered balance of histone proteins rather than to overproduction per se since overproduction of both pairs of histone proteins simultaneously had no effect on the rate of chromosome loss.[37] This result was not altogether surprising since a histone octomer, which is a major structural component of chromosomes, has a fixed stoichiometry of the four histone proteins. Nevertheless, these data led to the discovery of two unique genes, *MIF1* and *MIF2,* from a multicopy plasmid library, that in high copy number result in decreased fidelity of chromosome transmission. Deletions of *MIF1* were viable and had an increased rate of chromosome loss in mitosis. Thus, either increased levels of *MIF1* or the absence of *MIF1* altered the fidelity of chromosome transmission. *MIF2* is an essential gene, and its loss results in a cell-cycle-specific arrest as large unbudded cells, a phenotype expected of genes whose products are required for nuclear division.[38]

[36a] G. M. Cole, D. E. Stone, and S. I. Reed, *Mol. Cell. Biol.,* **10,** 510 (1990); M. Whiteway, L. Hougan, and D. Y. Thomas, *Mol. Cell. Biol.* **10,** 217 (1990).
[36b] E. Floor, *J. Mol. Biol.* **47,** 293 (1970).
[37] D. W. Meeks-Wagner and L. Hartwell, *Cell (Cambridge, Mass.)* **44,** 43 (1986).
[38] D. W. Meeks-Wagner, J. S. Wood, B. Garvik, and L. Hartwell, *Cell (Cambridge, Mass.)* **44,** 53 (1986).

A theme that emerges from the studies discussed in this section is that the fidelity of biological processes carried out by macromolecular assemblies is more sensitive to altered dosages of the components than are the processes themselves. Loss of fidelity may be generalizable as a screen for genes that are involved in a variety of interesting processes. For example, interactions between membrane vesicles and their target membranes are relatively specific. Increased levels of individual proteins that participate in controlling vesicle targeting may decrease the fidelity of targeting and result in mislocalization. Perhaps secretion of proteins otherwise destined for the vacuole would be a phenotype sensitive enough to detect decreased fidelity of membrane interactions. In this regard, the lethality of cells lacking YPT1 protein, which is required in the Golgi for membrane interactions, can be suppressed by overproduction of SLY2, SLY12, and SLY41 (D. Gallwitz, personal communication). It will be interesting to determine whether these substitute functions, which operate in place of a normal function, cause decreased fidelity of membrane interactions in wild-type cells. Of course, loss of fidelity of a process could be achieved in rather indirect ways, but if a gene identified by a loss-of-fidelity phenotype, when disrupted, proves to be essential for the process, then that gene is likely to be directly involved in the process.

Altering Gene Dosage to Understand Cell Structure

The actin cytoskeleton of yeast directs localized growth to daughter cells and controls the budding pattern of cells. Because there is only one actin gene in yeast, the different roles of actin in the cell must be mediated by actin-associated proteins. Studies of actin-associated proteins have led to purification of three major actin-binding proteins in yeast, one of which is likely to be myosin. Antibodies against the other two proteins, known as ABP67 and ABP85, were used to clone the genes encoding the proteins.[39] Although these proteins are candidates for structural components of the yeast cytoskeleton, the ability of actin to bind tightly to nonphysiologically relevant substrates, such as pancreatic DNase,[40] emphasizes the possibility of artifactual *in vitro* associations between proteins. Strong evidence for a role for ABP85 in the cytoskeleton comes from the observation that increased dosage of the ABP85 structural gene alters the pattern of actin staining in the daughter cells. Since changing the dosage of a putative actin-binding protein changes the *in vivo* properties of actin, that actin-binding protein is almost certainly an important component of the actin cytoskeleton. To generalize, in situations in which an assay exists for

[39] D. G. Drubin, K. G. Miller, and D. Botstein, *J. Cell Biol.* **107,** 2551 (1988).
[40] K. Zechel, *Eur. J. Biochem.* **110,** 343 (1980).

component B but not for A, potential interactions between A and B can be revealed if altering the dose of A results in a change detected by the assay for B.

A dramatic example of the consequences of protein overproduction on cell structure is the response of cells to overproduction of HMG-CoA reductase, the enzyme that catalyzes the rate-limiting step of the isoprene biosynthetic pathway. HMG-CoA reductase is an integral membrane protein of the endoplasmic reticulum, held in the membrane by seven hydrophobic α helices. The two isozymes of HMG-CoA reductase in yeast are encoded by the *HMG1* and *HMG2* genes. Tenfold overproduction of the HMG1 isozyme, a protein of low abundance, causes proliferation of a membrane assembly referred to as karmellae. Karmellae are stacked pairs of membrane bilayers that tightly surround the nucleus and contain HMG-CoA reductase. Karmellae may represent exaggerated intermediates between nuclear envelope synthesis and endoplasmic reticulum formation.[41] Formation of karmellae reflects a response of the cell to the level of this particular protein and not to increased activity of the enzyme, as mutant forms of the protein lacking a catalytic domain, when overproduced, still form karmellae (R. Wright, personal communication). Thus, these studies have revealed a hitherto unknown regulatory mechanism in the cell. By monitoring the level of a key protein, in this case HMG-CoA reductase, the cell senses the need to synthesize more of the membrane within which the protein resides.

Altering Gene Dosage to Analyze Rate-Limiting Steps

The activity of a biological pathway is commonly rate-limited at a single step, although the rate-limiting step can change under different conditions. The mating-type locus regulates expression of the α-factor structural gene at the transcriptional level. Thus, with respect to mating type, synthesis of α-factor mRNA is rate-limiting for α-factor secretion. However, once α-factor mRNA is synthesized, a posttranscriptional step becomes limiting for secretion of mature α-factor, as increased copy number of the *MFα1* gene, with a corresponding increase in *MFα1* mRNA, does not lead to increased secretion of mature α-factor. Synthesis and secretion of mature α-factor form a multistep process requiring the concerted action of a variety of posttranslational processing steps that occur at different locations in the cell. The *STE13* and *KEX2* genes encode proteases required for processing α-factor peptides from the α-factor precursor. Increased copy number of both *MFα1* and *STE13* in a wild-type cell

[41] R. Wright, M. Basson, L. D'Ari, and J. Rine, *J. Cell Biol.* **107**, 101 (1988).

results in increased production of α-factor.[42] Thus, use of elevated *MFα1* gene dosage overcame one rate-limiting step and led to the identification of the next rate-limiting step.

This analysis has practical importance. The α-factor leader is widely used to direct the secretion of heterologous proteins from yeast cells, and increased levels of STE13 protein may increase the yield of heterologous proteins. Success in secretion of heterologous proteins from yeast is somewhat idiosyncratic, and there is no reason to believe that STE13 would be limiting for all secreted proteins. However, screening multicopy-plasmid libraries for clones that increase the yield of a secreted protein could identify whatever gene product is limiting for the secretion of individual heterologous proteins.

Another example of the use of overproduction in the analysis of rate-limiting steps comes from the analysis of *cdc* mutants. The rate-limiting step in cell division is a point in the G_1 phase of the cell cycle known as START. *CDC28,* a homolog of a component of vertebrate maturation-promoting factor, encodes a protein kinase essential for initiation of a mitotic cell cycle in yeast. Temperature-sensitive alleles of *CDC28* cause G_1 arrest at START in cells grown at the restrictive temperature. Hence the *CDC28*-dependent step is rate-limiting for cell division. Two genes, isolated from a multicopy-plasmid library, that restore growth at high temperature in a *cdc*28 mutant are homologs of vertebrate cyclins. These results are thought to indicate that a class of cyclins are rate-limiting for the G_1-S transition in yeast.[43] More generally, these results underscore the ability of multicopy-plasmid suppression to identify limiting components of regulated steps.

Perspectives

The use of gene overexpression has proved valuable in genetic studies in a number of different ways. Although many clones identified from multicopy-plasmid libraries have proved not to contain the gene that corresponds to the mutant locus, one is impressed by the reliability with which these other genes are relevant to the gene being studied. From past experience one can predict a few areas in which protein overproduction can be expected to be particularly revealing. For example, protein kinases such as the calcium-dependent protein kinase (O. Fields, D. Levin, and J.

[42] D. Julius, L. Blair, A. Brake, G. Sprague, and J. Thorner, *Cell (Cambridge, Mass.)* **32,** 839 (1983).

[43] J. A. Hadwiger, C. Wittenberg, M. D. Mendenhall, and S. I. Reed, *Proc. Natl. Acad. Sci. U.S.A.* **86,** 6255 (1989).

Thorner, personal communication), STE7,[29] and CDC7 offer particularly fertile opportunities for multicopy-plasmid suppression studies. Similarly, the recent identification of a putative phosphatase that affects the expression of specific genes[44] increases the probability that altered balance between certain kinases and phosphatases would have revealing phenotypes. The existence of high-copy number, endogenous plasmids in a variety of yeasts[45] offers the opportunity for overproduction by use of multicopy plasmid vectors in organisms lacking the well-developed genetic map and methodologies of *Saccharomyces*.

Acknowledgments

I thank the members of my laboratory for many stimulating discussions of the ideas described in this chapter, and Jim Umen and Christine Guthrie for perceptive and useful comments on the manuscript. Research in my laboratory is supported by grants from the National Institutes of Health and the Lucille P. Markey Charitable Trust.

[44] K. T. Arndt, C. A. Styles, and G. R. Fink, *Cell (Cambridge, Mass.)* **56**, 527 (1989).
[45] F. C. Volkert, D. W. Wilson, and J. R. Broach, *Microbiol. Rev.* **53**, 299 (1989).

[17] Preparation of Clone Libraries in Yeast Artificial-Chromosome Vectors

By DAVID T. BURKE and MAYNARD V. OLSON

Introduction

Yeast has proved to be an excellent host in which to clone large fragments of exogenous DNA as yeast artificial chromosomes (YACs). Linear DNA molecules up to several hundred kilobase pairs in length can be constructed *in vitro,* transformed into host yeast cells, and propagated as faithful replicas of the source genomic DNA.[1] The YAC system has been successfully employed to assist in the physical mapping of the nematode genome,[2] to clone large centromeric regions from the fission yeast *Schizosaccharomyces pombe,*[3] to clone a variety of multicopy and single-copy human genes,[4,5] and to clone sequences adjacent to human telomeres.[6-9]

[1] D. T. Burke, G. F. Carle, and M. V. Olson, *Science* **236**, 806 (1987).
[2] A. Coulson, R. Waterston, J. Kiff, J. Sulston, and Y. Kohara, *Nature (London)* **335**, 184 (1988).

As described in this chapter, the procedures required to prepare and screen YAC libraries of sufficient complexity to provide good theoretical coverage of a complex genome have been reduced to practice. The techniques remain laborious relative to established genomic cloning methods. There is no doubt that λ and cosmid cloning remain the methods of choice for problems that require only the isolation of a few tens of kilobase pairs of genomic DNA. However, when the need arises to analyze regions 10 times larger, the appropriate comparison is between the difficulty of isolating a single YAC or carrying out a 10-cosmid walk. At best, such walks are extremely laborious, and, particularly in mammalian systems, there is a high failure rate. YAC cloning is also ideally suited to the analysis of functional units that are too large to isolate in a single cosmid. In such cases, single YAC clones may provide ideal material for gene-transfer experiments, an approach that has already been successfully applied to the *S. pombe* centromeres.[3]

The most widely used first-generation YAC vector has been pYAC4[1] (see restriction map in Fig. 1). This vector illustrates the basic principles of YAC construction. As is, pYAC4 can be transformed into either *Escherichia coli* or yeast and will replicate as a circular molecule in either host. However, pYAC4 is normally used to clone large inserts as linear yeast chromosomes. Five types of sequence modules, assembled into the framework of the classic *E. coli* plasmid vector pBR322,[10] provide the needed functions:

1. A *CEN* sequence. *CEN* sequences provide all the cis-acting information required to confer mitotic and meiotic centromere function on DNA molecules introduced into yeast.[11] The particular *CEN* sequence in pYAC4 is *CEN4*, which is derived from the natural centromere sequence present on yeast chromosome IV.[12]

---

[3] K. M. Hahnenberger, M. P. Baum, C. M. Plizzi, J. Carbon, and L. Clarke, *Proc. Natl. Acad. Sci. U.S.A.* **86,** 577 (1989).
[4] R. D. Little, G. Porta, T. Labella, G. Carle, D. Schlessinger, and M. D'Urso, *Proc. Natl. Acad. Sci. U.S.A.* **86,** 1598 (1989).
[5] B. H. Brownstein, G. A. Silverman, R. D. Little, D. T. Burke, S. J. Korsmeyer, D. Schlessinger, and M. V. Olson, *Science* **244,** 1348 (1989).
[6] H. C. Riethman, R. K. Moyzis, J. Meyne, D. T. Burke, and M. V. Olson, *Proc. Natl. Acad. Sci. U.S.A.* **86,** 6240 (1989).
[7] S. H. Cross, R. C. Allshire, S. J. McKay, N. I. McGill, and H. J. Cooke, *Nature (London)* **338,** 771 (1989).
[8] W. R. A. Brown, *Nature (London)* **338,** 774 (1989).
[9] J.-F. Cheng, C. L. Smith, and C. R. Cantor, *Nucleic Acids Res.* **15,** 6109 (1989).
[10] J. G. Sutcliffe, *Cold Spring Harbor Symp. Quant. Biol.* **43,** 77 (1978).
[11] M. Fitzgerald-Hayes, *Yeast* **3,** 187 (1987).
[12] C. Mann and R. W. Davis, *Mol. Cell. Biol.* **6,** 241 (1986).

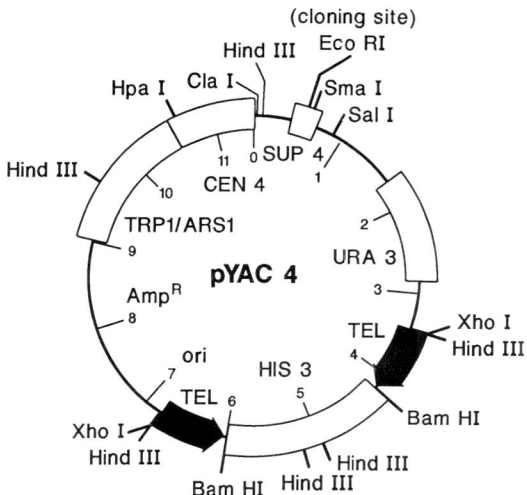

FIG. 1. Restriction map of the YAC vector pYAC4. A detailed description of the way in which the vector was constructed has been presented by Burke et al.[1]

2. An *ARS* sequence. DNA molecules introduced into yeast must contain an *ARS* (autonomous replication sequence) in order to replicate as autonomous molecules (i.e., not integrated into a natural yeast chromosome).[13] The particular *ARS* present in pYAC4 is *ARS1*, which is directly adjacent to the *TRP1* gene in the centromeric region of chromosome IV.[14]

3. The *TEL* sequences required for telomere formation. Following cleavage of pYAC4 with *Bam*HI and transformation into yeast, natural yeast telomeres form by the addition of simple sequences to the *Bam*HI ends. The *TEL* sequences, which are derived from *Tetrahymena,* can be regarded as "telomere-seeding" sequences: they do not themselves function as telomeres, but they seed the formation of functional telomeres at very high efficiency.[15]

4. An interruptible marker containing a cloning site. In pYAC4, the interruptible marker is in the short intervening sequence of the *SUP4-o* gene, an ochre-suppressing allele of a tRNA^{Tyr} gene. The cloning site is an *Eco*RI site, which is unique in the plasmid. Insertion of any large DNA fragment into the cloning site inactivates the ochre suppressor. Markers that are included in the host yeast strain allow phenotypic discrimination between transformants that display or lack ochre suppression.

[13] B. J. Brewer and W. L. Fangman, *Cell (Cambridge, Mass.)* **51,** 463 (1987).
[14] D. T. Stinchcomb, K. Struhl, and R. W. Davis, *Nature (London)* **282,** 39 (1979).
[15] A. W. Murray, T. E. Claus, and J. W. Szostak, *Mol. Cell. Biol.* **8,** 4642 (1988).

5. Two yeast selectable markers. The two selectable markers are *TRP1* and *URA3*, which are on opposite sides of the cloning site. Consequently, in the linear artificial chromosomes created from pYAC4, *TRP1* is at one end of the chromosome and *URA3* is at the other. These two genes are the wild-type alleles of the genes that are mutated in yeast strains bearing the auxotrophic markers *ura3* and *trp1;* selection involves complementation of these auxotrophic markers in suitable hosts.

Other features of pYAC4, aside from the *Amp*^{R} gene and *ori* sequence from pBR322 which allow selection in *E. coli,* are largely incidental. In particular, the yeast *HIS3* gene is in a "throwaway" fragment that is not present in the final YAC clones.

Additional vectors in the pYAC4 series include pYAC3 and pYAC55.[1] These vectors differ from pYAC4 only at the cloning site in the *SUP4* intervening sequence: pYAC3 contains a *Sna*BI cloning site for bluntended fragments, and pYAC55 contains a *Not*I site (pYAC55 is a newer name for the vector described as pYAC5 in Ref. 1; the name pYAC5 was retired from use because some isolates of pYAC5 are in circulation that do not have the specified structure). Further derivatives of these vectors include the following: pYAC-RC,[16] which contains a polylinker at the cloning site that can be cleaved with *Not*I, *Sac*II, *Sal*I, *Mlu*I, *Cla*I, and *Sna*BI; pYAC4-Neo,[17] which adds a module to pYAC4 that confers G418 resistance to mammalian cells; and pYACneo,[18] which provides both a G418-resistance module and a second *E. coli* replication origin that is useful in plasmid-rescue strategies.

In this chapter, we describe the use of pYAC4 for cloning large fragments of human DNA prepared by partial digestion with *Eco*RI. Straightforward modifications of these protocols allow the cloning of other types of fragments with alternative YAC vectors. Using the protocols described, we have obtained stable ligation mixtures in which titers are sufficiently high to prepare a library of 100,000 clones with an average insert size of 275 kb from a few hundred micrograms of high molecular weight human DNA. The expected sampling redundancy for single-copy sequences in such a library is nearly 10. Methods are also described for storing large collections of YAC clones, picked individually into the wells of microtiter plates in a form suitable for library screening. Sucrose density gradients are used for size fractionation either before ligation, after ligation, or at both steps of the procedure. Illustrative data are shown, indicating the effects of these choices on the size distribution of the YACs obtained.

---

[16] D. Marchuk and F. S. Collins, *Nucleic Acids Res.* **16,** 7743 (1988).
[17] H. Cooke and S. Cross, *Nucleic Acids Res.* **16,** 11817 (1988).
[18] C. N. Traver, S. Klapholz, R. W. Hyman, and R. W. Davis, *Proc. Natl. Acad. Sci. U.S.A.* **86,** 5898 (1989).

Isolation of High Molecular Weight Human DNA

High molecular weight DNA is prepared from circulating leukocytes that are harvested from whole blood by a modification of the method of Luzzatto.[19] The DNA is purified by a sucrose step-gradient procedure originally developed for the isolation of intact chromosomal DNA molecules from yeast spheroplasts.[20,21] Although this protocol involves only a one-step purification of a crude lysate, it produces DNA samples that are free of contaminating nucleases and readily cleaved by most restriction endonucleases.

Special Reagents. Unless otherwise indicated, reagents are prepared from the component stock solutions without adjusting the final pH. For example, if a reagent is described as containing 0.45 M EDTA (pH 9) and 10 mM Tris-HCl (pH 8), the implication is that stock solutions of EDTA at pH 9 and Tris-HCl at pH 8 are mixed and diluted to the specified final concentrations. EDTA stock solutions of 0.5 M are prepared from the disodium salt, and the pH is adjusted with NaOH. Tris-HCl stock solutions of 1 M are prepared from Tris base, and the pH is adjusted with HCl. Unless otherwise specified, concentrations expressed as percent refer to weight-to-volume ratios.

SCE: 1 M sorbitol, 0.1 M sodium citrate, 60 mM EDTA (final pH adjusted to 7 with HCl)

Lysis buffer: 0.5 M Tris-HCl (pH 9), 0.2 M EDTA (pH 8), 3% sodium N-lauroylsarcosinate

Sucrose solutions: Prepare to indicated concentration in 0.8 M NaCl, 20 mM Tris-HCl (pH 8), 10 mM EDTA (pH 8)

TE8: 10 mM Tris-HCl (pH 8), 1 mM EDTA (pH 8)

Protocol

1. Collect a 40-ml blood sample into a standard EDTA- or citrate-based anticoagulant, add 8 ml of 0.3% dextran (T500; Pharmacia, Piscataway, NJ) prepared in 0.9% NaCl, and incubate the suspension for 30 min at 37°. During this period, the erythrocytes settle out spontaneously while the leukocytes stay suspended. Harvest the leukocytes from the supernatant by gentle centrifugation (5 min at 1000 g) and rinse the pellet twice with 0.9% NaCl. Resuspend the final leukocyte pellet in 5 ml of SCE buffer.

2. Prepare a series of four 4-ml cell suspensions of progressively lower density by serial dilution: the first is simply a 4-ml aliquot of the SCE

---

[19] L. Luzzatto, *Biochem. Biophys. Res. Commun.* **2**, 402 (1960).
[20] M. V. Olson, K. Loughney, and B. D. Hall, *J. Mol. Biol.* **132**, 387 (1979).
[21] G. F. Carle and M. V. Olson, *Nucleic Acids Res.* **12**, 5647 (1984).

suspension described above; the three remaining samples are prepared by successive 5-fold dilutions of the remaining 1 ml of this suspension. Usually, the two middle dilutions yield 100–500 μg of DNA, which is in the optimum range for loading on one sucrose gradient using the protocol described below.

3. Lyse the 4-ml cell suspensions in a gently swirled (20 rpm) 250-ml Erlenmeyer flask by slowly dripping (1 ml/min) the cell suspension down the side of the flask into 7 ml of lysis buffer. Place the flask in a water bath at 68° for 15 min and then cool it quickly in an ice bath.

4. Gently pipette the viscous lysate onto the top of a 27-ml sucrose step gradient either by using the wide end of a 10-ml glass pipette or by decantation. The gradients are prepared by placing 12 ml of 20% sucrose solution in a centrifuge tube (2.5 × 8.9 cm) and then rapidly pouring in 12 ml of 15% sucrose solution. A 3-ml 50% sucrose solution pad is then underlaid onto the bottom of the tube using a 5-ml pipette.

5. Spin the gradients in a swinging-bucket rotor (Sorvall AH627 or equivalent) at 26,000 rpm for 3 hr at 20°.

6. Remove and discard the top 10 ml of each gradient with a pipette. This fraction should contain the bulk of the cellular material and should not be highly viscous. Obvious viscosity near the top of the tube indicates that the gradient was overloaded, in which case both the yield and the quality of the DNA are likely to be poor.

7. Remove the high molecular weight DNA by inserting the wide end of a 10-ml pipette to the bottom of the tube and gently withdraw 5–6 ml of the highly viscous solution. In successful separations, the leukocyte DNA is largely confined to the 50% sucrose pad.

8. Gently pipette the DNA-containing fraction into a cellulose dialysis bag (molecular weight cutoff 6000–8000) and dialyze it against TE8 at 4°. The volume of dialysis buffer should be at least 1 liter, and the buffer should be changed at least 3 times at intervals of 12 hr or more.

9. Concentrate the dialyzed sample by placing the dialysis bag on solid sucrose at 4° until the volume is reduced to approximately 4 ml, then clamp the bag so that the residual DNA solution is localized at one end. Continue the dialysis against TE8 to remove sucrose that enters the bag during the concentration step. If necessary, carry out additional cycles of concentration and dialysis until the sample volume has been reduced to less than 2 ml.

10. Assay the DNA sample by loading aliquots onto a 0.7 or 1.0% agarose gel and electrophoresing the samples at a relatively low voltage gradient (<1 V/cm). Under these conditions, molecules larger than 25 kb comigrate in a single sharp band whose intensity can be reliably compared with that of a band containing a known amount of bacteriophage λ DNA.

A single, optimally loaded sucrose gradient should yield approximately 250 μg of DNA.

11. Check the size distribution of the DNA by pulsed-field gel electrophoresis (see Fig. 2, lane 4, for a typical result).[22,23]

Partial Digestion of High Molecular Weight DNA

In order to minimize shearing, it is important to exercise special care when pipetting high molecular weight DNA. Disposable plastic pipette tips are cut with a razor blade to an inner diameter of 1.5–2 mm, and all mixing is done by gentle stirring using a pipette tip. Many of the samples are highly viscous and they often appear inhomogeneous. Consequently, they are difficult to mix, measure accurately, or load onto gels, and some irreproducibility in the behavior of replicate portions of the same sample is unavoidable.

Because the results of these digests can vary dramatically from sample to sample, it is important to carry out pilot reactions on each DNA preparation. Even when digestion conditions have been optimized for a particular preparation, some hedging is recommended when carrying out the preparative digestions. Large quantities of lymphocyte DNA are easily prepared; consequently, it is advisable to regard this material as expendable, carrying along several samples in parallel and discarding those that are under- or overdigested.

It is also important to carry out a control reaction containing restriction enzyme buffer and no enzyme. Samples must be rejected as unacceptably contaminated with endogenous nucleases if there is any change in the size distribution of the DNA during this procedure.

Protocol

1. For the pilot-scale reactions, add 0.001–10 units of *Eco*RI to 5 μg of DNA in a volume of 50 μl of standard *Eco*RI buffer. Incubate the samples at 37° for 10 min, stopping the reaction by placing the samples on ice and adding EDTA (pH 8) to a final concentration of 50 mM.

2. Assay the digestions by loading the entire samples onto a pulsed-field gel (Fig. 2). In the experiment shown, the range of digestion conditions suitable for preparing YACs averaging 200–300 kb in size is bracketed by the results with 0.05 and 0.5 units of *Eco*RI (lanes 6 and 7);

---

[22] G. Chu, D. Vollrath, and R. W. Davis, *Science* **234**, 1582 (1986).
[23] M. V. Olson, in "Genetic Engineering" (J. K. Setlow, ed.), Vol. 11, p. 183. Plenum, New York, 1989.

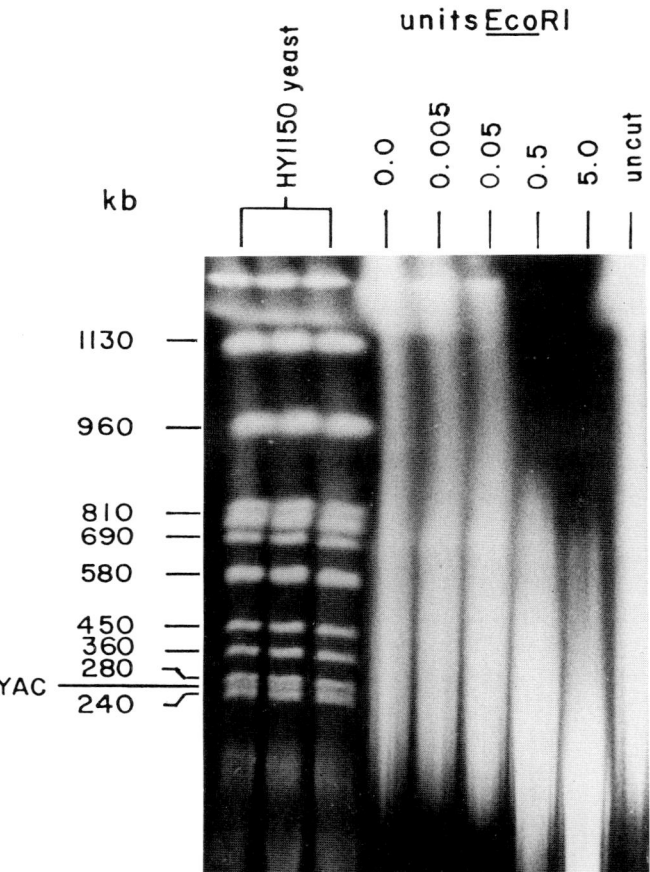

FIG. 2. Partial digestion of high molecular weight human DNA with EcoRI. The samples have been analyzed by pulsed-field gel electrophoresis using a CHEF (contour-clamped homogeneous electric field) apparatus,[22,23] and the gel was stained with ethidium bromide. The three leftmost lanes are DNA samples of the YAC-transformed yeast strain HY1150, which maintains an artificial chromosome of 250 kb. The rightmost lane is a sample of undigested human DNA that was prepared by gentle lysis of human leukocytes followed by purification on a sucrose step gradient. The five center lanes contain aliquots of this DNA that have been incubated in restriction enzyme buffer in the presence of increasing concentrations of EcoRI. The sizes for the natural yeast chromosomes are indicated at left (A. Link and M. V. Olson, unpublished, 1989). Electrophoresis was carried out using 1.0% agarose gels cast and run in 0.5× TBE.[21] The electrophoresis was performed at 13° and 6 V/cm (measured in the gel, see Ref. 23). The switching interval was 60 sec and the run time 24 hr.

basically, one is looking for the smallest amount of enzyme that causes an observable, dose-dependent decrease in the size distribution.

3. Scale up the reaction conditions that gave the desired size distribution so that the preparative reactions contain 50–100 μg for each set of conditions. Typically, three preparative digestions are performed to insure that at least one sample is digested to the desired size range. One reaction is run at the previously determined "optimal" enzyme concentration, one with 5-fold less enzyme, and one with 5-fold more enzyme. Stop the reactions as described above and assay 5-μg aliquots by pulsed-field gel electrophoresis.

4. Treat the digested samples with phenol equilibrated with 1 M Tris-HCl (pH 8). A volume of phenol equal to that of the aqueous sample is gently pipetted down the side of a 1.5-ml microcentrifuge tube, and the capped tube is laid on its side for 15 min and then placed upright for an additional 10 min. The tube is spun in a microcentrifuge for 5 sec, and the aqueous layer and interface are removed with a cutoff pipette tip and placed in a new tube.

5. Perform a chloroform treatment in a similar fashion to remove phenol from the aqueous phase.

6. Transfer the final aqueous phase to the bottom of a Schleicher and Schuell UH 100 collodion bag (Keene, NH) and dialyze it for 2–4 hr against 250 ml of TE8.

7. Assay 5-μg aliquots of the dialyzed sample on a pulsed-field gel to ensure that no change in the DNA size distribution occurred during Steps 4–6.

8. Depending on the strategy chosen (see Discussion), proceed either to the sucrose density gradient or the ligation step.

Ligation

The ligation step is straightforward, although the same precautions in handling the DNA that are outlined in the section on partial digestion of high molecular weight DNA should be observed. We have chosen to treat the vector rather than the insert with alkaline phosphatase and to use a sizable molar excess of vector. Consequently, the only function of the phosphatase treatment is to prevent the formation of concatenated vector fragments that would later be difficult to separate from the desired ligation products by size fractionation. Concatenated vector fragments would undoubtedly lead to the formation of some anomalous transformants with phenotypes indistinguishable from those of transformants containing authentic YACs. We have not experimented extensively with the alternative possibility of treating the inserts with phosphatase. In principle, such a step might reduce cloning artifacts arising from the ligation of unrelated insert fragments; however, in our experience, exposure of insert fragments to enzymes should be kept to an absolute minimum.

Protocol

1. Digest the pYAC4 cloning vector with *Bam*HI and *Eco*RI and then treat the digested DNA with calf intestinal alkaline phosphatase.[24] The dose of phosphatase is important and must be optimized for each batch of DNA and source of phosphatase. A series of enzyme levels should be tested and one selected that is approximately 2-fold above the minimum dose that eliminates detectable self-ligation of the vector, as assayed on conventional gels.

2. Mix 40–100 μg of insert DNA with an equal weight of prepared pYAC4 vector. Adjust the volume to 250 μl and the buffer composition to 50 mM Tris-HCl, 10 mM MgCl$_2$, 1 mM ATP, pH 7. Add 25 units T4 DNA ligase and incubate the sample at 15° for 8–12 hr.

3. Treat the reaction mixture with phenol and chloroform, as described in Steps 4 and 5 of the protocol for partial digestion of high molecular weight DNA.

4. Dialyze the sample against 10 mM Tris-HCl (pH 8), 5 mM EDTA (pH 8) in a UH 100 collodion bag.

5. Depending on the strategy chosen (see Discussion), proceed either to the sucrose density gradient or the transformation step.

Notes on Ligation

1. The ligation step can be assayed by separating the products on a pulsed-field gel and analyzing them by gel-transfer hybridization using the vector as a probe. However, the large amount of unligated vector tends to obscure the results of this assay. If the ligation mixture is size fractionated on a sucrose density gradient, better results can be obtained by carrying out gel-transfer hybridization on the same pulsed-field gel that is used to assay the fractions from the sucrose density gradient (see Step 9 of the sucrose density gradient protocol), as nearly all the unligated vector appears in different density gradient fractions than do the products of vector–insert ligations.

2. If one is carrying out the ligation using less than the recommended concentration of insert molecules (see Note 3 in the sucrose density gradient protocol), the concentration of vector molecules should be kept close to that recommended here. Otherwise, the fraction of insert molecules that are converted to potential YACs is likely to decrease.

[24] T. Maniatis, E. F. Fritsch, and J. Sambrook, "Molecular Cloning: A Laboratory Manual." Cold Spring Harbor Laboratory, Cold Spring Harbor, New York, 1982.

Sucrose Density Gradient Fractionation

We have used sucrose density gradients for size fractionation at two steps in the YAC-cloning procedure: immediately after the partial digestion of the insert DNA and following ligation. The relative merits of carrying out size fractionation at these two steps are addressed in the Discussion. The use of sucrose density gradients rather than pulsed-field gels for size fractionation allows relatively large amounts of DNA to be processed and obviates the need to extract the fractionated molecules from agarose.

The volume of the gradients can be either 17 or 34 ml depending on whether the 18-ml (1.6 × 10.2 cm) or 36-ml (2.5 × 8.9 cm) tubes are being used for a Sorvall AH627 swinging-bucket rotor. We are more likely to use the larger gradients when fractionating the partial digests and the smaller ones when fractionating ligation mixtures, but the procedure is the same in both cases. The volumes in the following protocol are appropriate for the 18-ml tubes; the minor changes appropriate to the larger tubes are specified in Note 1.

The gradients are exponential but not strictly isokinetic. They are prepared using the gradient marker described by Noll[25] using a closed mixing chamber with a volume only slightly larger than the volume of the gradient. The buffer for the sucrose solutions is the same as that described in the protocol for preparing high molecular weight DNA.

Protocol

1. Place 17 ml of 5% sucrose solution in the closed mixing chamber. Pump 17 ml of 25% sucrose solution into this chamber, displacing an equal volume of sucrose solution into the centrifuge tube.

2. Layer the DNA sample on top of the gradient in a volume of 250 μl or less. Bromphenol blue may be added to the sample to increase its visibility during loading.

3. Spin the gradient at 10,000 rpm for 18–24 hr at 20°.

4. Place a rubber stopper with an inlet for mineral oil in the top of the centrifuge tube.

5. Puncture the bottom of the tube with a 16-gauge hypodermic needle, taking care not to introduce air bubbles.

6. Displace 250-μl fractions into the wells of a 96-well microtiter dish by pumping oil into the top of the tube.

---

[25] H. Noll, *Nature (London)* **215**, 360 (1967).

7. Assay undiluted 50-μl aliquots of every fourth fraction on a pulsed-field gel in order to determine the size distribution of the DNA.

8. Pool gradient fractions containing DNA of the desired size range and dialyze and concentrate the DNA using a Schleicher and Schuell UH 100 collodion bag (MW cutoff 75,000). Transfer the sample with a cutoff plastic pipette tip to the bottom of the bag and concentrate it under reduced pressure against 250 ml of TE8 buffer. Since some DNA remains bound to the surface of the collodion bag, keep the volume of the sample in the bag under 0.5 ml at any one time in order to minimize the surface area of the bag that comes in contact with the DNA solution. A final concentration of 0.5–1.0 μg/μl is desirable. The DNA can be assayed as described in Step 10 of the protocol for isolating high molecular weight DNA.

9. Assay a 5-μg aliquot of the final pooled and concentrated sample by pulsed-field gel electrophoresis to determine the size distribution of the pooled DNA.

Notes on Sucrose Density Gradients

1. As much as 50–100 μg of DNA can be loaded into a single 17-ml gradient (or twice that on a 34-ml gradient); however, the fractionation is somewhat better if the gradients are less heavily loaded. For 34-ml gradients, double the volumes in Steps 1 and 2 and assay every eighth fraction in Step 7. Assay results for a 34-ml gradient are shown in Fig. 3.[26]

2. The concentration step is a common place to lose DNA or to suffer reductions in size. There appears to be variability from batch to batch in the collodion bags. Before committing valuable samples to this step, it is advisable to test the recovery of high-molecular-weight DNA using more easily prepared and assayed material such as yeast DNA solutions in which the smaller yeast chromosomes are intact. Such samples can be prepared by a method virtually identical to that described above for the isolation of leukocyte DNA.[20,21]

3. If the size of the DNA decreases observably during concentration or if unacceptable losses occur, consideration should be given to substituting simple dialysis for the concentration step. We have never encountered difficulties simply dialyzing high molecular weight DNA in collodion bags; when DNA has become degraded or has been lost, it has always been during concentration (i.e., dialysis with a large pressure drop across the membrane). The ligation and transformation steps can be carried out with lower concentrations of insert DNA than those recommended in step 8. The main penalty for adopting this course is that the number of transformants obtained per plate drops.

[26] G. F. Carle, M. Frank, and M. V. Olson, *Science* **232**, 65 (1986).

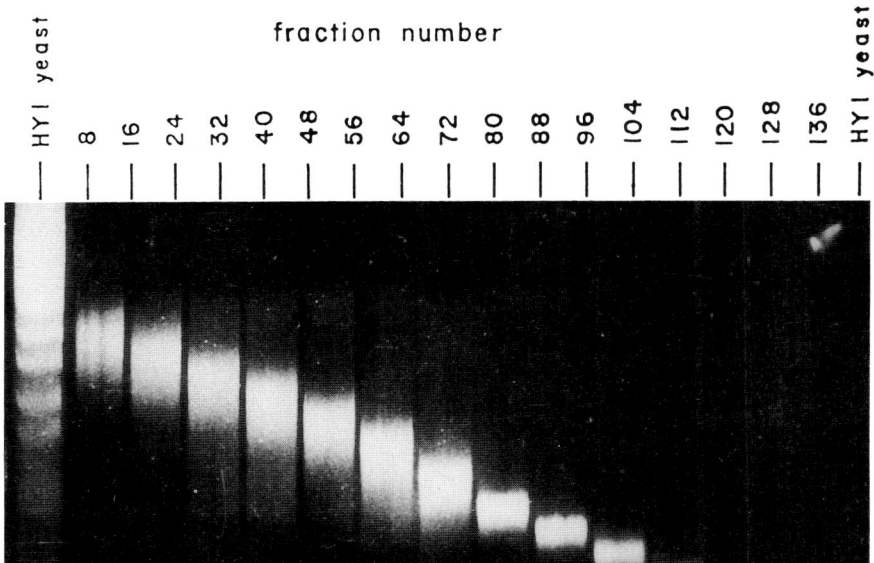

FIG. 3. Size fractionation of high molecular weight DNA by velocity sedimentation on a sucrose gradient. The fractions were analyzed by field-inversion gel electrophoresis (FIGE), and the gel was stained with ethidium bromide.[23,26] The two DNA marker lanes are samples of the YAC-transformed yeast strain HY1,[1] which maintains an artificial chromosome of 120 kb and for which chromosomal DNA sizes are indicated at left. In the gradient shown, the sample was 200 μg of sheared *Drosophila* DNA, and the gradient volume was 34 ml. Aliquots of every eighth fraction, of a total of 136, were assayed. This sample was used for illustrative purposes since a high fraction of the DNA is in the size range of effective fractionation. Similar fractionation is obtained using *Eco*RI partial digests of human DNA, but with these samples a much smaller fraction of the DNA is in the size range of interest; consequently, when the recommended gradient loading and assay conditions are employed, the signals on a gel of this type are much fainter than those shown here, and a significant amount of DNA pellets at the bottom of the tube. FIGE was carried out using a ramped program of switching times:[23,26] the forward interval was ramped from 3 to 60 sec while the reverse interval was ramped from 1 to 5 sec over a 12-hr period. Other FIGE conditions were standard.[26]

Transformation

Most experience with YAC cloning has involved the transformation of yeast spheroplasts, a technique discussed in [12], this volume. We have followed the protocol of Burgers and Percival[27] without modification. The yeast strain AB1380 has been used as the host.[1] The genotype of this strain is *MATa* ψ^+ *ura3 trp1 ade2-1 can1-100 lys2-1 his5* (note that the mating type is incorrectly specified as *MATα* in Ref. 1). The ψ^+ genotype indicates the presence of a cytoplasmic determinant that enhances suppression; the

[27] P. M. J. Burgers and K. J. Percival, *Anal. Biochem.* **163**, 391 (1987).

main consequence of this trait in AB1380 is that the strain tolerates strong ochre suppressors such as *SUP4-o* poorly.

In a typical experiment, for plating on a single standard petri plate (100-mm diameter), aliquots of the ligation mixtures containing 0.1–10 μg human DNA in less than 10 μl are added to 100 μl of the final spheroplast suspension, which contains 6×10^8 to 1.5×10^9 cells in a volume of 2 ml. No carrier has been employed. Transformation experiments should be controlled with a conventional yeast *CEN* plasmid. When 10 ng of YCp50, a *URA3 CEN* plasmid, is substituted for the ligation mixture in the above procedure, we typically obtain a few thousand transformants per plate. YAC vectors containing an intact *SUP4-o* gene should not be used as the control in transformation experiments because ψ^+ strains such as AB1380 transform poorly with these plasmids. The number of transformants obtained per plate with ligation mixtures is highly variable, depending on the success of the prior steps, but the range is generally 50–1000.

As recommended by Burgers and Percival, the spheroplasting is taken to 80–90% completion. A convenient assay for the percent completion is simply to use a microscope to observe the fraction of the cells that lyse when exposed to the same lysis buffer described in the protocol for isolating leukocyte DNA. We have had good success carrying out the spheroplasting with commercially available lyticase preparations (Sigma, St. Louis, MO, L8137).

There are no published protocols for preparing YAC clones with simpler transformation protocols such as the lithium acetate method.[28] One disadvantage of spheroplast transformation is that the transformants are obtained embedded in agar, which interferes with direct colony screening of the transformation plates. There is a report that this problem can be circumvented by plating the spheroplasts on an agar surface in the presence of calcium alginate.[18] However, in our hands there is a significant loss of transformation efficiency with this method.

Transformants are selected on a synthetic medium that lacks uracil (i.e., −Ura plates); these plates are prepared following standard recipes.[29] For reasons that are not well understood, selection on −Ura −Trp plates gives many fewer transformants than selection on −Ura plates, even though most of the transformants selected on the latter medium are Trp$^+$. Both the bottom and top agar in the −Ura transformation plates contain 1 *M* sorbitol and 2.5% agar. The success of the transformation protocol appears to be sensitive to the purity of the sorbitol. We have had good success using the FisherBiotech reagent (Fisher Scientific, Pittsburgh, PA).

---

[28] H. Ito, Y. Fukuda, K. Murata, and A. Kimura, *J. Bacteriol.* **153,** 163 (1983).
[29] F. Sherman, G. R. Fink, and J. B. Hicks, "Methods in Yeast Genetics." Cold Spring Harbor Laboratory, Cold Spring Harbor, New York, 1986.

Storage of Library

Primary transformants are picked to −Ura −Trp plates. In principle, the desired transformants are expected to form red colonies, as does AB1380 itself. The red color is a phenotype of *ade2* mutants. AB1380 contains an *ade2* ochre mutation so it forms red colonies unless it has been transformed with a plasmid containing a functional *SUP4-o* gene. Although the red color will ultimately develop on the primary transformation plates, this process is slow and erratic; the color assay is particularly problematic for colonies that are deeply embedded in the top agar. Consequently, we normally pick the primary transformants, irrespective of color, to −Ura −Trp plates as soon as they can be clearly distinguished. Red color development on these plates is favored if the adenine concentration is no higher than 10 mg/liter. It is worth picking all colonies that can be clearly distinguished since colony size on the transformation plates is unrelated to the subsequent growth characteristics of a clone.

Clones that give strong patches of red growth on the −Ura −Trp plates are added to the library. In all early experiments, the clones were picked into 800 μl of −Ura −Trp medium in 1.0-ml plastic tubes arranged in the same geometry as a microtiter plate (Bio-Rad Titertube Micro Test Tubes, Richmond, CA). More recent experience during the expansion of the library has suggested that better viability and adequate YAC maintenance are obtained by growing the clones in the standard rich medium YPD[29] at this stage.[5] The cultures are grown for 18 hr in YPD or for 2 days in selective medium at 30° by mounting the box of tubes on a roller drum at an angle that provides maximal surface area without spillage; the individual tubes are not capped, but a lid is placed on each box. Alternately, the cultures can be grown on a Bellco Mini Orbital Shaker (Vineland, NJ), which provides a rapid rotary motion with an orbit diameter of less than 2 mm. Using a 12-tip pipettor, 0.25 ml of 80% (v/v) glycerol is added to each culture, and the resultant cell suspension is stirred with the pipette tips; the same tips are then used to remove two 200-μl aliquots to the corresponding rows of two separate microtiter plates, which have 300-μl round-bottomed wells. The latter plates are sealed with clear plastic tape (Falcon MicroTest Film, Becton Dickinson, Cockeysville, MD), while the individual culture tubes in the original array are capped. This procedure yields three copies of the library for storage at −70°. We use the initial capped tubes as an archival copy of the library, and the two secondary copies are used to inoculate filters for colony hybridization.

Colony Screening

The colony-screening protocol involves growing the colonies on the surface of a nylon membrane, spheroplasting the yeast, lysing the spheroplasts with detergent, and denaturing the DNA with base. The following protocol closely follows that described by Brownstein et al.[5]

Protocol

1. If the clones are stored in liquid cultures in microtiter plates, use a multiprong replicator to stamp a replica of the clone array on a dry nylon membrane.

2. Place the membrane on the surface of a −Ura −Trp plate, taking care to avoid air bubbles between the membrane and the plate. If small numbers of colonies stored on Petri plates are to be screened, they can be transferred to the membrane at this stage, just as if one were inoculating a patch on the surface of the plate itself.

3. Incubate the plate at 30° for approximately 2 days or until vigorous growth is obtained.

4. Transfer the membrane to a thick paper filter, saturated with 2 mg/ml yeast lytic enzyme [ICN 152270, >70,000 units (U)/g], 1.0 M sorbitol, 0.1 M sodium citrate, 50 mM EDTA, and 15 mM dithiothreitol (pH of the enzyme buffer adjusted to 7). Incubate overnight at 30°.

5. Transfer the membrane to a paper filter saturated with 10% sodium dodecyl sulfate and let stand for 5 min at room temperature.

6. Transfer the membrane to a paper filter saturated with 0.5 M NaOH and let stand for 10 min.

7. Neutralize the membrane by transferring it to three successive paper filters saturated with 0.3 M NaCl, 30 mM sodium citrate, 0.2 M Tris HCl, pH 7.5, leaving it for 5 min each time.

8. Air dry the filters.

9. Carry out labeling of probes, hybridization, and autoradiography by standard methods.[24,30]

Notes on Colony Screening

1. Although there is some batch dependence, this protocol works well with most standard nylon membranes such as Hybond-N (Amersham, Arlington Heights, IL) and SUREBLOT (Oncor, Gaithersburg, MD).

2. For applications that involve screening large numbers of colonies stamped onto membranes from microtiter plates, Brownstein et al. recommend a richer medium than standard −Ura −Trp plates.[5] This medium, which is prepared from casein hydrolysate, still lacks uracil and tryptophan but results in better growth of lightly inoculated patches.

3. At Step 4, it is convenient to place the nylon membrane onto the wet

---

[30] A. P. Feinberg and B. Vogelstein, *Anal. Biochem.* **132**, 6 (1983).

filter paper in a plastic petri plate and then to seal the entire assembly in a plastic bag to avoid evaporation during the overnight incubation.

4. In choosing hybridization probes, it is important to avoid any contamination with the pBR322 sequences present in the YAC vectors.

Discussion

The effectiveness of these protocols is illustrated by data on two large ligation mixtures that have been used to develop a complex human YAC library. The first mixture, referred to as LM-I, was prepared by direct ligation of vector to the products of *Eco*RI partial digestion with no size fractionation before or after ligation. The second, referred to as LM-II, was prepared by ligation of vector arms to size-fractionated insert fragments, and an additional size-fractionation step was also incorporated following ligation. During the construction of LM-II, insert fragments were pooled from all fractions in which over 90% of the DNA appeared to be longer than 100 kb (e.g., fractions 1–48 in a gradient such as that shown in Fig. 3). After ligation, the ligation mixture was fractionated on a second, identical sucrose gradient with a more stringent low molecular weight cutoff: fractions were only pooled, in this case, if over 90% of the DNA appeared to be greater than 200 kb in size (e.g., 1–40 on a gradient such as that shown in Fig. 3).

Following test transformations with aliquots of the two ligation mixtures, Ura$^+$ transformants were selected and then picked to −Ura −Trp plates. For each of the ligation mixtures tested, essentially all of the colonies picked from the transformation plates grew on the −Ura −Trp plates. The only exceptions occurred when picking extremely small colonies, which are difficult to distinguish from defects in the agar. Of the colonies that grew, over 95% formed the expected red colonies. LM-I contained approximately 75 μg of human DNA in 150 μl. Using LM-I, a typical yield of transformants with the expected phenotype was 300 clones/μg, indicating the potential of producing a library with 45,000 clones. LM-II, which contained approximately 150 μg of human DNA in 150 μl, typically yielded transformants at a frequency of 500–1000/μg, indicating the potential of producing a library with 100,000 clones. Some of this material was used for pilot experiments in which products were not saved, but over 17,000 individual clones have actually been picked following more than 20 separate transformations with LM-II. These transformations have been carried out over a period of more than 1 year, during which time there has been no substantial change in either the titer of the ligation mixture or the size distribution of the YACs obtained (B. H. Brownstein, D. Schlessinger, and M. V. Olson, unpublished results, 1989). The screening of this library for single-copy human genes and the characterization of the resultant clones have been described by Brownstein et al.[5]

In typical transformations using either LM-I or LM-II, one or more artificial chromosomes in the size range 50–700 kb are present in over 95% of the clones. While most transformants contain a single YAC, 10–20% contain two or more. In some cases, the existence of two YACs in the same isolate simply reflects the picking of an impure clone from the primary transformation plates, whereas in other cases single cells actually contain artificial chromosomes of two different sizes.

The size distributions of the YACs obtained from the two ligation mixtures clearly demonstrate the effectiveness of size fractionation in eliminating small clones. Representative histograms are shown in Fig. 4; the data in Fig. 4A are for LM-I (no size fractionation), and the data in Fig. 4B are for LM-II (size fractionation before and after ligation). In the library constructed from LM-II, the average size of the YACs is estimated to be 275 kb. Given a YAC library with this average insert size, 11,000 clones would be required to provide "single-hit" coverage of the human genome (i.e., on the average, single-copy sequences would be represented once). In order to reduce the probability to less than 1% of missing a single-copy sequence because of inadequate sampling, a library must provide five-hit coverage, or 55,000 clones. The protocols presented here allow the construction of a library of that complexity from a single ligation reaction on the scale described.

The importance of size fractionation is demonstrated by the comparison between the size distributions of YACs obtained with and without this step. Simple theoretical considerations indicate that size fractionation is essential, regardless of how carefully the partial digestion is controlled. If a complex DNA sample is cleaved at some fraction of a set of randomly distributed sites, the expected size distribution is $(1/S_a) \exp(-S/S_a)$, where S_a is the average size of the fragments.[31] To the extent that *Eco*RI sites are randomly distributed and cleaved at equal rates, this equation predicts that when genomic DNA is partially digested to an average size of 400 kb, only 38% of the molecules are in the size range 200–600 kb, whereas 39% are under 200 kb. Consequently, the size distribution shown in Fig. 4A is approximately what would be expected even in the absence of size discrimination at the transformation step.

In choosing at which stage or stages to introduce size fractionation, it should be noted that size fractionation is likely to have different consequences before and after ligation. By depleting the samples of smaller fragments before ligation, size fractionation of the partially digested ge-

---

[31] S. S. Potter and C. A. Thomas, *Cold Spring Harbor Symp. Quant. Biol.* **42**, 1023 (1978).

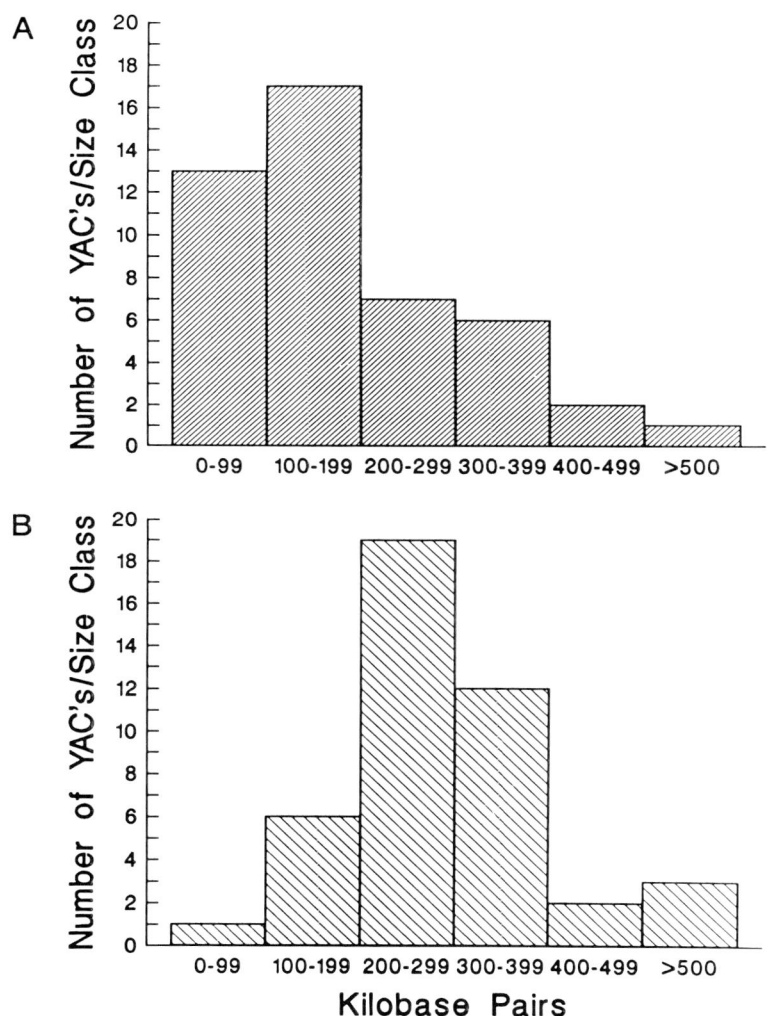

FIG. 4. Histograms showing the size distributions obtained by transformation of two YAC ligation mixtures. (A) Data on the sizes of the YACs present in 46 randomly picked clones from LM-I (no size fractionation). (B) Comparable data for the YACs present in 43 clones from LM-II (size fractionation both before and after ligation). Fifty clones were picked from each pool to develop the histograms that are shown; data are included only for clones that contained a single, pBR322-hybridizing YAC.

nomic DNA would be expected to reduce the number of ligation events that occur *in vitro* between unrelated insert fragments. Size fractionation after ligation, on the other hand, removes unligated vector molecules from the transformation samples, thereby reducing the potential for unwanted recombination events *in vivo* that could lead to artifactual Ura$^+$ transformants.

A final issue about YAC libraries concerns the relative merits of working with individually picked clones, as opposed to pools of clones. The approach described here presupposes hand picking of individual clones from the transformation plates and storage of these clones in individual wells of microtiter plates. This strategy has a number of obvious advantages (e.g., avoiding representational bias owing to different colony sizes on the transformation plates, allowing direct comparison of the results of colony screening for an unlimited number of probes), but it is laborious. A five-hit mammalian library would occupy nearly 600 microtiter plates. For some applications, it may be more desirable to carry out colony screening on the transformation plates themselves by relying on the calcium alginate transformation procedure to obtain colonies directly on the surface of the plates.[18]

Acknowledgments

This study was supported by a grant from the Monsanto Company. Bernard Brownstein and David Schlessinger provided helpful comments about the experience of the Washington University Center for Genetics in Medicine in using the ligation mixtures described in the Discussion to develop a large human YAC library. Harold Riethman, Takashi Imai, and Eric Green made helpful comments on the manuscript.

Section III
Making Mutants

[18] Classical Mutagenesis Techniques

By CHRISTOPHER W. LAWRENCE

Introduction

Generating mutants, to identify new genes and to study their properties, is the starting point for much of molecular biology. Forward mutations and metabolic suppressors obtained by reversion can provide powerful insights into the functions and relationships of normal gene products. Similarly, mutations and intragenic revertants provide the raw material for the analysis of gene product structure–function relationships. Reverse genetics and other methods based on recombinant DNA techniques are increasingly used for these purposes, and they are clearly the methods of choice where specific changes in specific genes or genetic sites are needed. Nevertheless, classical methods, in which cells are treated with mutagens, are in many circumstances likely to remain the chief means for inducing mutations because they require no prior knowledge of gene or product and are generally applicable: the user need only specify an appropriate alteration in phenotype. However, where selection for the desired strain is not possible, hunting for mutants can be extremely laborious and analyzing the material obtained even more so. Good planning, efficient mutagenesis, careful choice of strain, and effective mutant detection usually pay off in time and labor.

Choice of Mutagen and Dose

The best mutagens for most purposes are those that induce high frequencies of base-pair substitutions and little lethality. The widely used alkylating agents N-methyl-N'-nitro-N-nitrosoguanidine (MNNG) and ethylmethane sulfonate (EMS) fulfill these criteria but are highly specific in their action: they almost exclusively produce transitions at G · C sites.[1] For most purposes, such as forward mutagenesis, this specificity is unlikely to pose any problem, though it may be a disadvantage when, as in some kinds of reversion, specific mutations at specific sites are needed. Ultraviolet light (UV) is also a fairly efficient mutagen, and it has the advantage of producing a greater range of substitutions: most occur in runs of pyrimidines, particularly T–T pairs, and include both transitions and transversions.[2,3]

---

[1] S. E. Kohalmi and B. A. Kunz, *J. Mol. Biol.* **204**, 561 (1988).
[2] B. A. Kunz, M. K. Pierce, J. R. A. Mis, and C. N. Giroux, *Mutagenesis* **2**, 445 (1987).
[3] G. S. F. Lee, E. A. Savage, R. G. Ritzel, and R. C. von Borstel, *Mol. Gen. Genet.* **214**, 396 (1988).

UV also induces a significant frequency of frameshift mutations, almost exclusively of the single nucleotide deletion variety. Together, one or other of the chemical mutagens and UV are likely to satisfy most experimental needs, and it may be an advantage to induce different samples of mutants with each of these agents. Although base-pair substitution mutagens have most general utility, the alkylating acridine mustard ICR-170 {2-methoxy-6-chloro-9-[3-(ethyl-2-chloroethyl)aminopropylamino]acridine·2HCl} can be used when +1 frameshift mutations are required.[4-6] The majority of mutations induced by this agent are single G insertions in runs of two or more G's, with preference for runs of three or more G's.[6]

Choosing an optimal dose usually requires balancing the competing needs for a high mutation frequency, reasonably high survival, and avoidance of multiple mutations. The highest proportion of mutants per *treated* cell is usually found at doses giving 10 to 50% survival. The highest fraction of mutations per *surviving* cell most commonly requires a larger dose, but mutation frequencies often decline at very high doses. In any case, it is desirable to avoid doses that kill more than 99% of cells, because atypical resistant variants, which occur spontaneously in all cell populations, may be selected. In addition, multiple mutants become more common and may interfere with analysis.

Growth Conditions after Mutagen Treatment

After being treated with mutagens, cell cultures should be allowed to grow for several generations under nonselective or permissive conditions, to enhance the production and expression of mutations. With some mutagens, such as MNNG, mutations are thought to occur principally during S-phase replication, and unrepaired damage can continue to produce mutations in successive generations. With others, like UV, most mutations occur during G_1 excision repair synthesis. Growth is also required to promote dilution and turnover of gene products, to allow full expression of mutant or revertant phenotype. In addition, cells may require time to recover from mutagen damage, which can cause some cells to stop growing temporarily or to grow more slowly. Full recovery from mutagen damage is particularly important when mutagen enrichment procedures are used.

Various ways of accomplishing outgrowth of mutagenized cultures can be chosen, depending on experimental needs. Plating dilutions of treated cells on solid medium, to get colonies for screening, has the advantage that

[4] D. J. Brusick, *Mutat. Res.* **10,** 11 (1970).
[5] M. R. Culbertson, L. Charnas, M. T. Johnson, and G. R. Fink, *Genetics* **86,** 745 (1977).
[6] L. Mathison and M. R. Culbertson, *Mol. Cell. Biol.* **5,** 2247 (1985).

each induced mutation identified is of independent origin. Some of the desired mutations may occur as sectors in otherwise normal colonies, however, and therefore be hard to detect by some screening procedures. Outgrowth in liquid medium is convenient and allows segregation of pure mutant clones, but different mutant isolates may represent repeat copies of the same, rather than independent, events. If outgrowth in liquid medium is needed, independent mutations can be isolated by dividing the mutagenized culture before outgrowth and taking a single mutant from each subculture. When selective methods allow a large number of cells to be spread on each plate, as in the selection of prototrophs from an auxotrophic strain, outgrowth can be achieved by adding small amounts of the required nutrilite to the medium. In experiments of this kind it is usually advisable to spread no more than about 10^7 cells on each plate, since the efficiency with which revertants are recovered drops greatly at higher cell densities. Finally, it should be noted that all mutagens increase the frequency of rho^o petites, some, like ICR-170, to very high levels. It may therefore be useful to grow mutagenized cultures in medium containing a nonfermentable carbon source, such as glycerol, to avoid recovering such strains.

Choice of Strain

With many experimental species, it is customary to isolate mutations in a designated wild-type strain, but there is no such wild type of *Saccharomyces cerevisiae* in general laboratory use. However, many mutations have been isolated in the haploid strains S288C and A364A, both of which are obtainable from the Yeast Genetics Stock Center (Berkeley, CA). In addition, a pair of isogenic strains, X2180-1A and X2180-1B, of opposite mating type that are also isogenic with S288C are available. Mutants isolated from these strains can therefore be easily "cleaned up," that is, the mutation of interest placed in a nonmutagenized genetic background, by repeated crossing with one or other of this isogenic pair.

Although these strains are sometimes useful, in many instances it will be necessary to select a strain tailored to meet specific experimental needs: particular mutations for enrichment, selection, or analytical methods may be required. In this circumstance it is prudent to check that the strain performs satisfactorily with respect to mating, transformation, and, when crossed to other strains, sporulation, since some laboratory strains perform badly in these respects. Nonflocculent strains that give single cells directly or after brief sonication are also highly desirable. It should be noted that haploid yeast strains may carry additional copies of one or more chromosomes. Such aneuploidy may underlie the not uncommon failure to recover mutations at one locus, even though similar mutations at other loci are found readily. When exhaustive mutagenesis studies are planned, it may therefore be desirable to examine the parent strain for the presence of aneuploidy, by crossing it with strains carrying recessive markers and analyzing the tetrads. Aneuploidy is unlikely if 2:2 segregation for these markers is observed.

Mutant Enrichment Procedures

Although mutants can sometimes be selected, they more often can be isolated only by screening individual clones from mutagenized cell populations, a highly laborious process. Enrichment procedures, which increase the proportion of mutants, can sometimes be used to reduce this labor. Various procedures of this kind have been proposed,[7-10] but most depend on the same principle: the use of conditions that temporarily prevent mutant, but not nonmutant, growth and promote the selective killing of growing cells. The method using inositol starvation[9] to achieve selective killing is convenient and has been widely used. For good enrichment with mutagenized cultures, the cells must be grown without restriction for several generations, to allow mutant expression and promote recovery of damaged, but nonmutant, cells. To ensure the independent origin of the mutants eventually isolated, such outgrowth can be done on solid medium. Alternatively, a single mutant can be isolated from each of a series of liquid cultures.

Safety

Powerful mutagens are powerful carcinogens: their use and disposal require care. Chemical mutagens should be handled only in a hood, using the usual protective clothing. MNNG decomposes to release volatile diazomethane, a powerful carcinogen, and EMS is itself volatile. Handle open bottles only in a hood with the window closed as much as possible, and avoid inhaling the volatile materials. Keep a freshly made supply of 10% (w/v) sodium thiosulfate on hand, to deal with accidental spills. Germicidal UV is particularly damaging to the eyes, but it can also cause sunburn and skin cancer. UV tubes should be housed in a wood or metal structure, with screened ventilation louvres, painted matt black. Samples being irra-

[7] R. Snow, *Nature (London)* **211**, 206 (1966).
[8] B. S. Littlewood, *in* "Methods in Cell Biology" (D. M. Prescott, ed.), Vol. 11, p. 273. Academic Press, New York, 1975.
[9] S. A. Henry, T. F. Donahue, and M. R. Culbertson, *Mol. Gen. Genet.* **143**, 5 (1975).
[10] M. T. McCammon and L. W. Parks, *Mol. Gen. Genet.* **186**, 295 (1982).

diated can be observed through 6-mm-thick Lucite, which effectively blocks scattered UV.

Treatments with MNNG and EMS can be stopped, and the mutagens destroyed, by making the cell suspension 5% in sodium thiosulfate, using a filter-sterilized stock solution of this reagent. ICR-170, removed from cells by centrifugation, can be destroyed by making the solution 0.1 M in sodium hydroxide.

Methods

MNNG and EMS Mutagenesis

1. Inoculate 10 ml (or other appropriate volume) of liquid YPD medium with a freshly subcloned sample of the yeast strain to give approximately 1×10^6 cells/ml (just detectably turbid). Incubate overnight at 30° with vigorous shaking. In the morning, the culture should contain about 2×10^8 cells/ml.

2. Wash 2.5 ml of the overnight culture twice in 50 mM potassium phosphate buffer, pH 7.0, and resuspend in 10 ml of this buffer. Cell concentration should be 5×10^7 cells/ml. If sonication is needed, chill cells in ice and sonicate for 15 sec. A second cycle of chilling and sonication can be given, but further cycles are unlikely to be of benefit.

3a. For MNNG mutagenesis, add 40 μl of a solution of MNNG in acetone (10 mg/ml) to 10 ml of cells in a screw-cap glass tube, tighten the cap, and mix well. Carry out all operations in a hood, wear gloves and a laboratory coat, and avoid inhaling volatile substances. Incubate in the tightly capped tube at 30° without shaking for 60 min. The MNNG solution is made by dispensing (in a hood, with window lowered as much as possible) approximately 10 mg of MNNG into a capped, preweighed glass vial, followed by reweighing and the addition of a sufficient volume of acetone to bring the concentration to 10 mg/ml. Since MNNG is light sensitive, the vial should be wrapped in aluminum foil or otherwise darkened, and the mutagen handled in subdued light.

3b. For EMS mutagenesis, add 300 μl of EMS to 10 ml of cells in a screw-cap glass tube, tighten the cap well, and vortex vigorously: EMS is poorly miscible in the buffer. Incubate for 30 min at 30°. Carry out all operations in a hood, wear gloves and a laboratory coat, and avoid inhaling volatile substances. Most commercial samples of EMS contain contaminants that increase its toxicity but not mutagenicity: redistilled EMS is a significantly better mutagen.

4. Stop MNNG and EMS mutagenesis in the cell suspensions by adding, in a hood, an equal volume of a freshly made 10% (w/v) filter-sterilized

solution of sodium thiosulfate, mixing well, collecting the cells by centrifugation, and washing them twice with sterile water.

5. Incubate the cells in liquid medium or on plates as appropriate for the particular experimental needs. The mutagen doses suggested above kill 50 to 90% of the cells of most strains, but it is usually desirable to check the survival level of the particular strain used.

UV Mutagenesis

1. Inoculate 10 ml of liquid YPD medium with a freshly subcloned sample of the yeast strain to give approximately 1×10^6 cells/ml. Incubate overnight at 30° with vigorous shaking. In the morning, the culture should contain about 2×10^8 cells/ml.

2. Wash cells twice in sterile water, sonicate if necessary, and irradiate them either on plates or in suspension, according to need. To irradiate on plates, spread 200 μl of an appropriate dilution of the cell suspension on each plate, allow the liquid to be absorbed, and expose them, with lids removed, to 50 J/m^2 UV (or for an empirically determined time). Carry out the irradiation under illumination from "gold" fluorescent lights (e.g., F40G0), or very low light, and incubate the plates in the dark for at least 24 h, to avoid photoreactivation. To irradiate suspensions, 30–50 ml of washed cells in 0.9% (w/v) KCl are placed in a standard 9-cm petri dish, stirred vigorously and continuously with a magnetic mixer, and exposed to UV with petri dish lid removed. Depending on the cell concentration, most suspensions significantly absorb and scatter UV, and suspensions therefore usually need to be exposed to higher UV fluences than cells on plates, to achieve the same level of killing and mutagenesis. As a rough rule of thumb, fluences to suspensions, relative to plates, need to be equal at 10^6 cells/ml or less, 1.5-fold higher at 10^7 cells/ml, and 10-fold higher at 10^8 cells/ml. High cell concentrations are best given long UV exposures (several minutes) at low fluence rates, but even so the results are often less reproducible. Protect cells from photoreactivation as before.

A convenient source of UV can be made by enclosing G8T5 germicidal UV tubes in a box containing ventilation holes screened with matt-black painted panels to prevent the escape of direct or scattered radiation. Ventilation is needed as the relative output of the tube at 254 nm, the major effective wavelength, depends on tube temperature. A tube to sample distance of at least 50 cm is needed to give uniform radiation. Tightly woven metal mesh makes an excellent neutral filter, to reduce fluence rates.

A fluence of 50 J/m^2 kills about 50% of the cells in many strains, but it is often easier to determine exposure time empirically. Many commercial

UV meters underread fluence rates by as much as a factor of 2, because they are calibrated against collimated beams. If needed, fluence rates for specific circumstances can be determined by potassium ferrioxalate actinometry.[11]

Mutant Enrichment by Inositol Starvation

Selective killing of growing cells by starvation for inositol, and hence enrichment of mutants unable to grow in the particular conditions used, was first described for *Neurospora*[12] and has since been developed for yeast.[9] A necessary prerequisite is the presence of one or more *ino* mutations in the parent strain. Initial studies[9] used an *ino1-13 ino4-8* double mutant, but a single mutant containing the recently constructed *ino1* deletion/disruption[13] is likely to be more convenient.

1. Mutagenize cells by one of the methods described above. Inoculate the culture to be treated with a carefully subcloned strain: enrichment procedures indiscriminantly increase the frequency of all slow-growing cells, such as mitochondrial petites which can occur at high frequency in old cultures. If independent mutations at any given locus are needed, distribute aliquots of the mutagenized cells to different tubes before outgrowth, and carry out the procedure on each in parallel.

2. Allow the treated cells to recover from the mutagen damage and to express mutations by resuspending them in YPD or other appropriate medium at a concentration of about 5×10^5 cells/ml, grow the culture to no more than 10^7 cells/ml, and collect the cells by centrifugation: it is important to harvest exponential-phase cells.

3. Wash the cells in prewarmed prestarvation medium, resuspend in this medium at a concentration of between 1×10^4 and 1×10^6 cells/ml, and incubate for 3-4 hr under conditions that will stop the growth of the mutants for which enrichment is desired. If histidine auxotrophs are sought, for example, prestarvation medium is synthetic complete medium that contains inositol but not histidine. For temperature-sensitive mutations, prestarvation medium is any complete medium containing inositol, but the incubation is carried out at 35°.

4. Wash cells twice in prewarmed starvation medium and resuspend at a concentration of no more than 5×10^6 cells/ml. Starvation medium is prestarvation medium lacking inositol. Incubate for 24 hr at 35° to enrich for temperature-sensitive mutants; otherwise incubate at 30°. At 30°, cells

---

[11] J. Jagger, "Introduction to Research in Ultraviolet Photobiology," p. 137. Prentice Hall, Englewood Cliffs, New Jersey, 1967.
[12] H. E. Lester and S. R. Gross, *Science* **139**, 572 (1959).
[13] M. Dean-Johnson and S. A. Henry, *J. Biol. Chem.* **264**, 1274 (1989).

begin to die after 5–6 hr, and eventually only about 0.1–1% should remain viable.

5. Plate cells on any suitable permissive medium and incubate under permissive conditions to obtain well-separated colonies. Rich medium containing a nonfermentable carbon source (e.g., YPG) can be used at this stage if selection against petites is desirable.

6. Screen surviving clones for the desired mutant.

A second cycle of enrichment could be tried, but is usually not required when cells are mutagenized. Solid medium can be used in place of liquid for the starvation phase of the procedure, and surviving cells recovered from these plates by velveteen replication.[9,14] When this is done, the plating density needs to be adjusted to account not only for inositol-less death, but also for the fact that only about 10% cells are transferred by velveteen.

Inositol Starvation Medium per Liter

Ammonium sulfate	5 g
Potassium phosphate, monobasic	1 g
Magnesium sulfate	0.5 g
Sodium chloride	0.1 g
Calcium chloride	0.1 g
Boric acid	500 μg
Copper sulfate	40 μg
Potassium iodide	100 μg
Ferric chloride	200 μg
Manganese sulfate	400 μg
Sodium molybdate	200 μg
Zinc sulfate	400 μg
Biotin	2 μg
Calcium pantothenate	400 μg
Folic acid	2 μg
Niacin	400 μg
p-Aminobenzoic acid	200 μg
Pyridoxine hydrochloride	400 μg
Riboflavin	200 μg
Thiamin hydrochloride	400 μg
Dextrose	20 g

To prevent the selection of auxotrophs, the following nutrilites can be

[14] T. D. Fox, L. S. Folley, J. J. Mulero, T. W. McMullin, P. E. Thorsness, L. O. Hedin, and M. C. Costanzo, this volume [10].

added to the inositol starvation medium: adenine sulfate, uracil, L-arginine-HCl, L-histidine-HCl, L-methionine, L-tryptophan, each at 20 mg/liter; L-isoleucine, L-leucine, L-lysine-HCl, L-tyrosine, each at 30 mg/liter; L-phenylalanine, 50 mg/liter; L-valine, 150 mg/liter; L-aspartic acid, L-glutamic acid, each at 100 mg/liter; L-homoserine, L-threonine, each at 200 mg/liter; L-serine, 375 mg/liter. The last five nutrilites can often be omitted from the medium, since auxotrophs with these requirements are rare.

Synthetic prestarvation medium is inositol starvation medium plus 2000 µg of inositol. Starvation medium can also be made using Difco vitamin-free yeast nitrogen base (16.9 g/liter) but this provides 1% dextrose and also histidine, methionine, and tryptophan. YPD medium is 1% yeast extract, 2% peptone, and 2% dextrose. YPG medium is similar, but with 2% glycerol (v/v) replacing the dextrose.

[19] Targeting, Disruption, Replacement, and Allele Rescue: Integrative DNA Transformation in Yeast

By RODNEY ROTHSTEIN

Introduction

The ability to introduce exogenous DNA into microorganisms has been used extensively by investigators to manipulate the genomes of those organisms.[1] The availability of purified DNA fragments enabled yeast researchers in the late 1970s to introduce these fragments into yeast and for the first time achieve transformation reproducibly.[2,3] Hinnen, Hicks, and Fink[3] reported the transformation of yeast using a cloned *LEU2* DNA fragment isolated by Ratzkin and Carbon.[4] They transformed a nonreverting double mutation, *leu2-3,112*, and showed that yeast efficiently integrates circular DNA into the genome by a single homologous reciprocal exchange that results in a direct repeat of the target sequence. At the same time, Beggs showed that *LEU2* DNA, cloned in a plasmid that included sequences that allow the endogenous yeast plasmid, the 2-µm circle, to replicate autonomously, also transform yeast.[2] In addition, chromosomal

[1] S. A. Lacks, *in* "Genetic Recombination" (R. Kucherlapati and G. R. Smith, eds.), p. 43. American Society for Microbiology, Washington, D.C., 1989.
[2] J. D. Beggs, *Nature (London)* **275**, 104 (1978).
[3] A. Hinnen, J. B. Hicks, and G. R. Fink, *Proc. Natl. Acad. Sci. U.S.A.* **75**, 1929 (1978).
[4] B. Ratzkin and J. Carbon, *Proc. Natl. Acad. Sci. U.S.A.* **74**, 487 (1977).

sequences that support autonomous replication *(ARS)* were discovered by their ability to cause an approximately 1000-fold increase in the frequency of transformation.[5] Manipulation of these autonomously replicating circular plasmids led to the development of vectors that permitted the cloning, by complementation, of any yeast selectable marker (see [14] this volume).

In the absence of *ARS* sequences, DNA transformed into yeast cells integrates into the genome exclusively by homologous recombination. Thus, sequences modified *in vitro* can be used to precisely replace the resident chromosomal copy of any cloned gene.[6] Homologous recombination of transforming DNA can also be used to create null alleles by gene disruption of cloned yeast genes.[7,8] The frequency of homologous recombination after transformation can be stimulated by introducing a double-strand break within the yeast sequences on the plasmid.[9] Plasmids that contain more than one yeast homologous sequence can be integrated to the site of choice by introducing a double-strand break into that sequence on the plasmid.

The ability to direct integration of plasmids into any desired gene led to the development of methods to recover mutant chromosomal alleles by subsequently retrieving the sequence adjacent to the integrated plasmid as a circular plasmid in bacteria.[10] Next, it was shown that gapped plasmid molecules repair the missing information using the resident chromosomal allele as template.[11] This permits the rescue of mutant alleles by genetic recombination with plasmids that contain autonomous replication sequences. Methods to retrieve these plasmids in *Escherichia coli* are described in [21], this volume.

Utility of DNA Transformation

Transformation has been used to clone genes by genetic complementation,[12] to clone functional chromosomal components such as origins of replication,[13] centromeres,[14] and telomeres,[15] and to clone functional sup-

[5] K. Struhl, D. T. Stinchcomb, S. Scherer, and R. W. Davis, *Proc. Natl. Acad. Sci. U.S.A.* **76**, 1035 (1979).
[6] S. Scherer and R. W. Davis, *Proc. Natl. Acad. Sci. U.S.A.* **76**, 4951 (1979).
[7] D. Shortle, J. E. Haber, and D. Botstein, *Science* **217**, 371 (1982).
[8] R. J. Rothstein, this series, Vol. 101, p. 202.
[9] T. L. Orr-Weaver, J. W. Szostak, and R. J. Rothstein, *Proc. Natl. Acad. Sci. U.S.A.* **78**, 6354 (1981).
[10] J. I. Stiles, J. W. Szostak, A. T. Young, R. Wu, S. Consul, and F. Sherman, *Cell (Cambridge, Mass.)* **25**, 277 (1981).
[11] T. L. Orr-Weaver, J. W. Szostak, and R. J. Rothstein, this series, Vol. 101, p. 228.
[12] K. A. Nasmyth and S. I. Reed, *Proc. Natl. Acad. Sci. U.S.A.* **77**, 2119 (1980).
[13] C. S. Chan and B. K. Tye, *Proc. Natl. Acad. Sci. U.S.A.* **77**, 6329 (1980).
[14] L. Clarke and J. Carbon, *Nature (London)* **287**, 504 (1980).
[15] J. W. Szostak and E. H. Blackburn, *Cell (Cambridge, Mass.)* **29**, 245 (1982).

pressors identifying interacting genes.[16] Once isolated, yeast chromosomal sequences can be manipulated *in vitro* and reintroduced into cells either on an autonomously replicating plasmid or at their normal chromosomal location to permit a functional analysis. Null mutations of any cloned gene can be created by introducing an *in vitro*-generated deletion into the genome at the precise chromosomal position of the gene. If a gene or region is difficult to score phenotypically, a genetic marker can be inserted adjacent to the gene or within the gene, resulting in an easily scored phenotype. In addition, novel genotypes can be created in which either new genes are introduced into regions to develop specific assays[17] or novel chromosomal structures are created, such as shortened chromosomes useful for genetic analysis.[18] A recently devised technique uses these methods to determine the physical map distance between a marker and its telomere.[19] Finally, the strategies described for yeast sequences can also be applied to foreign DNAs cloned in yeast on yeast artificial chromosomes.[20]

In this chapter, some of the basic strategies used for manipulating the yeast genome using exogenous DNA are described. Directed integration of plasmid sequences as well as several methods for gene disruption used to construct null mutations are discussed. Next, gene replacement strategies based on pop-in/pop-out recombination and fragment-mediated gene conversion are described. Finally, plasmid gap repair used to rescue chromosomal alleles for DNA sequence and functional analysis is illustrated.

Materials

Yeast Strains and Media

Gene targeting and gene disruption can be performed in virtually any genetically marked yeast strain. The genetic markers for selection of transformation events in the host strain should exhibit low reversion frequencies (less than 10^{-8}). The most commonly employed auxotrophic markers are *ura3, leu2, his3*, and *trp1*, since double point mutations or low reverting alleles of these markers are available. In some cases, deletions have been engineered (e.g., *his3-Δ200*,[21] *trp1-Δ1*[22]) that can be used to decrease

---

[16] I. Herskowitz, *Nature (London)* **329**, 219 (1987).
[17] J. W. Wallis, G. Chrebet, G. Brodsky, M. Rolfe, and R. Rothstein, *Cell (Cambridge, Mass.)* **58**, 409 (1989).
[18] A. W. Murray and J. W. Szostak, *Nature (London)* **305**, 189 (1983).
[19] D. Vollrath, R. W. Davis, C. Connelly, and P. Hieter, *Proc. Natl. Acad. Sci. U.S.A.* **85**, 6027 (1988).
[20] V. Pachnis, L. Pevny, R. Rothstein, and F. Costantini, *Proc. Natl. Acad. Sci. U.S.A.* **87**, 5109 (1990).
[21] M. T. Fasullo and R. W. Davis, *Mol. Cell. Biol.* **8**, 4370 (1988).
[22] R. S. Sikorski and P. Hieter, *Genetics* **122**, 19 (1989).

recombination between the incoming plasmid DNA and the chromosomal site. These methods are not limited to the markers described above. Any marker for which a suitable genetic selection exists can be employed in targeting or gene disruption.

Standard genetic methods are used to grow yeast strains[23] (see [1] in this volume). Prototrophic transformants are selected on synthetic medium lacking the nutrient for which the marked strain is auxotrophic. Recipes for drug-containing media [e.g., media with 5-fluoroorotic acid (5-FOA), cycloheximide (CYH), or canavanine (CAN)] are found in [20] in this volume.

Yeast Shuttle Vectors

There is no rule to follow in choosing the vector for use in targeted integration. Plasmids based on pBR322[24] or pUC[25] are equally efficient in transformation. The yeast vector YIp5,[5] often used for pop-in/pop-out experiments, is shown in Fig. 1 along with some other useful cloning vectors for targeted integration. A series of multipurpose yeast vectors has also been described.[23]

Restriction Enzyme Digestion

Restriction enzymes used for targeting can leave either a 5' overhang, a 3' overhang, or blunt ends on the DNA. No substantial difference in transformation efficiencies has been noted. Standard digestion buffers and reaction conditions recommended by the suppliers are followed. The presence of residual restriction enzyme(s) in the transforming DNA does not appreciably inhibit or alter the transformation efficiencies.

Yeast Markers Used for Disruptions

Some commonly used selectable markers for disruption have been cloned in pUC-based vectors by John Hill[26] and are illustrated in Fig. 1. These vectors are available from our laboratory. Restriction sites that are unique or that lie outside the gene within the yeast DNA are underlined. These sites can be used to excise a functional fragment to use for gene disruption purposes.

---

[23] F. Sherman, G. R. Fink, and J. B. Hicks, "Methods in Yeast Genetics." Cold Spring Harbor Laboratory, Cold Spring Harbor, New York, 1986.
[24] F. Bolivar, R. L. Rodriguez, P. J. Greene, M. C. Betlach, H. L. Heynecker, and H. W. Boyer, *Gene* **2**, 95 (1977).
[25] C. Yanisch-Perron, J. Vieira, and J. Messing, *Gene* **33**, 103 (1985).
[26] J. Hill, personal communication (1988).

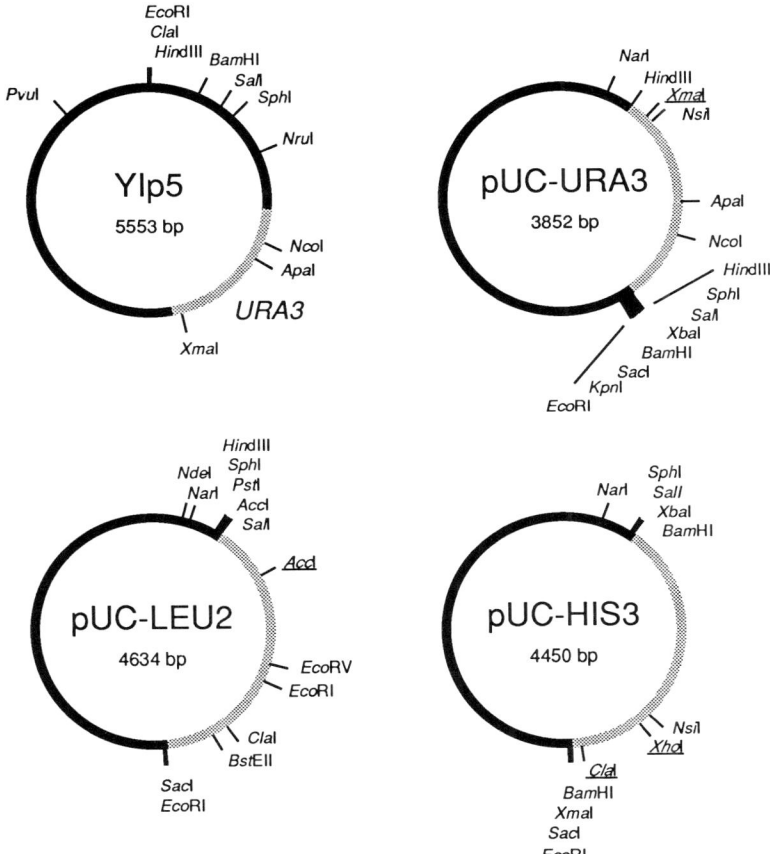

FIG. 1. Common yeast vectors. The four vectors shown can be used for targeted integration or for extracting genetic markers for gene disruption. Yeast sequences are indicated by stippled lines. Restriction sites within the yeast fragments that can be used to extract a functional DNA fragment for gene disruption are underlined.

Methods

Integration by Homologous Recombination

Exogenous plasmid DNA that contains a yeast sequence, in the absence of *ARS* sequences, integrates at the homologous chromosomal locus after DNA transformation.[3] Even plasmids that contain *ARS* sequences can integrate into the genome by homologous recombination. However, if the plasmid contains centromere sequences, integration creates a dicentric chromosome, and the event is selected against.[27] Plasmid integration in yeast is used to genetically mark a region to determine if the cloned fragment of interest is tightly linked to the original complemented mutation. It is also the first step of the pop-in/pop-out replacement technique described below.[6] Last, it is used to insert a stable single copy of a gene at a unique chromosomal site. Plasmid integration is most efficiently achieved by targeting the molecule with a double-strand break.[9] However, plasmids that contain yeast fragments with lengths less than 250 base pairs (bp) cannot be targeted easily.[28]

DNA Transformation with Circular Molecules: Nontargeted Integration. In Fig. 2, a circular plasmid molecule is illustrated that contains two yeast sequences, *AUXA* for selection and *YFG1* ("your favorite gene"). The recipient yeast strain is auxotrophic for *auxA* so that integration events can be selected. The recipient strain can be either auxotrophic or wild type for *YFG1*. After transformation, if there is no *ARS* on the plasmid, *AUXA* prototrophs arise from either a single reciprocal exchange between either *AUXA* or *YFG1* and the corresponding homologous sequence in the chromosome (e.g., Fig. 2C1) or a replacement of *auxA* with *AUXA* without any plasmid sequences integrated (likely by a gene conversion or a double crossover, Fig. 2C3). The relative ratio of integration versus replacement is fragment dependent. Colony hybridization using radioactively labeled bacterial plasmid sequences as a probe distinguishes the integration events from replacements.[29] Genomic blots or genetic linkage can be used to determine into which homologous sequence the plasmid has integrated. Since the restriction enzyme sites surrounding the two regions are different (indicated by arrows in Fig. 2), a genomic blot digested with either of these enzymes and probed with radioactively labeled plasmid DNA will generate a unique pattern for each integration.[30] If the yeast fragment on the plasmid contains a portion of a repetitive sequence such as a *Ty* element or a δ sequence,[31] then integration into the homologous repeated sequences may occur, resulting in what appear to be random integration events.[32] For example, the *Pst*I fragment of the *LEU2* gene in YEp13 contains a portion of a *Ty* element and its associated δ sequence and was shown to integrate at genomic *Ty* elements. To avoid this problem, the *Hpa*I–*Sal*I *LEU2* frag-

[27] C. Mann and R. W. Davis, *Proc. Natl. Acad. Sci. U.S.A.* **80,** 228 (1983).
[28] R. Rothstein, unpublished observations.
[29] R. Rothstein, in "DNA Cloning: A Practical Approach" (D. M. Glover, ed.), p. 45. IRL Press, Oxford, 1985.
[30] T. Maniatis, E. F. Fritsch, and J. Sambrook, "Molecular Cloning: A Laboratory Manual." Cold Spring Harbor Laboratory, Cold Spring Harbor, New York, 1982.
[31] J. D. Boeke, in "Mobile DNA" (D. Berg and M. Howe, eds.), p. 382. American Society for Microbiology, Washington, D.C., 1989.
[32] H. L. Klein and T. D. Petes, *Mol. Cell. Biol.* **4,** 329 (1984).

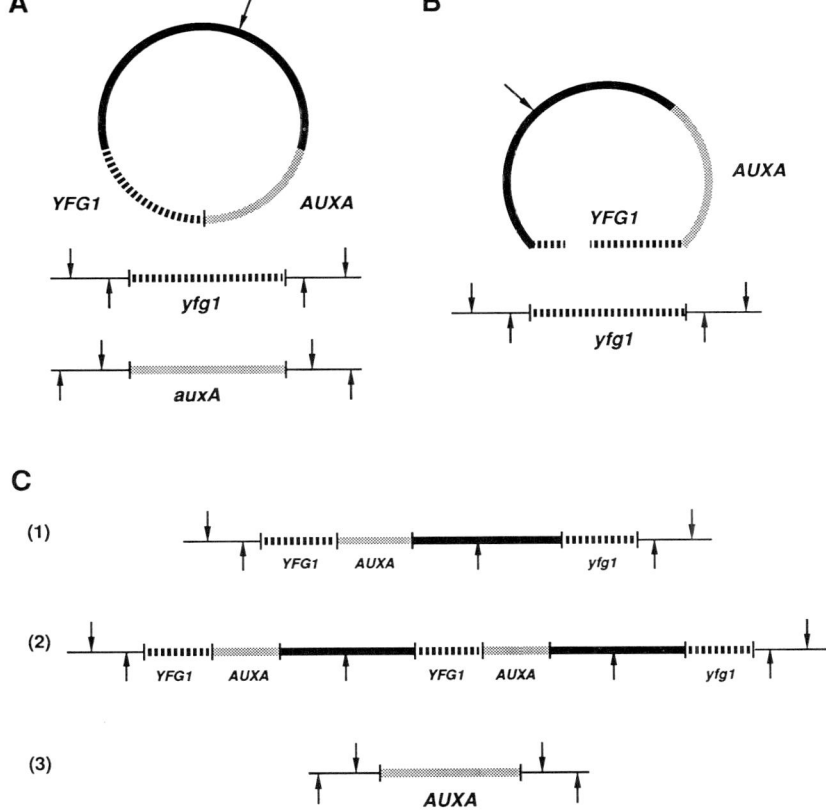

FIG. 2. Targeted integration in yeast. (A) A circular plasmid that contains two yeast sequences, *YFG1* (barred area) and *AUXA* (dotted area), can pair with homologous chromosomal sites, indicated as straight lines below the plasmid. The arrows surrounding each gene indicate restriction sites. The arrow on the plasmid and the upward arrows represent the same restriction site. The downward arrows represent unique chromosomal restriction sites that are not present on the plasmid. Note that the positions of the sites surrounding the two genes are different. (B) Targeted integration into *yfg1* is achieved by creating a double-strand break, indicated by the space in the dashed line of the circular plasmid. The two ends of the linear molecule pair with the chromosomal sequence as indicated. The double-strand break stimulates recombination, and plasmid integration events occur at *yfg1*. (C) Three kinds of events are detected after selection for *AUXA* prototrophs. (1) A single crossover event at *YFG1* leads to a direct repeat of *YFG1* flanking the plasmid. Genomic blots of DNA digested with a restriction enzyme that digests the plasmid once detect two bands with sizes characteristic for the integration site. Note that integration of the plasmid at *auxA* results in different sized restriction fragments (not shown). (2) A multiple integration event at *YFG1* leads to a third, plasmid-length, band after genomic blotting of DNA digested with a restriction enzyme that digests the plasmid once. The number of plasmids integrated can be determined by digesting genomic DNA with a restriction enzyme that does not digest the plasmid (downward arrow) and measuring the increased size of the parental band. (3) Replacement events occur at *auxA* when *AUXA* sequences from the plasmid replace the chromosomal mutant site, resulting in an *AUXA* prototroph.

ment in pUC-LEU2 (Fig. 1) is used since it does not have any repeated sequences.

Occasionally, multiple tandem integrations of the plasmid sequence occur.[9] These events can also be distinguished on genomic blots by digesting DNA from the transformant with a restriction enzyme that digests the plasmid once outside of the marker gene (indicated by the upward arrow). Hybridization with labeled plasmid DNA generates two bands for a single integration event. A multiple tandem array gives a third band equivalent in size to the plasmid (Fig. 2C2). The precise number of copies in the multiple tandem array can be determined by digesting with a restriction enzyme that fails to cut anywhere within the plasmid (indicated by the downward arrow) and measuring the increase in size of the parental fragment.

Transformation with Linear DNA Molecules: Targeted Integration. To ensure that the plasmid sequence integrates at the chromosomal location of interest, the integration event is targeted by introducing a double-strand break into the plasmid (Fig. 2B).[9] When the double-strand break is made within one of the two (or more) yeast sequences on a plasmid, the frequency of transformation increases 10- to 1000-fold compared to uncut circular plasmid without an *ARS* sequence. In addition, the double-strand break directs plasmid integration to the chromosomal region homologous to the cut sequence. An example of targeted integration is shown in Fig. 2B. To direct the integration event to *yfg1*, a unique double-strand break is introduced by restriction enzyme digestion within the *YFG1* sequence on the plasmid. The best position for the double-strand break is determined by factors that may influence the frequency of integration, such as the length of homology adjacent to the double-strand break or the position of the break relative to a chromosomal mutant site (see next section). After transformation with the linearized DNA and selection for *AUXA* prototrophs, plasmid integration events are readily detectable (Fig. 2C1, 2C2). In addition, at a lower frequency, replacement events at *auxA* may occur that have not integrated the plasmid sequence (Fig. 2C3). The distribution of these two kinds of events is dependent on factors such as the length of homology in the targeted sequence and the length of homology in the selectable marker and cannot be predicted *a priori*. As described above, integration events can be distinguished from replacements by performing yeast colony hybridization and genomic blots.

Variables that influence frequency of targeted integration. The role of the length of homology on the frequency of transformation has not been studied systematically. Anecdotal evidence suggests that longer homologies result in increased integration frequency. There is likely a minimum length of homology necessary for plasmid integration. In one case, cut sequences with 125 nucleotides of homology (37 nucleotides on one side and 88 on

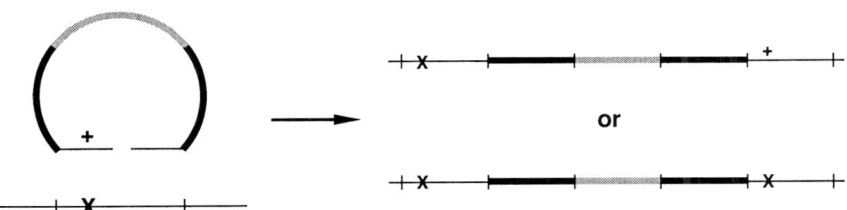

FIG. 3. Gene conversion of a site adjacent to a double-strand break. When a double-strand break is introduced adjacent to the wild-type information on a fragment, there are two possible outcomes after integration into a region that contains a mutation (X). The plasmid integrates, creating a duplication of the target sequence, with one wild-type copy and one mutant copy. Alternatively, the wild-type information on the plasmid copy is converted to mutant, giving rise to a duplication in which both sides of the duplication flanking the plasmid sequence contain mutations. See text for further discussion.

the other) did not integrate whereas targeted integration was obtained with a sequence containing 250 nucleotides of homology (40 and 210).[28]

The natural variability of different genetic regions for recombination also affects integration frequencies. For example, the *LEU2* region exhibits more than 100-fold lower integration frequencies when compared to the *HIS3* region.[28] These differences cannot be predicted in advance, so each case has to be tested independently. However, targeting generally increases significantly the probability of integration within a particular sequence over that obtained with circular, uncut plasmid DNA.

The apparent frequency of integration is also influenced by the position of the mutant site on the chromosome relative to the cut site on the integrating plasmid.[33] During integration, mutant sequences from the chromosome can replace the wild-type information on the incoming DNA strand, resulting in direct repeats that each contain the mutant chromosomal copy (Fig. 3). This is formally a gene conversion event, that is, the nonreciprocal transfer of genetic information. Orr-Weaver *et al.* showed that the frequency of gene conversion is highest when the break is closest to the mutant site.[33] Therefore, the position of the double-strand break affects the frequency at which wild-type information is converted to the mutant chromosomal sequence and affects the frequency of prototrophs recovered since a fraction of the integration events are lost to gene conversion.

The length of homology also influences the frequency of integration. To reduce the frequency of replacement events of the selectable marker and/or integration events that occur at exogenous sequences outside of the desired integration target, host strains that contain a deletion or rearrange-

[33] T. L. Orr-Weaver, A. Nicolas, and J. W. Szostak, *Mol. Cell. Biol.* **8**, 5292 (1988).

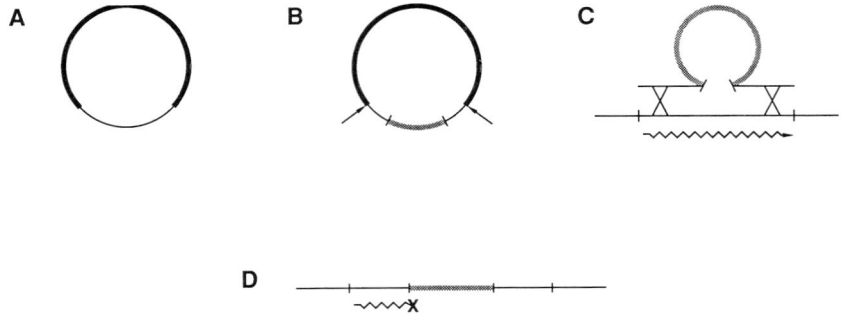

FIG. 4. One-step gene disruption. (A) To make a mutation in YFG1 (thin line), (B) a selectable marker (dotted area) is cloned into the middle of YFG1. The disrupted fragment is liberated from the parental plasmid using restriction enzymes (arrows). (C) The liberated DNA fragment is used to transform yeast. The homologous ends pair with the chromosome, and recombination results in a chromosomal gene replacement (D).

ment of the chromosomal copy of the selectable marker can be used. This reduces homology with the selectable marker and lowers the frequency of background replacements. For example, a plasmid linearized at YFG1, as illustrated in Fig. 2, integrates only at yfg1 if the selectable marker sequences (AUXA) have been deleted from the genome. The ura3-52 mutation is a Ty insertion that almost completely eliminates URA3 replacement events and reduces, but does not eliminate, integration events.[34]

Gene Disruption Techniques

One-Step Gene Disruption. One-step gene disruption or replacement results in a genetically stable disruption since no direct repeats are left flanking the insertion site.[8] The method requires a cloned gene and a restriction map of the fragment to identify a restriction site(s) for inserting a selectable genetic marker (Fig. 4A). Two kinds of disruptions can be constructed. One, an insertion, results from inserting a genetic marker into a region at a single restriction site within the gene of interest. The other, an insertion–deletion, results in a deletion of all or a portion of the gene after the insertion. When creating a null mutation, it is best to delete as large a portion of the gene as possible to avoid the possibility that the insertion–disruption leads to a fortuitously functional fusion.

The selectable marker fragment used for the disruption is cloned into the gene of interest with sufficient homology adjacent to the insertion point

[34] M. Rose and F. Winston, *Mol. Gen. Genet.* **193,** 557 (1984).

to permit homologous pairing with both sides of the chromosomal target sequence (Fig. 4B,C). The amount of DNA homology adjacent to the inserted gene should be greater than 500 bp whenever possible. Although no systematic study has been performed, the greater the length of homology, the more efficient the gene disruption. However, it is noteworthy that some gene disruptions have been successful with as few as 28 nucleotides of homology on one side of the insertion.[35]

Next, the disrupted fragment is liberated from the plasmid vector by cutting with restriction enzymes that generate a linear fragment that is homologous to the chromosome at both of its ends (Fig. 4B,C). It is not always necessary to completely remove plasmid sequences from both ends of the linear fragment. In fact, up to 4000 nucleotides of plasmid sequence can be left on either end or on both ends as long as the unpaired end sequences are not homologous to any sequences in the yeast genome. These nonhomologous ends do not significantly lower the frequency of successful gene disruptions. Similarly, it is not necessary to purify the liberated fragment from the vector sequences before transformation unless sequences on the vector can recombine with other sequences in the genome such as other plasmid sequences in the recipient cell.

Disruptions work equally well with yeast transformation procedures based on spheroplast formation,[2,3] Li^+ ions,[36] or electroporation.[28] As a general rule 1 to 10 μg of plasmid (0.1–1 pmol of a 15-kb plasmid) are used to transform 10^7 competent cells. Yields of transformants vary from 1 to 1000 transformants/μg/10^7 cells.

It is important to verify a successful gene disruption by a genomic blot since occasionally the disrupted copy on the plasmid may integrate adjacent to the wild-type genomic copy without replacing it. This leads to a duplication that still contains a functional gene and results in a misinterpretation of the true null phenotype.

When the null phenotype is unknown, gene disruption experiments should be performed in diploid cells to maintain one wild-type copy after replacement. The diploid is sporulated, and tetrads containing the four products of meiosis are dissected to determine the phenotype of the disruption (Fig. 5). When all four spores survive, the spores containing the disrupting marker also contain the null allele. If only two spores survive and neither contains the marker used for the disruption, the gene is either essential for germination, essential for growth, or both. To distinguish between these possibilities, a wild-type copy of the gene is cloned onto a centromere plasmid that contains the *URA3* gene and introduced into the

---

[35] J. Strathern, personal communication (1982).
[36] H. Ito, Y. Fukuda, K. Murata, and A. Kimura, *J. Bacteriol.* **153**, 163 (1983).

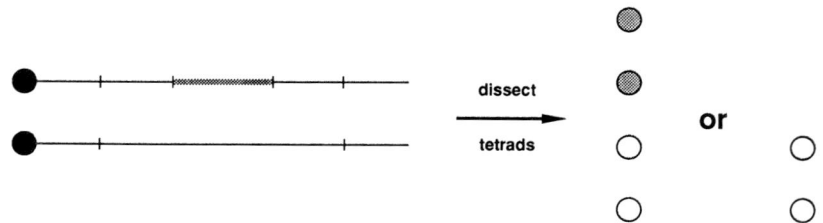

FIG. 5. Gene disruption in a diploid. When the disruption strategy shown in Fig. 4 is applied to a diploid, one of the two chromosomes becomes disrupted. After sporulation and tetrad dissection, there are two possible outcomes: If the gene is nonessential, four spores survive, two containing the disrupted allele (dotted circle) and two wild-type spores (open circle). Alternatively, if the disrupted gene is essential for germination, growth, or both, only the two wild-type spores survive.

heterozygous disruption strain by transformation. After sporulation and dissection, spores that contain the chromosomal gene disruption can be obtained since the wild-type plasmid copy can complement the defect. Next, plasmid loss is selected on 5-FOA medium which selects against the URA3-containing centromere plasmid.[37] If the gene is essential, cells containing the disrupted allele will fail to grow on 5-FOA medium. If the gene is essential only for germination, all of the cells will grow after selecting for plasmid loss.

Using standard gene disruption technology, a problem is encountered when multiple disruptions need to be analyzed, since the number of disruptions that can be created in a given strain is limited by the availability of useful selectable markers. Recently, a construct was designed to permit repeated disruptions without the loss of any selectable markers within the yeast strain.[38] This construct has a URA3 gene cloned between duplicated copies of a fragment from the Salmonella hisG gene. This URA3 "cassette" is used to disrupt the cloned gene of interest, and the disruption is transferred to the chromosome as described above. The hisG direct repeats that flank the URA3 gene can recombine to leave a single hisG fragment disrupting the gene of interest. Recombinants selected on 5-FOA medium are ura3 auxotrophs, and thus the entire procedure can be repeated.

Transposon mutagenesis in bacteria has been used to create multiple independent insertions within a DNA fragment. Several useful sets of bacterial transposons have been engineered to contain yeast selectable

[37] J. D. Boeke, F. Lacroute, and G. R. Fink, *Mol. Gen. Genet.* **197**, 345 (1984).
[38] E. Alani, L. Cao, and N. Kleckner, *Genetics* **116**, 541 (1987).

marker genes.[39-41] One method is described in [22], this volume. In conjunction with these methods, the location of a gene on a cloned DNA fragment can be determined. Independent transposon-induced disruptions of a yeast fragment are transformed into yeast cells to saturate the region with insertion–mutations. If the null phenotype of the gene is unknown (and may be lethal), each independent transposition-disrupted fragment must be individually transformed into a diploid. To avoid aberrant events and ensure replacement at the chromosomal locus, the disruptions are introduced into a diploid that has been singly interrupted in the region of interest with a marker different from the one used for the transposon-mediated disruptions.[42] After transformation and selection for the marker in the transposon, the colonies are screened for simultaneous loss of the first marker. Positive colonies are sporulated, and the four spores are dissected to assess the phenotype of the new transposon-generated disruptions.

Internal Fragment Disruption. A gene disruption can also be created by integrating a plasmid containing an internal segment of the gene into the homologous chromosomal copy of the gene.[7] The homologous reciprocal exchange between the internal fragment and the chromosome creates a disruption because, after integration, the two copies of the gene flanking the plasmid sequences are not full length (Fig. 6): one is truncated at the 3' end, and the other is truncated at the 5' end. The use of targeted integration by cutting uniquely in the internal fragment increases the frequency of the disruption event. This method is useful when a convenient internal fragment is available (at least 350 bp). However, it is necessary to select continuously for the maintenance of the integrated plasmid sequence, otherwise spontaneous recombination results in loss of the plasmid at a frequency between 10^{-4} and 10^{-3}, restoring the full-length gene. This high reversion frequency can often make it difficult to assess the precise phenotype of the disruption.

Gene Replacement Techniques

In the previous section, techniques for gene disruption were described that lead to a complete loss of function of the gene of interest. The selectable marker itself was used to create the mutation, resulting in either

---

[39] H. S. Seifert, E. Y. Chen, M. So, and F. Heffron, *Proc. Natl. Acad. Sci. U.S.A.* **83,** 735 (1986).
[40] M. Snyder, S. Elledge, and R. W. Davis, *Proc. Natl. Acad. Sci. U.S.A.* **83,** 730 (1986).
[41] O. Huisman, W. Raymond, K. U. Froehlich, P. Errada, N. Kleckner, D. Botstein, and M. A. Hoyt, *Genetics* **116,** 191 (1987).
[42] J. W. Wallis, W. L. Arthur, M. Rolfe, and R. Rothstein, in preparation.

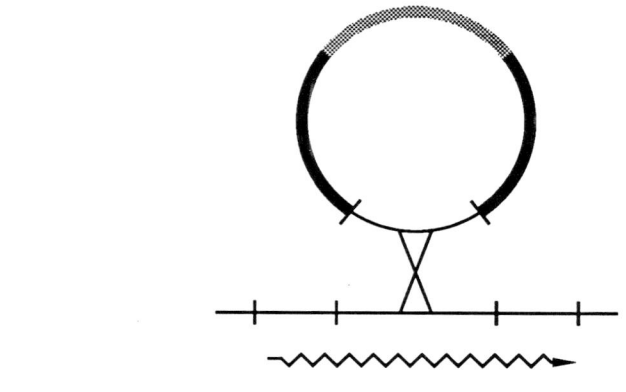

Fig. 6. Internal fragment disruption. An internal fragment of *YFG1* (thin line) is cloned into an integrating vector that contains a selectable marker (dotted area). Homologous recombination with the chromosomal locus results in a duplication that contains a mutated 3' fragment and a mutated 5' fragment; neither of which is functional.

an insertion or an insertion–deletion. In an exhaustive analysis of a gene, it is sometimes necessary to examine the phenotype of a large number of *in vitro*-generated constructs. The plasmid shuffle strategy described in [20], this volume, is one useful method for such an analysis. There are, however, occasions when the phenotype of a gene on a plasmid is different from the phenotype at its normal chromosomal location.[43,44] Methods for substituting any kind of *in vitro*-constructed mutation back into the chromosome are described below.

Pop-In/Pop-Out Replacement. The pop-in/pop-out replacement method, developed by Scherer and Davis,[6] involves two steps: plasmid integration using the *URA3* gene as a selectable marker and plasmid excision selecting against the *URA3* gene. When the method was first described, the drug ureidosuccinic acid was used to select against the *URA3* gene for the plasmid excision step. However, this selection does not work

[43] J. Abraham, J. Feldman, K. A. Nasmyth, J. N. Strathern, A. J. Klar, J. R. Broach, and J. B. Hicks, *Cold Spring Harbor Symp. Quant. Biol.* **47**, 989 (1983).
[44] A. H. Brand, L. Breeden, J. Abraham, R. Sternglanz, and K. Nasmyth, *Cell (Cambridge, Mass.)* **41**, 41 (1985).

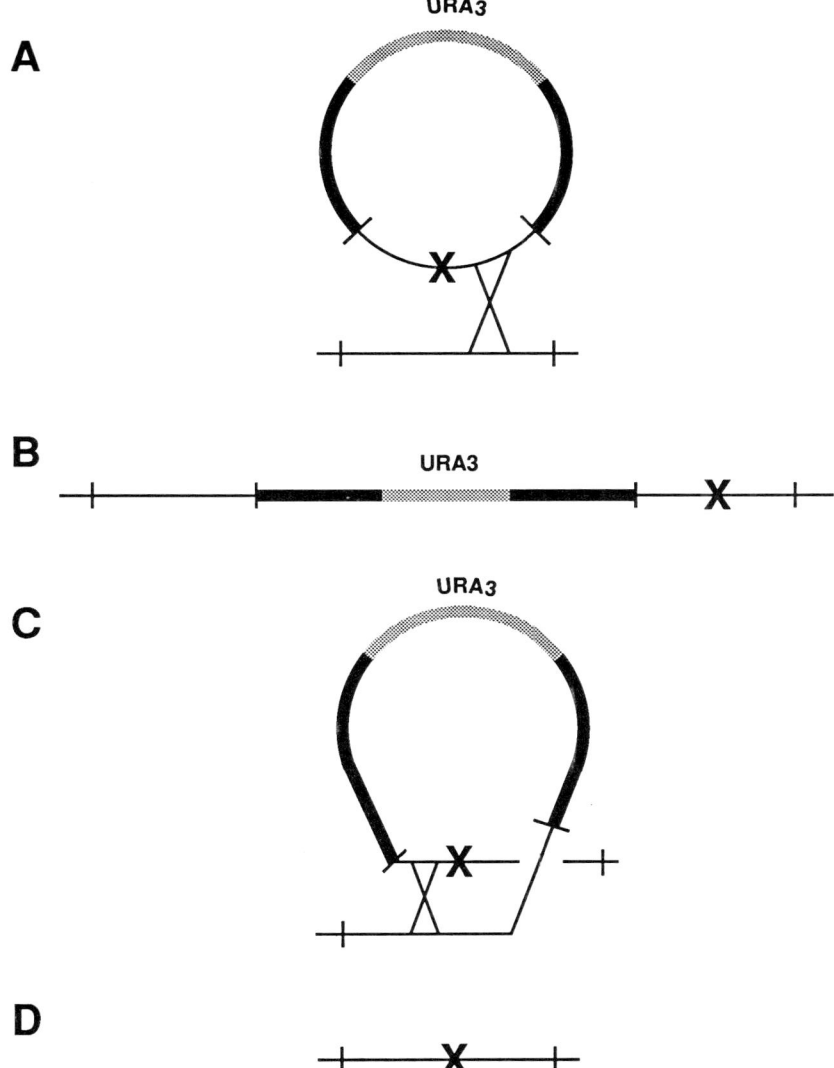

FIG. 7. Pop-in/pop-out allele replacement. (A) A mutation (X) is introduced in *YFG1* (thin line) and is cloned in an integrating vector that contains the *URA3* selectable marker (dotted area). (B) Integration of the circular molecule results in direct repeats of *YFG1* with one mutant copy and one wild-type copy. (C) Pairing of the homologous direct repeats and recombination can result in the loss of the plasmid sequence. This event is selected on 5-FOA-containing medium. (D) Crossovers that occur on the appropriate side of the mutant site replace the wild-type chromosomal site with the mutant sequence.

in all genetic backgrounds. Subsequently, the drug 5-FOA has been shown to work in virtually all genetic backgrounds tested.[37]

The basic strategy of the pop-in/pop-out method is as follows: a specific alteration is introduced into the gene of interest cloned in a *URA3*-based integrating vector (e.g., YIp5 or pUC18-URA3, see Fig. 7A). The alteration in the DNA can be a deletion, insertion, or a single base-pair change somewhere within the fragment. For example, a temperature-sensitive mutation or a suppressible mutation can be substituted for the wild-type copy as long as the mutant copy can be distinguished from the wild type (see below). Next, the plasmid is integrated into its chromosomal location by homologous recombination (Fig. 7B). This creates a duplication containing the wild-type copy and the mutant copy flanking the plasmid sequences. Finally, excision of the plasmid is selected using 5-FOA, and the 5-FOA-resistant colonies are screened for the mutant phenotype (see Fig. 7C,D).

The pop-in and pop-out events must occur on different sides of the alteration for the mutation to be retained after pop-out. It is best to leave as much homology on both sides of the mutant site as possible for the integration and excision events. For example, if the mutant site is asymmetrically positioned on the fragment, both the pop-in and the pop-out recombination events often occur on the same side of the mutation, and the altered site is not transferred to the chromosome. Targeted integration can be used to improve the probability of getting the pop-out to leave the mutant site in the chromosome. Whenever possible, the restriction site for targeting should be between the mutant site and the shortest stretch of homology. Although this does not guarantee that the pop-in crossover occurs in this position, it often does. This increases the probability that the pop-out crossover event will occur in the longer region of homology, resulting in a successful replacement. Finally, since the chromosomal sequences can sometimes replace the mutant sequences in the integrating plasmid by gene conversion (e.g., see Fig. 3) the integration event itself can sometimes lead to loss of the mutation on the incoming plasmid. Therefore, it is best to have a method to detect the presence of the altered allele (e.g., a genomic blot) to ensure that, after the pop-in, the strain can generate the desired allele replacement by a pop-out.

Conditional mutations in essential genes, generated *in vitro* and screened *in vivo* by plasmid shuffling (see [20] in this volume for a complete description), can be introduced at their chromosomal location by using a modification of the pop-in/pop-out strategy. A diploid is disrupted with *URA3* in one copy of the essential gene, deleting as much of the coding sequence as possible. The strain is transformed with the *in vitro*-generated mutant allele cloned into an integrating vector. Transformants

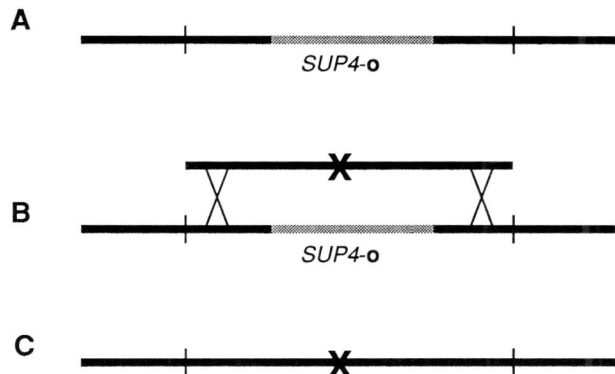

FIG. 8. Counterselection for allele replacement. (A) The gene of interest is first disrupted with a counterselectable marker (*SUP4*-o in this example, dotted area), removing as much of the wild-type sequence of the gene as possible. (B) Linear DNA that contains the altered allele (X) is cotransformed into the recipient strain along with a selectable plasmid. The double crossover occurs nonselectively in approximately 0.1 to 5% of the transformants. (C) Transformants that have lost the counterselectable marker contain the replaced allele.

that integrate into the *URA3*-disrupted copy are identified by genomic blots or by genetic linkage. Haploid progeny containing the mutated allele linked to the disruption are isolated after sporulation and dissection. The conditional mutant is selected as a 5-FOA-resistant cell *(ura3)* that simultaneously loses the integrating plasmid.

Use of Counterselectable Markers to Facilitate Direct Gene Replacement. Methods have been devised to facilitate repeated allele replacement at nonessential genes. First, a counterselectable marker is inserted at the chromosomal position of the gene (Fig. 8A). Next, linear DNA fragments containing the mutation of interest are transformed into the strain, where they recombine with the resident disrupted region (Fig. 8B) and result in the loss of the counterselectable marker (Fig. 8C). The *SUP4*-o gene, the *CAN1* gene,[45] or the *CYH2* gene[46] can be used as counterselectable markers. To use the *SUP4*-o gene, the genetic background must contain an ochre-suppressible allele, such as *can1-100*.[47] To use the *CAN1* gene, the strain may contain any allele of *can1*.[48] In either case, before the disruption with the counterselectable marker, the parent strain is resistant to canavanine, an arginine analog toxic to the cell. After introduction of the

[45] W. Hoffmann, *J. Biol. Chem.* **260,** 11831 (1985).
[46] N. F. Kaufer, H. M. Fried, W. F. Schwindinger, M. Jasin, and J. Warner, *Nucleic Acids Res.* **11,** 3123 (1983).
[47] K. Nasmyth, *Cell (Cambridge, Mass.)* **42,** 213 (1985).
[48] K. Struhl, *Gene* **26,** 231 (1983).

SUP4-o gene or the *CAN1* gene at the locus of interest, the strain becomes sensitive to canavanine owing to restored function of the *CAN1* gene, the arginine permease, and is ready for counterselection. After a successful replacement, the cell is again resistant to canavanine.

Replacements cannot be selected directly on canavanine-containing medium because the background level of revertants resistant to canavanine is too high ($\sim 10^{-6}$). Therefore, the most efficient way to substitute the mutated fragment into the chromosome is to cotransform the mutated linear molecule into the cell along with a circular plasmid that gives high-frequency transformation. The frequency of cotransformation varies from 0.1 to 5% of the total; after replica plating to canavanine medium, those transformants that have replaced the *SUP4*-o or *CAN1* gene can grow.

A similar strategy applies for the *CYH2* locus.[48] In this case, the starting strain is cycloheximide resistant *(cyh2^R)*, and the *CYH2* wild-type gene is cloned into the gene of interest. Cotransformants are screened for cycloheximide resistance.

Allele Recovery

It is often desirable to recover mutant chromosomal alleles for DNA sequence analysis. Rather than constructing a library from each mutant strain, the most common way to retrieve a chromosomal mutation is based on the ability of DNA ends to promote homologous recombination and takes advantage of the efficient repair of gapped molecules. Using procedures similar to targeting, it was found that yeast recombines gapped linear molecules that are missing information from the targeted sequence.[11] This requires that the two ends, homologous to the integration site, pair with the chromosome as illustrated in Fig. 9A. During the integration process, the gapped region is repaired using chromosomal sequences as template. Orr-Weaver and Szostak found that in the presence of an *ARS* sequence on the plasmid there are two possible outcomes for this repair event.[49] One results in the repair of the gap using chromosomal sequences as template, leading to an autonomously replicating plasmid that contains the mutant allele. The other is the integration of the plasmid after repair of the gap. Use of a centromere-based vector ensures that all of the gap repaired events are recovered as autonomously replicating plasmids. To rescue the mutant allele from such plasmids, total yeast DNA from the transformants is isolated (see [21] in this volume) and used to transform *Escherichia coli*.

This method has been modified to locate more precisely the region containing the mutant chromosomal allele.[11] For example, the plasmid

[49] T. L. Orr-Weaver and J. W. Szostak, *Proc. Natl. Acad. Sci. U.S.A.* **80**, 4417 (1983).

[19] TARGETING, DISRUPTION, REPLACEMENT, AND ALLELE RESCUE 299

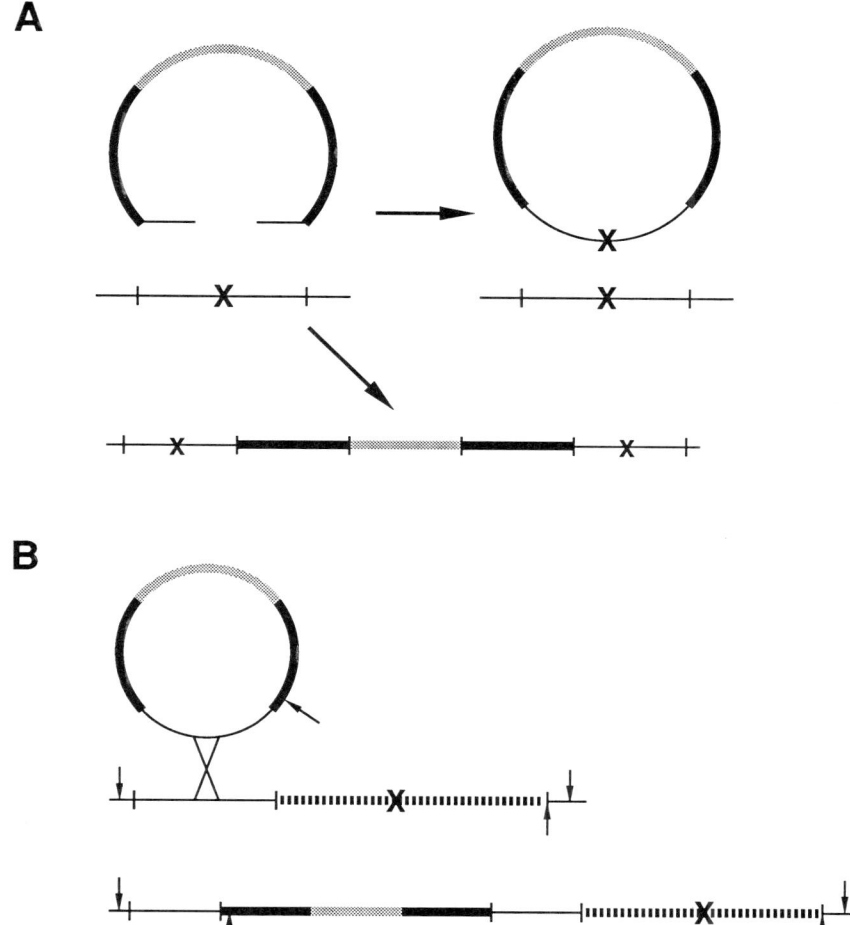

FIG. 9. Allele rescue strategies. (A) The circular gapped plasmid contains an *ars* sequence that permits it to replicate autonomously in yeast as a circle. When the gap spans the chromosomal mutant site (X), two kinds of events take place after repair of the gap. Either the gap is repaired and the plasmid integrates, resulting in a duplication of the mutant site on both sides of the vector, or, after gap repair, the circular plasmid replicates autonomously and contains the mutant chromosomal allele. In either case, the mutant allele can be rescued in *E. coli* as described in the text. (B) Integration of a plasmid adjacent to the gene of interest can also be used to rescue alleles. The integration event leads to plasmid sequences juxtaposed to the allele of interest. Appropriate restriction enzymes are used to cut either chromosomal DNA flanking the gene (downward arrows) or a restriction enzyme site that is present once in the plasmid and once near the gene (upward arrow). Circularization of these linear fragments by ligation and subsequent transformation in *E. coli* result in gene rescue as described in the text.

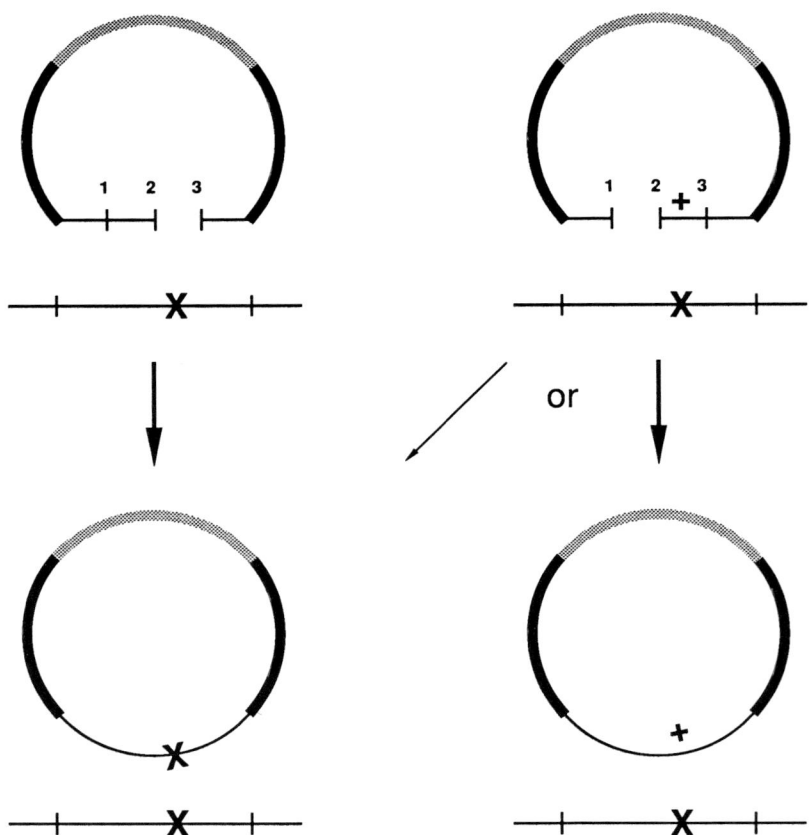

FIG. 10. Site mapping by plasmid gap repair. The position of mutant sites (X) can be mapped using restriction enzymes to create different gaps. A site maps within a gap (2–3) when only mutant transformants are recovered. If the mutant site maps outside of the gap (1–2), both wild-type and mutant plasmids are recovered. The frequency of wild-type gap-repaired plasmids is dependent on the distance from the wild-type site on the plasmid to the border of the gap (as shown in Fig. 3).

illustrated in Fig. 10 can be gapped at separate locations. If the chromosomal mutation is contained within the gap, then all transformants will exhibit the mutant phenotype. On the other hand, transformation with molecules gapped to the right or the left of the mutant chromosomal site will result in a mixed population of transformants expressing either the mutant or the wild-type phenotype of the chromosomal allele. The mixed population is due to the recovery of some transformants that are mutant as a result of gene conversion of the wild-type information on the incoming plasmid DNA by the mutant chromosomal allele (see Fig. 3). A similar method has been described for mapping plasmid-borne mutations.[50]

Gap repair cannot be used for plasmids that contain repetitive sequences since there are many alternative integration sites for the plasmid. A method that does not involve the use of gap repair can be used to recover mutant sites from such sequences. It requires the integration of a plasmid adjacent to the allele of interest, as illustrated in Fig. 9B.[10] After integration, chromosomal DNA is isolated (see [11] in this volume), and 100–300 ng is digested with a restriction enzyme that digests the DNA once near the end of the plasmid sequences and once near the end of the allele (Fig. 9B). The reaction is diluted to 1.0 ml and incubated with 1 mM ATP and 0.5–1.0 unit of T4 DNA ligase at 15° overnight. Transformation of this ligation mix (diluted with 2× transformation buffer) gives rise to between 10 and 50 transformants, depending on the competence of the bacteria.

Summary

The methods described in this chapter permit the manipulation of virtually any cloned yeast chromosomal sequence by virtue of the fact that DNA transformed into yeast integrates into the chromosome by homologous recombination. Furthermore, double-strand breaks in transforming DNA stimulate recombination and can be used to target integration events. This allows simple one-step gene disruption methods using yeast selectable markers. The availability of counterselectable markers makes it possible to replace chromosomal sequences with mutant alleles that cannot be directly selected. Finally, these same methods can be used to rescue chromosomal alleles on plasmids for subsequent molecular analysis.

Acknowledgments

I especially thank D. Shore for critical reading of the manuscript. I also thank A. Bailis, M. Fasullo, J. McDonald, and P. Verlander for their comments. I am an American Heart Association Established Investigator. This work was also supported by National Science Foundation Grant DCB 8703833, National Institutes of Health Grants GM34587 and CA21111, and grants from the Irma T. Hirschl Trust and the MacArthur Foundation.

---

[50] S. Kunes, H. Ma, K. Overbye, M. S. Fox, and D. Botstein, *Genetics* **115**, 73 (1987).

[20] *In Vitro* Mutagenesis and Plasmid Shuffling: From Cloned Gene to Mutant Yeast

By ROBERT S. SIKORSKI and JEF D. BOEKE

Introduction

Genetic analysis offers a powerful approach to the study of biological processes, and the relatively simple yeast, *Saccharomyces cerevisiae,* provides an excellent experimental system for such analysis. At the heart of this system are an impressive array of technological advantages (see other chapters in this volume) and the ability to uniquely exploit a method which has been termed "reverse genetics." Typically, in standard "forward genetics," one mutagenizes the entire organism and isolates mutants affected in the process of interest. These mutants are usually examined for interesting phenotypes (dominance or recessiveness, conditional lethality, etc.), and the corresponding wild-type genes are cloned (see [14] in this volume). The deduced amino acid sequence of these cloned genes often yields clues as to the function of the mutated loci (see, e.g., the isolation of *TOP3*).[1] Reverse genetics starts with a cloned gene and, through *in vitro* mutagenesis and transformation techniques, generates mutant yeast strains. The result is a set of strains that differ only at the loci selected for the mutagenesis. With this method one can create an unlimited number of mutant alleles from a single cloned gene. A collection of alleles allows one not only to probe the full spectrum of mutant phenotypes that can be produced from a cloned gene (some of which may not be anticipated) but allows one to screen for those alleles which possess the most "desirable" phenotypes. For example, an investigator studying kinetic aspects of DNA synthesis may need specifically those DNA polymerase mutants which most rapidly cease DNA synthesis.

The generation of mutant yeast strains from a cloned, nonessential (for vegetative growth) yeast gene is straightforward. To remove the wild-type gene product, an essential step for the analysis of recessive alleles, DNA at the wild-type locus can be deleted entirely from the genome (see [19] in this volume). A collection of mutant alleles can then be made and introduced into this host cell using replicating plasmid vectors. The generation of mutants in an essential yeast gene poses a problem in removing the wild-type allele, since deletion or inactivation of the gene results in an inviable

[1] J. W. Wallis, G. Chrebet, G. Brodsky, M. Rolfe, and R. Rothstein, *Cell (Cambridge, Mass.)* **58,** 409 (1989).

genotype. Two methods have been developed to circumvent this problem and thus allow the selective mutagenesis of essential yeast genes.

Principles of Methods

The first method, developed by Shortle et al.[2] uses homologous recombination to target in vitro mutagenized DNA to the chosen wild-type gene locus. By mutagenizing and targeting a truncated form of the gene, one can introduce mutations at the gene locus and destroy the wild-type gene in one step (see [19] in this volume for a further description of this method). Although this integration–disruption procedure has been used successfully, there are several drawbacks to the technique. One stems from the requirement for a truncated form of the gene; the entire coding sequence cannot be mutagenized in one experiment. Also, once mutants have been generated by the integrative technique, problems can arise owing to the repetitive nature of the mutated locus and the fact that the wild-type gene is not deleted. Simple recombination events, which have been shown to occur at a frequency of about 10^{-4},[3] can revert the locus to wild type. This genetic instability may make studies which require relatively stable mutations, such as suppressor analysis, difficult with these alleles as they are originally isolated. Finally, a variable but sometimes large fraction of the conditional mutants thus far generated by integration have been shown (unexpectedly) to map to loci other than the mutagenized gene.[2,4] An extreme example can be found in the mutagenesis of the yeast actin gene,[2] where 10 out of 11 temperature-sensitive isolates were unlinked to the *ACT1* locus. The exact cause of these induced mutations is unknown, but it makes extensive genetic linkage analysis of each putative mutant essential.

A second method for generating mutant alleles uses replicating yeast episomes as a means of exchanging the wild-type gene for mutant copies.[5-7] The basic scheme for the exchange, known as plasmid shuffling, is diagrammed in Fig. 1. In the first step, one copy of the gene of interest [termed your favorite gene *(YFG)* in this volume] is inactivated in a diploid, and a wild-type copy is propagated in the cell on an episome. This allows the generation of a haploid strain with a chromosomal null allele. Mutagenized copies of the gene are then introduced into this cell on a

---

[2] D. Shortle, P. Novick, and D. Botstein, *Proc. Natl. Acad. Sci. U.S.A.* **81**, 4889 (1984).
[3] E. Alani, C. Liang, and N. Kleckner, *Genetics* **116**, 541 (1987).
[4] C. Holm, T. Goto, J. C. Wang, and D. Botstein, *Cell (Cambridge, Mass.)* **41**, 553 (1985).
[5] J. D. Boeke, J. Trueheart, G. Natsoulis, and G. R. Fink, this series, Vol. 154, p. 164.
[6] C. Mann, J. Buhler, I. Treich, and A. Sentenac, *Cell (Cambridge, Mass.)* **48**, 627 (1987).
[7] M. Budd and J. L. Campbell, *Proc. Natl. Acad. Sci. U.S.A.* **84**, 2838 (1987).

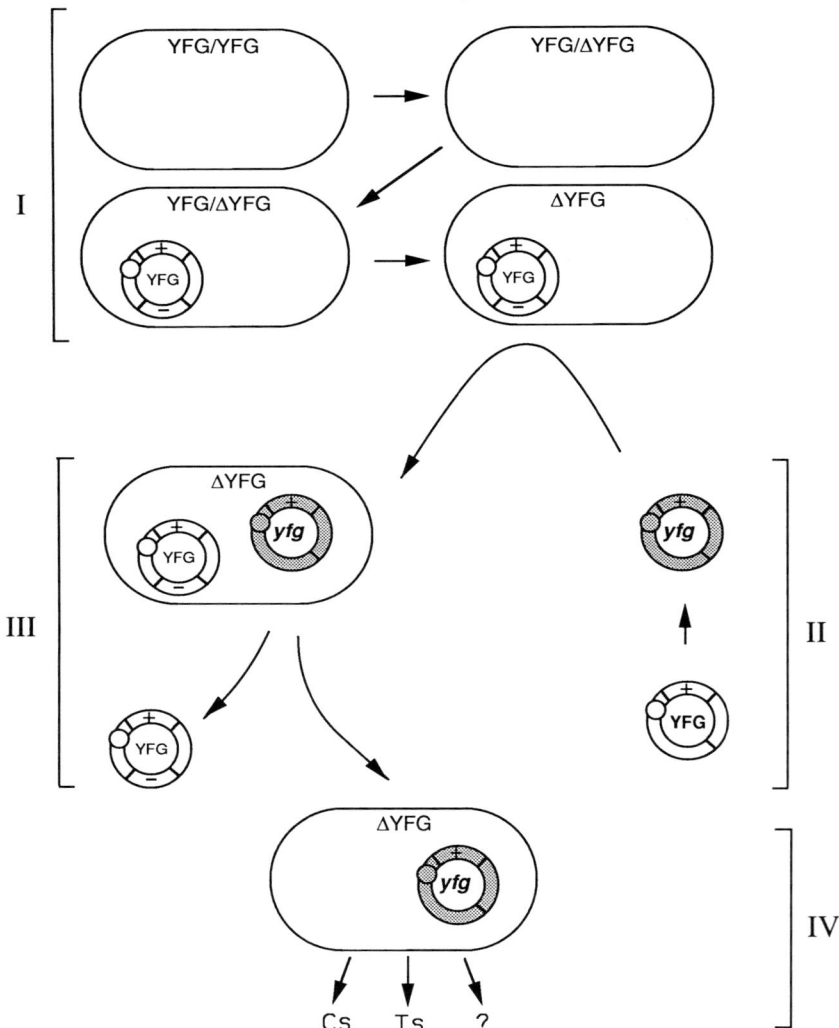

FIG. 1. Steps involved in plasmid shuffling. (I) Constructing the recipient yeast host strain. (II) *In vitro* mutagenesis. (III) Removing the wild-type gene by counterselection. (IV) Identifying the mutant alleles of interest. The (+) and (−) symbols refer to selectable and counterselectable functions, respectively. These may in some cases be provided by a single yeast gene. Cs, Cold sensitive for growth; Ts, temperature sensitive; ?, other conditional phenotypes. See text for details.

second episome and exchanged (or "shuffled") with the wild-type version. Unlike the integrative–disruptive procedure, the entire coding sequence can be mutagenized in one experiment. Also, if the chromosomal copy of the gene has been inactivated by a deletion which removes all of the sequence of the gene, the plasmid-borne mutant alleles cannot revert to wild type by homologous recombination. Any revertants that do arise will do so at the much lower frequency expected of single base-pair mutations. The percentage of unlinked temperature-sensitive lethal mutations generated by plasmid shuffling is apparently low, being, for instance, 2 in 7 in the case of *RPA190*[8] and 0 in 23 in the mutagenesis of *CDC23*.[9] More importantly, such unlinked mutations can be easily identified and discarded by checking the phenotype of the strain before shuffling (Step III in Fig. 1). At this stage, strains carrying recessive mutations in the gene of interest will be phenotypically wild type (since they are "covered" by the wild-type gene on a plasmid), whereas unlinked mutations will display their mutant phenotype.

Removal of the wild-type gene, *YFG* in our example, is the key step in any plasmid shuffling scheme. This can be accomplished by taking advantage of two factors. First, even relatively stable YCp episomes (such as YCp-*YFG*) are lost from a cell by missegregation or misreplication at a rate of 10^{-2} per generation. Second, compounds are available that prevent the growth of cells carrying specific yeast genes, and in the presence of such compounds these genes act as counterselectable markers, allowing one to directly select for cells which have lost this marker. By including one of these counterselectable markers on the same plasmid that contains the wild-type *YFG* gene, an investigator can select for cells that have lost the entire plasmid. Several negative selection schemes have been described, but only four make use of easily handled compounds that are readily available through commercial sources. The yeast genes in these schemes are *URA3, LYS2, CAN1,* and *CYH2.*

The *URA3* gene encodes orotidine-5'-phosphate decarboxylase, an enzyme required for the biosynthesis of uracil. Selection of *ura3* cells (cells which, e.g., have lost a *URA3*-based plasmid) is accomplished by plating the cells on media containing 5-fluoroorotic acid (5-FOA). This compound is apparently converted to a toxic product, 5-fluorouracil, by the action of the decarboxylase, killing *URA3* cells; *ura3* cells are resistant to 5-FOA. The 5-FOA negative selection procedure is very efficient and selective, and under appropriate conditions only one in several hundred 5-FOA-resistant

[8] M. Wittekind, J. Dodd, L. Vu, J. M. Kolb, J. Buhler, A. Sentenac, and M. Nomura, *Mol. Cell. Biol.* **8,** 3997 (1988).
[9] R. S. Sikorski and P. Hieter, in preparation (1990).

cells will be Ura⁺.[10] Quite conveniently, one can also positively select for *URA3* cells (and transformants with *URA3*-based vectors) by using uracil-free medium.

Similarly, the *LYS2* gene encodes α-aminoadipate reductase, an enzyme required for lysine biosynthesis. Yeast cells with wild-type *LYS2* activity will not grow on media containing α-aminoadipate (α-AA) as a primary nitrogen source.[11] High levels of α-AA are thought to cause the accumulation of a toxic intermediate, and *lys2* (as well as rarer *lys5*) mutants block the formation of this intermediate.[12] Although *lys5* mutants are α-AA resistant, they arise at a frequency much lower than that of loss of a centromere plasmid; therefore, they present no problem in plasmid shuffling. As is the case with *URA3*, the *LYS2* gene can also be selected in a positive fashion by using lysine-free medium.

The *CAN1* gene encodes an arginine permease.[13] This permease is also the sole route of entry for the arginine analog, canavanine, into yeast cells. In the absence of arginine, canavanine is readily incorporated into proteins with lethal consequences; therefore, *CAN1* (permease-producing) cells are sensitive to canavanine whereas *can1* (permease-deficient) cells are resistant. The sensitivity is dominant in that *CAN1/can1* cells also fail to grow on appropriate concentrations of canavanine. *CAN1* cells can be selected directly by transformation of the *CAN1* gene,[14] but the procedure is complicated by the fact that the host cell must be both *can1* and Arg⁻ for this to work. For routine use of *CAN1* in a negative selection scheme it is more convenient to include an additional selectable marker on the plasmid that can be used in selecting transformants [in pRS319 (Fig. 2), the *LEU2* gene is included].

The *CYH2* gene encodes the L29 protein of the yeast ribosome.[15] Cycloheximide, a drug which blocks polypeptide elongation during translation, prevents the growth of cells that contain the wild-type *CYH2* gene. Cycloheximide resistance results from a single amino acid change in the *CYH2* protein.[15] The lethality of the drug is dominant; cells containing both the sensitive (wild-type) and the resistant (mutant) *CYH2* alleles fail to grow on media containing cycloheximide. Therefore, the loss of a *CYH2*-containing plasmid can be selected directly if the host carries the

---

[10] J. D. Boeke, F. LaCroute, and G. R. Fink, *Mol. Gen. Genet.* **197**, 345 (1984).
[11] B. B. Chatoo, F. Sherman, D. A. Azubalis, T. A. Fjellstedt, D. Mehnert, and M. Ogur, *Genetics* **93**, 51 (1979).
[12] K. S. Zaret and F. Sherman, *J. Bacteriol.* **162**, 579 (1985).
[13] W. Hoffman, *J. Biol. Chem.* **260**, 11831 (1985).
[14] J. R. Broach, J. N. Strathern, and J. B. Hicks, *Gene* **8**, 121 (1979).
[15] A. F. Kaufer, H. M. Fried, W. F. Schwindinger, M. Jasin, and J. R. Warner, *Nucleic Acids Res.* **11**, 3123 (1983).

resistant allele chromosomally. No positive selection exists for *CYH2*, and another selectable marker must be used for the introduction of a *CYH2* vector. The *LEU2* gene has been linked to *CYH2* in the plasmid pRS318 (Fig. 2) for this purpose.

Why choose one counterselection scheme over the other? The *URA3*/5-FOA system is probably the best in most circumstances; only one gene is required for selection and counterselection, and the *URA3* gene is small and completely sequenced. *LYS2*/α-AA is the next best choice, but the gene itself is large and not yet completely sequenced. In *URA*$^+$ and *LYS*$^+$ strains, counterselection with a *CAN1*- or *CYH2*-containing plasmid is useful since the resistant alleles can be easily selected for in these strains.

The plasmid shuffling technique can be viewed as a series of steps from the generation of the proper yeast host strain to the phenotypic analysis of the resulting mutants. The following details these steps and the materials which will be required (plasmids, media, etc.). We have tried to list the variations that can be afforded at each step so that the entire procedure can be tailored to the specific needs of the investigator.

Materials

Yeast Strains

Preparing Host Strain for Plasmid Shuffling. A useful property of the plasmid shuffling approach is that any yeast strain of interest can be easily converted to a strain suitably marked for plasmid shuffling. The four shuffling strategies outlined above all require the presence of a counterselectable marker on the plasmid that also bears *YFG*. Some of these counterselectable markers are themselves nonselectable (*CYH2*, conferring cycloheximide sensitivity, which is dominant) or for all intents and purposes nonselectable (*CAN1*, conferring canavanine sensitivity, which is semidominant), whereas others are both selectable and counterselectable (*URA3* and *LYS2*).[10,11] For most applications, the *URA3* gene is probably the most convenient marker to use; many workers will already have *ura3* strains of suitable strain background in hand.

Selecting cyh2 Strains. Use of the cycloheximide selection will also require the presence of appropriate auxotrophies in the strain to allow selection of the plasmids in later steps. Yeast cells of the appropriate genotype are grown in 10 ml of YPD broth to saturation. Ten 0.3-ml aliquots are plated onto YPD agar plates supplemented with 10 µg/ml cycloheximide. Cycloheximide can be stored at −20° and added to medium as a 10 mg/ml stock solution in ethanol. After 2–4 days, a few colonies will appear, often on only a subset of the plates. Pick resistant

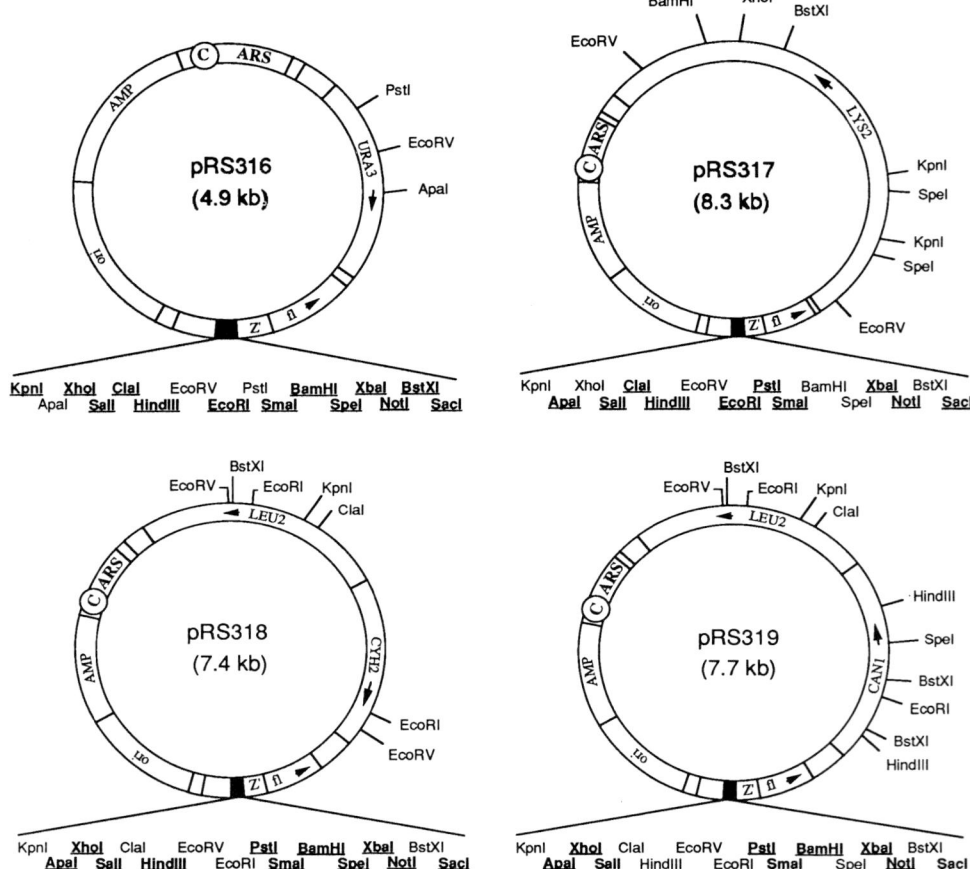

FIG. 2. Restriction maps of pRS shuttle vectors useful for plasmid shuffling. Note that all of the sites in the polylinker (derived from pBluescript) are not unique. The unique sites are underlined. C, Centromere; ARS, autonomously replicating sequence (yeast DNA replication origin); f1, filamentous phage origin of replication (for generating single-stranded DNA); Z', α complementing fragment of *lacZ*; *ori*, bacterial origin of DNA replication. pRS317 was made by digesting the plasmid pRSS56 [R. S. Sikorski and P. Hieter, *Genetics* **122**, 19 (1989)] with *Nde*I and filling in the overhang using Klenow polymerase. The *LYS2* gene was ligated into this backbone as a 4.5-kb *Hin*dIII–*Eco*RI fragment (both sites blunted with Klenow polymerase) isolated from YIp601 [D. A. Barnes and J. Thorner, *Mol. Cell. Biol.* **6**, 2828 (1986)]. A small *CEN/ARS* DNA cassette [isolated and prepared as in the construction of pRS316; see R. S. Sikorski and P. Hieter, *Genetics* **122**, 19 (1989)] was inserted into the *Aat*II site (blunt T4 DNA polymerase) of this plasmid to yield pRS317. For construction of pRS318, a 1.4-kb *Bam*HI fragment containing the sensitive allele of *CYH2* was first isolated from the plasmid pKC3 (kind gift of Karen Chapman, Johns Hopkins University School of Medicine, Baltimore, MD). The *CYH2* gene [A. F. Kaufer, H. M. Fried, W. F. Schwindinger,

colonies from a plate giving the smallest number of resistant colonies. Under these conditions only *cyh2* mutants will grow. (Growth in the presence of lower cycloheximide concentrations can result from mutations at a variety of other loci.) The *cyh2* genotype can be confirmed by observing the cycloheximide-sensitive phenotype on transformation with a *CYH2*-containing plasmid.

Selecting can1 Strains. Use of the canavanine selection will also require the presence of an appropriate auxotrophy in the strain to allow selection of the plasmid in later steps. Wash a 10-ml YPD overnight culture of the strain of interest and resuspend in 10 ml water. Plate 0.1 ml of the washed culture and 0.1 ml of 10- and 100-fold dilutions (in water) on SC − Arg agar plates (see [1] in this volume) containing 60 µg/ml canavanine. Canavanine can be autoclaved and stored at room temperature as an aqueous solution of 60 mg/ml. Canavanine SD plates (supplemented only with required compounds) can also be used for this selection, although growth of the resistant colonies will be slower than on SC − Arg plates. Arginine in the plates will allow growth of canavanine-sensitive cells and so must be omitted from the medium. Pick resistant colonies from a plate giving the smallest number of resistant colonies. Only *can1* mutants will form colonies under these conditions.

Selecting lys2 Strains. Follow the directions above for *can1* selection, except plate the cells on α-aminoadipic acid-containing medium (Table I). Unlike for the *can1* selection, the starting yeast strain may be Arg⁻.

Constructing ura3 Strains. In most cases, if a useful preexisting counterselectable marker is not present in the strain of interest, introducing a *ura3* mutation is desirable. The best markers for this purpose are probably the *ura3-52* mutation (a nonreverting Ty insertion in the *ura3* gene[16]) or

[16] M. Rose and F. Winston, *Mol. Gen. Genet.* **193**, 557 (1984).

M. Jasin, and J. R. Warner, *Nucleic Acids Res.* **11**, 3123 (1983)] on this fragment had been previously manipulated so that it is actually a hybrid between a *CYH2* promoter fragment and a *CYH2* cDNA (J. R. Warner, personal communication, 1988). The *Bam*HI fragment was blunted with Klenow polymerase and ligated to *Hpa*I-cut pRS315 to make pRS318. (The *Hpa*I site is just upstream of the *LEU2* promoter and does not interfere with functional *LEU2* expression.) To make pRS319, the *CAN1* gene [W. Hoffman, *J. Biol. Chem.* **260**, 11831 (1985)] was isolated as a 1.8-kb *Bam*HI-*Sal*I fragment from the plasmid pSH2 (kind gift of Paul Sutton, University of Illinois at Chicago), blunted with Klenow polymerase, and ligated to *Hpa*I-digested pRS315. The resulting *CAN1* gene is truncated such that the C-terminal 31 amino acids of the arginine permease protein have been deleted; however, this allele still confers canavanine sensitivity (P. Sutton, personal communication, 1989). Recombinants in the above constructions were identified by probing filter replicas of *Escherichia coli* colonies with radiolabeled probes made to the desired inserts, and positive clones were restriction mapped using all of the enzyme sites present in the polylinker.

TABLE I
MEDIA USED IN COMMON COUNTERSELECTION SCHEMES[a]

5-Fluoroorotic acid (URA3)		DL-α-Aminoadipic acid (LYS2)		Canavanine (CAN1)		Cycloheximide (CYH2)	
Solution component	Amount	Component	Amount	Component	Amount	Component	Amount
A. YNB [+ (NH$_4$)$_2$SO$_4$][b] Water	7 g 375 ml	YNB [− (NH$_4$)$_2$SO$_4$][b] Water	2 g 360 ml	YNB [+ (NH$_4$)$_2$SO$_4$][b] Water	7 g 400 ml	Yeast extract Peptone Water	10 g 20 g 400 ml
B. 20% Glucose	100 ml	20% Glucose	100 ml	20% Glucose	100 ml	20% Glucose	100 ml
C. 4% Agar	500 ml	4% Agar	500 ml	4% Agar	500 ml	4% Agar	500 ml
D. Uracil (2 mg/ml)	25 ml	Lysine (4 mg/ml) DL-α-Aminoadipate[c] Water Adjust pH to 6.0 with 1 M KOH (requires ∼12 ml)	7.5 ml 2 g 30 ml	—		—	
E. 5-Fluoroorotic acid	1 g			Canavanine (60 mg/ml)	1 ml	Cycloheximide (10 mg/ml)	1 ml

[a] Autoclave solutions A, B, and C separately. Cool these to about 50° and combine with filter-sterilized D and E. For 5-FOA medium, mix the 5-FOA powder with A, B, and D and filter sterilize. The powder will take some time to go into solution; heating to 50° will help.
[b] YNB is yeast nitrogen base (Difco). Note that YNB is available with (+) and without (−) added (NH$_4$)$_2$SO$_4$, a distinction which is important for LYS2 counterselection.
[c] A solution of α-AA is made and the pH adjusted separately (final volume ∼42 ml).

one of several constructed *ura3* deletion mutations.[5,17] These markers can be introduced into cells by transformation using the appropriate linearized transplacement vector. The following procedure has been used successfully by F. Winston (personal communication, 1989) to construct a *ura3-52* derivative of S288C. Two micrograms of plasmid pMRFW2[16] is digested to completion with *Bam*HI and *Sma*I (releasing a linear fragment containing *ura3-52*) and used to transform S288C strains by the lithium acetate method.[18] Following this, the transformed cells are resuspended in YPD at about 1×10^7 cells/ml and grown overnight to saturation. One-tenth milliliter of the culture gave rise to over 500 Ura⁻ colonies on 5-FOA medium; control cells that had not seen DNA gave only 18 colonies. Five of five colonies checked by Southern blot had incorporated the *ura3-52* mutation. This outgrowth period in YPD medium (to allow dilution of the *URA3* gene product) is apparently essential for good recovery of *ura3* cells on 5-FOA medium. 5-FOA-resistant cells should be checked for uracil auxotrophy and nonrevertibility and, by Southern blotting with a *URA3* gene probe, to ensure that the desired removal of the *URA3* gene has been effected.

Yeast Growth Media

Positive selection for the presence of a shuttle vector in yeast can be accomplished using standard SD or SC dropout media.[19] Counterselection against the presence of a shuttle vector will require the use of special media. Recipes for preparing media for use in conjunction with the *URA3, LYS2, CAN1,* and *CYH2* genes can be found in Table I. The required compounds can be purchased from the following vendors: 5-fluoroorotic acid (5-FOA) from SCM Specialty Chemicals, Gainesville, FL; DL-α-aminoadipic acid (α-AA), canavanine, and cycloheximide from Sigma, St. Louis, MO. Because 5-FOA is quite expensive if purchased in small amounts, it can be obtained in bulk quantities via the Genetics Society of America, Bethesda, MD, at a greatly reduced price. A relatively easy synthesis procedure[20] has also been used to make 5-FOA suitable for use in yeast genetics at a cost of approximately \$20/g.[21]

---

[17] Joachim Li, personal communication, 1989.
[18] H. Ito, Y. Funkuda, K. Murata, and A. Kimura, *J. Bacteriol.* **153**, 163 (1983).
[19] F. Sherman, G. R. Fink, and J. B. Hicks, "Methods in Yeast Genetics," Cold Spring Harbor Laboratory, Cold Spring Harbor, New York, 1979.
[20] S. N. Alam, T. K. Shires, and H. Y. Aboul-Enein, *Acta Pharm. Suec.* **12**, 375 (1975).
[21] R. L. Keil, Dept. of Biological Chemistry, Hershey Medical Center, Hershey, PA 17033, personal communication, 1988.

Yeast–Escherichia coli Shuttle Vectors

Two plasmid constructs and consequently two distinct shuttle vectors are needed for the plasmid shuffle. One construct will consist of a vector carrying the wild-type gene, and the other will consist of a vector carrying a "library" of mutant alleles. The vector requirements for each construct are different. The mutant-allele construct can be made using any of the various preexisting YCp, YEp, or YRp shuttle vectors (see [14] in this volume). The only stipulation is that the selectable marker on the vector be compatible with the genotype of the *E. coli* (to check level of mutagenesis) and yeast (to select transformants) host strains. The pRS series of shuttle vectors[22] are particularly desirable for *in vitro* mutagenesis because they are small and yield large amounts of plasmid DNA in *E. coli*. The backbone of these plasmids presents a minimal target for the mutagen, and the high yield of DNA in bacteria allows easy amplification of a mutant library (see below). Regarding the wild-type allele construct, the vector chosen must contain both appropriate selectable and counterselectable markers. The counterselectable marker will be used to select against cells harboring the wild-type gene, thus uncovering the phenotypes of the mutant alleles. Although several shuttle vectors containing counterselectable markers have been made, these vectors are usually quite bulky in size, contain a limited number of usable restriction sites, and are not well characterized. We have, therefore, made a new series of shuttle vectors specifically for the purpose of plasmid shuffling. These vectors (Fig. 2) are extensions of the pRS series of YCp plasmids which were made for more efficient cloning of DNA in yeast.[22]

pRS316 has been described previously.[22] It contains the *URA3* gene as both the selectable and counterselectable marker. The extensive polylinker with a blue/white screen for inserts makes cloning the wild-type gene into this vector, or its derivatives, quite simple. pRS317 is identical in design to pRS316 but contains the *LYS2* gene instead of *URA3*. Again, one gene serves as both the selectable and counterselectable marker. pRS318 is a *CYH2*-containing derivative of the *LEU2 CEN/ARS* vector pRS315.[22] pRS319 is a *CAN1*-containing derivative of pRS315.

Escherichia coli Strains

The overall level of mutagenesis of a plasmid can be gauged by quantifying the level of induced null mutations in the selectable marker of the plasmid. This testing is easily performed in *E. coli*, taking advantage of the fact that many yeast biosynthetic enzymes (including the products of

---

[22] R. S. Sikorski and P. Hieter, *Genetics* **122**, 19 (1989).

the *URA3, HIS3, TRP1,* and *LEU2* genes) are expressed in bacteria directly from a subcloned yeast DNA fragment. The expression levels are sufficient to complement growth requirements of certain *E. coli* auxotrophs, and null alleles can be readily identified by replica plating onto the appropriate minimal medium. A very useful *E. coli* strain is MH1066 [Δ*lacX74 hsr⁻ rpsL pyrF*::Tn5, *leuB600 trpC9830 galE galK*].[23,24] In this bacterial strain one can select for the function of the yeast *URA3, TRP1,* or *LEU2* genes by assaying for growth on minimal M9 medium[25] lacking the relevant nutrient (e.g., when testing for *URA3* function, add tryptophan and leucine but not uracil). Supplements are added to M9 at the following concentrations: uracil, 20 μg/ml; tryptophan, 30 μg/ml; and leucine, 40 μg/ml.

Methods

Step I: Construction of Recipient Yeast Host Strain

The first step in plasmid shuffling involves the construction of a special yeast strain, the recipient strain, in which only one functional copy of the essential gene of interest *(YFG)* is present. This copy has been manipulated such that it is located on an episomal plasmid vector. The strain also bears markers such as *cyh2, can1, ura3, leu2, trp1,* and *his3,* at least one of which must be counterselectable. This strain is usually made by first inactivating one chromosomal copy of *YFG* in a suitably marked homozygous a/α diploid with a deletion or deletion–insertion mutation, using one-step or two-step gene replacement strategies[26,27] (see also [19] in this volume). Sporulation of the resultant heterozygote should result in two viable and two inviable spores, the latter containing only a nonfunctional copy of *YFG*. By introducing (via transformation) a replicating plasmid such as a YCp vector containing the wild-type *YFG* gene (YCp-*YFG*) into the heterozygote *before* sporulation, some of the *YFG*Δ spores will be "rescued" for viability by the *YFG* gene on the episome. The vector chosen to carry the wild-type gene must be compatible with one of the negative selection schemes described in Step III. This new haploid strain with the desired genotype (*YFG*Δ, YCp-*YFG*) can serve as the recipient for plasmids containing mutations in *YFG*.

[23] M. N. Hall, L. Hereford, and I. Herskowitz, *Cell (Cambridge, Mass.)* **36**, 1057 (1984).
[24] M. N. Hall, personal communication, 1989.
[25] T. Maniatis, E. E. Fritsch, and J. Sambrook, "Molecular Cloning: A Laboratory Manual." Cold Spring Harbor Laboratory, Cold Spring Harbor, New York, 1982.
[26] R. J. Rothstein, this series, Vol. 101, p. 202.
[27] F. Winston, F. Chumley, and G. R. Fink, this series, Vol. 101, p. 211.

Step II: In Vitro Mutagenesis

The exact method chosen for the mutagenesis will depend largely on the existing knowledge of the structure or function of the gene in question. If the gene is well characterized and its function is known, then local, directed mutagenesis of chosen residues or domains may be in order. Several excellent protocols for local mutagenesis using mutant oligonucleotides,[28,29] gapped-duplex target DNA,[30] and nucleotide misincorporation[31] have been developed. The many strategies and applications of *in vitro* mutagenesis have been reviewed.[32] Because many of these methods work best with a single-stranded DNA template to serve as a target, yeast shuttle vectors which can be converted to single-stranded DNA by virtue of an f1 phage origin of DNA replication are quite useful.[22]

Often, however, the biochemical function of the selected gene (and/or its nucleotide sequence) may not be known, and a method for generating an assortment of mutations randomly throughout the coding sequence may be more informative. In this regard, mutagenesis using nitrous acid, bisulfite, or hydroxylamine is particularly useful. Nitrous acid produces a relatively wide variety of base modifications and, consequently, of point mutation types[33] and so may be the best mutagen if the aim is to obtain as wide a variety of mutant alleles as is possible. However, its use for mutagenizing yeast plasmids has not to our knowledge been reported. Ultraviolet light also produces a wide spectrum of mutant types[34] but may be less desirable owing to the fact that frameshift mutations are quite common. Bisulfite reacts with DNA *in vitro* causing deamination of cytosine residues and C to T transition mutations. Because bisulfite reacts preferentially with unpaired bases, the target DNA should be in single-stranded form. This limits its usefulness in generalized mutagenesis because a relatively narrow spectrum of mutation types is obtained if only one strand is mutagenized. Hydroxylamine will react with a double-stranded target DNA to create N^4-hydroxycytosine, which can base pair with adenosine, and results in both C to T and G to A transition mutations when doubled-stranded plasmid DNA is mutagenized. A variety of other mutations have been reported to occur with hydroxylamine under extended treatment condi-

[28] T. A. Kunkel, *Proc. Natl. Acad. Sci. U.S.A.* **82**, 488 (1985).
[29] M. J. Zoller and M. Smith, *DNA* **3**, 4789 (1984).
[30] W. Kramer, V. Drutsa, H. W. Jansen, B. Kramer, M. Pflugfelder, and H. J. Fritz, *Nucleic Acids Res.* **12**, 9441 (1984).
[31] D. Shortle, P. Grisafi, S. J. Benkovic, and D. Botstein, *Proc. Natl. Acad. Sci. U.S.A.* **79**, 1588 (1982).
[32] D. Shortle and D. Botstein, *Science* **229**, 1193 (1985).
[33] R. M. Meyers and T. Maniatis, *Science* **229**, 242 (1985).
[34] J. H. Miller, *J. Mol. Biol.* **182**, 45 (1985).

tions.[35] The simplicity of the procedure and the ability to utilize plasmid DNA directly make hydroxylamine mutagenesis an attractive method for the initial generation of mutants. We have obtained excellent results using the following hydroxylamine protocol, which has been modified from that of Busby et al.[35]

Hydroxylamine Mutagenesis Protocol

1. Make several milliliters of fresh hydroxylamine solution consisting of 1 M hydroxylamine (Sigma), 50 mM sodium pyrophosphate (pH 7.0), 100 mM sodium chloride, and 2 mM EDTA.

2. Add 10 μg of the target plasmid (10 μl of 1 mg/ml stock) to 500 μl of hydroxylamine solution.

3. Allow the reaction to proceed at 75°. The degree of mutagenesis can be titered by removing 100-μl aliquots at several different time points of incubation (0, 30, 60, 90, and 120 min) and stopping the reaction on ice.

4. The excess hydroxylamine is removed easily by dialysis on filters (type VS 0.025-μm filters; Millipore, Bedford, MA) floating on a large volume of TE or by gel filtration through a small Sephadex G-25 column. This step is essential, as hydroxylamine is inhibitory to bacterial transformation.

5. Transform 1–5 μl from each time point into bacteria to assess the degree of mutagenesis. The level of mutagenesis can be approximated by scoring null mutations in a nonessential plasmid-encoded function, usually the yeast selectable marker gene. The *E. coli* strain MH1066 is particularly useful here because, as mentioned previously, functional yeast *URA3, TRP1,* and *LEU2* genes complement the analogous bacterial mutations. For scoring, simply replica plate the bacterial transformants from rich medium to minimal medium that lacks the compound in question. For example, in scoring lesions in the *TRP1* gene of a pBR322-based plasmid, one would first select transformants on LB – Amp medium and then replica plate these to M9 medium lacking tryptophan. A productive mutagenesis should produce auxotrophs at a frequency of roughly 1–5%, depending on the exact gene tested and the total size of the plasmid.

6. The original mutagenized DNA can be amplified in *E. coli* in order to obtain enough DNA for transforming yeast. Collect and pool at least 10,000–20,000 bacterial colonies from a chosen time point of the original mutagenized DNA by scraping the transformation plates into Luria broth or buffer. Isolate plasmid DNA from this pool. Remember that the number of bacterial colonies will determine the maximum number of individual yeast mutants that one can obtain.

[35] S. Busby, M. Irani, and B. De Crombrugghe, *J. Mol. Biol.* **154**, 197 (1982).

7. The DNA is now ready to transform into the recipient yeast strain (see [12] in this volume). Try to obtain 200 to 500 transformants per normal petri dish (300 is optimal). Any more colonies than this will make replica-plate analysis difficult. Since in our hands the competency of the yeast cells can vary greatly from batch to batch, we have found it important to titer the competency before transforming and plating large amounts of the mutagenized DNA. Titered batches of cells made competent by the lithium acetate procedure[18] can be left in 0.1 M lithium acetate at 4° for at least 1 week without a significant change in the transformation frequency. When selecting for transformants, *do not* maintain selection for the original, unmutagenized plasmid. By allowing this plasmid to be lost freely, counterselection (see below) can be performed directly using replicas made from these transformants plates.

The above protocol has been used to introduce mutations into the *CDC23* gene.[9] A 10-kb YCp vector carrying the *LEU2* (2.2 kb) and *CDC23* (2 kb) genes was mutagenized such that 3% of Amp^R *E. coli* colonies were Leu$^-$. This degree of mutagenesis was sufficient to yield 23 Ts and 14 Cs *cdc23* mutant alleles after completing the plasmid shuffling procedure on 8500 yeast transformants. The nucleotide sequence obtained from 10 of these alleles revealed that 7 had received single, unique base-pair substitutions and 3 received multiple (usually 2) substitutions. All changes were C to T or G to A transitions expected from the mutagen. Thus, 3% knockout mutations in the selectable marker serves as a reasonable level for mutagenesis, at least in the case of *CDC23*.

A variation of the above strategy is to transform the mutagenized DNA directly into yeast.[36] This obviates removal of the hydroxylamine and bypasses the passaging through *E. coli*. All yeast transformants should be independent if they are plated directly onto solid medium. Care should be taken in the analysis of the resulting mutants when using this shortened protocol, however, since the mutagenic effects of hydroxylamine on yeast cells is not clear.

An alternative to chemical mutagenesis is "*in vivo*" mutagenesis using a mutator strain of *E. coli* such as *mutD5*.[37] In this strain, spontaneous mutations involving all possible base-pair changes are elevated by 10^3- to 10^4-fold, and the frequency of base-pair insertions is also increased.[38] DNA changes can be introduced into a plasmid by simply passaging the plasmid through a round of amplification in a *mutD5* strain. In one study utilizing

---

[36] M. D. Rose and G. R. Fink, *Cell (Cambridge, Mass.)* **48,** 1047 (1987).
[37] R. Fowler, G. Degnen, and E. Cox, *Mol. Gen. Genet.* **133,** 179 (1974).
[38] L. W. Enquist and R. A. Weisberg, *J. Mol. Biol.* **111,** 97 (1977).

mutD5 mutagenesis in conjunction with plasmid shuffling, the frequency of *URA3* knockout mutations in one passage was 1%.[39]

Step III: Removing Episomal Wild-Type Gene by Counterselection

The phenotype(s) of the mutagenized gene can be "uncovered" by counterselection against the original, unmutagenized episomal plasmid carrying *YFG*. This is performed by simply replica plating the yeast transformants obtained in Step II to medium containing the appropriate counterselective agent (5-FOA, α-AA, canavanine, or cycloheximide). These replicas are then incubated at two or more temperatures to allow for the isolation of conditional alleles (see below). The behavior of these colony replicas will indicate the phenotype of any mutation(s) that may have occurred in the mutagenesis.

Step IV: Identifying Mutant Alleles of Interest

For the essential gene *YFG* in the strain derived in Step I, the plasmid containing the counterselectable marker can be lost only if a second *YFG*-bearing plasmid is present to provide sufficient *YFG* function to allow growth. This feature provides the basis for identifying mutations in *YFG*. Mutants which cannot provide sufficient *YFG* function will not survive loss of the YCp-*YFG* plasmid and will not grow on media containing the compound used for negative selection of this plasmid. Mutants that provide *YFG* function conditionally do so only under appropriate conditions (selected by the investigator). Conditions such as low or high temperature, altered osmotic conditions,[40] growth on heavy-water containing medium,[41] and supersensitivity to toxic compounds[42] have been used successfully.

The conditional alleles of the essential gene are identified by replica plating to the appropriate counterselective medium under the chosen conditions, looking for those colonies which do not grow. Lack of growth can at times (especially in dense platings) be difficult to score, and we have found that the visible red color produced by an *ade2* mutation can make the identification much easier. Actively growing yeast cells with an *ade1* or *ade2* genotype form red colonies on solid medium containing limiting amounts of adenine (see Koshland and Hieter[43] for details). Nongrowing cells, even dense patches of nongrowing cells, are distinctly white on such a medium. Therefore, one can transfer many cells from a fully adenine-sup-

---

[39] F. Winston, personal communication, 1989.
[40] D. C. Hawthorne and J. Friis, *Genetics* **50**, 829 (1964).
[41] B. Bartel and A. Varshavsky, *Cell (Cambridge, Mass.)* **52**, 935 (1988).
[42] A. Hoyt, personal communication, 1989.
[43] D. Koshland and P. Hieter, this series, Vol. 155, p. 351.

plemented plate, on which all colonies are white, to a counterselection medium with limiting adenine. Nongrowing colonies (i.e., mutants) will remain white while growing colonies (i.e., wild type) will turn red.

Expected Results

It is not possible to predict the frequency with which temperature-sensitive mutants will be generated in any given plasmid shuffle. In some genes it may be difficult to generate the amino acid changes which will produce a protein that is both functional under one set of conditions and nonfunctional under another. Successful examples of previous plasmid shuffles can, however, serve as a guide for what to expect.[5-8] Consider two plasmid shuffles in which the hydroxylamine protocol was used for mutagenesis. Wittekind et al.[8] mutagenized a plasmid containing the gene for the largest subunit of RNA polymerase I, *RPA190,* and gauged the degree of mutagenesis by measuring the frequency of knockout mutations in *RPA190* itself (these did not shuffle under any condition). At a frequency of 4–5% knockout mutations they obtained 7 Ts alleles from 15,000 yeast transformants. Sequencing revealed that all of these were G to A or C to T transitions, as expected. In another shuffle, Budd and Campbell[7] created mutations in a *POL1*-containing plasmid. They measured the degree of mutagenesis by recording the frequency of both knockout mutations in the selectable marker *(TRP1)* on the plasmid vector and knockouts in the *POL1* gene itself. With a mutagenesis level sufficient to generate 1% Trp$^-$ (as judged in *E. coli*) and 10% *pol1*-plasmids, this particular shuffle yielded 12 Ts strains out of 1000 yeast transformants. Finally, Boeke et al.[5] reported a plasmid shuffle in which 11 nonconditional and 2 Ts mutants were obtained from 800 plasmids tested.

Conclusion

The plasmid shuffling scheme we have described is a rapid, four-step method for the generation of yeast strains that harbor mutations in a selected, essential gene. These new mutant alleles can be readily retrieved in *E. coli* for sequencing or further manipulation.

Acknowledgments

The plasmid vectors described in the manuscript were constructed in the laboratory of Dr. Philip Hieter, and we greatly appreciate his support and encouragement. R.S.S. is supported by a Medical Scientist Training Program grant from the National Institutes of Health and J.D.B. by grants from the NIH and the Searle Foundation.

[21] Recovery of Plasmids from Yeast into *Escherichia coli*: Shuttle Vectors

By JEFFREY N. STRATHERN and DAVID R. HIGGINS

Introduction

Many of the important techniques of molecular genetics applied to yeast utilize yeast–*Escherichia coli* shuttle vectors. The term shuttle vector reflects the ability to move the vector back and forth between the two hosts. Hence, a means of selecting the vector in both hosts is required as well as a means of propagating it in both organisms. Several quite successful systems have been developed based generally on the *E. coli* plasmid pBR322, a yeast metabolic gene (*LEU2, TRP1, URA3*, etc.), and a yeast *ARS* sequence derived from the chromosome (e.g., *ARS1*) or the *ARS* sequence of the endogenous plasmid 2-μm circle.[1-4] Low-copy-number vectors that incorporate sequences from yeast centromeric regions were also developed.[5] These tools opened the door to the cloning of yeast genes by complementation of yeast mutants. There followed an exciting era during which genes important to the yeast cell cycle, yeast life cycle, yeast metabolism and its regulation, DNA repair and recombination, protein synthesis, transcription, secretion, and organelle biology were cloned. That pleasant era continues. This volume includes multiple examples of clever shuttle vector constructions that can be used to clone your favorite gene *(YFG)*, regulate its expression, generate mutant alleles, and place them back in the yeast genome or fuse them to indicator proteins. Once a gene has been cloned by complementation, subsequent characterization and manipulation are greatly facilitated if it can be shuttled back into *E. coli*. This chapter is designed to give a simple protocol for this rather important step.

Having identified a DNA clone that contains *YFG*, it is often necessary to mutagenize it to define important sequences involved in the activity of the gene product. In some cases this is employed as a means of localizing the gene of interest within the clone, as in the use of transposon insertion mutagenesis described elsewhere in this volume. In other cases it is a component of our attempts to determine functional domains of the gene or

---

[1] J. D. Beggs, *Nature (London)* **275**, 104 (1978).
[2] J. R. Broach, J. N. Strathern, and J. B. Hicks, *Gene* **8**, 121 (1979).
[3] K. Struhl, D. T. Stinchcomb, S. Scherer, and R. W. Davis, *Proc. Natl. Acad. Sci. U.S.A.* **76**, 1035 (1979).
[4] M. Carlson and D. Botstein, *Cell (Cambridge, Mass.)* **28**, 145 (1982).
[5] M. Rose, P. Novick, J. H. Thomas, D. Botstein, and G. Fink, *Gene* **60**, 237 (1987).

gene product. These approaches involve mutagenizing the clone on a shuttle vector, transforming the mutagenized bank into the tester yeast strain, and then screening for plasmids that have lost the ability to complement the initial yeast mutation. Having identified such transformants, it is necessary to recover the mutant plasmids so that the nature of the mutation (e.g., the position of the transposon insertion) can be determined.

Once a clone of *YFG* has been obtained, it can be used to clone alleles that have been generated and characterized by classical means. This process could, of course, be done by merely using the clone as a homologous probe to a yeast clone bank (prepared from the mutant strain) present on plasmids[6] or in a bacteriophage.[7] However, the efficient level of homologous recombination in yeast makes possible simpler means of recovering mutant alleles based on gap repair or "retriever vectors." Retriever vectors capitalize on the fact that yeast are capable of faithfully repairing a double-strand gap in DNA via a homology-dependent recombination process. If a shuttle vector is engineered so that it has homology to sequences flanking both sides of *YFG* but is deleted for the gene, repair of the gap will result in copying the mutant allele from the chromosome into the plasmid (see [19] in this volume). These gap repair [or golden retriever (GR)] plasmids can be recovered directly into *E. coli* by the techniques described below.[8,9]

It is occasionally necessary to isolate additional genomic sequences surrounding a cloned gene. The physical techniques that have produced long regions of overlapping clones in *Drosophila*[10] and other higher eukaryotes are applicable to yeast. In addition, the fact that homologous recombination can be used to integrate plasmids allows the formation of strains from which flanking sequences can be readily cloned.[11] Integrative transformation results in the formation of a tandem duplication flanking the vector sequences. By using restriction endonucleases that do not cut within the duplication, combined with religation and transformation into *E. coli*, additional flanking sequences can be cloned. The cycle can be repeated using, as the site of integration, the sequences derived from the end of the novel flanking DNA.

In order to illustrate some of the important parameters of recovering

---

[6] M. Grunstein and D. Hogness, *Proc. Natl. Acad. Sci. U.S.A.* **72**, 3961 (1975).
[7] W. Benton and R. Davis, *Science* **196**, 180 (1977).
[8] J. B. Hicks, J. N. Strathern, A. J. S. Klar, and S. L. Dellaporta, *in* "Genetic Engineering" (J. Setlow and A. Hollander, eds.), p. 219. Plenum, New York, 1982.
[9] T. Orr-Weaver and J. Szostak, *Proc. Natl. Acad. Sci. U.S.A.* **80**, 4417 (1983).
[10] W. Bender, P. Spierer, and D. S. Hogness, *J. Mol. Biol.* **168**, 17 (1983).
[11] F. Winston, F. Chumley, and G. R. Fink, this series, Vol. 101, p. 211.

the shuttle plasmids from yeast into *E. coli* we report here an example of the use of these methods. In this experiment we have used a high-copy-number vector (YEp24), a low-copy *CEN* vector (YCp50), and a strain in which the shuttle vector (YIp5) is integrated. We report the efficiencies of plasmid recovery in two commonly used *E. coli* strains (DH5 and HB101) prepared by the calcium shock technique or obtained as commercially available competent cells. The results also demonstrate that the commonly used small-scale yeast DNA preparations contain inhibitors of *E. coli* transformation that can be removed.

Materials and Methods

Plasmids. The plasmids used in these experiments were YEp24,[12] YCp50,[13] and a derivative of YIp5. Plasmid YEp24 contains pBR322 sequences allowing for replication and ampicillin resistance selection in *E. coli*, the yeast *URA3* gene for selection in yeast, and the yeast 2-μm origin of replication for autonomous replication in yeast. This plasmid is stably maintained in yeast at a copy number of 30–50 per cell. The YCp50 plasmid contains pBR322 sequences and the yeast *URA3* gene, as described for YEp24, but also contains the yeast centromere sequence from chromosome IV (*CEN4*) and an autonomous replicating sequence (*ARS1*) derived from yeast chromosome IV near *TRP1* that allows for autonomous replication of the plasmid. This centromere-containing plasmid is stable in yeast even in the absence of selective pressure for the *URA3* gene. Plasmid YIp5 containing *MAT* DNA was integrated into the yeast chromosome III next to the *MAT* locus.

Yeast Strain. The haploid yeast strain JSS104-15B (*MATa his3-Δ200 leu2-Δ1 lys2-801 trp1-Δ1 tyr7-1 ura3-52*), used in these experiments, was constructed in this laboratory using standard yeast genetics techniques.

Yeast Transformations. Plasmids YEp24, YCp50, and YIp5 were introduced into yeast strain JSS104-15B using the lithium acetate method of yeast transformation[14] (see also [12] in this volume).

Bacterial Strains. The bacterial strains used in these experiments were *E. coli* HB101 (F$^-$ *hsdS20[r^-_B, m^-_B] supE44 ara14 galK2 lacY1 proA2 rpsL20[str^R] xyl5 leu mtlL λ^- recA13*) and DH5 (F$^-$ *endA1 hsdR17[r^-_k, m^+_k] supE44 thi-1 λ^- recA1 gyrA96 relA1*), both obtained from Bethesda Research Laboratories (Gaithersburg, MD).

[12] D. Botstein, S. C. Falco, S. E. Stewart, M. Brennan, S. Scherer, D. T. Stinchcomb, K. Struhl, and R. W. Davis, *Gene* **8,** 17 (1979).
[13] M. Rose, P. Novick, J. H. Thomas, D. Botstein, and G. Fink, *Gene* **60,** 237 (1987).
[14] H. Ito, Y. Fukuda, K. Murata, and A. Kimura, *J. Bacteriol.* **153,** 163 (1983).

Yeast DNA Preparations

Two methods were employed to isolate DNA from yeast strains containing plasmids. The first method is a modification of a protocol by Lorincz,[15] and similar to a method by Hoffman and Winston,[16] and uses glass beads to break the cells. A modification that we have added here is the addition of a purification step using commercially available Glassmilk[17] (Geneclean trademark of BIO101, Inc., La Jolla, CA). The details of the procedure are as follows.

Method 1: Quick Plasmid DNA Preparations from Yeast

1. Pellet cells from 1 ml of an overnight YEPD or SC − Ura culture in microcentrifuge tube (for media, see [1] in this volume).

2. Resuspend pellet in 200 μl of 100 mM NaCl−10 mM Tris-HCl, pH 8.0−1 mM EDTA−0.1% sodium dodecyl sulfate (SDS).

3. Add sterile glass beads (0.45 mm diameter) until just below the level of the liquid. Mix vigorously on vortex mixer for 1 min. Add 200 μl phenol and vortex for another 1 min. Spin in a centrifuge for 2 min and transfer the aqueous layer to a second microfuge tube. Add another 200 μl phenol and extract a second time. Transfer the aqueous layer to a new microfuge tube.

4. The DNA solution is treated with Glassmilk[17] (Geneclean, registered trademark of BIO101), per manufacturer's instructions.

5. Ethanol precipitate for at least 30 min in 300 mM sodium acetate at −20°. Centrifuge and wash pellet in 80% ethanol. Resuspend each dried pellet in 50 μl TE [10 mM Tris-HCl (pH 8.0)−1 mM EDTA].

6. Transform *E. coli*. In these experiments, 2.5 μl of DNA is used for each *E. coli* transformation.

The second method for yeast DNA preparations is one in general use in our laboratory to isolate yeast genomic DNA suitable for Southern blots. It is similar to a protocol described by Holm *et al.*[18]

Method 2: Yeast Genomic DNA Preparations

1. Start with a 10-ml YEPD or SC − Ura overnight culture.

2. Pellet cells by centrifugation. Resuspend the pellet in 1 ml water and transfer to a microfuge tube.

[15] A. Lorincz, *BRL Focus* **6**, 11 (1985).
[16] C. Hoffman and F. Winston, *Gene* **57**, 267 (1987).
[17] L. G. Davis, M. D. Dibner, and J. F. Battey, *in* "Basic Methods in Molecular Biology," p. 123. Elsevier, New York, 1986.
[18] C. Holm, D. W. Meek-Wagner, W. L. Fangman, and D. Botstein, *Gene* **42**, 169 (1986).

3. Spin 20–30 sec in a microfuge and resuspend pellet in 0.15 ml SZB (1 M sorbitol–100 mM sodium citrate–60 mM EDTA–0.5 mg/ml Zymolyase 60,000–100 mM 2-mercaptoethanol). Incubate at 37° for 30–40 min with occasional shaking. Check cells after 20 min for spheroplasting by adding a very small amount to a drop of 5% Triton X on a microscope slide and check for lysed cells. Zymolyase treatment is complete when more than 80% of the cells are spheroplasted (i.e., lysed).

4. Add 0.15 ml SDS–TE solution [2% SDS–0.1 M Tris-HCl (pH 8.0)–10 mM EDTA], vortex briefly, and incubate at 60–65° for 5–10 min.

5. Add 0.15 ml of 5 M potassium acetate solution, vortex briefly, and incubate on ice for 30–45 min.

6. Spin at 4° for 15 min; remove and save 0.3 ml supernatant. Discard pellet.

7. Add to the supernatant 0.2 ml of 5 M ammonium acetate and 1 ml prechilled 2-propanol. Incubate at −20° for 10 min to overnight.

8. Spin in a microfuge for 5 min; pour off supernatant. Rinse pellet with cold 80% ethanol. Dry pellet lightly under reduced pressure.

9. Redissolve pellet in 200 μl water, add 2 μl of 10 mg/ml RNase, and incubate at 37° for 10 min.

10. Add 4 μl of 5 M NaCl, 2 μl of 20 mg/ml pronase, and incubate at 37° for an additional 10 min.

11. Extract with 200 μl phenol, spin, and transfer supernatant (aqueous phase) to a new tube.

12. Ethanol precipitate supernatant. Spin and redissolve the pellet in 50 μl TE.

For plasmids YEp24 and YCp50, yeast DNA preparations were used directly for *E. coli* transformation, either with or without a Glassmilk purification step first (Tables I and II). In order to recover the integrated variant of YIp5, it is necessary to convert it to a circular form. The yeast genomic DNA was digested with *Bam*HI, phenol extracted, and ligated overnight, and the ligation mix was used to transform *E. coli* (see protocol in Table III). This resulted in the recovery of circular plasmid products (YIp5) derived from the yeast chromosomal DNA (Table III).

Bacterial Transformations

Bacterial strains transformed to ampicillin resistance using DNA prepared from yeast were either commercially prepared, frozen competent cells obtained from Bethesda Research Laboratories (BRL, Cat. No.

TABLE I
BACTERIAL TRANSFORMATIONS WITH QUICK PLASMID DNA PREPARATIONS FROM YEAST (METHOD 1)[a]

E. coli strain	Plasmid	Yeast growth	Total amp^R colonies	amp^R colonies/ 10^8 yeast cells
hb101	pBR322	—	726	—
(Ca shock)	YEp24	YEPD	56	2220
	YCp50	YEPD	12	480
	YEp24 (gc)	YEPD	116	4620
	YCp50 (gc)	YEPD	20	780
	pBR322 + YEp24	YEPD	198	—
	pBR322 + YEp24 (gc)	YEPD	702	—
dh5	pBR322	—	225	—
(Ca shock)	YEp24	YEPD	48	1920
	YCp50	YEPD	11	420
	YEp24 (gc)	YEPD	60	2400
	YCp50 (gc)	YEPD	9	360
	pBR322 + YEp24	YEPD	86	—
	pBR322 + YEp24 (gc)	YEPD	257	—
HB101	pBR322	—	623	—
	YEp24	YEPD	11	420
	YCp50	YEPD	2	60
	YEp24 (gc)	YEPD	14	540
	YCp50 (gc)	YEPD	23	900
	pBR322 + YEp24	YEPD	0	—
	pBR322 + YEp24 (gc)	YEPD	278	—
DH5	pBR322	—	226	—
	YEp24	YEPD	2	60
	YCp50	YEPD	0	0
	YEp24 (gc)	YEPD	17	660
	YCp50 (gc)	YEPD	11	420
	pBR322 + YEp24	YEPD	30	—
	pBR322 + YEp24 (gc)	YEPD	176	—

[a] *Escherichia coli* strains designated by lowercase letters indicate that the competent cells were made by CaCl$_2$ treatment. The strains in uppercase indicate BRL frozen, competent cells. The plasmids followed by the designation (gc) represent preparations that included the Geneclean step. In all transformations, 100 μl of cells was combined with 2.5 μl of yeast DNA and/or 1 μl of pBR322 DNA (pBR322 DNA is 0.01 ng/μl). All bacterial transformants were plated on LB containing 100 μg/ml ampicillin, 0.2 ml/plate, and were scored after 24 hr of incubation at 37°. YEPD refers to the growth medium of the yeast culture from which the DNA was isolated. YEPD is nonselective for either of the plasmids, although the cells reach a higher final density than when grown in selective media.

TABLE II
BACTERIAL TRANSFORMATIONS WITH YEAST GENOMIC DNA PREPARATIONS (METHOD 2)[a]

E. coli strain	Plasmid	Yeast growth	Total amp^R colonies	amp^R colonies/ 10^8 yeast cells
hb101	pBR322	—	141	—
(Ca shock)	YEp24	YEPD	348	696
	YCp50	YEPD	24	48
	YEp24	−Ura	0	0
	YCp50	−Ura	65	1290
	pBR322 + YEp24	YEPD	396	—
	pBR322 + YEp24	−Ura	363	—
dh5	pBR322	—	65	—
(Ca shock)	YEp24	YEPD	423	846
	YCp50	YEPD	9	18
	YEp24	−Ura	0	0
	YCp50	−Ura	74	1470
	pBR322 + YEp24	YEPD	551	—
	pBR322 + YEp24	−Ura	429	—
HB101	pBR322	—	1026	—
	YEp24	YEPD	42	84
	YCp50	YEPD	0	0
	YEp24	−Ura	0	0
	YCp50	−Ura	0	0
	pBR322 + YEp24	YEPD	19	—
	pBR32 + YEp24	−Ura	0	—
DH5	pBR322	—	557	—
	YEp24	YEPD	65	129
	YCp50	YEPD	0	0
	YEp24	−Ura	2	40
	YCp50	−Ura	0	0
	pBR322 + YEp24	YEPD	0	—
	pBR322 + YEp24	−Ura	0	—

[a] *Escherichia coli* strains designated by lowercase letters indicate that the competent cells were made by CaCl$_2$ treatment. The strains in uppercase indicate BRL frozen, competent cells. In all transformations, 100 μl of cells was combined with 2.5 μl of yeast DNA and/or 1 μl of pBR322 DNA (pBR322 DNA is 0.01 ng/μl). All bacterial transformants were plated on LB containing 100 μg/ml ampicillin, 0.2 ml/plate, and were scored after 24 hr of incubation at 37°. YEPD and −Ura refer to the growth medium of the yeast culture from which the DNA was isolated. YEPD is nonselective for either of the plasmids; −Ura is selective medium for the plasmid. The cells reach a higher final density in YEPD.

TABLE III
BACTERIAL TRANSFORMATIONS USING DNA PREPARATION METHODS 1 AND 2 FROM YEAST STRAIN CONTAINING INTEGRATED PLASMID[a]

Strain	Plasmid	DNA preparation method	Total amp^R colonies	amp^R colonies/ 10^8 yeast cells
hb101	pBR322	—	176	—
(Ca shock)	YIp5-18.3	1	3	120
	YIp5-18.3	2	7	14
	YIp5-18.3(gc)	1	4	160
	YIp5-18.3(gc)	2	9	18
dh5	pBR322	—	124	—
(Ca shock)	YIp5-18.3	1	2	80
	YIp5-18.3	2	2	4
	YIp5-18.3(gc)	1	4	160
	YIp5-18.3(gc)	2	5	10
HB101	pBR322	—	346	—
	YIp5-18.3	1	1	40
	YIp5-18.3	2	2	4
	YIp5-18.3(gc)	1	3	120
	YIp5-18.3(gc)	2	3	6
DH5	pBR322	—	196	—
	YIp5-18.3	1	2	80
	YIp5-18.3	2	0	0
	YIp5-18.3(gc)	1	1	40
	YIp5-18.3(gc)	2	2	4

[a] *Escherichia coli* strains designated by lowercase letters indicate that the competent cells were made by CaCl$_2$ treatment. The strains in uppercase indicate BRL frozen, competent cells. The plasmids followed by the designation (gc) represent preparations that included the Geneclean step. Genomic DNAs (10 µl) were digested with *Bam*HI, phenol extracted, and ethanol precipitated overnight. The *Bam*HI-digested DNAs were religated in 100 µl total volume overnight at 15° with 10,000 units of T4 DNA ligase. In all transformations, 100 µl of cells were combined with 2.5 µl of ligation mix or 1 µl of pBR322 DNA (pBR322 DNA is 0.01 ng/µl). All bacterial transformants were plated on LB containing 100 µg/ml ampicillin, 0.2 ml/plate, and were scored after 24 hr of incubation at 37°. All yeast cultures were grown in YEPD.

8260SA and 8262SA) or cells that were made competent by CaCl$_2$ shock by the method of Mandel and Higa[19] as described below.

Preparation and Transformation of Calcium-Shocked Escherichia coli

1. Dilute an overnight LB culture 1/100 in fresh LB or SOC (0.5 ml in 50 ml). Incubate at 37° with vigorous shaking. Monitor the OD$_{600}$ and harvest cells when OD$_{600}$ equals 0.5–1.0.

[19] M. Mandel and A. Higa, *J. Mol. Biol.* **53**, 154 (1970).

2. Resuspend cell pellet in 1/2 volume ice-cold 50 mM CaCl$_2$ (e.g., centrifuge 25 ml; add 12 ml of 50 mM CaCl$_2$). Keep on ice for at least 30 min.

3. Centrifuge CaCl$_2$-treated cells in a 4° rotor. Gently resuspend cells in 1/5 volume of ice-cold 50 mM CaCl$_2$. Keep cells on ice. Cells are usable for about 24 hr.

4. Aliquot 100 μl of cells for each transformation into a prechilled (on ice) Falcon 2059 tube. Add transforming DNA and incubate on ice for 30 min.

5. After a 30-min incubation on ice, heat-shock cells at 42° for 45–60 sec (in a water bath). After heat shock, return the tube(s) to ice for 2 min.

6. Add 1 ml of SOC medium and incubate at 37° for 1 hr with vigorous shaking. SOC contains Bacto-tryptone, 2%; yeast extract, 0.5%; NaCl, 10 mM; KCl, 2.5 mM; MgCl$_2$, 10 mM; MgSO$_4$, 10 mM; and glucose, 0.4%.

7. Plate 100–200 μl/plate on LB plus ampicillin.

8. Incubate plates at 37° overnight.

Bacterial DNA Minipreparations. Plasmid DNA was isolated from *E. coli* using the method of Birnboim and Doly,[20] as modified by Ish-Horowicz and Burke.[21]

Results

Yeast DNA was prepared from strains containing plasmid YEp24 or YCp50 or an integrated plasmid by the two methods outlined above. The yeast strains were grown overnight in either YEPD or synthetic medium lacking uracil (SC − Ura) to a density of approximately 10^8 and 10^7 cells/ml, respectively. The DNA was resuspended in a final volume of 50 μl, and 2.5 μl of it was used for each bacterial transformation. Note that in the glass bead method, the 50-μl final volume represents the DNA from 1.0 ml of a yeast culture, whereas in the genomic DNA preparation the 50 μl represents the DNA from 10 ml of cells. The data reported here for ampicillin-resistant bacterial colonies per 10^8 yeast cells have been adjusted for this difference.

The efficiency of transformation of each strain was monitored by using pBR322 DNA (obtained from BRL). The transformability of CaCl$_2$-shocked cells with pBR322 DNA is more variable. The data generated in Tables I, II, and III represent *E. coli* transformations on different days using different batches of both CaCl$_2$ and frozen competent cells. It is

[20] H. C. Birnboim and J. Doly, *Nucleic Acids Res.* **7**, 1513 (1979).
[21] D. Ish-Horowicz and J. F. Burke, *Nucleic Acids Res.* **9**, 2989 (1981).

worth noting that HB101 transformants grow noticeably better than DH5 transformants regardless of their source (i.e., $CaCl_2$ treated or from BRL).

The mixing experiment using yeast DNA and pBR322 DNA simultaneously gives an indication of the presence of anything in the yeast DNA preparation that inhibits the transformability of *E. coli*. If there is no inhibition of transformation with pBR322 DNA by the yeast DNA preparation, then the number of *E. coli* transformants in the mixed DNA reaction should yield bacterial transformants approximately equal to the sum of the two DNAs used separately. In general the BRL frozen competent cells are much more sensitive to "inhibitors" of transformation in the yeast DNA preparations than are the $CaCl_2$-treated cells.

Plasmid preparations for *E. coli* transformants were done and plasmids were confirmed by restriction site analysis.

Discussion

Comparison of Methods 1 and 2. Inspection of Table I (Method 1) and Table II (Method 2) suggests that both protocols for DNA preparations yield shuttle plasmids that can be recovered in *E. coli*.

Comparison of YEPD and Minimal Media. For the plasmids utilized here, the stability under nonselective growth (YEPD) conditions was fairly high so that the growth of the yeast culture selectively is not necessary.

Treatment with Glassmilk (Geneclean). In general, treatment of the yeast DNA with Glassmilk[17] (Geneclean) resulted in increased transformation frequency (1- to 6-fold). This was apparent in an increased number of transformants with the yeast DNA and a decrease in the inhibitory effect of the yeast DNA on the transforming potential of the control plasmid pBR322. These mixing experiments illustrate the importance of using clean DNA and suggest that the Glassmilk[17] (Geneclean) protocol is an effective and easy means to accomplish this goal.

Comparison of Calcium-Shocked and Commercial Competent Bacterial Cells. The commercially available bacterial cells, both DH5 and HB101, seemed particularly sensitive to the inhibition of transformation by something in the yeast DNA preparations. In several cases, plasmids that could be readily recovered in the calcium-shocked cells gave no transformants with the commercial cells. The mixing experiments with the control pBR322 plasmid demonstrate that this is a consequence of an inhibitor of transformation present in the yeast DNA.

Comparison of DH5 and HB101. In our hands HB101 transforms at least as well as DH5 and grows better.

Recovery of Integrated Plasmids. Both Methods 1 and 2 (Table III) work for the recovery of integrated plasmids from the genome.

Conclusions

1. The glass bead method is better for plasmid recovery on a per yeast cell basis than the genomic preparation.

2. There is something in the yeast DNA preparations (both methods) that inhibits *E. coli* transformation to ampicillin resistance with pBR322 DNA. (A) This inhibition is greater on BRL frozen competent cells than on $CaCl_2$-treated cells, and we therefore recommend for plasmid recovery from yeast that competent *E. coli* cells made by the $CaCl_2$-shock method be used. (B) Inhibition of *E. coli* transformation by the yeast preparations can be reduced somewhat by the addition of a Glassmilk[17] (Geneclean) step into the glass bead protocol.

3. At least for plasmids YEp24, YCp50, and an integrated plasmid (YIp5) it is not necessary to grow the yeast cells in medium that selects for the plasmid.

Acknowledgments

Many members of our laboratory, past and present, contributed to the evolution of these protocols. We thank Patti Hall for preparation of the manuscript. Research was sponsored by the National Cancer Institute, U.S. Department of Health and Human Services, under Contract No. N01-C0-74101 with ABL. The contents of this publication do not necessarily reflect the views or policies of the Department of Health and Human Services, nor does mention of trade names, commercial products, or organizations imply endorsement by the U.S. government.

[22] Shuttle Mutagenesis: Bacterial Transposons for Genetic Manipulations in Yeast

By MERL F. HOEKSTRA, H. STEVEN SEIFERT, JAC NICKOLOFF, and FRED HEFFRON

Introduction

The ability to rapidly isolate and characterize mutations physically linked to a transposable element has greatly facilitated research in bacterial genetics. Prokaryotic transposons offer high single-hit mutation frequencies, a selectable marker for strain construction, and a means for recovery of the mutated gene. Efforts to extend the advantages of bacterial genetics approaches to yeast genetics has resulted in the development of two mutagenesis systems: shuttle mutagenesis[1] (SM) and transplacon[2] transposons. Both systems utilize elements that are defective or miniature versions of bacterial transposons. These transposons contain bacterial and yeast selectable markers and are capable of transposition only when functions normally encoded by the wild-type transposon are supplied in trans. Bacterial transposons specifically designed for use with *Saccharomyces cerevisiae* have been based on Tn*3*[1] and Tn*10*.[2,3] The minitransposons in SM are defective derivatives of Tn*3* that are designed to be inserted into yeast genes carried on plasmid vectors. Transposons based on Tn*10* find their greatest utility in mutagenizing λ[2] or cosmid clones but can also be used for plasmid mutagenesis. In addition, Ty retrotransposons have also been designed for use as yeast mutagens[4] (see [23] in this volume).

The utility of shuttle mutagenesis and transplacon transposons is severalfold. The ability of these chimeric transposable elements to insert at many places in cloned yeast genes and the ability to select for insertion of the disrupted gene into the homologous region of the yeast chromosome have aided the analysis of several genes.[1-5] The minitransposons have been specifically employed for a variety of purposes, including delineation of gene boundaries,[1-3] rapid introduction of selectable markers for transplacement experiments,[1-3] gene-counting experiments,[5] mapping antigenic coding regions,[2] production of fusion proteins,[1-3] production of regulatory mutants with transposon-borne regulatory elements (see below), and DNA sequencing using oligonucleotide primers homologous to the ends of the transposons (Ref. 3; H. S. Seifert, M. So, R. Ajioka, M. Vito, and F. Heffron, unpublished; M. F. Hoekstra, E. Chiaou, F. Heffron, and R. M. Liskay, unpublished).

The Tn*3*- and Tn*10*-based mutagenesis systems have distinct advantages and disadvantages that should be considered during experimental design. The differences are based on the mechanisms of transposition in each system, the means for introducing the transposon into the mutagenesis protocol, the manner in which the particular transposon derivatives have been constructed, and target sequence specificity differences. For example, since the transposons are carried on an F derivative, the Tn*3* system involves simple bacterial matings for initiating mutagenesis and

---

[1] H. S. Seifert, E. Y. Chen, M. So, and F. Heffron, *Proc. Natl. Acad. Sci. U.S.A.* **83**, 735 (1986).
[2] M. Snyder, S. Elledge, and R. W. Davis, *Proc. Natl. Acad. Sci. U.S.A.* **83**, 730 (1986).
[3] O. Huisman, W. Raymond, K.-U. Froehlich, P. Errada, N. Kleckner, D. Botstein, and M. A. Hoyt, *Genetics* **116**, 191 (1987).
[4] J. D. Boeke, H. Xu, and G. R. Fink, *Science* **239**, 280 (1988).
[5] M. Goebels and T. Petes, *Cell (Cambridge, Mass.)* **46**, 983 (1986).

selecting transposition products and thus requires little "hands-on" time in mutagenesis. Mutagenesis with mini-Tn*10* elements requires the preparation of bacteriophage lysates which are then used to introduce the element using phage λ. With the Tn*3*-based shuttle mutagenesis system, transposition can be directed preferentially to the target molecule (for efficient selection of molecules containing insertions). Because the chromosome of *Escherichia coli,* as well as any molecule containing the 38-base pair (bp) inverted repeat of Tn*3*, is immune to Tn*3* transposition, insertions are directed to the plasmid carrying the yeast DNA. This initial transposition event results in a cointegrate between the donor plasmid and the yeast clone. Thus, all of the transposition products can be isolated by selecting for the transfer of the cointegrate into a secondary strain by conjugation. In contrast, Tn*10* can freely transpose into both the *E. coli* chromosome and the target plasmid. This generates a high background of false positives unless transposition products are selected by transforming or infecting a second *E. coli* strain. The process of isolating DNA or phage and introducing it into a secondary strain are more labor intensive and less efficient that conjugation. On the other hand, transposition immunity conferred by the 38-bp terminal repeat of Tn*3* limits the use of the SM system.

Many vectors are derivatives of pBR322, a molecule containing the β-lactamase gene along with the 38-bp end from Tn*3*. These vectors are immune to Tn*3* transposition but susceptible to insertion by Tn*10*. Hence, to use the Tn*3* shuttle mutagenesis system, subcloning into a Tn*3*-free vector is required. With the Tn*10*-based systems subcloning should not be necessary. Finally, a greater variety of genetic constructs have been generated to use in SM compared to the Tn*10* systems (see Discussion). This allows latitude as to which yeast background is used and the type of genetic analysis that can be performed.

In this chapter we concentrate on the methods and application of the Tn*3* shuttle mutagenesis system and discuss several previously undescribed elements and cloning vectors. In addition to the applicability of this system to *Escherichia coli* and *Saccharomyces cerevisiae,* the system has been successfully employed with other organisms including *Salmonella typhimurium* (S. Libby and F. Heffron, unpublished observation), *Neisseria gonorrhoeae,*[6] and *Schizosaccharomyces pombe* (G. Smith, personal communication). The system should be amenable for use with many other transformable organisms but may be less useful for organisms containing large introns in most genes.

[6] W. Carbonetti, V. I. Sinaud, H. S. Seifert, M. So, and P. F. Sparling, *Proc. Natl. Acad. Sci. U.S.A.* **85,** 6841 (1988).

Materials and Methods

Shuttle Mutagenesis Procedure

The shuttle mutagenesis procedure is summarized in Fig. 1. First, your favorite gene *(YFG)* is cloned into a Tn3-free vector [e.g., the pHSS series (Fig. 2)]. A vector containing *YFG* (pHSS-*YFG*) is transformed into a pLB101-containing strain. The resulting strain contains the pHSS-*YFG* target molecule and the pLB101 transposase-producing plasmid. The transformant is then mated with a strain containing the conjugal plasmid pOX38 into which a defective minitransposon has been previously placed (pOX38::m-Tn3). Plasmid pOX38 acts as a delivery vehicle for the transposon. Transconjugants contain cointegrates between pOX38::mini-Tn3 and pHSS-*YFG*. The cointegrate is then resolved by conjugation with an F⁻ strain that expresses the Cre protein from phage *P1*. The Cre enzyme is able to catalyze a site-specific recombination/resolution event at *loxP* sites carried on the Tn3-derived minitransposons. The products formed from cointegrate resolution are the original pOX38::mini-Tn3 and the target plasmid with a minitransposon insertion (pHSS-*YFG*::m-Tn3).

Bacterial Strains and Media

Escherichia coli strains (Table I) are routinely grown on LB medium (1% tryptone, 0.5% yeast extract, 1% NaCl, solidified with 2% agar). Other rich media like SOB or YT[7] can also be used. Kanamycin (Km), chloramphenicol (Cm), carbenicillin (Carb), streptomycin (Sm), and spectinomycin (Sp) are used at final concentrations of 40, 30, 200, 100 µg/ml. (Ampicillin at 50–100 µg/ml can be used in place of carbenicillin.) We routinely use carbenicillin because of its stability when incorporated into media or stored in solution. In addition, carbenicillin reduces the problem of satellite colony formation that is often observed with ampicillin.

Mutagenesis Protocol

Step-by-step details of the shuttle mutagenesis procedure from subcloning a DNA fragment to returning a mutagenized insert to yeast are as follows.

1. To initiate shuttle mutagenesis, the appropriate DNA fragment is cloned into the multicloning site of any of the vectors shown in Fig. 2, Table II, or described below. Multicloning sites are flanked by *Not*I or *Sfi*I sites to facilitate linearization of the DNA for transplacement experiments in yeast.

[7] J. H. Miller, "Experiments in Molecular Genetics." Cold Spring Harbor Laboratory, Cold Spring Harbor, New York, 1972.

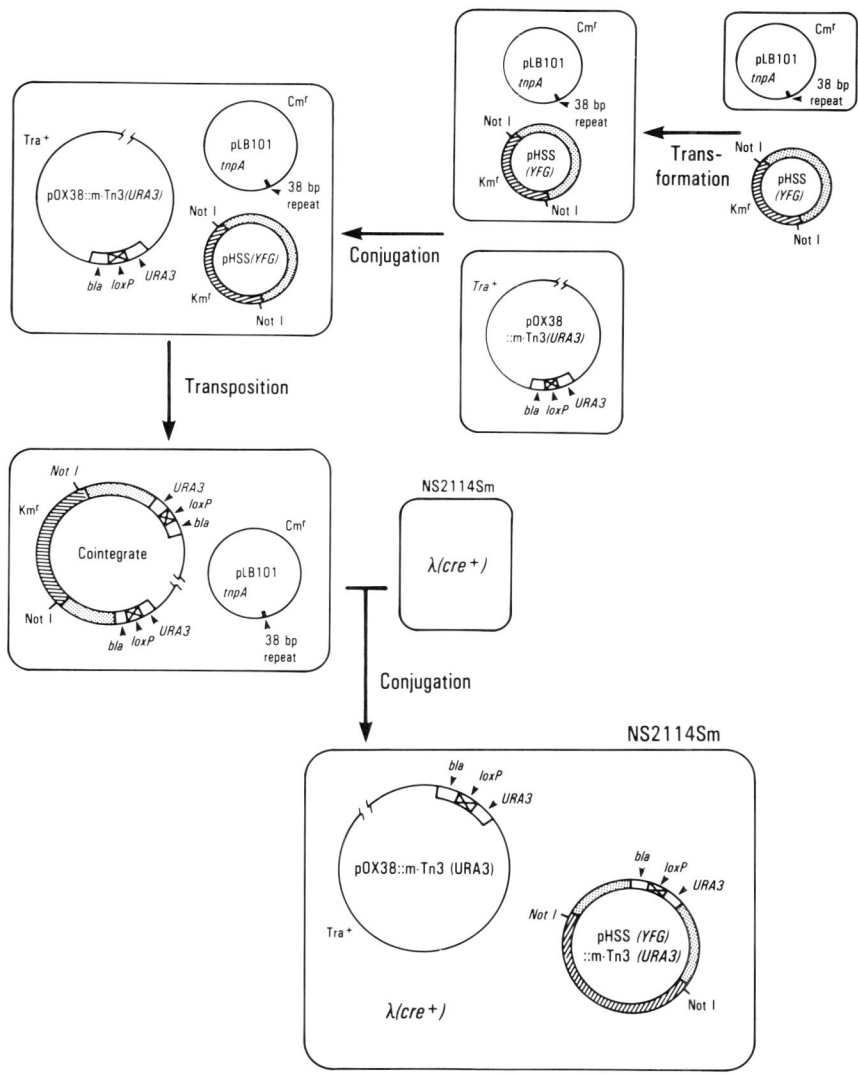

FIG. 1. The shuttle mutagenesis procedure. A kanamycin-resistant plasmid containing a subcloned insert (pHYSS*YFG*) is transformed into a pLB101-containing strain. Plasmid pLB101 contains a Tn*3* 38-bp terminal repeat and is immune to transposition. The resulting transformant is mated with a mini-Tn*3*-containing strain, the transconjugant is selected on the appropriate medium, and transposition is allowed to occur. Transposition resolution proceeds by mating the cointegrate into NS2114Sm, which expresses the Cre resolvase, and allowing for site-specific recombination to produce an insertion in the desired target region (pHSS*YFG*::m-Tn*3*).

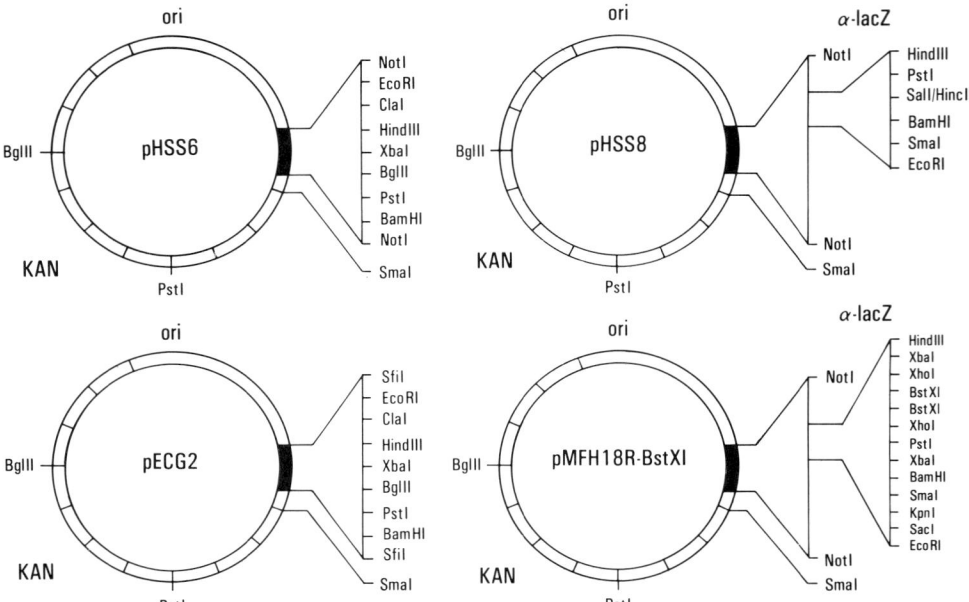

FIG. 2. Physical and genetic maps of Tn3-free target vectors. These vectors contain a ColE1 replicon *(ori)* and the kanamycin resistance determinant from Tn5 *(KAN)*. The vectors contain various multicloning sites (dark box). Vectors pHSS6 and pECG2 contain the multiple cloning site (MCS) from pLink322,[8] pHSS8 contains the MCS from pUC9,[9] and pMFH18R-*Bst*XI has the MCS from pTZ18R-*Bst*XI.[10]

TABLE I
Escherichia coli Strains

Strain	Plasmid	Genotype and relevant properties
RDP146	—	F$^-$ *recA1* (Δ*lac-pro*)*rpsE* (spectinomycin resistant)
	pLB101	Chloramphenicol-resistant pACYC184 derivative which constitutively expresses Tn3 transposase[11]
	pOX38::m-Tn3	Conjugative F factor derivative[12] which carries mini-Tn3
NS2114Sm	—	F$^-$ *recA rpsL* (streptomycin resistant, contains a λ-*cre* lysogen)

[8] T. Maniatis, E. F. Fritsch, and J. Sambrook, "Molecular Cloning: A Laboratory Manual." Cold Spring Harbor Laboratory, Cold Spring Harbor, New York, 1982.
[9] C. Yanish-Perron, J. Vierra, and J. Messing, *Gene* **33**, 103 (1985).
[10] Invitrogen, San Diego, California.
[11] C.-J. Huang, F. Heffron, J.-S. Twu, R. H. Schloemer, and C.-H. Lee, *Gene* **41**, 23 (1986).
[12] M. S. Guyer, *J. Mol. Biol.* **126**, 347 (1978).

TABLE II
VECTORS FOR USE WITH SHUTTLE MUTAGENESIS[a]

Vector	Cloning sites for transplacement	Other features
pHSS4	EcoRI, BamHI	
pHSS6	EcoRI, ClaI, HindIII, XbaI, BamHI	
pHSS8	EcoRI, SalI, BamHI, HindIII	Blue/white
pHSS9	EcoRI, ClaI, HindIII, XbaI, BamHI	M13 ori
pHSS11	EcoRI, ClaI, HindIII, XbaI, BglII, BamHI	Lacks BglII site in KAN
pHSS12	EcoRI, ClaI, HindIII, XbaI, BglII, BamHI	Lacks BglII and SmaI sites in KAN
pHSS13	EcoRI, SalI, BamHI, HindIII	Blue/white
pHSS19	EcoRI, SacI, SmaI, XmaI, BamHI, SalI, SphI, HindIII	Blue/white
pHSS20	EcoRI, SacI, BglII, BamHI, SalI, SphI, HindIII	Blue/white
pHSS21	EcoRI, SacI, XhoI, BamHI, SalI, SphI, HindIII	Blue/white
pECG2	EcoRI, ClaI, HindIII, XbaI, BamHI	MCS flanked by SfiI sites
pMFH18R-BSTX1	EcoRI, SacI, KpnI, BamHI, XbaI, XhoI, BstXI, HindIII	Blue/white

[a] Multiple cloning sites (MCS), with the exception of pECG2, are flanked by NotI sites for linearization of mutagenized inserts. The MCS in pMFH18R-BstXI contains two XhoI and BstXI sites.

2. The cloned DNA is introduced into RDP146(pLB101), selecting for Cm and Km. Either fresh or frozen CaCl$_2$ competent cells can be used for transformation. RDP146(pLB101) is amenable to transformation by the method of Hanahan[13] or Dagert and Ehrlich.[14] Frozen competent RDP146(pLB101) cells of 10^7-10^8 transformants/μg efficiency can be stored for several months. Strain RDP146(pLB101) should also be amenable to electroporation methods for introducing DNA. High competence is required at this step only when transforming gene banks for mutagenesis. Quick-preparation DNA samples are usually of sufficient quality for use in this transformation step.

3. RDP146(pOX38::m-Tn3) is mated with the transformants from Step 2, selecting for conjugal transfer of the transposon-borne antibiotic resistance gene, the target plasmid, and the transposase-producing plasmid by plating the mating mixture on medium containing Km/Carb/Cm. The details of bacterial mating are described elsewhere.[7] We routinely grow an overnight culture of both strains to be mated in the presence of antibiotic selection, dilute 1/100 into fresh medium without antibiotics, grow at 37°

[13] D. Hanahan, J. Mol. Biol. **166**, 557 (1983).
[14] M. Dagert and S. D. Ehrlich, Gene **6**, 23 (1979).

to mid-log phase (about $3-5 \times 10^8$ cell/ml), and mix the strains at a donor to recipient ratio of 1:2. The mating mixture is allowed to sit for at least 15 min at 37° before plating at $10^{-1}-10^{-4}$ dilutions. As controls we routinely plate 0.1 ml of donor and recipient separately. Alternative mating procedures that employ replica plating or physical mixing of strains on the surface of an LB plate with an inoculating loop also work effectively.

4. Tn*3* transposition is maximal at 30°. To allow transposition of mini-Tn*3*, a culture from Step 3 is grown at 30° with antibiotics. Our preference is to incubate the colonies on the plates from Step 3 at 30° for 4 hr to 2 days (but not longer), wash colonies off the surface of selection plates with a small volume of LB, dilute to early log phase (about 5×10^7 cells/ml), and grow at 37° for mating with *E. coli* strain NS2114Sm. This method also appears to result in the isolation of a large number of independent insertions since each colony from Step 3 produces at least one independent transposition. It is important that the strains not be held at 30° for extended periods. Prolonged growth at this temperature results in overproduction of transposase and a concomitant increase in the frequency with which adjacent deletions are isolated.

5. The transposition intermediate is resolved by mating the cointegrate into NS2114Sm. As in Step 3, an overnight of the culture from Step 4 and an overnight culture of NS2114Sm are diluted separately into fresh LB lacking antibiotics, grown to mid-log phase, mated, and plated onto medium containing Km/Carb/Sm. Resolution of the cointegrate intermediate occurs after transfer into NS2114Sm. Cre-mediated resolution can take place by intermolecular as well as an intramolecular recombination, but the intramolecular reaction is more efficient. Therefore, most of the plasmid isolates will be monomeric, but a small proportion will comprise every assortment of the two plasmids contained in the strain. It is recommended that matings with NS2114Sm be limited to 15 min as longer mating periods can result in colonies containing more than one transposon insertion.

6. Individual transconjugants can be screened by restriction endonuclease analysis of small-scale DNA preparations, or pools of DNA isolated from transconjugants can be transformed into yeast and screened for specific phenotypes. We have successfully isolated plasmid DNA from NS2114Sm with the boiling method[8] and the alkaline extraction method.[8] Digestion of DNA preparations with *Not*I (or *Sfi*I, depending on the cloning vector or minitransposon) will determine which insertions occurred in the vector as well as ascertain if any transposon-mediated deletions have taken place. On occasion we have noted problems with unusually high levels of endonuclease contamination in quick-preparations from NS2114Sm. Treatment with self-digested pronase at 65° for 30 min, fol-

lowed by ammonium acetate/2-propanol precipitation in the presence of 1 mM phenylmethylsulfonyl fluoride (PMSF) and Sepharose CL-6B spin dialysis, often remedies this problem. Others have noted that two phenol extractions help to reduce endonuclease problems. When endonuclease contamination is found to be a significant problem, undigested quick-preparation samples should be retransformed into other common *E. coli* strains like DH5, LE392, or HB101 prior to plasmid analysis.

7. Mating procedures for use of the epitope insertion element (EIE) transposon (Fig. 3m; see Discussion) are similar to those described above for elements derived from m-Tn*3*, with the exception that antibiotic selections for the EIE differ. We recommend, as a first step, the subcloning of the gene of interest into pECG2 (Fig. 2). This plasmid allows transplacement of *Sfi*I-linearized fragments containing EIE inserts back into yeast (the EIE contains two *Not*I sites). Plasmids are transformed into RDP146 (pLB101) as described in Step 2. These transformants are mated with RDP146(pOX38::mTn*3*EIE) as described in Step 3, but the mating mixture is plated on Cm/Km/tetracycline plates. The optimum concentration for tetracycline has been found to be 5 μg/ml. As in Step 5, resolution of cointegrates is facilitated by mating into NS2114Sm. This mating mixture is plated on medium containing Km/Sm/tetracycline. The use of 5 μg/ml tetracycline has been optimized for selecting single EIE insertions. Higher concentrations of tetracycline selects for multiple insertions, and lower concentrations result in the isolation of deletion products. In fact, removing tetracycline selection allows the isolation of resolved products that have deleted the entire inner portion of the element and results in a small, 135-bp insertion (see Discussion).

8. Mutagenized inserts have been successfully returned to yeast by either spheroplast[15] or lithium salt-mediated transformation[16] or by electroporation. As noted by others, there is great yeast strain variability in transformation efficiency, and the method of choice for returning a mutagenized insert is largely dependent on this factor. For routine transplacements, however, we recommend the use of 5 μg of linearized fragments per single lithium transformation experiment.

Discussion

Shuttle mutagenesis extends prokaryotic transposon mutagenesis to eukaryotes such as *S. cerevisiae*. The system is rapid and powerful, requiring only a subcloning into a Tn*3*-free vector, a bacterial transformation,

[15] A. Hinnen, J. Hicks, and G. Fink, *Proc. Natl. Acad. Sci. U.S.A.* **75,** 1929 (1978).
[16] H. Ito, Y. Fukada, K. Murata, and A. Kimura, *J. Bacteriol.* **153,** 163 (1983).

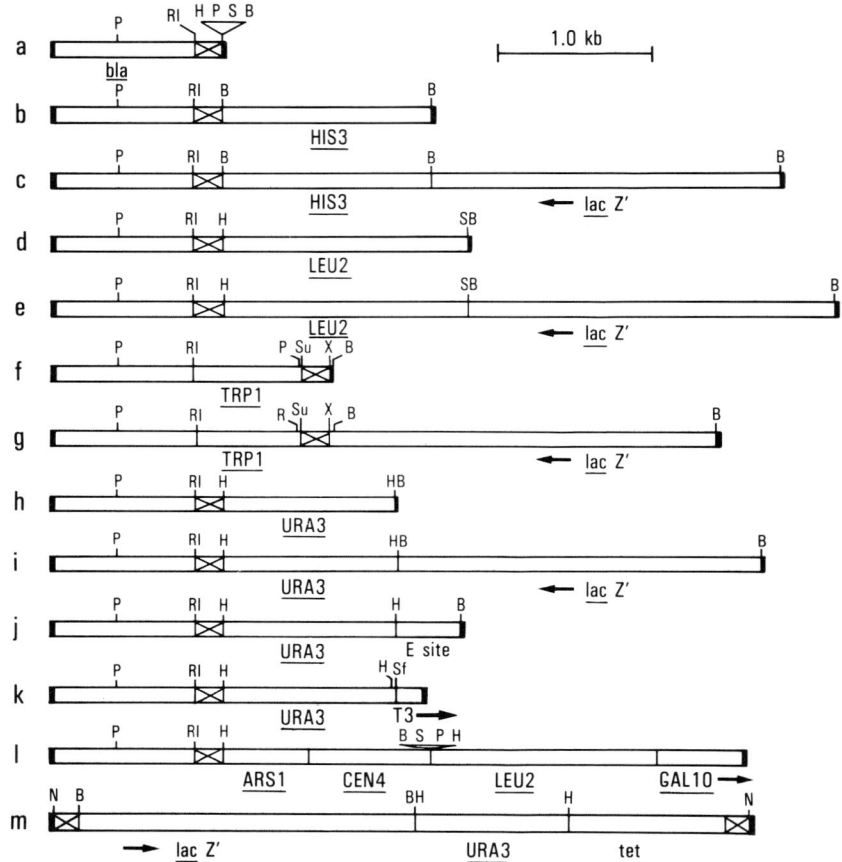

FIG. 3. Shuttle mutagenesis transposons. Various SM transposons derived from m-Tn*3* are shown. All elements are flanked by Tn*3* 38-bp repeats (dark ends) and carry *loxP* sites (crossed box) for cointegrate resolution. With the exception of m-Tn*3*(EIE), which confers tetracycline resistance, all m-Tn*3*s shown carry the β-lactamase gene (bla). (a) m-Tn*3*; (b) m-Tn*3(HIS)*; (c) m-Tn*3(HIS lac)*; (d) m-Tn*3(LEU2)*; (e) m-Tn*3(LEU2 lac)*; (f) m-Tn*3(TRP1)*; (g) m-Tn*3(TRP1 lac)*; (h) m-Tn*3(URA3)*; (i) m-Tn*3(URA3 lac)*; (j) m-Tn*3(URA3/E* site); (k) m-Tn*3(URAST)* (l) m-Tn*3*(flying promoter); (m) m-Tn*3*(EIE). Restriction sites: P, *Pst*I; RI, *Eco*RI; H, *Hin*dIII; S, *Sal*I; B, *Bam*HI; Su, *Sau*3AI; X, *Xho*I; Sf, *Sfi*I; N, *Not*I; and T3, T3 polymerase promoter.

two subsequent matings (Fig. 1), and transfer of one disrupted target plasmid to another host prior to analysis. As such, it does not require the extensive time and effort commitment demanded by *in vitro* manipulations.

Figure 2 shows the physical and genetic maps of four Tn*3*-free vectors compatible with the SM system. These vectors all contain the kanamycin resistance determinant from Tn5 and various multicloning sites (MCS). Other vectors compatible with the SM system are listed in Table II. These vectors include pHSS9–pHSS13 and pHSS19–pHSS21. These latter vectors are all based on pHSS6 or pHSS8 (Fig. 2). Plasmid pHSS9 contains the M13 phage origin or replication inserted in the SmaI site of pHSS6 that is located outside of the MCS. Plasmids pHSS11 and pHSS13 are derivatives of pHSS6 and pHSS8, respectively, and lack the BglII site within the kanamycin resistance gene. Vector pHSS12 is a derivative of pHSS11 that also lacks the SmaI site in the kanamycin resistance gene. Vector pHSS19 is pHSS12 with the MCS from pUC19 replacing the pUC9 MCS. Vectors pHSS20 and pHSS21 are pHSS19 with the SmaI site in the MCS converted to a BglII or XhoI site, respectively. Plasmids pHSS8, pHSS11, pHSS12, pHSS19–pHSS21, and pMFH18R-BstXI all afford α complementation blue/white insert screening on X-Gal-containing media.

Shuttle mutagenesis minitransposons appropriate for use in transformation experiments with S. cerevisiae are shown in Fig. 3b–m. All transposons are derivatives of the prototype β-lactamase element m-Tn3 (Fig. 3a) and contain either the HIS3, LEU2, TRP1, or URA3 genes. Derivatives of these transposons for making translational fusions between yeast genes and E. coli β-galactosidase are also shown in Fig. 3. These elements contain a truncated β-galactosidase gene from pMC1871[17] in frame with the one open reading frame of the 38-bp terminal repeat. It should be noted that the final recipient strain for the SM procedure, NS2114Sm, is lac+. Consequently, the direct identification of fusions on indicator media with this recipient is not feasible. To overcome this problem, we routinely prepare quick-preparation plasmid DNA from NS2114Sm conductants and transform a lac− strain such as DH5α to kanamycin resistance and screen for a lac+ phenotype on X-Gal-containing media.

Other minitransposons shown in Fig. 3 include m-Tn3(URA3/E site) (Fig. 3j), and m-Tn3(URAST) (Fig. 3k), m-Tn3(flying promoter) (Fig. 3l), and m-Tn3(EIE) (Fig. 3m). Minitransposon URA3/E site contains the E site region from HMRa[18] that has been implicated in transcriptional control.[19] This element has been successfully employed to silence the ADE1 gene in a SIR4-dependent manner.[20]

The URAST element contains, in addition to the URA3 gene, an SfiI recognition site and a T3 polymerase promoter. The SfiI site can be used to

---

[17] M. J. Casadaban, A. Martinez-Arias, S. K. Shapira, and J. Chou, this series, Vol. 100, p. 293.
[18] J. Abraham, K. A. Naysmyth, J. A. Strathern, A. J. S. Klar, and J. B. Hicks, J. Mol. Biol. **176**, 307 (1984).
[19] A. H. Brand, L. Breeden, J. Abraham, R. Sternglanz, and K. A. Nasmyth, Cell (Cambridge, Mass.) **41**, 41 (1985).
[20] C. Davies, M. F. Hoekstra, F. Heffron, and B. Errede, J. Cell Sci. (Suppl. 13E) (1989).

physically map the location of insertions relative to known and mapped *Sfi*I and *Not*I sites in the yeast genome, while the T3 promoter can be used for the production of riboprobes through regions into which the element has hopped.

A different kind of minitransposon that contains a selectable marker *(LEU2)*, an origin of replication *(ARS2)*, and a centromere sequence *(CEN4)* [m-Tn*3(LEU2/ARS/CEN)*; not shown] can be used to introduce into yeast a stable, self-replicating plasmid molecule, as a result of transposition insertion, rather than a transplaced mutation. This element can be used for the reintroduction of mutagenized inserts of large yeast subclones. For example, when delineating the size of a gene, insertion and deletion analysis based on complementation is often performed. By using a self-replicating insertion, transplacement is not required.

The so-called flying promoter transposon (Fig. 3l) is a derivative of m-Tn*3(LEU2/ARS/CEN)* and contains the *GAL10* inducible promoter reading out one end of the element. The flying promoter construct was designed to provide inducible expression of cloned yeast genes. This minitransposon has been used to drive the expression of a *RAD3::lacZ* fusion construct in a galactose-dependent fashion (M. F. Hoekstra and F. Heffron, unpublished observation). Another version of the flying promoter (not shown) has been constructed that contains an ATG initiation codon adjacent to the *GAL10* promoter and in frame with the open reading frame that spans the Tn*3* terminal repeat (flying start). This element could be used to express regions into which the flying promoter has inserted.

The epitope insertion element (EIE; Fig. 3m) allows insertion of a small, defined DNA segment throughout a gene after transposition and resolution. This 135-bp insertion results from Cre-mediated deletion of the bulk of the element and consists of two 38-bp ends of Tn*3*, a *Not*I site, and a *loxP* site. In-frame translation of this small DNA segment with the rest of the gene results in the possible production of an epitope that could serve as an antigenic tag for the target protein. The minitransposon contains a *lacZ* translational fusion gene in frame with an open reading frame through a Tn*3* 38-bp end and a *loxP* site to select insertions that are in frame with the target coding sequence. We have isolated in-frame EIE inserts into the yeast *HO* gene by examining mutagenized isolates for *lacZ* expression in *E. coli*. Many of these unresolved inserts were found to express β-galactosidase in yeast as well (M. F. Hoekstra, D. G. Burbee, J. Singer, E. Chiaou, E. Mall, and F. Heffron; submitted, 1990). The choice of yeast or *E. coli* for screening EIE fusions for β-galactosidase activity will depend on whether a specific yeast promoter is active in *E. coli*.

The EIE element also contains *URA3* and *tet*^{R}, which have both positive and negative selections in *S. cerevisiae* and *E. coli*, respectively. Immediately inside each end of the EIE are directly repeated *lox*P site-specific recombination sites from phage *P1*. The *lacZ, URA3*, and *tet*[R] genes can be deleted by site-specific recombination catalyzed by the *cre* recombinase. The Cre enzyme functions both in *E. coli* and yeast,[21] and for use with the EIE system in yeast we have constructed a derivative of pBS49[21] that contains a *GAL1*::Cre fusion on a *LEU2/ARS1/CEN4* vector. The presence of both the unresolved EIE (Ura$^+$) and *GAL1*::Cre (Leu$^+$) can thus be selected in yeast. Addition of galactose to yeast strains containing this *GAL1*::Cre fusion construct results in the functional expression of the recombinase, and, after Cre-mediated recombination, two 38-bp ends of Tn*3* and a *loxP* site are left as resolution products.

The Cre protein stimulates precise deletion of the EIE between the *loxP* sites and results in the loss of *URA3, tet*[R], and *lacZ*. The deletion event is selected by plating yeast cells onto 5-fluoroorotic acid-containing medium. The recombination leaves a 135-bp insert that can be detected by examining for a *Not*I site that is part of the resolution product. We have found that homologous intrachromosomal recombination can also generate the appropriate deletion but recommend the use of *GAL1*::Cre for selection of precise resolved EIE inserts. Potentially, EIE insertions can be used to map domains in a multifunctional protein or for strain construction through the introduction of multiple mutations in different genes by sequential transformation and resolution of EIE insertions.

In addition to their utility as mutagens for the rapid production of insertions into cloned yeast genes, these shuttle mutagenesis elements can be used directly as mobile priming sites for DNA sequence analysis. Oligonucleotide primers homologous to sequences just within the 38-bp repeats (5' CTC ATG ACC AAA ATC CC 3' and 5' GGA TTC CCC TTA ACG 3') can be used to sequence across the left and right ends of the minitransposons. This allows the determination of at least 400 bp of DNA sequence surrounding each minitransposon insertion. Consequently, one can determine the specific location of an insertion in a known sequence or, by this approach, overcome the requirement of subcloning or constructing nested deletions when determining a new sequence. This approach has been used to determine the site of insertion in the gonococcal pilin gene (H. S. Seifert, M. Vito, R. Ajioka, F. Heffron, and M. So; submitted, 1989) as well as to determine the sequence of a yeast DNA repair function (M. F. Hoekstra, E. Chiaou, F. Heffron, and R. M. Liskay, unpublished observation).

In summary, shuttle mutagenesis has been designed to efficiently and rapidly produce a large number of insertions into cloned genes with a minimum number of manipulations. Insertions can be transplaced into yeast to produce chromosomal gene replacements. Transposons can be

---

[21] B. Sauer, *Mol. Cell. Biol.* **7**, 2087 (1987).

engineered for many experimental needs. For example, in combination with yeast gene libraries constructed in Tn3-free vectors, random insertions can be scattered throughout the genome[5] and used to screen for specific phenotypes linked to the insertion mutations. Transposons like the flying promoter could also be used in this type of gene bank mutagenesis approach to screen for inducible misexpression or overexpression phenotypes.

In summary, shuttle mutagenesis systems can facilitate the mapping and sequencing of cloned genes, and they offer many opportunities for rapid and large-scale genetic manipulations. In addition to its applicability for *S. cerevisiae,* the shuttle mutagenesis system is amenable to any organism with a DNA-mediated transformation system. With the increasing interest in fungal systems other than bakers' yeast, generally applicable mutagenesis systems based on *E. coli* transposons should remain central to rapid experimentation.

Acknowledgments

We are deeply indebted to Chris Davies for discussion on *E* site regulation of *ADE1* and to Mike Liskay, Maja Vito, Rich Ajioka, and Maggie So for communication of results and helpful discussions. Portions of this work were supported by a grant from the Lucille P. Markey Charitable Trust (M.F.H.), by National Institutes of Health Grant AI20978 (F.H.), National Science Foundation Grant DMB-8217002 (F.H.), and a Damon Runyon–Walter Winchell Cancer Fund Fellowship DGR-847 (H.S.S.). M.F.H. is a Lucille P. Markey Scholar in Biomedical Sciences.

[23] Ty Mutagenesis in *Saccharomyces cerevisiae*

By DAVID J. GARFINKEL and JEFFREY N. STRATHERN

It has been shown in such diverse organisms as bacteria, maize, and *Drosophila* that insertional mutagenesis mediated by transposable elements is an effective means of tagging genes. Using the native yeast transposon Ty, we have developed a method for transposon mutagenesis in *Saccharomyces cerevisiae.* This chapter covers the basic protocol used to generate high levels of Ty transposition in yeast (referred to as transposition induction) and ways to identify relevant Ty-induced mutations. Further information on the biology of Ty elements and their properties as insertional mutagens can be found in a recent review.[1]

[1] J. D. Boeke, *in* "Mobile DNA" (D. E. Berg and M. M. Howe, eds.), p. 335. American Society for Microbiology, Washington, D.C., 1989.

Induced Ty Mutagenesis with Marked Ty Elements

Why Use Ty Mutagenesis?

The advent of genetically marked Ty elements that transpose at high frequency has made Ty mutagenesis a useful tool in yeast genetics.[2] Ty mutagenesis is useful because it physically and genetically tags a chromosomal site with a unique sequence. This insertion tag offers the following practical advantages: (1) Mutations made by insertion of a marked Ty element into a gene permit the rapid cloning of that gene into *Escherichia coli* without any intervening steps. (2) Once a gene or locus is tagged with a marked Ty, it can be rapidly mapped to a chromosome or chromosome arm by hybridization to chromosomes separated by electrophoresis. (3) Mutations caused by Ty elements have useful phenotypes. Insertions within the coding region are usually stable to reversion and are usually null alleles. Ty element insertions in the promoter region of yeast genes have varied and interesting effects on gene expression. These mutations often result in altered gene regulation and increased expression of the adjacent cellular gene. Thus, Ty insertions can result in classes of mutants not readily obtained using mutagens that cause point mutations

Marked Ty Elements and pGTy Plasmids

When a genetically marked Ty1 element (Ty1-H3) is fused to the controllable yeast *GAL1* promoter on a high-copy plasmid (pGTy1-H3), addition of galactose induces high levels of transposition of both the galactose-regulated Ty1-H3 element and genomic Ty elements.[2] When cells are propagated on glucose-containing media, the Ty transposition system is repressed. Ty elements can be genetically and physically marked with foreign sequences without severely lowering the transposition level. Foreign markers ranging in size from a 40-base pair (bp) *lacO* fragment to selectable genes of at least 2.7 kilobases (kb) have been used successfully.[2,3] The ability to regulate high levels of Ty transposition and to genetically tag a particular element to distinguish its movement from that of other elements forms the basis of the Ty mutagenesis system. Because the pGTy vector is a plasmid that can be lost by segregation, cells which have experienced a marked Ty transposition can be identified as those which have lost the marker carried on the backbone of the plasmid but which

---

[2] J. D. Boeke, D. J. Garfinkel, C. A. Styles, and G. R. Fink, *Cell (Cambridge, Mass.)* **40,** 491 (1985).
[3] E. Jacobs, M. Dewerchin, and J. D. Boeke, *Gene* **67,** 259 (1988).

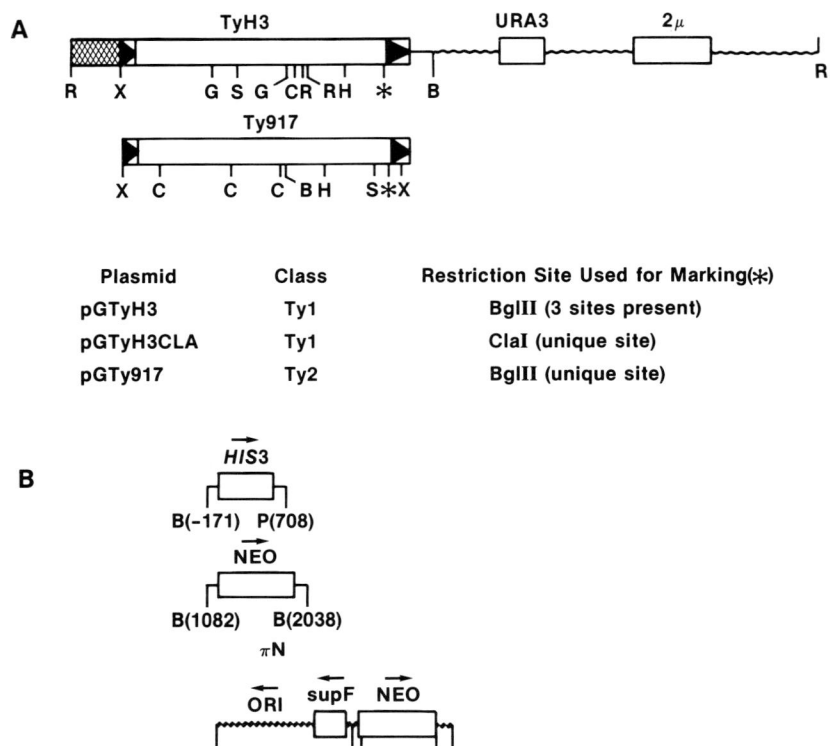

FIG. 1. Restriction maps of the pGTy1-H3 and pGTy2-917 plasmids (A) and marker genes (B) (maps not drawn to scale) and features useful for cloning marked Ty2-917πN insertions. Boxed segments represent yeast or bacterial genes; arrows indicate the direction of transcription. The marker genes and the Ty elements are drawn in the same transcriptional orientation. The wavy line represents pBR322 sequences. Restriction sites are abbreviated as follows (not all restriction sites are shown): B, BamHI; C, ClaI; G, BglII; H, HindIII; P, PstI; R, EcoRI; S, SalI; Sm, SmaI; X, XhoI; Xb, XbaI. (A) Organization of pGTy1-H3 and pGTy2-917. The cross-hatched box represents the GAL1 promoter sequence. The boxed arrows are the Ty long terminal repeat (LTR) sequences. In the pGTy plasmids, either Ty1-H3 or Ty2-917 is fused to the GAL1 promoter at a XhoI site within the 5′ LTR. The Ty1-H3 segment is derived from a Ty transposition into plasmid pNN162. The Ty1-H3CLA segment is derived from pGTy1-H3 using oligonucleotide-directed mutagenesis. The Ty2-917 segment is derived from plasmid phis917 (kindly provided by G. Fink). The segment containing the 2-μm plasmid origin is shown. The asterisk represents the BglII restriction site just inside the 3′ LTR where the marker genes are inserted. (B) Marker genes used to construct the plasmids pGTy1-H3HIS3, pGTy1-H3NEO, and pGTy2-917πN. The numbers in parentheses indicate the nucleotide position in the HIS3 gene, or in the Tn903 element for the NEO gene. The plasmid πN contains the Tn903 NEO gene. The arrows above the HIS3, NEO, and supF genes indicate the direction of transcription. The arrow above the origin of replication (ORI) in the plasmid πN indicates the direction of DNA replication. We have cloned insertions

retain the marker carried within the Ty element. Our experience suggests that most such cells have had a novel transposition of the marked Ty into a chromosomal site. In the absence of the plasmid, the marked transpositions can be followed in subsequent genetic and molecular analyses.

Several marked Ty1 and Ty2 elements have been shown to be transposition competent when expressed on a pGTy plasmid.[4] We created a derivative of pGTy1-H3 that makes it easier to insert a variety of foreign sequences (Fig. 1).[5] Since the plasmid pGTy1-H3 contains sites for most common restriction endonucleases, including three *Bgl*II sites, inserting a polylinker in the permissive *Bgl*II site at nucleotide 5561 is of little value. Instead, we have constructed the plasmid pGTy1-H3CLA, which contains a unique *Cla*I restriction site placed at the *Bgl*II site (at nucleotide 5561). A *Cla*I site is used because we could remove the only *Cla*I site in the plasmid without changing the Ty protein sequence and because a *Cla*I adaptor plasmid pCLA12 is available as an intermediate vector.[6] This plasmid is useful because it contains the polylinker array from pUC12 flanked by *Cla*I sites. Virtually any segment of DNA can be converted to a fragment with *Cla*I ends and then inserted into the *Cla*I site of pGTy1-H3CLA.

The second pGTy plasmid contains the Ty2-917 element. This element is the causative agent of the *his4-917* promoter mutation.[7] We chose Ty2-917 as an additional mutagen for several reasons. Ty2-917 belongs to the Ty2 structural class of Ty elements, whereas Ty1-H3 is a Ty1 element. Although Ty1 and Ty2 elements are clearly related, they persist as separate retrotransposon families in yeast. It is possible that Ty1-H3 and Ty2-917 have different functional properties, such as different insertion site specificities or different mutagenic effects on target genes.[8] Furthermore, the

[4] M. J. Curcio, N. J. Sanders, and D. J. Garfinkel, *Mol. Cell. Biol.* **8**, 3571 (1988).
[5] D. J. Garfinkel, M. F. Mastrangelo, N. J. Sanders, B. K. Shafer, and J. N. Strathern, *Genetics* **120**, 95 (1988).
[6] S. H. Hughes, J. J. Greenhouse, C. J. Petropoulos, and P. Sutrave, *J. Virol.* **61**, 3004 (1987).
[7] G. S. Roeder, P. J. Farabaugh, D. T. Chaleff, and G. R. Fink, *Science* **209**, 1375 (1980).
[8] G. S. Roeder, A. B. Rose, and R. E. Perlman, *Proc. Natl. Acad. Sci. U.S.A.* **82**, 5428 (1985).

---

from strains containing either the Ty917πND or the Ty917πNI transpositions using *Pst*I or *Sac*I. *Sac*I cleaves once at position 4850, allowing one to isolate the 3′ flanking sequence (with respect to the direction of Ty transcription) regardless of the πN orientation. The only *Pst*I site in Ty2-917πN is located immediately adjacent to one of the *Bgl*II sites that bracket the πN miniplasmid. Depending on the orientation of the πN plasmid within Ty917, yeast sequences on either side of the integration site have been recovered in *E. coli*. Ty2-917πND transpositions yield πN plasmids containing almost all of the marked Ty and genomic sequences 5′ to the transposition, and Ty2-917πNI transpositions yield plasmids containing the 3′ Ty LTR and 3′ flanking sequences. Alternatively, enzymes that do not cleave within Ty2-917πN can be used for cloning. These include *Bsp*MII, *Mlu*I, *Nco*I, *Nsi*I, *Pvu*I, and *Stu*I.

pGTy2-917 plasmid is easy to tag with foreign sequences because it contains a single BglII site in the correct position. pGTy plasmids that are easy to manipulate can also simplify the construction of sophisticated Ty element vectors, which stably amplify and express genes in yeast.[3]

Two criteria are important in designing marker genes for Ty elements. First, as expected for an element that transposes through an RNA intermediate, sequences that serve as transcriptional terminators inhibit transposition of the element (P. Rogan, unpublished results, 1989). Therefore, the 5' and 3' untranslated sequence of the marker gene should be trimmed as much as possible without destroying gene function. Second, Ty transposition is sensitive to extensive inverted or direct repeats.[9] Although no marked transpositions were obtained with a Ty element carrying a neomycin phosphotransferase gene *(NEO)* flanked by several hundred bases of inverted repeat,[9] removal of the inverted repeat restores Ty*NEO* transposition.[10] Ty elements containing markers with direct repeats lose one copy of the direct repeat at a high frequency during transposition.[11] If another gene is placed between the direct repeats, this segment as well as one copy of the repeat is lost.

We have marked Ty elements with a truncated yeast *HIS3* gene (pGTy1-H3*HIS3*) or the *E. coli* miniplasmid πN (pGTy2-917πND and pGTy2-917πNI carry the miniplasmid in either orientation) (Fig. 1). We also use a marked version of pGTy1-H3 or pGTy2-917 containing the *NEO* gene. In addition, Ty1-H3 has also been successfully tagged with *TRP1*.[9] The yeast *HIS3* or *TRP1* genes present in Ty1-H3 allow direct selection in *his3* or *trp1* yeast strains and *hisB* or *trpC E. coli* strains. The *NEO* gene confers dominant resistance to the antibiotic G418 in yeast[12] and to neomycin and kanamycin in *E. coli*. The Ty2-917πN element can be used to recover any Ty917-induced mutation directly, since it contains sequences required for selection and replication in *E. coli* (see below).

The majority of the pGTy plasmids have derived from the *URA3*-based, *GAL1* expression vector pCGE329 (kindly provided by J. Schaum and J. Mao). The functional definition of a cell that has had a marked transposition is that retention of the marker carried by the Ty is not dependent on retention of the plasmid. The *URA3* gene on the plasmid backbone facilitates the identification of cells that have lost the plasmid because the Ura⁻ phenotype can be selected as resistance to 5-fluoroorotic acid (5-FOA).[13] Other highly expressed promoters, such as *ADH2* and

[9] J. D. Boeke, H. Xu, and G. R. Fink, *Science* **239**, 280 (1988).
[10] C. M. Joyce and N. D. F. Grindley, *J. Bacteriol.* **158**, 636 (1984).
[11] H. Xu and J. D. Boeke, *Proc. Natl. Acad. Sci. U.S.A.* **84**, 8553 (1987).
[12] A. Jimenez and J. Davies, *Nature (London)* **287**, 869 (1980).
[13] J. D. Boeke, F. Lacroute, and G. R. Fink, *Mol. Gen. Genet.* **197**, 345 (1984).

PGK1, have been used to drive Ty element expression,[14,15] but transcription cannot be regulated to the same degree as with *GAL1*.

Strain Requirements

The major requirements of any strain used for transposition induction is that it be Gal$^+$ and contain the relevant mutant alleles for plasmid selection and tracking the Ty marker. We use two tests for determining the strength of galactose induction. The first involves growing the strains on galactose indicator plates containing the dye bromthymol blue. This medium contains 1% (w/v) yeast extract, 2% (w/v) peptone, 2% (w/v) agar, 2% (w/v) galactose, and bromthymol blue at a final concentration of 80 μg/ml. Galactose and bromthymol blue [prepared as stock solutions of 20% (w/v) and 4 mg/ml, respectively, and filter sterilized] are added to the medium after autoclaving. Plates are stored in the dark. We obtain the best results if relatively few cells are inoculated onto the plates. Strong Gal$^+$ strains turn the agar yellow. Alternatively, minimal SD medium (see below) supplemented with the required nutrients, 2% galactose, and 5 μg of ethidium bromide per milliliter of agar can be used to test strains. Aqueous stock solutions of ethidium bromide (1–10 mg/ml) are filter sterilized and stored in the dark at 4°. Gal$^+$ cells grow on this medium, whereas Gal$^-$ cells do not grow, or grow weakly. The second assay is to compare the level of β-galactosidase produced in various strains containing a p*GAL1-lacZ* expression plasmid pCGS286 (kindly provided by J. Schaum and J. Mao). Standard biochemical methods are used to detect β-galactosidase activity in yeast. Our strongest *GAL* strains yield 10,000–20,000 units of activity.

The chromosomal markers used for plasmid selection and tracking Ty transposition should be nonreverting. In the case of the Ty markers, it is preferable if the chromosomal locus is completely deleted. Most of our mutagenesis studies have been done with Ty1-H3*HIS3*. For this marker, there are two chromosomal *his3* deletions in common use. One is the *his3Δ1* allele, which is an internal deletion, and the other is *his3Δ200*, which is a complete deletion of the gene.[16] *his3Δ200* is especially useful because the chromosomal deletion completely covers the functional *HIS3* segment present in Ty1-H3. Use of the *his3Δ200* deletion eliminates the possibility of homologous recombination events between the plasmid and chromosomal *HIS3* loci. Another marker combination is the yeast *TRP1*

---

[14] J. Mellor, M. H. Malim, K. Gull, M. F. Tuite, S. M. McCready, T. Finnsysesn, S. M. Kingsman, and A. J. Kingsman, *Nature (London)* **318**, 583 (1985).

[15] F. Muller, K.-H. Bruhl, K. Freidel, K. V. Kowallik, and M. Ciriacy, *Mol. Gen. Genet.* **207**, 421 (1987).

[16] K. Struhl, *Nucleic Acids Res.* **13**, 8587 (1985).

gene and a complete chromosomal gene deletion *trp1Δ1*. There is one cautionary note about using *trp1Δ1*. *TRP1* is tightly linked to *GAL3* on chromosome IV. *gal3* mutants have a 72- to 96-hr induction lag, whereas *GAL3* strains take only a few minutes to adapt to galactose. It is likely that *trp1Δ1* deletes most of the *GAL3* promoter and therefore influences galactose induction.[17] The presence of *trp1Δ1* does not severely reduce the level of marked transpositions if strains are grown on galactose for at least 5 days (see below).

A dominant marker that has no specific strain requirement is the Tn*903 NEO* gene.[12] Presently, resistance to the antibiotic G418 can only be scored on YEPD plates.[18] We use the standard recipe for YEPD (per liter, it is composed of 10 g yeast extract powder, 20 g peptone, 20 g dextrose, and 20 g agar) and add filter-sterilized G418 just before pouring the plates. Stock solutions of G418 are made at 25–50 mg/ml and stored at $-20°$ (concentrations refer to the active drug concentration as given by the vendor). The plates are very stable if stored at 4°. The levels of endogenous resistance to G418 vary among yeast strains. Most of our strains are sensitive to G418 concentrations in the range of 50–100 μg/ml, but each strain should be tested individually. Occasionally, a strain has a high endogenous resistance to several hundred micrograms of G418 per milliliter. Strains containing Ty*NEO* transpositions are routinely scored at 100–400 μg/ml of G418. In crosses, the high and low endogenous resistance phenotype segregates as a multigenic trait. Ty*NEO* transpositions can be followed in crosses if colonies are replica plated to YEPD containing different G418 concentrations.

Transposition induction occurs in haploid or diploid cells (D. Garfinkel, unpublished results, 1987). The presence of the *GAL1* promoter apparently overrides normal repression of Ty transcription in **a**/α diploids (reviewed by J. Boeke).[1] This feature should prove useful in searches for Ty-induced mutations that are lethal in haploid cells (see below).

Assaying Ty Transposition

The first step in the transposition assay is to introduce a pGTy plasmid into a suitable strain. For simplicity, we discuss Ty elements marked with *HIS3* or *NEO* on a *URA3*-based pGTy plasmid. We introduce the plasmid into yeast cells by transformation (refer to [25] in this volume for proto-

---

[17] W. Bajwa, T. E. Torchia, and J. E. Hopper, *Mol. Cell. Biol.* **8**, 3439 (1988).
[18] T. D. Webster and R. C. Dickson, *Gene* **26**, 243 (1983).

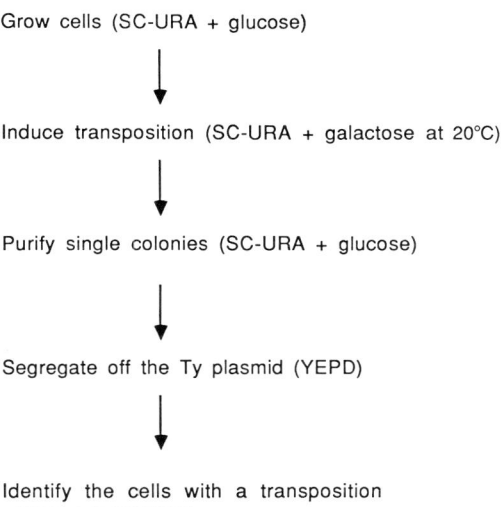

FIG. 2. Flow sheet for transposition of a marked Ty element into a chromosome.

cols), but other techniques (electroporation, bacterial conjugation) can also be used.

Transformant colonies should be ready to pick after 2–3 days of incubation at 30° (refer to Fig. 2 for a flow sheet of this procedure). Six transformants are purified by streaking for single colonies on SC − Ura (refer to [1], this volume, for composition of media). Single colonies are then streaked to SC − Ura, and replica plated to SC − His or YEPD (G418) to check for the presence of the *HIS3* or *NEO* gene on the pGTy plasmid. Only Ura$^+$, His$^+$, or G418^R transformants are picked for further study. For the purposes of mutagenesis, we recommend testing several transformants for transposition.

Optimum transposition levels are obtained by streaking cells from a fresh patch grown on SC − Ura to SC − Ura plus galactose solid medium. SC − Ura (galactose) is made the same way as SC − Ura except galactose is substituted for glucose to a final concentration of 2% (100 ml of a 20% filter-sterilized stock solution per liter of medium). Streak for single colonies on SC − Ura (galactose). Incubate the plates at 20° for 5 days, as Paquin and Williamson found that the rate of spontaneous Ty transpositions is 100-fold higher at 15–20° than at 30°,[19] which is the optimum growth temperature for *S. cerevisiae*. The efficiency of Ty1-H3*NEO* trans-

[19] C. E. Paquin and V. M. Williamson, *Science* **226**, 53 (1984).

position using the pGTy system drops about 5-fold at 30°, and is practically undetectable at 37° (D. Garfinkel, unpublished results, 1990).

At this point, most of the single colonies that form are relatively small, although some larger colonies also appear. It is our experience that cells taken from large colonies as well as cells taken from confluent areas of growth have fewer marked transpositions. Small colonies are streaked for single colonies on SC − Ura containing glucose to abolish transcription from the *GAL1* promoter and to select cells that retain pGTy1-H3*HIS3* for the entire galactose induction. Colonies are then grown on nonselective YEPD plates to allow segregation of the plasmid. Ura⁻ segregants are identified by replica plating to SC − Ura or to medium containing 5-FOA.

As mentioned above, 5-FOA can be used as a positive selection for cells that have lost a functional *URA3* gene. In this application, hundreds of independent segregations can be done with minimal effort on a few plates. SC + 5-FOA plates are prepared the same way as SC − Ura plates with the following modifications. Solid uracil (50 mg/liter of medium) and 5-FOA (800 mg) is added to the 2× SC − Ura solution, allowed to dissolve, and filter sterilized. Less 5-FOA may be used; however, this results in more background growth of Ura⁺ cells. Alternatively, supplemented SD plates can be used for 5-FOA selection. 5-Fluoroorotic acid is very expensive, so yeast researchers periodically organize special purchases of this compound through the Genetics Society of America [GSA Administrative Office, 9650 Rockville Pike, Bethesda, MD 20814; (301)-571-1825]. The last purchase was from PCR Inc. (Gainesville, FL).

Transpositions are scored after 1 day at 30° if *HIS3* is the Ty marker, after 2 days if *NEO* is the marker. The transposition efficiency is defined as the total number of His⁺ or G418^R, Ura⁻ segregants divided by the total number of colonies analyzed. This assay gives an underestimate of the transposition frequency in that some of the cells have multiple insertions of the marked element. In our hands, at least half of the cells that retain the pGTy plasmid (Ura⁺) for the entire induction period have at least one marked Ty transposition. Representative transposition efficiencies for several elements are presented in Table I.

Transposition also occurs at about the same efficiency in liquid cultures containing SC − Ura (galactose). Marked transpositions are detected earlier if the cells are pregrown in a nonrepressing carbon source such as glycerol (3% final concentration) or raffinose (2% final concentration), and then diluted about 20- to 40-fold into SC − Ura (galactose) liquid medium. Cells are plated for single colonies on SC − Ura and processed as described above. Under these conditions, maximum transposition frequencies are observed after about 16 hr of incubation with aeration at 20°.

TABLE I
TRANSPOSITION EFFICIENCY OF MARKED Ty ELEMENTS

Ty element[a]	Background[b]	Marker length (nucleotides)	Transposition efficiency (%)[c]	Copies per genome[d]
Ty1-H3*HIS3*	*SPT3*	750	26/52 (50)	1.9 (28/15)
Ty1-H3*NEO*	*SPT3*	956	40/45 (89)	2.7 (78/29)
Ty1-H3*NEO*	*spt3-101*	956	39/47 (83)	1.5 (39/26)
Ty2-917*NEO*	*SPT3*	956	46/48 (96)	6.8 (314/46)
Ty2-917πND	*SPT3*	1830	42/38 (88)	2.3 (16/7)
Ty2-917πNI	*SPT3*	1830	40/46 (87)	1.0 (4/4)
Ty2-917*NEO*	*spt3-101*	956	45/45 (100)	ND

[a] Transcription of the Ty element and the marker gene is in the same direction for all of the marked elements except Ty2-917πNI. In Ty2-917πNI, Ty2-917 and *NEO* are transcribed in opposite directions.

[b] Isogenic *spt3-101* derivatives of *SPT3* strains are constructed by transplacement using the *spt3-101* integrating plasmid pFW33 (kindly provided by F. Winston). *spt3-101* is a frameshift mutation in the *SPT3* gene [F. Winston, K. J. Durbin, and G. R. Fink, *Cell (Cambridge, Mass.)* **39**, 675 (1984)].

[c] Transposition efficiency is the number of G418^R or His$^+$, Ura$^-$ segregants divided by the total number of Ura$^-$ segregants analyzed.

[d] Measured as the number of bands hybridizing with an appropriate probe on a genomic Southern blot. The fraction in parentheses is the total number of bands counted, divided by the total number of His$^+$ or G418^R, Ura$^-$ colonies analyzed. ND, Not determined.

Experimental Approach

We present a general scheme to generate transposition-induced mutations using Ty1-H3*HIS3* in a *his3Δ200* background. Our standard conditions optimize the total number of marked transpositions per cell during one cycle of growth on SC − Ura (galactose) (Table I). However, there are many variables that can be manipulated to optimize transposition induction for a particular mutant hunt. These include temperature, length of induction, type of pGTy plasmid, cell ploidy, and other genotypic requirements. To obtain even more marked transpositions per cell, induction plates can be replica plated to fresh SC − Ura (galactose) plates. Three rounds of induction results in 10–20 marked transpositions per cell (R. Nash and B. Futcher, personal communication, 1989).[9] However, multiple genetic crosses are then required to isolate the relevant marked insertion (see below).

Mutations can be caused by unmarked Ty elements and by other spontaneous events as well as by Ty1-H3*HIS3*. The background of unwanted mutations depends on several factors, including the length of galactose induction, the particular screen or selection, and whether cells without a marked transposition are still present in the population. In the limited number of tests we have done, it appears that the highest fraction of Ty-induced mutations occurs in selections for gene activation.

After transposition induction the cells can be handled in different ways. In instances where a selection exists for a particular mutant class, the induction plates can be replica plated directly and the resulting mutant colonies can be analyzed for marked transpositions. Any His$^+$, Ura$^-$ colony has a potential Ty1-H3*HIS3*-induced mutation in the target gene.

An alternative is to first enrich for cells containing at least one Ty1-H3*HIS3* insertion and then perform the specific mutant selection or screen. We have been able to collect thousands of Ty1-H3*HIS3* insertions by making use of the powerful 5-FOA selection and the instability of the pGTy plasmid under transposition-inducing conditions. The *URA3*-based pGTy plasmids are reasonably stable when cells are propagated on glucose. However, during galactose induction many cells lose the plasmid despite selection for plasmid maintenance. About 10% of the cells retain the pGTy plasmid after standard transposition induction on plates. The transposition efficiency in these cells is relatively high (Table I). In contrast, among those cells that have lost the plasmid during the induction period, about 5–25% have at least one Ty1-H3*HIS3* transposition (D. Garfinkel, unpublished results, 1988). Only cells containing a Ty1-H3*HIS3* transposition and lacking the plasmid are able to grow on SC + 5-FOA plates that also lack histidine (5-FOA − HIS). 5-FOA − HIS medium is prepared the same way as SC + 5-FOA, except a nutrient mix is used that does not contain histidine. Cells from the galactose-induction plates can be directly replica plated to 5-FOA − HIS, or cells taken from the galactose plates can be suspended in water, diluted, and plated on 5-FOA − HIS. We have also constructed transposition libraries by making permanent stocks [15% glycerol (v/v) stored at −70°] from the 5-FOA − HIS cultures.

The background level of His$^+$ cells is very low and presumably represents other recombinational events. *GAL1*-promoted Ty1-H3*HIS3* expression results in more than a 1,000-fold increase in colony formation on 5-FOA − HIS plates compared with controls that are not grown in galactose or contain mutations in *TYB* (M. J. Curcio and D. Garfinkel, unpublished results, 1989).

Genetic Characterization of Mutants

A useful first step in the characterization of mutants is to cross putative Ty1-H3*HIS3*-induced mutants with an appropriate strain, then perform tetrad or random spore analysis (refer to [1]–[3] in this volume). Examples of this are presented below. The cross indicates if the mutant contains multiple marked transpositions and whether the mutant phenotype is linked to a Ty1-H3*HIS3* insertion. Like other mutagenesis techniques, Ty-induced mutants should be backcrossed several times to minimize the possibility of carrying an unwanted event.

Standard complementation tests to establish allelism and dominance relationships can be done with Ty-induced mutants. However, the influence of cell type on some Ty-induced mutations (the ROAM phenotype) should be kept in mind.[20] These mutations result in constitutive gene expression in *MAT*a, α, a/a, or α/α cells, but they are repressed in a/α cells. For example, some mutations may appear recessive in a/α diploids and dominant in a/a or α/α diploids. Because the mutant phenotype may be suppressed or altered in a/α diploids, the mutations should be characterized in diploids with an a or α phenotype. There are several genetic tricks to accomplish this that can be easily incorporated into mutant analysis (refer to [8] in this volume). For example, if mutagenesis is done in an α strain, *MAT*a and *mat*a testers can be used in the complementation analysis. In the resulting diploids, ROAM mutants are recessive in the first case (*MAT*a/*MAT*α) and dominant in the second (*mat*a/*MAT*α) because a/α repression does not occur.

Examples of Ty Mutagenesis

Mutating Specific Targets: LYS2 and LYS5 Genes

To analyze Ty1-H3*HIS3*-induced mutations at a specific chromosomal target, strain DG662 [*MAT*α *his3Δ200 ura3-167 trp1Δ1 leu2Δ GAL* (pGTy1-H3*HIS3*)] was grown in the presence of galactose, the resulting colonies were replica plated to selective medium, and 150 L-α-aminoadipate-resistant colonies were picked for further analysis. This selection yields mostly *lys2* mutants (≥90%) and a few *lys5* mutants,[21] and it has been used successfully to recover Ty insertions at *LYS2* in normal cells (1–5% are Ty-induced mutations)[22,23] and in transposition-induced cells (30–40% are Ty-induced).[2] Of the 60 His⁺, L-α-aminoadipate-resistant

---

[20] B. Errede, T. S. Cardillo, F. Sherman, E. Dubois, J. Deschamps, and J. M. Wiame, *Cell (Cambridge, Mass.)* **22**, 427 (1980).
[21] B. B. Chattoo, F. Sherman, D. A. Azubalis, T. A. Fjellstedt, D. Mehvert, and M. Ogur, *Genetics* **93**, 51 (1979).
[22] H. Eibel and P. Philippsen, *Nature (London)* **307**, 386 (1984).
[23] G. Simchen, F. Winston, C. A. Styles, and G. R. Fink, *Proc. Natl. Acad. Sci. U.S.A.* **81**, 2431 (1984).

mutants, 52 fail to complement a *lys2* tester strain. DNA from 17 of the 52 mutants (33%) show alterations that are indicative of Ty insertions at *LYS2*. Ty1-H3*HIS3* transpositions cause 6 of 17 (35%) Ty-induced *lys2* mutants and 6 of 52 (11.5%) total *lys2* mutants; the rest are caused by unmarked Ty elements or other events. Of the 60 His$^+$, L-α-aminoadipate-resistant mutants we isolated, 8 fail to complement a *lys5* tester strain, and half of these mutants contain multiple Ty1-H3*HIS3* transpositions. In the absence of a *LYS5* hybridization probe, we have looked at the segregation of the Ty1-H3*HIS3* transpositions and the *lys5* mutation by tetrad analysis.

Each mutation caused by the marked Ty1-H3*HIS3* element should carry a functional *HIS3* gene genetically linked to the new mutation. Two types of strains are presented as examples: one in which the marked Ty is the only copy of Ty1-H3*HIS3* in the genome and another in which there are two copies of Ty1-H3*HIS3* in the genome. In three different Ty1-H3*HIS3*-induced mutants at *LYS2* (*lys2-941*, *-956*, and *-923*), in which the marked insertion is the only copy in the genome, *HIS3* segregates as a gene tightly linked to *lys2* (Table II). Multiple unlinked Ty1-H3*HIS3* transpositions should assort independently during meiosis. As a result, the ratio of His$^+$ and His$^-$ segregants should increase as the number of unlinked Ty1-H3*HIS3* transpositions increases. For example, if two unlinked copies of Ty1-H3*HIS3* are present in the genome, the ratio of His$^+$ and His$^-$ progeny should be 3:1; if three copies are present, the ratio should be 7:1. To test these predictions, we show the segregation pattern of a *lys2* mutant that appears to contain two Ty1-H3*HIS3* transpositions: one

TABLE II
TETRAD ANALYSIS OF SINGLE Ty1-H3*HIS3* TRANSPOSITIONS

Target gene	Gene pair	Ascus type[a]		
		PD	NPD	T
LYS2	*lys2-941/HIS3*	20	0	0
LYS2	*lys2-956/HIS3*	18	0	0
LYS2	*lys2-923/HIS3*	14	0	0
LYS5[b]	*lys5-973/HIS3*	18	0	0

[a] PD, Parental ditype; NPD, nonparental ditype; T, tetratype. Only tetrads with four viable spores were included. These asci showed 2:2 segregation for both markers.

[b] Represents the tetrads from two different His$^+$, Lys$^-$ segregants.

TABLE III
TETRAD ANALYSIS OF Ty1-H3*HIS3*-INDUCED *lys2* OR *lys5* MUTANTS THAT CONTAIN ADDITIONAL MARKED TRANSPOSITION

Target gene	Tetrads analyzed[a]	Spore phenotype				His$^+$: His$^-$ [b]
		His$^+$, Lys$^+$	His$^+$, Lys$^-$	His$^-$, Lys$^+$	His$^-$, Lys$^-$	
LYS2	13	12	26	14	0	38 : 14 (2.7 : 1)
LYS5	32	33	64	31	0	97 : 31 (3.1 : 1)

[a] Only tetrads with four viable spores are included.
[b] The total number of His$^+$ and His$^-$ spores present.

at *LYS2* and another elsewhere in the genome (Table III). In 13 tetrads, the ratio of His$^+$ and His$^-$ segregants approaches 3 : 1 (38 : 14), but there are no His$^-$, Lys$^-$ segregants present (Table III). These results suggest there are two unlinked Ty1-H3*HIS3* insertions in the genome, and one of these has mutated *LYS2*. When crossed with a suitable strain, seven of the eight *lys5* mutants are not marked by Ty1-H3*HIS3* because His$^-$, Lys$^-$ segregants appeared among the progeny. One cross shows a different segregation pattern (Table III). Even though two *HIS3* genes segregate, there is an association between the *lys5* mutation and one copy of the *HIS3* gene. No Lys$^-$, His$^-$ progeny appear in the cross. Subsequent analysis of a segregant that contains only the Ty1-H3*HIS3*-induced *lys5* mutation *(lys5-973)* confirms the linkage of the His$^+$ and Lys$^-$ phenotypes, as well as their 2 : 2 segregation (Table II). The ability to tag the *LYS5* gene with Ty1-H3*HIS3* demonstrates that we can identify a marked mutation by genetic analysis alone.

Other Target Genes

Spontaneous Ty1- and Ty2-induced mutations in at least 12 different genes have been described since Ty elements were discovered in 1979 (reviewed by J. Boeke).[1] Since the *GAL1*-promoted Ty transposition system was developed, Ty-induced mutations have been used to tag a variety of additional genes. A Ty1-H3*NEO* element has been used to tag the *WHI3* gene, which is involved in determining cell size (R. Nash and B. Futcher, personal communication, 1989). In the continuing search for new genes involved in mating, Ty1-H3*HIS3*-marked *ste* mutants have been isolated using the α-pheromone resistance selection (M. Mastrangelo, K. Weinstock, D. Garfinkel, and J. Strathern, unpublished results, 1988). Previously identified *ste* genes, such as the α-pheromone receptor *(STE2)*, are present in the collection. Interestingly, four **a**-specific sterile mutations are the result of Ty insertions in the repressed mating cassette $HML\alpha$ (see below). Ty1-H3*HIS3* has also been used to tag the *KLA1* gene [which is required for potassium uptake (C. Ko and R. Gaber, personal communication, 1989)], genes important for the response to heat shock (K. Kawagami, personal communication, 1988), and random auxotrophs (D. Garfinkel, unpublished results, 1988). Taken together, these results indicate that Ty transposition events occur at a variety of loci.

The screen for auxotrophs is useful because of the large number of genes that might be mutable. Ty1-H3*HIS3*-induced auxotrophs affecting several different pathways appear at a frequency of about 0.1% from a pool of 10,000 colonies containing at least one Ty1-H3*HIS3* insertion. These include mutants in adenine, arginine, inositol, leucine, and methionine/cysteine biosynthesis. If one assumes there are 100 mutable loci in the screen, the average target gene is 2.5 kilobases (kb), and there is one marked element inserted randomly per cell, mutants should have been recovered at a frequency of about 1–2%. This 10- to 20-fold discrepancy in frequency may reflect a target-site specificity for Ty1-H3*HIS3*. Recent work on Ty target-site preference supports this idea.[24,25]

Chromosomal Manipulations of Ty-Induced Mutations

An advantage of Ty mutagenesis is that a mutation is physically and genetically tagged with a unique sequence. This feature coupled with the development of electrophoretic systems that separate yeast chromosomes should allow any mutation caused by a marked transposition to be assigned to a chromosome.[26] We have tested this application for two singly marked insertions at *LYS2*.[5] Standard protocols for preparation of yeast chromosomes and CHEF (contour-clamped homogeneous electric field) electrophoresis have been used in the analysis (refer to [4], this volume, on physical mapping and chromosome separation). Southern analysis using either a *HIS3*- or a *LYS2*-specific probe shows that *HIS3* sequences have been inserted in the chromosome that carries *LYS2* (chromosome II).

A similar approach with mutations caused by Ty elements inserted into unknown genes should establish the chromosomal location of the novel mutations. The position of the gene on the chromosome can be further localized using the chromosome fragmentation technique of Vollrath *et al.*[27] (see [4], this volume). To direct the fragmentation vector (YCF) to the

[24] G. Natsoulis, W. Thomas, M.-C. Roghmann, F. Winston, and J. D. Boeke, *Genetics* **123**, 269 (1989).
[25] C. M. Wilke, S. H. Keidler, N. Brown, and S. W. Liebman, *Genetics* **123**, 655 (1989).
[26] G. Chu, D. Vollrath, and R. W. Davis, *Science* **234**, 1582 (1986).
[27] D. Vollrath, R. W. Davis, C. Connelly, and P. Heiter, *Proc. Natl. Acad. Sci. U.S.A.* **85**, 6027 (1988).

site of the Ty-induced mutation, the YCF vector should contain the Ty marker gene.

Marked Ty Insertions as Portable Genetic Markers

Even though many new yeast genes are being identified and mapped, there are large regions of yeast chromosomes that are poorly characterized. We have established a collection of yeast strains that contain unselected Ty1-H3*HIS3* insertions throughout the yeast genome. These insertions should serve as portable genes to establish linkage with other chromosomal markers. The chromosome profile of 12 of these strains is presented in Fig. 3. The available evidence suggests that Ty1-H3*HIS3* transposes to many yeast chromosomes and that Ty insertions act as genetic markers in crosses.[4,7,28] As a result, linkage data obtained using a marked Ty should not be greatly distorted.

Cloning Ty-Induced Mutations

There are several ways to isolate a Ty-induced mutation and/or the wild-type gene. Any recessive Ty-induced mutation can be cloned by complementation using standard techniques (also see [14], this volume).[29] Marked Ty element insertions can be isolated from a clone bank made from a mutant strain by using the Ty marker as a probe in colony or plaque hybridizations. Such flanking sequences can then be used to isolate the wild-type gene. The latter approach has been used to clone Ty1-H3*HIS3* insertions from the *HML* locus (K. Weinstock, unpublished data, 1989). We have also constructed *URA3*-based integrating vectors containing the *HIS3* or *NEO* genes to aid in cloning Ty-induced mutations caused by Ty*HIS3* and Ty*NEO* elements. The methods used to clone adjacent sequences by this approach have been previously described.[30]

To facilitate the isolation of the Ty–mutant allele junction fragment, we have developed a Ty element that carries an origin of replication and a selectable marker that functions in *E. coli*. The present marked elements are derived from Ty2-917 and carry the miniplasmid πN (Fig. 1). To determine if Ty2-917πN-induced mutations can be cloned directly, we have used the πN replicon present on Ty2-917 to recover several random Ty2-917πN transpositions in *E. coli*. The following is a general protocol used to recover marked Ty2-917πN insertions. It is similar to protocols for cloning yeast mutations by "eviction."[30]

Large quantities of yeast DNA can be prepared by the procedure of

---

[28] H. L. Klein and T. D. Petes, *Mol. Cell. Biol.* **4**, 329 (1984).
[29] M. D. Rose, this series, Vol. 152, p. 481.
[30] F. Winston, F. Chumley, and G. R. Fink, this series, Vol. 101, p. 211.

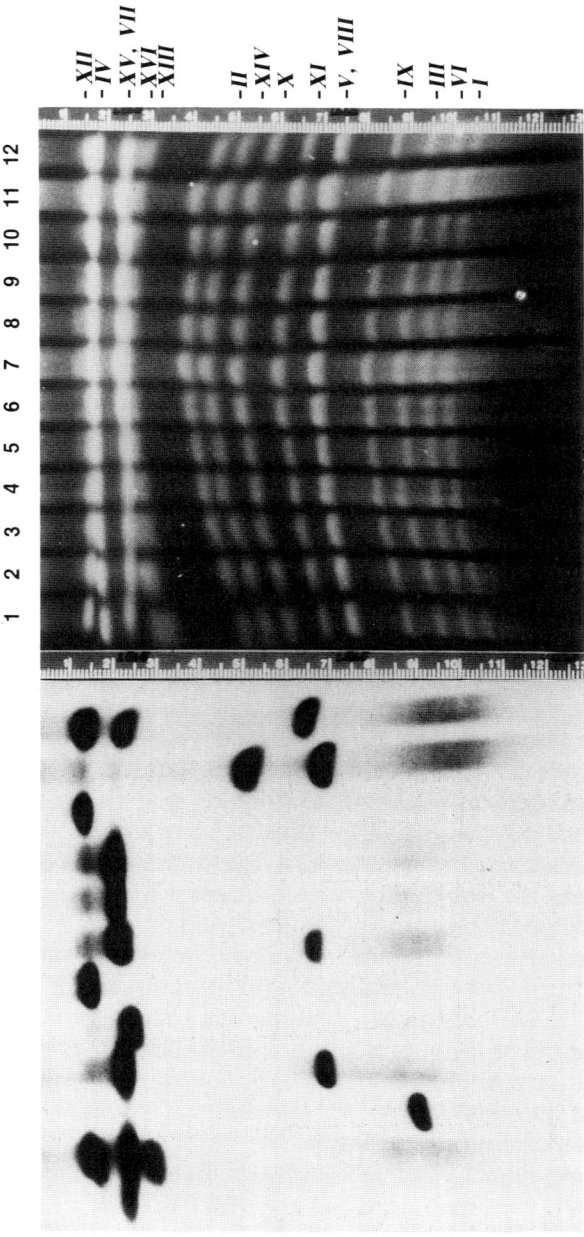

Cryer et al.[31] Yeast DNA can also be prepared from 10- to 40-ml cultures in quantities large enough for a cloning experiment (refer to [11], this volume).[29,30] Yeast DNA is digested with a restriction endonuclease that will yield plasmids with flanking sequences to one or both sides of the insertion (Fig. 1). After digestion, the enzyme is inactivated by heating the digestion reaction or by phenol extraction, followed by ethanol precipitation. The DNA is ligated using T4 DNA ligase under conditions that promote recircularization. Typically, 2 μg of digested DNA is ligated in a total volume of 200 μl overnight at 15°. The ligated DNA is phenol extracted, ethanol precipitated, and suspended in 100 μl of DNA buffer. Aliquots of 5, 10, or 20 μl are introduced into 0.3 ml of competent *E. coli* strain HB101, and the cells from an entire transformation are plated on one or two LB plates containing 25 μg/ml kanamycin. The transformation and media are prepared as described by Maniatis et al.[32] Between 10 and 100 small transformant colonies resistant to kanamycin appear after 24 hr at 37°. We observed that bacteria containing the original πN or pGTy2-917πN plasmids also form small colonies. It is possible that the expression of the PiAN7 *supF* gene on a high-copy plasmid is deleterious to cells. Similar TyπN elements that lack the *supF* gene are being constructed.

We have been able to recover plasmids from several different single insertions. The recovered plasmids are accurate circular forms of the genomic DNA and the predicted segment of the starting Ty element, as judged by Southern and restriction enzyme analyses. In one case, we did not recover all the insertions present in the strain. It is possible some insertions will be difficult to rescue by this technique, especially if fragments are larger than 20 kb or if the genes recovered are toxic in *E. coli*.

Demonstration that the flanking sequence obtained by cloning contains the Ty-induced mutation, and in fact is the desired gene, can be done by a combination of molecular and genetic techniques. If possible, flanking sequences on both sides of the insertion should be used as hybridization

[31] D. R. Cryer, R. Eccleshall, and J. Marmur, *Methods Cell Biol.* **12**, 39 (1975).
[32] T. Maniatis, E. F. Fritsch, and J. Sambrook, "Molecular Cloning: A Laboratory Manual." Cold Spring Harbor Laboratory, Cold Spring Harbor, New York, 1982.

FIG. 3. Unselected transpositions of Ty1-H3*HIS3* into various yeast chromosomes. Yeast chromosomal DNA was prepared from strains carrying independent, unselected Ty1-H3*HIS3* transpositions (lanes 1–12). The DNA is separated on a 1% agarose gel using the CHEF electrophoretic system. The ethidium bromide staining pattern of the gel is shown at right. The DNA was UV-nicked, transferred to Hybond-N, and hybridized with a *HIS3*-specific probe. Autoradiograms of the resulting blots are shown at left. The position of different chromosomes is shown alongside the gel. The parental strain contains the *his3Δ200* deletion.

probes to isolate a set of clones from the various yeast gene banks that span the original insertion (refer to [14] in this volume).[29] The clones obtained from the banks must fulfill two criteria: some clones should complement a recessive Ty-induced mutation, and the cloned segments should result in a restriction map that is colinear with the region surrounding the Ty-induced mutation. If the Ty-induced mutation is dominant or results in a ROAM phenotype, a gene bank from the mutant strain can be made in a centromere-based vector and assayed for a dominant mutant phenotype in a suitable recipient strain.

The most serious problems with directly cloning Ty-induced mutations are genomic rearrangements, such as Ty-mediated deletions, and complex Ty insertions. There is one example of a marked mutation, as defined by tight genetic linkage, that is the result of a deletion event involving Ty1-H3*HIS3* and another Ty or solo δ (K. Kawagami, B. Shafer, D. Garfinkel, and J. Strathern, unpublished results, 1988). Consequently, the genomic flanking sequence is not adjacent to or within the wild-type gene, but rather is from a deletion breakpoint. Recently, complex Ty insertions have also been obtained at the *CAN1*[25,33] and *HML* loci (K. Weinstock, M. Mastrangelo, B. Shafer, D. Garfinkel, and J. Strathern, unpublished results, 1989). The insertion at *CAN1* appears to result from the concurrent integration of two Ty elements. The *HML* insertions are multimeric Ty insertions that contain up to 100 kb of Ty DNA. In either case, obtaining the flanking genomic DNA adjacent to these insertions is much more complicated than with a single Ty insertion. However, the available data suggest that aberrant Ty insertions occur rather infrequently at most loci (comprising less than 10% of the total number of Ty insertions). In addition, the relevant gene can still be isolated using standard techniques for cloning by complementation.[29]

Use of spt3 Mutants

Using the pGTy system, Ty-induced mutations are caused both by native chromosomal elements and by marked elements. It would be advantageous to reduce the background created by unmarked Ty elements. In the study summarized here, 12% of the *lys2* (6/52) and *lys5* (1/8) mutants are caused by Ty1-H3*HIS3* and 21% (11/52) of the *lys2* mutants are caused by unmarked chromosomal Ty elements. It should be possible to virtually eliminate the chromosomal Ty transpositions by inducing transposition in an *spt3* mutant background. The *SPT3* gene was originally

[33] C. M. Wilke and S. W. Liebman, *Mol. Cell. Biol.* **9**, 4096 (1989).

isolated as an extragenic suppressor of Ty-induced mutations.[34] It is required for transposition of chromosomal elements, but transposition of the *GAL1*-promoted Ty elements is relatively unaffected (Table I).[35] However, *spt3* mutants also affect diploid formation and sporulation.[36] These pleiotropic defects in normal yeast physiology may have unforeseen consequences in various mutant searches.

Tagging Essential Genes

It is likely that Ty-induced mutations will occur in essential genes. Because it is possible to induce transposition in sporulation-competent diploids, Ty-induced recessive lethal mutations may be maintained in the heterozygous state. If Ty transpositions can be limited to about 1 Ty1-H3*HIS3* element per cell, one can use random spore analysis to detect mutants that are His$^+$ as diploids but are unable to give rise to His$^+$ haploid cells. A standard protocol for random spore analysis uses two recessive drug resistance markers, such as resistance to canavanine and cycloheximide. Diploids heterozygous for these unlinked markers are sensitive to both drugs. However, 25% of the haploid spores are resistant to both and can be selected for by plating on the appropriate medium. The probability that a marked Ty-induced mutation would be linked to one of the drug resistance loci is assumed to be low but can be addressed by adding a third drug resistance marker. Only doubly drug resistant, haploid segregants that are His$^-$ can be derived from the parental diploid if Ty1-H3*HIS3* transposed into an essential gene. Tetrad analysis can then be performed on the parental His$^+$ diploid to verify the nature of the mutation. This technique can conceivably be used to generate a library of marked Ty-induced mutations that is analogous to the currently available "*ts* bank."

Acknowledgments

Research sponsored by the National Cancer Institute, U.S. Department of Health and Human Services, under Contract No. N01-CO-74101 with BRI. The contents of this publication do not necessarily reflect the views or policies of the Department of Health and Human Services, nor does mention of trade names, commercial products, or organizations imply endorsement by the U.S. Government. We are grateful to A. Arthur, J. Boeke, G. Fink, R. Fishel, and several anonymous reviewers for helpful comments on the manuscript; to members of our laboratories and other yeast researchers for contributing unpublished data; to P. Farabaugh for the sequence of Ty2-917; and to P. Hall and L. Summers for expert secretarial assistance.

[34] F. Winston, D. T. Chaleff, B. Valent, and G. R. Fink, *Genetics* **107**, 179 (1984).
[35] J. D. Boeke, C. A. Styles, and G. R. Fink, *Mol. Cell. Biol.* **6**, 3575 (1986).
[36] F. Winston, K. J. Durbin, and G. R. Fink, *Cell (Cambridge, Mass.)* **39**, 675 (1984).

[24] Transformation of Yeast Directly with Synthetic Oligonucleotides

By RICHARD P. MOERSCHELL, GOUTAM DAS, and FRED SHERMAN

Introduction

Specific changes can be introduced into DNA by using synthetic oligonucleotides with any of a number of methods. The most versatile methods for producing highly specific changes require single-stranded *Escherichia coli* vectors containing the target sequence and a short synthetic oligonucleotide containing the desired alterations; the synthetic oligonucleotide is hybridized to the target sequence, the complementary strand is synthesized *in vitro* by extending the synthetic oligonucleotide, and the duplex vectors containing the mismatches are introduced and amplified in *E. coli*.[1] Highly efficient recoveries of the desired alterations have been achieved by preferentially destroying the original parental strand.[2] In addition, random or multiple changes can be produced in a short region with mixed oligonucleotides.[3-5] However, identification of the desired alterations requires repetitive sequencing, and the complete set of desired alterations is often not recovered.

We have described a convenient procedure for producing specific alterations of genomic DNA by transforming yeast directly with synthetic oligonucleotides.[6] This procedure is easily carried out by transforming a defective mutant and selecting for at least partially functional revertants. The oligonucleotide should contain a sequence that would correct the defect and produce additional alterations at nearby sites. Alternatively, defective mutants could be selected with methods such as those using chlorolactate for selecting *cyc1* mutants,[7] 5-fluoroorotic acid for selecting *ura3* mutants,[8] α-aminoadipic acid for selecting *lys2* mutants,[9] methyl

---

[1] M. J. Zoller and M. Smith, this series, Vol. 154, p. 329.
[2] T. A. Kunkel, J. D. Roberts, and R. A. Zakour, this series, Vol. 154, p. 367.
[3] D. E. Hill, A. R Oliphant, and K. Struhl, this series, Vol. 155, p. 558.
[4] K. Singh, J. G. Tokuhisa, E. S. Dennis, and W. J. Peacock, *Proc. Natl. Acad. Sci. U.S.A.* **86**, 3733 (1989).
[5] S. A. Goff, S. R. Short Russell, and J. F. Dice, *DNA* **6**, 381 (1987).
[6] R. P. Moerschell, S. Tsunasawa, and F. Sherman, *Proc. Natl. Acad. Sci. U.S.A.* **85**, 524 (1988).
[7] S. B. Baim, D. F. Pietras, D. C. Eustice, and F. Sherman, *Mol. Cell. Biol.* **5**, 1839 (1985).
[8] J. D. Boeke, F. Lacroute, and G. R. Fink, *Mol. Gen. Genet.* **197**, 345 (1984).
[9] B. B. Chattoo and F. Sherman, *Genetics* **93**, 51 (1979).

```
              1              16
              |  1  2  3  4   |
         (Met)Thr-Glu-Phe-Lys-Ala-Gly-
         ATA ATG ACT GAA TTC AAG GCC GGT     CYC1⁺
                       ↓↓ TC → A
         ATA ATG ACT GAA TA- AAG GCC GGT     cyc1-31
          ↓   ↓↓  ↓↓↓ ↓↓↓ ↓↓
         ATA ATA ATG TCT CCA TTG GCC GGT
             (Met) Ser-Pro-Leu-Ala-Gly-      CYC1-880
```

```
         229   (CH3)3      240
          |76    |        79 |
         -Asn-Pro-Lys-Lys-Tyr-Ile-Pro-
         AAC CCA AAG AAA TAT ATT CCT     CYC1⁺
                       ↓ -T
         AAC CCA AAG AAA TA- ATT CCT     cyc1-812
          ↓                 ↓
         AAC GCA AAG AAA TAT ATT CCT     CYC1-813
         -Asn-Ala-Lys-Lys-Tyr-Ile-Pro-
                    |
                  (CH3)3
```

FIG. 1. Amino acid sequences of two regions of the iso-1-cytochromes c and the corresponding DNA sequences of the *CYC1* alleles. The *cyc1-31* and *cyc1-812* mutants lack iso-1-cytochrome c because of the frameshift and TAA nonsense mutations. The *CYC1-880* and *CYC1-813* functional mutants were obtained by transforming the *cyc1-31* and *cyc1-812* defective mutants, respectively, with a 40-mer and 50-mer. The oligonucleotides corrected the defects and produced amino acid replacements at the indicated nearby sites. Amino acid residues and nucleotides that differ from the *CYC1⁺* normal sequence are in bold type. Nucleotides of the *CYC1-880* and *CYC1-813* transformants that differ, respectively, from the *cyc1-31* and *cyc1-812* alleles are underlined and indicate mismatches. The excised aminoterminal methionine residues are denoted in parentheses. The ϵ-N-trimethyllysine residue at amino acid position 77 is also shown.

mercury for selecting *met15* mutants,[10] and canavanine sulfate for selecting *can1* mutants.[11,12] If no selective system is available, cotransformation could be used to enrich for the proportion of transformants (see below), and colony hybridization could be used to detect desired transformants among unselected colonies. However, transformation of yeast directly with synthetic oligonucleotides is ideally suited for producing a large number of specific alterations that change a completely nonfunctional allele to at least a partially functional form.

[10] A. Singh and F. Sherman, *Genetics* **81**, 75 (1975).
[11] A. M. Srb, *C. R. Lab. Carlsberg, Ser. Physiol.* **26**, 363 (1955).
[12] J. M. Wiame, J. Bechet, M. Mousset, and M. de Deken, *Arch. Int. Physiol. Biochim.* **70**, 766 (1962).

With oligonucleotides in hand, a single worker has produced replacements of 15 different amino acids at one site in less than 2 working days. The high fidelity of the process was revealed by the correct sequence of over 60 altered iso-1-cytochromes c, including 4 that arose by 11 mismatches (see Fig. 1) (S. Tsunasawa, R. P. Moerschell, and F. Sherman, unpublished results, 1989). Furthermore, because the method results in a single altered chromosomal copy, the phenotype and relative activity of the mutant allele can be readily assessed *in vivo.*

The major disadvantage of the convenient transformation of yeast directly with oligonucleotides is a requirement for a selection system and the inability to generate both functional and completely nonfunctional alleles with the same system. Furthermore, in many cases, it is unknown whether the alteration will produce a functional or nonfunctional allele. If only a few altered forms are required and if a selective system has to be constructed, it would be advisable to use conventional systems where alterations are made *in vitro* and amplified in *E. coli.*

This procedure is illustrated with the two mutants *cyc1-31* and *cyc1-812* (Fig. 1) that are ideally suited for this method and that have been used to generate missense mutants having various single and multiple amino acid substitutions of iso-1-cytochrome c. Although different systems will vary, the selection procedure used with *cyc1* mutants allow recovery of altered iso-1-cytochrome c with less than 1% of the normal activity. The *cyc1-31* mutant arose spontaneously, whereas the *cyc1-812* mutant was specifically constructed for our studies by the general method[2] using *E. coli* vectors (G. Das and F. Sherman, unpublished results, 1988). Each of the mutants has an in-frame TAA nonsense codon and a deleted base pair, resulting in a complete deficiency of iso-1-cytochrome c and an extremely low reversion frequency. Because the amino-terminal region encompassing up to Gly-11 is dispensable,[13] presumably any iso-1-cytochrome c altered within this region should be recoverable. As many missense mutants have at least partial function,[14] numerous replacements within the internal portion of the molecule also have been generated. This method has proved invaluable for investigating amino-terminal processing and has been used to produce numerous amino acid replacements in the evolutionarily invariant region encompassing the *cyc1-812* mutation (Fig. 1).

Methods

Synthetic Oligonucleotides

Although any of the commercially available instruments can be used for synthesizing oligonucleotides, we have exclusively used the Applied

Biosystems 380A DNA Synthesizer (Foster City, CA) using the 0.2 μmol fast cycle set for automatic removal of the trityl group. Chemicals used for the syntheses can be purchased from Applied Biosystems, Biosearch (San Rafael, CA), and other vendors.

The oligonucleotides are removed from the synthesizer; dried in a heated SpeedVac (Hicksville, NY); redissolved in 2 ml of 30% ammonium hydroxide; incubated at 55° for 4 to 6 hr; dried again in a heated Speed-Vac; redissolved in 0.5 ml of deionized distilled water; and transferred to a 1.5-ml microcentrifuge tube. Insoluble material is removed by centrifuging at 15,000 g for 10 min and transferring the solution to a fresh tube. Three extractions are performed with 1 ml of water-saturated n-butanol. Two further extractions are performed with regular n-butanol. The solution is again dried in a heated SpeedVac. The pellet is redissolved in 0.1 ml of 0.5 M ammonium acetate, then precipitated at room temperature for 5 min with the addition of 95% (v/v) ethanol. The oligonucleotide precipitate is centrifuged (15,000 g, room temperature) for 5 min. The pellet is rinsed once with 1 ml of 95% ethanol, dried under reduced pressure, and redissolved in a sufficient volume of deionized distilled water to make the oligonucleotide concentration approximately 30 mg/ml. The concentration can be determined from the absorbancy of the solution at 260 nm. The purity of the oligonucleotides is analyzed by electrophoresis with a 0.75-mm-thick, 12% acrylamide gel and by visualization after staining the gel with a 170 ng/ml ethidium bromide solution for 10 min. Oligonucleotide preparations with a substantial proportion of failure sequences should not be used.

Yeast Transformation

Yeast can be transformed with synthetic oligonucleotides using standard procedures.[15] The lithium acetate[16] procedure yields higher frequencies of transformants and is more convenient than the spheroplasting method. For transformation of lithium acetate-treated cells, 10 μl or less of a solution typically containing 200 μg of a 40-mer is added to 0.2 ml of a cell suspension typically containing 3×10^7 viable cells. The cells are transformed and washed according to the method of Itoh et al.[16] After the final wash, the transformed cells are suspended in 0.5 ml of a solution containing 10 mM Tris-HCl and 1 mM EDTA (pH 7.5), and the suspen-

---

[13] F. Sherman, J. W. Stewart, M. Jackson, R. A. Gilmore, and J. H. Parker, *Genetics* **77**, 255 (1974).
[14] D. M. Hampsey, J. F. Ernst, J. W. Stewart, and F. Sherman, *J. Mol. Biol.* **201**, 471 (1988).
[15] J. C. Schneider and L. Guarente, this volume [25].
[16] H. Itoh, Y. Fukuda, K. Murata, and A. Kimura, *J. Bacteriol.* **153**, 163 (1983).

sion is plated on a medium that selects for the transformants or allows for their easy detection. Because completely deficient *cyc1* mutants do not utilize nonfermentable carbon sources and grow poorly on complete media, *CYC1* transformants having at least 1% activity form distinct colonies over a lawn of untransformed cells after 3 days of incubation on YPD medium (1% Bacto-yeast extract, 2% Bacto-peptone, and 2% dextrose).

Factors Affecting Transformation Frequencies

Cotransformation

The number of transformants obtained with oligonucleotides can be greatly increased by transforming concomitantly with conventional plasmids such as those containing the *URA3* selectable marker. By selecting the Ura$^+$ transformants, the proportion of oligonucleotide transformants was approximately 100-fold higher. Although such enrichments have not proved to be useful when the system allows for the direct selection, cotransformation increases the frequencies from approximately 10^{-5} to about 10^{-3} per recovered cell, allowing detection of the desired transformants by testing unselected populations by colony hybridization.

Oligonucleotide Concentrations

Experiments with the *cyc1-31* mutant revealed that the frequencies of transformation increase with higher amounts of oligonucleotide and that saturating levels were not attained with either the spheroplasts or the lithium acetate-treated cells. The increased frequencies of transformants per amount of oligonucleotide with increasing concentrations of the oligonucleotide suggest that the oligonucleotide may also have a protective function (Fig. 2). These results also indicate that the lithium acetate procedures yield higher frequencies of transformants. Although 50 μg of an oligonucleotide is generally sufficient for obtaining transformants with the lithium acetate procedure, we recommend using 200 μg per reaction because high amounts of oligonucleotides are generally produced by commercial instruments and because excesses are usually discarded.

Oligonucleotide Length

Two series of oligonucleotides having two and five mismatches, respectively, were used to establish the relationship between transformation frequencies and length of oligonucleotides.[6] The results (Fig. 2) indicated an optimum size of about 50 nucleotides. Although coupling during oligonucleotide synthesis is generally efficient, long oligonucleotides inherently

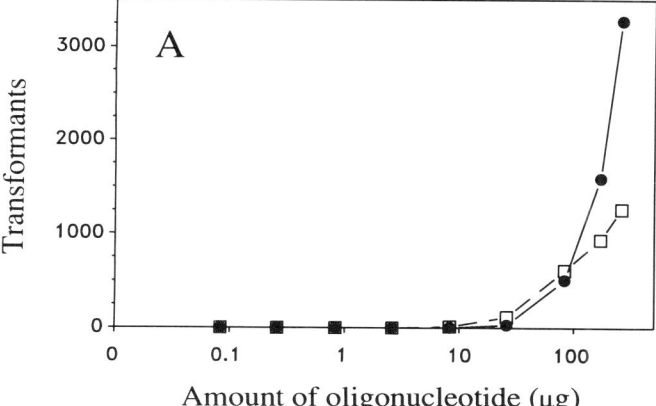

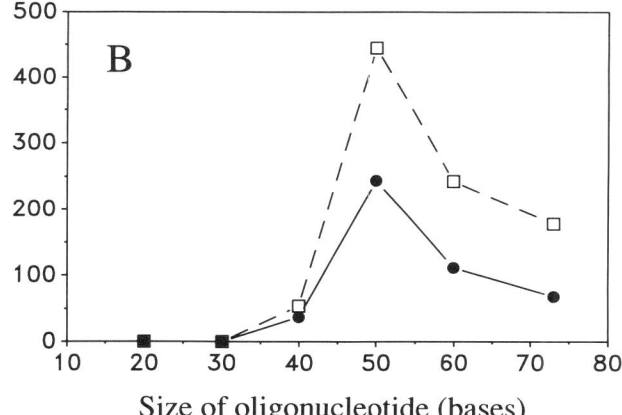

FIG. 2. Relationships between transformation frequencies and concentrations (A) and lengths (B) of oligonucleotides. (A) The *cyc1-31* mutant was transformed by the lithium acetate procedure with various concentrations of a 50-mer having the normal sequence. The data are presented as the number of transformants obtained with a particular amount (●) or normalized frequencies per 100 μg (□). (B) Frequencies of transformants obtained by transforming the *cyc1-31* mutant with the spheroplast method, using 85 μg of a series of *CYC1-793* oligonucleotides having various lengths and generating the sequence (Met)Ser-Glu-Phe-Leu-Ala-. The results are shown as the frequencies per reaction tube (●) and frequencies per nanomole (□).

have a lower percent yield than short oligonucleotides. It is not known why oligonucleotides longer than 50-mers are less effective, nor whether this diminution reflects an inherent property of the transformation process or is perhaps due to poorer quality of longer oligonucleotides. Because shorter oligonucleotides are more economical to produce, we recommend using 40-mers even for generating alterations with numerous mismatches. Transformation frequencies, using 200 μg of a 40-mer, have ranged from approximately 10 to over 2000 per plate, depending on the degree of mismatches and on which DNA strand the oligonucleotide is homologous to.

Strand Dependency

Transformation of the *cyc1-31* mutant with a series of oligonucleotides revealed that oligonucleotides homologous to the mRNA (sense oligonucleotides) had an approximately 100-fold higher transformation frequency than oligonucleotides complementary to the mRNA (antisense oligonucleotides). This strand specificity was noted with 40-mers having a 95% similarity with the target allele. Oligonucleotides with less similarity to the target sequence, and therefore yielding lower numbers of transformants, did not demonstrate this strand specificity.[6] Furthermore, it is unclear whether the differences between sense and antisense oligonucleotides apply to the other regions of the gene. Possibly, antisense oligonucleotides may be inhibiting translation of the *CYC1* message in the transformants, thereby inhibiting their development. Alternatively, these particular sense oligonucleotides may be incorporated more efficiently owing to the nature of their sequence.

Oligonucleotide Mismatches

Comparisons between two 40-mers having two and five mismatches, respectively, showed approximately 10-fold higher transformation frequencies of the *cyc1-31* allele with the more homologous oligonucleotide. Even though transformation frequencies decline with more mismatches, we have easily obtained altered iso-1-cytochromes *c* with 40-base-long oligonucleotides having 11 mismatches (Fig. 1).

Concluding Remarks

Although the method is rapidly carried out with few steps, a 40-mer is typically used, and the method is therefore more costly than conventional methods with *E. coli* vectors, which usually use shorter oligonucleotides. In principle, this and other procedures could be carried out with mixed oligonucleotides for generating a predetermined distribution of mutations. The specific alteration could be conveniently identified by retrieving and sequencing the transformants by methods using the polymerized chain reaction. However, transformants arising by multiple mismatch would be recovered only rarely. It has been our experience that multiple mutants and specifically desired alterations are best constructed separately with single oligonucleotides.

Acknowledgments

This work was supported by U.S. Public Health Service Research Grant R01 GM12702.

Section IV
Biochemistry of Gene Expression

[25] Vectors for Expression of Cloned Genes in Yeast: Regulation, Overproduction, and Underproduction

By JANE C. SCHNEIDER and LEONARD GUARENTE

Often it is necessary to separate a gene from its own promoter in order to control expression of its product in different conditions or to produce the product to a greater or lesser degree than normal. This chapter discusses yeast transcription cassettes of several types used for the expression of cloned genes. The first type includes vectors with regulatable promoters; *GAL, PGK,* and *ADH1* respond to the carbon source, and *PHO5* responds to the inorganic phosphate concentration. These vectors allow the expression of the cloned gene to be turned on and off by changes in the growth medium (or, in some specially constructed strains, by temperature changes) (see also [26], this volume). Another class creates a protein fusion to an amino-terminal signal sequence so that the protein will be secreted from the cell (see also [34], this volume). A third kind of cassette makes protein fusions of the cloned gene to the coding region for another protein, such as β-galactosidase, or to a small antigenic epitope. Hence, proteins for which there is no antibody or convenient assay available can be tagged with a heterologous enzymatic activity or antigen (see also [35], this volume). Two families of plasmids which are useful for introducing cloned genes into yeast are also described.

Requirements for Transcription Cassettes

Sites Required for Gene Expression

The amount of gene transcription depends on the nature of the flanking sequences. Cis-acting sites near yeast genes are responsible for determining the position of initiation and termination of the message as well as regulation of mRNA levels in response to environmental signals. Activation of transcription is mediated by upstream activation sites (UASs), which are analogous to enhancers found near metazoan genes.[1] In most cases, these sites bind to regulatory proteins which, under appropriate conditions for the particular gene, transmit to RNA polymerase II the signal to initiate transcription. In addition, some genes have sites independent of the UAS which interact with repressors; their effect is to reduce the extent of activation.

[1] L. Guarente, *Cell (Cambridge, Mass.)* **52**, 303 (1988).

Transcription of a gene also depends on a TATA box sequence in the 5′ flanking region. These sites bind TFIID and general transcription factors[2] and are necessary to elicit mRNA initiation by RNA polymerase II at sites found 60–120 base pairs (bp) downstream.[3] Although many mRNA initiation sites fit one of two consensus sequences, deletion of the original site results in initiation at another nearby sequence, without affecting transcriptional levels. In most cases, UASs function as well with heterologous TATA boxes and mRNA initiation regions as they do with their original ones. Thus, the UAS of a gene may be replaced with that of a heterologous one in order to confer a different regulatory regimen. UAS elements may be combined as well, placing gene expression under two regulatory systems.

It is preferable to use a cassette without an initiation codon so that the introduced gene is translated from its own AUG; otherwise, a few amino acids from the cassette will be attached to the amino terminus. Consideration should be given to the subsequent location in the resulting transcript of the ATG initiation codon. In the great majority of cases, the AUG nearest the 5′ end of the message is used to initiate protein synthesis.[4] Thus, if fortuitous ATG sequences from the cloned gene are introduced upstream of the actual initiation codon in the gene fusion, they will probably interfere with proper translation of the gene. When cloning into cassettes without ATG initiation codons (−ATG vectors), the introduced gene supplies the flanking sequence along with its cognate initiation codon. The identity of the sequence flanking the AUG initiation codon may affect the efficiency of translational initiation and thus gene expression. Dobson et al.[5] found, in a compilation of several yeast genes, that position −3 from the ATG was always occupied by an A residue. In addition, Kozak[6] found that changing this base in a rat gene from C to A resulted in a 15-fold increase in expression. However, yeast genes may not be as sensitive as those of mammalian cells to this flanking sequence (I. Griff and L. Guarente, unpublished observations, 1986). Identity of the flanking sequence can probably be ignored without major consequences, unless maximum expression of the gene is required (see Ref. 7). The −ATG vectors pLGSD5 and YEp62 (see below) are favored since they add only a few extra amino

[2] S. Hahn, S. Buratowski, P. Sharp, and L. Guarente, *Proc. Natl. Acad. Sci. U.S.A.* **86,** 5718 (1989).
[3] L. Guarente, *Annu. Rev. Genet.* **21,** 425 (1987).
[4] M. Kozak, *Cell (Cambridge, Mass.)* **34,** 971 (1983).
[5] M. J. Dobson, M. F. Tuite, N. A. Roberts, A. J. Kingsman, and S. M. Kingsman, *Nucleic Acids Res.* **10,** 2625 (1982).
[6] M. Kozak, *Nature (London)* **308,** 241 (1984).
[7] M. F. Tuite, M. J. Dobson, N. A. Roberts, R. M. King, D. C. Burke, S. M. Kingsman, and A. J. Kingsman, *EMBO J.* **1,** 603 (1982).

acids to the amino terminus. When creating a protein fusion, care must be taken to ensure that the reading frame has been preserved. It occasionally happens that a construction yields an out-of-frame fusion owing to contaminating exonuclease activities in enzyme preparations. Therefore, the fusion junction should be checked by sequencing.

The presence of a transcription termination site is necessary for 3'-end formation of the transcript. These sites are generally found near the actual 3' end of the message, within a few hundred base pairs downstream of the translation termination codon. Many transcription cassettes do not carry an obvious termination region; however, examination of transcripts produced from these fusions shows that the messages terminates either at a site within the 2-μm sequences, at the terminators of other yeast genes carried on the plasmid, or at fortuitous sites within the plasmid (e.g., *CYC1-lacZ* transcribed from pLGΔ312 terminates in the *Amp^R* gene.[8]) A few cases, though, show that a termination sequence located a few hundred base pairs downstream of the open reading frame results in a higher level of expression.[9,10] The extra sequences present in a transcript which is longer than necessary may expose it to a greater chance of endonucleolytic attack. In addition, a termination sequence near the end of the gene may prevent read-through transcription from downstream genes.[11] For example, the same sequence serves as a terminator for *CYC1* and for a downstream gene, *UTR1,* which is transcribed from the opposite strand; removal of this sequence causes *CYC1* transcription to decrease, possibly owing to interference caused by the lack of termination of polymerases transcribing the downstream gene. Thus, although a terminator close to the 3' end of the coding sequence is not necessary, its presence may lead to higher expression.

Origin of Replication

Most plasmids are based on the 2-μm circle. Occasionally, however, it may be necessary to create a vector with a higher or lower copy number, or one with greater stability. For example, cloning vectors for libraries often use a low-copy ARSCEN vector such as YCp50[12] to avoid cloning a related gene which happens to complement a mutation when overexpressed. Here we summarize salient properties of available vectors.

[8] B. I. Osbourne and L. Guarente, *Genes Dev.* **2,** 766 (1988).
[9] G. Ammerer, this series, Vol. 101, p. 192.
[10] J. Mellor, M. J. Dobson, N. A. Roberts, M. F. Tuite, J. S. Emtage, S. White, D. A. Lowe, T. Patel, A. J. Kingsman, and S. M. Kingsman, *Gene* **24,** (1983).
[11] K. S. Zaret and F. Sherman, *Cell (Cambridge, Mass.)* **28,** 563 (1982).
[12] M. Johnston and R. W. Davis, *Mol. Cell. Biol.* **4,** 1440 (1984).

2-μm Vector. The level of expression of a fusion is, of course, influenced by the copy number and stability of the vector. Many vectors contain the origin of the 2-μm circle, a multicopy endogenous yeast episome. The part present in most yeast episomal vectors is a 2.2-kilobase pair (kb) *Eco*RI (e.g., YEp13) or 2.1-kb *Hin*dIII fragment from the B form containing an *ARS* sequence, which serves as an origin of replication in yeast, and the *REP3* locus, which is required in cis for partitioning during cell division (for review, see Rose and Broach[13]). Other gene products necessary for replication and maximum stability of the 2-μm plasmids (*REP1, REP2,* and *FLP*) are provided in trans from endogenous 2-μm circles present in most yeast strains (such a strain is referred to as [cir$^+$]; strains lacking the 2-μm circle are [cir$^{\circ}$]). The 2-μm origin present on a plasmid results in an average copy number of 10–40 per cell; there tends to be a large variation in copy number from cell to cell.[14] Although vectors based on the 2-μm circle origin are fairly stable in the absence of selection, cells which have lost the plasmid arise at a rate of about 2–6% per generation;[14] thus, the strain must be grown in a selectable medium to maintain the plasmid.

Very High Copy 2-μm Origins. A special class of 2-μm vectors has a very high copy number of 100 to 200 per cell. The basis for the high copy number is a *LEU2* allele *(leu2-d)* which complements a leucine auxotrophy poorly. The vector pJDB219 contains an entire 2-μm circle (B form), the *leu2-d* marker, and unique cloning sites[15] into which exogenous genes may be introduced. Under selection for growth in the absence of leucine, cells with higher copy numbers, which presumably compensate for poor expression of the *leu2-d* gene, have a growth advantage[16] (however, a plasmid derived from pJDB219, YEpsec1, apparently exists at high copy number without leucine selection[17]). For maximum stability and integrity of the plasmid, [cir$^{\circ}$] strains should be used to propagate pJDB219-derived plasmids. Owing to incompatibility between pJDB219 and the endogenous 2-μm circle, the vector is less stable in [cir$^+$] strains.[14] In addition, rearrangement between pJDB219 and endogenous 2-μm has been observed.[13] (Rose and Broach[13] present a method for creating [cir$^{\circ}$] strains by overexpressing *FLP*.) The stability of pJDB219 in a [cir$^{\circ}$] strain is very high even without continued selection for expression of *leu2-d;* after 100 generations, only about 20% of the cells had lost the plasmid.[14] Plasmids that contain

---

[13] A. B. Rose and J. R. Broach, this series, Vol. 185, p. 234.
[14] A. B. Futcher and B. S. Cox, *J. Bacteriol.* **157**, 283 (1984).
[15] J. D. Beggs, in "Molecular Genetics in Yeast," (D. von Wettstein, ed.), Vol. 16, p. 383. Alfred Benzon Symposium, Copenhagen.
[16] E. Erhart and C. P. Hollenberg, *J. Bacteriol.* **156**, 625 (1983).
[17] C. Baldari, J. A. H. Murray, P. Ghiara, G. Cesareni, and C. L. Galeotti, *EMBO J.* **6**, 229 (1987).

the *leu2-d* marker but only part of 2-μm (e.g., pJDB207) demonstrate the same stability and replicate at a high copy number as long as the cell maintains endogenous 2-μm circles to supply missing trans-acting gene products.

Centromeric Vectors. Centromeric vectors contain an *ARS* origin of replication and one of several centromere loci that have been cloned. Vectors with only an *ARS* replicate at high copy but are extremely unstable; YRp7, containing only *ARS,* can be lost within two generations when grown without selection.[18] Even when selection is maintained, the plasmid may not be present in more than 80% of the cells.[19] Vectors carrying only *ARS* were widely used because of their high copy number; the instability and difficulty of retaining the plasmid in a large proportion of cells, however, make the choice of other origins more appropriate. The centromeric vectors exist at a low copy number (1 to 2 per cell) and are generally more stable than those based on the 2-μm circle origin; 97% of cells retained a plasmid with *CEN3* after 20 to 30 generations.[20] When diploids made by crossing strains with and without plasmids are sporulated, most tetrads (50–80%) segregate the plasmid to two of the four spores. However, a substantial proportion of the tetrads segregate plasmids to all four spores, or to none.[18,20] These results imply that the copy number and segregation of centromeric plasmids are not as stringently controlled as those of chromosomes.

Integration. The most stable way to maintain introduced genes is by integration of the entire plasmid into the chromosome by homologous recombination. The gene is then present in only one copy (in haploids) but is very stable; plasmid loss occurs mainly through a very low-frequency recombination, a consequence of looping out the integrated sequences. Continuous selection to maintain the gene is not generally required unless expression is deleterious to the cell. An altered gene can be integrated at the locus of the gene with no chance of looping out by the gene replacement method.[21] Alternatively, the gene and an adjacent selectable marker may be cloned into the coding sequence of a nonessential yeast gene, such as *HO*.[22] Transformation of a linear DNA fragment containing the gene, the marker, and flanking *HO* homology on either side stably integrates the

---

[18] M. Fitzgerald-Hayes, L. Clarke, and J. Carbon, *Cell (Cambridge, Mass.)* **29,** 235 (1982).
[19] R. A. Hitzeman, F. E Hagie, H. L. Levine, D. V. Goeddel, G. Ammerer, and B. D. Hall, *Nature (London)* **293,** 717 (1981).
[20] L. Clarke and J. Carbon, *Nature (London)* **287,** 504 (1980).
[21] R. J. Rothstein, this series, Vol. 101, p. 202.
[22] R. Yocum, "Genetic Engineering of Industrial Yeast." Butterworth, Boston, Massachusetts, 1986.

gene into the genome at the *HO* locus. For a discussion of integration, see [19] in this volume.

Selectable Markers

The selectable marker used is a matter of convenience; occasionally, markers need to be changed to suit the background of a strain. (See [20] in this volume for methods to create *lys2* and *ura3* strains.)

Most laboratory strains have a lesion in either *URA3* or *LEU2;* many plasmids thus carry these genes in order that transformation events may be selected by growth in the absence of uracil or leucine, respectively. Certain situations may create a need for other markers, however. Complicated strain constructions which call for deleted chromosomal loci, integrated fusions, and plasmids all at the same time can easily use up the available markers. In addition, a particular strain may not have appropriate lesions for use with certain plasmids. Table I[23-26] presents a list of selectable markers which have been cloned, along with the vector and restriction fragment on which they reside. The most popular markers are genes encoding amino acid biosynthetic enzymes; most strains have at least one of these mutations, and transformation events are easily selected by omitting the amino acid from the medium. The *URA3-hisG* marker deserves specific mention; if this marker is used to select for integration of a plasmid, it can be subsequently deleted, leaving the rest of the plasmid integrated, so that the *URA3* marker can be reused in further transformations. The *URA3* gene is flanked by direct repeats of the bacterial gene, *hisG;* recombination between these repeats results in looping out of the *URA3* gene in a proportion of cells, an event for which selection is available.[23] In addition to the amino acid synthetic genes, there are two dominant markers which encode resistance to drugs that inhibit yeast growth (hygromycin B for Hgm^R [24,25] or tunicamycin for TUN^R [26]). These may be useful in cases where there are no markers available. In both cases, transformation events have been selected directly by treatment with the growth inhibitor, with a transformation efficiency near that obtained for *URA3* selection. In the experiments reported, all drug-resistant colonies obtained also became Ura$^+$ owing to the *URA3* carried on the plasmid; spontaneous drug resistance was therefore not a problem.

If there are no vectors available with appropriate replication origins or selectable markers, a gene fusion to the desired promoter can be con-

---

[23] E. Alani, L. Cao, and N. Kleckner, *Genetics* **116,** 541 (1987).
[24] L. Gritz and J. Davies, *Gene* **25,** 179 (1983).
[25] K. R. Kaster, S. G. Burgett, and T. D. Ingolia, *Curr. Genet.* **8,** 353 (1984).
[26] J. Rine, W. Hansen, E. Hardeman, and R. W. Davis, *Proc. Natl. Acad. Sci. U.S.A.* **80,** 6750 (1983).

gene into the genome at the *HO* locus. For a discussion of integration, see [19] in this volume.

Selectable Markers

The selectable marker used is a matter of convenience; occasionally, markers need to be changed to suit the background of a strain. (See [20] in this volume for methods to create *lys2* and *ura3* strains.)

Most laboratory strains have a lesion in either *URA3* or *LEU2;* many plasmids thus carry these genes in order that transformation events may be selected by growth in the absence of uracil or leucine, respectively. Certain situations may create a need for other markers, however. Complicated strain constructions which call for deleted chromosomal loci, integrated fusions, and plasmids all at the same time can easily use up the available markers. In addition, a particular strain may not have appropriate lesions for use with certain plasmids. Table I[23-26] presents a list of selectable markers which have been cloned, along with the vector and restriction fragment on which they reside. The most popular markers are genes encoding amino acid biosynthetic enzymes; most strains have at least one of these mutations, and transformation events are easily selected by omitting the amino acid from the medium. The *URA3-hisG* marker deserves specific mention; if this marker is used to select for integration of a plasmid, it can be subsequently deleted, leaving the rest of the plasmid integrated, so that the *URA3* marker can be reused in further transformations. The *URA3* gene is flanked by direct repeats of the bacterial gene, *hisG;* recombination between these repeats results in looping out of the *URA3* gene in a proportion of cells, an event for which selection is available.[23] In addition to the amino acid synthetic genes, there are two dominant markers which encode resistance to drugs that inhibit yeast growth (hygromycin B for Hgm^R [24,25] or tunicamycin for TUN^R [26]). These may be useful in cases where there are no markers available. In both cases, transformation events have been selected directly by treatment with the growth inhibitor, with a transformation efficiency near that obtained for *URA3* selection. In the experiments reported, all drug-resistant colonies obtained also became Ura⁺ owing to the *URA3* carried on the plasmid; spontaneous drug resistance was therefore not a problem.

If there are no vectors available with appropriate replication origins or selectable markers, a gene fusion to the desired promoter can be con-

---

[23] E. Alani, L. Cao, and N. Kleckner, *Genetics* **116**, 541 (1987).
[24] L. Gritz and J. Davies, *Gene* **25**, 179 (1983).
[25] K. R. Kaster, S. G. Burgett, and T. D. Ingolia, *Curr. Genet.* **8**, 353 (1984).
[26] J. Rine, W. Hansen, E. Hardeman, and R. W. Davis, *Proc. Natl. Acad. Sci. U.S.A.* **80**, 6750 (1983).

TABLE I
MARKERS, VECTORS, AND RESTRICTION FRAGMENTS

Marker	Vector	Restriction fragment	Sequence ref.
URA3	YEp24[a]	1.2 kb HindIII or 1.1 kb HindIII–SmaI	b
URA3–hisG	pNKY51[23]	3.8 kb BamHI–BglII	
LEU2	YEp13[c]	2.2 kb SalI–XhoI	d
LYS2	pDA6200[e]	4.6 kb EcoRI–HindIII 4.8 kb EcoRI–ClaI	
HIS3	YEp6[f]	BamHI–BamHI	
HIS4	pPB54 (cited by Hahn et al.[g])	1.8 kb SacI–BstEII	h
TRP1	YRp7[f]	1.5 kb EcoRI	i
ARG4	pCL1[j]	3.0 kb HindIII–HindIII	
Hgm^R	pIT219[25]	BamHI partial	24
TUN^R	pJR41[26]	2.3 kb EcoRI	

[a] M. Carlson and D. Botstein, Cell (Cambridge, Mass.) 28, 145 (1982).
[b] M. Rose, P. Grisafi, and D. Botstein, Gene 29, 113 (1984).
[c] J. R. Branch, J. N. Strathern, and J. B. Hicks, Gene 8, 121 (1979).
[d] A. Andreadis, Y. Hsu, M. Hermodson, G. Kohlhaw, and P. Schimmel, J. Biol. Chem. 259, 8059 (1984).
[e] D. A. Barnes and J. Thorner, Mol. Cell. Biol. 6, 2828 (1986).
[f] K. Struhl, D. T. Stinchcomb, S. Scherer, and R. W. Davis, Proc. Natl. Acad. Sci. U.S.A. 76, 1035 (1979).
[g] S. Hahn, J. Pinkham, R. Wei, R. Miller, and L. Guarente, Mol. Cell. Biol. 8, 655 (1988).
[h] T. F. Donahue, P. J. Farabaugh, and G. R. Fink, Gene 18, 47 (1982).
[i] G. Tschumper and J. Carbon, Gene 10, 157 (1980).
[j] C. Hsiao and J. Carbon, Gene 15, 157 (1981).

structed in a vector from which the fusion can subsequently be removed on a restriction enzyme fragment. The fusion can be then subcloned into a yeast cloning vector with the desired attributes. Parent et al.[27] have compiled an extensive list of yeast cloning vectors. In addition, sets of yeast vectors containing multiple unique cloning sites as well as other useful sequences for plasmid manipulation are discussed below (see section on Plasmid Shuttle Vectors).

Regulatable Promoters

Regulated Expression with GAL Promoter

The regulation of the genes required for metabolism of galactose has been extensively studied by biochemical, genetic, and molecular biological

[27] S. A. Parent, C. M. Fenimore, and K. A. Bostian, Yeast 1, 83 (1985).

methods (for review, see Ref. 28). There are two regulatory proteins, GAL4 and GAL80, which calibrate transcription of the structural genes: *GAL2*, a permease; *GAL1*, a kinase; *GAL7*, a transferase; *GAL10*, an epimerase, and *MEL1*, a galactosidase. *GAL11* is possibly an additional regulatory gene, as mutations in this gene reduce the level of induction.[29] The *GAL3* gene product is thought to produce the intracellular inducer from galactose. When the inducer is present, the GAL4 protein binds to sites within the UAS and activates transcription. When no inducer is present, in lactate medium, for instance, GAL80 binds to the carboxyl terminus of GAL4, masking the activation domain. If cells are growing on glucose medium, expression is repressed in several ways in addition to absence of inducer; the actions of repressors at sites between the UAS and the TATA box possibly account for the interference with binding of GAL4 to the UAS, and inhibition of galactose uptake and inducer synthesis as well. Addition of glucose to cells growing in galactose thus causes immediate repression of transcription, allowing the regulation of the *GAL* UAS to be manipulated. The UAS of the galactose genes contains one or more conserved 17-bp palindromic sequences to which the GAL4 protein binds. Only one of the palindromes is necessary for induction by GAL4 in galactose, although different levels of gene transcription can be obtained by the use of various numbers and combinations of the palindromes, which vary in strength.[30]

This tight regulation of expression by carbon source makes the divergent UAS of the *GAL1* and *GAL10* genes highly suitable for manipulating the expression of cloned genes. In addition, the extremely strong induction of transcription in galactose is useful for expressing high levels of a protein. For example, expression of galactokinase *(GAL1)* is induced 1000-fold in galactose from barely detectable levels in glucose,[31,32] resulting in 0.8% of total cell protein.[33] However, proteins which are overexpressed to this extent can affect viability of the cell. Because of the regulated nature of the system, cells containing the gene fusion can be easily propagated in a noninducing carbon source, then the gene can be induced by addition of galactose.

For the regulated expression of a cloned gene by the GAL system, two options exist. First, the UAS of the divergently transcribed *GAL1* and *GAL10* genes is present on a 365-bp *Sau*3A–*Dde*I fragment of pSc4812. This fragment contains sites sufficient for maximum galactose induction

[28] M. Johnston, *Microbiol. Rev.* **51**, 458 (1987).
[29] Y. Suzuki, Y. Nogi, A. Abe, and T. Fukasawa, *Mol. Cell. Biol.* **8**, 4991 (1988).
[30] E. Giniger and M. Ptashne, *Proc. Natl. Acad. Sci. U.S.A.* **85**, 382 (1988).
[31] T. P. St. John and R. W. Davis, *J. Mol. Biol* **152**, 285 (1981).
[32] L. Guarente, R. R. Yocum, and P. Gifford, *Proc. Natl. Acad. Sci. U.S.A.* **79**, 7410 (1982).
[33] T. Fukasawa, K. Obonai, T. Segawa, and Y. Nogi, *J. Biol. Chem.* **255**, 2705 (1980).

and tight glucose repression[31,32] (see Fig. 1). No TATA box sequences or mRNA initiation sites are present on this fragment. Thus, it may simply be inserted in either orientation upstream of the TATA box to place a cloned gene under *GAL* UAS control. The cognate promoter does not have to be removed; GAL regulation will operate in addition to normal regulation. This is a convenient method for overproduction of a protein if the approximate location of the 5' end of the gene is known and convenient restriction sites are present. The UAS can be placed quite far upstream (at least 1 kb) from the TATA box and still activate transcription. Second, vectors exist into which an open reading frame, either with or without its cognate ATG initiation codon, can be placed. These contain a variety of origins, selectable markers, and cloning sites (see Table II).

Example of Overproduction of Toxic Protein. HAP1 is a trans-activator which mediates transcriptional activation of the *CYC1* gene in response to heme by binding to sites in *CYC1* UAS1.[3] The HAP1 open reading frame was inserted into the *Sma*I–*Sal*I fragment of pKP151 [creating pKP(SD5-HAP1)] with the intention of overproducing the gene to facilitate purification of large quantities of the protein. Continual growth of the cells in galactose, however, was not possible, presumably because of the toxic effects on the transcription of other genes caused by overproduction of an

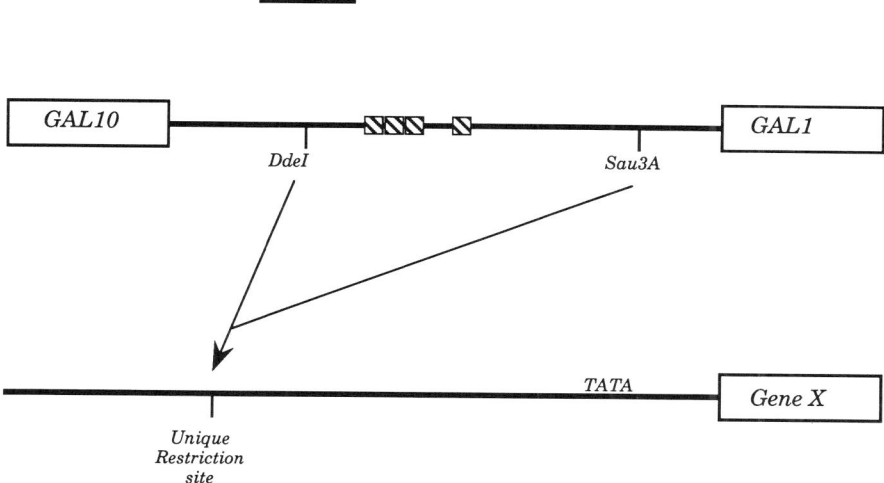

FIG. 1. Upstream activation site for the divergently transcribed *GAL1* and *GAL10* genes. Striped boxes indicate the 17-bp conserved palindromes which constitute the GAL4 binding sites.[28] These can be conveniently removed on a *Sau*3A–*Dde*I restriction fragment and inserted into the promoter of another gene.

TABLE II
Vectors with Promoter Cassettes

UAS	Vector	ATG	Cloning sites	Marker	Origin	Comments	Ref.
GAL1–GAL10	pBM150	No	EcoRI BamHI	URA3	CEN4 ARS1	EcoRI is 4–6 nucleotides and BamHI is 56 nucleotides downstream of mRNA initiation site	12
	pYEp51	No	SalI to BamHI, HindIII or BclI	LEU2	2 μm		a
	pYEp52	No	HindIII to HindIII or BclI	LEU2	2 μm		a
	pLGSD5	Yes	BamHI	URA3	2 μm	Contains the 356-bp GAL UAS upstream of CYC1 TATA and mRNA initiation regions; can create lacZ fusion protein	32
	pLGSD5-ATG	No	BamHI	URA3	2 μm	7 bp containing the ATG removed from pLGSD5 (referred to as G1)[b]	b
	pKP151	No	Replace BamHI to SalI or SmaI to SalI	URA3	2 μm	Derived from pLGSD5-ATG by removing lacZ sequences	c
	YEp61	Yes	SmaI	LEU2	2 μm	Can create lacZ fusion protein	a
	YEp62	Yes	SalI, BamHI or SmaI	LEU2	2 μm	Can create lacZ fusion protein	a
PHO5	pAM82	No	XhoI	LEU2	2 μm	UAS and mRNA leader can be removed as an EcoRI–XhoI cassette	d

Promoter	Plasmid		Site	Marker	Origin	Comments	Ref.
ADH1	pYE4	No	EcoRI	TRP1	2 μm	TRP1 mRNA terminator downstream of the cloning site	e
	pAAH5	No	HindIII	LEU2	2 μm	ADH1 mRNA terminator downstream of cloning site	9
	pMA56	No	EcoRI	TRP1	2 μm	Message terminates in 2-μm sequences	9
	pAH9, pAH10, pAH21	Yes	HindIII	LEU2	2 μm	HindIII cloning site present in all three reading frames	9
PGK	pMA230	Yes	BamHI	LEU2	2 μm	BamHI 33 bp downstream of the ATG	7
	pMA91	No	BglII	LEU2	2 μm	PGK mRNA terminator downstream of cloning site	10
	YEp1PT	No	EcoRI	TRP1	2 μm	Message terminates in 2-μm sequences	f

[a] J. R. Broach, Y. Li, L. C. Wu, and M. Jayaram, in "Experimental Manipulation of Gene Expression" (M. Inouye, ed.). Academic Press, New York, 1983.
[b] L. Guarente, this series, Vol. 101, p. 181.
[c] K. Pfeifer, Ph.D. Thesis, Massachusetts Institute of Technology, Cambridge, Massachusetts (1988).
[d] A. Miyanohara, A. Toh-e, C. Nosaki, F. Hamada, N. Ohtomo, and K. Matsubara, Proc. Natl. Acad. Sci. U.S.A. **80**, 1 (1983).
[e] R. A. Kramer, T. M. DeChiara, M. D. Schaber, and S. Hilliker, Proc. Natl. Acad. Sci. U.S.A. **81**, 367 (1984).
[f] R. A. Hitzeman, D. W. Leung, L. J. Perry, W. T. Kohr, H. L. Levine, and D. V. Goeddel, Science **219**, 620 (1983).

TABLE III
INDUCTION OF HAP1 FROM THE *GAL* UAS[a]

1. Cells of strain BWG1-7a carrying pKP(SD5-HAP1) are grown to saturation in rich media, YEP-lactate (1% yeast extract, 2% Bacto-peptone, 2% lactate)
2. Minimal media, YMM-lactate (0.67% yeast nitrogen base without amino acids, 2% lactate, 0.004% amino acid supplements), is inoculated with the saturated culture to an OD_{600} of 0.1 and incubated overnight at 30°
3. When the OD_{600} reaches 1, galactose [Sigma Cat. No. G0750 (St. Louis), filter sterilized] is added to 0.2%. Incubation is continued at 30° for 6–15 hr, then cells are harvested for preparation of crude protein extracts. The OD_{600} generally reaches 6–7 by this time

[a] From K. Pfeifer, Ph.D. Thesis, Massachusetts Institute of Technology, Cambridge, Massachusetts (1988).

activator protein. Therefore, the cells were grown in a noninducing carbon source to mid-log phase, then supplemented with galactose in order to induce HAP1 transcription from the *GAL* UAS; the protocol is presented in Table III.

Several points about this procedure bear discussion. First, it is useful to have an analogous vector with the promoter fused to an easily assayed reporter gene in order to monitor the extent of transcriptional induction. In this instance a *GAL* UAS-*lacZ* fusion was used (pLGSD5, see below). The induction of *lacZ* was immediate, but it took 6 hr to reach a maximum after addition of galactose, much longer than the 30 min that has been reported.[34] This difference may be due to strain variations or differences in culture media. Second, the noninducible carbon source should not be glucose. Owing to repression of the *GAL* UAS by glucose, initial induction of *lacZ* is not observed until at least 2 hr after galactose addition to cells growing in glucose.[35] The lack of expression in lactate medium, by contrast, is due only to lack of the inducer, not to repression. For cells which are petite, and therefore unable to grow on lactate, raffinose can be used as the noninducing carbon source. At least 12 hr was required after addition of galactose to reach maximum induction in this case. Third, the amount of HAP1 protein isolated from cells was about 1000-fold higher than that from genomic HAP1, as assayed by its ability to bind to DNA fragments carrying the *CYC1* UAS1. However, posttranscriptional control may not allow a particular protein to be overexpressed to this extent, even though mRNA levels are highly induced.

Use of GAL Promoter to Curtail Protein Expression. The strong glucose repression of the *GAL* UAS has been used to curtail expression of essential

[34] B. G. Adams, *J. Bacteriol.* **111**, 308 (1972).
[35] K. Pfeifer, Ph.D. Thesis, Massachusetts Institute of Technology, Cambridge, Massachusetts (1988).

genes for which functionless mutants are not possible. Thus, the carbon source can be used to regulate expression of the fusion gene in much the same way that temperature shifts have been used to control activity of temperature-sensitive alleles. With this method Deshaies et al.[36] showed that a particular family of HSP70-related proteins was required for targeting of protein precursors to the secretory pathway and mitochondria. They deleted the chromosomal loci of all members of the family and supplemented the strain with one of the genes, *SSA1*, on a plasmid under control of the *GAL* UAS. After addition of glucose to galactose-grown cells, they showed that accumulation of the HSP70 protein declined severely after 6 hr. Concurrently, the precursors, prepro-α-factor (a secreted protein) and preF$_1$β (a mitochondrially localized protein), accumulated to higher levels than normal, thereby implicating HSP70 in protein secretion.

Similarly, the *GAL* UAS was used to control expression of the H4 histone in a cell with a deletion of endogenous H4 genes. The mRNA for H4 was gone completely within 2 hr after addition of glucose to cells growing on galactose. Washing the cells and reinoculation into galactose media caused reinduction of H4 mRNA synthesis. The elimination of synthesis of H4 by addition of glucose during synchronization of cells in G$_1$ led to nucleosome depletion.[37] This shutoff of H4 caused the activation of transcription of several genes by a pathway independent of the activators.[38]

Regulation by Inorganic Phosphate

PHO5 Regulation. Another gene which is tightly regulated is *PHO5*. This gene encodes an inducible acid phosphatase which is secreted from the cell when extracellular levels of inorganic phosphate are low. Transcription of the gene is highly induced under these conditions, yet the message is undetectable when inorganic phosphate is added to the medium.[39] As with the *GAL* UAS, these properties make the promoter very useful for controlling the expression of cloned genes. Since some petite strains do not grow well in galactose, this system may be a useful substitute when a total elimination of gene expression is required. In addition, a special strain has been constructed (see below) which makes the *PHO5* UAS temperature inducible. Temperature induction may be more convenient than galactose addition for *GAL* UAS-regulated genes, especially with petite strains, since raffinose is expensive (see above).

[36] R. J. Deshaies, B. D. Koch, M. Werner-Washburne, E. A. Craig, and R. Schekman, *Nature (London)* **332**, 800 (1988).
[37] U. Kim, M. Han, P. Kayne, and M. Grunstein, *EMBO J.* **7**, 2211 (1988).
[38] M. Han and M. Grunstein, *Cell (Cambridge, Mass.)* **55**, 1137 (1988).
[39] D. T. Rogers, J. M. Lemire, and K. A. Bostian, *Proc. Natl. Acad. Sci. U.S.A.* **79**, 2157 (1982).

A set of repressors and activators, defined by mutation, regulate the transcription of *PHO5*. *PHO2* and *PHO4* are required for activation; *PHO80* and *PHO85* are required for repression. In addition, the *PHO84* gene product is required for import of the corepressor, inorganic phosphate.[40] A conserved dyad sequence present four times in the UAS of *PHO5* is also found in the UAS of other coregulated genes; deletion analysis of the promoter and subcloning of the dyads established that each site can mediate phosphate regulation independently, although some are stronger than others.[41] Binding of PHO4 to two of the conserved dyads has been observed *in vitro;* PHO2 binds to a region between the two dyads.[42] Strangely, although the PHO4 binding site is necessary for activation in response to low phosphate, the PHO2 binding site can be removed without greatly reducing transcription. *PHO2* turns out to be the same as *BAS1,* which is needed to maintain the basal level of *HIS4* transcription.[43] Vogel et al.[42] suggest that the PHO2 protein may act at other levels in addition to DNA binding. Although the *PHO80* and *PHO85* gene products are required for repression in the presence of inorganic phosphate, the mechanism of their action is not yet clear.

Phosphate- and Temperature-Regulated Expression of Interferon by PHO5 Upstream Activation Site. A gene for interferon (IFN) was cloned into the *Eco*RI site of pYE4. High expression of rIFN-αD was obtained by growing the cells on media lacking inorganic phosphate.[44] No interferon was produced when inorganic phosphate was added to the medium, until levels had been depleted by cell growth. Thus, interferon was produced under phosphate control. Because prompt induction of expression by phosphate depletion is awkward, a special strain was devised by taking advantage of mutants in the phosphate regulatory pathway. As a result, synthesis from this promoter was induced in response to a temperature change, instead of phosphate levels. The strain carries a loss-of-function mutation in the repressor gene, *PHO80,* and a temperature-sensitive activator *pho4*^{ts}. At the permissive temperature (23°) the *pho80* mutant allows activation of the *PHO5* UAS by *pho4*^{ts} regardless of inorganic phosphate levels. At restrictive temperatures, the *pho4*^{ts} activator, and thus the *PHO5* UAS, is not functional. In this strain, interferon was produced within a few hours after a shift from the higher to the lower temperature. The levels produced, however, were about 10–20% those in the wild-type strain,

---

[40] Y. Oshima, "The Molecular Biology of the Yeast *Saccharomyces*—Metabolism and Gene Expression." Cold Spring Harbor Laboratory, Cold Spring Harbor, New York, 1982.
[41] H. Rudolph and A. Hinnen, *Proc. Natl. Acad. Sci. U.S.A.* **84,** 1340 (1987).
[42] K. Vogel, W. Horz, and A. Hinnen, *Mol. Cell. Biol.* **9,** 2050 (1989).
[43] K. T. Arndt, C. Styles, and G. R. Fink, *Science* **237,** 874 (1987).
[44] R. A. Kramer, T. M. DeChiara, M. D. Schaber, and S. Hilliker, *Proc. Natl. Acad. Sci. U.S.A.* **81,** 367 (1984).

A set of repressors and activators, defined by mutation, regulate the transcription of *PHO5*. *PHO2* and *PHO4* are required for activation; *PHO80* and *PHO85* are required for repression. In addition, the *PHO84* gene product is required for import of the corepressor, inorganic phosphate.[40] A conserved dyad sequence present four times in the UAS of *PHO5* is also found in the UAS of other coregulated genes; deletion analysis of the promoter and subcloning of the dyads established that each site can mediate phosphate regulation independently, although some are stronger than others.[41] Binding of PHO4 to two of the conserved dyads has been observed *in vitro;* PHO2 binds to a region between the two dyads.[42] Strangely, although the PHO4 binding site is necessary for activation in response to low phosphate, the PHO2 binding site can be removed without greatly reducing transcription. *PHO2* turns out to be the same as *BAS1*, which is needed to maintain the basal level of *HIS4* transcription.[43] Vogel et al.[42] suggest that the PHO2 protein may act at other levels in addition to DNA binding. Although the *PHO80* and *PHO85* gene products are required for repression in the presence of inorganic phosphate, the mechanism of their action is not yet clear.

Phosphate- and Temperature-Regulated Expression of Interferon by PHO5 Upstream Activation Site. A gene for interferon (IFN) was cloned into the *Eco*RI site of pYE4. High expression of rIFN-αD was obtained by growing the cells on media lacking inorganic phosphate.[44] No interferon was produced when inorganic phosphate was added to the medium, until levels had been depleted by cell growth. Thus, interferon was produced under phosphate control. Because prompt induction of expression by phosphate depletion is awkward, a special strain was devised by taking advantage of mutants in the phosphate regulatory pathway. As a result, synthesis from this promoter was induced in response to a temperature change, instead of phosphate levels. The strain carries a loss-of-function mutation in the repressor gene, *PHO80,* and a temperature-sensitive activator *pho4*^{ts}. At the permissive temperature (23°) the *pho80* mutant allows activation of the *PHO5* UAS by *pho4*^{ts} regardless of inorganic phosphate levels. At restrictive temperatures, the *pho4*^{ts} activator, and thus the *PHO5* UAS, is not functional. In this strain, interferon was produced within a few hours after a shift from the higher to the lower temperature. The levels produced, however, were about 10–20% those in the wild-type strain,

---

[40] Y. Oshima, "The Molecular Biology of the Yeast *Saccharomyces*—Metabolism and Gene Expression." Cold Spring Harbor Laboratory, Cold Spring Harbor, New York, 1982.
[41] H. Rudolph and A. Hinnen, *Proc. Natl. Acad. Sci. U.S.A.* **84**, 1340 (1987).
[42] K. Vogel, W. Horz, and A. Hinnen, *Mol. Cell. Biol.* **9**, 2050 (1989).
[43] K. T. Arndt, C. Styles, and G. R. Fink, *Science* **237**, 874 (1987).
[44] R. A. Kramer, T. M. DeChiara, M. D. Schaber, and S. Hilliker, *Proc. Natl. Acad. Sci. U.S.A.* **81**, 367 (1984).

possibly because the *pho4*ᵗˢ is not fully functional at the permissive temperature.

Overexpression in Glucose

Regulation of ADH1 and PGK Upstream Activation Sites. The genes coding for the fermentative enzyme alcohol dehydrogenase I *(ADH1)* and the glycolytic enzyme 3-phosphoglycerate kinase *(PGK)* are expressed at very high levels, greater than 1% of total RNA each, in glucose media.[45] These promoters are thus widely used for high-level expression of cloned yeast or foreign genes. *ADH1* is often considered to be a "constitutive" gene; however, expression is repressed when the cells are growing on a nonfermentable carbon source such as lactate. The extent of this repression may be as slight as 2-fold[46] or as much as 10-fold.[47] As with variation in expression of *GAL* UAS, this difference may be due to strain variations. Likewise, *PGK* is repressed 20- to 30-fold when cells are growing on acetate, a nonfermentable carbon source.[7]

Expression of Viral Protein and Interferons in Yeast Using ADH1 and PGK Upstream Activation Sites. Yeast has been used as an alternative to *Escherichia coli* for the expression of foreign genes. A hepatitis antigen, HBsAg (hepatitis B surface antigen), and a variety of interferons have been successfully produced at high levels from the *ADH1* and *PGK* transcription vectors. The *ADH1* promoter from pMA56 was used to express the mature form of HBsAg.[48] This protein normally contains a secretion signal sequence; in this case, the fusion junction with the *ADH1* leader was placed in the middle of the signal so that the protein would not be exported from the cell. A fortuitous ATG at the 5′ end of the coding region for the mature protein was used as an initiation codon. This construction produced 2–5 μg of HBsAg from a 200-ml culture; the HBsAg synthesized in yeast had the formed particles similar to those purified from human tissue culture cells, as judged by cross-reaction to antibodies, molecular weight, density, sedimentation rate, and electron microscopy. Since the yeast-synthesized protein was not exported, however, it was not glycosylated, unlike that from tissue culture.

Tagging Proteins

Fusions to a gene such as *lacZ* may be used to "tag" proteins for which no convenient assay or antibodies are available. (See [35] in this volume

---

[45] M. J. Holland and J. P. Holland, *Biochemistry* **17**, 4900 (1978).
[46] D. R. Beier and E. T. Young, *Nature (London)* **300**, 724 (1982).
[47] C. L. Denis, J. Ferguson, and E. T. Young, *J. Biol. Chem.* **258**, 1165 (1983).
[48] P. Valenzuela, A. Medina, W. J. Rutter, G. Ammerer, and B. D. Hall, *Nature (London)* **398**, 347 (1982).

for a discussion of epitope tagging.) If the coding sequence of the cloned gene is inserted (minus the termination codon) into a site upstream of *lacZ*, it is possible to create a bifunctional protein consisting of the introduced gene at the amino-terminal half and *lacZ* at the carboxyl-terminal half. (β-Galactosidase does not retain activity if a protein is fused to its carboxyl terminus.) These fusions have been used in our laboratory to follow a protein during purification by β-galactosidase assays[49] or for intracellular localization by indirect immunofluorescense with antibodies against β-galactosidase.[50,51]

Plasmid Shuttle Vectors

Often a fusion cassette is not present on a vector appropriate for propagation in the desired yeast strains. In this case, the fusion can be subcloned into another yeast vector with different markers (see [20] in this volume). In addition, Elledge and Davis[52] have developed a series of centromeric vectors (pUN) designed so that a colony sectoring assay can be used to analyze mutants in essential genes. These vectors are similar to the pBluescript-based vectors described by Sikorski and Boeke ([20], this volume), except they carry ochre suppressor genes (*SUP4* or *SUP11*) for the colony-sectoring assay. To analyze mutants, a chromosomal lesion in an essential gene is introduced into a strain with an *ade2-101* ochre mutation. The *ade2-101* mutation causes accumulation of a colored intermediate, so the cells are pink. Viability of the strain is maintained through complementation of the gene lesion by a functional copy on a pUN plasmid. Cells containing the pUN plasmid are white owing to suppression of *ade2-101* by *SUP4* or *SUP11*. This strain is transformed with a second plasmid carrying mutants in the essential gene and a different selectable marker. If the second copy of the gene is functional, the original plasmid (with the suppressor marker) is free to segregate. In this case, plasmid loss leads to pink sectors on a white background. On the other hand, if the new plasmid fails to complement, no pink sectors will be observed, as cells which lose the suppressor plasmid are nonviable. This method may be useful to generate large numbers of mutants, including temperature-sensitive ones, since complementation can be observed directly, without the need for selection against the original plasmid.[52]

---

[49] S. Hahn and L. Guarente, *Science* **240**, 317 (1988).
[50] M. N. Hall, L. Hereford, and I. Herskowitz, *Cell (Cambridge, Mass.)* **36**, 1057 (1984).
[51] J. L. Pinkham, J. T. Olesen, and L. P. Guarente, *Mol. Cell. Biol.* **7**, 578 (1987).
[52] S. J. Elledge and R. W. Davis, *Gene* **70**, 303 (1988).

[26] Vectors for Constitutive and Inducible Gene Expression in Yeast

By MARK SCHENA, DIDIER PICARD, and KEITH R. YAMAMOTO

Introduction

The development of plasmid vectors for expression of cloned DNA sequences in yeast has aided in establishing *Saccharomyces cerevisiae* as a model system for cellular, molecular, and genetic studies. Vectors that mediate high-level constitutive gene expression in yeast commonly employ the upstream activator sequences (UASs) and promoters from yeast genes encoding metabolic enzymes such as alcohol dehydrogenase I *(ADH1)*[1] and 3-phosphoglycerate kinase (PGK).[2] Other plasmids have been described that allow inducible gene expression in yeast. For example, the UAS and promoter region from the yeast *GAL10* gene[3-5] permits selective expression of fused sequences when cells are grown on nonfermentable carbon sources supplemented with galactose. In a second example, the UAS and promoter of the yeast *PHO5* gene confer strongly induced expression of linked sequences in media depleted of inorganic phosphate.[6,7] These vectors are maintained at a high copy number in yeast by the 2-μm origin of replication and by the presence of dominant selectable markers such as *URA3, LEU2,* or *TRP1*.[5]

Many of the available expression systems, however, possess certain limitations. Thus, although the *ADH1* and *PGK* vectors are generally considered to be "constitutive," expression from these promoters is actually repressed as much as 10-fold[8] and 30-fold,[2] respectively, on nonfermentable carbon sources. In the case of inducible expression vectors, induction generally involves drastic alterations in growth conditions, such as carbon source changes, which have highly pleiotropic effects on cellular

---

[1] G. Ammerer, this series, Vol. 101, p. 192.
[2] M. F. Tuite, M. J. Dobson, N. A. Roberts, R. M. King, D. C. Burke, S. M. Kingsman, and A. J. Kingsman, *EMBO J.* **1,** 603 (1982).
[3] L. Guarente, R. R. Yocum, and P. Gifford, *Proc. Natl. Acad. Sci. U.S.A.* **79,** 7410 (1982).
[4] M. Johnston and R. W. Davis, *Mol. Cell. Biol.* **4,** 1440 (1984).
[5] J. C. Schneider and L. Guarente, this volume [25].
[6] A. Miyanohara, A. Toh-e, C. Nozaki, F. Hamada, N. Ohtomo, and K. Matsubara, *Proc. Natl. Acad. Sci. U.S.A.* **80,** 1 (1983).
[7] R. A. Kramer, T. M. DeChiara, M. D. Schaber, and S. Hilliker, *Proc. Natl. Acad. Sci. U.S.A.* **81,** 367 (1984).
[8] C. L. Denis, J. Ferguson, and E. T. Young, *J. Biol. Chem.* **258,** 1165 (1983).

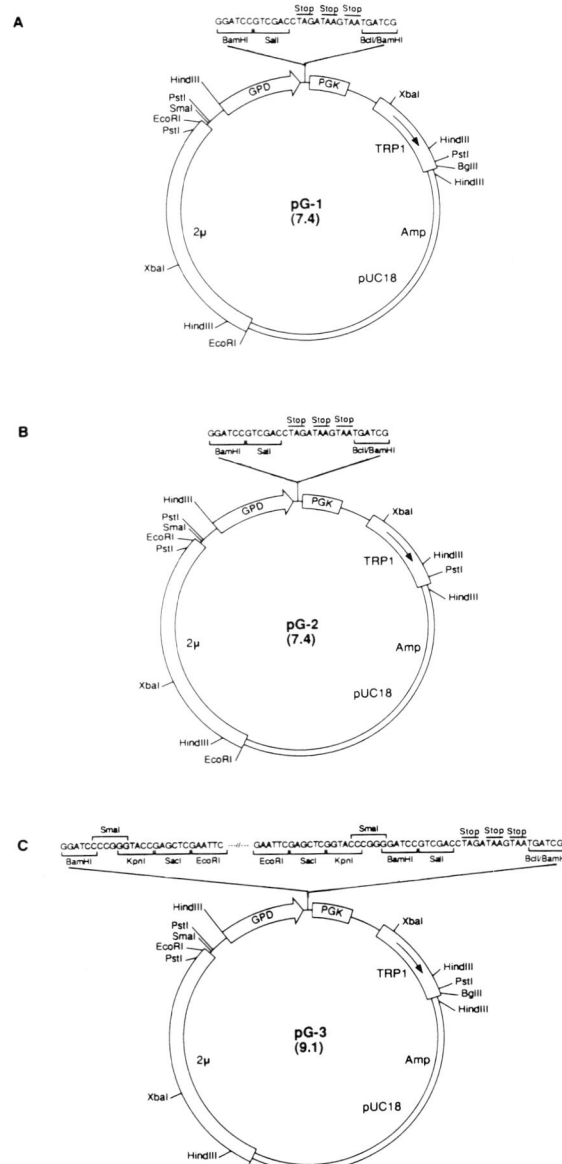

FIG. 1. Constitutive expression plasmids. (A) Plasmid pG-1. Thick boxes represent the yeast *GPD* promoter, the yeast *PGK* transcription terminator and polyadenylation signal, the yeast *TRP1* gene, and the yeast 2-μm origin of replication. The thin box indicates the bacterial origin of replication and ampicillin resistance gene from pUC18. Thin lines between the 2-μm region and the *GPD* promoter and between the *PGK* terminator and the *TRP1* gene

metabolism; moreover, the levels of expression achieved on induction are often lower than those obtained with constitutive promoters. Finally, many of the available yeast expression vectors lack convenient cloning sites and replicate poorly in *Escherichia coli.*

Here we describe two novel vector systems for constitutive and inducible gene expression in *Saccharomyces cerevisiae* that appear to remedy some of these problems. In particular, we have constructed a series of glyceraldehyde-3-phosphate dehydrogenase (GPD) promoter-based vectors that allow convenient, high-level constitutive gene expression in yeast. In addition, we have developed a hormone-inducible expression vector whose low basal promoter activity is strongly enhanced by the addition of glucocorticoids to yeast cells expressing the glucocorticoid receptor; importantly, steroid hormones are gratuitous inducers of gene expression in yeast having little or no effect on the expression of endogenous genes. The purpose of this chapter is to provide a practical discussion on the features and use of these plasmids.

Constitutive Expression

We have constructed three new plasmids (pG-1, pG-2, and pG-3) that direct high-level constitutive gene expression in yeast. These vectors, derived from plasmids originally developed for expression of the rat glucocorticoid receptor cDNA in yeast,[9,10] contain the very efficient yeast glyceraldehyde-3-phosphate dehydrogenase promoter.[11,12] In addition, each plasmid contains the yeast *TRP1* gene and 2-μm origin of replication and the ampicillin resistance gene and prokaryotic origin of replication from pUC18.

Using unique restriction sites present in a polylinker in each construct, cloned genes or cDNAs can be readily inserted downstream of the GPD promoter (Fig. 1); transcripts initiate approximately 20 nucleotides up-

[9] M. Schena and K. R. Yamamoto, *Science* **241**, 965 (1988).
[10] M. Schena, L. P. Freedman, and K. R. Yamamoto, *Genes Dev.* **3**, 1590 (1989).
[11] G. A. Bitter, this series, Vol. 152, p. 673.
[12] G. A. Bitter and K. M. Egan, *Gene* **32**, 263 (1984).

---

denote pBR322 sequences. The nucleotide sequence of the polylinker is shown, including restriction sites and translation termination codons. Arrows indicate the direction of transcription from the *GPD* promoter and of the *TRP1* gene, respectively. The approximate size of the plasmid (in kilobases) is given in parentheses. (B) Plasmid pG-2. Plasmid pG-2 is identical to pG-1 except that the *Bgl*II site downstream of the *TRP1* gene has been eliminated (see text). (C) Plasmid pG-3. Plasmid pG-3 is identical to pG-1, except that a fragment containing additional restriction sites has been inserted into the pG-1 polylinker at the *Bam*HI site. The dashed line between the *Eco*RI sites in the polylinker denotes 1.7 kb of spacer DNA from the *lac* operon that has been inserted into the *Bam*HI site (see text).

TABLE I
RESTRICTION SITES PRESENT IN CONSTITUTIVE
(pG-1, pG-2, pG-3) AND INDUCIBLE (p2UG)
EXPRESSION VECTORS[a]

Enzyme	pG-1	pG-2	pG-3	p2UG
BamHI	1	1	2	1
BglII	1	0	1	0
EcoRI	2	2	4	3
HindIII	4	4	4	2
KpnI	0	0	2	1
NcoI	0	0	0	1
PstI	3	3	3	2
SacI	0	0	2	1
SalI	1	1	1	0
SmaI	1	1	1	1
XbaI	2	2	2	1
XhoI	0	0	0	1

[a] Number of restriction sites present in each vector for 12 commonly used restriction enzymes. See Figs. 1 and 3 for map locations.

stream of the proximal BamHI site in each vector.[13,14] As this "leader sequence" does not introduce an upstream AUG codon, translation initiation depends on the presence of a start codon within the inserted sequences; translation then proceeds either to a termination codon within the insert or to a cluster of termination codons in all three reading frames immediately distal to the polylinker. Polyadenylation and termination of the transcripts are conferred by sequences from the 3' untranslated region of the yeast phosphoglycerate kinase gene that resides downstream of the polylinker.

Plasmids pG-1 and pG-2 (Fig. 1A,B) are identical except that pG-2 lacks a BglII site located downstream of the TRP1 gene in pG-1, thus enabling introduction of a unique BglII site into the pG-2 polylinker if needed. Plasmid pG-3 is also similar to pG-1, except that it contains a 1.7-kilobase (kb) insert at the BamHI site in the pG-1 polylinker, which provides additional novel cloning sites (Fig. 1A,C). Table I lists restriction sites present in each vector for 12 commonly used restriction enzymes.

[13] A. M. Musti, Z. Zehner, K. A. Bostian, B. M. Paterson, and R. A. Kramer, Gene 25, 133 (1983).
[14] J. P. Holland, L. Labieniec, C. Swimmer, and M. J. Holland, J. Biol. Chem. 258, 5291 (1983).

To compare the levels of expression obtained with these constructs to standard yeast expression systems, a 2.8-kb cDNA fragment encoding the 795-amino acid rat glucocorticoid receptor[15] was inserted into the BamHI site in plasmid pG-1 and into the BamHI site of plasmid pLGSD5,[5] to yield plasmids pG-N795 and pSD-N795, respectively. Expression plasmid pLGSD5 contains the GAL10 UAS fused to the CYC1 TATA region, as well as the yeast URA3 gene and 2-μm origin of replication, and thus directs high-level expression of inserted sequences when transformants are grown on minimal galactose medium. Yeast strain BJ2168[16,17] was transformed with pG-N795, pG-1, or pSD-N795 according to the method of Ito et al.,[18] and transformants were selected and propagated on minimal glucose or galactose medium lacking tryptophan or uracil.[19] Extracts prepared from these cultures were assessed for expression of the glucocorticoid receptor protein (denoted N795) by immunoblotting, using a receptor-specific monoclonal antibody.

The N795 protein was detected in extracts from both glucose- and galactose-grown transformants carrying pG-N795 and in extracts from galactose-grown transformants containing pSD-N795 (Fig. 2, lanes 3–5). Importantly, expression from pG-N795 in cells grown on glucose or galactose was 5- and 2-fold higher, respectively, relative to that observed in galactose-induced cells containing pSD-N795 (Fig. 2, compare lanes 4 and 5 to lane 3). Reconstitution experiments with purified glucocorticoid receptor indicate that the N795 product constitutes approximately 0.1% of the soluble protein in BJ2168 transformants propagated on minimal glucose media (data not shown). Levels of N795 approaching 1% of the soluble protein have been achieved with pG-N795 in yeast strains with shorter generation times than BJ2168; in general, both strain background and insert characteristics appear to influence expression level (data not shown).

Growth of pG-N795 transformants in rich media allows strains to be "cured" of the expression plasmids at a rate of about 0.1% per generation. The pUC18 origin of replication facilitates efficient shuttling of the plasmids between yeast and E. coli;[10] plasmid yields from minilysate preparations are approximately 10 μg/ml bacterial culture.

---

[15] R. Miesfeld, S. Rusconi, P. J. Godowski, B. A. Maler, C. Okret, A.-C. Wikström, J. Å. Gustafsson, and K. R. Yamamoto, *Cell (Cambridge, Mass.)* **46**, 389 (1986).
[16] E. W. Jones, *Genetics* **85**, 23 (1977).
[17] P. K. Sorger and H. R. B. Pelham, *EMBO J.* **6**, 3035 (1987).
[18] H. Ito, Y. Fukuda, K. Murata, and A. Kimura, *J. Bacteriol.* **153**, 163 (1983).
[19] F. Sherman, G. R. Fink, and J. B. Hicks, "Methods in Yeast Genetics." Cold Spring Harbor Laboratory, Cold Spring Harbor, New York, 1986.

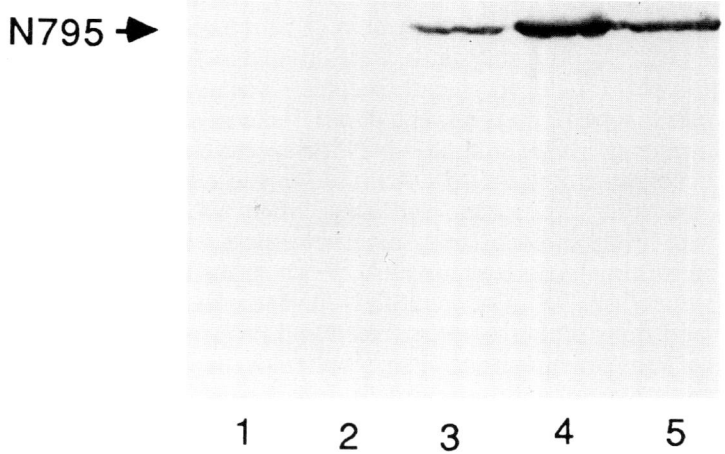

FIG. 2. Constitutive expression in yeast. Shown is an immunoblot of yeast extracts fractionated by polyacrylamide gel electrophoresis, transferred to nitrocellulose, and probed with a glucocorticoid receptor-specific monoclonal antibody [B. Gametchu and R. W. Harrison, *Endocrinology* **114,** 274 (1984)] and with a goat anti-mouse IgG conjugated to alkaline phosphatase (Bio Rad, Richmond, CA). Expression plasmids pG-1, pG-N795, and pSD5-N795 (see text) were transformed into yeast strain BJ2168 (from E. Jones, *MATa pep4-3 prc1-407 prb1-1122 ura3-52 trp1 leu2*) according to Ito et al.,[18] and extracts were prepared as described[10] from cells grown in minimal yeast medium.[19] Each lane contains 10 μg of total yeast protein prepared from logarithmic BJ2168 cultures transformed with pG-1 (lane 1), pSD5-N795 (lanes 2, 3), or pG-N795 (lanes 4, 5), grown for 36 hr in minimal glucose medium (lanes 1, 2, 4) or in minimal galactose medium (2% galactose, 3% glycerol, 2% ethanol) (lanes 3, 5). The arrow indicates the migration position of the N795 receptor protein (88 kDa).

Inducible Expression

Mammalian steroid receptors maintain their activities as conditional transcriptional regulators when expressed in heterologous species such as *Saccharomyces cerevisiae*[9,20] and in *Schizosaccharomyces pombe*.[21] These findings prompted the development of a glucocorticoid-inducible yeast expression vector termed p2UG.[21] Plasmid p2UG contains three tandem 26-base pair (bp) glucocorticoid response elements (GREs) fused upstream of the yeast *CYC1* promoter region; in cells containing p2UG and a second plasmid (pG-N795) that encodes the glucocorticoid receptor, the addition of glucocorticoids results in specific binding of receptor to GREs within p2UG and transcriptional enhancement of the *CYC1* promoter. Thus, as

[20] D. Metzger, J. H. White, and P. Chambon, *Nature (London)* **334,** 31 (1988).
[21] D. Picard, M. Schena, and K. R. Yamamoto, *Gene* **86,** 257 (1990).

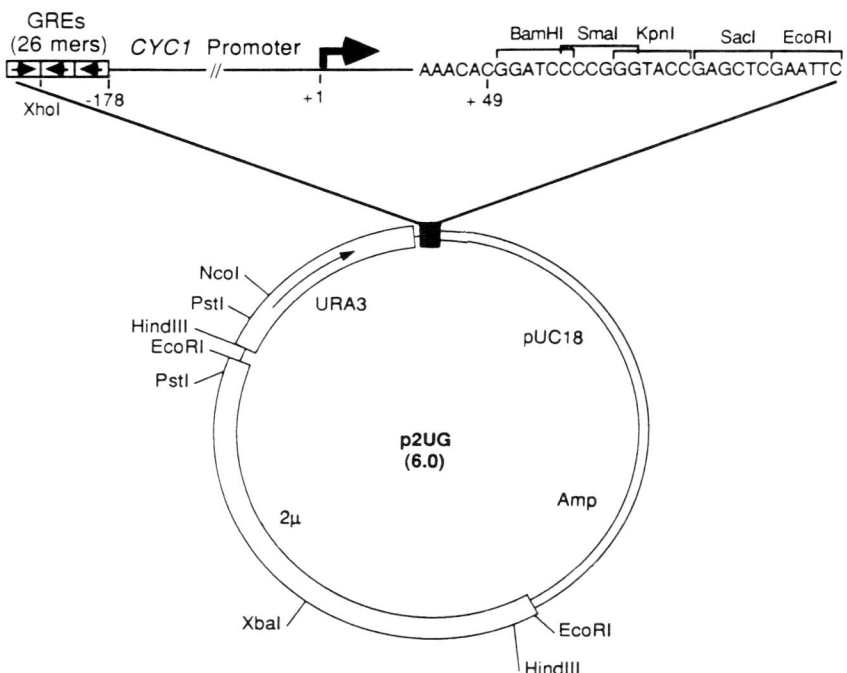

FIG. 3. A hormone-inducible expression vector. Shown is a diagram of plasmid p2UG.[21] Thick boxes represent the yeast 2-μm origin of replication and the yeast *URA3* gene. The arrow indicates the direction of transcription of the *URA3* gene. The thin box indicates the bacterial origin of replication and the ampicillin resistance gene from pUC18. The solid box (detailed above the plasmid diagram) depicts the yeast *CYC1* TATA region [L. Guarente and T. Mason, *Cell (Cambridge, Mass.)* **32**, 1279 (1983)] fused to three copies of a 26-bp GRE oligonucleotide in the designated orientations[9,21] derived from the rat tyrosine aminotransferase gene [H.-M. Jantzen, U. Strähle, B. Gloss, F. Stewart, W. Schmid, M. Boshart, R. Miksicek, and G. Schütz, *Cell (Cambridge, Mass.)* **49**, 29 (1987)] and positioned 178 bp upstream of the 5′-most cap site [S. Hahn, E. T. Hoar, and L. Guarente, *Proc. Natl. Acad. Sci. U.S.A.* **82**, 8562 (1985)]. The nucleotide sequence of the polylinker containing the designated restriction sites is also shown.

in mammalian cells,[22] expression of sequences inserted into a polylinker downstream of a glucocorticoid-regulated promoter in yeast is strongly dependent on expression of the glucocorticoid receptor and on addition of steroid hormone to the yeast culture medium. Plasmid p2UG also contains the yeast *URA3* gene and 2-μm origin of replication (Fig. 3); selected restriction sites in p2UG are listed in Table I.

[22] K. R. Yamamoto, *Annu. Rev. Genet.* **19**, 209 (1985).

To determine the levels of expression from p2UG, the bacterial chloramphenicol acetyltransferase (CAT) gene was inserted at the BamHI site within the polylinker to generate p2UGCAT.[21] Plasmids p2UGCAT and pG-N795 (see above) were cotransformed into yeast strain BJ2168 and selected and propagated on minimal glucose media deficient of uracil and tryptophan.[19] Treatment of these cultures with the glucocorticoid deoxycorticosterone at a level of 10 μM resulted in a 50- to 100-fold increase in CAT expression compared to uninduced cultures (Table II; see also Ref. 21). This level of enhancement agrees well with the magnitude of β-galactosidase induction observed using a closely related reporter plasmid;[9] moreover, the level of CAT expression obtained on induction of p2UGCAT was equivalent to that observed when CAT was expressed from the GPD promoter in the constitutive pG-CAT construct (Table II).

An important feature of the p2UG/pG-N795 inducible expression system is that steroid hormones are gratuitous inducers in yeast; they have little or no effect on the expression of endogenous genes. Thus, GRE-

TABLE II
HORMONE-INDUCIBLE CAT EXPRESSION[a]

Experiment	Plasmids	CAT activity		Induction ratio
		−DOC	+DOC	
1	pG-1 + p2UGCAT	300	n.d.	—
	pG-N795 + p2UGCAT	1,100	57,000	52
	pG-CAT	60,000	n.d.	—
2	pG-N795 + p2UGCAT	450	30,000	67
	pG-N795 + p2UGCAT	300	31,000	103
	pG-N795 + p2UGCAT	400	30,000	75

[a] Yeast strain BJ2168 (see legend to Fig. 2) was transformed[18] with plasmids p2UGCAT, pG-1, or pG-N795 (see text) together with plasmid pG-CAT, which was constructed by inserting the CAT gene from p2UGCAT into pG-1. Yeast transformants were selected and propagated on minimal media deficient in the appropriate amino acids.[19] Extracts were prepared according to Jones et al. [R. H. Jones, S. Moreno, P. Nurse, and N. C. Jones, Cell (Cambridge, Mass.) **53**, 659 (1988)] from cells grown in the absence (−) or presence (+) of 10 μM deoxycorticosterone (DOC) (Sigma St. Louis, MO, Cat. No. D-6875; prepared as a 10 mM stock in 100% ethanol and stored at −20°). CAT activities were measured according to M. J. Sleigh [Anal. Biochem. **156**, 251 (1986)], using 10–70 ng of extract protein incubated with substrate for 1 hr at 37°. Results from two experiments are shown; Experiment 2 presents data from three independent transformants. CAT activity is expressed as ^{3}H-acetylated chloramphenicol (cpm)/[extract protein (μg) × reaction time (min)]. n.d., Not determined.

linked sequences can be specifically and strongly induced without general metabolic perturbations. Moreover, induction kinetics are rapid (studies in mammalian cells reveal promoter activation with a $t_{1/2}$ of 7–9 min after hormone addition,[23] and intermediate levels of induction can be achieved simply by titrating the levels of hormone in the culture medium between 1 nM and 10 μM (Ref. 21; M. J. Garabedian and K. R. Yamamoto, unpublished results, 1989). In principle, any gene or cDNA that contains a translation initiation sequence and a stop codon can be inserted into p2UG and faithfully expressed in yeast in a hormone-dependent manner.

Vector Constructions

Plasmid pG-1 was derived from pG-D[10] by deleting the 1.7-kb BamHI fragment encompassing a glucocorticoid receptor cDNA insert. Plasmid pG-D was constructed from pGPD-556a[9] by deleting the 5.5-kb EcoRI fragment of yeast 2-μm DNA, followed by replacing the pBR322 sequences with pUC18. These deletions were accomplished by a quadruple ligation of the following pGPD-556a and pUC18 fragments: a BglII to EcoRI fragment encompassing pUC18; an EcoRI to XbaI fragment of yeast 2-μm plasmid; an XbaI to SacI fragment encompassing the remainder of 2-μm, the GPD promoter, and the glucocorticoid receptor cDNA; and a SacI to BglII fragment encompassing the PGK terminator and the yeast TRP1 gene. The BglII site was introduced into pUC18 by adding a BglII linker to the PstI polylinker site rendered blunt-ended with T4 DNA polymerase.

Plasmid pG-1 thus contains the following sequences: the 650-bp TaqI fragment of the GPD promoter,[13] the 3' border of which lies at position −24 relative to the +1 ATG initiation codon;[11] the 38-bp BamHI to BclI polylinker sequence from pSV7d;[15] the 380-bp BglII to HindIII fragment of the transcription termination and polyadenylation region of the yeast PGK gene;[24] the 850-bp EcoRI to BglII fragment of the yeast TRP1 gene;[25] the 2700-bp BglII to EcoRI fragment of pUC18;[26] and the 2246-bp EcoRI to EcoRI fragment of the B-form of the yeast 2-μm plasmid.[27,28] The nucleotide sequence of the GPD promoter and additional cloning details are given by Bitter.[11]

---

[23] D. S. Ucker and K. R. Yamamoto, *J. Biol. Chem.* **259**, 7416 (1984).
[24] R. A. Hitzeman, F. E. Hagie, J. S. Hayflick, C. Y. Chen, P. H. Seeburg, and R. Derynck, *Nucleic Acids Res.* **10**, 7791 (1982).
[25] G. Tschumper and J. Carbon, *Gene* **10**, 157 (1980).
[26] C. Yanisch-Perron, J. Vieira, and J. Messing, *Gene* **33**, 103 (1985).
[27] J. L Hartley and J. E. Donelson, *Nature (London)* **286**, 860 (1980).
[28] A. B. Rose and J. R. Broach, this series, Vol. 185, p. 234.

Plasmid pG-2 was generated by the blunt-end ligation of plasmid pG-1 linearized with BglII. Plasmid pG-3 was generated by inserting a 1.7-kb fragment into the BamHI site of pG-1; this fragment corresponds to a HinII fragment of the *lac* operon,[29] which was inserted via EcoRI linkers into a plasmid bearing inverted pUC19 polylinkers (Bob DuBridge, Genentech) and liberated by digestion with BamHI. Plasmid p2UG[21] was derived from parent plasmid pSX26.1[9] by substituting a polylinker for *CYC1* sequences downstream of position +49 as well as the β-galactosidase coding region; furthermore, the pBR322 sequences were replaced with pUC18.

Acknowledgments

We thank C. Peterson for plasmids, E. Jones for yeast strains, and R. Harrison for the BUGR-1 antibody. We are also grateful to members of the Guthrie laboratory for communication of results, and to members of the Guthrie, Herskowitz, and Yamamoto laboratories for discussions. This work was supported by grants from the National Institutes of Health and the National Science Foundation; postdoctoral fellowship support (D.P.) was from the Swiss National Research Foundation.

[29] J. E. Mott, J. Van Arsdell, and T. Platt, *Nucleic Acids Res.* **12**, 4139 (1984).

[27] Preparation of High Molecular Weight RNA

By KARL KÖHRER and HORST DOMDEY

The preparation of high molecular weight RNA of good quality is an essential step in the analysis of gene expression in yeast. At first glance, preparing RNA seems to be a rather simple procedure; however, it requires some special precautions owing to the ubiquitous presence of RNA-degrading enzymes. For this reason it is important to wear gloves during the isolation and preparation procedure, and it is essential to separate all chemicals, tubes, and tips used for RNA preparation from commonly used materials. All glassware has to be baked for 2 hr at 200°; commercially available sterile plastic tubes or tips do not have to be specially pretreated (e.g., autoclaving), because they seem to be free of ribonucleases. Since the degradation of RNA by RNases is a time-dependent reaction, it is also advisable to work as fast as possible. Working according to these guidelines, it is possible to isolate high molecular weight RNA from yeast cells without adding any ribonuclease inhibitors like RNasin or vanadyl–ribonucleoside complex.

High molecular weight RNA can be isolated from haploid or diploid yeast cells as well as from ascospores. The conditions under which the yeast cells are grown (complete or selective media) do not influence the preparation procedure, although higher yields of RNA are generally obtained from cells grown in rich medium.

Isolation of Total RNA from Vegetative Yeast Cells

Several possibilities have been described to break the yeast cell wall, for example, by enzymatic digestion with Zymolyase, Lyticase, or Glusulase.[1,2] Because of the short half-life of some RNAs, however, it is often essential to carry out the initial steps of the isolation procedure very rapidly. Therefore, generally a combination of mechanical and chemical breakage of the cells is preferred. In our opinion the method of choice is to break the cells by vigorous shaking in the presence of hot phenol.

Materials

Sodium acetate buffer: 50 mM sodium acetate, 10 mM EDTA; adjusted to pH 5.0 with acetic acid
10% SDS (sodium dodecyl sulfate)
Phenol: 500 ml liquified phenol equilibrated with 500 ml sodium acetate buffer (stored at 4° in brown glass bottles)
ANE buffer: 10 mM sodium acetate, 100 mM NaCl, 1 mM EDTA, pH 6.0
Chloropane: 50% liquified phenol, 50% chloroform, 0.5% 8-hydroxychinoline, equilibrated with ANE buffer and stored at 4° in brown glass bottles
Chloroform–isoamyl alcohol (24:1)
3 M Sodium acetate, pH 5.3
Ethanol
DEPC-treated water: 0.1% DEPC (diethyl pyrocarbonate) in distilled water; allow to stand overnight, then autoclave

Procedure

Logarithmically growing yeast cells are the best source for the preparation of total RNA. Therefore, an overnight culture is diluted in the morning with fresh prewarmed medium and grown for another one or two generations to an optical density at 600 nm (OD$_{600}$) of 1–3. It is most convenient to isolate RNA from a liquid yeast culture, 100 to 400 ml

[1] O. Necas, *Bacteriol. Rev.* **35**, 149 (1971).
[2] J. H. Scott and R. Schekman, *J. Bacteriol.* **142**, 414 (1980).

volume; however, the RNA can also be isolated from several liters of yeast culture by adjusting the respective volumes. Generally, about 2-4 mg of total RNA can be isolated from 1 g of yeast cells, which corresponds to about 100 ml of logarithmically growing cells. Before starting the isolation, a dry ice-ethanol bath is prepared, and the phenol is prewarmed to 65°. A centrifuge with the chosen rotor is precooled, and a water bath or, if available, a water bath shaker is heated to 65°.

The cultured cells (2-4 g) are collected by centrifugation (2500 g, 5 min, 4°) in a precooled rotor, resuspended in 10 ml sodium acetate buffer, and transferred to 50-ml plastic tubes. If the amount of cells is higher, a multifold of buffer is added, and the cell suspension is aliquoted in 10-ml portions to several tubes, rather than using different plastic tubes taking bigger volumes. One-tenth volume of 10% SDS and 1.2 volumes phenol (65°) are added to the cell suspension and immediately transferred to the water bath shaker set at maximum speed (e.g., 400 strokes/min in a New Brunswick gyratory water bath shaker) for 4 min at 65°. Alternatively, it is also possible to agitate the tubes in a 65° water bath over a total period of 5 min with intermittent (every 30 sec) short vortexing at room temperature. Subsequently, the sample is cooled down quickly to room temperature in the dry ice-ethanol bath. The sample should not get too cold, because the SDS will precipitate. The organic and aqueous phases are separated by a 10-min centrifugation (2500 g) at room temperature. The lower organic phase is removed with a sterile pipette, leaving the interphase and the pellet of unbroken cells and cell debris in the tube. If the organic and aqueous phases are not well separated, more hot phenol should be added and the tubes mixed and centrifuged again.

To the remaining aqueous phase (including interphase and cell debris) the same volume of prewarmed phenol (65°) is added, and the extraction procedure is repeated as described above. After separation of the organic and aqueous phases, the upper aqueous phase is transferred to a new 50-ml plastic tube. One volume of Chloropane is added, the sample is vortexed for 2 min, and the two phases are again separated by centrifugation. The aqueous phase is transferred to a new tube and extracted once with 1 volume chloroform-isoamyl alcohol (24:1). After phase separation, the upper aqueous phase is transferred to a new tube, 1/10 volume of 3 M sodium acetate (pH 5.3) and 3 volumes ethanol are added, and, after the sample is briefly vortexed, the RNA is precipitated at −20° for several hours.

The precipitated RNA is collected by centrifugation (2500 g, 4°) for at least 15 min, and the RNA pellet is washed with 70% ethanol, dried briefly under reduced pressure, and dissolved in an appropriate amount of water (e.g., in 600 μl). (Sometimes it is necessary to heat the sample for a short time in order to get the RNA completely dissolved.) Aliquots (300 μl) of the RNA solution are transferred to 1.5-ml Eppendorf tubes, 1/10 volume of 3 M sodium acetate (pH 5.3) and 3 volumes ethanol are added, and the tubes are mixed and left at $-70°$ for 20 min. The RNA is pelleted by centrifugation (15,000 g) at 4° for 15 min. The RNA pellet is rinsed with 70% ethanol, dried briefly, and dissolved to a final concentration of 10 $\mu g/\mu l$ in DEPC-treated water.

General Comments

The above-described hot phenol extraction procedure yields high molecular weight RNA with very little (almost no) contaminating DNA. RNA prepared by this method is perfect for any enzymatic or chemical manipulation, such as Northern blot, S1, RNase H, primer extension, RNA sequence, or RNA secondary structure analyses, as well as for *in vitro* translations.

A different phenol-based extraction procedure of total RNA has been described in the Cold Spring Harbor manual, "Methods in Yeast Genetics."[3] The main differences from the above-described hot phenol method are the use of glass beads, the temperature (room temperature), and the pH of 7.4. Although this method yields higher amounts of RNA, the isolated RNA is contaminated with considerable amounts of DNA. In order to obtain the higher yields of RNA, but without contaminating DNA, the above-described hot phenol extraction procedure can be carried out in the presence of glass beads. The use of glass beads is indispensible if RNA is prepared from ascospores (see below).

An alternative procedure for the isolation of intact high molecular weight RNA, which is based on the nuclease-denaturing capacity of guanidinium salts, is described in [28], this volume.

Isolation of RNA from Yeast Ascospores

The yeast ascospore wall is much more rigid than the cell wall of vegetative cells and therefore much more resistant to mechanical breakage. Consequently, it is essential to use glass beads in addition to hot phenol in order to efficiently crack the ascospore wall. This method can, of course, also be used for vegetative cells.

Additional Materials

    Glass beads with an approximate size of 0.4 mm (40 mesh), treated once with 1 M HCl, washed several times with water, and autoclaved

---

[3] F. Sherman, G. R. Fink, and J. B. Hicks, "Methods in Yeast Genetics." Cold Spring Harbor Laboratory, Cold Spring Harbor, New York, 1986.

Procedure

Yeast ascospores can be enriched and isolated according to the method described in [9], this volume. The spores are washed with water, up to 2 g of spores are suspended in 2.5 ml sodium acetate buffer (pH 5.0), and transferred to a 50-ml plastic tube, and 11 g of glass beads in 3 ml hot phenol are added. Subsequently, the samples are alternately vortexed at top speed for 30 sec and kept for 30 sec in the 65° water bath for a total of 3 min.

Centrifugation of the samples (2500 g, 10 min, 4°) leads to the separation of the organic phase, with the glass beads at the bottom of the tube, and the upper aqueous phase. The aqueous phase is transferred to a new tube and extracted once with Chloropane and once with chloroform-isoamyl alcohol (24:1). The RNA is precipitated with ethanol and dissolved in DEPC-treated water as described above.

Isolation of Poly(A)$^+$ RNA

Only 1.3–1.4% of the total yeast RNA is poly(A)$^+$ RNA.[4] The excess of nonpolyadenylated rRNA and tRNA very often results in cross-hybridization with one of these abundant RNA species, especially when degenerate or heterologous probes are used. Also, in order to detect low-abundance poly(A)$^+$ RNAs by Northern blot, S1, or primer extension analyses, an enrichment of poly(A)$^+$ RNA may be essential. Furthermore, for the generation of cDNA libraries it is also necessary to remove the nonpolyadenylated ribosomal and transfer RNAs. In order to enrich poly(A)$^+$ RNA from a preparation of total RNA, either oligo(dT)-cellulose or poly(U)-Sephadex column chromatography is used. This procedure results in an efficient removal of tRNAs, rRNAs, and also contaminating traces of DNA.

Materials

Oligo(dT)-cellulose
Low-salt buffer (LSB): 10 mM Tris-HCl (pH 7.5), 0.2% SDS, 1 mM EDTA
High-salt buffer (HSB): 10 mM Tris-HCl (pH 7.5), 0.2% SDS, 5 mM EDTA, 0.5 M NaCl
0.1 M NaOH
5 M NaCl
3 M Sodium acetate, pH 5.3
Ethanol
DEPC-treated water: 0.1% DEPC (diethyl pyrocarbonate) in distilled water; allow to stand overnight, then autoclave

[4] E. Kraig and J. E. Haber, *J. Bacteriol.* **144**, 1098 (1980).

Procedure

For preparation of the oligo(dT)-cellulose column, 0.6–0.8 ml (wet volume in water) of oligo(dT)-cellulose is pipetted into a 15-ml column, washed with several column volumes of sterile water, washed with 10 ml of 0.1 M NaOH, and subsequently washed with sterile water until the pH of the effluent is neutral. The column is equilibrated with 50 ml HSB.

The ethanol-precipitated total RNA (up to 20 mg) is dissolved in 2.5 ml LSB, incubated at 65° for 3 min, and cooled in an ice bath to room temperature. One-tenth volume of 5 M NaCl is added, and the RNA solution is applied to the column. For good binding of the poly(A)$^+$ RNA, the flow rate has to be adjusted to around 200 μl/min (or 1 drop/5 sec) and should be controlled from time to time. The effluent is collected, heated again to 65°, cooled, and reapplied to the column. Subsequently, the column is washed with at least 100 ml HSB to remove most of the rRNA and tRNA. The bound poly(A)$^+$ RNA is eluted with 8 ml LSB, which is added in 0.5-ml portions. For efficient removal of rRNA and tRNA a second passage through the column is recommended. To do so, NaCl is added to a final concentration of 0.5 M, and the sample is loaded on the freshly equilibrated (HSB) column.

The eluted poly(A)$^+$ RNA is precipitated at $-20°$ overnight with 1/10 volume of 3 M sodium acetate (pH 5.3) and 3 volumes ethanol. A second ethanol precipitation helps to get rid of traces of SDS. The RNA is dissolved to a final concentration of 2 μg/μl in DEPC-treated water and stored at $-70°$.

General Comments

It is also possible to use poly(U)-Sephadex instead of oligo(dT)-cellulose for the column chromatography. Although we routinely use oligo(dT) cellulose, poly(U)-Sephadex is occasionally preferred by others, because the Sephadex-coupled poly(U) stretch is longer than the oligo(dT) attached to the cellulose.

Another procedure to enrich poly(A)$^+$ RNAs is paper affinity chromatography using poly(U) filter paper which is prepared by poly(U) inactivation of diazothiophenyl paper.[5,6] Since here the enrichment of poly(A)$^+$ RNA is not as efficient as with oligo(dT) chromatography (even when the RNA is applied twice to the paper), we recommend this procedure only if a number of poly(A)$^+$ RNA preparations have to be carried out simultaneously.

[5] D. Werner, Y. Chemla, and M. Herzberg, *Anal. Biochem.* **141**, 329 (1984).
[6] D. Wreschner and M. Herzberg, *Nucleic Acids Res.* **12**, 1349 (1984).

Evaluation of Quality of Prepared RNA

The best and fastest way to analyze the quality of the isolated RNA is gel electrophoresis in a nondenaturing 1–1.5% TBE-agarose gel (TBE: 89 mM Tris–borate, 89 mM boric acid, 2 mM EDTA). Different RNA dilutions are coelectrophoresed in an agarose minigel alongside DNA size markers for a short period (20 min, 5 V/cm). High-quality total RNA or poly(A)$^-$ RNA show two distinct bands (28 and 18 S ribosomal RNA) comigrating with DNA fragments of 1 and 0.7 kilobase pairs (kbp) length, respectively (Fig. 1). In addition, a smear of low molecular weight RNA comigrating with 50–200 base pair (bp) long DNA fragments will be visible. The quality of the RNA preparation can then be estimated by comparing the amount of high molecular weight RNA (the two 28 and 18 S ribosomal RNA bands) and the amount of low molecular weight RNA (tRNA, 5 S RNA, and degraded RNA). The amount and intensity of the high molecular weight RNA should be at least twice the amount and intensity of low molecular weight RNA. The application of different RNA

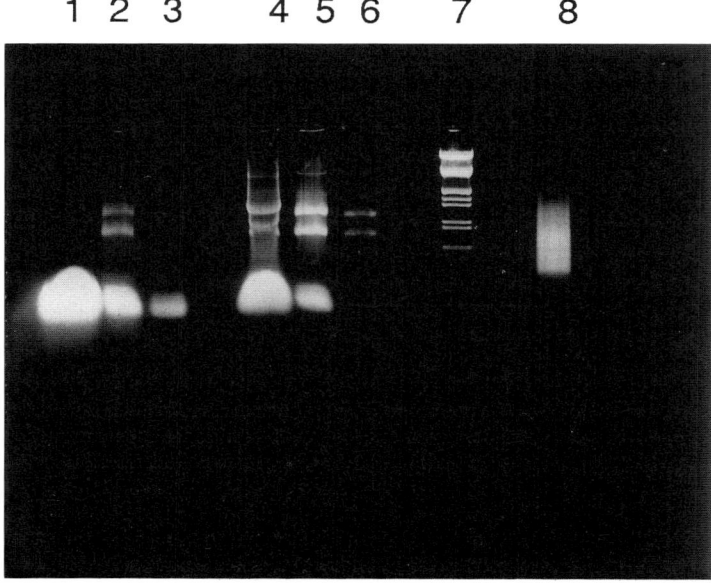

FIG. 1. TBE-agarose (1.2%) gel electrophoreses of yeast RNA under native conditions. Lanes 1–3: 50 μg (lane 1), 10 μg (lane 2), and 2 μg (lane 3) partially degraded total yeast RNA. Lanes 4–6: 50 μg (lane 4), 10 μg (lane 5), and 2 μg (lane 6) good quality total RNA. Lane 7: DNA size markers (λ DNA digested with EcoRI and HindIII). Lane 8: 5 μg poly(A)$^+$ RNA.

dilutions is very important because overloaded gels will give the wrong impression of most of the RNA being degraded (Fig. 1, lane 4).

Using such native agarose gels it is also possible to detect traces of contaminating DNA in RNA preparations. With DNase that is absolutely free of RNase activity (Boehringer Mannheim, Mannheim, FRG) it is possible to get rid of the DNA. If total RNA from ascospores is analyzed, in addition to the two ribosomal RNAs (18 and 28 S rRNA) an intermediate processing form which accumulates during sporulation and which migrates between these two bands will be detected.

The first step in the evaluation of the quality of the poly(A)$^+$ RNA is to determine the amount of enriched poly(A)$^+$ RNA. Therefore, one should measure the optical density at 260 nm (25 units A_{260} = 1 mg) to calculate the amount of collected RNA. Generally approximately 1% of the total RNA isolated from vegetative yeast cells is recovered as poly(A)$^+$ RNA. A smaller fraction is obtained if the RNA was isolated from ascospores. We also recommend determining the ratio of the optical densities at 260 nm and 280 nm, since only if this ratio is between 1.7 and 2.0 one can be sure that the isolated poly(A)$^+$ RNA is free of any proteins and phenol. If high-quality poly(A)$^+$ RNA is separated on a nondenaturing TBE-agarose gel, a smear ranging between DNA size markers of 2 kbp and 500 bp, but no bands corresponding to the ribosomal or transfer RNA, should be visible (Fig. 1, lane 8).

[28] Preparation and Analysis of Low Molecular Weight RNAs and Small Ribonucleoproteins

By JO ANN WISE

The repertoire of functions ascribed to low molecular weight RNAs has expanded immensely in recent years. In addition to their familiar roles in protein synthesis, various species of small RNA participate in premessenger RNA splicing, processing of tRNA precursors, and protein secretion; future research will undoubtedly add to this list. Several classes, including transfer (t) RNAs, small nuclear (sn) RNAs, and small cytoplasmic (sc) RNAs, can be prepared and analyzed by similar procedures. Although these RNAs are far more stable than messenger RNA because of extensive intramolecular hydrogen bonding, it is still necessary to observe the precautions described in [27], this volume, to avoid breakdown by endogenous nucleases. In addition to extraction of RNA from yeast cells, gel and blot techniques commonly used to analyze small RNAs are described. Total RNA preparations generally provide the starting point for these protocols; however, in some cases it is desirable to enrich for the RNA(s) of interest by subfractionating the cells before extraction. It is also worth noting that these procedures are applicable to other species such as the distantly related fission yeast *Schizosaccharomyces pombe* (see, e.g., Ref. 1), except that provision must be made for differences in the cell wall if spheroplasts are to be prepared.

Extraction of Total RNA

As noted in [27] in this volume, several methods can be used to break or permeabilize the yeast cell wall to allow extraction of nucleic acids. For routine preparation of total RNA, we employ mechanical breakage with glass beads in the presence of the chaotropic agent guanidine thiocyanate to obtain high yields of good quality RNA.

Materials

Diethyl pyrocarbonate (DEPC)-treated water: Add DEPC to 0.1% (the limit of solubility), shake vigorously, and autoclave to remove residual DEPC (DEPC is an alkylating agent which destroys RNase activity)

Glass beads: 250–300 μm, baked for at least 3 hr at 350°F (160°C) to destroy RNase activity

RNase-free tubes: Cell harvesting and extractions should be performed in disposable 50-ml screw-cap polypropylene tubes, available from several manufacturers; handle the tubes with gloves and take care not to touch the rims

GTE buffer: 4 M guanidine thiocyanate, 100 mM Tris-HCl, pH 7.6, 10 mM EDTA, pH 8.0

4 M Sodium acetate, pH 5.5 (DEPC-treated)

Phenol: 500 ml liquefied phenol (purchase high quality or redistill) equilibrated with water; many people equilibrate against buffer (e.g., see [27] in this volume), but this may increase DNA contamination

Chloropane: 50% liquefied redistilled phenol containing 0.5% hydroxyquinoline, 49% chloroform, 1% isoamyl alcohol (v/v/v)

Chloroform–isoamyl alcohol (24:1)

Procedure. As for preparation of high molecular weight RNA, it is best to harvest cells from a liquid culture in late logarithmic phase. The procedure given below is for 100 ml of culture at an OD_{600} of 1–3 (1 OD = 3 ×

[1] P. Brennwald, X. Liao, K. Holm, G. Porter, and J. A. Wise, *Mol. Cell. Biol.* **8**, 1580 (1988).

10^7 cells/ml), from which a yield of approximately 2–4 mg of total RNA can be expected. If a larger or smaller amount of RNA is desired, the volumes of all reagents should be adjusted accordingly.

Before beginning the isolation, be certain that all required materials and buffers are on hand and at the desired temperature, since it is important to proceed rapidly through the steps in order to isolate intact RNA. We often set up the tubes for the organic extractions prior to harvesting cells. The first extraction is performed in a heated (60°–65°) water bath, preferably one that shakes. The 5.8 S rRNA is inefficiently released from 25 S rRNA if the initial extraction is performed at temperatures lower than 60°. Prewarming the phenol helps to ensure efficient release of small RNAs that may be involved in hydrogen bonding or other tight associations with higher molecular weight structures.

The cells are collected by centrifugation (2500 g, 5 min), preferably at 4° in a precooled rotor, but an unrefrigerated centrifuge may be used if this is more convenient. We generally wash away residual medium by resuspending the cells in 20 ml of cold, sterile DEPC-treated water and collecting again by centrifugation. After discarding the supernatant, add to the tube a volume of RNase-free glass beads approximately equal to the volume of the cells, using a spatula dipped in ethanol and flamed to destroy RNases. Finally, add an equal volume of cold GTE buffer (~0.5 ml). Vortex the mixture at high speed for 45–50 sec and cool for 10–15 sec on ice; repeat 3–4 times. Dilute the suspension by adding DEPC-treated water to 10 times the volume of GTE buffer. Add 1/10 volume of 4 M sodium acetate buffer (pH 5.5) followed by a volume of hot phenol equal to that of the aqueous phase. The sodium acetate may also be added later in the procedure, just before precipitating with ethanol. Vortex for about 15 sec and transfer the tube to the shaking water bath (we generally place at a slant in a beaker of water). Shake at high speed for 10 min, vortexing for 15 sec every 2–3 min. The extraction time and/or the frequency of vortexing should be increased if a gyratory water bath is unavailable. To facilitate phase separation, add chloroform–isoamyl alcohol to 1/5 the volume of the aqueous phase before centrifugation. After spinning at 2500 g for 10 min at room temperature, both layers will probably be cloudy, and they will be separated by a flocculent white mass of protein and cell debris. If the aqueous and organic phases cannot be clearly distinguished (e.g., if the amount of starting material was higher than assumed), add more phenol and more water, vortex briefly, and spin again.

The aqueous (top) layer is removed to a new tube containing an equal volume of cold chloropane, carefully avoiding the flocculent material; at every step, yield should be sacrificed in favor of purity. Vortex for 1–2 min at top speed, spin for 5 min at 2500 g, and remove the aqueous phase to a new tube containing chloropane. Vortex, spin, and transfer the top phase to a tube containing chloroform–isoamyl alcohol. After the chloroform extraction (and sometimes earlier in the preparation), both layers should be clear, and there should be no material at the interphase. Add 1/10 volume of 4 M sodium acetate (pH 5.5) if you did not do so earlier in the preparation and precipitate the RNA by adding 2.5–3 volumes of 100% ethanol. Cover the tube, mix by vortexing or inverting, and store at $-20°$ for several hours (we generally leave overnight, but this is unnecessary). You should observe opalescence immediately owing to the large amount of RNA.

The precipitated RNA is collected by centrifugation for 15–30 min at 4° at at least 2500 g. Wash the RNA pellet with 5–10 ml of 70% ethanol (made with DEPC-treated water), dry briefly under reduced pressure, and dissolve in an appropriate amount of water (~ 500 μl to yield a 4–8 mg/ml RNA solution). Store frozen at $-20°$, preferably in aliquots so that the entire sample need not be thawed for each use.

Preparation of *in Vivo* ^{32}P-Labeled RNA

We generally employ ortho[^{32}P]phosphate, since efficient labeling of RNA to quite high specific activities can be achieved. Alternative protocols involving incorporation of radioactive precursors such as ^{3}H- or ^{14}C-labeled uracil are described in [30], this volume.

Procedure. A fresh overnight culture is diluted 1:100 in YEPD and grown to an OD_{600} of 0.3. The cells are then collected by centrifugation and resuspended in the same volume of low-phosphate YEPD (a recipe is provided in [30], this volume). The size of the culture will depend on its intended purpose. For example, to obtain sufficient incorporation to analyze 5' end structures of individual species present at only 200 copies/cell (see, e.g., Ref. 2), 50–100 ml of cells and 10–20 mCi of ^{32}P are required. To simply examine the gel electrophoretic pattern of RNA from cells harboring mutations or grown under different conditions, a small (≤ 10 ml) culture will do.

Published protocols specify a 1- to 3-hr wait before addition of ^{32}P (see, e.g., Ref. 3). We have found little difference in the labeling of metabolically stable species such as snRNAs if the starvation for inorganic phosphate is omitted; however, if one wants to examine short-lived species such as precursors to tRNAs or snRNAs, this waiting period may be more critical. Carrier-free ortho[^{32}P]phosphate (an inexpensive preparation supplied in

[2] J. A. Wise, D. Tollervey, D. Maloney, H. Swerdlow, E. J. Dunn, and C. Guthrie, *Cell (Cambridge, Mass.)* **35**, 743 (1983).

[3] T. Etcheverry, D. Colby, and C. Guthrie, *Cell (Cambridge, Mass.)* **18**, 11 (1979).

HCl is suitable) is added to the culture at a concentration of at most 0.25 μCi/ml. The time of incubation after adding label will depend on the experimental goals. We generally allow at least 2 generation times (2–4 hr at 30°) if stable species are to be examined, whereas cells are harvested after no longer than 30 min to optimize labeling of unstable RNAs.

RNA can be prepared from labeled cells as described above except that in general we perform only one chloropane extraction after the hot phenol and then proceed to the ethanol precipitation. This RNA must be used rather quickly to avoid decay of the ^{32}P, and thus need not be as clean as unlabeled RNA, which often undergoes repeated cycles of freezing and thawing. An alternative method for preparing low molecular weight RNAs from small (≤20 ml) cultures is to add hot phenol directly rather than spinning out the cells. This has the added advantage of minimizing the length of time that the investigator is exposed to radioactivity. At the end of the labeling period, simply add an equal volume of prewarmed phenol to the culture and extract small RNAs by shaking vigorously at 60°–65° for 15–30 min. Separate the phases by centrifugation as described above and reextract the aqueous phase with chloropane. Add 1/10 volume of 4 M sodium acetate and precipitate with 2.5–3 volumes of ethanol.

General Comments

An alternative to *in vivo* labeling is to postlabel individual species or mixtures of RNA using RNA ligase and [5'-^{32}P]pCp.[4] A caveat is that, since wide variations are observed in the efficiency with which each RNA is labeled by this procedure, reliable conclusions cannot be drawn about the relative amounts of different species within a population (cf. Refs. 2 and 5).

Isolating Nuclear and Cytoplasmic RNA

To determine in which cell compartment the RNA of interest resides, or to enrich for a species if its localization is known, it is often desirable to fractionate cells before extracting RNA.

Preparation of Nuclear RNA

Materials

SB: 1.2 M sorbitol, 10 mM EDTA (pH 8.0), 10 mM potassium phosphate (pH 7.5), 0.1% 2-mercaptoethanol (add just before use)

---

[4] T. E. England, A. G. Bruce, and O. C. Uhlenbeck, this series, Vol. 65, p. 65.
[5] N. Riedel, J. A. Wise, H. Swerdlow, A. Mak, and C. Guthrie, *Proc. Natl. Acad. Sci. U.S.A.* **83**, 8097 (1986).

Zymolyase (100T)
HMC buffer: 25 mM HEPES (pH 7.6), 5 mM magnesium acetate, 0.5 M sucrose
HMS buffer: 25 mM HEPES (pH 7.6), 5 mM magnesium acetate, 0.25 M sucrose
Solutions for the preparation of RNA (see above)

Procedure. Because the ribonucleoprotein particles in which snRNAs reside are small and easily diffuse out of nuclei, it is unwise to employ protocols which involve time-consuming centrifugation over Percoll or sucrose gradients for their isolation. An additional reason to minimize manipulations between cell breakage and RNA extraction is that yeast nuclei are relatively fragile. The following procedure results in retention of the vast majority of trimethylguanosine-capped RNAs within the nuclear fraction.[5] This protocol is designed for 400 ml of cells and can be scaled up or down with the caveat that the small quantity of RNA in the nucleus may prevent efficient precipitation if the amount of starting material is too low (≤ 1 μg/ml). Recoveries are generally lower than for total RNA extraction from intact cells, largely owing to losses during the lysis and first centrifugation steps.

In order to isolate intact nuclei, cell walls must first be removed to allow gentle lysis. Commercially available Zymolyase preparations[6] may be used since they are relatively free of RNases. Grow the cells to an OD$_{600}$ of at most 0.6–0.8 (overgrowth will lead to inefficient spheroplast formation) and harvest by centrifugation for 5 min at 2500 g at room temperature. Wash in 20 ml of freshly made 0.5% 2-mercaptoethanol, centrifuge, and discard the supernatant. Resuspend the cells in 20 ml of SB and transfer to a flask or bottle in which good aeration can be achieved. Add 5–10 mg of Zymolyase and shake gently for 15–20 min at 30°. The rate of spheroplast formation is strain specific and is best monitored microscopically (cells become spherical and lose their buds). A crude measure is lysis in water: after sufficient digestion, cells will burst to give a clear solution relative to the same dilution made in SB. Harvest the spheroplasts by spinning at 2500 g at room temperature or 4° for 1 min (the shorter spin is to avoid lysis owing to spheroplast fragility). Wash with 15 ml of HMC buffer and recentrifuge; a smaller second pellet means that the cells have probably been overdigested. (If the loss is too great, it may be desirable to start over.) Resuspend spheroplasts in 12 ml HMS buffer containing 0.1% nonidet P-40 (NP-40) and transfer to a prechilled glass (Dounce) homogenizer.

---

[6] J. H. Scott and R. Schekman, *J. Bacteriol.* **142**, 414 (1980).

Lyse the cells on ice using several strokes with a loose pestle ("loose" versus "tight" is defined by the manufacturer), then switch to a tight pestle to remove cytoplasmic components (e.g., ribosomes) adhering to the nuclei; the solution should clear during this step. Lysis can be monitored with a phase-contrast microscope by the appearance of nuclei, spheres approximately 1/10 the size of cells which contain dense dots; empty cell "ghosts," distorted in appearance, may also be present. The next step depends on whether quantitative recovery or purity of the nuclei is most important. For highest purity, carry out a 5-min spin at 2500 g to remove unlysed cells; however, up to 20–30% of the nuclei may be lost as well, in part because of aggregation. To quantitate RNAs in the different cell fractions, it is advisable to proceed immediately to pelleting the nuclei. Layer the lysate onto a 5-ml cushion of HMC using a plastic pipette (nuclei stick to glass) and spin at 8000 g at 4° for 5 min, preferably in a swinging-bucket rotor. The cloudy top layer is the cytoplasmic fraction, from which RNA may be extracted if desired; hold on ice until the nuclei are ready to be processed. The cushion should remain clear and can be discarded. The nuclear pellet should be white; a brown color indicates the presence of unlysed cells. Cell "ghosts" and other membraneous debris are not sufficiently dense to penetrate the sucrose cushion. We sometimes wash the nuclei in 10 ml HMS and centrifuge again through an HMC cushion, although it is not clear that this improves purity significantly, and it provides additional time for losses owing to leakage and lysis. Resuspend the nuclei in 10 ml HMS.

Before extracting RNA, dilute all samples with at least an equal volume of DEPC-treated water; it may be necessary to dilute the nuclear fraction further to ensure good phase separation during the phenol extraction. Add 1/10 volume of 4 M sodium or potassium acetate (pH 5.5), EDTA to 20 mM final concentration, and an equal volume of prewarmed phenol, then proceed with the extractions as described above for total RNA.

Preparation of Small Cytoplasmic RNAs

Small cytoplasmic RNAs relatively free of contamination with ribosomal RNAs or their breakdown products can be obtained from a postribosomal supernatant.

Procedure. Carry out cell lysis as described above for the preparation of nuclear RNA, but discard the nuclear fraction. Since small cytoplasmic ribonucleoprotein particles may be associated with polysomes or other larger structures in salt-labile interactions [e.g., signal recognition particle (SRP); see Ref. 1 and references therein], extracting the cytoplasmic fraction with high salt before pelleting the ribosomes may be desirable. In this case, add 1/10 volume of 5 M potassium acetate (pH 7.5) and mix by inverting several times; you should observe clarification of the solution. Hold on ice for 15–20 min, mixing gently every 2–3 min. Carefully layer the cytoplasmic fraction over 7 ml HMC in an ultracentrifuge tube; if appropriate, the cushion buffer should be brought to 0.5 M potassium acetate to avoid mixing. Centrifuge for 3 hr at 140,000 g (calculated at average rotor radius) at 4°. The pellet is light brown and can be resuspended for rRNA extraction if desired (it will probably be necessary to dislodge it manually). Small RNAs and RNPs are in the cloudy supernatant. We generally avoid the filmy white material (lipids) at the top of the tube and discard the cushion as well. Extract RNA as described above.

To specifically enrich for small ribonucleoproteins (RNPs) that are associated with polysomes and microsomes, we perform the high-speed spin at low salt and then extract the pellet with salt. After removal of the supernatant (including the cushion), resuspend the 140,000 g pellet in 10 ml of HMS plus 0.5 M potassium acetate (pH 7.5) using a glass rod. Incubate on ice for 15 min with occasional gentle agitation, layer over a 7-ml HMC cushion containing 0.5 M potassium acetate, and pellet the large material as above. The supernatant will yield small RNPs relatively free of tRNA contamination.

Size Fractionation on DEAE Resins

It is sometimes desirable to enrich for small RNAs, for example, to lower the background on gels or blots. Because high molecular weight RNAs are bound very tightly to DEAE resins, we often include a chromatographic or batch elution step in our protocols. This is also an effective way to deplete polyphosphates formed during ^{32}P labeling (see [30] in this volume), which also stick tenaciously.

Materials

DEAE-cellulose: DE-52, preswollen
DEAE-Sephacel: Pharmacia (Piscataway, NJ); wet particle size 40–150 μm
Wash buffer: 0.3 M NaCl, 10 mM Tris-HCl (pH 7.4) (autoclave but do not treat Tris buffers with DEPC)
Elution buffer: 1.0 M NaCl, 10 mM Tris-HCl (pH 7.4)

Procedure. For radioactive RNA, we generally use DE-52 columns poured in disposable plastic pipettes. A 4 ml bed volume is required per 100 ml of culture at an OD$_{600}$ of 0.3 (~0.2 g cells). The resin, equilibrated in wash buffer, is poured into the column and allowed to settle. The RNA is diluted with an equal volume of wash buffer and loaded. Most of the label should bind. Wash the column extensively (at least 3 column volumes); there should be a substantial decline in radioactivity in sequential fractions. Small RNAs will be released by elution buffer after approximately 1 column volume has been added; pool the hottest three 1-ml fractions. Dilute the RNA 1:1 with DEPC-treated water and ethanol precipitate at $-20°$ overnight.

For nonradioactive samples, we use batch elution from DEAE-Sephacel, equilibrated in 100 mM Tris-HCl (pH 7.5). After binding the RNA, the resin is washed extensively with the same buffer used for DEAE-cellulose and the RNA eluted as above. Note that since the tenacity with which an RNA is bound to the column is proportional to its size (i.e., number of phosphates), losses of the larger snRNAs, e.g., snR20, which is 1175 nucleotides (nt) long[5] may be observed with this procedure.

Gel Electrophoresis

Because of the complexity of the population of both tRNAs and snRNAs in yeast, one-dimensional gels often provide inadequate resolution. We have therefore adopted a semidenaturing two-dimensional system[7] for analysis of RNAs labeled either *in vivo* or *in vitro*. For examples, see Refs. 1–3 and 5.

Procedure. The polyacrylamide content of the first dimension will depend on the molecular weight range to be analyzed; most yeast snRNAs (~100–600 nt) are optimally resolved on a 6% gel, while tRNAs (70–95 nt) are best displayed on a 10% gel. We generally use 4 M urea, but higher concentrations may also be employed. The RNA is loaded into a slot about 1 cm wide and electrophoresed in 1 × TBE buffer[8] for the desired time (see Ref. 8 for a chart of polynucleotide mobilities relative to commonly used marker dyes). The portion of the first-dimension lane of interest is then excised with a clean razor blade or scalpel and placed perpendicular to the direction of migration between two glass plates separated by spacers. The second-dimension gel (20% polyacrylamide/4 M urea) is then poured around the first-dimension slice, taking care to avoid trapping bubbles. To make pouring easier, we generally run this gel bottom to top; thus, the gel slice is positioned approximately 2 cm from the bottom. Because of the high concentration of acrylamide, these gels must be run slowly (~150–200 V) to avoid heating; at this voltage, it is not necessary to perform the

[7] A. Fradin, H. Gruhl, and H. Feldmann, *FEBS Lett.* **50**, 185 (1975).
[8] T. Maniatis, E. F. Fritsch, and J. Sambrook, "Molecular Cloning: A Laboratory Manual." Cold Spring Harbor Laboratory, Cold Spring Harbor, New York, 1982.

electrophoresis in a cold room or to use a cooling device. Include a slice of the first dimension which contains dye to allow the progress of the gel to be monitored. Note that for a 40-cm gel, it will take 2–3 days to achieve good resolution of snRNAs. We generally change the buffer each day to compensate for evaporation.

Northern Blot Analysis

RNAs of the size discussed in this chapter are only poorly resolved on agarose gels. However, passive blotting from polyacrylamide is extremely inefficient. Both problems can be solved by electrophoretically transferring the RNA from a high-resolution gel.

Materials

Transfer buffer: 50 mM sodium acetate, pH 5.5; store at 4°
Transfer apparatus
Nylon membrane

Procedure. We generally employ fully denaturing (7 M urea) polyacrylamide gels for Northern blotting, since a primary use is to determine accurately the sizes of rare RNAs. The gel can be stained with ethidium bromide and photographed beside a ruler to record the locations of abundant species of known size, namely, 5 and 5.8 S rRNAs. Use a clean razor blade to cut away excess acrylamide and notch the gel for purposes of orientation. Wash the gel 2 times, 30 min each, at room temperature with transfer buffer, shaking gently. This step removes ethidium bromide and urea, which inhibit binding to the filter. We generally use GeneScreen, but other nylon-based membranes may also be employed. Cut a filter to fit the gel using RNase-free techniques and slowly (to avoid trapping bubbles) immerse in transfer buffer. Shake gently for 20–30 min.

Several electrophoretic transfer chambers are available commercially (e.g., Bio-Rad, Richmond, CA). The gel is immobilized during the blotting procedure by a plastic holder. The blot "sandwich" is set up as follows: porous cushion; two squares of Whatman 3 MM paper; gel; membrane; two additional squares of Whatman 3 MM paper; second porous cushion. The cushions and 3 MM sheets should be thoroughly soaked in buffer before assembly to ensure even conductivity during the transfer. Clamp the holder, place in the transfer chamber, and slowly cover with buffer, avoiding bubbles. Circulate the buffer during the transfer using a magnetic stirrer to avoid localized heating. Run overnight (16 hr) at 7 V (~120–150 mA) in a cold room (4°); excessive voltage can cause the RNA to bypass the filter. Carefully disassemble and mark the positions of lanes if desired.

Bake the blot for 2 hr at 80° under reduced pressure or cross-link the RNA to the filter by UV irradiation (10 min on each side on a 254-nm transilluminator). Hybridization conditions will depend on the probe to be used.[2,9]

Concluding Remarks

The complete characterization of a newly isolated RNA molecule requires in-depth structural analysis beyond the scope of this chapter. A useful initial classification tool is to determine the 5' end structure and whether it contains base or sugar modifications (see Refs. 1 and 2 and references therein). The complete sequence is best obtained after cloning the gene, particularly if the RNA is rare in the cell. However, direct RNA sequence analysis may be desirable or essential in some cases; modern methods are described in a recent volume on RNA processing in this series.[10] This volume also includes techniques for purifying and analyzing the ribonucleoprotein particles in which many small RNAs are housed.

Acknowledgments

I would like to thank Christine Guthrie, in whose laboratory these protocols were developed, and colleagues Pat Brennwald and Greg Porter for helpful comments on the manuscript. Research in the author's laboratory is supported by grants from the National Institutes of Health (GM38070) and the National Science Foundation (DCB 88-16325).

[9] B. Patterson and C. Guthrie, *Cell (Cambridge, Mass.)* **49**, 613 (1987).
[10] J. E. Dahlberg and J. N. Abelson (eds.), this series, Vol. 180.

[29] Measurement of mRNA Decay Rates in *Saccharomyces cerevisiae*

By ROY PARKER, DAVID HERRICK, STUART W. PELTZ, and ALLAN JACOBSON

Introduction

Several different experimental procedures can be used to measure the decay rates of individual mRNAs in the yeast *Saccharomyces cerevisiae*. Transcripts can be labeled *in vivo* and the rate of mRNA decay determined either from the disappearance of specific mRNAs during a chase (pulse-chase) or from the kinetics of the initial labeling (approach to steady state). Alternatively, transcription can be inhibited, and decay rates derived, by determining the relative abundance of individual mRNAs after such inhibition. Although the *in vivo* labeling techniques can provide accurate measures of mRNA decay rates, they have several disadvantages (see Table I), including a requirement for large quantities of radioactively labeled nucleic acid precursors, poor signal-to-noise ratios with mRNAs that are transcribed at low rates, a dependence on accurate pool size measurements, and a failure to provide information on mRNA integrity during the course of an experiment. Given these disadvantages, the emphasis of this chapter is on protocols which measure mRNA decay rates in yeast subsequent to transcriptional inhibition. Information on *in vivo* labeling techniques is available in papers published previously[1-4] (see also [30] in this volume).

In all of the following protocols mRNA synthesis is inhibited, either in general or for specific genes, and, at various times after such inhibition, the abundance of particular mRNAs is monitored by simple techniques (e.g., Northern blotting). Inhibition of total mRNA synthesis is accomplished by the use of either specific drugs or a temperature-sensitive RNA polymerase II mutant. In contrast to *in vivo* labeling procedures, these approaches are straightforward, require minimal amounts of radioactive material, are applicable to mRNAs of any abundance class or transcription rate, and they provide information on mRNA integrity in parallel with the quantitation of mRNA decay rates. The potential disadvantages of these protocols include the nonspecific side effects of drugs used to inhibit transcription and the potential loss of labile turnover factors in the absence of ongoing transcription. However, for most of the mRNAs which we have analyzed, the decay rates obtained by methods dependent on transcriptional inhibition are consistent with (a) those obtained by pulse–chase analysis[3] and (b) comparisons of the steady-state levels of mRNAs with identical transcription rates but different decay rates.[5] For the analysis of the decay rate of a single mRNA, an alternative approach utilizes the regulated *GAL1* promoter to selectively repress the synthesis of a specific transcript. This method, however, requires specific plasmid constructions.

It should be noted that no one method is perfect and that, for a specific situation, one protocol may be more suitable than another. The relative advantages and disadvantages of each procedure are summarized in Table I.

[1] C. H. Kim and J. R. Warner, *J. Mol. Biol.* **165,** 79 (1983).
[2] R. Losson and F. Lacroute, *Proc. Natl. Acad. Sci. U.S.A.* **76,** 5134 (1979).
[3] D. Herrick, R. Parker, and A. Jacobson, *Mol. Cell. Biol.* **10,** 2269 (1990).
[4] R. S. Zitomer, D. L. Montegomery, D. L. Nichols, and B. D. Hall, *Proc. Natl. Acad. Sci. U.S.A.* **76,** 3627 (1976).
[5] R. Parker and A. Jacobson, submitted for publication.

TABLE I
COMPARISON OF METHODS FOR MEASUREMENT OF mRNA DECAY RATES

Method	Advantages	Disadvantages
Transcriptional inhibition		
Drugs (e.g., thiolutin, phenanthroline)	Simple to execute Requires small amount of radioactive materials mRNA integrity monitored in parallel with mRNA decay Several mRNAs can be analyzed simultaneously; blots can be reprobed for additional mRNAs Homologous mRNAs (e.g., endogenous and chimeric mRNAs) can be analyzed in same experiment	Other cellular pathways (e.g., protein synthesis) may be affected Thiolutin is not commercially available Inhibition of transcription may deplete cell of labile turnover factors
Temperature-sensitive RNA polymerase II mutant	Same advantages as above RNA polymerase II is specifically inhibited	Requires special strain Possible complications of heat shock, although not supported by available data (see ref. 3) Inhibition of transcription may deplete cell of labile turnover factors
GAL promoter	Minimal perturbation of cells Simple to execute (after gene fusion has been constructed) Requires small amount of radioactive materials	Requires construction of promoter fusion Comparisons limited to mRNAs under GAL control Change in carbon source may affect decay of some mRNAs
In vivo labeling		
Pulse-chase and approach to steady state	Minimal perturbation of cells	Labor intensive Requires substantial amount of radioactive materials Difficult to obtain efficient chase Dependent on knowledge of pool sizes Poor signal-to-noise ratio for mRNAs with low transcription rates Difficult to detect possible decay products Difficult to analyze chimeric mRNAs

Growth of Cells

Cultures of yeast cells in the exponential phase of growth, started by dilution from an overnight broth culture or by inoculation from a fresh single colony, are used for all of the procedures described here. These cultures are grown at the appropriate temperature (see below), in either rich (YEP) or synthetic media (see [1] in this volume), until the cell density is between 5×10^6 and 1×10^7 cells/ml (OD_{600} 0.5 to 1.0). At this time one of the following protocols is used to determine mRNA decay rates.

Inhibition of Transcription with Thiolutin or 1,10-Phenanthroline

Two inhibitors which can be used to block transcription *in vivo* in yeast are thiolutin[6] (obtained from Pfizer, Groton, CT), or 1,10-phenanthroline[7] (obtained from Sigma, St. Louis, MO). Thiolutin is prepared as a stock solution of 2 mg/ml in dimethyl sulfoxide (DMSO), and 1,10-phenanthroline is prepared as a stock solution of 100 mg/ml in ethanol (store at $-20°$). To inhibit transcription in cell cultures, thiolutin is added to a final concentration of 3 μg/ml (inhibiting mRNA transcription to $<5\%$ of wild-type levels[3,6]). Likewise, 1,10-phenanthroline, added to a final concentration of 100 μg/ml, inhibits transcription to approximately 10% of wild-type levels.[7] A standard experiment is outlined below.

1. Grow a 250-ml culture of cells to mid-log phase.
2. Add the transcriptional inhibitor (e.g., 375 μl of thiolutin stock solution).
3. Immediately after addition of the drug, remove and quickly harvest an aliquot of cells (usually 25–30 ml). By definition, this is the t_0 sample. Rapid harvesting of cells is accomplished by brief centrifugation (15 sec at 6,000–10,000 rpm in the SS34 rotor), removal of the medium supernatant by aspiration, and rapid freezing of the cell pellet in crushed dry ice. Alternatively, cells can initially be concentrated into smaller volumes prior to the inhibition of transcription (e.g., 200 ml of culture into 20 ml of fresh medium) and 4 ml of culture harvested at each time point in two 2-ml Eppendorf tubes. In this case, cells need only be pelleted for 10 sec, followed by aspiration of the supernatant and quick freezing.
4. Remove additional aliquots of cells at various times. Since the half-lives of most yeast mRNAs are between 3 and 45 min,[3] useful time points for a preliminary experiment are 5, 10, 20, 30, 40, 50, and 60 min. Once an

---

[6] A. Jimenez, D. J. Tipper, and J. Davies, *Antimicrob. Agents Chemother.* **3,** 729 (1973).
[7] T. C. Santiago, I. J. Purvis, A. J. E. Bettany, and A. J. P. Brown, *Nucleic Acids Res.* **14,** 8347 (1986).

approximate half-life is determined, different time points can be employed in subsequent experiments to define the actual decay rate more precisely. For example, if the half-life of an mRNA is approximately 8 min, time points useful for a more accurate assessment of the decay rate are 3, 6, 9, 12, 15, 18, and 21 min.

5. Isolate RNA from the frozen cell pellets using either of the RNA isolation procedures described in [27] or [28], this volume.

6. Determine the amount of the transcript of interest at each time point by Northern blotting, dot blotting, or RNase protection. Changes in mRNA levels with time should be quantitated by densitometry of autoradiographs (within the linear response range of the film) or, preferably, by direct counting of the blot in a two-dimensional blot analyzer (such machines are available from Betagen, Waltham, MA, or AMBIS Systems, San Diego, CA). Methods for the analysis of the data obtained are discussed below (see section on Analysis of mRNA Decay Rates).

Inhibition of Transcription Using Temperature-Sensitive RNA Polymerase II Mutants

In principle, the protocol with RNA polymerase II mutants is essentially the same as that outlined above for experiments utilizing transcriptional inhibitors. In this case, however, mRNA synthesis is inhibited by rapidly raising the temperature of mutant cells which harbor a thermolabile RNA polymerase II. The mutant strains which we have used with success in our laboratory are Y260 *(MATa ura3-52 rpb1-1)* and Y262 *(MATα ura3-52 his4-519 rbp1-1)*.[8] With these strains, transcription of mRNAs is reduced to less than 10% of wild-type levels within 2 min of a shift from 24° to 36°.[8] A standard procedure is outlined below, and an example of the results of a typical experiment is shown in Fig. 1.

1. Grow a 125 ml culture of cells containing the *rpb1-1* allele to midlog phase at 24°. (As discussed above, cells can be concentrated 5–10 times prior to the temperature shift to facilitate the harvesting of time points in smaller volumes.)

2. Shift cells to 36° rapidly by the addition of an equal volume of medium (125 ml) preheated to 48° and place the flask of cells in a 36° shaking water bath.

3. Immediately remove and harvest (see previous section, Step 3) an aliquot of cells for the t_0 time point.

4. Continue incubating cells in a 36° shaking water bath. Remove and

[8] M. Nonet, C. Scafe, J. Sexton, and R. Young, *Mol. Cell. Biol.* **7**, 1602 (1987).

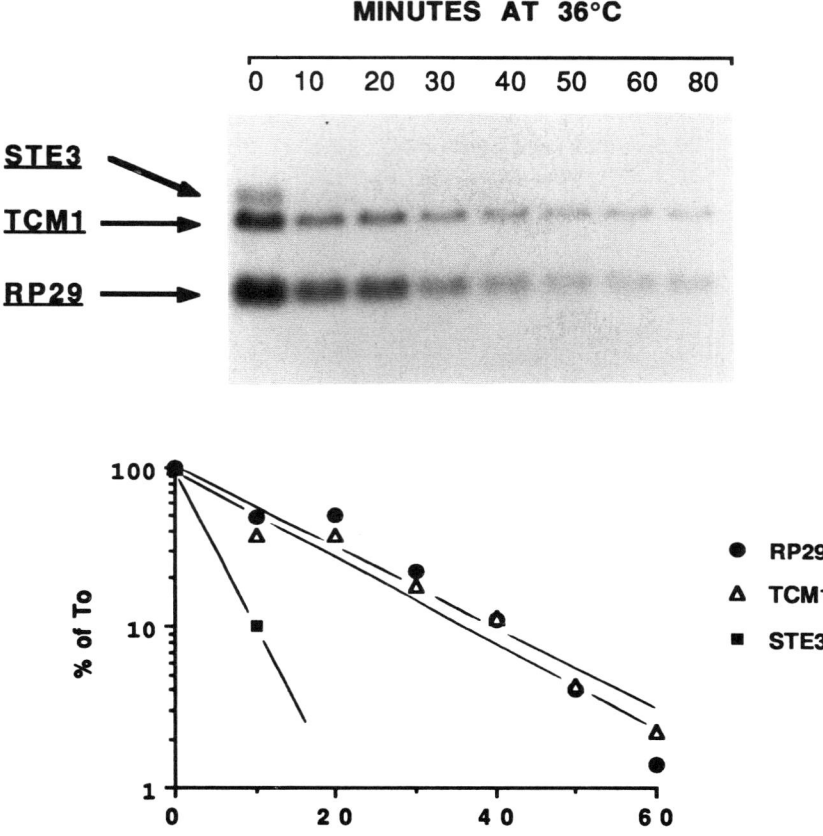

FIG. 1. Decay of mRNAs following inhibition of transcription in a strain with a thermolabile RNA polymerase II. The top part shows an autoradiogram of a Northern blot of RNA isolated from cells at different times after a shift from 24° to 36°. The blot has been hybridized to probes specific for the mRNAs encoded by the *STE3*, *TCM1*, and *RP29* genes. The bottom part is a semi-log plot of the fraction of each mRNA remaining at different times after the temperature shift. mRNA levels were quantitated with a Betascope Blot Analyzer and expressed as the percentage of the t_0 sample.

harvest additional aliquots of cells at appropriate times (see previous section, Step 4).

5. Isolate RNA as described in other chapters of this volume.

6. Quantitate the amount of the transcript of interest at each time point by Northern blotting, dot blotting, or RNase protection (see previous section, Step 6). Methods for the analysis of the data obtained are discussed below.

Use of *GAL* Promoters

By expressing a gene under the control of the repressible *GAL1* promoter, it is possible to inhibit the transcription, and monitor the decay, of a single mRNA species.[9] As a first step in this procedure, the gene of interest is fused to either the *GAL1* or *GAL10* upstream activation site such that the normal transcriptional initiation and termination sites are utilized. Construction of such regulated genes is described elsewhere (see [25] in this volume). To monitor mRNA decay rates using the *GAL1* promoter, transcription is first induced in the presence of galactose and then repressed in the presence of glucose. Such repression inhibits *GAL*-specific transcription to less than 10% of induced levels within 5 min.[10] A standard experiment is described below.

1. Grow 250 ml of cells in YEP containing 2% galactose to mid-log phase at 30°.

2. Harvest the cells by centrifugation and resuspend in 250 ml of YEPD containing 2% glucose. Immediately remove and harvest a cell aliquot (t_0 time point).

3. Continue incubation in a 30° shaking water bath. Remove and harvest additional aliquots of cells at appropriate times (see above).

4. Isolate RNA as described in other chapters of this volume.

5. Quantitate the amount of the transcript of interest at each time point by Northern blotting or RNase protection (see above). Methods for the analysis of the data obtained are discussed below.

Analysis of mRNA Decay Rates

Assuming first-order kinetics, a simple mathematical analysis allows the expression of mRNA decay as a function of time. The decrease in transcript concentration ([mRNA]) with respect to time, in the absence of mRNA synthesis, can be expressed in Eq. (1):

$$d[\text{mRNA}]/dt = -k[\text{mRNA}] \tag{1}$$

where k represents the rate constant for decay. Integration and rearrangement of Eq. (1) allows the concentration of mRNA at any time to be expressed as a function of time, as described in Eq. (2):

$$\ln([\text{mRNA}]_t/[\text{mRNA}]_0) = -kt \tag{2}$$

where $[\text{mRNA}]_0$ is the mRNA concentration at time zero. Plotting the

---

[9] H. G. Nam and H. M. Fried, *Mol. Cell. Biol.* **6**, 1535 (1986).
[10] R. T. Surosky and R. E. Esposito, personal communication.

percentage of RNA remaining (compared with time zero) with respect to time on semi-log axes should yield a straight line, the slope of which is dependent on the mRNA half-life. (Biphasic decay kinetics have been observed for a small number of yeast mRNAs.[3,11] The molecular basis for biphasic decay is not understood, but the phenomenon may reflect incomplete inhibition of transcription or heterogeneity in the cell population.)

As an example we have shown an RNA blot and data analysis from an experiment in which transcription was inhibited using the temperature-sensitive RNA polymerase II allele (Fig. 1). This Northern blot was probed for the mRNAs encoded by the *STE3, RP29,* and *TCM1* genes and was quantitated directly using a two-dimensional blot analyzer. The amount of mRNA present at each time point, expressed as a percentage of the 0 min time point, was plotted on semilog axes (see Fig. 1, bottom). mRNA half-lives can be determined by linear regression analysis and visual determination of the point at which the best-fit line crosses the 50% intercept (e.g., in Fig. 1, the $t_{1/2}$ for the *STE3* mRNA is 2.5–3.0 min and the $t_{1/2}$ for the *TCM1* and *RP29* mRNAs is 10–12 min). Alternatively, mRNA half-lives can be obtained from the slope of the best-fit line, using the simple relationship $T_{1/2} = \ln 2/k$ (where $-k$ is the slope of the line).

The results presented in Fig. 1 also illustrate two important advantages of the use of Northern blots for the analysis of mRNA decay: (1) simultaneous hybridization with more than one probe provides information on multiple mRNA species (provided that the mRNAs are sufficiently different in molecular weight), and (2) the integrity of the RNA isolated at each time point is readily monitored.

Measurement of Decay Rates of Chimeric mRNAs

Recent experiments demonstrate that unstable mRNAs contain sequences or structures which dictate their instability. Identification of such "instability elements" in yeast has been facilitated by an analysis of the decay rates of chimeric transcripts comprised of segments derived from stable and unstable mRNAs.[5,11] Essential to these analyses is a direct comparison of the decay rate of the chimeric mRNA with its endogenous "parental" mRNAs. Using the transcriptional inhibition protocols described here, this is feasible if the chimeric and endogenous mRNAs are separable on a Northern blot[5,11] or if probes are designed such that an RNase protection assay distinguishes the protected fragments complementary to the different mRNAs.[5] Moreover, a comparison of the steady-state levels of chimeric and endogenous mRNAs at t_0 provides an internal

---

[11] R. Parker and A. Jacobson, *Proc. Natl. Acad. Sci. U.S.A.* **87**, 2780 (1990).

control for the experiment.[5] Since a chimeric mRNA will utilize the same promoter as one of its parental mRNAs, the two mRNAs are presumably transcribed at approximately the same rate. To a first approximation, differences in the steady-state levels should, therefore, be consistent with differences in the decay rates.

Acknowledgments

Most of the methods described here were developed with the support of a grant (GM27757) to A.J. and a postdoctoral fellowship (GM11479) to R.P. from the National Institutes of Health. We thank David Munroe, Janet Donahue, and Laura Steel for their advice and assistance.

[30] Labeling of RNA and Phosphoproteins in *Saccharomyces cerevisiae*

By JONATHAN R. WARNER

The labeling of macromolecules *in vivo* is an important part of many types of experiments. In some cases it is necessary only to incorporate as much radioactivity as possible. In other cases it is necessary to incorporate radioactivity under controlled conditions, so that one can, for instance, compare the transcriptional or translational activity of cells under different experimental conditions. In yet other cases it is important to be able to provide a pulse of label followed by a chase with nonradioactive precursors in order to follow the fate of a particular species of macromolecule. We describe methods available for such labeling and potential pitfalls that may be encountered. Methods for preparation of RNA ([27] and [28]) and protein ([35]) are described elsewhere in this volume.

[^{32}P]Phosphate: RNA, Protein

Most media contain inorganic phosphate that competes with added $^{32}PO_4$ for uptake into the cell. Rubin[1] described a medium, adapted from one used for *Escherichia coli,* in which the inorganic phosphate is removed by precipitation with Mg^{2+} at high pH. The cells grow rather well on the organic phosphate present, but they avidly take up inorganic $^{32}PO_4$. The recipe is as follows: Dissolve 10 g yeast extract (Difco, Detroit, MI) and

[1] G. M. Rubin, *J. Biol. Chem.* **248,** 3860 (1973).

20 g peptone (Difco) in 920 ml water. With vigorous stirring, add 10 ml of 1 M MgSO$_4$ followed by 10 ml concentrated NH$_4$OH. Allow the phosphate salts to precipitate at room temperature for 30 min. Filter twice through a Büchner funnel fitted with Whatman 3 MM paper. Adjust to pH 5.8 with concentrated HCl. Autoclave 25 min. Add 40 ml sterile 50% dextrose. Grow cells to log phase in this medium. We generally label with 150 μCi/ml of carrier-free ^{32}PO$_4$ for periods from 5 to 60 min.

To our knowledge, no one has performed careful experiments to determine the uptake of phosphate from this medium under different experimental conditions. Its use is primarily to prepare highly radioactive RNA, DNA, or protein molecules for further analysis. Because the [^{32}P]phosphate is rapidly converted to polyphosphates, it is not feasible to "chase" under these conditions except over a long period of time. Furthermore, the polyphosphates confound attempts to measure the incorporation of ^{32}P into macromolecules by the usual precipitation with trichloroacetic acid. Nevertheless, one can expect several million counts per minute per milliliter of culture incorporated into RNA, and several thousand into any given phosphoprotein.

Organic Precursors: General Considerations

It is frequently useful to label macromolecules with their immediate precursors, uracil or adenine in the case of nucleic acids, or amino acids in the case of proteins. In our experience it is always more effective to label a cell prototrophic for the precursor rather than auxotrophic. In the latter case, one must grow the cells in the presence of unlabeled precursor to reach the experimental condition. That precursor will compete with the labeled precursor unless it is removed. To do so, however, the cells must be shifted from the growth medium to a starvation medium. This shocks them in at least 3 ways, by the manipulations of centrifugation or filtration, by the new "unconditioned" medium, and by starvation for the required nutrient. Although controls for such perturbations are possible, they are tedious and may be unconvincing.

In our hands[2] cells prototrophic for uracil, adenine, or any of several amino acids, for example, leucine, lysine, or methionine, incorporate added radioactive precursors avidly when growing in synthetic medium (SC), containing, per liter, 6.7 g yeast nitrogen base (without amino acids) (Difco), 20 g glucose, 50 mg required amino acids, and 20 mg required purines or pyrimidines. In some cases addition of 10 g succinic acid and 6 g of NaOH (per liter) improves growth by buffering the medium.

[2] J. R. Warner, *J. Biol. Chem.* **246**, 447 (1971).

Labeling of RNA

Bases or Nucleosides as Precursors. ^{14}C- or ^{3}H-labeled uracil is a useful precursor for the labeling of nucleic acids. Alternatively, labeled uridine, substantially less expensive, can be employed. Adenine can be used as well, though in our hands the signal-to-noise ratio is poorer. For either precursor, there are three potential drawbacks.

1. Uracil, for example, is incorporated into RNA as UTP and CTP. Therefore, a number of enzymatic reactions must ensue between the uptake of the uracil into the cell and its utilization for transcription. Although the incorporation of radioactive uracil sometimes appears linear from an early time, this is in fact a delusion arising from the countervailing effects of two nonlinear events: (a) the convex labeling profile of unstable RNAs, namely, the fact that unstable RNAs are synthesized and *degraded* [the amount of radioactivity in unstable RNAs after a short pulse is out of proportion to their representation in bulk RNA; for example, poly(A)$^+$ mRNA makes up 3–5% of total RNA but as much as 30% of a 2-min pulse], and (b) the approach to maximum value of the specific activities of UTP and CTP, which can take up to 15 min.[3] These considerations may be of little import in most instances, but if one is interested in the rate of synthesis of a molecule under different experimental conditions, one must be careful about the equilibration of the labeled precursor with the nucleoside triphosphate precursor pools. In one example,[4] it was found that starvation for an amino acid effectively prevented the uptake of exogenous uracil into the precursor pools that were maintained by the reutilization of nucleotides resulting from the turnover of RNA within the cell.

2. If one desires to measure carefully the flow of precursor into an RNA species, as, for example, when measuring the approach to equilibrium labeling of an unstable species, it is necessary to measure continuously the specific activity of both the UTP and CTP pools. Alternatively, the UTP pool can be measured and the specific activity of the UMP in RNA determined. A method for doing so is detailed in Ref. 3.

3. Once the uracil has been taken into the cell and phosphorylated, it rarely leaves the cell. Therefore, a pulse–chase can be carried out only over a considerable length of time. Realistically, chase times of less than 10–15 min are impractical.

In practice,[3] if one labels a culture at 10^7 cells/ml in synthetic medium lacking uracil for 5 min at 23° with 200 μCi/ml of [^{3}H]uracil (19 Ci/mmol), the yield is about 10^5 cpm/μg RNA. The yield of RNA is generally

---

[3] C. H. Kim and J. R. Warner, *J. Mol. Biol.* **165**, 79 (1983).
[4] R. W. Shulman, C. E. Sripati, and J. R. Warner, *J. Biol. Chem.* **252**, 1344 (1976).

20 μg/10⁷ cells. Therefore, 1 ml of culture will yield 2×10^7 cpm of [³H]RNA. The radioactivity found in RNA is generally proportional to the amount of radioactivity added to the culture.

Methionine as a Source of Labeled Methyl Groups in RNA. One of the most useful ways to label many RNAs, especially ribosomal RNA, is through the methyl group of methionine. [*methyl*-³H]Methionine is converted readily to S-adenosylmethionine (SAM), which donates the ³H-methyl in the methylation of nucleic acids. Many of the problems described above are not present when methionine is used as a precursor. In our experience, labeling a 1-ml culture of 10⁷ cells in the minimal medium described above with 100 μCi of [*methyl*-³H]methionine for 3–5 min yields several thousand counts per minute incorporated into ribosomal RNA precursors.[5] Similar protocols have been useful in studying the m⁷G cap at the 5′ end of mRNAs[6] and its metabolism.

A useful feature of labeling RNA with methionine is that the uptake of radioactivity into the SAM precursor pool is far less sensitive to environmental perturbation than is that of nucleotides.[4] Furthermore, the pools of SAM are rapidly saturated (a half-time of less than 2 min) and even more rapidly chased (a half-time of about 0.5 min).[7] The actual measurement of the specific activity of the SAM pool is rather simple if the cells have been prelabeled with [¹⁴C]adenine. (See Ref. 7 for details.)

Methionine can therefore be particularly useful for pulse–chase experiments, for example, to study the processing of ribosomal precursor RNA. A 1-ml culture in methionine-free medium is labeled with 60 μCi of [*methyl*-³H]methionine for 1–5 min. If needed, cold methionine can be added to 0.5 mg/ml in order to chase the label. RNA is prepared from the cells (see [27] in this volume), and the RNA from 0.2–0.4 ml of culture is fractionated on a 1.5% agarose denaturing gel, containing 2.2 M formaldehyde and 10 mM sodium phosphate. To detect the RNA by autoradiography, the gel is soaked in EN³HANCE (NEN–Du Pont, Boston, MA), dried, and exposed to X-ray film at $-80°$ without an intensifying screen.[8]

A more quantitative analysis is possible if the cells have been uniformly labeled with a purine or pyrimidine precursor to assist in bookkeeping.[5] We generally use 0.05 μCi [¹⁴C]uracil, diluted with 10 μg cold uracil, per milliliter. In this case the gel must be sliced, solubilized, and analyzed in a scintillation counter.[5]

The one potential drawback of the labeling of RNA with methyl groups is uncertainty over the degree of methylation of RNA in a given experi-

[5] S. A. Udem and J. R. Warner, *J. Mol. Biol.* **65**, 227 (1972).
[6] C. E. Sripati, Y. Groner, and J. R. Warner, *J. Biol. Chem.* **251**, 2898 (1976).
[7] J. R. Warner, S. A. Morgan, and R. W. Shulman, *J. Bacteriol.* **125**, 887 (1976).
[8] M. R. Underwood and H. M. Fried, *EMBO J.* **9**, 91 (1990).

mental condition. This can usually be assessed by analysis of an alkali digest of the RNA, since the bulk of the methylation in ribosomal RNA is at the 3'-hydroxyl, to yield an alkali-resistant dinucleotide. The dinucleotides can be readily separated from mononucleotides on a DEAE-Sepharose column in 7 M urea (see Ref. 4 for details).

Run-On Transcription: Labeling with UTP

Although there is much interest in the regulation of transcription of genes in yeast, transcription is rarely measured. Rather, one measures the concentration of mRNA or its products. This is largely due to the problems of incorporating sufficient radioactivity and of correcting for pool specific activity. As an alternative, investigators working with mammalian cells have developed the procedure of run-on transcription. Nuclei are prepared and incubated with radioactive nucleoside triphosphates. RNA polymerase molecules which were actively transcribing in the cells continue transcribing for a few hundred nucleotides. The radioactive RNA produced appears to be a faithful representation of the transcription that was going on in the cell.[9,10]

An analogous method was developed for *Saccharomyces cerevisiae* by Jerome and Jaehning.[11] However, to avoid the difficulty and duration of the procedure for preparing yeast nuclei, we have modified the approach by using a detergent to permeabilize the membrane of whole cells, with cell walls remaining intact.[12] The rigid cell wall protects the permeabilized cells during the procedure, which takes only a few minutes. These will now incorporate nucleoside triphosphates for a brief period. A useful feature is that the cells are also permeable to α-amanitin. RNA can then be prepared from the permeabilized cells for analysis, for example, by slot-blot hybridization to cloned gene fragments.

The procedure is as follows.[12] Pour a culture containing about 3×10^7 cells onto one-half volume of crushed ice. Do all subsequent steps on ice until the incubation. Centrifuge for 5 min at 5000 rpm; pour off supernatant. Suspend cells in 5 ml of TMN (10 mM Tris, pH 7.4, 100 mM NaCl, 5 mM MgCl$_2$) at 0°. Centrifuge for 5 min at 5000 rpm; aspirate supernatant. Suspend cells in 0.95 ml of cold water. Add 50 μl of 10% (w/w) *N*-lauroylsarcosine (sodium salt) [Sigma (St. Louis, MO) L-5125]. Leave on ice for 15 min. Transfer to an Eppendorf tube. Spin for 2 min in the cold. Aspirate supernatant.

[9] J. Weber, W. Jelinek, and J. E. Darnell, Jr., *Cell (Cambridge, Mass.)* **10**, 611 (1977).
[10] G. S. McKnight and R. D. Palmiter, *J. Biol. Chem.* **254**, 9050 (1979).
[11] J. F. Jerome and J. A. Jaehning, *Mol. Cell. Biol.* **6**, 1633 (1986).
[12] E. A. Elion and J. R. Warner, *Mol. Cell. Biol.* **6**, 2089 (1986).

Suspend permeabilized cells in 120 μl of reaction mix:

Component	Final concentration
Tris, pH 7.9	50 mM
KCl	100 mM
MgCl$_2$	5 mM
MnCl$_2$	1 mM
Dithiothreitol	2 mM
ATP	500 mM
GTP	250 mM
CTP	250 mM
Phosphocreatine*	10 mM
Creatine phosphokinase*	12 ng/μl
[α-^{32}P]UTP (800 Ci/mmol)	1 μCi/μl

*These are not essential but seem to increase incorporation slightly.

Incubate at 25° for 5–10 min. (The reaction is usually finished in 3–5 min.) Transfer 2 μl into cold 5% trichloroacetic acid (TCA) to determine incorporation. Add 1 ml of cold TMN containing 50 μM UTP. Spin, then aspirate and carefully discard the radioactive supernatant. Suspend cells and prepare RNA. (See [27] and [28] in this volume.) Under the best conditions we get about 1 cpm/cell. Frequently the yield is only one-third of that.

Acknowledgments

Research in the author's laboratory is supported by grants from the National Institutes of Health (GM25532 and CA13330) and the American Cancer Society (MV-323S).

[31] Tackling the Protease Problem in *Saccharomyces cerevisiae*

By ELIZABETH W. JONES

Introduction

The yeast *Saccharomyces cerevisiae* contains a large number of proteases that are located in various compartments (cytosol, vacuole, mitochondria, endoplasmic reticulum, and Golgi complex, at least) and membranes (vacuole, endoplasmic reticulum, Golgi complex, and plasma, at least) of the cell. These include endoproteinases, carboxypeptidases, aminopeptidases, and dipeptidylaminopeptidases.[1-5] Some of the proteases pose significant impediments to analysis of biochemical processes and/or purification of proteins and can generate artifacts concerning the activity, structure, and, even, intracellular location of proteins.

Of the many cellular proteases, the lumenal vacuolar proteases probably comprise the major source of problems. Found soluble within the vacuole are endoproteinases A and B (PrA and PrB), carboxypeptidases Y and S (CpY and CpS), aminopeptidase I (the 600 kDa species, ApI; called LAPIV by Trumbly and Bradley[6]), and aminopeptidase yscCo (ApCo).[1,2] Polypeptide inhibitors of PrA, PrB, and CpY are located in the cytosol.[1,2] Salient characteristics of the vacuolar proteases are summarized in Table I. Also listed are the relevant structural genes, where known, and inhibitors for the enzymes. Of the vacuolar proteases, protease B (PrB) is thought to be the greatest source of protease problems for several reasons. John Pringle some time ago compiled a list of protease artifacts, most of which were caused by PrB.[7] The pH optimum of PrB is near neutrality, unlike that of PrA, the other major vacuolar endoproteinase. Conditions that activate PrB and free it from its cytosolic inhibitor I_B [heat and sodium dodecyl sulfate (SDS), see below][7-12] are built into most biochemical analyses. Finally, from the fact that strain EJ101, which on recent retest (by Jones, upon receipt from Dave Engelke) proved to be of genotype α his1 prb1-1122 prc1-126, not α trp1 pro1-126 prb1-112 pep4-3 prc1-126,[13] has proved to have significant utility for purposes of enzyme purification, I infer that elimination of PrB activity can eliminate some protease problems.

An additional source of protease problems is Zymolyase, for commercial preparations of Zymolyase contain substantial amounts of a protease that will catalyze hydrolysis of azocoll (E. W. Jones, unpublished). Users of procedures that employ spheroplasts will need to bear this in mind.

In this chapter, I describe conditions and procedures that affect the

[1] E. W. Jones, *Annu. Rev. Genet.* **18**, 233 (1984).
[2] T. Achstetter and D. H. Wolf, *Yeast* **1**, 139 (1985).
[3] T. Achstetter, C. Ehmann, and D. H. Wolf, *Arch. Biochem. Biophys.* **207**, 445 (1981).
[4] T. Achstetter, C. Ehmann, and D. H. Wolf, *Arch. Biochem. Biophys.* **226**, 292 (1983).
[5] T. Achstetter, O. Emter, C. Ehmann, and D. H. Wolf, *J. Biol. Chem.* **259**, 13334 (1984).
[6] R. Trumbly and G. Bradley, *J. Bacteriol.* **156**, 36 (1983).
[7] J. R. Pringle, *Methods Cell Biol.* **12**, 149 (1975).
[8] E. W. Jones, *Genetics* **85**, 23 (1977).
[9] D. H. Wolf and C. Ehmann, *FEBS Lett.* **92**, 121 (1978).
[10] G. S. Zubenko, A. P. Mitchell, and E. W. Jones, *Proc. Natl. Acad. Sci. U.S.A.* **76**, 2395 (1979).
[11] J. Schwenke, *Anal. Biochem.* **118**, 315 (1981).
[12] R. E. Ulane and E. Cabib, *J. Biol. Chem.* **251**, 3367 (1976).
[13] R.-J. Lin, A. J. Newman, S.-C. Cheng, and J. Abelson, *J. Biol. Chem.* **260**, 14780 (1985).

TABLE I
VACUOLAR PROTEASES[a]

Enzyme	Abbreviation	Structural gene	Type	Inhibitors
Proteinase A	PrA	*PEP4*	Aspartic protease; endoproteinase	Pepstatin
Proteinase B	PrB	*PRB1*	Serine protease– subtilisin family; endoproteinase	DFP, PMSF, PCMB, Hg^{2+}, chymostatin, antipain
Carboxypeptidase Y	CpY	*PRC1*	Serine carboxypeptidase	DFP, PMSF, TPCK, PCMB
Carboxypeptidase S	CpS	*CPS1*	Metallo(Zn^{2+}) carboxypeptidase	EDTA
Aminopeptidase I	ApI	*LAP4*?	Metallo(Zn^{2+}) aminopeptidase	EDTA, PCMB, nitrilotriacetic acid, bestatin
Aminopeptidase yscCo	ApCo	Unknown	Metallo(Co^{2+}) aminopeptidase	EDTA, Zn^{2+}

[a] DFP, Diisopropyl fluorophosphate; PMSF, phenylmethylsulfonyl fluoride; PCMB, 4-chloromercuribenzoic acid; TPCK, L-1-*p*-tosylamino-2-phenylethyl chloromethyl ketone; EDTA, ethylenediaminetetracetic acid.

levels and activities of the vacuolar proteases and present both genetic and biochemical methods for coping with protease problems. For many purposes, a combination of a mutant strain and an inhibitor cocktail may provide the optimum solution.

Effects of Growth Stage, Medium Composition, and Genotype on Enzyme Levels. Levels of activity of all of the vacuolar proteases except CpS increase as cells approach stationary phase.[4,5,14-19] For PrB, enzyme activity and antigen are undetectable in log-phase cells growing on YEPD.[19,20] A small increase in activity occurs at the diauxic plateau. The largest in-

[14] T. Saheki and H. Holzer, *Biochim. Biophys. Acta* **384**, 203 (1975).
[15] A. Klar and H. Halvorson, *J. Bacteriol.* **124**, 803 (1975).
[16] J. Frey and K. H. Röhm, *Biochim. Biophys. Acta* **527**, 31 (1978).
[17] D. H. Wolf and C. Ehmann, *FEBS Lett.* **91**, 59 (1978).
[18] T. Achstetter, C. Ehmann, and D. H. Wolf, *Biochem. Biophys. Res. Commun.* **109**, 341 (1982).
[19] C. M. Moehle, M. W. Aynardi, M. R. Kolodny, F. J. Park, and E. W. Jones, *Genetics* **115**, 255 (1987).
[20] V. L. Nebes and E. W. Jones, unpublished (1989).

crease, to a level at least 100 times that of log-phase cells, occurs as the cells enter stationary phase.[19,21] These increases as a function of growth stage are presumed to reflect a release from glucose repression, for all of the enzymes save CpS are expressed at higher levels when acetate or lactate serve as carbon sources (ApCo was not tested).[6,14,16,21-24] For CpY, ApI, and PrB, increased enzyme levels are correlated with increased levels of mRNA.[21,23]

Levels of all of the vacuolar proteases increase on provision of a poor nitrogen source like valine or proline[22,25,26] or Cbz-Gly-Leu for CpY and CpS (PrB did not respond; other enzymes were not tested.)[17] Nitrogen starvation, under conditions conducive or not to sporulation, results in increased levels of proteases (CpS was not examined).[4,15,22,27] The *snf2* and *snf5* mutations[28,29] result in high, constitutive levels of PrB activity, whether or not glucose is present in the medium.[22,24] Levels of the other proteases were not examined in the mutants. The *snf1*, *snf3*, *snf4*, *snf6*, and *hex2* mutations were without effect on PrB levels.[22,24] Some investigators have suggested that the presence of peptone in the medium induces higher levels of vacuolar proteases. What evidence there is suggests the reverse, for protease levels on minimal medium seem higher than on complex media.[16,22,30] No systematic study has been reported, however.

Activation of Proteases. When cells are broken open, PrA, PrB, and CpY, at least, complex with their corresponding polypeptide inhibitors to form inactive complexes.[31-36] Incubation of cell extracts at low pH (4–5) will activate the proteases, apparently by hydrolysis of the polypeptide inhibitors.[31-38] Addition of SDS to crude extracts activates PrB, as does

---

[21] C. M. Moehle, Ph.D. Thesis, Carnegie Mellon University, Pittsburgh, Pennsylvania (1988).
[22] R. J. Hansen, R. L. Switzer, H. Hinze, and H. Holzer, *Biochim. Biophys. Acta* **496**, 103 (1977).
[23] B. Distel, E. J. M. Al, H. F. Tabak, and E. W. Jones, *Biochim. Biophys. Acta* **741**, 128 (1983).
[24] C. M. Moehle and E. W. Jones, *Genetics* **124**, 39 (1990).
[25] D. J. Klionsky, L. M. Banta, and S. D. Emr, *Mol. Cell. Biol.* **8**, 2105 (1988).
[26] C. M. Moehle, C. K. Dixon, and E. W. Jones, *J. Cell Biol.* **108**, 309 (1989).
[27] H. Betz and U. Weiser, *Eur. J. Biochem.* **62**, 65 (1976).
[28] L. Neigeborn and M. Carlson, *Genetics* **108**, 845 (1984).
[29] E. Adams, L. Neigeborn, and M. Carlson, *Mol. Cell. Biol.* **6**, 3643 (1986).
[30] T. R. Manney, *J. Bacteriol.* **96**, 403 (1968).
[31] H. Betz, H. Hinze, and H. Holzer, *J. Biol. Chem.* **249**, 4515 (1974).
[32] T. Saheki, Y. Matsuda, and H. Holzer, *Eur. J. Biochem.* **47**, 325 (1974).
[33] T. Saheki and H. Holzer, *Biochim. Biophys. Acta* **384**, 203 (1975).
[34] E. P. Fischer and H. Holzer, *Biochim. Biophys. Acta* **615**, 187 (1980).
[35] J. F. Lenney and J. M. Dalbec, *Arch. Biochem. Biophys.* **129**, 407 (1969).
[36] J. F. Lenney, *J. Bacteriol.* **122**, 1265 (1975).
[37] J. F. Lenney, *J. Biol. Chem.* **221**, 919 (1956).
[38] J. F. Lenney and J. M. Dalbec, *Arch. Biochem. Biophys.* **120**, 42 (1967).

TABLE II
REPRESENTATIVE INHIBITOR COCKTAILS[a]

		Strain	
Source	Cocktail	Normal	Deficient
Badaracco et al.[b]	0.1 mM EDTA 5 mM 2-Mercaptoethanol 1% Me$_2$SO	Wild type	—
Jong et al.[c]	1 mM EDTA 1 mM PMSF 2 µg/ml Pepstatin A 1 mM EGTA 1 mM Benzamidine 1 µg/ml Leupeptin	Wild type (A364A)	BJ926 (called PEP4D[c])
Johnson et al.[d]	10 mM EDTA 2 mM Benzamidine 1 mM PMSF	—	BJ926 (called *PEP4D*[c])
Olesen et al.[e]	1 mM EDTA 1 mM PMSF 1 µg/ml Leupeptin 1 µg/ml Pepstatin	Wild type	—
Davis et al.[f]	1 mM EDTA 1 mM PMSF 1 mM Benzamidine or 1 µg/ml Pepstatin A	Wild type (X2180-1A)	"pep4-3"
Lue and Kornberg[g]	1 mM EDTA 1 mM PMSF 2 µM Pepstatin A 0.6 µM Leupeptin	—	BJ926
Lue et al.[h]	1 mM EDTA 1 mM PMSF 2 µM Pepstatin A 0.6 µM Leupeptin 2 µg/ml Chymostatin 2 mM Benzamidine	—	BJ926
Aris and Blobel[i]	1 mM ε-Aminocaproic acid 5 µg/ml Aprotinin 1 mM p-Aminobenzamidine 1 µg/ml Chymostatin 5 µg/ml Pepstatin 250 µM PMSF 50 µM p-Chloromercuri- phenylsulfonic acid	—	BJ2168

[a] The strains used are indicated as normal or (protease) deficient. Protease-deficient strains (made by us and sent to the Yeast Genetics Stock Center) for which genotypes are given at the end of the chapter are identified by BJ numbers.

increasing the temperature.[7-12] The process of sample preparation (boiling) for denaturing electrophoresis activates PrB in the run up to maximum temperature, providing the opportunity for proteolysis and artifact generation. The effects can be readily seen by comparing the protein signatures of wild-type and *prb1* strains or of extracts of wild-type cells with and without preincubation with phenylmethylsulfonyl fluoride (PMSF) prior to denaturation. The most striking effects are seen for the largest polypeptides. For a more extensive discussion of the contribution of denaturation to accelerated proteolysis, see Pringle.[7]

Protease Inhibitor Cocktails. In Table II are given a number of different "cocktails" that have been used in various purification procedures. In several cases cocktails were used even though the starting strain was BJ2168 or BJ926, strains that have greatly reduced levels of PrA, PrB, CpY, and ApI [as well as RNase(s) and the repressible alkaline phosphatase] because they carry *pep4-3, prb1-1122,* and *prc1* mutations. In the various cocktails, EDTA will obviously inhibit metalloproteases; mercurials will inhibit PrB and CpY. Pepstatin A will inhibit carboxylproteases including PrA.[36] Chymostatin inhibits some serine proteases including PrB,[36] and PMSF reacts covalently with active site serines of serine proteases[39,40] like PrB[12,41] and CpY.[42] The target(s) for leupeptin is unclear, for, although it inhibits some serine and thiol proteases, it does not inhibit PrA, PrB, or CpY.[36,41] The target for benzamidine is likewise unclear. It inhibits trypsin, plasmin, and thrombin,[43-45] apparently through its resemblance to arginine.[46] No protease with trypsinlike specificity has been described in yeast. The data presented by Achstetter *et al.*[5] supply possible candidates,

[39] A. M. Gold and D. Fahrney, *Biochemistry* **3**, 783 (1964).
[40] A. M. Gold, *Biochemistry* **4**, 897 (1965).
[41] E. Kominami, H. Hoffschulte, and H. Holzer, *Biochim. Biophys. Acta* **661**, 124 (1981).
[42] R. Hayashi, Y. Bai, and T. Hata, *J. Biochem.* **77**, 1313 (1975).
[43] J. W. Ensink, C. Shepard, R. J. Dudl, and R. H. Williams, *J. Clin. Endocrinol. Metab.* **35**, 463 (1972).
[44] S. L. Jeffcoate and N. White, *J. Clin. Endocrinol. Metab.* **38**, 155 (1974).
[45] M. Mares-Guia and E. Shaw, *J. Biol. Chem.* **240**, 1579 (1965).
[46] M. Krieger, L. M. Kay, and R. M. Stroud, *J. Mol. Biol.* **83**, 209 (1974).

---

[b] G. Badaracco, L. Capucci, P. Plevani, and L. M. S. Chang, *J. Biol. Chem.* **258**, 10720 (1983).
[c] A. Y. S. Jong, R. Aebersold, and J. L. Campbell, *J. Biol. Chem.* **260**, 16367 (1985).
[d] L. M. Johnson, M. Snyder, L. M. S. Chang, R. W. Davis, and J. L. Campbell, *Cell (Cambridge, Mass.)* **43**, 369 (1985).
[e] T. N. Davis, M. S. Urdea, F. R. Mesiarz, and J. Thorner, *Cell (Cambridge, Mass.)* **47**, 423 (1986).
[f] J. Olesen, S. Hahn, and L. Guarente, *Cell (Cambridge, Mass.)* **51**, 953 (1987).
[g] N. F. Lue and R. D. Kornberg, *Proc. Natl. Acad. Sci. U.S.A.* **84**, 8839 (1987).
[h] N. F. Lue, A. R. Buchman, and R. D. Kornberg, *Proc. Natl. Acad. Sci. U.S.A.* **86**, 486, (1989).
[i] J. P. Aris and G. Blobel, *J. Cell Biol.* **107**, 17 (1989).

however. Benzamidine might significantly improve polypeptide integrity even when the enzyme source is a multiply protease-deficient strain like BJ926, and, although I cannot document the finding, I include it for its possible utility in designing experiments. Possibly ϵ-aminocaproic acid, aprotinin, and *p*-aminobenzamidine also are targeted to these enzymes, for they inhibit some of the same serine proteases as are inhibited by benzamidine. PrB is not inhibited by aprotinin (Trasylol).[41] Soybean inhibitor inhibits PrB, but at concentrations 100 times that shown in Johnson *et al.*[47] What its target here might be is unknown. EGTA will inhibit Ca^{2+}-dependent proteases, including the *KEX2* protease.[48] And $NaHSO_3$, which is known to inhibit PrA and PrB,[38] is thought to enhance histone stability in nuclear preparations.[45,49]

When using inhibitors, it is important that they be included in all buffers at all steps. Because proteases can be cryptic, through association with their inhibitors, and because the purification steps may aid dissociation, protease activity can be generated during a purification. (See the final warning section for *in vitro* maturation of the PrB precursor present in *pep4* mutant extracts as a source of PrB activity during a purification.) For PMSF there is the added complication that the compound is quite unstable, particularly if the buffer is above pH 7.[50] If use of a protease-deficient mutant and/or use of PMSF is precluded for some reason, fairly extensive analyses of potential inhibitors are available.[36,38,41]

Protease-Deficient Mutants

Most of the genetic analysis has concentrated on the two endoproteinases, PrA and PrB, and the two carboxypeptidases, CpY and CpS. Plate tests or microtiter well tests have been developed that allow one to test directly for activity of PrB, CpY, CpS, and aminopeptidases and indirectly for activity of PrA in colonies. In this chapter I first present the plate and well tests for assaying protease activities of colonies, followed by assays that allow quantitation of activity levels in cell-free extracts. Guidelines for designing useful protease-deficient strains and a list of strains that we have sent to the Yeast Genetics Stock Center are provided (see Comments). A cautionary note in the form of a summary of the problems that have been known to surface at least once during use of protease-deficient strains is also given.

[47] L. M. Johnson, M. Snyder, L. M. S. Chang, R. W. Davis, and J. L. Campbell, *Cell (Cambridge, Mass.)* **43**, 369 (1985).
[48] R. S. Fuller, A. Brake, and J. Thorner, *Proc. Natl. Acad. Sci. U.S.A.* **86**, 1434 (1989).
[49] J. Thorner, personal communication (1989).
[50] G. T. James, *Anal. Biochem.* **86**, 574 (1978).

Genetic Analyses

Requirements of Tests

Several requirements must be met to succeed in assaying activities of particular intracellular enzymes in colonies. The first condition is that the enzyme gain access to the externally supplied substrate. This is accomplished either by permeabilizing the cells in a colony with a solvent or by establishing conditions that result in lysis of cells. The second requirement is that, where lysis is employed, conditions be established that free the enzyme from its naturally occurring, intracellular, polypeptide inhibitor.[1,2] This is accomplished, for PrB, by including SDS in the overlay. The third requirement is to find a substrate or condition that tests for one enzyme activity only. How this is accomplished for each enzyme will be given in the procedure for that enzyme.

Protease B

Protease B activity in colonies can be assayed using an overlay test. Initially we developed a procedure for cells grown on YEPD plates (20 g Difco Bacto-peptone, 10 g Difco yeast extract, 20 g dextrose, 13–20 g agar, according to brand, per liter) that necessitated use of a lysis mutation to cause release of intracellular protease B from cells. We have since realized that growth of cells on YEPG plates (20 g Difco Bacto-peptone, 10 g Difco yeast extract, 50 g glycerol, 13–20 g agar, according to brand, per liter) obviates use of a lysis mutation, since some lysis occurs when cells are grown on this medium. The substrate is particulate Hide Powder Azure (HPA Calbiochem, La Jolla, CA), for PrB is the only protease in *Saccharomyces cerevisiae* that catalyzes cleavage and solubilization of this substrate.[51]

HPA Overlay Test for PrB Activity[10,19]

Principle. Protease B, which is freed from cells by lysis and from its inhibitor by the SDS present in the overlay, solubilizes the particles of Hide Powder Azure in the overlay, uncovering the colony and surrounding it with a clear halo. Mutant colonies remain covered.

Reagents

Sodium dodecyl sulfate (SDS), 20% (w/v) in 0.1 M Tris-HCl, pH 7.6
Cycloheximide, sterile solution at 5 mg/ml

---

[51] R. E. Ulane and E. Cabib, *J. Biol. Chem.* **249**, 3418 (1974).

Penicillin G-streptomycin: Use a sterile solution containing 5000 units/ml penicillin base and 5000 µg/ml streptomycin base

0.6% Agar, molten, held at 50°

Hide Powder Azure (Calbiochem): Hide Powder Azure is pulverized by homogenization (VirTis homogenizer) of 200 ml of a slurry (100 mg/ml 95% ethanol) in a 500-ml flask for 5 min at 40,000 rpm or by sonication of a slurry of the same proportions. Aliquots containing 50–100 mg are transferred to sterile 13 × 100 mm tubes, centrifuged (5 min at 1650 g), and the supernatants are discarded. The pellets are washed with 2.5–3 ml of sterile water, repelleted, and the supernatants discarded

Procedure. Add 0.2 ml cycloheximide solution, 0.2 ml penicillin–streptomycin solution, and 0.1 ml of 20% SDS to a hide powder azure pellet, vortex, then add 4 ml of molten agar. Vortex to mix resuspend particles. Streaks or replica plates of cells grown for 2 days at 30° on YEPG agar are overlaid with the molten agar cocktail, with pouring along the length of the stripes rather than across. After the agar solidifies, the plates are incubated at 34°–36° for 8 hr to 2 days, depending on the properties of the strains. (Some strains lyse well and are difficult to score at later times.) Plates can (and should) be incubated upside down, so long as the medium will absorb the moisture in the overlay. It is important that a seal *not* form between the lid and base of the petri dish. If a seal forms, the test simply does not work for unknown reasons. Even wild-type strains will fail to form a proper halo. Thus, the use of freshly poured plates is not advised.

Utility. The HPA overlay test works very well for following mutations in the PrB structural gene, *PRB1,* and less well for pleiotropic mutations like the *pep* mutations. The *pep4::HIS3* insertion mutation present in strains like BJ3501 and BJ3505, the BJ5400 series, and the BJ5600 series can be scored very easily in this test, since it causes a much tighter Prb$^-$ phenotype than does the *pep4-3* mutation.[52] In using *prb1* mutations or in constructing strains, be aware that, although *prb1* homozygotes sporulate, the asci may be very small (the size of a normal spore) and that superimposition of heterozygosity for *pep4* in such *prb1* homozygotes may prevent sporulation.[53]

Carboxypeptidase Y

Carboxypeptidase Y activity in colonies can be assessed by using an overlay test that relies on the esterolytic activity of the enzyme. The

[52] C. A. Woolford and E. W. Jones, unpublished observations (1988).
[53] G. S. Zubenko and E. W. Jones, *Genetics* **97,** 45 (1981).

substrate is N-acetyl-DL-phenylalanine β-naphthyl ester (APE), cleavage of which, in colonies anyway, is catalyzed only by CpY.

APE Overlay Test for CpY Activity[8]

Principle. Dimethylformamide present in the initial overlay permeabilizes cells on the surface of colonies. CpY within cells catalyzes cleavage of the ester. The produce β-naphthol reacts nonenzymatically with the diazonium salt Fast Garnet GBC to give an insoluble red dye: Cpy$^+$ colonies are red; Cpy$^-$ colonies are yellow or pink.

Reagents

N-Acetyl-DL-phenylalanine β-naphthyl ester (Sigma, St. Louis, MO): Make a solution 1 mg/ml dimethylformamide
0.6% Agar, molten, held at 50°
Fast Garnet GBC (Sigma)
0.1 M Tris-HCl, pH 7.3 – 7.5

Procedure. Replica plate strains to thick YEPD plates (40 – 45 ml/100-mm plate). Grow 3 days at 30°. To form the overlay mix, add 2.5 ml of the ester solution to 4 ml molten agar in a 13 × 100 mm tube. Vortex or cover with Parafilm and invert 3 or 4 times until the schlieren pattern disappears. After the bubbles exit, pour the contents over the surface of colonies or stripe (along, not across, stripes; colonies must be covered). After 10 min (or after the agar is hard), carefully flood the surface of the agar with 4.5 – 5 ml of a solution of Fast Garnet GBC (5 mg/ml 0.1 M Tris-HCl, pH 7.3 – 7.5). Do not tear the agar overlay during the flooding. The Fast Garnet GBC solution must be made immediately before use, for diazonium salts are very unstable in solution. We use Fast Garnet GBC from Sigma and store it in the freezer. Watch the color develop and pour off the fluid when Cpy$^+$ colonies turn red (a few to several minutes). If color development takes longer than 5 – 10 min, use a fresh bottle of Fast Garnet GBC. Do the test at room temperature. The color is not stable, but is more stable if the plates are placed in the cold in the dark. Although diazonium salts other than Fast Garnet GBC could, in principle, be used, Fast Garnet GBC is usually preferred because it is not a zinc salt and, thus, is less inhibitory to enzyme activity.

Utility. The APE test has wide utility for following many mutations that reduce protease activity so long as CpY activity is among the activities reduced as a consequence of the mutation. The APE test can be used for following mutations in the CpY structural gene, *PRC1*, and for following the pleiotropic *pep* mutations, including *pep4-3*.[8] If an *ade1* or *ade2* mutation is segregating in the cross, the red pigmentation problem can be circumvented by growing cells on YEPG or on YEPD supplemented with 100 μg/ml adenine sulfate. Petites give aberrant phenotypes in the test (but see well test below).

PEP4 is the structural gene for the PrA precursor.[54,55] PrA activity is essential for proper maturation of several vacuolar hydrolases, including PrB, CpY, one or more RNase species, the 600-kDa vacuolar aminopeptidase, ApI, and the repressible species of alkaline phosphatase.[56-58] We have cloned and sequenced several of the pleiotropic *pep4* mutations (including *pep4-3*), as well as the allelic *pra* mutations[59] that are not fully pleiotropic.[60] Pleiotropic *pep4* mutations like *pep4-3* usually prove to be nonsense mutations that totally eliminate protease A activity and greatly reduce levels of all hydrolase activities, including CpY, that require PrA activity for maturation.[59] For this reason, one can follow many mutations that eliminate protease A activity (the pleiotropic *pep4* mutations) by means of the APE test. Indeed, the more devastating the effect of the *pep4* mutation on protease A activity, the easier it is to follow by the APE test. Of course, *pep4* mutations, because of their effects on hydrolase maturation, result in reduction or elimination of several protease activities simultaneously, including PrA, PrB, CpY, and ApI (but not CpS; the Co^{2+}-dependent aminopeptidase was not tested) as well as RNase and alkaline phosphatase activity.[57]

In working with pleiotropic *pep4* mutations (in crosses or in constructing alleles), be aware that *pep4* homozygotes do not sporulate[53] and that the mutations show phenotypic lag.[61] Strains showing phenotypic lag will appear to be Pep⁺, even though genetically they carry the *pep4* allele. The lag is physiological in origin, not genetic, and is apparently a reflection of the ability of active PrB packaged within the spore to continue to activate its own as well as the CpY precursor. This positive feedback loop results in

[54] C. A. Woolford, L. B. Daniels, F. J. Park, E. W. Jones, J. N. Van Arsdell, and M. A. Innis, *Mol. Cell. Biol.* **6**, 2500 (1986).
[55] G. Ammerer, C. Hunter, J. Rothman, G. Saari, L. Valls, and T. Stevens, *Mol. Cell. Biol.* **6**, 2490 (1986).
[56] B. A. Hemmings, G. S. Zubenko, A. Hasilik, and E. W. Jones, *Proc. Natl. Acad. Sci. U.S.A.* **78**, 435 (1981).
[57] E. W. Jones, G. S. Zubenko, and R. R. Parker, *Genetics* **102**, 665 (1982).
[58] B. Mechler, M. Müller, H. Müller, and D. H. Wolf, *Biochem. Biophys. Res. Commun.* **107**, 770 (1982).
[59] E. W. Jones, C. A. Woolford, C. M. Moehle, J. A. Noble, and M. A. Innis, in "Proceedings UCLA Symposium, Cellular Proteases and Control Mechanisms" (T. E. Hugli, ed.), p. 141. Alan R. Liss, New York, 1989.
[60] E. W. Jones, G. S. Zubenko, R. R. Parker, B. A. Hemmings, and A. Hasilik, in "Alfred Benzon Symposium" (D. von Wettstein, J. Friis, M. Kielland-Brandt, and A. Stenderup, eds.), Vol. 16, p. 182. Munksgaard, Copenhagen, 1981.
[61] G. S. Zubenko, F. J. Park, and E. W. Jones, *Genetics* **102**, 679 (1982).

continued production of CpY (detected in the APE test) long after division would have diluted away active PrA. To circumvent phenotypic lag, streak spore clones for single colonies on YEPD plates (quadrants suffice). Do the APE test. Stab negative colonies through the agar overlay and make a new master plate. Then carry out the usual analyses for other markers. If this procedure is followed, the overlays should be made with sterile solutions and the colonies should be stabbed soon after the test, for the cocktail will kill cells.

Well Test for CpY Activity[54]

Principle. Dimethylformamide present in the solution permeabilizes cells. Cleavage of the amide bond in *N*-benzoyl-L-tyrosine *p*-nitroanilide (BTPNA) to give the yellow product *p*-nitroaniline is catalyzed only by CpY. This test works for petites as well as for grandes, and it is unaffected by *ade1* and *ade2* mutations.

Reagents

   N-Benzoyl-L-tyrosine *p*-nitroanilide (Sigma): Make a solution 2.5 mg/ml dimethylformamide
   0.1 M Tris-HCl, pH 7.5: Use a sterile solution

Procedure. Mix 4 volumes of buffer to 1 volume BTPNA solution. Distribute 0.2 ml into wells in a 96-well microtiter test plate. Cells are transferred into the solution by rotating an applicator stick in the fluid after dipping the sterile stick into a colony grown on a YEPD plate. (Alternatively, a 48-prong replicator can be used, taking care to adjust individual wells for colonies that may not transfer effectively with the technique.) Cover the wells and incubate overnight at 34°–37°. Cpy$^+$ cells give yellow fluid in the wells; Cpy$^-$ cells produce no color.

Utility. This test can be used to follow *prc1* mutations as well as pleiotropic *pep* mutations, like *pep4*, that result in CpY deficiency.

Carboxypeptidase S

Carboxypeptidase S activity in colonies can be assessed by using a well test that incorporates a coupled assay which detects release of free leucine from the blocked dipeptide carbobenzoxyglycyl-L-leucine (Cbz-Gly-Leu). CpY, which also catalyzes this cleavage, is inactivated by preincubation with PMSF, which reacts covalently to inactivate serine proteases like CpY.

Principle. The principle of the coupled assay, devised by Lewis and

Harris[62] and adapted by Wolf and Weiser[63] is shown in the following reactions:

$$N\text{-Cbz-Gly-Leu} + H_2O \xrightarrow{\text{carboxypeptidase}} \text{leucine}$$

$$\text{Leucine} + O_2 \xrightarrow{\text{L-amino-acid oxidase}} \text{keto acid} + NH_3 + H_2O_2$$

$$H_2O_2 + o\text{-dianisidine} \xrightarrow{\text{peroxidase}} \text{oxidized dianisidine}$$

Oxidized dianisidine is dark brown in color.

The amino acid generated must be a substrate for the L-amino-acid oxidase employed. Typically snake venoms serve as sources. For *Crotalus adamanteus* (Eastern diamondback rattlesnake) venom, leucine, isoleucine, phenylalanine, tyrosine, methionine, and tryptophan are good substrates, arginine, valine, and histidine are poor substrates, and the other amino acids are not oxidized. We have successfully used venoms for *Crotalus atrox* (Western diamondback rattlesnake) and *Bothrops atrox* (a viper). Cps⁺ colonies give brown fluid in the wells; Cps⁻ cells do not. Cells are permeabilized and CpY is inactivated by preincubation of cells in a solution containing Triton X-100 and PMSF.

Reagents

0.1% Triton X-100, made 1 mg/ml in PMSF
0.2 M Potassium phosphate, pH 7.0
50 mM MnCl$_2$
N-Carbobenzoxyglycyl-L-leucine
Horseradish peroxidase, type I (Sigma)
L-Amino-acid oxidase, type VI (Sigma; actually crude dried venom from *Crotalus atrox*) or type II (Sigma; dried venom from *Bothrops atrox*)
o-Dianisidine dihydrochloride

Procedure. Cells are transferred into 50 μl/well of the Triton X-100/PMSF solution in a 96-well microtiter test plate by applicator stick or multiprong replicator (see CpY procedure) after growth on a YEPD plate. Cover and let sit 2 hr at room temperature. Add to each well 150 μl of the substrate mix made in the following proportions: 1 ml buffer, 10 μl MnCl$_2$, 3.22 mg Cbz-Gly-Leu (final concentration, 10 mM), 0.2 mg peroxidase, 0.4 mg amino acid oxidase type VI or 1.2 mg type II, and 0.4 mg dianisidine dihydrochloride. Cover and incubate at 37° for 17–18 hr. (We

---

[62] W. H. P. Lewis and H. Harris, *Nature (London)* **215**, 351 (1967).
[63] D. H. Wolf and U. Weiser, *Eur. J. Biochem.* **73**, 553 (1977).

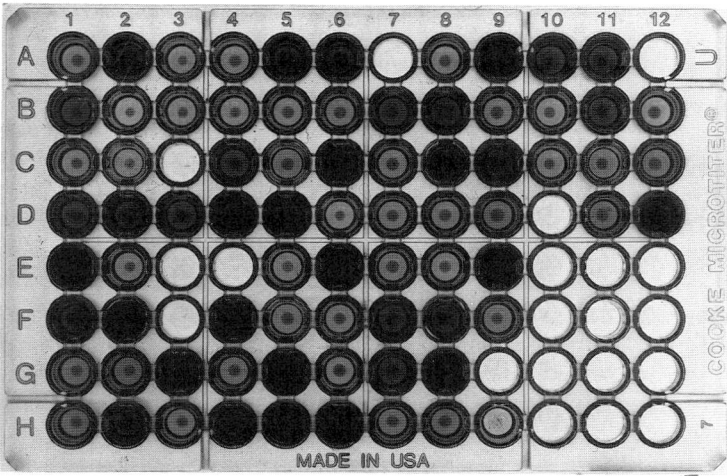

FIG. 1. Scoring for the presence or absence of carboxypeptidase S activity in meiotic segregants using the well test for carboxypeptidase S activity. Parents of the cross were of genotype α *leu2 trp1 prc1-407* (well 9E) and **a** *ura3-52 leu2 prc1-407 dut1-1* (well 9F). The *dut1-1* mutation results in failure to produce carboxypeptidase S activity. Spores of a given tetrad are in the following columns: 1 A–D; 2 A–D; 13 E–H; 14 E–H; etc. The scoring for CpS activity for tetrads 1 and 2 is, respectively, (1) −+−+, (2) +−−+. Blank wells such as 3C correspond to segregants that did not grow. Well 9H contains the reaction mixture but no cells. The Cps$^+$ phenotype and ability to use Cbz-Gly-Leu as a nitrogen source cosegregated in all 20 tetrads.

have not investigated other concentrations of venoms but know that these work.)

An example of this test for 20 tetrads from a cross is shown in Fig. 1. One parent of this cross carried mutations in *PRC1*, the structural gene for CpY, and *DUT1*, a gene required for production of CpS activity. The *prc1-407 dut1-1* strain was isolated based on its inability to use Cbz-Gly-Leu as a nitrogen source.[64] (Either CpY or CpS activity alone is sufficient to allow utilization of the dipeptide as a nitrogen source.) The diploid was homozygous for *prc1-407* and heterozygous for *dut1-1*. In this cross, segregants possessing CpS activity should retain the ability to use Cbz-Gly-Leu as a nitrogen source and give oxidation of *o*-dianisidine; those lacking CpS activity should do neither. The two characteristics cosegregated in all tetrads, as expected.

Utility. This test appears to work well for following mutations that result in CpS deficiency, whether or not the strain lacks CpY activity. We

[64] D. H. Wolf and C. Ehmann, *J. Bacteriol.* **147**, 418 (1981).

have used it particularly to follow *dut1-1*,[65] a mutation that might be allelic to *cps1*.[64]

Protease A

We have found no general plate test to directly assess protease A activity that is satisfactory, although preliminary tests indicate that it may be possible to adapt the fluorescence-based assay given in the biochemistry section of this chapter for a well test. (A plate test necessitating that strains carry a lysis mutation has been described.[66]) However, since total loss of function for protease A results in failure to activate a set of vacuolar hydrolase precursors, including that of CpY, many mutations in the protease A structural gene, *PEP4*, can be followed using the indirect test (APE test, see under carboxypeptidase Y) that detects the esterolytic activity of CpY, the processing and activation of which is dependent on protease A. This latter test is satisfactory for following the pleiotropic *pep4* mutations (Pra$^-$ Prb$^-$ Cpy$^-$ . . .) that are of most utility in biotechnological applications.

Aminopeptidases

Trumbly and Bradley[6] isolated mutants defective in one or more aminopeptidases by sequential mutagenesis. An agar overlay test was used that incorporated leucine β-naphthylamide (LBNA) as substrate. Cleavage releases β-naphthylamine. In fairly extensive but unpublished work, I isolated and partially characterized mutants by similar procedures but used several different amino acid β-naphthylamides as substrate.[67] I describe both procedures below.

Principle. Dimethylformamide present in the overlay permeabilizes cells on the surface of colonies. Aminopeptidases catalyze cleavage of the naphthylamide. The product β-naphthylamine reacts nonenzymatically with the diazonium salt Fast Garnet GBC to give an insoluble red dye. *Caution:* β-Naphthylamine is a carcinogen.

Trumbly and Bradley Method[6]

Reagents

Leucine β-naphthylamide: Make a solution 50 mg/ml dimethylformamide

---

[65] G. S. Zubenko, Ph.D. Thesis, Carnegie Mellon University, Pittsburgh, Pennsylvania (1981).
[66] B. Mechler and D. H. Wolf, *Eur. J. Biochem.* **121,** 47 (1981).
[67] E. W. Jones, unpublished (1973-1979).

0.8% Molten agar, held at 50°
Fast Garnet GBC

Procedure. Replica plate or streak strains on YEPD plates. Grow 2–3 days. Mix LBNA solution, agar, and Fast Garnet GBC in the proportion 0.02 ml, 1 ml, and 1 mg, respectively, and cover colonies with the solution. Red color should develop within 30 min and be stable for 2–3 days.

Jones Method[67]

Reagents

Methionine β-naphthylamide, leucine β-naphthylamide, alanine β-naphthylamide, threonine β-naphthylamide: Make solutions 3 mg naphthylamide/ml dimethylformamide
0.6% molten agar, held at 50°
Fast Garnet GBC
0.1 M Tris-HCl, pH 7.3–7.5

Procedure. Replica plate or streak strains on YEPD plates. Grow 3 days at 30°. Mix 4 ml of molten agar with 2.5 ml of naphthylamide solution in a 13 × 100 mm tube. Vortex until the schlieren pattern disappears. After the bubbles exit, pour the contents over the surface of colonies or stripes (along, not across the stripes; colonies must be covered). After 10 min (or after the agar is hard), carefully flood the surface of the agar with 4.5–5 ml of a solution of Fast Garnet GBC (5 mg/ml 0.1 M Tris-HCl, pH 7.3–7.5). Do not tear the agar overlay during flooding. The Fast Garnet GBC solution must be made immediately before use, for diazonium salts are very unstable in solution. We use Fast Garnet GBC from Sigma and store it in the freezer. Watch the color develop and pour off the fluid when wild-type colonies are red (several minutes to 1 hr, depending on the substrate and/or genotype). If color development takes more than 5–10 min for the wild type, replace your bottle of Fast Garnet GBC. The test is done at room temperature. Fast Garnet GBC is the preferred diazonium salt since it does not contain zinc, which inhibits some enzyme activities.

Utility. When leucine β-naphthylamide is used as substrate, activity in colonies is apparently the sum of four different aminopeptidase activities.[6] Whether ApCo is among them is unclear, but this seems unlikely. Whether segregation of the *lap* mutations other than *lap1 (lap2–lap4)*[6] can be followed in the presence of wild-type alleles for the other three genes has not been determined.

Analysis of cleavage patterns for several naphthylamides across column profiles or in activity gels for mutant and wild-type extracts indicates that different enzymes have different cleavage specificities.[4,16,67] One amino acid naphthylamide can be used to follow segregation of one mutation and a second naphthylamide to separately follow segregation of a second mutation. Obviously the particular substrate for each must be judiciously chosen. Because the relationship to *lap1–lap4* of the mutations I isolated and studied is not known, I cannot provide more guidance for substrate choice. For wild-type colonies, the rates of cleavage of naphthylamide decline in the order methionine > alanine > leucine > threonine. The last three are better for scoring segregation of mutations than is methionine β-naphthylamide.

The structural gene for the vacuolar ApI (probably *LAP4*) has recently been cloned and sequenced.[68] It should now be possible to generate an insertion mutation within the gene and follow segregation of the inserted gene.

Biochemical Analyses

Growth of Cells and Preparation of Extracts. For most assays of cell-free extracts for protease activities we employ extracts prepared according to the following protocol. Cells are grown to stationary phase in YEPD at 30° with vigorous shaking (usually 48–52 hr for our conditions of inoculation), harvested by centrifugation, washed once with distilled water, and resuspended in 2 ml of 0.1 M Tris-HCl, pH 7.6, per gram of cells. The cells are broken (3 min) with 0.45-mm glass beads [40:60 to 50:50 (v/v) glass beads to cell suspension] in a Braun homogenizer (Braun, Melsungen, Germany) without CO_2 cooling. After centrifugation for 30 min at 35,000 g in the cold, the supernatant is removed to a fresh tube and placed on ice.

Enzymatic Assays

Protease A

Two assays for protease A activity are described: the first based on release of peptides from hemoglobin, the second a fluorescence assay based on cleavage of a peptide.

Principle. The most commonly used assay for protease A activity measures the release of tyrosine-containing acid-soluble peptides from acid-denatured hemoglobin. Protease A is apparently the only protease to catalyze the reaction at acid pH. The procedure is based on that of Lenney et al.[69]

---

[68] Y.-H. Chang and J. A. Smith, *J. Biol. Chem.* **264,** 6979 (1989).

[69] J. Lenney, P. Matile, A. Wiemken, M. Schellenberg, and J. Meyer, *Biochem. Biophys. Res. Commun.* **60,** 1378 (1974).

Reagents

2% Acid-denatured hemoglobin: Dissolve 2.5 g hemoglobin (Sigma) in 100 ml distilled water; dialyze against 3 changes of 3 liters of water in the cold. Bring the pH to 1.8 with 1 N HCl. After 1 hr of incubation with stirring at 35°, bring the pH to 3.2 with 1 M NaOH and adjust the volume to 125 ml. Aliquots can be stored frozen for years

0.2 M glycine-HCl, pH 3.2
1 N Perchloric acid
0.5 M NaOH
2% Na_2CO_3 in 0.1 M NaOH
1% $CuSO_4 \cdot 5H_2O$
2% Sodium or potassium tartrate
Folin and Ciocalteu's phenol reagent (diluted 1:1 with water)

Procedure. The reaction mixture consists of 2 ml of a hemoglobin solution (prepared by mixing equal volumes of the 2% hemoglobin, pH 3.2, described above and 0.2 M glycine, pH 3.2) and 0.1 ml of cell-free extract (1–2 mg protein/incubation), with incubation at 37°. At 0, 15, and 30 min, 0.4-ml samples are removed to 0.2 ml of 1 N perchloric acid on ice, and the tubes are shaken briefly. After centrifugation at 1650 g for 5 min, 0.1 ml of each sample is removed to 0.1 ml of 0.5 M NaOH. Tyrosine-containing peptides in the neutralized 0.2-ml sample are determined with the Folin reagent according to Lowry *et al.*[70] To each 0.2-ml sample add 1 ml of a reagent consisting of 2% Na_2CO_3 in 0.1 M NaOH, 1% $CuSO_4 \cdot 5H_2O$, and 2% sodium or potassium tartrate (100:1:1) (mix just before use). Incubate at room temperature for at least 10 min. Add 0.1 ml of the diluted phenol reagent and vortex immediately. After 30 min, determine the absorbance at 750 nm.

Definition of Unit and Specific Activity. One unit corresponds to 1 μg tyrosine per minute. The A_{750} of 1 μg tyrosine is 0.058 in the Lowry assay performed as described. Using the change in absorbance for a 30-min incubation, the conversion to micrograms Tyr/minute/milligram protein is made by the following calculation:

$$\frac{\Delta A_{750}}{30} \times \frac{1}{0.058} \times \frac{2.1}{0.1} \times \frac{6}{4} \times \frac{1}{0.1 \text{ (mg protein/ml extract)}} =$$

$$(181)(\Delta A_{750})/(\text{mg protein/ml extract})$$

An abbreviated protease A assay can be used for segregants of crosses known to be segregating a *pra* mutation, where distinction between + and

---
[70] O. Lowry, N. Rosebrough, A. Farr, and R. Randall, *J. Biol. Chem.* **193**, 265 (1951).

— is all that is sought. It can also be used for mutant screens if a 0 min point is added. Fifty microliters of extract is added to 1 ml of the hemoglobin solution (equal volumes of 2% hemoglobin and 0.2 M glycine, pH 3.2). Incubate at 37° for 30 min. Remove a sample to perchloric acid and work up as described above.

An alternate, much more sensitive assay for protease A is available. We find it more satisfactory than the hemoglobin assay with respect to linearity, reproducibility, etc. It was developed for assaying renin[71] but has been used for assaying protease A.[72] We have adapted it for use in crude extracts.

Principle. Protease A will catalyze cleavage at the Leu-Val bond of the octapeptide N-succinyl-L-arginyl-L-prolyl-L-phenylalanyl-L-histidyl-L-leucyl-L-leucyl-L-valyl-L-tyrosine-7-amido-4-methylcoumarin. After removal of the valine and tyrosine residues by aminopeptidase M, the fluorescence of 7-amino-4-methylcoumarin can be determined at 460 nm after excitation at 380 nm. PMSF is included to covalently react with and inactivate PrB, a serine protease that could also catalyze cleavage of the peptide.

Reagents

McIlvaine's buffer, 0.2 M Na$_2$HPO$_4$, 0.1 M citric acid; adjust pH to 6.0 with NaOH or HCl

N-Succinyl-L-arginyl-L-prolyl-L-phenylalanyl-L-histidyl-L-leucyl-L-leucyl-L-valyl-L-tyrosine-7-amido-4-methylcoumarin (Sigma, MW 1300), 0.325 mg/ml dimethylformamide (0.25 mM)

7-Amino-4-methylcoumarin (Sigma, MW 175): Make a 0.4 mg/ml solution and serially dilute this 1/100 × 1/40 (1/4000) to make a stock solution for the standard curve

PMSF (Sigma), 0.2 M in 95% ethanol (34.8) mg/ml)

Aminopeptidase M (Sigma L-0632), 0.5 mg/ml in McIlvaine's buffer (we have not explored whether cheaper and apparently more active preparations like L-9876 or L-1503 will work, as they should, since the original protocol for renin calls for only 50 mU[23])

Procedure. Mix 90 or 85 μl McIlvaine's buffer, pH 6.0, 10 μl peptide solution, 10 or 15 μl crude extract (pretreat 1 ml of extract with 5 μl of 0.2 M PMSF for 2 hr at room temperature). Incubate 15 min at room temperature. Immerse the tube in boiling water for 5 min to stop the reaction. Cool. Add 5 μl aminopeptidase M. After a 90-min incubation at room temperature, add 1.61 ml buffer (to dilute the reaction 15-fold) and centrifuge the tubes at 1650 g for 5 min. Remove the supernatant and

---

[71] K. Murakami, T. Ohsawa, S. Hirose, K. Takada, and S. Sakakibara, *Anal. Biochem.* **110**, 232 (1981).

[72] H. Yokosawa, H. Ito, S. Murata, and S.-I. Ishii, *Anal. Biochem.* **134**, 210 (1983).

determine the fluorescence in a fluorimeter, with excitation at 380 nm and emission at 460 nm. A standard curve is constructed for 7-amino-4-methylcoumarin. The diluted stock (0.1 μg/ml) is mixed with buffer as given below and the fluorescence is determined.

Stock (μl)	Buffer (ml)	Total (nmol)
350	1.375	0.2
175	1.550	0.1
88	1.637	0.05
44	1.681	0.025
22	1.703	0.0125
11	1.714	0.00625

For experimental samples, read nanomoles in the sample from the standard curve (the volumes for the standard and the experimental sample are the same). The samples and standards must be read in exactly the same way (same slit width, etc.). Prepare the standard curve the same day.

Definition of Unit and Specific Activity. We have been using 1 unit to equal 1 nmol/min and the specific activity to be units/milligram protein.

Crude extracts for this assay procedure are made as follows. The volume of cells sampled is 2500 ml/Klett unit. Pellet the cells, wash with water, and freeze the pellet. Resuspend the pellet in 1.5 ml of 0.1 M Tris-HCl, pH 7.6, and transfer into a small Braun homogenizer tube (40–50% full of glass beads). Add buffer to the top (fill tube completely) to prevent foaming. Homogenize 3 min at room temperature. Transfer to a long centrifuge tube and centrifuge 20 min at 25,000 g in the cold. Transfer the supernatant to a 1.5-ml microcentrifuge tube. (The small homogenizer tubes are about 35 mm high and are cut down from 12 × 75 mm tubes.)

We have preliminary evidence that suggests that this assay can be adapted for use in a well test for genetic analyses.

Protease B

Protease B is an endoproteinase that will solubilize particulate substrates like Hide Powder Azure and Azocoll. It is apparently the only enzyme in the yeast cell that can do so.[73]

Principle. Protease B catalyzes hydrolysis of the peptide bonds in Azocoll, resulting in release of the trapped red dye. Absorbance of the dye is read at 520 nm.[10]

[73] E. Juni and G. Heym, *Arch. Biochem. Biophys.* **127**, 89 (1968).

Reagents

Azocoll (Calbiochem)
1% Triton X-100
0.1 M Tris-HCl, pH 7.6
20% SDS in 0.1 M Tris-HCl, pH 7.6

Procedure. Use 20 mg Azocoll for each 0.54 ml of solution made by mixing the three listed solutions in the proportion 0.125 ml Triton X-100, 0.375 ml buffer, 40 μl SDS. Once the Azocoll is thoroughly wetted, 0.54 ml of the suspension is transferred with a wide-bore pipettor (cut 1/4 inch off a 1-ml pipette tip) to a tube. Two-tenths milliliter of extract of a suitable dilution is added, and the tube is placed in a 37° constant temperature block. At 1-min intervals each tube is removed and shaken gently (do not vortex) to resuspend the Azocoll, then replaced in the block. Avoid leaving Azocoll on the tube walls. At the end of the 15-min incubation, the tubes are plunged into ice and 3.5 or 2 ml of ice-cold distilled water is added. Tubes are immediately centrifuged for 3–5 min at 1650 g, and the supernatants are removed to fresh tubes. The absorbance of these supernatants is relatively stable. Absorbance is read at 520 nm. Dilutions are chosen such that the kinetics are linear with time and protein concentration. Best results are obtained if the ΔA_{520} is less than 0.3. When comparing different strains, the extracts should be diluted such that the protein concentrations are similar for all strains. For extracts made from stationary-phase wild-type cells, about 0.1 mg extract protein/assay is appropriate. Correction is made for a blank lacking extract. A 0 min time point is needed for *ade2* mutant strains.

Definition of Unit and Specific Activity. One unit of protease B activity is defined as a change in absorbance at 520 nm of 1.0 per minute for the 0.74-ml reaction mixture as assayed at 37°. For a reaction stopped with 3.5 ml of water and run for 15 min, the conversion to units/mg protein is

$$\frac{\Delta A_{520}}{15} \times \frac{4.24}{0.74} \times \frac{1}{0.2 \text{ (mg protein/ml extract)}} =$$

$$(1.91)(\Delta A_{520})/(\text{mg protein/ml extract})$$

Carboxypeptidase Y

Carboxypeptidase Y will catalyze cleavage of esters, amides, and peptides. An assay based on its amidase activity can be employed for kinetic analyses[74] or modified to a fixed-time point assay.[8]

---

[74] S. Aibara, R. Hayashi, and T. Hata, *Agric. Biol. Chem.* **35**, 658 (1971).

Principle. Carboxypeptidase Y will catalyze the hydrolysis of N-benzoyl-L-tyrosine p-nitroanilide to give the yellow product p-nitroaniline. Production can be followed by absorbance at 410 nm.

Reagents

N-Benzoyl-L-tyrosine p-nitroanilide (Sigma) (6 mM): Dissolve 2.43 mg in 1 ml dimethylformamide (DMF)
0.1 M Tris-HCl, pH 7.6
1 mM HgCl$_2$
20% (w/v) SDS in 0.1 M Tris-HCl, pH 7.6

Procedure. One-tenth milliliter of 6 mM BTPNA in dimethylformamide is added to a tube containing 0.40 ml of 0.1 M Tris-HCl, pH 7.6, and 0.1 ml of extract at 37°. After 30 min, 1.5 ml of 1 mM HgCl$_2$ is added to stop the reaction. If the extract being assayed has low activity (as is typical for wild-type strains), 0.2 ml of 20% SDS, pH 7.6, is added, and, after vortexing, the tubes are incubated at 70° until solubilization of the protein, as evidenced by clearing, ensues. Absorbance at 410 nm is determined. Permeabilized cells can be used as an enzyme source in this assay. We have used cells permeabilized with 0.1–0.2% Triton X-100 or with 10–20% DMF.

Definition of Unit and Specific Activity. One unit of activity corresponds to 1 μmol p-nitroaniline produced per minute, assuming a molar absorbance of 8800. Corrections for absorbance owing to substrate and protein are made. The conversion to units/mg protein is

$$\frac{\Delta A_{410}}{30} \times \frac{2.3}{8800} \times \frac{10^3}{0.1 \text{ (mg protein/ml extract)}} =$$

$$(0.087)(\Delta A_{410})/(\text{mg protein/ml extract})$$

Carboxypeptidase Y levels can also be determined using the kinetic assay described below for carboxypeptidase S, but using Cbz-Phe-Leu as the substrate, since 95% of the hydrolytic activity toward this peptide is apparently due to CpY.[17]

Carboxypeptidase S

Carboxypeptidase S will catalyze hydrolysis of dipeptides that are blocked at the amino terminus.

Principle. Free leucine released from a peptide by CpS catalysis is oxidized by L-amino-acid oxidase. Reduction of the product hydrogen peroxide by horseradish peroxidase is coupled to oxidation of o-dianisidine, yielding brown oxidized dianisidine. The absorbance at 405 nm is followed. The kinetic assay was developed by Wolf and Weiser.[63]

Reagents

0.2 M Potassium phosphate buffer, pH 7.0, containing 0.5 mM MnCl$_2$
L-amino-acid oxidase type I (Sigma)
Horseradish peroxidase type I (Sigma)
Carbobenzoxyglycyl-L-leucine (20 mM), 6.44 mg/ml 0.2 M potassium phosphate buffer, pH 7.0 [or carbobenzoxy-L-phenylalanyl-L-leucine (15 mM) in 0.2 M potassium phosphate buffer, pH 7.0, for CpY]
o-Dianisidine dihydrochloride 2 mg/ml water
0.2 M PMSF in 95% ethanol

Procedure. A solution is made in the proportion 1 ml phosphate buffer–MnCl$_2$, 0.25 mg L-amino-acid oxidase, and 0.4 mg peroxidase. To 0.5 ml of this is added 0.5 ml of the peptide solution followed by 50 μl of the *o*-dianisidine solution and 50 μl of dialyzed extract. The mixture is incubated at 25° and the absorbance followed at 405 nm. To render the assay specific for CpS when Cbz-Gly-Leu is the substrate, extracts are preincubated for 2 hr at 25° with 0.1 mM PMSF to inactivate CpY. Use Cbz-Phe-Leu as the substrate if CpY activity is to be measured with the assay.

Definition of Unit and Specific Activity. One unit corresponds to production of 1 nmol L-leucine per minute; 0.1 μmol of leucine corresponds to a change in absorbance of 0.725 for this procedure. Specific activity is expressed as nanomoles L-leucine/minute/milligram extract protein.

Aminopeptidases

Quantitative assays for aminopeptidase activity of extracts exist but do not distinguish among the several activities.[4,16]

Principle. Aminopeptidases will catalyze hydrolysis of amino acid *p*-nitroanilides to give the yellow *p*-nitroaniline. Production is monitored by absorbance at 405 nm.[4]

Reagents

Amino acid *p*-nitroanilide, 2 mM in 4 mM H$_2$SO$_4$
0.2 M Tris-HCl, pH 7.5

Procedure. Add extract to buffer. Start the reaction by mixing this with an equal volume of the substrate solution and incubate at 37°. Follow absorbance at 405 nm. Frey and Röhm[16] add Zn^{2+} to 50 μM to activate ApI. Achstetter *et al.*[4] and Frey and Röhm[16] used various amino acid *p*-nitroanilides (PNA), Trumbly and Bradley[6] used leucine PNA, and Chang and Smith[68] used methionine PNA.

Utility. A great many aminopeptidases are present in cells. No substrate differentiates among them in crude extracts.[4,16] The difference between the activity toward leucine PNA measured in the presence of 50 μM Zn^{2+} and that measured in the presence of only 1 nM Zn^{2+} (by use of nitrilotriacetic acid)[16] is taken as the activity of ApI.[16]

Comments

Designing the Most Useful Protease-Deficient Strain. In all known cases the use of a protease-deficient strain has eased protease problems. Strains bearing the pleiotropic *pep4-3* (UGA) mutation have greatly reduced, but (except for PrA) not zero, levels of PrA, PrB CpY, and ApI [as well as RNases(s) and the repressible alkaline phosphatase]. Strains like EJ101 *(α his1 prb1-1122 prc1-126)* that carry *prb1-1122* have also been used successfully in some purifications. I recommend use of a double mutant that carries mutations both in the *PEP4* gene and the *PRB1* gene. Whether a mutation in the CpY gene, *PRC1*, is needed has not been determined, but, in most cases, it probably would not be required.

We have constructed and lodged the following strains in the Yeast Genetics Stock Center (MCB/Biophysics and Cell Physiology, 102 Donner Laboratory, University of California, Berkeley, CA 94720) for use by interested parties.

BJ926	α *trp1* ± *prc1-126 pep4-3 prb1-1122 can1 gal2* / **a** ± *his1 prc1-126 pep4-3 prb1-1122 can1 gal2*
BJ1984	≡ 20B-12 α *trp1 pep4-3 gal2*
BJ1991	α *leu2 trp1 ura3-52 prb1-1122 pep4-3 gal2*
BJ1995	α *leu2 trp1 ura3-52 prb1-1122 pep4-3 gal2*
BJ2168	**a** *leu2 trp1 ura3-52 prb1-1122 pep4-3 prc1-407 gal2*
BJ2407	α *leu2 trp1 ura3-52 prb1-1122 prc1-407 pep4-3 gal2* / **a** *leu2 trp1 ura3-52 prb1-1122 prc1-407 pep4-3 gal2*
BJ3501	α *pep4::HIS3 prb1-Δ1.6R his3-Δ200 ura3-52 can1 gal2*
BJ3505	α *pep4::HIS3 prb1-Δ1.6R his3-Δ200 lys2-801 trp1-Δ101 (gal3) ura3-52 (gal2) can1*
BJ5457	α *ura3-52 trp1 lys2-801 leu2Δ1 his3Δ200 pep4::HIS3 prb1Δ1.6R can1 GAL*
BJ5458	α *ura3-52 trp1 lys2-801 leu2Δ1 his3Δ200 pep4::HIS3 prb1Δ1.6R can1 GAL*
BJ5459	**a** *ura3-52 trp1 lys2-801 leu2Δ1 his3Δ200 pep4::HIS3 prb1Δ1.6R can1 GAL*
BJ5460	**a** *ura3-52 trp1 lys2-801 leu2Δ1 his3Δ200 pep4::HIS3 prb1Δ1.6R can1 GAL*

BJ5461	**a** *ura3-52 trp1 lys2-801 leu2Δ1 his3Δ200 pep4::HIS3 prb1Δ1.6R can1 GAL*
BJ5462	α *ura3-52 trp1 leu2Δ1 his3Δ200 pep4::HIS3 prb1Δ1.6R can1 GAL*
BJ5464	α *ura3-52 trp1 leu2Δ1 his3Δ200 pep4::HIS3 prb1Δ1.6R can1 GAL*
BJ5465	**a** *ura3-52 trp1 leu2Δ1 his3Δ200 pep4::HIS2 prb1Δ1.6R can1 GAL*
BJ5622	α *ura3-52 trp1 lys2-801 leu2Δ1 his3Δ200 pep4::HIS3 prb1Δ1.6R can1 GAL* / **a** *ura3-52 trp1 lys2-801 leu2Δ1 his3Δ200 pep4::HIS3 prb1Δ1.6R can1 GAL*
BJ5623	α *ura3-52 trp1 lys2-801 leu2Δ1 his3Δ200 pep4::HIS3 prb1Δ1.6R can1 GAL* / **a** *ura3-52 trp1 lys2-801 leu2Δ1 his3Δ200 pep4::HIS3 prb1Δ1.6R can1 GAL*
BJ5624	α *ura3-52 trp1 leu2Δ1 his3Δ200 pep4::HIS3 prb1Δ1.6R can1 GAL* / **a** *ura3-52 trp1 leu2Δ1 his3Δ200 pep4::HIS3 prb1Δ1.6R can1 GAL*
BJ5626	α *ura3-52 trp1 his3Δ200 pep4::HIS3 prb1Δ1.6R can1 GAL* / **a** *ura3-52 leu2Δ1 his3Δ200 pep4::HIS3 prb1Δ1.66 can1 GAL*
BJ5627	α *ura3-52 trp1 his3Δ200 pep4::HIS3 prb1Δ1.6R can1 GAL* / **a** *ura3-52 leu2Δ1 his3Δ200 pep4::HIS3 prb1Δ1.6R can1 GAL*
BJ5628	α *ura3-52 leu2Δ1 his3Δ200 pep4::HIS3 prb1Δ1.6R can1 GAL* / **a** *ura3-52 trp1 his3Δ200 pep4::HIS3 prb1Δ1.6R can1 GAL*

The *prb1-1122* and *pep4-3* alleles are nonsense mutations (UAA[75] and UGA,[59] respectively). The *pep4::HIS3* mutation is an insertion of a *Bam*HI fragment bearing *HIS3* into the *Hin*dIII site in *PEP4*.[54] The *prb1-Δ1.6R*

---

[75] G. S. Zubenko, A. P. Mitchell, and E. W. Jones, *Genetics* **96,** 137 (1980).

mutation is a deletion of a 1.6-kilobase *Eco*RI fragment internal to the *PRB1* gene.[19] The *trp1-Δ101* deletion inactivates the adjacent *GAL3* gene.[76] **a**/**a** and α/α derivatives of diploids BJ5622–5624 will be sent to the Stock Center.

Cautions to Users of Protease-Deficient Strains. The *pep4-3* and *prb1-1122* mutations are nonsense mutations (UGA and UAA, respectively). Investigators employing nonsense suppressor-bearing plasmids must bear this in mind. *pep4-3*-bearing strains accumulate suppressors of the *pep4* mutation on continued subculturing.[57,77,78] Stress in the form of overproduction of toxic proteins may exacerbate the problem and result in selection of revertants[77] or reactivation of phenotypic lag.[77,78]

While attempting to purify the 40-kDa precursor to PrB (see Moehle *et al.*[26] for PrB maturation pathway) from a *pep4*-bearing strain, C. M. Moehle found that the precursor became activated to PrB of mature size during one column purification step.[21] We have no information on whether the activation was autocatalytic or due to the activity of another protease. Nevertheless, it seems clear that strains of *pep4 PRB1* genotype that contain PrB precursor at levels comparable to levels of PrB in wild-type strains may not be ideal starting strains for biochemical analyses and/or purifications, since the PrB precursor constitutes a potential reservoir for production of PrB during a purification. This provides an additional reason for using *prb1 pep4* double mutants.

[76] P. A. Hieter, personal communication (1988).
[77] J. R. Shuster, A. Randolph, and C. George-Nacimento, *J. Cell. Biochem.* **9C** (Suppl.), 111 (1985).
[78] E. W. Jones, unpublished (1984–1988).

[32] Structural and Functional Analysis of Yeast Ribosomal Proteins

By H. A. RAUÉ, W. H. MAGER, and R. J. PLANTA

Structural Analysis

Ribosomal proteins of yeast have been separated and characterized predominantly by various types of two-dimensional gel electrophoresis, all of which are modifications of the procedures developed by Kaltschmidt and Wittmann[1] (basic urea/acidic urea) and Mets and Bogorad[2] (acidic urea/sodium dodecyl sulfate) respectively, for *Escherichia coli* ribosomal proteins. In our hands best results have been obtained using the Kaltschmidt and Wittmann procedure for small subunit and the Mets and Bogorad procedure for large subunit ribosomal proteins. Both separations can be carried out conveniently and rapidly by employing a minigel system developed in our laboratory.[3]

Isolation of Ribosomes and Ribosomal Subunits

Yeast cells are grown in a fermenter to mid-log phase ($A_{550} \sim 3$) on a synthetic medium containing, per liter, 11.3 ml lactic acid, 14 ml sodium lactate, 1 g glucose, 5 g yeast extract (Difco, Detroit, MI), 2 g $MgSO_4 \cdot 7H_2O$, 6 g $(NH_4)_2HPO_4$, and 0.2 ml antifoam (Serva). The pH of the medium should be adjusted to 4.5 by addition of KOH. Growth is stopped by pouring the culture over crushed ice, and cells are harvested by centrifugation at 4°, preferably in a continuous-flow rotor (e.g., the Beckman Instruments JCF-Z rotor). Cells can be stored in convenient portions at $-70°$. The average yield is about 4 g/liter. Harvesting at late log phase increases the yield but makes it more difficult to break open the cells.

All subsequent manipulations have to be carried out at 0–4°. Glassware should be baked at 120° for 4 hr in an oven and buffers autoclaved at 110° for 15 min to inactivate nucleases, throughout the procedures described in this and the following sections. If ribosomal subunits are to be isolated, cells (16–20 g) are resuspended in an equal volume of ice-cold twice-concentrated dissociation buffer [buffer A: 20 mM Tris-HCl, pH 7.4, 16 mM $MgCl_2$, 1.0 M KCl, 12 mM 2-mercaptoethanol (added after autoclaving the rest of the buffer), 0.2 mM EDTA] by vortexing. Glass beads (2.5 times the weight of the cells, diameter 0.5 mm) are added, and the cells are broken by vigorously shaking the mixture in a Braun shaker (2 times, 1 min each time). During this procedure temperature has to be carefully controlled by cooling the shaking vessel with expanding CO_2 gas. Freezing of the mixture should be avoided since it results in a considerable reduction in the yield of subunits. Glass beads are removed by filtration under suction on a glass filter. The resulting cell homogenate is centrifuged for 15 min at 10,000 rpm in a Beckman 60Ti rotor at 4°. The supernatant is removed using a Pasteur pipette, avoiding the white upper layer containing lipid material, and recentrifuged at 20,000 rpm for 30 min. The clear yellowish upper two-thirds of the supernatant is again removed using a

---

[1] E. Kaltschmidt and H. G. Wittmann, *Anal. Biochem.* **36**, 401 (1970).
[2] L. J. Mets and L. Bogorad, *Anal. Biochem.* **57**, 200 (1974).
[3] G. H. P. M. Bollen, W. H. Mager, and R. J. Planta, *Mol. Biol. Rep.* **8**, 37 (1981).

Pasteur pipette and the A_{260} (1 cm path length) determined. Inclusion of material from the cloudy lower part of the homogenate should be avoided. Although the supernatant containing the subunits can be stored at $-70°$, best results are obtained if the material is used without freezing. The supernatant should be left at 4° for at least 2–3 hr prior to sucrose gradient centrifugation, however, to allow complete dissociation of the ribosomes.

Ribosomal subunits are separated by layering about 350 A_{260} units of the supernatant on a 10–40% linear sucrose gradient in 1 × dissociation buffer. Up to 2 ml of the supernatant can be applied to a single SW28 gradient without significant loss of resolution of the subunit peaks. The gradients are centrifuged for 17 hr at 4° and 20,000 rpm in a Beckman SW28 rotor. Large and small subunits are pooled by continuously monitoring the A_{280} of the gradient (the amount of material is too high to allow monitoring at 260 nm). The OD pattern usually shows a small shoulder of undissociated 80 S ribosomes. The pooled subunit fractions are dialyzed overnight at 4° against 2 times 200 volumes of standard buffer (buffer B: 10 mM Tris-HCl, pH 7.4, 10 mM MgCl$_2$, 50 mM KCl, 6 mM 2-mercaptoethanol). The dialysis tube should be tightly closed since osmotic pressure within the tube increases considerably during this procedure. The dialyzed subunits are stored in portions at $-70°$.

For isolation of intact ribosomes the yeast cells are suspended in buffer A containing 0.1 M instead of 1.0 M KCl. Ribosomes can be pelleted from the purified cell homogenate by centrifugation at 50,000 rpm for 2 hr at 4° in a Beckman 50H rotor.

Protein Extraction

Protein is prepared from purified ribosomes by resuspending the ribosome pellet in 10 mM Tris-HCl containing 6 mM 2-mercaptoethanol and extraction with 67% acetic acid/30 mM MgCl$_2$.[4] Stored subunits stored are first dialyzed against this buffer. The precipitated rRNA is removed by centrifugation and the protein solution dialyzed overnight at 4° against 5% acetic acid. Dialysis tubing should be boiled for 10 min in either 1% NaHCO$_3$ or 10% trichloroacetic acid (TCA) to inactivate nucleases and then thoroughly rinsed with distilled water. The dialyzed protein solution is either lyophilized or stored in 5% acetic acid at $-20°$.

Two-Dimensional Gel Electrophoresis in Minigel System

For the Kaltschmidt and Wittmann procedure lyophilized ribosomal proteins are dissolved in sample buffer containing 8 M deionized urea,

[4] S. J. S. Hardy, C. G. Kurland, P. Voynow, and G. Mora, *Biochemistry* **8**, 2897 (1969).

2 mM dithiothreitol (DTT), 25% (w/v) sucrose, and 0.1% fuchsin to a final concentration of 20 μg/μl. From 60 to 100 μg of protein is used per sample. The first-dimension gel is prepared in 90 × 2.7 mm cylindrical tubes which should be carefully cleaned with concentrated H_2SO_4 and a detergent, rinsed with 70% ethanol, and dried. The tubes are tightly sealed with Parafilm and filled from the bottom using a syringe with a long needle to avoid the inclusion of air bubbles. First-dimension gel solution contains 6 M deionized urea, 4% (w/v) acrylamide, 0.13% (w/v) bisacrylamide, 18 mM EDTA, 0.4 M Tris, and 3 ml/liter TEMED. The pH is adjusted to 8.6 with boric acid. The tops of the gels should be kept flat by overlayering with a small volume of isobutanol. After polymerization the isobutanol is removed by thoroughly rinsing with distilled water. Electrophoresis is carried out at 220 V (4 mA/gel) at 4° for 2 hr for separation of the acidic (pI < 8.6) and 2.5 hr for separation of the basic (pI > 8.6) ribosomal proteins. The electrophoresis buffer contains 0.12 M Tris–borate, pH 8.6, 6 M deionized urea, 6 mM EDTA.

Gels are removed from the tubes by pressing distilled water around the outside of the gel using a syringe with a long, thin needle, then soaked for at least 10 min in unpolymerized second-dimension gel solution [6 M deionized urea, 18.6% (w/v) acrylamide, 0.51% (w/v) bisacrylamide, 5.4% (v/v) glacial acetic acid, 50 mM KOH, 6 ml/liter TEMED]. Second-dimension gels are poured between two 100 × 100 mm glass plates (carefully cleaned in the same way as the first-dimension tubes) separated by 2.7 mm spacers. About 1 cm free space should be left at the top. After polymerization, the first-dimension gel is polymerized on top of the slab with the aid of the second-dimension gel. Air bubbles between the first- and second-dimension gels should be absolutely avoided. Electrophoresis is carried out for 3.5 hr at 250 V (maximum 50 mA/gel) and 16° using 0.19 M glycine/1.5% (v/v) glacial acetic acid, pH 4.0, as running buffer. The gels are stained in a solution containing 0.12% (w/v) Coomassie Brilliant Blue R250 (CBB), 45% (v/v) methanol, 9% (v/v) acetic acid for 1 hr. Destaining is carried out by gently shaking the gel in 25% (v/v) methanol containing 7% acetic acid. The amount of solution as well as the time needed for destaining can be kept to a minimum by including a piece of polyurethane which efficiently absorbs the CBB.

The same minigel system can also be used for separation of basic ribosomal proteins according to the procedure of Mets and Bogorad.[2] In this case the sample buffer consists of 10 mM bis-Tris–acetate, pH 4.2, containing 8 M deionized urea, 1% 2-mercaptoethanol, and 0.1% fuchsin. The first-dimension gel solution contains 4% (w/v) acrylamide, 0.66% (w/v) bisacrylamide, 8 M deionized urea, 57 mM bis-Tris–acetate, pH 5.0, and 1 ml/liter TEMED. A discontinuous buffer system is used. The electrophoresis buffer in the upper compartment consists of 10 mM bis-Tris–acetate, pH 3.8, 5 mM 2-mercaptoethanol, whereas the buffer in the lower compartment is composed of 0.179 M potassium acetate, pH 5.0. First-dimension electrophoresis is at 50 V for 15 min and then 120 V for 90–120 min (2 mA/gel) at 4°. The first-dimension gels are soaked for 30 min in 40 mM bis-Tris–acetate, pH 6.0, 6 M deionized urea, 1% sodium dodecyl sulfate (SDS). The second dimension consists of both a separation and a stacking gel. The separation gel [12.5% (w/v) acrylamide, 0.25% (w/v) bisacrylamide, 0.1 M bis-Tris–acetate, pH 6.75, 1 ml/liter TEMED] is prepared 1 day in advance. About 2 cm free space should be left at the top, which is kept flat by overlayering with upper second-dimension running buffer. Before pouring the stacking gel the buffer is carefully removed and the top of the gel dried by blotting with filter paper. About 1 cm of stacking gel [4% (w/v) acrylamide, 0.066% (w/v) bisacrylamide, 6 M deionized urea, 40 mM bis-Tris–acetate, pH 6.0, 0.2% SDS, 2 ml/liter TEMED] is polymerized on top of the separation gel. The first-dimension gel is then polymerized on top of the stacking gel as described above using stacking gel solution. Electrophoresis starts at 50 V for 30 min and is continued for 3–4 hr at 80–100 V, using 50 mM bis-Tris–acetate, pH 6.5, 0.07% (w/v) 2-(N-morpholino)ethanesulfonic acid (MES), 0.2% SDS, 0.02% (v/v) thioglycolic acid in the upper and 20 mM bis-Tris–acetate, pH 6.75, in the lower compartment.

Correlation between Various Nomenclatures for Yeast Ribosomal Proteins

The use of different two-dimensional gel systems for characterizing yeast ribosomal proteins has led to various numbering systems which, unfortunately, have not yet been fully correlated. Table I lists the currently available correlation between nomenclatures based on the three most frequently used two-dimensional gel electrophoresis systems.[1,3,5–7] Also included are the molecular weights of the proteins, either calculated from the known amino acid sequence (see Table II) or determined by gel electrophoretic analysis.[7,8]

Sequences of Yeast Ribosomal Proteins

Over the last few years a large number of genes for yeast ribosomal proteins have been cloned and analyzed, allowing deduction of the com-

---

[5] E. Otaka and S. Osawa, *Mol. Gen. Genet.* **181,** 176 (1981).
[6] J. A. Warner and C. G. Gorenstein, in "Methods in Cell Biology" (D. Prescott, ed.), Vol. 20, p. 45. Academic Press, New York, 1978.
[7] S. Michel, R. R. Traut, and J. C. Lee, *Mol. Gen. Genet.* **191,** 251 (1983).
[8] T. Kruiswijk and R. J. Planta, *Mol. Biol. Rep.* **1,** 409 (1974).

TABLE I
Correlation of Ribosomal Protein Nomenclature[a]

40 S subunit proteins				60 S subunit proteins			
A	B	C	MW ($\times 10^{-3}$)[b]	A	B	C	MW ($\times 10^{-3}$)[b]
S0	—	YS1, 2?	31.0	L1a	rp3?	YL3	38.5
S1	—	YS1, 2?	35.5	L1b	—	—	20.0
S2	rp14	YS8	26.5	L1c	rp4?	YL31	13.5
S3	rp13	YS3	28.5	L2	rp2	YL2	**39.1**
S4	rp12	YS5	28.0	L3	rp1	YL1	**43.5**
S6	—	YS7	27.5	L4	rp6	YL5	28.0
S7	rp5	YS6	27.0	L5	rp8	YL6	27.5
S10	rp9	YS4	**26.8**	L6	rp11	YL8	26.5
S11	rp40	YS10?	22.0	L7	—	YL9	26.5
S12	rp30	YS10?	24.0	L8	rp25	YL7, 11?	24.5
S13	rp21	YS11	21.5	L9	—	YL7, 11?	26.0
S14	rp19	YS9?	24.0	L10	rp16	YL12, 13?	27.5
S15	—	YS9?	24.0	L11	—	—	?
S16	—	YS16	14.5	L12	—	YL12, 13?	24.5
S16a	rp55	—	**15.8**	L13	rp15?	YL10	26.5
S17	—	YS17, 18, 19?	15.5	L15	—	YL23	**17.8**
S18	rp41	YS12	20.5	L16	rp39	YL22	**19.7**
S19	—	YS13	15.5	L17	rp18?	YL16	23.0
S20	rp42	YS21	14.5	L17a	—	YL32	**14.5**
S21	rp52	YS17, 18, 19?	15.5	L18	—	YL18	22.5
S22	—	YS17, 18, 19?	15.0	L18a	—	YL18?	22.5
S23	—	—	?	L19	—	YL19	22.0
S24	rp50	YS22	**14.5**	L20	—	YL17	23.0
S25	—	YS17, 18, 19?	15.0	L21	rp22	YL15	22.5
S26	—	YS25	9.7	L22	rp33?	YL20	20.5
S26a	—	—	?	L23	rp33?	YL14	23.5
S27	rp61	YS20	13.5	L24	—	YL31	?
S28	rp37	YS14	17.5	L25	—	YL25	**15.7**
S29	—	—	?	L26	—	YL28	14.0
S30	—	—	?	L29	rp44	YL24	**16.7**
S30a	—	—	?	L30	rp29	YL21	**17.6**
S31	rp45	YS23	**12.0**	L31	—	YL29	16.5
S32	—	YS28	< 10.0	L32	rp73	YL38	**11.4**
S32a	—	—	?	L33	—	YL33	14.5
S32b	—	—	?	L34	—	YL36	**12.9**
S33	—	YS27	7.6	L35	—	YL30	16.0
S34	—	YS29?	?	L36	—	YL34	13.0
S35	—	YS24?	?	L37	rp47	YL37	11.0
S36	—	YS29?	< 10.0	L38	—	YL26	16.0
S37	—	YS24?	**8.7**	L39	—	YL39	10.5
—	rp51	—	**15.8**	L40	—	—	10.0
—	rp59	—	**14.7**	L41	—	YL27	**11.8**

(Continued)

TABLE I (Continued)

40 S subunit proteins				60 S subunit proteins			
A	B	C	MW ($\times 10^{-3}$)[b]	A	B	C	MW ($\times 10^{-3}$)[b]
				L42	—	YL42	< 10.0
				L43	—	YL35	9.8
				L44	—	YL44	**10.7**
				L44'	—	—	10.7
				L45	—	—	10.9
				L46	—	YL40, 41?	6.2
				L47	—	YL40, 41?	3.3
				—	rp28	—	20.5
				—	—	YL43	?
				—	(A0)[c]	—	26.2
				—	(A1)[c]	—	18.7

[a] From (A) Kruiswijk and Planta,[8] (B) Warner and Gorenstein,[6] and (C) Otaka Osawa.[5]

[b] Molecular weights as determined by SDS–gel electrophoresis.[8] Boldface numbers indicate molecular weights calculated from amino acid sequence of the protein.

[c] Genes for these acidic proteins were recently isolated and sequenced. [K. Mitsui and K. Tsurugi, *Nucleic Acids Res.* **16**, 3573 (1988).]

plete amino acid sequences of these proteins. These data, together with those on complete or partial (usually amino-terminal) sequences derived by classic amino acid sequence analysis, have been compiled in Table II.[9-42]

[9] E. Otaka, K. Higo, and T. Itoh, *Mol. Gen. Genet.* **195**, 544 (1984).
[10] E. Otaka, K. Higo, and S. Osawa, *Biochemistry* **21**, 4545 (1982).
[11] R. J. Leer, M. M. C. van Raamsdonk-Duin, C. M. T. Molenaar, H. M. A. Witsenboer, W. H. Mager, and R. J. Planta, *Nucleic Acids Res.* **13**, 5027 (1985).
[12] C. M. Molenaar, L. P. Woudt, A. E. M. Jansen, W. H. Mager, and R. J. Planta, *Nucleic Acids Res.* **12**, 7345 (1984).
[13] R. J. Leer, M. M. C. van Raamsdonk-Duin, P. Kraakman, W. H. Mager, and R. J. Planta, *Nucleic Acids Res.* **13**, 701 (1985).
[14] K. Suzuki and E. Otaka, *Nucleic Acids Res.* **16**, 6223 (1988).
[15] T. Itoh, E. Otaka, and K. A. Matsui, *Biochemistry* **24**, 7418 (1985).
[16] R. T. M. Nieuwint, C. M. T. Molenaar, J. H. Van Bommel, M. M. C. van Raamsdonk-Duin, W. H. Mager, and R. J. Planta, *Curr. Genet.* **10**, 1 (1985).
[17] R. J. Leer, M. M. C. van Raamsdonk-Duin, P. J. Schoppink, M. T. E. Cornelissen, L. H. Cohen, W. H. Mager, and R. J. Planta, *Nucleic Acids Res.* **11**, 7759 (1983).
[18] E. Özkaynak, D. Finley, M. J. Solomon, and A. Varshavsky, *EMBO J.* **6**, 1429 (1987).
[19] J. L. Teem and M. Rosbash, *Proc. Natl. Acad. Sci. U.S.A.* **80**, 4403 (1983).
[20] N. Abovich and M. Rosbash, *Mol. Cell. Biol.* **4**, 1871 (1984).
[21] J. C. Larkin, J. R. Thompson, and J. L. Woolford, Jr., *Mol. Cell. Biol.* **7**, 1764 (1987).

TABLE II
AMINO ACID SEQUENCES OF YEAST RIBOSOMAL PROTEINS

Protein	Sequence[c]	Ref.
S3	VALISKkRKL VANGVFYAQ L...	9
S7	ARGPKKHLKR LAAPHHxLLL NLLGGYAP.....................................	10
S10[a]	KLNISYPVNG SQKTFEIDDE HRIRVFFDKR IGQEVDGEAV GDEFKGYVFK ISGGNDKQGF PMKQGVLLPT RIKLLLTKNV SCYRPRRDGE RKRKSVRGAI VGPDLAVLAL VIVKKGEQEL EGLTDTTVPK RLGPKRANNI RKFFGLSKED DVRDFVIRRE VTKGEKTYTK APKIQRLVTP QRLQRKRHQR ALKVRNAQAQ REAAAEYAQL LAKRLSERKA EKAEIRKRRA SSLKA	10, 11
S13	PRAPRTYSKT YSTPKRPYQS NRL...	10
S14	GISRNSRHKRS ATGAKRAQFR KRRKFQLGR QPANTKIGAK RIHsVRTkGG................	9
S16[a]	P_AGVSVRDVAA QDFINAYASF LQRQGKLEVP GYVDIVKTSS GNEMPPQDAE GWFYKRAASV ARHIYMRKQV GVGKLNKLYG GAKSRGVRPY KHIDASGSIN RKVLQALEKI GIVEISPKGG RRISENGQRD LDRIAAQTLE EDE	10, 12
S24[a]	TRSSVLADAL NAINNAEKTG KRQVLIRPSS KVIIKFLQVM QKHGYIGEFE YIDDHRSGKI VVQLNGRLNK CGVISPRFNV KIGDIEKWTA NLLPARQFGY VILTTSAGIM DHEEARRKHV SGKILGFVY	10, 13
S26[a]	MENDKGQLVE LYVPRKCSAT NRIIKADDHA SVQINVAKVD EEGRAIPGEY VTYALSGYVR SRGESDDSLN RLAQNDGLLK NVWSYSR	14, 15
S27a	GRMHSAGKGI SSSAIPYSRN APAGFKLSSE CVIExIVKYA...........................	10
S31[a,b]	mPPKQLSKA AKAAAALAGG KKSKKKWSKK SMKDRAQHAV ILDQEKYDRI LKEVPTYRYV SVSVLVDRLK IGGSLARIAL RHLEKEGIIK PISKHSKQAI YTRATASE	9, 16
S33[a]	mDNKTPVTLA KVIKVLGRTG SRGGVTQVRV EFLEDTSRTI VRNVKGPVRE NDILVLMESE REARRLR	17
S36	AHENVGFSH...	9

[32]   YEAST RIBOSOMAL PROTEINS   461

S37[a]	GKKRKKKVYT TPKKIKHKHK KVKLAVLSYY KVDAEGKVTK LRRECSNPTC GAGVFLANHK DRLYCGKCHS VYKVNA					9, 18
rp51[a]	mGRVRTKTVK RASKALIERY YPKLTLDFQT NKRLCDEIAT IQSKRLRNKI AGYTHLMKR IQKGPVRGIS FKLQEEERER KDQYVPEVSA LDLSRSNGVL NVDNQTSDLV KSLGLKLPLS VINVSAQRDR RYRKR$_N$					19, 20
rp59[a]	mSNVVQARDN SQVFGVARIY ASFNDTFVHV TDLSGKETIA RVTGGMKVKA DRDESSPYAA MLAAQDVAAK CREVGITAVH VKIRATGGTR TKTPGPGGQA ALRALARSGL RIGRIEDVTP VPCDSTRKKG GRRGRRL					21
L1a	AFQKDAKSSA YSSRFQYPFR RRREGKTDYY......(~164)........ EELADDDEER FSELFKGYLA DDIDADSLED IYTSAHEAIR ADPAFKPTEK KFTKEQYAAE SKKYRQTKLS KQERAARVAA KIAALAGQQ					22, 23
L1c	APNTSRKQKI AKTFTVNVSS PTENGVFNPA SYAKYLINHI KVQGAVGNLG N.........					10
L2[a]	mSRPQVTVHS LTGEATANAL PLPAVFSAPI RPDIVHTLFT SVNKNKRQAY AVSEKAGHET SAESWGTGRA VARIPRVGGG GTGRSGQGAF GNMCRGGRMF APTKTWRKWN VKVNHNEKRY ATASAIAATA VASLVLARGH RVETIPEIPL VVSTDLDSIQ KTKEAVAALK AVGAHSDLLK VLKSKLRAG KGKYRNRRWT QRRGPLVVYA EDNGIVKALR NVPSVETANV ASLNLLQLAP SAHLGRFVIW TEAAFTKLDQ VWGSETVASS KVGYTLPSHI ISTSDVTRII NSSEIQSAIR PAGQATQKRT KVLKKNPLKN KQVLLRLNPY AKVFAAEKLG SKKAEKTGTK PAAVFTETLK HD					24
L3[a]	SHRKYEAPRH GHLGFLPRKR AASIRARVKA FPKDDRSKPV ALTSFLGYKA GMTTIVRDLD RPGSKFHKRE VVEAVTVVDT PPVVVVGVG YVETPRGLRS LTTVWAEHLS DEVKRRFYKN WYKSKKKAFT KYSAKYAQDG AGIERELARI KKYASVVRVL VHTQIRKTPL AQKKAHLAEI QLNGGSISEK VDWAREHFEK TVAVDSVFEQ NEMIDAIAVT KGHGFEGVTH RWGTKKLPRK THRGLRKVAC IGACHPAHVM WSVARAGQRG YHSRTSINHK IYRVGKGDDE ANGATSFDRT KKTITPMGGF VHYGEIKNDF IMVKGCIPGN RKRIVTLRKS LYTNTSRKAL EEVSLKWIDT ASKFGKGRFQ TPAEKHAFMG TLKKDL					25, 26
L5	GRVIRNQRKG AGSIFTSHTR LRQGAAKLRT LNYAQRHGYI..............					9
L6	AAEKILTPES QLKKSKAQQK TAQQVAAERA ARKAANKExR.............					9
L8	MKYIQTQQQI EVPQGVTVSI KSRIVKVVGP RGTLTKN̲LxH.............					9

(Continued)

TABLE II (Continued)

Protein	Sequence[c]	Ref.
L13	GAYKYLEELQ RKKQSDVLRF LQRVRVGEYR QKNVIHRAAR PTR............	10
L15[a,b]	MPPKFDPNEV KYLYLRAVGG EVGASAALAP KIGPLGLSPK KVGEDIAKAT KEFKGIKVTV QLKIQNRQAA ASVVPSASSL VITALKEPPR DRKKDKNVKH SGNIQLDEII EIARQMRDKS FGRTLASVTK EILGTAQSVG CRVDFKNPHD IIEGINAGEI EIPEN	9, 27
L16[a]	mS[A]KAQNPMR DLKIEKLVLN ISVGESGDRL TRASKVLEQL SGQTPVQSKA RYTVRTFGIR RNEKIAVHVT VRGPKAEEIL ERGLKVKEYQ LRDRNFSATG NFGFGIDEHI DLGIKYDPSI GIFGMDFYVV MNRPGARVTR RKRCKGTVGN SHKTTKEDTV SWFKQKYDAD VLDK	28, 29
L17	TAQQAPKXYP SQNVAAPKKT RKAV............	9
L17a[a]	mSGNGAQGTK FRISLGLPVG AIMNCADNSG ARNLYIIAVK GSGSRLNRLP AASLGDMVMA TVKKGKPELR KKVMPAIVVR QAKSWRRRDG VFLYFEDNAG VIANPKGEMK GSAITGPVGK ECADLWPRVA SNSGVVV	30
L20	ARYGATSTNP AKSASARGSY LRVSFKNTRQ TAQAINGQL............	9
L25[a]	mAPSAKATAA KKAVVKGTNG KKALKVRTSA TFRLPKTLKL ARAPKYIASKA VPHYNRLDSY KVIEQPITSE TAMKKVEDGN ILVFQVSMKA NKYQIKKAVK ELYEVDVLKV NTLVRPNGTK KAYVRLTADY DALDIANRIG YI	30, 31
L26	AGLKDVVTRQ YTINLHKRLH GVSFKKRAPR AVKQIKKFAK LHMGTNNVRL............	9
L29[a]	mPSRFTKTRK HRGHVSAGKG RIGKHRKHPG GRGMAGGEHH HRINMDKYHP GYFGKVGMRY FHKQQAHFWK PVINLDKLWT LIPEDKRDQY LKSASKETAP VIDTLAAGYG KILGKGRIPN VPVIVKARFV SKLAEEKIRA AGGVELIA	32
L30[a]	mKVEIDSFSG AKIYPGRGTL FVRGDSKIFR FQNSKSASLF KQRKNPRRIA WTVLFRKHHK KGITEEVAKK RSRKTVKAQR PITGASLDLI KERRSLKPEV RKANREEKLK ANKEKKAEK AARKAEKAKS AGTQSSKFSK QQAKGAFQKV ATTSR	33
L32[a]	mAPVKSQESI NQKLALVIKS GKYTLGYKST VKSLRQGKSK LIIIAANTPV LRKSELEYYA MLSKTKVYF QGGNNELGTA VGKLFRVGVV SILEAGDSDI LTTLA	34

L33	AKQSLDVSSN	RxKARKAYFTA	PSSQRRVLLS	APLSKQLRA................	9
L34[a]	mAGLKDVVTR KRGVKGVEYR	EYTINLHKRL LRLRISRKRN	HGVSFKKRAP EEEDAKNPLF	RAVKEIKKFA SYVEPVLVAS	KLHMGTDDVR AKGLQTVVVE	LAPELNQAIW EDA	35
L37	AESHRLYVKG	KHLSYQRSKR	VNDPN.............		9
L39	AVKTGIAIGL	NKGKKVTQMT	PAPKISYKKG	AASDRTKFVR................	9
L41	VNVPTKRKTY AKTTKKVVLR	CKGKTCRKHT LECVKCKTRA	QHKVTQYKAG QLTLKRHFEL	KASLFAQGRK GGEKKQKGQA	RYDRKQSGFG LQF	GQTKPVFHKK	36, 37
L43	GKGTPSFGKR GRMRYLKHVS	HNKSHTLCNR RRRFLNGFQT	CGRRSFHVQK GSASKASA	KTCSSCGYPA	AKTRSHNWAA	KAKRRHTTGT	37
L44[a]	mKYLAAYLLL LAAVPAAGPA	NAAGNTPDAT SAGGAAAASG	KIKAILESVG DAAAEEEKEE	IEIEDEKVSS EAAEESDDDM	VLSALEGKSV GFGLFD	DELITEGNEK	38, 39
L44'[a]	mSDSIISFAA HNAGPVAGAG	FILADAGLEI AASGAAAAGG	TSDNLLTITK DAAAEEEKEE	AAGANVDNVW EAAEESDDDM	ADVYAKALEG GFGLFD	KDLKEILSGF	38
L45[a]	MKYLAAYLLL KKFATVPTGG	VQGGNAAPSA ASAAAGAAGA	ADIKAVVASV AAGGDAAEEE	GAEVDEARIN KEEEAKEESD	ELLSSLEGKG DDMGFGLFD	SLEEIIAEGQ	26, 38, 40
L46[a]	AAQKSFRIKQ	KMAKAKKQNR	PLPQWIRLRT	NNTIRYNAKR	RNWRRTKMNI		13, 26
L47	MRAKWRKKRT	RRLKRKRRKV	RARSK............		9
rp28[a]	mGIDHTSKQH PVSVSRIARA CITLDQLAVR SKGFKV	KRSGHRTAPK LKQEGAANKT APKGQNTLIL	SDNVYLKLLV VVVVGTVTDD RGPRNSREAV	KLYTFLARRT ARIFEFPKTT RHFGMGPHKG	DAPFNKVVLK VAALRFTAGA KAPRILSTGR	ALFLSKINRP RAKIVKAGGE KFERARGRRR	12
YL43	AKSKNHTAHN	QTRKAHRNGI	KKPKTYKYPS	LKGVDPKFR............	9
L?[a,d]	IIEPSLKALA	SKYNCDKSVC	RKCYARLPPR	ATNCRKRKCG	HTNQLRPKKK	LK	18

(Continued)

TABLE II (Continued)

Protein	Sequence[c]	Ref.
A0[a]	mGGIREKKAE YFAKRLEYLE EYKSLFVVGV DNVSSQQMHE VRKELRGRAV VLMGKNTMVR RAIRGFLSDL PDFEKLLPFV KGYVGFVFTN EPITEIKNVI VSNRVAAPAR AGAVAPEDIW VRAVNTGMEP GKTSFFQALG VPTKIARGTI EIVSDVKVVD AGNKVGQSEA SLLNLLNISP FTFGLTVVQV YDNGQVFPSS ILDITDEELV SHFVSAVSTI ASISLAIGYP TLPSVGHTLI NNYKDLLAVA IAASYHYPEI EDLVDRIENP EKYAAAAPAA TSAASGDAAP AEEAAEEEE ESDDDMGFGL FD	41
A1[a]	mSTESALSYA ALILADSEIE ISSEKLLTLT NAANVPDENI WADIFAKALD GQNLKDLLVN FSAGAAAPAG VAGGVAGGEA GEAEAEKEEE EAKEESDDDM GFGLFD	42

[a] Sequence deduced from nucleotide sequence of the gene. If the amino-terminal Met residue is shown as a capital or if no Met residue is shown at the amino terminus, its presence or absence has been established by amino acid sequence analysis. If the amino-terminal Met residue is shown in lowercase type, its presence or absence has not been established. Two residues shown at the same position indicate a difference between proteins derived from duplicate genes.
[b] Amino-terminal sequencing of the isolated protein[10] identified a sequence starting at the arrow.
[c] N indicates either asparagine or aspartic acid; Q indicates glutamine or glutamic acid. Residues shown in small type have not been unambiguously identified. x indicates an unidentified residue.
[d] Protein is synthesized as part of a fusion protein with ubiquitin.[18]

Homologies between Yeast Ribosomal Proteins and Ribosomal Proteins from Other Organisms

Visual inspection of the two-dimensional gel electrophoretic patterns of ribosomal proteins from yeast and other eukaryotes suggests a considerable degree of homology. Both structural and functional studies have indeed provided compelling evidence for homology between individual ribosomal proteins from yeast on the one hand and mammals (predominantly rat) on the other. Moreover, several cases of clear homology, both functional and structural, with prokaryotic ribosomal proteins have been brought to light. Table III contains a compilation of the homologies so far uncovered.[43-81]

---

[22] R. N. Nazar, M. Yaguchi, G. E. Willick, C. F. Rollin, and C. Roy, *Eur. J. Biochem.* **102**, 573 (1979).
[23] M. Yaguchi, C. F. Rollin, C. Roy, and R. N. Nazar, *Eur. J. Biochem.* **139**, 451 (1984).
[24] C. Presutti, A. Lucioli, and I. Bozzoni, *J. Biol. Chem.* **263**, 6188 (1988).
[25] L. Schultz and J. D. Friesen, *J. Bacteriol.* **155**, 8 (1983).
[26] E. Otaka, K. Higo, and T. Itoh, *Mol. Gen. Genet.* **191**, 519 (1983).
[27] G. Puccierelli, M. Remacha, and J. P. G. Ballesta, *Nucleic Acids Res.* **18**, in press, (1990).
[28] R. J. Leer, M. M. C. van Raamsdonk-Duin, W. H. Mager, and R. J. Planta, *FEBS Lett.* **175**, 371 (1984).
[29] J. L. Teem, N. Abovich, N. F. Käufer, W. F. Schwindinger, J. R. Warner, A. Levy, J. Woolford, R. J. Leer, M. M. C. van Raamsdonk-Duin, W. H. Mager, R. J. Planta, L. Schultz, J. D. Friesen, H. Fried, and M. Rosbash, *Nucleic Acids Res.* **12**, 8295 (1984).
[30] R. J. Leer, M. M. C. van Raamsdonk-Duin, M. J. M. Hagendoorn, W. H. Mager, and R. J. Planta, *Nucleic Acids Res.* **12**, 6685 (1984).
[31] L. P. Woudt, W. H. Mager, J. G. Beek, G. M. Wassenaar, and R. J. Planta, *Curr. Genet.* **12**, 193 (1987).
[32] N. F. Käufer, H. M. Fried, W. F. Schwindinger, M. Jasin, and J. R. Warner, *Nucleic Aicds Res.* **11**, 3123 (1983).
[33] G. Mitra and J. R. Warner, *J. Biol. Chem.* **259**, 9218 (1984).
[34] M. D. Dabeva and J. R. Warner, *J. Biol. Chem.* **262**, 16055 (1987).
[35] P. J. Schaap, C. M. T. Molenaar, W. H. Mager, and R. J. Planta, *Curr. Genet.* **9**, 47 (1984).
[36] T. Itoh and B. Wittmann-Liebold, *FEBS Lett.* **96**, 399 (1978).
[37] T. Itoh, K. Higo, E. Otaka, and S. Osawa, in "Genetics and Evolution of RNA Polymerase, tRNA and Ribosomes" (S. Osawa, H. Ozeki, and H. Uchida, eds.), p. 609. Univ. of Tokyo Press, Tokyo, 1980.
[38] M. Remacha, M. T. Sáenz-Robles, M. D. Villela, and J. P. G. Ballesta, *J. Biol. Chem.* **263**, 9094 (1988).
[39] K. Mitsui and K. Tsurugi, *Nucleic Acids Res.* **16**, 3575 (1988).
[40] T. Itoh, *Biochim. Biophys. Acta* **671**, 16 (1981).
[41] K. Mitsui and K. Tsurugi, *Nucleic Acids Res.* **16**, 3573 (1988).
[42] K. Mitsui and K. Tsurugi, *Nucleic Acids Res.* **16**, 3574 (1988).
[43] K. Lechner, G. Heller, and A. Böck, *J. Mol. Evol.* **29**, 20 (1989).
[44] Y.-L. Chan and I. G. Wool, *J. Biol. Chem.* **263**, 2891 (1988).
[45] J.-L. Lalanne, M. Lucero, and J.-M. le Moullec, *Nucleic Acids Res.* **15**, 4490 (1987).
[46] J. B. Lott and G. A. Mackie, *Gene* **65**, 31 (1988).
[47] C. E. Sripati and M. Cuny, *Eur. J. Biochem.* **162**, 669 (1987).

TABLE III
HOMOLOGY BETWEEN RIBOSOMAL PROTEINS FROM YEAST AND OTHER ORGANISMS

Yeast	Mammals[a]	Prokaryotes Eubacteria[b]	Prokaryotes Archaebacteria[c]	Others[d]
S7	S4 (R)[10e]		ORFc (M.v.)[43j]	
S10	S6 (R, M, H)[44-46]			S7 (T. t.)[47i]
S13			S9 (H.c.)[10e]	
S16			S12 (H.m.)[48]	
S24		S8 (E.c.)[48]	S16 (H.m.)[48]	
S26	S21 (R)[15]	S6 (E.c.)[49]		S28 (S.p.)[15]
S27a			S11 (H.c.)[10e]	
S37	S27a (R)[50]			
rp51	S17 (R, Ha)[51,52]			
rp59	S14 (H, Ha)[53-55]	S11 (E.c.)[21,54]	S19 (H.m.)[48]	S14 (D.m.)[56]
L1a	L5 (R)[57,58]	L5 (E.c.)[51]	L13 (H.c.)[22g]	
		L18 (E.c.)[22,23g]		
L2				L1 (X.l., D.m.)[24,59]
L3	L4 (R)[26e]			
L5		L2 (E.c.)[60h]		
L13				L12 (S.p.)[26e]
L15		L11 (E.c.)[27,61i]	L11 (H.c.)[62]	
L23	L19 (R)[51]			L15 (S.p.)[26e]
L25		L23 (E.c.)[63i]	L25 (H.m.)[48]	L25 (C.u.)[31]
L29	L27' (M)[64]			L29 (N.c.)[65]
L32	L30 (R, M)[51,66]		ORF1 (M.v.)[43j]	
L34	L31 (R)[67]			
L41	L36a (R);[68] L44 (H)[69]			
L43	L37 (R)[70]	L34 (E.c.)[70]		L27 (S.p.)[26e]
L44 ⎫[f]				
L44' ⎬		L7/L12 (E.c.)[72]		rpA1 (D.m.)[73]
L45 ⎭	P2 (R, H)[71,74]			L40c (S.p.)[75]
A1	P1(H)[74]			
A0	P0 (H)[74]	L10 (E.c.)[76]	L10 (H.c.; M.v.)[76,77]	
L46	L39 (R)[78]		L46e (S.s.)[79]	L36 (S.p.)[26e]
rp28	L18 (R)[80]			L14 (X.l.)[81]

[a] R, rat; M, mouse; H, human; Ha, hamster.
[b] E.c., *Escherichia coli.*
[c] H.c., *Halobacterium cutirubrum*; H.m., *Halobacterium marismortui*; M.v., *Methanococcus vanniellii*; S.s., *Sulfolobus solfataricus.*
[d] T.t., *Tetrahymena thermophila*; S.p., *Schizosaccharomyces pombe*; D.m., *Drosophila melanogaster*; X.l., *Xenopus laevis*; C.u., *Candida utilis*; N.c., *Neurospora crassa*
[e] Based on comparison of amino-terminal sequence.
[f] 40-60% mutual homology. Homologs occur in all organisms studied so far.
[g] Functional homology (5 S rRNA binding); partial structural homology.
[h] Immunologically related.
[i] Functional homology established.
[j] Identified in transcribed gene clusters equivalent to *E. coli* S10, *str*, and *spc* operons.

Functional Analysis

In bacteria, analysis of the role of ribosomal proteins in determining the structure and function of ribosomes has benefited enormously from our ability to take prokaryotic ribosomal subunits apart and reassemble the purified components into biologically active particles. As far as eukaryotes are concerned, however, such complete reconstitution has not (yet) been accomplished. At the moment, reconstitution of eukaryotic ribosomal subunits is confined to reassociation of only a few ribosomal proteins,

[48] M. Kimura, E. Arndt, T. Hatakeyama, T. Hatakeyama, and I. Kimura, *Can. J. Biochem.* **35**, 195 (1989).
[49] E. Otaka, T. Itoh, and T. Kumazaki, *J. Mol. Evol.* **23**, 337 (1986).
[50] K. L. Redman and M. Rechsteiner, *Nature (London)* **33**, 438 (1989).
[51] A. Lin,, Y-L. Chan, R. Jones, and I. G. Wool, *J. Biol. Chem.* **262**, 14343 (1987).
[52] I-T. Chen, A. Dixit, D. D. Rhoads, and D. J. Roufa, *Proc. Natl. Acad. Sci. U.S.A.* **83**, 6907 (1986).
[53] D. D. Rhoads, A. Dixit, and D. J. Roufa, *Mol. Cell. Biol.* **6**, 2774 (1986).
[54] T. Tanaka, K. Ishikawa, and K. Ogata, *FEBS Lett.* **202**, 295 (1986).
[55] D. D. Rhoads and D. J. Roufa, *Mol. Cell. Biol.* **5**, 1655 (1985).
[56] S. J. Brown, D. D. Rhoads, M. J. Stewart, B. van Slyke, I-T. Chen, T. K. Johnson, R. E. Denell, and D. J. Roufa, *Mol. Cell. Biol.* **8**, 4314 (1988).
[57] Y-L. Chan, A. Lin, J. McNally, and I. G. Wool, *J. Biol. Chem.* **262**, 12879 (1987).
[58] S. Tamura, Y. Kuwana, T. Makayama, S. Tanaka, T. Tanaka, and K. Ogata, *Eur. J. Biochem.* **168**, 83 (1987).
[59] F. Rafti, G. Gargiulo, A. Manzi, C. Malva, and F. Graziani, *Nucleic Acids Res.* **17**, 456 (1989).
[60] G. Schmid, O. Strobel, M. Stöffler-Meilicke, G. Stöffler, and A. Böck, *FEBS Lett.* **177**, 189 (1984).
[61] T. T. A. L. El-Baradi, V. C. H. F. de Regt, S. W. C. Einerhand, J. Teixido, R. J. Planta, J. P. G. Ballesta, and H. A. Raué, *J. Mol. Biol.* **195**, 909 (1987).
[62] L. C. Shimmin and P. P. Dennis, *EMBO J.* **8**, 1225 (1989).
[63] T. T. A. L. El-Baradi, H. A. Raué, V. C. H. F. de Regt, E. C. Verbree, and R. J. Planta, *EMBO J.* **4**, 2101 (1985).
[64] P. Belhumeur, G. D. Paterna, G. Boileau, J. M. Claverie, and D. Skup, *Nucleic Acids Res.* **15**, 1019 (1987).
[65] C. A. Kreader and J. E. Heckman, *Nucleic Acids Res.* **15**, 9027 (1987).
[66] L. M. Wiedemann and R. P. Perry, *Mol. Cell. Biol.* **4**, 2518 (1984).
[67] T. Tanaka, Y. Kuwano, T. Kumazaki, K. Ishikawa, and K. Ogata, *Eur. J. Biochem.* **162**, 45 (1987).
[68] M. J. Gallagher, Y.-L. Chan, A. Lin, and I. G. Wool, *DNA* **7**, 269 (1988).
[69] M. S. Davies, A. Henney, W. H. J. Ward, and R. K. Craig, *Gene* **45**, 183 (1986).
[70] A. Lin, J. McNally, and I. G. Wool, *J. Biol. Chem.* **258**, 10664 (1983).
[71] A. Lin, B. Wittmann-Liebold, J. McNally, and I. G. Wool, *J. Biol. Chem.* **257**, 9189 (1982).
[72] E. Otaka, T. Ooi, T. Kumazaki, and T. Itoh, *J. Mol. Evol.* **22**, 342 (1985).
[73] S. Qian, J.-Y. Zhang, M. A. Kay, and M. Jacobs-Lorena, *Nucleic Acids Res.* **15**, 987 (1987).
[74] B. E. Rich and J. A. Steitz, *Mol. Cell. Biol.* **7**, 4065 (1987).
[75] M. Beltrame and M. E. Bianchi, *Nucleic Acids Res.* **15**, 9089 (1987).

dissociated from the subunits by various types of treatment (the "split proteins"), with the resulting ribonucleoprotein "core" particle.[82-84] Even such partial reconstitution has, except in an isolated case,[82] not been exploited for functional studies on the members of the split protein fraction.

One important aspect of ribosomal protein function is the ability of a number of ribosomal proteins to bind individually and specifically to one or another of the rRNA species. As demonstrated by numerous studies in *E. coli*, the identification of these so-called primary binding ribosomal proteins and their recognition sites on the various rRNA species provides valuable information on the structure and assembly of the ribosomal subunits.[85,86] Moreover, ribosomal protein–rRNA interactions were shown to play a central role in the regulation of ribosome biogenesis in prokaryotes.[87] In contrast to the situation in bacteria, binding of eukaryotic ribosomal proteins to rRNA is still a largely unexplored subject, in particular where high molecular mass rRNA is concerned. Yeast ribosomes constitute an excellent system for this type of study for two reasons. First, large amounts of ribosomes and ribosomal subunits can easily be obtained, and, second, essential structural information is available on all four rRNAs as well as a large number of the ribosomal proteins from analysis of the genes. The availability of the cloned genes also opens the possibility to study ribosomal protein–rRNA interactions using *in vitro* synthesized components. This obviates the need for laborious purification of individual ribosomal protein species and allows predetermined structural alterations to be introduced into either the ribosomal protein or its recognition site on the rRNA to study structure–function relationships.

[76] C. Ramirez, L. C. Shimmin, C. H. Newton, A. T. Matheson, and P. P. Dennis, *Can. J. Microbiol.* **35**, 234 (1989).
[77] A. K. E. Köpke, G. Baier, and B. Wittmann-Liebold, *FEBS Lett.* **247**, 167 (1989).
[78] A. Lin, J. McNally, and I. G. Wool, *J. Biol. Chem.* **259**, 487 (1984).
[79] C. Ramirez, K. A. Louie, and A. T. Matheson, *FEBS Lett.* **250**, 416 (1989).
[80] K. R. Gayathri-Devi, Y.-L. Chan, and I. G. Wool, *DNA* **7**, 157 (1988).
[81] E. Beccari and P. Mazetti, *Nucleic Acids Res.* **15**, 1870 (1987).
[82] F. Juan-Vidales, F. Sánchez-Madrid, M. Sáenz-Robles, and J. P. G. Ballesta, *Eur. J. Biochem.* **136**, 275 (1983).
[83] A. Vioque, J. A. Pinto-Toro, and E. Palaçian, *J. Biol. Chem.* **237**, 6477 (1982).
[84] F. Conquet, J.-P. Lavergne, A. Paleologue, J.-P. Reboud, and A.-M. Reboud, *Eur. J. Biochem.* **163**, 15 (1987).
[85] H. A. Raué, T. T. A. L. El-Baradi, and R. J. Planta, *Biochim. Biophys. Acta* **826**, 1 (1985).
[86] H. A. Raué, J. Klootwijk, and W. Musters, *Prog. Biophys. Mol. Biol.* **52**, 77 (1988).
[87] R. L. Gourse, R. A. Sharrock, and M. Nomura, *in* "Structure, Function and Genetics of Ribosomes" (B. Hardesty and G. Kramer, eds.), p. 766. Springer-Verlag, Berlin, 1985.

Preparation of Ribosomal Core Particles and Core Proteins

"Stripping" of ribosomal subunits using solutions of chaotropic salts such as KCl, NH_4Cl, or LiCl removes a portion of the ribosomal proteins from the subunit. The set of ribosomal proteins that remains associated with the rRNA will be enriched for primary binding ribosomal proteins and can be isolated as a ribonucleoprotein complex. In this manner attention can be focused on a limited number of potential candidates while at the same time a partial purification of these candidates is effected.

Our studies on stripping of yeast ribosomal subunits have concentrated on the 60 S particle of the yeast *Saccharomyces carlsbergensis* (strain S74 of the British National Collection of Yeast Cultures).[63,88] Treatment with various concentrations of LiCl was found to give the best results. By carefully controlling the concentration of this salt, five different types of core particles have been obtained, the largest containing six and the smallest only a single ribosomal protein.[88] Further studies on the latter ribosomal protein as well as one of the other five have shown that these proteins bind specifically to the 26 S rRNA at evolutionarily conserved sites.[61,63,89]

Stripping of 60 S Ribosomal Subunits. All manipulations should be carried out at 4° using sterilized glassware and buffers. Five milliliters (total 25 A_{260} units) of a 60 S subunit preparation (or 12 A_{260} units of a 40 S subunit preparation) are carefully thawed and centrifuged at low speed to remove aggregates. The sample is then dialyzed for 3 hr against 50 volumes of standard buffer containing the appropriate concentration of LiCl, taking into account the volume of the sample. Dialysis was found to give much better results than adding concentrated LiCl solution to the subunit preparation since the latter procedure caused considerable precipitation. The dialyzed subunit preparation is centrifuged for 15 min at 12,000 rpm in a Beckman 50H rotor to remove small amounts of precipitated material. The core particles are then collected by centrifugation for 4 hr at 50,000 rpm in a Beckman SW50 rotor. After centrifugation, the supernatant is removed by pipette, and the tube is carefully dried. The pellet containing the core particles is resuspended in 10 mM Tris-HCl, pH 7.4, containing 6 mM 2-mercaptoethanol, first using a glass rod to loosen the pellet and then by rotating the tube for several hours at 4° on a roller bench. The core particles can then be used for further analysis. Protein is extracted as

[88] T. T. A. L. El-Baradi, H. A. Raué, V. C. H. F. de Regt, and R. J. Planta, *Eur. J. Biochem.* **144**, 393 (1984).
[89] T. T. A. L. El-Baradi, V. C. H. F. de Regt, R. J. Planta, K. H. Nierhaus, and H. A. Raué, *Biochimie* **69**, 939 (1987).

described above for ribosomes and ribosomal subunits. Starting with 25 A_{260} units of 60 S particles, sufficient material is obtained for two or three analyses of protein composition by two-dimensional gel electrophoresis (see above). The procedure can easily be scaled up to obtain larger amounts of ribosomal protein for further purification and functional or structural analysis. We have used this procedure to isolate protein L25 for *in vitro* rRNA binding studies.[63,88] The protein is either lyophilized or stored in 5% acetic acid at $-20°$. In both cases it should retain its full rRNA-binding capacity for up to 2 years. Repeated freezing and thawing should be avoided, however.

Using 60 S subunits from *S. carlsbergensis* strain S74, LiCl concentrations up to 0.4 M did not have any effect on the protein composition of the treated particles. Treatment with 0.5 M LiCl removed the majority of the ribosomal proteins, leaving six species (L4, L8, L10, L12, L15, and L25) associated with the 26 S rRNA. Five of these six ribosomal proteins could be removed sequentially by further increasing the LiCl concentration in steps of 0.1 M, leaving only protein L25 associated with the 26 S rRNA after treatment with 1.0 M LiCl.[88] Although we have not carried out extensive experiments on 60 S subunits from other yeast strains, our data indicate that the degree of stripping caused by a specific concentration of LiCl is to a certain extent strain specific. When using yeast strains other than *S. carlsbergensis* S74, the above concentrations should, therefore, be considered merely as a guidance. So far our attempts to use this procedure for obtaining core particles from 40 S subunits of either *S. carlsbergensis* or *S. cerevisiae* have not been successful. Treatment of 40 S subunits with various concentrations of LiCl, KCl, or NH_4Cl either had no measurable effect at all or resulted in dissociation of all ribosomal protein species from the 17 S rRNA (this laboratory, 1988, unpublished results).

Analysis of Ribosomal Protein – rRNA Interactions

Analysis of the interaction between a ribosomal protein and an rRNA molecule consists of three main steps: (1) positive identification of the ribosomal protein as a primary binding species; (2) identification of the binding site for the ribosomal protein on the rRNA; and (3) identification of the rRNA-binding domain in the ribosomal protein. Each of these steps can be carried out using either components purified from ribosomes or components synthesized *in vitro*.

Nitrocellulose Filter Binding. The nitrocellulose filter binding technique is based on the ability of nitrocellulose filters to retain a protein – RNA complex while free RNA passes through the filter. This technique can therefore be used to determine whether a specific ribosomal protein does interact with an rRNA species. Failure of the rRNA to be retained on the filter does not necessarily indicate the absence of interaction with the protein, however.[90] Filter binding experiments also allow identification of the binding site on the rRNA, either by using specific *in vitro* synthesized rRNA fragments or by analyzing the rRNA fragments retained after filtration of a ribosomal protein–rRNA complex subjected to prior limited digestion with a ribonuclease (see below).

When a purified ribosomal protein is available, its rRNA-binding capacity can be tested by the following procedure. (Techniques for *in vivo* labeling of rRNA species are described elsewhere in this volume [30]. We have routinely labeled cells with ^{32}P by growing them in low-phosphate medium[91] to which 50 μCi/ml of ^{32}PO$_4^{3-}$ had been added.) Uniformly ^{32}P-labeled 17 S or 26 S rRNA (10,000–50,000 cpm as determined by Cerenkov counting) is incubated with increasing amounts of the purified protein in 300 μl (total volume) binding buffer (20 mM Tris-HCl, pH 7.4, 20 mM MgCl$_2$, 300 mM KCl, 6 mM 2-mercaptoethanol) for 15 min at 37°. Since high molecular mass rRNA dissolves very slowly in binding buffer owing to the high salt concentration, it is dissolved in buffer containing no KCl and the KCl concentration adjusted afterward. After incubation, the mixture is filtered through a nitrocellulose membrane filter (pore width 0.45 μm, diameter 1.2 cm; Sartorius, Göttingen, FRG) by a short low-speed centrifugation (3000 rpm) in a table-top centrifuge using a swing-out rotor. The filter is washed 3 times with 0.5 ml ice-cold binding buffer in the same way. The amount of radioactivity on the filter as well as that in the combined filtrates is quantitated by Cerenkov counting. From these data the percent radioactivity retained on the filter is calculated and plotted as a function of the amount of protein added to the incubation mixture. Binding of the protein to the rRNA results in a typical saturation curve which normally reaches a plateau at about 40–60% retention of the rRNA. By using progressively shorter *in vitro* synthesized fragments of the rRNA species containing the binding site it is possible to define the boundaries of this binding site.[89]

In our experiments with ribosomal protein L25, the protein stored in 5% acetic acid is dialyzed at 4°, first for 3 hr against 20 mM Tris-HCl, pH 7.4, 20 mM MgCl$_2$, 6 mM 2-mercaptoethanol and subsequently for 15 hr against the same buffer containing 300 mM KCl. This two-step dialysis is necessary to avoid precipitation of the protein which occurs when the KCl is present in the first buffer. Lyophilized L25 can be redissolved by adding

---

[90] J. Schwartzbauer and G. R. Craven, this series, Vol. 59, p. 583.
[91] R. C. Brand, J. Klootwijk, C. P. Sibum, and R. J. Planta, *Nucleic Acids Res.* 7, 121 (1979).

a small amount of binding buffer and shaking overnight in an Eppendorf shaker at 4°. After clarification of the solution by low-speed centrifugation it should be stored in small aliquots at −20° to avoid repeated freezing and thawing.

For unknown reasons preparations of *in vivo* labeled high molecular mass rRNA may suddenly show very high blanks in this type of experiment (up to 100% retention in the absence of ribosomal protein). This problem can often, though not always, be cured by again extracting the rRNA solution with phenol saturated with binding buffer (without KCl), containing 0.2% SDS and redissolving the rRNA as described above. We have not encountered this phenomenon using *in vitro* synthesized shorter fragments of rRNA.

Identification of Ribosomal Protein-Binding Site by RNase Protection. Digestion of a ribosomal protein–rRNA complex with a ribonuclease under limiting conditions will result in complete fragmentation of the regions of the rRNA not in contact with the protein, while the ribosomal protein-binding site will to a large extent be protected from the action of the enzyme. If the conditions are carefully chosen, these protected fragments will remain associated with the protein. Consequently they can be isolated by nitrocellulose filtration and characterized by sequencing. This procedure has been successfully employed to identify the binding sites for yeast ribosomal proteins L25 and L15 on yeast 26 S rRNA as well as related rRNA species from other organisms.[61,63,89]

Complex formation between the purified ribosomal protein and the ^{32}P-labeled rRNA is effected as described above using an amount of ribosomal protein sufficient to reach the plateau value of the saturation curve. After formation of the complex a sample containing about 10^6 cpm (Cerenkov) is mixed with 40 µg of unlabeled carrier RNA and incubated in a total volume of 300 µl for 15 min at 37° with 100 units of RNase T1 (Boehringer, Mannheim, Germany). The incubation mixture is then filtered through a nitrocellulose filter and washed exactly as described above. The bound RNA fragments are recovered by suspending the filter in 10 mM Tris-HCl, pH 7.4, 10 mM MgCl$_2$, 0.2% SDS (400 µl) and extracting twice with an equal volume of buffer saturated with phenol. Recovery in this step usually is about 50% of the amount of activity bound to the filter. The combined water layers are dialyzed overnight at 4° against distilled water and lyophilized. The lyophilized fragments are dissolved in 90 mM Tris–borate, pH 8.3, 1 mM EDTA, 6 M deionized urea, 10% glycerol, 0.1% bromphenol blue, loaded on a 12.5% polyacrylamide gel (20 × 40 × 0.05 cm) in the same buffer (without glycerol), and separated by electrophoresis until the marker dye has migrated about 30 cm. The fragments are visualized by overnight autoradiography at −70° using an intensifying screen. They are recovered by cutting out the appropriate bands and incubating them for 18–24 hr at room temperature in 400 μl of a solution containing 0.5 M ammonium acetate, 10 mM magnesium acetate, 0.1 M EDTA, 0.1% SDS. After removal of the gel slice(s), the fragments are precipitated by addition of $\frac{1}{10}$ volume of 3 M sodium acetate and 3 volumes ethanol and leaving the mixture at $-70°$ for at least 2 hr. No carrier should be added. The precipitate is collected by centrifugation for 15 min in an Eppendorf centrifuge, dissolved in 50 μl distilled water, and lyophilized.

We have also used the "native" L25–26 S rRNA complex, obtained by treating 60 S ribosomes isolated from ^{32}P-labeled cells with 1.0 M LiCl, in this type of experiment.[63] In this case RNase treatment was carried out exactly as described above immediately after dialysis of the subunits against the LiCl solution without removing the split proteins or changing the buffer.

The protected fragments recovered from the gel can be sequenced after 5'-end labeling with [γ-^{32}P]ATP using standard enzymatic sequencing techniques.[92] Comparison of the sequences obtained with that of the complete rRNA identifies the nature and location of the binding site. The uniform labeling of the rRNA fragments does not interfere with the sequencing since the amount of 5'-end label exceeds that of the uniform labeling by a factor 10 or more.

In Vitro Synthesis of rRNA Fragments. The binding sites of most primary binding ribosomal proteins occupy only a relatively small region of the rRNA in question.[61,85,89] Thus, it is possible to synthesize *in vitro* fragments of rRNA that retain full capability to interact with a specific ribosomal protein. The use of such *in vitro* transcripts opens the possibility for detailed analysis of structure–function relationships in the ribosomal protein binding site using the full potential of recombinant DNA technology.[61,89,93] Moreover, it allows easy access to the corresponding rRNA regions of other organisms for studies on evolutionary conservation of ribosomal protein–rRNA interactions.[61,89]

Appropriate rDNA fragments can be cloned for *in vitro* transcription by standard techniques in commercially available vectors under control of either the SP6 or T7 promoter (e.g., the pSP or pGEM vectors available from Promega, Madison, WI). In particular cases it may be necessary to use a cloning strategy that avoids the presence at the 5' end of the transcript of "foreign" nucleotides derived from vector sequences. The 3' end of the transcript is defined by cutting the recombinant vector with a restriction

---

[92] H. Donis-Keller, *Nucleic Acids Res.* **8**, 3133 (1980).
[93] C. L. Thomas, R. J. Gregory, G. Winslow, A. Muto, and R. A. Zimmermann, *Nucleic Acids Res* **16**, 8129 (1988).

enzyme either within or downstream of the rDNA insert so run-off transcription will occur.

In vitro transcription is carried out in 50-μl incubation mixtures using a commercially available transcription system (Promega) essentially as recommended by the manufacturer. Ten micrograms of cut template is added to the mixture. When radioactively labeled transcript is required, the concentration of cold UTP in the incubation mixture is decreased from 0.5 mM to 12 μM, and 10 μCi/μl [α-^{32}P]UTP (specific activity 3000 Ci/mmol) is added. Incubation is for 30 min at 40°. The transcript is purified by extraction with phenol/chloroform [1:1 (v/v), containing 10 ml/liter isoamyl alcohol and 0.1% (w/v) 8-hydroxyquinoline], ethanol precipitation, electrophoresis on a denaturing 6% polyacrylamide gel, and recovery from the gel exactly as described above for the RNase protection experiment. Unlabeled transcript is detected by fluorescence under UV light after soaking the gel for 15 min in a solution of 0.5 μg/ml ethidium bromide in distilled water. The purified transcript is again extracted with phenol/chloroform, precipitated, and redissolved in binding buffer as described above. The yield is routinely 2–5 μg/μg template.

In Vitro Synthesis of Ribosomal Proteins and Protein Fragments. Yeast ribosomal proteins, because of their low solubility, are difficult to purify and handle, which constitutes a severe impediment for detailed studies on their rRNA-binding properties. In order to circumvent this problem we have prepared yeast ribosomal proteins by *in vitro* transcription of cloned genes and *in vitro* translation of the resulting mRNA. We have established that at least two primary binding species synthesized in this way retain the ability to interact with rRNA. Moreover, we have used this approach successfully in delineating the rRNA binding domain of ribosomal protein L25 by testing the capacity of various amino- and carboxyl-terminal fragments of this protein to recognize the L25 binding site on 26 S rRNA.[94]

The template for *in vitro* translation is obtained by *in vitro* transcription of a ribosomal protein gene cloned under control of either the SP6 or T7 promoter and purification of the transcript exactly as described above for the rRNA fragments. The purified, usually unlabeled, transcript is dissolved in 10 mM Tris-HCl, pH 7.5, 0.1 mM EDTA. In cloning the ribosomal protein gene, attention should be paid to the following two points: (1) the intron, which is present in a large number of yeast ribosomal protein genes, should be removed, and (2) the presence in the transcript of AUG codons upstream of the authentic translation initiation codon (which may be introduced by transcription of vector sequences) should be

---

[94] C. A. Rutgers, P. J. Schaap, J. van't Riet, C. L. Woldringh, and H. A. Raué, *Biochim. Biophys. Acta,* in press (1990).

avoided. We have not found it necessary to cap the *in vitro* transcripts in order to obtain satisfactory template activity in the *in vitro* translation system.

In vitro translation is carried out in a commercial wheat germ lysate (Promega) again essentially according to the manufacturer's protocol. Incubation is for 60 min at 25° in a 100-μl incubation mixture containing 2–5 μg of template, 1 mM of each of 19 amino acids (excluding the one supplied in labeled form), 50 μl wheat germ lysate, and either [^{35}S]methionine (10 μCi/μl; specific activity 300–600 μCi/mmol) or [^{3}H]leucine (1 μCi/μl; specific activity 169 μCi/mmol). After incubation, free label is removed by applying the mixture to a 1-ml Sephadex G-25 column in 10 mM Tris-HCl, pH 7.4, 10 mM MgCl$_2$, 6 mM 2-mercaptoethanol and centrifuging. The eluate is stored at $-20°$ after adding glycerol to a final concentration of 15%. It contains no labeled products other than the one specified by the added template and can be used for binding studies without further purification.

Specific ribosomal protein fragments can be prepared by manipulating the cloned gene in such a way that a transcript encoding the desired portion of the protein is obtained. Amino-terminal protein fragments are obtained via run-off transcription after cutting within the gene with an appropriate restriction enzyme or by introducing 3'-terminal deletions by means of controlled *Bal*31 digestion. It should be noted that polypeptides translated from this kind of template, lacking a translation stop codon, may still be partially esterified with a tRNA at their carboxyl terminus.[95] Although we have not encountered any difficulties that might be due to partial esterification, such problems can easily be avoided by inserting a synthetic oligonucleotide linker containing an in-frame stop codon downstream of the remaining coding region. In the same way an AUG start codon has to be added in front of the coding region remaining after 5'-terminal deletion in order to obtain carboxyl-terminal fragments of the ribosomal protein. Using the above procedures, we have shown that at least 8 amino acids can be removed from the carboxyl terminus of L25 and at least 42 from the amino terminus without affecting the rRNA binding of the protein.[94]

Binding Assay for in Vitro Synthesized Ribosomal Protein. rRNA binding of the *in vitro* prepared ribosomal protein or protein fragments is assayed by incubating 2–5 μl of the mixture containing the labeled product with 0.5–1 μg of rRNA (either intact rRNA or *in vitro* synthesized fragment) in 100 μl of 10 mM Tris-HCl, 20 mM MgCl$_2$, 6 mM 2-mercaptoethanol. KCl is not included in the incubation mixture since it causes severe precipitation, probably because of the presence of components from

---

[95] M. T. Haeuptle, R. Frank, and B. Dobberstein, *Nucleic Acids Res.* **14**, 1427 (1986).

the wheat germ lysate. Independent filter binding experiments using purified L25 have shown, however, that the absence of KCl does not affect the binding of this protein to the rRNA.[94] After clarifying the incubation mixture by low-speed centrifugation, it is layered on a 10–40% sucrose gradient in the same buffer and centrifuged at 10° in a Beckman SW41 rotor. Speed and duration of this centrifugation are determined by the molecular weight of the rRNA (fragment) used. After centrifugation, the distribution of ^{35}S or ^{3}H across the gradient is analyzed by liquid scintillation counting.

Identification of Nuclear Localization Signals of Ribosomal Proteins

Since ribosome assembly is predominantly a nuclear process, ribosomal proteins have to be imported into the nucleus. Thus, they are likely to contain within their amino acid sequences one or more nuclear localization signals (NLS) ensuring active nuclear import.[96] The nature and localization of the NLS can be identified by analyzing the intracellular distribution of fusion proteins consisting of various parts of the ribosomal protein attached to a reporter protein that normally remains in the cytoplasm. Such analysis is carried out by means of indirect immunofluorescence using an antibody directed against the reporter protein, usually *E. coli* β-galactosidase.[97,98]

Construction of Fusion Genes. Fusions between fragments of a ribosomal protein gene and the *E. coli lacZ* gene are constructed by standard recombinant DNA techniques. We have used pLGSD5[99] as the vehicle for introducing the fusion genes into yeast cells since in this vector the fusion gene is placed under control of the inducible *GAL* promoter. In this way growth problems owing to the possible toxicity of the fusion protein can be avoided. The *Bam*HI site located closely downstream of the AUG start codon of the *lacZ'* gene of pLGSD5 is used for fusing the ribosomal protein gene fragment to this gene either directly or via a linker. In-frame fusions are easily scored by the blue coloration of the colony when cells are plated on X-Gal plates containing 0.1 M NaH$_2$PO$_4$/H$_3$PO$_4$, pH 7.0, 6.7 g/liter yeast nitrogen base (YNB), 2% (w/v) galactose, and 1 mg/ml 5-bromo-4-chloro-3-indolylgalactoside (BCIG). Indirect immunofluorescense can be carried out using the procedures described in [51] and [52] in

[96] P. A. Silver and M. N. Hall, in "Protein Transfer and Organelle Biosynthesis" (A. C. Das and P. W. Robbins, eds.), p. 749. Academic Press, New York, 1988.
[97] R. B. Moreland, H. G. Nam, M. Hereford, and H. M. Fried, *Proc. Natl. Acad. Sci. U.S.A.* **82**, 6561 (1985).
[98] P. J. Schaap, J. van't Riet, C. L. Woldringh, N. Nanninga, and H. A. Raué, manuscript in preparation.

this volume. In addition to inspection of the cells with a standard fluorescence microscope, we have analyzed the distribution of the fusion proteins in three dimensions using confocal scanning laser microscopy (CSLM).[99] This technique allows one to record fluorescence in a series of "optical coupes" of the same cell.

Strain Specificity

Although the techniques described in this chapter were primarily developed in studies on ribosomal proteins from the yeast strain *S. carlsbergensis* S74, most should be applicable to any of the standard laboratory yeast strains currently in use. This is true in particular for the *in vitro* techniques, which we have successfully used in analyzing the interaction of yeast ribosomal proteins with heterologous (prokaryotic and vertebrate) rRNAs, as well as prokaryotic ribosomal proteins with yeast and vertebrate rRNAs.[61,88,89] As far as the *in vivo* methods are concerned, a certain degree of strain specificity was observed for LiCl stripping of ribosomal subunits, as discussed in the section describing this technique. With respect to the *in vivo* analysis of nuclear localization signals of yeast ribosomal proteins, we have not observed any strain specificity using ribosomal protein genes derived from *S. carlsbergensis* S74 and several *S. cerevisiae* strains in the construction of fusions and expression of the fusion genes in several standard laboratory strains of *S. cerevisiae*.

[99] J. V. Kilmartin and A. E. M. Adams, *J. Cell Biol.* **98,** 992 (1984).

[33] High-Expression Vectors with Multiple Cloning Sites for Construction of *trpE* Fusion Genes: pATH Vectors

By T. J. KOERNER, JOHN E. HILL, ALAN M. MYERS, and ALEXANDER TZAGOLOFF

Introduction

Molecular cloning technology frequently allows the characterization of a gene that determines a specific phenotype, in the absence of any knowledge regarding the polypeptide coded for by that gene. Nucleotide sequence analysis of such a gene describes a polypeptide, termed a "genetically identified protein," for which limited or no information is available concerning its function. This situation could arise, for example, when (1) cloned genes are selected by complementation of a defined mutation, (2) cDNA clones are selected on the basis of high abundance at specific developmental stages, or (3) genes are isolated by transposon tagging. To progress from genetic to biochemical characterization of the particular cellular function under investigation, the cloned gene often is introduced into a bacterial host for production of the genetically identified protein.

Foreign proteins can be produced in bacteria either in the native form or as a hybrid protein in which the foreign polypeptide, or a portion thereof, is fused to a polypeptide normally expressed in the host cell. Production of native polypeptides offers the advantage that the genetically identified protein may be isolated in a biochemically active form. However, there are often difficulties in producing large amounts of foreign proteins in their native form in bacterial hosts, and identification of the proteins may not be possible in the absence of previous biochemical characterization. Hybrid proteins using bacterial promoters and genes, on the other hand, usually are produced in relatively large amounts regardless of the nature of the fused polypeptide because both the transcription and translation signals are indigenous to the host cell. In this case, the foreign polypeptide is produced as a "passenger" at the carboxyl terminus of the hybrid. The presence of the host polypeptide sequence in the hybrid can stabilize the hybrid protein. Although hybrid proteins frequently are not suitable for biochemical analysis, they are extremely useful as antigens for the production of antibodies that can aid in characterization of the genetically identified protein.

This chapter describes the structure and applications of a series of *Escherichia coli* plasmids, the pATH plasmids, designed for the production of proteins from any cloned DNA sequence that contains an open reading frame (ORF).[1] The cloned DNA sequences are fused in-frame to the *trpE* gene of *E. coli,* which codes for anthranilate synthase.[2] Thus, the hybrid protein produced contains the amino-terminal 323 residues of anthranilate synthase followed by the translation product specified by the cloned DNA. The advantages of pATH plasmids for production of hybrid proteins are as follows: (1) pATH plasmids are maintained in *E. coli* at high copy number. (2) The plasmids are relatively small [3.8 kilobase pairs (kbp)], and their complete sequence is known. (3) Multiple cloning sites (MCS) are present following codon 323 of the *trpE* gene, for easy construction of in-frame

---

[1] bp, Base pair; IAA, indoleacrylic acid; MCS, multiple cloning site; nt, nucleotide; ORF, open reading frame; SDS, sodium dodecyl sulfate; SDS–PAGE, SDS–polyacrylamide gel electrophoresis; W, tryptophan.
[2] C. Yanofsky, T. Platt, I. P. Crawford, B. P. Nichols, G. E. Christie, H. Horowitz, M. VanCleemput, and A. M. Wu, *Nucleic Acids Res.* **9,** 6647 (1989).

gene fusions. Different pATH plasmids carry MCS in each of the three registers of the translational reading frame. (4) Stop codons exist in all three registers and are very close to the MCS. (5) Transcription is controlled by the *E. coli trp* operon promoter. This promoter is transcribed at very high levels under inducing conditions. (6) The anthranilate synthase amino terminus provides stability to the hybrid protein. (7) The hybrid proteins usually are produced in *E. coli* in an insoluble form, which greatly facilitates purification of the hybrid.[3]

Production of Hybrid Proteins Using pATH Plasmids

The first step in producing anthranilate synthase hybrid proteins using the pATH series is to choose the particular plasmid and restriction sites for construction of the gene fusion. The nucleotide sequence of the ORF to be expressed will reveal restriction enzyme recognition sites that can be fused in-frame to *trpE'* by ligation into the appropriate pATH vector. The ORF is then cloned into pATH, and recombinant plasmids are selected and characterized by restriction mapping.

Cells containing the appropriate plasmid are grown to mid-log phase in synthetic medium supplemented with tryptophan (W). The culture is then diluted by a factor of 10 into the same medium lacking tryptophan. Depletion of the corepressor tryptophan causes induction of the operon by preventing binding of the *trp* repressor to the operator, and depletion of charged tRNA^{Trp} eliminates transcriptional attenuation.[4] After several hours the tryptophan analog indoleacrylic acid (IAA) is added to the culture, as an additional means of relieving repression and attenuation.[5-7]

The induced culture is harvested, and a total cell lysate is prepared by lysozyme treatment followed by combined osmotic and detergent lysis. DNA is then fragmented by sonication, after which the total lysate is

[3] D. G. Kleid, D. Yanasura, B. Small, D. Dowbenko, D. M. Moore, M. J. Grubman, P. D. McKercher, D. O. Morgan, B. H. Robertson, and H. L. Bachrach, *Science* **214**, 1125 (1981).

[4] C. Yanofsky and I. P. Crawford, in "*Escherichia coli* and *Salmonella typhimurium:* Cellular and Molecular Biology" (F. C. Niedhardt, J. L. Ingraham, K. Brooks Low, B. Magasanik, M. Schaechter, and H. E. Umbarger, eds), p. 1453. American Society for Microbiology, Washington, D.C., 1987.

[5] D. E. Morse, R. D. Mosteller, R. F. Baker, and C. Yanofsky, *Nature (London)* **223**, 40 (1969).

[6] D. E. Morse, R. D. Mosteller, and C. Yanofsky, *Cold Spring Harbor Symp. Quant. Biol.* **34**, 725 (1969).

[7] S. French, K. Martin, T. Petterson, R. Bauerle, and O. L. Miller, *Proc. Natl. Acad. Sci. U.S.A.* **82**, 4638 (1985).

separated into soluble and insoluble fractions by centrifugation. To detect the hybrid protein each fraction is analyzed by sodium dodecyl sulfate–polyacrylamide gel electrophoresis (SDS–PAGE). The hybrid usually is found as the most prevalent protein in the insoluble fraction and can easily be identified based on abundance, apparent molecular weight, and absence in an uninduced control culture (Fig. 1).

Since the anthranilate synthase hybrid proteins are most often found in the insoluble cell fraction, their further purification depends on solubilization with denaturing agents. SDS–PAGE or gel-exclusion chromatography

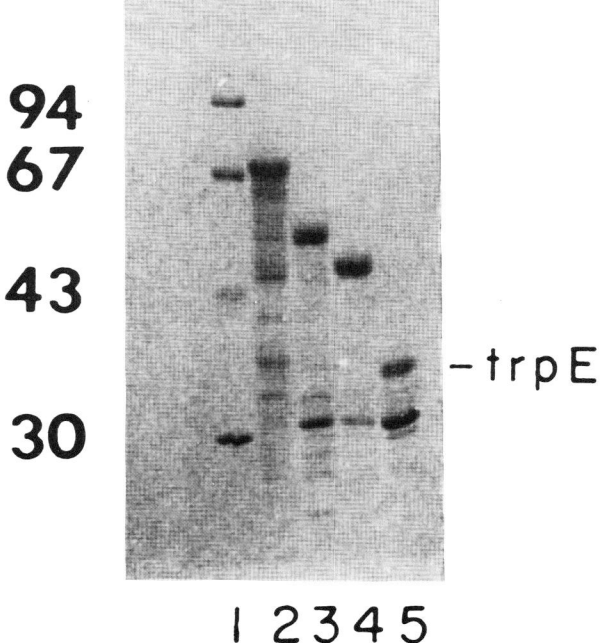

FIG. 1. SDS–PAGE analysis of anthranilate synthase hybrid proteins. *Escherichia coli* strains carrying various pATH plasmids were subjected to the protocol described in the text (see Induced Expression of Hybrid Proteins by Tryptophan Starvation). The insoluble protein fraction was prepared from each strain (see Extraction of Hybrid Protein) and 20 μg was solubilized in 1× loading buffer and applied to a SDS–12% (w/v) polyacrylamide gel (0.75 mm thick). Staining with Coomassie blue reveals that by far the most prevalent protein in the insoluble fractions is the anthranilate synthase hybrid protein produced from the pATH plasmid. Lane 1, Molecular weight ($\times 10^{-3}$) standards. Lane 2, pATH3 with an open reading frame of 909 nt ligated at the *Sal*I site. Lane 3, pATH1 with an open reading frame of 600 nt ligated at the *Bam*HI site. Lane 4, pATH11 with an open reading frame of 309 nt ligated at the *Sac*I site. Lane 5, pATH1 with no insert. The band migrating slightly slower than the molecular weight standard of 30,000 is presumed to be stained β-lactamase. The amount of this protein produced varies among the plasmid constructions.

are suitable for further purification prior to use of the hybrid protein as an antigen.

pATH1

The pATH plasmids were constructed by modifying plasmid pKRS101.[8] pATH1 is a 3779-base pair (bp) plasmid composed of (1) a 1391-bp fragment of *E. coli* DNA containing the *trp* operon promoter and most of the first gene in that operon, *trpE*,[1] (2) a MCS following codon 323 of *trpE*, and (3) a segment of pBR322[9] containing the β-lactamase gene *(bla)* and the ColE1 replication origin *(ori)* of *E. coli* (Fig. 2). Nucleotide 1 is the first nucleotide of the *trp* fragment, located at the beginning of a *Pvu*II site 261 nucleotides (nt) upstream of the transcription start site. The coding region of *trpE'* begins at nt 423. The *trp* fragment extends to nt 1391, the third nucleotide of *trpE* codon 323. The terminating point of the *trp* fragment was originally a *Bgl*II recognition sequence. In pATH1 the MCS[10] comprises nt 1392 to 1455, the last nucleotide of the *Hin*dIII site. The plasmid backbone begins at nt 1456 and extends to nt 3779, containing the region of pBR322 from the *Cla*I site, through *bla* and *ori,* to the *Pvu*II site. The *bla* coding region comprises nt 1696 through 2556. The *Eco*RI site of the pBR322 backbone was altered prior to the construction of pKRS101 by filling in and ligation.[8]

The region of pBR322 serving as the pATH1 backbone is similar to that used as the backbone of the pUC plasmids.[11] Both pUC and pATH have eliminated the *rop* gene, which encodes for a negative regulator of ColE1 replication.[12,13] Thus, pATH plasmids accumulate in *E. coli* at high copy levels.

Other pATH Plasmids

The additional pATH plasmids described in this chapter are virtually identical to pATH1. Each vector has different sequences in the MCS region. The nucleotide sequence of the MCS present in each pATH plasmid is shown in Fig. 3, and Table I[14] lists the register in the anthranilate

[8] K. R. Spindler, D. S. E. Rosser, and A. J. Berk, *J. Virol.* **49**, 132 (1984).
[9] F. Bolivar, R. L. Rodriquez, P. J. Greene, M. C. Betlach, H. L. Heynecker, H. W. Boyer, J. H. Crosa, and S. Falcow, *Gene* **2**, 95 (1977).
[10] J. Messing, R. Crea, and P. H. Seeburg, *Nucleic Acids Res.* **9**, 309 (1981).
[11] J. Vieria and J. Messing, *Gene* **19**, 259 (1982).
[12] J. Davison, *Gene* **28**, 1 (1984).
[13] G. Cesareni, M. Cornelissen, R. M. Lacatena, and L. Castagnoli, *EMBO J.* **3**, 1365 (1984).
[14] C. Yanisch-Perron, J. Vieria, and J. Messing, *Gene* **33**, 103 (1985).

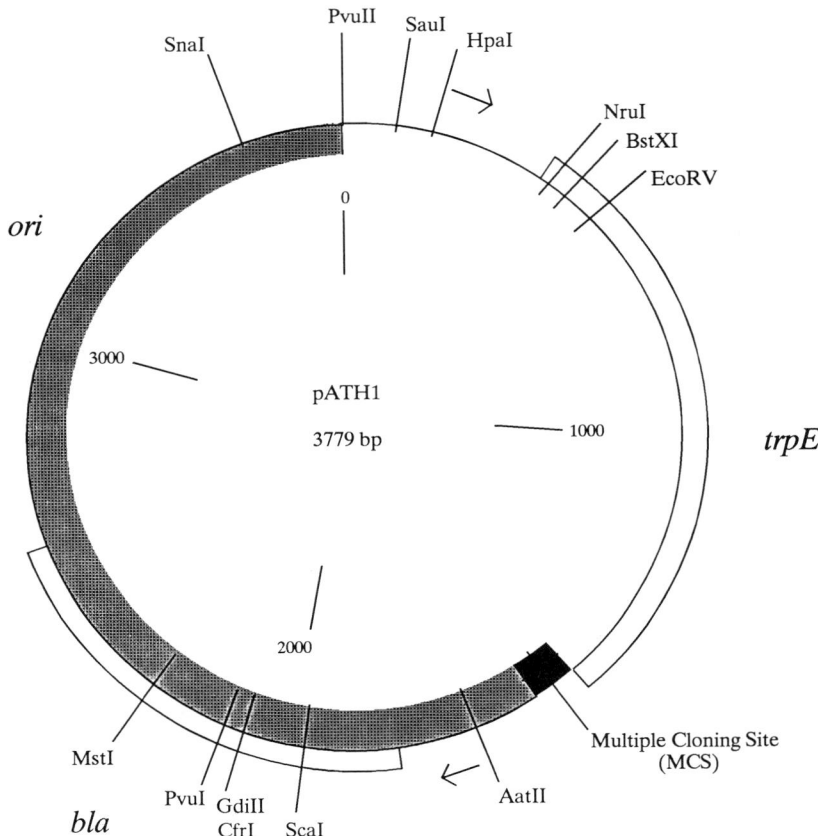

FIG. 2. Schematic description of pATH1. Numbering of the nucleotides begins with the *Pvu*II recognition sequence at the pBR322/*trp* junction. The numbering continues along the genomic *E. coli trp* segment. This is followed by the MCS (see Fig. 3). The juncture with pBR322 is at the *Hin*dIII site. Downstream is the *bla* gene, which confers ampicillin resistance, and the *ori* segment. The synthetic MCS is denoted by the solid box. The gray area inside the circle delimits the segment derived from pBR322. The open reading frames of *trpE'* and *bla* are shown as open boxes. Arrows indicate the direction of transcription. Refer to the text for detailed description of the vector components. The recognition sites of restriction enzymes which cut the vector once are indicated. Additional unique sites in the MCS are listed in Fig. 3.

synthase translational reading frame of each restriction enzyme recognition site. The *Pst*I site present in the *bla* gene of pATH1, pATH2, and pATH3 has been removed from all subsequent pATH vectors. The *Ava*II fragment in pATH was replaced with the *Ava*II fragment of pUC8; thus, the *Pst*I recognition sequence is present twice in pATH1, pATH2, and pATH3 but is unique in the MCS of pATH10 through pATH23.

```
pATH 1       Sac I         Sac I         BamH I        Sal I         Hind III
1386         ========      ++++++++      ========      ++++++++      ++++++++
GAG ATC CCC GGG CGA GCT CGA ATT CGA GCT CGC CCG GGG ATC CTC TAG AGT CGA CCT GCA GCC CAA GCT TAT CGA TGA TAA GCT GTC AAA CAT GAG AAT TAA TTC TTG AAG
             =======       ========      ++++++++      ++++++++      ++++++++      ++++++++
             Sma I         EcoR I        Sma I         Xba I         Pst I         Cla I

pATH 2       BamH I        Sal I         Hind III
             ========      ========      ========
GAG ATC CCC GGG GAT CCT CTA GAG TCG ACC TGC AGC CCA AGC TTA TCG ATG ATA AGC TGT CAA ACA TGA GAA TTA ATT CTT GAA GAC GAA
                           ++++++++      ========
             Sma I         Xba I         Pst I         Cla I

pATH 3                     BamH I        Sal I         Hind III
                           =======       =======       =======
GAG ATC CCC CCG AAT TCG GGG GGA TCC TCT AGA GTC GAC CTG CAG CCC AAG CTT ATC GAT GAT AAG CTG TCA AAC ATG AGA ATT AAT TCT TGA AGA CGA
             =======       =======       ++++++++      =======
             EcoR I        Xba I         Pst I         Cla I

pATH 10                    Sac I         BamH I        Sal I         Hind III
                           =======       ========      ========      ========
GAG ATC CCC CGG AAT TCG AGC TCG CCC GGG GAT CCT CTA GAG TCG ACC TGC AGC CCA AGC TTA TCG ATG ATA AGC TGT CAA ACA TGA GAA TTA ATT CTT GAA GAC
             =======       =======       =======       =======       =======
             EcoR I        Sma I         Xba I         Pst I         Cla I

pATH 11                    Sac I         BamH I        Sal I         Hind III
                           =======       =======       =======       =======
GAG ATC CCC CCG GAA TTC GAG CTC GCC CGG GAA TCC TCT AGA GTC GAC CTG CAG CCC AAG CTT ATC GAT GAT AAG CTG TCA AAC ATG AGA ATT AAT TCT TGA AGA CGA
             =======       =======       =======       =======       =======
             EcoRI         Sma I         Xba I         Pst I         Cla I

pATH 20                    Sac I         Sma I         Xba I         Pst I         Hind III
                           =======       =======       =======       =======       =======
GAG ATC CCC CCG AAT TGG GAA TTC GAG CTC GGT ACC CGG GGA TCC TCT AGA GTC GAC CTG CAG GCA TGC AAG CTT
             =======       =======       =======       =======       =======
             EcoR I        Kpn I         BamH I        Sal I         Sph I

pATH 21                    Sph I         Sal I         BamH I        Kpn I         EcoR I
                           =======       =======       ++++++++      ++++++++      ++++++++
GAG ATC CCC CCG AAT TGG GAA GCT TGC ATG CCT GCA GGT CGA CTC TAG AGG ATC CCC GGG TAC CGA GCT CGA ATT CCC
             =======       =======       ++++++++      +++++++       ++++++++
             Hind III      Pst I         Xba I         Sma I         Sac I

pATH 22                    Sac I         Sma I         Xba I         Pst I         Hind III
                           =======       =======       =======       =======       =======
GAG ATC CCC CCG AAT TGG AGG GGG AAT TCG AGC TCG GTA CCC GGG GAT CCT CTA GAG TCG ACC TGC AGG CAT GCA AGC TTN
             =======       =======       =======       =======       =======
             EcoR I        Kpn I         BamH I        Sal I         Sph I

pATH 23                    Hind III      Pst I         Xba I         Sma I         Sac I
                           =======       =======       =======       =======       =======
GAG ATC CCC CCG AAT TGG AGG GGG AAG CTT GCA TGC CTG CAG GTC GAC TCT AGA GGA TCC CCG GGT ACC GAG CTC GAA TTC
                           =======       =======       =======       =======       =======
                           Sph I         Sal I         BamH I        Kpn I         EcoR I
```

FIG. 3. Nucleotide sequence of the multiple cloning sites of the pATH vectors. Beginning with nucleotide 1386 (the start of codon 322 of *E. coli* anthranilate synthase), the multiple cloning site region of each pATH vector is given. They are presented as triplets representing the translated codons. ====== represents readily usable restriction sites. Restriction sites that are downstream of a stop codon or that are not unique sites in the vector (see Table I) are denoted by ++++++. The stop codons in register are underlined.

TABLE I
pATH VECTORS: POSITION OF RESTRICTION ENZYME CLEAVAGE SITES IN READING FRAME[a]

Enzyme	pATH1	pATH2	pATH3	pATH10	pATH11	pATH20	pATH21	pATH22	pATH23
BamHI	2	3	1	3	1	1	2S	3	1
ClaI	3S	1	2	1	2	?	?	?	?
EcoRI	2	—	3	3	1	1	2S	3	1
HindIII	2S	3	1	3	1	1	2	3	1
KpnI	—	—	—	—	—	2	3S	1	2
PstI	3MS	1M	2M	1	2	2	3	1	2
SacI	3M	—	—	1	2	2	3S	1	2
SalI	2S	3	1	3	1	1	2	3	1
SmaI	3M	3	—	3	—	1	3S	3	2
SphI	—	—	—	—	—	2	3	1	2
XbaI	2S	3	1	3	2	2	2S	3	1
XmaI	1M	1	—	1	—	2	1S	3	3

[a] pATH vectors were designed to facilitate expression of an open reading frame (ORF). Most of the restriction enzyme recognition sequences in the multiple cloning site (MCS) are represented in each of the three registers of translation. For each vector, the number (1, 2, or 3) represents the number of nucleotides remaining in the codon after cutting with the indicated restriction enzyme. The cleavage position in the ORF should be determined by the same criterion and matched with a vector in Table I. S denotes that the site is not usable as the 5' juncture because it is downstream from a stop codon that is in the XbaI recognition sequence. M indicates that the recognition sequence for that enzyme is present more than once in the vector although in several cases only one other site exists. ? indicates that the position has not been determined. The pATH vectors can be distinguished as follows. The first vectors constructed (pATH1, pATH2, and pATH3) have the bla gene from pBR322 which contains a PstI site. Starting with pATH10 a region of the bla gene containing the PstI site has been replaced with the corresponding region of the bla gene from pUC8, which lacks this PstI site. Thus, the PstI site in the MCS of pATH10 through pATH23 is unique. The MCSs, or multiple cloning sites, for pATH20 through pATH23 were constructed by inserting, in either direction, a cassette corresponding to the MCS of pUC18.[14]

Usage of pATH Plasmids as Cloning Vectors

Standard laboratory procedures can be utilized[15-17] with the following exceptions. Cells containing pATH plasmids should at *all times* be grown on a medium supplemented with tryptophan (LB + AW) to prevent expression of the *trp* operon promoter. Large-scale plasmid isolations can be performed using standard procedures such as cesium chloride equilibrium density gradient centrifugation. Chloramphenicol amplification is not necessary. We recommend that the plasmids be stored as DNA solutions at $-20°$. These precautions are noted because long-term storage of pATH plasmids in cells sometimes results in the selection of promoter mutations that no longer express the *trp* operon; we presume that expression of the operon from a high-copy plasmid is deleterious to cell growth.

The multiple cloning sites in pATH plasmids are not located within a selectable marker or an indicator gene such as the *lacZ'* sequence of the pUC plasmids.[11] Therefore, cloning in the pATH plasmids is facilitated by treatment of the vector with alkaline phosphatase prior to ligation to the insert DNA. We include 30–50 units of calf alkaline phosphatase (Boehringer-Mannheim, Indianapolis, IN), in the restriction enzyme digestion used to prepare the vector. The linearized vector is then purified by agarose gel electrophoresis. Under these conditions the background of recircularized vector is very low, so that ligation products containing the insert are the predominant plasmids obtained after transformation of the ligation mixture into *E. coli*. Recombinant plasmids are identified by minipreparations of plasmid from random transformants.

Materials and Methods

Strains

Escherichia coli strain RR1 (*proA2 leuB6 galK2 xyl-5 mtl-1 ara-14 rpsL20 supE44 hsdS* λ^-) is used routinely in our experiments. Many other strains have been used successfully. The faster a strain doubles, the better it serves as a "factory" for overproduction of the hybrid polypeptide. It has been noted that *recA* strains give poor yields.

Solutions

$10 \times$ M9 salts (per liter): 60 g $Na_2HPO_4 \cdot H_2O$, 30 g KH_2PO_4, 5 g NaCl, 10 g NH_4Cl; this solution can be autoclaved if desired

---

[15] T. Maniatis, E. F. Fritsch, and J. Sambrook, "Molecular Cloning: A Laboratory Manual." Cold Spring Harbor Laboratory, Cold Spring Harbor, New York, 1982.
[16] D. Hanahan, *J. Mol. Biol.* **166**, 557 (1983).
[17] B. Perbal, "A Practical Guide to Molecular Cloning." Wiley, New York, 1984.

10 mg/ml Thiamin B_1, 10 mg/ml tryptophan (W), 20 mg/ml ampicillin, 1 M $MgSO_4$, and 0.5 M $CaCl_2$ are prepared and sterilized by filtration through a 0.45-μm cellulose acetate membrane. The tryptophan stock solution is stored at 4° in a foil-wrapped container for up to 2 weeks. Indoleacrylic acid (IAA), 2 mg/ml, is prepared in 95% ethanol (not absolute ethanol); 40% (w/v) glucose is prepared and sterilized by autoclaving

1 M Tris-HCl, pH 7.5, 0.5 M EDTA, 5 M NaCl, and 10% (w/v) Nonidet P-40 are prepared for the isolation procedure. Just prior to use, the lysozyme is weighed out and added to the ice-cold buffered solution described later

Solutions necessary for SDS–PAGE are as described elsewhere[18]

Media

LB + AW (per liter): 5 g Difco Bacto-yeast extract, 10 g Difco Bacto-tryptone, 5 g NaCl, 10 mM Tris-HCl, pH 8.0; after autoclaving add 2 ml/liter of tryptophan stock solution and 2 ml/liter of ampicillin stock solution. The ampicillin can be raised to a final concentration of 50–100 μg/ml

LB + AW plates: 15 g Difco Bacto-agar added to 1 liter of LB + AW

Modified M9 (per 100 ml): Mix and autoclave 10 ml 10× M9 salts, 90 ml water, 0.5 g Difco Bacto-casamino acids. (Note: Casamino acids do not contain tryptophan.) After autoclaving add 0.1 ml of 1 M $MgSO_4$, 20 μl of 0.5 M $CaCl_2$, 0.5 ml of 40% glucose, 0.1 ml of 10 mg/ml thiamin B_1, 0.2 ml of 10 mg/ml tryptophan, and 0.2 ml of 20 mg/ml ampicillin

Modified M9 − W: Omit tryptophan addition from the modified M9 medium described above

Induced Expression of Hybrid Proteins by Tryptophan Starvation

pATH-transformed cells to be used for expression are patched onto an LB + AW plate the day before performing the procedure. A loop of cells is inoculated from the plate into 10 ml of modified M9 in a 50-ml flask. Grow the culture for about 2–4 hr with shaking at 37° to midlogarithmic phase (A_{600} of 0.2–0.4). Next, the entire 10-ml culture is poured into 100 ml of modified M9 − W, in a 2-liter flask. The large ratio of flask volume to culture volume provides a great amount of aeration to the culture, which is very important for good yields of the hybrid polypeptide. The culture is grown at 37° with vigorous shaking for 1–2 hr. The culture is treated with 0.5 ml of IAA stock, and incubation is continued for an additional 4 hr. At this point the cells can be harvested to begin prepara-

---

[18] U. K. Laemmli, *Nature (London)* **227**, 680 (1970).

tion of the hybrid protein. Alternatively, the cells can be stored at 4° overnight prior to beginning the isolation procedure.

The control(s) used to evaluate expression of the hybrid polypeptide can include the following: an induced culture of the chosen *E. coli* strain not transformed with pATH (note: remember not to add ampicillin to these cultures); an induced culture transformed with the parental pATH vector; and an uninduced culture of the recombinant-transformed strain. The difference in expression owing to the addition of IAA to tryptophan-starved cells is not always dramatic, and, therefore, we do not suggest comparing cultures with and without treatment with IAA.

Extraction of Hybrid Protein

Induced cells are harvested by centrifugation for 5 min at approximately 3300 g (5000 rpm in a Sorvall GSA rotor), resuspended in 10 mM Tris-HCl, pH 7.5, and pelleted again. The cell pellet from each 100-ml culture is resuspended in 20 ml of 50 mM Tris-HCl, pH 7.5, 5 mM EDTA, and 3 mg/ml lysozyme and kept on ice for 2 hr. Next, 1.4 ml of 5 M NaCl is added, and the cells are mixed by inversion; the final NaCl concentration of 0.3 M will cause the cells to lyse. To complete the lysis, 1.5 ml of 10% Nonidet P-40 is added, and the solution is mixed by inversion. The resulting lysate will be extremely viscous owing to release of chromosomal DNA. To shear DNA and reduce the viscosity, the solution is treated by sonication. Sonication is at approximately 350 W and is administered in 30-sec bursts until the solution is able to drop freely through a Pasteur pipette (usually three separate bursts of sonication are required). This procedure decreases the possibility of foaming, which is detrimental to isolation of the hybrid polypeptide.

A portion of the total cell lysate is saved for further analysis. The remainder of the lysate is separated into soluble and insoluble fractions by centrifugation for 10 min at 9000 g (10,000 rpm in a Sorvall SS34 rotor). The soluble cell fraction is also retained for further analysis. The insoluble pellet is washed once in 20 ml of 10 mM Tris-HCl, pH 7.5, 1 M NaCl and once in 10 mM Tris-HCl, pH 7.5. Finally, the insoluble pellet from each 100-ml induction culture is resuspended in 1 ml of 10 mM Tris-HCl, pH 7.5. Since the pellet fraction will not dissolve, the material is dispersed by brief hand homogenization with a Teflon pestle (Potter–Elvehjem) tissue grinder.

Analysis of Protein Samples by Denaturing Electrophoresis

To detect the presence of the hybrid protein, each cell fraction (total, soluble, and insoluble) is analyzed by SDS–PAGE using the Laemmli gel system.[18] Protein concentrations of the fractions are determined by standard assays, and the desired amount of protein is diluted with water and 4× Laemmli loading buffer to a final concentration of 1× loading buffer (the high concentration of SDS in this buffer will solubilize the pelleted protein fraction). The final concentration of protein in the sample to be applied to the gels ranges from 1 to 10 mg/ml, and between 10 and 100 μg is loaded per lane. An example of such an analysis is shown in Fig. 1. Typically, the bulk of anthranilate synthase hybrid protein is found in the insoluble protein fraction; however, we are aware of several hybrid proteins expressed from pATH plasmids where significant amounts are found in the soluble fraction.

Purification of Hybrid Polypeptides for Antigenic Use

Further purification of the hybrid protein can be accomplished by scaling up the electrophoresis procedures. The proteins can be electroeluted from the acrylamide gels prior to inoculation of the animals for preparation of antibodies; alternatively, the entire strip of gel containing the hybrid protein can be crushed in adjuvant and used to inoculate animals. We have also used gel-exclusion chromatography to partially purify hybrid proteins expressed from pATH plasmids.

Potential Pitfalls with pATH Plasmids

pATH plasmids have been used to produce genetically identified proteins from many different organisms. In most cases, the use of these plasmids is straightforward and reliable. However, in some cases a gene fusion which appears by all criteria to be in the correct orientation and reading frame fails to express a hybrid protein. We believe that mutations in the *trp* promoter are a frequent cause of such failures. To avoid this problem, we strongly recommend that pATH plasmids be prepared only from cultures known to express at high levels and that they be stored as purified DNA solutions. Furthermore, we recommend that, prior to beginning construction of a gene fusion, the individual pATH plasmid to be used be tested for the ability to express the truncated anthranilate synthase. To do so, the chosen pATH plasmid is freshly transformed from a stored DNA stock into *E. coli,* then expressed as described above. The insoluble protein fraction from the pATH transformant is compared to the same fraction from a mock expression of a strain lacking the plasmid. The presence of a prominent protein at a molecular weight of approximately 37,000, which is not seen in the control culture, indicates the *trp* promoter is efficiently working in that plasmid.

Another potential problem using pATH plasmids is that large hybrid proteins (>90,000 molecular weight) are not expressed as efficiently as shorter hybrids. If production of a large hybrid polypeptide is desired, we recommend that in addition a shorter ORF that codes for a polypeptide of 60,000 or less also be fused to *trpE'*.

Additional Comments

Antisera raised against hybrid polypeptides produced from pATH plasmids routinely recognize the native form of the genetically identified protein. The specificity of these antisera is sufficient for their use in most applications. However, in the typical case where the hybrid proteins are insoluble, the only means of further purification is based on size separations in the presence of denaturing agents such as SDS. Therefore, the antigen preparations may be contaminated with other proteins, giving rise to antibodies that react with proteins other than the genetically identified protein of interest. Thus, in cases where specificity of the antiserum is critical, it may be advisable to use affinity purification methods either to remove antibodies specific for the anthranilate synthase portion of the hybrid protein or to collect those antibodies specific for the genetically identified protein. In the former case, an affinity column can be prepared containing the total insoluble protein fraction from a pATH induction where no ORF has been fused to *trpE'*.[19] In the latter case, an affinity column can be prepared containing the genetically identified protein as part of a different hybrid protein, for example, fused to β-galactosidase.[20-23]

The pATH vectors can be used even when the sequence of the ORF is not known. If the boundaries of the ORF are known and some restriction map has been constructed, a restriction site mapping within the boundaries can be chosen. This fragment would then be ligated into three appropriate pATH vectors. For example, if a *Bam*HI site mapped within the ORF and a *Xba*I site was known to be 3' of the *Bam*HI site, the fragment could be cloned into pATH1, pATH2, and pATH3 to test each register of translation.

Although the pATH vectors have been generally utilized for production of antigenic material, they can be employed for other applications. The pATH vectors have also proved useful in mapping epitopes.[24] In addition it

---

[19] N. Segev, J. Mulholland, and D. Botstein, *Cell (Cambridge, Mass.)* **52,** 915
[20] B. K. Haarer and J. R. Pringle, *Mol. Cell. Biol.* **7,** 3678 (1987).
[21] J. R. Pringle, R. A. Preston, A. E. M. Adams, T. Stearns, D. G. Drubin, B. K. Haarer, and E. W. Jones, *Methods Cell Biol.* **31,** 357 (1989).
[22] H. B. Kim, B. K. Haarer, and J. R. Pringle, *J. Cell Biol.* (in press).
[23] J. S. Johnson and A. M. Myers, unpublished observations (1989).
[24] C. H. Gross and G. F. Rohrmann, *BioTechniques* **8,** 196 (1990).

is possible that enzymatic or other biochemical characteristics of the ORF may be expressed in the hybrid polypeptide. The attributes of the pATH series can be compared to the numerous cloning vectors that are available.[25] Despite the potential problems, the pATH vectors have been used successfully by many laboratories. They have proved, through many field tests, to be a rapid and efficient means of generating genetically identified proteins for use in production of antibodies.

Availability of Plasmids and Sequences

The pATH plasmids (1, 2, 3, 10, 11, 20, 21, 22, and 23) have been deposited with the American Type Culture Collection (ATCC) using the identifiers 37695 through 37703. The nucleotide sequences for the pATH vectors (1, 2, 3, 10, and 11) have been submitted to the GenBank/EMBL Databank with the following accession numbers: M32985, M33624, M33622, M33623, and M33625.

Nomenclature

As these vectors were designed to be steady and simple vehicles for producing large amounts of the desired polypeptides, it was desired that the name also be simple and straightforward. This vector series was designed by John Hill, T. J. Koerner, and Alexander Tzagoloff to be the route taken by investigators to produce large quantities of the desired polypeptide. The name for this series of vectors came from description of their function: these plasmids are amenable for making Trp hybrids.

Acknowledgments

We wish to acknowledge Dr. Carol Dieckmann for helpful discussions early in the development of this project. We greatly appreciate the assistance of Dr. Katherine Spindler in supplying us with the vector pKRS101 and advice during the preparation of the manuscript. This work was supported by Research Grant GM25250 from the National Institutes of Health, U.S. Public Health Service.

---

[25] P. H. Pouwels, B. E. Enger-Valk, and W. J. Brammer (eds.), "Cloning Vectors: A Laboratory Manual." Elsevier, Amsterdam, 1985–1988.

[34] Production of Proteins by Secretion from Yeast

By DONALD T. MOIR and LANCE S. DAVIDOW

Introduction

Tools and techniques perfected over the past few years permit the production of a variety of proteins in the bakers' yeast *Saccharomyces cerevisiae*. Both homologous and heterologous proteins may be produced in the cytoplasm or directed through the secretory pathway. This latter approach, secretion of proteins by yeast, is the focus of this chapter. For a variety of reasons, secretion is the preferred route of production for many proteins. For example, correct folding of proteins with disulfide bonds appears to occur more readily in the secretory pathway, glycosylation of proteins coincides with secretion, and secretion of proteins removes them from the bulk of the yeast proteins and proteases, often resulting in better stability and higher initial purity of the desired product. Generally, secretory proteins and integral membrane proteins, which normally pass through the secretory pathway in their native hosts, are produced more faithfully in yeast by directing them into the secretory pathway. In all cases characterized so far, specific activities are normal, reflecting accurate folding and disulfide bond formation, and the proper Asn residues receive N-linked glycosylation.[1-5]

The aim of this chapter is to outline the methods available for secreting proteins from *Saccharomyces cerevisiae*. We briefly describe some of the vectors, promoters, and secretion signals which have been used successfully in our laboratory and in others. We focus mainly on methods which we have used to identify and manipulate mutations affecting secretion and glycosylation in yeast. These mutations have proved useful for increasing the secreted yield of several mammalian proteins from yeast, for reducing

---

[1] A. J. Brake, J. P. Merryweather, D. G. Coit, U. A. Heberlein, F. R. Masiarz, G. T. Mullenbach, M. S. Urdea, P. Valenzuela, and P. J. Barr, *Proc. Natl. Acad. Sci. U.S.A.* **81**, 4642 (1984).
[2] R. A. Smith, M. J. Duncan, and D. T. Moir, *Science* **229**, 1219 (1985).
[3] D. T. Moir and D. R. Dumais, *Gene* **56**, 209 (1987).
[4] G. Loison, A. Findeli, S. Bernard, M. Nguyen-Juilleret, M. Marquet, N. Riehl-Bellon, D. Caravallo, L. Guerra-Santos, S. W. Brown, M. Courtney, C. Roitsch, and Y. Lemoine, *Bio/Technology* **6**, 72 (1988).
[5] L. M. Melnick, B. G. Turner, P. Puma, B. Price-Tillotson, K. A. Salvato, D. R. Dumais, D. T. Moir, R. J. Broeze, and G. C. Avgerinos, *J. Biol. Chem.* **265**, 801 (1990).

the complexity of the carbohydrate applied to secretory glycoproteins, and for understanding the mechanism of secretion.[6]

Mutant yeast host strains have proved valuable because many mammalian secretory proteins are not secreted efficiently from wild-type yeast strains.[7,8] The glycosylation pattern of proteins remaining within the cell suggests that these proteins encounter a rate-limiting step somewhere between the endoplasmic reticulum (ER) and Golgi.[3,9] We describe here detailed protocols for screening mutagenized yeast cultures for mutants which secrete many mammalian proteins more efficiently—so-called supersecreting mutants. These same protocols have proved useful for constructing production strains by aiding in the identification of transformants and spore progeny which secrete maximal amounts of the desired protein. Finally, these methods have been used to screen for variant proteins with greater or lesser activity than the parental form. In all of these cases, a key feature of these methods is the facility with which large numbers of yeast colonies can be screened.

In summary, secretion from yeast is an attractive method for producing many proteins both because of the facility with which genetic manipulations and fermentation can be carried out and because of the fidelity of posttranslational modifications. A variety of proteins have been secreted in biologically active form from yeast in yields ranging from tens of micrograms to over 50 mg per liter of culture. Examples in this chapter illustrate some of the tools required for constructing secretion vectors. In addition, we also provide screening methods which can be applied to large numbers of colonies to identify those secreting higher levels of a desired protein or those secreting variant proteins with desired properties.

Methods

Promoters

A variety of regulated and constitutive promoters have been described. The most useful have been derived from highly expressed genes such as those encoding the glycolytic enzymes triose-phosphate isomerase *(TPI1),*[10] phosphoglycerate kinase *(PGK1),*[11] and glyceraldehyde-3-phos-

---

[6] H. R. Rudolph, A. Antebi, G. R. Fink, C. M. Buckley, T. E. Dorman, J. LeVitre, L. S. Davidow, J. Mao, and D. T. Moir, *Cell (Cambridge, Mass.)* **58**, 133 (1989).

[7] R. A. Hitzeman, D. W. Leung, L. J. Perry, W. J. Kohr, H. L. Levine, and D. V. Goeddel, *Science* **219**, 620 (1983).

[8] J. F. Lemontt, C.-M. Wei, and W. R. Dackowski, *DNA* **4**, 419 (1985).

[9] S. Elliot, J. Griffin, S. Suggs, E. P. Lau, and A. R. Banks, *Gene* **79**, 167 (1989).

[10] T. Alber and G. Kawasaki, *J. Mol. Appl. Genet.* **1**, 419 (1982).

[11] R. Hitzeman, F. E. Hagie, J. S. Hayflick, C. Y. Chen, P. H. Seeburg, and R. Derynck, *Nucleic Acids Res.* **10**, 7791 (1982).

phate dehydrogenase *(TDH3)*.[12] These promoters may be considered virtually constitutive since variations in gene expression from them are generally less than an order of magnitude when cells are grown glycolytically versus gluconeogenically.[13] These three promoters from glycolytic genes direct roughly equal levels of gene transcription, consistent with the fact that each produces a glycolytic enzyme at about 1% of the soluble protein in an untransformed cell.

An equally "strong" regulated promoter is that from the galactokinase gene *(GAL1)*.[14] Gene expression from the *GAL1* promoter is induced by the presence of galactose but repressed by the presence of glucose. In the repressed, uninduced state, during growth in the absence of galactose and in the presence of glucose, transcription is 1000-fold lower than in the fully induced, derepressed state during growth in galactose-containing medium.[14] Thus, *GAL1* is not only a "strong" promoter, but it is also very tightly regulated. This feature is extremely useful when expression of a foreign gene or overexpression of a yeast gene is deleterious to growth of the yeast cell. Cells can be grown in the absence of expression, and then transcription can be turned on as cells reach stationary phase.[15]

Since protein secretion by yeast is a growth-associated process, we have chosen to use constitutive promoters whenever possible so that the entire growth phase is productive. This approach involves a simple one-carbon source fermentation and is compatible with continuous fermentation processes which are more productive for a given fermentor size. However, if the gene product is toxic to yeast, then it may be necessary to divide the overall process into a growth phase and a production phase. Some degree of cell growth will be necessary during the production phase to accommodate secretion, and the proper balance between length of growth phase and length of production phase must be determined empirically.

In yeast, the sequence preceding the AUG of the mRNA for a gene is not particularly critical for translation efficiency. Translation of yeast messages is maximal when A is found at position -3; however, in contrast to the case with mammalian genes, translation of yeast messages is affected no more than 2-fold by the context 5' to the initiator AUG.[16] Similarly, the distance and sequence context between the TATA element and the mRNA

---

[12] J. P. Holland and M. J. Holland, *J. Biol. Chem.* **255**, 2596 (1980).

[13] D. G. Fraenkel, in "The Molecular Biology of the Yeast *Saccharomyces*: Metabolism and Gene Expression" (J. N. Strathern, E. W. Jones, and J. R. Broach, eds.), p. 1. Cold Spring Harbor Laboratory, Cold Spring Harbor, New York, 1982.

[14] M. Johnston and R. W. Davis, *Mol. Cell. Biol.* **4**, 1440 (1984).

[15] J. C. Fieschko, K. M. Egan, T. Ritch, R. A. Koski, M. Jones, and G. A. Bitter, *Biotechnol. Bioeng.* **29**, 1113 (1987).

[16] J. M. Clements, T. Laz, and F. Sherman, in "Yeast Genetic Engineering" (P. J. Barr, A. J. Brake, and P. Valenzuela, eds.), p. 65. Butterworth, Boston, Massachusetts, 1989.

TABLE I
3' SEQUENCE OF YEAST PROMOTER REGIONS FUSED TO *SUC2* SECRETION SIGNAL

Yeast gene	Portion of promoter sequence[a]	Expression–secretion plasmid[b]	Ref.
TPI	... TATAACTACAAAAAACTATAAGCTCCATG	pCGS681	3
	... TATAACTACAAAAAACACATACATAAACTAAAA		10
PGK	... CTTTTTACAACAAATATAAAACAACCATG	pCGS740	5
	... CTTTTTACAACAAATATAAAACA		11
TDH3	... TTTCGAATAAACACACATAAATAAACAAA	pCGS795	c
	... TTTCGAATAAACACACATAAATAAACAAA		12
SUC2	... AAAAGCTTTTCTTTTCACTAACGTATATG	pCGS370	d
	... AAAAGCTTTTCTTTTCACTAACGTATATG		e
GAL1	... TCAAGGAGAAAAAACCCCGGATCGGA[f]	pCGS471	2
	... TCAAGGACAAAAAACTATA		14

[a] Between 0.8 and 1.2 kilobases (kb) of upstream sequence is present in these expression–secretion vectors. The sequence of the 3' ends, including regions altered by addition of oligonucleotide linkers, are shown. Unless otherwise noted, these sequences are fused directly to the ATG of the *SUC2* secretion signal coding sequence shown in Table II. For comparison, the sequence of the promoter as found in the naturally occurring yeast gene is shown directly below the promoter sequence present in each expression vector.
[b] Examples of expression plasmids carrying these promoter–secretion signal fusions.
[c] J. Mao, K. Hsiao, L. Melnick, D. T. Moir, and L. S. Davidow, unpublished observations (1988).
[d] D. T. Moir, J. Mao, M. J. Duncan, R. A. Smith, and T. Kohno, in "Developments in Industrial Microbiology," Vol. 26 (L. Underkofler, ed.), p. 75. Society for Industrial Microbiology, Arlington, Virginia, 1985.
[e] R. Taussig and M. Carlson, *Nucleic Acids Res.* **11,** 1943 (1983).
[f] Unlike the case for the other vectors shown, in this vector, the *GAL1* promoter is fused to the invertase secretion signal at the *Hin*dIII site in the *SUC2* promoter region (the fourth A in the *SUC2* sequence shown above).

start are not particularly critical for transcription initiation in yeast.[17] Thus, there is considerable flexibility in the design of expression vectors. A few promoter–secretion signal gene junctions which we have used successfully for secretion of several heterologous proteins including bovine prochymosin, bovine growth hormone, human urokinase-type plasminogen activator (u-PA), and human α_1-antitrypsin are shown in Table I.

Vectors

Both autonomously replicating and integrating vectors are available for maintenance of genes in yeast. Autonomously replicating vectors make use of replication origins from the 2-μm plasmid or from the yeast chromo-

[17] W. Chen and K. Struhl, *EMBO J.* **4,** 3273 (1985).

somes *(ARS)*. Integrating vectors lack a replication origin but carry at least one yeast gene segment to provide homology for integration. Autonomously replicating vectors are quite useful for rapid evaluation of promoter–gene constructions. However, maintenance of these plasmids requires that cells containing these vectors be grown constantly under selective pressure. Usually, this requires growth in a synthetic defined medium, resulting in slower growth and a significantly lower yield of cells than can be achieved with growth in a rich complex medium. Considerable experimentation is often required to optimize growth of a particular yeast strain in a synthetic medium, and the resulting medium is not universally useful for other strains.

A useful alternative for larger scale growth of yeast and production of a desired gene product is to integrate the expression units stably into the yeast chromosomes.[2,18] Cells may then be grown without selective pressure in a richer medium such as YPD, and cell yields together with accompanying product yields typically approach levels 10-fold higher than those obtained in an unoptimized minimal medium. Current techniques permit precise targeting of integrating vectors to desired regions of the yeast chromosomes (see [19] in this volume).

Examples of the two types of integrating vectors which we have used successfully are shown in Fig. 1. Copies of pCGS740 can be integrated at the *PGK1, SUC2,* and *LEU2* loci. We and others have found that such integrated vectors are stable for up to 60 generations.[18] This has proved suitable for the production of 40-liter cultures secreting a variety of heterologous proteins in our laboratory. The transcriptional unit in pCGS732 can be integrated at any of the 20–30 copies of the Ty element found throughout the yeast chromosomes and possibly at solo δ elements as well. In a single transformation, we have obtained strains carrying up to eight tandemly integrated copies of the pCGS732 transcriptional unit at a single Ty locus. Individual transformants vary both in the site of integration as well as in the number of tandem units integrated. From our experience with u-PA, the number of tandemly integrated transcriptional units correlates well with the level of protein product obtained (L. Melnick and L. Davidow, unpublished observations, 1989).

Surprisingly, many secreted proteins are produced at higher levels when the transcriptional unit is integrated in the yeast chromosomes than when it is carried on an autonomously replicating vector. There are several potential explanations for this phenomenon. First, the copy number of integrated vectors placed at multiple locations can approach that of most

---

[18] T. S. Lopes, J. Klootwijk, A. E. Veenstra, P. C. van der Aar, H. van Heerikhuizen, H. A. Raue, and R. J. Planta, *Gene* **79,** 199 (1989).

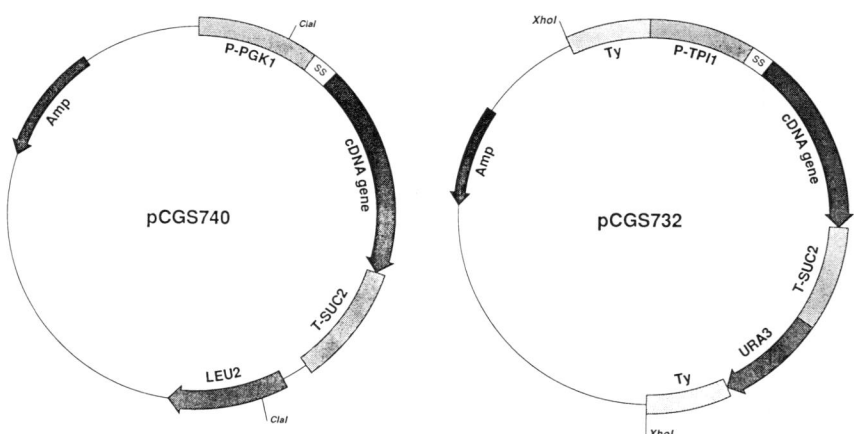

FIG. 1. Two types of integrating u-PA expression–secretion vectors. Vector pCGS740, which is based on YIp5, is shown at left. Integration is directed to the yeast *PGK1* or *LEU2* locus by partial restriction with endonuclease *Cla*I (sites are shown) prior to transformation. In vector pCGS732 shown at right, most of the yeast transposable element Ty has been replaced with the transcriptional unit (P-*TPI1*-SS-cDNA-T-*SUC2*) and selectable marker *(URA3)*, leaving about 800–1000 base pairs (bp) of Ty on either end.[5] Integration is directed to Ty and possibly solo δ elements by complete restriction with *Xho*I prior to transformation. P, Promoter; T, terminator; thin lines including *Amp* represent pBR322; thick lines represent yeast genes and mammalian cDNA sequences; arrows indicate the direction of transcription of genes.

autonomously replicating vectors. Second, for unknown reasons, secretion of heterologous proteins such as calf prochymosin and human u-PA is more efficient when transcription occurs from integrated vectors than when it occurs from autonomously replicating vectors.[2,5] Third, the growth advantage in richer media permits production of more cell mass, and, since secretion generally parallels growth, this results in production of more of the desired protein.

Finally, besides integration, there is another approach which appears to provide vector stability even during growth in complex media. Two groups have built vectors which supply necessary genes to their respective host strains, and, in these cases, the host strains cannot bypass the requirements by utilizing components from outside the cell. Loison *et al.*[19] placed the *URA3* gene on an expression vector and inserted it into a *ura3 fur1* double mutant which cannot utilize exogenous uracil. Similarly, Thim *et al.*[20]

---

[19] G. Loison, M. Nguyen-Juilleret, S. Alouani, and M. Marquet, *Bio/Technology* **4**, 433 (1986).

[20] L. Thim, M. T. Hansen, K. Norris, I. Hoegh, E. Boel, J. Forstrom, G. Ammerer, and N. P. Fiil, *Proc. Natl. Acad. Sci. U.S.A.* **83**, 6766 (1986).

placed the *Schizosaccharomyces pombe* triose-phosphate isomerase gene on an expression vector and inserted it into a *tpi1 Saccharomyces cerevisiae* mutant. Since the *S. pombe TPI* gene is not expressed well in *S. cerevisiae*, viable transformants must have many copies of the vector. This latter approach has been used successfully for secretion of human proinsulin variants.[20]

Secretion Signals

A number of heterologous and homologous secretion signals have been used successfully to direct proteins into the yeast secretory pathway. The cell permits considerable flexibility in the secretion signal sequence, as judged by the work of Kaiser *et al.*[21] Despite the variety of permissible functional secretion signals, most constructions have used either the secretion signal from yeast α-factor or that from yeast invertase. In our hands, these two signals work with equivalent efficiency for secretion of calf prochymosin.[2] The mating factor-α signal has been somewhat more popular, mainly because of the convenient naturally occurring HindIII site providing access for fusion of other structural genes to the end of the DNA encoding the α-factor prosegment.[22]

We have engineered a convenient NcoI access site into the invertase signal DNA and have used that signal most extensively in our work. No processing other than signal peptidase cleavage appears necessary for removal of the invertase signal. In contrast, processing of α-factor fusions requires the action of two proteases, the *KEX2* gene product to remove the prosegment and the *STE13* gene product to remove the initial Glu-Ala-Glu-Ala segment.[22] We have had very few problems with proteases in production of a variety of heterologous proteins fused to the *SUC2* signal, whereas undesired proteolytic clipping of β-endorphin and α-interferon secreted by fusion to the *MFα1* signal has been well documented.[23] The DNA sequences of two *SUC2* signals carrying convenient NcoI sites in two reading frames are shown in Table II.

Yeast Colony Screens

Application of the above engineering principles permits secretion of most heterologous and homologous proteins at levels of between 1 and

---

[21] C. A. Kaiser, D. Preuss, P. Grisafi, and D. Botstein, *Science* **235**, 312 (1987).
[22] J. Kurjan and I. Herskowitz, *Cell (Cambridge, Mass.)* **30**, 933 (1982).
[23] G. A. Bitter, K. K. Chen, A. R. Banks, and P.-H. Lai, *Proc. Natl. Acad. Sci. U.S.A.* **81**, 5330 (1984).

TABLE II
DNA AND TRANSLATED PROTEIN SEQUENCE OF *SUC2* SECRETION SIGNAL CODING REGION FUSED
TO YEAST PROMOTERS WITH ACCESS IN TWO READING FRAMES

```
                                                                       NcoI
  M   L   L   Q   A   F   L   F   L   L   A   G   F   A   A   K   I   S   A↓
 ATG CTT TTG CAA GCT TTC CTT TTC CTT TTG GCT GGT TTT GCA GCC AAA ATA TCT GC C ATG G..


 ATG CTT TTG CAA GCT TTC CTT TTC CTT TTG GCT GGT TTT GCA GCC AAA ATA TCT GCC CAT GG.
                                                                          ↑
                                                                         NcoI
```

10 mg per liter of culture. Additional techniques have proved useful for boosting secreted levels of relatively large proteins ($M_r > 40{,}000$) to over 10 mg per liter.[24] In particular, mutagenesis coupled with screens for overproducing colonies have yielded strains secreting higher levels of calf prochymosin, bovine growth hormone, and human u-PA.[2,5] There are at least three useful applications for these colony screening methods for secreted proteins. First, they permit isolation of stable mutant strains of yeast with beneficial secretion properties. Second, they permit identification of the most productive integrants following integrative transformation with DNA carrying the desired expression unit. Third, they may be used to screen the products of random mutagenesis for more or less active variants of the enzyme whose activity is measured in the screen.

Detailed descriptions of two such screening procedures are given below. These are based on measurement of the activity of the proteins secreted from colonies on petri plates. More general methods can also be devised to measure the amount of various antigens secreted from colonies on plates by adapting the antibody detection method of Rothman *et al.*[25]

Screening Yeast Colonies for Secretion of Calf Prochymosin. Screening has been carried out with cells containing the invertase secretion signal–calf prochymosin fusion gene on the autonomously replicating plasmid pCGS514 (ATCC 20753). This plasmid consists of a *TPI*-promoted *SUC2* secretion signal fusion to bovine prochymosin in a 2-μm-based vector

---

[24] D. T. Moir, in "Yeast Genetic Engineering" (P. J. Barr, A. J. Brake, and P. Valenzuela, eds.), p. 215. Butterworth, Boston, Massachusetts, 1989.
[25] J. H. Rothman, C. P. Hunter, L. A. Valls, and T. H. Stevens, *Proc. Natl. Acad. Sci. U.S.A.* **83**, 3248 (1986).

(pCGS40 of Goff et al.[26]). Supersecreting mutations were mapped quickly to the chromosomes, as opposed to the promoter or the secretion signal, by curing the cells of plasmid and retransforming with a new copy of the expression unit. The following procedure was used to obtain cells carrying supersecreting mutations in the *SSC1* gene (now known as the *PMR1* gene[6]) and at a variety of other loci.[2]

1. Cells are mutagenized with ethyl methanesulfonate (EMS) following the procedure of Sherman et al.[27]

2. About 200 to 300 mutagenized colonies are spread on each of several minimal medium (SD) plates and incubated at 30° for 3 days. (Note: If cells are grown longer, higher levels of pepstatin may be required in Step 6 below. Use of SD medium[28] is critical here because it is poorly buffered, resulting in acid activation of prochymosin as cells grow and secrete it.)

3. Colonies are removed from the surface of the agar by blotting with sterile 9.0-cm Whatman #1 filter paper circles. Most of the cells are transferred to the first filter, particularly if the filter is removed before it is completely saturated with moisture from the agar plate. The mirror image replica is marked for orientation purposes and saved by placing it colony-side up on a SD agar plate at 4°. Viable cells are recovered from these colonies after mutants are identified (see Step 9 below). The surface of each plate is blotted with additional filters to remove all traces of the yeast colonies, and these blots are discarded.

4. Milk, the substrate for chymosin, is prepared as follows. Twelve grams of nonfat dry milk powder is dissolved in 100 ml of 10 mM CaCl$_2$ by stirring for 5 min and allowing the solution to stand for 1 hr before use.

5. A solution of 0.5% agarose in 10 mM CaCl$_2$ is melted and held at 50° until needed.

6. Pepstatin (2 mg/ml stock solution in dimethyl sulfoxide) is added to the milk solution to a final concentration of 5 μg/ml. The addition of pepstatin dampens the activity of chymosin secreted by wild-type cells, permitting easier detection of larger amounts of chymosin secreted by desired mutant cells. (Note: It may be necessary to adjust the pepstatin concentration, depending on the basal level of prochymosin secretion by the particular strain being screened and the age and size of the colonies.

---

[26] C. G. Goff, D. T. Moir, T. Kohno, T. C. Gravius, R. A. Smith, E. Yamasaki, and A. Taunton-Rigby, *Gene* **27**, 35 (1984).

[27] F. Sherman, G. R. Fink, and J. B. Hicks, "Methods in Yeast Genetics," p. 9. Cold Spring Harbor Laboratory, Cold Spring Harbor, New York, 1986.

[28] F. Sherman, G. R. Fink, and J. B. Hicks, "Methods in Yeast Genetics," p. 164. Cold Spring Harbor Laboratory, Cold Spring Harbor, New York, 1986.

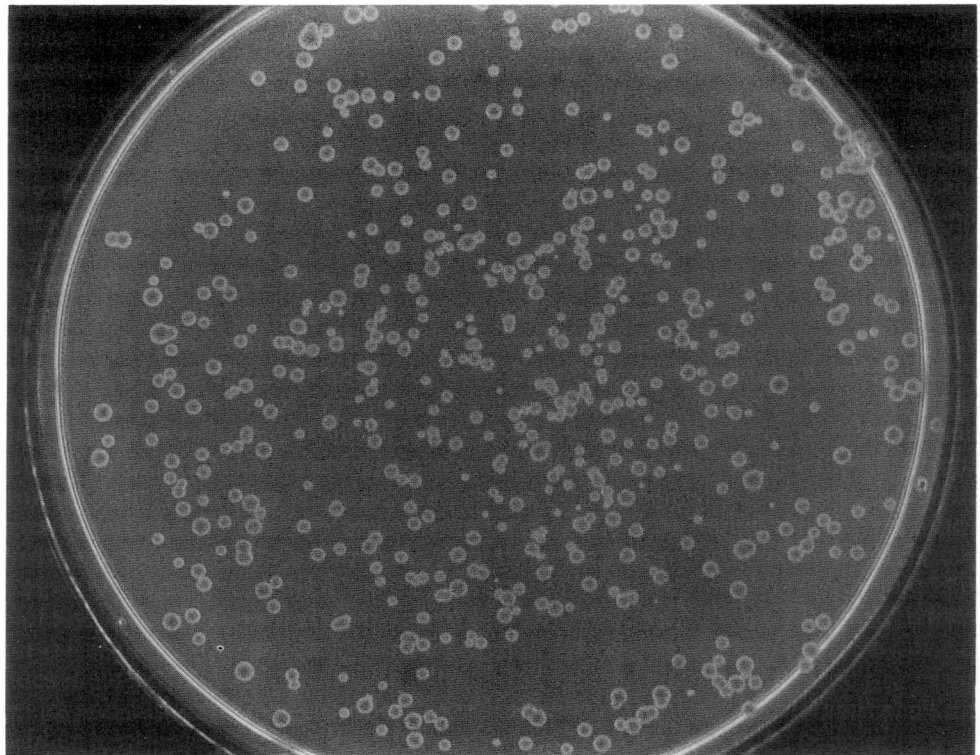

FIG. 2. Yeast colony screen for secretion of bovine prochymosin. A petri plate containing yeast colonies carrying an autonomously replicating vector directing the production and secretion of bovine prochymosin is shown at left. The same plate about 3 hr after removal of the colonies by blotting with sterile filter paper and application of an agarose overlay containing milk and pepstatin is shown at right. The size of the milk "clots" and the speed with which they develop indicate the amount of prochymosin secreted by the colony formerly occupying that location.

Pepstatin concentrations of 0.5 μg/ml have little or no effect while concentrations as high as 40 μg/ml inhibit clot formation almost completely.) The recommended 5 μg/ml level allows a faint clot to form over virtually every colony after 3 hr of incubation at room temperature, and this permits convenient orientation of the plate and the filter replica.

7. Four milliliters of milk and 4 ml of agarose are mixed in a 14 × 150 mm test tube and poured over the surface of petri plate from which prochymosin secreting yeast colonies have been removed by blotting. Timing and temperature are important to prevent premature solidification of the agarose and curdling of the milk.

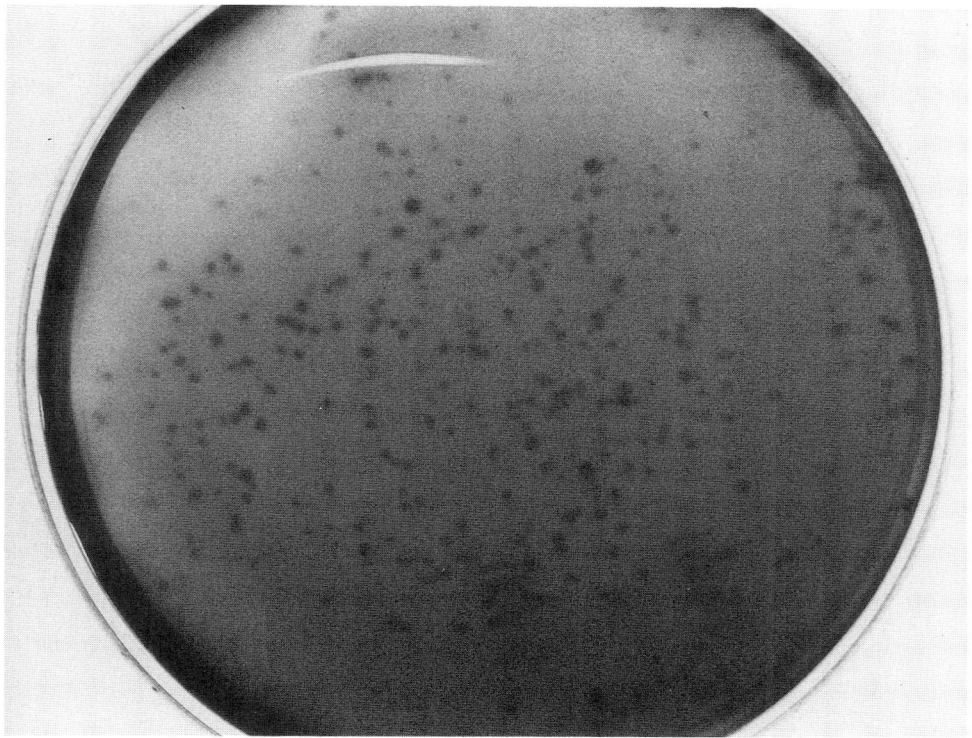

Fig. 2. *(Continued)*

8. The plates are incubated at room temperature or at 30°. After 1 hr, clots begin to form over a few of the removed colonies; after 3 hr, the clots intensify, and clots form over the former locations of virtually all of the colonies (see Fig. 2). There is variability between plates and even between areas on the same plate, so supersecretors are picked by their performance in relation to neighboring colonies.

9. The locations of intense or rapidly forming clots are noted on the backs of the petri plates. The corresponding colonies are picked from the filter paper replica (see Step 3), streaked, and retested.

By using this method, we obtained, from a total of 120,000 colonies screened, 39 mutant colonies in at least 3 different complementation groups which secreted from 3 to 10 times the amount of prochymosin secreted by the wild-type starting strain.[2] Several of these mutant strains carried mutations in the *PMR1* gene. Recently, we cloned and characterized the wild-type *PMR1* gene by complementation of a sporulation defect in a *pmr1-1/pmr1-1* homozygous diploid and discovered that it apparently encodes a Ca^{2+}-pumping P-type ATPase.[6] That this gene product functions in the secretory pathway is further demonstrated by the observed truncation of glycosylation of invertase and the suppression of *ypt1-1* and several *sec* mutants by the null allele of *PMR1*.[6,29] The patterns of suppression and glycosylation suggest that the *PMR1* gene product may function in the ER to Golgi transition or in early Golgi.

The utility of strains carrying mutations in the *PMR1* gene extends far beyond secretion of calf prochymosin. We have observed significant enhancement of secretion of several heterologous gene products, including human urokinase-type plasminogen activator, human tissue-type plasminogen activator, and bovine growth hormone, when *pmr1* strains are used as hosts instead of wild-type strains.[24] Strains carrying mutations in the *PMR1* gene should be useful for secreting many other proteins which encounter a rate-limiting step between the ER and Golgi.

Screening Yeast Colonies for Secretion of Human u-PA. The petri dish colony assay for functional u-PA secretion, as in the case of the prochymosin plate assay, has been used to find the most productive integrative transformants as well as to find desired new host mutants that secrete u-PA more efficiently. In addition, we have used *in vitro* mutagenesis of u-PA genes on plasmids to find mutant forms of u-PA with either decreased or increased activity compared with the parental form of the molecule, as described below.

1. DNA of plasmid pCGS721 directing production and secretion of u-PA[5] is mutagenized with hydroxylamine.[30]

2. For each experiment, a time course of treatment with hydroxylamine is investigated. Mutagenized vector from time points yielding Ura^- transformants of an *E. coli pyrF* strain at a frequency of 5% is selected for building a library of mutagenized u-PA genes. At this mutagenesis frequency, 12 of 13 completely sequenced mutant u-PA genes encoding u-PA with decreased activity contain single nucleotide changes.

3. Following the determination of the correct time point for hydroxylamine treatment of vector DNA, a larger aliquot of the mutagenized DNA is used to transform *E. coli*. At least 100,000 ampicillin-resistant *E. coli* colonies are obtained per library of mutagenized parental u-PA vector. These colonies are washed off the ampicillin plates and used to prepare plasmid DNA.

[29] A. Antebi, personal communication (1989).
[30] H. C. M. Nelson and R. T. Sauer, *Cell (Cambridge, Mass.)* **42**, 549 (1985).

4. The mutagenized plasmid DNA preparation is then used to transform yeast host strain CGY1585 (MATα *leu2 ura3 pmr1*)[6] to uracil prototrophy by the lithium acetate method of Ito et al.[31] Yeast transformant colonies are washed off the uracil-deficient plates and stored frozen in 15% glycerol in several tubes for subsequent screening. The stored, transformed CGY1585 cells are plated at 50–100 colonies per plate on nylon 66 filters (Schleicher and Schuell, Keene, NH, #00160) which had been placed on top of agar plates containing supplemented synthetic medium lacking uracil.[28] The plates are incubated for 3–4 days until the colony sizes are approximately 2 mm in diameter. (Note: In screens for cells producing u-PA variants with increased activity, starting with cells producing u-PA of low activity, 150–200 colonies can be screened easily on each plate since the background consists of colonies producing very small zones of fibrin lysis.)

5. Filters are then transferred to plates containing synthetic media and the following salts (in g/liter), which are necessary to obtain maximal secretion of u-PA: KH_2PO_4 (7.5), NaCl (3.0), $(NH_4)_2HPO_4$ (20), and $NH_4H_2PO_4$ (20). The salts are made up and autoclaved separately at double strength (in 500 ml) and then are combined with 50 ml of 20× yeast nitrogen base without amino acids (Difco), 50 ml of 40% glucose, 400 ml of 4% agar, and all necessary nutritional supplements prior to pouring plates. Incubation is overnight at 25°–30°.

6. To assay the activity of the secreted u-PA, filters are transferred to agar plates containing bovine fibrin for overnight incubation at 37°. [Note: Plasmid pCGS721 directs the production and secretion of single-chain u-PA (scu-PA) which must be converted to the two-chain form (tcu-PA) in order to detect its plasminogen activator activity with maximum sensitivity. scu-PA is at least 100-fold less active than tcu-PA *in vitro*.[32] Fortunately, scu-PA is converted to tcu-PA quantitatively and rapidly in this system, either by the action of bovine thrombin added to polymerize the fibrinogen or by traces of bovine plasmin or other proteases in the plates.] The active human tcu-PA formed in the assay can activate traces of bovine plasminogen (presumably contained in the relatively crude fibrinogen used in the plates) to create a zone of clearing in the turbid fibrin network in the agar (see Fig. 3). The clear zones centering on the yeast colonies on the filters have been visualized either by lifting up one edge of a filter with a forceps and examining the fibrin–agar surface directly or by shining a strong light on the back of a plate with the filter remaining on the surface. scu-PA secreted by a yeast colony 2–3 mm in diameter can generate

---

[31] H. Ito, Y. Fukuda, K. Murata, and A. Kimura, *J. Bacteriol.* **153**, 163 (1983).
[32] V. Ellis, M. F. Scully, and V. V. Kakkar, *J. Biol. Chem.* **262**, 14998 (1987).

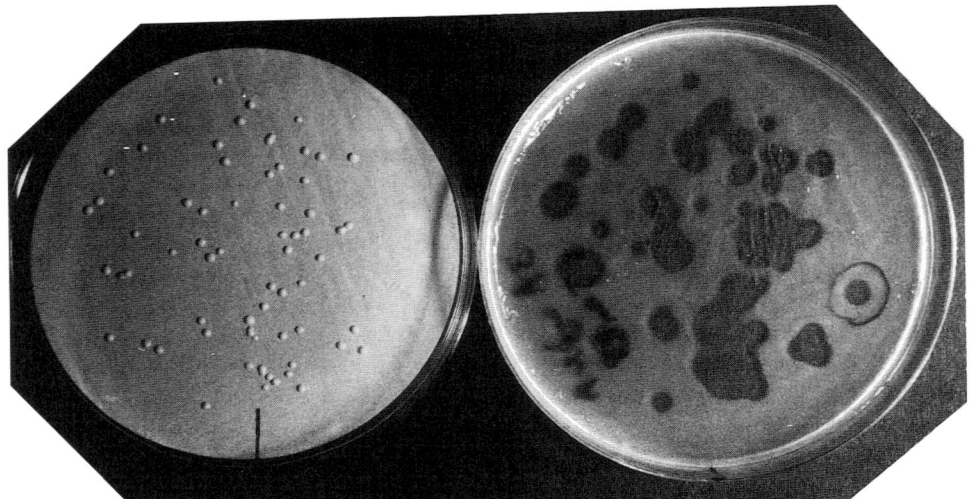

FIG. 3. Yeast colony screen for secretion of human urokinase-type plasminogen activator (u-PA). (Left) A nitrocellulose filter containing yeast colonies grown by placing the filter on the surface of a nutrient agar petri plate. (Right) Zones of clearing in a turbid fibrin–agar plate resulting from incubation of the nitrocellulose filter, colony-side up, on the surface of the fibrin–agar plate for about 12 hr. Human u-PA secreted through the nitrocellulose activates trace amounts of plasminogen in the plate, thereby causing dissolution of the turbid fibrin network in local areas under the colonies. The colony corresponding to the circled zone of clearing is not visible because it was removed for further study.

cleared zones of 12 mm diameter following overnight incubation at 37°.

To prepare the fibrin plates, three reagent solutions are made. (1) The fibrinogen reagent contains 1% (w/v) Miles Pentex (Kankakee, IL) bovine fibrinogen, 75% clottable (Cat. No. 82-021-5), dissolved in sterile water plus 1/4 volume sterile 10× phosphate buffer [containing, in g/liter, $NaH_2PO_4 \cdot H_2O$ (26.9), Na_2HPO_4 (43.3), and NaCl (58.5), pH of 7.0]. Extensive stirring with a sterile stirring bar is needed to dissolve the fibrinogen. To avoid bacterial contamination, ampicillin (25 µg/ml) and/or tetracycline (10 µg/ml) are added. (2) The thrombin reagent contains 1000 units of Miles Pentex bovine thrombin (20 µg/1000 units, Cat. No. 82-036) plus 10 ml of sterile 1× phosphate buffer. (3) The agar reagent contains sterile 2% agar maintained in a 65° water bath. To generate a fibrin plate, a 5-ml drop of the fibrinogen reagent and a 0.1 ml drop of the thrombin reagent are pipetted into different locations of a petri dish. Then a 5-ml drop of agar is added to a third location in the dish, and the three drops are mixed by shaking the plate by hand for approximately 5 sec on a table surface, allowing the fibrinogen to polymerize and the agar to solidify uniformly.

7. Yeast colonies demonstrating the desired mutant phenotype, in this case smaller zones of clearing, are restreaked for single colonies and retested for their u-PA phenotype on filters as well as in a liquid culture assay. In the latter assay, 5-ml cultures (synthetic complete minus uracil growth medium) of the yeast strains to be tested are set up in roller tubes. Following rapid overnight growth at 30°, the cultures are harvested by centrifugation and resuspended in the salts-supplemented medium SMII (Table III) containing the appropriate nutritional supplements (in this case, leucine) to allow u-PA accumulation in the culture broth. Following overnight growth in secretion medium, 20-μl aliquots of clarified broths from

TABLE III
COMPONENTS OF SMII SECRETION MEDIUM[a]

Ingredient	Amount
Component I: Autoclaved in 430 ml	
KH_2PO_4	7.5 g
NaCl	3.0 g
$(NH_4)_2HPO_4$	20.0 g
$NH_4H_2PO_4$	20.0 g
Component II: Autoclaved in 400 ml	
Inositol	300 mg
Vitamin B_1	52 mg
Vitamin B_6	15 mg
Calcium d-pantothenate	7.5 mg
Yeast extract (Difco)	3.75 g
Glutamic acid	5 g
Citric acid	5 g
Component III: Autoclaved in 20 ml	
$MgSO_4$	350 mg
$ZnSO_4$	30 mg
$CuSO_4$	4.5 mg
Component IV: Filter sterilized in 10 ml	
Biotin	10 mg
Component V: Filter sterilized in 10 ml	
$FeNH_4SO_4$	60 mg
Component VI: Autoclaved in 10 ml	
$CaCl_2$	150 mg
Component VII: Dissolved in 20 ml	
NaOH	4 g
Component VIII: Autoclaved in 100 ml	
Glucose	40 g

[a] All components are mixed in order. The final volume is 1 liter. The pH should be approximately 5.4 before addition of NaOH and 6.5 afterward.

cultures (as well as from tcu-PA standards used for calibration purposes) are placed in 5-mm cylindrical wells cut into fibrin plates with a cork borer. Zones of fibrinolysis are measured following incubation at 37° for between 3 and 17 hr, depending on the activity of the molecules being tested. This assay shows a nearly logarithmic relationship between u-PA concentration and the diameter of fibrinolytic zone formed. The same culture broths are used for dot blots to measure the amount of u-PA immunological cross-reacting material present and are also processed for Western blot analysis.

To recover the plasmids carrying mutant u-PA genes from yeast cells, we perform a rapid glass-bead breakage and phenol extraction to prepare total yeast DNA.[33] The yeast DNA is then used to transform *Escherichia coli* to ampicillin resistance. The recovered plasmids in *E. coli* are checked for the correct restriction pattern. Minipreparation DNA from several *E. coli* transformants resulting from DNA of a single yeast strain is used to retransform the yeast host strain to verify that the plasmid is the actual cause of the phenotypic change observed in u-PA activity. Plasmids deemed to carry a u-PA mutation are prepared in large scale for DNA sequencing. In a recent search for less active u-PA variants in our laboratory, the DNA sequences of u-PA genes from 13 mutants producing less active u-PA were determined. The mutations occurred throughout the molecule (L. Davidow, D. Dumais, and D. Moir, unpublished observations, 1989).

We have also used this same procedure to isolate more active revertants of specific u-PA variants exhibiting less than wild-type activity. We examined a total of 40,000 yeast transformants in each mutant hunt because, at the plasmid mutagenesis frequency used, this number of transformants yielded five or more repeat isolations of verified true revertants. Several apparent revertants were shown to be false positives consisting of "petite" (mitochondrial respiration deficient) strains. On filters, petites were found to secrete a higher percentage of the u-PA produced than do wild-type yeast strains. We now eliminate this class of colonies by testing all apparent revertant colonies for growth on a medium containing a nonfermentable carbon source (acetate growth medium). Recently, Kaisho *et al.*[34] have reported a similar increase in expression of a foreign protein by petite yeast.

Summary and Conclusion

To summarize, a variety of stable vectors and efficient promoters and secretion signals are available in yeast for engineering the secretion of any

---

[33] F. Winston, personal communication (1987).
[34] Y. Kaisho, Y. Koji, and N. Kazuo, *Yeast* **5**, 91 (1989).

protein of interest. Since secretion is growth-associated, we have favored the use of constitutive promoters and moderate copy number integrated vectors. This is because (1) heterologous gene expression from very high copy number vectors is frequently deleterious to growth[35] and (2) delaying gene expression until after the most rapid cell growth phase is cumbersome on a large scale. Methods are available for dividing the total process into growth and production/secretion phases, but they appear worthwhile only when expression of the engineered protein compromises growth significantly.

Even with these useful tools, it is frequently helpful to enlist the aid of mutant host strains in order to maximize secretion of a desired protein. Mutations in the *PMR1* gene have proved effective in a number of different cases. Moreover, it is possible to identify new host strains tailored to specific needs by applying activity screens to mutagenized colonies growing on petri plates.

Finally, colony screens such as the ones described here for active secreted enzymes are useful for routine strain construction. For example, they may be applied to identify the most productive strain from a large number of clones following a transformation or genetic cross. In addition, these screens may be used for characterizing the products of random mutagenesis of the gene encoding the secreted enzyme. The resulting structure–function information can be used to identify regions of the enzyme involved in different activities and to build new enzymes with different characteristics.

Acknowledgment

The authors gratefully acknowledge the work of many present and former colleagues at Collaborative Research, Inc., whose efforts have helped to build the understanding conveyed in this summary. In particular, Robert A. Smith and Margaret J. Duncan developed the chymosin colony screen described here, Dennis Dumais, Larry Melnick, and Bonnie Price-Tillotson assisted in the development of the u-PA colony screen, and Jay Raina developed the u-PA secretion medium. We also acknowledge the generous financial support of Dow and Sandoz for work on prochymosin and u-PA, respectively.

---

[35] J. F. Ernst, *DNA* **5**, 483 (1986).

[35] Epitope Tagging and Protein Surveillance

By PETER A. KOLODZIEJ and RICHARD A. YOUNG

Introduction

Epitope tagging is the process of fusing a set of amino acid residues that are recognized as an antigenic determinant to a protein of interest.[1-3] Tagging a protein with an epitope allows the surveillance of the protein with a specific monoclonal antibody. This approach can elucidate the size of a tagged protein as well as its abundance, cellular location, posttranslational modifications, and interactions with other proteins.[1-4] In addition, epitope tagging allows the protein to be purified in the absence of a functional assay.

The epitope tagging approach offers significant advantages over the use of antibodies generated directly against the protein of interest. The tagged protein can be monitored with a well-characterized monoclonal antibody whose immunoprecipitation capabilities and spectrum of cross-reactivity with nontagged proteins are already known. Proteins that associate with a tagged protein can be identified in a manner not possible with antibodies generated against the protein of interest; such associated proteins will be absent in immunoprecipitates from extracts prepared from cells lacking the tagged protein.[3] The use of independently raised antibodies cannot provide a comparable negative control. The intracellular location of epitope-tagged proteins can be identified in immunofluorescence experiments in a similarly controlled manner.[1] Finally, the epitope tagging approach may be particularly useful for discriminating among similar gene products.

The yeast *Saccharomyces cerevisiae* is an organism of choice for epitope tagging because the altered protein can be tested rapidly for normal function. The limitation of the epitope tagging approach is the potentially deleterious effect that the additional amino acids may have on protein function. Fortunately, such effects can be diagnosed rapidly in yeast by testing the ability of the tagged allele to complement a null allele. Moreover, introducing the epitope into various portions of the protein can provide results that corroborate those obtained from a single epitope tag. This chapter describes the epitope tagging method, two particularly useful

---

[1] S. Munro and H. Pelham, *Cell (Cambridge, Mass.)* **48,** 899 (1987).
[2] J. Field, J. Nikawa, D. Broek, B. MacDonald, L. Rodgers, I. A. Wilson, R. Lerner, and M. Wigler, *Mol. Cell. Biol.* **8,** 2159 (1988).
[3] P. A. Kolodziej and R. Young, *Mol. Cell. Biol.* **9,** 5387 (1989).
[4] D. Finley, B. Bartel, and A. Varshavsky, *Nature (London)* **338,** 394 (1989).

monoclonal antibodies, procedures for labeling yeast proteins *in vivo* and immunoprecipitation of labeled proteins, and some results that have been obtained using this approach with yeast RNA polymerase II.

Epitope Addition

Principles

The amino or carboxyl terminus of the protein is typically chosen as a tagging site because the ends of proteins are more likely to be accessible to the antibody and to be susceptible to modification without affecting function. Addition of the epitope coding sequence to the protein coding sequence is accomplished with minimal effort and minimal disruption of flanking sequences by oligonucleotide-mediated site-directed mutagenesis of the corresponding gene. Alternatively, the additional amino acids can be introduced via the polymerase chain reaction (PCR).[5]

Epitope Addition Methods

The sequences used to tag various proteins are shown in Fig. 1. Both of these epitope coding sequences have been optimized for codon usage in yeast and marked via the inclusion of a restriction site. The restriction site facilitates screening for the tagged gene. Once the epitope has been added, wild-type and modified copies of the gene should be tested for the ability to complement a mutant allele, either by direct transformation or plasmid shuffle.[6]

Tools for site-directed mutagenesis are available commercially (from Bio-Rad, Richmond, CA, and Amersham, Arlington Heights, IL, for example). In most methods, the gene to be altered is cloned into either an M13 vector or a yeast shuttle vector bearing a single-stranded bacteriophage origin of replication.[7-9] An oligonucleotide is synthesized that contains epitope coding sequences flanked by 15–17 nucleotides that are complementary to DNA on each side of the insertion site. Single-stranded DNA is produced, and the oligonucleotide is annealed to the single-stranded DNA form of the vector. The oligonucleotide serves as a primer for *in vitro* synthesis of the complementary DNA strand. Selection against

---

[5] R. Saiki, D. Gelfand, S. Stoffel, S. Scharf, R. Higuchi, G. Horn, K. Mullis, and H. Erlich, *Science* **239**, 487 (1988).
[6] J. D. Boeke, J. Trueheart, G. Natsoulis, and G. R. Fink, this series, Vol. 154, p. 164.
[7] C. Yanisch-Perron, J. Vieira, and J. Messing, *Gene* **33**, 103 (1985).
[8] S. Elledge and R. Davis, *Gene* **70**, 303 (1988).
[9] R. Sikorski and P. Hieter, *Genetics* **122**, 19 (1989).

9E10 epitope   GAACAA<u>AAGCTT</u>ATTTCTGAAGAAGACTTG (<u>Hind</u>III)

12CA5 epitope   TACCCATAC<u>GACGTC</u>CCAGACTACGCT (<u>Aat</u>II)

FIG. 1. Nucleotide sequences encoding epitopes. These sequences are typically flanked by 15–17 nucleotides that are complementary to DNA on each side of the insertion site. Underlined sequences indicate restriction enzyme recognition sites.

the template strand allows efficient recovery of mutants. For example, the template strand can be made sensitive to nucleases either by incorporation of uracil during replication in a dut^- ung^- strain[10] or by specific nicking *in vitro*.[11]

Yeast centromere vectors that contain both M13 and yeast replication origins facilitate efficient and iterative mutagenesis.[8,9] They dispense with the need to subclone mutagenized DNA from M13 phage vectors into yeast shuttle vectors; moreover, double-stranded DNA is easier to prepare from them than from M13 phage vectors. These vectors require a helper phage[12] in order to replicate as single-stranded DNA molecules in *Escherichia coli*. Single-stranded vector DNA may be more difficult to obtain with an insert size greater than 4 kilobases (kb), and yields may vary with the helper phage and *E. coli* strain used. M13KO7, a helper phage that allows kanamycin selection of infected cells, gives good results (available from Pharmarcia, Piscataway, NJ), as does VCS M13 (Stratagene Cloning Systems, La Jolla, CA). Currently available dut^- ung^- strains are $recA^+$, so, if these are used, care should be taken to assure that rearrangements do not occur during the mutagenesis.

Antibodies

Useful Antibodies

Two antibodies are described here, one (12CA5) that recognizes an epitope derived from the influenza hemagglutinin protein[13] and a second (9E10) that is specific for a portion of the human c-*myc* gene.[14] The 12CA5 and 9E10 cell lines were obtained from Dr. Ian Wilson (Scripps Institute, La Jolla, CA), and Dr. J. Michael Bishop (University of California at San

[10] T. Kunkel, J. Roberts, and R. Zakour, this series, Vol. 154, p. 367.
[11] J. Sayers, W. Schmidt, and F. Eckstein, *Nucleic Acids Res.* **16**, 791 (1988).
[12] M. Russel, S. Kidd, and M. Kelley, *Gene* **45**, 333 (1986).
[13] I. Wilson, H. Niman, R. Houghten, A. Cherenson, M. Connolly, and R. Lerner, *Cell (Cambridge, Mass.)* **37**, 767 (1984).
[14] G. Evan, G. Lewis, G. Ramsay, and J. M. Bishop, *Mol. Cell. Biol.* **5**, 3610 (1985).

Francisco), respectively. Both of the antibodies are highly specific in Western blot procedures[15] at concentrations of 1–10 μg/ml. Of these two antibodies, 12CA5 has the highest avidity. It can bind its epitope even in 0.4 M ammonium sulfate. The amino acid sequence recognized by the monoclonal antibody 12CA5 is YPYDVPDYA, and that recognized by 9E10 is EQKLISEEDL.

Methods and Materials for Hybridoma Cell Culture

The growth medium for the 12CA5 cell line consists of Dulbecco's modified Eagle's medium (DMEM, high glucose), 10% fetal calf serum (FCS), 2 mM L-glutamine, 1 mM sodium pyruvate, 0.1 mM hypoxanthine, and 16 μM thymidine, and is supplemented from 100× stocks (available from Gibco, Grand Island, NY) with penicillin, streptomycin, nonessential amino acids, and vitamins, with the pH adjusted to 7.5 with NaOH. The growth medium for the 9E10 cell line consists of RPMI, 0.3% sodium bicarbonate, 125 μg/ml gentamicin, and 10% FCS, pH adjusted to 7.5 with NaOH. Procedures for tissue culture and ascites production are described elsewhere.[16] These cell lines grow especially well if they are passaged at high cell density (10^5 cells/ml and split 1 : 1). For large volume tissue culture, 12CA5 cells can be grown in spinner bottles at 37°. After the cultures have grown to saturation and the medium has turned orange or yellow, a yield of about 5–10 μg/ml 12CA5 antibody is obtained. Ascites fluid provides a thousandfold more concentrated source of antibody. We inoculate BALB/c mice with 10^6 cells/mouse. Larger numbers of cells produce solid tumors rather than ascites fluid.

Initial Examination of Epitope-Tagged Protein

The ability of the tagged protein to complement a null allele should be tested. The investigator should carefully examine the modified strain to ensure that the epitope addition does not alter its growth characteristics.

The presence of the modified protein in the cell should be investigated initially by Western blot analysis. Cell extracts can be prepared by trichloroacetic acid (TCA) extraction of cells[17] or by disrupting cells with glass beads (see below) in Laemmli sample buffer[18] using a vortex mixer, followed by heating to 65° for several minutes. The cell debris is removed by

---

[15] H. Towbin, T. Staehelin, and J. Gordon, *Proc. Natl. Acad. Sci. U.S.A.* **76**, 4350 (1979).
[16] E. Harlow and D. Lane, "Antibodies: A Laboratory Manual." Cold Spring Harbor Laboratory, Cold Spring Harbor, New York, 1988.
[17] G. Reid and G. Schatz, *J. Biol. Chem.* **257**, 13056 (1982).
[18] U. K. Laemmli, *Nature (London)* **227**, 680 (1970).

a 5-min centrifugation in a microcentrifuge, and a portion of the supernatant subjected to sodium dodecyl sulfate–polyacrylamide gel electrophoresis (SDS–PAGE). Techniques for separation of proteins by electrophoresis transfer to nitrocellulose, and identification of specific polypeptides with antibodies are described elsewhere.[15] Promega (Madison, WI) manufactures a secondary antibody–alkaline phosphatase conjugate kit useful for the detection of both the 12CA5 and 9E10 antibodies on Western blots of yeast extracts.

Protein Surveillance

This section provides methods for labeling yeast proteins *in vivo*, for making yeast extracts, and for analytical immunoprecipitation of the epitope-tagged protein.

Media, Materials, and Buffers

YPD, SD, and YNB-DO are described elsewhere.[19] Low-sulfate (LSM) and low-phosphate (LPM) media contain 200 μM ammonium sulfate and 30 μM potassium phosphate (monobasic), respectively.[20] Low-sulfate medium is SD medium in which chloride salts replace sulfate salts. No-sulfate medium is SD medium which lacks the 200 μM ammonium sulfate.

Triton X-100 and Tween 20 were obtained from Bio-Rad, protein A-agarose from Repligen (Cambridge, MA), glass beads (425–600 μm), aprotinin, and other chemicals from Sigma (St. Louis, MO), ammonium sulfate, Polymin P (50% solution), and glycerol from BRL (Gaithersburg, MD), protease inhibitors from Chemicon, (Temecula, CA) [^{35}S]methionine (>600 mCi/mmol), $^{35}SO_4$ (in water, 1200–1400 Ci/mmol), and $^{32}PO_4$ (in water, 8500–9120 Ci/mmol) from New England Nuclear (Boston, MA) or Amersham. Glass beads should be washed with several bead volumes of concentrated HCl, then with distilled water, then with 1 M Tris base, and finally with water until the pH is neutral. Dry them by baking in a vacuum oven.

The buffers used in procedures described in this section are buffers A, B, and C. Buffer A (lysis buffer) contains 10% glycerol, 20 mM HEPES, pH 7.9, 10 mM EDTA, 1 mM dithiothreitol (DTT), phosphatase inhibitors (1 mM NaN$_3$, 1 mM NaF, 0.4 mM NaVO$_3$, 0.4 mM Na$_2$VO$_5$, 0.1 mg/ml phosvitin), and protease inhibitors [5 mM benzamidine, 1 mM phenylmethylsulfonyl fluoride (PMSF), 0.5 mg/ml bovine serum albumin (BSA), 10 μg/ml each aprotinin, antipain, chymostatin, leupeptin, and pepstatin

---

[19] F. Sherman, G. R. Fink, and J. B. Hicks, "Laboratory Course Manual for Methods in Yeast Genetics." Cold Spring Harbor Laboratory, Cold Spring Harbor, New York, 1986.
[20] D. Julius, R. Schekman, and J. Thorner, *Cell (Cambridge, Mass.)* **36**, 309 (1984).

A]. Lysis buffers may also contain various divalent cations in millimolar amounts in lieu of EDTA. EDTA lysis tends to reduce background in immunoprecipitations. Buffer A should also contain 100 mM to 0.9 M ammonium sulfate. For an initial experiment, a range of ammonium sulfate concentrations is suggested in order to determine the optimal concentration for extraction of the antigen. Buffer B contains 5% glycerol, 20 mM HEPES, pH 7.9, and 10 mM EDTA. Buffer C is 20 mM HEPES, pH 7.9, 10 mM DTT, 5 mM benzamidine, 1 mM PMSF, 0.5 mg/ml BSA, 10 μg/ml of each of aprotinin, antipain, chymostatin, leupeptin, and pepstatin A, and 1% lithium dodecyl sulfate.

In Vivo Labeling of Yeast Proteins

The following procedure works well for RNA polymerase II, a moderately abundant enzyme found in about 10^4 molecules per cell. Shorter labeling times may be required if the tagged protein has a short half-life or for a pulse–chase experiment. The key feature of the procedure is the initial incubation in high concentrations of [^{35}S]methionine. If the protein has a short half-life, its specific activity will be diluted by endogenous synthesis of methionine. Incubation times are somewhat arbitrary and are based largely on a need to ensure that the yeast cells do not exhaust nutrients in the culture medium during the labeling procedure.

1. Inoculate an overnight YPD culture with a single yeast colony.

2. The next day, dilute the overnight culture 1:1000 into low-sulfate medium (LSM) at 30°. Grow the culture to an OD$_{600}$ of 0.5 to 1.0 units. For immunoprecipitation of RNA polymerase II, 5 × 10^7 cells (equivalent to 5 ml at 1 OD$_{600}$) are labeled per immunoprecipitation.

3. Harvest the cultures by centrifugation at 2000 to 3000 rpm for 5 min in a clinical centrifuge at room temperature.

4. Resuspend the cells in 1 ml of no-sulfate medium and transfer them to a screw-capped Eppendorf tube.

5. Spin briefly, remove the supernatant, and vortex the cell pellet in 1 mCi of ^{35}S label per 5 × 10^7 cells and the volume of 5× no-sulfate medium required to make the suspension 1×. For 40 μl of [^{35}S]methionine (~1 mCi), 12 μl of 5× no-sulfate medium is added.

6. After 5 min, dilute the samples with LSM to 1 mCi/ml, transfer the cultures to a plastic screw-capped Erlenmeyer flask, and incubate with shaking at 30° for 20 min. For ^{35}SO$_4$ labeling, the ammonium sulfate concentration in LSM is reduced to 20 μM.

7. Add 5 volumes of LSM prewarmed to 30° and continue growth for 90 min.

8. Harvest the cells by centrifugation as in Step 3.

To label with $^{32}PO_4$, cells are grown overnight in low-phosphate medium (30 μM KH_2PO_4), resuspended in no-phosphate medium (1 ml/5 × 10^7 cells), and transferred to a 125-ml plastic screw-capped Erlenmeyer flask, and 1 mCi/ml $^{32}PO_4$ is added to the culture. After a 20-min incubation with shaking, the culture is diluted 10-fold with low-phosphate medium and grown for an additional 2-6 hr. Cultures can also be labeled overnight in the presence of 100-250 μCi $^{32}PO_4$ in low-phosphate medium. If the culture is initially at 10^6 cells/ml, the cells can be harvested in mid- to late-log phase 12 hr later. Strains that grow poorly can be grown on richer medium and then resuspended in low-phosphate medium the day or a few hours before the experiments. In order to minimize radiation exposure, the cells should be lysed in a screw-capped Eppendorf tube surrounded by a plexiglass thimble (available from the M.I.T. Radiation Safety Office, Cambridge, MA).

Preparation of Cell Extracts

All steps are performed at 4°.

Cell Lysis (Native Conditions)

1. Harvest 5 × 10^7 labeled yeast cells by centrifugation and resuspend the cell pellet in 1 ml of distilled water (4°).

2. Transfer the cell suspension to a screw-capped Eppendorf tube, spin the cells down quickly, and resuspend them in 2 pellet volumes (100-200 μl) of buffer A containing 100 mM ammonium sulfate at 4°. In an initial experiment, a range of ammonium sulfate concentrations should be tried in order to determine the optimal concentration for extraction.

3. Add 50 μl of glass beads and vortex the samples for seven 30-sec periods, alternating with 30 sec of incubation on ice.

4. Centrifuge for 5 min in a microfuge and remove the supernatant to a fresh tube.

5. Repeat Step 4.

6. Often it is desirable to precipitate labeled proteins by the addition of ammonium sulfate. This can be conveniently done by adding the cell lysate to an appropriate amount of finely crushed ammonium sulfate, vortexing briefly, and rotating the sample end over end for 1 hr. The presence of BSA in buffer A provides sufficient carrier protein to prevent pellet loss. After 1 hr, the proteins are pelleted by centrifugation in a microfuge for 20 min and the supernatant is carefully and completely discarded. For examination of RNA polymerase II subunits, 250 μl of lysate is added to 0.1 g finely ground ammonium sulfate.

7. If ammonium sulfate fractionation is performed, resuspend the pellet in 600 µl of buffer A containing 50 mM ammonium sulfate, 5% nonfat dry milk, 1% Triton X-100, and 0.1% SDS. The detergent may be omitted, but the use of the nonfat dry milk blocks much of the nonspecific cross-reactivity seen in crude extracts and will aid in the visualization of poorly labeled, low abundance antigens. Spin this buffer for 10 min in a microfuge before using it to resuspend the pellet. This step will remove aggregated milk proteins. If ammonium sulfate fractionation is not performed, dilute into buffer A containing sufficient Triton X-100, SDS, and nonfat dry milk so that the final buffer contains 50 mM ammonium sulfate, 1% Triton X-100, 0.1% SDS, and 5% nonfat dry milk.

8. Samples can be frozen on dry ice for later use.

Cell Lysis (Denaturing Conditions). Immunoprecipitation of denatured proteins can produce less background than immunoprecipitation of native proteins and can be used as a positive control for a native immunoprecipitation experiment.

1. Cells are harvested, washed, and vortexed with glass beads as described above, but in 2 pellet volumes of buffer C at room temperature.

2. The extract is then boiled for 5 min, centrifuged in a microfuge for 5 min, diluted 10-fold into buffer A containing 50 mM ammonium sulfate and 1% Triton X-100, and centrifuged again for 15 min. Immunoprecipitation is then performed as described for native proteins.

Immunoprecipitation

To obtain the best signal-to-noise ratio, the immunoprecipitation procedure should be coupled with a simple initial fractionation step, such as an ammonium sulfate cut or Polymin P precipitation. All steps are performed at 4°.

Procedure

1. Calculate the volume of buffer A necessary to add to the crude cell extract in order to bring the final ammonium sulfate concentration to 50 mM. Add sufficient 10% Triton X-100/1% SDS to this volume of buffer A so that, after addition of cell extract, the sample will contain 1% Triton X-100/0.1% SDS. Add the crude cell extract to this mixture.

2. Proteins that bind protein A-agarose must be removed from the extract. Add 4 µl protein A-agarose beads (Repligen) equilibrated in buffer A containing 50 mM ammonium sulfate, 1% Triton X-100, and 0.1% SDS and rotate the sample end over end for 30 min. Thaw an aliquot of antibody and clear it by centrifugation for 5 min in a microfuge. Spin the

extract sample briefly to pellet the protein A-agarose beads, and transfer the supernatant to a fresh Eppendorf tube.

3. Add 1 μg of antibody to the supernatant (if added as ascites, a titration should be performed to determine the optimal amount) and incubate for 1 hr with end-over-end rotation.

4. Spin the samples for 15 min in a microfuge, transfer the supernatant to a fresh tube, add 6 μl protein A-agarose beads, and incubate for 1 hr with mixing.

5. Centrifuge the samples briefly to collect the immunoprecipitate. Remove the supernatant and freeze at $-70°$. Remove the last few microliters of supernatant from the beads with a fine-tipped 200-μl Pipetman and wash the beads twice with buffer B containing 400 mM ammonium sulfate and once with buffer B containing 50 mM ammonium sulfate. If 9E10 is the immunoprecipitating antibody, lower the salt concentration in the first 2 washes to 100 mM. Detergents can be added to buffer B if more stringent wash conditions are desired. To wash the beads effectively, invert the tube briefly so that the beads are resuspended and centrifuge briefly to collect the beads. Aspirate the wash buffer carefully.

6. After removing all of the wash buffer from the beads with a fine-tipped 200-μl Pipetman, heat the beads for 10 min at 65° in 10 μl of Laemmli loading buffer.[18] Samples are subjected to SDS-PAGE, and fluorography is performed by soaking in glacial acetic acid for 15 min, followed by 20% 2,5-diphenyloxazole (PPO) in acetic acid (w/w) for 15 min, and finally water until the gel turns white from precipitation of the PPO. Exposures are made at $-70°$, preferably without an intensifying screen, on Kodak XAR-5 film. Intensifying screens increase background in fluorography experiments.

Troubleshooting

There are a number of sources of background in the immunoprecipitations. Excessive background may be reduced by altering antibody concentration or incubation time. With any given antibody sample, a titration should be performed to determine optimal conditions. Signal quality does not seem to be improved with room temperature or overnight incubations, and these conditions may enhance proteolysis. The presence of detergent during the antibody incubation can increase the signal-to-noise ratio. However, the presence of detergent during the wash steps does not appear to reduce background. Increasing the number of washes fails to increase the signal-to-noise ratio and tends to reduce the signal. Inclusion of a simple fractionation step prior to immunoprecipitation probably is the most effective strategy for reducing background.

Preparative Immunochromatography

The epitope-tagged protein can be purified by immunochromatography on a preparative scale, under conditions that are likely to lead to retention of biochemical activity.[3,21] This approach may be particularly useful when a precise biochemical assay for a protein does not exist or when the protein is difficult to purify. When additional chromatography is required, the epitope tag permits identification of the fractions containing the protein.

Application to RNA Polymerase II

Epitope tagging has provided a particularly useful tool for the investigation of the subunits that associate to form RNA polymerase II. The epitope-tagged enzyme can be rapidly purified, the subunits defined by their ability to assemble with the tagged subunit, and the relative stoichiometry of subunits can be estimated.[22] In addition, the posttranslational modifications of the subunits can be identified and mutations studied for their effects on enzyme assembly and stability.[3,22]

Fractionation with Polymin P

Polymin P acts to precipitate nucleic acids found in the whole cell extract along with bound proteins.[23] Polymin P fractionation of a crude yeast extract permitted the immunoprecipitation of pure RNA polymerase II. All steps in the following protocol are performed at 4°.

1. Lyse the labeled cells using glass beads in 200 μl buffer A containing 0.35 M ammonium sulfate.

2. Remove cell debris and beads by centrifugation in a microfuge for 5 min. Transfer the supernatant to a new tube and dilute with 500 μl buffer A (to a salt concentration of 100 mM). Incubate the sample for 10 min on ice.

3. Centrifuge the sample for 20 min in a microfuge. Transfer the supernatant to a fresh tube and dilute it with 300 μl buffer A (to 70 mM ammonium sulfate). Add 5 μl of 10% Polymin P (adjust the pH of the Polymin P supplied by BRL to pH 7.9 with NaOH) per milliliter of supernatant. Vortex and incubate for 10 min on ice.

4. Centrifuge the sample for 20 min and discard the supernatant. Extract the Polymin P pellet with 250 μl buffer A plus 0.25 M ammonium

[21] C. Schneider, R. Newman, D. Sutherland, U. Asser, and M. Greaves, *J. Biol. Chem.* **257**, 10766 (1982).
[22] P. A. Kolodziej, N. Woychik, S.-M. Liao, and R. A. Young, *Mol. Cell. Biol.* **10**, 1915 (1990).
[23] G. Bitter, *Anal. Biochem.* **128**, 294 (1983).

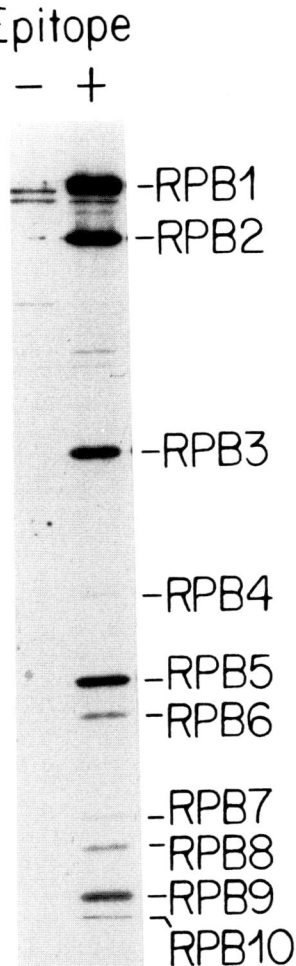

Fig. 2. Immunoprecipitation of *Saccharomyces cerevisiae* RNA polymerase II. The RPB3 protein is a subunit of RNA polymerase II, an enzyme composed of 10 subunits. Lane 1: 12CA5 immunoprecipitate from a Polymin P-fractionated extract prepared from a strain lacking the 12CA5 epitope. Lane 2: 12CA5 immunoprecipitate from a Polymin P-fractionated extract prepared from a strain expressing a 12CA5 epitope-tagged RPB3 protein.

sulfate by vortexing the samples briefly and then rotating them end over end for 20 min.

5. Centrifuge the sample for 20 min, transfer the supernatant to a new tube, and add 0.4 g finely crushed ammonium sulfate per milliliter of

supernatant. Incubate for 40 min with end-over-end rotation. Centrifuge the sample for 20 min in a microfuge. Discard the supernatant and carefully remove the remaining liquid using a 200-μl Pipetman with yellow tip. Dissolve the pellet by resuspension in 200 μl of buffer A plus 50 mM ammonium sulfate, 1% Triton X-100, and 0.1% SDS.

Immunoprecipitation of RNA polymerase II is performed as described above. The results are shown in Fig. 2.

Summary

The epitope tagging approach offers advantages of economy, universality, and precision over the use of antibodies raised directly against a protein of interest. The latter strategy promises a potentially greater diversity of reagents and obviates the need to modify the protein, but it may not yield sufficiently high-affinity, abundant, or specific antibodies. The major uncertainty in an epitope-tagging strategy, namely, the ability of the altered protein to function *in vivo,* is readily resolved in yeast by testing complementation of a null allele by the modified gene. Modification of the protein is easily accomplished by addition of the epitope coding sequence to the gene via oligonucleotide-mediated site-directed mutagenesis. The uniqueness of the epitope in the genome and the use of the monoclonal antibody assure a high-affinity, specific, and abundant antibody. Unrelated but identically modified proteins can be immunoprecipitated and affinity purified under the same conditions. Only extraction conditions and possibly a simple initial fractionation step need vary. Moreover, otherwise identical but differentially tagged proteins can be separated. Even proteins completely defective in an essential *in vivo* function can be purified and studied. Finally, polypeptides coprecipitating with the protein of interest are normally difficult to distinguish from those merely cross-reactive with the antibody used. As an alternative to defining a complex of proteins using a battery of antibodies, complexes are defined as a set of immunoprecipitable polypeptides present only in extracts containing the modified protein.

Acknowledgments

We thank Jeffrey Field, Gail Fieser, Gerry Fink, David Miller, Linda Rodgers, Adam Antebi, Joan Park, and Dan Finley for helpful discussions and Carolyn Carpenter for assistance with the preparation of the manuscript.

[36] Reverse Biochemistry: Methods and Applications for Synthesizing Yeast Proteins *in Vitro*

By KEVIN STRUHL

Introduction

Reverse biochemistry is a recent and relatively general approach for determining the function of proteins encoded by cloned genes and for analyzing the relationship between protein structure and function. The key feature of reverse biochemistry involves the synthesis of a desired protein from the cloned gene by transcription and translation *in vitro*. In contrast, classical biochemistry involves the identification and purification of proteins from crude cell-free extracts derived from living organisms. As is evident from the name, the relationship of reverse and classical biochemistry closely parallels the relationship between reverse and classical genetics.

The reverse biochemical approach has a number of attractive features. First, proteins can be synthesized *in vitro* in a short time, typically a few hours.[1,2] Moreover, the resulting products are radiopure and can be labeled to very high specific activities [up to 10^9 counts per minute (cpm)/μg protein]. In contrast, protein purification by classical procedures is usually very laborious and time consuming, and it is difficult to obtain radioactively labeled proteins that are functionally active. Second, any desired mutant protein can be generated and then analyzed simply by modifying the DNA template.[3,4] This is considerably easier than the isolation and purification of mutant proteins from living organisms, especially in cases where the mutant protein lacks activity. Third, because it is easy to follow their fate, radiolabeled proteins and their mutated derivatives can be very useful as substrates to assay for factors that modify,[5–7] directly interact with,[8,9] localize, or transport the protein.[5,7,10–12] Moreover, such *in vitro*

---

[1] D. Stueber, I. Ibrahimi, D. Cutler, B. Dobberstein, and H. Bujard, *EMBO J.* **3**, 3143 (1984).
[2] D. A. Melton, P. A. Krieg, M. R. Rebagliati, T. Maniatis, K. Zinn, and M. R. Green, *Nucleic Acids Res.* **12**, 7035 (1984).
[3] I. A. Hope and K. Struhl, *Cell (Cambridge, Mass.)* **46**, 885 (1986).
[4] I. A. Hope and K. Struhl, *Cell (Cambridge, Mass.)* **43**, 177 (1985).
[5] D. Baker, L. Hicke, M. Rexach, M. Schleyer, and R. Schekman, *Cell (Cambridge, Mass.)* **54**, 335 (1988).
[6] G. Hawlitschek, H. Schneider, B. Schmidt, M. Tropschug, F.-U. Hartl, and W. Neupert, *Cell (Cambridge, Mass.)* **53**, 795 (1988).
[7] J. A. Rothblatt and D. I. Meyer, *Cell (Cambridge, Mass.)* **44**, 619 (1986).
[8] C. J. Brandl, *Proc. Natl. Acad. Sci. U.S.A.* **86**, 2652 (1989).

synthesized proteins are well suited for low-resolution structural analysis such as proteolytic mapping of domains and thermodynamic stability.[13] Fourth, the ability to cosynthesize functional proteins of different sizes can greatly facilitate the determination of the subunit structure (e.g., monomer, dimer)[14] of a protein.

Despite all the advantages listed above, it is important to note the limitations of reverse biochemistry. Most importantly, the approach is suitable only for the analysis of individual proteins for which the genes have been cloned. It is clearly also inappropriate for identifying new biochemical functions or for carrying out complex biochemical processes such as DNA replication, transcription, translation, and splicing. In addition, there is always the possibility that *in vitro* synthesized proteins may differ in some structurally or functionally significant way from proteins produced *in vivo*. Finally, *in vitro* synthesis can generate only relatively small amounts of proteins (likely maximum about 1 μg) and hence is unsuitable for high-resolution structural analyses such as X-ray crystallography or NMR.

Principle of Method

The availability of a cloned gene makes it possible to synthesize the encoded protein by *in vitro* transcription and translation. In the first step of the procedure (Fig. 1A), the desired protein-coding sequences are cloned into a vector containing a promoter for SP6 or T7 RNA polymerase.[15] The protein-coding sequence can be obtained from a cDNA clone, or, for the vast majority of yeast genes that lack introns, it can be generated from cloned genomic DNA. In accord with the rules for efficient translation in eukaryotic organisms, it is essential that the correct initiation codon is the 5'-proximal AUG in the mRNA to be synthesized.[16,17] In certain situations, the initial cDNA or genomic clone might have to be modified to optimize the lengths of the 5'- or 3'-untranslated sequences adjacent to the

---

[9] N. F. Lue, D. I. Chasman, A. R. Buchman, and R. D. Kornberg, *Mol. Cell. Biol.* **7**, 3446 (1987).
[10] W. Hansen, P. D. Garcia, and P. Walter, *Cell (Cambridge, Mass.)* **45**, 397 (1986).
[11] A. P. G. M. vanLoon, A. W. Brandli, and G. Schatz, *Cell (Cambridge, Mass.)* **44**, 801 (1986).
[12] M. Mueckler and H. F. Lodish, *Cell (Cambridge, Mass.)* **44**, 629 (1986).
[13] I. A. Hope, S. Mahadevan, and K. Struhl, *Nature (London)* **333**, 635 (1988).
[14] I. A. Hope and K. Struhl, *EMBO J.* **6**, 2781 (1987).
[15] P. A. Krieg and D. A. Melton, this series, Vol. 155, p. 397.
[16] M. Kozak, *Nucleic Acids Res.* **15**, 8125 (1987).
[17] A. M. Cigan and T. F. Donahue, *Gene* **59**, 1 (1987).

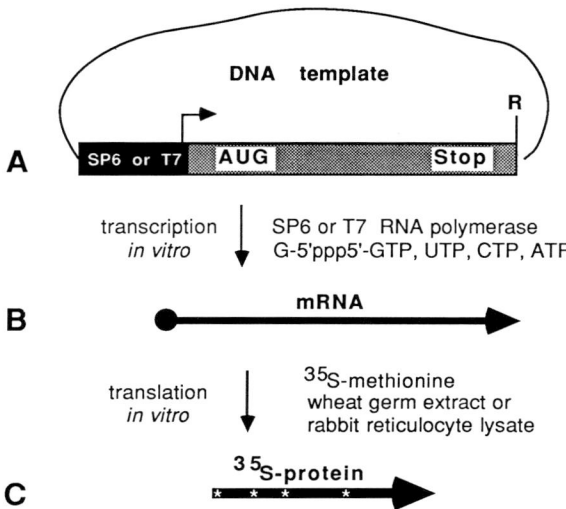

FIG. 1. Scheme for synthesizing proteins *in vitro*. (A) The template DNA contains a promoter for bacteriophage SP6 or T7 RNA polymerase (black bar; arrow indicates RNA start site) fused to a DNA fragment encoding the desired protein (gray bar with AUG and stop codons indicated) and is generally linearized by cleaving with a restriction endonuclease (R) just downstream of the protein-coding region. (B) The messenger RNA contains a 5' cap (black circle) that is produced by transcription *in vitro*. (C) The protein labeled with ^{35}S at methionine residues (asterisks) is produced by translation *in vitro*.

protein-coding region. Given an initial DNA template that can be transcribed and translated *in vitro* to yield reasonable amounts of a desired protein, the wide array of standard recombinant DNA manipulations can be employed to generate any mutant template of interest from which any desired mutant protein can be synthesized.

Second, messenger RNA encoding the protein is produced by transcribing the DNA template with the appropriate bacteriophage RNA polymerase (Fig. 1B). This *in vitro* transcription step is performed at high concentrations of ribonucleotide triphosphates in order to obtain large amounts of full-length RNA. However, unlike standard transcription reactions, 90% of the GTP in the reaction mixture is replaced by a "capped" GTP analog such as diguanosine-GTP. In this way, 90% of the RNA synthesized contains a 5'-capped structure that is typical of eukaryotic mRNAs and is important for efficient translation.[18] Because SP6 and T7 RNA polymerases are highly specific for initiation only from their respec-

[18] P. A. Krieg and D. A. Melton, *Nucleic Acids Res.* **12**, 7057 (1984).

tive promoters,[19,20] the RNA synthesized is a single species that encodes the desired protein.

Third, the essentially pure mRNA is translated *in vitro* using wheat germ extracts or reticulocyte lysates (Fig. 1C). In most cases, translation is performed in the presence of [^{35}S]methionine so that the protein is synthesized as a radiolabeled species. Most importantly, because the translation reaction is programmed with a pure mRNA species, the only radioactively labeled species should be the protein of interest. However, it is important to remember that, although the protein synthesized *in vitro* is radiopure, it represents only a very small percentage of the protein present in the translation extract. In some cases, these unlabeled proteins in the translation extract can influence the properties of the radiolabeled protein of interest.

Once the protein is synthesized, it can be used for a wide variety of purposes. In most cases, the products of the translation reaction can be used directly without further purification. This chapter is not intended, however, to be a complete survey of the large number of current and potential applications for *in vitro* synthesized proteins. Nevertheless, the last section discusses some specific uses of reverse biochemistry such as determining whether a cloned gene encodes a specific DNA-binding protein and analyzing the structure and function of such specific DNA-binding proteins.

Methods

Design and Preparation of DNA Template

The first step is to subclone protein-coding sequences of interest into a plasmid vector that contains a promoter for bacteriophage SP6 or T7 RNA polymerase. The major consideration in designing the DNA template is that the mRNA generated by the bacteriophage RNA polymerase should permit efficient synthesis of the desired protein. Since *in vitro* translation reactions are almost always carried out in extracts from eukaryotic organisms (typically wheat germ, rabbit reticulocytes, and yeast), this means that the mRNA produced from the DNA template should fit the rules for efficient translational initiation in eukaryotes.[16,17] Obviously, synthesis of the desired protein requires that the uninterrupted protein-coding sequence be cloned in the correct orientation downstream of the bacterio-

---

[19] E. T. Butler and M. Chamberlin, *J. Biol. Chem.* **257**, 5772 (1982).
[20] M. Chamberlin and T. Ryan, in "The Enzymes," (P. Boyer, ed.), Vol. 7, p. 87. Academic Press, New York, 1972.

phage promoter. In addition, it is essential that the correct initiation codon is the 5'-proximal AUG in the mRNA to be synthesized because initiation (or reinitiation) from internal AUG codons is generally very inefficient.[21] As for most eukaryotic mRNAs, the AUG initiation codon should be relatively close (25-100 bases) to the 5' end of the message.[16]

Because most yeast genes lack introns, protein-coding sequences are typically obtained from cloned genomic DNA. However, to ensure that the AUG initiation codon is properly situated, it is usually necessary to remove excess 5'-flanking nontranslated sequences from the initial cloned DNA. This is accomplished by cleavage at a fortuitous restriction site, sequential deletion by conventional Bal31 or exonuclease III-S1 nuclease procedures, or oligonucleotide-directed deletion. For some genes, synthesis of the encoded protein may be inefficient because the AUG initiation codon resides in a poor sequence context or is inaccessible owing to the secondary structure of the mRNA. In such cases, a reasonable way around the problem is to replace the normal 5'-untranslated sequences (and possibly the first few translated codons) with the equivalent region from an efficiently translated protein.

Given an appropriately designed template, plasmid DNA is prepared either by centrifugation in cesium chloride gradients or by an alternative procedure that yields relatively high-quality DNA; crude minipreparation-grade DNA may not be adequate.[15] Such DNA is then cleaved with a restriction endonuclease that cuts just downstream of termination codon [ideally 50-200 base pairs (bp)] and does not cut within the protein-coding region. Restriction sites in the polylinker of the cloning vector are often useful for this purpose, although if the site is too close (less than 50 bases) to the termination codon, the mRNA may be translated less efficiently. Such cleavage, though not essential, will usually result in higher molar amounts of the desired mRNA by minimizing the length of the transcript. It is desirable to avoid restriction enzymes that generate 3' overhanging ends as these sometimes lead to in vitro transcription artifacts. Following cleavage, the DNA is purified by phenol extraction and then resuspended in Tris-EDTA (TE) buffer; at this stage the DNA should be as free of ribonucleases as possible. Typically, 10 μg of DNA is cleaved and resuspended in a volume of 50 μl; this is enough for 10 in vitro transcription and translation reactions.

Preparing mRNA by in Vitro Transcription

Detailed descriptions of in vitro transcription reactions involving bacteriophage SP6 or T7 RNA polymerase have been presented elsewhere.[15] For

[21] F. Sherman, J. W. Stewart, and A. M. Schweingruber, *Cell (Cambridge, Mass.)* **20,** 215 (1980).

the purpose of *in vitro* translation, the only significant difference is that 90% of the GTP in the reaction mixture is replaced by diguanosine-GTP so that 5′ capping of the RNA occurs concurrently with transcription. A typical transcription reaction is carried out in a volume of 25 µl and contains the components listed below. Reactions involving SP6 RNA polymerase should be set up at room temperature to avoid precipitation of the DNA template by spermidine. If desired, a small amount of labeled ribonucleotides (10 µCi) can be included to monitor the reaction.

Reaction Mix

  8 µl Water (treated with diethyl pyrocarbonate)
  5 µl DNA (total 1 µg)
  5 µl 5× Ribonucleoside triphosphates [5 mM each ATP, CTP, UTP, diguanosine triphosphate (G-5′ppp5′-GTP); 0.5 mM GTP]
  2.5 µl 10× Transcription buffer (400 mM Tris-HCl, pH 7.5; 60 mM MgCl$_2$; 100 mM dithiothreitol; 1 mg/ml bovine serum albumin)
  2.5 µl 10 mM Spermidine (for SP6 RNA polymerase only; otherwise add water)
  1 µl (30–60 units) Pancreatic ribonuclease inhibitor (RNasin)
  1 µl (5–20 units) SP6 or T7 RNA polymerase

Transcription reactions are incubated at 40° (SP6 reactions) or 37° (T7 reactions) for 60 min. The reaction is terminated by extraction first with 25 µl buffered phenol, then twice with 2-butanol. The resulting mixture is brought to a final concentration of 2 M ammonium acetate, precipitated with ethanol, and washed once with ethanol. The RNA is resuspended in TE buffer containing 2 M ammonium acetate, reprecipitated with ethanol, washed once, and resuspended in 10 µl TE buffer. The RNA should be translated immediately or quick-frozen on dry ice and stored at −70°.

The most important consideration for a successful transcription reaction is the absence of ribonucleases throughout the procedure. A single ribonucleolytic cleavage in an RNA molecule will prevent the synthesis of the desired protein. With reasonable precautions, the transcription of full-length mRNA is rarely a problem because there are few efficient termination sequences for the bacteriophage RNA polymerases. When in doubt, the amount and quality of the RNA synthesized can be analyzed by electrophoresis in formaldehyde gels or by acid precipitation (if radiolabeled precursors are included in the reaction).

Preparing Protein by in Vitro Translation

For reasons of convenience, translation reactions are often carried out with commercially available kits that use either wheat germ extracts or rabbit reticulocyte lysates. Such commercial kits, though relatively expensive, are generally recommended because the preparation of active translation extracts is not routine for most investigators and can be problematic. Procedures for making large quantities of active extracts from wheat germ[22] or rabbit reticulocytes[23] are well established, however, and they significantly reduce the cost and variability of commercial preparations.

When using commercial kits, simply add the products of the *in vitro* transcription reaction (1 – 10 μl of mRNA) to the appropriate reagents and follow the directions of the manufacturer; typical reactions are carried out in 30-μl volumes at room temperature for 30 – 60 min. In general, 1 – 2 μl of RNA is sufficient to produce the maximum amount of protein, although this may vary depending on the specific transcript. For each new preparation of translation extract, it is important to perform a control reaction that lacks added RNA. For most applications, the reaction mixture should contain 15 μCi of [^{35}S]methionine [1400 Ci/mmol) in order to radiolabel the protein. In some cases, however, unlabeled methionine is added so that the synthesized proteins are unlabeled or of lower specific activity. Proteins should be used as soon as possible, although they are generally stable at 0° – 4° for 1 week. The relative instability of *in vitro* synthesized proteins primarily reflects the presence of proteases in the crude translation extracts. Protein preparations can be quick-frozen on dry ice and stored at −70° for longer periods of time. For this purpose, the protein should be divided into small aliquots prior to freezing; individual aliquots should be thawed only once, and then discarded after use.

To measure the amount of [^{35}S]methionine incorporated into protein, add 1 μl of the translation products to 50 μl of 0.1 M NaOH and incubate for 15 min at 37°. Then add 1 ml of 10% trichloroacetic acid, incubate for an additional 15 min on ice, and collect the acid-precipitated protein on glass fiber filters. By comparing the amount of incorporated [^{35}S]methionine in the sample to the control lacking added RNA, it is possible to estimate the amount of protein synthesized. Since wheat germ extracts lack any endogenous methionine, the specific activity of the protein can be directly calculated from the specific activity of the [^{35}S]methionine.[22] Reticulocyte lysates contain variable amounts of endogenous methionine, thus making the calculation more difficult.[23] In order to calculate the molar amount of protein synthesized, it is necessary to know the number of methionine residues per monomer product.

To determine the quality of the protein synthesized, 1 to 3 μl of the translation reaction should be analyzed by sodium dodecyl sulfate (SDS) – polyacrylamide gel electrophoresis; the ^{35}S-labeled proteins are easily visu-

[22] A. H. Erickson and G. Blobel, this series, Vol. 96, p. 38.
[23] R. J. Jackson and T. Hunt, this series, Vol. 96, p. 50.

alized by fluorography and autoradiography. Ideally, the synthesis should yield a single band in 1–4 hr, the molecular weight of which corresponds to that expected for the protein of interest (Fig. 2). For some proteins (especially DNA-binding proteins), the mobility of the *in vitro* synthesized product is much slower than expected, and the apparent molecular weight can be strongly influenced by the gel conditions.[4] Further confirmation that the correct protein is being synthesized can be obtained by cleaving the template DNA at various positions within or just beyond the protein-coding sequences and examining the proteins synthesized[4] (Fig. 2).

Even in successful syntheses, minor translation products are often observed. For bands corresponding to lower molecular weight products, the most common reasons are premature termination of translation, aberrant initiation at internal AUG codons, or proteolysis. Bands representing higher molecular weight species are less common and may be due to inefficient termination at the correct stop codon or posttranslational modification. Some *in vitro* synthesis reactions yield a set of distinct products of similar molecular weight, which may arise from protein modification.

Many mRNAs can be translated efficiently either in wheat germ extracts or in rabbit reticulocyte lysates. In general, it is believed that wheat germ extracts initiate translation somewhat better than reticulocyte lysates but are more prone to premature termination (or degradation of the mRNA template). Thus, wheat germ extracts may be preferred for shorter proteins, whereas reticulocyte lysates may be the best choice for longer proteins. However, other nonpredictable factors can influence the translation efficiency for any particular protein. Wheat germ extracts contain no exogenous methionine and hence generate proteins with somewhat higher specific activity than those produced in reticulocyte lysates, which contain variable amounts of exogenous methionine. A final consideration is that the activity of proteins synthesized *in vitro* may vary owing to differences in modification or the amounts of critical cofactors or auxiliary proteins. Thus, the choice of translation extract can have profound effects on the amount and/or functional properties of protein synthesized.

For particular applications, it may be highly advantageous to use *in vitro* translation extracts from *Saccharomyces cerevisiae* itself[10,24] (see [37] and [46] in this volume). Obviously, such extracts must be employed in studies of the yeast translational machinery itself or in studies of processes that are intimately connected with translation such as some cases of protein transport. One major benefit of employing yeast extracts is that proteins are synthesized in a homologous system that is more conducive for correct posttranslational modifications and more likely to contain auxil-

---

[24] K. Moldave and E. Gasior, this series, Vol. 101, p. 644.

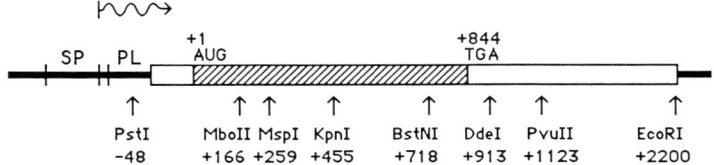

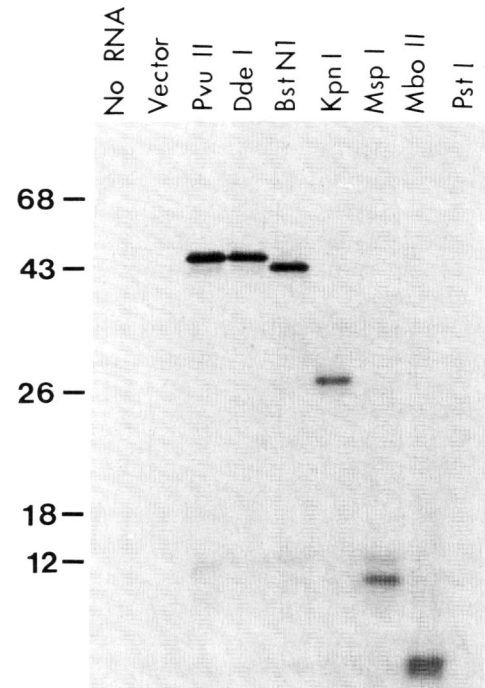

FIG. 2. Analysis by electrophoresis in denaturing gels of proteins synthesized *in vitro*. Shown above is a diagram of the template containing a bacteriophage SP6 promoter (SP) fused to a DNA fragment (open bar) encoding a protein (shaded region) with sites for restriction endonucleases as indicated. Shown below is an autoradiogram in which proteins were synthesized from template DNA cleaved with the indicated enzymes and examined by SDS–PAGE [mobilities of molecular weight markers ($\times 10^{-3}$) indicated at left]. Cleavage within the protein-coding region results in synthesis of proteins truncated at their carboxyl termini. (From Ref. 3.)

iary proteins necessary for function. In this sense, the proteins synthesized can be analyzed under conditions that are as physiologically significant as possible. Another powerful aspect is the potential to synthesize proteins in extracts from yeast strains that contain essentially any mutation of interest. Such a fusion between reverse biochemistry and classical and molecular genetics could provide a new way to examine how specific gene products affect the structure and function of a protein of interest.

The main disadvantage of *in vitro* translation extracts from yeast is that they are not commercially available, and there is considerably less experience in generating and using them. At this time, it is difficult to assess whether yeast extracts are equally suitable for efficient synthesis of a wide range of proteins. For making yeast extracts, it is probably advantageous to utilize *pept4* strains that lack the major proteases.[25] Another disadvantage of using a homologous translation system is that the extract itself will probably contain the protein of interest (unless the extract is generated from the appropriate mutant strain); this might complicate the interpretation of an experiment. Also, the presence of auxiliary proteins in the extract might make it difficult to determine if the protein of interest is directly or indirectly involved in a particular process.

Applications

Analysis of DNA-Binding Proteins

Reverse biochemistry is often used for determining whether a cloned gene encodes a specific DNA-binding protein. For this purpose, DNA-binding activity can be detected by incubating the labeled protein with appropriate DNA fragments, then separating the protein–DNA complexes from unbound protein and unbound DNA by electrophoresis in native acrylamide gels.[4] This DNA-binding assay is essentially the reverse of the standard mobility shift assay in which unlabeled proteins are examined for their ability to retard the mobility of a ^{32}P-labeled DNA fragment.[26-28] However, the "reverse assay" has the advantage that the fate of the ^{35}S-labeled protein is followed directly, thus making it possible to examine individual DNA fragments for both specific and nonspecific binding.

DNA fragments to be tested are prepared by restriction endonuclease cleavage of plasmid DNA followed by phenol extraction. Such fragments can be tested simultaneously, or they can be purified individually by

[25] E. W. Jones, *Annu. Rev. Genet.* **18**, 233 (1984).
[26] M. Garner and A. Revzin, *Nucleic Acids Res.* **9**, 3047 (1981).
[27] M. Fried and D. Crothers, *Nucleic Acids Res.* **9**, 6505 (1981).
[28] F. Strauss and A. Varshavsky, *Cell (Cambridge, Mass.)* **37**, 889 (1984).

agarose or acrylamide gel electrophoresis. Ideally, the specific binding site should reside on a fragment with a length between 50 and 500 bp and which differs from that of any other fragment generated by restriction cleavage. A typical DNA-binding reaction is carried out in a volume of 15 μl and contains the components indicated below. Of course, in order to interpret the results, it is essential to carry out parallel control reactions that contain no DNA and that contain nonspecific DNA lacking a binding site.

Reaction Mix

5 μl Water
3 μl 5× Binding buffer (100 mM Tris, pH 7.4; 250 mM KCl; 15 mM MgCl$_2$; 5 mM EDTA; 500 μg/ml gelatin)
5 μl DNA [DNA fragments each at 9 nM; equivalent to 0.5 μg of a 5-kilobase (kb) molecule]
1 μl 10 mg/ml Double-stranded poly(dI–dC)
1 μl ^{35}S-Labeled protein

After incubation at room temperature for 20 min, 5 μl of loading buffer (1× binding buffer containing 20% glycerol, 1 mg/ml each of xylene cyanol FF and bromphenol blue) is added, and the resulting mixture is immediately loaded on a 5% nondenaturing polyacrylamide gel. The polyacrylamide gel can either be of standard composition (30:0.8 acrylamide and bisacrylamide in 90 mM Tris–borate buffer, pH 8.3) or of low percentage and low ionic strength (40:0.5 acrylamide and bisacrylamide in 10 mM Tris–acetate buffer). Electrophoresis is performed under conditions in which the gel does not heat up above room temperature and until the bromphenol blue is near the bottom of the gel (a few hours). After electrophoresis, the gel is fixed in a solution of 45% ethanol, 10% acetic acid for 1 hr at room temperature, treated with En^{3}Hance (NEN–Du Pont, Boston, MA) for 1 hr, dried down, and autoradiographed.

As is the case with the conventional band-shift assay, the conditions for carrying out the binding reaction and electrophoresis can have an enormous impact on whether a particular protein–DNA interaction can be detected. The relative amounts of radiolabeled protein, specific DNA fragment, nonspecific DNA fragments, and bulk carrier DNA are particularly important because the presence of labeled protein means that both specific and nonspecific protein–DNA complexes are observable. Other important parameters that can be varied include the concentration of salt and divalent cations, temperature, the composition of the gel, and the ionic strength of the gel buffer.

The mobility of an individual protein depends on its charge-to-mass ratio, a property that varies greatly among proteins and that is strongly affected by pH. Thus, the band corresponding to free protein can appear anywhere on the gel, and its location is strongly affected by the precise gel conditions. Under normal electrophoresis conditions, many radiolabeled proteins will migrate in the wrong direction owing to their positive charge at the pH of gel. This unpredictability emphasizes the necessity for proper controls to distinguish between bands corresponding to free protein and those corresponding to protein-DNA complexes (the unbound DNA is unlabeled; hence its mobility is irrelevant).

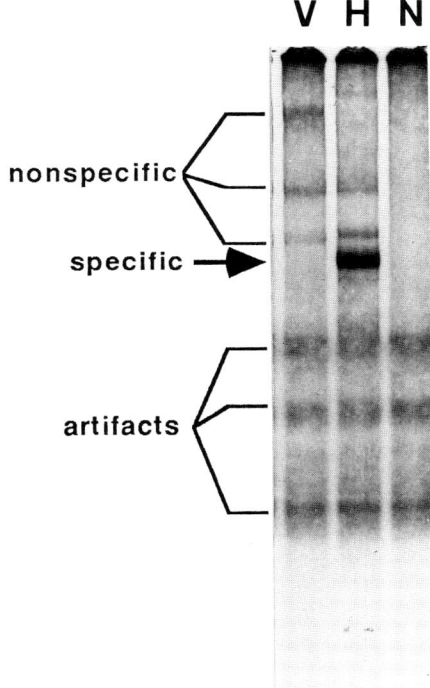

FIG. 3. Analysis of DNA-binding activity by a reverse mobility-shift assay. ^{35}S-Labeled protein was incubated in the presence of restriction fragments of plasmid DNAs that did (H) or did not (V) contain a specific binding site or in the absence of any DNA (N), and the resulting reaction products were subjected to electrophoresis in native acrylamide gels. Fragments containing specific binding sites generate dark bands (arrow), whereas fragments lacking such sites generate light bands owing to low-affinity nonspecific binding; the distinct mobilities of these bands reflect differences in the sizes of the DNA fragments. Artifact bands observed even in the absence of DNA are probably due to aberrant *in vitro* translation products.

Protein-DNA complexes are detected by the presence of new bands that do not appear in the control reaction containing protein alone (Fig. 3). Complexes involving a specific binding site should have significantly higher band intensities than those containing mutant binding sites or unrelated DNA sequences. Nonspecific, low-affinity interactions with DNA are characteristic of specific DNA-binding proteins and are observed as low-intensity bands. In considering the differences in band intensities between specific and nonspecific complexes, it is important to remember that a 400-bp DNA fragment containing a single specific binding site will have nearly 400 nonspecific binding sites. For such fragments, band intensities differing by a factor of 5 can reflect relative binding affinities that differ by a factor of 2000. The distinction between specific and nonspecific interactions can be enhanced by competition experiments in which the protein is incubated with a mixture of different sized DNA fragments.

The mobilities of protein–DNA complexes are influenced by a number of factors in addition to the electrophoretic conditions. First, complexes to larger DNA fragments migrate more slowly than complexes to smaller fragments, thus making it possible to simultaneously assay different sized DNA fragments.[4] Second, the mobility is affected by the amount of carrier DNA such as poly(dI–dC) in the reaction (C. R. Wobbe and K. Struhl, unpublished observations). In the absence of carrier DNA, the complex migrates very slowly because it contains nonspecific DNA-binding proteins from the translation extract in addition to the ^{35}S-labeled protein. As these nonspecific DNA-binding proteins are displaced by competition from the target fragment by increasing concentrations of carrier DNA, the protein–DNA complex migrates further in the gel. Third, the mobility of a protein–DNA complex is strongly influenced by the molecular weight of the protein component.[3,14] In general, complexes involving larger proteins migrate more slowly, although other factors such as shape and charge of the protein might also be involved.

Although the reverse band-shift assay just described is convenient, other methods for analyzing protein–DNA interactions can be applied to proteins synthesized *in vitro*. For example, DNA-binding activity can be assayed by the standard mobility shift assay using ^{32}P-labeled DNA,[26–28] in which case the protein does not have to be radiolabeled. Alternatively, protein–DNA complexes can be detected by immunoprecipitation.[29,30] *In vitro* synthesized proteins are also suitable for higher resolution analyses such as DNase I footprinting or methylation interference.

---

[29] V. Giguere, S. M. Hollenberg, M. G. Rosenfeld, and R. M. Evans, *Cell (Cambridge, Mass.)* **46,** 645 (1986).

[30] A. D. Johnson and I. Herskowitz, *Cell (Cambridge, Mass.)* **42,** 237 (1985).

Mapping Functional and Structural Domains

The region(s) of a protein necessary and sufficient for function can be mapped by generating appropriate deletions of the original DNA template, synthesizing the encoded mutant proteins, and testing them for the property of interest.[3,4] Initial mapping experiments typically involve a series of amino- or carboxyl-terminal deletions of the protein. However, proteins containing internal deletions or single amino acid substitutions can easily be tested, as can chimeric proteins. Since the amount and quality of the radiopure proteins can be easily assessed by SDS–PAGE, the failure of a mutant protein to carry out a particular function is not due to trivial reasons such as degradation. This approach has been used to localize protein regions required for DNA binding,[3,4] transcriptional activation *in vitro*,[31] ligand interactions,[29] and protein–protein associations.[8,9] However, it should be generally applicable to any functional property for which there is an appropriate assay. In general, it is unnecessary to purify the radiolabeled protein away from the vast excess of unlabeled proteins in the translation extract.

Because *in vitro* synthesized proteins are radiopure, they are well suited for mapping structural domains by partial proteolytic cleavage.[13] Unstructured regions in a protein are cleaved much more readily than independent structural domains.[32-34] In a typical experiment, a series of 10-μl reactions are performed in which a constant amount of radiolabeled protein is incubated with varying amounts of a given protease under appropriate conditions for 20 min.[13] Initially, it is useful to vary the protease concentration over a wide range (3–6 orders of magnitude) because a structurally informative amount will depend greatly on the specific protease and on the experimental conditions (a very rough average of a useful protease concentration is 10^{-4} units/μl of ^{35}S-labeled protein). The digestion products are then diluted with 20 μl of SDS sample buffer containing 500 μg/ml bovine serum albumin, boiled for 5 min, and examined by SDS–PAGE and autoradiography.

Protease cleavage sites can be mapped by cleaving a series of terminally deleted proteins.[13] For example, a band corresponding to the carboxyl-terminal region of the protein will be observed in amino-terminally deleted derivatives, whereas the band corresponding to the amino terminus will

---

[31] K. E. Vrana, M. E. A. Churchill, T. D. Tullius, and D. D. Brown, *Mol. Cell. Biol.* **8,** 1684 (1988).
[32] R. R. Porter, *Biochem. J.* **73,** 119 (1959).
[33] H. Jacobsen, H. Klenow, and K. Overgaard-Hansen, *Eur. J. Biochem.* **45,** 623 (1974).
[34] C. O. Pabo, R. T. Sauer, J. M. Sturtevant, and M. Ptashne, *Proc. Natl. Acad. Sci. U.S.A.* **76,** 1608 (1979).

become progressively shorter (Fig. 4). Moreover, the presence of a given band in a set of proteolytically cleaved deletion proteins is indicative of the existence and boundaries of an independent structural domain. Further structural information can be obtained by using appropriate deletion and point mutants and a variety of proteases and by varying the cleavage conditions (e.g., temperature, ionic strength, concentration of denaturants). In interpreting the results, it is important to remember that protein fragments must contain at least one methionine residue in order to be visualized and that relative band intensities must be normalized to the number of methionines in each fragment.

Determining Subunit Structure

The availability of truncated but functional proteins can be useful for determining the subunit structure of a protein. Specifically, two derivatives of a given protein that differ in size are cosynthesized by carrying out *in vitro* transcription and translation starting from an equimolar mixture of DNA templates.[14] If, for example, the protein is a dimer, the cosynthesized mixture will generate three distinct species in a 1:2:1 molar ratio, with the intermediate species representing a heterodimer. In such an experiment, monomer proteins will generate two equimolar species, whereas tetramers will yield five species in a 1:4:6:4:1 molar ratio.

The principle of using different sized proteins to determine stoichiometry can be applied to a variety of methods for detecting the protein species

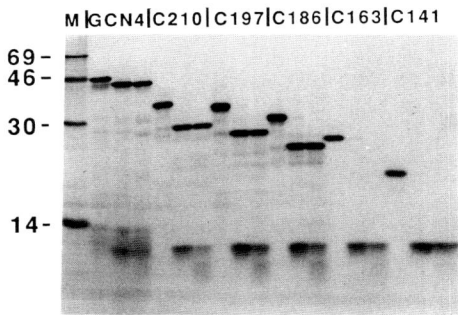

FIG. 4. Analysis of protein structure by partial proteolysis. Full-length GCN4 protein or derivatives containing the indicated number of carboxyl-terminal amino acids were incubated with 0, 10^{-4}, or 10^{-3} units of chymotrypsin for 30 min at 37° and analyzed by SDS-PAGE [molecular weight markers ($\times 10^{-3}$) indicated at left]. The arrow represents a stable proteolytic fragment that is present in all the derivatives and hence maps to the carboxyl terminus; the amino-terminal proteolytic fragments of the different derivatives vary in molecular weight.

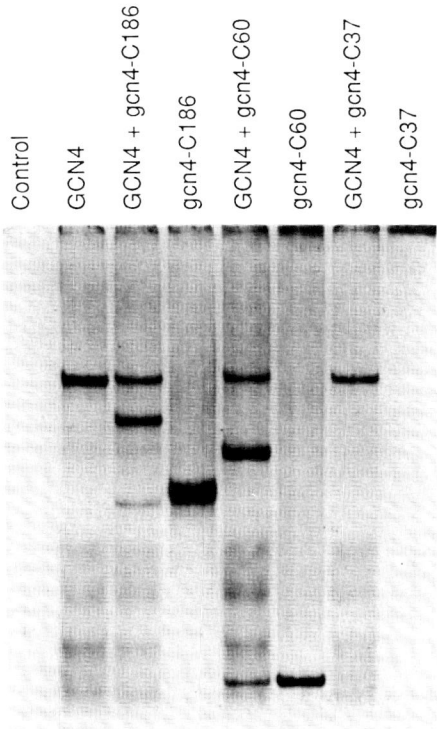

FIG. 5. Determining subunit structure by cosynthesis of differently sized proteins. Wild-type GCN4 protein and derivatives containing the indicated number of carboxyl-terminal amino acids were synthesized individually or in combination, incubated with a DNA fragment containing a specific binding site, and analyzed by electrophoresis in a native acrylamide gel (see Fig. 2). Complexes dependent on the individual proteins have different electrophoretic mobilities owing to their different molecular weights. New complexes with intermediate mobilities reflect heteromeric species; the existence of three bands in a 1:2:1 molar ratio indicates that the protein binds DNA as a dimer. (From Ref. 14.)

of interest. In the case of the reverse band-shift DNA-binding assay described above, protein–DNA complexes involving different-sized protein derivatives have different electrophoretic mobilities; thus, homo- or heteromeric complexes can be distinguished easily[14] (Fig. 5). A more general method, however, would be glutaraldehyde cross-linking followed by SDS gel electrophoresis to examine the protein species.[35]

[35] A. J. Joachimiak, R. L. Kelley, R. P. Gunsalus, and C. Yanofsky, *Proc. Natl. Acad. Sci. U.S.A.* **80**, 668 (1983).

[37] In Vitro Protein Synthesis

By MICHAEL J. LEIBOWITZ, FRANCIS P. BARBONE, and DENISE E. GEORGOPOULOS

Introduction

Translation in a cell-free system directly determines the ability of a messenger RNA (mRNA) molecule to encode a protein product. Although *Saccharomyces cerevisiae* has been widely used as an experimental organism in molecular genetics and biochemistry, extracts from other eukaryotes have been much more popular for cell-free translation studies. The yeast translation system described here is an alternative to the more widely used systems derived from wheat germ,[1] rabbit reticulocytes,[2] and ascites tumor cells.[3] Commercial availability of extracts from the first two sources makes them accessible to all laboratories. However, the relative genetic tractability of *S. cerevisiae* makes it an ideal eukaryote for the study of translational regulatory mechanisms, which have been well characterized in yeast at the level of the whole organism.[4] The translational regulation of gene expression has been extensively studied in the rabbit reticulocyte lysate system. However, the unique regulatory mechanisms in this differentiated anucleate cell are probably quite different from those present in most other cell types.[5] Molecular characterization of the genes encoding ribosomal RNA, ribosomal proteins (as reviewed in Ref. 6), and factors involved in translational initiation[7,8] and elongation[9] in yeast make this an exciting system in which to study the biochemistry of translation.

In addition to the system presented in this chapter, several yeast cell-free translation systems which are dependent on added messenger RNA (mRNA) have been described.[10,11] Tuite *et al.*[12] modified the method of

[1] G. E. Roberts and B. M. Paterson, *Proc. Natl. Acad. Sci. U.S.A.* **70,** 2330 (1973).
[2] H. R. B. Pelham and R. J. Jackson, *Eur. J. Biochem.* **67,** 247 (1976).
[3] M. B. Mathews and A. Korner, *Eur. J. Biochem.* **17,** 328 (1970).
[4] G. R. Fink, *Cell (Cambridge, Mass.)* **45,** 155 (1986).
[5] S. Ochoa and C. de Maro, *Annu. Rev. Biochem.* **48,** 549 (1979).
[6] J. Warner, *Microbiol. Rev.* **53,** 256 (1989).
[7] C. Keierleber, M. Wittekind, S. Qin, and C. S. McLaughlin, *Mol. Cell. Biol.* **6,** 4419 (1986).
[8] M. Altman, C. Handschin, and H. Trachsel, *Mol. Cell. Biol.* **7,** 998 (1987).
[9] F. Herrera, J. A. Martinez, N. Moreno, I. Sadnik, C. S. McLaughlin, B. Feinberg, and K. Moldave, *J. Biol. Chem.* **259,** 14347 (1984).
[10] M. F. Tuite and J. Plesset, *Yeast* **2,** 35 (1986).
[11] K. Moldave and E. Gasior, this series, Vol. 101, p. 644.
[12] M. F. Tuite, J. Plesset, K. Moldave, and C. S. McLaughlin, *J. Biol. Chem.* **255,** 8761 (1980).

Gasior et al.[13] to prepare lysates capable of mRNA-dependent protein synthesis from homogenized spheroplasts of Saccharomyces cerevisiae. A 27,000 g supernatant prepared from crude lysates was centrifuged at 100,000 g for 30 min to remove polysomes and an unknown particulate inhibitory factor which prevented translation initiation in vitro. Sephadex G-25 chromatography subsequently removed endogenous amino acid pools, and micrococcal nuclease treatment improved mRNA dependence by eliminating endogenous RNA. This method works best with extracts prepared from the prototrophic diploid strain SKQ2n, but it has also been successfully used with other strains.[10] Spheroplast preparation and lysis can be replaced by vigorous vortex lysis with glass beads after prolonged preincubation in 1 M sorbitol followed by medium containing 0.4 M $MgSO_4$,[10] with subsequent steps as described.[13] Although micrococcal nuclease treatment was used in the methods cited above, spheroplast extracts have been shown to initiate translation of exogenous RNA even without micrococcal nuclease treatment. However, in such extracts the background arising from endogenous mRNA was high, so that definitive identification of the translation product of the added RNA (human leukocyte interferon, in this case) required immunodetection.[14] Another method based on spheroplast lysis but without ultracentrifugation, gel filtration, or micrococcal nuclease digestion was also reported to yield extracts with considerable dependence on added mRNA.[15]

Hofbauer et al.[16] described a cell extract prepared by gentle glass bead lysis requiring neither prolonged pretreatment before lysis nor ultracentrifugation. This chapter is based on our previous modification of this method.[17] This method produces active extracts from various haploid and diploid yeast strains which are stable frozen at $-70°$. We prefer this method for its reproducibility and ease of preparation. The results obtained with it indicate properties similar to those described for the other systems.

Materials

Strains. The haploid nonkiller yeast strain M654 (**a** *trp5 arg4 his4 lys1 ade2 met1 leu2 gal2*) is used as a source of translation extract in all experiments described here, but comparable results have been obtained with various auxotrophic and prototrophic haploid and diploid strains.

[13] E. Gasior, F. Herrera, I. Sadnik, C. S. McLaughlin, and K. Moldave, *J. Biol. Chem.* **254**, 3965 (1979).
[14] P. K. Chanda and H. H. Kung, *Proc. Natl. Acad. Sci. U.S.A.* **80**, 2569 (1983).
[15] E. Szczesna and W. Filipowicz, *Biochem. Biophys. Res. Commun.* **92**, 563 (1980).
[16] R. Hofbauer, F. Fessl, B. Hamilton, and H. Ruis, *Eur. J. Biochem.* **122**, 199 (1982).
[17] I. Hussain and M. J. Leibowitz, *Gene* **46**, 13 (1986).

The diploid prototrophic type 1 killer strain A364A × S7 (a/α *ade1/ + ade2/ + ura1/ + tyr1/ + his7/ + lys2/ + gal1/gal1* [KIL-k1]) is used as a source of poly(A)$^+$ RNA and killer virions.

Media and Cells. Cultures for isolation of killer virions are grown to stationary phase (5 days) in 1% (w/v) yeast extract (Difco), 2% (w/v) peptone (Difco), and 5% (v/v) ethanol. Cell-free extracts for translation are prepared from logarithmically growing cells ($A_{650\ nm}$ 1.0) in 1% (w/v) yeast extract, 1% (w/v) peptone, and 2% (w/v) dextrose. For adenine auxotrophs, adenine sulfate dihydrate (0.4 g/liter) should be added to media. Aerobic growth is in Erlenmeyer flasks (less than half full) shaken at 150 revolutions/min at 28°.

Chemicals and Enzymes. Creatine kinase from Sigma (St. Louis, MO) is stored in 30 mM HEPES–KOH, pH 7.4, containing 50% glycerol at −20°. [^{35}S]Methionine (~800 Ci/mmol), "translation grade" (shipped in 50 mM Tricine, pH 7.4, 10 mM 2-mercaptoethanol, under nitrogen), is from New England Nuclear (Boston, MA); Sephadex G-10 from Pharmacia (Piscataway, NJ); and micrococcal nuclease from Worthington Biochemicals (Freehold, NJ). Placental RNase inhibitor (RNasin) is from Promega Biotech (Madison, WI). All nucleotides are from P-L Biochemicals (Pharmacia). Oligo(dT)-cellulose, type 3, is from Collaborative Research (Bedford, MA).

All laboratory ware and solutions are sterilized by autoclaving or ultrafiltration.

RNA. Total poly(A)$^+$ RNA is prepared from yeast cells as previously described.[17,18] Killer virions are purified as described[19] from strain A364A × S7 and are used to catalyze synthesis of m transcript derived from the M double-stranded (ds) RNA segment of the viral genome; m transcript is purified by oligo(dT)-cellulose chromatography.[18] Sindbis virus 26 S mRNA is purified from infected primary chick embryo fibroblasts as described.[20] The killer virus m transcript prepared *in vitro* appears to be uncapped,[21] whereas yeast mRNAs in the poly(A)$^+$ RNA preparation[22] and Sindbis virus mRNA synthesized in vertebrate cells[23] have "type 0" caps, with the structure m^{7}G(5')pppA or m^{7}G(5')pppG. Sindbis virus intracellular RNA species also contain some caps with the 5'-terminal hypermethylated bases m$_2^{2,7}$G and m$_3^{2,2,7}$G.[23]

[18] E. M. Hannig, D. J. Thiele, and M. J. Leibowitz, *Mol. Cell. Biol.* **4**, 101 (1984).
[19] D. J. Thiele, E. M. Hannig, and M. J. Leibowitz, *Mol. Cell. Biol.* **4**, 92 (1984).
[20] V. Stollar, T. E. Shenk, and B. D. Stollar, *Virology* **47**, 122 (1972).
[21] J. Bruenn, L. Bobek, V. Brennan, and W. Held, *Nucleic Acids Res.* **8**, 2985 (1980).
[22] C. E. Sripati, Y. Groner, and J. R. Warner, *J. Biol. Chem.* **251**, 2898 (1976).
[23] D. T. Dubin, V. Stollar, C.-C. HsuChen, K. Timko, and G. M. Guild, *Virology* **77**, 457 (1977).

Methods

Preparation of Translation Extract

Cultures are chilled in an ice-water bath, and cells are collected by centrifugation in a GSA rotor in a Sorvall–Du Pont RC-5B centrifuge (5000 rpm, 5 min, 4°). Cells are washed twice with water and twice with homogenization buffer (8.5% mannitol, 30 mM HEPES–KOH, pH 7.4, 100 mM potassium acetate, 2 mM magnesium acetate, 2 mM dithiothreitol) and resuspended in homogenization buffer (2–3 ml/g of cells) containing 0.5 mM phenylmethylsulfonyl fluoride (PMSF). Phenylmethylsulfonyl fluoride enhances translational activity of the extracts, presumably by inactivating proteases. Cells are broken using ice-cold acid-washed glass beads (0.45 mm diameter, 3–4 g/ml of cell suspension) in a capped centrifuge tube (35 ml) as described.[24] The tube is shaken by hand in a vertical motion through a 50-cm path in a 4° cold room at a frequency of 2 cycles/sec for three 1-min periods of breakage with 1-min chillings on ice in between. This method releases approximately the same amount of protein per gram of cells as does more violent breakage by vortexing with glass beads or lysis in a French pressure cell, which yield extracts failing to show mRNA dependence.

Glass beads and unbroken cells are removed by centrifugation at 30,000 g for 15 min. The supernatant is chromatographed on a Sephadex G-10 column (45 × 1.1 cm) equilibrated with homogenization buffer without mannitol to remove endogenous amino acids and other small molecules. Excluded fractions with the highest $A_{260 \text{ nm}}$ values are pooled; the resulting solution routinely has a protein concentration of about 5–10 mg/ml assayed as described.[25] This spectrophotometric protein assay is based on the proportionality of protein concentration to the difference in absorbance at 233 and 224 nm; contaminating nucleic acids and reducing agents do not interfere. For storage, multiple aliquots of extracts taken either before or after the Sephadex column are flushed with nitrogen gas, quick-frozen in dry ice/ethanol, and stored at −70° in snap-top microcentrifuge tubes. Extracts are stable for at least 1 year. Repeated freezing and thawing are avoided.

Micrococcal Nuclease Titration

Immediately prior to use, extracts are thawed on ice and treated with micrococcal nuclease (2–10 μg/ml) in the presence of 1 mM CaCl$_2$ at 20°

---

[24] B. Lang, G. Burger, I. Doxiadis, D. Y. Thomas, W. Bandlow, and F. Kaudewitz, *Anal. Biochem.* **77**, 110 (1977).

[25] W. E. Groves, F. C. Davis, Jr., and B. M. Sells, *Anal. Biochem.* **22**, 195 (1968).

for 2-10 min. Reactions are terminated by adding EGTA, pH 7.4, to 2.3 mM, after which translation is assayed in the presence and absence of added poly(A)$^+$ RNA in order to determine the nuclease concentration and digestion time yielding optimum message dependence. These titrations are required with each new extract and lot of nuclease. Nuclease-treated extracts are used within 5-10 min. Freshly nuclease-treated extracts are more active than extracts stored at $-70°$ after nuclease digestion.

Translation Reaction

Assay Conditions. Translation reactions are run containing the following components, including those contributed by the cell-free extract: 50 μM of each of 19 nonradioactive amino acids (excluding methionine), 12.5-18.5 μCi of [^{35}S]methionine (~800 Ci/mmol), 56 mM HEPES-KOH, pH 7.4, 160 mM potassium acetate, 1.2 mM magnesium acetate, 3.9 mM dithiothreitol, 0.5 mM ATP, 0.1 mM GTP, 20 mM creatine phosphate, 0.5 mM each of spermidine and putrescine hydrochlorides, 2 mM glucose 6-phosphate, yeast tRNA (40 μg/ml), creatine kinase (20 μg/ml), 0.6 mM CaCl$_2$ and 1.4 mM EGTA (both unavoidably contributed by the micrococcal nuclease-treated extract), placental ribonuclease inhibitor (200 units/ml), translational extract (~6 mg/ml), and exogenous mRNA. The indicated concentrations of potassium and magnesium are optimal for translation of total yeast poly(A)$^+$ RNA. The optima may vary with different RNA templates. Concentrated stock solutions used in the reaction are routinely neutralized to keep the final pH at about 7.4. Incubations are routinely at 20° for 120 min, with shorter incubations used to determine the rate of protein synthesis.

Quantitation of Translation. Active templates such as yeast poly(A)$^+$ RNA[17] or capped mRNA species stimulate translation sufficiently to easily quantitate protein synthesis by incorporation of [^{35}S]methionine or other radioactive amino acids into trichloroacetic acid (TCA)-precipitable material. Reactions are stopped by spotting aliquots (5 μl of reaction) onto Whatman 3 MM filter disks, and hot TCA-insoluble radioactivity is determined by a modification of the method of Mans and Novelli.[26] Disks are immersed in 10% (weight/volume) TCA containing 0.1 M methionine at 0°, rinsed once with 5% TCA, boiled in 5% TCA for 10 min, and then washed 3 times each with 5% TCA (10 min/wash) and with ethanol (5 min/wash). Disks are air dried, and radioactivity is determined by liquid scintillation spectrometry. Since low molecular weight polypeptides may be partially soluble in hot TCA, translation may alternatively be quantitated by stopping reactions by digestion with pancreatic ribonuclease (200

---

[26] R. J. Mans and G. D. Novelli, *Biochem. Biophys. Res. Commun.* **3**, 540 (1960).

μg/ml) at 37° for 15 min followed by TCA precipitation and washing at 0°.

In order to quantitate incorporation programmed by less active templates, such as uncapped m transcript of killer virus of yeast,[17] or to determine incorporation of amino acids into specific polypeptide products, reactions may be stopped by quick freezing on dry ice or by digestion with pancreatic ribonuclease as above and the proteins analyzed by electrophoresis on 12.5% (w/v) polyacrylamide–sodium dodecyl sulfate (SDS) slab gels with 5% polyacrylamide stackers.[27] After electrophoresis, the gel is fixed in 40% (v/v) methanol, 10% (v/v) acetic acid, 5% (v/v) glycerol for at least 1 hr (can be left overnight), rinsed 3 times with deionized water, and dried under reduced pressure with heat (Hoefer slab gel drier). Radioactive proteins are visualized by autoradiography using XAR-5 film (Kodak) with intensifying screens (Cronex Lightning-Plus) at $-70°$.

We have found that this method requires less processing of the gel and results in sharper autoradiographic bands than does fluorography,[28] although fluorography may be more sensitive for rapid detection of very faint bands. Presumably, either method could be used to detect ^{14}C-labeled proteins, although ^{3}H-labeled proteins can only be detected in a reasonable time using fluorography. For gel analysis of ^{35}S-labeled peptides, ^{14}C-methylated polypeptide molecular weight markers (New England Nuclear) are used, including phosphorylase b (97,000), ovalbumin (46,000), carbonate dehydratase (30,000), lactoglobulin A (18,300), and cytochrome c (12,300). Following autoradiography, specific polypeptide products may be quantitated by excision from the dried gel and liquid scintillation spectrometry of specific bands, or by densitometric scanning of the autoradiogram.

As previously reported,[17] in some reactions with [^{35}S]methionine as a substrate, low levels of incorporation into a 48,000 polypeptide independent of added RNA were noted. This incorporation is apparently enzyme catalyzed and is reduced or eliminated by using "translation grade" [^{35}S]methionine.

Results

As previously reported,[17] the yeast system is active in the translation of exogenous capped mRNA molecules and is inhibited by the addition of various cap analogs, including m^{7}G(5')pppA and m^{7}G(5')pppG. Sindbis virus RNA from infected primary chick embryo fibroblasts programs the synthesis of the 32,000 capsid protein (gel not shown, but a comparable gel has been published[17]) in the yeast translation system. As shown in Fig. 1,

[27] V. Laemmli, *Nature (London)* **227**, 680 (1970).
[28] W. M. Bonner and R. A. Laskey, *Eur. J. Biochem.* **46**, 83 (1974).

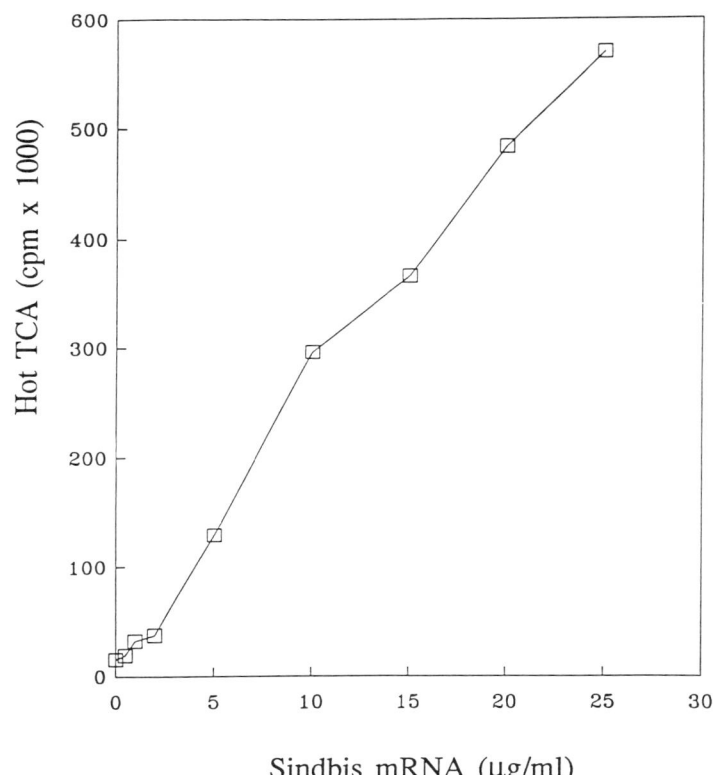

FIG. 1. Dose–response curve for Sindbis virus mRNA translation. The indicated amounts of Sindbis virus mRNA (containing low levels of contaminating ribosomal RNA) were added to the yeast translation system, and translation reactions were incubated for 2 hr at 20° as described in the text. Quantitation of protein synthesis by [^{35}S]methionine incorporation was assayed by determination of hot TCA-insoluble radioactivity on 5-μl aliquots. No incorporation above that seen in a TCA blank (a disk without sample subjected to the same TCA treatment) was observed in the absence of the translational extract.

this RNA preparation results in a dose-dependent stimulation of protein synthesis, which can be quantitated by measurement of TCA-precipitable radioactivity. This dose dependence may not be seen at higher concentrations of some mRNAs, possibly owing to inhibitory substances present as impurities in these preparations.

Figure 2 demonstrates that the rate of RNA-stimulated protein synthesis seen in this system declines beyond 2 hr, as seems to be the case in all

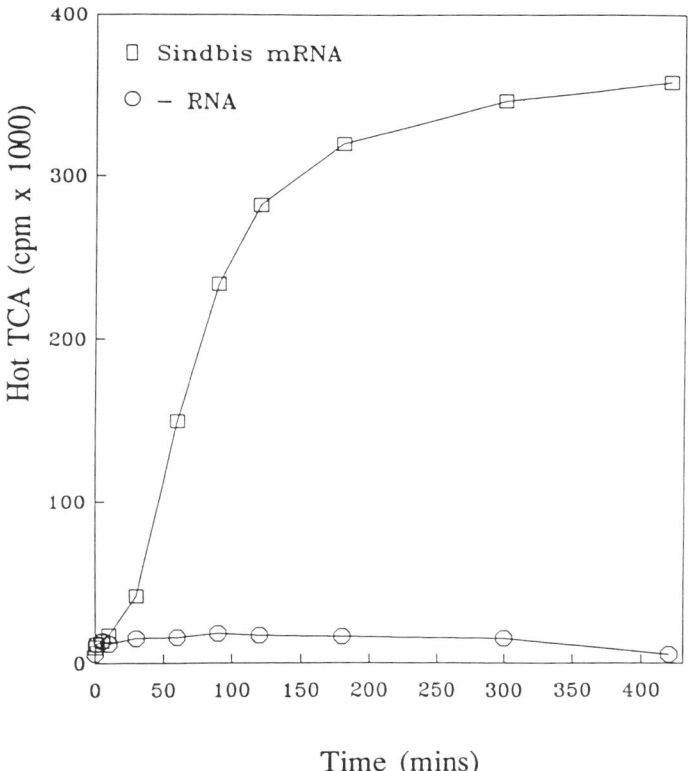

Time (mins)

FIG. 2. Time course for translation of Sindbis virus mRNA. Translation reactions were incubated for the indicated times, as in Fig. 1, in the presence of Sindbis virus mRNA (10 μg/ml). A control reaction without exogenous RNA is also shown.

other message-dependent translation systems. As described for the *Escherichia coli* and wheat germ systems, longer reactions could presumably be run using a continuous flow apparatus to deliver buffer and substrates (amino acids, ATP, and GTP) to the reaction.[29]

When a less active uncapped mRNA, such as the m transcript of killer virus of yeast, was added to the cell-free translation system, minimal or no stimulation of amino acid incorporation into TCA-precipitable material was detected. However, gel electrophoretic analysis of the reaction prod-

[29] A. S. Spirin, V. I. Baranov, L. A. Ryabova, S. Y. Ovodov, and Y. B. Alakhov, *Science* **242,** 1162 (1988).

ucts demonstrates that incorporation into a 32,000 polypeptide (preprotoxin) is programmed by this RNA in a dose-dependent fashion.

Discussion

The method of preparing cell-free translation extracts of yeast described here retains the advantages of previously described systems; it can be applied to many strains of yeast and is reproducible and simple. Application of this method to cells with genetic alterations in the components of the translational machinery should prove extremely useful for investigation of the mechanism and regulation of protein synthesis. In yeast, both classical and recombinant genetics can be used to determine the structure–function relationship of gene products making up the translational apparatus, unlike the more widely used systems from metazoan eukaryotes. The extensive genetic and biochemical characterization of protein secretion in yeast[30-32] indicates the potential for developing an *in vitro* system coupling translation and secretion processes derived from yeast cells.

As discussed elsewhere in this volume,[33] the yeast transcriptional apparatus has been studied extensively. Although extracts of yeast cells support accurate initiation at RNA polymerase II promoters,[34] there has been relatively little attention paid to the *in vitro* synthesis and processing of mRNA for use as a template for translation *in vitro*, except in the killer virus system. The fact that killer virus is in the cytoplasmic compartment,[35] and thus may be expressed by a coupled transcription and translation process, suggests the possibility of directly coupling these two processes *in vitro*.[36] However, since nuclear gene transcripts must be processed and transported to the cytoplasm prior to translation, direct coupling of transcription and translation is not likely to be a physiological process. Although DNA-dependent transcription and translation are not coupled *in vivo*, these reactions could presumably be coupled *in vitro* and might be useful in analyzing the ability of different transcripts of a single gene to compete as translational templates. Such a coupled system might also be

[30] D. T. Moir and L. S. Davidow, this volume [34].
[31] A. Franzusoff, J. Rothblatt, and R. Schekman, this volume [45].
[32] P. D. Garcia, W. Hansen, and P. Walter, this volume [46].
[33] N. F. Lue, P. M. Flanagan, R. J. Kelleher III, A. M. Edwards, and R. D. Kornberg, this volume [38].
[34] N. F. Lue and R. D. Kornberg, *Proc. Natl. Acad. Sci. U.S.A.* **84,** 8839 (1987).
[35] M. Dihanich, E. van Tuinen, J. D. Lambris, and B. Marshallsay, *Mol. Cell. Biol.* **9,** 1100 (1989).
[36] F. P. Barbone and M. J. Leibowitz, unpublished results (1990).

useful for directly comparing the coding capacities of different DNA fragments or recombinant plasmids.

Acknowledgments

We thank Dr. V. Stollar for providing Sindbis virus and primary chick embryo fibroblasts for preparation of Sindbis virus RNA for these experiments. This work was supported in part by Army Research Office Grant 24931-LS and by the New Jersey Center for Advanced Biotechnology and Medicine.

[38] RNA Polymerase II Transcription *in Vitro*

By NEAL F. LUE, PETER M. FLANAGAN, RAYMOND J. KELLEHER III, ALED M. EDWARDS, and ROGER D. KORNBERG

Introduction

The first observation of accurately initiated transcription by RNA polymerase II in yeast extracts[1] depended on the following: (1) The preparation of nuclear (rather than whole cell) extracts, from nuclei isolated in the presence of a high polymer. Nuclear components are concentrated and inhibitors are removed during the isolation of nuclei. A polymer prevents the loss of transcription factors from the nuclei, presumably through a macromolecular "crowding" effect.[2] (2) The use of acetate rather than chloride salts in the transcription reaction. Chloride inhibits initiation by yeast RNA polymerase II, whereas it is the major anion in other transcription systems. (3) An assay procedure in which specific transcripts are enriched before detection. Enrichment is essential because of a high background of extraneous transcription in crude extracts.

Yeast nuclear extracts that support initiation by RNA polymerase II also exhibit responses to a number of gene activator proteins,[3-5] and these extracts may show effects of negative regulators as well. The extracts may prove useful for studies of other nuclear transactions, such as RNA polymerase I and III transcription, mRNA splicing and polyadenylation, and

[1] N. F. Lue and R. D. Kornberg, *Proc. Natl. Acad. Sci. U.S.A.* **84,** 8839 (1987).
[2] A. P. Minton, *Mol. Cell. Biochem.* **55,** 119 (1983).
[3] A. R. Buchman, N. F. Lue, and R. D. Kornberg, *Mol. Cell. Biol.* **8,** 5086 (1988).
[4] N. F. Lue, A. R. Buchman, and R. D. Kornberg, *Proc. Natl. Acad. Sci. U.S.A.* **86,** 486 (1989).
[5] D. I. Chasman, J. Leatherwood, M. Carey, M. Ptashne, and R. D. Kornberg, *Mol. Cell. Biol.* **9,** 4746 (1989).

DNA replication. We therefore devote most attention in this chapter to the method of preparing nuclear extracts and associated concerns. We describe one of several assay procedures that provides the simplest and most sensitive way of determining the quality of an extract for initiation by RNA polymerase II; methods of assessing other aspects, such as the precision of initiation and the dependence on sequences surrounding the start site, may be found elsewhere.[1,6]

Preparation of Nuclear Extracts

Yeast nuclei are isolated following lysis of spheroplasts, produced by degradation of cell walls with crude yeast lytic enzyme. Although procedures are available for the purification of yeast nuclei[7] (see also [53] in this volume), we have preferred to manipulate the isolated nuclei as little as possible, to avoid losing or damaging nuclear components. We simply centrifuge lysed spheroplasts to separate nuclear and cytosolic fractions. The nuclear pellet is treated with 0.5 M ammonium sulfate, to help dissociate transcription factors from DNA, and soluble proteins are concentrated by precipitation with ammonium sulfate. The precipitate is dissolved and dialyzed, and the resulting nuclear extract is stored at $-196°$.

A major concern throughout the preparation and subsequent use of yeast nuclear extract is with damage caused by proteolysis. Although we have prepared extracts from a protease-deficient strain of yeast, the inclusion of protease inhibitors in all solutions is probably more important. It is also essential to work in the cold, as rapidly as possible. Ironically, the main source of proteolytic activity in the procedure described here is the yeast lytic enzyme added to convert the cells to spheroplasts. This necessitates extensive washing of the spheroplasts prior to lysis. The use of purified glucanases[8] instead of crude lytic enzyme would presumably alleviate the problem.

We have varied parameters of the procedure such as the density of the cell culture (A_{600} values between 2 and 8) and the concentrations of ammonium sulfate used to extract (0.5–0.9 M) and precipitate (50–75% of saturation) nuclear proteins, with little effect on activity in the initiation of RNA polymerase II transcription. Such variation may be important when extracts are prepared for study of other activities.

[6] N. F. Lue, P. M. Flanagan, K. Sugimoto, and R. D. Kornberg, *Science* **246**, 661 (1989).
[7] L. D. Schultz, *Biochemistry* **17**, 750 (1978).
[8] J. H. Scott and R. Schekman, *J. Bacteriol.* **142**, 414 (1980).

Procedure

A culture (3 liters) of *Saccharomyces cerevisiae* strain BJ926 (α/a *trp1/+ prc1 − 126/prc1 − 126 pep4 − 3/pep4 − 3 prb1 − 1122/prb1 − 1122 can1/can1;* from Dr. E. Jones, Carnegie-Mellon University, Pittsburgh, PA) in YPD [1% yeast extract/2% Bacto-peptone/2% glucose (w/v)] is grown at 30° to an A_{600} value of 6–8 (measured at a dilution of 1:40 in a Hewlett-Packard 8451A diode array spectrophotometer). The cells are harvested by centrifugation in a Beckman JA-10 rotor at 6000 rpm for 10 min at 4°. The cell pellet (about 30 g) is suspended in 200 ml of 50 mM Tris-HCl, pH 7.5/30 mM dithiothreitol, shaken slowly (about 20 rpm) for 15 min at 30°, centrifuged as before, and resuspended in 30 ml of YPD containing 1 M sorbitol. Solid (40 mg) yeast lytic enzyme (100,000 units/mg, ICN Biochemicals, Costa Mesa, CA) is added, and the suspension is shaken slowly at 30° until the A_{600} on dilution 1:100 in 1% sodium dodecyl sulfate (SDS) is less than 10% of the starting value. This takes about 60 min. Digestion is stopped by the addition of 300 ml of ice-cold YPD/1 M sorbitol, and the spheroplasts are collected by centrifugation in a Beckman JA-10 rotor at 5000 rpm for 5 min at 4°. The spheroplasts are washed 3 times in 300 ml of ice-cold YPD/1 M sorbitol, suspended in 800 ml of YPD/1 M sorbitol, and shaken slowly for 30 min at 30° to allow recovery from the physiological effects of the previous treatments.

The spheroplasts are collected and washed once in 300 ml of 1 M sorbitol, then suspended in 200 ml of 18% (w/v) Ficoll 400 (Pharmacia, Piscataway, NJ) /10 mM Tris-Cl, pH 7.5/20 mM KCl/5 mM MgCl$_2$/3 mM dithiothreitol/1 mM EDTA/0.5 mM spermidine/0.15 mM spermine plus protease inhibitors (1 mM phenylmethylsulfonyl fluoride, 2 μM pepstatin A, 0.6 μM leupeptin, 2 μg/ml chymostatin, and 2 mM benzamidine). Lysis is effected with 2 strokes of a motor-driven Teflon/glass homogenizer (Potter–Elvehjem tissue grinder, 55 ml capacity, Wheaton, Millville, NJ) at 400 rpm. Unlysed spheroplasts and cell debris are removed by centrifugation in a Beckman JA-10 rotor twice at 6000 rpm for 10 min and then 3 times at 5000 rpm for 5 min at 4°.

Nuclei are recovered from the final supernatant by centrifugation in a Sorvall SS34 rotor at 13,000 rpm for 30 min at 4° and are resuspended in 15 ml of 100 mM Tris–acetate, pH 7.9/50 mM potassium acetate/10 mM MgSO$_4$/3 mM dithiothreitol/2 mM EDTA/20% glycerol plus protease inhibitors (as above). Ammonium sulfate (3 M, neutralized) is added over about 10 min with stirring to a final concentration of 0.5 M. The suspension is stirred for 30 min at 4° and centrifuged in a Beckman SW60 rotor at 40,000 rpm for 1 hr at 0°. The supernatant is adjusted to 69% of saturation with ammonium sulfate by the addition of the solid salt (0.35 g/ml), stirred for 20 min at 0°, and centrifuged in a Beckman SW60 rotor at 25,000 rpm for 20 min at 0°. The pellet is dissolved at a concentration of 20 mg of protein/ml (about 4 ml total volume; protein determined by the method of Bradford[9] with bovine serum albumin as standard) in 20 mM HEPES–KOH, pH 7.6/10 mM MgSO$_4$/5 mM dithiothreitol/10 mM EGTA/20% glycerol (w/v) plus protease inhibitors (as above) and dialyzed against the same buffer until the conductivity is less than 20 mS/cm. The resulting nuclear extract is stored in liquid nitrogen, and it can be thawed and refrozen many times.

Assay for Initiation by RNA Polymerase II

The most convenient assay for accurately initiated transcription employs a template with an initiation site and sequence downstream devoid of guanosine residues.[10] ^{32}P-Labeled products of transcription are digested with RNase T1, which cleaves after G residues, sparing only the transcripts of the G$^-$ region, which are revealed by gel electrophoresis and autoradiography (Fig. 1). This procedure eliminates a high background of extraneous RNA products which would otherwise obscure those of interest.

The particular templates we have used contain either a G$^-$ form of the yeast *CYC1* promoter or the adenoviral major late promoter in front of a 377-base pair (bp) G$^-$ sequence.[6] In both cases there are multiple specific transcription starts within the G$^-$ region, resulting in two major bands in the autoradiogram (Fig. 1). Occasionally an additional band is observed slightly above the major ones, which results from spurious initiation upstream and readthrough of the G$^-$ region. Inhibition by α-amanitin is indicative of transcription by RNA polymerase II and not polymerases I or III.

Most parameters of the assay have been optimized, but the amount of nuclear extract used should be varied. The dependence of the reaction on the amount of extract is nonlinear (Fig. 1), and the threshold for obtaining significant transcription differs from one preparation of extract to another.

Procedure

The reaction mixture (25 μl) contains 50 mM HEPES–KOH, pH 7.3/70 mM potassium acetate/5 mM magnesium acetate/3.3 mM MgSO$_4$/2.5 mM dithiothreitol/5 mM EGTA/4 mM phosphoenolpyruvate/6.6% glycerol/0.4 mM each of ATP and CTP/4 $\mu$$M$ UTP/20 μCi of [α–^{32}P]UTP (600 Ci/mmol)/0.2 unit of Inhibit-ACE (5 Prime–3 Prime, Inc., West Chester, PA)/0.3 μg of template [3 kilobases (kb) DNA, either linear or

---

[9] M. M. Bradford, *Anal. Biochem.* **72**, 248 (1976).
[10] M. Sawadogo and R. G. Roeder, *Proc. Natl. Acad. Sci. U.S.A.* **82**, 4394 (1985).

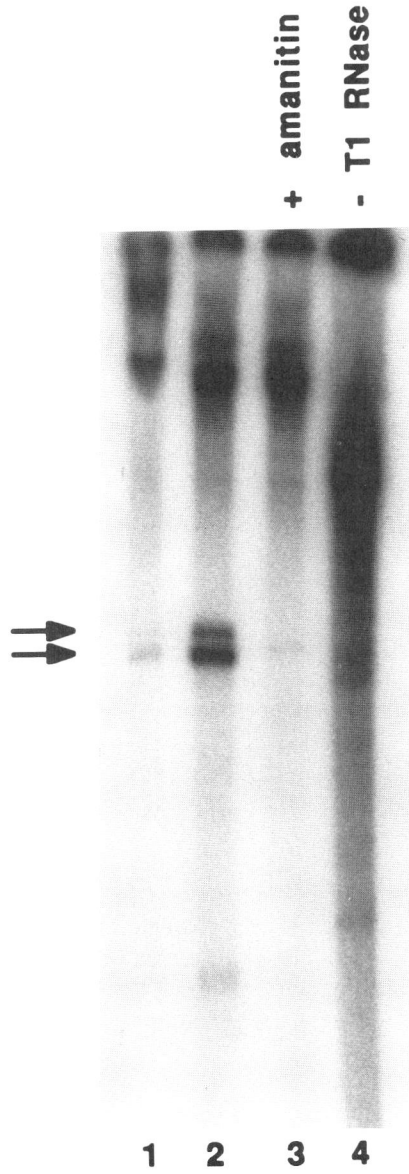

FIG. 1. Transcription of the yeast *CYC1* promoter fused to a G⁻ sequence with yeast nuclear extract. The detailed structure of the template, pΔCG⁻, is given elsewhere.[6] The procedure was as described in the text, with 2 μl (lane 1) and 4 μl (lanes 2, 3, and 4) of nuclear extract (18 mg of protein/ml), except that α-amanitin (10 μg/ml) was included in the reaction analyzed in lane 3 and T1 RNase was omitted during processing of the sample in lane 4. Arrows indicate accurately initiated transcripts.

circular supercoiled], and nuclear extract (50–100 μg of protein). The mixture is incubated for 1 hr at 20°, followed by treatment with RNase T1 (8 units in 200 μl of 10 mM Tris-Cl, pH 7.5/300 mM NaCl/5 mM EDTA) for 10 min at 22°, treatment with 12 μl of 10% SDS and 100 μg of proteinase K for 20 min at 30°, and precipitation with 0.6 ml of ethanol and 15 μg of carrier tRNA for 10 min on dry ice. The precipitate is collected in a microcentrifuge at top speed for 10 min, washed with 70% ethanol, dried, dissolved in 20 μl of 80% formamide/0.1 × TBE[11] buffer/ 0.01% xylene cyanol FF/0.01% bromphenol blue, and analyzed in a 1.5 mm thick 7% polyacrylamide–7 M urea gel (12 × 18 cm) in TBE buffer.[11] The gel is prerun for 30 min at 130 V, electrophoresis is at 30 V until the xylene cyanol has run off the bottom (about 2.5 hr), and the gel is then washed for at least 20 min in water to remove urea and unincorporated label, dried, and autoradiographed for about 30 min with an intensifying screen.

[11] T. Maniatis, E. F. Fritsch, and J. Sambrook, "Molecular Cloning: A Laboratory Manual." Cold Spring Harbor Laboratory, Cold Spring Harbor, New York, 1982.

[39] Direct Sequence and Footprint Analysis of Yeast DNA by Primer Extension

By JON M. HUIBREGTSE and DAVID R. ENGELKE

Introduction

For many applications direct sequence analysis of single-copy yeast genes in total genomic DNA is faster than first cloning the genes and, in addition, can eliminate the possibility of introducing cloning artifacts. One obvious application is the characterization of gene sequences after selection of putative mutants. Other possibilities include fine structure characterization of DNA rearrangements and confirmation of the lack of artifactual mutations after gene replacement.

The two approaches used to directly sequence genomic DNA are modifications of the standard chemical[1] and dideoxynucleotide chain-termination[2] strategies used for cloned DNA. In the first case the positions of base-specific chemical cleavages are determined relative to an adjacent

[1] A. Maxam and W. Gilbert, this series, Vol. 65, p 499.
[2] F. Sanger, S. Nicklen, and A. R. Coulson, *Proc. Natl. Acad. Sci. U.S.A.* **74**, 4951 (1979).

restriction endonuclease cleavage by transferring total genomic DNA from a denaturing polyacrylamide sequencing gel to a membrane support.[3] An internally radiolabeled probe is then annealed adjacent to the restriction site (i.e., indirect end labeling). In the second approach radiolabeled oligonucleotide primers are hybridized to unique sites on denatured genomic DNA and extended with a DNA polymerase in the presence of dideoxynucleotide chain terminators.[4-6] Until recently the chemical methods had the advantage that the more highly labeled probes gave stronger signals and could be hybridized under stringent conditions not tolerated in the DNA polymerase extension reactions. Both of these factors are necessary for obtaining clear signals from genomic DNA more complex than that of yeast, but we chose to develop the chain-termination methods primarily because of their technical simplicity and because the relatively small yeast genome allows oligonucleotide primers to hybridize uniquely and give acceptable signal strengths. After testing a variety of DNA-dependent and RNA-dependent DNA polymerases, we found that modified T7 DNA polymerase[7] and *Thermus aquaticus (Taq)* DNA polymerase[8] reproducibly gave the best results.

In this chapter, we describe protocols for direct sequencing using both the T7 and *Taq* DNA polymerases, since either might be superior for a given application or target sequence. We find that the T7 enzyme often gives good results, but the stringency of probe hybridization is limited by the necessity of using the enzyme at less than 50° and of adding new enzyme for each round of primer annealing and extension. Only about one-half of the 15- to 17-nucleotide-long probes we have tested anneal to unique sites under these conditions. The *Taq* enzyme, on the other hand, tolerates stringent hybridization conditions with longer oligonucleotides and can be used more easily in multiple rounds of extension using an automated thermal cycler. Another major advance in genomic DNA sequencing has been the ability to greatly amplify the target region relative to the DNA as a whole using the polymerase chain reaction.[9-12] We have

[3] G. M. Church and W. Gilbert, *Proc. Natl. Acad. Sci. U.S.A.* **81**, 1991 (1984).
[4] J. M. Huibregtse and D. R. Engelke, *Gene* **44**, 151 (1986).
[5] J. M. Huibregtse, D. R. Engelke, and D. J. Thiele, *Proc. Natl. Acad. Sci. U.S.A.* **86**, 65 (1989).
[6] J. D. Axelrod and J. Majors, *Nucleic Acids Res.* **17**, 171 (1989).
[7] S. Tabor and C. C. Richardson, *Proc. Natl. Acad. Sci. U.S.A.* **84**, 4767 (1987).
[8] F. C. Lawyer, S. Stoffer, R. K. Saiki, K. Myambo, R. Drummond, and D. H. Gelfand, *J. Biol. Chem.* **264**, 6427 (1989).
[9] R. K. Saiki, S. Scharf, F. Faloona, K. B. Mullis, G. T. Horn, H. A. Erlich, and N. Arnheim, *Science* **230**, 1350 (1985).
[10] C. Wong, C. E. Dowling, R. K. Saiki, R. G. Higuchi, H. A. Erlich, and J. J. Kazazian, Jr., *Nature (London)* **330**, 384 (1987).

not usually found this to be necessary for our purposes, and therefore we avoid the extra steps involved; however, the sequencing conditions described below are compatible with a preamplified DNA template.

The ability to examine single-copy sequences in genomic DNA with single-nucleotide resolution has applications in addition to obtaining sequence information. We have been particularly interested in characterizing native chromatin protein–DNA interactions at high resolution to guide and verify *in vitro* studies of gene transcription. "Footprinting" methods derived from both indirect end labeling[13,14] and primer extension[5,6,15] have been developed in which intact cells or nuclei are treated with base modification or DNA cleavage reagents, followed by deproteinization and determination of the sensitive sites. By comparing the patterns of cleavage or modification to those of DNA alone, presumptive sites of bound protein can be deduced. This capability is especially useful when testing models built around the behavior of DNA-binding proteins *in vitro* or when testing whether a genetically identified trans-acting component affects DNA–protein interactions. For these applications we find that both the T7 and *Taq* DNA polymerases can give high-resolution footprints. Although this probing method should be useful for any cleavage or base modification reagent that blocks the extension reaction (e.g., dimethyl sulfate or UV light), the second section below gives sample protocols only for DNase I cleavage of chromatin in fresh cell lysates.

DNA Sequencing

The following section describes a protocol for isolation of total yeast DNA suitable for sequencing and protocols for dideoxynucleotide chain-termination sequencing using modified T7 and *Thermus aquaticus* DNA polymerases (see Fig. 1).

Isolation of Yeast DNA

1. Grow 100-ml liquid cultures to mid-log phases (OD_{600} 0.5–1.0).
2. Harvest cells by centrifugation (3,000 g, 4 min).
3. Resuspended cells in 5 ml of 40 mM EDTA, 90 mM 2-mercap-

---

[11] D. R. Engelke, P. A. Hoener, and F. S. Collins, *Proc. Natl. Acad. Sci. U.S.A.* **85,** 544 (1988).
[12] M. A. Innis, D. H. Gelfand, J. J. Sninsky, and T. J. White (eds.), "PCR Protocols." Academic Press, San Diego, 1989.
[13] G. M. Church, A. Ephrussi, W. Gilbert, and S. Tonegawa, *Nature (London)* **313,** 798 (1985).
[14] E. Giniger, S. Varnum, and M. Ptashne, *Cell (Cambridge, Mass.)* **40,** 767 (1985).
[15] J. M. Huibregtse and D. R. Engelke, *Mol. Cell. Biol.* **9,** 3244 (1989).

toethanol. Transfer to a 40-ml Oak Ridge-type centrifuge tube. Incubate at room temperature for 5 min.

4. Collect cells by centrifugation (3,000 g, 5 min).

5. Resuspend cells in 5 ml of 1 M sorbitol, 1 mM EDTA, 3 mM dithiothreitol (DTT), and 2 mg/ml Zymolyase 20,000 (ICN, Costa Mesa, CA or Seikagaku, St. Petersburg, FL). Incubate at 30° for 40 min with gentle swirling.

6. Collect spheroplasts by centrifugation (3,000 g, 4 min).

7. Resuspend in 2 ml of 100 mM NaCl, 10 mM Tris (pH 7.4), 10 mM MgCl$_2$, 5 mM 2-mercaptoethanol. Add 2 ml of 50 mM Tris (pH 7.4), 1 M NaCl, 2% sodium dodecyl sulfate (SDS), 50 mM EDTA. Mix by inverting. Incubate at 45° for 10 min.

8. Extract the aqueous phase twice with phenol/chloroform. Add an equal volume of phenol/chloroform (1:1), mix by inversion, centrifuge at 10,000 g, remove the aqueous phase to a new tube, and repeat.

9. Precipitate the nucleic acids by adding an equal volume of 2-propanol and mixing gently. Set at room temperature for 10 min, centrifuge at 10,000 g, and pour off the supernatant. Add 2 ml of cold 75% ethanol and swirl gently for a few seconds. Pour off the supernatant and air-dry the pellet.

10. Resuspend the pellet in 1 ml of 10 mM HEPES (pH 7.9), 0.1 mM EDTA, 50 μg/ml DNAse-free RNase A. Incubate at 37° at least 4 hr or overnight.

11. Reprecipitate the DNA by adding 0.1 volume of 3 M sodium acetate and 2.5 volumes of ethanol. Let sit at room temperature for 10 min. Centrifuge at 10,000 g for 10 min, wash the pellet with cold 75% ethanol, dry the pellet, and resuspend it in 0.5 ml of 10 mM HEPES (pH 7.9), 0.1 mM EDTA, 20 μg/ml RNase A. It may take up to 1 hr for DNA to go into solution. The yield should be 1–2 mg of DNA, as measured by the OD$_{260}$.

Genomic Sequencing Using Modified T7 DNA Polymerase

Buffers and Nucleotide (N) Mixes

10× Buffer: 0.2 M HEPES (pH 7.5), 0.1 M MgCl$_2$, 0.5 M NaCl
10× dNTP mix: 3 mM each dATP, dTTP, dCTP, and dGTP
10× Dideoxy (dd)NTP mixes: 0.3 mM ddNTP, 3.0 mM each dATP, dTTP, dCTP, and dGTP

Procedure. For each sequencing reaction, combine the following in a microcentrifuge tube: 3 μl of 10× buffer, 1.5 μl of 0.1 M DTT, 3 μl of 10×

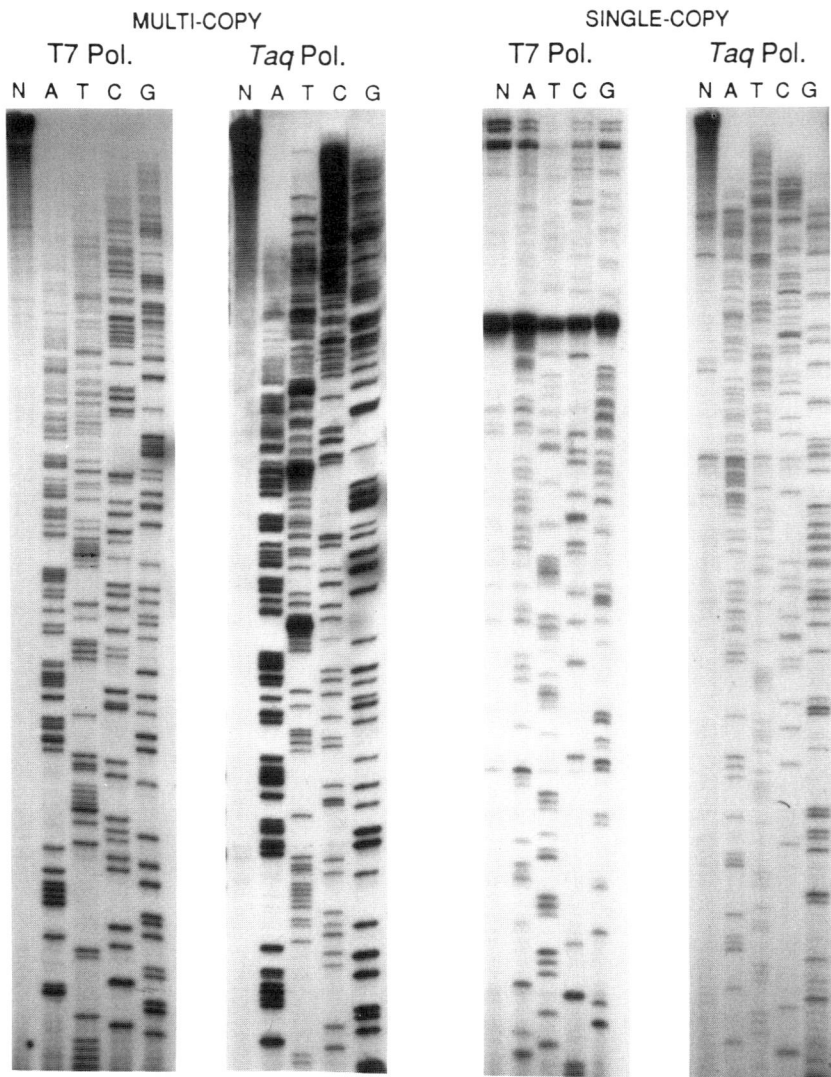

Fig. 1. Genomic DNA sequencing of multicopy and single-copy genes using modified T7 and *Taq* DNA polymerases. Yeast DNA was probed with a radiolabeled oligonucleotide (a 21-mer) that hybridized upstream of the *CUP1* gene [J. M. Huibregtse, D. R. Engelke, and D. J. Thiele, *Proc. Natl. Acad. Sci. U.S.A.* **86,** 65 (1989)], which is reiterated chromosomally at a copy number of approximately 10–20, or with an oligonucleotide (a 17-mer) that hybridized within the tRNA^{Ser} (UCG) gene, which is present as a single copy [M. V. Olson, G. S. Page, A. Sentenac, P. W. Piper, M. Worthington, R. B. Weiss, and B. D. Hall, *Nature (London)* **291,**

dNTP or ddNTP mix, 20-40 μg yeast genomic DNA,* 1-2 μl ^{32}P-end-labeled oligonucleotide,† and water to give total volume of 29 μl.

1. Heat at 95°-100° for 5 min.
2. Centrifuge for a few seconds in the microfuge to bring down liquid that has condensed on the sides of the tube.
3. Place at 37°-44° for 6 min.‡
4. Add 1 μl of modified T7 DNA polymerase diluted in 1× buffer.§
5. Continue incubation at 37°-44° for 4 min.
6. Additional rounds of primer extension may be done by repeating Steps 1-5.‖
7. Stop the reaction by addition of 4 μl of solution containing 1% SDS, 0.1 M EDTA, and 1 mg/ml proteinase K. Incubate for 10 min at 44°.
8. Precipitate DNA with 2.5 volumes of ethanol. Freeze on dry ice, then microfuge for 12 min, remove the supernatant, wash the pellet with cold 75% ethanol, and dry the pellet. Resuspend in 4 μl formamide con-

---

*In general, the more DNA used the stronger the signal, with the constraint being how much DNA can be loaded in a lane of a sequencing gel without significantly diminishing the resolution. In some cases treating the DNA with a restriction endonuclease that cleaves outside of the region of interest may help reduce background owing to polymerase pausing.

†About 200,000-500,000 counts per minute (cpm), specific radioactivity of approximately 3000-6000 Ci/mmol.

‡The optimal hybridization temperature should be determined for each oligonucleotide. About one-half of the 16- to 21-nucleotide oligomers that we have tested hybridize uniquely between 37° and 44°. Longer oligonucleotides (24- to 30-mers) have sometimes given unacceptable background priming within this temperature range. It is also possible to use a hybridization temperature greater than 44° and then place the reaction at 44° when the enzyme is added. (Modified T7 DNA polymerase will work well up to only about 45°.)

§Equivalent to about 1-3 units of United States Biochemical's Sequenase (Cleveland, OH).

‖Typically 3-5 rounds of primer extension is sufficient. The exposure times are usually 2-3 days when sequencing single-copy genes (Kodak XAR5 film, with intensifying screen). The number of rounds of primer extension necessary to obtain a suitable signal will vary with the individual oligonucleotide, as well as with the copy number of the gene of interest.

---

464 (1981)]. Three and five rounds of primer extension were performed on the multicopy gene with the T7 and *Taq* DNA polymerases, respectively. Five and 10 rounds of primer extension were performed on the single-copy gene with the T7 and *Taq* DNA polymerases, respectively. The primer annealing temperature for all four sets of reactions was 44°. N lanes represent primer extensions in the presence of dNTPs only, whereas lanes A, T, C, and G represent the corresponding dideoxy sequencing reactions.

taining 10 mM NaOH and 1 mM EDTA. Heat at 95°–100° for 2 min prior to loading on a sequencing gel.[16#]

Genomic Sequencing Using Taq DNA Polymerase

Buffers and Nucleotide Mixes

10× Buffer: 0.1 M Tris (pH 8.5), 0.5 M KCl, 15 mM MgCl$_2$
10× dNTP mix: 1 mM each dATP, dTTP, dCTP, and dGTP
10× ddATP mix: 4 mM ddATP, 40 μM dATP, 1 mM each dCTP, dGTP, dTTP
10× ddTTP mix: 4 mM ddTTP, 50 μM dTTP, 1 mM each dATP, dCTP, dGTP
10× ddCTP mix: 4 mM ddCTP, 0.33 mM dCTP, 1 mM each dGTP, dATP, dTTP
10× ddGTP mix: 2 mM ddGTP, 0.1 mM dGTP, 1 mM each dCTP, dATP, dTTP

Procedure. For each sequencing reaction, combine the following in a 0.5-ml microfuge tube: 3 μl of 10× buffer, 3 μl of 10× dNTP or ddNTP mix, 20–40 μg yeast genomic DNA, 1–2 μl ^{32}P-end-labeled oligonucleotide (see above), 1 μl *Taq* DNA polymerase diluted in 1× buffer [equivalent to 1–2 units of Cetus (Perkin-Elmer, Norwalk, CT) *Taq* DNA polymerase], and water to give total volume of 30 μl.

1. Place a drop of mineral oil (about 35 μl) over the solution to prevent evaporation. Multiple rounds of denaturation, hybridization, and primer extension are carried out automatically in a thermal cycler, or manually by moving the reactions between water baths (see below).

2. After the reactions are completed, the lower aqueous phases of the reactions are removed to new tubes using a pipetman set at 25 μl, with care being taken to transfer as little of the mineral oil as possible. Four microliters of 1% SDS, 0.1 M EDTA, 1 mg/ml proteinase K is added, and the reactions are incubated for 10 min at 44°. The DNA is then ethanol precipitated and loaded on a sequencing gel.

#This protocol should be compatible with any denaturing polyacrylamide gel system used for DNA sequencing. The obligatory ^{32}P radiolabel and use of intensifying screens, when necessary, limit resolution on unmodified Sanger and Coulson gels to 150–250 nucleotides.

---

[16] F. Sanger and A. R. Coulson, *FEBS Lett.* **87,** 107 (1978).

Notes. For shorter oligonucleotides (16- to 21-mers) a three-step cycle is used: denaturation at 94°, 2 min; primer annealing at 37°–55°, 3–6 min; primer extension at 72°, 3 min. The primer annealing temperature and time should be optimized for each primer. For longer oligonucleotides (>21-mers) a two-step cycle is often feasible, with primer annealing and extension both performed at 72°: denaturation at 94°, 2 min; primer annealing and extension at 72°, 6 min. Typically 10–12 cycles of primer extension will give an acceptable signal for single-copy genes (exposure times 1–3 days, Kodak XAR film, with intensifying screen).

Chromosomal Footprinting

When probing chromatin for stable nucleoprotein complexes one would ideally like to examine single-copy genes in their normal chromosomal location to avoid potential problems with gene dosage and artifactually inactive loci. For the purpose of showing well-defined footprints consistent with previously characterized DNA–protein interactions,[15,17] however, analysis is shown of the DNase I sensitivity of tRNA genes on multicopy plasmids (10–15 copies per cell), using both modified T7 and *Taq* DNA polymerases. The patterns of protection for these genes on plasmids is entirely consistent with the single-copy chromosomal patterns at the *SUP53* locus[15] but of higher resolution owing to an increased signal-to-background ratio. The footprinting data for a tRNA$_3^{Leu}$ gene *(SUP53)* and three variants containing intragenic promoter mutations (C_{19}, AAA_{10-12}, TTT_{23-25}, G_{56})[15] are presented in Fig. 3. Titrations of DNase I are routinely performed for each cell lysate so that the appropriate level of digestion can be selected by visualizing an aliquot on agarose gels before continuing with primer extension. A typical digestion profile is presented in Fig. 2 and discussed below. The experimental result and interpretations, which are described in greater detail elsewhere, can be summarized as follows: (1) the wild-type gene has a stable complex protecting the transcription initiation site and upstream region, with the internal promoters occupied relatively poorly by transcription factor IIIC, (2) mutations in either internal promoter that severely reduce transcription also eliminate the upstream complex, suggesting that formation but not maintenance of the upstream interactions requires the interaction of TFIIIC with the internal promoters.

We have also analyzed DNA digestion patterns produced by other nucleases (micrococcal nuclease, restriction endonucleases) and base mod-

[17] G. A. Kassavetis, D. L. Riggs, R. Negri, L. H. Nguyen, and E. P. Geiduschek, *Mol. Cell. Biol.* **9**, 2551 (1989).

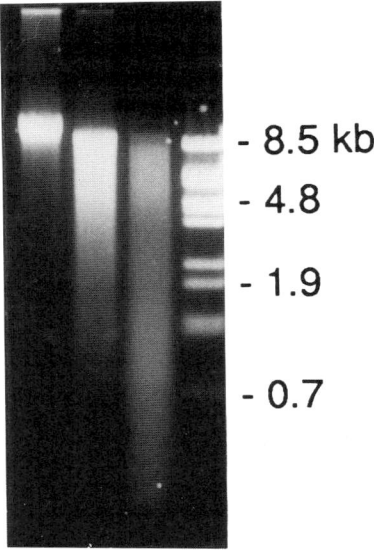

FIG. 2. Agarose gel electrophoresis of yeast DNA treated with (−) or with (+, ++) DNase I in cell lysates. The final concentrations of DNase I in the lysate digestions were 35 and 70 μg/ml, respectively. Approximately 10 μg of yeast DNA was loaded per lane. The 35 μg/ml digestion (+ lane) would, in this case, represent the more optimal degree of digestion for primer extension analysis. The DNA size markers (far right lane) are BstEII-digested DNA (New England Biolabs, Beverly, MA).

ification patterns produced by dimethyl sulfate and ultraviolet light (unpublished data). Protection of selected sites relative to others nearby in restriction endonuclease partial digests is most useful for estimating site occupation, since measurements involve only single-band intensities. The partial digests and primer extension analysis are done as shown below for DNase I, except that after exposure of the gels to X-ray film the individual bands are excised and subjected to liquid scintillation counting. For actual footprinting, we favor DNase I because the characteristic intensified cleavages and large, obvious regions of protection have facilitated identification of previously unsuspected complexes and have suggested that some sites are occupied only part of the time.[15]

Procedure

1. Grow 250-ml liquid cultures in YPD or synthetic media to mid-log phase (OD_{600} 0.5–1.0).

2. Harvest cells by centrifugation (3,000 g, 4 min).

3. Resuspend cells in 7.5 ml of 40 mM EDTA, 90 mM 2-mercaptoethanol. Transfer to a 40-ml Oak Ridge-type centrifuge tube. Let sit at room temperature for 5 min.

4. Collect cells by centrifugation (3,000 g, 4 min).

5. Resuspend cells in 8 ml of 1 M sorbitol, 1 mM EDTA, 3 mM DTT, and 2 mg/ml Zymolayse 20,000 (ICN or Seikagaku).* Incubate at 30° for 30–40 min with gentle swirling.

6. Collect spheroplasts by centrifugation (3,000 g, 4 min).

7. Resuspend rapidly but gently in 1–2 ml lysis buffer† [100 mM NaCl, 10 mM Tris (pH 7.4), 10 mM MgCl$_2$, 5 mM 2-mercaptoethanol, 0.075% Nonidet P-40 (NP-40)].

8. Transfer to a small (5–7 ml) Dounce homogenizer. Treat with 5–10 strokes of a loose pestle to resuspend cells evenly.

9. Immediately‡ remove 300-μl aliquots to 1.5-ml microfuge tubes which contain 30 μl of lysis buffer containing either no DNase I or 0.2–1.2 mg/ml DNase I.§

10. Incubate reactions at room temperature for 5 min. Stop reactions by adding 330 μl of solution containing 50 mM Tris (pH 7.4), 1 M NaCl, 2% SDS, 50 mM EDTA. Shake the tube vigorously several times to mix.

11. Incubate at 45° for 10–15 min.

12. Purify DNA by extracting twice with an equal volume of phenol/chloroform (1:1).

13. Precipitate the nucleic acids by adding an equal volume of 2-propanol and mixing gently. Set at room temperature for 10 min, centrifuge

---

* In some cases it may be advantageous for the sorbitol solution to contain media components (yeast nitrogen base, amino acids, carbon source, specific metabolites or inducers) to ensure that the spheroplasts remain metabolically active or that certain genes of interest remain induced or repressed (J. M. Huibregtse and D. R. Engelke, unpublished observations; P. B. Zhou and D. J. Thiele, personal communication, 1989).

† The lysis buffer may have to be altered slightly for study of some systems. For example, the NP-40 detergent, which helps to lyse the cells, might disrupt certain features of chromatin. Too high a salt concentration might have a similar effect.

‡ The purpose of probing chromatin structure in the first minutes after lysis is to minimize any time-dependent rearrangements that might occur after lysis or during the isolation of nuclei. In the cases of the DNase I protection upstream of tRNA genes, we find that the footprint is slowly lost over a period of about 30 min after cell lysis (J. M. Huibregtse and D. R. Engelke, unpublished observations).

§ Since the degree of digestion varies with the degree of spheroplasting and on the amount of cells used, it is best to conduct DNase I digestions at about three different concentrations within this range (see footnote ||).

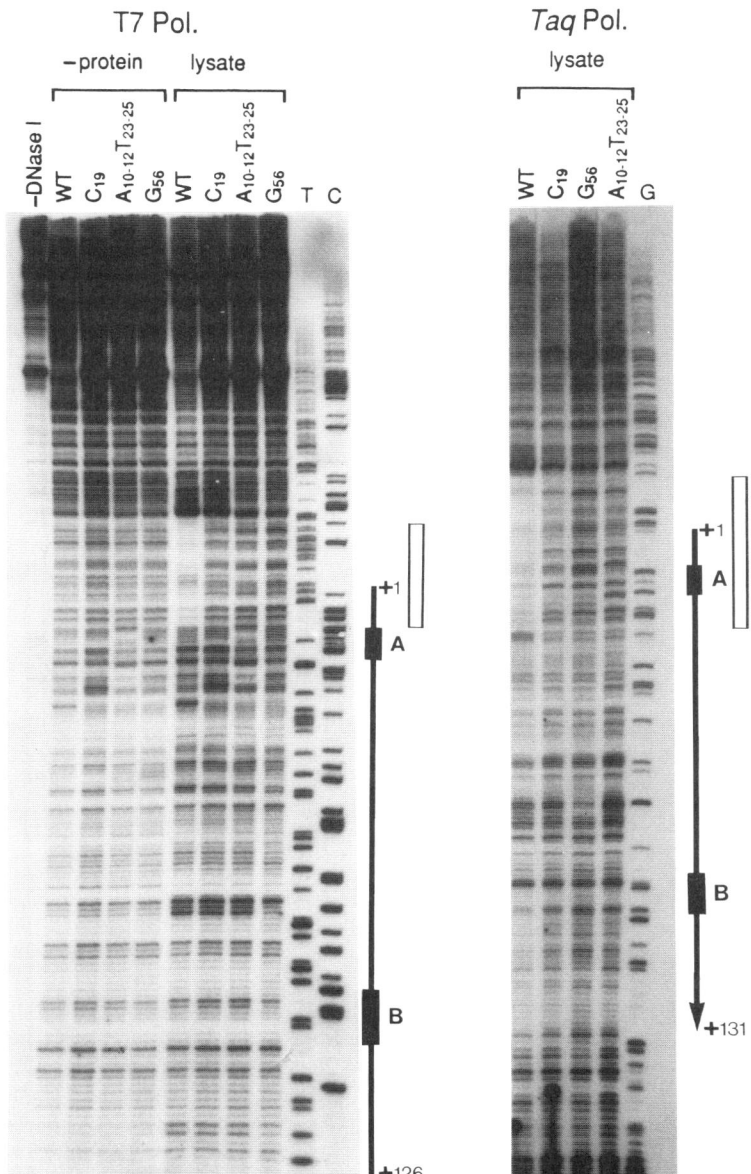

FIG. 3. Chromosomal footprinting of wild-type tRNA$_3^{Leu}$ and *SUP53* variants on multicopy plasmids. Cell lysates (lysate lanes) from YEp13 transformants containing either the wild-type (WT) gene or the indicated *SUP53* variants were treated with DNase I, and the digestion products were probed by primer extension using either modified T7 (T7 Pol.) or *Taq* DNA polymerase (*Taq* Pol.). The oligonucleotide primer (a 16-mer) hybridized approxi-

for 5 min in a microfuge, pour off the supernatant, wash the pellet with 0.5 ml 75% cold ethanol, and air- or vacuum-dry the pellet.

14. Resuspend the pellet in 150 μl of 10 mM HEPES (pH 7.9), 0.1 mM EDTA, 50 μg/ml DNase-free RNase A. Incubate at 37° for at least 4 hr or overnight.

15. Reprecipitate DNA by adding 0.1 volumes of 3 M sodium acetate and 2.5 volumes of ethanol. Incubate at room temperature for 10 min, centrifuge for 10 min in a microfuge, remove the supernatant, wash the pellet with 75% ethanol, and dry the pellet. Resuspend the DNA in 75 μl of 10 mM HEPES (pH 7.9), 0.1 mM EDTA, 20 μg/ml RNase A. Read the OD$_{260}$. Each reaction yield about 250–500 μg of DNA.

16. Analyze the DNase I cleavage pattern by primer extension in the presence of dNTPs as described above for genomic sequencing.[||]

Controls for primer extension analysis of DNase I cleavage patterns include the following: (1) Genomic sequencing lanes provide exact markers for DNase I cleavage (lanes T, C, and G, Fig. 3). More importantly, they provide a control for unique hybridization of the oligonucleotide during primer extension. The DNA used for sequencing should therefore be from the experiment of interest (the −DNase I sample). (2) By performing primer extension analysis with dNTPs on genomic DNA that has not been treated with DNase I (−DNase I lane, Fig. 3), the degree of polymerase pausing over the region of interest can be assessed. If pausing over a particular region is significant, interpretation of the DNase I cleavage pattern is difficult or impossible. In addition the −DNase I lane serves

---

[||] In order to determine which reactions will be most useful for DNase I cleavage pattern analysis, run about 3 μl of each reaction on a 0.8% agarose gel (Fig. 2). DNA from reactions not treated with DNase I should run as a wide band greater than 10 kilobase pairs (kbp). DNA from DNase I-treated reactions should show a smear from about 10 kbp downward, with the average sized fragments decreasing in size with increasing DNase I concentration. Reactions that give a smear from 10 to 4 kbp will probably be most useful. Over- or underdigested DNA will give less primer extension signal over the region of interest. There is generally a small amount of approximately 10 kbp DNA in reactions that are treated with even a very high DNase I concentration. This represents DNA from cells that were not lysed and does not generally pose a problem in analysis of DNase I cleavage patterns.

---

mately 50-bp downstream of the gene. DNase I digestion patterns of purified genomic DNA from each transformant are shown for the modified T7 polymerase extensions (−protein lanes). The −DNase I lane represents primer extension of undigested DNA using dNTPs, whereas lanes T, C, and G represent dideoxy sequencing lanes using undigested DNA. The position and orientation of the tRNA gene are indicated by the black bars, with the position of the A and B box internal promoters indicated. The open box represents the region spanning the transcription initiation site and upstream region that is clearly protected from DNase I digestion on the wild-type gene [J. M. Huibregtse and D. R. Engelke, *Mol. Cell. Biol.* **9**, 3244 (1989)]. All of the internal promoter mutations shown eliminate this DNase I protection.

as a control for any nonspecific DNA breakdown during isolation. (3) To assess the sequence-dependent specificity of DNase I over the region of interest it is necessary to perform primer extension on genomic DNA that has been digested after purification (−protein lanes, Fig. 3). This is most easily done by isolating a large amount of DNA, dissolving it in the lysis buffer described above, and treating aliquots with increasing amounts of DNase I. The reactions are stopped by the addition of EDTA to 25 mM, and the DNA is ethanol precipitated and resuspended in 10 mM HEPES (pH 7.9), 0.1 mM EDTA. Portions of these samples can then be run on an agarose gel as described above to determine which digestion was optimal. The DNase I cleavage pattern of this DNA is compared with that of an equivalent amount of DNA that was treated with DNase I in cell lysate (lysate lanes, Fig. 3)

Acknowledgments

This work was supported by National Science Foundation Research Grant DMB89-01559.

Section V
Cell Biology

[40] Immunofluorescence Methods for Yeast

By JOHN R. PRINGLE, ALISON E. M. ADAMS, DAVID G. DRUBIN, and BRIAN K. HAARER

Introduction

The major attraction of yeast as an experimental organism is clearly its susceptibility to the methods of classical and molecular genetics. However, it is also clear that effective use of yeast for study of cell biological problems requires that satisfactory morphological (both light and electron microscopic) and biochemical (including cell fractionation) methods also be available. Among the most powerful of light microscopic techniques is immunofluorescence, in which the exquisite specificity of antibodies is used to localize molecules of interest within cells; the antibodies are rendered visible by attachment of a fluorophore, which emits visible fluorescent light on excitation at a shorter wavelength. In this chapter, we provide protocols for the application of immunofluorescence procedures to yeast. Other fluorescence microscopic procedures for work with yeast are described elsewhere.[1-3] Table I lists suppliers of reagents and other items mentioned in this chapter.

It should perhaps be stressed that immunofluorescence and other light microscopic techniques play a role that is separate from but equal to the role of electron microscopy. Although in some situations the greater resolving power of the electron microscope is clearly essential to obtain the needed structural information (see Refs. 4 and 5 for discussions of modern electron microscopic methods for yeast), in other situations the necessary information can be obtained more easily, more reliably, or both, by light microscopy. The potential advantages of light microscopic approaches derive from these facts: (1) they can be applied to lightly processed or (in some cases) living cells, (2) much larger numbers of cells can be examined than by electron microscopy (note especially the great labor involved in visualizing the structure of whole cells by serial-section methods), and (3)

---

[1] J. R. Pringle, this volume [52].
[2] A. E. M. Adams and J. R. Pringle, this volume [51].
[3] J. R. Pringle, R. A. Preston, A. E. M. Adams, T. Stearns, D. G. Drubin, B. K. Haarer, and E. W. Jones, *Methods Cell Biol.* **31,** 357 (1989).
[4] B. Byers and L. Goetsch, this volume [41]; M. W. Clark, this volume [42].
[5] R. Wright and J. Rine, *Methods Cell Biol.* **31,** 473 (1989).

TABLE I
SUPPLIERS OF REAGENTS AND OTHER ITEMS MENTIONED IN THIS CHAPTER

Name of company	Address	Telephone number
Accurate Chemical & Scientific Corp.	300 Shames Dr. Westbury, NY 11590	800/645-6264 or 516/433-4900
Acufine, Inc.	5441 N. Kedzie Ave. Chicago, IL 60625	312/539-8700
Bio-Rad Laboratories	1414 Harbour Way South Richmond, CA 94804	800/227-5589 or 800/227-3259
BRL (Bethesda Research Laboratories)	P.O. Box 6009 Gaithersburg, MD 20877	800/638-8992 or 800/638-4045
Cappel (see Organon Teknika Corp.)		
Cel-Line Associates, Inc.	P.O. Box 35 Newfield, NJ 08344	609/697-4590
Du Pont NEN Products	549 Albany St. Boston, MA 02118	800/551-2121
E-Y Laboratories, Inc.	P.O. Box 1787 San Mateo, CA 94401	800/821-0044 or 415/342-3296
Flow Laboratories, Inc.	7655 Old Springhouse Rd. McLean, VA 22102	800/368-3569
ICN Immunobiologicals	P.O. Box 1200 Lisle, IL 60532	800/348-7465
Jackson ImmunoResearch Labs, Inc.	P.O. Box 9, 872 W. Baltimore Pike West Grove, PA 19390	800/367-5296
Leitz (see Wild Leitz)		
Lumicon Corp.	2111 Research Dr. #5 Livermore, CA 94550	415/447-9570
Molecular Probes, Inc.	4849 Pitchford Ave. Eugene, OR 97402	503/344-3007
Omega Optical, Inc.	3 Grove St., P.O. Box 573 Brattleboro, VT 05301	802/254-2690
Organon Teknika Corp.	100 AKZO Avenue Durham, NC 27704	800/523-7620
Pharmacia LKB Biotechnology, Inc.	800 Centennial Ave., P.O. Box 1327 Piscataway, NJ 08855-1327	800/922-0318
Polysciences, Inc.	400 Valley Rd. Warrington, PA 18976-2590	800/523-2575
Ted Pella, Inc.	P.O. Box 2318 Redding, CA 96099-2318	800/237-3526
Vector Laboratories, Inc.	30 Ingold Rd. Burlingame, CA 94010	800/227-6666 or 415/697-3600
Wild Leitz USA, Inc.	24 Link Dr. Rockleigh, NJ 07647	201/767-1100 or 800/654-4488
Zeiss (Carl Zeiss, Inc.)	One Zeiss Drive Thornwood, NY 10594	914/747-1800
Zymed Laboratories, Inc.	52 S. Linden Ave. South San Francisco, CA 94080	800/874-4494 or 415/871-4494

some structures (e.g., the cytoplasmic microtubules[6-8]) have simply been easier to see by light microscopy than by electron microscopy.

General Remarks

The Broad Utility of Immunofluorescence in Yeast

Before immunofluorescence methods had been applied to yeast, there was considerable pessimism as to how well these methods (which were developed for large, flat, wall-less animal cells) would work on yeast (which are small, round, and heavily walled). However, it is now clear from the steadily expanding list of successful applications that this pessimism was unwarranted. Immunofluorescence on yeast has been used to visualize the intranuclear and cytoplasmic microtubules;[6-21] apparent components of the spindle-pole body;[22,23] actin[7,24-28] and associated proteins;[25,27] the 10-nm filaments in the mother–bud neck;[21,29] apparent constituents of the cytoplasm,[30,31] plasma membrane,[32-34] and cell wall;[35] proteins that localize to sites of cell-surface growth in budding[36] and mating[37] cells; proteins of the nucleus,[38-46] including those of the nucleolus[44,47-51] and nuclear envelope;[49,52-56] proteins apparently of the endoplasmic reticulum,[54,55] Golgi (or trans-Golgi network?),[57,58] and secretory vesicles;[59] and proteins of the vacuole,[60,61] peroxisomes,[62,63] and mitochondria.[64,65] Although some

[6] A. E. M. Adams and J. R. Pringle, *J. Cell Biol.* **98**, 934 (1984).
[7] J. V. Kilmartin and A. E. M. Adams, *J. Cell Biol.* **98**, 922 (1984).
[8] C. W. Jacobs, A. E. M. Adams, P. J. Szaniszlo, and J. R. Pringle, *J. Cell Biol.* **107**, 1409 (1988).
[9] L. Pillus and F. Solomon, *Proc. Natl. Acad. Sci. U.S.A.* **83**, 2468 (1986).
[10] J. H. Thomas and D. Botstein, *Cell (Cambridge, Mass.)* **44**, 65 (1986).
[11] J. Hašek, J. Svobodová, and E. Streiblová, *Eur. J. Cell Biol.* **41**, 150 (1986).
[12] J. Hašek, I. Rupeš, J. Svobodová, and E. Streiblová, *J. Gen. Microbiol.* **133**, 3355 (1987).
[13] M. D. Rose and G. R. Fink, *Cell (Cambridge, Mass.)* **48**, 1047 (1987).
[14] M. Han, M. Chang, U.-J. Kim, and M. Grunstein, *Cell (Cambridge, Mass.)* **48**, 589 (1987).
[15] T. C. Huffaker, J. H. Thomas, and D. Botstein, *J. Cell Biol.* **106**, 1997 (1988).
[16] M. Snyder and R. W. Davis, *Cell (Cambridge, Mass.)* **54**, 743 (1988).
[17] P. J. Schatz, F. Solomon, and D. Botstein, *Genetics* **120**, 681 (1988).
[18] B. A. Guthrie and W. Wickner, *J. Cell Biol.* **107**, 115 (1988).
[19] W. S. Katz and F. Solomon, *Mol. Cell. Biol.* **8**, 2730 (1988).
[20] F. Matsuzaki, S. Matsumoto, and I. Yahara, *J. Cell Biol.* **107**, 1427 (1988).
[21] H. B. Kim, B. K. Haarer, and J. R. Pringle, *J. Cell Biol.* (in press).
[22] M. D. Rose, personal communication (1989).
[23] M. Snyder, personal communication (1989).
[24] P. Novick and D. Botstein, *Cell (Cambridge, Mass.)* **40**, 405 (1985).
[25] D. G. Drubin, K. G. Miller, and D. Botstein, *J. Cell Biol.* **107**, 2551 (1988).
[26] A. E. M. Adams, D. Botstein, and D. G. Drubin, *Science* **243**, 231 (1989).
[27] H. Liu and A. Bretscher, *Cell (Cambridge, Mass.)* **57**, 233 (1989).

of these successful applications have involved staining of spheroplasts, isolated organelles, or cell sections, many others have involved staining of whole cells whose shape and spatial organization have been preserved by fixation prior to the removal of the cell wall. In practice, of course, the detectability of particular antigens will depend on their abundance, degree of localization, and retention of antigenicity after fixation, as well as on the

---

[28] P. Novick, B. C. Osmond, and D. Botstein, *Genetics* **121**, 659 (1989).
[29] B. K. Haarer and J. R. Pringle, *Mol. Cell. Biol.* **7**, 3678 (1987).
[30] C. Wittenberg, S. L. Richardson, and S. I. Reed, *J. Cell Biol.* **105**, 1527 (1987).
[31] J. L. Celenza and M. Carlson, *Science* **233**, 1175 (1986).
[32] J. L. Celenza, L. Marshall-Carlson, and M. Carlson, *Proc. Natl. Acad. Sci. U.S.A.* **85**, 2130 (1988).
[33] L. Popolo, R. Grandori, M. Vai, E. Lacaná, and L. Alberghina, *Eur. J. Cell Biol.* **47**, 173 (1988).
[34] J. Lisziewicz, A. Godany, D. V. Agoston, and H. Küntzel, *Nucleic Acids Res.* **16**, 11507 (1988).
[35] M. Watzele, F. Klis, and W. Tanner, *EMBO J.* **7**, 1483 (1988).
[36] M. Snyder, *J. Cell Biol.* **108**, 1419 (1989).
[37] J. Trueheart, J. D. Boeke, and G. R. Fink, *Mol. Cell. Biol.* **7**, 2316 (1987).
[38] M. N. Hall, L. Hereford, and I. Herskowitz, *Cell (Cambridge, Mass.)* **36**, 1057 (1984).
[39] R. L. Last and J. L. Woolford, Jr., *J. Cell Biol.* **103**, 2103 (1986).
[40] M. W. Clark and J. Abelson, *J. Cell Biol.* **105**, 1515 (1987).
[41] R. B. Moreland, G. L. Langevin, R. H. Singer, R. L. Garcea, and L. M. Hereford, *Mol. Cell. Biol.* **7**, 4048 (1987).
[42] P. A. Silver, A. Chiang, and I. Sadler, *Genes Dev.* **2**, 707 (1988).
[43] I. Uno, T. Oshima, and T. Ishikawa, *Exp. Cell Res.* **176**, 360 (1988).
[44] T.-H. Chang, M. W. Clark, A. J. Lustig, M. E. Cusick, and J. Abelson, *Mol. Cell. Biol.* **8**, 2379 (1988).
[45] C. N. Giroux, M. E. Dresser, and H. F. Tiano, *Genome* **31**, 88 (1989).
[46] B. Weinstein and F. Solomon, personal communication (1989).
[47] A. Y.-S. Jong, M. W. Clark, M. Gilbert, A. Oehm, and J. L. Campbell, *Mol. Cell. Biol.* **7**, 2947 (1987).
[48] J. P. Aris and G. Blobel, *J. Cell Biol.* **107**, 17 (1988).
[49] E. C. Hurt, A. McDowall, and T. Schimmang, *Eur. J. Cell Biol.* **46**, 554 (1988).
[50] C. H. Yang, E. J. Lambie, J. Hardin, J. Craft, and M. Snyder, *Chromosoma* **98**, 123 (1989).
[51] K. Armstrong, J. R. Broach, and M. D. Rose, personal communication (1989).
[52] E. C. Hurt, *EMBO J.* **7**, 4323 (1988).
[53] J. P. Aris and G. Blobel, *J. Cell Biol.* **108**, 2059 (1989).
[54] R. Wright, M. Basson, L. D'Ari, and J. Rine, *J. Cell Biol.* **107**, 101 (1988); R. Wright and J. Rine, personal communication (1989).
[55] M. D. Rose, L. M. Misra, and J. P. Vogel, *Cell (Cambridge, Mass.)* **57**, 1211 (1989).
[56] L. Davis and G. R. Fink, *Cell (Cambridge, Mass.)* **61**, 965 (1990).
[57] N. Segev, J. Mulholland, and D. Botstein, *Cell (Cambridge, Mass.)* **52**, 915 (1988).
[58] K. Redding, A. Franzusoff, and R. Fuller, personal communication (1989).
[59] B. Goud, A. Salminen, N. C. Walworth, and P. J. Novick, *Cell (Cambridge, Mass.)* **53**, 753 (1988).
[60] T. H. Stevens, C. Roberts, C. Raymond, and P. Kane, personal communication (1989).

properties of the particular antisera used. Moreover, when spatial resolution rather than simply detection is at issue (e.g., in double-label experiments), the inherent limit of resolution of the light microscope may also be a factor.

The Need for a Flexible, Empirical Approach

The protocols provided below have worked well in a variety of specific applications. However, there is no guarantee that they will work without complications in all other specific applications. Indeed, experience to date makes clear that the procedures that are necessary and sufficient to get good results vary from case to case. We have noted below such complications and useful alternative protocols of which we are aware, but others seem certain to emerge with further experience. Thus, a flexible, empirical approach is required; this approach must incorporate controls sufficient to establish the specificity of staining and the biological meaningfulness of the results in each particular case.

Choice of Strains and Growth Conditions

Choices of strain and growth conditions are obviously constrained by the biological problem to be studied. Investigators should be aware, however, that these choices may affect the results obtained with immunofluorescence methods. The most obvious and general problem is that cell wall removal and permeabilization for immunofluorescence are usually more difficult with nonexponential-phase cells (e.g., cells approaching, in, or just leaving stationary phase; sporulating cells; arrested cell-cycle mutants), as discussed further below. However, more subtle effects have also been noted. For example, immunofluorescence localization of the *SAC6* 67-kDa actin-binding protein[25] was more satisfactory when the cells had been grown in defined medium than in rich medium. In addition, if other things are equal, visualization of various structures is generally better in bigger than in smaller cells. Thus, diploids are generally preferable to haploids, and tetraploids (when available) are better yet. In addition, structures that are present only transiently (e.g., during a short period of the cell cycle) may be more convincingly visualized if an appropriately enriched or synchronized population is examined.[21,66]

---

[61] R. A. Preston, personal communication (1989).
[62] R. Thieringer and P. Lazarow, personal communication (1989).
[63] D. McCollum and S. Subramani, personal communication (1989).
[64] S. McConnell and M. Yaffe, personal communication (1989).
[65] R. Azpiroz and R. A. Butow, personal communication (1989).
[66] V. Berlin and G. R. Fink, personal communication (1989).

Fluorophores

A wide variety of fluorescent probes is now available,[3,67,68] some of which are useful in immunofluorescence and related techniques because they can be coupled covalently to antibodies and other proteins such as lectins and avidin (see also below).[67-70] Of these, the workhorses for immunofluorescence have been fluorescein isothiocyanate (FITC) and tetramethylrhodamine isothiocyanate (TRITC), because of their convenient spectral properties and ease of coupling to antibodies. In general, FITC and TRITC are equally satisfactory; although FITC photobleaches more rapidly, this problem is easily solved for most purposes by the use of chemical antibleaching agents (see below). Investigators should also be aware that additional fluorophores, such as other rhodamine derivatives (notably Texas Red) and the phycobiliproteins,[67,71-79] can be of great value in certain applications. Some useful information about these and other relevant fluorescent probes is collected in Table II. A point to note is that the possibility of using two or more fluorophores in double- or multiple-label experiments depends on their spectral resolution at the level of excitatory wavelengths, emission wavelengths, or (preferably) both; inadequate spectral resolution can lead to troublesome "crossover" fluorescence (see also below). Investigators should also be aware that work continues on developing additional fluorophores designed to be easy to use and to have improved spectral properties [i.e., molar absorption and quantum yield (hence brightness); resistance to photobleaching; and spectral resolution].[67,68,70] For example, a recent attractive addition is 7-amino-4-methylcoumarin-3-acetic acid (AMCA) (Table II), whose blue fluorescence allows

[67] R. P. Haugland, "Handbook of Fluorescent Probes and Research Chemicals." Catalogue of Molecular Probes, Inc., Eugene, Oregon, 1989.

[68] Y.-L. Wang and D. L. Taylor (eds.), "Fluorescence Microscopy of Living Cells in Culture, Part A" (*Methods Cell Biol.* **29**). Academic Press, San Diego, California, 1989.

[69] R. P. Haugland, in "Excited Stages of Biopolymers" (R. F. Steiner, ed.), p. 29. Plenum, New York, 1983.

[70] A. S. Waggoner, in "Applications of Fluorescence in the Biomedical Sciences" (D. L. Taylor, A. S. Waggoner, R. F. Murphy, F. Lanni, and R. R. Birge, eds.), p. 3. Alan R. Liss, New York, 1986.

[71] V. T. Oi, A. N. Glazer, and L. Stryer, *J. Cell Biol.* **93**, 981 (1982).

[72] R. R. Hardy, K. Hayakawa, D. R. Parks, and L. A. Herzenberg, *Nature (London)* **306**, 270 (1983).

[73] A. N. Glazer and L. Stryer, *Trends Biochem. Sci.* **9**, 423 (1984).

[74] D. R. Parks, R. R. Hardy, and L. A. Herzenberg, *Cytometry* **5**, 159 (1984).

[75] L. L. Lanier and M. R. Loken, *J. Immunol.* **132**, 151 (1984).

[76] R. Festin, B. Björklund, and T. H. Tötterman, *J. Immunol. Methods* **101**, 23 (1987).

[77] D. C. Nguyen, R. A. Keller, J. H. Jett, and J. C. Martin, *Anal. Chem.* **59**, 2158 (1987).

[78] J. C. White and L. Stryer, *Anal. Biochem.* **161**, 442 (1987).

[79] R. S. Lewis and M. D. Cahalan, *Science* **239**, 771 (1988).

TABLE II
SOME FLUOROPHORES USEFUL IN IMMUNOFLUORESCENCE MICROSCOPY OF YEAST[a]

Fluorophore	Excitation wavelength[b] (nm)	Emission wavelength[b] (nm)	Appropriate filter sets[c]	
			Leitz	Zeiss
FITC[d]	490–495	525	I2/3 or L3	487709 or 487710
TRITC[e]	540–552	570	N2 or M2	487714 or 487715
Texas Red[f]	590–596	615–620	N2	487700 or 487714
AMCA[g]	350	450	A or A2	487701 or 487702
B-Phycoerythrin[h]	540–565	576	N2 or M2	487714 or 487715
C-Phycocyanine[h]	605–620	645–650	N2?	487700
Calcofluor[i]	340–360	400–440	A	487702
DAPI[j]	340–365	450–488	A	487701 or 487702
Acridine orange[j]	470–503	523–650	I2/3 or K3	487709 or 487711
Hoechst 33258[j]	365–374	472–480	A or B2	487702 or 487718

[a] Additional information about these and many other fluorescent probes can be found in the Molecular Probes handbook (R. P. Haugland, "Handbook of Fluorescent Probes and Research Chemicals." Molecular Probes, Inc., Eugene, Oregon, 1989).

[b] The approximate maxima for excitation and emission are given. However, it should be noted that both processes actually occur over a more-or-less broad band of wavelengths, and the positions of the maxima may vary somewhat with the chemical environment.

[c] The filter information is provided simply for general guidance. In some applications, other standard or nonstandard filter combinations will be equally or more appropriate for use with Zeiss or Leitz microscopes, and other types of fluorescence microscopes will have their own versions of appropriate filter sets. Note that the Zeiss filter numbers given are for the older series microscopes; filters for the new Axio line are the same except for a 9 instead of a 7 in the fourth place (e.g., 487909 or 487910 for FITC).

[d] Fluorescein 5-isothiocyanate. The spectral data shown refer to conjugates at pH ≥ 8 (Haugland, *op. cit.*).

[e] Tetramethylrhodamine 5 (and 6)-isothiocyanate. The effect of pH on the spectral properties is not clear.

[f] Trademark name of Molecular Probes, Inc., for a rhodamine X sulfonyl chloride. Spectral data refer to conjugates and are reportedly pH independent near pH 7. Texas Red offers a comparable quantum yield (hence brightness) to TRITC and has better spectral resolution from FITC (thus minimizing crossover fluorescence in double-label experiments). A limitation for the use of Texas Red has been the failure of the microscope manufacturers to offer filter sets optimal for its use.

[g] 7-Amino-4-methylcoumarin-3-acetic acid, recently made available by Jackson ImmunoResearch in a variety of antibody conjugates. It should prove useful in double- and triple-label experiments with FITC and TRITC.

[h] Two of a number of phycobiliproteins that have great promise as fluorescent labels because of their spectral properties (intense fluorescence, reasonable photostability, widely separated excitation and emission maxima). See the references given in the text. The large size of these molecules may possibly limit their usefulness in labeling intracellular structures.

[i] Used in the staining of cell wall chitin (J. R. Pringle, this volume [52]), sometimes in conjunction with immunofluorescence (see text).

[j] DAPI (4',6-diamidino-2-phenylindole dihydrochloride), acridine orange, and Hoechst 33258 are used in the staining of nuclear and mitochondrial DNA, sometimes in conjunction with immunofluorescence (see text). Note that acridine orange stains RNA heterochromatically [M. E. Dresser and C. N. Giroux, *J. Cell Biol.* **106**, 567 (1988)], accounting for its wide range of emission wavelengths.

triple-label immunofluorescence using FITC and TRITC as the other fluorophores.

Direct and Indirect Immunofluorescence

Most immunofluorescence experiments done with yeast and other cell types have utilized *indirect* immunofluorescence. That is, the fluorophore ultimately visualized is present not on the *primary* antibodies (i.e., the antibodies that recognize the antigen of interest) but on *secondary* antibodies that recognize and bind to the primary antibodies. (For example, to visualize actin, one might use rabbit antiactin primary antibodies and fluorophore-conjugated goat anti-rabbit-IgG secondary antibodies.) This approach offers some amplification of signal (several secondary antibody molecules, each with its complement of fluorophores, can bind to one primary antibody molecule) as well as convenience (the fluorophore-conjugated secondary antibodies, each of which may be used with a variety of different primary antibodies, can be purchased ready-made from commercial suppliers). However, despite these advantages, *direct* immunofluorescence, in which the fluorophore is conjugated directly to the primary antibodies, may sometimes be preferable.[7,80] The coupling reactions with conventional fluorophores such as FITC and TRITC are not difficult.[80,81] Although the signal is likely to be weaker than in indirect immunofluorescence, the signal-to-noise ratio may actually be better (thus facilitating video enhancement of the weak signal; see below), as nonspecific binding of secondary antibodies appears to be a frequent source of troublesome "background" fluorescence. Moreover, direct immunofluorescence can greatly facilitate double-label experiments, which are otherwise difficult if the primary antibodies available are all derived from the same type of animal (see also below).

Microscopy and Photomicroscopy

Microscopes and Objectives

Conventional Microscopes. In modern epifluorescence microscopes, the excitatory illumination is directed onto the specimen through the objective lens. The backscattered excitatory illumination that reenters the objective is then absorbed by a suppression filter that is an integral part of the packet that also supplies the excitatory illumination. [The booklets "Fluorescence

[80] J. C. Talian, J. B. Olmsted, and R. D. Goldman, *J. Cell Biol.* **97,** 1277 (1983).
[81] J. J. Haaijman, in "Immunohistochemistry" (A. C. Cuello, ed.), p. 47. Wiley, Chichester, England, 1983.

Microscopy" (by E. Becker; distributed by Leitz) and "Worthwhile Facts about Fluorescence Microscopy" (by H. M. Holz; distributed by Zeiss) provide helpful, simple introductions to this and other aspects of the hardware for fluorescence microscopy.] Epifluorescence microscopes thus offer major advantages over older fluorescence microscopes in terms of simplicity of operation (there is no need to worry about alignment, focusing, and eventual de-oiling of a separate condenser, and all the filters are put into position by a single manipulation), safety (there is no danger of accidentally leaving out a suppression filter and thus searing the eyes with UV excitatory illumination), brightness of object fluorescence (the excitatory illumination is focused intensely on the area of the object that is being observed), increased spectral resolution (the more intense excitatory illumination means that a narrower band of excitatory wavelengths can be used), and longevity of the sample (areas of the object that are not being examined are not illuminated, and hence do not suffer photobleaching).

Excellent results can be obtained with any of the high-quality epifluorescence microscopes presently on the market. It is important to realize that the vast majority of applications in yeast cell biology do not require a top-of-the-line microscope with its costly features designed to provide maximum stability, versatility, and semiautomatic operation. In particular, the same quality optics are available in the less costly versions of the research microscope lines of the major manufacturers. Nor are large numbers of objectives required. As a result, although it is not hard to spend over $50,000 on a microscope system, it is possible to do very well while spending much less. For example, although we have been very pleased by the performance of the Leitz Orthoplan microscope, we have also been very satisfied by the performance of the less expensive Zeiss Axioscop, which was set up for bright-field, differential interference contrast (DIC, or Nomarski), and fluorescence work on yeast. Comparable results may well be obtainable at still lower cost using instruments from Leitz (Laborlux or Diaplan series microscopes), Nikon, or Olympus.

Confocal Scanning Microscopes. Confocal scanning microscopes are expensive to set up and not yet widely available, and they are certainly not essential for all applications. Nonetheless, investigators should be aware of the great potential benefits offered by this latest major advance in light microscope technology.[82-85] In confocal scanning microscopes, the object

[82] A. Boyde, *Science* **230**, 1270 (1985).
[83] G. J. Brakenhoff, H. T. M. van der Voort, E. A. van Spronsen, W. A. M. Linnemans, and N. Nanninga, *Nature (London)* **317**, 748 (1985).
[84] G. J. Brakenhoff, H. T. M. van der Voort, E. A. van Spronsen, and N. Nanninga, *Ann. N.Y. Acad. Sci.* **483**, 405 (1986).
[85] J. G. White, W. B. Amos, and M. Fordham, *J. Cell Biol.* **105**, 41 (1987).

is viewed not as a whole, but as a series of points on which both illumination and detection are tightly focused. A two-dimensional scan of the object thus produces an optical section, whereas a series of such scans allows a three-dimensional image to be constructed and presented on a video screen or as stereo-pair micrographs. These systems offer an increase in potential resolution, and the digitally stored information is readily available for further computer analysis. However, the greatest advantage of the confocal scanning systems appears to be in fluorescence applications, where a major problem is often the swamping of the in-focus fluorescence signal by noise in the form of fluorescence from out-of-focus portions of the object. By taking a thin (~0.5 μm) optical section, the confocal scanning microscope can sharply reduce the contribution of out-of-focus fluorescence, producing in some applications a dramatic increase in the sharpness and contrast of the ultimate images.[85] One potential liability of confocal microscopy is that the intensely focused illumination can lead to especially rapid photobleaching, producing misleading images under some circumstances. Confocal microscopes have been applied successfully in fluorescence work with yeast.[53,58,86,87]

Lamps. Optimal illumination (hence proper alignment of the light source) is critical to high-quality fluorescence microscopy. Another consideration is that the brightness and quality of the fluorescence image deteriorate as the mercury lamp begins to fail, especially for fluorophores requiring UV excitation. In addition, explosions of these lamps (which can occur as the quartz gets brittle with age) are reputedly unpleasant to experience, even if successfully contained by the lamp housings (which are designed to meet this challenge). Thus, although the mercury lamps are expensive, it is generally a false economy to continue using them after they have reached their stated lifetime or after their brightness begins to decrease or fluctuate noticeably.

Filters. Microscope manufacturers provide a choice of prepackaged sets of excitation and suppression filters that are appropriate for use with the most commonly used fluorophores (see Table II and the Becker and Holz books, mentioned above). Alternatively, special filter sets can be assembled for particular purposes (e.g., certain double-label experiments; see below) using the filters available from the microscope manufacturers or from other suppliers such as Omega Optical.

Field Diaphragm. Epifluorescence microscopes are equipped with a field diaphragm (iris) that controls the area of the object that is illuminated. To reduce background fluorescence and photobleaching of parts of the

---

[86] J. L. Allen and M. G. Douglas, personal communication (1989).

[87] C. Copeland and M. Snyder, personal communication (1989).

specimen that have not yet been examined, the field diaphragm should always be adjusted so that the area being illuminated is no larger than the field of view. Indeed, following the same arguments, it is sometimes helpful to stop this diaphragm down further so that only the cell(s) of immediate interest is illuminated.

Objectives. The objective lens chosen must obviously provide appropriate magnification (for yeast, almost always 40–100×). However, several additional important considerations also influence the choice of an appropriate objective for immunofluorescence work. First, in epifluorescence microscopy, both the intensity of the excitatory illumination and the efficiency of collection of the emitted fluorescence are proportional to the square of the numerical aperture (NA) of the objective. Thus, for objectives of a given magnification, the brightness of the fluorescence image is proportional to the fourth power of the NA. In photomicroscopy, the effect is even greater, as a brighter image allows shorter exposure times and thus less photobleaching. As the optical resolving power of an objective also increases with the NA, it is clear that there are strong incentives to use high-NA (hence oil immersion) objectives. On the other hand, it is not uncommon to find that optical resolution is not limiting and that an object is actually too brightly stained. In this situation, use of an objective of reduced NA can facilitate both observation (by reducing the brightness to a comfortable level and reducing the rate of photobleaching) and photography (by allowing more flexibility in the setting of exposure times). This is most conveniently accomplished by using an objective in which the NA is actually adjustable by means of an iris diaphragm. Moreover, we have the strong impression that stopping down the iris in such an objective can actually increase contrast, though we are unsure whether this visual effect is reproduced in photomicrographs.

Second, other things being equal, the brightness of the fluorescence image is inversely proportional to the square of the total magnification (objective magnification times eyepiece magnification). Thus, if image brightness is a limiting factor for observation or photography, use of a lower magnification should be considered. Third, in phase-contrast and dedicated DIC objectives, there are unavoidable losses of both excitatory and emitted light in the objectives because of the phase rings or built-in interference-contrast prisms, resulting in less bright fluorescence images. (This problem should be less severe with DIC than with phase-contrast objectives.) Thus, in choosing an objective, one must weigh the advantages of obtaining a maximally bright fluorescence image with a bright-field objective against the advantages (often considerable) of obtaining companion phase-contrast or DIC images without having to change the objective. In some new microscopes, this problem has been solved for DIC by placing the necessary prisms in a special slider (so they can be moved out of position when not in use) and using ordinary bright-field objectives. Finally, the expense of objectives is determined in part by the width of the flat field that they provide. Although the uniformity of focus provided by a wide flat field is sometimes convenient for photography, it is rarely if ever essential in yeast work. Comparably good correction for chromatic and spherical aberrations and equally good fluorescence properties are available in objectives that are substantially less expensive than those offering the largest flat fields.

Slides and Immobilization of Cells

Immunofluorescence is best carried out on cells attached by polylysine to multiwell slides, as described below. This approach allows efficient staining and washing using small volumes of solutions, and the cells are immobilized for eventual photomicroscopy. We have used slides from Flow Laboratories (Cat. No. 60-408-05) or Polysciences (Cat. No. 18357). Some lot-to-lot variability has occurred in the wettability of the glass wells of these slides; on several occasions, the polylysine solution has beaded up rather than spread properly. A variety of similar slides (with which we have no personal experience) are also available from Cel-Line and other suppliers.

Bleaching Problem

All fluorophores undergo photobleaching to some extent, although the rates vary considerably with the fluorophore, the chemical environment, and the wavelength and intensity of the excitatory illumination. To minimize photobleaching prior to observation, we routinely perform all manipulations involving fluorophores under minimal illumination and keep the samples in the dark (aluminum foil is convenient) during incubations and storage. In addition, immunofluorescence preparations are routinely mounted in a medium that contains the chemical antibleaching agent *p*-phenylenediamine (see recipe below). This agent greatly retards the photobleaching of several important fluorophores (notably fluorescein, rhodamine, and their derivatives) and may even intensify their initial fluorescence.[88-90] The photobleaching of both fluorescein and rhodamine conjugates can also be retarded by *n*-propyl gallate,[91] but we have no personal experience with this agent. Photobleaching can also sometimes be

---

[88] G. D. Johnson and G. M. de C. Nogueira Araujo, *J. Immunol. Methods* **43**, 349 (1981).
[89] G. D. Johnson, R. S. Davidson, K. C. McNamee, G. Russell, D. Goodwin, and E. J. Holborow, *J. Immunol. Methods* **55**, 231 (1982).
[90] J. L. Platt and A. F. Michael, *J. Histochem. Cytochem.* **31**, 840 (1983).
[91] H. Giloh and J. W. Sedat, *Science* **217**, 1252 (1982).

alleviated by using nonstandard filter sets that provide a different range of excitatory wavelengths.

Film

Exposure time is often the limiting factor in getting good immunofluorescence photomicrographs. At the same time, using too fast a film can result in an undesirably grainy micrograph. Thus, it is important to choose an appropriate combination of film and developing conditions.

Black-and-White Film. We have taken many satisfactory photomicrographs using standard Kodak Tri-X Pan film (ASA 400), usually "pushing" the development to ASA 1600 using Diafine developer (Acufine Corp.; available in many local photography stores). However, it is clear that the slightly more expensive Kodak T-MAX film (ASA 400) gives a finer-grain negative and can be pushed to ASA 1600 or 3200 by extending developing times with the special T-MAX developer (Kodak). (In our experience to date, the best compromise between sensitivity and graininess was achieved at ASA 1600.) Recently, we have also had good results with the still more sensitive Kodak T-MAX p3200. Even better in its combination of a fine grain size with high sensitivity is the somewhat more expensive hypersensitized Kodak Technical Pan 2415 film (Lumicon),[92,93] developed according to the supplier's instructions.

Color Film. Color slides are effective for seminars, especially when double-label experiments are to be shown. We have routinely used Kodak Ektachrome 400 with satisfactory results. It often helps to push the development to higher effective ASA by extending the developing time. On the rare occasions when color prints are needed, they can be prepared from the color slides either by making an internegative or by a direct printing process.

The Importance of Taking Many Photographs

Even with the best equipment and the best hands, the outcome of photomicroscopy is somewhat uncertain. Moreover, once a good preparation is available, the cost (both in materials and in investigator time) of taking additional photographs is generally small in relation to the total cost of doing the research. Thus, we strongly recommend that investigators routinely take many more photographs than they expect to need. This allows later documentation of the range of structures observed and also provides many choices from which to select the best micrographs for presentation. Along the same lines, it is convenient to keep two separate

---

[92] E. Schulze and M. Kirschner, *J. Cell Biol.* **102,** 1020 (1986).
[93] E. Schulze and M. Kirschner, *J. Cell Biol.* **104,** 277 (1987).

cameras loaded with black-and-white and color film, respectively, so that particularly favorable images can be captured in both formats.

Antibodies

Sources of Primary Antibodies

Sometimes useful primary antibodies can be obtained from colleagues or commercial sources. More often, however, the desired primary antibodies must be prepared by the investigator. This requires tactical decisions as to what type of antibodies (polyclonal, monoclonal, or oligopeptide-directed) is to be prepared, as well as to the nature of the immunogen (protein purified from yeast, fusion protein purified from *Escherichia coli*, etc.) to be used. Full consideration of these issues and of the details of raising antibodies is beyond the scope of this chapter; novices are referred to the relevant volumes in this series,[94-97] as well as to their local immunologists.

Type of Antibodies. It is clear that monoclonal and oligopeptide-directed antibodies can offer unexcelled specificity, which is essential for some purposes (e.g., distinguishing between the intracellular locations of two closely related antigens[98-101]). However, it should not be forgotten that such antibodies can sometimes recognize an epitope shared by two or more related or unrelated antigens,[102-104] so that uncritical acceptance of their specificity is not warranted. Moreover, for most immunofluorescence purposes, it appears that properly purified (monospecific) polyclonal antibodies are equally good or even superior reagents, as well as being (usually) easier and less expensive to prepare. The potential superiority of polyclonal antibodies derives from the fact that, in most cases, the serum will contain multiple different antibody species that recognize different determinants on the antigen of interest. Thus, although any one antigenic determinant

---

[94] H. Van Vunakis and J. J. Langone (eds.), this series, Vol. 70.
[95] J. J. Langone and H. Van Vunakis (eds.), this series, Vol. 73.
[96] J. J. Langone and H. Van Vunakis (eds.), this series, Vol. 92.
[97] J. J. Langone and H. Van Vunakis (eds.), this series, Vol. 121.
[98] H. S. Phillips, K. Nikolics, D. Branton, and P. H. Seeburg, *Nature (London)* **316**, 542 (1985).
[99] J. F. Bond, J. L. Fridovich-Keil, L. Pillus, R. C. Mulligan, and F. Solomon, *Cell (Cambridge, Mass.)* **44**, 461 (1986).
[100] G. Piperno, M. LeDizet, and X. Chang, *J. Cell Biol.* **104**, 289 (1987).
[101] Y. Deng and B. Storrie, *Proc. Natl. Acad. Sci. U.S.A.* **85**, 3860 (1988).
[102] R. M. Pruss, R. Mirsky, M. C. Raff, R. Thorpe, A. J. Dowding, and B. H. Anderton, *Cell (Cambridge, Mass.)* **27**, 419 (1981).
[103] J. V. Kilmartin, B. Wright, and C. Milstein, *J. Cell Biol.* **93**, 576 (1982).
[104] S. J. Elledge and R. W. Davis, *Mol. Cell. Biol.* **7**, 2783 (1987).

may be fixation sensitive, or inaccessible to antibody in the native structure, it is unlikely that all will be. Moreover, a single molecule of a macromolecular antigen should be able to bind several different antibody molecules, thus increasing the ultimate strength of the immunofluorescence signal.

Type of Immunogen. The choice of immunogen will in most cases be conditioned by whether the original identification of an antigen of interest was biochemical or genetic. In the latter case, there are usually several ways to approach the acquisition of an immunogen that can be used to raise antibodies specific for that gene product. Although every case should be evaluated on its own merits, it is worth noting that the use of fusion proteins frequently offers advantages. Given modern vector systems (notably for constructing fusions of the gene of interest to the *E. coli lacZ* and *trpE* genes), the necessary constructions at the DNA level are usually straightforward, especially if the sequence of the gene of interest is known. The fusion proteins are then produced in large amounts by *E. coli* and can usually be purified sufficiently to serve as immunogens by simple one- or two-step procedures. As any contaminating proteins derive from *E. coli* rather than yeast, they are less likely to elicit antibodies that will subsequently give confusing results. Moreover, the fusion proteins themselves then provide convenient bases for affinity purification.[21,29,54,57] Especially valuable is the fact that two quite different fusion proteins can usually be prepared (nearly) as cheaply as one. This increases the probability of eliciting a strong immune response and allows a particularly effective affinity purification, in which antibodies raised against one fusion protein are purified on a matrix composed of the other.[21,29]

The Importance of Affinity Purification

The serum of a mammal or bird is an extremely complex reagent. Numerous examples make it clear that the serum of an immunized animal may contain, in addition to the antibodies specifically elicited by the immunization procedure, other antibodies that react with antigens in the organism of interest. For example, preimmune sera from many rabbits contain antibodies that react with some component of mammalian spindle poles.[105,106] In yeast work, Payne and Schekman[107] observed that the sera of rabbits immunized with purified clathrin heavy chain also contained antibodies (apparently recognizing carbohydrate structures) that were a problem in immunoblotting experiments unless they were first removed by

---

[105] J. A. Connolly and V. I. Kalnins, *J. Cell Biol.* **79**, 526 (1978).
[106] B. W. Neighbors, R. C. Williams, Jr., and J. R. McIntosh, *J. Cell Biol.* **106**, 1193 (1988).
[107] G. S. Payne and R. Schekman, *Science* **230**, 1009 (1985).

preadsorption of the sera with whole cells. Similarly, attempts to immunolocalize the yeast *CDC12* gene product were at first thoroughly obfuscated by antibodies, present in the preimmune sera of all rabbits tested (~20, including both males and females), that reacted with a component (apparently chitin) of the cell wall.[29] In one rabbit, the titer of this cell wall-reactive antibody was boosted approximately 50-fold during the course of immunization with fusion protein purified from *E. coli,* so that the comparison of preimmune to immune sera was initially quite misleading. Lillie and Brown[108] also observed that the comparison of preimmune to immune sera is not always an adequate test of the specificity of an immunofluorescence result. One of the immune sera produced a striking staining pattern of spots and blotches that was not apparent with the preimmune serum used at comparable concentrations; however, affinity purification showed clearly that this staining pattern was not due to the antibodies recognizing the protein that had been used as immunogen. (Indeed, such staining patterns have proved to be a rather common artifact in attempts to do immunofluorescence with insufficiently purified sera.) Another apparently artifactual result that we have observed on several occasions is immunofluorescence staining of the mitotic spindle when using antisera raised against proteins that are almost certainly not associated with the spindle.[109] In one case, the relevant antibodies were already detectable in the preimmune serum, but in other cases they were not. Lillie and Brown[108] also observed that the preimmune sera of various rabbits frequently recognized a variety of yeast proteins in immunoblotting experiments, and that the pattern of proteins recognized could change on a time scale of weeks even in the absence of any intentional immunization of the rabbits.

These results demonstrate clearly that it is very dangerous to take seriously an immunofluorescence result until there is good evidence that the antibodies used are in fact monospecific for the antigen of interest. Immunoblotting experiments can provide important evidence for such monospecificity, but it should be noted that this test is not foolproof; for example, antibodies reactive with cell wall carbohydrates or nucleic acids presumably would not be detected by testing antisera on blots of cellular proteins. Therefore, careful affinity purification should *always* precede attempts to determine the intracellular localization of an antigen by immunofluorescence. When possible, the affinity purification should use a matrix other than the original immunogen (e.g., antibodies raised against one fusion protein can be purified using a different fusion protein), to minimize the chances that antibodies recognizing a contaminating antigen in

---

[108] S. H. Lillie and S. S. Brown, *Yeast* **3,** 63 (1987).
[109] S. K. Ford, B. Rahe, and J. R. Pringle, unpublished results (1988, 1989).

the original immunogen preparation will be affinity purified right along with the antibodies recognizing the antigen of interest. In any case, the success of the affinity purification should be monitored by performing both immunoblotting and immunofluorescence experiments using both the purified and "depleted" fractions (see below). Another powerful control that should be applied whenever possible (i.e., whenever the protein studied is nonessential) is to perform immunofluorescence on a strain carrying a null mutation in the structural gene for the protein of interest.

Affinity Purification on Nitrocellulose Blots

The use of blots of electrophoretically separated proteins as affinity-purification matrices was pioneered by Olmsted.[110] In comparison with column purification methods (see below), this approach offers simplicity and facilitates effective work with low-abundance antigens and their corresponding antibodies. Although the yields from blot purification are lower, they are sufficient for numerous immunofluorescence experiments, the screening of λgtll libraries, etc. The original procedure of Olmsted and some subsequent successful applications[111] have utilized diazotized paper; this may possibly offer some advantage in terms of the number of times a given blot can be reused. However, for most purposes, it appears simpler and at least as satisfactory to use nitrocellulose blots.[29,80,108,112] These blots also can be reused many times; in one case, a blot used more than 100 times is still giving satisfactory affinity purification.[113]

Purification Protocol. We have consistently had satisfactory results with the following protocol. After the initial preparation of the blots, all steps are performed at room temperature except as noted.

1. Preparation of blots. Separate proteins by sodium dodecyl sulfate (SDS)–polyacrylamide gel electrophoresis and transfer electrophoretically to nitrocellulose paper using standard procedures.[114,115] Allow the blot to dry without rinsing out the residual transfer buffer.

2. Identification of protein bands. Immediately after preparation of the blot or whenever desired (the dried blots are stable for months if not years), visualize the protein bands of interest on the blot by staining with Ponceau

---

[110] J. B. Olmsted, *J. Biol. Chem.* **256**, 11955 (1981).
[111] K. J. Green, R. D. Goldman, and R. L. Chisholm, *Proc. Natl. Acad. Sci. U.S.A.* **85**, 2613 (1988).
[112] D. E. Smith and P. A. Fisher, *J. Cell Biol.* **99**, 20 (1984).
[113] S. H. Lillie, personal communication (1988).
[114] H. Towbin, T. Staehelin, and J. Gordon, *Proc. Natl. Acad. Sci. U.S.A.* **76**, 4350 (1979).
[115] W. N. Burnette, *Anal. Biochem.* **112**, 195 (1981).

S.[116,117] Immerse the blot in 0.2% Ponceau S (Sigma, St. Louis, MO, Cat. No. P-3504) in 0.3% trichloroacetic acid (TCA) for approximately 10 min, then destain with several changes of distilled water until bands are visible (typically 5–10 min), using gentle agitation at each step. At this point, the band(s) of interest can be cut out using a clean razor blade and the remaining stain removed by washing with several changes of phosphate-buffered saline (PBS) (to minimize the chance of interference with the antigen–antibody reaction). The strip of blot can then be used immediately for affinity purification or dried and stored indefinitely. Alternatively, the Ponceau-stained blot can be dried and stored until needed; the stain is then removed by washing with PBS as above just before proceeding with the affinity purification.

3. "Blocking" (to minimize nonspecific binding of antibodies). Incubate the strip in 3–10 ml blocking solution (5% nonfat dry milk in PBS) for 50 min with gentle agitation in a plastic petri dish or other suitable container.

4. Washing. Wash the strip 3 times, using 3–10 ml PBS and 5 min of gentle agitation per wash.

5. Incubation with antiserum. Place a piece of Parafilm on the bottom of a Petri dish and lay the nitrocellulose strip on top (drained, but wet; protein side up). Carefully layer the crude serum (undiluted or diluted several fold with PBS) or IgG fraction [prepared by chromatography on protein A-Sepharose (Pharmacia Cat. No. 17-0963-03), protein G-Sepharose (Pharmacia Cat. No. 17-0618), or Affi-Gel Protein A (Bio-Rad Cat. No. 153-6153), following the manufacturer's instructions] on top of the nitrocellulose, using about 200 μl for a 1 × 3 cm strip. Place on shaker for 2–3 hr, shaking fast enough to see the antibody solution move back and forth. It may help to tape a wet Kimwipe to the top of the petri dish to prevent excessive evaporation.

6. Removal of "depleted fraction." Lift the strip slowly from one corner while removing liquid with a Pipetman; save this "depleted fraction" below 4° as a control on the effectiveness of the affinity purification.

7. Washing. Wash the strip 3 times, using 3–10 ml PBS and 10 min of gentle agitation per wash.

8. Elution of purified antibodies. Place the drained, but wet, nitrocellulose strip on a fresh piece of Parafilm in a petri dish. Layer on 200 μl of low-pH buffer (0.2 M glycine, 1 mM EGTA, pH 2.3–2.7) and shake as in Step 5 for 10–20 min. (The optimal pH and time are probably compromises between elution and inactivation of the antibodies of interest, and

[116] L. S. B. Goldstein, R. A. Laymon, and J. R. McIntosh, *J. Cell Biol.* **102**, 2076 (1986).
[117] N. Moreau, N. Angelier, M.-L. Bonnanfant-Jais, P. Gounon, and P. Kubisz, *J. Cell Biol.* **103**, 683 (1986).

may vary from case to case.) Remove the liquid as in Step 6, quickly neutralize either by adding an equal volume of cold 100 mM Tris base or by adding 3 N NaOH (~4–4.5 μl) to bring to approximately pH 7 (check by spotting 1-μl aliquots onto pH paper), and store below 4°. In some cases, a significant amount of additional purified antibody may be recovered by repeating the elution procedure.

9. Washing and storing filter. Wash the strip 3 times with PBS, as above, and store below 4° in PBS or after drying for reuse.

Comments. As with other protocols, the conditions necessary to give optimal results will almost certainly vary from case to case. The following considerations may be helpful.

1. Although our own experience has been with blots of proteins separated by SDS–polyacrylamide gel electrophoresis, it seems likely that blots of proteins separated by other means (e.g., urea–gel electrophoresis or two-dimensional gel electrophoresis) would also form satisfactory matrices for affinity purification. In addition, spotting purified native proteins onto nitrocellulose might provide satisfactory matrices in some cases; this approach could be especially useful in purifying antibodies that do not react well with denatured proteins.

2. Although we have had good success to date with the Ponceau S method of visualizing the protein bands of interest prior to excision, it seems likely that in some cases this stain or the associated exposure to TCA may be inimical to subsequent antigen–antibody reactions. In such a case, the bands of interest can be located by staining outside lanes and/or a central lane. Staining for 1 min in an aqueous solution containing 0.1% (w/v) amido black, 25% (v/v) 2-propanol, and 10% (v/v) acetic acid, followed by destaining for 20 min in the same solution without the amido black, has given satisfactory results. Also, in some cases involving complex protein mixtures, it may be necessary to use antibody staining (e.g., with the relevant primary antibody plus ^{125}I-labeled protein A or peroxidase-coupled secondary antibody) to locate the protein of interest prior to excision of the appropriate region of the blot. In such a case, however, special caution would be needed to ensure that the antibodies purified really recognized only the protein of interest, and not also one or more comigrating proteins.

3. Although nonfat dry milk seems particularly effective as a blocking agent, other proteins or protein mixtures [e.g., bovine serum albumin (BSA) or gelatin solutions] may also give satisfactory results.

4. Although to date we have had excellent success in eluting the affinity-purified antibodies with the low-pH buffer described above, several lines of evidence suggest that this protocol will not always give an acceptable yield of active antibodies. First, in attempts to purify antibodies

specific for the mammalian microtubule-associated protein tau using affinity columns, elution with low-pH buffers consistently failed to release active antibodies.[118] However, elution with 4.5 M MgCl$_2$ did release active antibodies.[119] Second, when affinity columns were used to purify antibodies specific for the yeast *CDC46* gene product, the desired antibodies from the serum of one rabbit were eluted effectively with 4.5 M MgCl$_2$; on the other hand, with the serum from a second rabbit, 4.5 M MgCl$_2$ was not effective, but the antibodies were eluted by a subsequent wash with low-pH buffer.[120] Third, in attempts to use blots of purified yeast profilin to purify profilin-specific antibodies, little active antibody was recovered after eluting with low-pH buffer;[121] apparently the antibodies remained bound to the profilin (as opposed to being eluted but inactivated). Surprisingly, however, good yields of profilin-specific antibodies were obtained when the same protocol was used with a blot of *trpE*-profilin fusion protein as the affinity-purification matrix. Given these precedents, we suggest that if low pH buffer does not seem to be eluting the desired antibodies satisfactorily in a given case, other elution regimens (such as one using 4.5 M MgCl$_2$; see also below) should be tried before the procedure is abandoned.

5. In some cases, better results seem to be obtained if the desired antibodies are carried through two cycles of affinity purification. For example, we significantly improved the signal-to-noise ratio observed both in immunoblots and in immunofluorescence with *CDC10*-specific antibodies by carrying the antibodies through two cycles of affinity purification, one on a blot of *trpE-CDC10* fusion protein, the other on a blot of *lacZ-CDC10* fusion protein.[122]

6. At least in some cases, blots of the appropriate protein can be used to concentrate as well as purify the antibodies of interest. To do this, simply incubate the nitrocellulose strip (or strips) in a relatively large volume (e.g., 1–2 ml) of serum and then elute with a smaller volume of low-pH buffer. It may help in such applications to increase the incubation times. In this case, be sure to keep the environment humid; it may also help to conduct the longer incubations at 4° rather than room temperature.

Affinity Purification on Columns

Affinity purification on columns is somewhat more laborious and requires more antigen than does affinity purification on blots, but it offers higher yields of purified antibodies. (If large amounts of antibodies are

[118] D. G. Drubin, unpublished results (1989).
[119] S. R. Pfeffer, D. G. Drubin, and R. B. Kelly, *J. Cell Biol.* **97**, 40 (1983).
[120] K. Hennessy, personal communication (1989).
[121] S. S. Brown, personal communication (1989).
[122] H. B. Kim and J. R. Pringle, unpublished results (1989).

required, it may be less work in the long run to do a single large preparation by affinity column than to do repeated small preparations using blots.)

Purification Protocol. Protocols such as the following have been effective in a variety of specific applications;[25,30,57,119] except as noted, all steps are conducted at room temperature.

1. Prepare buffer T (50 mM Tris-HCl, pH 7.4) and buffer M (4.5 M MgCl$_2$, 0.1% BSA in buffer T; prepare by adding 91.5 g of MgCl$_2 \cdot$6H$_2$O to 40 ml buffer T containing 2.5 mg/ml BSA).

2. Prepare a 0.5- to 1-ml column containing 0.2–2 mg of antigen coupled to CNBr-activated Sepharose 4B (Pharmacia Cat. No. 17-0430-01) according to the protocol supplied by Pharmacia.

3. Wash the column with 15 ml of 6 M guanidine-HCl, then equilibrate with buffer T by passing 25 ml through the column.

4. Wash the column with 20 ml buffer M, then wash again with 50 ml buffer T.

5. Run 5–30 ml crude serum or purified IgG fraction (see above) over the column in 2 hr or longer. Save the flowthrough (depleted fraction) as a control to determine how efficiently the antibodies of interest were bound.

6. Wash the column successively with 20 ml buffer T, 40 ml of 1.0 M guanidine-HCl, and 20 ml buffer T.

7. Elute purified antibodies with buffer M. Collect six or more 1-ml fractions and dialyze immediately (separately or after pooling) against 1 liter PBS for at least 3 hr. To assay antibody in the fractions, dilute 1 μl of each fraction (before dialysis) with 10 μl PBS and spot on nitrocellulose. When the filter is dry, probe it with ^{125}I-labeled protein A or with enzyme-conjugated secondary antibody.

8. After dialyzing against PBS (Step 7), dialyze for approximately an additional 12 hr against PBS containing 35% (v/v) glycerol, then store at $-20°$. The solution should not freeze because of the glycerol, and the antibodies are stabilized by the BSA.

9. Flush the column with buffer T containing 0.02% NaN$_3$ and store below 4°.

Comments. (1) The Affi-Gel supports (Bio-Rad Cat. Nos. 153-6046 and 153-6052) should provide a satisfactory alternative to CNBr-activated Sepharose. (2) Although the elution of purified antibodies from affinity columns with 4.5 M MgCl$_2$ is generally effective, it is not always so (see comments above in relation to affinity purification on blots). Other elution regimens that are reportedly effective in certain applications are as follows: high-pH buffer (50 mM diethylamine-HCl, pH 11.5); high-pH buffer containing 10% dioxane; low-pH buffer (0.2 M glycine, 1 mM EGTA, pH 2.3–2.7); and low-pH buffer containing 10% dioxane. See Bio-Rad Bulletin 1099 for additional discussion. (3) As in the case of purification on blots (see above), better results may sometimes be obtained by using more than one cycle of affinity purification. For example, Segev *et al.*[57] used two cycles of preadsorption on a column of *trpE* protein plus affinity purification on a column of *trpE-YPT1* fusion protein to prepare Ypt1p-specific antibodies.

Secondary Antibodies

High-quality, fluorophore-conjugated secondary antibodies are available from a variety of sources. We have worked mainly with antibodies obtained from Cappel (now Organon Teknika), Accurate, Zymed, Jackson ImmunoResearch, and Sigma, but products from other suppliers may be equally good. Our experience is mixed as regards the use of non-affinity-purified or affinity-purified secondary antibodies. On some occasions, the non-affinity-purified products have appeared to give a brighter signal, with no particular problem of background fluorescence. On other occasions, the additional expense of the affinity-purified products has apparently been repaid by the diminution of background fluorescence. (Recall that the ability to localize an antigen of interest, and to produce convincing photomicrographs documenting this localization, depends not just on the signal but on the signal-to-noise ratio.) Even with affinity-purified secondary antibodies, absolute specificity for the primary IgG of interest should not be assumed without trial; for example, some goat anti-rabbit IgG sera cross-react also with rat IgG, causing potential problems in double-label immunofluorescence experiments[25] (see also below).

Secondary antibodies are also available with a choice of fluorophores (see Table II and text, above). The commercial availability of biotin-conjugated secondary antibodies should also be noted; these can be used in conjunction with fluorochrome-conjugated avidin or streptavidin in schemes to amplify immunofluorescence signals (see below).

Fixation

Standard Fixation Protocol

For most purposes (some exceptions and caveats are noted below), the following fixation procedure may be used. (1) Add concentrated formaldehyde solution directly to the cells in growth medium to a final concentration of 3.7–5% (w/v) formaldehyde. (2) After 2–30 min at room temperature, recover the cells by centrifugation and resuspend in phosphate-buffered formaldehyde [40–100 mM potassium phosphate, pH 6.5, containing 0.5 mM MgCl$_2$ and 3.7–5% (w/v) formaldehyde]. (3) After

2–4 hr at room temperature, wash the cells free of formaldehyde with the appropriate solution and process them for immunofluorescence as described below.

Comments on Fixation Protocol

Importance of Rapid Fixation. Some features of cell structure can change surprisingly rapidly when cells are subjected to the stress of harvesting. For example, the asymmetric distribution of actin in budding cells[6,7,24] is much more evident when cells are fixed by adding fixative directly to the culture medium than when cells are first harvested by centrifugation and then resuspended in fixative;[3,25] the difference appears to reflect a response to energy limitation during harvesting.[3,28] Rapid changes in mitochondrial and vacuolar morphology have also been observed.[3] Thus, we recommend that cells ordinarily be fixed initially by adding fixative directly to the growth medium. At least with formaldehyde, this appears to stop life processes very rapidly, even in rich medium.[123]

Choice of Fixative. Nearly all immunofluorescence done to date with yeast has utilized formaldehyde-fixed cells. However, it should be noted that there will almost certainly be applications in which formaldehyde is not the optimal fixative. First, formaldehyde clearly gives insufficient preservation of cellular fine structure to be adequate for electron microscopy; there may be applications in which this limitation is also a factor in light microscopy. In this regard, it should be noted that a protocol involving fixation with 3% (w/v) paraformaldehyde plus 1% (w/v) glutaraldehyde (and incorporating $NaBH_4$ treatments to reduce the otherwise intolerable background fluorescence) has been used successfully with yeast[7] (see also Ref. 93), although the results in this particular case were not actually better than those obtained with formaldehyde. Alternatively, we have also had success with a fixation protocol employing a short (~3 min) fixation with a low concentration (0.025%) of glutaraldehyde in the growth medium, followed by further fixation (~30 min) with 3.5% formaldehyde in PBS;[109,124] with this protocol, $NaBH_4$ treatment was not necessary. Second, some antigens (or particular antigenic determinants) may be sensitive to fixation by formaldehyde. Whether for this or other reasons, the fixation conditions yielding optimal immunofluorescence results with different antigen–antibody combinations can vary widely, as shown clearly by work

---

[123] J. R. Pringle and J.-R. Mor, *Methods Cell Biol.* **11,** 131 (1975).
[124] L. B. Chen, S. Rosenberg, K. K. Nadakavukaren, E. S. Walker, E. L. Shepherd, and G. D. Steele, *in* "Hybridoma Technology in the Biosciences and Medicine" (T. A. Springer, ed.), p. 251. Plenum, New York, 1985.

on animal cells.[102,124-126] Indeed, Chen et al.[124] recommend screening each new monoclonal antibody on cells fixed by each of 10 different procedures, some of which avoid aldehyde fixation altogether. In our own experience, one batch of antibodies recognizing the CDC11 product gave very poor immunofluorescence results with the standard fixation protocol, but behaved well when we used either shorter (20–30 min) formaldehyde fixation or any of several alternative fixation protocols.[109,124] Third, there is at least one example in yeast work in which formaldehyde fixation had to be followed by additional treatments with methanol and acetone in order to obtain good immunofluorescence[25] (see also below); possibly an alternative approach to the initial fixation would also have yielded good results.

Deterioration of Formaldehyde Stocks. Concentrated formaldehyde stocks deteriorate with time, as evidenced by the accumulation of precipitate (polymerization product) at the bottom of the bottle. Although the presence of some of this precipitate does not appear to be a problem, we are uncertain how to gauge exactly when a stock should be replaced; thus, we recommend buying small bottles of 37% formaldehyde, removing solution as needed from the top of the undisturbed bottle, and replacing the bottle as soon as polymerization becomes extensive. Better results may possibly be obtained with "ultrapure E.M. grade" formaldehyde (Polysciences Cat. No. 4018), methanol-free formaldehyde (Ted Pella), or fresh formaldehyde prepared by heat depolymerization of paraformaldehyde.[127]

Resuspension in Phosphate-Buffered Formaldehyde. Resuspension in buffered fixative was incorporated into our standard fixation protocol based in part on general principles (the idea that cell structure would be preserved better if the pH stayed near physiological values despite the chemical action of the fixative) and in part on the routine buffering of fixatives by electron microscopists.[4,5] However, we are not certain how often the buffering really matters or whether any of the parameters of our particular buffered fixative are really critical. Indeed, we have often omitted this step (conducting the entire fixation right in growth medium) without obvious detriment to the results. A compromise that also seems effective is to add concentrated buffer to the growth medium at the same time the fixative is added.[7,58]

Duration of Fixation. In some immunofluorescence applications, the duration of fixation appears to be important. For example, in one case, 20–30 min of formaldehyde fixation gave much better results than did

[125] A. M. Gown and A. M. Vogel, *J. Cell Biol.* **95**, 414 (1982).
[126] D. E. Greenwalt and I. H. Mather, *J. Cell Biol.* **100**, 397 (1985).
[127] M. A. Hayat, "Fixation for Electron Microscopy." Academic Press, New York, 1981.

longer fixations (see above).[109,128] In contrast, in another case (involving microtubule staining with a particular antitubulin antibody), 2–3 hr of fixation gave more satisfactory results than did less than 1 hr fixation.[129] In other cases, however, cells have been fixed for as little as a few minutes or as long as several days without detectable effect on the results. In particular, satisfactory results have generally been obtained after storing cells up to 12 hr at 4° in formaldehyde-containing growth medium or in phosphate-buffered formaldehyde after about 2 hr of fixation at room temperature. In addition, we have successfully used cells fixed for 2 hr at room temperature, then washed into solution A (see below) and stored 24 hr at 4° before removal of cell walls (see below), as well as cells stored over 2 days at 4° after the cell wall-removal step. However, we do not yet know how frequently such variations yield satisfactory results.

Immunofluorescence Procedures

Solutions

Solution A: Mix 1 M K_2HPO_4 with 1 M KH_2PO_4 to obtain a solution at pH 6.5. Dilute with water to 40 mM. Add $MgCl_2$ to 0.5 mM and sorbitol to 1.2 M

Solution B: Mix 1 M K_2HPO_4 with 1 M KH_2PO_4 to obtain a solution at pH 7.5. Dilute with water to 100 mM. Add sorbitol to 1.2 M

Solution C: Mix together 4 ml of 1 M Tris-HCl, pH 9.0, 4 ml of 0.1 M disodium EDTA, pH 8.0, 10 ml of 2 M NaCl, and 2 ml water. Add 120 μl of 2-mercaptoethanol just before use

Solution D: Make phosphate–citrate buffer, pH 5.8 (contains 22.32 g KH_2PO_4 and 9.41 g sodium citrate per liter), then mix 1:1 with 2 M NaCl

Solution E: Dissolve 180 g sorbitol in 250 ml phosphate–citrate buffer, pH 5.8 (as in solution D). Dilute to 1 liter with water

Solution F: Add 10 mg KH_2PO_4 to 90 ml water and titrate to pH 7.4 with 0.1 N KOH. Dilute to 100 ml with water. Add 0.85 g NaCl, 0.1 g BSA, and 0.1 g NaN_3

Polylysine stock: Dissolve 10 mg polylysine (MW > 300,000; Sigma Cat. No. P-1524 or comparable) in 10 ml water. This solution can

---

[128] A liability of the shorter fixation was that the cells were more fragile and tended to be mangled during preparation for immunofluorescence; thus, the best results in this case were actually obtained with the sequential glutaraldehyde/formaldehyde fixation protocol described above.

[129] P. Schatz, personal communication (1989).

be stored for several months at $-20°$, with multiple freeze-thaw cycles, without obvious detriment

Phosphate-buffered saline (PBS): Dissolve 160 g NaCl, 4 g KCl, 22.8 g Na_2HPO_4, and 4 g KH_2PO_4 in water to a final volume of 1 liter. Adjust to pH 7.3 with 10 N NaOH. Dilute 20× with water prior to use

Mounting Medium. Our standard mounting medium contains p-phenylenediamine to retard photobleaching (see above). As this chemical is reportedly carcinogenic, we prefer to weigh it out infrequently; thus, we make up mounting medium in large batches as follows: (1) Dissolve 100 mg p-phenylenediamine (Sigma Cat. No. P-6001 or comparable) in 10 ml PBS and adjust to pH 9 if needed. (Stir vigorously at room temperature to facilitate dissolving the p-phenylenediamine.) (2) To this solution, add 90 ml glycerol and stir until homogeneous. (3) If desired for DNA staining, add 2.25 μl of fresh DAPI stock solution (1 mg/ml 4′,6-diamidino-2-phenylindole dihydrochloride in water). Note that this is frequently useful and rarely harmful, so that we almost always include this step. (Another dye, Hoechst 33258, is sometimes used instead of DAPI to stain the DNA in immunofluorescence preparations.[16,49,52]) (4) Store at low temperature in the dark. The full stock can be stored in a bottle at $-20°$ (it stays liquid) and samples removed as needed without warming. However, the solution does gradually deteriorate (time scale of a few months), developing a dark color and producing apparent fluorescence artifacts. Thus, we recommend storing multiple small aliquots of the mounting medium in capped microcentrifuge tubes at $-70°$ and retrieving these one at a time as needed. Under these conditions the mounting medium appears to stay good for many months if not years.

Standard Immunofluorescence Protocols

The following protocols are written for standard immunofluorescence on whole cells. Appropriate modifications can adapt them for use with spheroplasts, isolated organelles, or cell sections.

Permeabilization. Spin down approximately 2×10^8 fixed cells (see above) and wash twice with the solution appropriate to one of the following cell wall-removal protocols. For ordinary growing cells, we have generally found either protocol i or ii to be satisfactory. For cells with walls more resistant to digestion (e.g., stationary-phase cells or arrested cell-cycle mutants), protocol ii or iii has generally given better results.

i. Wash cells with solution A and resuspend in 1 ml solution A containing 10 μl of 2-mercaptoethanol and 55 μl Glusulase (Du Pont NEN

Cat. No. NEE-154). Incubate for 2 hr at 36° with gentle agitation (e.g., in a roller drum).

ii. Wash cells with solution B and resuspend in 1 ml solution B containing 2 μl of 2-mercaptoethanol and 20 μl of a Zymolyase stock [1 mg/ml Zymolyase 100T (ICN Immunobiologicals Cat. No. 32093-1)[130] in water]. Incubate 30 min at 37° with gentle agitation.

iii. Wash cells with water, then incubate in 1 ml solution C for 10 min at room temperature. Spin down and wash once with solution D, then twice with solution E. Resuspend cells in 1 ml solution A containing 10 μl of 2-mercaptoethanol, 110 μl Glusulase, and 22 μl of a Zymolyase stock [27 mg/ml Zymolyase 20T (ICN Immunobiologicals Cat. No. 32092-1)[130] in water]. Incubate 30 min at 37° with gentle agitation.

After digestion by any of these procedures, spin cells down at low speed and wash once with solution A or B, then resuspend gently in 1 ml solution A or B. Note that (1) this final wash may not be necessary, unless the cells are going to be stored for a time before proceeding; (2) once washed and resuspended, the cells can be stored for some time before proceeding (at least for some antigens; see above), so long as they are kept cold; (3) even though the cells are fixed, once their walls are removed they lose their shapes with rough handling (hard centrifugation, vortexing), which should therefore be avoided; (4) although additional treatments (e.g., with acetone and methanol or detergent) do not appear to be necessary for permeabilization per se, such treatments do facilitate the visualization of some antigens, presumably by contributing to fixation and/or denaturation of the antigen (see below).

Preparation of Slides. Put approximately 10 μl of polylysine stock solution in each well of a multiwell slide (see above). After 5–10 sec, aspirate the solution off and air-dry. Wash each well 3 times with drops of water that are removed by aspiration. Air-dry completely. Normally, the slide is now ready for use. However, if background fluorescence proves to be a problem (this may vary with different batches of slides and antibodies), it often helps to wash the slides more extensively. To do this, take the slide after polylysine treatment and drying and place it in distilled water in a capped plastic tube. Shake this or place it in a roller drum for about 10 min, then air-dry completely. Slides can be prepared at least several hours before use.

Staining of Cells with Antibodies. The following steps describe standard indirect immunofluorescence. The obvious modifications would be neces-

---

[130] In protocols calling for Zymolyase, it may be possible to substitute Sigma Lyticase (Cat. Nos. L8012, L8137, and L5263) with similar efficacy and less expense; however, we have not checked this systematically.

sary to use direct immunofluorescence (see above) or "sandwiching" methods (see below).

1. Place 10 μl of cell suspension (see above) in each well. After about 10 sec, aspirate off the fluid and allow the slide to air-dry. Check the slide microscopically to ensure that the cells have retained their shapes and are at a suitable density and not clumped.

2. For most antigens, the cells can now be reacted directly with the primary antibodies (Step 3). However, for some antigens (or at least for some antigen–antibody combinations; e.g., actin and the antiactin antibodies that have been used to date), additional treatment for fixation and/or denaturation is essential before treatment with antibodies.[25] Use fresh methanol and acetone that have been chilled to −20°. Immerse the wells of the slide in methanol at −20° for 6 min, then in acetone at −20° for 30 sec, then air-dry completely.

3. Place 5–10 μl of primary antiserum [diluted as appropriate with solution F (see above) or with PBS containing 1 mg/ml BSA (PBS–BSA)] in each well. In some cases, eventual background fluorescence may be reduced by incubating briefly with solution F or PBS–BSA alone before the incubation with primary antiserum. Use a control antiserum in one well per slide (or per batch of slides) as needed.

4. Incubate the slide at room temperature in a moist environment (e.g., in a petri dish with a wet Kimwipe in it) for 0.5–1.5 hr. (The exact time required may vary with different antigen–antibody combinations.)

5. Aspirate off the primary antiserum and wash cells ~10 times with solution F or PBS–BSA by placing a drop of solution in each well and aspirating it off after a few seconds. Do not let the wells dry out completely during these washes; that is, put in a new drop of solution as soon as the old drop is aspirated off. We generally wash the first well about 7 times, leave it under a drop of wash solution while we wash the remaining wells approximately 7 times apiece, then return to the beginning and wash each well 3 more times, again leaving each well under wash solution. Then these final washes are removed and replaced immediately by the secondary antibody solution.

6. Place 5–10 μl of fluorophore-conjugated secondary antiserum (diluted as appropriate in solution F or PBS–BSA) in each well and incubate at room temperature in a moist environment for 0.5–1.5 hr (see note at Step 4). Note that it is important to conduct these manipulations in low light and to incubate in the dark, to avoid photobleaching of the fluorophore.

7. Aspirate off the secondary antiserum and wash approximately 10 times with solution F or PBS–BSA (as described in Step 5; remove the last drop of wash solution immediately before adding mounting medium).

Note that in some applications (e.g., screening of large numbers of antisera), it is feasible and convenient to conduct staining with a method using larger puddles of solution that cover several or all wells at one time. The slides can then be washed either using similar puddles of wash buffer or by simply immersing the slides in wash buffer. However, it should be noted that once wells have been "fused" in this way they cannot subsequently be treated separately without danger of cross-contamination of solutions.

Mounting of Slides. Place a drop of mounting medium on the slide (one small drop per four wells is adequate, and too much causes problems such as floating coverslips or mixing of immersion oil with mounting medium) and cover with coverslips. View immediately or after storage at $-20°$ in the dark. The images deteriorate little or not at all (opinions differ) during storage under these conditions for days or even weeks. If the coverslips are sealed around the edges with clear nail polish, slides can be stored at $-20°$ in the dark for months or years without gross deterioration of immunofluorescence images.

The Value of a Positive Control

Even in experienced hands, the results of immunofluorescence are not always equally successful, especially when applying the method to a new strain or to cells grown under nonstandard conditions. (The vagaries of permeabilization appear to be one major source of variable results.) A positive control should be included routinely each time that immunofluorescence is done. One convenient and effective control is provided by the Kilmartin monoclonal antitubulin antibody YOL1/34; this antibody is commercially available (Accurate Cat. No. MAS078; purchase the supernatant form and use it at 1:100 to 1:500 dilution), is derived from a rat cell line (making it convenient for double-label experiments with rabbit antibodies), and gives a well-characterized pattern of staining of both cytoplasmic and intraorganellar structures.[7,8,15,103] However, it should be noted that this particular control is not a panacea: for example, conditions that routinely give excellent antitubulin immunofluorescence have required modification to give acceptable antiactin immunofluorescence (see above).[25] Thus, as in other areas, a flexible, empirical approach is required for the selection and evaluation of appropriate positive controls.

Detection of Nonabundant Antigens

It is clear that immunofluorescence is capable of localizing a wide variety of antigens of unequal abundance in the yeast cell. However, it is also clear that sufficiently nonabundant antigens will be difficult or impossible to localize. We mention briefly here some approaches that may be helpful in the borderline cases.

Video Enhancement of Signals. Recent spectacular results with animal cells and subcellular systems *in vitro* show clearly the potential power of electronic enhancement of the optical signals.[68,131-136] However, it should be noted that in immunofluorescence of whole yeast cells, immunodetection is likely to be limited more often by the ratio of signal (from the object of interest) to noise (contributed mainly by nonspecific background fluorescence) than by the strength of the signal per se. Thus, simple amplification of the optical signals (which will amplify the nonspecific background as well as the signal of interest) will probably not be useful in many cases. However, the video systems can also improve signal-to-noise ratios in some situations. Although much of the technology is directed toward reducing noise from within the optical and video systems themselves, the systems can also enhance image contrast electronically, seek discrete structures using algorithms such as those for thresholding and edge detection, and play tricks such as electronically subtracting an out-of-focus image from an in-focus image. It remains unclear how generally useful such approaches will be in practical work with yeast. The only successful applications of which we are aware are the studies of *Schizosaccharomyces pombe* nuclei and chromosomes by Yanagida and co-workers,[137] and even these workers have relied primarily on conventional fluorescence micrographs. It should be noted that use of direct rather than indirect immunofluorescence should sometimes facilitate the application of video enhancement methods by improving the starting ratio of signal to nonspecific background fluorescence. Confocal microscopy also has great promise in this regard[85] (and see above).

Overexpression of Gene Products and of Fusion-Gene Products. It seems clear that overexpression of a gene product (by introducing the gene on a high-copy plasmid or linking it to a strong promoter) should in some cases push a weak immunofluorescence signal above the threshold of detectability. However, it also seems clear that overexpression will sometimes lead to mislocalization of the gene product, perhaps accompanied by a general

---

[131] R. D. Allen, D. G. Weiss, J. H. Hayden, D. T. Brown, H. Fujiwake, and M. Simpson, *J. Cell Biol.* **100**, 1736 (1985).

[132] R. D. Vale, B. J. Schnapp, T. Mitchison, E. Steuer, T. S. Reese, and M. P. Sheetz, *Cell (Cambridge, Mass.)* **43**, 623 (1985).

[133] T. Horio and H. Hotani, *Nature (London)* **321**, 605 (1986).

[134] S. Inoué, "Video Microscopy." Plenum, New York, 1986.

[135] P. J. Sammak and G. G. Borisy, *Nature (London)* **332**, 724 (1988).

[136] D. L. Taylor and Y.-L. Wang (eds.) "Fluorescence Microscopy of Living Cells in Culture, Part B" (*Methods Cell Biol.* **30**). Academic Press, San Diego, California, 1989.

[137] M. Yanagida, K. Morikawa, Y. Hiraoka, S. Matsumoto, T. Uemura, and S. Okada, *in* "Applications of Fluorescence in the Biomedical Sciences" (D. L. Taylor, A. S. Waggoner, R. F. Murphy, F. Lanni, and R. R. Birge, eds.), p. 321. Alan R. Liss, New York, 1986.

cellular pathology, so that the immunofluorescence results obtained may be misleading. For example, Wright et al.[54] were unable to immunolocalize hydroxymethylglutaryl (HMG)-CoA reductase in normal cells but were successful with cells that overproduced the enzyme. However, the overproduced enzyme localized to a distinctly abnormal structure, namely, a set of stacked circumnuclear membranes not present in normal cells. Thus, although the results were interesting in a variety of ways, they did not answer the question of the normal localization of HMG-CoA reductase. Similarly, Clark and Abelson[40] observed that cells overproducing tRNA ligase under GAL10 control yielded a detectable nuclear staining with antiligase antibodies under fixation and permeabilization conditions with which normal or uninduced cells yielded no detectable signal. Although this observation was useful in helping to verify the results obtained with wild-type cells using other fixation and permeabilization conditions, detailed examination of the immunofluorescence and immunoelectron microscopy results suggested that tRNA ligase was partially mislocalized in the overproducing cells.

In another case, overexpression of KAR1[13] yielded an immunofluorescence signal (a small dot) with Kar1p-specific antibodies that was not detectable when normal cells were examined.[22] However, the overexpressing cells were arrested in the cell cycle and dying, and the dots of fluorescence were not consistently localized with respect to the spindle poles, so that the significance of the results was not clear. More satisfying results were obtained using an alternative approach that may also be useful with other nonabundant antigens.[22] Cells overexpressing a KAR1–lacZ fusion protein (which was not lethal to the cells) were examined using anti-β-galactosidase antibody. Under these conditions, immunofluorescence revealed a dot of staining that associated consistently with the spindle poles in such a way as to suggest strongly that the bona fide localization of the KAR1 product had been revealed.

In summary, if immunofluorescence on normal cells yields no detectable signal, overexpression of the gene product of interest or of an appropriate fusion protein is probably worth a try. However, any results obtained must be interpreted with considerable caution.

"Sandwiching" Methods. Schulze and Kirschner[93] described a method in which up to four successive layers of secondary antibodies were built up on a single primary antibody; in particular, microtubules containing biotinylated tubulin were reacted with rabbit antibiotin antibody, then successively with goat anti-rabbit-IgG, rabbit anti-goat-IgG, goat anti-rabbit-IgG, and rabbit anti-goat-IgG. Two, three, or all four of the layers of secondary antibodies utilized fluorescein-conjugated antibodies. This method served both to amplify the immunofluorescence signal over what was obtained

with a single layer of fluorophore-conjugated secondary antibodies and to block the microtubules containing biotinylated tubulin from reaction with antitubulin antibodies during a subsequent incubation (used to reveal microtubules that did not contain biotinylated tubulin). In yeast work, this approach has been used successfully to visualize the intracellular localization of the *CDC46* gene product;[120] each layer of antibodies was applied using standard procedures. Similar methods have been used to localize the *KEX2* product in normal cells[58,138] and to localize the *SIR2* product in cells overproducing this protein.[139] In each of the cases cited, conventional immunofluorescence procedures yielded no reliably detectable signal.

The major potential problem with this powerful approach is that nonspecific "background" binding of antibodies at any stage will be amplified in the subsequent stages, with deleterious consequences to the signal-to-noise ratio. Thus, it is desirable to use the cleanest possible antibodies,[140] and it may be necessary to invest some effort in optimizing the dilutions at which the various layers of secondary antibodies are applied. Another variable is the number of layers of fluorophore-conjugated secondary antibodies to be used. Schulze and Kirschner[93] reported that beyond two such layers their signal-to-noise ratio actually decreased; however, Hennessy[120] has had good success with three such layers, and Redding *et al.*[58] successfully used four layers of antibodies in which only the antibodies of the final layer were fluorophore-conjugated.

A related approach to signal amplification involves the use of a biotin-conjugated primary or secondary antibody followed by fluorophore-conjugated avidin or streptavidin[98,126,141,142] (reagents available from Accurate, BRL, Molecular Probes, Vector, Zymed, and other sources). This approach has been used successfully to localize the *SPA1* and *SPA2* gene products in mitotic and meiotic yeast cells.[23] Further amplification can be attempted by adding a biotinylated carrier such as BSA, or biotinylated antiavidin or antistreptavidin antibodies, after the first avidin or streptavidin treatment, then adding another layer of fluorophore-conjugated avidin or streptavidin.[98] A common problem with this general approach is background due to endogenous biotin-containing macromolecules, other avi-

---

[138] A detailed protocol is available from K. Redding and R. Fuller, Department of Biochemistry, B400 Beckman Center, Stanford, California 94305 (telephone 415/723-5872).

[139] L. Pillus and J. Rine, personal communication (1989).

[140] Redding and Fuller strongly recommend preadsorbing all antibodies to be used against fixed cells (before or after their walls have been digested) in order to remove potential wall-reactive antibodies that would otherwise bind nonspecifically and then be amplified.

[141] D. A. Fuccillo, *BioTechniques* **3**, 494 (1985).

[142] T. Kobayashi, T. Sugimoto, T. Itoh, K. Kosaka, T. Tanaka, S. Suwa, K. Sato, and K. Tsuji, *Diabetes* **35**, 335 (1986).

din- or streptavidin-binding substances, or biotin-binding proteins in the cells. The limited experience to date suggests that this problem is also significant in yeast work but can be alleviated using the same methods that have worked in other systems, notably, preliminary incubation of the cells with avidin or streptavidin, followed by incubation with free biotin.[23,58,120,143,144]

Other Methods. Clark and Abelson[40] reported that removal of cell walls prior to fixation allowed them to detect the nuclear localization of tRNA ligase in normal (i.e., nonoverproducing) cells, which was not possible if the cells were fixed first as in the usual protocol. Although this approach may also be useful in other cases, it should be used with caution because of the danger of rearrangement of cell constituents during the prolonged incubations prior to fixation (see comments above on the importance of rapid fixation). The use of fluorophores with superior quantum yields (e.g., phycobiliproteins; see Table II) may also be of some value in detecting nonabundant antigens.

Double-Label Immunofluorescence

In attempting to compare the intracellular localizations of two cellular constituents, it is frequently valuable to perform double-label experiments so that these localizations can be compared in the same individual cells. This is relatively straightforward if primary antibodies from different types of animals are available or if the available antibodies are of different immunoglobulin classes (e.g., an IgG and an IgM[145]); in either case, appropriate secondary antibodies can effect the necessary discrimination. Thus, for example, one could incubate the cells (sequentially or simultaneously) with rabbit antiactin and rat antitubulin, then apply FITC-conjugated goat anti-rabbit-IgG and TRITC-conjugated goat anti-rat-IgG. Two potential problems should be noted; both are more troublesome if the antigens of interest appear to colocalize than if they show obviously different localizations. First, commercially available secondary antibodies are not always entirely specific for their target IgGs; for example, some goat anti-rabbit-IgG antibodies cross-react with rat IgG.[25] Fortunately, in this case at least, the rat-reactive component could be effectively removed by preadsorption with immobilized rat IgG. Second, illumination intended for the fluorophore that absorbs and emits at shorter wavelengths often results also in some detectable fluorescence from the fluorophore that absorbs and emits at longer wavelengths ("crossover" fluorescence). For example, illumina-

---

[143] G. S. Wood and R. Warnke, *J. Histochem. Cytochem.* **29**, 1196 (1981).
[144] R. C. Duhamel and D. A. Johnson, *J. Histochem. Cytochem.* **33**, 711 (1985).
[145] P. A. W. Edwards, I. M. Brooks, and P. Monaghan, *Differentiation* **25**, 247 (1984).

tion for FITC generally results also in some visible TRITC fluorescence from a double-labeled sample. If this is a problem, it can usually be solved either by choosing different fluorophores (e.g., Texas Red instead of TRITC; see Table II) or by using additional (or different) filters to circumscribe the exciting wavelengths, the emitted wavelengths that are allowed to pass through to the eyepieces, or both. (See the lists of filters available from the microscope manufacturers or from Omega Optical.)

Double-label immunofluorescence is more difficult if both of the available primary antibodies are of the same immunoglobulin class and are derived from the same type of animal. One solution is to resort to direct immunofluorescence, conjugating an appropriate fluorophore directly to one or both primary antibodies (see above). Thus, for example, J. Kilmartin achieved excellent double labeling with rat anti-yeast-actin and anti-yeast-tubulin antibodies using an approach adapted from that of Hynes and Destree;[146] cells were incubated successively with the rat antitubulin, TRITC-conjugated goat anti-rat-IgG, excess unlabeled rat IgG (to block residual binding sites on the goat anti-rat-IgG), and FITC-conjugated rat antiactin.[7] In related approaches that may allow stronger signals to be obtained (see above), one or both primary antibodies can be biotinylated or one can be biotinylated and the other dinitrophenylated.[98,145,147] Appropriate sequences of incubations with fluorophore-conjugated avidin, streptavidin, or antidinitrophenyl-group antibodies, plus appropriate blocking solutions (free biotin or unlabeled IgGs), then allow the double-label fluorescence to be visualized.

Another approach to double-labeling that has been used successfully at least once[58] is based on the "sandwiching" method of Schulze and Kirschner[93] (see also above). A sequence of five incubations was used: (1) one rabbit primary antibody; (2) Fab [*not* F(ab')$_2$] fragment (crucial to eliminate the effects of antibody bivalency) of goat anti-rabbit-IgG; (3) the other rabbit primary antibody; (4) mouse anti-rabbit-IgG; and (5) FITC-labeled rabbit anti-mouse-IgG plus Texas Red-labeled rabbit anti-goat-IgG.[138]

Finally, it should be noted that multiple-label immunofluorescence (localizing three or more antigens simultaneously) is possible at least in some situations using presently available fluorophores.[72,74,75] Such approaches may be further facilitated by new fluorophores recently made available (e.g., AMCA; see Table II) or presently under development.

---

[146] R. O. Hynes and A. T. Destree, *Cell (Cambridge, Mass.)* **15**, 875 (1978).
[147] J. B. Miller, M. T. Crow, and F. E. Stockdale, *J. Cell Biol.* **101**, 1643 (1985).

Detection of Specific Structures

In performing immunofluorescence on some new protein of interest, it is frequently important to be able to localize that protein relative to previously characterized ones in double-label experiments (or, less satisfactorily, in parallel-labeling experiments). For example, such experiments can be invaluable in defining rigorously what compartment of the cell the new protein is in, or in defining how it localizes relative to known proteins or structures during the cell cycle. With these considerations in mind, we offer the following comments on what appear, at present, to be the best ways to visualize particular cellular constituents in the context of immunofluorescence experiments. We also note three explicit caveats. First, there is no guarantee that all of the potential reagents referred to below will be available at any given time. (Note in particular that any given antiserum may become limiting, and therefore extremely precious, to the investigators who generated it.) Second, in some areas, our current views as to what is the most appropriate reagent may be rapidly superceded. Finally, for some purposes, certain organelles may be best visualized by using vital stains on living cells.[3]

Cell Surface

Cell Wall. Until recently, attempts to visualize specific cell wall components in cells that had also been prepared for immunofluorescence staining of intracellular structures had seemed stymied by the need to remove the cell wall in order to permeabilize the cells to antibodies. However, it has recently been found[148,149] that reduced durations of cell wall digestion can yield cells that are adequately permeabilized to antibodies, yet retain enough cell wall structure that bud scars are held in place and can be visualized with Calcofluor.[1] It seems likely that a similar approach would allow immunofluorescence staining of intracellular structures in conjunction with staining of cell wall components using appropriate antibodies (e.g., see Ref. 35) or visualization of cell wall growth zones using concanavalin A.[3]

Plasma Membrane. Although several proteins have been identified that appear to localize to the plasma membrane by immunofluorescence,[32-34] it is not clear that any of these provides a satisfactory general marker for colocalization experiments on cells or cell fractions. Antibodies against the

---

[148] R. Palmer, M. Koval, and D. Koshland, *J. Cell Biol.* **109**, 3355 (1989).
[149] R. Palmer and D. Koshland, personal communication (1989).

plasma membrane ATPase would seem to have great promise in this regard.

Cytoskeletal Elements

Microtubules and Spindle-Pole Body. Although a variety of antitubulin antibodies have been used successfully to visualize the intranuclear and cytoplasmic microtubules in yeast,[6-21] the most generally useful one still appears to be the monoclonal YOL1/34,[103] which has among its several virtues that of being commercially available (see above, The Value of a Positive Control). In cells in which the microtubules have not been destroyed by mutation or drug treatment, tubulin-specific antibodies also provide the best way to visualize the spindle-pole bodies. In cells that have lost their microtubules, it may be possible to visualize the spindle-pole bodies using antibodies that recognize the *KAR1*[13,22] or *SPA1*[23] products (or related fusion proteins).

Actin System. Actin and various associated proteins have been visualized successfully using several different antibodies.[7,24-28,150] It should also be noted that staining of actin with fluorophore-conjugated phalloidin[2] sometimes offers a simpler and equally effective way to visualize the actin cytoskeleton, particularly in double-label experiments.

Neck Filaments. Antibodies specific to the *CDC3, CDC10, CDC11,* and *CDC12* products all appear to decorate the neck filaments in budded cells.[21,29,151] It remains unclear whether the staining by these antibodies of former and future budding sites in unbudded cells always reflects the presence of filaments per se at the sites of staining.

Membrane-Bounded Organelles

Nucleus. The routine inclusion of DAPI or Hoechst 33258 in the mounting media used for immunofluorescence preparations allows the nuclear DNA to be visualized in conjunction with immunofluorescence using FITC, TRITC, or both. However, it should be noted that the localization of the nuclear DNA is not always an adequate guide to the boundaries of the nucleus as a whole or of its specific, non-DNA components.[10,148,149] In this regard, monoclonals[48,51] and human autoantibodies[44,50] that stain the nucleolus,[152] as well as monoclonal antibodies (MAb 414,[53,153]306,[56]

[150] B. K. Haarer, S. H. Lillie, A. E. M. Adams, V. Magdalen, W. Bandlow, and S. S. Brown, *J. Cell Biol.* **110**, 105 (1990).

[151] S. K. Ford, H. B. Kim, B. K. Haarer, and J. R. Pringle, unpublished results (1989).

[152] It may also be possible to visualize the nucleolus by heterochromatic staining with acridine orange [S. Royan and M. K. Subramaniam, *Proc. Indian Acad. Sci., Sect. B* **51B**, 205 (1960); M. E. Dresser and C. N. Giroux, *J. Cell Biol.* **106**, 567 (1988)].

[153] L. I. Davis and G. Blobel, *Cell (Cambridge, Mass.)* **45**, 699 (1986).

and RL1[87,154]) and a high-titer polyclonal antiserum[22,55] that stain components of the nuclear envelope, should be very useful. The monoclonal antibody 8C5 (generated by S. Benzer and co-workers against *Drosophila* heads) also appears to give a general nuclear staining in yeast.[46] Staining of microtubules (see above) of course also allows localization of the nuclear envelope associated with the spindle-pole bodies.

Endoplasmic Reticulum. Polyclonal sera raised against the *KAR2* product[55] appear to stain nonnuclear-envelope elements of the endoplasmic reticulum (ER) as well as the nuclear envelope even when used at high dilutions.[22] At present, this seems the best approach to visualization of the ER, as antibodies specific to HMG-CoA reductase (another presumed ER protein) have given detectable immunofluorescence signals only in cells overproducing the protein (and demonstrably abnormal thereby).[54]

Golgi Complex. At present, there remains considerable uncertainty about the structure and localization of the Golgi in normal yeast cells, so that the significance of the staining patterns obtained with various antibodies remains somewhat unclear. However, it is encouraging that antibodies against the *YPT1* gene product appear to stain the Golgi in mammalian cells as well as in yeast,[57] and that antibodies specific to the *SEC7* and *KEX2* proteins appear to stain the same structure[58] (which, from genetic and biochemical evidence, ought to be the Golgi complex as a whole, a particular subcompartment of the Golgi, or a Golgi-related compartment such as the trans-Golgi network).

Secretory Vesicles. Secretory vesicles have been visualized using antibodies specific for the *SEC4* gene product.[59] It seems likely that antibodies recognizing the products of other "late *SEC* genes" may also be useful in this regard.

Vacuoles. Although good immunofluorescence visualization of vacuoles has been achieved both with antibodies recognizing carboxypeptidase Y and with antibodies recognizing the membrane protein dipeptidylaminopeptidase B,[60,61] at present it appears that the most satisfactory general reagent may be antibodies specific for the vacuolar membrane protein alkaline phosphatase.[60,155] Monoclonal antibodies specific for subunits of the vacuolar membrane ATPase also appear promising in this regard.[60]

Peroxisomes. Successful immunofluorescence visualization of peroxisomes (both in oleic acid-induced and in glucose-grown cells, and verified by immunoelectron microscopy) has been achieved using both antibodies raised against total *Candida tropicalis* peroxisomal proteins[156] and anti-

---

[154] C. M. Snow, A. Senior, and L. Gerace, *J. Cell Biol.* **104**, 1143 (1987).
[155] D. J. Klionsky and S. D. Emr, *EMBO J.* **8**, 2241 (1989).
[156] G. M. Small, T. Imanaka, H. Shio, and P. B. Lazarow, *Mol. Cell. Biol.* **7**, 1848 (1987).

bodies specific for thiolase (generated by W. Kunau).[62,63] It seems likely that antibodies specific for peroxisomal catalase would also be useful in this regard in yeast, as they have been in mammalian cells.[157]

Mitochondria. Good immunofluorescence visualization of mitochondria has been achieved using both a polyclonal antiserum raised against citrate synthase (by P. Srere)[65] and monoclonal and polyclonal antibodies raised against several outer membrane proteins.[64,158,159]

Acknowledgments

We thank the numerous colleagues who have participated in the development of the ideas and procedures described above, provided us with unpublished information, or commented on the manuscript. These include, but are not limited to, J. Allen, K. Armstrong, R. Azpiroz, V. Berlin, D. Botstein, J. Broach, S. Brown, R. Butow, C. Copeland, L. Davis, M. Douglas, G. Fink, S. Ford, R. Fuller, C. Giroux, S. Gould, K. Hennessy, T. Huffaker, E. Jones, J. Kilmartin, H. Kim, W. Kunau, P. Lazarow, S. Lillie, N. Martin, D. McCollum, S. McConnell, P. Novick, B. Page, L. Pillus, R. Preston, K. Redding, S. Reed, J. Rine, C. Roberts, M. Rose, P. Schatz, M. Snyder, F. Solomon, T. Stearns, T. Stevens, S. Subramani, P. Takasawa, R. Thieringer, A. Waggoner, B. Weinstein, M. Welsh, R. Wright, and M. Yaffe. Unpublished work from J.R.P.'s laboratory was supported by National Institutes of Health Grant GM31006. Other unpublished work cited above was done while A.E.M.A. and D.G.D. were in D. Botstein's laboratory at the Massachusetts Institute of Technology, supported by NIH Grants GM21253 and GM18973 and American Cancer Society Grant MV 90, and by postdoctoral fellowships from the Burroughs Wellcome Fund of the Life Sciences Research Foundation (to A.E.M.A.) and Helen Hay Whitney Foundation (to D.G.D.).

[157] M. J. Santos, T. Imanaka, H. Shio, G. M. Small, and P. B. Lazarow, *Science* **239**, 1536 (1988).
[158] H. Riezman, R. Hay, S. Gasser, G. Daum, G. Schneider, C. Witte, and G. Schatz, *EMBO J.* **2**, 1105 (1983).
[159] D. Vestweber, J. Brunner, A. Baker, and G. Schatz, *Nature (London)* **341**, 205 (1989).

[41] Preparation of Yeast Cells for Thin-Section Electron Microscopy

By BRECK BYERS and LORETTA GOETSCH

Technical difficulties in the preparation of *Saccharomyces cerevisiae* for electron microscopy have contributed significantly to suspicion among cell biologists about the evolutionary status of the organism. In actuality, although cytological studies of yeast may have lagged somewhat behind those in plants and animals, the organization of several yeast organelle systems was clearly delineated in excellent early work and found to be quite similar to that in other eukaryotes.[1,2] These and subsequent studies[3,4] have provided a basis for effectively utilizing microscopic techniques to take advantage of the exquisite genetics of yeast in addressing fundamental issues of cell biology.

Principles

Yeast cells present the electron microscopist with a variety of challenges that have been met by the modification of standard protocols suitable for higher eukaryotes. A principal difficulty is the unusually high density of the cell, a property which renders both nuclei and cytoplasm stained in the usual manner with such uniformly high electron density that organelles are difficult to distinguish from the ribosome-rich cytoplasmic ground substance. A partial solution is achieved by enzymatic removal of the cell wall after an initial fixation in glutaraldehyde, before further fixation (postfixation) and embedding. Removing the wall not only facilitates permeation of the embedding resin but also permits the cell to expand slightly, conferring differences in density that provide for variation in visual contrast. The walls of cells fixed during logarithmic growth are easily digested with appropriate enzymes without making any special provision at the time of glutaraldehyde fixation, but the walls of meiotic cells and certain types of vegetative cells, such as those approaching stationary phase, are refractory to digestion and require a "pretreatment" procedure. Use of these methods (detailed below) permits visualization of microtubules, certain filament systems, and many other features, but still fails to reveal membrane systems clearly. As an alternative, membrane systems can be accentuated nicely by special methods, such as fixation with permanganate (which is no longer used) or the more recently applied technique of postfixation in an osmium tetroxide–ferrocyanide solution,[5] but the spindle is then poorly seen and other structures, such as chromatin, become ill-defined. No single method has been described to date that addresses both requirements fully, so the investigator must opt for the method that favors structures of primary concern.

The following procedures are intended to complement the wealth of information available on various aspects of electron microscopy in text-

[1] C. F. Robinow and J. Marak, *J. Cell Biol.* **29,** 129 (1966).
[2] P. B. Moens and E. Rapport, *J. Cell Biol.* **50,** 344 (1971).
[3] B. Byers and L. Goetsch, *J. Bacteriol.* **124,** 511 (1975).
[4] D. Zickler and L. W. Olson, *Chromosoma* **50,** 1 (1975).
[5] R. Wright, M. Basson, L. D'Ari, and J. Rine, *J. Cell Biol.* **107,** 101 (1988).

books on technique, such as those by Hayat[6] and Robinson et al.[7] Following sample preparation by the means detailed in the present protocols, the textbook procedures are suitable for the thin-sectioning, staining, and electron microscopic examination of yeast cells. In our own work, we favor the use of thin Formvar films borne on wire loops (Ref. 6, pp. 134–141) to recover and mount ribbons of serial sections. General procedures for immunoelectron microscopy,[8,9] as well as many other electron microscopic techniques, are suitable for yeast cells (see [40], [42], and [50]–[52] in this volume).

Methods

Materials

Enzymes. Glusulase, an enzyme solution adequate for removal of yeast cell walls, is available from Du Pont NEN (Boston, MA). Other workers report successful use for this purpose of other enzymes, such as lyticase, which is available from Sigma (St. Louis, MO).

Chemicals. Primary fixation is in glutaraldehyde, available as a 50% (w/v) stock solution (biological grade) from Polysciences, Inc. (400 Valley Road, Warrrington, PA 18976). Postfixation is with osmium tetroxide, which we obtain in crystalline form from Polysciences. Low-viscosity "Spurr" resin kits and BEEM capsules for embedding cells are also obtained from this supplier.

Reagents for Standard Procedures

1 M $MgCl_2$
1 M $CaCl_2$
1 M Tris(hydroxymethyl)aminomethane, made up to pH 9.0 according to Trizma instructions
0.1 M Ethylenediaminetetraacetic acid
40 mM Phosphate–magnesium buffer: 40 mM K_2HPO_4–KH_2PO_4 (pH 6.5), 0.5 mM $MgCl_2$
Phosphate-buffered glutaraldehyde: 2% (w/v) glutaraldehyde diluted from 50% (w/v) stock in 40 mM phosphate–magnesium buffer; the mixture is centrifuged and then Millipore-filtered (0.54-μm filter) to remove precipitate

[6] M. A. Hayat (ed.), "Principles and Techniques of Electron Microscopy, Volume 8: Biological Applications." Van Nostrand-Reinhold, Princeton, New Jersey, 1978.
[7] D. G. Robinson, U. Ehlers, R. Herken, B. Herrmann, F. Mayer, and F.-W. Schurmann, "Methods of Preparation for Electron Microscopy." Springer-Verlag, Berlin, 1987.
[8] J. M. Polak and I. M. Varndell (ed.), "Immunolabeling for Electron Microscopy." Elsevier, Amsterdam, 1984.
[9] J. DeMay, *J. Neurosci. Methods* **7**, 1 (1983).

0.2 M Phosphate–citrate buffer: 0.17 M KH$_2$PO$_4$ and 30 mM sodium citrate (weighed out) to yield pH 5.8

0.1 M Sodium acetate adjusted with glacial acetic acid to pH 6.1

4% Osmium tetroxide, made up in glass-distilled water (can be stored at room temperature in a hood for several months)

1% (w/v) Uranyl acetate in water (stored in the dark)

Pretreatment solution: 0.2 M tris(hydroxymethyl)aminomethane (from 1 M stock adjusted to pH 9.0) and 20 mM EDTA (from 0.1 M stock adjusted to pH 8.0), plus addition of 0.1 M 2-mercaptoethanol immediately before use

0.2 M Cacodylate–calcium buffer: 0.2 M sodium cacodylate and 10 mM CaCl$_2$ adjusted to pH 6.5 with concentrated HCl

Cacodylate-buffered glutaraldehyde: 3% (w/v) glutaraldehyde prepared by dilution from a 50% (w/v) stock into a 1:1 dilution of 0.2 M cacodylate–calcium buffer

Standard Procedures for Fixing and Embedding Vegetatively Grown Cells

1. Cells to be fixed should be grown only to the early stage of logarithmic growth, generally under conditions that permit rapid doubling. Use about 10^8 cells for each sample that is to be prepared for electron microscopy.

2. Rinse the cells briefly in water or the buffer to be used for fixation, using either centrifugation or Millipore filtration to recover them from medium or washing solutions. For this and all subsequent steps, 0.5–1.0 ml of the stated solution is sufficient for fixative solutions; we use 2–4 ml for washes and dehydration steps. We use 10-ml glass test tubes for all steps. All treatments are done at room temperature unless specified otherwise.

Excess growth medium in the sample to be fixed generally appears to interfere with fixation, possibly by competing with the cells for reaction with the fixative (glutaraldehyde). Where there is a requirement for such rapid initiation of fixation that the complete culture (including growth medium) must be mixed directly with the fixative, leave the cells in the medium/fixative mixture only briefly (5 min) before transfer to a fresh change of fixative.

3. Suspend cells in phosphate-buffered glutaraldehyde and incubate for about 30 min. If desired for convenience, the samples can then be transferred to ice and stored at 0° overnight (or even for several days) before further processing.

4. Rinse cells twice in 0.1 M phosphate–citrate buffer (pH 5.8) and resuspend in this buffer containing a 1/10 dilution of Glusulase. Incubate until walls have been removed (usually about 2 hr), as indicated by loss of

wall refractility under phase-contrast microscopy. (Excessively dense cells seen on electron microscopy often result from failure of wall removal.)

5. For postfixation, wash the cells twice in 0.1 M sodium acetate (pH 6.1), transfer them (in a fume hood) to a 2% osmium tetroxide fixation solution generated by mixing equal volumes (0.25 ml per sample) of 0.1 M sodium acetate buffer and the 4% osmium tetroxide stock solution. Incubate for 15 min.

6. Rinse cells with distilled water and transfer to 1% aqueous uranyl acetate for 60 min of incubation in the dark.

7. Wash cell twice in distilled water and dehydrate by transferring them (5 min each step) through two changes of 95% ethanol and two changes of 100% ethanol.

8. Pellet the cells, resuspend them in Spurr resin (a low-viscosity embedding mixture that permeates yeast cells well), and permit infiltration of the resin for 30 min. Some workers prefer to transfer cells from 100% ethanol to a 1:1 (v/v) solution of Spurr resin in absolute ethanol and permit infiltration to occur gradually by placing the uncapped vessel overnight on a slowly rotating platform, so the ethanol can evaporate slowly. For either method, the Spurr resin is a thoroughly mixed solution of 10 g vinylcyclohexene dioxide, 4 g diglycidyl ether of polypropylene glycol, 26 g nonenylsuccinic anhydride, and 0.4 g dimethylaminoethanol. [This is Recipe B (hard) recommended in the reagent kit from Polysciences.]

9. After either method of infiltration, transfer the cells to fresh Spurr resin by pelleting and resuspension. Then pipette the suspension into a BEEM embedding capsule and concentrate the sample in the narrow bottom of the capsule by centrifugation (~1000 g). We use large corks perforated with several capsule-sized holes as multisample centrifuge adaptors in a table-top centrifuge. Insert paper labels with designations written in pencil into the capsules. The capsules are then incubated overnight at 65° to harden. (If trapped gases are found to have formed air bubbles, subject the liquid resin to moderately reduced pressure, e.g., from an aspirator, before or during the 65° incubation in subsequent experiments.)

10. Cut thin sections and stain with lead citrate for electron microscopy as described in standard textbooks. We stain only for 15–30 sec to avoid the beam-damage artifacts that result from overstaining.

Modified Procedures for Stationary-Phase Cells or Other Cell Types with Walls Stable to Digestion

1. Harvest about 10^8 cells to be fixed.
2. Wash the cells twice in water (1 min each).
3a. Fix the cells by resuspension and incubation in phosphate-buf-

fered glutaraldehyde for 30 min (at room temperature, here and in following steps). As in the standard procedure, one may store the cells overnight on ice at this point.

3b. To condition the cells for subsequent wall removal, resuspend and incubate them in freshly prepared (see Reagents) pretreatment solution for 10 min.

4–10. Proceed with wall removal, postfixation, and embedding as in the standard procedure.

Modified Procedures for Sporulating Cells

1. Bring cultures to desired stages of sporulation in liquid sporulation medium (such as 1% potassium acetate) at approximately 2×10^7 cells/ml.

2. Harvest, wash once in water, and pretreat cells in pretreatment solution with 1 M NaCl added (to guard against osmotic damage) for 5 min (at room temperature).

3. Wash cells twice in 0.7 M sorbitol and transfer to cacodylate-buffered glutaraldehyde for 30 min.

4–10. Proceed with wall removal, postfixation, and embedding as in the standard procedures.

Osmium Tetroxide – Ferrocyanide Fixation Procedures for Visualization of Membrane Structures

The following procedures were adapted by Wright et al.[5] from methods of McDonald.[10]

1. Grow at least 10^8 cells to early log phase in supplemented minimal medium and harvest by centrifugation.

2. Prefix by resuspending in 10 ml of a prefixation solution consisting of 2% glutaraldehyde in 0.1 M cacodylate (pH 6.8), 1 mM $MgCl_2$, and 1 mM $CaCl_2$; incubate on ice for 30 min.

3. Wall removal, if desired, is accomplished by pelleting the cells, resuspending them in 10 ml TMS [50 mM Tris-HCl (pH 7.5), 5 mM $MgCl_2$, 1.4 M sorbitol, and 0.44% (v/v) 2-mercaptoethanol], adding 1500 units lyticase, and incubating at room temperature with occasional shaking for 10–15 min, or until walls are digested.

4. Wash the cells by resuspension in 10 ml of 0.1 M cacodylate buffer and pelleting; repeat resuspension and pelleting.

5. Postfix by resuspension in 5 ml of a freshly prepared, ice-cold solution of 0.5% osmium tetroxide and 0.8% potassium ferrocyanide and

[10] K. McDonald, *J. Ultrastruct. Res.* **86**, 107 (1984).

incubate on ice for 5 min; pellet (~5 min) and resuspend the pellet 5 min longer in fresh postfixation solution as the original solution darkens in color (total postfixation time, ~15 min).

6. Wash cells 3 times in distilled water by pelleting and resuspension and perform *en bloc* staining (if desired for added contrast) by resuspending the cells in 1% aqueous uranyl acetate for 30–60 min at room temperature.

7–10. Dehydrate and embed according to the standard procedures.

[42] Immunogold Labeling of Yeast Ultrathin Sections

By MICHAEL W. CLARK

Introduction

The popularity of yeast as an experimental organism for studies in cell biology has been tempered by the difficulties encountered in ultrastructural localization of proteins within the yeast cell. A high-resolution cellular map of the location of a protein of interest is necessary to comprehend fully its role in the life of the cell. Cytochemical and immunofluorescence-based optical microscopic techniques on whole mount yeast cells, in both *Saccharomyces cerevisiae*[1-3] and *Schizosaccharomyces pombe*,[4] have provided much useful information on the subcellular and suborganellar localization of many proteins, such as actin, tubulin, and SSB-1 (a yeast nucleolar protein). Still, the small size of yeast (a haploid *S. cerevisiae* cell is 3–5 μm in length) and the limit of resolution of the light microscope (0.25–0.5 μm)[5] restrict the amount of structural data that can be obtained by optical methods.

To ascertain the exact suborganellar location of a protein, the resolving power of electron microscopy is required. Electron microscopic studies using immunogold labeling of ultrathin sections of yeast are beginning to appear in the literature.[6-9] These immunoelectron microscopic (IEM)

[1] A. E. M. Adams and J. R. Pringle, *J. Cell Biol.* **98**, 934 (1984).
[2] J. V. Kilmartin and A. E. Adams, *J. Cell Biol.* **98**, 922 (1984).
[3] A. Y.-S. Jong, M. W. Clark, M. Gilbert, A. Oehm, and J. L. Campbell, *Mol. Cell. Biol.* **7**, 2947 (1987).
[4] M. I. Hagan and J. S. Hyams, *J. Cell Biol.* **89**, 343 (1988).
[5] M. Spencer, *in* "Fundamentals of Light Microscopy." Cambridge Univ. Press, New York, New York, 1982.
[6] M. W. Clark and J. Abelson, *J. Cell Biol.* **105**, 1515 (1987).
[7] E. van Tuinen and H. Riezman, *J. Histochem. Cytochem.* **35**, 327 (1987).

studies have been very informative, but they all demonstrate that there are specific difficulties in the procedure which arise from some unusual characteristics of the yeast cell. In this chapter, these problems are discussed, and workable protocols for alleviating them are provided.

In planning IEM mapping in yeast, a number of variables must be weighed before deciding on which fixation and embedding regimen to follow. Because of the relative harshness of many of the electron microscopic procedures, certain compromises must be made between preservation of cellular ultrastructure and retention of the affinity of an antibody for the fixed and embedded protein. Many of these difficulties can only be resolved by empirical means. Various combinations of fixation protocols and embedding resins have to be tried to determine the optimal conditions for a particular protein. Below are protocols that I have used successfully. Comments following each step should be used as a guide by the investigator in choosing which protocol to follow. In the final section of this chapter, I describe a data collection and representation method for successful IEM of very low copy number proteins. Thus, even if one can obtain only a very low immune labeling efficiency for a protein with a given antibody, the cellular location of the protein can still be determined.

Materials

Listed below are the reagents, equipment, and sources from which to obtain these items necessary to reproduce the IEM procedure.

Fixatives

Glutaraldehyde, 70%, EM grade (Polysciences, Warrington, PA, Cat. No. 1201)
Paraformaldehyde, EM grade, powder (Polysciences, Cat. No. 0380)
Osmium tetroxide (OsO_4), 4% solution (Polysciences, Cat. No. 972A)
Acrolein, distilled, EM grade (Polysciences, Cat. No. 0016)

Embedding Resins

LR White (Polysciences, Cat. No. 17411)
LR Gold (Polysciences, Cat. No. 17412)
Poly/Bed 812 Embedding kit (Polysciences, Cat. No. 8792)
Araldite 502, epoxy kit (Polysciences, Cat. No. 2600)
Spurr low-viscosity embedding kit (Polysciences, Cat. No. 1916)

---

[8] R. Wright, M. Basson, L. D'Ari, and J. Rine, *J. Cell Biol.* **107**, 101 (1988).
[9] G. S. Payne, D. Baker, E. van Tuinen, and R. Schekman, *J. Cell Biol.* **106**, 1453 (1988).

Heavy Metal Stains

Ammonium heptamolybdate (Polysciences, Cat. No. 0085)
Uranyl acetate (Polysciences, Cat. No. 0379)
Vanadyl sulfate (Polysciences, Cat. No. 1310)

Colloidal Gold Secondary Antibody Conjugates

20 nm Colloidal gold/anti-rabbit IgG conjugates (E-Y Labs, San Mateo, CA, Cat. No. GAF-012)
20 nm Colloidal gold/anti-human IgG conjugates (E-Y Labs, Cat. No. GAF-311)

Proteins

Pentex, bovine serum albumin (Miles Diagnostics, Elkhart, IN, Cat. No. 81-001-3)
Protein A-Sepharose CL-4B (Sigma Chemical Co., St. Louis, MO, Cat. No. P3391)
β-Glucuronidase, Type H-2 (Sigma, Cat. No. G-2887)
Human autoimmune serum to DNA (Sigma, Cat. No. ANA-H)
Zymolyase 100T (ICN ImmunoBiologicals, Lisle, IL, Cat. No. 320931)

Other Chemicals

Sodium metaperiodate (Mallinckrodt, St. Louis, MO, Cat. No. 1139)
Hydrogen peroxide (H_2O_2) (30%) (Mallinckrodt, Cat. No. 5240)
Propylene oxide (Polysciences, Cat. No. 00236)
Freon 113, trichlorodifluoromethane (Ted Pella, Redding, CA, Cat. No. 17388)

Equipment

Slimbar hexagonal mesh, nickel grids (400 HH) (SPI, West Chester, PA, Cat. No. 2240N)
Multiple grid staining unit (Polysciences, Cat. No. 7332)
Teflon-coated forceps, #4 (SPI, Cat. No. 504T)
Embedding capsules, conical tips, 00 size (Polysciences, Cat. No. 0294)
Adjustable-temperature drying oven (Polysciences, Cat. No. 8416A)
12 × 17 inch Digitizing tablet (Jandel, Sausalito, CA, Cat. No. 3062)
Sigma scan (Jandel)
Multiwell depression plates (Coors, Golden, CO 60429)
Nalgene filter units, 0.45 μm (Cat. No. 245-0045)

Cell Fixation and Embedding

Choice of Fixatives

The accepted fixation and prestaining procedure for visualization of yeast cell ultrastructure is that described in [41], this volume. This procedure provides good morphological fixation while yielding highly contrasted cellular organelles. The distinct staining of the organelles is obtained by treating the cells with uranyl acetate in addition to osmium tetroxide (OsO_4) postfixation prior to infiltration and resin embedding. Unfortunately, excessive use of heavy metals (osmium and uranium) has been shown to mask epitopes recognized by some antibodies.[10] Although there is no effective way to avoid this problem with uranyl acetate prestaining, methods have been developed that allow the use of the OsO_4 for tissue fixation. Treating the thin section with a saturated solution of sodium metaperiodate will restore the immune reactivity of many proteins in OsO_4 postfixed tissues (for details of this procedure, see Appendix D at the end of this chapter). As this procedure shows much variability, empirical determination of the proper procedure for a given protein will be necessary.

In some cases, it may be necessary to omit entirely heavy metal fixation. The elimination of the OsO_4 postfixation, however, will cause a loss of distinct membrane images and a reduction of ultrastructural detail; thus, a modification of the initial fixation procedure is required to compensate. This cannot be done simply by increasing the concentration of glutaraldehyde, as using 3–4% glutaraldehyde to fix cells can cause a large reduction in protein antigenicity.[11] Presumably this diminished reactivity with the antibody is caused by the bifunctional cross-linking of the protein that results from the dialdehyde nature of the glutaraldehyde molecule. Glutaraldehyde should not be totally removed from the fixation procedure though, as its presence is still necessary for good cellular preservation. The concentration of glutaraldehyde should be greatly reduced to 0.25–0.5%. Another type of aldehyde fixative that has less deleterious affects must be added to supplement the cross-linking ability of the glutaraldehyde. Formaldehyde, at a concentration of 3–4%, fulfills these criteria.[11] A combination fixative of 4% formaldehyde and 0.5% glutaraldhyde provides a good compromise between morphological preservation and retention of antigenicity. A stock solution of formaldehyde should be made just before use from paraformaldehyde (see Appendix A at end of chapter for this procedure). This precaution is taken to prevent the presence of any aldehyde oxidation by-products that would interfere with fixation and epitope

---

[10] M. Bendayan and M. Zollinger, *J. Histochem. Cytochem.* **31**, 101 (1983).
[11] J. DeMay, *J. Neurosci. Methods* **7**, 1 (1983).

recognition. This combination fixative, formaldehyde/glutaraldehyde, can be used alone or in conjunction with OsO_4 postfixation.

The initial penetration of the yeast cell wall has to be rapid and thorough to guarantee proper morphological preservation. Formaldehyde has the advantage of faster penetration into the cell than glutaraldehyde.[12] Another aldehyde, acrolein, is known for its very rapid penetration abilities.[13] Smith and Keefer[13] have reported that a combination of 2% acrolein and 0.25% glutaraldehyde allows short fixation times (1–2 hr) while giving very good ultrastructural preservation of mammalian tissues and good retention of antigenicity. This fixation works well for yeast (M. W. Clark, personal observation), and thus seems the ideal fixative for yeast IEM. Acrolein/glutaraldehyde fixatives should be considered for any experiments in IEM of yeast. Caution should be taken with acrolein, however, as it is volatile and a lacrimator. Always handle acrolein, as well as OsO_4, in the fume hood.

Choice of Embedding Resins

A wide range of embedding media are suitable for IEM, each having advantages and disadvantages. Listed below are some of the most commonly used resins.

Water-Soluble Resins: LR White and LR Gold. The two embedding resins LR White and LR Gold are water soluble and thus do not require extraction of the fixed cells with organic solvents before the infiltration process. Cellular membranes thus remain intact even without OsO_4 stabilization. Also, both LR White and LR Gold have more gentle hardening regimens than the epoxy resins. As a result, preservation of the antigenic epitopes on many proteins is improved. LR White can be hardened by the addition of a chemical accelerator. This chemical curing can even be done with the sample on ice. LR Gold can be photopolymerized at temperatures down to $-25°$. The problem of thermal denaturation of proteins by the high-temperature hardening required for epoxy resins is thus completely eliminated by using these two resins.

These resins do present some problems. LR White and LR Gold have sectioning characteristics different from the epoxy resins and thus require variation in knife angle. Thin sections of LR White and LR Gold do not have the same refractive properties as epoxy thin sections; visualization of

[12] M. A. Hayat, in "Principles and Techniques of Electron Microscopy," Vol. 1. Van Nostrand-Reinhold, Princeton, New Jersey, 1970.
[13] P. F. Smith and D. A. Keefer, *J. Histochem. Cytochem.* **30**, 1307 (1982).

the thin section on the surface of the water thus proves difficult. The ultramicrotome lighting must be adjusted appropriately. Other, more annoying, characteristics of LR White and LR Gold are more difficult to overcome. Thin sections of these resins show a relatively low contrast between the cellular components and the resin. Lightly stained cell samples are, at times, difficult to distinguish from the resin in the electron microscope. Also, the high water content of these water-soluble resins can cause the thin sections to distort when exposed to the electron beam. This distortion is presumably due to subliming of the water trapped in the thin section. The distortion of the resins under the electron beam causes a great deal of section movement, which makes it difficult to focus and can cause aberration of the cellular morphology. Despite these drawbacks, these two resins do provide great accessibility of the antibody to the protein. This is an attribute not to be overlooked when working with yeast. These resins can be quite useful for some applications

Epoxy Resins. Spurr's low viscosity resin, araldite-epoxy, and Poly/Bed-812 resins all have the disadvantage of requiring organic solvent extractions of fixed cells to remove the water from cellular constituents before infiltration and embedding. Such treatment of nonosmicated cells causes loss of the membrane lipid components. Also, the harsh hardening process, usually thermal curing, can cause a reduction in antibody reactivity with some proteins. If either of these treatments causes a loss of antigenicity, then the LR resins should be tried. Another difficulty encountered with the epoxy resins, particularly Poly/Bed-812, is the very smooth surface of thin sections. This smooth section surface renders much of the cellular protein inaccessible to the antibody. Kellenberger and co-workers[14] determined that the epoxy resins allow the antibody to penetrate to a depth of only 1.0 nm into the section. Since most ultrathin sections are from 20 to 30 nm thick, usually less than 5% of the cellular constituents are sampled by the antibody on any one thin section. This accessibility problem can be overcome though by various section etching techniques (see Appendices B and D at the end of this chapter).

The major advantages of the epoxy resins are the high contrast between the resin and cell constituents and the section stability under the electron beam. In the method detailed below, the cell thin sections are only lightly stained with heavy metal after the immunolabeling procedures so as not to obscure the colloidal gold particles. A high contrast between the epoxy resins and the cell components is thus necessary. The other attribute of

---

[14] E. Carlemalm, C. Colliex, and J. Kellenberger, *in* "Advances in Electronics and Electrophysics" (P. W. Hawkes, ed.), p. 280. Academic Press, New York, London, 1984.

epoxy resin sections is their stability under the electron beam. Because of the need in yeast IEM to collect large sets of electron micrographs (see section on Data Collection), the stability of the epoxy resin ultrathin section becomes invaluable. In the procedure reported in this chapter, Poly/Bed-812 (Polysciences) was used as the embedding resin.

Cell Growth

For most strains of yeast, it is optimal to collect cell cultures at a density of less than 0.5 OD_{600}. At this early stage of growth, the cell walls appear to be more readily permeable to the fixative and are more easily removed by enzymatic digestion. The cell wall must be removed to allow adequate penetration of the embedding resin during the infiltration step of the embedding procedure. Another important consideration in the growth of strains for immune mapping procedures, using both optical and electron microscopy, is the use minimal media. Yeast cells grown on limited nutrient media tend to contain large vacuoles that can take up almost the entire cellular volume and dramatically distort the cytoplasm and nucleus. If growth on minimal medium is required to retain marked plasmids, the strains can be grown to 0.5 OD_{600} in minimal medium then diluted with an equal volume of rich medium containing double the nutrient ingredients. This culture should be grown for another doubling time before fixation. This short growth period in rich medium will reduce the vacuole size greatly and cause less than 1% loss of even a 2-μm plasmid.

Fixation and Embedding Procedures

The following fixation procedure, which does not use OsO_4 postfixation, has been used successfully for both low (less than 500 copies per cell) and high (greater than 500 copies per cell) copy number proteins. If the protein of interest is a high copy number protein and a high affinity antibody is available, the paraformaldehyde/glutaraldehyde fixation and OsO_4 postfixation should yield good results. This was the case for IEM mapping of the overproduced yeast hydroxymethylglutaryl-CoA reductase protein.[8] For low copy number proteins, however, the most gentle treatments are required to retain as much antigenicity and accessibility of the protein to the antibody as possible. The procedures listed below are adjusted to provide such characteristics to the ultrathin sections.

Fixation and Embedding of Yeast Cells

The technique described below is the combined and modified procedures of Byers and Goetsch[15] and Zickle and Olson.[16]

Fixation

1. Grow the yeast culture to an approximate density of 0.5 OD_{600}.

2. Collect the cells rapidly by filtration through a 0.45-μm Nalgene filter unit.

3. While on the filter unit, wash the cells once with 10 ml of cold distilled water, then once with 10 ml of 0.1 M potassium phosphate, pH 7.

4. Resuspend the cells in 10 ml of 0.1 M potassium phosphate and transfer the suspension to a 15-ml Corex tube. Centrifuge in a Sorvall SS34 rotor at 5000 rpm for 5 min.

5. Pour off the supernatant and resuspend the washed cell pellet in 4% paraformaldehyde/0.5% glutaraldehyde in 0.1 M potassium phosphate, pH 7, buffer, 1–1.5 ml. Agitate in the cold room overnight.

6. Spin the fixed cells in a microcentrifuge about 15 sec in the cold.

7. Wash the cells once in distilled water at room temperature. (A washing refers to resuspending the pelleted cells in a solution then pelleting the cells again by centrifugation and pouring off the supernatant.)

8. Resuspend the cells in 1 ml of a buffer containing 20 mM Tris-HCl, pH 8, 5 mM disodium EDTA, 25 mM dithiothreitol (DTT), 1 M sorbitol and incubate for 10 min at 30°.

9. Wash the cells once in 0.1 M potassium phosphate–citrate, pH 5.8, 1.2 M sorbitol.

10. Resuspend the fixed cells in 0.1 M potassium phosphate–citrate, pH 5.8, 1.2 M sorbitol to an OD_{600} of 10–20 and add 0.1 volume of β-glucuronidase (Sigma, Cat. No. G-2887) and 0.1 volume of 5 mg/ml Zymolyase 100T (ICN ImmunoBiologicals Cat. No. 320931). Incubate this mixture at 30° for 2 hr. Shake the tube occasionally to keep the cells in suspension.

11. After this digestion, microcentrifuge the cells about 1 min and wash once with 0.1 M potassium phosphate, pH 7.

12. For cells to be used for IEM, go directly into the ethanol dehydration series below.

13. For normal cytological examination, the cells need much more contrast and should be postfixed with 2% OsO_4 in 0.1 M potassium phosphate, pH 7, at 6° for 1 hr, with agitation.

14. Wash the cells in cold distilled water.

15. Prestain the cells with 2% aqueous uranyl acetate for 1 hr at room temperature.

---

[15] B. Byers and L. Goetsch, *J. Bacteriol.* **124**, 511 (1975).
[16] D. Zickle and L. W. Olson, *Chromosoma* **50**, 1 (1975).

16. Wash the cells in distilled water once and then go to the ethanol dehydration series.

Dehydration and Embedding in Poly/Bed 812

1. Ethanol dehydration series: Microcentrifuge the fixed cells 15 sec between each step. Do these incubation steps at 6° until the propylene oxide step.
   a. Wash with 50, 70, 80, and 95% ethanol for 5 min each.
   b. Wash with 100% ethanol, 4 times for 5 min each.
   c. Wash with 100% propylene oxide, 2 times for 15 min each.
2. Infiltration
   a. Resuspend the fixed and dehydrated cells in 1:1 propylene oxide:Poly/Bed 812 (Polysciences) and agitate this solution overnight at room temperature. The next day microcentrifuge the suspension for 2 min to pellet the cells.
   b. Resuspend the pelleted cells in 100% freshly prepared Poly/Bed 812 and let the cells settle in the resin with the tube caps open for 6-8 hr. This treatment allows any propylene oxide left in the cells to diffuse out and evaporate. This suspension needs up to 5 min of microcentrifuging to pellet the cells.
   c. Resuspend the pelleted cells in 100% freshly prepared Poly/Bed 812, about 0.5 ml. Put a drop or two of the cell suspension in the tip of a dust-free, oven-baked BEEM capsule, and fill the remaining space of the capsule with fresh Poly/Bed 812.
   d. To harden the resin use the following three-step curing procedure: (1) 35° overnight, (2) 45° for 8 hr, and (3) 60° for overnight.

This procedure gives the cells time to settle to the tip of the capsule and facilitates sectioning.

Electron Microscope Grids and Sectioning

Section from a small block face (less than 0.1 mm) and collect the ultrathin sections on nickel, 400 mesh, slimbar hexagonal grids. The nickel grids are used to prevent any interaction of the colloidal gold particles of the secondary antibody with the EM grid as might happen if a copper EM grid were used. Furthermore, the hexagonal nature of these grids provides good support of the ultrathin section, while the slimbar configuration yields a wide viewing area for each thin section. The slimbars have the added advantage of allowing the edges of the thin sections to wrap around the slimbars, which securely anchors the thin section to the grid. To aid in thin-section attachment to the grids, rinse the grids in Freon 113 for 1 min (sonication is not necessary), pour off the excess Freon 113, and then invert the beaker containing the grids onto a piece of filter paper in a small petri dish. As the Freon evaporates the grids will fall from the beaker wall. These grids can then be used to pick sections from the surface of the water. It is convenient to do this for a batch of grids just before sectioning.

Before proceeding to the immunostaining, cell components can be made more accessible to the antibody by removing some surface resin from the sections using Luft's procedure (see Appendix B at the end of this chapter).

Preparation of Immunostaining Solutions and Antibodies

All buffers used in this procedure should be prepared from chemical stocks of reagent grade or better. These solutions are filtered through a 0.45-μm filter to remove small particulates that would interfere with colloidal gold visualization. The fixation buffer and the heavy metal stains used in the final steps of the "on-section" staining procedure are not only filtered but also microcentrifuged for 10 min just prior to use. The final centrifugation further removes aggregates and stain precipitates that might form after the stains have been prepared.

The primary antibody used for the "on-section" immunogold labeling procedure should be, at least, the IgG fraction of the antiserum. This IgG fraction can be purified on a protein A-Sepharose column by standard procedures.[17] The IgG fraction is used at a concentration of 2.5 – 25 μg/ml. Affinity-purified antibodies give the best results for the "on-section" staining and should be used when possible. Small quantities of affinity-purified antibody can be prepared from nitrocellulose filters to which proteins separated by denaturing electrophoresis have been electrotransferred, using the methods of Olmsted.[18] The affinity-purified antibody can be used for staining at a concentration of 0.05 – 0.5 μg/ml.

Secondary IgG – colloidal gold conjugates can be obtained from a number of commercial sources. Colloidal gold conjugates from E-Y Labs and Janssen Lifesciences Products (Piscataway, NJ) are used in experiments reported here, but materials from other companies can be used. It is necessary to preadsorb the antibody conjugates first with whole yeast cells to eliminate any yeast cell wall reactivity. Reactivity with yeast cell wall components has been found in almost 60% of the secondary antibodies examined from a variety of companies, so this procedure is done routinely. To preadsorb the antibody, yeast cells are washed in sterile distilled water 5

[17] D. Winkleman, L. Kahn, and J. Lake, *Proc. Natl. Acad. Sci. U.S.A.* **79,** 5188 (1982).
[18] J. B. Olmsted, *J. Biol. Chem.* **256,** 11955 (1981).

times and then resuspended in sterile distilled water at a concentration of 1–2 OD_{600}. About 1.5 hr before the secondary antibody conjugate is needed, the conjugate is diluted to a concentration of 0.5 μg/ml in the staining buffer (see below) and then 1 ml is added to 50 μl of the yeast cell solution. This solution is kept on ice for 1 hr. Just prior to use, the IgG–colloidal gold conjugate dilution is centrifuged to remove the yeast cells and any large aggregates of colloidal gold. For gold conjugates 20–25 nm in diameter, centrifuge the solution at $700 \times g$ (2100 rpm in a Sorvall HB-4 rotor) for 15 min. For gold particles 10–15 nm in diameter, centrifuge the solution at $2000 \times g$ (3500 rpm in a Sorvall HB-4 rotor) for 15 min. After this centrifugation use only the upper 0.75 ml of the solution.

Immunostaining

All antibody incubations, grid fixations, and heavy metal staining are done in an acid-washed, autoclaved, opaque multiwell depression plate held in a clear plastic box. This set-up keeps everything together, visible, and in a covered chamber.

The buffer used for the antibody staining procedure contains 20 mM Tris-HCl, pH 7.5, 500 mM NaCl,* and 0.1% BSA. After removing the surface resin by Luft's procedure (see Appendix B at the end of this chapter), block the nonspecific binding sites on the section by incubation in the above buffer containing BSA at a concentration of 8%. Pass this solution through a Millipore filter before use. Incubate the grids with thin sections attached in this solution for at least 10 min at room temperature.

Primary Antibody Staining

1. Microcentrifuge all reagents for 10 min just before use to remove any aggregates or debris. The colloidal gold solution which has a size dependency is centrifuged as described above.

2. Using the ranges described above under preparation of solutions, titrate each antibody to determine the dilution that gives the best IEM staining. It is best to use as little antibody as possible because background staining is a major problem with this technique, especially when working with a low copy number protein. For low copy number proteins, the sections should be incubated with the antibody overnight at 6°. For high copy number proteins, the primary antibody incubation can usually be

---

* 200 mM NaCl can be used, but the higher salt concentration helps prevent nonspecific binding of the antibodies to the resin surface. This salt concentration also helps break up aggregates that can form in the colloidal gold solutions.

done at room temperature for 1 hr. Of course, these incubation times will depend on the nature of the antibody.

3. Take the grids, with thin sections attached, directly from the blocking solution and place them in the primary antibody solution. Teflon-coated forceps are used for the grid transfer, since uncoated forceps could introduce heavy metals into the antibody and colloidal gold solutions.

Secondary Anti-IgG Antibody–Colloidal Gold Conjugates

1. After the primary antibody incubation, hold the grid in forcep tips and pass buffer over the grid, about 2 ml per grid (jet wash), then place the grid into the colloidal gold–antibody conjugate solution.

2. About 2 hr before using the colloidal gold–antibody conjugate, treat the solutions as follows: dilute the colloidal gold–antibody conjugate to about 0.5 μg/ml in antibody buffer and add the yeast cells prepared for the preadsorption (see section on preparation of immunostaining solutions and antibodies). After the preadsorption, these solutions must be centrifuged: for gold particles of 20–25 nm, centrifuge the dilution in an HB-4 Sorvall rotor at 2100 rpm (700 × g) for 15 min; for gold particles of 10–15 nm, centrifuge in the HB-4 rotor at 3500 (2000 × g) for 15 min. This will remove any large aggregates of gold and the yeast cells from the preadsorption step.

3. Incubate the grids in the gold–antibody conjugate for 1 hr at room temperature.

4. The next rinse is done in an EM Multiple grid staining unit (Polysciences, Cat. No. 7332), which will hold up to 24 grids at one time.

    a. Remove the staining unit cover, place the unit into the cover of a petri dish, and add antibody buffer to cover the unit. Remove any bubbles from the grid slots.

    b. Place the immunostained grids into the slots, replace the cover, and place the unit into a 150-ml beaker filled with antibody buffer without BSA. Mix the solution with a stir bar for 3–5 min.

    c. Remove the staining unit from the beaker, remove the cover, and place the unit gently into the cover of a petri dish containing antibody buffer minus BSA.

Fixation of Antibodies and Heavy Metal Staining.

1. Remove the grids from the staining unit slots and place them onto a drop of 1% glutaraldehyde in antibody buffer minus BSA (centrifuge the buffer first for 10 min in a microcentrifuge). Incubate for 10 min at room temperature.

2. Heavy metal staining (microcentrifuge both stains 10 min before

use): After the fixation, jet wash the grids with distilled water, then proceed as follows:
   a. Place the grids in 1% aqueous uranyl acetate for 5 min in a covered petri dish. Keep the uranyl acetate stain in the dark.
   b. Rinse the grids with distilled water.
   c. Place them in 1% vanadatomolybdate for 15 min (see Appendix C).*
   d. Jet wash the grids with distilled water and dry on filter paper.

The resulting sections are now ready for viewing in a transmission electron microscope. The grids can be stored indefinitely at room temperature.

Data Collection

Because of the low copy number of many of the tRNA and mRNA splicing components that I have been investigating (~0.5–5 copies of the protein were accessible to the antibody on each ultrathin section of the nucleus), it was necessary to develop a data collection technique that could depict the location of a protein with mapping data derived from a large number of cell sections. The strategy I developed uses the membrane of a cellular organelle as a reference point. The shortest distance between that point and the gold particle, both in and out of the organelle, is determined. The distances are then plotted, in two dimensions, on a histogram. For this procedure, electron micrographs of individual cells are taken at random. These micrographs are enlarged and printed, then subjected to the measurement process and histogram representation. This technique proved successful for IEM mapping of tRNA ligase (a yeast tRNA splicing protein),[6] RNA11[19] and Sm antigens (M. W. Clark, unpublished results) (yeast mRNA splicing components), and yeast heat-shock transcription factor (M. W. Clark, unpublished result). By this procedure, these nuclear components were found to be located in the periphery of the yeast nucleus within 200 nm of the nuclear envelope.

Examples of this histogram representation for two different nuclear components are shown in Figs. 1 and 2. Immunoblot analysis of total yeast proteins with a human autoimmune serum (#238) showed that this serum recognizes a high copy number yeast protein of 55 kDa (Fig. 1A, see black asterisk). This protein can be seen by immunofluorescence on whole

---

[19] T.-H. Chang, M. W. Clark, A. J. Lustig, M. E. Cusick, and J. Abelson, *Mol. Cell. Biol.* **8**, 2379 (1988).

---

* Vanadyl stains less densely than lead, leading to lower contrast but better visualization of gold particles. Vanadyl also does not tend to precipitate as readily as does lead. A lead precipitate can look like a colloidal gold particle.

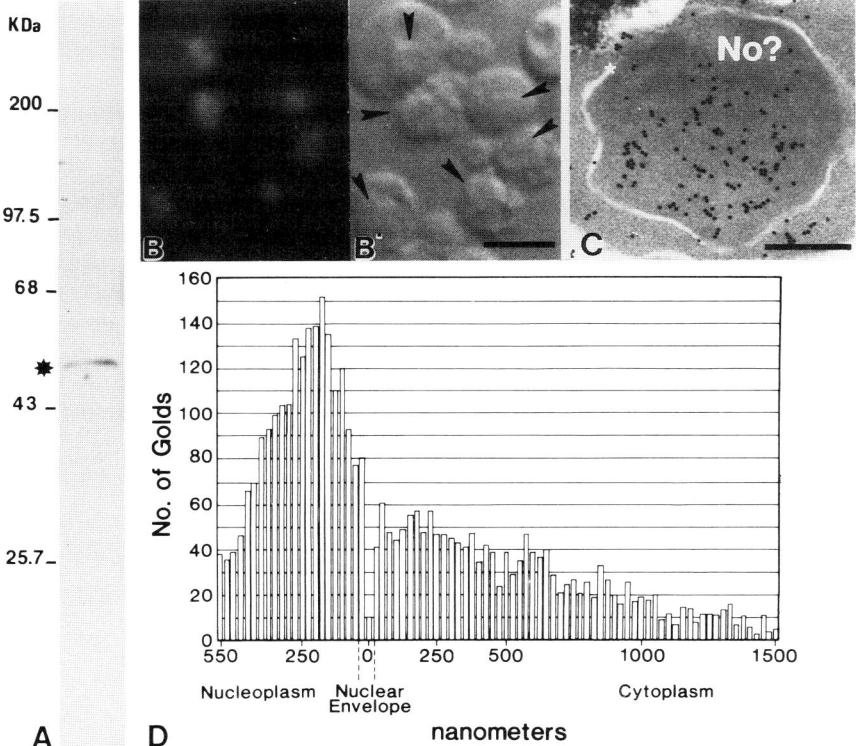

FIG. 1. Immune localization using a human autoimmune serum (#238). (A) Immunoblot of 100 μg of total yeast proteins stained with a 1/500 dilution of the #238 serum. The black asterisk indicates the 55-kDa protein. (B and B') Optical immunofluorescence microscopy on whole mount yeast cells: (B) shows cells viewed under the FITC excitation wavelength; (B') shows the same cells viewed by Nomarski differential interference optics. The arrowheads indicate the yeast nucleus. Bar: 5 μm. (C) An ultrathin section of a yeast cell showing the nucleus stained for IEM by the #238 antiserum. Colloidal gold particles of 20 nm were used in this labeling. NO? indicates area not stained by the #238 serum, which could be the yeast nucleolus. The white asterisk indicates the perinuclear space between the inner and outer membranes of the nuclear envelope, which is slightly swollen by the fixation procedures. Bar: 0.5 μm. (D) Histogram of data collected from 22 cell thin sections demonstrating the location of the 55-kDa protein with which the #238 serum reacts.

mount yeast cells to stain the majority of the nucleoplasm (Fig. 1B). In IEM labeling of yeast sections, the #238 serum stains the majority of the nuclear area (Fig. 1C). The nuclear area not stained by the serum is most likely the nucleolus (Fig. 1C, NO?). A histogram of data collected from 22 cell sections readily reflects the major nucleoplasmic staining pattern (Fig. 1D). A limitation of this data representation procedure is that the absence

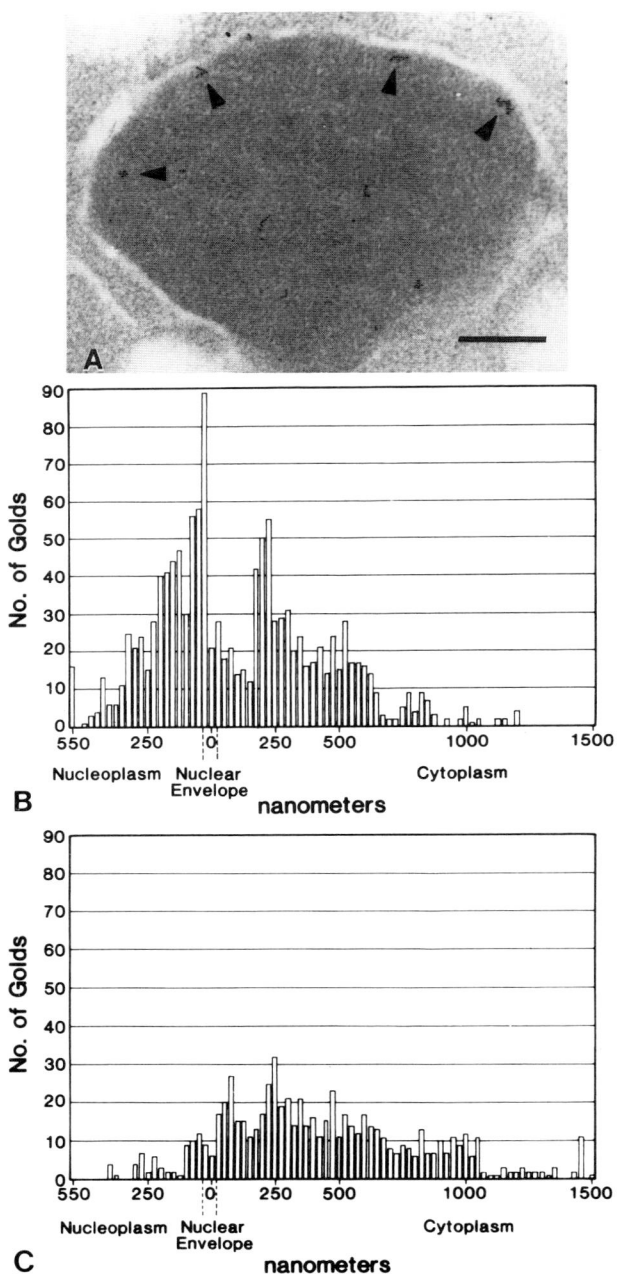

of staining in the nucleolus is not demonstrated. This limitation can be circumvented by changing the reference point for data collection.[6]

Another human autoimmune serum (Sigma), which reacts with B-DNA, stains the yeast nucleoplasm with a low efficiency in IEM labeling experiments (Fig. 2A, arrowheads). The histogram from 58 labeled thin sections shows that the DNA available to the antibody (Fig. 2B) appears to be in a more peripherial location in the yeast nucleus than the 55-kDa protein stained by the #238 serum. Major DNA staining starts in the nucleoplasm within about 350 nm from the nuclear envelope and increases toward the nuclear envelope. The peak of DNA staining is seen at the nuclear envelope. This predominant nuclear staining could reflect the telomere attachments of the 16 yeast chromosomes to the nuclear envelope. Figure 2C shows the control experiment for the anti-DNA staining. The 77 sections represented in this histogram were stained with the anti-DNA antibody that had been preadsorbed with double-stranded DNA. The nucleoplasmic and nuclear envelope staining were almost completely removed by this preadsorption treatment while the cytoplasmic staining remained about the same.

These results indicate that histogram representation of IEM labeling data provides detailed information on nuclear component placement in yeast. Also, collection of the labeling information in this manner provides an opportunity to examine the mapping results in a more quantitative manner than could be accomplished simply from visual inspection of each electron micrograph. Table I gives the data quantification of the experiments discussed above. Labeling intensities can easily be calculated using the preadsorbed antibody as background staining and then subtracting this from the experimental data sets.

Similar data collection and representation techniques have been used successfully in IEM investigation of rabbit muscle actin[20] and an *Escherichia coli* cytoplasmic membrane protein.[21] This method can be per-

[20] H. Shida, M. Shida, and R. Ohga, *J. Electron. Microsc.* **36**, 361 (1987).
[21] B. K. Ghosh, K. Owens, R. Pietri, and A. Peterkosky, *Proc. Natl. Acad. Sci. U.S.A.* **86**, 849 (1989).

FIG. 2. IEM localization of DNA in the yeast nucleus revealed by a human autoimmune serum that reacts with DNA. (A) Ultrathin section of a yeast cell stained with the anti-DNA serum and a 10 nm colloidal gold–secondary antibody conjugate. The arrowheads indicate clusters of 10 nm gold particles at the nuclear periphery. The white asterisk indicates the perinuclear space between the inner and outer membrane of the nuclear envelope, which is slightly swollen by the fixation procedure. Bar: 0.25 μm. (B) Histogram of data collected from 58 cell thin sections stained with the anti-DNA serum showing a peripheral location of the yeast DNA. (C) Histogram of data collected from 77 cell thin sections stained with the anti-DNA serum that was first preadsorbed with double-stranded DNA to remove the anti-DNA activity of the serum.

TABLE I
QUANTITATION OF IMMUNOELECTRON MICROSCOPY DATA[a]

Measured from nuclear envelope	#238 Serum	Anti-DNA	Anti-DNA plus dsDNA
Golds in nucleoplasm	2005 (53.2)	489 (38.6)	63 (9)
Golds in nuclear envelope	91 (2.4)	118 (9.3)	15 (2)
Golds in cytoplasm	1676 (44.4)	659 (52.1)	624 (89)
Total golds counted	3772	1266	702
Number of cells counted	22	58	77

[a] Numbers in parentheses represent the percentages of total cellular gold particles found within the indicated region.

formed with relatively little scientific hardware. Electron micrographs can be measured with a ruler or by using a digitizing tablet connected to a personal computer containing a measurement program, such as Sigma Scan (Jandel). The measurements can thus be stored directly in the computer. Both procedures have been used for the studies described in this chapter.

Problems with Immunoelectron Microscopic "On-Section" Labeling Technique

The difficulties with the "on-section" immunolabeling technique used for yeast are many. The long fixation times and drastic treatment of the cells sometimes required for infiltration and embedding can be extremely traumatic for certain proteins. Also, the length of time required for preparation of cells for sectioning (4–5 days) can be a hardship for the investigator, especially when examining an array of time points for a protein induction experiment. Moreover, the relatively low accessibility of cell constituents to the antibodies, even with the most permeable resins, usually prevents the antibody from sampling more than 10% of the cell components. In this chapter I have reported methods devised to overcome these and other difficulties pertaining to IEM on yeast. Arguably, the data collection and representation technique is the most useful. This procedure overcomes many of the problems that arise from low-efficiency antibody labeling and small sampling size caused by sectioning. Still, the histogram representation of immunolabeling data gives a static, two-dimensional depiction of the dynamic, three-dimensional cell interior. This problem can be partially remedied by using different reference points for the data collection.

It may be possible to solve many of the problems with yeast IEM through the use of ultrathin cryosectioning techniques.[22] Cryoultramicrotomy uses cells that are rapidly frozen in liquid nitrogen, which prevents any morphological changes that occur during initial collection and fixation steps. The frozen cells are sectioned and the sections collected on plastic-coated grids. The interior of the cell is thus exposed, and a short, gentle fixation with an aldehyde fixative can be done directly on the grid. The cell section is then ready for the immunogold staining procedure. The advantage of this sectioning technique is that the cells are not subject to enzymatic digestion to remove the cell wall. Furthermore, antibodies can penetrate the entire depth of the section because no embedding resin is used. Therefore, cryoultrathin sections should provide a more accurate localization of yeast proteins.

So far cryoultramicrotomy has had only limited use on yeast cells, so many of the immune procedures on frozen yeast sections have not been thoroughly worked out. Furthermore, most investigators do not have access to a cryoultramicrotome. Thus, at this time, the immunogold labeling of yeast ultrathin sections is the most reliable and reproducible technique for high-resolution localization of proteins in the yeast cell. If one keeps in mind the reservations discussed here, IEM mapping of yeast proteins will provide an adequate depiction of the cellular location for the protein of interest.

Appendix A: 8% Paraformaldehyde Stock Solution[12]

1. Suspend 0.8 g of paraformaldehyde in 10 ml of distilled water.
2. Heat the solution to 60°.
3. Add a few drops of 0.5 M NaOH to the solution until it clears.
4. Cool and dilute an appropriate amount of the solution 1:1 with 2× fixation buffer, then add the 70% glutaraldehyde to make 0.5%.

Appendix B: Removal of Epoxy Resins from Ultrathin Sections*

This procedure is used to remove surface epon from sections before doing the "on-section" staining procedure, thus increasing the surface area available to the antibody.

1. Prepare a solution of 7 pellets of KOH in 100 ml of *absolute* ethanol. Make this solution fresh each time it is used.
2. Treat the sections on grids for 5–10 min with the above solution.

[22] G. Griffiths, K. Simons, G. Warren, and K. T. Tokuyasu, this series, Vol. 96, p. 466.

* John Luft, personal communication, 1975.

3. Rinse the grids with TBS buffer to bring down the pH, then immediately go to the 8% BSA blocking solution. After this treatment of the thin sections do not let them dry out until after the heavy metal staining step.

Caution: It is critical that the KOH solution be free of water because KOH in water is a very caustic reagent, whereas KOH in alcohol is quite mild and will not drastically disrupt cytological detail. After removing the resin, the sections cannot be air dried. The staining time should be reduced after KOH treatment as KOH leaves the epon matrix quite permeable to stains.

Appendix C: Vanadium Stain for Grid Staining*

Vanadium is probably the lightest metal used as an electron microscopy stain. Vanadium solutions were originally introduced as an alternate to lead hydroxide during the 1960s. The staining solution is artifact-free, simple to prepare, and acts as an effective general stain. Vanadium stains cytoplasmic organelles, collagen fibrils, and glycogen granules intensely; it stains ribosomes adequately but less intensely than does lead. Other cytoplasmic constituents are lightly stained and present a nongranular appearance not easily obtained with other stains. Vanadatomolybdate is quite soluble in water and is not precipitated by normal atmospheric gases.

To make the vanadatomolybdate staining solution mix the following:
  Vanadyl sulfate (1% aqueous), 20 ml
  Ammonium heptamolybdate (1% aqueous), 80 ml

The blue vanadyl sulfate solution and the ammonium molybdate (colorless) produce a dark, purple-black solution when mixed; this should be stirred until oxidation produces a clear, yellow solution. Adjust the pH of the vanadatomolybdate to between 5.5 and 6.0 with 0.1 N NaOH. The vanadatomolybdate solution can be stored at room temperature and used for 2 to 3 months. Use the solution directly to stain the ultrathin sections.

Appendix D: Alternative Epoxy Section Etching Methods

M. Bendayan and M. Zollinger[10] found that epoxy resin thin sections treated with a saturated aqueous solution of sodium metaperiodate for 1 hr increased antibody labeling of both glutaraldehyde- and glutaraldehyde/OsO_4-fixed cells. The metaperiodate etches the resin surface and the tissue only slightly. A H_2O_2 etching procedure can also be used but causes more drastic etching of the tissue. In this procedure, sections are incubated in 10% H_2O_2 for 10 min.

Acknowledgments

I thank Dr. S. Hoch for providing the #238 human autoimmune serum. I thank J. Abelson for providing laboratory space and support during this work.

---

* Modified from M. A. Hayat, *in* "Positive Staining for Electron Microscopy." Van Nostrand-Reinhold, Princeton, New Jersey, 1975.

[43] Analysis of Mitochondrial Function and Assembly

By MICHAEL P. YAFFE

Introduction

Yeast is an extremely valuable organism for the study of mitochondrial function and assembly. The ability of wild-type yeast to grow either aerobically or anaerobically has allowed isolation of numerous mutants defective in aerobic growth.[1] These mutants define several hundred genes, some in the mitochondrial genome and many more encoded in the nucleus, and these genes encode proteins essential for mitochondrial respiration. The characterization of such mutant strains, combined with the purification and analysis of many mitochondrial enzymes and electron carriers, has contributed to a detailed understanding of mitochondrial metabolism and oxidative phosphorylation.[2] The study of these processes has been expanded more recently by the development of powerful molecular genetic techniques that have enabled researchers to isolate and characterize genes encoding a number of mitochondrial proteins (see [10] in this volume). The analysis and manipulation of these genes allow researchers to examine the role of individual polypeptides.

Much current mitochondrial research focuses on the assembly of the organelle. Mitochondria proliferate by the growth and division of preexisting mitochondria, and major components of mitochondrial growth are the synthesis of polypeptides encoded in the mitochondrial genome and the import of nuclear-encoded proteins from the extramitochondrial cytoplasm.[3] Again, yeast has provided an excellent experimental system for the investigation of these essential processes. Studies of both nuclear and mitochondrial mutants are leading to new insights into the expression of the mitochondrial genome. Additionally, three types of experiments using yeast have revealed many facets of the process of mitochondrial protein import. The first of these approaches involves *in vitro* assays employing isolated mitochondria and has illuminated many of the key steps in mitochondrial protein import.[4-9] A second approach has been the cloning,

[1] A. Tzagoloff and A. M. Myers, *Annu. Rev. Biochem.* **55,** 249 (1986).
[2] A. Tzagoloff, "Mitochondria." Plenum, New York, 1982.
[3] G. Attardi and G. Schatz, *Annu. Rev. Cell Biol.* **4,** 289 (1988).
[4] S. M. Gasser, G. Daum, and G. Schatz, *J. Biol. Chem.* **257,** 13034 (1982).
[5] P. Bohni, S. Gasser, C. Leaver, and G. Schatz, *in* "The Expression and Organization of the Mitochondrial Genome" (A. M. Kroon and C. Saccone, eds.), p. 423, North-Holland Biomedical, Amsterdam, 1980.

analysis, and manipulation of genes encoding imported mitochondrial proteins.[10-12] This type of investigation has revealed targeting signals of many imported polypeptides. A third approach has involved the isolation and characterization of mutant yeast defective in mitochondrial protein import.[13] Such studies have led to the identification of specific components of the import apparatus.[14,15]

The experimental procedures described below should allow the analysis of many aspects of mitochondrial function and assembly. The approaches focus on the analysis of mitochondria in intact cells, the isolation of functional mitochondria, and the assay of mitochondrial function and protein import in cell-free systems. These procedures can be employed to characterize the fate of a specific mitochondrial protein or to analyze the effect of a mutation on mitochondrial function or on protein import.

Analysis of Mitochondria in Intact Cells

Mitochondrial Function

The simplest way to analyze mitochondrial function is to examine the growth of yeast on various nonfermentable carbon sources. A classic diagnostic test is to plate cells on YPDGE [1% Bacto-yeast extract, 2% Bacto-peptone, 0.1% glucose, 3% (v/v) glycerol, 2% ethanol, 2% Bacto-agar], a rich medium containing the mitochondrial substrates glycerol and ethanol and only a small amount of glucose. Cells with complete mitochondrial function grow well on this medium while many mutants affecting mitochondrial activities grow to only small (*petite*) colonies. Growth can also be examined on rich (1% Bacto-yeast extract, 2% Bacto-peptone) or minimal (0.67% yeast nitrogen base) media containing only 2% ethanol, 3% glycerol, or 2% lactate. Growth on one but not another of these carbon sources may reveal defects either in uptake or in a specific metabolic pathway. Some (so-called) wild-type strains will not grow in minimal media with

---

[6] F.-U. Hartl, J. Ostermann, B. Guiard, and W. Neupert, *Cell (Cambridge, Mass.)* **51**, 1027 (1987).
[7] W.-J. Chen and M. G. Douglas, *J. Biol. Chem.* **262**, 15605 (1987).
[8] M. Eilers and G. Schatz, *Nature (London)*, **322**, 228 (1986).
[9] M. Schleyer and W. Neupert, *Cell (Cambridge, Mass.)* **43**, 339 (1985).
[10] E. C. Hurt, B. Pesold-Hurt, and G. Schatz, *EMBO J.* **3**, 3149 (1984).
[11] M. G. Douglas, B. L. Geller, and S. D. Emr, *Proc. Natl. Acad. Sci. U.S.A.* **81**, 3983 (1984).
[12] T. Hase, U. Muller, H. Riezman, and G. Schatz, *EMBO J.* **3**, 3157 (1984).
[13] M. P. Yaffe and G. Schatz, *Proc. Natl. Acad. Sci. U.S.A.* **81**, 4819 (1984).
[14] C. Witte, R. E. Jensen, M. P. Yaffe, and G. Schatz, *EMBO J.* **7**, 1439 (1988).
[15] R. E. Jensen and M. P. Yaffe, *EMBO J.* **7**, 3863 (1988).

nonfermentable carbon sources, so it is essential to analyze mutant growth with respect to that of the parental strain.

A second approach to assess the function of mitochondria over a short time period (e.g., after the shift of a temperature-sensitive, mutant strain to the nonpermissive temperature) is to examine cellular protein synthesis during growth on a nonfermentable carbon source such as lactate. Protein synthesis is almost entirely dependent on mitochondrially produced ATP during growth on this carbon source. We used such an analysis to demonstrate that mitochondrial energy metabolism is not affected (in the short run) by a temperature-sensitive mutation that blocks import of proteins into mitochondria.[13]

To perform this analysis cells are grown on semisynthetic medium[16] (containing, per liter, 3 g Bacto-yeast extract, 1 g KH_2PO_4, 1 g NH_4Cl, 0.5 g $CaCl_2 \cdot 2H_2O$, 0.5 g NaCl, 0.6 g $MgCl_2 \cdot 6H_2O$) with 2% lactate (the pH of the medium is adjusted to 5.5 with NaOH) to an A_{600} of 2-4. Cells are collected by centrifugation and resuspended in the same medium to 10 A_{600}/ml. The cells are then preincubated under appropriate conditions (e.g., 37°) for up to 2 hr. Labeling of proteins is initiated by addition of [^{35}S]methionine (10 μCi to 1 ml cells). Aliquots (0.1 ml) are removed periodically during labeling and added to 2 ml of 5% trichloroacetic acid (TCA). It is convenient to remove aliquots every 1-2 min. The precipitated proteins are collected on glass-fiber filters (Whatman GF/A), and the filters are washed with 3-5 ml of 5% (w/v) TCA and then 3-5 ml of 95% ethanol. Radioactivity on filters is analyzed by liquid scintillation counting. The apparent rate of protein synthesis can be determined from the slope of the curve of the counts per minute per aliquot versus the time of incubation with radiolabeled methionine. The addition of KCN (to 1 mM) or other inhibitors of mitochondrial energy metabolism during labeling will abolish the increase in incorporation over time and demonstrate that the labeling is dependent on mitochondrial function.

Mitochondrial Protein Import

Mitochondrial protein import can be analyzed in intact cells by isotope-tracer (pulse–chase) experiments. One advantage of such analysis is that one can assess import in the complete cellular system without disrupting structures or altering concentrations of essential components. Additionally, import can be examined in cells grown on glucose or any other carbon source, and import into mitochondria of cells with *petite* mutations may be evaluated. This approach takes advantage of the shift in molecular

---

[16] G. Daum, P. C. Bohni, and G. Schatz, *J. Biol. Chem.* **257**, 13028 (1982).

size on processing of many precursors during or following their import into mitochondria.[5] The approach involves (1) labeling of cells with [^{35}S]methionine or $^{35}SO_4^{2-}$, (2) extraction of cellular proteins, (3) immunoprecipitation of specific mitochondrial proteins, and (4) analysis of immunoprecipitates by sodium dodecyl sulfate (SDS)–polyacrylamide gel electrophoresis and fluorography.

Cell Labeling. Cells are grown to an OD_{600} of 2–4 on semisynthetic medium (see above) supplemented with 2% lactate, 2% glucose, 2% galactose, or other carbon source. Cells are collected by centrifugation (3000 g, 5 min) and resuspended in labeling media to 10 OD_{600}/ml. Cells are labeled either in semisynthetic medium plus 1% glucose with [^{35}S]methionine or in sulfate-free, minimal glucose medium (0.67% sulfate-free, yeast nitrogen base, 1% glucose) with $^{35}SO_4^{2-}$. Other labeling media and protocols should also be adequate. Cells are incubated 5 min or longer in the medium before addition of labeled precursor. Next, 20–100 μCi of [^{35}S]methionine or $^{35}SO_4^{2-}$ is added per 10 OD of cells. The cells are incubated an additional 5 min. This labeling often results in the incorporation of 1000 to 5000 cpm in relatively abundant mitochondrial proteins (e.g., β-subunit of the F_1-ATPase or citrate synthase); however, labeling efficiency is rather strain-specific. Labeling is stopped by adding either unlabeled methionine to 2 mM or cycloheximide (from a fresh stock of 10 mg/ml in 50 mM potassium phosphate, pH 6.0) to 0.1 mg/ml. The incubation is then continued for a chase period. To examine the rate of import of a mitochondrial protein, aliquots (usually 1 ml) are removed at the end of the pulse and periodically during the chase.

One can perform a variation of the pulse–chase protocol to examine only posttranslational import of precursors into mitochondria. This is particularly useful for studying the import of proteins whose import is so rapid that only a very small pool of precursor is normally present in the cell. For such an analysis, precursors are accumulated by biosynthetically labeling proteins in the presence of 20 μM carbonyl cyanide *m*-chlorophenylhydrazone (CCCP, added from a 2 mM stock in ethanol), an uncoupler of the mitochondrial membrane potential. Cells are labeled as described above, except that labeling is for 10 min. 2-Mercaptoethanol is then added to 0.05% to inactivate the CCCP. Cycloheximide or methionine is added also, as described above, to block the further incorporation of label into precursor proteins. The incubation is continued, aliquots (1 ml) are removed intermittently, and proteins are extracted as described below. Generally, a 10- to 15-min incubation with 2-mercaptoethanol is required before the CCCP is inactivated fully and the potential, on which import depends, is restored.

Protein Extraction. Proteins are extracted from the cells by the rapid procedure described by Yaffe and Schatz.[13] To perform this extraction, the 1-ml aliquots are added directly to 0.15 ml of a freshly made solution of 1.85 M NaOH with 7.4% 2-mercaptoethanol in 15-ml conical polypropylene centrifuge tubes. The mixture is held on ice for 10 min, then 0.15 ml of 50% TCA is added to precipitate the proteins. The samples are held for an additional 10 min on ice, and the precipitate is collected by centrifugation for 5 min in a clinical-style benchtop centrifuge at top speed. The supernatant is carefully removed by aspiration and discarded. The pellet is washed by adding 1.5 ml of ice-cold acetone and vortexing. It is not necessary to thoroughly resuspend the pellet in the acetone, as the function of this wash is to remove residual TCA. The pellet is collected by centrifugation as described above, and the acetone supernatant is removed carefully by aspiration. Proteins are extracted from the pellet by adding 0.5 ml of 5% SDS and heating in a boiling water bath for 5 min. The samples are centrifuged once again for 5 min to remove insoluble material. The supernatant, containing the extracted proteins, is carefully removed to a 50-ml conical polypropylene tube (e.g., Falcon 2098) and diluted 50-fold with sterile TNET [1% Triton X-100, 150 mM NaCl, 50 mM Tris-HCl (pH 7.4), 5 mM EDTA]. Specific, labeled proteins are then immunoprecipitated from this solution.

Immunoprecipitation and Analysis. Immunoprecipitation and gel electrophoretic analysis of proteins can be carried out by standard procedures.[17,18] Antibodies against a variety of mitochondrial proteins can be used, and often the same labeled samples can be treated successively with different antisera. Table I lists some representative mitochondrial proteins for which import has been examined *in vivo* by pulse–chase analyses.

Isolation of Mitochondria

Mitochondria are isolated from a homogenate of yeast spheroplasts by differential centrifugation. The procedure closely follows that described by Daum *et al.*[16] and is optimized to yield organelles which can carry out respiratory functions, maintain structural integrity, and import polypeptide precursors *in vitro*.

Preparation of Spheroplasts. Cells may be grown on a variety of media; however, optimal yields of mitochondria are obtained with growth on a semisynthetic medium supplemented with 2% lactate and 0.05% glucose

---

[17] S. W. Kessler, *J. Immunol.* **115**, 1617 (1975).
[18] M. G. Douglas and R. A. Butow, *Proc. Natl. Acad. Sci. U.S.A.* **73**, 1083 (1976).

TABLE I
YEAST MITOCHONDRIAL PROTEINS USEFUL FOR
IMPORT STUDIES *in Vivo*

Protein	Location[a]	Apparent molecular size (kDa)		Ref.
		Precursor	Mature	
F_1-ATPase, β-subunit	IM[b]	56	54	c
Citrate synthase	M	50	47	d
Alcohol dehydrogenase isoenzyme III	M	43	39	e
Cytochrome c_1	IM	37	31	f
Cytochrome b_2	IMS	68	58	f, g
Cytochrome-c peroxidase	IMS	39.5	33.5	g

[a] M, Matrix; IM, inner membrane; IMS, intermembrane space.
[b] Peripheral membrane protein on the matrix side of IM.
[c] M.-L. Maccecchini, Y. Rudin, G. Blobel, and G. Schatz, *Proc. Natl. Acad. Sci. U.S.A.* **76**, 343 (1979).
[d] M. P. Yaffe and G. Schatz, *Proc. Natl. Acad. Sci. U.S.A.* **81**, 4819 (1984).
[e] D. Pilgram and E. T. Young, *Mol. Cell. Biol.* **7**, 294 (1987).
[f] S. M. Gasser, A. Ohashi, G. Daum, P. C. Bohni, J. Gibson, G. A. Reid, T. Yonetani, and G. Schatz, *Proc. Natl. Acad. Sci. U.S.A.* **79**, 267 (1982).
[g] G. A. Reid, T. Yonetani, and G. Schatz, *J. Biol. Chem.* **257**, 13068 (1982).

(described above). Mitochondria have also been purified from cells grown in semisynthetic media with glycerol, galactose, or raffinose as carbon sources. These latter two sugars allow the growth of *petite* strains without causing repression (as does glucose) of expression of many mitochondrial proteins. However, some strains will not grow on galactose as *petites*. Cells are grown with substantial aeration (e.g., 700 ml medium in a 2-liter Erlenmeyer flask shaking on a rotary platform at 325 rpm or in a microfermentor with high aeration). The standard growth temperature is 30°, although satisfactory mitochondria have been prepared from cells grown at 18°–37°. Cells are grown to an A_{600} of 2–4.

Cells are collected by centrifugation at 3000 g for 5 min at room temperature. They are resuspended in water to approximately 10% of the original volume of the culture. All resuspensions of cells and spheroplasts are performed by stirring with a glass rod which has a rubber bulb fitted over the tip contacting the cells. The cells are collected again by centrifugation in preweighed buckets. After discarding the supernatant from this spin, the wet weight of the cells is determined. This weight is used as the basis for all resuspensions described below.

Cells are resuspended in 0.1 M Tris–SO_4, pH 9.3, to which dithiothreitol has been freshly added to 10 mM. Cells are resuspended to 0.1–0.3 g/ml. They are incubated with gentle shaking for 10 min at 30° or 20 min at 23° and then collected by centrifugation as above. The cells are next washed once by resuspension in a solution of 1.2 M sorbitol, 20 mM potassium phosphate, pH 7.4 (spheroplasting buffer), and recentrifugation.

The cells are resuspended to 0.1 g/ml in spheroplasting buffer to which Zymolyase 20K (ICN Immunobiologicals, Lisle, IL) has been added (2.5 mg/g cells) immediately prior to the resuspension. Cells are incubated for 30–45 min at 30° or 60–90 min at 23° with gentle shaking. One can check for the completeness of spheroplasting by examining under the microscope a small amount of the cell suspension diluted 1:10 with water. Spheroplasts should readily lyse under these conditions. Following the incubation with Zymolyase, the spheroplasts are collected by centrifugation and washed 3 times by resuspension in spheroplasting buffer (5–10 ml/g cells; without zymolyase) and recentrifugation.

Mitochondrial Isolation. All subsequent steps are carried out at 0–4°. Spheroplasts are resuspended in mitochondrial isolation buffer (MIB) composed of 0.6 M mannitol, 20 mM HEPES–KOH, pH 7.4, 0.5 mM phenylmethylsulfonyl fluoride (PMSF) [added immediately prior to use from a 1 M stock in dimethyl sulfoxide (DMSO)] to a final concentration of 0.5 g cells/ml MIB. Spheroplasts are broken in a Dounce homogenizer with 15 strokes using the B pestle. The homogenate is diluted 2-fold with MIB and centrifuged at 3000 g for 5 min at 0°–4°. The supernatant from this spin is held on ice after addition of more PMSF (1 μl of the 1 M stock to 10 ml supernatant). The pellet is resuspended in MIB and homogenized a second time as described for the spheroplasts above. This second homogenization releases additional mitochondria from the pellet, which contains many unbroken cells. The homogenate is diluted 2-fold and centrifuged at 3000 g for 5 min. The pellet contains unbroken cells, nuclei, and some mitochondria and is usually discarded. The supernatant from this spin is combined with the first supernatant and then centrifuged at 9500 g, for 10 min. The pellet comprises the crude mitochondrial fraction. The supernatant is discarded unless postmitochondrial supernatant, microsomal, cytosolic, or other light fractions are desired.

Mitochondria are purified further from the crude fraction by successive washes with MIB. Immediately prior to each wash the PMSF is added to the resuspension buffer. This protease inhibitor is omitted from the last two washes if the mitochondria will be treated with proteases as part of a later procedure. Mitochondrial pellets are resuspended by blowing gently in and out with a Pasteur pipette fitted with a rubber bulb. The crude mitochondrial pellet is resuspended in 5–10 ml MIB per original gram of cells. The suspension then is centrifuged at 9500 g for 10 min. This is repeated for a total of 4 washes. Prior to the last spin, the supernatant is centrifuged at 2000 g for 3 min. The resulting pellet contains precipitated PMSF and some proteinaceous aggregates and is discarded. The supernatant fraction is spun for a final run at 9500 g for 10 min. The pellet is resuspended in a minimal volume of MIB, usually 50 μl/g of original cells.

The mitochondrial yield can be determined by measuring mitochondrial protein with any of the convenient protein assays (e.g., Lowry, Bradford, BCA). An extremely convenient assay of mitochondrial protein involves the solubilization of proteins in SDS and the determination of protein content by ultraviolet absorption. To perform this procedure 10 μl of the final mitochondrial preparation is added to 1 ml of 0.6% SDS. The mixture is heated at 95° for 4 min, and the absorbance at 280 nm is measured against a 0.6% SDS blank. For mitochondria isolated from cells grown on lactate, an A_{280} value of 0.21 is equivalent to an original mitochondrial protein concentration of 10 mg/ml.

Mitochondria can be purified further by banding them in a density gradient. First, a Percoll (Pharmacia, Piscataway, NJ) gradient is formed by centrifuging a 10 ml solution of 20% Percoll in MIB at 15,000 rpm for 12 min in a Sorvall SS34 rotor. Next, 0.5 ml of mitochondrial suspension (10 mg/ml) is layered on the top of the gradient, which is then spun for 10 min at 12,000 rpm in a Sorvall HB-4 (swinging-bucket) rotor. The turbid, mitochondrial band (the major band in the gradient) is removed with a Pasteur pipette and diluted 10-fold with MIB, and the mitochondria are reisolated by centrifugation at 9500 rpm for 10 min in the SS34 rotor. The pelleted mitochondria are gently resuspended in a minimal volume of MIB. In some strains vacuoles largely copurify with mitochondria during differential centrifugation but can be separated by the Percoll gradient. The vacuoles appear as a white, upper (less dense) band while the pink or orange lower band comprises the mitochondrial fraction.

Functional Assays of Isolated Mitochondria

Assays for the characterization of mitochondrial function have been described extensively (see this series, Vols. X and LV). Some assays evaluate the presence or activity of individual enzymes, whereas others measure multienzyme processes such as electron transport, oxidative phosphorylation, or mitochondrial protein synthesis. Described below are tests we frequently use to characterize isolated mitochondria, particularly in comparing mitochondria from mutant strains with those from wild-type, parental strains.

One of the most useful and direct analyses of mitochondrial function is the determination of respiratory rate and the respiratory coupling index (RCI; also called the respiratory control ratio, RCR). In this assay mitochondria are incubated with respiratory substrates, and the rate of oxygen utilization (the respiratory rate) is measured with an oxygen electrode. The oxygen consumption is dependent on the availability of ADP, and the addition of a small amount of ADP induces a rapid utilization of oxygen. The ratio of the rate of oxygen consumption after addition of ADP (state 3 respiration) to the rate after exhaustion of the added ADP (state 4 respiration) is the RCI. The value of this ratio provides a measure of the intactness or integrity of the mitochondrial inner membrane. Uncoupled mitochondria display a ratio of 1. Additionally, the phosphorylating capacity (P/O or ADP/O ratio) of a mitochondrial preparation can be determined by measuring the amount of oxygen consumed after addition of a known amount of ADP. The oxygen electrode apparatus and its use to characterize isolated mitochondria have been described thoroughly.[19] The conditions described below are appropriate for the assay of yeast mitochondria.

Isolated mitochondria are incubated at a final concentration of 250–500 μg/ml in a reaction mixture containing 0.6 M mannitol, 20 mM HEPES–KOH (pH 7.4), 10 mM potassium phosphate (pH 7.4), 2 mM MgCl$_2$, 1 mM EDTA, 5 mg/ml bovine serum albumin (BSA). Oxygen uptake can be monitored at a variety of temperatures (18°–37°). The background respiratory rate is first measured for 1–2 min. Respiratory substrates, either Tris–succinate or Tris–malate plus Tris–pyruvate, are added to 10 mM (the substrates are prepared as 0.5 M stocks of the appropriate acid, and the pH is adjusted to 7.4 with Tris base). Other respiratory substrates such as D-lactate, ethanol, or α-ketoglutarate can also be used. The rate of oxygen uptake is measured again for 1–2 min. Next, an aliquot of ADP is added to a final concentration of 100 μM (e.g., for a 1-ml reaction volume, add 2 μl of 50 mM ADP). The new respiratory rate (state 3) is measured. Measurement continues until the slower rate (state 4) resumes following the exhaustion of the ADP. The RCI is calculated by dividing the slope of the curve in state 3 by that from state 4 respiration. Common values for intact yeast mitochondria are 2–3.5. The P/O ratio is calculated as the number of nanomoles of ADP added divided by the amount of oxygen consumed during state 3 respiration. For yeast mitochondria this ratio is commonly 1.6–1.8 for respiration on succinate and 1.8–2.0 for respiration on malate plus pyruvate. Even with the best yeast

---

[19] R. W. Estabrook, this series, Vol. 10, p. 41.

mitochondrial preparations, these ratios only approach 2.0 rather than the 3.0 found with mammalian mitochondria, because yeast mitochondria do not contain a phosphorylation site for the NADH-linked substrates.

A second useful and easy test of mitochondrial function is the determination of mitochondrial membrane potential by examining the uptake of the fluorescent dye 3,3′-dipropylthiocarbocyanine iodide [diS-C_3-(5), Molecular Probes, Eugene, OR]. This dye is taken up by mitochondria in a potential-dependent manner,[20,21] and uptake can be measured with a fluorescence spectrophotometer or fluorometer. A typical assay mixture contains 0.6 M mannitol, 20 mM potassium pohosphate (pH 7.4), 10 mM $MgCl_2$, 0.5 mM EDTA, 1 mg/ml BSA (fatty acid-free), Tris–malate and/or Tris–succinate at 5 mM, mitochondria at a final concentration of 100–200 µg/ml, and the dye at 2 µM. The dye should be added from a 2 mM stock solution in DMSO. All components except the mitochondrial substrates, malate and succinate, are mixed together in a stirred cuvette. Excitation of the sample is at 620 nm, and emission is measured at 670 nm. After the emission from the sample has stabilized, the substrates are added and changes in the fluorescence recorded. The addition of uncouplers (e.g., CCCP to 0.2 µM) or various mitochondrial inhibitors (e.g., 1 mM KCN and 50 µg/ml oligomycin) will demonstrate that changes in fluorescence are due to the generation of membrane potential.

The analysis of mitochondrial protein synthesis examines the functioning of a different set of mitochondrial proteins. This process can be measured by assaying the incorporation of a labeled amino acid into proteins by isolated mitochondria.[22] Isolated mitochondria (100 µg) are incubated in an assay volume of 0.2 ml in a mixture containing 0.6 M mannitol, 20 mM potassium phosphate (pH 7.4), 150 mM KCl, 10 mM sodium succinate, 10 mM sodium malate, 10 mM α-ketoglutarate, 50 µM GDP, 10 mM $MgCl_2$, 1 mM ATP, and 100 µg/ml cycloheximide. After a 5- to 15-min incubation, labeling is initiated by adding 25 µCi of [^{35}S]methionine (1000 Ci/mmol). Incubation is continued for 30 to 45 min. The optimal incubation temperature is 30°; however, assays at other temperatures are possible. At the end of the incubation period, unlabeled methionine is added to a final concentration of 10 mM, and incubation is continued for an additional 10 min. Mitochondria are next reisolated by layering the incubation mixture on 1 ml of 25% sucrose solution (4°) containing 10 mM unlabeled methionine and spinning for 10 min in a microcentrifuge at 4°. The mitochondrial pellet is resuspended in SDS sample buffer and heated at 95° for 4 min. Labeled proteins are resolved by electrophoresis in a 12%

---

[20] P. J. Sims, A. S. Waggoner, C. H. Wang, and J. F. Hoffman, *Biochemistry* **13**, 3315 (1974).
[21] A. S. Waggoner, this series, Vol. 55, p. 689.
[22] A. Ohashi and G. Schatz, *J. Biol. Chem.* **255**, 7740 (1980).

polyacrylamide gel in the presence of SDS and detected by fluorography. The most abundant products of mitochondrial protein synthesis are cytochrome oxidase subunits I, II, and III, with apparent molecular weights of 40,000, 27,300, and 25,000, respectively.

Protein Import into Isolated Mitochondria

The ability of isolated mitochondria to import precursor proteins[23] allows one to study many aspects of the import process in a defined, *in vitro* system. Such a system has proven essential for identifying major steps in the import pathway, assessing the energy requirements of protein import, and analyzing the state of precursors during their translocation into yeast mitochondria.[4,6,8,24-26]

Early *in vitro* studies involved incubation of mitochondria with a mixture of labeled precursors synthesized by the translation of total cellular mRNA.[4,23] Following import, mitochondria were reisolated, proteins were extracted in SDS, and specific, labeled proteins were immunoprecipitated. The immunoprecipitated proteins were then resolved by SDS–PAGE and detected by fluorography. Although this procedure remains adequate for characterizing import, it requires tedious immunoprecipitation steps, and the low degree of labeling of many proteins demands lengthy exposures of the dried gel to photographic film. The cloning of genes for many mitochondrial proteins and the availability of *in vitro* transcription vectors now allows the synthesis of individual protein precursors with high specific radioactivity by *in vitro* transcription and translation.[10] Use of such precursors eliminates the immunoprecipitation step and long exposure times for detection of labeled protein.

Precursor proteins are synthesized by the *in vitro* transcription of cloned genes[27,28] followed by the translation of the mRNA in an *in vitro* protein synthesis system such as rabbit reticulocyte lysate.[29] The translation is carried out in the presence of [^{35}S]methionine or another labeled amino acid. Any procedure and set of reagents which yield a substantial amount of highly labeled protein should be satisfactory for the production

---

[23] M.-L. Maccecchini, Y. Rudin, G. Blobel, and G. Schatz, *Proc. Natl. Acad. Sci. U.S.A.* **76**, 343 (1979).

[24] S. M. Gasser, A. Ohashi, G. Daum, P. C. Bohni, J. Gibson, G. A. Reid, T. Yonetani, and G. Schatz, *Proc. Natl. Acad. Sci. U.S.A.* **79**, 267 (1982).

[25] W.-J. Chen and M. G. Douglas, *Cell (Cambridge, Mass.)* **49**, 651 (1987).

[26] M. Eilers, W. Oppliger, and G. Schatz, *EMBO J.* **6**, 1073 (1987).

[27] D. A. Melton, P. A. Krieg, M. R. Rebagliati, T. Maniatis, K. Zinn, and M. R. Green, *Nucleic Acids Res.* **12**, 7035 (1984).

[28] D. Stueber, I. Ibrahimi, D. Cutler, B. Dobberstein, and H. Bujard, *EMBO J.* **3**, 3143 (1984).

[29] H. R. B. Pelham and R. J. Jackson, *Eur. J. Biochem.* **67**, 247 (1976).

of precursors. Vectors containing the T_7, SP6, and T_5 promoters have all been used for efficient production of mRNA. We routinely employ a commercial rabbit reticulocyte lysate system (Promega, Madison, WI), but successful synthesis of mitochondrial precursor proteins has also been described with wheat germ lysate[30] and yeast lysate[31] translation systems. Table II lists some precursor proteins which have been produced by *in vitro* transcription–translation and used in import assays.

Mitochondrial Preparation. Mitochondria used for import assays are prepared from cells grown on semisynthetic medium supplemented with 2% lactate by spheroplast formation, homogenization, and differential centrifugation as described above. Mitochondria can also be isolated from cells grown on other carbon sources as the experiments dictate; however, *in vitro* import into *petite* (p^-) mitochondria has been demonstrated only under highly specialized import conditions.[32]

Frozen mitochondria can also be used for import assays. In this case mitochondria are frozen for storage and are thawed as described by Murakami *et al.*[30] with minor modifications. The freshly isolated mitochondria are adjusted first to a protein concentration of 30 mg/ml. One volume of an ice-cold solution of 20% DMSO, 20 mg/ml BSA, 0.6 M mannitol, 20 mM HEPES–KOH (pH 7.4), and 0.5 mM PMSF is added dropwise. Aliquots (100 μl) are frozen in liquid nitrogen and stored at $-70°$. Immediately prior to use in an import assay, frozen aliquots are thawed at room temperature and diluted with 1 ml of ice-cold washing buffer containing 0.6 M mannitol, 20 mM HEPES–KOH (pH 7.4), 0.1% BSA, 0.5 mM magnesium acetate, and 0.5 mM PMSF. Samples are centrifuged for 2 min in a microcentrifuge at 4° and then resuspended gently in MIB (see section on mitochondrial preparation above).

Import Reaction. Mitochondria (200 μg) are incubated in an import mixture (200 μl total volume) consisting of 3–15 μl of lysate containing the labeled precursor, 0.6 M mannitol, 20 mM HEPES–KOH (pH 7.4), 1 mM ATP, 1 mM MgCl$_2$, 5 mM phosphoenolpyruvate, 2 units pyruvate kinase (Boehringer-Mannheim, Indianapolis, IN), 40 mM KCl, 5 mM methionine, and 3 mg/ml BSA. It is often convenient to make a stock "energy mix" of ATP, MgCl$_2$, and phosphoenolpyruvate at 5 to 15 × concentration in 0.6 M mannitol. The pH of this stock solution is adjusted to approximately 7, and aliquots are stored frozen at $-20°$. To set up the assay, all components except lysate and mitochondria are mixed together first. An appropriate amount of a 1 M mannitol solution (warm to 37° to get the sugar into solution at this concentration) is added to compensate for

[30] H. Murakami, D. Pain, and G. Blobel, *J. Cell Biol.* **107**, 2051 (1988).
[31] K. Verner and M. Weber, *J. Biol. Chem.* **264**, 3877 (1989).
[32] A. S. Lewin, L. J. Wells, and D. K. Norman, in "Mitochondria 1983" (R. J. Schweyen, K. Wolf, and F. Kaudewitz, eds.), p. 343. de Gruyter, Berlin, 1983.

TABLE II
YEAST MITOCHONDRIAL PRECURSORS PRODUCED BY
in Vitro TRANSCRIPTION AND TRANSLATION

Precursor	Target[a]	Molecular size (kDa)		Promoter	Ref.
		Precursor	Mature		
F_1-ATPase					
β-Subunit	M	56	54	T7	c
α-Subunit	M	61	58	T7	d
Citrate synthase	M	50	47	SP6	e
PUT2 protein[b]	M	64	61	SP6	f
Alcohol dehydrogenase III	M	43	39	SP6	g
Cytochrome oxidase					
Subunit IV	IM	17	14	T5	h
Subunit V	IM	15	12.5	SP6	i
ADP/ATP carrier	IM	34	34	T7	j
Iron-sulfur protein	IMS	27	25	T7	k
Cytochrome b_2	IMS	68	58	T7	l

[a] Submitochondrial destination of protein: M, matrix; IM, inner membrane; IMS, intermembrane space.
[b] Δ^1-Pyrroline-5-carboxylate dehydrogenase.
[c] W.-J. Chen and M. G. Douglas, *Cell (Cambridge, Mass.)* **49**, 651 (1987).
[d] M. Takeda, W.-J. Chen, J. Saltzgaber, and M. G. Douglas, *J. Biol. Chem.* **261**, 15126 (1986).
[e] R. E. Jensen and M. P. Yaffe, *EMBO J.* **7**, 3863 (1988).
[f] H. Murakami, D. Pain, and G. Blobel, *J. Cell. Biol.* **107**, 2051 (1988).
[g] D. Pilgram and E. T. Young, *Mol. Cell. Biol.* **7**, 294 (1987).
[h] E. C. Hurt, B. Pesold-Hurt, and G. Schatz, *EMBO J.* **3**, 3149 (1984).
[i] M. G. Cumsky, personal communication (1989).
[j] C. Smagula and M. G. Douglas, *J. Biol. Chem.* **263**, 6787 (1988).
[k] A. P. G. M. Van Loon and G. Schatz, *EMBO J.* **6**, 2441 (1987).
[l] F.-U. Hartl, J. Ostermann, B. Guiard, and W. Neupert, *Cell (Cambridge, Mass.)* **51**, 1027 (1987).

components which do not contain 0.6 *M* mannitol (e.g., the labeled lysate) so that the final concentration will be 0.6 *M*. The lysate is then added, the solution is mixed gently, and the assay is begun by adding the mitochondria. The standard import temperature is 30°, but import has been performed at a variety of temperatures between 12° and 37°. Some precursors are imported poorly at extreme temperatures (e.g., pre-$F_1\beta$ is not imported well at 37°). An average incubation time is 20–30 min, but it is often useful to perform import for various times so that the time course of the reaction can be documented.

At the end of the import incubation, samples are loaded onto ice-cold, 1-ml cushions composed of 25% sucrose, 20 mM HEPES-KOH (pH 7.4), and 2 mM EDTA in 1.5-ml microcentrifuge tubes. The tubes are spun for 10 min at 4° in a microcentrifuge. A small amount of the supernatant (50 μl) from the top of the tube is saved, and the remaining supernatant is carefully removed by aspiration and discarded. The mitochondrial pellet is gently resuspended in 10-25 μl of 0.6 M mannitol, 20 mM HEPES-KOH (pH 7.4) and analyzed as described below.

Evaluation of Protein Import. There are four principal criteria to evaluate the import of a precursor protein into isolated mitochondria: (1) association (sedimentation) of the labeled protein with mitochondria, (2) proteolytic processing of a precursor into mature form, (3) energy dependence of the import process, and (4) delivery of the imported protein to the correct submitochondrial compartment. Usually, a single test is insufficient to prove the correct import of a polypeptide, but the satisfaction of several criteria suggests that the *in vitro* assay often reflects the import process occurring in the cell.

The first two criteria may be evaluated simply by recovering the mitochondria from the incubation (import) mixture and analyzing the extracted, labeled proteins by SDS-polyacrylamide gel electrophoresis and fluorography. Mitochondria can be recovered by pelleting through a cushion of 25% sucrose as described above. Additionally (or as an alternative), the mitochondria can be washed by resuspending in MIB containing 0.2 M KCl and repelleting the organelles. The mitochondria are resuspended in a small volume (10-25 μl) of 0.6 M mannitol, 20 mM HEPES-KOH (pH 7.4) and then mixed with SDS sample buffer. The mixture is heated at 95° for 4 min, and the proteins are resolved by electrophoresis on a polyacrylamide gel in the presence of SDS. Labeled proteins are detected by fluorography of the dried gel. The amount of specific, labeled protein recovered with the mitochondria should be compared with the amount of precursor added to the import incubation by running a measured amount of precursor in one lane of the gel. Recoveries of 5-30% are common and vary greatly with the time and temperature of the import incubation and with the nature of the precursor. The recovery of labeled protein with mitochondria, by itself, is usually insufficient as a demonstration of precursor import since substantial amounts of many precursors bind (often nonspecifically) to the mitochondrial surface and will pellet in sedimentation gradients with the organelle.

The conversion of a labeled precursor to mature-sized protein can be evaluated by the same analysis used to examine the association of labeled proteins with mitochondria following import. Since this conversion is dependent on a matrix-localized protease,[5] processing suggests the successful delivery of the precursor into the matrix compartment. Most proteins targeted to the mitochondrial matrix, inner membrane, and intermembrane space are processed by this enzyme.[33] In analyzing this processing, the amount of mature protein or the ratio of precursor to mature forms is determined by examining the bands on the fluorogram using densitometric scanning or other quantitative methods (e.g., elution and quantitation of silver grains from the fluorogram as described by Suissa).[34] The mobilities of authentic precursor and mature species can be identified by running a sample of the labeled precursor and a sample of the precursor processed to mature form, respectively. The mature-form protein standard is prepared by first resuspending isolated mitochondria in a solution of 1% Triton X-100, 50 μM $ZnCl_2$, 50 μM $MnCl_2$, 1 mM PMSF, 1 mM aprotinin, 1 μg/ml leupeptin, 100 μg/ml α_2-macroglobulin (Boehringer-Mannheim), 20 mM HEPES–KOH (pH 7.4) to a mitochondrial protein concentration of 10 mg/ml and then incubating 10 μl of this mixture with 5 μl of lysate containing the labeled precursor. Incubation is for 15 min at 30°. Alternatively, a mature-form standard may be generated by labeling cells for a number of generations with $^{35}SO_4^{2-}$, extracting cellular proteins, immunoprecipitating the specific polypeptide with the appropriate antibody, and running the precipitate in one lane of the gel as described by Maccecchini et al.[23]

A third criterion for correct import is the demonstration of energy dependence. Import of proteins into the mitochondrial matrix or inner membrane and the import of most proteins into the intermembrane space require a potential across the mitochondrial inner membrane.[4,35] The addition of an uncoupler to the incubation mixture eliminates the import of these proteins and can be used in conjunction with other analyses to demonstrate that the association, submitochondrial delivery, and proteolytic processing of a precursor required energized, coupled mitochondria. Import is abolished by addition to the incubation mixture of valinomycin to 1 μg/ml (KCl is already present in the assay mixture) or CCCP to 0.2 μM. In the presence of these uncouplers, many precursors will bind to the mitochondrial surface and may be recovered with pelleted mitochondria; however, these precursors will not be processed to mature form and will be susceptible to exogenous protease digestion.

Authentic protein import requires the faithful delivery of a precursor to its correct submitochondrial compartment. Although the complete fractionation of the small quantity of mitochondria used in a typical import

---

[33] R. Hay, P. C. Bohni, and S. M. Gasser, *Biochim. Biophys. Acta* **779**, 65 (1984).
[34] M. Suissa, *Anal. Biochem.* **133**, 511 (1983).
[35] M. Schleyer, B. Schmidt, and W. Neupert, *Eur. J. Biochem.* **125**, 109 (1982).

assay is extremely difficult, the subcellular location of a labeled, imported protein can be inferred by examining its accessibility to exogenously added proteases after the mitochondria are subjected to various treatments. The simplest test involves incubating the mitochondria with protease following import to demonstrate that precursors have, indeed, been taken up by the organelles. Following the import incubation and reisolation (as described above) the resuspended mitochondria (10 μl) are mixed gently with 10 μl of 500 μg/ml proteinase K or with 10 μl of 500 μg/ml trypsin. Stock protease solutions are prepared in 0.6 M mannitol, 20 mM HEPES–KOH (pH 7.4). The mitochondria are treated with protease at 0° (in an ice-water bath) for 20–30 min. At the end of this period, proteinase K digestion is terminated by addition of PMSF to 1 mM, and trypsin is inhibited by addition of trypsin inhibitor and PMSF to final concentrations of 2.5 mg/ml and 1 mM, respectively. SDS sample buffer is added, the samples are heated immediately at 95° for 4 min, and the proteins are analyzed as described above. An important control for protease accessibility involves protease digestion in the presence of 0.5% Triton X-100. This detergent disrupts both mitochondrial membranes yet allows proteolysis, demonstrating that the labeled protein exists (somewhere in the organelle) in a protease-sensitive form.

To further determine the submitochondrial location of an imported protein one can disrupt the outer membrane by osmotic shock and then test the susceptibility of the protein to protease treatment. Digestion of the protein suggests that it resided in the intermembrane space (or was embedded in the membrane but exposed to the intermembrane space), whereas continued protection implies that the protein had been transported beyond the inner membrane. The outer membrane is ruptured by adding 5 volumes of ice-cold 20 mM HEPES–KOH (pH 7.4) and holding the mixture on ice for 10 min. Samples are then treated with protease as described above. After terminating the protease digestion by addition of inhibitors, SDS sample buffer may be added directly to the mixture and the labeled proteins analyzed by electrophoresis and fluorography.

More extensive fractionation is possible only if one begins with a greater amount of mitochondria than that used in a typical import assay. To accomplish such a fractionation the import incubation is scaled-up 10-fold, so that 2 mg of mitochondria and a total volume of 2 ml are employed. Following the import incubation, the mitochondria are recovered by centrifugation and resuspended in 0.2 ml of 0.6 M mannitol, 20 mM HEPES–KOH (pH 7.4). The outer membrane is broken as described above, and mitoplasts are recovered by spinning for 10 min in a microcentrifuge at 4°. The supernatant from this spin comprises the intermembrane space fraction. The mitoplasts can be fractionated further into membranes and soluble matrix by resuspending them in a solution of 0.6 M sucrose, 3 mM MgCl$_2$, 20 mM HEPES–KOH (pH 7.4), 3 mM ATP, sonicating for 1 min at 0°, and pelleting the membranes in an airfuge (Beckman) for 20 min at 100,000 g. The purity of the fractions can be characterized by subjecting a portion of each fraction to immune-blot analysis[36] using antibodies against specific marker proteins. Useful markers are cytochrome b_2 for intermembrane space,[16] citrate synthase for matrix,[37] and cytochrome c_1 for inner membrane.[38] This fractionation procedure has been used to characterize the targeting of fusion proteins to various submitochondrial locations.[39]

The assay of protein import into the mitochondrial outer membrane presents special problems since outer membrane proteins are initially synthesized at their mature size and their import does not require a potential across the inner membrane. Furthermore, the import of only one outer membrane protein, the 29-kDa porin, has been characterized thoroughly in an *in vitro* import assay.[40,41] During incubation with mitochondria the porin precursor (synthesized *in vitro*) is converted from a trypsin-sensitive form to a trypsin-resistant one; however, the trypsin-resistant character remains even when the mitochondrial membranes are dissolved with detergents. It is unclear if this conversion to protease insensitivity represents true protein import. Further characterization of outer membrane import awaits the identification and analysis of other precursors for which import can be confirmed by several independent criteria.

Acknowledgments

I am grateful to Jeff Schatz for many contributions and leadership in the study of yeast mitochondria. Many of the procedures described here were developed or perfected in his laboratory. I thank Chris Wills, Leslie Stewart, and Barbara Smith for critical reading of the manuscript.

[36] A. Haid and M. Suissa, this series, Vol. 96, p. 192.
[37] H. Riezman, R. Hay, C. Witte, N. Nelson, and G. Schatz, *EMBO J.* **2**, 1113 (1983).
[38] A. Ohashi, J. Gibson, I. Gregor, and G. Schatz, *J. Biol. Chem.* **257**, 13042 (1982).
[39] A. P. G. M. Van Loon, A. W. Brandli, and G. Schatz, *Cell (Cambridge, Mass.)* **44**, 801 (1986).
[40] K. Mihara, G. Blobel, and R. Sato, *Proc. Natl. Acad. Sci. U.S.A.* **79**, 7102 (1982).
[41] S. Gasser and G. Schatz, *J. Biol. Chem.* **258**, 3427 (1983).

[44] Methods for Studying the Yeast Vacuole

By CHRISTOPHER J. ROBERTS, CHRISTOPHER K. RAYMOND, CARL T. YAMASHIRO, and TOM H. STEVENS

Introduction

The yeast vacuole is the equivalent of lysosomes in animal cells and the vacuoles of plant cells. It is an acidic compartment that contains a large number of hydrolases.[1-3] Use of the genetic techniques available in yeast, together with biochemical and cell biological approaches, has shed light on many aspects of the vacuole, including (1) the structure and biosynthesis of several vacuolar proteins,[1,4-6] (2) the localization determinants on vacuolar proteins that are necessary and sufficient for proper targeting to the vacuole,[7-9] (3) the identification of genes necessary for vacuolar protein sorting and organelle biogenesis (*VPS* genes, vacuolar protein sorting)[10-15] and acidification of the lumen of the vacuole,[13,16,17] (4) the composition and biochemical characteristics of the vacuolar H^+-ATPase known to be necessary for the generation of the acidic interior of the vacuole,[18-20] and (5) the behavior of the vacuole during the cell cycle.[21-24b]

---

[1] J. H. Rothman and T. H. Stevens, *in* "Protein Transfer and Organelle Biogenesis" (R. Das and P. Robbins, eds.), p. 159. Academic Press, San Diego, 1988.
[2] D. J. Klionsky, L. M. Banta, J. S. Robinson, and S. D. Emr, *in* "Molecular Biology of Intracellular Protein Sorting and Organelle Assembly" (R. Bradshaw, L. McAllister-Henn, and M. Douglas, eds.), p. 173. Alan R. Liss, New York, 1988.
[3] J. H. Rothman, C. T. Yamashiro, P. M. Kane, and T. H. Stevens, *Trends Biochem. Sci.* **14,** 347 (1989).
[4] B. Mechler, H. H. Hirsch, H. Müller, and D. H. Wolf, *EMBO J.* **7,** 1705 (1988).
[5] C. M. Moehle, C. K. Dixon, and E. W. Jones, *J. Cell Biol.* **108,** 309 (1989).
[6] C. J. Roberts, G. Pohlig, J. H. Rothman, and T. H. Stevens, *J. Cell Biol.* **108,** 1363 (1989).
[7] L. A. Valls, C. P. Hunter, J. H. Rothman, and T. H. Stevens, *Cell (Cambridge, Mass.)* **48,** 887 (1987).
[8] L. M. Johnson, V. A. Bankaitis, and S. D. Emr, *Cell (Cambridge, Mass.)* **48,** 875 (1987).
[9] D. J. Klionsky, L. M. Banta, and S. D. Emr, *Mol. Cell. Biol.* **8,** 2105 (1988).
[10] J. H. Rothman and T. H. Stevens, *Cell (Cambridge, Mass.)* **47,** 1041 (1986).
[11] V. A. Bankaitis, L. M. Johnson, and S. D. Emr, *Proc. Natl. Acad. Sci. U.S.A.* **83,** 9075 (1986).
[12] J. S. Robinson, D. J. Klionsky, L. M. Banta, and S. D. Emr, *Mol. Cell. Biol.* **8,** 4936 (1988).
[13] L. M. Banta, J. S. Robinson, D. J. Klionsky, and S. D. Emr, *J. Cell Biol.* **107,** 1369 (1988).
[14] J. H. Rothman, I. Howald, and T. H. Stevens, *EMBO J.* **8,** 2957 (1989).
[15] V. Dulic and H. Riezman, *EMBO J.* **8,** 1349 (1989).
[16] J. H. Rothman, C. T. Yamashiro, C. K. Raymond, P. M. Kane, and T. H. Stevens, *J. Cell Biol.* **109,** 93 (1989).

The purpose of this chapter is to describe techniques used in the study of the vacuole. We focus attention on the procedures used in our laboratory, and, where possible, we try to describe (or at least reference) the methods used by others. The chapter is divided into three sections. In the first section, methods for visualizing vacuoles by light and fluorescence microscopy are outlined. The next section describes a method for purification of vacuoles, marker enzyme assays used to assess yield and purity, and activities of the purified organelle that have been described. The final section describes two methods used in our laboratory for assessing the fidelity of protein sorting to the vacuole. Unless otherwise noted, all reagents used in these procedures are available from Sigma Chemical Co. (St. Louis, MO).

Visualization of Yeast Vacuoles by Fluorescence Microscopy

General Remarks

The yeast vacuole is a large, dynamic organelle which can be detected by light microscopy. It is most prominent when cells are viewed using differential interference-contrast optics (commonly referred to as Nomarski optics). However, the vacuoles of many cells in a wild-type population are not readily visualized by Nomarski optics. In this section, we outline procedures for more reliable detection of vacuoles using vital staining with fluorescent dyes in conjunction with fluorescence microscopy. In addition, we discuss immunofluorescence techniques for observation of vacuolar antigens in fixed cells.

Two important issues regarding the visualization of yeast vacuoles are the specificity of the stains used and the morphology of vacuoles observed under various conditions. The vital stains discussed below have been shown to specifically label the region which coincides with the vacuole as

---

[17] R. A. Preston, R. F. Murphy, and E. W. Jones, *Proc. Natl. Acad. Sci. U.S.A.* **86,** 7027 (1989).
[18] E. Uchida, Y. Ohsumi, and Y. Anraku, *J. Biol. Chem.* **260,** 1090 (1985).
[19] P. M. Kane, C. T. Yamashiro, J. H. Rothman, and T. H. Stevens, *J. Cell Sci. Suppl.* **11,** 161 (1989).
[20] P. M. Kane, C. T. Yamashiro, and T. H. Stevens, *J. Biol. Chem.* **264,** 19236 (1989).
[21] A. Wiemken, P. Matile, and H. Moor, *Arch. Mikrobiol.* **70,** 89 (1970).
[22] L. H. Hartwell, *Proc. Natl. Acad. Sci. U.S.A.* **66,** 352 (1970).
[23] L. S. Weisman, R. Bacallao, and W. Wickner, *J. Cell Biol.* **105,** 1539 (1987).
[24a] L. S. Weisman and W. Wickner, *Science* **241,** 589 (1988).
[24b] C. K. Raymond, P. J. O'Hara, G. Eichinger, J. H. Rothman, and T. H. Stevens, *J. Cell Biol.*, in press, (1989).

seen by Nomarski optics.[23,25,26] These same stains fail to label particular mutants of yeast which are defective in the assembly of the vacuole.[13,15,26] With respect to immunofluorescence applications on fixed cells, it is essential to demonstrate that the antibodies used are specific for the vacuolar protein of interest.

The morphology of the vacuole is highly variable and sensitive to many factors, including growth conditions and the genetic background of a particular strain. The vacuole is visible at all stages of the cell cycle and becomes apparent in the bud soon after emergence.[23,24b] Under conditions of rapid growth, most cells of the yeast strains commonly used in our laboratory exhibit several small vacuolar compartments which assume a variety of shapes and sizes. Other conditions, such as transfer to glucose-free buffers or approach to stationary phase, appear to cause these smaller vacuoles to coalesce into a large, roughly spherical vacuole. What constitutes normal vacuolar morphology has not been resolved.[23,24b,25]

This section is limited to the application of fluorescence microscopy to the study of yeast vacuole. For a comprehensive overview of fluorescence microscopy application in yeast cell biology, see Refs. 25 and 27.

Vital Staining

A variety of fluorescent probes have been used to label vacuoles in living yeast cells.[25] Among these are membrane-impermeant compounds, including the sulfonate dyes (e.g., Lucifer Yellow[26,28]) and fluorescein derivatives,[13,17,23,29] lysosomotropic dyes, such as quinacrine,[13,16,23] and endogenously produced fluorophores that accumulate in the vacuoles of several yeast mutants, such as in *ade2* mutants.[13,23] Although all of the molecules specifically label vacuoles, little is known about the mechanisms by which they are concentrated in these organelles.

Three protocols for vital staining of yeast vacuoles are given below. The first and second utilize the fluorescein derivatives 5- (and 6-)carboxy-2′,7′-dichlorofluorescein diacetate (CDCFDA; Molecular Probes, Eugene, OR; Cat. No. C-369) and 6-carboxyfluorescein diacetate (CFDA; Molecular Probes, Inc., Eugene, OR; Cat. No. C-195), which are relatively easy to use and which stain the vacuoles of many different strains reliably (Fig. 1A,B).[17,25] The third protocol describes the use of quinacrine, a weakly

---

[25] J. R. Pringle, R. A. Preston, A. E. M. Adams, T. Stearns, D. G. Drubin, B. K. Haarer, and E. W. Jones, *Methods Cell Biol.* **31**, 357 (1989).
[26] H. Riezman, *Cell (Cambridge, Mass.)* **40**, 1001 (1985).
[27] J. R. Pringle, this volume [52].
[28] V. Dulic, M. Egerton, I. Elguindi, S. Raths, B. Singer, and H. Riezman, this volume [48].
[29] R. A. Preston, R. F. Murphy, and E. W. Jones, *J. Cell Biol.* **105**, 1981 (1987).

basic dye which accumulates in low-pH compartments, such as the vacuole.[23] Quinacrine accumulation has been used to screen the *vps* mutants for those which are defective in acidification of the vacuole.[13,16] However, the amount of dye taken up in different yeast strains can vary, and thus quinacrine is a less reliable vital stain than CDCFDA or CFDA.

Solutions for Vital Staining of Yeast Cells

CDCFDA (1000×): 10 mM, prepared in dimethylformamide; stable for months at $-20°$

CFDA (1000×): 10 mM, prepared in dimethylformamide; stable for months at $-20°$

20 × Phosphate-buffered saline (PBS): Dissolve 160 g NaCl, 22.8 g Na_2HPO_4, and 4 g KH_2PO_4 in 1 liter of water; adjust the pH to 7.3 with NaOH

Phosphate-buffered YEPD: YEPD buffered with 50 mM Na_2HPO_4, pH 7.6

Staining with CDCFDA. Suspend approximately 1×10^7 cells/ml from a fresh plate or actively growing liquid culture in SD minimal broth supplemented with the appropriate nutrients plus 50 mM citric acid, pH 5.0, and 5–10 μM CDCFDA. Incubate with shaking at 30° for 10–30 min. Spot 8 μl on a concanavalin A (Con A)-coated microscope slide (prepare Con A-coated slides by spreading 10 μl of 1 mg/ml Con A onto a slide and allowing to air dry; the Con A coating causes the cells to adhere to the slide). Cover the cells with a coverslip and view immediately. The CDCFDA-derived fluorescence can be viewed with standard fluorescein optics (Fig. 1A,B).

Staining with CFDA. Staining cells with CFDA is performed as with CDCFDA, except the cells are incubated in SD plus nutrients, 50 mM citrate, pH 3.0, and 5–10 μM CFDA, and the incubation should be extended to 30 min at 30° (Fig. 1C,D).

Notes. Staining may also be performed in buffered YEPD broth, but cells will not stick to Con A-coated slides. The cells will adhere if they are washed with PBS plus 2% glucose after labeling. In our hands and as reported by others,[25] the composition of the wash buffer greatly influences vacuolar morphology. Washing with buffers containing glucose yields many cells with several vacuoles, whereas cells washed in solutions free of glucose predominantly possess a single large vacuole (Fig. 1).

The fluorescent compounds which accumulate in yeast vacuoles during CDCFDA and CFDA staining are most likely carboxy-2',7'-dichlorofluorescein and carboxyfluorescein (CF), respectively, or a polymerization product of these compounds.[25] These compounds photobleach rapidly and also leak out of the vacuole with a $t_{1/2}$ of about 30 min at 30°; therefore, it

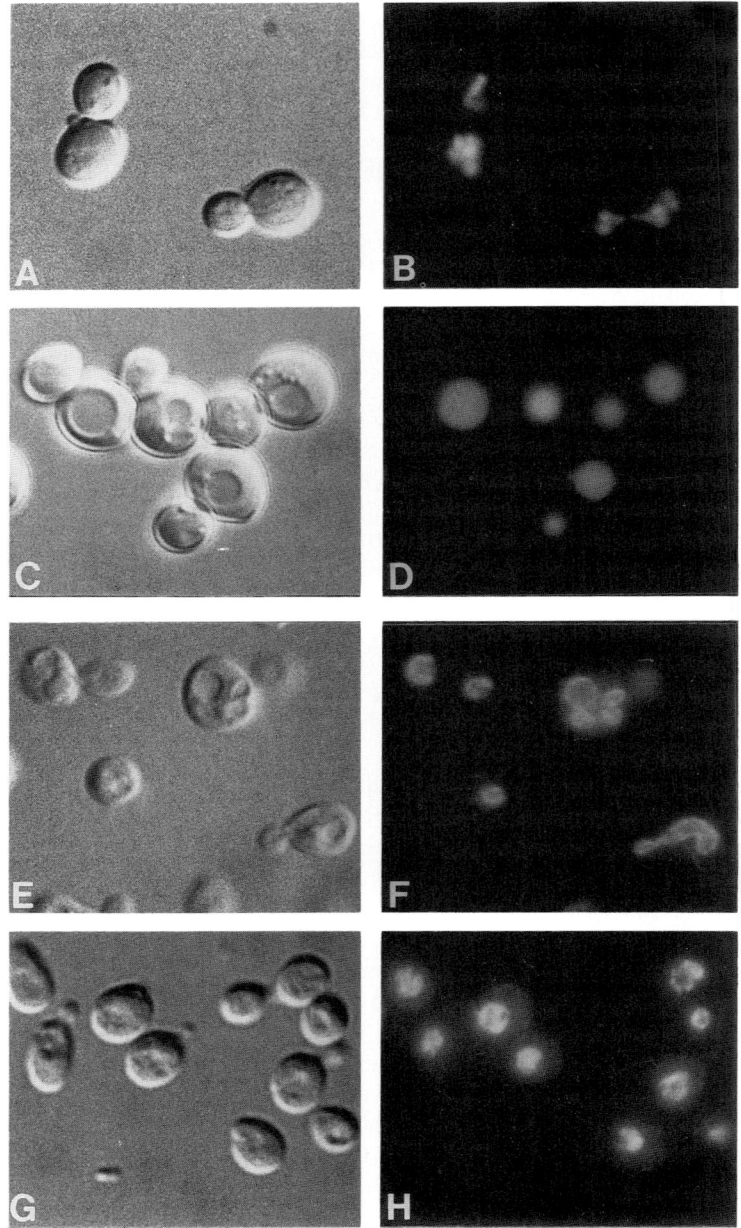

is necessary to view the cells immediately after staining. CF undergoes pH-dependent changes in both the absorption and fluorescence emission spectra. This property has been exploited to determine vacuolar pH and to screen for mutants defective in maintenance of a low vacuolar pH.[17] CDCFDA and CFDA have similar staining properties, but CDCFDA can be used over a wider range of pH with shorter labeling periods, and the fluorescence signal from this compound is brighter than CFDA.

Staining with Quinacrine. Harvest ~5 × 10⁶ log-phase yeast cells, resuspend in 500 μl of phosphate-buffered YEPD containing 200 μM quinacrine in a microfuge tube, and incubate for 5 min at room temperature. Harvest the cells by a 5-sec centrifugation, and wash once in 500 μl of minimal media or 2% glucose, either of which should be buffered to pH 7.6 with 50 mM Na$_2$HPO$_4$. Resuspend in 100 μl of the same solution and apply 10 μl to a Con A-coated microscope slide. Quinacrine fluorescence can be viewed using standard fluorescein fluorescence wavelengths.

Notes. Quinacrine-labeled cells may be viewed up to 15 min after labeling. As with the CDCFDA-labeling protocol, the entire procedure can be carried out in phosphate-buffered YEPD if immobilizing the cells to slides is not critical.

Immunofluorescence

Immunofluorescence techniques have been applied successfully to the study of the yeast vacuole (Fig. 1E–H).[6,9] In this section we outline the procedures used in our laboratory, with emphasis placed on the techniques that have been most critical for success, including increasing the specificity of antibodies, treatment of fixed cells with sodium dodecyl sulfate (SDS), and enhancement of weak signals by antibody amplification.

We typically use rabbit polyclonal antibodies directed against yeast antigens that have been expressed in and purified from *Escherichia coli*. Subsequent affinity purification of these antibodies has been essential for immunofluorescence experiments. Affinity columns are prepared by at-

---

FIG. 1. Fluorescence detection of the yeast vacuole in living and fixed cells. (A) and (B) show vacuoles in living cells stained with CDCFDA and applied directly to slides without washing. (C) and (D) show cells stained with CFDA and washed in glucose-free buffer. Indirect immunofluorescence microscopy of fixed yeast cells is shown in (E)–(H). Fixed cells were labeled with either rabbit anti-alkaline phosphatase antibody (E, F) or anti-CPY antibody (G, H) followed by fluorescein-conjugated goat anti-rabbit antibody. Alkaline phosphatase is a vacuolar membrane protein, and CPY is a soluble vacuolar protein. Cells were photographed through Nomarski optics (A, C, E, and G) or by fluorescein fluorescence (B, D, F, and H).

taching the original immunogen to CNBr-activated Sepharose beads, and affinity purification of antibodies is carried out essentially as described.[25,30,31] We have found that two successive purifications of the antibodies (double-affinity purification) results in optimal antibody titer and specificity.

Successful antigen detection in immunofluorescence experiments is best monitored by comparison of the staining pattern observed in wild-type cells to that of a deletion strain, that is, a strain in which the gene encoding the protein of interest has been deleted, if this is possible. If a background signal is seen in the deletion strain then further purification of the antibody is necessary. We have accomplished this by adsorbing small amounts of affinity-purified antibodies to deletion strain cells that have been fixed and prepared as described below. Apparently this procedure improves the signal-to-noise ratio by removing nonspecific antibodies.

We use well-established techniques for indirect immunofluorescence detection of proteins in yeast.[25,32] A protocol for treatment of spheroplasted fixed cells with antibodies on polylysine-coated slides is described below. Also, we describe a signal amplification procedure that has been used successfully for immunolocalization of nonabundant antigens. This protocol has worked well using polyclonal and monoclonal antibodies for staining of both soluble and membrane-associated vacuolar antigens. These include dipeptidyl aminopeptidase B (DPAP B),[6] carboxypeptidase Y (CPY), alkaline phosphatase, proteinase B (PrB), and the 60- and 69-kDa subunits of the vacuolar H^+-ATPase, using polyclonal or monoclonal antibodies (Fig. 1E–H; C. Raymond, I. Howald, and T. Stevens, unpublished results).

Solutions for Immunofluorescence Microscopy of Yeast Cells

TEB: 200 mM Tris-HCl, pH 8.0, 20 mM EDTA, 1% 2-mercaptoethanol; prepare immediately before use

SPM: 1.2 M sorbitol, 50 mM potassium phosphate, pH 7.3, 1 mM $MgCl_2$

Fixative (4% formaldehyde): To prepare 50 ml, add 2 g paraformaldehyde to 50 ml of water and heat. Add 350 μl of 6 N NaOH and stir until the solution clears. Remove from heat and add 0.68 g of KH_2PO_4 and 50 μl of 1 M $MgCl_2$ (the pH of the solution should be approximately 6.5). Always prepare immediately before use

Polylysine: 1 mg/ml polylysine (M_r 400,000) in water; stable for months at 4°

---

[30] C. Wittenberg, S. L. Richardson, and S. I. Reed, *J. Cell Biol.* **105**, 1527 (1987).
[31] J. P. Aris and G. Blobel, *J. Cell Biol.* **107**, 17 (1988).
[32] J. V. Kilmartin and A. E. M. Adams, *J. Cell Biol.* **98**, 922 (1984).

Mounting medium: To prepare 5 ml, mix 4.5 ml of 100% glycerol, 500 μl 10 mg/ml p-phenylenediamine (in 1× PBS, pH 9), and 0.1 μl DAPI (4′,6′-diamidino-2-phenylindole, 1 mg/ml in water; DAPI stains nuclei with fluorescence in the blue range, and is routinely included). Mounting medium should be stored at −20° or −80° in the dark; discard the solution when it turns dark brown

PBS–BSA: 5 mg/ml bovine serum albumin (BSA) in 1× PBS plus 5 mM sodium azide as preservative

Adsorption of Affinity-Purified Antibodies to Fixed Yeast Cells. Harvest approximately 1×10^{10} cells lacking the gene product of interest and prepare the cells for immunofluorescence as described below, except scale up the procedure 50- to 100-fold. Collect the cells by centrifugation and resuspend in 20 ml of PBS–BSA. Gently shake at room temperature for 30 min, then divide the suspension into two tubes and collect the cells. Resuspend one-half of the cells in 500 μl of PBS–BSA and add 20–50 μl of approximately 1 mg/ml affinity-purified antibody. Incubate at room temperature for 1 hr with gentle agitation. Centrifuge, remove the antibody solution, and use it to resuspend the other half of the fixed cells. Wash the first cell pellet with 500 μl PBS–BSA and save the solution. Repeat the 1 hr incubation with the second cell pellet. Remove the fixed cells from the adsorbed antibody preparation by centrifugation, wash the second cell pellet with the 500 μl PBS–BSA wash from above, pool the supernatant solutions, and repeat the centrifugation once more to remove any remaining debris. Use immediately or store at 4°. Most adsorbed antibody preparations are stable at 4° for months. Although nonspecific antibodies are removed by this procedure, substantial losses in specific antibody titer may occur. The optimal antibody dilution for immunofluorescence experiments must be determined empirically, but dilutions of 1 to 10 are typically used in our laboratory.

Fixation of Cells for Immunofluorescence Microscopy. Grow 10 ml of yeast cells in YEPD or selective medium to a density of about 1×10^7 cells/ml. Add 1.2 ml of 37% formaldehyde (commercially available) to the culture and incubate for 1 hr at 30°. Harvest the cells by centrifugation and resuspend in 2 ml of 1× fixative. Shake gently at room temperature or 30° for about 16 hr (we routinely fix cells overnight at either temperature with good results). Harvest the cells, resuspend in 1 ml TEB, and shake gently at 30° for 10 min. Centrifuge and resuspend the pellet in 1 ml SPM. Remove the cell walls by adding 25 μl of Glusulase (Du Pont NEN Products, Boston, MA) and 15 μl of 10 mg/ml Zymolyase 100T (ICN Immunobiologicals, Lisle, IL; freshly prepared in SPM) and shaking gently at 30° for 30–60 min. Pellet the spheroplasted fixed cells, wash once in 1 ml of 1.2 M sorbitol, and treat with SDS by resuspending in 500 μl of 1.2 M sorbitol and adding 500 μl 1.2 M sorbitol plus 4% SDS. After a 2-min incubation, harvest the cells by centrifugation at slow speed (e.g., 30 sec at 4000 g) and wash gently 2 times with 1-ml aliquots of 1.2 M sorbitol. Resuspend in 1–2 ml of 1.2 M sorbitol. The cells are now ready to be applied to a microscope slide.

Notes. As discussed previously, the morphology of the vacuole is influenced by the way the cells are treated prior to fixation. The protocol just described involves fixing the cells directly in glucose-containing medium and thus many of the cells will contain several vacuoles. Also, cells with small buds will exhibit vacuolar segregation structures, in which the vacuole appears as an extended tubular structure that extends from the mother cell into the bud[24b] (Fig. 1E,F). Modifying the procedure to include washing cells in glucose-free buffers prior to fixation will result in most of the cells containing a single, large vacuole (Fig. 1G,H).

We fix cells for much longer periods of time than conventional protocols,[25] and this has greatly improved the morphological preservation of cells and the subcellular structures within them. Spheroplasted fixed cells are quite fragile and should be treated gently. Damaged cells have a squashed, fuzzy appearance when viewed under the microscope. We generally use fixed cells immediately after preparation and have little experience with their stability in storage.

SDS treatment is required in order to achieve staining with certain antibodies (e.g., polyclonal rabbit anti-alkaline phosphatase), but the staining with other antibodies deteriorates with extended SDS treatment (e.g., rabbit polyclonal or mouse monoclonal anti-60-kDa vacuolar H^+-ATPase subunit). The optimal concentration of SDS (1–5%) and time of exposure (0.5–5 min) must be worked out empirically. Treatment of fixed cells with SDS makes the cells very adherent to polylysine-coated microscope slides (see Antibody Staining of Fixed Cells below). This adherence is a distinct advantage as most untreated cells can be lost during the extensive washes required for antibody treatment. Carefully controlled experiments have shown that SDS treatment does not change the distribution of vacuolar staining patterns observed, nor does it exert adverse effects on cell morphology.

Antibody Staining of Fixed Cells. Prepare polylysine-coated slides by placing 10 μl of 1 mg/ml polylysine in each well of an 8-well multitest slide (Flow Laboratories, Inc., McLean, VA; Cat. No. 60-408-05). Make sure the solution covers the entire glass surface of the well. After 1 min, aspirate the wells, wash each 5–6 times with 20 μl of water, aspirate again, and allow the slides to dry. Apply 40 μl of a freshly prepared fixed cell suspension to each well and allow to settle for 10 min. Aspirate the fluid and wash each well 3 times with 20 μl of PBS–BSA. Apply 20 μl of PBS–BSA and incubate for 30 min at room temperature in a humid chamber (a petri dish containing a damp paper towel works well). *Never* allow the cells to dry out.

Aspirate the wells and apply 10 μl of the primary antibody solution diluted appropriately in PBS–BSA. A 1 to 10 dilution of adsorbed primary antibody is typically used. Incubate the slide in a humid chamber for 1 hr at room temperature. Aspirate the primary antibody solution and wash each well 6 times with PBS–BSA. Apply the secondary antibody solution and repeat the incubation as above (for amplification of signal, see below). After removal of the last antibody solution, wash the wells at least 9 times. Place small drops of mounting medium in the wells, cover with a 24 × 60 mm coverslip, and seal the ends of the coverslip to the slide with fingernail polish. Slides can be viewed immediately or stored at −20° for weeks (Fig. 1E–H).

Notes. In cases where the antigen being detected is abundant, fluorochrome-conjugated second antibody is applied directly. For proteins of low abundance, this two-step incubation procedure often yields a signal that is too weak to detect above the inevitable slight background fluorescence. To improve the signal-to-noise ratio, an antibody amplification procedure has been used,[24b,25] which involves decorating the primary antibody with goat anti-rabbit IgG, followed by treatment with rabbit anti-goat IgG, followed finally by goat anti-rabbit IgG conjugated to the appropriate fluorochrome. In theory each bound antibody binds more than one antibody in the subsequent incubation, thus amplifying a weak signal. In practice, one observes a greater increase in vacuolar versus background fluorescence; however, the procedure is limited by the specificity of the primary antibody. All incubations with secondary antibodies should be done as described above. This procedure has given us better amplification of signals than procedures that use biotinylated second antibodies followed by fluorochrome-conjugated streptavidin that have been described elsewhere.[25]

We have had good results using secondary antibodies, reconstituted to 1 mg/ml, from both Jackson ImmunoResearch Laboratories, Inc. (West Grove, PA) and Organon Teknika–Cappel Laboratories (West Chester, PA). Fluorochrome-conjugated secondary antibodies are used at 1 to 100 dilutions. When the amplification procedure is used, 1 to 1000 dilutions of intermediate secondary antibodies are used.

Purification of Vacuoles

Several methods are available for purification of the yeast vacuole, all of which take advantage of the low buoyant density of the vacuole.[33,34] Our method of choice, which is a modified version of the procedure used by

Kakinuma et al.,[35] involves enzymatic removal of the cell walls, osmotic lysis of the spheroplasts (without lysis of the vacuoles), followed by flotation of the vacuoles on a discontinuous Ficoll gradient. Other methods use polybase (e.g., DEAE-dextran) induced lysis of spheroplasts[11,36] or lysis of spheroplasts under isotonic conditions followed by density gradient centrifugation.[33,34] The procedure that follows has consistently given an 10–20% yield of vacuoles and low contamination of other organelles.

Solutions

Zymolyase solution: 50 mM Tris-HCl, pH 7.7, 1 mM EDTA, 50% glycerol, 400 units (U)/ml Zymolyase 100T (4 mg/ml; ICN Immunobiologicals)

Buffer A: 10 mM 2-(N-morpholino)ethanesulfonic acid (MES)/Tris, pH 6.9, 0.1 mM MgCl$_2$, 12% Ficoll 400

Buffer B: 10 mM MES/Tris, pH 6.9, 0.5 mM MgCl$_2$, 8% Ficoll 400

Buffer C (1×): 10 mM MES/Tris, pH 6.9, 5 mM MgCl$_2$, 25 mM KCl

Cell Growth and Spheroplast Formation

Grow yeast cells to mid-log phase and collect 4×10^{10} cells (i.e., 4000 OD$_{600}$ units) by centrifuging at 4400 g for 5 min. Wash the cells once with distilled water at room temperature, and convert the cells to spheroplasts by resuspending in 100 ml of 1 M sorbitol and then adding 1 ml of Zymolyase solution. Shake the culture gently at 30° for 60–90 min and check for completion of spheroplasting by diluting aliquots 10-fold into both buffer A and 1 M sorbitol; after about 5 min at room temperature, the buffer A suspension should be much less opaque than the sorbitol suspension owing to lysis of the spheroplasts.

Spheroplast Lysis

Collect the spheroplasts by centrifuging at 2200 g for 5 min, and wash twice in 1 M sorbitol. Lyse the spheroplasts by resuspending the final pellet in 25 ml of buffer A and homogenizing at 0°. Centrifuge the lysate at 2200 g for 10 min at 4° to remove any unlysed spheroplasts, and save the supernatant. Save 2% of the spheroplast lysate for assessment of yield and purity.

---

[33] A. Wiemken, in "Methods in Cell Biology" (D. M. Prescott, ed.), Vol. 12, p. 99. Academic Press, London, 1975.

[34] A. Wiemken, M. Schellenberg, and K. Urech, *Arch. Microbiol.* **123**, 23 (1979).

[35] Y. Kakinuma, Y. Ohsumi, and Y. Anraku, *J. Biol. Chem.* **256**, 10859 (1981).

[36] M. Dürr, T. Boller, and A. Wiemken, *Arch. Microbiol.* **105**, 319 (1975).

Vacuole Flotation

All subsequent manipulations are carried out at 0°–4°. Transfer the supernatant to a polyallomer tube of appropriate size for a Beckman SW 28 rotor. Carefully overlay with about 13 ml of buffer A and centrifuge at 60,000 g for 30 min. Collect the white wafer floating on top of the Ficoll and homogenize in 6 ml of buffer A. Transfer the suspension to a polyallomer tube appropriate for a Beckman SW 41 Ti rotor. Overlay with 6 ml buffer B, then centrifuge at 60,000 g for 30 min. Collect the white wafer at the top of the tube and resuspend in a small volume (0.2–1 ml) of 2× buffer C. Homogenize the purified vacuolar vesicles by passing the suspension up and down through a micropipette tip several times, then add an equal volume of 1× buffer C. Homogenization in buffer C causes fragmentation of the vacuoles into vacuolar vesicles, which may result in some loss of lumenal content. Store the vacuolar vesicles at −80°

Notes. The procedure just described usually yields 1–3 mg of vacuolar protein, corresponding approximately to a 10–20% recovery of vacuolar vesicles as determined by vacuolar membrane marker assays (see below). The yield obtained may vary depending on the yeast strain used.

It is important to proceed through the purification protocol as quickly as possible to avoid extensive loss of enzymatic activity and protein integrity (the vacuole, as noted above, contains a large proportion of the cellular hydrolases). Also, it is crucial to keep the sample cold during the last two steps of the purification, including precooling the swinging buckets and rotor for the ultracentrifuge runs.

Collection of the floated vacuolar wafers can be very difficult, especially for strains that give low yields of vacuoles. We generally use a spoon-shaped stainless steel spatula that has been prewetted in buffer A to scoop up the vacuoles.

Determination of Yield: Vacuolar Enzyme Marker Assays

There is an abundance of hydrolytic activities which are specific to the vacuole and are easily assayed. Several of these enzyme markers are soluble proteases such as carboxypeptidase Y, proteinase A (PrA), and proteinase B; the assays for these enzymes are described elsewhere in this volume.[37] Our laboratory commonly measures three vacuolar membrane-associated activities for determination of yield: H^+-ATPase, dipeptidyl aminopeptidase B, and α-mannosidase. Protein determination is performed according

---

[37] E. W. Jones, this volume [31].

to Lowry et al.[38] except that 2% SDS is included in order to release all the lumenal proteins.

Vacuolar H^+-ATPase. The activity of the vacuolar H^+-ATPase is monitored by a coupled enzyme–ATP regeneration assay system as described by Lötscher et al.[39] with several modifications. Vacuolar vesicles (5–100 μl) are assayed in 1 ml of a solution containing 25 mM Tris–acetate, pH 7.0, 25 mM KCl, 5 mM MgCl$_2$, 2 mM phosphoenolpyruvate, 2 mM ATP, 0.5 mM NADH, 30 units of L-lactate dehydrogenase, and 30 units of pyruvate kinase at 30°. The assay is performed by adding the vacuolar vesicles directly to a cuvette containing the other components, then immediately observing the change in absorbance at 340 nm using the time drive mode on a spectrophotometer. Absorbance readings are linear up to an A_{340} value of 3.0. The molar extinction coefficient for NADH (ϵ) is 6.22 mM^{-1} cm^{-1}, and depletion of NADH is directly correlated to ATP hydrolysis. Specific activity corresponds to micromoles ATP hydrolyzed per minute per milligram protein. The vacuolar H^+-ATPase is insensitive to 2 mM sodium azide and 0.1 mM sodium vanadate, distinguishing it from the mitochondrial and plasma membrane ATPases, respectively. The yeast vacuolar H^+-ATPase,[20] like other vacuolar H^+-ATPases,[40] is specifically inhibited by the macrolide antibiotic bafilomycin A$_1$ at very low concentrations ($I_{50} = 0.6$ μg bafilomycin A$_1$/mg vacuolar vesicles);[20] unfortunately, this drug is not commercially available at this time.

Dipeptidyl Aminopeptidase B. The following assay was adapted from Suarez Rendueles et al.[41] To a 200-μl sample containing 20–100 μl of vacuolar vesicles and 0.5% Brij 58, add 250 μl of HEPES/Tris, pH 7.0, and 50 μl of 3 mM alanylprolyl-p-nitroanilide (Bachem Biochemicals, Switzerland; prepare in 25% methanol and store at 4°). Incubate at 37° for 1 hr, and stop the reaction by adding 500 μl of 5% ZnSO$_4$ and 100 μl of 7.5% Ba(OH)$_2 \cdot$8H$_2$O (must boil to dissolve; store in a flask with a soda lime tube trap at 25°). Pellet the particulate matter in a clinical centrifuge at full speed for 5 min, and then read the absorbance of the optically clear supernatant at 405 nm. Readings are linear up to an A_{405} value of 0.7.

For whole cell lysate determinations, cells must first be collected and then washed in 0.9% NaCl. Freeze the washed pellet at $-80°$ for at least 10 min. After thawing on ice, resuspend the cells in 50 mM potassium acetate, pH 5.0 (use 5 μl per OD$_{600}$ unit of cells). Add 5% Brij 58 to a final

---

[38] O. H. Lowry, N. J. Rosebrough, A. L. Farr, and R. J. Randall, *J. Biol. Chem.* **193**, 265 (1951).
[39] H. Lötscher, C. deJong, and R. A. Capaldi, *Biochemistry* **23**, 4128 (1984).
[40] E. J. Bowman, A. Siebers, and K. Altendorf, *Proc. Natl. Acad. Sci. U.S.A.* **85**, 7972 (1988).
[41] M. P. Suarez Rendueles, J. Schwencke, N. Garcia Alvarez, and S. Gascon, *FEBS Lett.* **131**, 296 (1981).

concentration of 1% and leave on ice for 30 min. Increase the sample volume to 200 µl in 0.5% Brij 58 and assay as described above. Yeast cells contain two dipeptidyl aminopeptidases, DPAP A and DPAP B.[42] DPAP A is heat stable and is localized to the secretory pathway, where it processes the mating pheromone, α-factor. DPAP B is situated in the vacuolar membrane[6,43] and is sensitive to heat treatment. To distinguish between the two DPAP activities, divide the sample to be assayed in half and incubate one tube at 60° for 15 min, followed by 5 min on ice. DPAP B activity corresponds to the difference between the A_{405} readings of the heat-treated and untreated samples.

α-Mannosidase. α-Mannosidase activity is assessed as described.[44] Mix 0.6 ml of 0.1 M MES/NaOH, pH 6.5, plus 0.2% Triton X-100 and 0.2 ml of 4 mM p-nitrophenyl-α-D-mannopyranoside with 25–200 µl of vacuolar vesicles and adjust the final reaction volume to 1 ml with water. Incubate at 37° for 1 hr and stop the reaction by adding 1 ml of 0.5 M glycine/Na$_2$CO$_3$, pH 10.0. Measure the absorbance at 400 nm. Readings are linear up to an A_{400} of 1.5. The molar extinction coefficient for p-nitrophenol is 16 mM^{-1} cm^{-1}.

Assessment of Purity: Nonvacuolar Marker Enzyme Assays

The extent of contamination of the isolated vacuoles by nonvacuolar proteins can be determined by assaying marker enzymes of other subcellular compartments. We routinely check glucose-6-phosphate dehydrogenase (cytoplasm), NADPH–cytochrome-c reductase (endoplasmic reticulum), azide-sensitive ATPase (mitochondria), and vanadate-sensitive ATPase (plasma membrane). We have not checked for nuclear contamination, although we expect that it is very low. More information on marker assays for different subcellular compartments is provided elsewhere in this volume.[45]

Cytoplasm: Glucose-6-Phosphate Dehydrogenase. Glucose-6-phosphate dehydrogenase is a commonly measured cytoplasmic activity.[46] Resuspend the sample in 1.6 ml of 50 mM Tris-HCl, pH 7.6, and 0.2 ml of 5 mM NADP$^+$ in 1% NaHCO$_3$. Incubate at room temperature for 5 min, then read the A_{340} of one-half the mix ($t = 0$). Immediately add 0.1 ml of 6.67 mM glucose 6-phosphate to the other half of the mix to initiate the

---

[42] D. Julius, L. Blair, A. Brake, G. F. Sprague, Jr., and J. Thorner, *Cell (Cambridge, Mass.)* **32**, 839 (1983).
[43] C. Bordallo, J. Schwencke, and M. S. Suarez Rendueles, *FEBS Lett.* **173**, 199 (1984).
[44] D. J. Opheim, *Biochim. Biophys. Acta* **524**, 121 (1978).
[45] A. Franzusoff, J. Rothblatt, and R. Schekman, this volume [45].
[46] G. W. Löhr and H. D. Waller, *in* "Methods of Enzymatic Analysis" (H. U. Bergmeyer, ed.), 2nd Ed., Vol. 2, p. 636. Academic Press, New York, 1974.

reaction. Incubate at room temperature for 15 min then read the A_{340}. Readings are linear up to an A_{340} of 0.5. The molar extinction coefficient for NADH is 6.22 mM^{-1} cm^{-1}. The level of contamination by glucose 6-phosphate in purified vacuoles is typically very low (i.e., <0.5% of the total activity).

Endoplasmic Reticulum: NADPH–Cytochrome-c Reductase. The NADPH–cytochrome-c reductase activity is measured to assess the yield of endoplasmic reticulum.[47] The reaction solution consists of 0.1 mM NADPH, 70 μM cytochrome c, 0.4 mM KCN, 1 μM flavin mononucleotide, and 50 mM potassium phosphate, pH 7.4. Add 1–10 μl of vacuolar membranes and incubate at 30° for 30 min. Stop the reaction by incubating the tubes on ice. Measure the absorbance at 550 and 540 nm. The value of ΔE ($= A_{550} - A_{540}$) is linear up to 0.4. The molar extinction coefficient is $\epsilon = 19.1$ mM^{-1} cm^{-1}. The level of contamination in purified vacuoles by NADPH–cytochrome c activity is typically less than 1% of the total activity.

Mitochondria (Azide-Sensitive H^+-ATPase) and Plasma Membrane (Vanadate-Sensitive H^+-ATPase): We measure the azide- and vanadate-sensitive ATPase activities as markers for the mitochondria and plasma membrane, respectively. The mitochondrial and plasma membrane ATPases are normally assayed at pH 8.0 and 6.5, respectively;[48] however, the degree of contamination of these activities can be determined at pH 7.0 by comparing the activity present in the starting material and the purified vacuoles. The extent of contamination by these two enzymes is usually less than 0.5% of the total activity.

Golgi Apparatus. Enzymatic markers for the Golgi apparatus in yeast are less well characterized than markers for other compartments owing to the lack of a purification procedure for yeast Golgi membranes. However, *in vitro* assays have been described for the Golgi enzyme mannosyltransferase I,[49] which catalyzes the transfer of mannose units from GDP-mannose to the core glycosyl residues of certain secreted mannoproteins. Also, assays for the Kex2 protease[50,51] and DPAP A,[42] which are membrane-bound proteases involved in the maturation of the α-factor mating pheromone, have been described. These proteases are thought to reside in a later Golgi compartment than mannosyltransferase I.[50,52] We do not know the extent to which mannosyltransferase I or the Kex2 protease copurifies with vacuoles in the purification protocol described above; however, essentially

---

[47] S. Kubota, Y. Yoshida, H. Kumaoka, and A. Furumichi, *J. Biol. Chem.* **81**, 197 (1977).
[48] R. Serrano, *Mol. Cell. Biochem.* **22**, 51 (1978).
[49] T. Nakajima and C. E. Ballou, *Proc. Natl. Acad. Sci. U.S.A.* **72**, 3912 (1975).
[50] D. Julius, R. Schekman, and J. Thorner, *Cell (Cambridge, Mass.)* **36**, 309 (1984).
[51] R. S. Fuller, A. Brake, and J. Thorner, *Proc. Natl. Acad. Sci. U.S.A.* **86**, 1434 (1989).
[52] K. W. Cunningham and W. T. Wickner, *Yeast* **5**, 25 (1989).

no DPAP A activity (heat-stable DPAP activity, as opposed to the heat-labile activity of DPAP B) is found in our purified vacuoles.

Other Activities Assayed Using Purified Vacuolar Vesicles

A number of activities have been characterized using purified vacuolar membranes. As mentioned before, a vacuolar H^+-ATPase activity has been described that utilizes the energy of ATP hydrolysis to pump protons into the vacuole, causing it to be acidic.[35] The proton pumping activity of the vacuolar H^+-ATPase can be monitored using acridine orange[53] or quinacrine[54] fluorescence quenching methods. Assays on purified vacuolar vesicles have shown that the proton gradient is used to drive several H^+/substrate antiport systems for the transport of calcium[54] and amino acids[55,56] into the vacuole. Also, an ion channel of the vacuolar membrane has been identified which conducts K^+ and other monovalent cations and is presumed to play a role in the formation and regulation of the osmotic or electrical potential difference across the vacuolar membrane.[57]

Methods for Assessing the Fidelity of Protein Sorting to the Vacuole

The search for mutations that perturb vacuolar protein sorting has uncovered at least 40 *VPS* genes that are necessary for this process.[10–12,14,15] All *vps* mutants secrete a substantial portion of the soluble vacuolar protease CPY as well as other vacuolar proteins. Two plate assays have been used in our laboratory for assaying extracellular vacuolar hydrolases. The first detects extracellular CPY activity, whereas the second is a colony immunoblotting technique that detects extracellular vacuolar antigens. These simple assays should be useful to the researcher who wishes to assess whether vacuolar protein sorting is compromised in a particular mutant of interest.

Carboxypeptidase Y Activity Filter Assay

When CPY is mislocalized, because of either overproduction[58] or a *vps* mutation,[10,14] it is secreted as the inactive (proCPY) zymogen. However, a portion of the mislocalized proCPY is converted to an enzymatically active

---

[53] S. Gluck, S. Kelly, and Q. Al-Awquati, *J. Biol. Chem.* **257,** 9230 (1982).
[54] Y. Ohsumi and Y. Anraku, *J. Biol. Chem.* **258,** 5614 (1983).
[55] Y. Ohsumi and Y. Anraku, *J. Biol. Chem.* **256,** 2079 (1981).
[56] T. Sato, Y. Ohsumi, and Y. Anraku, *J. Biol. Chem.* **259,** 11505 (1984).
[57] Y. Wada, Y. Ohsumi, M. Tanifuji, M. Kasai, and Y. Anraku, *J. Biol. Chem.* **262,** 17260 (1987).
[58] T. H. Stevens, J. H. Rothman, G. S. Payne, and R. Schekman, *J. Cell Biol.* **102,** 1551 (1986).

form by an unidentified periplasmic protease, and so extracellular CPY activity can be used as an indication of vacuolar protein missorting. The assay described below is based on the observation that CPY catalyzes the peptide bond cleavage of the N-blocked dipeptide N-carbobenzyloxy-L-phenylalanyl-L-leucine (CBZ-Phe-Leu)[59] and is the major enzyme in yeast responsible for this activity.[60]

Grow patches of the desired yeast strains on plates containing minimal medium. Replica plate fresh patches onto 11-cm Whatman No. 1 filters (Whatman, Inc., Clifton, NJ) that have been soaked in a freshly prepared reaction cocktail containing 0.2 M potassium phosphate, pH 7.0, 1 mM CBZ-Phe-Leu, 1 mg/ml L-amino-acid oxidase, 0.2 mg/ml o-dianisidine (from a freshly prepared stock solution), and 7 μg/ml horseradish peroxidase. Ten milliliters of reaction cocktail is enough for 4 filters. Keep the filters moist in plastic wrap and incubate at room temperature. Begin scoring the filters after 10 min, and keep checking periodically for 2 hr. Patches that are secreting CPY will turn dark pink. Leave the patches at room temperature overnight and score again in the morning.

Notes. Always include control patches on each filter to be assayed, including a wild-type strain and a *vps* mutant (if available). Also, some wild-type strains may give substantial levels of background activity owing to cell lysis; therefore, a strain lacking CPY *(prc1)* is a good control for the nonenzymatic background activity.

Colony Immunoblotting

Extracellular vacuolar proteins can be detected by a colony immunoblotting technique, in which a nitrocellulose filter is overlayed onto growing yeast colonies.[61] Proteins that are secreted adhere to the filter, and they can be detected using the appropriate antibody and standard Western blotting procedures. We routinely use this protocol to monitor the secretion of CPY[7] and PrA.[61]

Replica plate fresh yeast patches onto a YEPD or minimal medium plate, carefully overlay a prewetted 85-mm nitrocellulose filter (0.45 μm) onto the patches, and incubate the plate at 30° for 12–24 hr. Peel the filter off, rinse with distilled water to remove cells, and detect the secreted proteins using standard Western blotting techniques. We generally detect the primary antibodies by staining the filters with goat anti-rabbit antibody conjugated to alkaline phosphatase (Promega Corp., Madison, WI), fol-

---

[59] R. W. Kuhn, K. A. Walsh, and H. Neurath, *Biochemistry* **13**, 3871 (1974).
[60] D. H. Wolf and U. Weiser, *Eur. J. Biochem.* **73**, 553 (1977).
[61] J. H. Rothman, C. P. Hunter, L. A. Valls, and T. H. Stevens, *Proc. Natl. Acad. Sci. U.S.A.* **83**, 3248 (1986).

lowed by color development using reagents from Bio-Rad Laboratories (Richmond, CA).

Notes. As with the CPY activity filter assay, it is important that fresh patches be used for colony immunoblotting, as older patches will give a higher background. Also, be sure to include the proper controls on each plate. Colony immunoblotting may also be used as a control for cell lysis by using antibody to a cytoplasmic protein, such as phosphoglycerate kinase (C. Vater and T. Stevens, unpublished results).

As a general tool for monitoring mislocalization of vacuolar proteins, the CPY activity filter assay is of greater utility than colony immunoblotting because all of the reagents needed for the assay are commercially available. These assays are an effective means for qualitatively assessing whether a particular strain is secreting vacuolar hydrolases; however, a quantitative assessment of mislocalization requires immunoprecipitations of CPY or PrA from intracellular and extracellular fractions of radiolabeled mid-log phase cells growing in culture.[10-12,14]

The localization of a CPY–invertase fusion protein has also been used to monitor vacuolar protein targeting.[11,12] Strains containing a deletion of the invertase structural gene *(SUC2)* that mislocalize this fusion protein exhibit extracellular invertase activity, which is easily monitored by whole cell or plate assay.

Acknowledgments

We thank Rob Preston for helpful discussions concerning fluorescent labeling of vacuoles, Kevin Redding and Robert Fuller for suggesting antibody amplification as a means for detecting nonabundant antigens by immunofluorescence, Harrison Howard for expert assistance with the microscopy and photography, Patricia Kane, Carol Vater, and Karen Moore for comments on the manuscript, and to the past and present members of the Stevens laboratory for helping us develop the techniques described in this chapter.

[45] Analysis of Polypeptide Transit through Yeast Secretory Pathway

By ALEX FRANZUSOFF, JONATHAN ROTHBLATT, and RANDY SCHEKMAN

Introduction

The secretory pathway in the yeast *Saccharomyces cerevisiae* is one of several systems employed by the cell to ensure correct localization and compartmentation of proteins and lipids. Assembly of the endoplasmic reticulum (ER) nuclear envelope, Golgi apparatus, vacuole, and plasma membrane depend on a progression of events that begins by biosynthetic insertion of macromolecules into and across the ER membrane. Mitochondria and peroxisomes recruit their constituents directly from the cytosol. The biogenesis of proteins that become localized to organelles has been an active area of investigation for some time, and consequently a number of techniques are available for the analysis of membrane and protein intermediates in transport. A complete review of such techniques is well beyond the scope of this chapter. Two simple operations, differential centrifugation of membrane fractions and radiolabeling of biosynthetic intermediates, are discussed in sufficient detail to permit a preliminary study of the localization pathway employed for a typical secretory protein. A more detailed account of the many techniques of examining yeast glycoprotein localization is presented elsewhere.[1]

Yeast Organelle Fractionation

A scheme for the complete resolution of the major yeast organelles by subcellular fractionation has not been established. This is not because yeast membranes are inherently difficult to fractionate, rather, the system simply has not received the dedicated attention that has allowed such resolution to be achieved with mammalian tissues. Certain organelles, such as the nucleus,[2] ER,[3] plasma membrane,[4] vacuole,[5] mitochondria,[6,7] secretory vesi-

---

[1] J. Rothblatt and R. Schekman, *in* "Methods in Cell Biology" (A. M. Tartakoff, ed.), Vol. 32, p. 3. Academic Press, San Diego, 1989.
[2] J. Aris and G. Blobel, *J. Cell Biol.* **107**, 17 (1988).
[3] J. Rothblatt and D. I. Meyer, *EMBO J.* **5**, 1031 (1986).
[4] J. Tschopp and R. Schekman, *J. Bacteriol.* **156**, 222 (1983).
[5] A. Wiemken, *in* "Methods in Cell Biology" (D. M. Prescott, ed.), Vol. 12, p. 99. Academic Press, New York, 1975.

cles,[8,9] and peroxisomes,[10] may be isolated from appropriate strains. Additional information relevant to these goals can be found in [43], [44], and [46] in this volume.

In spite of the lack of experience in achieving complete membrane fractionation, it is usually possible to assign a subcellular location and pathway through the use of monospecific antibody directed against a protein of interest. First, a simple scheme for membrane fractionation is presented that may be used to achieve a preliminary assessment of the location of a protein.

Spheroplast Formation and Lysis

Yeast organelles are fragile and easily broken by the shear forces necessary to break intact yeast cells. Removal of the yeast cell wall generates spheroplasts, which are easily broken under conditions that preserve the integrity of subcellular organelles. Spheroplasts may be formed by different commercial preparations of lytic enzyme, most of which (e.g., Glusulase, Zymolyase) are contaminated with proteases and nucleases that can wreak havoc with intracellular constituents. We have developed a simple procedure for obtaining purified lytic enzyme (lyticase) from the soil bacterium, *Oerskovia xanthineolytica*.[11] Lyticase is particularly valuable for large-scale isolation, where the commercial enzyme is prohibitively expensive. For small scale (≤ 100 mg cell pellet), however, zymolyase may be used as long as precaution is taken to remove spheroplasts from the lytic enzyme by at least two washes with isosmotic buffer.

Spheroplasts are formed from cells grown to early to mid-logarithmic phase (2–4 OD_{600} units/ml; $\sim 10^7$ cells/OD_{600} unit) in YPD growth medium [1% yeast extract, 2% Bacto-peptone, and 2–5% glucose (all w/v)]. After centrifugation at 3000 g for 5 min (all steps usually at room temperature), cells are washed once in distilled water and resuspended at 20 OD_{600}/ml in 0.1 M Tris–SO_4, pH 9.4, 10 mM dithiothreitol (DTT) (or to 50 mM 2-mercaptoethanol) for 10 min. This incubation loosens the outer mannoprotein layer to allow subsequent attack of the underlying $\beta(1 \rightarrow 3)$-glucan layer by the lytic enzyme. Cells are then washed in 1.2 M sorbitol

---

[6] A. W. Linnane and H. B. Lukens, in "Methods in Cell Biology" (D. M. Prescott, ed.), Vol. 12, p. 285. Academic Press, New York, 1975.
[7] G. Daum, P. C. Böhni, and G. Schatz, *J. Biol. Chem.* **257**, 13028 (1982).
[8] C. Holcomb, T. Etcheverry, and R. Schekman, *Anal. Biochem.* **166**, 328 (1987).
[9] N. C. Walworth and P. N. Novick, *J. Cell Biol.* **105**, 163 (1987).
[10] J. M. Goodman, C. W. Scott, P. N. Donahue, and J. P. Atherton, *J. Biol. Chem.* **259**, 8485 (1984).
[11] J. Scott and R. Schekman, *J. Bacteriol.* **142**, 414 (1980).

and resuspended to 50 OD_{600}/ml in spheroplasting buffer (1.2 M sorbitol, 10 mM potassium phosphate, pH 7.2). Lyticase (Fraction II, 10–25 units/OD_{600} unit cells) or Zymolyase 5000 (5μg/OD_{600} unit cells) is added, and spheroplasts form during incubation with gentle shaking at 25° for temperature-sensitive strains or 30° for other strains.[11] The formation of metabolically active spheroplasts is ensured by including 0.5 × YP (1% yeast extract, 2% Bacto-peptone) in the spheroplasting buffer.[8] Spheroplasting is monitored by dilution of an aliquot (~10 μl or ~0.4 OD_{600} unit cells) into 1 ml water. Complete spheroplasting, which should occur within 45 min, results in a greater than 10-fold drop in OD_{600} in the dilution lysis assay. Spheroplasts are recovered by centrifugation through a cushion of 0.8 M sucrose, 1.5% (w/v) Ficoll 400, 20 mM 2-(N-morpholino)ethanesulfonic acid (MES), pH 6.5, for 10 min at 4000 g at 4°. Spheroplast stability is enhanced at pH 5.5–6.5.

Efficient spheroplast lysis is achieved with a motor-driven Potter–Elvehjem tissue grinder fitted with a serrated Teflon pestle (0.1–0.15 mm clearance). The spheroplast pellet is resuspended to 50 OD_{600} unit cell equivalents/ml of chilled lysis buffer (0.7 M sorbitol, 20 mM MES, pH 6.5, 5 mM MgCl$_2$). Protease inhibitors, such as phenylmethylsulfonyl fluoride (1 mM), pepstatin, leupeptin, chymostatin, and aprotinin (20 μg/ml each), are added to the lysis buffer just before use. A few drops of lysis buffer are added to the spheroplast pellet, which is mixed on ice with the aid of a rubber Pasteur pipette bulb to obtain an even suspension. The diluted suspension is transferred to the cooled homogenizer, and spheroplasts are disrupted with three 1-min bursts using a pestle connected to a Black and Decker drill (Model 7193) set at approximately 1500 rpm. Spheroplast lysis is monitored qualitatively by examination in a phase-contrast microscope. Lysis should be at least 75% complete. A few more bursts (two or three) with the homogenizer may be tolerated to achieve this efficiency of lysis. A hand-driven Dounce-type homogenizer [stainless steel Dura-grind; Wheaton Scientific (Millville, NJ), clearance 12.7 μm] provides an acceptable alternative. The number of homogenizer passes required for efficient lysis is quite variable. The efficiency of lysis is increased by reducing the sorbitol concentration in the lysis buffer by a factor of 2.

Gentler, but somewhat less efficient, lysis is achieved by agitation with glass beads [0.5 mm, Biospec Products (Bartlesville, OK)]. Acid-washed beads (0.25 g) are mixed with 0.6 ml lysis buffer and 200 OD_{600} cell equivalents of spheroplasts in a round-bottomed tube. The mixture is agitated on a vortex mixer set at half-maximal force for 5 periods of 10 sec each with 15-sec intervals on ice. The supernatant fraction is removed to a fresh tube, and the beads are washed with enough fresh lysis buffer to bring the homogenate to 100 OD_{600} cell equivalent units/ml. This procedure yields about 50% lysis.[8]

All lysis procedures leave a fraction of intact cells or spheroplasts remaining in the homogenate. Most unlysed cells and large cell wall debris are removed by centrifugation of the lysate at 650 g for 3 min at 4°. The supernatant fraction is removed to a fresh tube for further manipulation.

Evaluation of Membrane Association

The crude homogenate may be used to analyze the disposition of a protein in the cytosol, on the cytosolic face of a membrane, or enclosed within an organelle. Proteins in the first two locations generally are readily degraded by exogenous protease, whereas proteins sequestered within an organelle are degraded only in the presence of detergent. Procedures for this analysis, and the appropriate controls, are described elsewhere.[1,12,13] It is important to note that centrifugation of the homogenate to form a membrane pellet may cause cytoplasmically exposed membrane proteins to become latent to protease attack because of membrane aggregation and the consequent occlusion of membrane surfaces.[12] For this reason, protease accessibility experiments should be performed with a fresh unfractionated homogenate.

Many different procedures have been described for the enrichment of yeast organelles from a homogenate (Table I).[14-25] It is beyond the scope of this review to present a detailed analysis of each method; however, each organelle has distinctive characteristics that allow reasonable enrichment.

Nuclei, endoplasmic reticulum, and vacuoles tend to sediment at the lowest centrifugal forces. Nuclei are further enriched by density sedimentation; aside from cell wall fragments, the nucleus is the densest organelle in a yeast homogenate.[2] ER membrane tends not to vesiculate into microsomes as is seen with mammalian tissues that are broken by homogeniza-

[12] R. Feldman, M. Bernstein, and R. Schekman, *J. Biol. Chem.* **262,** 9332 (1987).
[13] R. Deshaies and R. Schekman, *J. Cell Biol.* **105,** 633 (1987).
[14] J. P. Aris and G. Blobel, *J. Cell Biol.* **108,** 2059 (1989).
[15] R. Wright, M. Basson, L. D'Ari, and J. Rine, *J. Cell Biol.* **107,** 101 (1988).
[16] M. D. Rose, L. M. Misra, and J. R. Vogel, *Cell (Cambridge, Mass.)* **57,** 1211 (1989).
[17] R. Fuller and J. Thorner, personal communication (1989).
[18] A. Wiemken and M. Dürr, *Arch. Microbiol.* **101,** 45 (1974).
[19] C. J. Roberts, G. Pohlig, J. H. Rothman, and T. H. Stevens, *J. Cell Biol.* **108,** 1363 (1989).
[20] C. L. Holcomb, W. J. Hansen, T. Etcheverry, and R. Schekman, *J. Cell Biol.* **106,** 641 (1988).
[21] G. R. Willsky, *J. Biol. Chem.* **254,** 3326 (1979).
[22] G. Schatz, *Z. Naturforsch. Teil B* **18,** 145 (1963).
[23] R. W. Parish, *Arch. Microbiol.* **105,** 187 (1975).
[24] M. Skoneczny, A. Chelstowska, and J. Rytka, *Eur. J. Biochem.* **174,** 297 (1988).
[25] M. Veenhuis, M. Mateblowski, W. H. Kunau, and W. Harder, *Yeast* **3,** 77 (1987).

TABLE I
PROCEDURES FOR ENRICHMENT OF YEAST ORGANELLES

Organelle	Sedimentation and buoyant density[a] properties	Immunological markers[c]	Ref.
Nucleus	Rapid sedimentation	Nuclear envelope: p110/p95 (m)	14
	High density	Nucleolus: p38	2
Endoplasmic reticulum	Rapid sedimentation	HMG-CoA reductase (m)	15, 16, 26
	Low density, 1.06 g/cm³ (P)	Binding protein (BiP)	
Golgi apparatus	Slow sedimentation	Kex2 endopeptidase (m)	17
	Low density		
Vacuole	Rapid sedimentation	Carboxypeptidase Y	5, 18
	Low density, 1.03 g/cm³ < V < 1.075 g/cm³ [b]	Dipeptidyl aminopeptidase B (m)	19
Secretory vesicles	Slow sedimentation	Acid phosphatase, Mg²⁺-ATPase (m)	8, 9, 20
	High density, 1.08–1.09 g/cm³ (P)		
Plasma membrane	Variable sedimentation	Mg²⁺-ATPase (m)	4
	High density, 1.15, 1.17 g/cm³ (R); 1.22 g/cm³ (S)		21
Mitochondria	Intermediate sedimentation	IMM,[d] F₁,β-ATPase (m)	22, 7, 6, 20
	Variable density, 1.06, 1.08 g/cm³ (P); 1.16–1.18 g/cm³ (S)	OMM,[d] porin (m)	
Peroxisomes	Intermediate sedimentation	Catalase	23
	High density, 1.22 g/cm³ (S); 1.08–1.09 g/cm³ (F)	β-Oxidation bifunctional enzyme (enoyl-CoA hydratase/L-3-hydroxy-acyl-CoA dehydrogenase)	24, 25

[a] Particle buoyant density in P (Percoll, Pharmacia), S (sucrose), F (Ficoll, Sigma), or R (Renografin, Squibb) density gradients.
[b] Density of vacuoles (V) relative to 0.6 M sorbitol (1.03 g/cm³) and 0.6 M sucrose (1.075 g/cm³) after flotation in an isotonic density gradient.
[c] m, Membrane-associated protein; all others are lumenal or matrix constituents.
[d] IMM, Inner mitochondrial membrane; OMM, outer mitochondrial membrane.

tion. Crude ER fractions that sustain secretory protein translocation may be enriched by differential centrifugation.[3] Further purification by density sedimentation may rely on the increased density created by ribosomes bound to the ER membrane.[26] Vacuole density is most dramatically influenced by the osmotic pressure of the lysis buffer. With a modest osmotic shock during spheroplast homogenization, the vacuole swells without lysis. Vacuoles behave like osmometers and can be isolated by flotation up a density and osmotic gradient.[5]

Mitochondria and plasma membrane fragments tend to sediment at g forces somewhat higher than those required for the largest organelles. In fractions that are depleted of nuclei and ER, intact mitochondria can be further enriched at dense positions in a sedimentation equilibrium gradient.[6] Plasma membrane fragments do not seal to form closed vesicles, and they have a different density than intact mitochondria. Alternatively, mitochondria are selectively precipitated by mild acid (pH 5).[27] The behavior of plasma membrane fragments can be controlled by coating spheroplasts with concanavalin A prior to lysis.[28] Large plasma membrane sheets are generated which sediment rapidly from the crude homogenate. Elution of the lectin with α-methylmannoside allows more slowly sedimenting plasma membrane vesicles to be generated.

Peroxisomes are induced in *S. cerevisiae* grown on oleic acid;[25] however, the isolation of intact peroxisomes from this source has not yet been reported. Another very rich source of peroxisomes are the methylotropic yeasts *Hansenula polymorpha, Pichia pastoris,* and *Candida boidinii.* Growth of these yeasts on methanol allows peroxisome induction up to 30% of the cell mass. The large size and high buoyant density of peroxisomes in *C. boidinii* have been exploited to obtain highly purified fractions.[10] The organelle is quite fragile so lysis conditions must be carefully controlled.

Golgi membranes and secretory vesicles constitute the slowest sedimenting membranes in a homogenate. Golgi membranes are retained in the supernatant fraction after centrifugation at 17,000–25,000 g for 15 min and are collected in the pellet fraction after sedimentation at 100,000 g for 60 min. Further enrichment has not been reported, although it should be possible to separate the smooth-membrane Golgi from the ribosome-bound ER on a density gradient. Secretory vesicles constitute an insignificant fraction of membranes in a wild-type cell. Secretory mutants that accumulate mature secretory vesicles have been used as a source of this material.[8,9] The homogeneous small size, high density, and electropho-

---

[26] M. Marriot and W. Tanner, *J. Bacteriol.* **139,** 565 (1979).
[27] A. Franzusoff and V. P. Cirillo, *J. Biol. Chem.* **258,** 3608 (1983).
[28] G. A. Scarborough, *J. Biol. Chem.* **250,** 1106 (1975).

retic mobility of secretory vesicles have been used to achieve substantial purification.

Marker Enzyme Analysis

Marker enzyme assay is the traditional means of assessing membrane fractionation. In practice, membrane yield and purification based on an enzyme assay can be misleading since it requires a stable enzyme whose activity is not influenced by agents used to achieve fractionation. This problem is most serious given the high concentrations of solutes or particles used in density gradient fractionation.

Fortunately, virtually every yeast membrane and lumenal space can now be detected by immunoblot of fractions using monospecific antibodies directed against organelle-specific marker proteins. The technique is sensitive, quantifiable, and not dependent on retention of marker enzyme activity. Table I lists the organelles and marker proteins for which immunoblotting-quality antibodies have been prepared. Some problems are encountered in the use of immunoblotting to examine membrane fractionation. Certain membrane protein antigens, particularly those that span the membrane multiple times, aggregate irreversibly when exposed to sodium dodecyl sulfate (SDS) at high temperature. The vanadate-sensitive plasma membrane ATPase is a good example. This protein is efficiently solubilized, and migrates according to its expected size on SDS-polyacrylamide gels, only when membranes are solubilized with SDS at a moderate temperature, such as 50°.

Most applications for immunoblotting are not quantitative. To give reliable data that allow antigen yield and purification to be assessed, antibody must be in excess and the secondary detecting agent [^{125}I-labeled protein A or horseradish peroxidase (HRP) secondary antibody] must be measured over a linear range of exposure. A test should be performed with each antigen diluted to achieve a proportional response on the immunoblot.

Marker proteins that are transported between distinct compartments (e.g., proteins in the secretory pathway) may appear in multiple fractions. Usually some form of covalent modification (glycosylation, proteolytic maturation) may be used to distinguish transport intermediates, and thus the compartments in which they reside (see Cell Disruption and Immune Detection of Transit Forms). However, some proteins, such as the plasma membrane vanadate-sensitive ATPase, sustain no detectable changes in SDS-PAGE mobility during biogenesis.[20] Pulse-kinetic radiolabeling followed by immunoprecipitation from isolated fractions may be used to identify transit compartments. Independent physical marking procedures, such as catalytic ^{125}I-iodination of the plasma membrane prior to rupture of spheroplasts, may be used to distinguish the plasma membrane from its precursor membranes.[8]

Analysis of Protein Transport by Radiolabeling Wild-Type and *sec* Mutant Cells

Arrest of Protein Transport and Radiolabeling Transit Intermediates

The rapid rate of protein transport through the secretory pathway makes it difficult to define certain intermediate steps. For this reason, *sec* mutations which block specific stages in the pathway have been particularly useful in defining the biogenetic process of many different localized proteins. Morphological and biochemical analysis of the *sec* mutations has shown that all the major secretory, vacuolar, and plasma membrane proteins accumulate in precursor forms and intermediate organelles when cells are incubated at a restrictive growth temperature.[29]

More than 30 genes have been defined by the isolation of *sec* mutations. Five stages in the pathway may be distinguished by analysis of the mutants (Table II).[30-40] Protein translocation from the cytosol into the ER is blocked in a group of mutants that cause unglycosylated precursors to accumulate in forms that resemble the primary translation products.[13,35,36] Another two mutations block core glycosylation, but not ER membrane penetration, of secretory polypeptides and cause signal peptide-processed forms to accumulate in the ER lumen.[12,30] Conversely, *sec11* affects signal peptidase and causes glycosylated but signal peptide-unprocessed secretory precursors to accumulate within the ER.[31] Thirteen *SEC* genes are required for transport of core-glycosylated and signal peptide-processed forms from the ER to the Golgi apparatus.[32-34] Transport within the Golgi

---

[29] R. Schekman, in "Annual Review of Cell Biology" (G. Palade, B. Alberts, and J. Spudich, eds.), Vol. 1, p. 115. Annual Reviews Inc., Palo Alto, California, 1985.
[30] M. Bernstein, F. Kepes, and R. Schekman, *Mol. Cell. Biol.* **9**, 1191 (1989).
[31] P. Böhni, R. Deshaies, and R. Schekman, *J. Cell Biol.* **106**, 1035 (1988).
[32] P. Novick, S. Ferro, and R. Schekman, *Cell (Cambridge, Mass.)* **25**, 461 (1981).
[33] A. P. Newman and S. Ferro-Novick, *J. Cell Biol.* **105**, 1587 (1987).
[34] H. D. Schmitt, M. Puzicha, and D. Gallwitz, *Cell (Cambridge, Mass.)* **53**, 653 (1988).
[35] J. A. Rothblatt, R. J. Deshaies, S. Sanders, G. Daum, and R. Schekman, *J. Cell Biol.* **109**, 2641 (1989).
[36] J. Toyn, A. R. Hibbs, R. Sanz, J. Crowe, and D. I. Meyer, *EMBO J.* **7**, 4347 (1988).
[37] S. Ferro-Novick, P. Novick, C. Field, and R. Schekman, *J. Cell Biol.* **98**, 35 (1984).
[38] P. Novick, C. Field, and R. Schekman, *Cell (Cambridge, Mass.)* **21**, 205 (1980).
[39] N. Segev, J. Mulholland, and D. Botstein, *Cell (Cambridge, Mass.)* **52**, 915 (1988).
[40] T. Stevens, B. Esmon, and R. Schekman, *Cell (Cambridge, Mass.)* **30**, 439 (1982).

TABLE II
ANALYSIS OF PROTEIN TRAFFIC

Defective step in secretion	Mutations	Diagnostic features of glycoproteins	Ref.
Translocation into ER	sec61, 62, 63; ptl1	Unmodified, cytoplasmic precursor	13, 35, 36
Signal peptide cleavage	sec11	Partially processed lumenal precursor	12, 30, 31, 37
Core oligosaccharide addition	sec53, 59		
ER to Golgi transport	sec12, 13, 16, 17, 18, 19, 20, 21, 22, 23; bet1, 2; ypt1	Core-glycosylated intermediate	32, 33, 34, 38, 39
Vacuole/secretory vesicle formation	sec7, 14	Outer chain-glycosylated intermediate; endoproteolytic processing blocked	32, 38, 40
Exocytosis	sec1, 2, 3, 4, 5, 6, 8, 9, 10, 15	Outer chain-glycosylated intermediate; endoproteolytic processing completed	32, 38

and sorting of secretory and vacuolar proteins require the action of one or two Sec products.[32] Finally, 10 *SEC* genes are required for the transport or fusion of mature secretory vesicles with the plasma membrane in the bud portion of a growing cell.[32]

In the simplest case, the biogenesis of a protein may be evaluated in parallel wild-type and *sec* mutant cultures radiolabeled after a brief preincubation of cells at 37°. Unfortunately, many proteins of interest are subject to metabolic regulation, and changes in growth medium often must precede incubation at the *sec* restrictive temperature. Derepression of the secreted form of invertase begins within 10 min of cell transfer to medium containing low glucose. High-level expression of the sulfate permease, galactose permease, or regulated acid phosphatase requires anywhere from 2 to 12 hr of derepression in specific growth medium. Such changes in growth medium may severely reduce the expression of constitutively exported proteins. For example, the conditions of low glucose optimal for invertase synthesis in *sec* mutant cells depress synthesis of α-factor precursor and of the plasma membrane ATPase.[20] Obviously, these problems must be dealt with individually.

Metabolic labeling with ^{35}S is the most convenient means of detecting protein species. $^{35}SO_4^{2-}$ or [^{35}S]methionine may be used; each has advantages. $^{35}SO_4^{2-}$ is inexpensive but generally requires higher levels of isotope to achieve the desired protein labeling, and efficient uptake into cells occurs only when the growth medium is adjusted to low concentrations of sulfate.[1] [^{35}S]Methionine is more expensive but impure preparations [e.g., Tran ^{35}S-label, ICN Radiochemicals (Costa Mesa, CA)] suffice for cell labeling. The advantages of [^{35}S]methionine are that lower levels of radioactivity can be used to achieve significant labeling, cells need not be starved for sulfate during radiolabeling, and shorter periods of radiolabeling may be used to detect short-lived protein intermediates.

Cultures are grown in minimal (MV) medium[41] in which all the sulfate salts are replaced by chloride salts, then supplemented with 100–200 μM $(NH_4)_2SO_4$, 2% (w/v) D-glucose, and auxotrophic components (except methionine or cysteine). A permissive temperature of 25° is used to prepare an overnight culture which generally (depending on the generation time of the strain) is inoculated at 0.025–0.05 OD_{600}/ml (measured in a Zeiss PMQII spectrophotometer) in a sterile Erlenmeyer flask. Typically, a 50-ml overnight culture suffices for a complete radiolabeling experiment. Once cells have reached mid-logarithmic phase (OD_{600} 0.2–0.5), growth conditions may be altered for derepression of synthesis of a desired protein, for introduction of a drug that interferes with protein modification (such as

---

[41] L. J. Wickerham, *J. Bacteriol.* **52**, 293 (1946).

tunicamycin), or for imposition of a *sec* mutant block. In a typical experiment, 12 OD_{600} units of cells are harvested by centrifugation at 3000 g (top speed in a clinical centrifuge) for 2 min at room temperature. Cells are resuspended to 3 OD_{600} units/ml and divided into two portions of 2 ml in 13 × 100 mm sterile tubes (larger volumes require flasks for proper aeration). Tubes are incubated with agitation in gyrotory water baths set at 24° and 37°. The duration of the temperature shift necessary for a completely restrictive secretory block ranges from 10–20 min at 37° for most secretory mutants to 1–2 hr for translocation-defective mutants (e.g., *sec61*).[13] Lengthy preincubations require a cell dilution adjustment for the absorbance increase.

[^{35}S]Methionine labeling is initiated with 25 μCi/OD_{600} unit cells. A labeling period of 30 min usually is sufficient to detect production of the mature form of a localized protein. To detect transit precursors, 5 min or less may be necessary. In a pulse–chase regimen, an aliquot (1 ml) is transferred to a chilled tube containing 4 ml of 10 mM sodium azide. A concentrated chase solution [100 mM $(NH_4)_2SO_4$, 0.3% L-cysteine, and 0.4% L-methionine] is diluted 100-fold into the remaining culture, and incubations are continued, with aliquots removed into ice-cold azide at timed intervals.

Cell Disruption and Immune Detection of Transit Forms

Detection of the molecular form of a transit intermediate or mature species may be achieved by immunoprecipitation of an antigen solubilized from a crude cell lysate or subcellular fraction. For immunoprecipitation of an antigen from a crude cell lysate, rapid lysis by agitation with glass beads followed by heating in SDS gel sample buffer is preferred. Radiolabeled cells are collected by centrifugation and washed in chilled 10 mM sodium azide. Cell pellets (3 OD_{600} units) are mixed with 100–200 mg of acid-washed glass beads (Biospec Products) and diluted with 200 μl of Laemmli sample buffer (2% SDS, 10% glycerol, 2% 2-mercaptoethanol, 62.5 mM Tris-HCl, pH 6.8, 0.01% bromphenol blue).[42] The amount of beads is adjusted to just match the volume of the cell–lysis buffer mixture. Lysis is achieved by agitation on a vortex mixer at top speed for 90 sec. Samples are heated to 95° for 5 min (or at lower temperatures for certain membrane proteins.[20,43] and cooled. Secreted proteins may be resolved into extracellular, periplasmic, and intracellular forms by collecting the culture fluid in the radiolabeled sample and converting cells to spheroplasts as described earlier. Medium, spheroplast supernatant, and spheroplast

[42] U. K. Laemmli, *Nature (London)* **227**, 680 (1970).
[43] K. Blumer, J. Reineke, and J. Thorner, *J. Biol. Chem.* **263**, 10836 (1988).

pellet fractions are adjusted with 10× Laemmli buffer and heated as above. Subcellular fractions prepared from spheroplasts are handled similarly.

Reagents for Immunoprecipitation

IP dilution buffer: 1.25% (v/v) Triton X-100, 190 mM NaCl, 6 mM EDTA, 60 mM Tris-HCl, pH 7.4

IP buffer: 1% Triton X-100, 0.2% (w/v) SDS, 150 mM NaCl, 5 mM EDTA, 50 mM Tris-HCl, pH 7.4

Urea wash buffer: 1% Triton X-100, 0.2% SDS, 2 M urea, 250 mM NaCl, 5 mM EDTA, 50 mM Tris-HCl, pH 7.4

High-salt wash buffer: 1% Triton X-100, 0.2% SDS, 50 mM NaCl, 5 mM EDTA, 50 mM Tris-HCl, pH 7.4

Detergent-free wash buffer: 150 mM NaCl, 5 mM EDTA, 50 mM Tris-HCl, pH 7.4

10% (v/v) IgG Sorb (fixed *Staphylococcus aureus* cells; The Enzyme Center, Inc., Malden, MA) in IP buffer

20% (v/v) Protein A-Sepharose CL-4B (Pharmacia Biochemicals, Piscataway, NJ) in IP buffer

Four volumes (~800 μl) of IP dilution buffer is added to the cooled SDS-denatured sample which is transferred, without the glass beads, to a 1.5-ml microcentrifuge tube. Fifty microliters of 10% IgG Sorb is added to the sample, and the mixture is incubated at room temperature with continuous rotation. Optionally, approximately 25 μl of unlabeled "delete" extract (300–400 OD$_{600}$ units of cell equivalents/ml), prepared as above from a yeast strain with a chromosomal deletion of the protein of interest (or from a *MAT*a strain for immunoprecipitation of α-factor), is added together with the IgG Sorb to the sample. The IgG Sorb suspension is included to adsorb any material that might bind nonspecifically to protein A, and to form a tight pellet together with the yeast cell debris collected during a clearing centrifugation. After 10 min the samples are centrifuged in a microcentrifuge for 5 min, and the supernatant fractions are transferred to fresh tubes.

A preliminary experiment is performed to titrate the amount of antibody per OD$_{600}$ unit of cell equivalents required to immunoprecipitate all of the protein of interest. The optimum amount of antibody is added, and immunoprecipitation is continued for 12 hr at 4° with continuous rotation. For every microliter of antiserum used, 5 μl of 20% protein A-Sepharose suspension is added and the mixture rotated 2 hr at room temperature. The Sepharose beads are allowed to settle or are centrifuged briefly in a microcentrifuge, and the supernatant solution is aspirated with a 22-

gauge hypodermic needle into radioactive waste or transferred to a fresh tube for immunoprecipitation of a different protein. Sepharose beads are washed twice with 1 ml of IP buffer, twice with urea wash buffer, and once with high-salt wash buffer. Finally, the sample is washed with 1 ml of detergent-free wash buffer, and the supernatant is aspirated with a 25-gauge needle which does not draw up the beads. The sample is then prepared for SDS–PAGE or reimmunoprecipitated to reduce the level of contaminating protein species. For SDS–PAGE, sample buffer containing DTT or 2-mercaptoethanol is added and the sample heated as before. Beads are removed by sedimentation.

The various transit forms of α-factor provide examples of each stage in the secretory pathway that can be evaluated with *sec* mutants. In wild-type cells the 19-kDa primary translation product is rapidly translocated from the cytosol into the ER where signal peptide cleavage[44] and core glycosylation[45] create a 62-kDa transit form. Untranslocated intact precursor and translocated signal peptide-processed precursors accumulate in *sec61* and *sec53* mutants, respectively.[12,13] Core-glycosylated precursor moves to the Golgi apparatus with a $t_{1/2}$ of 1–2 min but is retained in the ER indefinitely in mutants such as *sec18*.[45] Within the Golgi apparatus α-factor precursor oligosaccharides are elongated by outer chain carbohydrate, and the glycosylated intermediate is proteolytically matured in what may be sequential compartments.[45,46] These events are so rapid that a highly glycosylated intact precursor is not detected in wild-type cells. The intermediate is trapped within the Golgi apparatus in the *sec7* mutant, which displays a heterogeneous collection of α-factor precursor molecules ranging from approximately 30–110 kDa in size. Mature α-factor is formed prior to its export: the 1.5-kDa peptide accumulates within mutant cells, such as *sec1*, that block secretory vesicle fusion with the plasma membrane.[45]

Acknowledgments

We thank Peg Smith for preparing the manuscript. The work in this laboratory is supported by grants from the National Institutes of Health (GM26755 and GM36881). J.R. has been supported by a Postdoctoral Fellowship from the NIH (GM11791) and a Senior Postdoctoral Fellowship from the American Cancer Society, California Division.

---

[44] M. G. Waters, E. A. Evans, and G. Blobel, *J. Biol. Chem.* **263**, 6209 (1988).
[45] D. Julius, R. Schekman, and J. Thorner, *Cell (Cambridge, Mass.)* **36**, 309 (1984).
[46] A. Franzusoff and R. Schekman, *EMBO J.* **8**, 2695 (1989).

[46] *In Vitro* Protein Translocation across Microsomal Membranes of *Saccharomyces cerevisiae*

By PABLO D. GARCIA, WILLIAM HANSEN, and PETER WALTER

The ability to assay the translocation of secretory proteins across yeast microsomal membranes *in vitro*[1-3] has allowed us to gain new insights into the molecular nature of the process.[4-11] In addition, different secretory proteins were found to have different requirements for translocation. Some can be translocated after protein synthesis has terminated, that is, posttranslationally (e.g., prepro-α-factor[1-3]), whereas others require the presence of microsomal vesicles during translation, that is, cotranslationally (e.g., preinvertase[10]). We describe here the preparation of all components required for co- or posttranslational translocation assays.

Preparation of Reagents for Translocation Assays

In Vitro Transcription of DNA Sequences Encoding Secretory Proteins

General Comments. We have used both bacteriophage SP6 and T7 RNA polymerases to obtain transcripts which contain cap analogs at their 5' ends and are efficiently translated in several cell extracts. This has been possible owing to the ability of both polymerases to utilize an analog diguanosine 5'-triphosphate (G5'ppp5'G), of the cap structure found in the 5' end of eukaryotic mRNA, as the first nucleotide during the initiation of RNA synthesis.[12] The *in vitro* transcription conditions described below

---

[1] W. Hansen, P. D. Garcia, and P. Walter, *Cell (Cambridge, Mass.)* **45**, 397 (1986).
[2] G. Waters and G. Blobel, *J. Cell Biol.* **102**, 1543 (1986).
[3] J. A. Rothblatt and D. I. Meyer, *EMBO J.* **5**, 1031 (1986).
[4] J. A. Rothblatt, J. R. Webb, G. Ammerer, and D. I. Meyer, *EMBO J.* **6**, 3455 (1987).
[5] G. Waters, W. J. Chirico, and G. Blobel, *J. Cell Biol.* **102**, 2629 (1986).
[6] R. J. Deshaies and R. Schekman, *J. Cell Biol.* **105**, 633 (1987).
[7] H. Murakami, D. Pain, and G. Blobel, *J. Cell Biol.* **107**, 2051 (1988).
[8] W. J. Chirico, M. G. Waters, and G. Blobel, *Nature (London)* **332**, 805 (1988).
[9] R. J. Deshaies, B. D. Koch, M. Werner-Washburne, E. A. Craig, and R. Schekman, *Nature (London)* **332**, 800 (1988).
[10] W. Hansen and P. Walter, *J. Cell Biol.* **106**, 1075 (1988).
[11] M. A. Poritz, V. Siegel, W. Hansen, and P. Walter, *Proc. Natl. Acad. Sci. U.S.A.* **85**, 4315 (1988).
[12] P. A. Krieg and D. A. Melton, this series, Vol. 155, p. 397.

yield a 5- to 100-fold higher translational activity in the yeast system (depending on the translation extract and the particular DNA template used) compared to the same mRNAs made in the absence of the cap analog.

The coding sequences of several secretory proteins were subcloned in plasmid vectors containing the bacteriophage SP6 RNA polymerase promoter [pSP64 and pSP65, Promega Biotec (Madison, WI)] or the T7 RNA polymerase promoter (pGEM1 and pGEM2, Promega Biotec). In most cases, these plasmids produced transcripts active in translation. However, the 5'-untranslated sequences of some genes (e.g., bovine preprolactin) diminished the translational activity of the mRNAs drastically. We have overcome this difficulty by subcloning these genes into the pSP64T vector (a generous gift of D. Melton, Harvard University), thus replacing the natural 5'-untranslated sequences with the 5'-untranslated sequences of *Xenopus β*-globin mRNA. Plasmid templates are routinely linearized with a restriction endonuclease downstream of the protein coding sequences. By delimiting the extent of the DNA to be transcribed, a better yield of transcripts per DNA template molecule is obtained.

Transcription with Bacteriophage SP6 RNA Polymerase. This procedure was adapted from the protocol described by Krieg and Melton.[12]

Reagents. All solutions are prepared with diethyl pyrocarbonate (DEPC)-treated water, frozen in liquid nitrogen, and stored in small aliquots at $-80°$, unless indicated otherwise.

$5\times$ SP6 Transcription buffer: 200 mM Tris-HCl, pH 7.5, 30 mM MgCl$_2$, 10 mM spermidine

100 mM Dithiothreitol (DTT)

Ribonuclease inhibitor: 40,000 units/ml (Promega Biotec or Boehringer-Mannheim Biochemicals, Indianapolis, IN), store at $-20°$

DNA template, linearized plasmids at 0.5 mg/ml, store at $-20°$

SP6 bacteriophage RNA polymerase at 10–20 units/μl (Promega Biotec or Boehringer-Mannheim Biochemicals), store at $-20°$

$5\times$ Ribonucleotides mixture (rNTPs): 2.5 mM each ATP, UTP CTP, and G(5')ppp(5')G, 0.5 mM GTP [PL-Biochemicals (Milwaukee, WI) or Sigma (St. Louis, MO)]

Phenol, saturated and equilibrated to about pH 7.0 with TE buffer [10 mM Tris-HCl, pH 8.0, 1 mM EDTA (ethylenediaminetetraacetic acid)]; store at 4°

Chloroform–isoamyl alcohol: 24 parts of chloroform/1 part isoamyl alcohol; store at room temperature

3 M Sodium acetate, store at room temperature

Transcription Reactions. For a 20-μl transcription reaction, 4.5 μl of DEPC-treated water, 4 μl of the rNTPs mixture, 4 μl of the DNA template, and 1 μl of the SP6 RNA polymerase are mixed at room temperature in a sterile tube. The reaction is incubated at 40° for 60 min and terminated by the addition of 50 μl of phenol, 50 μl of chloroform–isoamyl alcohol, and 30 μl DEPC-treated water. After vortexing and centrifugation, the aqueous phase is reextracted with phenol–chloroform, followed by extraction with chloroform–isoamyl alcohol. Nucleic acids are precipitated by addition of 5 μl of 3 M sodium acetate and 150 μl of 100% ethanol. After centrifugation, the pellet is dissolved in 50 μl of DEPC-treated water, frozen in liquid nitrogen, and stored at $-80°$. Transcripts prepared and stored under these conditions remained active for more than 1 year, even after several freeze–thaw cycles.

Transcription with Bacteriophage T7 RNA Polymerase. This procedure was adapted from previously published methods.[13]

Reagents. All solutions are the same as that indicated for SP6 polymerase transcription, with the following exceptions.

10× T7 Transcription buffer: 400 mM Tris-HCl, pH 8.0, 80 mM MgCl$_2$, 50 mM DTT, 40 mM spermidine; store at $-80°$

5× T7 Ribonucleotides mixture (rNTPs): 5.0 mM each ATP, CTP, and UTP, 2.5 mM each GTP and G(5′)ppp(5′)G; store at $-80°$

Transcription Reactions. A 20-μl transcription reaction contains 8 μl of DEPC-treated water, 2 μl of 10× T7 transcription buffer, 4 μl of T7 rNTPs, 4 μl of digested DNA (0.5 mg/ml), 1 μl of RNase inhibitor, and 1 μl of T7 RNA polymerase (20 units/μl). The reaction is incubated at 37° for 1 hr, and the nucleic acids are extracted and stored as described for the SP6 transcriptions.

Preparation of Yeast Cell Extracts

General Comments. Yeast cytoplasmic translation extracts and microsomal membrane fractions competent for protein translocation are prepared separately. Each procedure is optimized starting at the cell lysis step. The procedure to obtain translation extracts was adapted from Gasior *et al.*[14] Similar procedures for the preparation of yeast microsomal vesicles were independently developed in the laboratories of Blobel[2] and Meyer.[15]

---

[13] P. Davanloo, A. H. Rosenberg, J. J. Dunn, and F. W. Studier, *Proc. Natl. Acad. Sci. U.S.A.* **81**, 2035 (1984).

[14] E. Gasior, F. Herrera, I. Sadnik, C. S. McLaughlin, and K. Moldave, *J. Biol. Chem.* **254**, 3965 (1979).

[15] J. A. Rothblatt and D. I. Meyer, *Cell (Cambridge, Mass.)* **44**, 619 (1986).

Cell Growth and Spheroplast Preparation

Media and Solutions. All solutions are stored sterile at room temperature.

YEPD Medium[16]: 10 g/liter yeast extract, 10 g/liter peptone; the solution is autoclaved and 20 g/liter of glucose is added from a 50% (w/v) autoclaved stock solution

Sorbitol buffer: 1.4 M sorbitol, 50 mM potassium phosphate, pH 7.5, 40 mM 2-mercaptoethanol or 10 mM DTT; the solution is autoclaved prior to the addition of 2-mercaptoethanol or DTT, which is added immediately before use

YM5–Mg medium: 2× YM5 medium[17] is prepared as follows: 2 g of yeast extract and 4 g of peptone are dissolved in 500 ml of distilled water, and the solution is autoclaved. Adenine and uracil are added from individually sterilized 100× solutions to a final concentration of 10 mg/liter. From sterile 10× solutions yeast nitrogen base is added to 6.7 g/liter, succinic acid to 10 g/liter, sodium hydroxide to 6 g/liter, and glucose to 10 g/liter. Sterile water is added to 1 liter, and the medium is stored at 4° for up to 2 weeks. Five hundred milliliters of YM5–Mg medium is prepared by mixing 250 ml of the 2× YM5 solution, 200 ml of sterile 1 M MgSO$_4$, and 50 ml of sterile water.

Procedure. Saccharomyces cerevisiae strains which contain reduced levels of vacuolar proteases because of point mutations[18] or deletions in the *PEP4* gene[19] are used as the starting material for both cytoplasmic and microsomal extracts. For both preparations, the cells are grown in 4 liters of YEPD medium to a density of 1–2 OD$_{600}$/ml. Cells are collected by centrifugation (5 min, 3000 rpm, Sorvall RC-5B centrifuge). The pelleted cells are washed with ice-cold distilled water and resuspended in 200 ml of the sorbitol buffer at room temperature. Zymolyase 5000 (Kirin Brewery, Tokyo, Japan) powder is directly dissolved in the cell suspension to a final concentration of 50 µg/ml, and the suspension is incubated at room temperature for 1 hr. The efficiency of spheroplast formation is monitored by measuring the decrease in turbidity, caused by cell lysis, after dilution of a small aliquot of the suspension with 100 volumes of distilled water. The spheroplasts are harvested by centrifugation as indicated before and al-

---

[16] R. Mortimer and D. Hawthorne, *in* "The Yeast" (A. Rose and J. Harrison, eds.), Vol. 1, p. 385. Academic Press, New York, 1969.
[17] L. Hartwell, *J. Bacteriol.* **93**, 1662 (1967).
[18] B. Hemmings, G. Zubenko, A. Haslik, and E. Jones, *Proc. Natl. Acad. Sci. U.S.A.* **78**, 435 (1981).
[19] G. S. Payne, T. B. Hasson, M. S. Hasson, and R. Schekman, *Mol. Cell. Biol.* **7**, 3888 (1987).

lowed to resume growth by resuspending them in 500 ml of YM5–Mg medium followed by incubation at room temperature for 90 min with gentle shaking. The spheroplast culture is transferred to an ice-water bath and allowed to cool for 10 min. All further procedures are performed at 4°. The spheroplast culture is underlayed in a 1-liter centrifuge bottle with 200 ml of ice-cold sorbitol buffer. Spheroplasts are collected by centrifugation as indicated above and washed once with 200 ml of ice-cold sorbitol buffer. After the wash, the pelleted spheroplasts are immediately resuspended in the lysis buffer corresponding to the particular extract preparation.

Preparation of Extracts for in Vitro Protein Synthesis

Solutions. The lysis and gel filtration buffers are made immediately before use. The other solutions are stored as indicated.

> Lysis buffer: 20 mM HEPES (N-2-hydroxyethylpiperazine-N'-2-ethanesulfonic acid)/KOH, pH 7.5, 100 mM potassium or ammonium acetate, pH 7.5, 2 mM magnesium acetate, 2 mM DTT, 0.5 mM phenylmethylsulfonyl fluoride (PMSF)
> Gel filtration buffer: 20 mM HEPES/KOH, pH 7.5, 100 mM potassium or ammonium acetate, pH 7.5, 2 mM magnesium acetate, 2 mM DTT, 0.5 mM PMSF, 20% (v/v) glycerol
> 1 M Calcium chloride, store at room temperature
> Micrococcal nuclease, 10,000 units/ml; store at $-20°$
> 200 mM EGTA [ethylene glycol bis(β-aminoethyl ether) $N,N,N'N'$-tetraacetic acid], adjusted with NaOH to pH 8.0; store at room temperature

Procedure. Spheroplasts from a 4-liter culture are resuspended in 15 ml of lysis buffer, and the suspension is homogenized with 10 strokes in a motor-driven Potter homogenizer. The homogenate is centrifuged at 27,000 g for 15 min in a Beckman Ti 50.2 rotor at 4°, and the supernatant is collected and centrifuged for an additional 30 min at 100,000 g in the same rotor. The resulting supernatant (typically 12–15 ml) contains mainly soluble cytoplasmic components. This extract is passed over an 80-ml Sephadex G-25 column equilibrated in gel filtration buffer. Fractions of 3 ml are collected. The void volume fractions with absorbances of 20 A_{260}/ml or higher are pooled (typically 9–12 ml). Calcium chloride and micrococcal nuclease are added to final concentrations of 1 mM and 300 units/ml, respectively. The extract is transferred to a 20° water bath and incubated for 15 min. The nuclease digestion is terminated by the addition of EGTA to a final concentration of 2 mM, and the extract is transferred to an ice-water bath. The extract is cleared by centrifugation at 15,000 rpm in a Beckman JA20 rotor for 10 min at 4°, and the supernatant is frozen in liquid nitrogen in 100-μl aliquots. These aliquots can be stored at −80° for up to 2 years without loss of activity.

Preparation of Microsomes Active in Protein Translocation

Solutions. All solutions are made immediately before use.

Lysis buffer: 20 mM HEPES/KOH, pH 7.5, 500 mM sucrose, 1 mM DTT, 3 mM magnesium acetate, 1 mM EGTA, 1 mM EDTA, 100 units/ml aprotinin, 0.5 mM PMSF, 2 μg/ml each pepstatin A, chymostatin, antipain, and leupeptin (Sigma)

EDTA buffer: 20 mM HEPES/KOH, pH 7.5, 250 mM sucrose, 50 mM EDTA, 1 mM DTT

Membrane buffer: 20 mM HEPES/KOH, pH 7.5, 250 mM sucrose, 1 mM DTT

Percoll cushion solution: 30% (v/v) Percoll (Pharmacia, Piscataway, NJ) in lysis buffer

Procedure. Spheroplasts prepared from a 4-liter culture are resuspended in 15 ml of lysis buffer and homogenized with 10 strokes in a motor-driven Potter homogenizer. Lysis is checked by light microscopy. The homogenate is centrifuged in half-filled tubes in a Beckman JS-13 swinging-bucket rotor at 8000 rpm for 10 min, and the resulting supernatant is collected. The pellet is resuspended in another 15 ml of the lysis buffer, homogenized again, and centrifuged under the same conditions. Both supernatants should contain about the same amount of microsomal vesicles. They are pooled and recentrifuged as above. Five-milliliter fractions of this supernatant (total volume of 25–30 ml) are transferred into 24-ml centrifuge tubes and are underlayed with 18 ml of the Percoll cushion solution. The samples are centrifuged at 29,000 rpm (76,000 g) for 1 hr in a Beckman Ti 50.2 rotor. Two turbid bands are reproducibly observed in the Percoll gradient, of which the upper band contains the endoplasmic reticulum vesicles.[1] These bands are collected and pooled, resulting in about 20–25 ml of a crude microsomal fraction.

Calcium chloride and micrococcal nuclease are added to a final concentrations of 1 mM and 300 units/ml respectively, and the solution is incubated at 20° for 15 min. The nuclease reaction is stopped by the addition of EGTA to a final concentration of 2 mM; the sample is transferred to an ice-water bath and diluted with 1-volume of ice-cold EDTA solution. After 15 min at 0°, the extract is distributed into several tubes and centrifuged at 43,000 rpm in a Beckman Ti 70 rotor for 1 hr. Under these conditions, the remaining Percoll from the previous gradient forms a transparent pellet, and the microsomal membranes form a compact band above it. The membrane band is collected with a Pasteur pipette and resuspended in 0.5 ml of membrane buffer using a Dounce homogenizer. A 10-μl aliquot of this suspension is diluted into 1 ml of 1% SDS, and the absorbance at 280 nm is determined. The final concentration of the microsomal preparation is adjusted to 25 A_{280}/ml by the addition of membrane buffer. Aliquots are frozen in liquid nitrogen and stored at $-80°$.

Protein Translation and Translocation Assays

Translocation of proteins across the yeast microsomal membranes can be assayed both co- and posttranslationally,[1-3] depending on whether the microsomal membranes are added before or after synthesis of the secretory protein. The following protocol allows the use of the same salts compensation buffers (see below) for both types of assays.

Solutions. Small aliquots of all solutions are stored frozen at $-80°$, unless indicated otherwise. Aliquots are thawed slowly in an ice-water bath before use. The aliquots can be refrozen (including translation and microsomal extracts) in liquid nitrogen up to 4 times without any apparent loss of activity.

E-Mix: 5.0 mM ATP, 0.4 mM GTP, 87.5 mM creatine phosphate, 1.0 mM each of the 19 amino acids (excluding methionine), 11.0 mM putrescine, 0.5 mg/ml yeast tRNA, 1.0 mg/ml creatine phosphate kinase, 2.5 mCi/ml [^{35}S] methionine (1000 Ci/mmol), 1,000 units/ml ribonuclease inhibitor

Yeast translation extract, prepared as indicated above

Yeast microsomal membranes, prepared as indicated above

mRNA, usually an *in vitro* transcript prepared as indicated above; natural mRNAs should be titrated for optimal translational efficiency

Compensation buffer: 100 mM HEPES/KOH, pH 7.5, 1.2 M potassium acetate, pH 7.5, 25 mM MgCl$_2$, 16 mM DTT, 80 mM S-adenosylmethionine

Yeast microsomal membrane buffer, prepared as indicated above

30% (w/v) Trichloroacetic acid (TCA), store at room temperature

Procedure. The following components are mixed in an ice-water bath: 3 μl DEPC-treated distilled water, 6 μl yeast translation extract, 2 μl compensation buffer, 1 μl mRNA solution, 4 μl E-mix, and 4 μl of microsomal membrane buffer, if translations are made in the absence of microsomal membranes, or 4 μl of yeast microsomal membranes, if protein translocation is to be assayed in the presence of ongoing protein synthesis. The components are mixed and the reactions incubated at $20°$ for 1 hr. The reactions are stopped by transfer to an ice-water bath, and 20 μl of the TCA solution is added if no further test is desired. When protein translocation is assayed posttranslationally 4 μl of microsomal membranes and 1 μl of 25 mM cycloheximide are added after 1 hr of translation. The incubation at 20° is continued for another 1 hr, and 25 μl of the TCA solution is added to stop the reaction. After centifugation, the protein pellets are dissolved in sample buffer and subjected to denaturing electrophoresis followed by autoradiography of the gel.

In a typical experiment, translation products are visible as prominent bands after 6 to 16 hr of exposure in the absence of fluorography. Translocation across the microsomal membranes is assessed by the shift in apparent molecular weight caused by either signal peptidase cleavage and/or core glycosylation of the translation products. At the indicated membrane concentration, about 60–80% translocation is obtained for prepro-α-factor in a co- or posttranslational reaction.[1-3] In contrast, less than 10% of preinvertase is translocated, and translocation is only observed cotranslationally.[10] All translocated products are found protected from externally added proteases under standard assay conditions,[1-3] owing to their location within sealed microsomal vesicles.

[47] Analysis of Glycoproteins from *Saccharomyces cerevisiae*

By P. ORLEAN, M. J. KURANDA, and C. F. ALBRIGHT

Introduction

Many *Saccharomyces cerevisiae* proteins are modified with carbohydrate; proteins that have entered the secretory pathway can be modified by the attachment of oligosaccharides to asparagine residues (N-glycosylation) or to serine or threonine residues (O-mannosylation). By analogy with higher eukaryotic cells, cytoplasmic and nuclear proteins may also be modified by the attachment of GlcNAc to serine or threonine residues. In this chapter we present several methods to analyze proteins that are potentially N-glycosylated or O-mannosylated. We expect that some of these techniques will also be applicable to the analysis of proteins in the cytoplasm and nucleus which potentially contain O-linked GlcNAc residues.

N-Glycosylation begins in the lumen of the endoplasmic reticulum (ER) with the transfer of a preassembled core oligosaccharide from a lipid-linked intermediate [dolichol pyrophosphate (Dol-PP)-GlcNAc$_2$Man$_9$Glc$_3$] to acceptor asparagine residues in the sequence Asn-X-Ser/

Thr.[1] The N-linked oligosaccharide is then modified by the action of mannosidases and mannosyltransferases. Analyses of bulk cell wall mannoprotein and of purified proteins showed that the resulting structures can be divided into two broad classes based on size: $Man_{8-14}GlcNAc_2$ and $Man_{>50}GlcNAc_2$. The latter structures are formed by extensive addition of mannose in the Golgi apparatus.[2-4] An Asn-X-Ser/Thr site in a given protein need not necessarily be used; studies with invertase indicate that in a given yeast N-glycosylated protein, some potential acceptor asparagines are always glycosylated, some receive carbohydrate in only a proportion of molecules of a given protein, and some are never glycosylated.[3]

O-Mannosylation is also initiated in the ER lumen with the transfer of a mannose residue from Dol-P-Man to serine or threonine in nascent protein. Further mannose residues are attached to this monosaccharide, yielding a linear oligosaccharide chain of 2 to 5 mannose residues.[1] These types of structures have been found on bulk cell wall mannoprotein,[2] a cell surface agglutinin,[5] chitinase,[6] and human insulin-like growth factor expressed in yeast.[7] The signals to initiate O-mannosylation are not known; the only obvious feature of known O-mannosylated yeast proteins is the presence of unusually large amounts of serine and threonine residues.[6]

The type and extent of glycosylation can be assessed by perturbing glycosylation in various ways that affect the apparent size of the protein. For example, apparent protein size can be decreased by treatment of yeast cells with the antibiotic tunicamycin, which inhibits N-glycosylation, or by treatment of the protein with enzymes such as endoglycosidase H (Endo-H), which specifically cleaves N-linked carbohydrate chains. Further, the extent of glycosylation can change as a result of defects in the glycosylation apparatus; some mutations, such as *sec53* and *dpm1*, affect the formation of oligosaccharide precursors, while others, such as *sec18*, trap secretory proteins in the endoplasmic reticulum, thereby preventing the action of processing enzymes later in the secretory pathway.

The carbohydrate moieties of glycoproteins can also be analyzed. One approach involves labeling yeast glycoproteins with radioactive sugars. The

---

[1] W. Tanner and L. Lehle, *Biochim. Biophys. Acta* **906**, 81 (1987).
[2] C. E. Ballou, "The Molecular Biology of the Yeast *Saccharomyces:* Metabolism and Gene Expression." Cold Spring Harbor Laboratory, Cold Spring Harbor, New York, 1982.
[3] V. A. Reddy, R. S. Johnson, K. Biemann, R. S. Williams, F. D. Ziegler, R. B. Trimble, and F. Maley, *J. Biol. Chem.* **263**, 6978 (1988).
[4] F. D. Ziegler, F. Maley, and R. B. Trimble, *J. Biol. Chem.* **263**, 6986 (1988).
[5] P. Orlean, H. Ammer, M. Watzele, and W. Tanner, *Proc. Natl. Acad. Sci. U.S.A.* **83**, 6263 (1986).
[6] M. J. Kuranda, *Mol. Cell. Biol.*, submitted.
[7] K. Hard, W. Bitter, J. P. Kamerling, and J. F. G. Vliegenthart, *FEBS Lett.* **248**, 111 (1989).

labeled oligosaccharides can then be cleaved from the protein and analyzed by chromatography.

We illustrate these techniques using invertase and chitinase as model glycoproteins. Invertase, the product of the *SUC2* gene, is a model N-glycosylated protein; it is extensively glycosylated and contains on average 9 or 10 carbohydrate chains attached at Asn-X-Ser/Thr sites.[3,4] Chitinase, the product of the *CTS1* gene, is a model O-mannosylated protein which is modified with short mannose chains (Man_2-Man_5). A 170-amino acid region of the peptide near the carboxyl terminus of the protein contains more than 50% serine and threonine and includes the majority of the potential glycosylation sites.[6]

Analysis of Glycosylation State of Proteins

In this section, we describe procedures to determine whether a protein contains N-linked or O-linked oligosaccharides and ways to estimate the number of N-linked chains per protein molecule. These procedures require a means to isolate the protein of interest, and we assume that either specific antiserum or a purification scheme is available. If known, the amino acid sequence of a protein is first examined to see whether it contains any Asn-X-Ser/Thr sites which could potentially be N-glycosylated. If one or more are present, we begin our analysis by investigating the occurrence and extent of N-glycosylation. This is done by treating the protein with Endo-H or isolating the protein synthesized in cells that have been treated with tunicamycin. A decrease in apparent protein size following both treatments indicates the native protein is N-glycosylated. Endo-H specifically cleaves normally processed N-linked oligosaccharides in yeast between the GlcNAc residues and will remove all such chains from SDS-denatured glycoproteins, while tunicamycin prevents N-glycosylation by blocking the synthesis of the lipid-linked precursor oligosaccharide.

Since there are neither specific glycosidases that remove O-linked mannooligosaccharide chains nor inhibitors of O-mannosylation, we use temperature-sensitive yeast mutants to infer the presence of O-linked oligosaccharides. Phosphomannomutase is defective in the *sec53* mutant,[8] and consequently the formation of GDPmannose and lipid-linked precursor saccharides is impaired. Proteins synthesized in these cells at the restrictive temperature typically contain 5–10% of the normal level of core N-linked oligosaccharides[9,10] and little, if any, O-linked mannose. Dolichol phos-

[8] F. Kepes and R. Schekman, *J. Biol. Chem.* **263**, 9155 (1988).
[9] S. Ferro-Novick, P. Novick, C. Field, and R. Schekman, *J. Cell Biol.* **98**, 35 (1984).
[10] D. Julius, R. Schekman, and J. Thorner, *Cell (Cambridge, Mass.)* **36**, 309 (1984).

phate mannose synthase (EC 2.4.1.83) is defective in *dpm1* cells. Proteins synthesized in these cells at the restrictive temperature receive truncated N-linked saccharides ($Man_5GlcNAc_2$) and very little O-linked mannose.[11] No alteration in electrophoretic mobility by tunicamycin or Endo-H treatment and an alteration by the *sec53* and *dpm1* mutations are preliminary indications of O-mannosylation. The presence of O-linked mannose can be confirmed by direct analysis of the carbohydrate, as outlined in the following section.

Mobility Shifts of N-Glycosylated Protein Invertase

When yeast cells are induced to synthesize invertase, three classes of invertase protein are seen: (1) a cytoplasmic form, (2) an ER form, and (3) a secreted form (Fig. 1a). The cytoplasmic form (60 kDa), which is produced constitutively, lacks a signal sequence and is consequently unglycosylated. The ER form is a mixture of proteins (80–86 kDa[12]) which differ in the number of core oligosaccharide chains that are attached. The secreted form of the protein is heterogeneous in size (100–150 kDa), resulting from the elongation of core oligosaccharide chains in the Golgi apparatus. Treatment with Endo-H removes the bulk of the N-linked carbohydrate, leaving only single GlcNAc residues attached to asparagine; the Endo-H-treated form shows a mobility slightly slower than that of the cytoplasmic form (about 63 kDa; Fig. 1a). Tunicamycin prevents N-glycosylation completely, resulting in an invertase protein with the same mobility as the cytoplasmic form. In *sec53* cells, addition of oligosaccharides is virtually abolished at 37°;[9] consequently, the protein shows a mobility only slightly lower than that of the cytoplasmic form. In *dpm1* cells at 37°, a truncated oligosaccharide ($Man_5GlcNAc_2$) is transferred to the protein (Fig. 1a). Although this structure can serve as an alternative acceptor for outer chain extension,[12] most of the truncated chains transferred to invertase in the *dpm1* strain at 37° are not in fact extended, and most of the invertase made has an apparent size between 66 and 69 kDa. Endo-H has only a slight effect on invertase made in the *dpm1* strain since the truncated chains formed are Endo-H-resistant; if outer chains are attached to these structures, they remain Endo-H-resistant.[12] The small mobility shift on Endo-H treatment presumably reflects the facts that, in the mutant, a small amount of $Dol-PP-GlcNAc_2Man_9Glc_3$ is made and that a few of these chains are transferred to protein, yielding Endo-H-sensitive structures.

[11] P. Orlean, *Mol. Cell. Biol.*, in press (1990).
[12] T. C. Huffaker and P. W. Robbins, *Proc. Natl. Acad. Sci. U.S.A.* **80**, 7466 (1983).

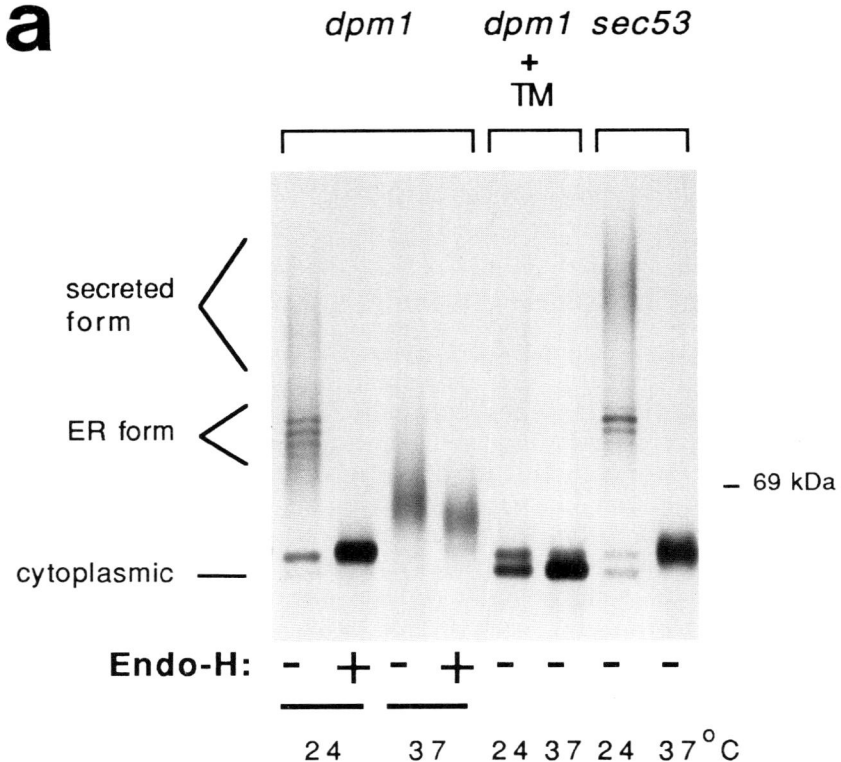

FIG. 1. (a) Response of invertase to conditions affecting glycosylation. The electrophoretic mobility of invertase was examined in the temperature-sensitive glycosylation mutants *dpm1* and *sec53*. Invertase synthesis was induced at 24°, and yeast proteins were then radiolabeled with [^{35}S]methionine. Radiolabeled invertase was purified by immunoprecipitation and analyzed by denaturing electrophoresis in gels containing 7% acrylamide. For radiolabeling at the restrictive temperature, the strains were shifted to 37° for 15 min before the addition of radiolabel. For tunicamycin treatment, the drug was added to cultures 5 min before the addition of radiolabel. Endo-H digestions were carried out after immunoprecipitation. The normal glycosylated forms of invertase are represented by the forms radiolabeled in the *dpm1* and *sec53* strains at 24°. (The 63-kDa band radiolabeled at 24° is not due to invertase for it was also immunoprecipitated from extracts of cells in which the *SUC2* gene had been deleted. The cytoplasmic form of invertase was also radiolabeled in the *dpm1* mutant at 37° and was visible after longer exposure of the fluorograph.) (b) Response of chitinase to conditions affecting glycosylation. The chitinase protein was immunoprecipitated from the extracts used in Fig. 1a. The positive control for both treatments was invertase, immunoprecipitated from the same cells. The normal glycosylated form of chitinase is represented by the forms radiolabeled in the *dpm1* and *sec53* strains at 24°.

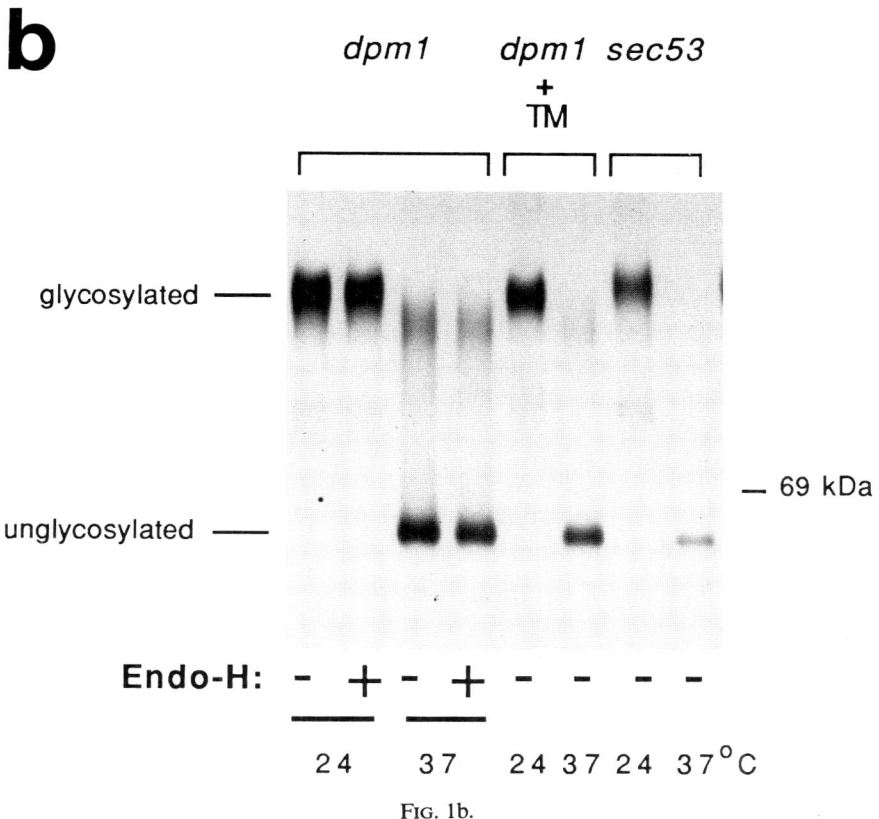

FIG. 1b.

Mobility Shifts of O-Mannosylated Protein Chitinase

Fully glycosylated chitinase migrates with an apparent size of 130 kDa, and neither Endo-H treatment nor incubation of the cells in the presence of tunicamycin resulted in any shift in molecular mass since there are no N-linked oligosaccharides (Fig. 1b). Tunicamycin treatment was short (we use a 5-min preincubation), since prolonged incubation causes an inhibition of the incorporation of mannose into O-linked carbohydrate.[13] Indeed, this is seen in the *dpm1* mutant, which is slightly leaky; some higher molecular mass material is made at the restrictive temperature, but in the presence of tunicamycin less of this material is made. Chitinase radiolabeled in both the *sec53* and *dpm1* mutants at the restrictive temperature

[13] P. Orlean and W. Tanner, in "Microbial Cell Wall Synthesis and Autolysis." Elsevier, Amsterdam, 1984.

shows a large mobility increase and migrates with an apparent size of 60 kDa, the size predicted for the unglycosylated protein (Fig. 1b). This is the expected result, since both mutations prevent synthesis of Dol-P-Man, the donor of the first mannose residue that is O-linked to protein. The exact contribution of O-linked mannose chains to the apparent molecular weight of chitinase is hard to estimate because extensive O-mannosylation may result in aberrant electrophoretic mobility. This may be the case with the *KEX2* protein.[14]

Estimation of Number of N-Linked Chains on Glycoprotein

Once it is established that a protein contains N-linked carbohydrate, it is possible to estimate the number of N-linked chains per protein molecule; this can be done using Endo-H and the *sec18* mutant.[15] N-Linked oligosaccharides made in *sec18* cells at the restrictive temperature have the common structure $Man_8GlcNAc_2$. This homogeneity allows the number of N-linked chains to be estimated, since each core oligosaccharide contributes about 2000 to the molecular weight. Alternatively, partial digestions with Endo-H of the protein synthesized in *sec18* cells at the restrictive temperature can generate a ladder of polypeptides, each differing in size by a single N-linked oligosaccharide chain. A similar glycosylation ladder generated *in vivo* can sometimes be seen when invertase is biosynthetically labeled (Fig. 2). An indication of the amount of mannose added during later stages along the secretory pathway is obtained by comparing the sizes of the major intermediates accumulating at the *sec18* block with that of the fully processed protein.

Alternative Techniques

The enzyme *N*-glycanase (peptide *N*-glycosidase F) cleaves N-linked chains between the asparagine and GlcNAc residues,[16,17] and it has been used as an alternative to Endo-H in at least one study.[18] Further, treatment with anhydrous HF, which cleaves O-glycosidic bonds between neutral sugars, would remove the bulk of both N- and O-linked carbohydrate. Deglycosylated protein was subsequently recovered in one case,[19] but in another study the treatment degraded the protein.[20]

[14] R. S. Fuller, A. Brake, and J. Thorner, *Proc. Natl. Acad. Sci. U.S.A.* **86,** 1434 (1989).
[15] R. Schekman and P. Novick, in "The Molecular Biology of the Yeast *Saccharomyces:* Metabolism and Gene Expression." Cold Spring Harbor Laboratory, Cold Spring Harbor, New York, 1982
[16] F. K. Chu, *J. Biol. Chem.* **261,** 172 (1986).
[17] A. Haselbeck, and W. Hosel, *Topics Biochem. (Boehringer, Mannheim)* **8** (1988).
[18] R. I. Feldman, M. Bernstein, and R. Schekman, *J. Biol. Chem.* **262,** 9332 (1987).
[19] M. Watzele, F. Klis, and W. Tanner, *EMBO J.* **7,** 1483 (1988).
[20] J. Frevert and C. E. Ballou, *Biochemistry* **24,** 753 (1985).

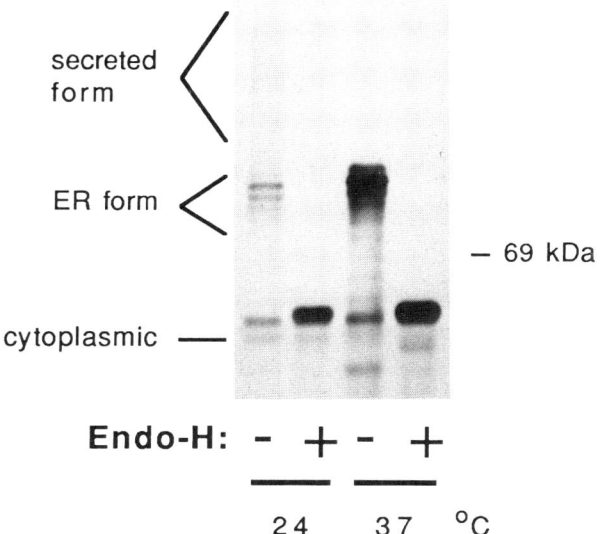

FIG. 2. Invertase synthesized in the *sec18* mutant. The conditions used in Fig. 1a were used to examine invertase synthesized in *sec18* cells at 24° and 37°.

Certain lectins, such as concanavalin A, will bind carbohydrates, and kits that are based on the use of lectins, or which use carbohydrate-specific antibodies, are becoming available commercially (Boehringer-Mannheim, Indianapolis, IN; Genzyme Corporation, Boston, MA). These could be used to detect the presence of carbohydrate on immunoprecipitated or purified proteins.

Analysis of Carbohydrate

The procedures described in the previous section allow us to obtain a reasonable estimate of the extent of N-glycosylation of a protein and the average number of chains per molecule. Further analysis of the N-linked carbohydrate would involve determining which Asn-X-Ser/Thr sites are glycosylated, the size of the attached chains, and how often a site is used in a population of molecules of a given protein. Such analysis has been performed on invertase and involved cleavage of the glycoprotein into peptides, separation and identification of those that contain carbohydrate, and analysis of the composition and structure of the individual oligosaccharides. Such a complex analysis, however, is beyond the scope of this chapter, and the reader is referred to the studies by Trimble and co-workers.[3,4]

The methods we presented earlier allow the presence of O-linked carbohydrate to be inferred. In the absence of specific inhibitors of O-mannosylation and glycosidases specific for O-linked mannooligosaccharides, it is necessary to demonstrate directly the presence of O-linked sugars. In this section, we present a protocol for analysis of the O-linked mannooligosaccharides on an individual protein. This procedure could be adapted for the study of proteins suspected to contain O-linked GlcNAc residues. The first step in this analysis is specific radiolabeling of the carbohydrate portion of the protein, followed by chemical cleavage of the O-linked sugars and analysis of the released oligosaccharides.

Radiolabeling Carbohydrate Portion of Yeast Glycoproteins: General Considerations

The carbohydrate of N-glycosylated proteins can be specifically radiolabeled with mannose and glucosamine, and the carbohydrate on O-mannosylated proteins can be labeled only with mannose. O-Linked GlcNAc-containing proteins could be radiolabeled with glucosamine. [2-^{3}H] Mannose (typical specific activity 14 Ci/mmol) is used since the label is lost as tritiated water if mannose is catabolized. To label GlcNAc *in vivo*, [1-^{14}C]glucosamine (typical specific activity 54 mCi/mmol) is used as precursor;[21] this is rapidly converted to UDPGlcNAc *in vivo*. Although glucosamine is not catabolized, about two-thirds of this sugar is incorporated into chitin, a cell wall polysaccharide. Both mannose and glucosamine enter yeast cells via the glucose transport systems. Therefore, to obtain a reasonable concentration of radiolabel within the cell, the amount of glucose in the medium must be kept to the minimum required to support cell growth. Longer radiolabeling can be done in the presence of 0.5% sucrose using a concentration of [2-^{3}H]mannose of 0.1–1.0 mCi/ml, while pulse labeling is done in 0.05–0.1% glucose with higher concentrations of radiolabel (2.5–20 mCi/ml[22]). Radiolabeling with [1-^{14}C]glucosamine can be done at at least 2 μCi/ml,[21] and the rate of glucosamine uptake is reported to be increased at pH 7 to 8.[23] The K_m for even the "high affinity" monosaccharide transport system is relatively high at 1.0 mM for glucose,[24] and the K_m reported for mannose uptake is 27 mM.[25] Therefore, the intracellular concentration of precursor and the amount of radiolabel in glycoproteins will increase in proportion to the amount of radiolabel in the medium for all practical concentrations of radiolabeled precursor.

[21] P. Orlean, E. Arnold, and W. Tanner, *FEBS Lett.* **184**, 313 (1985).
[22] T. C. Huffaker and P. W. Robbins, *J. Biol. Chem.* **257**, 3203 (1982).
[23] M. Burger and L. Hejmova, *Folia Microbiol. (Prague)* **6**, 80 (1961).
[24] L. F. Bisson and D. G. Fraenkel, *Proc. Natl. Acad. Sci. U.S.A.* **80**, 1730 (1983).
[25] A. Kotyk, *Folia Microbiol. (Prague)* **12**, 121 (1967).

Analysis of O-Linked Mannooligosaccharides from Chitinase

We illustrate the analysis of an O-mannosylated protein using chitinase as a model protein and an electroblotting technique developed in this laboratory.[6] Analysis of chitinase suggested the presence of O-linked mannose, and this possibility was supported by the fact that the protein could be radiolabeled with [2-^{3}H]mannose. When *S. cerevisiae* is radiolabeled in rich medium, [2-^{3}H]mannose-labeled chitinase is secreted into the medium, and some is also found associated with the cells.[6] Cell-associated chitinase was purified by immunoprecipitation, and the secreted form by adsorption to chitin. When the purified proteins were separated by sodium dodecyl sulfate–polyacrylamide gel electrophoresis (SDS–PAGE), transferred to nitrocellulose, and visualized with tritium-sensitive film, a major band of fully glycosylated chitinase was observed in each case (Fig. 3). To characterize the oligosaccharides further, regions of the nitrocellulose filter corresponding to the [2-^{3}H]mannose-labeled chitinase were excised using the autoradiogram as a template, and the filter-bound chitinase was treated with mild alkali. This treatment, β-elimination, renders the O-linked oligosaccharides soluble by cleaving the peptide–oligosaccharide bond; the protein, however, is degraded during β-elimination, hence the method cannot be used to prepare deglycosylated protein. Analysis of the released oligosaccharides by paper chromatography revealed oligosaccharides 2 to 5 resi-

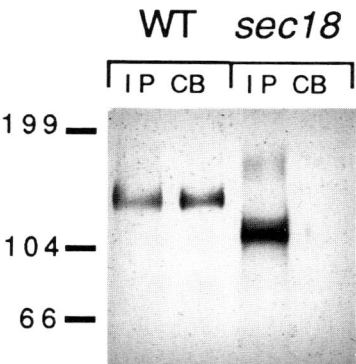

FIG. 3. Autoradiogram of [2-^{3}H]mannose-labeled chitinase transferred to nitrocellulose. Wild-type (DBY1315) and *sec18* cells were radiolabeled with [2-^{3}H]mannose at 37°. Cell-associated chitinase was extracted from the radiolabeled cells and isolated by immunoprecipitation (IP), and secreted chitinase was isolated from the culture medium by binding to chitin (CB). The [2-^{3}H]mannose-labeled proteins were separated by denaturing electrophoresis, then transferred to a nitrocellulose filter by electroblotting, after which the radiolabeled forms of chitinase were visualized using tritium-sensitive film. The *sec18* strain secreted fully glycosylated chitinase at 24°, but chitinase secretion into the medium was completely blocked at 37°.

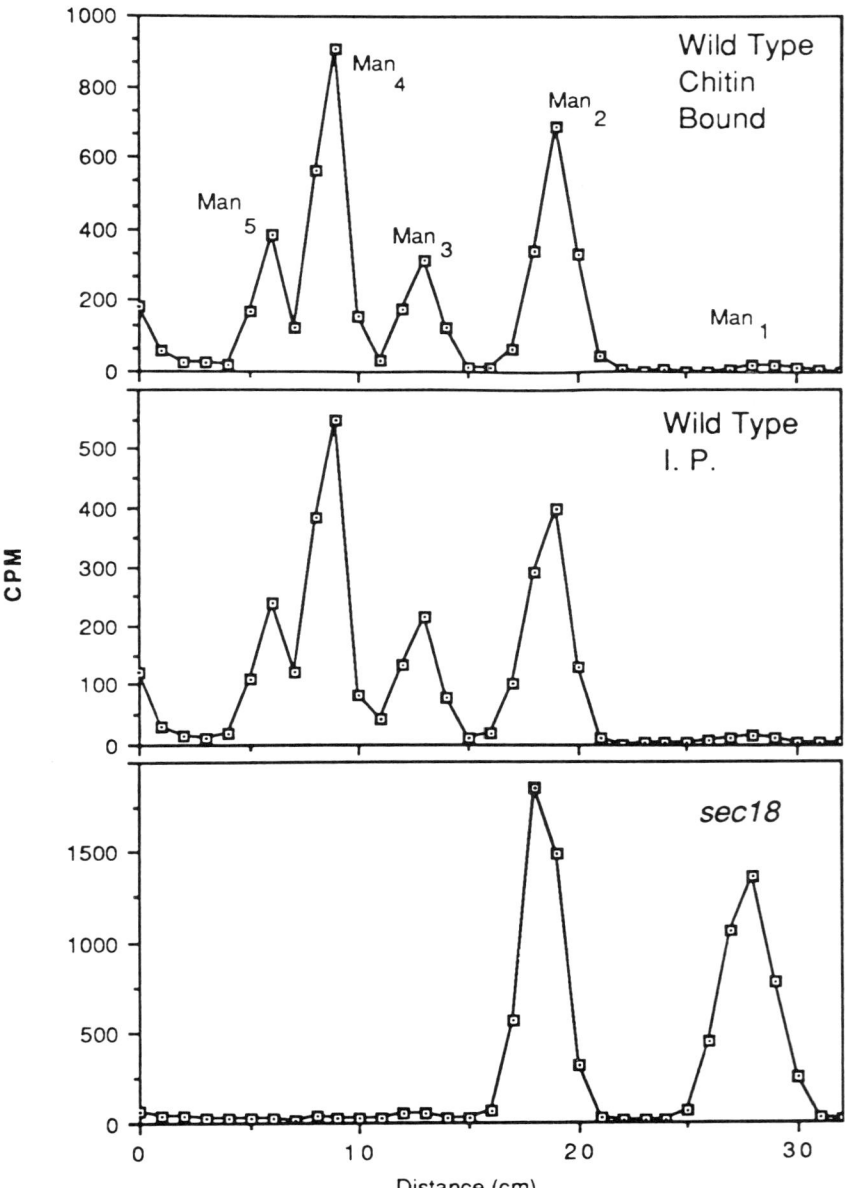

FIG. 4. Mannooligosaccharides released from chitinase by β-elimination. Regions of the nitrocellulose filter from Fig. 3 containing radiolabeled chitinase were cut out using the autoradiogram as template, and the pieces were treated with 0.1 M NaOH. All filter-bound

dues in size (Fig. 4). Similar profiles are observed for oligosaccharides released by β-elimination of total cell wall material and for other known O-mannosylated proteins.[5,26] Any glycopeptides or N-linked oligosaccharides remain at the origin of the chromatogram in the solvent system used. In the case of chitinase, very little radioactivity remains at the origin of the chromatogram, indicating that chitinase contains mainly, if not exclusively, O-linked carbohydrate. Digestion of β-eliminable oligosaccharides with α-mannosidase yielded only mannose, confirming that the oligosaccharides contain only mannose. As a further test of this technique, we analyzed the O-linked mannose chains of chitinase synthesized in *sec18* cells at the restrictive temperature. Under these conditions, a lower molecular weight form accumulates in the cells (Fig. 3), and no chitinase is recovered from the medium. β-Elimination of this material released only mannose and mannobiose (Fig. 4). The decreased chain length is due to the fact that extension of these saccharides cannot take place in this mutant, and this is consistent with the smaller apparent molecular weight of the chitinase produced under these conditions.

Methods

Materials. D-[2-^{3}H]Mannose (14 Ci/mmol) was from American Radiolabeled Chemicals (St. Louis, MO), and Tran[^{35}S] label, used for methionine labeling, was from ICN Radiochemicals (Costa, Mesa, CA). ^{3}H-Sensitive film (Ultrofilm, LKB/Pharmacia Piscataway, NJ) was from Cambridge Instruments (Deerfield, IL) (^{3}H-sensitive film is also offered by Amersham, Arlington Heights, IL). Nitrocellulose (Grade BA83, 0.2-μm pore size) was from Schleicher and Schuell (Keene, NH). Protein A-Sepharose was supplied by LKB/Pharmacia.

Strains. sec18 and *sec53* have been described.[9,15] The *dpm1* strain [*dpm1::LEU2 trp1 ura3-52 lys2,* harboring pDM8 (a derivative of YCp410[27] containing the *DPM1* gene] was obtained by *in vitro* mutagenesis of the *DPM1* gene on pDM8, followed by selection of temperature-

[26] A. Haselbeck and W. Tanner, *FEBS Lett.* **158,** 335 (1983).
[27] H. Ma, S. Kunes, P. J. Schatz, and D. Botstein, *Gene* **58,** 201 (1987).

---

radioactivity was solubilized by this treatment. The solubilized material was neutralized, then submitted to paper chromatography to separate the β-eliminable mannooligosaccharides. (Top) β-Eliminable oligosaccharides from chitinase isolated from the culture supernatant of wild-type cells by adsorption to chitin. (Center) Oligosaccharides from cell-bound chitinase extracted from the radiolabeled cells and isolated by immunoprecipitation. (Bottom) β-Eliminable mono- and disaccharides from cell-associated chitinase radiolabeled in *sec18* at 37°.

sensitive alleles using the plasmid shuffling technique.[11,28] The three strains were transformed with the plasmid pRB58, which contains the *SUC2* gene.[29] Strain DBY1315[30] was also used for [2-^{3}H]mannose labeling of chitinase.

Reagents and Media

Tunicamycin: A stock solution (1-2 mg/ml) is made by dissolving the drug (from Calbiochem, San Diego, CA) in 20 μl of 0.1 *M* NaOH, then diluting the solution with water and adding HCl to give a pH of 8.0

Endo-H: This enzyme was prepared in our laboratory but is also commercially available

Endo-H buffer: 50 m*M* sodium citrate, pH 5.5, containing 10 m*M* NaN$_3$

α-Mannosidase: Jack bean mannosidase (Sigma) is desalted using a Centricon 30 microconcentrator (Amicon, Danvers, MA) and taken up in 50 m*M* sodium citrate, pH 4.5

Phosphate-buffered saline (PBS): 0.8% NaCl, 0.02% KCl, 0.12% Na$_2$HPO$_4$, 0.02% KH$_2$PO$_4$ (all w/v)

Breakage buffer: PBS containing 0.5% (w/v) SDS and 0.1% (v/v) 2-mercaptoethanol

Solubilizing buffer: 10 m*M* Tris-HCl, pH 6.8, containing 1% (w/v) SDS and 2% (v/v) 2-mercaptoethanol

Chitin: Purified chitin powder (Sigma) is heated in 1% SDS containing 1% 2-mercaptoethanol to 70° for several hours, then washed extensively with water; the washed chitin is stored as a 50% (v/v) suspension in water

Minimal medium: 0.67% (w/v) Bacto-yeast nitrogen base (without amino acids), supplemented where appropriate to complement strain auxotrophies

YPS medium: 1% Bacto-yeast extract, 2% Bacto-peptone, and 0.6% sucrose (all w/v); sucrose stocks are filter-sterilized to avoid the generation of glucose by hydrolysis during autoclaving

Radiolabeling, Immunoprecipitation, and Endo-H Treatment

Cultures are grown to logarithmic phase at 24° in methionine-free minimal medium containing 5% glucose, washed once in medium containing 0.1% glucose, and incubated in low glucose medium at 24° for

---

[28] J. D. Boeke, J. Truehart, G. Natsoulis, and G. R. Fink, this series, Vol. **154**, p. 164.
[29] M. Carlson and D. Botstein, *Cell (Cambridge, Mass.)* **28**, 145 (1982).
[30] M. J. Kuranda and P. W. Robbins, *Proc. Natl. Acad. Sci. U.S.A.* **84**, 2585 (1987).

10 min to induce synthesis of invertase. At least 1×10^7 cells per sample are needed for easy handling. For radiolabeling experiments at the restrictive temperature, *sec18, sec53,* and *dpm1* cells are incubated for 15 min at 37°, and, when used, tunicamycin (final concentration 10 µg/ml) is added to cultures 5 min before the addition of [^{35}S]methionine. After radiolabeling for 30 min, cells are harvested, washed twice with cold 10 mM NaN$_3$, and resuspended in 50 µl breakage buffer in a 1.5-ml microcentrifuge tube. Glass beads are added to just below the surface of the liquid, and the tube is vortexed for 30 sec and placed in boiling water for 3 min. Vortexing and boiling are repeated twice, after which the broken cell slurry is removed using a drawn-out Pasteur pipette. The beads are washed twice with breakage buffer and the washings added to the slurry, yielding a final volume of about 100 µl. The slurry is then centrifuged in a microfuge for 5 min and the supernatant liquid removed. Centrifugation is repeated to ensure that no particulate material remains. After preclearing with protein A-Sepharose, invertase and chitinase are immunoprecipitated from the extracts by standard procedures. After binding to protein A-Sepharose, the immune complexes are washed twice with 0.1 M Tris-HCl, pH 7.6, containing 2.0 M urea, 0.2 M NaCl, and 1.0% Triton X-100, then twice with 1.0% 2-mercaptoethanol.[31] The immunoprecipitated proteins are then released by boiling in 35 µl of solubilizing buffer, after which 35 µl Endo-H buffer is added to the extract. Endo-H (4 mU[32]) is added to one-half of the sample, while the other half is left untreated. The samples are incubated overnight at 37°, after which SDS–PAGE sample buffer is added and the samples boiled for 3 min. Material extracted from 2.5 to 5×10^6 cells, radiolabeled with [^{35}S]methionine (200 µCi/ml; specific activity 1 Ci/mmol), is loaded per lane for SDS–PAGE, and radiolabeled proteins are visualized by fluorography.

[^{3}H]Mannose Labeling

Cultures are grown at 24° as above but in YPS medium. One-milliliter cultures containing 10^7 cells are then shifted to 37° for the required preincubation period at the nonpermissive temperature. Disposable 15-ml centrifuge tubes for labeling are prepared prior to the experiment by drying 1 mCi [2-^{3}H]mannose in the bottom of tubes with a stream of N$_2$. [2-^{3}H]Mannose incorporation is initiated by transferring the temperature-shifted cultures to prewarmed labeling tubes, which are then shaken at the

---

[31] B. Goud, A. Salminen, N. C. Walworth, and P. J. Novick, *Cell (Cambridge, Mass.)* **53**, 753 (1988).
[32] A. L. Tarentino, R. B. Trimble, and F. Maley, this series, Vol. 50, p. 574.

appropriate temperature for 90 min. The cells are harvested by centrifugation and the medium collected for subsequent isolation of chitinase. Extracts are prepared from the cells, and chitinase is immunoprecipitated from them as described above.

Isolation of Secreted Chitinase by Chitin Binding

[2-^{3}H]Mannose-labeled chitinase for use as a control O-mannosylated protein is isolated as follows. Chitin suspension (50 µl) is added to the culture supernatant (1 ml in a 1.5-ml microfuge tube) remaining after [2-^{3}H]mannose labeling in medium containing yeast extract and Bactopeptone, and the tubes are then rocked overnight at 4°. The chitin pellet is collected by centrifugation and washed 4 times with 1.5 ml PBS. Radiolabeled chitinase is then eluted by boiling in 75 µl of solubilizing buffer.

Autoradiography and Analysis of O-Linked Oligosaccharides

Mannose-labeled immunoprecipitates or secreted chitinase samples from the equivalent of 10^7 cells are initially separated by SDS–PAGE. Following electrophoresis, the separated proteins are transferred to nitrocellulose by electroblotting [TE Series Transphor Electrophoresis Unit (Hoefer Scientific, San Francisco, CA), using 25 mM Tris, 192 mM glycine, 10% methanol/100 mA/16 hr]. Filters are washed briefly in distilled water and then air-dried on Whatman 3 MM paper. Tritium-sensitive film is placed in direct contact with the membrane for 24 hr at room temperature. The resulting autoradiogram is then used as a template to excise regions of the membrane containing labeled glycoproteins. The approximately 4 × 7 mm portions are cut into smaller pieces and incubated overnight at room temperature in 100 µl of 0.1 M NaOH. The solution is removed, neutralized by the addition of 10 µl of 10% acetic acid, and lyophilized. The residue is dissolved in 25 µl of 50% ethanol and the base-released oligosaccharides separated by descending paper chromatography on Whatman No. 1 paper with ethyl acetate/butanol/acetic acid/water (3:4:2.5:4, by volume) as solvent. A standard sugar mixture (10 µl of a stock containing 3, 2, and 2 mg/ml, respectively, of raffinose, maltose, and mannose) is applied and run in parallel, and the sugars are detected by silver staining.[33] Elution positions of the di- and trisaccharide standards coincide approximately with those of the mannose oligomers Man$_2$ and Man$_3$. Lanes to be assayed for radioactivity are left unstained and cut into 1 × 2 cm strips. Radiolabeled saccharides are then eluted from chromatogram segments by adding 1 ml of water to the scintillation vials containing the strips. Aqueous scintillation fluid is added prior to counting.

[33] W. E. Trevelyan, D. P. Procter, and J. S. Harrison, *Nature (London)* **166**, 444 (1950).

Acknowledgments

We are grateful to Dr. R. B. Trimble for the gift of anti-invertase serum. We thank Dr. Phillips W. Robbins for his interest in this work, and for reviewing the manuscript. This work was supported by Grants GM-31318 and CA-26712 from the National Institutes of Health to P. W. Robbins.

[48] Yeast Endocytosis Assays

By VJEKOSLAV DULIC, MARK EGERTON, IBRAHIM ELGUINDI, SUSAN RATHS, BIRGIT SINGER, and HOWARD RIEZMAN

Introduction

Endocytosis is the process whereby cells internalize portions of their own plasma membrane and macromolecules from the external environment. The functions of endocytosis are many, including a nutritive function (uptake of iron, cholesterol, vitamin B_{12}), removal of unwanted molecules from the plasma membrane and external environment (asialoglycoproteins, foreign agents such as viruses), and a possible role in hormone response as many polypeptide hormones and their receptors are cleared from the cell surface by endocytosis.[1] Two different types of endocytosis can be followed, receptor-mediated and fluid-phase endocytosis. Receptor-mediated endocytosis is a specific process involving recognition of the ligand by cell surface receptors and is therefore saturable with respect to ligand concentration. In fluid-phase endocytosis, substances are internalized without binding to the cell surface, and therefore the process is not saturable with respect to the marker measured. Both processes require energy and involve vesiculation of the plasma membrane. In most examples of receptor-mediated endocytosis, the receptor–ligand complexes are concentrated into coated invaginations of the plasma membrane that subsequently bud off of the membrane. After removal of the coat, composed of a complex set of proteins including clathrin,[2] the resulting endocytic vesicles fuse with a cellular compartment termed the early endosome. The receptors and ligands are further transported by a vesicular carrier to a compartment called the late endosome or prelysosomal compartment. In this organelle, sorting of macromolecules to the lysosome, to the trans-

---

[1] J. L. Goldstein, M. S. Brown, R. G. W. Anderson, D. W. Russell, and W. Schneider, *Ann. Rev. Cell Biol.* **1**, 1 (1985).
[2] B. M. F. Pearse, *EMBO J.* **6**, 2507 (1987).

Golgi network, or back to the cell surface occurs.[3] The process of endocytosis is very complex, and attempts to use a genetic approach in mammalian cells have not been very fruitful. *Saccharomyces cerevisiae* would offer an excellent system to develop a genetic approach to the study of endocytosis. With the aid of genetics one should be able to identify the proteins involved in endocytosis and investigate the functions of endocytosis in the physiology of the yeast cell.

In yeast, several markers have been introduced to follow endocytosis. They can be divided into two groups, specific markers that bind to a cell surface receptor and nonspecific markers. Among the nonspecific markers that have been introduced are virus particles, α-amylase, fluorescein isothiocyanate dextran,[4] and Lucifer Yellow CH.[5] We prefer the latter as it is the nonspecific marker that has been most convincingly demonstrated to be taken up by endocytosis. Uptake of Lucifer Yellow is time-, temperature-, and energy-dependent, and it is nonsaturable. It accumulates in the vacuole with time, but the vacuolar concentration of Lucifer Yellow never reaches the external concentration. Lucifer Yellow has been used to screen for mutants that are defective in accumulation of endocytic content in the vacuole.[6] Many mutants can be screened at once as the assay is simple to perform.

An alternative mechanism for Lucifer Yellow uptake could be direct transport across the plasma membrane to the cytoplasm with subsequent direct transport from cytoplasm to the vacuole. We consider this alternative unlikely because of the fact that many secretion mutants[7] are defective in accumulation of Lucifer Yellow in the vacuole at the restrictive temperature.[5] In particular, *sec18*, for which the wild-type gene product or mammalian homolog has been shown to be directly involved in the membrane fusion events of secretion and endocytosis,[8,9] leads to defects in accumulation of Lucifer Yellow in the vacuole. This points more likely to a delivery mechanism involving vesicular flow. The absolute proof for endocytic uptake of Lucifer Yellow has not yet been obtained, but it would involve the identification of a vesicular intermediate in the transport of this compound to the vacuole. This is very difficult to do because of the small size

[3] G. Griffiths, B. Hoflack, K. Simons, I. Mellman, and S. Kornfeld, *Cell (Cambridge, Mass.)* **52**, 329 (1988).
[4] M. Makarow, *EMBO J.* **4**, 1861 (1985).
[5] H. Riezman, *Cell (Cambridge, Mass.)* **40**, 1001 (1985).
[6] Y. Chvatchko, I. Howald, and H. Riezman, *Cell (Cambridge, Mass.)* **46**, 355 (1986).
[7] P. J. Novick, C. Field, and R. Schekman, *Cell (Cambridge, Mass.)* **21**, 205 (1980).
[8] D. W. Wilson, C. A. Wilcox, G. C. Flynn, E. Chen, W. J. Kuang, W. J. Henzel, M. R. Block, A. Ullrich, J. E. Rothman, *Nature (London)* **239**, 355 (1989).
[9] R. Diaz, L. S. Mayorga, P. J. Weidman, J. E. Rothman, and P. D. Stahl, *Nature (London)* **339**, 398 (1989).

of Lucifer Yellow, which allows the fluorescent molecule to leak out of vesicles during purification or during chemical fixation procedures.

A special word of caution should be introduced concerning FITC-dextran. Even though FITC-dextran accumulation was claimed to occur by endocytosis,[4] several lines of new evidence suggest that this is not the case. The accumulation of FITC-dextran is sharply pH-dependent,[10] and FITC-dextran is taken up into ATP-depleted cells.[11] This is not consistent with an endocytic mechanism because ATP has been shown to be necessary for several steps of endocytosis, including the initial uptake step.[9,12] It has been reported that impurities in commercial FITC-dextran are actually accumulated in the vacuole and not FITC-dextran itself.[13] The most likely mechanism of vacuole labeling by FITC-dextran is that small fluorescent impurities in the product are transported across the plasma membrane through the cytoplasm to intracellular compartments. This problem is much less likely to exist for Lucifer Yellow, owing to the characteristics of its energy-dependent uptake and vacuolar labeling. Vacuolar labeling with FITC-dextran but not Lucifer Yellow is blocked by azide in the presence of glucose (H. Riezman, unpublished observations, 1987).[5] Azide, in the presence of glucose, does not have a large effect on ATP levels, but it inhibits many plasma membrane transport systems in yeast by destroying the electrochemical potential across this membrane.[14] Thus, although commercial FITC-dextran is conveniently used to visualize the vacuole by fluorescent microscopy, it should not be considered as a marker of endocytosis.

The only specific marker of endocytosis by yeast cells described to date is the pheromone α-factor. α-Factor binds to the α-factor receptor on **a** cells. The receptor is the product of the *STE2* gene.[15] There are approximately 10^4 receptors per **a** cell, with a K_D of about $10^{-8} M$.[16] On binding, α-factor is rapidly internalized[6,17] concomitant with the disappearance of cell surface receptor activity. It has not yet been directly shown that the α-factor receptor is internalized, but this is most likely the case. Once internalized, α-factor is degraded,[6] presumably in the vacuole because

[10] H. Riezman, Y. Chvatchko, and V. Dulic, *Trends Biochem. Sci.* **11**, 325 (1986).
[11] M. Makarow and L. T. Nevalainen, *J. Cell Biol.* **104**, 67 (1987).
[12] E. Smythe, M. Pypaert, J. Lucocq, and G. Warren, *J. Cell Biol.* **108**, 843 (1989).
[13] R. A. Preston, R. F. Murphy, and E. W. Jones, *J. Cell Biol.* **105**, 1981 (1987).
[14] T. G. Cooper, in "The Molecular Biology of the Yeast *Saccharomyces:* Metabolism and Gene Expression" (J. N. Strathern, E. W. Jones, and J. R. Broach, eds.), p. 399. Cold Spring Harbor Laboratory, Cold Spring Harbor, New York, 1982.
[15] D. D. Jenness, A. C. Burkholder, and L. H. Hartwell, *Cell (Cambridge, Mass.)* **35**, 521 (1983).
[16] D. D. Jenness, A. C. Burkholder, and L. H. Hartwell, *Mol. Cell. Biol.* **6**, 318 (1986).
[17] D. D. Jenness and P. Spatrick, *Cell (Cambridge, Mass.)* **46**, 345 (1986).

degradation is dependent on the *PEP4* gene product.[18] Thus, using α-factor one can distinguish three distinct events: binding to cell surface receptors, internalization of the pheromone, and degradation of internalized α-factor. Recently, we have been able to detect a vesicular intermediate in the transport of α-factor to the vacuole. The α-factor carried in this vesicle is protected from protease digestion by the membrane and floats in density gradients.[18] The identification of this vesicular intermediate provides evidence for the delivery of α-factor to the vacuole via an endocytic mechanism. α-Factor internalization is to date the most reliable way to quantitatively assess the earlier stages of endocytosis in yeast. An inability to degrade internalized α-factor could reflect, depending on the circumstances, the absence of vacuolar hydrolase activity or a defect in transport of α-factor to the vacuole.

Lucifer Yellow: A Nonspecific Marker for Endocytosis

Lucifer Yellow carbohydrazide (Fluka, Buchs, Switzerland) (LY) is a small fluorescent molecule that has been used as a marker of fluid-phase endocytosis in mammalian systems.[19,20] It has several characteristics which make it a suitable marker to study endocytosis: (1) it is a highly hydrophilic molecule and is thus incapable of diffusion across biological membranes; (2) it is nontoxic to cells even at concentrations much higher than those required to label endocytic compartments; (3) it is a pure fluorophore with a very high quantum yield, resulting in high fluorescence after uptake by cells; (4) its quantum yield is independent of pH over a wide range; (5) it is resistant to bleaching and therefore permits extended viewing. Results obtained with LY are best interpreted when combined with the results from α-factor uptake and degradation assays.

Internalized LY accumulates in the yeast vacuole and can easily be visualized by fluorescence microscopy.[5] Unfortunately, owing to limitations in the resolution of yeast organelles by light microscopy, it has not yet been possible to visualize internalized LY in any intermediate endocytic compartment. In addition, a strong labeling of the cell wall can, in some strains, obscure the visualization of the internalized LY and cause a high background in quantitation experiments.

Visualization of Cell-Associated Lucifer Yellow

Cultures of yeast cells are grown to mid-logarithmic phase ($1-2 \times 10^7$ cells/ml) in either YPUAD (1% yeast extract, 2% peptone, 30 mg/ml each

---

[18] B. Singer and H. Riezman, *J. Cell Biol.* **110,** 1911 (1990).
[19] J. A. Swanson, B. D. Yirinec, and S. C. Silverstein, *J. Cell Biol.* **100,** 851 (1985).
[20] D. K. Miller, E. Griffiths, J. Lenard, and R. A. Firestone, *J. Cell Biol.* **97,** 1841 (1983).

of uracil and adenine, 2% glucose) or minimal selective medium[21] (see below) if plasmid-bearing strains are being studied. Approximately 10^7 cells are harvested by centrifugation, resuspended in 90 µl of fresh medium, and transferred to a 1.5-ml microcentrifuge tube. Ten microliters (one can add more) of a 40 mg/ml LY solution (in water) is added and the cultures incubated at the desired temperature (usually 24–37°) for the desired time (30 min–2 hr). After incubation the cells are harvested by centrifugation (1 min at low speed in a benchtop microfuge), washed 3 times in 1 ml ice-cold buffer (50 mM succinate–NaOH, pH 5.0, 20 mM NaN$_3$), and finally resuspended in 10 µl buffer. An equal volume of cells (usually 3–4 µl) is mixed with 1.6% low melt agarose (precooled to 45°) and mounted on a microscope slide that can be kept on ice until observation under the microscope.

Lucifer Yellow CH can be visualized by fluorescence microscopy using FITC optics. We use the Zeiss Axiophot microscope equipped with the following filters: excitation 450–490 nm, FT510, LP520. The LY fluorescence can be photographed using Kodak TMAX 400 film pushed to 1600 ASA with approximately 20- to 50-sec exposures. The exposure time should be the same for samples that are to be compared. It is always advisable to photograph the same cells using Nomarski optics. A defect in accumulation of LY in the vacuole can only be proven in cells that contain vacuoles visualized by Nomarski optics.

Quantification of Cell-Associated Lucifer Yellow

One would often like to know whether a cell is defective in uptake or subsequent delivery of endocytic contents to the vacuole. In this case it is desirable to quantitate the uptake rate of LY. To do this, Lucifer Yellow is internalized in an experiment identical to that described above except scaled up 10-fold. After incubation the cells are diluted into 5 ml ice-cold buffer (50 mM succinate–NaOH, pH 5.0, 100 mM NaCl, 10 mM MgCl$_2$, 20 mM NaN$_3$), immediately collected by vacuum filtration on a 0.45-µm nitrocellulose filter, and washed 8 times with 2 ml cold buffer. The cells are eluted off the filter into a microfuge tube with 1 ml buffer and harvested by centrifugation. The supernatant is removed and the cells placed on ice until all samples are ready to be processed. The cells are resuspended in 1 ml of 50 mM Tris-HCl, pH 7.5, 10 mM 2-mercaptoethanol, 2000 units (U) lyticase/ml.[22] After 15 min at 37° (complete lysis), 50 µl is removed and precipitated with 10% trichloroacetic acid (TCA). The protein content is determined by a modification of the Lowry assay.[5] Fifty microliters of 10% sodium dodecyl sulfate (SDS) is then added to the lysed cells, and LY

[21] L. J. Wickerham, *J. Bacteriol.* **52**, 293 (1946).
[22] J. Scott and R. Schekman, *J. Bacteriol.* **142**, 414 (1980).

is quantified using a Kontron SFM spectrofluorometer with excitation at 426 nm and emission at 550 nm calibrated with a standard curve of LY. Background fluorescence is determined by treating cells without LY in the same manner: a negative control that has to be included in the same experiments is incubation conditions at 0°. In addition, incubation of mutants defective in endocytosis (we use a secretion mutant, e.g., *sec1*) or cells treated with 20 mM each NaN_3 and NaF can be used. It is important to take several early time points in the quantification of Lucifer Yellow uptake. For every three molecules of Lucifer Yellow taken up at 30°, two are subsequently lost back to the medium. A steady-state uptake rate is reached in about 10 min at 30°.[5]

The ability to quantify Lucifer Yellow accumulation is very strain dependent. As Lucifer Yellow binds to the yeast cell wall, only strains that accumulate a large amount of Lucifer Yellow in the vacuole show significant differences in cell-associated Lucifer Yellow between 30° and 0°. Cells that tend to lyse, as many mutants do at the nonpermissive temperature, give abnormally high apparent uptake and cannot be quantified in this manner. It may be feasible to adapt the above protocol to quantify Lucifer Yellow uptake even in strains with a high ratio of wall to vacuole labeling by adding sodium azide and fluoride (see above), converting the cells to spheroplasts, and then quantifying the Lucifer Yellow that cosediments with the spheroplasts.

α-Factor: A Specific Marker for Endocytosis

Following the binding, uptake, and degradation of α-factor requires the use of a well-defined, radioactively labeled pheromone. Methods have been published for the biosynthetic labeling of α-factor with $^{35}SO_4$.[15,23] Furthermore, chemical synthesis has made available at least two other means by which radioactive α-factor of sufficient specific activity for use in these assays can be obtained.[16,24] The method that we currently use is a modification of the method for biosynthetic labeling of α-factor published previously by Jenness *et al.*[15]

Preparation of ^{35}S-Labeled α-Factor

A *Matα* strain (RH449: *Matα his4 leu2 ura3 lys2 bar1*) is transformed with plasmid pDA6300[25] (contains both the *MFα1* and *STE13* genes on a *LEU2*, 2-μm-based plasmid, kindly provided by Jeremy Thorner, University of California, Berkeley). Individual colonies are picked and grown to stationary phase in minimal selective medium, and the culture supernatant

[23] K. J. Blumer, J. E. Reneke, and J. Thorner, *J. Biol. Chem.* **263,** 10836 (1988).
[24] S. K. Raths, F. Naider, and J. M. Becker, *J. Biol. Chem.* **263,** 17333 (1988).
[25] D. A. Barnes, Ph.D. Dissertation, University of California, Berkeley, California (1985).

is monitored by the halo assay[15] for α-factor production. Plates for the halo assay are prepared by adding 1 ml of a stationary YPUAD culture of RC898 (Mata leu1 trp5 ade2 can1 sst1 sst2, kindly provided by Russell Chan, Pioneer Hi-Bred International, Johnston, IA) to 250 ml of YPUAD medium containing 0.8% agar at 47°. Five milliliters is poured per petri dish (8.5-cm diameter). Aliquots (5 μl) from culture supernatants are pipetted directly onto the halo plates and incubated overnight at 30°. Transformants showing at least a 10-fold increase in pheromone production (~2-cm diameter halo) are then selected for biosynthetic labeling. We have noticed that α-factor production can decrease in transformants that have been kept on plates for extended periods of time.

For labeling, 4×10^7 cells are harvested from an exponentially growing culture in minimal selective medium, washed with sterile water, and resuspended in 25 ml of minimal selective medium without sulfate and leucine using 2% glucose (analytical grade, BDH, Poole, England) as a carbon source (see below). Twenty-five millicuries of carrier-free $H_2{}^{35}SO_4$ (New England Nuclear, Boston, MA) is added, and the culture is incubated at 30° in a rotary shaker at 250 rpm for 8 hr. After determining the extent of incorporation by analyzing the amount of ^{35}S present as TCA-precipitable counts (>80% TCA-precipitable counts expected), the culture is centrifuged for 5 min at 3500 rpm to remove the cells. The supernatant is collected, and 2-mercaptoethanol, p-tosyl-L-arginine methyl ester, and EDTA are added to final concentrations of 1, 1, and 0.2 mM, respectively. At this point, if necessary, the supernatant can be frozen at −20° overnight. However, we routinely continue with the purification by application of the supernatant to a CG-50 column.

The CG-50 column is prepared by washing a 10-ml volume of Amberlite CG-50 dry resin (100–200 mesh, Serva Fine Biochemicals, Heidelberg, Germany) successively with 3 volumes of 3 N HCl, 1 mM 2-mercaptoethanol, followed by 3 volumes of water, then 3 volumes of 0.01 N HCl, 80% ethanol, 1 mM 2-mercaptoethanol, and again 3 volumes of water, after which the resin is equilibrated with 0.1 M acetic acid, 1 mM 2-mercaptoethanol. This solution and all subsequent solutions are purged with argon before use. The slurry containing the CG-50 resin is then poured into a Bio-Rad (Richmond, CA) Econo-column (0.7-cm i.d., 15 cm length) such that the bed volume is roughly 3.5 ml. The culture supernatant from the 8-hr labeling is loaded onto the resin, and the column is washed with at least 25 bed volumes of 50% ethanol, 1 mM 2-mercaptoethanol. The ^{35}S-labeled α-factor is eluted with 0.01 N HCl, 80% ethanol, 1 mM 2-mercaptoethanol, and fractions are collected in siliconized Eppendorf microfuge tubes. After determining which fractions contain α-factor activity by the halo assay, the active fractions are pooled and concentrated using a Speed-Vac centrifuge.

Further purification of ^{35}S-labeled α-factor is achieved using high-performance liquid chromatography (HPLC). All HPLC buffers are flushed with argon and kept under helium. Under isocratic conditions (29.6% acetonitrile, 0.025% trifluoroacetic acid), at a flow rate of 1 ml/min, labeled α-factor elutes approximately 6.5 min after injection onto a C_{18} reversed phase column (30 cm × 3.9 mm i.d.; 10-μm particle size). Absorbance is monitored at 220 nm, and 1-μl aliquots of the 0.5-ml fractions collected in siliconized Eppendorf tubes (containing 10 μl of 0.5% Triton X-100, 0.5 mg/ml hemoglobin to prevent nonspecific absorption of the α-factor to the tubes) are screened for radioactivity by scintillation counting. Typically, two major peaks of radioactivity are eluted. They are dried down in the Speed-Vac and resuspended to their original volume with 0.01 N HCl, 0.2 mM EDTA, 1 mM dithiothreitol. The major peak (retention time 6.5 min) coelutes with chemically synthesized tridecapeptide under identical conditions. Furthermore, thin-layer chromatography of both synthetic tridecapeptide and the major peak of radioactivity on silica gel 60 plates (Merck, Darmstadt, Germany), using the solvent system n-butanol:propionic acid:water (50:25:35, v/v/v), and autoradiography reveal one major spot of radioactivity which comigrates with the ninhydrin-positive spot of the synthetic standard (R_f 0.27). The minor peak of radioactivity which elutes at approximately 4.5 min under the HPLC conditions described above is most likely the methionine sulfoxide form of the native pheromone and migrates with an R_f of 0.21 in the thin-layer system. This species also appears to a very slight degree in the 6.5-min peak after HPLC purification and increases on extended storage. Although both peaks of radioactivity are biologically active, binding and uptake of the minor peak are less when compared to those of the major peak, suggesting this to be a modified form of lower affinity. Once purified, ^{35}S-labeled α-factor peak fractions are resuspended in 0.01 N HCl, 0.2 mM EDTA, 1 mM dithiothreitol, pooled, and aliquoted into siliconized microfuge tubes for storage in liquid nitrogen.

We routinely obtain between 6×10^7 and 1×10^8 cpm total in the ^{35}S-labeled α-factor peaks from one preparation. Approximately 75% of the counts which elute after the application of the 0.01 N HCl, 80% ethanol, 1 mM 2-mercaptoethanol eluate reside in the two peaks described above after HPLC. Based on the halo assay using a standard curve with known α-factor concentrations, we calculate the specific radioactivity of our ^{35}S-labeled α-factor preparations to vary between 5 and 10 Ci/mmol. Between 40 and 90% of the radioactivity in the most active fractions binds specifically to **a** cells. At 30° in pulse–chase experiments (see below), greater than 90% of the specifically bound radioactivity is internalized.

We have found a number of factors which may affect incorporation of $^{35}SO_4^{2-}$ and the recovery of ^{35}S-labeled α-factor using this method. First, siliconization of the tubes in which the ^{35}S-labeled α-factor is collected is very important but is not always completely successful in preventing some of the radioactivity from sticking to the tubes. In fact, we have noted batch differences between microfuge tubes obtained from the same supplier. Second, we find the quality of glucose used in the labeling medium to be an important variable. Sulfate contamination differs among commercially available glucose and can reduce ^{35}S incorporation significantly if present in sufficient amounts. We use D-glucose supplied by BDH Chemicals Ltd. which has a maximum sulfate impurity level of 5 ppm. Finally, since the α-factor is overproduced, fresh transformants are always taken for labeling and α-factor production to ensure the maximum amount of ^{35}S incorporation into the tridecapeptide species.

Minimal Selective Medium

Sugar and salts	g/liter
Glucose	20
KH_2PO_4	0.875
K_2HPO_4	0.125
NaCl	0.1
$(NH_4)_2SO_4$	0.5
$MgCl_2$	0.5
$CaCl_2$	0.1

(For sulfate-free minimal medium, NH_4Cl is substituted for NH_4SO_4.)

Vitamins	mg/liter
(Vitamins are filter-sterilized and added to the medium after it is autoclaved.)	
Biotin	0.002
Pantothenic acid	4.0
myo-Inositol	20
Niacin	4.0
p-Aminobenzoate	2.0
Thiamin	4.0
Riboflavin	2.0

Trace elements	M
H_3BO_3	10^{-8}
$CuCl_2$	10^{-8}
KI	10^{-7}
$FeCl_3 \cdot 6H_2O$	5×10^{-8}
$ZnCl_2$	7×10^{-8}

Optional supplements	mg/liter
Adenine	40
Uracil	20
Leucine	30
Histidine	30
Tryptophan	30
Tyrosine	30
Lysine	30

Binding of Radiolabeled α-Factor to Mata Cells

In some cases one would like to determine the extent of cell surface expression and activity of the α-factor receptor. This may be necessary in the analysis of receptor mutants. In order to measure the binding step alone, cells are incubated in presence of radioactive α-factor under conditions that minimize nonspecific binding, α-factor degradation, and internalization. To minimize nonspecific interactions of α-factor with cells, we measure binding in YPUAD medium or in 50 mM potassium phosphate, pH 6, containing 1% bovine serum albumin (BSA). YPUAD medium is preferred unless defined conditions are necessary, for example, to determine ion requirements. Extracellular α-factor degradation is minimized by using strains that are *bar1* mutants. The *BAR1* gene product is an **a** cell-specific secreted protease that degrades α-factor.[26] Endocytosis and secretion can be blocked by performing the binding step at 0° or by including 20 mM NaN_3 and 20 mM NaF.

a cells are harvested from a log-phase culture (usually $1-2 \times 10^7$ cells/ml) in YPUAD, washed twice with medium, and resuspended in ice-cold YPUAD. Cells are diluted to the desired concentration (10^8 to 2×10^9 cells/ml), and purified biosynthetically ^{35}S-labeled α-factor is added to 10^4 cpm per 100 μl assay. After incubation at 0° with gentle shaking, 100-μl aliquots (in duplicate) are withdrawn and diluted into 20 ml of

---

[26] V. L. MacKay, S. K. Welch, M. Y. Insley, T. R. Manney J. Holly, G. C. Saari, and M. L. Parker, *Proc. Natl. Acad. Sci. U.S.A.* **85,** 55 (1988).

ice-cold 50 mM potassium phosphate, pH 6.0. Cells are then collected on GF/C filters (Whatman) set on a filtration apparatus connected to a vacuum pump. This assembly can be set up in a cold room. The filters are washed twice 5 ml phosphate buffer each time, transferred into scintillation vials, dried at 80° for 1 hr, and counted for radioactivity in a β-counter. The results are plotted as counts per minute bound versus cell concentration of time of incubation, depending on the desired experiment.

In order to determine the extent of nonspecific binding, the assay is performed under identical conditions using equal quantities of α cells. If an accurate number of cell surface receptors and their affinity for α-factor is needed, then **a** cells at 1×10^9 cells/ml are incubated with a mixture of ^{35}S-labeled α-factor and varying amounts of unlabeled synthetic α-factor (10^{-9} to 10^{-6} M) for 60 min at 0°. Cells are then collected and counted to determine bound radioactivity as above. Data can be plotted according to the method of Scatchard.[27] In our strains the intercept of the line on the abscissa gives about 8000 molecules of α-factor bound per cell. The slope of the curve gives an equilibrium dissociation constant, K_D of $5-7 \times 10^{-9}$ M.

Assay of α-Factor Internalization

As mentioned in the Introduction, the best way to assess quantitatively the earliest stages of endocytosis in yeast is to follow the uptake of radioactive α-factor by **a** cells. These experiments are based on results showing that the cell surface-bound α-factor can be dissociated from the cells by short incubation in an acidic buffer.[6] The amount of pheromone that is resistant to the "acid wash" increases in a time-, temperature-, and energy-dependent fashion consistent with the hypothesis of receptor-mediated endocytosis. An alternative way to measure ligand-induced α-factor receptor clearing is described by Jenness and Spatrick.[17] Here, we describe our standard assay to measure α-factor internalization.

Cultures of **a** *bar1* cells are grown to exponential phase ($1-2 \times 10^7$ cells/ml) in YPUAD medium at 24° (for temperature-sensitive mutants) or 30°. Usually, 100 ml of such a culture should be sufficient for one uptake experiment. The cells are harvested by centrifugation in a tabletop Sorvall centrifuge (3000 rpm, 5-10 min), washed twice with YPUAD, and resuspended in ice-cold YPUAD to $0.5-1.0 \times 10^9$ cells/ml. The internalization of α-factor can be monitored by using two approaches that differ in their initial steps. In the pulse-chase experiment, the binding of radioactive α-factor to **a** cells is achieved by shaking the cells for 1 hr at 0° (ice bath) with 10^5 cpm of radioactive α-factor/10^9 cells at $0.5-1 \times 10^9$ cells/

[27] G. Scatchard, *Ann. N. Y. Acad. Sci.* **51**, 660 (1948).

ml. Under these conditions 40–90% of the radioactive pheromone can be bound. Unbound α-factor is removed by centrifugation at 0° in a refrigerated Sorvall centrifuge, and the cells are quickly resuspended (0.5–1.0 × 10^9 cells/ml) in prewarmed YPUAD (usually in 50-ml Falcon tubes) and incubated with shaking at the desired temperature. After various incubation times (e.g., 0, 2.5, 5, 7.5, 10, 15, 30, 45, and 60 min) aliquots of 100 μl are removed in duplicate and diluted into 10–20 ml of either ice-cold 50 mM potassium phosphate buffer, pH 6.0, to determine total cell-associated radioactivity, or 50 mM sodium citrate–HCl, pH 1.1, to remove surface-bound α-factor, thus giving a measure of pheromone internalization. The samples that are washed with pH 1.1 buffer are kept on ice for a minimum of 15 min to allow complete dissociation of externally bound α-factor. The cells are then collected on GF/C filters mounted on a filtering apparatus and washed 3 times with 5 ml of the appropriate ice-cold buffer. The filters are then transferred into scintillation vials and processed as described in the binding assay. When ^{3}H-labeled α-factor is used, scintillation fluid is added only after solubilizing the cells for several hours in 1 ml/vial of Soluene-350 (Packard Instrument Int., Zurich, Switzerland) as internalized α-factor is more readily quenched than surface-bound pheromone. The yellow cell-solubilization product, which can interfere with counting, is almost instantaneously bleached by addition of 80–100 μl of 30% hydrogen peroxide.

In a typical experiment, after all of the surface-bound α-factor is internalized (15–30 min), one can notice at elevated temperatures (30°–37°) a decrease of about 20% of both the total cell-associated and internalized radioactivity. This decrease does not occur in mutant strains that cannot degrade α-factor, such as *pep4* or *end1*.[28]

Internalization in the continuous presence of radioactive α-factor can be used in situations when mutant strains require growth at the nonpermissive temperature in order to express the mutant phenotype. In this case, the radioactive pheromone can be added directly to prewarmed cells (usually 15–30 min to express the mutant phenotype) suspended in YPUAD. The rest of the protocol is the same as for the pulse–chase experiments. In some cases (e.g., uptake experiments in the continuous presence of ^{35}S-labeled α-factor at 37°), not all of the bound α-factor is internalized even after 60 min of incubation. Although we do not understand this effect, this observation should be kept in mind, especially in cases where comparisons between strains are to be made. The best way to avoid any confusion regarding a possible endocytic defect attributed to a given mutant is to compare the isogenic mutant and wild-type strains under the same experimental conditions (if possible using mutants which can also be assayed in

---

[28] V. Dulic and H. Riezman, *EMBO J.* **8**, 1349 (1989).

the presence of a *CEN* plasmid containing the wild-type gene) at the permissive and nonpermissive temperatures.

Degradation of α-Factor

Another assay used in the analysis of α-factor endocytosis is to measure the extent of pheromone degradation. In wild-type cells, α-factor prebound to cells at 0° is rapidly degraded on internalization at 30°.[6] This assay measures the arrival of the internalized α-factor to the active hydrolytic compartment, presumably the vacuole. In order to characterize the endocytic pathway of α-factor, it is therefore necessary to study both internalization and degradation of the pheromone. Experimental conditions that allow α-factor uptake but prevent transport to the vacuole, indicated by a lack of degradation, could help to identify the possible transport steps. The state of internalized ^{35}S-labeled α-factor is monitored by extracting the cell-associated radioactivity and analyzing it by thin-layer chromatography and autoradiography.

Binding of ^{35}S-labeled α-factor, internalization, and washing in a pH 1.1 and pH 6 buffer are done using either the *pulse–chase* or *continuous presence* protocol, depending on the desired experiment. After being washed on nitrocellulose filters (type HA, Millipore), the cells are eluted from the filters with 1 ml of 50 mM potassium phosphate, pH 6.0, and are collected in 1.5-ml Eppendorf tubes. The cells are then pelleted by a 1-min high-speed centrifugation, resuspended in 150 μl of extraction buffer containing 40% (v/v) methanol, 0.025% (v/v) trifluoroacetic acid, 0.008% (v/v) 2-mercaptoethanol, and 0.06% (v/v) acetic acid, and subjected to 5 cycles of freezing in liquid N_2 and thawing at room temperature. We find that by this permeabilization procedure we can recover 30–80% of the cell-associated or internalized radioactivity in the supernatant after a 5-min centrifugation at high speed. This supernatant, which should contain at least 2000 cpm, can be analyzed directly by thin-layer chromatography on preparative silica gel 60 plates (20 × 20 cm). The solvent system we use is butanol:propionic acid:water (50:25:35, v/v/v). After chromatography (8–9 hr) the plates are dried, sprayed with En^{3}Hance (Du Pont, Boston, MA) and subjected to fluorography at −70°. Externally bound and internalized undegraded α-factor comigrate with the authentic intact pheromone (see α-factor production; a radioactive α-factor standard should always be included in the chromatography). On internalization at 30° in wild-type cells, degradation products become visible after 15 min.[6,28] The major degradation products migrate more slowly than intact α-factor as a smear. Depending on the experimental conditions and on the strains used, we see one to three major degradation products appearing with different kinetics during the course of the experiment; since ^{35}S-labeled α-factor is susceptible to oxidation (see α-factor production), one cannot exclude the possibility that some of the newly appearing spots are oxidized derivatives of breakdown products. We quantify both the disappearance of intact α-factor and the appearance of degradation products by scanning the X-ray films using a computing densitometer (Molecular Dynamics, Sunnyvale, CA).

Acknowledgments

The authors wish to thank Maryse Moya and Bettina Zanolari for their efforts to improve the α-factor purification and degradation assay and Gaby Häusermann for her help with the manuscript.

[49] Inducing and Assaying Heat-Shock Response in *Saccharomyces cerevisiae*

By CHARLES M. NICOLET and ELIZABETH A. CRAIG

A nearly universal response of organisms to an increase in temperature or other stresses is the induction of a set of proteins referred to as heat-shock or stress proteins (hsp). For general reviews, see Refs. 1–4. In this chapter we provide information concerning the methods we commonly use in our laboratory to induce and monitor the heat-shock response in the yeast *Saccharomyces cerevisiae*.

Standard Protocol for Inducing Heat-Shock (Stress) Response

Typically, induction of the heat-shock response involves shifting a culture in early log phase ($4-10 \times 10^6$ cells/ml, corresponding to an OD_{600} of 0.4–1.0) growing at 23° to 37° or 39° (see, e.g., Refs. 5 and 6). A shift to 37° does not always give a maximal induction of hsp but is often used in experiments since *S. cerevisiae* grows quite well at this temperature. A shift to 39°, a marginally permissive temperature for *S. cerevisiae*, gives a more consistent pattern of induction. For maximal induction, it is important

[1] E. A. Craig, *Crit. Rev. Biochem.* **18**, 239 (1985).
[2] M. Bienz and H. R. B. Pelham, *Adv. Genet.* **24**, 31 (1987).
[3] S. Lindquist and E. A. Craig, *Annu. Rev. Genet.* **22**, 631 (1988).
[4] J. E. Rothman, *Cell (Cambridge, Mass.)* **59**, 591 (1989).
[5] M. Werner-Washburne, D. E. Stone, and E. A. Craig, *Mol. Cell. Biol.* **7**, 2568 (1987).
[6] R. E. Susek and S. Lindquist, *Mol. Cell. Biol.* **9**, 5265 (1989).

that the temperature change occur as rapidly as possible. If a small amount of cells is to be heat-shocked, a portion of the 23° culture is removed and transferred to a prewarmed flask or tube in a 37° or 39° water bath. Alternatively, an entire flask can be transferred to a prewarmed water-bath. If a large volume needs to be heat-shocked (>50 ml) one should transfer the culture to a much larger prewarmed flask to maximize the heat transfer, thereby ensuring a rapid upshift. These protocols are effective in rich as well as defined media using a variety of carbon sources.

Assaying Induction of Heat-Shock Protein Genes

The heat-shock response is manifested at a number of levels that are easily assayed. These include a change in the pattern of mRNA accumulation and protein synthesis, as well as a change in the physiological state of the heat-shocked cell. In analyzing RNA and protein it is important to harvest cells such that metabolic activity is arrested rapidly, so that induction of the heat-shock response does not occur during the harvesting period.

RNA Levels. The levels of mRNA are most easily investigated by Northern analysis. We have found the time course of induction of a number of heat-shock genes to be very similar; a substantial increase is noted at 5 min, levels peak at 15 min, and by about 60 min the levels of mRNA are returning to pre-heat-shock values. This general induction pattern has been observed for the heat-inducible members of the stress seventy family (*SSA1, SSA3, SSA4, SSC1,* and *KAR2*) as well as at least one gene outside this family, *STI1*.[7] We have not intensively investigated the kinetics of mRNA accumulation when the cells are briefly heat-shocked and returned to a low temperature.

A load of 2–4 µg of total RNA per lane is sufficient to determine the heat-shocked level of the genes we have examined. Therefore, one can obtain from a small volume of cells (<1.5 ml) enough RNA for multiple Northern blot analyses. When such small volumes are taken, cells are pipetted into 1.5-ml microcentrifuge tubes, the tubes spun for 30 sec in a microfuge, the supernatant removed with a drawn-out Pasteur pipette, and the tube placed in a dry ice/ethanol bath. In this circumstance, approximately 1 min has transpired between the removal of the sample from the culture and freezing. If larger volumes need to be harvested, the culture can be rapidly cooled prior to centrifugation by placing ice in the centrifuge tube prior to the addition of the culture media. The pellet is quickly frozen in dry ice/ethanol, 15- or 50-ml plastic disposable tubes are suitable for this

---

[7] C. M. Nicolet and E. A. Craig, *Mol. Cell. Biol.* **9**, 3638 (1989).

purpose. It is not necessary to rinse the pellet, as we have found no adverse effects on RNA integrity from the small amount of culture fluid that remains.

A number of *HSP* genes can be used as probes to monitor heat-shock induction. In our laboratory we typically use *SSA1* or *SSA4* sequences. *SSA1* is used because it is the best characterized of the stress seventy genes. It has a substantial basal level of expression; the amount of *SSA1* mRNA increases approximately 10-fold after a heat shock. *SSA4* is used because it has a very low basal level of expression but is induced very strongly upon heat shock. Also, *SSA4* seems to be very sensitive to perturbations which slightly stress the cell. We have observed induction of detectable *SSA4* expression under conditions of stress when induction of some other HSP was not seen. Appropriate plasmids to use for such experiments are SSA1P3 (formerly called YG100P3,[8] containing the entire *SSA1* structural gene plus flanking sequences carried on a *Pst*I fragment in pBR322) and SSA4H[5] (containing the entire *SSA4* structural gene plus flanking sequences carried on a *Hin*dIII fragment in pBR322). It should be noted that when using probes including the *SSA1* coding region, the *SSA2* mRNA will also be detected. These two genes are 96% identical at the nucleotide level, preventing distinction between them even by stringent hybridization and washing conditions. Work from our laboratory utilizing strains with disrupted *SSA1* or *SSA2* genes has demonstrated that the observed heat induction results from an increase in *SSA1* rather than *SSA2* RNA levels.

Protein Synthesis. The use of protein labeling allows a more global observation of the proteins that are synthesized in response to heat shock. This method has the advantage over Northern analysis in that one can simultaneously observe a number of heat-shock proteins, including proteins for which the corresponding genes are not available. We (and others) have found that maximal labeling of heat-shock proteins (with concomitant decrease in labeling of other proteins) occurs in the interval between 15 to 30 min following temperature upshift. This corresponds closely to the peak levels of mRNA for these proteins (15 min). Efficient labeling can be carried out by the following method: A culture of early log cells in synthetic medium lacking methionine at 23° is shifted to 39°. Fifteen minutes after the temperature upshift, 1 ml of culture is removed to a prewarmed microfuge tube containing 30–100 μCi of ^{35}S (we have always had good results with the Trans ^{35}S-label from ICN Biomedicals, Costa Mesa, CA). Labeling is carried out for 15 min in the closed microfuge tube. Cells are then spun and rinsed once (in contrast to collection for RNA), and the pellet is frozen in dry ice/ethanol for future workup. This method

---

[8] E. A. Craig and K. Jacobsen, *Cell (Cambridge, Mass.)* **38**, 841 (1984).

of labeling results in no detectable increase in synthesis of heat-shock proteins in cells maintained at 23°.

The samples can be analyzed by either one- or two-dimensional polyacrylamide gel electrophoresis (PAGE). Although the resolution is not so great in the one-dimensional gel, it is often sufficient to determine if a large induction of hsp has occurred. Figure 1 shows a typical *in vivo* labeling experiment, with the proteins separated by one-dimensional PAGE in the presence of sodium dodecyl sulfate (SDS); induced proteins at approxi-

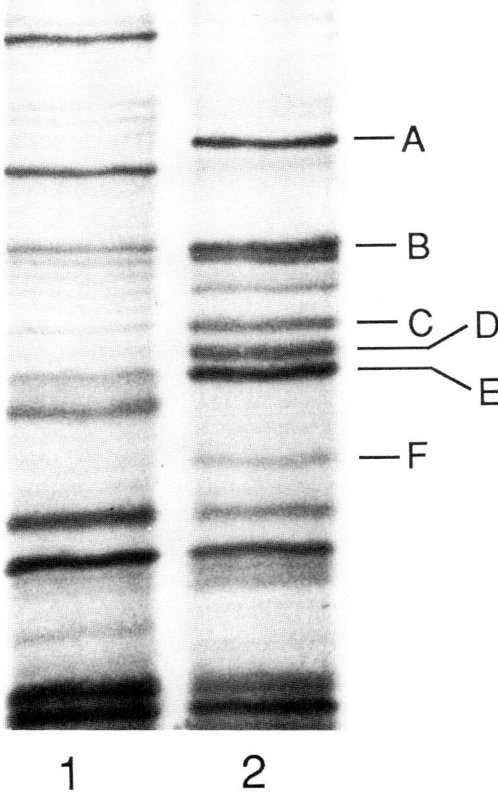

FIG. 1. One-dimensional SDS–PAGE analysis of heat-shocked *Saccharomyces*. Cells in early log phase (5×10^6 cells/ml) grown in synthetic medium lacking methionine were shifted from 23° to 39°. After 15 min, 1 ml of culture was removed to a prewarmed microfuge tube containing 100 μCi [^{35}S]methionine (obtained from a protein hydrolyzate) and labeling carried out for 30 min at 39°. For the control, labeling was carried out for 15 min at 23°. Qualitatively similar results are obtained if heat-shock labeling is carried out under a variety of regimens, including 15–20, 15–30, and 30–45 min after heat shock. Lane 1, Control; lane 2, heat shock. A, Hsp104; B, Hsp90; C, Sti1p + Kar2p; D, Ssa3p + Ssa4p; E, Ssa1p; F, Hsp58.

mately 70, 83 (hsp in this size range are broadly denoted as Hsp90 to account for organismal variations in size), and 104 kDa are easily detected. However, because of the comigration of several bands (e.g., Ssa1p and Ssa2p), under certain conditions induction is not detected by one-dimensional PAGE. In these circumstances two-dimensional analysis is required to observe the induction of one or more heat-shock proteins. Figure 2 shows a portion of a two-dimensional gel that includes the 70- to 85-kDa region. The induced proteins detected include Ssa1p, Ssa3p, Ssa4p, Ssc1p, Hsp90, and Sti1p. Ssb1p and Ssb2p serve as a good indicator of the decrease in synthesis of many proteins that occurs after heat shock.

Thermotolerance. Another easily assayable manifestation of the heat-shock response is the acquisition of thermotolerance by heat-shocked cells. Although this is an indirect analysis, the development of thermotolerance does correlate well, in general, with the induction of heat-shock proteins. Thermotolerance is the phenomenon whereby preexposure of cells to a nonlethal heat shock allows them to survive temperatures which would otherwise be lethal. This effect is quite dramatic. Yeast cells shifted from 23° to 37°, then exposed briefly to 52°, show four to five logs greater resistance to killing than do cells which have not been preexposed to 37°. Since cells approaching stationary phase also become thermotolerant (perhaps by induction of some hsp by a regulatory system different from that utilized by stress), care should be taken to carry out assays for induction of thermotolerance during exponential growth at low cell density.

There are two relatively simple procedures for assaying thermotolerance. The first, which is more quantitative, involves growing cells in YPD at 23° to early log phase (OD_{600} 0.4–0.8). The culture is then split (a small culture-to-flask volume is desirable since in these circumstances the final temperature is reached more rapidly), and one-half is transferred to 37° for 50 min. Both cultures are then shifted to 52° for 12 min, after which cells are diluted, plated onto YPD solid medium, and incubated for 2 days at 23° in order to determine the number of viable cells.

The second method is a simple plate assay. Cells from a fresh culture are streaked onto YPD plates and incubated for 6 hr at 23° to allow cells to enter exponential growth. The plates are then sealed into a plastic bag, incubated in a 38° water bath for 90 min, and then transferred to a 56° bath for 8 or 16 min. The plates are then incubated for 1–2 days at 23° and growth observed. The exact time of incubation at 56° that gives the best differentiation between the tolerant and nontolerant states must be determined empirically for each strain background.

Induction of Heat-Shock Protein–lacZ Fusions. The use of heat-shock promoter–*lacZ* fusion gene vectors as a reporter system is perhaps the most convenient of the assays discussed, but it is the least informative in terms of the true *in vivo* kinetics of the response. Two types of fusions are

A) WILD TYPE 23°C

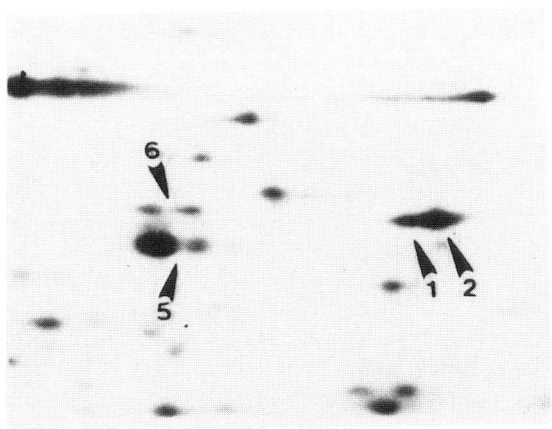

B) WILD TYPE ↑ 39°C

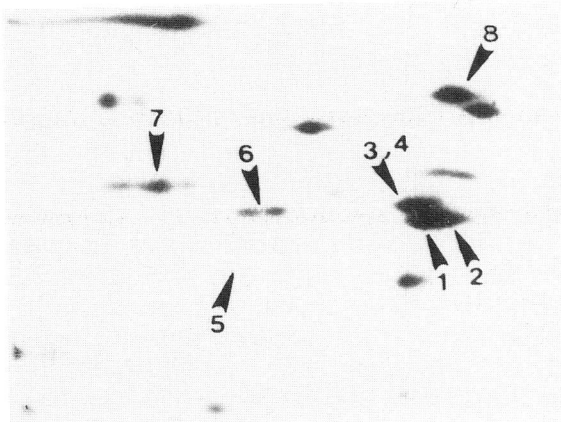

FIG. 2. Two-dimensional gel analysis of heat-shocked *Saccharomyces*. (Adapted from Werner-Washburne *et al.*[5]) The acidic portion of the gel is oriented toward the right. 1, Ssa1p; 2, Ssa2p; 3, Ssa3p; 4, Ssa4p; 5, Ssb1p/Ssb2p doublet; 6, Ssc1p doublet; 7, Sti1p; 8, Hsp90. Kar2p (GRP78/BiP) is tentatively identified as an elongated spot just above and slightly to the right of Ssa2p, as determined by its molecular weight, induction pattern, cross-reactivity with anti-Hsp70 antibodies, and ability to bind ATP.

available: transcriptional fusions, in which defined promoter fragments replace the *CYC1* upstream activating sequence (UAS) in a *CYC1* promoter–*lacZ* fusion vector, and transcriptional–translational fusions, in which the first few amino acids of the heat-shock gene are joined in frame to the *lacZ* protein coding sequence. When using transcriptional fusions, one is measuring only the transcriptional activation conferred by the fragment inserted into the *CYC1* promoter. With the translational fusion, one can measure transcriptional activation as well as translational components of regulation which may be dependent on the 5' end of the mRNA.

These types of fusions are convenient to use, since β-galactosidase activity is readily assayed in yeast cells. Cells can be treated with various agents or grown in different temperature regimens and the inducibility of the promoter easily determined. One caveat to this approach is that the overall kinetics of a heat-shock response are not precisely represented by the amount of β-galactosidase activity. This is probably due to the fact that there is a time lag between synthesis of β-galactosidase monomers and assembly of enzymatically active tetramers, and the stability of β-galactosidase, once synthesized, may lead to artificially elevated levels at later heat-shock time points. Nevertheless, such fusions are an invaluable resource for the researcher investigating or assaying heat shock. Cell harvesting and β-galactosidase assay procedures have been thoroughly described elsewhere.[9] Several fusions which we have found particularly useful are described below.

Transcriptional Fusions. The vector pZJHSE2-26 contains a 26-base pair core heat-shock element, HSE2, from the *SSA1* promoter in a *CYC1*–*lacZ* fusion vector. It has a relatively high basal activity but responds to heat shock, inducing approximately 8-fold by 75 min after temperature upshift. The vector pZJHSE2-137 contains a larger fragment from the *SSA1* promoter and has a very low level of basal activity, as compared to pZJHSE2-26. The 137-base pair promoter fragment inserted in place of the *CYC1* UAS contains the HSE2 sequence but also a repressing sequence responsible for the low basal activity. pZJHSE2-137 shows a 200-fold induction after temperature upshift. A more thorough description of these two transcriptional fusion plasmids can be found in Refs. 10 and 11. Both plasmids are derivatives of pLG669Z.[12] They have a 2-μm origin of replication, and are thus multicopy, and carry the *URA3* gene as a selectable metabolic marker.

[9] L. Guarente, this series, Vol. 101, p. 181.
[10] M. R. Slater and E. A. Craig, *Mol. Cell. Biol.* **7**, 1906 (1987).
[11] H. O. Park and E. A. Craig, *Mol. Cell. Biol.* **9**, 2025 (1989).
[12] L. Guarente and M. Ptashne, *Proc. Natl. Acad. Sci. U.S.A.* **78**, 2199 (1981).

Translational Fusion. The vector pWB215, constructed by Will Boorstein (University of Wisconsin, Madison), contains the entire promoter from the *SSA4* gene, which might be considered the archetypal Hsp70 gene from *S. cerevisiae.* At 23°, the promoter directs a barely detectable level of *lacZ* synthesis, but when maximally induced it shows a very strong induction, of the order of 100-fold. pWB215 is a centromeric plasmid, present in one or two copies per cell, harboring the *TRP1* gene as a metabolic marker.

The *SSA4-lacZ* fusion is particularly useful to assay conditions in which the heat-shock response might just be "tickled" a bit, but not fully induced. We have seen conditions in which this fusion shows an approximately 10-fold induction, indicating the cell has been stressed, yet *SSA1-lacZ* fusions show essentially no change. The sensitivity of this promoter makes it useful for detecting conditions in which the heat-shock response has been inappropriately induced by some laboratory manipulation. One needs to be aware of the sensitivity of the stress response, since in many laboratory manipulations it is very difficult to avoid stress induction caused by perturbation of the culture.

Acknowledgments

We thank W. R. Boorstein and H. O. Park for critical reading of the manuscript.

[50] Nucleolar-Specific Positive Stains for Optical and Electron Microscopy

By Michael W. Clark

Introduction

The elliptical, almost spherical nature of the *Saccharomyces cerevisiae* cell presents a special problem to the yeast cell biologist. What method should be used to determine the orientation of the cell in a localization experiment? As described in the chapter on immunogold labeling of yeast ultrathin sections ([42], this volume) and in this chapter, the yeast nuclear components appear to have a defined order with respect to each other. Proper investigation of this higher order arrangement of nuclear components requires a detailed correlation of cellular orientation in relation to nuclear component orientation. The bud on the cell does provide a frame of reference for positioning features in the mother cell in relation to the daughter. The bud scar, which can be stained specifically with a fluorescent dye, Calcofluor, has also been used for orienting the position of the new bud relative to the old ones.[1] Although this procedure is useful for determining the polarity of a yeast cell, it does not convey much more information. Moreover, some immunofluorescence microscopy preparations, which require extensive enzymatic digestions, will remove the bud-scar material required for Calcofluor staining. The only other common feature used to determine yeast cellular orientation has been the spindle pole apparatus, which can be visualized by immunofluorescent staining with antitubulin antibodies.[2] Changes in spindle pole morphology during the cell cycle can, however, cause some confusion in placement of cellular components. Here I propose the nucleolus as a potential marker for yeast cell orientation. The nucleolus is a cellular component that is ubiquitous in eukaryotic cells and that does not extensively change its morphology during the cell cycle in *S. cerevisiae*.[3]

The *S. cerevisiae* nucleolus, the site of rRNA synthesis and preribosome assembly, has been described as a dense crescent that occupies about one-third to one-half of the nuclear volume.[4] The yeast nucleolus is closely applied to the inner membrane of the nuclear envelope.[5] This juxtaposition of the nucleolus with the nuclear envelope suggests an actual physical attachment to some feature of the envelope. Indicative of such a physical attachment to the nuclear envelope is the fact that the nucleolus appears to have a specific orientation in the nucleus and in the cell. The nucleolus has been observed to maintain a specific position opposite to that of the emerging bud (M. W. Clark, personal observation, 1989; see Fig. 1). The nucleus is known to maintain specific orientation in the cell during cell growth. The spindle pole body remains on the side of the nucleus underlying the emerging bud.[5] Double-label immunofluorescence microscopy experiments with antitubulin and antibodies to nucleolar proteins have confirmed both of these observations (Ref. 6; M. W. Clark, personal observation). The nucleolus was always seen to be opposite of the spindle pole body. These and other data on the immune mapping of other nuclear

[1] J. R. Pringle, K. Coleman, A. Adams, S. Lillie, B. Haarer, C. Jacobs, J. Robinson, and C. Evans, in "Molecular Biology of the Cytoskeleton" (G. G. Borisy, D. W. Cleveland, and D. B. Murphy, eds.), p. 193. Cold Spring Harbor Laboratory, Cold Spring Harbor, New York, 1984.
[2] J. V. Kilmartin and A. E. Adams, *J. Cell Biol.* **98**, 922 (1984).
[3] C. F. Robinson and J. Marak, *J. Cell Biol.* **29**, 129 (1966).
[4] W. W. Sillevis-Smitt, J. M. Vlak, I. Molenaar, and T. H. Rozijin, *Exp. Cell Res.* **80**, 313 (1973).
[5] B. Byers, in "Molecular Biology of the Yeast *Saccharomyces*: Life Cycle and Inheritance" (J. N. Strathern, E. W. Jones, and J. R. Broach, eds.), p. 59. Cold Spring Harbor Laboratory, Cold Spring Harbor, New York, 1981.
[6] C. H. Yang, E. J. Lambie, J. Hardin, J. Craft, and M. Snyder, *Chromosoma* in press (1989).

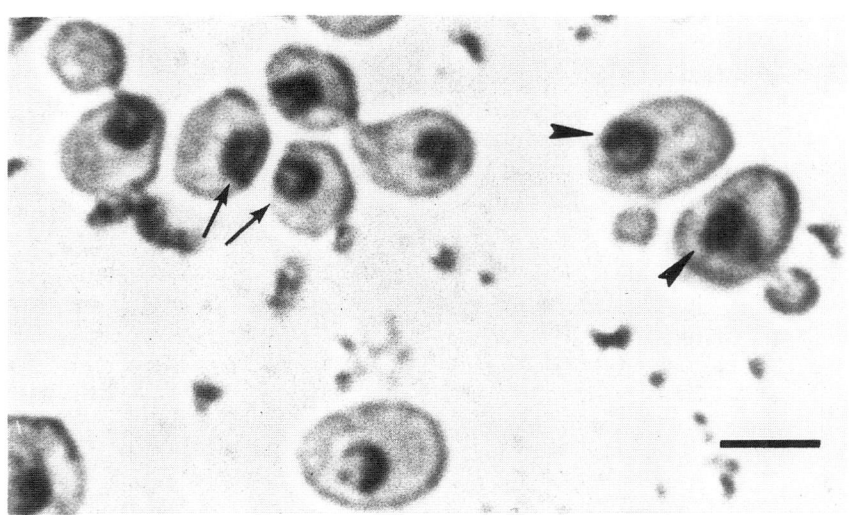

FIG. 1. Whole-mount, haploid yeast cells in which the nucleolus has been contrasted with the nucleolar-specific silver staining procedure. Arrowheads indicate the yeast nucleolus positioned opposite of the growing bud, showing a crescent-shaped nucleolus. Arrows indicate nucleoli showing the halo-like staining pattern. Bright-field illumination. Bar: 5 μm.

RNA processing components (see [42] in this volume) indicate that a high order of structural arrangement exists in the yeast nucleus. The specific orientation of the nucleolus in the nucleus and in the cell makes the nucleolus the perfect nuclear structure to utilize as a frame of reference in studies of not only nuclear organization but also the arrangement of cytoplasmic components.

A number of antibodies are now available that cross-react with yeast nucleolar components, such as SSB-1,[7] fibrillarin,[6,8] p38,[9,10] p90.[6] These antibodies can all be used in double-label immunofluorescence and immunoelectron microscopic experiments to map proteins relative to the yeast nucleolus (for immunomicroscopic methods, see [42], [51], and [52],

[7] A. Y.-S. Jong, M. W. Clark, M. Gilbert, A. Oehm, and J. L. Campbell, *Mol. Cell. Biol.* **7**, 2947 (1987).
[8] T.-H. Chang, M. W. Clark, A. J. Lustig, M. E. Cusick, and J. Abelson, *Mol. Cell. Biol.* **8**, 2379 (1988).
[9] J. P. Aris and G. Blobel, *J. Cell Biol.* **107**, 17 (1988).
[10] E. C. Hurt, A. McDowall, and T. Schimmang, *Eur. J. Cell. Biol.* **46**, 554 (1988).

this volume). Still, there are situations in which antibody mappings are difficult, as in some temperature-sensitive *(ts)* mutants, in growth curve experiments where there are many samples, or when a third label is required. For these cases, I describe in this chapter nucleolar-specific positive, cytochemical staining protocols that can be used easily and reproducibly in studies of yeast cellular morphology.

Materials

All the materials necessary for these procedures can be obtained from general chemical supply houses or from microscopic suppliers such as Polysciences (Warrington, PA).

Light Microscopy

The first cytochemical technique described here utilizes silver binding under acidic conditions to selectively stain certain nucleolar proteins. This technique was developed for studying the nucleolar organizing regions (NOR) on mammalian chromosomes.[11] A number of nucleolar-specific silver-binding proteins have been described for mammalian cells, particularly, nucleolin, also known as C23 (100–100 kDa),[12] B23 (36 kDa) (although for some investigators B23 is not a silver-binding protein),[12,13] and AgNOR (40 kDa).[14] Although the actual mechanisms of the nucleolar specificity of silver binding is still unclear, the phenomenon is reproducible. Nucleolar-specific silver stains have been used quite readily in conjunction with immunofluorescence microscopy to map the nucleolar-specific silver-binding proteins in relation to known nucleolar proteins.[15] I have modified the mammalian procedure slightly so it can be used for the yeast cell.[7] In attempting to silver stain the yeast nucleolus, a problem arose with nonspecific staining of the yeast vacuole. This vacuolar staining can be avoided by first incubating the cells in a Schiff base reagent. Thiebaut and co-workers[16] showed this treatment to reduce nonspecific staining in mammalian cells. The entire yeast nucleolar silver staining procedure is as follows.

[11] W. M. Howell, *in* "The Cell Nucleus" (H. Busch, ed.), Vol. 11, p. 89. Academic Press, New York, 1982.
[12] M. A. Lischwe, K. Smetana, M. O. J. Olson, and H. Bush, *Life Sci.* **25,** 701 (1979).
[13] D. L. Spector, R. L. Ochs, and H. Bush, *Chromosoma* **90,** 139 (1984).
[14] H. R. Hubbell, L. I. Rothblum, and T. C. Hsu, *Cell Biol. Int. Rep.* **3,** 615 (1979).
[15] R. L. Ochs, M. A. Lischwe, W. H. Spohn, and H. Busch, *Biol. Cell (1980)* **54,** 123 (1985).
[16] F. Thiebaut, J. P. Rigaut, and A. Reith, *Stain Technol.* **59,** 181 (1984).

Nucleolar-Specific Silver Staining Procedure for Whole-Mount Yeast Cells

Pretreatment of Cells to Remove Nonspecific Staining. This first step of the procedure is modified from Thiebaut and co-workers.[16]

A. Fix the cells in 3.7% formaldehyde in 0.1 M potassium phosphate, pH 7, buffer and mount them on glass slides as for immunofluorescence (see [40], this volume).

B. Apply 0.1–0.2 ml of Schiff reagent (see Appendix A at the end of this chapter) to the mounted cells. Incubate the slides at room temperature for 2 hr.

C. Wash the cells once with 2 ml of distilled water.

Silver Staining. This is the slightly modified procedure of Hernandez-Verdun and co-workers.[17]

A. Dissolve 1 g of $AgNO_3$ in 2 ml of distilled water.
B. Apply 0.1 ml of the 50% $AgNO_3$ solution onto the pretreated cells.
C. Incubate the slides at 55° in a closed moist chamber for 2 hr.
D. Rinse the slides once with 2 ml of distilled water.
E. Silver development
  1. Apply 0.1 ml of ammoniacal silver (see Appendix B at the end of this chapter) to the cells and let sit 1–2 min at room temperature.
  2. Rinse the slides with 0.2 ml of distilled water. (This rinse is only to remove excess ammoniacal silver from around the cells and to prevent extensive silver precipitation. If the cells are rinsed too much it will remove the ammoniacal silver from the cell, and the nucleolus will not stain. If this situation occurs, reapply the ammoniacal silver solution.)
  3. Apply 0.1 ml of the 3% formaldehyde developer (see Appendix B) to the cells.
  4. Watch the development under a light microscope at low magnification. When the nucleolar morphologies are visible, stop the development by rinsing with distilled water. (A precipitate might form during the development. Be careful, and if the precipitate forms very rapidly with the addition of the developer, rinse the slide with 0.1 ml of distilled water and reapply the developer.)
  5. At this stage these cells can be rinsed in buffer and subjected to the immunofluorescence staining procedure (see [40], this volume). The silver staining of the yeast nucleoli is stable for months.

---

[17] D. Hernandez-Verdun, H. J. Bourgeois, and C. A. Bouteille, *Chromosoma* **79**, 349 (1980).

6. After the staining is complete, mount the slides in 90% buffer glycerol and attach a cover slip (#1) with clear nail polish around the edges.

Figure 1 shows haploid yeast cells stained by the nucleolar-specific silver-binding procedure viewed with bright-field optical microscopy. Phase-contrast and Nomarski differential interference optics are adequate for viewing the nucleolar-specific silver staining, except that the contrast seen with the bright-field optics is somewhat reduced. The nucleolus can be seen as a single nuclear structure; this stucture appears as a cap that fits on the nucleus, taking up about one-third to one-half the nuclear volume. The diploid yeast cell nucleolus appears also as a single nuclear cap structure (M. W. Clark, personal observation). The cells shown in Fig. 1 were harvested in the early stages of logarithmic growth (0.3–0.4 OD_{600}). Cells examined from later stages of growth show a slightly smaller but similar caplike structure for the nucleolus. Figure 2 is a diagram of this nucleolar morphology, showing the orientation-dependent appearance of this structure. In a majority of the cells (~70%) (Fig. 2A; also see Fig. 1, arrowheads), the nucleolus is seen as the characteristic crescent, while in the remaining 30% of the cells the entire nucleus is covered (Fig. 2B; also see

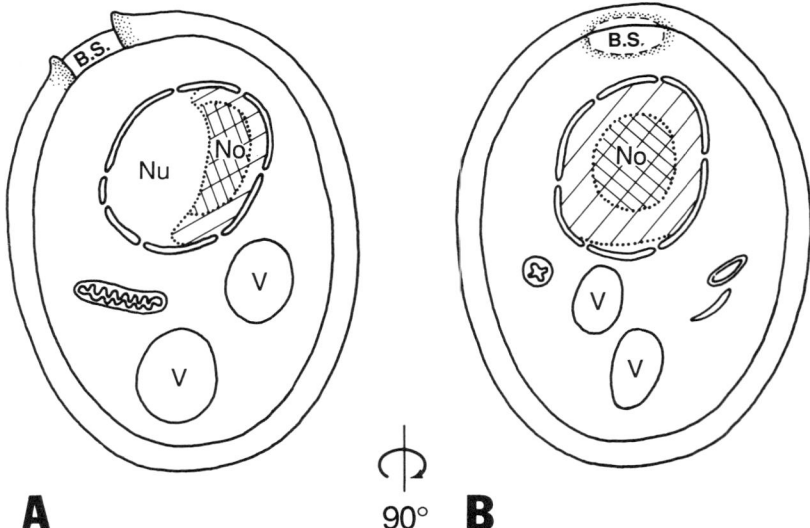

FIG. 2. Diagrammatic representation of two views of the yeast nucleolus. The nucleolus is shown as the cross-hatched areas. (A) The crescent pattern of the nucleolus seen in 70% of cells. (B) Turning of the cell 90° to produce the entire or halo-like pattern of the nucleolus seen in 30% of cells. No, Nucleolus; Nu, nucleoplasm; V, vacuole; B.S., bud scar.

Fig. 1, arrows). The latter staining pattern sometimes appears as a halo-like pattern. Such a characteristic staining pattern lead us first to suspect the yeast SSB-1 protein to be a nucleolar protein.[7] SSB-1 has since been shown to be not only a nucleolar protein but also a yeast nucleolar-specific silver-binding protein (M. W. Clark, unpublished result). The nucleoli of all yeast cell strains tested have reacted with the nucleolar-specific silver stain. The crescent or entire nuclear staining pattern was seen in all cells. Lack of nucleolar silver staining usually results from incomplete dispersal of the staining reagents over the slide surface.

As mentioned in the Introduction, the nucleolus has a specific orientation relative to the growing bud. Figure 1 (arrowheads) shows the nucleolus on the opposite side of the emerging bud. During mitotic cell growth the nucleolus will rotate slightly toward the bud pole of the cell. Then, as the nucleus elongates into the daughter cell, the nucleolus also elongates with the nucleus. At no time during this process does the nucleolus totally disassemble. The duplicate nucleoli simply separate with the duplicate nuclei.

A note of caution should be kept in mind when attempting to use the silver-binding procedure with immunofluorescence. It must be remembered that the silver developer has 3% formaldehyde and will further fix the cells. This treatment might interfere with some antibody-protein interaction. There is an alternative nucleolar-specific cytochemical procedure utilizing bismuth instead of silver. Bismuth staining of the nucleolus is a more gentle procedure and might be attempted.

Bismuth Staining of Nucleolus

Locke and Huie[18] introduced a procedure that will predominantly stain the cell nucleolus on formaldehyde-fixed tissue. This nucleolar-specific staining technique works as reported on yeast cells. The staining is very light compared to the silver staining procedure. Such a light staining of the nucleolus might be an attribute in some immunofluorescence experiments. Bismuth staining fades within a few hours. This instability may limit the usefulness of this technique. All reagents for the bismuth nucleolar-specific staining can be obtained from Polysciences. The protocol described below is the Locke and Huie procedure[18] modified by Gas and co-workers.[19]

Staining Solutions

Solution A, bismuth oxynitrate: Dissolve 400 mg sodium tartrate in 10 ml of 1 N NaOH (freshly made), then add dropwise to 200 mg of

[18] M. Locke and P. Huie, *Tissue Cell* **9**, 347 (1977).
[19] N. Gas, G. Inchauspe, M. Azum, and B. Stevens, *Exp. Cell Res.* **151**, 447 (1984).

bismuth oxynitrate ($BiONO_3 \cdot H_2O$, also called bismuth subnitrate or basic nitrate)

Solution B, Triethanolamine buffer: 0.2 M triethanolamine-HCl (TEA), pH 7.0

Bismuth staining solution: Add solution B to solution A in a 2:1 to 5:1 ratio and readjust the pH to 7.0–7.2 with 12 N HCl (this staining solution is unstable and should be used within 1–2 hr)

Staining Schedule

1. Fix the cells in 0.1 M sodium cacodylate buffer, pH 7.2, containing 4% paraformaldehyde/0.5% glutaraldehyde for 1 hr at room temperature or several hours in the cold. (After fixation, enzymatically remove the cell walls as you would for the immunofluorescence procedure, see [40] in this volume.)
2. Wash the cells in 50 mM sodium cacodylate-HCl, pH 7.2 (with 10% sucrose) for 30 min.
3. Then wash the cells in 0.1 M TEA, pH 7.0, for 30 min. These fixed cells are now mounted on polylysine-coated glass microscope slides.
4. Stain the mounted cells with the bismuth staining solution for 1 hr at room temperature.
5. Rinse the mounted cells with 0.2 ml of 0.1 M TEA, pH 7.0. Do this rinse 3 times for 10 min each.
6. React the mounted cells with 2% ammonium sulfide in the 0.1 M TEA, pH 7.0, buffer to reveal the bound bismuth as Bi_2S_3. The cells are then washed with distilled water.
7. At this point the cells can be stained for immunofluorescence (see [40] in this volume). Remember, these cells have not been through the methanol/acetone extractions called for in most immunofluorescence procedures; thus, the antibody solutions will require the addition of detergent to remove the membrane lipids of the cell and allow the antibodies to penetrate.

Electron Microscopy

Many cellular morphology studies will require higher resolution protein mapping than can be obtained by optical microscopy, so the electron microscope must be utilized. Two reasonably simple nucleolar-specific staining procedures for electron microscopy are described below. In the first procedure, a silver staining is done before the cells are infiltrated and embedded in the epoxy resin.

Nucleolar-Specific Silver Staining

Pretreatment of Cells to Remove Nonspecific Staining[16]

A. Fix the cells in 4% paraformaldehyde/0.5% glutaraldehyde in buffer and remove the cell walls by enzymatic digestion as described in [42] this volume.

B. Immerse the fixed cells in Schiff reagent (see Appendix A) for 0.5 hr at room temperature.

C. Wash the cells in distilled water. (A wash step consists of centrifuging cells to a pellet then resuspending the pelleted cells in the solution, followed by another centrifugation.)

Silver Stain[17]

A. Resuspend the fixed cells in a 25% aqueous silver solution and incubate at 55° for 5–10 min. Watch for a precipitate to form; if a precipitate forms, centrifuge the cells, wash them once in distilled water, and reapply the 25% silver.

B. Rinse the cells in distilled water.

C. Resuspend the cells in the ammoniacal silver solution (see Appendix B) and incubate for 1–2 min at room temperature.

D. Rinse the cells in distilled water.

E. Resuspend the cells in the 3% formaldehyde developer (see Appendix B), for 1–2 min at room temperature. Watch for a yellowish color to form; once it becomes obvious, stop the development by rinsing in distilled water.

F. Dehydrate the stained cells through an ethanol series and embed in Poly/Bed 812 resin or another resin of choice (see [41] or [42] in this volume).

This procedure will yield cell sections which have a nucleolus with granular silver throughout. These thin sections are perfectly adequate for most immunoelectron microscopic (IEM) mapping of nucleoplasmic and cytoplasmic proteins (see [42] in this volume). A problem does arise if you wish to use the cell sections for immune localization of a nucleolar protein with colloidal gold–antibody conjugates. The silver grains and the colloidal gold particles can look quite similar. An alternative to the silver staining procedure is also available. A procedure developed by Barnard[20] to visualize the ribonucleoproteins (RNP) in ultrathin sections can be used.

---

[20] M. Bernard, *J. Ultrastruct. Res.* **27**, 250 (1969).

This heavy metal staining procedure for the nucleolus is done after the IEM staining procedure ([42] in this volume) for a nucleolar protein.

Ribonucleoprotein Staining of Ultrathin Sections for Visualization of Yeast Nucleolus

1. For aldehyde fixation of the cells, 1% glutaraldehyde, 4% formalin/ 0.5% glutaraldehyde, or 2% acrolein/0.25% glutaraldehyde (all w/v) is required. Never use OsO_4; for yeast, 1% glutaraldehyde fixation works best.
2. The resins Poly/Bed 812 or LR White (Polysciences) work the best.
3. The section thickness only effects the Na_2EDTA exposure time: the thicker the section, the longer the exposure.
4. Between each step of the staining procedure described in the table below rinse the grids with 1–2 ml of distilled water.

Staining	Poly/Bed 812	LR White
a. 0.5% Aqueous uranyl acetate	2–3 min	2–3 min
b. 0.2 M Na_2EDTA	0.5–1 hr	0.5 hr
c. 2.6% Lead citrate (Reynolds)	1 min	1 min

5. a. Uranyl acetate stains the deoxyribonucleoprotein (DNP) and ribonucleoprotein (RNP).
   b. The Na_2EDTA preferentially removes the uranyl acetate from DNP faster than RNP.
   c. The lead citrate stain then enhances the remaining uranyl acetate: RNP, stain darker; DNP, stain lighter.

Figure 3 shows a thin section of a yeast cell stained with the RNP procedure. Small RNP structures can be seen in the periphery of the yeast nucleus within about 300 nm of the nuclear envelope (Fig. 3, brackets). These stained regions are most likely the regions of mRNA[8] and tRNA[21] transcription and processing that have been described by Abelson and co-workers. The nucleolus will be recognized by its darker staining over the rest of the nucleoplasm (Fig. 3, No). Martin and co-workers[22] have used this RNP staining technique on ultrathin sections of mammalian cells to demonstrate by immune localization the association of mammalian mRNA splicing components, the SM antigens and hnRNP proteins, with nuclear ribonucleoprotein complexes.

[21] M. W. Clark and J. Abelson, *J. Cell Biol.* **105**, 1515 (1987).
[22] S. Fakan, G. Leser, and T. E. Martin, *J. Cell Biol.* **98**, 358 (1984).

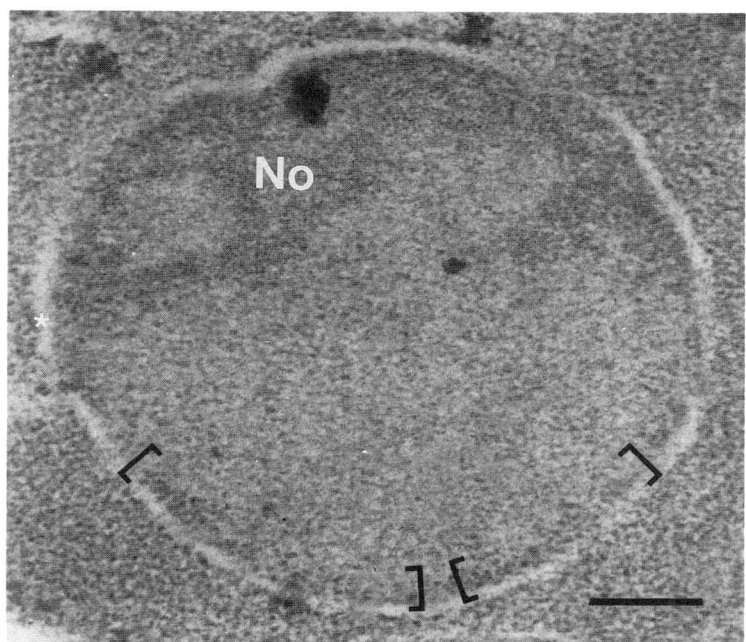

FIG. 3. Ultrathin section of a yeast nucleus that has been stained with the Bernard RNP staining technique to demonstrate the yeast nucleolus. The cell has been fixed with 1% glutaraldehyde, dehydrated through ethanol, and embedded in Poly/Bed 812 resin. Brackets show the RNP staining of the nuclear pheriphery. No indicates the position of the nucleolus, the white asterisk, the perinuclear lumen. Bar: 0.5 μm.

The Bernard RNP staining procedure has the benefit of being performed directly on the ultrathin sections. Thus, it is much easier to control the nucleolar staining density and reproduce the experiments. In contrast, the nucleolar-specific silver staining for electron microscopy is a preembedding procedure and, therefore, can present problems of over- or understaining of the nucleolus. This problem can be remedied only by terminating the experiment and starting again. The Bernard RNP staining procedure may be more compatible with the experimental design. Still, the nucleolar-specific silver stain provides more contrast of certain regions of the nucleolus[13] and might be of use in some investigations.

As more components of the yeast nucleus are identified, it will be necessary to know where these specific nuclear components are located relative to the rest of the yeast cell. The techniques described in this chapter provide methods for obtaining such data. Although, at present, the data are limited for the locations of specific yeast nuclear components,[5-10,21] these data demonstrate a very specific arrangement of components in the nucleus. These results can be interpreted to mean that there is a high order of structural arrangement occurring in the yeast nucleus. In view of the fact that these nuclear components are involved in major cellular processes, such as tRNA and mRNA transcription, RNA processing, and ribosome biogenesis, a highly developed structural organization for the yeast nucleus may be required for proper nuclear function. Further investigation of the structural arrangement of the yeast nucleus, at high resolution, will be required to elucidate whether the spatial ordering of events of nuclear processes affects the viability of yeast.

Appendix A: Modified Schiff Reagent, Decolorized Basic Fuchsin (Fuchsin–Sulfurous Acid)[23]

A. Basic fuchsin: Bring 200 ml of distilled water to a boil and stir in 1 g of the dye. Cool the solution to 50°, filter, then add HCl (see below).

B. Hydrochloric acid: Prepare 20 ml of 1 N HCl. Add to the dye solution and allow to cool to 25°. Then add sodium bisulfite.

C. Sodium bisulfite: Use 1 g of anhydrous sodium bisulfite. Add to the acidified dye solution.

D. Stir the red solution slowly at room temperature for 12–24 hr. The solution should turn to a pale yellow color when ready for use. Keep this solution in a dark, tightly sealed bottle. If the stain becomes red or dark orange, or if the dye precipitates, discard it.

Appendix B: Ammoniacal Silver Solution and Formaldehyde Developer

Prepare these solutions just prior to use.

Ammoniacal silver solution:[17] Combine 2.5 ml distilled water and 2.5 ml NH_4OH. Add 1.5 g $AgNO_3$ slowly to the above solution. If the solution turns brown, discard it.

3% Formaldehyde developer:[17] Combine 91.9 ml distilled water and 8.1 ml 37% formaldehyde. Add solid sodium acetate to make the pH 7, then add concentrated formic acid (1 or 2 drops) to make the pH 5–6.

Acknowledgments

I thank J. Abelson for providing laboratory space and support that allowed me to do this work.

---

[23] A. E. Galiger and E. N. Kozloff, *in* "Essentials of Practical Microtechnique," Lea & Febiger, Philadelphia, Pennsylvania, 1971.

[51] Staining of Actin with Fluorochrome-Conjugated Phalloidin

By ALISON E. M. ADAMS and JOHN R. PRINGLE

Introduction

Yeast cells contain actin[1] and a variety of actin-associated proteins.[2-4] These proteins are arranged in a network of cytoplasmic fibers and cortical patches (which may represent sites of attachment of fibers to the cell membrane).[1,2,4-6] This network displays distinctive, highly asymmetric arrangements in relation to sites of morphogenetic activity in both vegetative and mating (or shmooing) cells; in particular, the cortical patches cluster at, and the cytoplasmic fibers tend to impinge on, sites of new cell-surface deposition. Thus, visualization of actin or the associated proteins provides a graphic and sensitive indication of the morphogenetic activity of the cell. Although such visualization can be achieved by immunofluorescence,[7] it is sometimes convenient to use a simpler procedure utilizing phalloidin, a bicyclic heptapeptide (MW ~800) toxin produced by the mushroom *Amanita phalloides*.

Phalloidin interacts specifically with yeast[8] and other [9] actins; the interaction appears to be specific for polymerized (F) rather than unpolymerized (G) actin. The coupling of fluorochromes to phalloidin thus provides a quick and convenient means of visualizing the actin cytoskeleton in various types of cells; in some cases, staining of living as well as fixed cells has been achieved.[10,11] In yeast, phallotoxin staining has been accomplished to date only with fixed cells; this staining reveals patterns of localization very similar to those seen by immunofluorescence,[1,5] except that at least some

---

[1] J. V. Kilmartin and A. E. M. Adams, *J. Cell Biol.* **98**, 922 (1984), and references cited therein.
[2] D. G. Drubin, K. G. Miller, and D. Botstein, *J. Cell Biol.* **107**, 2551 (1988).
[3] V. Magdolen, U. Oechsner, G. Müller, and W. Bandlow, *Mol. Cell. Biol.* **8**, 5108 (1988).
[4] H. Liu and A. Bretscher, *Cell (Cambridge, Mass.)* **57**, 233 (1989).
[5] A. E. M. Adams and J. R. Pringle, *J. Cell Biol.* **98**, 934 (1984).
[6] P. Novick and D. Botstein, *Cell (Cambridge, Mass.)* **40**, 405 (1985).
[7] J. R. Pringle, A. E. M. Adams, D. G. Drubin, and B. K. Haarer, this volume [40].
[8] C. Greer and R. Schekman, *Mol. Cell. Biol.* **2**, 1270 (1982).
[9] J. Vandekerckhove, A. Deboben, M. Nassal, and T. Wieland, *EMBO J.* **4**, 2815 (1985), and references cited therein.
[10] E. Wulf, A. Deboben, F. A. Bautz, H. Faulstich, and T. Wieland, *Proc. Natl. Acad. Sci. U.S.A.* **76**, 4498 (1979).
[11] H. Faulstich, H. Trischmann, and D. Mayer, *Exp. Cell Res.* **144**, 73 (1983).

antibodies appear to stain the cytoplasmic fibers more brightly, relative to the cortical patches, than do the phalloidin probes. A particular advantage of phallotoxin staining in work with yeast is that the low molecular weight stain (in contrast to antibodies) can enter cells whose walls have not been digested. This facilitates double-label experiments in which the distribution of actin and the sites of incorporation of new cell wall material are visualized simultaneously.[1,5]

We have had good results with both the fluorescein and tetramethylrhodamine derivatives of phalloidin[10-12] (Molecular Probes[13] Cat. Nos. F-432 and R-415, respectively). Each of the following protocols has worked well with a variety of strains (wild-type and mutant) grown under a variety of conditions. For reasons that are not fully clear (but seemingly involving difficulties in getting effective washes), we have had less success with a third protocol in which formaldehyde-fixed cells were attached to a multiwell slide with polylysine and then stained without removing the cell walls. As with other microscopy methods, staining of bigger cells is generally more informative and facilitates photomicroscopy; thus, diploids are generally more satisfactory than haploids, and tetraploids (when available) are better yet. The importance of rapid fixation should also be stressed; like many other aspects of cell structure, the actin network rearranges rapidly when the cells are subjected to stresses such as the loss of an energy source during washes with glucose-free buffer.[14]

Rapid Staining

The following procedure is quick, easy, and readily combined with staining of cell wall components using fluorochrome-conjugated concanavalin A (Con A)[14] or Calcofluor.[15] All steps are conducted at room temperature.

1. Fix cells by adding concentrated formaldehyde solution directly to the growth medium to a final concentration of 3.7–5%. After a few minutes, the cells can be recovered by centrifugation and resuspended in buffered fixative,[7,14] but this does not appear necessary in most cases. Fix for 0.5–3 hr before proceeding. If desired, the cells can be stained prior to fixation with fluorochrome-conjugated Con A or after fixation with Calcofluor.

[12] T. Wieland, T. Miura, and A. Seeliger, *Int. J. Pept. Protein Res.* **21**, 3 (1983).
[13] Molecular Probes, Inc., P.O. Box 22010, 4849 Pitchford Avenue, Eugene, OR 97402; telephone (503) 344-3007.
[14] J. R. Pringle, R. A. Preston, A. E. M. Adams, T. Stearns, D. G. Drubin, B. K. Haarer, and E. W. Jones, *Methods Cell Biol.* **31**, 357 (1989).
[15] J. R. Pringle, this volume [52].

2. Collect approximately 10^7 cells by centrifugation, wash 2 or 3 times in water or phosphate-buffered saline (PBS),[7] then resuspend in 100 μl PBS.

3. Stain cells with 0.2–1.5 μM fluorochrome-conjugated phalloidin for 30–90 min in the dark. [Note that the fluorochrome-conjugated phalloidins are presently supplied by Molecular Probes as 3.3 μM stock solutions in methanol. We have achieved good staining both (a) by simply adding 6–80 μl of this stock solution to the 100 μl of cell suspension and (b) by first drying down the stock solution in a Speed-Vac concentrator, redissolving at a higher concentration in PBS, then diluting appropriately into the cell suspension.]

4. Wash the cells 5 times with PBS by centrifugation, resuspend in a drop of p-phenylenediamine-containing mounting medium (the p-phenylenediamine retards photobleaching),[7] and observe using the standard fluorescein or rhodamine filter sets.[7] Although we have not tried these variations, it may be possible to immobilize the washed cells with polylysine[7] or Con A[14] immediately before mounting.

Staining in Conjunction with Immunofluorescence

The following method is designed specifically for combining phalloidin staining of actin with immunofluorescence localization of some other component of interest.

1. Follow the immunofluorescence protocol[7] until the washes following treatment with secondary antibody have been completed. [One caution should be noted: the methanol and acetone treatments necessary to visualize actin (and presumably some other proteins) by immunofluorescence[7] appear to destroy the ability of actin to interact with phalloidin.]

2. Immediately following removal of the last wash, add to each well 6–10 μl of a solution of fluorochrome-conjugated phalloidin (the 3.3 μM stock supplied by Molecular Probes diluted 5- to 15-fold with PBS or with 40 mM potassium phosphate, 0.5 mM $MgCl_2$, pH 6.5).

3. Stain for 5–30 min (depending on the phalloidin concentration) at room temperature in the dark.

4. Remove the phalloidin solution by aspiration, wash several times with the buffer used in Step 2, and mount the cells in p-phenylenediamine-containing mounting medium.[7] (In working with the lower concentrations of fluorochrome-conjugated phalloidin, the washes are not essential.)

[52] Staining of Bud Scars and Other Cell Wall Chitin with Calcofluor

By JOHN R. PRINGLE

Introduction

Shortly before bud emergence, a ring of chitin is formed in the *Saccharomyces cerevisiae* cell wall; the bud then emerges within the confines of this chitin ring, which remains in place as the bud grows.[1-6] Subsequently, a chitin-rich "primary septum" forms within the confines of the chitin ring as an early step in the separation of mother and daughter cells.[7] After division, the chitin ring and septum remain on the mother cell as the craterlike "bud scars."[8] For reasons that remain somewhat unclear,[5] the chitin-rich structures of the *S. cerevisiae* cell wall can be stained with high specificity using the fluorescent dye Calcofluor White M2R New [Chem. Index 40622; also known as Fluorescent Brightener 28 (Sigma, St. Louis, MO, Cat. No. F6259) or Cellufluor (Polysciences,[9] Cat. No. 17353)], as first reported by Hayashibe and Katohda.[1] Staining with Calcofluor is thus very useful (1) in distinguishing mother from daughter cells, distinguishing mother cells of different ages, and monitoring the success of procedures designed to separate these different classes of cells; (2) in determining the effects of genetic background, mating type, specific genetic lesions, and environmental conditions on the localization of budding sites; and (3) in

---

[1] M. Hayashibe and S. Katohda, *J. Gen. Appl. Microbiol.* **19**, 23 (1973).
[2] O. Seichertová, K. Beran, Z. Holan, and V. Pokorný, *Folia Microbiol. (Prague)* **18**, 207 (1973).
[3] E. Cabib and B. Bowers, *J. Bacteriol.* **124**, 1586 (1975).
[4] B. F. Sloat and J. R. Pringle, *Science* **200**, 1171 (1978).
[5] J. R. Pringle, R. A. Preston, A. E. M. Adams, T. Stearns, D. G. Drubin, B. K. Haarer, and E. W. Jones, *Methods Cell Biol.* **31**, 357 (1989).
[6] More complete documentation of this and the other assertions in this paragraph can be found in Ref. 5. In particular, Ref. 5 considers at length the issue of the degree to which Calcofluor staining is specific for chitin, discusses other means of visualizing chitin and other cell wall constituents, and provides additional references illustrating the various uses to which Calcofluor staining has been put.
[7] R. L. Roberts, B. Bowers, M. L. Slater, and E. Cabib, *Mol. Cell. Biol.* **3**, 922 (1983).
[8] A scar (the "birth scar") is also left on the daughter cell and can be visualized by scanning electron microscopy [L. T. Talens, M. Miranda, and M. W. Miller, *J. Bacteriol.* **114**, 413 (1973)]. However, the birth scar contains little or no chitin and has been difficult or impossible to visualize by fluorescence methods.
[9] Polysciences, Inc., 400 Valley Road, Warrington, PA 18976-2590; telephone (800)-523-2575.

assessing the effects of mutations, drugs, and physiological effectors that induce abnormal patterns of cell wall deposition.

Staining Procedures

Calcofluor is easily used according to the following protocols.

1. Make a 1 mg/ml stock solution of Calcofluor in water (be sure to get all of the dye into solution; this may require several hours of stirring at room temperature) and store at 4°. It is best to keep the Calcofluor in the dark (use a brown and/or foil-covered bottle), both while dissolving it and during subsequent storage. This stock usually remains good for several weeks or longer but occasionally deteriorates more rapidly, as manifested by the appearance of overt precipitate in the stock or of brightly fluorescent specks in stained cell preparations. (Note that some protocols in circulation call for dissolving the Calcofluor in phosphate buffers. However, this approach appears to be less satisfactory than simply using water, perhaps because of reduced solubility of the dye in such buffers.)

2. Simply to observe chitin rings and bud scars, collect live or formaldehyde-fixed[10] cells by centrifugation. It may sometimes help to wash once with water at this stage, but this is not generally necessary. Resuspend the cells in Calcofluor solution (the undiluted stock or a 2- to 10-fold dilution of this) and incubate at room temperature for approximately 5 min. Depending on the strain and the growth conditions (the relevant variables are not well understood[5]), it may be necessary to adjust the Calcofluor concentration, the staining time, or both in order to get good differential staining of the chitin rings and bud scars. Wash the cells 2–5 times with water by centrifugation, mount on a slide, and observe using UV excitatory illumination (e.g., Leitz filter set A or Zeiss filter set 487702). Optimal photomicroscopy requires immobilization of the cells, which we have generally accomplished (with cells mounted in water or buffer) simply by allowing the slide to dry enough that the cells are gently squeezed between the slide and coverslip. This has the additional advantage (particularly helpful for mutants with complex morphologies) of squeezing the cells more nearly into one focal plane, thus allowing more of the structure to be captured in a single photograph. Alternatively, it should be possible to immobilize Calcofluor-stained cells with concanavalin A,[5] low melting point agarose,[5] or polylysine.[11]

3. It is sometimes useful to observe nuclei and bud scars simultaneously. Although several methods have been used, it seems that the best

[10] A. E. M. Adams and J. R. Pringle, this volume [51].
[11] J. R. Pringle, A. E. M. Adams, D. G. Drubin, and B. K. Haarer, this volume [40].

results are obtained with the least effort by simply staining fixed cells with Calcofluor and washing as just described, then mounting them on a slide in 4′,6′-diamidino-2-phenylindole (DAPI)-containing mounting medium.[11] Staining of the nuclear and mitochondrial DNA is good after 30 min, and both Calcofluor and DAPI are visible using the same filter set (see above).

4. It is sometimes useful to observe chitin rings and actin simultaneously. This can be accomplished by staining with Calcofluor and washing as described above, then staining with fluorochrome-conjugated phalloidin using standard procedures.[10]

5. It is sometimes useful to perform immunofluorescence on cells that have also been stained with Calcofluor. This can be done by staining fixed cells with Calcofluor and washing as described above, then preparing them for immunofluorescence by standard procedures.[11] However, the procedures normally used to permeabilize the cells to antibodies destroy enough of the cell wall that most bud scars become detached from their cells of origin.[12] Thus, the original spatial relationships are lost, and the double-label images are not very informative. Recently, it has been found that a moderation of the usual cell wall digestion conditions can produce cells with walls that retain sufficient structural integrity to hold the bud scars in place, yet are permeable to antibodies. Either a short digestion of the wall preceding a relatively short fixation[13] or a normal long fixation followed by a short digestion of the wall[14] was effective.

Comments

Photobleaching. Photobleaching of Calcofluor is not normally a problem, in part because the staining is so bright to begin with. However, photobleaching is not always negligible, and it may be a problem in some applications. I am not certain of the factor(s) responsible for the variable photobleaching that various laboratories have observed, but one factor appears to be the intensity and wavelengths of the excitatory illumination obtained with different microscopes and filter sets. In limited trials, p-phenylenediamine did not seem to retard appreciably the moderate rate of photobleaching observed with our Leitz Orthoplan microscope using filter set A.

Vital Staining. Cells are brightly stained after growth for approximately 2 hr in medium containing 25–500 µg/ml Calcofluor.[15,16] The staining

[12] B. K. Haarer and J. R. Pringle, *Mol. Cell. Biol.* **7**, 3678 (1987).
[13] R. Palmer, M. Koval, and D. Koshland, *J. Cell Biol.* **109**, 3355 (1989).
[14] R. Palmer and D. Koshland, personal communication (1989).
[15] M. V. Elorza, H. Rico, and R. Sentandreu, *J. Gen. Microbiol.* **129**, 1577 (1983).
[16] C. Roncero, M. H. Valdivieso, J. C. Ribas, and A. Durán, *J. Bacteriol.* **170**, 1950 (1988).

appears still to be specific for chitin, but it is distinctly abnormal in pattern, reflecting the fact that Calcofluor produces severe perturbations of cell wall biogenesis, culminating in complete growth arrest at the higher concentrations. Thus, any use of Calcofluor as a vital stain should proceed with appropriate caution.

Specificity of Staining. The degree to which the Calcofluor staining of the *S. cerevisiae* cell wall is really specific for chitin per se, as opposed to structural features of the cell wall at the base of the bud and in the bud scar, has long been controversial.[5] For many of the applications of Calcofluor staining, it does not matter how this controversy is resolved. However, it can be an important issue when the dyes are used for studies of cell wall biogenesis. As Calcofluor is manifestly not specific for chitin in all contexts (e.g., it stains both cellulose in plant and algal cell walls and some unknown component of *Schizosaccharomyces pombe* cell walls), abnormal Calcofluor staining should be regarded as indicative, but not proof, of abnormal chitin deposition.

[53] Isolation of Yeast Nuclei

By JOHN P. ARIS and GÜNTER BLOBEL

Introduction

It is generally believed to be more difficult to isolate nuclei from yeast than from higher eukaryotic cells, such as cultured cells or animal tissues. For example, a convenient method is available for preparing nuclei from rat liver.[1] However, a number of methods for the purification of yeast nuclei have been available for quite a while. Effective methods were described by Rozijn and Tonino (1964) and by May (1971) for *Saccharomyces carlsbergensis,* and these provided the basis for recent methods applicable to *Saccharomyces cerevisiae,* including the one presented in this chapter.[2,3] These two methods, among others, were reviewed in this series by Duffus in 1971.[4]

During the past few years a number of methods for the isolation of yeast nuclei have been described. This proliferation of methods has re-

[1] G. Blobel and V. R. Potter, *Science* **154,** 1662 (1966).
[2] T. H. Rozijn and G. J. M. Tonino, *Biochim. Biophys. Acta* **91,** 105 (1964).
[3] R. May, *Z. Allg. Mikrobiol.* **11,** 131 (1971).
[4] J. H. Duffus, *in* "Methods in Cell Biology" (D. M. Prescott, ed.), Vol. 11, p. 77. Academic Press, New York, 1975.

sulted from an increasingly keen desire to study various processes and structures occurring within the nucleus in this unicellular eukaryote. We mention some of these methods to provide the reader with an experimental context for the methodology presented in this chapter, as well as to cite alternative methods. The reader is also referred to a recent chapter by Lohr which contains a detailed discussion of various considerations relevant to the purification of yeast nuclei.[5]

Virtually all procedures used currently for purifying nuclei begin with the enzymatic removal of the yeast cell wall to yield spheroplasts. Typically, spheroplasts are lysed in hypotonic solution to release nuclei. However, nuclei isolated by French press disruption of spheroplasts have been used to study DNA uptake.[6] The majority of methods employ a medium for hypotonic lysis containing 18 or 20% Ficoll 400 and 20 mM potassium phosphate, pH 6.5. The use of Ficoll in hypotonic buffer, attributable to May, acts to prevent lysis of vacuoles, which reduces proteolysis during purification.[3] Nuclei in Ficoll buffer are separated from other cellular constituents by a number of means: differential centrifugation,[5,7] Ficoll cushions,[8,9] Percoll gradients,[10] step gradients of glycerol and Ficoll,[11-14] Ficoll step gradients,[15] and sucrose step gradients.[16] An alternative to hypotonic lysis is the lysis of spheroplasts with buffer containing nonionic detergent. An unpublished procedure of P. Baum and J. Thorner (University of California, Berkeley) has been used by a number of investigators for the isolation and study of residual nuclear structures.[17-19]

In this chapter we describe a method for isolating nuclei from *Saccharomyces cerevisiae* that includes some features of previously described methods.[7-9] This method may also be employed to isolate nuclei from other fungi, such as *Schizosaccharomyces pombe* and *Neurospora crassa*. In addition, a method for the fractionation of yeast nuclei is described. We

[5] D. Lohr, *in* "Yeast, A Practical Approach" (I. Campbell and J. H. Duffus, eds.), p. 125. IRL Press, Oxford and Washington, D.C., 1988.
[6] E. Tsuchiya, S. Shakuto, T. Miyakawa, and S. Fukui, *J. Bacteriol.* **170,** 547 (1988).
[7] U. Wintersberger, P. Smith, and K. Letnansky, *Eur. J. Biochem.* **33,** 123 (1973).
[8] K. Mann and D. Mecke, *FEBS Lett.* **122,** 95 (1980).
[9] S. Ruggieri and G. Magni, *Physiol. Chem. Phys.* **14,** 315 (1982).
[10] G. J. Ide and C. A. Saunders, *Curr. Genet.* **4,** 85 (1981).
[11] C. Szent-Gyorgyi and I. Isenberg, *Nucleic Acids Res.* **11,** 3717 (1983).
[12] J. F. Jerome and J. A. Jaehning, *Mol. Cell. Biol.* **6,** 1633 (1986).
[13] J. L. Allen and M. G. Douglas, *J. Ultrastruct. Mol. Struct. Res.* **102,** 95 (1989).
[14] J. F. Kalinich and M. G. Douglas, *J. Biol. Chem.* **264,** 17979 (1989).
[15] J. P. Aris and G. Blobel, *J. Cell Biol.* **107,** 17 (1988).
[16] E. C. Hurt, A. McDowall, and T. Schimmang, *Eur. J. Cell Biol.* **46,** 554 (1988).
[17] J. A. Potashkin, R. F. Zeigel, and J. A. Huberman, *Exp. Cell Res.* **153,** 374 (1984).
[18] L. C. Wu, P. A. Fisher, and J. R. Broach, *J. Biol. Chem.* **262,** 883 (1987).
[19] M. N. Conrad and V. A. Zakain, *Curr. Genet.* **13,** 291 (1988).

have used the methods described in this chapter in order to prepare a panel of monoclonal antibodies against the yeast nucleus, as well as in the study of protein components of the yeast nucleus and their homologs in mammalian cells.[15,20,21]

Media, Solutions, and Reagents

The volumes listed below are sufficient for the preparation of nuclei from 12 liters of yeast culture. Prepare the media, solutions, and reagents the day before the preparation of nuclei. All buffers and solutions are stored at 0–4°. Sterilized media are kept at approximately 25°.

Media

YPD medium, 12 liters: 1% (w/v) Bacto-yeast extract, 2% (w/v) Bacto-peptone, 2% (w/v) glucose; autoclave 2 liters in each of six 6-liter flasks for 30 min and cool to 25 –30°

SPH medium, 500 ml: 1 M sorbitol, 1% (w/v) glucose, 0.2% (w/v) Bacto-yeast nitrogen base (complete), 0.2% (w/v) Bacto-casamino acids (certified), 25 mM HEPES (free acid), 50 mM Tris base; autoclave as above; the initial pH will be approximately 8.2, and the pH after sterilization will be about 8.0

Solutions and Reagents

2× MES–Tris stock, 100 ml: 1.10 g MES [2-(N-morpholino)ethanesulfonic acid, free acid], 0.45 g Tris base, 4 g Ficoll 400; the pH should be approximately 6.5; add Ficoll slowly with rapid stirring

Cushion buffer, 50 ml, and wash buffer, 100 ml: 1.5 M sorbitol (cushion) or 1.2 M sorbitol (wash), 2% (w/v) Ficoll 400, 25 mM MES–Tris, pH 6.5; dissolve the appropriate amount of sorbitol in 25 ml (cushion) or 50 ml (wash) of 2× MES–Tris stock buffer and add double-distilled water to obtain the 1× concentrations listed here

2× PO_4–Mg stock, 200 ml: 2.5 ml 1 M K_2HPO_4, 5.5 ml 1 M KH_2PO_4, 0.4 ml 1 M $MgCl_2$; the pH should be 6.50 ± 0.02 pH units

50% Ficoll buffer, 200 ml: 50% Ficoll 400, 20 mM potassium phosphate, pH 6.45, 1 mM $MgCl_2$

To prepare 50% Ficoll buffer, dissolve 100 g of Ficoll 400 in 100 ml of 2× PO_4–Mg stock buffer and add double-distilled water to obtain the 1× concentrations listed here. Add the Ficoll slowly with continuous stirring

[20] J. P. Aris and G. Blobel, *J. Cell Biol.* **108**, 2059 (1989).
[21] R. Henríquez, G. Blobel, and J. P. Aris, *J. Biol. Chem.* **265**, 2209 (1990).

and heat the solution to 50°–60° in a covered beaker. Begin dissolving the Ficoll in the morning; it requires approximately 8 hr. The 50% Ficoll buffer will serve as a stock solution for the 30 and 40% Ficoll buffers, and as well will be used directly in the nuclei preparation. The pH of the 50% Ficoll stock should be 6.60 ± 0.02. The addition of the protease inhibitor cocktails will reduce the pH by about 0.15 unit to 6.45. It is important that the Ficoll solutions be at a pH value of 6.45 ± 0.02.

Lysis buffer, 200 ml: 20% (w/v) Ficoll 400, 20 mM potassium phosphate, pH 6.45, 1 mM MgCl$_2$; combine 80 ml of the 50% Ficoll stock buffer, 60 ml of the 2× PO$_4$–Mg stock, and 60 ml double-distilled water

30% Ficoll buffer and 40% Ficoll buffer, 50 ml each: 30 or 40% (w/v) Ficoll 400, 20 mM potassium phosphate pH 6.45, 1 mM MgCl$_2$; dilute either 30 or 40 ml of the 50% Ficoll stock buffer with the appropriate volumes of the 2× PO$_4$–Mg stock buffer and double-distilled water.

Protease Inhibitor Cocktails (PIC)

1000× PIC-D in dimethyl sulfoxide (DMSO), 1 ml: 88 mg Phenylmethylsulfonyl fluoride (PMSF) (0.5 M), 5 mg pepstatin A, 1 mg chymostatin, approximately 950 μl DMSO

1000× PIC-W in water, 1 ml: 208 mg p-Aminobenzamidine (1 M), 131 mg ε-Aminocaproic acid (1 M), 5 mg aprotinin, 1 mg leupeptin, approximately 700 μl water

Dilute both PIC solutions 1/1000 in buffer immediately prior to use. Add PIC-D with rapid stirring below the surface of the liquid.

Growth of Yeast

The protease-deficient haploid *Saccharomyces cerevisiae* strain BJ2168 is the yeast for which this method has been optimized (BJ2168, **a** *prcl-407 prbl-1122 pep4-3 leu2 trpl ura3-52*). For our standard preparation, 12 liters of BJ2168 is grown overnight at 30° in YPD medium using an incubator–shaker. Begin growing BJ2168 2 days before the nuclei preparation by inoculating an overnight culture (5 ml) from a fresh YPD plate. On the day before, dilute the overnight culture into a 50-ml culture volume such that the OD$_{600}$ value will be 1–2 at a convenient time (6–9 PM). In the preparation below, an OD$_{600}$ value of 1 is equivalent to ~4 × 10^7 cells/ml. At this density the yeast are in the mid-log growth phase. Calculate the appropriate dilution assuming a doubling time of about 100 min (usually 1/100). Check the doubling time of BJ2168 in the 50-ml culture. Calculate the volume from the 50-ml starter culture to be added to each of the six 2-liter volumes of YPD to give an OD_{600} of 1 after the overnight growing period. Add starter culture to the 2-liter volumes of YPD equilibrated to 30° and agitate at a moderate speed (200–250 rpm).

Preparation of Spheroplasts and Isolation of Nuclei

Method

1. Harvest the yeast at an OD_{600} of approximately 1 by placing the flasks on ice and centrifuging the culture in a cold rotor for 5 min at about 4000 g (e.g., Sorvall GS-3 rotor at 5000 rpm). If the OD_{600} is below 1, begin collecting the yeast without delay. If the OD_{600} is greater than 1, begin centrifugation after the yeast have been chilled to about 20°. If a Pellicon cell concentrator (Millipore, Bedford, MA) is available, it is convenient to concentrate the yeast before centrifugation. Place cell pellets on ice. All steps in this method are conducted at 0°–4°, *except* for the lyticase digestion and lysis of spheroplasts.

2. Resuspend and pool the yeast cells with about 180 ml of SPH medium. Resuspend the yeast cell pellets with a glass stirring rod and/or pipetting. Transfer the yeast to six 40-ml centrifuge tubes (tare two of the tubes). Spin for 5 min at approximately 3000 g (e.g., Sorvall SS34 tubes and rotor at 5000 rpm). Remove the supernatant. The yeast pellets should weigh about 3.0 g each.

3. To digest the cell wall, prepare 27 ml of SPH medium for each 3.0-g yeast pellet as follows: stir in 10,000 units of lyticase (Sigma, St. Louis, MO) and add 60 μl of a 1 M dithiothreitol (DTT) solution (2 mM final concentration). The lyticase may not completely dissolve. Resuspend each yeast pellet in 27 ml with a stirring rod, and place the centrifuge tubes in a water bath at 30° for 1 hr. Stir with the stirring rod every 5 min. Do not cap tubes. The yeast should grow actively and liberate CO_2 during the last 30 min of digestion. To ensure optimum purity and yield of nuclei, spheroplasting and lysis should be monitored microscopically during the preparation. Compared to undigested yeast, the spheroplast sample after 1 hr of digestion should show all spherical cells, extensive clumping, and about 5% lysis (liberated organelles will be visible).

4. Chill the tubes on ice for 5 min. Freshly add PIC solutions to the cushion buffer with stirring. Pipette 5 ml of cushion buffer below the SPH medium in the tubes. Centrifuge in a swinging-bucket rotor for 5 min at 6000 g (e.g., Sorvall HB-4 rotor at 6000 rpm).

5. Resuspend each of the pellets with approximately 15 ml of the wash

buffer containing freshly added PIC. Resuspend the spheroplasts carefully with gentle pipetting. Spin for 5 min at about 3000 g as in Step 2.

6. To lyse spheroplasts, add 27 ml of lysis buffer containing freshly added PIC, at ambient temperature, to each pellet. Quickly disperse the pellet with a stirring rod and pour into a 40-ml Dounce homogenizer. Use 20 strokes with a loose-fitting pestle, which should take about 3–4 min. Return the lysate to the centrifuge tube and chill on ice for at least 10 min. After lysis, the spheroplasts should be completely disrupted and the clumps dispersed.

7. Centrifuge the lysate in a swinging-bucket rotor for 5 min at 13,000 g (e.g., Sorvall HB-4 rotor at 9000 rpm). Pour the supernatant into a second centrifuge tube, leaving most of the loose pellet behind. Centrifuge again at 13,000 g for 10 min. A large, tight pellet should be visible. Place the tubes on ice for at least 10 min.

8. Set up the Ficoll step gradients during the previous steps. To pour the step gradients, warm the 50% Ficoll buffer to 35°–45°, add the PIC with stirring, and pipette 5 ml into the bottom of each of six Beckman SW28 ultraclear tubes. Place the tubes on ice (it is convenient to use a separate ice bucket for the gradients). Warm the 40% Ficoll buffer, add the PIC, and slowly pipette 5 ml on top of the chilled 50% Ficoll layers. Repeat this process with the 30% Ficoll buffer. Transfer approximately 25 ml of supernatant from each of the tubes from the 13,000 g spin to each of the SW28 tubes containing the chilled three-step gradients. Do not transfer any of the pellet. Top off the SW28 tubes with cold 1× PO_4–Mg buffer plus PIC if necessary. Centrifuge the gradients in a Beckman SW28 rotor at 18,000 rpm (58,400 g) at 2° for 60 min.

9. Nuclei occur in the 40% layer and at the 30–40% and 40–50% interfaces (bands are usually visible). A layer of white film should be visible at the top of the gradient. An optically dense layer should be visible at the 20–30% interface. A small pellet, with a halo appearance, should occur at the bottom. Carefully remove the film at the top of the gradient and about half of the 20% Ficoll load volume (aspiration works well). Collect nuclei with a 10-ml pipette connected to an automatic pipette pump. Slowly insert the pipette tip through the 20% layer and the 20–30% interface to a position just below the 40–50% interface at, or near, the wall of the tube. Keep a bubble at the pipette tip to prevent contamination. Slowly pipette 8–10 ml while manually rotating the tube slowly and keeping the pipette tip in position. This will require 1–2 min because the cold Ficoll suspension is viscous. Remove the pipette, wipe its exterior, and collect nuclei in a beaker on ice.

10. Repeat the Ficoll step gradient with four Ficoll gradients as in Step 8. Dilute approximately 48 ml of nuclei with 42 ml of cold 1× PO_4–Mg

buffer plus PIC (prepared from the 2× PO_4-Mg stock) using gentle stirring. After centrifugation, collect the nuclei from the 40–50% interface as above. Pool nuclei and stir gently with a glass rod to make a uniform suspension. The nuclei may be collected for experimental purposes by 10-fold dilution in cold 1× PO_4-Mg buffer plus PIC followed by centrifugation for 10 min at 10,000 g. For storage, aliquot into 1 to 10-ml volumes as a Ficoll suspension, freeze in liquid nitrogen, and place at −70°.

Results

A simple way to evaluate the isolation of nuclei is to view the nuclei directly in the Ficoll suspension by light microscopy. Nuclei appear dark, with the crescent-shaped nucleolus sometimes evident (Fig. 1c). Membrane(s) associated with nuclei may be visible (Fig. 1c, lower left). Electron microscopy reveals that nuclei isolated from two Ficoll gradients are substantially free of contamination (Fig. 1a,b). We have routinely observed smooth membrane in contact with nuclei and believe this may reflect regions of intimate contact between the nucleus and vacuole, which are often seen in micrographs of yeast cells and spheroplasts (not shown). The protein-to-DNA mass ratio of this preparation is approximately 20:1 as determined by standard methods.[22,23] The protein concentration of nuclei in Ficoll is typically 2–4 mg/ml. Relatively pure nuclei may be obtained after one Ficoll gradient (Fig. 1d). The membrane band above the 20% Ficoll layer, termed fraction L, and the band at the 20–30% Ficoll interface, termed fraction H, consist primarily of vacuoles, membrane vesicles, and ruptured spheroplasts (Fig. 1e,f). The pellet contains nuclei, ruptured spheroplasts, and cell wall fragments (Fig. 1g).

The best method to evaluate purity of the nuclei is sodium dodecyl sulfate–polyacrylamide gel electrophoresis (SDS–PAGE), which is convenient and allows proteolysis to be assessed. The fractions from the Ficoll step gradient, including the L, H, and pellet fractions described above, and the 20% Ficoll layer, termed fraction S, may be compared (Fig. 2). Pure nuclei will generate an electrophoretic profile with characteristic features: histone proteins are prominent, high molecular weight proteins are well represented, and a protein of molecular weight 38,000 is visible (Fig. 2). The 38K protein, yeast fibrillarin, is a nucleolar component that is perhaps the most easily discernable among the nuclear proteins.[15,21] We use standard methods for SDS–PAGE and immunoblotting.[15] To eliminate gel artifacts caused by the Ficoll, nuclear proteins are precipitated with 9

---

[22] O. H. Lowry, N. J. Rosebrough, A. L. Farr, and R. J. Randall, *J. Biol. Chem.* **193**, 265 (1951).
[23] G. Ceriotti, *J. Biol. Chem.* **214**, 59 (1955).

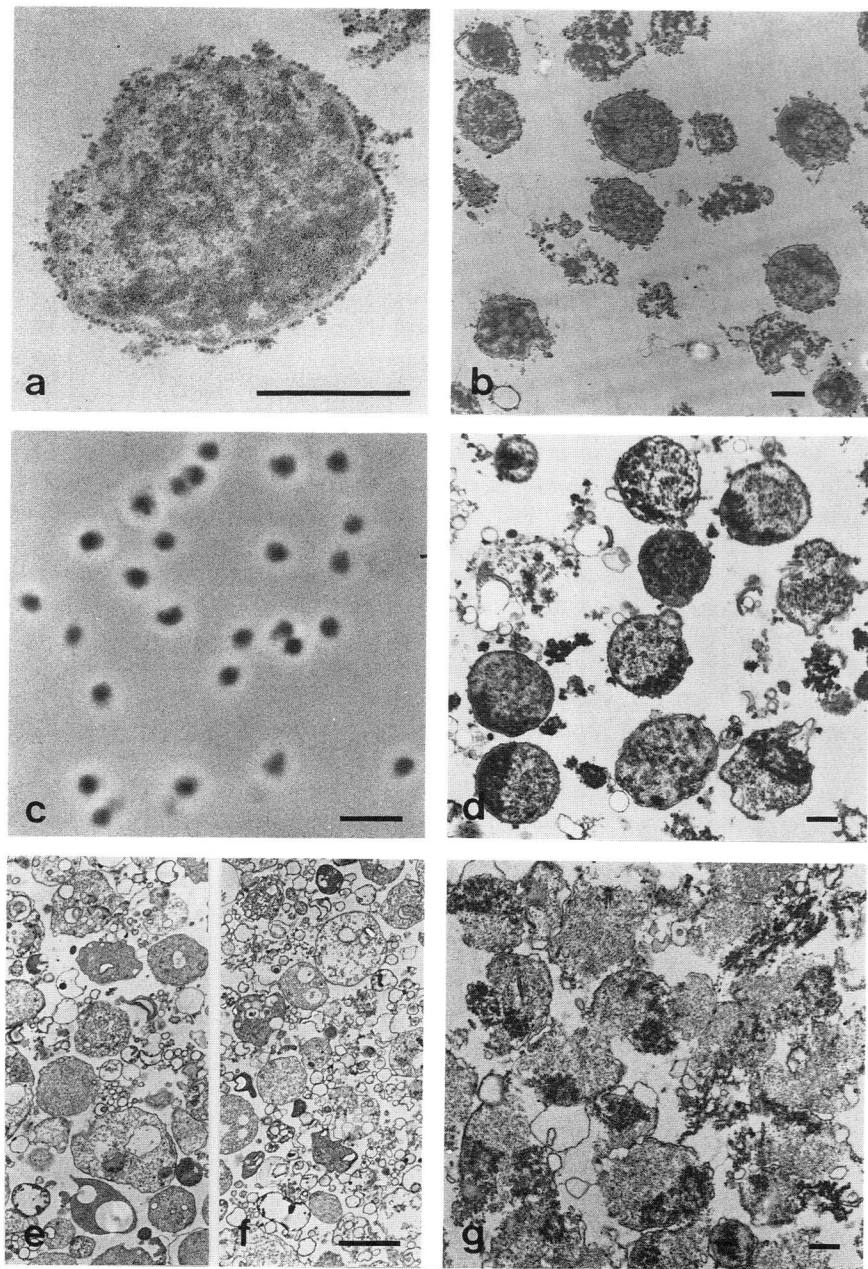

volumes of 10% (w/v) trichloroacetic acid (TCA) before dissolving in electrophoresis sample buffer. The D77 monoclonal antibody specific for yeast fibrillarin has been previously described.[15]

This method achieves high purity, but in so doing sacrifices yield. With BJ2168, the yield of nuclei after one Ficoll gradient is approximately 5%, based on chemical determination of DNA.[23] After two gradients, the yield is reduced to about 2.5%. In our experience it is difficult to achieve complete release of nuclei from lysed spheroplasts. Nuclei appear to remain tethered to, or trapped in, osmotically compromised but partially intact spheroplasts (darker in appearance owing to loss of birefringence). Concentrations of DTT higher than 2 mM in Step 3 do not appear to improve formation or lysis of spheroplasts. We have had limited experience with the use of a Polytron tissue disrupter for spheroplast lysis, as suggested by M. Rout and J. Kilmartin,[24] and obtained a modest improvement in the yield of morphologically intact nuclei, as assessed electron microscopy (EM). In general, however, one 12-liter preparation yields approximately 150 mg of total nuclear protein after two Ficoll gradients, which is sufficient for many experiments.

Nuclei from Different Yeast Strains and Other Fungi

To compare the yield and purity of nuclei from different *S. cerevisiae* strains, nuclei were isolated from the following: BJ2168 (**a** *prc1-407 prb1-1122 pep4-3 leu2 trp1 ura3-52*); BJ926 (**a**/α *prc1-407 prb1-1122 pep4-3 can1 gal2*); AMR1 (**a** *ade2-1 ura3-1 his3-11, 15 trp1-1 leu2-3,112 can1-100 nat1-5::LEU2*); and RS190 (**a** *ade2-1 ura3-1 his3-11,15 trp1-1 leu2-3,112 can1-100 top1-8::LEU2*). The four yeast cultures were harvested at 2.5–3.0 × 10^7 cells/ml, and nuclei were taken from the first Ficoll gradient. Nuclei after one Ficoll gradient are relatively pure (Fig. 3). However, the yields of nuclei obtained show variation over a range of approximately 4-fold (Fig. 3). We note that this method was developed using BJ2168, which gives the highest yield. This suggests that obtaining maximal yield of nuclei from other strains may require some minor modifications.

Nuclei have also been isolated from the fission yeast *Schizosaccharomyces pombe* (wild-type strain 926) and a cell wall-less slime mutant of

[24] Michael Rout, personal communication (1989).

FIG. 1. Microscopy of yeast nuclei and Ficoll gradient fractions. BJ2168 nuclei isolated after two Ficoll gradients viewed by electron (a, b) or light (c) microscopy. BJ926 nuclei after one Ficoll gradient (d). BJ926 fractions from the Ficoll gradient: L (e), H (f), and pellet (g) fractions (see text). Bars in (a), (b), (d), and (g): 0.5 µm. Bars in (c) and (f): 5 µm. Electron microscopy was done according to standard methods.

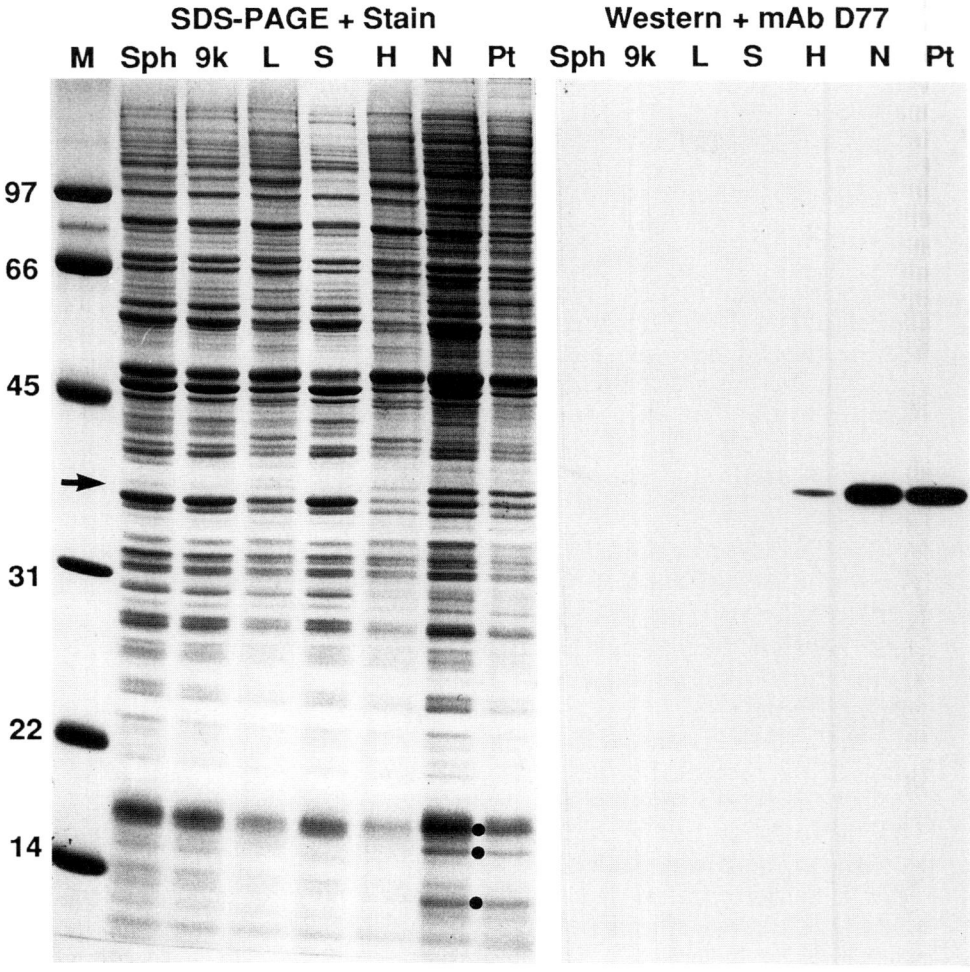

FIG. 2. Isolation of yeast nuclei. Samples from different steps in the isolation of nuclei from strain BJ926 were analyzed by SDS–PAGE and Western blotting: spheroplasts (Sph), supernatant loaded onto the Ficoll gradient (9k), fractions from the Ficoll gradient (L, S, H, see text), nuclei (N), and the pellet from Ficoll gradient (Pt). Molecular weight markers (M) are shown with M_r values ($\times 10^{-3}$). The gel was stained with Coomassie blue, and the Western blot was probed with MAb D77, which is specific for yeast fibrillarin, a 38K nucleolar protein (*arrow*). The portion of each total sample (from 2 liters of yeast culture and from one Ficoll gradient) that was loaded on the gel is given in percent. Histones are indicated (*dots*).

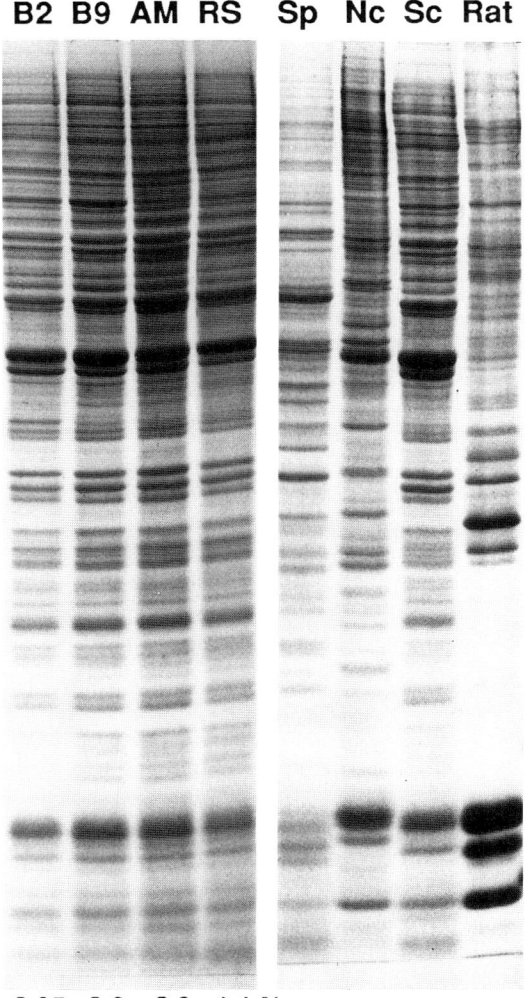

FIG. 3. Nuclei isolated from different strains of *S. cerevisiae* and other fungi. Nuclei from yeast strains BJ2168 (B2), BJ926 (B9), AMR1 (AM), and RS190 (RS) were electrophoresed and stained as in Fig. 2. The portion of nuclei (from 2 liters of yeast culture and from one Ficoll gradient) that was loaded on the gel is given in percent. Nuclei from *S. pombe* (Sp), *N. crassa* (Nc), and *S. cerevisiae* (Sc) are compared to nuclei from rat liver (Rat).

Neurospora crassa (*fz sg os-1*) (Fig. 3). *Schizosaccharomyces pombe* was cultured in NBYG medium (1% Bacto-nutrient broth, 1% Bacto-yeast extract, 1% glucose) and digested with Novozyme enzyme (Novo BioLabs, Danbury, CT) at 5 mg/ml in NBYG medium plus 1 M sorbitol and 50 mM MES–KOH, pH 6.5, at 30°, using a procedure similar to that described above for *S. cerevisiae*. *Neurospora crassa* was grown in NBYSM (1% Bacto nutrient broth, 1% Bacto-yeast extract, 2% sucrose, 0.3 M mannitol, pH 7.5) at 30°. Recent work by K. Bauer in our laboratory indicates that nuclei may be prepared from *S. pombe* with only minor modifications of the method for budding yeast.[25]

Fractionation of Nuclei

An inseparable companion to a method for the isolation of nuclei is a method for the fractionation of nuclei. A typical fractionation protocol for nuclei is to sequentially wash nuclei, remove chromatin with a DNase digestion, extract the resultant nuclear envelopes with salt, or salt and detergent, and extract the salt-treated envelopes further, for example, with urea. The yeast nuclear envelope fraction differs from that obtained from animal cell nuclei because the yeast nucleolus is large and extensively connected to the nuclear envelope. Thus, a DNase digestion results in a combined nuclear envelope–nucleolus fraction. The following method may be used on an analytical or preparative scale.

Method

1. All steps below are done at 0°–4°, except for the DNase digestion and urea extraction. To all buffers used below, add the PIC solutions and DTT (to 1 mM final concentration, from a 1 M DTT stock solution). Add nuclei to 9 volumes of PSM buffer (20 mM potassium phosphate, pH 7.0, 0.25 M sucrose, 0.1 mM MgCl$_2$). Thaw nuclei in Ficoll in an ice-water bath, or on ice, and freeze again in liquid nitrogen. Centrifuge for 10 min in a microcentrifuge ($\sim$ 12,000 g) for small volumes or in a swinging-bucket rotor (e.g., Sorvall HB-4) at 10,000 g for large volumes.

2. Remove the wash supernatant and resuspend the pellet in the same volume of PSM buffer. Resuspend the pellet with homogenization, or brief bath sonication in the case of small pellets. For each 10 μg of starting nuclear protein add 1 μg of DNase I (from a 10 mg/ml solution in 50% glycerol, 20 mM potassium phosphate, pH 7.0, 1 mM MgCl$_2$, 1 mM DTT; store for <1 month at $-20°$). Place in an ambient temperature water bath for 10 min. Add EDTA to 1 mM final concentration (from a 0.5 M

[25] Katharina Bauer, unpublished results (1990).

Na$_2$EDTA stock of pH 8) and hold at ambient temperature 5 min. Chill on ice and centrifuge as in Step 1. The resultant pellet is the nuclear envelope–nucleolus fraction.

3. Resuspend the pellet as in Step 2 in PSE buffer (20 mM potassium phosphate, pH 7.0, 0.25 M sucrose, 1 mM EDTA). Use one-half of the previous volume. Add an equal volume of PEN buffer (20 mM potassium phosphate, pH 7.0, 1 mM EDTA, 2 M NaCl), and mix well. Hold on ice for 10 min. Centrifuge as in Step 1.

4. Resuspend the pellet as in Step 2 in TUE buffer (50 mM Tris-HCl, pH 8.0, 8 M urea, 1 mM EDTA). Use one-tenth of the volume used in Step 1. Hold at ambient temperature for 10 min. Centrifuge at 20° in a Beckman TL100 tabletop ultracentrifuge at approximately 400,000 g for 20 min.

Results

This fractionation procedure may be evaluated by checking the fractionation of a few easily surveyed proteins (Fig. 4). The majority of the histones should be present in the DNase supernatant. The nucleolar protein yeast fibrillarin should be associated with the nuclear envelopes, but it should be released from the envelopes by salt extraction. A high concentration of salt with detergent solubilizes virtually all nuclear proteins. We have used standard methods for SDS–PAGE and Western blotting in Fig. 4. Samples are prepared for electrophoresis by precipitation of proteins in supernatants with 20% TCA (final concentration). For effective TCA precipitation, dilute the salt to less than 0.5 M and urea to below 2.0 M and extend the times of the incubation on ice and the centrifugation.

Additional Notes and Variations

We have observed that different *S. cerevisiae* strains fall in the range of 3–5 × 10^7 cells/ml at an OD$_{600}$ value of 1 and that OD$_{600}$ values from different spectrophotometers may vary about 2-fold. However, if the amount of yeast obtained at Step 2 of the nuclei isolation method is lower than suggested above (by up to 50%), the preparation may be continued nevertheless. Digest the cells with proportionally less SPH medium with lyticase and DTT. The volumes for the remainder of the preparation need not be changed. Lower starting weights of cells may be viewed as advantageous, in that nuclei of modestly higher purity are usually obtained. On the other hand, we have isolated nuclei from up to 50% greater amounts of yeast by increasing the concentration of lyticase and DTT. The higher yield but lower purity of the nuclei may be acceptable if the nuclei will be subjected to further fractionation and/or purification.

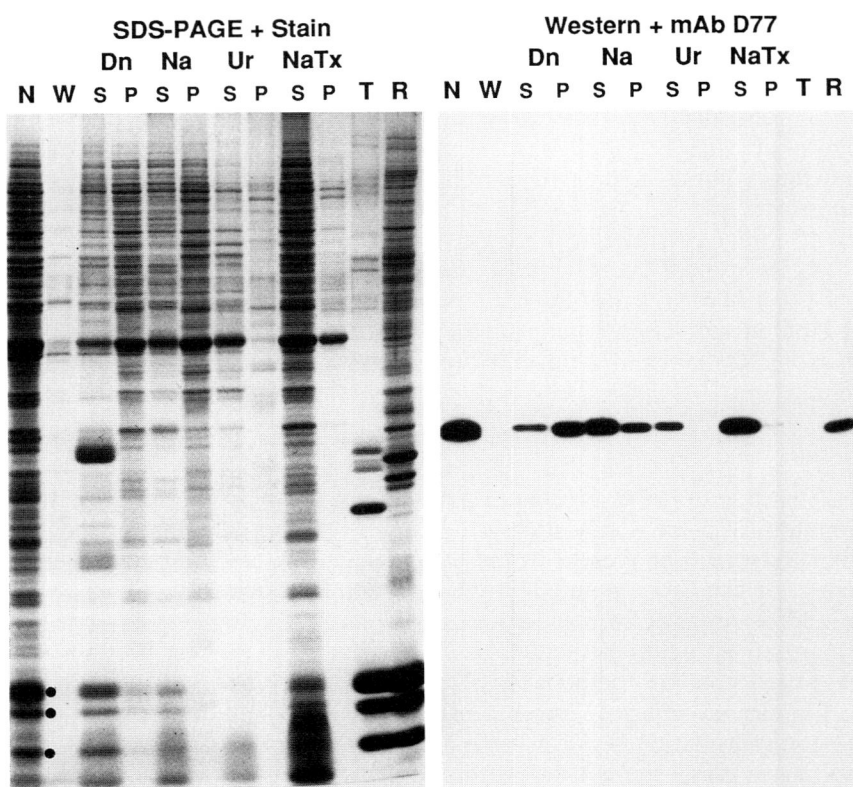

FIG. 4. Fractionation of nuclei. Nuclear fractions were analyzed by SDS–PAGE and Western blotting as described in Fig. 2. Nuclei from BJ2168 (N) were washed, resulting in a supernatant (W) and pellet, which was digested with DNase (Dn) and centrifuged to give a supernatant (S) and pellet (P). See text for details. The DNase pellet was extracted with 1 M NaCl (Na) and centrifuged. The NaCl pellet was extracted with 8 M urea (Ur) and centrifuged. Nuclei were also extracted with 1 M NaCl with 1% Triton X-100 detergent (NaTx) and centrifuged. Nuclear envelopes from turkey erythrocytes (T) and rat liver nuclei (R) serve as controls. Histones are indicated (dots).

The conditions that we have described for production of spheroplasts result in digestion of the cell wall during the first 30-min interval and continued growth, or "recovery," of the cells in the second 30-min interval. The pH drops from around 8 to about 6 during the 60-min incubation. Omission of DTT during the lyticase digestion, lysis of spheroplasts at $0°-4°$, or lysis buffer containing 5 mM MgCl$_2$, 5 mM spermine, or 5 mM spermidine causes incomplete spheroplast lysis and reduces the yield of nuclei by over 50%. The reader may wish to consider the use of other enzymes, such as oxalyticase, to remove the cell wall. The relative merits of four common enzymes used in preparing spheroplasts, Glusulase, lyticase, oxalyticase, and Zymolyase, have been examined by Lohr.[5]

The method for preparing yeast nuclei is conveniently carried out using six culture flasks, six centrifuge bottles, and sets of six centrifuge tubes. Thus, nuclei from up to six different yeast strains may be conveniently isolated simultaneously with this procedure. Alternatively, six different growth conditions, or conditions for nuclear isolation itself, may be processed in parallel.

The method for the fractionation of nuclei above should be viewed as a set of initial guidelines. Achieving selective protein solubilization usually requires the comparison of a number of extraction conditions. Salts other than NaCl, such as KCl or $MgCl_2$, may be tried. The extraction with salt may be substituted, preceded, or combined with an extraction with detergent. Triton X-100, or another detergent, may be added to a final concentration of 0.1–1% [from a 20% (w/v) stock solution]. When using salt and detergent together it is advisable to try a range of salt concentrations.

Acknowledgments

Eleana Sphicas provided expert assistance in the preparation and evaluation of the samples analyzed by electron microscopy. The rat liver nuclei were purified by Rubén Henríquez, and the turkey erythrocyte nuclear envelopes were prepared by Howard Worman. J.A. wishes to thank Randolph Addison, Ed Ching, Rubén Henríquez, Jim Kaput, Teri Mélèse, Mike Rout, and Gerry Waters for many contributions and suggestions regarding the isolation of yeast nuclei. We thank Katharina Bauer for insightful discussions on the preparation of nuclei from S. pombe. Rolf Sternglanz kindly provided yeast strains AMR1 and RS190. We thank Laura Davis and Rubén Henríquez for critically reading the manuscript.

[54] Analysis of Chromosome Segregation in Saccharomyces cerevisiae

By James H. Shero, Michael Koval, Forrest Spencer, Robert E. Palmer, Philip Hieter, and Douglas Koshland

Introduction

The yeast Saccharomyces cerevisiae is an excellent organism for the study of mitotic and meiotic chromosome segregation because it is possible to isolate mutations that effect the fidelity of this process. In this chapter we present two methods for analyzing chromosome segregation in these mutants. In the first method, digital imaging microscopy is used on individual live cells to analyze the segregation of fluorescently labeled chromosomes. In the second method, we describe the construction of artificial chromosomes which have been specifically designed for analysis of the fidelity of chromosome segregation in mitosis and meiosis.

Following Nuclear DNA Movement during Mitosis by Digital Imaging Microscopy

The analysis of chromosome movement by standard phase-contrast microscopy is not possible in *Saccharomyces cerevisiae* because its chromosomes are not directly visible with phase optics. However, chromosomal DNA can be observed by fluorescence microscopy when cells are stained with the DNA-specific dye, DAPI (2,6-diamidinophenylindole). Cells grow at normal rates in the presence of DAPI, even when they are exposed to low levels of incident light (such as room light). However, when cells are exposed to the high levels of incident light normally used to visualize DAPI-stained DNA, the cells stop dividing. Thus, to maintain cell viability and mitosis, it is desirable to lower the intensity of the excitation light source as much as possible, while continuing to produce a detectable fluorescence image. This goal can be accomplished by using low-light level digital imaging microscopy (DIM) to increase image contrast from the low-intensity fluorescence images produced by samples illuminated with highly attenuated incident excitation light.[1]

Reagents and Equipment

YPD medium: 10 g yeast extract, 20 g peptone, 0.9 liter water; autoclave 30 min, then add 100 ml of sterile 20% glucose
DAPI (2,6-diamidinophenylindole)
Low gelling temperature agarose derived from agar
Petroleum jelly
Microscope slides (3 × 1 inch)
Coverslips (22 × 22 mm)
Microscope equipped with epifluorescence optics
Low-light level imaging device
Image processing system
Electronic shutters (optimal)

Method

To prepare cells for microscopic analysis, complete medium is inoculated with a 1/1000 volume of fresh saturated culture. Cells are grown at

[1] S. Inoue, "Video Microscopy." Plenum, New York, 1986.

23° to a density of 7×10^6 cells/ml in complete YPD medium. Then DAPI is added to a final concentration of 2.5 µg/ml. Thirty minutes after addition of the stain, 18 µl of DAPI-stained culture is added to 6 µl of 0.6% low gelling temperature agarose made by boiling the agarose in YPD to melt it and then cooling the solution to 37° in a water bath. Then 15 µl of the mixture of cells plus agarose is placed on a flat slide and covered with a No. 1, 22-mm square coverslip. To prevent the sample from dehydrating, the edges of the coverslip are sealed with petroleum jelly. Petroleum jelly is placed in a syringe with a large-gauge needle. The needle is heated in a flame just enough to melt the jelly. The molten petroleum jelly is extruded from the needle by pushing the plunger of the syringe and then applied to the edges of the coverslip. Once on the coverslip, the petroleum jelly instantaneously solidifies, forming a water-soluble seal.

Changes in nuclear DNA staining pattern are observed using a microscope equipped with epifluorescence optics suitable for DAPI-stained samples. A low-light level imaging device, such as a silicon-intensified tube (SIT) camera or a multichannel plate intensifier, is coupled to the microscope and used to obtain video images. Incident excitation light from a 100-W mercury arc lamp is attenuated to only 1–3% of its normal intensity using neutral density filters. Typically, both phase and fluorescence images of the cells are obtained at 2- to 5-min intervals. Computer-controlled electronic shutters can be used to ensure that excitation light illuminates the sample only during image collection (~0.5–1.0 sec/exposure). Sixteen to thirty-two video images are digitized to 8-bit precision and then boxcar-averaged using an image processing system to increase the image signal-to-noise ratio.[2] Following each time-lapse series, the microscope is defocused, and a image of background fluorescence is obtained. Contrast of each fluorescence image in the time-lapse series is improved by subtraction of background from each fluorescence image followed by histogram equalization.[2]

Specific Example. Twenty-five milliliters of YPD medium was inoculated with 1/1000 volume of a saturated culture of DK208 (*MATa/MATα leu2/LEU2 his7/his7 hom3/HOM3 ADE2/ade2 ADE3,/ade3 can1/can1 sap3/sap3 CYC2/cyc2 cdc16/cdc16*), which was grown overnight at 23° to a density of 7×10^6 cells/ml. DAPI was added to a final concentration of 2.5 µg/ml. Thirty minutes after the addition of the stain, 15 µl of cells in medium was mounted onto the slide as described above. Changes in nuclear DNA staining pattern were observed with a 40× Zeiss Neofluar oil-immersion objective on a Zeiss Universal microscope equipped with epifluorescence optics suitable for DAPI-stained samples. A KS-1380

---

[2] R. Gonzales and P. Wintz, "Digital Image Processing." Addison-Wesley, Reading, Massachusetts, 1987.

image intensifier (Videoscope International, Ltd., Washington, DC) coupled to a Newvicon camera (DAGE-MTI, Michigan City, IN) was used to obtain video images with incident light from a mercury arc lamp attenuated to only 3% of its normal intensity using a combination of 1.0 and 0.7 neutral density filters.

Phase and fluorescence images were obtained at 2-min intervals using an IS-68K computer (Integrated Solutions, Inc., San Jose, CA) to control Uniblitz SD-122B electronic shutters (A. W. Vincent Associates, Inc., Rochester, NY) and image collection as follows. First, the shutter between the mercury arc lamp (fluorescence excitation light) was opened to illuminate the sample. Then, 16 video frames were digitized and boxcar-averaged using an IP-512 image processing system (Imaging Technology, Inc., Woburn, MA) controlled by the IS-68K computer. The excitation shutter was then closed, and a shutter between the tungsten incandescent lamp and the sample was opened to allow acquisition of a digital phase image. Intervals between each image in the time-lapse series were timed by the computer. Contrast was improved by removing background fluorescence from each fluorescence image followed by histogram equalization as described above. Recently, we have found that equal if not better images can be obtained with a Hewlett Packard Vectra QS/20 computer and a Image-1/AT image processing system (Universal Imaging Corporation, Media, PA) which is significantly less expensive than the computer and image processing system described above.

The time-lapse images of exponentially growing cells are presented in Fig. 1. Phase images of the cell taken at the beginning (A) and end (C) of the observation period show that the cell continued to grow, as evidenced by the increase in the bud size. Fluorescence images of these same cells (B and D) show that during the observation period othe nuclear DNA segregated, whereas fluorescence images of the cell taken at intermediate time points (F–K) display the segregation of the nuclear genome. Rates of chromosome movement can be determined by simply measuring the distance between the segregating genomes as a function of time.[3] In addition, positional changes of the nuclear DNA relative to the cell periphery can be observed by superimposing the phase image at each time interval on the corresponding fluorescence image.[3] Mutants which affect chromosome segregation may affect either the kinetics of chromosome movement or the position of nuclear DNA within the cell.[3]

Comments on Digital Imaging Microscopy

Visualization of nuclear DNA movement can be optimized by the choice of cell type and cell density. Diploids are advantageous over hap-

---

[3] R. E. Palmer, M. Koval, and D. Koshland, *J. Cell Biol.* **109**, 3355 (1989).

loids simply because the larger cell size and greater nuclear DNA content. Optimal cell density is between 3×10^6 and 7×10^6 cells/ml. This range of density is sufficient to enable the simultaneous observation of 2 to 4 cells in a given field. In addition, the spatial relationship of one cell to another provides a very sensitive metric for cell position which allows discrimination of chromosome movement from gross cell movement.

The choice of fluorescent dye, necessary for the visualization of nuclear DNA, is also an important parameter in this methodology. DAPI has several attractive features. Cells can grow at normal exponential rates in the presence of concentrations of DAPI as high as 25 µg/ml, provided the cells are not exposed to high levels of light, indicating that the presence of the dye in the medium is not toxic to yeast. Since background fluorescence is low with DAPI, the signal-to-noise is inherently high and can be significantly boosted by DIM. Thus, one can visualize DAPI-stained DNA using very low intensity excitation light, an essential feature of this methodology. Finally, the intensity of DAPI stain remains constant over several hours of time-lapse observation. The most serious drawback to DAPI is that it stains mitochondrial DNA as well, which can obscure features of nuclear DNA segregation (Fig. 1). This problem can be minimized by occasionally focusing through the cell to select a plane of focus that is optimized for nuclear DNA. It may be possible to double label DAPI-stained cells with a vital stain for mitochondria such as rhodamine 123 [4] to distinguish mitochondrial DNA from nuclear DNA. Alternatively, ρ^o strains which lack mitochondrial DNA may be used. A ρ^o strain can be constructed from any ρ^+ strain by adding ethidium bromide (10 µg/ml final) to an exponential culture of the ρ^+ strain and growing the cells to saturation in the dark.[5]

We have also used the dye Hoechst 33342 (8.5 µg/ml). The major asset of this dye is that it stains mitochondrial DNA poorly. However, it has several major drawbacks. Hoechst 33342 inhibits cell growth at a stage prior to medial nuclear division. Also, nuclear DNA staining spontaneously fades with time, even when cells are grown in the dark and in the presence of Hoechst 33342. Thus, the use of Hoechst 33342 has been restricted to the analysis of nuclear DNA movement in cells recovering from arrest at medial nuclear division because these cells are past the fluorophore-induced growth inhibition block.

Microscopy

Several parameters in the microscopy methodology were optimized to view nuclear DNA movement. The most important parameter is the amount of incident and emitted fluorescent light to which the cells are

[4] L. Johnson, M. Walsh, B. Bockus, and L. Chen, *J. Cell Biol.* **88**, 526 (1981).
[5] P. Slonimski, G. Perrodin, and J. Croft, *Biochem. Biophys. Res. Commun.* **30**, 232 (1968).

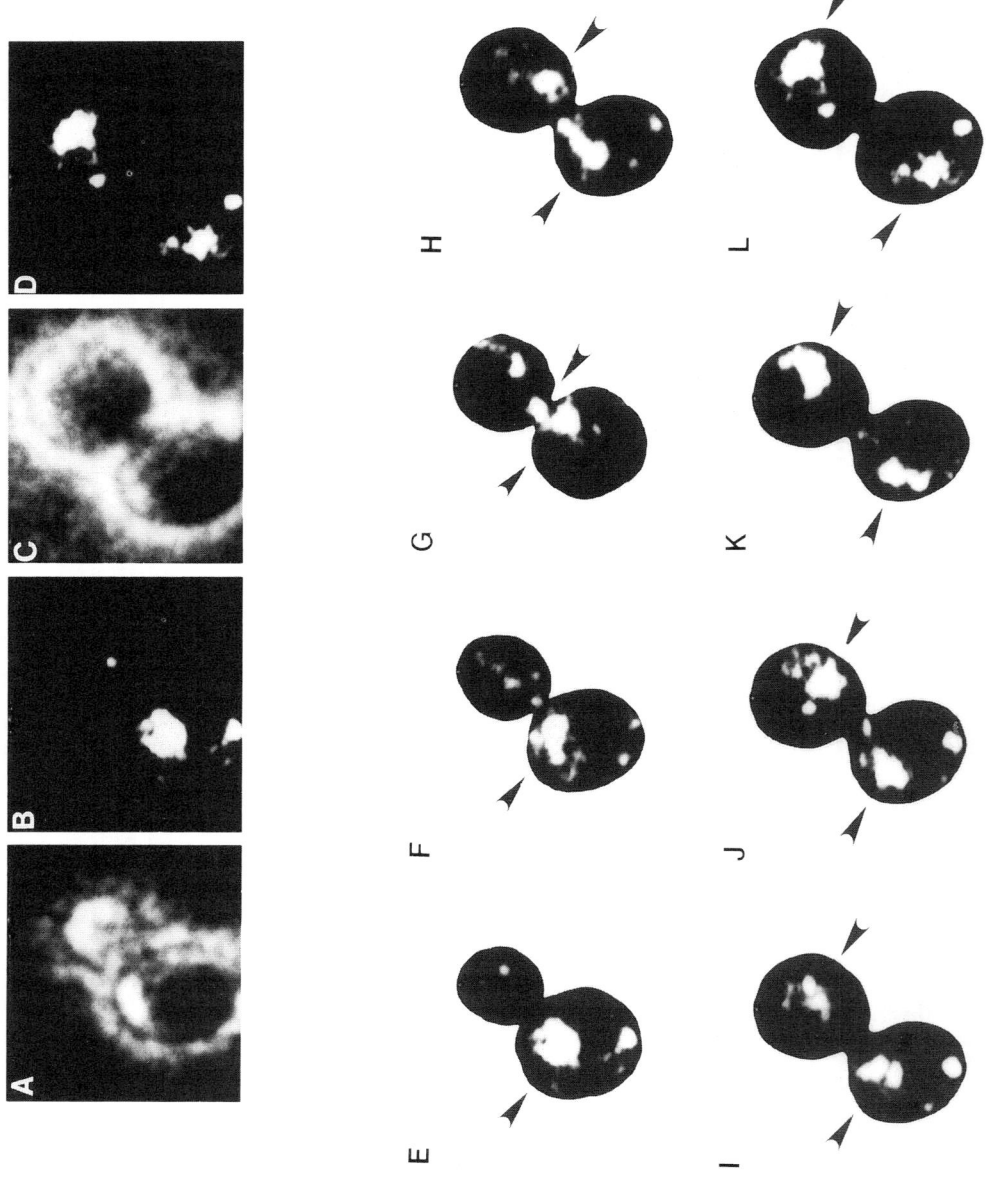

exposed because chromosome movement is destroyed by either brief exposure to high intensity light or prolonged exposure to low intensity light. A combination of neutral density filters (1.7–2.0) is used to reduce the incident light from a 100-W mercury arc lamp to 1–3% of its normal intensity. A test group of cells is first examined under fluorescence optics so that camera gain and black level can be optimized to produce maximum image contrast. When an image processor is used, this process can be simplified by the use of a software tool, such as a line histogram function, which allows monitoring of values for digitized camera output. The gain and black level should be adjusted so that the brightest regions of the field are digitized to values near 255 (in the 8-bit case), while the darkest regions should register near 0. Avoid the use of automatic gain or black level, since this will result in significant variation in output (drift) between images in the time-lapse series. The camera optimization process usually requires exposing the cells continuously for several minutes to incident excitation light which results in cell death. Thus, new fields are surveyed by phase-contrast microscopy to select an experimental group of cells. Note that the incandescent lamp should also be attenuated to allow phase image collection with the camera settings used for fluorescence image acquisition. The phase image is also used to position cells for image collection. In this way, cells are not exposed to incident light from the mercury lamp until the actual observation period begins. In addition, by having a computer control the exposure of the cells to incident light, the cells are exposed to light for the minimum amount of time required to collect an image (usually < 1 sec/exposure). Manual shutter control may also be feasible.

The quality of images collected during an experiment depends on the image remaining in focus and being stationary. Nuclear DNA is distinguished from mitochondrial DNA by its size and morphology. However, nuclear DNA is capable of moving out of the plane of focus, owing to the depth of spherical yeast cells. Therefore, it is important during image collection to occasionally focus through the image to unambiguously iden-

---

FIG. 1. Time-lapse analysis of chromosome segregation in *Saccharomyces cerevisiae* by contrast-enhanced video microscopy. Images A and C are phase-contrast images taken at the beginning (0 min) and end (60 min) of image collection, respectively. Images B and D correspond to the fluorescence images of A and C, respectively. Images E through L are taken at the following time points (minutes): E, 0; F, 20; G, 22; H, 26; I, 30; J, 38; K, 48; and L, 60. In order to view the movement of DNA relative to the periphery of the cell, a mask of the periphery of the cell was made from the phase images corresponding to these fluorescence images. These masks were superimposed on the fluorescence images. Arrows point to the nuclear DNA. Additional small dots of fluorescence are observed near the bulk of the nuclear DNA. These dots are likely to be mitochondrial DNA, though a fraction of them may represent small clusters of chromosomal DNA. The diameter of the cell is 10 μm.

tify nuclear DNA and to select a plane of focus that optimizes the brightness of nuclear DNA. The mounting of the cells on the slide has a significant influence on the success of the procedure. Mounting the cells in agarose is very important for holding the cell stationary during image collection. In the absence of agarose, cells remain stationary for only short periods of time. Sealing of the edges of the coverslip to prevent dehydration is also important because cell growth is impaired as they become dehydrated.

Mitotic Analysis of Chromosome Segregation in *Saccharomyces cerevisiae* Using Chromosome Fragments

Of critical importance in the analysis of mitotic chromosome segregation is the use of a reproducible and sensitive test system able to detect and quantitate minor changes in the fidelity of chromosome segregation. The ideal test chromosome for monitoring mitotic chromosome stability would have the following features. It would contain a genetic marker that would provide a simple but sensitive means for detecting both decrease and increase in its ploidy. Presence of zero, one, or two copies would have no effect on normal cell growth, ensuring that observed changes in its ploidy are not biased by selection for or against cells with different numbers of this chromosome. The mitotic stability of the test chromosome would mimic the high fidelity of mitotic transmission of endogenous chromosomes to ensure that it would be useful in detecting subtle changes in chromosome transmission. Finally, the genotype of the test chromosome and the strain that harbors it would be amenable to classical and recombinant DNA genetic methods to optimize the use of the test chromosome in the analysis of both cis and trans mutations that alter chromosome stability.

Small artificial circular minichromosomes or endogenous chromosomes have been used to monitor the fidelity of chromosome transmission. Minichromosomes and endogenous chromosomes have some of the virtues of an ideal test chromosome. For example, visual assays for monitoring the ploidy of minichromosomes have been developed, and changes in their ploidy have little effect on cell growth.[6-8] However, the high basal loss rates of minichromosomes ($> 1/100$ cell divisions) makes analysis of subtle changes in fidelity of chromosome transmission difficult. Additionally, functions involved in proper endogenous chromosome maintenance may be different or lacking entirely (e.g., telomere function). Use of an endogenous chromosome to analyze mitotic chromosome function leads to an

---

[6] P. Hieter, C. Mann, M. Snyder, and R. Davis, *Cell (Cambridge, Mass.)* **40**, 381 (1985).
[7] D. Koshland, J. Kent, and L. Hartwell, *Cell (Cambridge, Mass.)* **40**, 393 (1985).
[8] D. Koshland and P. Hieter, this series, Vol. 155, p. 351.

improvement in sensitivity, since the basal level of chromosome loss in *S. cerevisiae* is approximately 10^{-5}.[9,10] However, the genetic markers commonly used on endogenous chromosomes only allow for the detection of chromosome loss but not gain. Furthermore, a change in the ploidy of an entire chromosome will cause a change in the dosage of many genes, which may cause slow growth or other unwanted secondary phenotypes. Our goal was to design a convenient test chromosome which would retain the useful features of mini- and endogenous chromosomes while eliminating many of the drawbacks.

Principle of Method

Transformation of a strain with the appropriate linearized vector can create a new strain that contains a novel nonessential chromosome (henceforth referred to as a chromosome fragment). The structure and genotype of a chromosome fragment make it ideal for monitoring the fidelity of mitotic chromosome segregation.[11–13] The long arm of the chromosome fragment contains a portion of an endogenous chromosome. The length of the endogenous sequence in the chromosome fragment can be manipulated to ensure that the chromosome fragment is sufficiently long to be replicated and segregated with high fidelity. In addition, endogenous sequences can be chosen such that they are devoid of genes for which dosage affects viability or cell growth rate. The short arm of the chromosome fragment contains two genes embedded in pBR322 sequences. One gene is a selectable marker used during the initial transformation and in subsequent strain constructions. The second gene, either *SUP11* or *ade3-2p,* provides a color marker which can be used in a visual colony assay to monitor chromosome stability and ploidy;[6,7] considerations in deciding which colony color assay system to use have been described.[8] Since the color marker is embedded in pBR322, which bears no homology to yeast genomic DNA, the color marker cannot be lost by mitotic recombination between the chromosome fragment and endogenous chromosomes. Therefore, loss of the color marker equals chromosome loss. The color marker allows one to use the sectoring assay qualitatively to monitor the stability of the chromosome fragment during colony growth (G). Subtle differences

[9] L. Hartwell, S. Dutcher, J. Wood, and B. Garvik, *Recent Adv. Yeast Mol. Biol.* **1**, 28 (1982).
[10] M. Esposito, D. Maleas, K. Bjornstad, and C. Brushi, *Curr. Top. Genet.* **6**, 5 (1982).
[11] J. Hegemann, J. Shero, G. Cottarel, P. Philippsen, and P. Hieter, *Mol. Cell. Biol.* **8**, 2523 (1988).
[12] F. Spencer, C. Connelly, S. Lee, and P. Hieter, in "Cancer Cells" (T. Kelly and B. Stillman, eds.), Vol. 6, p. 441. Cold Spring Harbor Laboratory, Cold Spring Harbor, New York, 1988.
[13] F. Spencer, S. Gerring, C. Connelly, and P. Hieter, *Genetics* **124**, 237 (1990).

in chromosome fragment loss rates (2- to 3-fold) are detectable. With the color assay, the number of copies of the chromosome fragment per cell can be determined. Therefore, increases or decreases in the ploidy of the chromosome fragment can be easily monitored. Finally, the rate of increase or decrease in ploidy can be assayed quantitatively by fluctuation analysis or by measuring the frequency of half-sectored colonies.[8,11] The chromosome fragments described below are useful tools for studying the effects of *CEN* DNA mutants[11] as well as trans-acting mutants[12-15] on mitotic chromosome transmission fidelity.

Method 1: Chromosome Fragment Strain Construction

We have found it convenient to use a 125-kilobase (kb) chromosome fragment, CFIII(*D8B*.d), as a standard for the assay of mitotic chromosome transmission. This fragment contains a short arm with a color marker and a selectable marker as described above. Its long arm contains most of the left arm of chromosome III distal to a sequence *D8B* which is located 8 kb from the centromere.[16] This fragment has a loss rate of approximately 2 in 10^4 cell divisions. In addition, it results in disomy for less than 1% of the yeast genome and has no apparent detrimental effect on cell growth or endogenous chromosome fidelity.[11,13] Furthermore, unlike an entire chromosome III disomy, this chromosome fragment does not carry the *MAT* locus and therefore does not produce nonmating spores in crosses. This feature simplifies the use of classical genetic methods for introducing the chromosome fragment into a mutant strain of interest (see below).

Strains carrying a CFIII(*D8B*.d) chromosome fragment marked with *SUP11* or *ade3-2p* and a gene mutation can be generated by several methods using standard techniques.[17] A chromosome fragment can be generated *de novo* in the mutant strain of interest by transforming it with linearized plasmid pJS2 (*SUP11*) or pDK352 (*ade3-2p*) (Fig. 2A). A schematic depiction of the *in vivo* conversion of a linearized yeast chromosome fragmentation vector to a stably maintained chromosome fragment is shown in Fig. 2B. This method requires that the mutant strain carry the appropriate auxotrophies for selecting transformants and for monitoring chromosome stability with one of the two color assay systems. Verification

[14] C. Holm, T. Stearns, and D. Botstein, *Mol. Cell. Biol.* **9**, 159 (1989).
[15] V. Lundblad and J. W. Szostak, *Cell (Cambridge, Mass.)* **57**, 633 (1989).
[16] C. Newlon, unpublished.
[17] M. Rose, F. Winston, and P. Hieter, "Methods in Yeast Genetics: Course Manual." Cold Spring Harbor Laboratory, Cold Spring Harbor, New York, 1988.

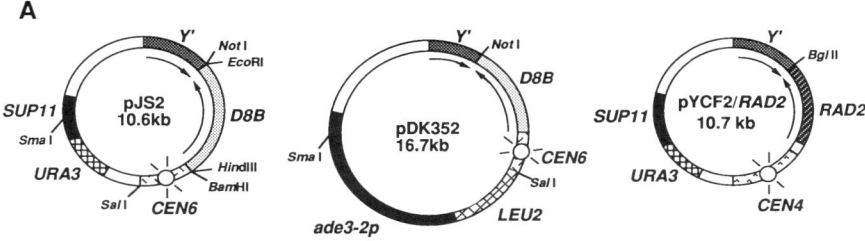

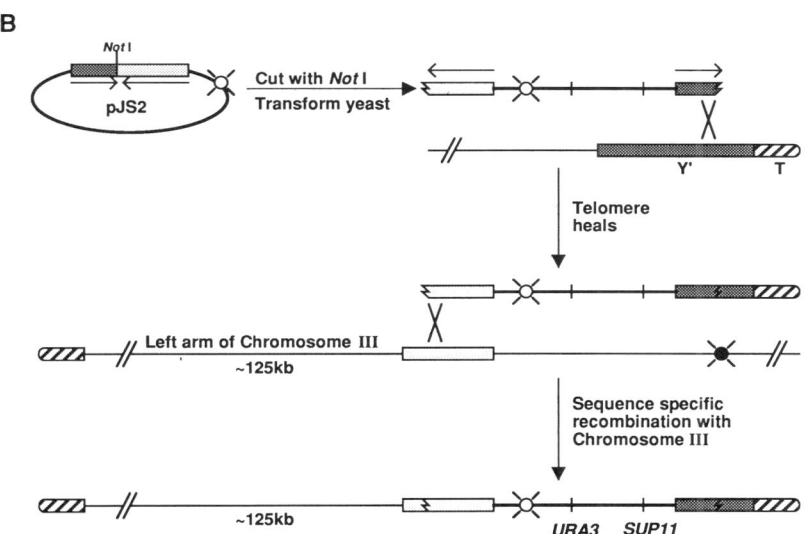

FIG. 2. *In vivo* generation of chromosome fragments. (A) Vectors for generating chromosome fragments. pJS2 was constructed by cloning a 1.16-kb *Bam*HI to *Sal*I *CEN6* fragment into pYCF5.[11] pYCF5 carries *URA3* (for selecting transformants), *SUP11* (for monitoring mitotic stability of chromosome fragments generated by transformation), *Y'* sequences (for efficient conversion of one end of the transformed DNA to a telomere), and a unique DNA sequence (*D8B*, from the left arm of chromosome III) for *in vivo* addition of most of the left arm of chromosome III to the plasmid DNA. pYCF2/*RAD2*[12] is a vector with similar characteristics but with *RAD2* targeting sequences for the *in vivo* addition of a 90-kb portion of the right arm of chromosome VII. pDK352 was constructed from pJS2 by cloning a 7-kb *Sma*I to *Sal*I *ade3-2p LEU2* fragment in place of the *SUP11 URA3* portion of pJS2. This vector is suitable for generating fragments that can be monitored using the *ade3-2p* color assay system. (B) Schematic of chromosome fragmentation. pJS2 plasmid DNA (or derivatives containing mutant *CEN6* sequences) is linearized with *Eco*RI or *Not*I and Ura+ transformants selected. One end of the linear plasmid molecule contains sequences (*Y'*) that are efficiently repaired to a functional telomere; the other end contains a unique DNA sequence (*D8B*), the genomic position of which is located on the left arm of chromosome III 8 kb from the centromere. Pink Ura+ transformants each possess a 125-kb chromosome fragment (bottom), in which all of the chromosome III sequences distal to the unique sequence *D8B* have been added to the transforming plasmid.

of the presence of a chromosome fragment is efficiently accomplished by comparing the new electrophoretic karyotype[18] to a known standard. The identity of the newly generated chromosome fragment may be further verified by Southern blot analysis, probing with a distal genomic sequence [e.g., *LEU2* DNA should hybridize a CFIII(*D8B*.d)]. Alternatively, a cloned mutant allele (if available) can be used to replace the wild-type copy in a strain already containing a *SUP11* or *ade3-2p* marked chromosome fragment by one-step gene replacement (Ref. 19; e.g., see Ref. 15) or a two-step replacement strategy (Ref. 20; e.g., see Ref. 14). Existing chromosome fragment-bearing strains containing various auxotrophies for transformant selection are listed in Table I. Finally, a strain containing both a mutation of interest and a chromosome fragment can be generated from spores obtained from a cross of appropriate haploid strains.

Comments on Strain Construction

Each of the methods for introducing a chromosome fragment into a strain of interest has certain virtues and drawbacks. Use of transformation either to construct a chromosome fragment *de novo* in a mutant strain or to introduce a mutant allele into a fragment-containing strain maintains the isogenic relationship of experimental and control strains. However, the newly generated chromosome fragment or newly introduced mutant allele must be characterized in the new background. Standard genetic crosses may often provide the most efficient method for constructing a strain with a chromosome fragment and a mutation of interest but may often sacrifice isogenicity.

There are several helpful hints pertaining to the method of generating chromosome fragments *de novo*. The color of colonies on the primary transformation plate can be used to identify the appropriate chromosome fragment-containing strains. For example, transformation of the linearized fragmenting *SUP11* vector into diploid strains will yield four colony colors on the selective plate. The investigator should choose pink colonies for further analysis. White-, red-, and copper-colored colonies should not be picked. White transformants are likely to be those transformants with more than 1 copy of *SUP11*. Red transformants carry a highly unstable plasmid resulting from recirculatization of the plasmid *in vivo* or incomplete restriction digestion *in vitro* (weak *ARS* activity is associated with the *SUP11* sequences). Petite transformants are copper-colored on minimal medium with limiting adenine.

[18] G. Carle and M. Olson, *Proc, Natl. Acad. Sci. U.S.A.* **82**, 3756 (1985).
[19] R. Rothstein, this series, Vol. 101, p. 202.
[20] F. Winston, F. Chumley, and G. R. Fink, this series, Vol. 101, p. 211.

TABLE I
YEAST STRAINS AND PLASMIDS FOR MITOTIC ANALYSIS OF CEN MUTATIONS

Mutation	URA3 SUP11 system		LEU2 ade3-2p system	
	Plasmid	Yeast strain[d]	Plasmid	Yeast strain
CEN6 (wild type)[a]	pJS2	YPH281	pDK352	BP5001-6-4 CF352
CEN4 (wild type)[a]	pJS3	YPH186		
CEN6 (ML)[b]	pJS27	YPH301		
CDEI(7-A)[a]	pJS41	YPH282		
CDEI(7-C)[a]	pJS42	YPH283		
CDEI(7-G)[a]	pJS43	YPH384	pDK354	BP5003-6-1 CF354
CDEI(8-T)[a]	pJS51	YPH285		
CDEI(8-A)[a]	pJS52	YPH286		
CDEI(8-C)[a]	pJS53	YPH287	pDK353	BP5002-5-1 CF353
CDEIΔ[c]	pJS108	YPH435		
CDEIII(2-A,3-G)[a]	pJS10	YPH288		
CDEIII(8-T)[a]	pJS11	YPH289		
CDEIII(8-T,9-G)[a]	pJS12	YPH290		
CDEIII(14-A)[a]	pJS13			
CDEIII(14-G)[a]	pJS14			
CDEIII(14-T)[a]	pJS15			
CDEIII(15-T)[a]	pJS16	YPH291	pDK355	DK4513-121 CF355
CDEIII(15-A)[a]	pJS17	YPH292		
CDEIII(15-C)[a]	pJS18	YPH293		
CDEIII(23-T,24-G)[a]	pJS19	YPH294		
CDEIII(24-G)[a]	pJS20	YPH295		
CDEIII(24-G,25-T)[a]	pJS21	YPH296		
CDEIII(21-C)[a]	pJS22	YPH297		
CDEIII(21-C,19∇ 20-TA)[a]	pJS23			
CDEIII(19∇ 20-T)[a]	pJS24	YPH298	pDK356	BP5004-1-2 CF356
CDEIII(19∇ 20-G)[a]	pJS25	YPH299	pBP100	
CDEIII(17Δ25)[a]	pJS26	YPH300	pBP101	DK4513-121 CF101
CDEII (Δ41 bp)[c]	pJS109	YPH429	pBP102	DK4513-121 CF102
CDEII (+45 bp)[c]	pJS110	YPH433		
CDEII (+86 bp)[c]	pJS111	YPH431	pBP103	

[a] J. Hegemann, J. Shero, G. Cottarel, P. Philippsen, and P. Hieter, *Mol. Cell. Biol.* **8**, 2523 (1988).
[b] G. Cottarel, J. Shero, P. Hieter, and J. Hegemann, *Mol. Cell. Biol.* **9**, 3342 (1989).
[c] L. Panzeri, L. Landonio, A. Stotz, and P. Philippsen, *EMBO J.* **4**, 1867 (1985).
[d] Yeast strains have the following genotypes: YPH281–301, *MATa/MATα ura3-52/ura3-52 lys2-801/lys2-801 ade2-101/ade2-101 trp1Δ1/trp1Δ1* + CFIII(D8B.d); YPH429, 431,433,435: *MATa ura3-52 lys2-801 ade2-101 trp1Δ1 his3Δ200 leu2Δ1* + CF III(D8B.d); DK4513-121, BP5001-6-4: *MATa leu2-3,112 his3 ade2 ade3 can1 sap3* + CFIII(D8B.d); BP5003-6-1, 5002-5-1, 5004-1-2: *MATa leu2-3.112 his7 ade2 ade3 can1 sap3* + CFIII(D8B.d).

Infrequently, the chromosome fragments obtained are not of the expected size on pulsed-field gels, indicating that they do not contain the desired structure. These undesired events occur because the plasmid vector sequences can occasionally target incorrectly. This situation most commonly arises when the chromosome fragment sought is highly unstable or is associated with deleterious secondary phenotypes. For example, in recovering *CEN* mutations which cause severe functional impairment, fragmentation events are occasionally recovered that target to chromosome VI instead of chromosome III (see Ref. 11 for details). These fragments presumably occur via degradation of the *D8B* sequence, exposing the *CEN6* sequence for recombination. The genomic origin of chromosome fragments can be unambiguously determined by hybridization of labeled probes to Southern blots of pulsed-field gels.

Manipulation of Chromosome Fragment Sequences

An additional feature of chromosome fragments, such as CFIII(*D8B*.d), is that one can choose which endogenous sequences are present on the long arm, which selectable and color markers are on the short arm, and what centromere sequences are present. This flexibility greatly facilitates the use of chromosome fragments to analyze chromosome transmission.

Changing Endogenous Sequences on the Long Arm. To determine if an effect of a mutation on chromosome transmission is a general property of all chromosomes or specific to a particular chromosome fragment, it is important to check the effect of the mutant on different chromosome fragments. Chromosome fragments containing endogenous chromosome sequences of choice on the long arm can be generated *de novo* by substituting different targeting sequences in the fragmentation vector.[21] In particular, we have constructed the vector YCF2/*RAD2* (Fig. 2A) which contains the *RAD2* gene as the recombination target and used this new fragmentation vector to generate a chromosome VII fragment in which the long arm is composed of sequences distal to *RAD2*. This CFVII(*RAD2*.d) chromosome fragment is approximately 90 kb, has stability similar to *D8B* distal, and has no apparent detrimental effects on cell growth or endogenous chromosome fidelity.[12] The *RAD2* fragmentation vector or strains harboring the CFVII(*RAD2*.d) chromosome fragment are available.

Changing Markers on the Short Arm. An additional feature of the system is the capacity to manipulate the selectable marker and color marker present on the short arm of a chromosome fragment. This feature accommodates various genetic backgrounds or other constraints. These changes may be made by altering the fragmentation vector used to produce

[21] S. L. Gerring, C. Connelly, and P. Hieter, this volume [4].

the chromosome fragment. For example, digestion of the fragmentation vector, pJS2, with *Sma*I and *Sal*I removes the *URA3 SUP11* sequences. Ligation of the vector sequences with a *Sma*I/*Sal*I *LEU2 ade3-2p* cassette derived from pDK351 produces a fragmentation vector, pDK352, in which the *URA3* and *SUP11* color marker has been replaced with *LEU2* and the *ade3-2p* system. Alternatively, it is also possible to substitute selectable markers on existing chromosome fragments. A series of YMC (yeast marker change) plasmids have been constructed which allow exchange via one-step gene replacement[19] of a marker on the short arm with any of seven different markers (see Table II). The YMC plasmids were constructed by inserting restriction fragments containing a marker gene (e.g., *HIS3*) into the *Pvu*II site of pBR322. Digestion of these plasmids with the appropriate combination of restriction enzymes (as indicated in Table II) releases a restriction fragment containing the selectable marker flanked by unique pBR322 sequences. When a pJS2- or YCF2/*RAD2*-derived chromosome fragment-containing strain is transformed with 3 μg of a YMC restriction fragment, the linearized transforming DNA bears appropriate pBR322 sequence similarity at the termini to allow one-step replacement of the existing selectable marker with the YMC selectable marker. Transformants are colony purified and checked for simultaneous loss of the target chromosome fragment marker and the presence of the chromosome fragment (on pulsed-field gels). It should be noted that, with the exeptions of replacements from *URA3 SUP11* to *TRP1 SUP11* or *HIS3 SUP11*, all marker changes are associated with loss of the *SUP11* marker for visual monitoring of the chromosome fragment.

Changing Centromere Structure. The substitution of the wild-type centromere on the fragmenting vector by a mutant centromere sequence enables phenotypic characterization of the effects of centromere DNA

TABLE II
YEAST MARKER CHANGE PLASMIDS

YMC plasmid	Enzymes used for linearizing prior to one-step gene replacement
YMC.*TRP1 SUP11*	*Eco*RI and *SA*lI
YMC.*HIS3 SUP11*	*Sal*I and *Cla*I
YMC.*URA3 SUP11*	*Eco*RI and *Sal*I
YMC.*HIS3*	*Sal*I and *Cla*I
YMC.*LEU2*	*Hin*dIII and *Bam*HI
YMC.*TRP1*	*Sal*I and *Cla*I
YMC.*LYS2*	*Sal*I and *Cla*I
YMC.*URA3*	*Eco*RI and *Sal*I

alteration on the behavior of the chromosome fragment in cis (see Ref. 11 and below) or assay for a synergistic effect of an additional mutation in trans.[22] Fragmentation vectors containing mutant centromere sequences and yeast strains containing chromosome fragments carrying mutant centromeres have been constructed (Table I), and they can be used in strain construction via the methods outlined above.

Method 2: Quantitation of Chromosome Stability

Quantitation of the rate of loss and gain of chromosome fragments can be accomplished by counting half-sectored colonies or by fluctuation analysis as described previously for minichromosomes.[8,11] In these methods, clonal populations of cells derived from a single cell containing one copy of the chromosome are plated nonselectively, and the resulting colonies are scored for changes in color which reflect changes in chromosome ploidy. However, quantitating chromosome loss by these methods is a tedious procedure because a large number of indicator plates are required to screen for rare colonies that have lost (or gained) the ploidy marker.

Using fluctuation analysis to quantitate chromosome loss would be much easier if one could identify rare cells that have lost the chromosome fragment by a direct selection for growth. Recently, we have worked out methods to select for cells lacking a *URA3* chromosome fragment by using the drug 5-fluoroorotic acid (5-FOA).[23] In our experience, direct plating of logarithmically growing colonies on 5-FOA-containing medium results in a leaky lethal phenotype, allowing Ura$^+$ cells to go through several cell divisions before arresting growth. It was necessary to eliminate the phenotypic lag to make this selection useful for quantitation of chromosome loss. A starvation incubation (4 hr at 4° with shaking) in the presence of 5-FOA (1 μg/ml in water) prior to plating solves this problem. The pretreatment slows cell growth so that subsequent plating on 5-FOA (1 μg/ml) results in death of Ura$^+$ cells at the single-cell stage, and an accurate count of cells which have lost the gene prior to plating can be obtained. We note that the 5-FOA pretreatment will also result in the death some Ura$^-$ cells when they are incubated for too long (e.g., 16 hr, 4°). Therefore, this method for quantitating chromosome fragment loss should be used with due caution.

Specific Example: Analysis of Mutant Centromere DNA Sequences Using Chromosome Fragments

A chromosome fragment, CFIII(*D8B*.d), was constructed using pJS24 (Table I) by transformation of YPH49.[11] pJS24 carries the centromere

---

[22] R. Palmer and D. Koshland, unpublished results (1988).
[23] J. D. Boeke, J. Trueheart, G. Natsoulis, and G. R. Fink, this series, Vol. 154, p. 164.

mutation CDEIII(19▽ 20-T).[11] (For a description of the technique, see above and Ref. 21.) The electrophoretic karyotypes of colony-purified Ura⁺ transformants were screened for the presence of the 125-kb CFIII(*D8B*.d) by pulsed-field gel electrophoresis. The colony color phenotypes in the absence of selection showed that the CDEIII(19▽ 20-T) mutation resulted in an obvious decrease in chromosome fragment stability (Fig. 3B). This phenotype was analyzed quantitatively (see Ref. 11 and Table III), scoring chromosome loss in small clonal populations by fluctuation analysis. (For technical details, see the above discussion.) This strategy is generally applicable to the analysis of any centromere mutation.

Analysis of Chromosome Segregation in Meiosis Using Yeast Artificial Chromosomes

Analysis of chromosome segregation in meiosis requires a pair of homologous test chromosomes which incorporate all the features described above for mitotic analysis plus additional features important for meiotic analysis. The test chromosomes must be nonessential for spore viability and must be genetically marked to monitor the fidelity of segregation of homologs during meiosis I and of sister chromatids during meiosis II. Since recombination between homologs contributes to their proper segregation in meiosis I, the test chromosomes should contain markers to assess the occurrence or absence of meiotic crossovers. In addition, sequence similarity between the test chromosomes and endogenous chromosomes should be kept to a minimum since the test chromosome and the endogenous

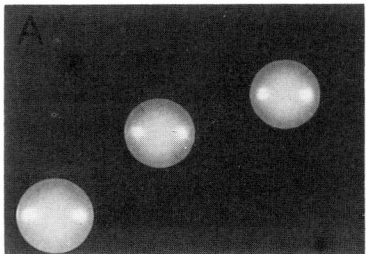
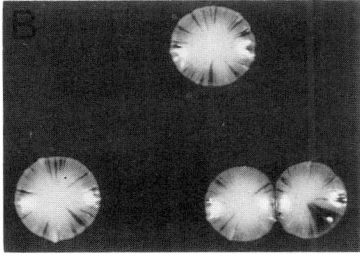

FIG. 3. Colony sectoring phenotypes reflecting chromosome fragment stability. The parent diploid yeast strain is homozygous for the *ade2-101* mutation and therefore gives rise to red colonies. (A) Transformation with a linearized pJS2 vector results in the presence of mitotically stable *SUP11*-marked chromosome fragments. On nonselective medium, strains generate homogeneously pink colonies, which show, on average, less than one visible red sector per colony. The loss rate of this telocentric chromosome fragment is 1.9×10^{-4}. (B) shows increased rates of loss of the chromosomal fragment caused by mutations within the *CEN6* sequence. The loss rate of the chromosome fragment for CDEIII(19▽ 20-T) is 1.5×10^{-3}.

TABLE III
FLUCTUATION ANALYSIS OF YPH298: CDEIII
(19▽ 20-T)[a]

Test colony	Colony size	Red segregants
1	5450	45
2	6620	38
3	6810	35
4	5460	27
5	3895	27
6	5205	25
7	4900	22
8	5620	15
9	4860	10
10	3840	7
Average	5260	
Median		26
Mean		7.9

[a] Chromosome fragment loss rate = $7.9/5260 = 1.5 \times 10^{-3}$.

chromosome may interact when extensive regions of sequence similarity exist. This pairing apparently interferes with the proper meiotic segregation of both the endogenous and test chromosomes (see below).

Various test chromosomes have been used to examine the effect of a *CEN* DNA mutation (in cis) or a gene mutation (in trans) on chromosome segregation during meiosis. The use of endogenous chromosomes for this purpose is limited because a nondisjunction event in meiosis I results in two nonviable spores. In addition, meiosis II errors result in one nonviable spore product, and the type of segregation error (nondisjunction or chromosome loss) cannot be easily determined.[24] Meiotic analysis of cis- or trans-acting mutations using plasmids is limited by the high basal level of missegregation events.[25] Moreover, circular minichromosomes may not be an accurate indicator of endogenous chromosome behavior because recombination between circular homologs can lead to the formation of dicentric molecules which are structurally and genetically unstable in meiosis. To circumvent this problem, Dawson *et al.*[26] constructed linear artificial chromosomes 61 kb in length that consisted of phage λ DNA for most of their lengths. Although these artificial chromosomes were able to

[24] A. Gaudet and M. Fitzgerald-Hayes, *Genetics* **121,** 477 (1989).
[25] S. Cumberledge and J. Carbon, *Genetics* **117,** 203 (1987).
[26] D. Dawson, A. Murray, and J. Szostak, *Science* **234,** 713 (1986).

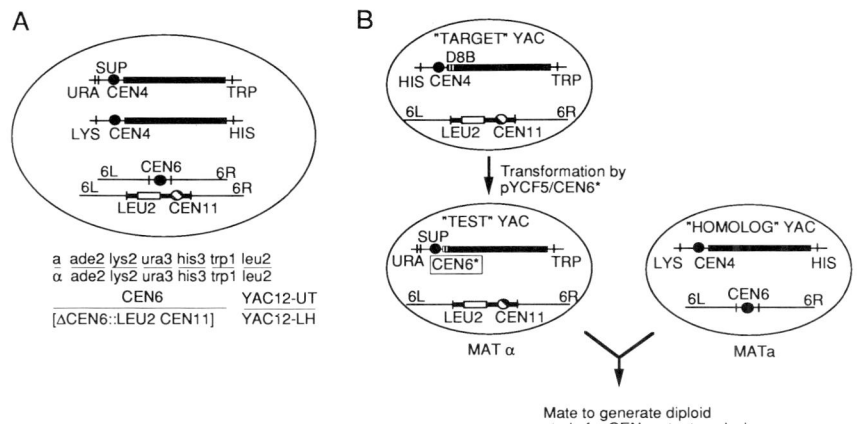

Fig. 4. Yeast strains useful for analysis of chromosome segregation in meiosis. (A) Schematic depiction of YPH607. Meiotic sister spores are unambiguously scored using the *LEU2* marker because one copy of chromosome VI contains a *LEU2 CEN11* replacement of *CEN6*. The strain contains two differentially marked YACs which contain an identical 350-kb exogenous human DNA segment. YAC12-UT is marked with *URA3 SUP11* (on the short arm) and *TRP1* (on the long arm). YAC12-LH is similarly marked with *LYS2* (short arm) and *HIS3* (long arm). (B) Scheme for meiotic analysis of *CEN* DNA mutations. The *CEN4* sequence of the *D8B CEN4 HIS3* pBR322 YAC in the "target" strain (YPH603) can be replaced with *CEN6 URA3 SUP11* DNA using NotI/PvuI restriction fragments from plasmids listed in Table I. The strain containing the resultant "test" YAC (*D8B CEN6* URA3 SUP11* pBR322) can then be mated to the "homolog" YAC strain (YPH604) and the diploid used for subsequent meiotic analysis.

disjoin in meiosis I more efficiently than small circular minichromosomes, they still exhibited an error rate of approximately 10%. Initial experiments characterizing the behavior of chromosome fragments in meiosis indicated that they possess a high background of nondisjunction owing to interaction with the homologous endogenous chromosomes (see below). Subsequently, we have devised a strategy using a yeast artifical chromosome (YAC) containing a large segment (350 kb) of exogenous DNA[27] which eliminates most of the shortcomings of previous methods.

Principle of Method

To study chromosome segregation in meiosis, we constructed a diploid strain (YPH607) containing two differentially marked YACs, YAC12-UT and YAC12-LH (Fig. 4A). The short arm of YAC12-UT is identical in structure to the chromosome fragments described above, with the *URA3*

[27] D. Burke, G. Carle, and M. Olson, *Science* **236,** 806 (1987).

and *SUP11* markers embedded in pBR322 sequences. These markers are therefore tightly linked to the centromere. The long arm contains a 350-kb segment of exogenous human DNA and the yeast *TRP1* marker (also embedded in pBR322 sequences), which is tightly linked to the distal telomere. To make YAC12-LH from YAC12-UT, the *URA3 SUP11* cassette was replaced with *LYS2* using the YMC.*LYS2* marker change plasmid described above. Similarly, the distal *TRP1* marker was replaced with *HIS3*. By using a large segment of human DNA, the YACs are of sufficient size to ensure that they are segregated with high fidelity during mitotic growth. In addition, the human DNA provides extensive sequence similarity between the two differentially marked YAC "homologs" to allow meiotic pairing and recombination, while limiting similarity to endogenous yeast chromosomes. The diploid strain contains a *LEU2 CEN11* replacement of *CEN6* on one copy of chromosome VI.[28] In this configuration, recombination cannot occur between *LEU2* and its centromere, and sister spores are thus unambiguously marked. The long arm markers of the two YACs allow one to monitor recombination. The combination of the YAC centromere-linked markers and the *LEU2 CEN11* marker on chromosome VI allows one to unambiguously monitor meiosis I nondisjunction, meiosis I precocious sister chromatid separation, meiosis II nondisjunction, and meiotic chromosome loss (depicted in Fig. 5).

Method

Construction of Strains Containing Artificial Chromosomes. The introduction of the 365-kb YAC into different strain backgrounds can be conveniently accomplished by standard genetic crosses. Alternatively, a genetic mutation of interest can be substituted by gene replacement into a YAC-containing background. A third possibility is to directly transform the YAC into the appropriate strain using whole chromosomal DNA and a spheroplast transformation protocol in the presence of polyamines.[29]

The two-YAC system can be used to study the effects of trans-acting mutations on meiotic segregation. A cloned mutant allele can be introduced into the two haploid strains, YPH602 and YPH604 (Table IV), by gene replacement techniques[19] and the appropriate diploid (homozygous for the mutation and carrying the YAC homologs) obtained by mating. Alternatively, the mutation can be introduced by transformation into the diploid YPH607, the resultant heterozygote sporulated, and haploid spore clones of appropriate genotypes picked and mated for subsequent meiotic analysis.

The two-YAC system can also be used to study the effects of *CEN* DNA

---

[28] J. Hegemann, J. Shero, and P. Hieter, unpublished results (1989).
[29] C. Connelly, M. McCormick, J. Shero, and P. Hieter, in preparation.

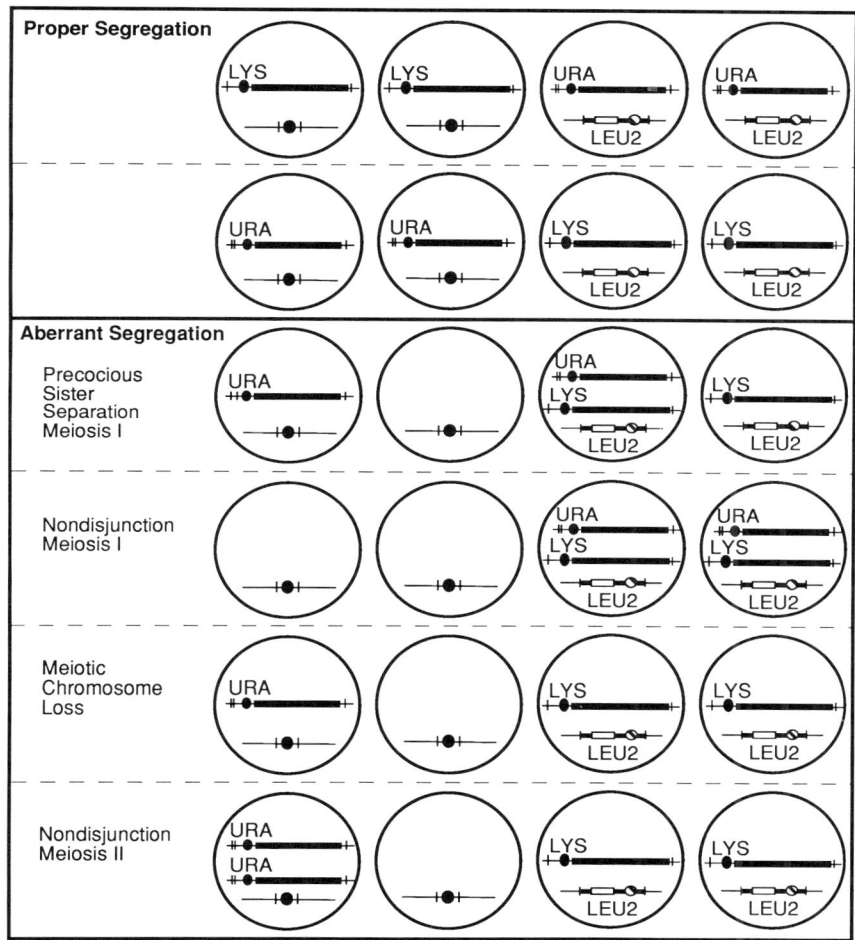

FIG. 5. Classes of spore products in YPH607 tetrads. The distribution of the relevant centromere-linked markers are diagrammed for normal and aberrant segregations of the *URA3*-marked YAC in meiosis. Segregation errors associated with the *LYS2*-marked YAC are also possible (not shown). Meiosis II nondisjunction and meiotic chromosome loss events (both yielding 1+:3− tetrads) are distinguished by mating *SUP11*-containing spore colonies to *ade2-101* testers and scoring the resultant diploid color (white indicates nondisjunction has occurred, pink indicates chromosome loss).

mutations (in cis) on meiotic chromosome segregation. Figure 4B depicts a scheme for construction of strains carrying "test" YACs with *CEN* DNA mutations. To facilitate introduction of *CEN* mutations onto YAC12-UT, a "target" YAC strain (YPH603) was constructed. The target YAC strain was designed to contain a *D8B CEN4 HIS3* pB322 recombination target

TABLE IV
GENOTYPES OF STRAINS USED FOR MEIOTIC ANALYSIS

Strain	Genotype
YPH602	MATα ura3-52 lys2-801 ade2-101 his3-Δ200 trp1-Δ1 leu2-Δ1 [ΔCEN6::LEU2 CEN11] + YAC12-UT (URA3 SUP11 CEN4 human insert TRP1)
YPH603	MATα ura3-52 lys2-801 ade2-101 his3-Δ200 trp1-Δ1 leu2-Δ1 [ΔCEN6::LEU2 CEN11] + "target" YAC (HIS3 CEN4 D8B human insert TRP1)
YPH604	MATa ura3-52 lys2-801 ade2-101 his3-Δ200 trp1-Δ1 leu2-Δ1 + YAC12-LH (LYS2 CEN4 human insert HIS3)
YPH605	MATα/MATa ura3-52/ura3-52 lys2-801/lys2-801 ade2-101/ade2-101 his3Δ200/his3Δ200 trp1Δ1/trp1Δ1 leu2-Δ1/leu2-Δ1 CEN6/[ΔCEN6::LEU2 CEN11] + CFIII(D8B.d) URA3 SUP11 + CFIII(D8B.d) HIS3
YPH606	MATα/MATa ura3-52/ura3-52 lys2-801/lys2-801 ade2-101/ade2-101 his3Δ200/his3Δ200 trp1Δ1/trp1Δ1 leu2-Δ1/leu2-Δ1 CEN6/[ΔCEN6::LEU2 CEN11] + YAC12-UT (URA3 SUP11 CEN4 human insert TRP1) + YAC12-LT (LYS2 CEN4 human insert TRP1)
YPH607	MATα/MATa ura3-52/ura3-52 lys2-801/lys2-801 ade2-101/ade2-101 his3Δ200/his3Δ200 trp1Δ1/trp1Δ1 leu2-Δ1/leu2-Δ1 CEN6/[ΔCEN6::LEU2 CEN11] + YAC12-UT (URA3 SUP11 CEN4 human insert TRP1) + YAC12-LH (LYS2 CEN4 human insert HIS3)
YM259	MATa ura3-52 ade2-101 his3Δ200 tyr1
YM260	MATα ura3-52 ade2-101 his3Δ200 tyr1

on the short arm in which the centromere can be replaced (by one-step gene replacement) using any of the pJS2 or pDK352 derivative plasmids listed in Table I. Plasmid digested with NotI and PvuI is used to transform YPH603, selecting Ura$^+$. The resultant structure on the short arm is *D8B CEN6* URA3 SUP11*. This "test" YAC-containing strain can then be mated to the YAC12-LH "homolog"-containing strain (YPH604) for subsequent meiotic analysis.

Quantitative Analysis of Meiotic Chromosome Segregation. Sporulation and tetrad dissection of the YAC-containing strains are carried out using standard techniques.[17] Meiosis I behavior of the YAC is scored in reference to *LEU2* prototrophy (see Figs. 4A and 5). The diploid strains contain *leu2Δ1* deletions[30] at the endogenous loci on chromosome III and replacement of one copy of *CEN6* with a *LEU2 CEN11* cassette.[28] In this

[30] R. Surosky and B.-K. Tye, *Genetics* **119,** 273 (1988).

configuration, *LEU2* perfectly marks sister spores because homologous recombination between *LEU2* and its centromere cannot occur.

Nondisjunction and loss events in meiosis II can be distinguished by following the behavior of the *URA3 SUP11*-containing YAC. This is accomplished by mating spore colonies to *ade2-101* tester strains (YM259, *MAT*a; YM260, *MAT*α). The diploids resulting from these matings will be white if the original spore contained two copies of the YAC containing *SUP11* (the result of nondisjunction) or pink if it contained one copy (the result of chromosome loss).

Specific Examples

Meiotic Segregation Errors Associated with Chromosome Fragments. Chromosome fragments, useful for the analysis of mitotic segregation, proved to be inappropriate for the analysis of chromosome segregation in meiosis. YPH605, a diploid strain containing two differentially marked chromosome fragment "homologs" [CFIII(*D8B*.d) *URA3 SUP11* and CFIII(*D8B*.d) *HIS3*] was sporulated and dissected to determine the fidelity of meiotic disjunction. Ten percent of the four viable spore tetrads showed meiosis I nondisjunction of the two chromosome fragments; 2% showed meiosis II errors (Table V). Furthermore, 12 of 169 tetrads (7%) contained two viable spores which were red, *his3, ura3,* and nonmaters. There data can be explained by cosegregation of the two normal copies of chromosome III to one daughter cell and the two chromosome fragments to the other during meiosis I. Apparently, the homology between the chromosome fragments and the left arm of endogenous chromosome III interferes with proper disjunction of chromosome III homologs at a significant rate. This result is consistent with previously published data.[30] Thus, the chromosome fragments were not appropriate for meiotic analysis because they interacted significantly with chromosome III and led to an unacceptably high background of missegregation and inviable spores.

Artificial Chromosomes as Marker Chromosomes for Meiotic Segregation. In contrast to chromosome fragments, the behavior of homologous yeast artificial chromosomes in meiosis makes them well suited for use as marker chromosomes in meiotic analysis. To test meiotic chromosome transmission fidelity of the 365-kb human YAC12, strain YPH606 was constructed to contain two differentially marked copies. One YAC (YAC12-UT) was marked with *URA3 SUP11* on the short arm; the second YAC (YAC12-LT) was a derivative of YAC12-UT in which the *URA3 SUP11* marker was changed to *LYS2* using YMC.*LYS2* (as described above). Meiotic analysis of this strain revealed 100% correct disjunction in meiosis I and 98% correct disjunction in meiosis II for both the *URA3 SUP11*- and the *LYS2*-marked YACs (Table V). These results demon-

TABLE V
MEIOTIC ANALYSIS

Strain	No. of tetrads analyzed (4 viable spores)	Proper segregation[a]	Meiosis I nondisjunction	Meiosis I precocious separation	Meiosis II nondisjunction plus meiotic chromosome loss
YPH279 (CFs)	154	131 (86%)	16 (10%)	0 (0%) CF.URA 0 (0%) CF.HIS3	5 (3%) CF.URA3 2 (1%) CF.HIS3
YPH603 (YACs)	51	50 (98%)	0 (0%)	0 (0%) YAC.URA3 0 (0%) YAC.LYS2	1 (2%) YAC.URA3 1 (2%) YAC.LYS2

[a] Proper segregation refers to the *CEN*-linked markers on the two artificial chromosomes segregating away from each other in meiosis I and properly disjoining in meiosis II.

strate that the YAC is able to segregate properly with a high degree of fidelity in both meiosis I and meiosis II. We conclude that these YAC-containing strains provide the necessary low background for analysis of subtle alterations in meiotic chromosome segregation.

Comments

Spore Dissection. We find that the YPH strains listed in Table IV and strains isogenic to them will give approximately 95% spore viability if the following conditions are used: (1) Zymolyase treatment should be kept to the minimum time needed to generate tetrads that can be separated with a glass needle. Our experience is that treatment in 200 μg/ml Zymolyase 20T in 1 M sorbitol for 8 min is usually sufficient. However, this time may vary with the batch of Zymolyase used and is best determined empirically. (2) Sporulated strains can be stored at 4° for approximately 1 week without drastic reduction in viability. Plates stored longer will have progressively lower spore viability. In general it is preferable to dissect sporulated strains as soon as possible.

Testing Marker-Changed Artificial Chromosomes. From our experience, when novel marker-change configurations are introduced into YACs, potential aberrant segregational behavior owing to cis effects should first be ruled out. The original strain for meiotic analysis was constructed using the YMC.*HIS3* plasmid to change the *URA3 SUP11* YAC short arm markers to *HIS3*. The data showed that the *HIS3* YAC disjoins from its homolog (the original *URA3 SUP11* YAC) with high fidelity in meiosis I (57/57 tetrads). However, the *HIS3* YAC demonstrated aberrant segregation in meiosis II (6/57 tetrads). The reason for this problem is unknown, but it may be due to the transcriptional orientation of the *HIS3* gene in relation to the centromere (transcription is toward the *CEN* sequence).[31] The YMC.*LYS2* plasmid used to convert *URA3 SUP11* to *LYS2* results in a transcriptional orientation away from the centromere.

Conclusion

In this chapter we describe cytological and genetic methods to study the fidelity of chromosome segregation in mitosis and meiosis. Application of these methods should contribute to our understanding of the cis- and trans-acting determinants required to achieve the observed accuracy of these processes.

---

[31] A. Hill and K. Bloom, *Mol. Cell. Biol.* **7**, 2397 (1987).

[55] Genetic Screens and Selections for Cell and Nuclear Fusion Mutants

By VIVIAN BERLIN, JULIE A. BRILL, JOSHUA TRUEHEART, JEF D. BOEKE, and GERALD R. FINK

Introduction

Haploid cells, confronted with cells of opposite mating type, leave the mitotic cell division cycle and enter into conjugation (Fig. 1), the name given to the mating events which produce the diploid yeast zygote. On exposure to mating pheromone, cells arrest at G_1 in the cell cycle.[1-3] Although pheromone-arrested cells fail to divide mitotically, they continue to grow, forming elongated surface projections that give the cell a pear or shmoo shape.[4,5] Cells of opposite mating type appear to stick together at the tip of these projections. This sticking may be promoted by mating type-specific agglutinins expressed on the surface of the cells.[6-8] Once contact is established between pairs of **a** and α cells, they become intimately associated such that they are no longer separable by sonication.[2] The intervening cell wall at the site of cell–cell contact is dissolved, exposing the plasma membranes of each parent cell which then undergo fusion to make a single continuous membrane around the cellular contents.[5,9] During wall dissolution and membrane fusion, the cell wall junctions on the outer periphery become continuous, presumably by local cell wall synthesis. Immediately after membrane fusion, the haploid nuclei fuse, beginning at the spindle pole bodies.[9] The net result of conjugation is that two cells, each with a haploid nucleus, fuse to form a dumbbell-shaped zygote containing a single diploid nucleus. The zygote subsequently produces diploid vegetative cells by budding.

Conjugation is conveniently divided into two processes: signal transduction and fusion. Signal transduction is the process by which the mating pheromone secreted by a cell of one mating type stimulates mating func-

---

[1] E. Bucking-Throm, W. Duntze, and L. H. Hartwell, *Exp. Cell Res.* **76**, 99 (1973).
[2] L. H. Hartwell, *Exp. Cell Res.* **76**, 111 (1973).
[3] L. E. Wilkinson and J. R. Pringle, *Exp. Cell Res.* **89**, 175 (1974).
[4] P. N. Lipke, A. Taylor, and C. E. Ballou, *J. Bacteriol.* **127**, 610 (1976).
[5] M. Osumi, C. Shimoda, and N. Yanagashima, *Arch. Microbiol.* **97**, 27 (1974).
[6] R. Betz, W. Duntz, and T. R. Manney, *FEMS Microbiol. Lett.* **4**, 107 (1978).
[7] G. Fehrenbucher, K. Perry, and J. Thorner, *J. Bacteriol.* **134**, 893 (1978).
[8] Y. Kawanabe, K. Yoshida, and N. Yanagashima, *Plant Cell Physiol.* **20**, 423 (1979).
[9] B. Byers and L. Goetsch, *J. Bacteriol.* **124**, 511 (1975).

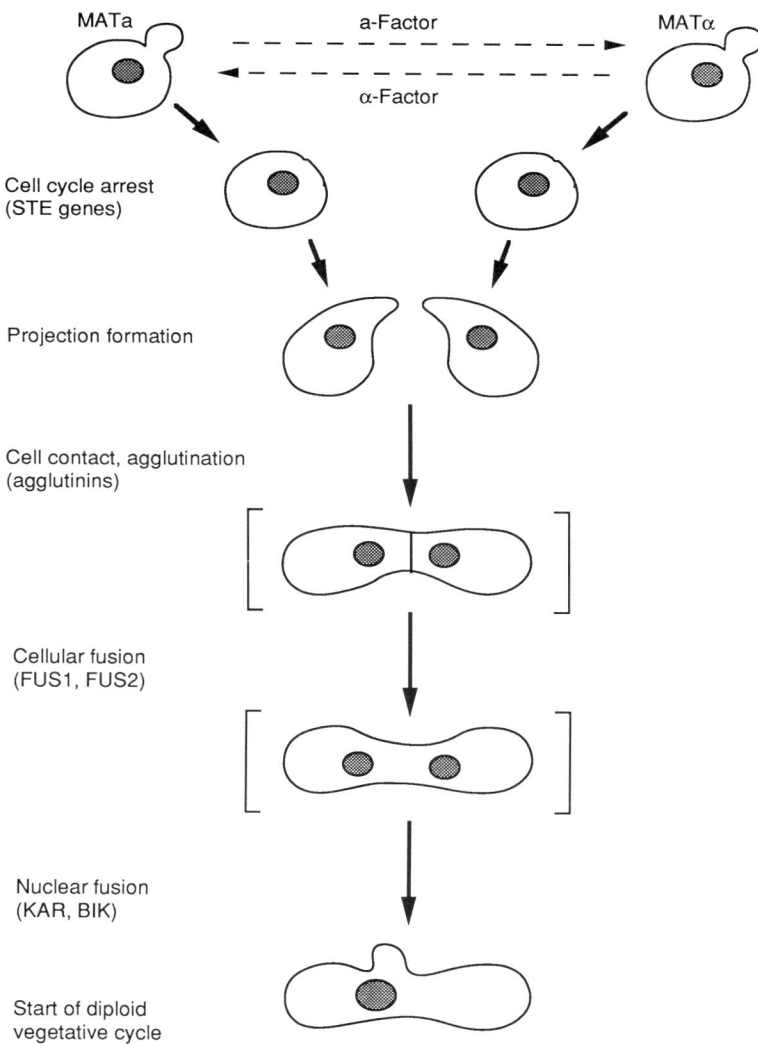

FIG. 1. The conjugation process in *Saccharomyces cerevisiae*. *MAT*a and *MAT*α cells, when mixed on solid medium, mutually stimulate each other via secreted peptide pheromones to arrest growth in G_1, induce transcription of mating-specific genes, and form projections in the direction of prospective mating partners. Contact usually occurs at the tip of the projections and leads to a rapid reorganization of the cell surface and mixing of cytoplasmic contents. At the same time, a signal is generated that prevents triparental interactions. Nuclear fusion follows cellular fusion, and the diploid cell commences vegetative growth.

tion in a cell of the opposite mating type. Thus, signal transduction includes both those functions required to arrest mitotic growth as well as those required for induction of the conjugation-specific genes. The mating pheromones (a-factor and α-factor), the pheromone receptors, the trimeric G protein that transfers the signal, and the trans-acting DNA binding proteins that induce conjugation-specific genes are all part of the signal transduction process.[10] The fusion pathway involves all of the functions required for joining the two isolated cells into a single zygote. The events in this pathway involve agglutination, cell wall fusion, plasma membrane fusion, and finally nuclear fusion. This chapter focuses on methods for obtaining and analyzing mutants in the fusion part of the conjugation pathway.

General Methods

Assays of Cell and Nuclear Fusion

Most fusion mutants are identified by their failure to form diploids. Failure of a haploid to form diploids when presented with a cell of opposite mating type could result from mutations in either signal transduction or fusion. Several tests have been developed to distinguish between mutations in signal transduction and those in fusion. Since signal transduction is a prerequisite for subsequent conjugation events, mutations that prevent signal transduction often fail to induce conjugation-specific genes or to manifest any of the morphological changes associated with conjugation. For example, *MAT*a *ste2* mutants lack the α-factor receptor and therefore fail to respond to α cells: they do not form projections, induce conjugation-specific genes, agglutinate, or fuse with cells of opposite mating type.[11] In actual practice a mutation in signal transduction is identified by two criteria: failure to respond to pheromone (no cell cycle arrest or morphological changes) and failure to induce transcription of *FUS1,* a gene that is massively induced by mating pheromones.[12] Cells that fail to form diploids, yet respond to pheromone and induce *FUS1,* are putative fusion mutants. These mutants can be analyzed by the following techniques in order to determine at what point in the fusion process they are blocked.

Signal Reception and Transduction. Pheromone production and sensitivity (responsiveness) can be assayed to distinguish mutants blocked in

[10] F. Cross, L. H. Hartwell, C. Jackson, and J. H. Konopka, *Annu. Rev. Cell Biol.* **4,** 429 (1988).
[11] D. D. Jenness, A. C. Burkholder, and L. H. Hartwell, *Cell (Cambridge, Mass.)* **35,** 521 (1983).
[12] J. Trueheart, J. D. Boeke, and G. R. Fink, *Mol. Cell. Biol.* **7,** 2316 (1987).

cell or nuclear fusion from mutants with a defect in the signal transduction pathway. If signal transduction and transmission are normal in a putative mutant, then the strain should be inhibited by pheromone and secrete its own pheromone. Inhibition by the pheromone can be tested by growing the *MAT*a strain in question at 30° in liquid YPD medium titrated to pH 4.0 with HCl. α-Factor is added to logarithmically growing cells (5 μ*M* final concentration), and cells are incubated for 2 hr at 30°, fixed (see below), and examined microscopically. Cells sensitive to α-factor arrest their growth in G_1 (unbudded) and form projections. To examine pheromone secretion, putative mutants are tested for their ability to arrest growth of a lawn of cells of the opposite mating type. The most sensitive test for pheromone secretion uses lawns of *MAT*a *sst1*(L3284) or *MAT*α *sst2* (F747) cells. The *sst* mutations confer a heightened sensitivity to pheromone production[13-16] and make the test easy to read. Patches of mutant and wild-type control strains are replica plated onto a dilute lawn of the desired *sst* strain and incubated at 24° for 1 or 2 days. Ability to secrete mating pheromone is judged by the diameter of the halo of growth inhibition of the lawn.

Cytological Analysis: Nuclear Staining. The position of the nucleus in fused cells can be examined directly by fluorescence microscopy of newly formed zygotes stained with the DNA-specific dye mithramycin[17] or DAPI (4′,6′-diamidino-2-phenylindole). Wild-type zygotes usually have a single nucleus because nuclear fusion rapidly follows cell fusion. Fusion mutants often show two or more nuclei. For DAPI staining, mixtures of cells (3×10^6 of each parent) are mated for 3.5 hr at 30° on sterile nitrocellulose filters placed aseptically on YPD plates. Filters are then placed in tubes with 0.85% (w/v) saline and vortexed to dislodge cells, and cells are collected by centrifugation. Alternatively, matings are performed by mixing cells of opposite mating type, pelleting the cells, resuspending them in 0.5 ml of YPD and spreading them on small YPD plates (60×15 mm). Cells are fixed in 1 ml of Carnoy fixative (methanol–glacial acetic acid, 3:1) at room temperature for 45 min and washed 4 times with 0.85% saline. Cells resuspended in 0.85% saline are stained with DAPI (1 μg/ml) for 45 min at room temperature, washed 4 times with 0.85% saline, sonicated briefly to disperse clumps, and then examined by fluorescence microscopy (see [40] in this volume).

---

[13] G. F. Sprague, Jr., and I. Herskowitz, *J. Mol. Biol.* **153**, 305 (1981).
[14] R. K. Chan and C. A. Otte, *Mol. Cell. Biol.* **2**, 11 (1982).
[15] E. Ciejek and J. Thorner, *Cell (Cambridge, Mass.)* **18**, 623 (1979).
[16] C. Dietzel and J. Kurjan, *Mol. Cell. Biol.* **7**, 4169 (1987).
[17] M. L. Slater, *J. Bacteriol.* **126**, 1339 (1976).

Cytological Analysis: Septum Formation. A smooth isthmus forms between the two lobes of the zygote. Mutations in cell fusion often lead to a characteristic septum at the junction between the lobes. These structures are easily identified when observing zygotes by phase-contrast microscopy or by Nomarski optics. Since the septum poses a physical block to nuclear fusion, unfused nuclei can be detected by staining zygotes with DAPI as described in the preceding test.

Cytoduction. Normal zygotes rarely produce haploid buds. Mutants defective for nuclear fusion undergo cell fusion normally but bud off cytoductants, haploid exconjugants containing the nucleus of one parent and the cytoplasm from both. These cytoductants can be identified because both the nuclei and the cytoplasm of each parent can be marked genetically. Figure 2 illustrates a standard cytoduction assay in which ρ is the cytoplasmic marker and *cyh2, can1,* or *nys*^{R}, recessive mutations conferring drug resistance, are the nuclear markers. Parent 1 is an X$^-$ *can1*^{R} *nys*^{R} petite ([ρ^o]) strain and parent 2 is a Y$^-$ [ρ^+] strain containing the wild-type alleles *CAN1* and *NYS*^{S} conferring drug sensitivity (X$^-$ and Y$^-$ are complementary auxotrophic markers). The two strains are mated for 4–5 hr on YPD, and then the mixture is plated onto synthetic complete medium minus arginine (canavanine competes with arginine for incorporation into proteins) containing 3% glycerol, 0.1% glucose, nystatin (2 μg/ml), and canavanine (60 μg/ml). (If parent 1 were *cyh2*, cytoductants would be selected on complete medium containing 3% glycerol, 0.1% glucose and 10 μg/ml of cycloheximide.) If cell fusion is normal but nuclear fusion is defective, one will obtain X$^-$ *can1 nys*^{R} [ρ^+] haploids. These are cytoductants because they contain the nuclear genotype of parent 1 and the mitochondrial genotype contributed by parent 2. This selective medium prohibits growth of the parental strains and diploid progeny; canavanine and nystatin inhibit the growth of parent 2 and diploid progeny, and glycerol does not support the growth of petite strains (i.e., parent 1). The total number of cells and diploids are determined by plating dilutions of the mating mixture onto YPD and minimal media, respectively. The efficiency of nuclear fusion is expressed as the ratio of cytoductants to diploids: the lower the efficiency of nuclear fusion, the higher the cytoductant to diploid ratio. The cytoductant to diploid ratio for most wild-type strains is approximately 1×10^{-4} to 1×10^{-3}.

Clonal Analysis of Zygotes by Micromanipulation. Wild-type haploid cells fuse to form zygotes that bud off diploid cells. Fusion mutants may form zygotes in which the cells fuse but the nuclei do not (heterokaryons) or prezygotes in which neither the cells nor nuclei fuse. If cell fusion is normal but nuclear fusion is defective, then zygotes will bud off cytoduc-

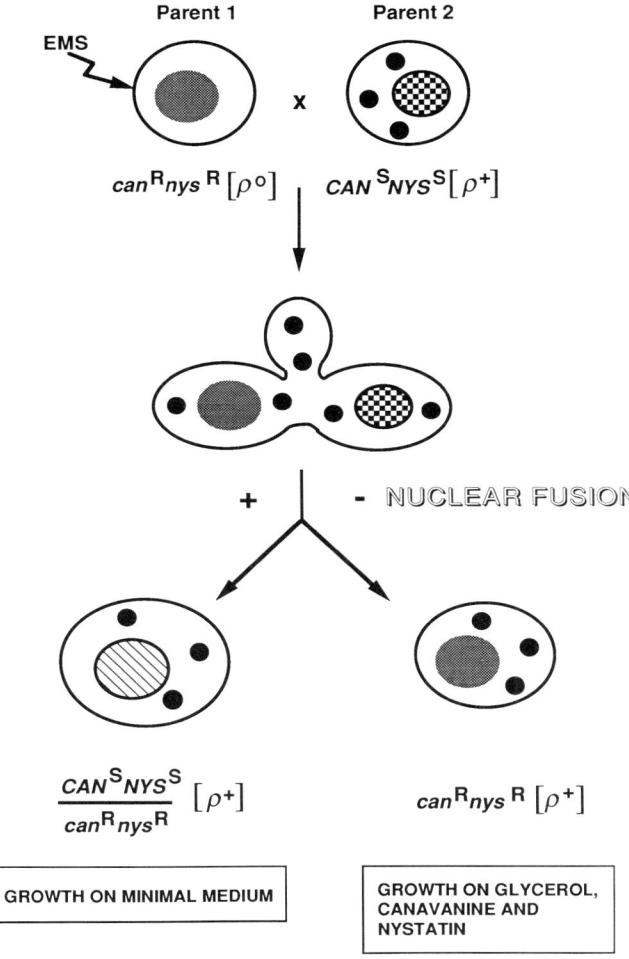

FIG. 2. Strategy for the isolation of mutants defective in nuclear fusion. Parent 1 is mutagenized and mated with parent 2. Kar⁻ mutants undergo cell fusion without concomitant nuclear fusion, producing transient heterokaryons. The products of such a mating, Can^R Nys^R $[\rho^+]$ cytoductants containing the nuclear genotype of parent 1 and the cytoplasmic genotypes of parents 1 and 2, are able to grow on selective medium containing glycerol, canavanine, and nystatin, whereas the parental strains and diploids cannot. Prototrophic diploids can grow on minimal medium, whereas cytoductants and parental strains cannot.

tants (see above) containing a haploid nucleus from one parent and the cytoplasmic contribution of both parents. If the cells adhere but the walls or membranes fail to fuse, then zygotes will bud off haploid cells that have the nuclear and cytoplasmic constitution of one of the original parents. The latter are not cytoductants since the cytoplasms failed to mix.

To perform clonal analysis of zygotes, mate putative mutants with wild-type or mutant tester strains for 3–4 hr. Individual zygotes are isolated by micromanipulation and grown for 3 days at 30° to form colonies. The composition of a zygotic colony is determined by resuspending each colony in water and plating approximately 100–200 cells on YPD medium. The phenotypes of the resulting single colonies are tested to determine whether they are diploid or haploid and, if haploid, whether they are cytoductants or exconjugants with an unmixed cytoplasm.

Isogenic MATa and MATα Strains Containing Fusion Mutations

Many of the diagnostic tests for fusion mutations require isogenic MATa and MATα derivatives containing the same fusion mutation. Outcrossing to obtain the opposite mating type can introduce genes that modify the expression of the original mutation and make diagnosis of the phenotype difficult. Outcrosses can be avoided if the mating type is switched by using a plasmid-borne HO gene. For this purpose, mutants are transformed with the plasmid, pSB283, carrying the HO gene under control of the GAL1 promoter. This plasmid, derived from YCp50-HO (gift of R. Jensen and I. Herskowitz; see [8] in this volume), contains LEU2 in addition to URA3, enabling selection in a variety of strains (J. Trueheart, unpublished results). leu2-3, 112 or ura3-52 strains carrying this plasmid are grown overnight in SC minus leucine (SC − Leu) or minus uracil (SC − Ura) medium containing 0.1% glucose. Cells are then washed and diluted into medium containing 2% galactose (YPGal). After incubation for 2–4 hr at 30° in YPGal medium, cells are plated for single colonies on medium containing glucose (YPD). Colonies are tested for mating type by performing matings with MATa and MATα tester strains: colonies are replica plated to mating tester lawns on YPD plates, incubated approximately 4 hr to overnight at 30°, and then replica plated to minimal medium to select for prototrophic diploids. Colonies are also replica plated to SC − Leu or SC − Ura medium to identify isolates that have lost pSB283.

Complementation and Dominance Tests

Dominance and complementation tests are carried out in a diploid heterozygote. The dominance or complementation test for fusion ability cannot be carried out in MATa/MATα heterozygotes because strains with this configuration of the mating type locus are unable to mate. The following procedures permit the creation of diploids that are competent to mate.

1. Construction of strains homozygous at MAT by mitotic recombination: Several thousand $MATa/MAT\alpha$ diploids are plated on YPD medium. Those that become homozygous for the MAT locus by mitotic recombination are identified by their ability to mate with $MATa$ or $MAT\alpha$ tester strains.

2. Construction of $MAT\alpha/mat$a1 or $MAT\alpha/mat$a$_0$ diploids: Cells carrying a defective MATa1 gene or a deletion of the MAT locus (mata$_0$) mate like $MATa$ cells; however, the resulting $MAT\alpha/mat$a1 or $MAT\alpha/mat$a$_0$ diploids are capable of mating with a $MATa$ cell.[18] The mata1 mutation is introduced into strains by performing a cross to a $kar1$-1 strain.[19] Such a cross must be performed since $MAT\alpha/mat$a1 strains are unable to sporulate. The $kar1$-1 mutation inhibits nuclear fusion (as described in a later section), producing heterokaryons when the $kar1$-1 strain is mated with the $MAT\alpha/mat$a1 strain. The mating mixtures are plated onto sporulation medium, and asci containing separate clusters of four spores and two spores are dissected. The two clusters of spores are the products of meiosis of the diploid and haploid nuclei of the heterokaryon. The spores from the cluster of four which mate like $MATa$ cells contain the mata1 mutation. The $MAT\alpha/mat$a1 strain can also be induced to sporulate by transforming the diploid with a plasmid containing the $MATa$ gene.

mata$_0$ strains are constructed by disrupting the MAT locus according to the method of Rothstein.[20] Strains are transformed with a Hindlll fragment from pSB284 (derived from pAK2, gift of A. Klar; J. Trueheart, unpublished results, 1987) which contains a deletion of the MAT locus into which the $LEU2$ gene is inserted. Integration of the Hindlll fragment at the MAT locus by a double crossover event (gene conversion) replaces the wild-type allele with the disrupted allele.

Dominance and complementation tests are performed by examining the mating efficiency of $MAT\alpha/mat$a$_0$ (or $MAT\alpha/mat$a1) diploids. For example, in the dominance test, a $MAT\alpha/mat$a$_0$ diploid is constructed between a fusion mutation and a wild-type strain. In the complementation test, a $MAT\alpha/mat$a$_0$ diploid is constructed between two fusion mutations. The heterozygous diploids are compared with a wild-type $MAT\alpha/mat$a$_0$ diploid for ability to mate with either a $MATa$ wild-type cell or fusion mutant. The actual mating test is performed by replica plating patches of the $MAT\alpha/mat$a$_0$ or $MAT\alpha/mat$a1 diploids to $MATa$ lawns of both mu-

[18] Y. Kassir and G. Simchen, *Genetics* **82**, 187 (1976).
[19] A. Klar, *Genetics* **94**, 597 (1980).
[20] R. J. Rothstein, this series, Vol. 101, p. 202.

tant and wild-type tester strains on YPD medium. These plates are incubated for 4 hr at 30°, then replica plated to minimal medium to select for growth of prototrophic progeny.

Isolation of Mutants with Bilateral Block in Cell or Nuclear Fusion

Rationale

The design of a scheme to isolate fusion mutations depends on whether the mutations sought are unilateral or bilateral. A bilateral mutation is one that manifests a fusion defect only if both parents contain the same mutation (e.g., mutant × mutant → few or no diploids; mutant × wild type → diploids). A unilateral mutation is one which manifests a fusion defect even when one of the parents in the cross is wild type (e.g., mutant × wild type → few or no diploids). Since several fusion mutants obtained serendipitously (*fus1* or *fus2*[12] and *bik1*[12,21]) have a bilateral defect, we decided to develop a strategy for systematically isolating mutants which exhibit a bilateral defect in cell or nuclear fusion during mating.

Identification of a bilateral fusion mutant requires a system that can produce a clone consisting of both *MATa* and *MATα* cells carrying the same mutation so that large numbers of such clones can be screened to identify those where mating has failed to occur within the clone. Certain *HO/HO* strains have the desired properties. On sporulation, *HO/HO* strains yield haploid spores which germinate and undergo mating-type switching to produce *MATa* and *MATα* cells. If no mutation affecting mating has occurred, the *MATa* and *MATα* cells within each ascospore clone will mate to produce a stable, nonmating *MATa/MATα* diploid early in the growth of the colony. If a mutation which causes a bilateral block in cell or nuclear fusion occurs, then the *MATa* and *MATα* cells within the clone will not mate with each other or will do so at a reduced frequency. These mutant clones can be identified because they will mate with tester strains of either mating type whereas nonmutant clones (diploid) will mate with neither mating type (Fig. 3). Unwanted mutations such as those which cause a failure to respond to mating pheromone (e.g., *ste*) or which prevent mating-type switching (e.g., *HO* → *ho*) would be easily identified because they would produce a nonmating phenotype or preferential mating to one mating type, respectively. Thus, the bilateral fusion defect provides an easily distinguishable phenotype: failure to diploidize within the clone and ability to mate with both *MATa* or *MATα* wild-type testers.

[21] V. Berlin, C. A. Styles, and G. R. Fink, *J. Cell Biol.*, in press (1990).

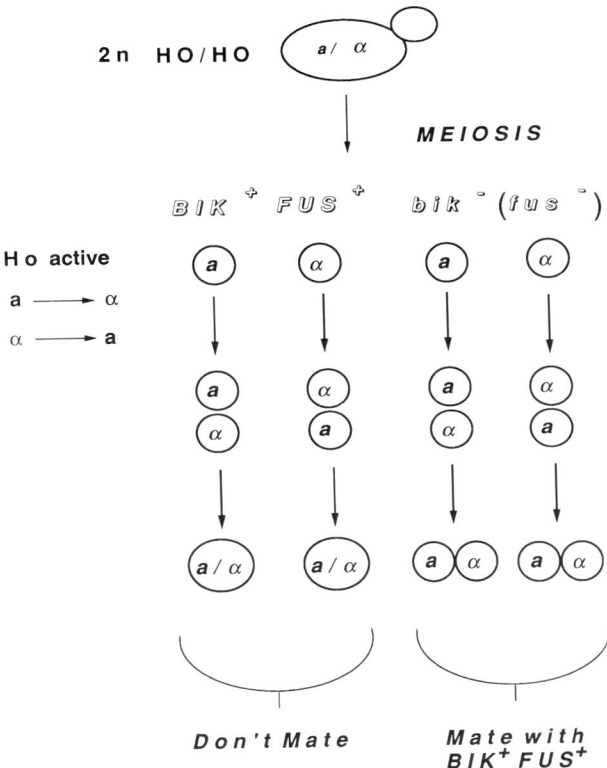

FIG. 3. Strategy for isolating mutants with a bilateral block in cell or nuclear fusion. An *HO* strain is mutagenized, sporulated, and plated for random spores on YPD medium. Wild-type cells (Fus⁺ and Bik⁺) form a/α diploid colonies unable to mate with **a** and α tester lawns. Fus⁻ or Bik⁻ colonies, unable to form diploids owing to a bilateral block in cell or nuclear fusion, respectively, can mate with both *MAT*a and *MAT*α tester lawns.

Mutant Screen

The strains used for the genetic screens and selections in this and in subsequent sections are listed in Table I. The media have been described previously.[22]

The *HO/HO* strain F764 was mutagenized with ethylmethane sulfonate (EMS).[23] Approximately 10^7 to 10^8 cells were spread on GNA plates

---

[22] F. Sherman, G. R. Fink, and C. W. Lawrence, "Methods in Yeast Genetics." Cold Spring Harbor Laboratory, Cold Spring Harbor, New York, 1979.
[23] G. R. Fink, this series, Vol. 17A, p. 59.

TABLE I
List of *Saccharomyces cerevisiae* Strains

Strain	Genotype	Source
7117-1A	*MATa ade2-1*	Lab Collection
F2	*MATα ade2-1*	R. K. Mortimer
F242	*MATa lys2*	F. Sherman
F243	*MATα lys2*	F. Sherman
F581	*mata1 trp1-1 ade8 ade6? his2 ilv3 ura1 MAL2*	?
F747	*MATα sst2-4 his6 ura1 cry1 (met can1 cyh2 rme1?)*	J. Thorner
F762	*MATa ura3-52 trp1Δ1*	J. Thomas
F763	*MATα ura3-52 trp1Δ1*	J. Thomas
F764	*MATa/MATα HO/HO his5/his5 ade5/ade5 ura4/ura4 met3/met3 met14/met14*	R. Jensen
GF4836-8C	*MATa leu1 thr1 [ρ⁺]*	Lab Collection
JY289	*mata1 fus2-1 ura1 trp1 ade*	J. Trueheart
JY290	*mata1 fus1-483 ura1 trp1 ade*	J. Trueheart
JY318	*MATa fus1-483 fus2-1 ura3-52 lys2-801 GAL⁺*	J. Trueheart
L461	*MATα his4 ade2 can1 nysᴿ [ρ⁰]*	J. Conde
L1543	*MATα lys9*	J. Boeke
L1544	*MATa lys9*	J. Boeke
L1545	*MATa lys9 his4Δ453*	J. Boeke
L1546	*MATα lys9 his4Δ453*	J. Trueheart
L2745	*MATα ade2-1 fus1-483 fus2-1*	J. Trueheart
L2746	*MATa ade2-1 fus1-483 fus2-1*	J. Trueheart
L3284	*MATa sst1-3 ura3-52 his6 lys2Δ201 trp1Δ1*	J. Trueheart
L3285	*MATα trp1Δ1 leu2-3,112 ade2*	J. Trueheart
L3286	*MATa ura3-52, leu2-3,112 ade2-1*	J. Trueheart
L3404	*MATa ura3-52 trp1Δ1 ade? ura4 fus3-1 (B7)*	V. Berlin
L3405	*MATα ura3-52 trp1Δ1 ade5 met lys2-801 fus3-1 (B7)*	V. Berlin
L3519	*mata₀::LEU2 leu2-3,112 ura3-52 his4-34 lys2*	E. Elion
L3520	*mata₀::LEU2 leu2-3,112 ura3-52 lys2 fus1Δ1*	E. Elion
L3522	*mata₀::LEU2 leu2-3,112 ura3-52 lys2 his4-34 fus2Δ3*	E. Elion
L3871	*mata₀::LEU2 leu2-3,112 ura3-52 lys2 his4-34 fus3-1*	J. Brill
VB07-7B	*MATα trp1Δ1 his5 met fus3-1 (B7)*	V. Berlin
VB18-16A	*MATa trp1Δ1 his5 (B1)*	V. Berlin
VB18-16D	*MATα trp1Δ1 ura3-52 (B1)*	V. Berlin
VB32-5A	*MATα ura3-52 trp1Δ1 met (B15)*	V. Berlin
VB32-5C	*MATa ura3-52 his5 (B15)*	V. Berlin
VB34-2C	*MATα ura3-52 trp1Δ1 met (B17)*	V. Berlin
VB34-2D	*MATa ura3-52 trp1Δ1 met (B17)*	V. Berlin

and were grown for 2 days at 30°. Cells were replica plated to sporulation medium supplemented with adenine, uracil, histidine, and methionine and then incubated at room temperature until cultures sporulated. Asci were scraped off the sporulation plates, digested with Glusulase, and sonicated.[22]

The random spores were diluted and spread onto YPD plates to an estimated density of 150 to 200 cells per plate, and the resulting colonies were replica plated to lawns of MATa and $MAT\alpha$ tester strains to perform matings. Mating plates were incubated at 30° for 4 hr, and diploids were selected by replica plating the mating colonies onto YNB medium. Colonies which formed diploids with both mating tester strains were considered putative mutants. Putative mutants were outcrossed to a MATa or $MAT\alpha$ ho strain, F762 or F763, respectively. ho mutant segregants from the cross were identified by pairwise mating of MATa and $MAT\alpha$ segregants. MATa and $MAT\alpha$ segregants unable to mate with each other but able to mate with mating testers were backcrossed to F762 or F763 and analyzed further.

Results

Of the 7250 colonies screened, approximately 0.8% or 58 colonies appeared to form diploids when mated with both MATa and $MAT\alpha$ strains. After these colonies were streaked to obtain pure cultures, only 24 of the 58 putative mutants exhibited the bimating phenotype in mating tests performed with the MATa or $MAT\alpha$ ho strain, F762 or F763, respectively. The diploids formed in matings with F762 and F763 were sporlated to isolate ho mutant segregants.

Putative ho mutant segregants were backcrossed to the wild-type ho strain F762 or F763 to examine segregation of the mating defect. Four mutants (B1, B7, B15, B17) showed 2:2 segregation for a bilateral defect in diploid formation. Ascospore segregants (listed in Table I as VB18, VB32, VB34, L3404, and L3405 strains) were tested for their mating ability under a number of conditions. B1 mutant ascospores exhibited reduced diploid formation when mated at 34° compared to matings at 24° with MATa and $MAT\alpha$ wild-type strains, F242 and F243. B7 mutant ascospore segregants showed a reduced frequency of diploid formation when mated with the *fus1 fus2* double mutant (L2745 or L2746) and with the *bik1 fus1* double mutant (L1545 or L1546) compared to matings with wild-type control strains, F2 and 7117-1A or L1543 and L1544, respectively.

The assessment of whether the mutants were blocked in cell or nuclear fusion was made by examining mating mixtures stained with DAPl by phase-contrast and fluorescence microscopy. Mating partners of the B7 mutant consistently displayed the characteristic septum produced as a consequence of a block in cell fusion. Zygotes with septa and unfused nuclei were observed occasionally in mating mixtures of B1, B15, and B17 mutants.

To determine whether B7 was in the same complementation group as *fus1* or *fus2*, *MATα/mat*a1 diploids were constructed which had the following genotypes: *fus1*/B7, *fus2*/B7, *FUS+*/B7. These strains were constructed by mating VB07-7B with JY290, JY289, and F581, respectively. Complementation analysis was performed by mating the *MATα/mat*a1 diploids with a *MAT*a *fus1 fus2* tester strain (L2746). B7 was able to complement the mating defects of both the *fus1* and *fus2* strains, indicating that B7 corresponds to a third gene required for cell fusion, designated *FUS3*. The *MATα/mat*a1 *FUS+*/B7 heterozygote mated as efficiently with L2746 as the *FUS+/FUS+* control strain, indicating that the B7 mutation, renamed *fus3-1*, is recessive.

Comments

Although many putative mutants were identified in the initial screen, few of these emerged as promising candidates on subsequent testing. A major difficulty was encountered in scoring the bilateral mating defect in the initial outcross to a wild-type *ho* strain. The initial outcross yields *HO BIK* (or *HO FUS*) and *HO bik* (or *HO fus*) segregants which have non-mating and bimating phenotypes, respectively, and *ho BIK* (or *ho FUS*) and *ho bik* (or *ho fus*) segregants (*MAT*a or *MAT*α) which can mate to *MAT*a or *MAT*α wild-type tester strains. Spores which mated as **a** and α were mated to each other to distinguish the *ho BIK* and *ho bik* segregants from each other. Since the pairwise matings were performed with spores which frequently contained the same auxotrophies, mating tests could not be done by selection for prototrophic diploids. Instead the mating ability of each spore had to be judged by microscopic analysis. The inability to score each member of a tetrad identically in the initial outcross confounded the segregation analysis.

The need to isolate *ho* derivatives of each mutant by tetrad analysis could be avoided by isolating the mutants in a haploid strain containing the *HO* gene under control of the *GAL1* promoter. On galactose-containing medium, an *ho* strain containing the *GAL1::HO* fusion on a *CEN* plasmid (see [8] in this volume) undergoes mating-type switching and diploid formation. Once the fusion mutant is identified, the *GAL1::HO* plasmid can be easily segregated off to produce the desired *ho* derivative. However, since *HO* is no longer under mating-type control when fused to the *GAL1* promoter, *MAT*a/*MAT*α diploids continue to undergo mating-type switching, producing *MAT*a/*MAT*a and *MAT*α/*MAT*α diploids. These *MAT*a/*MAT*a and *MAT*α/*MAT*α diploids give the bimating phenotype of the desired mutant and, therefore, confound the analysis. This problem was not circumvented by using a shorter induction time on galactose prior to shutting off mating-type switching by addition of glucose to the medium. One possible solution was to use *HO* under the control of its own promoter, which would shut off *HO* expression as soon as the cell became *MATa/MATα*. Unfortunately, a haploid strain with the *HO* gene on a plasmid undergoes mating-type switching but produces nonmutant clones which contain many mating cells. Perhaps *MAT* repression of *HO* is partially relaxed when the *HO* gene is resident on a plasmid (R. Jensen, personal communication, 1986). A potentially useful modification of the *HO* scheme is to disrupt the *HO* gene in putative mutants prior to backcrossing to simplify the segregation analysis.

Isolation of Mutants Defective in Cellular Fusion

Rationale

The *fus1* and *fus2* mutations already identified[12] can be used to obtain new mutations defective in the cell fusion pathway. The idea is to find mutants that mate poorly with a *fus1 fus2* double-mutant strain but mate normally with a wild-type strain. In principle, mutations causing this phenotype could result from new mutations in *fus1* or *fus2*, in positive regulators of *fus1* and *fus2*, or in new genes in which the function is evident only in crosses by a cell with compromised mating ability (the *fus1 fus2* strain).

A screen using a mixed lawn (Fig. 4) was employed to identify haploid colonies which mate more efficiently with wild-type than with *fus1 fus2* cells. The screen relies on the color phenotype of *ade2* mutants. Colonies containing the *ade2* mutation are red, owing to the accumulation of an intermediate in the pathway for adenine biosynthesis, whereas ADE^+ colonies are white. Colonies of a *MATα ade2* strain mate equally well with *MATa ade2* and *MATa fus1 fus2 ADE^+* cells in a mixed lawn. However, on medium containing limiting quantities of adenine, *ade2/ADE^+* cells will grow faster than cells requiring adenine, producing diploid colonies which are white or slightly pink. A *MATα ade2* Fus⁻ colony, unable to form diploids with the ADE^+ *fus1 fus2* strain, will mate preferentially with the *ade2* Fus⁺ strain, producing *ade2/ade2* colonies which are deep red in color and easily visible against the background of white colonies, even on a crowded plate.

Mutant Screen

L3285 cells were mutagenized with ethylmethane sulfonate[23] to a survival of 9%. Approximately 500 viable cells plated on each of 79 YPD plates containing 0.3 mM adenine were incubated for 3 days at 30°. A

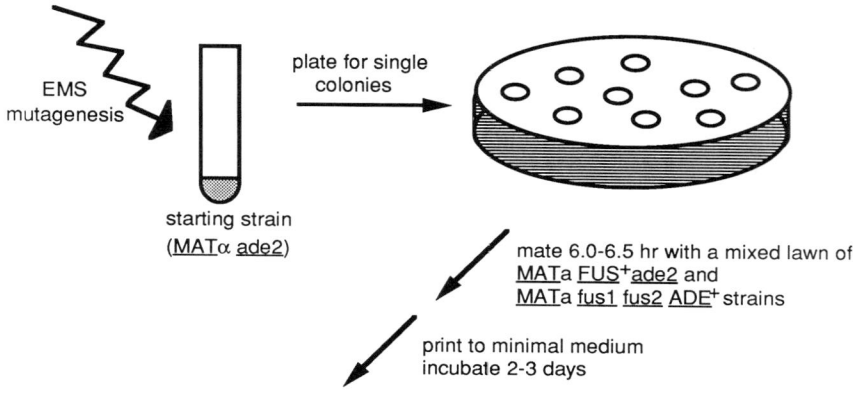

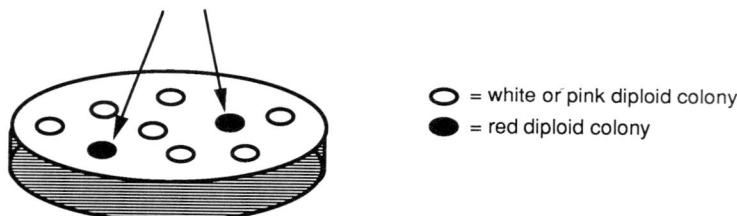

FIG. 4. Screen to isolate new *fus* mutants. A mutagenized culture of L3285 cells (*MATα ade2*) is plated for single colonies on YPD medium. Colonies are replica mated for 6.0–6.5 hr with a mixed tester lawn of L3286 (*MATa Fus⁺ ade2*) and JY318 (*MATa fus1 fus2 Ade⁺*) cells and printed to minimal medium limiting for adenine. The resulting diploid patches are examined 2–3 days later for color, and those with the most intense red color are selected for further study.

mixed lawn was prepared by plating 10^7 cells of each of two strains, L3286 and JY318, on YPD plus adenine. These lawns, grown overnight, were replica plated to fresh replicas of the mutagenized colonies and mated for 6.0–6.5 hr on YPD plus adenine. The plates were printed to minimal medium containing 0.1 mM adenine, and after 2–3 days the diploid colonies were scored for the intensity of the red color.[24]

Results

Approximately 40,000 colonies were screened by this method. Three hundred fifty putative mutants were retested for their Fus phenotype by

[24] J. Trueheart, Ph.D. Thesis, Massachusetts Institute of Technology, Cambridge, Massachusetts (1987).

4-hr replica matings to individual Fus⁺ (F762) and *fus1 fus2* (JY318) lawns. A final set of approximately 50 candidates that formed diploids efficiently with F762 but not with JY318 was selected for further analysis.

Complementation tests were performed to determine whether the Fus⁻ mutants were new alleles of *FUS1, FUS2,* or *FUS3* or whether they defined new genes. For this purpose, diploids able to mate were constructed by mating *MAT*α putative mutant strains with *mat*a1 Fus⁺, *fus1-483,* and *fus2-1* strains (F581, JY290, and JY289) or with *mat*a$_0$ Fus⁺, *fus1Δ1, fus2Δ3,* and *fus3-1* strains (L3519, L3520, L3522, and L3871). The *MAT*α/*mat*a1 and *MAT*α/*mat*a$_0$ diploids were then tested for their ability to mate with Fus⁺ and *fus1 fus2* strains. Three mutants failed to complement the mating defect of *fus1-483,* six failed to complement the mating defect of *fus2-1,* and two failed to complement the mating defect of *fus3-1* (described in the preceding section). Many of the putative mutants were examined for septum formation during self-matings. Each mutant was obtained in both mating types using the *HO* plasmid as described under General Methods. When examined under the microscope, the Fus⁻ mutants that were designated *fus1, fus2* and *fus3* on the basis of the complementation test showed the septum characteristic of fusion mutants. Those putative mutants that were not *fus1, fus2,* or *fus3* also did not produce septa during zygote formation. This group was examined for the ability to secrete mating pheromones by printing them to plates spread with a dilute culture of *MAT*a *sst1* or *MAT*α *sst2* cells. These putative mutants produced halos of growth arrest of the *sst* lawns equivalent to that produced by the wild-type control strain, indicating the former were secreting normal amounts of pheromone. These strains were also tested for α-factor sensitivity. All responded to pheromone by arresting growth and making projections. Thus, there is a group of strains that appeared positive in the screen which, on more careful testing, fail to give clear evidence for a fusion defect.

Comments

The screen described here allows easy detection of potential fusion mutants. Of the 44 putative new mutants in cellular fusion, three are new alleles of *FUS1,* seven are new alleles of *FUS2,* and two are new alleles of *FUS3.* Although *fus3-2* and *fus3-3* exhibit a bilateral defect in mating, the primary defect in these mutants appears to be a failure to arrest in G$_1$, suggesting an involvement of *FUS3* in the process of pheromone signal transduction. In fact, a null allele of *FUS3* is sterile and therefore required for transmission of the mating pheromone signal.[25] This screen may there-

---

[25] E. A. Elion, P. L. Grisafi, and G. R. Fink, *Cell (Cambridge, Mass.)* **60,** 649 (1990).

fore be useful in identifying a new class of mutants defective early in mating.

The majority of the mutants isolated by the *fus* screen do not appear to have a defect in cellular fusion. They may represent mutations leading either directly or indirectly to partial sterility. This procedure clearly yields new *fus1*, *fus2*, and *fus3* alleles. Further analysis will be required to determine whether any other genes in the cell fusion pathway can be obtained.

Isolation of Mutants Defective in Nuclear Fusion

Rationale

In matings of wild-type strains nuclear fusion occurs with greater than 99% efficiency. The strategy illustrated in Fig. 2 utilizes cytoductant formation as a selection for mutants with a unilateral defect in nuclear fusion. Cytoductants are produced if cell fusion occurs without concomitant nuclear fusion (see General Methods). The procedure involves matings between strains containing cytoplasmic and nuclear markers permitting the selection of cytoductants on medium which supports the growth of neither the parental strains nor diploid progeny. The advantage of this strategy is that it identifies mutants where cytoplasmic mixing between the two parents has occurred and therefore avoids sterile and cell fusion mutants.

Mutant Selection Strategy

The isolation of karyogamy-defective mutants has been described previously.[26,27] A *can1 nys^R* [ρ^o] strain (L461) was mutagenized with ethylmethane sulfonate.[23] After growth for 2 or 3 generations, mutagenized L461 was mixed with a Can^S Nys^S [ρ^+] strain (GF4836-8C). Aliquots (0.2 ml) of the mating mixtures containing 10^8 cells/ml of L461 and 10^9 cells/ml of GF4836-8C were spread on plates containing YPD medium. After 20–50 hr at 30°, cells from the mating plates were resuspended in water, diluted 1:10, and applied to plates containing selective medium, namely, YNB medium buffered with 0.1 M citrate phosphate (pH 6.5), containing 0.3 mM histidine, 0.15 mM adenine, 3% glycerol, 0.1% glucose, 60 μg/ml canavanine, and 2 μg/ml nystatin. This medium selects for the growth of cytoductants which are *can1 nys^R* and [ρ^+], that is, those containing the nuclear genotype of L461 and mitochondrial genotype of GF4836-8C. Colonies which grew after 5 to 15 days at 30° were tested for their nuclear and cytoplasmic traits. Those which were His$^-$ Ade$^-$ Can^R Nys^R [ρ^+] were tested further. His$^-$ Ade$^-$ Can^R Nys^R [ρ^+] cytoductants were

[26] J. Conde and G. R. Fink, *Proc. Natl. Acad. Sci. U.S.A.* **73**, 3651 (1976).
[27] J. Polaina and J. Conde, *Mol. Gen. Genet.* **186**, 253 (1982).

converted to [ρ°] by ethidium bromide treatment[22] and were crossed again to GF4836-8C to repeat the selection for cytoductants. Strains which produced cytoductants at a higher frequency than a control cross between the unmutagenized strain L461 × GF4836-8C were considered putative karyogamy-defective or Kar⁻ mutants.

Several of the putative mutants were clearly defective in nuclear fusion by two criteria. First, micromanipulation of individual zygotes from Kar⁻ mutants by wild-type crosses gave rise to zygotic colonies composed of haploid cells with the phenotype of either parental strain or a mixture of both. Second, staining mating mixtures with mithramycin[17] revealed zygotes with unfused nuclei. Mutants which exhibited a Kar⁻ phenotype by these two tests were backcrossed to Kar⁺ cells (GF4836-8C) to determine whether the Kar⁻ phenotypes segregated as single Mendelian traits. Segregants from the cross were grown on plates containing YPD medium and replica plated to a lawn of Kar⁺ cells of opposite mating type containing complementary auxotrophic markers. After 3–4 hr of incubation at 30°, mating mixtures were replica plated to minimal medium to select for growth of prototrophic diploid colonies. Kar⁻ segregants were distinguished from Kar⁺ segregants because the former produce diploids at a lower frequency than the latter when mated to a wild-type strain.

The Kar mutants were tested for dominance to Kar⁺ and for complementation with the other Kar⁻ mutants. Complementation and dominance tests were performed by constructing diploids homozygous for one mating type by the procedures described in General Methods. For the dominance and complementation tests Kar⁺/Kar⁻ and *karX/karY* diploids, respectively, were mated with a Kar⁺ haploid containing complementary auxotrophies. The diploid under analysis is said to have a Kar⁺ phenotype if it forms prototrophic progeny at a high frequency in a cross with a Kar⁺ haploid strain of the appropriate mating type.

Results

Karyogamy-defective mutants were isolated by selecting mutants which produce cytoductants at a high frequency when mated with a wild-type strain. From 11 independent mutagenesis experiments 22 mutants were isolated which exhibited 2:2 segregation for a karyogamy defect. These mutants fell into at least three complementation groups, designated *kar1, kar2,* and *kar3*. The *kar1* and *kar2* mutants were recessive, whereas the *kar3* mutants were semidominant.[26,27]

Comments

In preliminary attempts to isolate karyogamy-defective mutants by selecting for a high level of cytoductant formation, a strain containing one recessive drug resistance marker was used. A majority of the colonies which grew on selective medium were not cytoductants but prototrophic diploids which had become drug-resistant by mitotic recombination (producing diploids homozygous for drug resistance). The frequency of drug-resistant diploids arising by mitotic recombination was dramatically reduced by the use of a strain containing two recessive drug resistance markers. With this modification two classes of colonies grew on selective medium: bona fide cytoductants (the majority) and prototrophic diploids (possibly multiple aneuploids which had become drug resistant). The former were easily distinguished from the latter by the presence of auxotrophic markers.

The *kar1, kar2,* and *kar3* mutants were isolated using the selection for cytoductants. All three mutants form zygotes (i.e., undergo cell–cell fusion) at a normal frequency but undergo nuclear fusion in less than 10% of the zygotes. The consequence of a failure in nuclear fusion is a high cytoductant to diploid ratio. For example, the cytoductant to diploid ratio is approximately 4 for *kar1-1* × *KAR1* matings versus 1×10^{-3} for matings between two *KAR1* strains. Evidence suggests the *KAR1* gene may specify a component of the spindle pole body,[28] the site of initiation of nuclear fusion during zygote formation.[9] *KAR2* is the yeast homolog of mammalian BiP/GRP78, a protein of the endoplasmic reticulum possibly involved in the assembly of multimeric secreted proteins.[29] *KAR3* shows homology to kinesin, a microtubule motor protein.[30]

Conclusions

Genetic screens and selections are described for isolating mutants defective in cell or nuclear fusion. The selection for karyogamy mutants successfully yields unilateral *kar1, kar2,* and *kar3* mutants. New alleles of *FUS1, FUS2,* and *FUS3* were isolated by the screens for bilateral mutants. These screens and selections will prove useful in the identification of additional genes required for cell and nuclear fusion.

---

[28] M. D. Rose and G. R. Fink, *Cell (Cambridge, Mass.)* **48**, 1047 (1987).
[29] M. D. Rose, L. M. Misra, and J. R. Vogel, *Cell (Cambridge, Mass.)* **57**, 1211 (1989).
[30] P. B. Meluh and M. D. Rose, *Cell (Cambridge, Mass.)* **60**, 1029 (1990).

Section VI

Fission Yeast

[56] Molecular Genetic Analysis of Fission Yeast *Schizosaccharomyces pombe*

By SERGIO MORENO, AMAR KLAR, and PAUL NURSE

The fission yeast *Schizosaccharomyces pombe* is proving increasingly attractive as an experimental system for investigating problems of eukaryotic cell and molecular biology. Many of the powerful molecular genetic procedures developed for *Saccharomyces cerevisiae* can now be applied to *S. pombe*. In this chapter we describe a range of techniques concerned with classical and molecular genetics, cell biology, and biochemistry which can be used with *S. pombe* and are in routine operation in our laboratories.

Schizosaccharomyces pombe is a simple unicellular eukaryote with a genome size of 14 megabases (Mb), about 4 times that of the *Escherichia coli*. Despite being an Ascomycete fungus like *S. cerevisiae*, fission yeast is not closely related to budding yeast. Protein comparisons between homologous genes in the two yeasts have revealed identities of between 60 and 90% in amino acid type and position, values close to those found in similar comparisons between yeast and mammalian genes. Such divergence means that if a function is conserved between the two yeasts it is likely that an equivalent function will also be found in other eukaryotes. Also, sequence comparisons of homologous genes in the two yeasts are useful for identifying those regions of proteins which are conserved and probably essential for function. Evolution has provided us with an extensive mutagenesis experiment to examine the effects of specific amino acid changes on protein functions. Comparison of gene sequences has suggested that *S. pombe* may be slightly more similar to mammalian cells than is *S. cerevisiae*. Certain features such as cell cycle, chromosome structure, and RNA splicing are likely to be more similar between mammalian cells and *S. pombe* than mammalian cells and *S. cerevisiae*, but this need not be the case for all problems of cell and molecular biology. The fission yeast has been used successfully to study mating type, recombination, translation, RNA splicing, chromosome structure, meiosis, mitosis, and cell cycles. Most of these topics are covered in *The Molecular Biology of the Fission Yeast* (A. Nasim, P. Young, and B. F. Johnson, eds.), published by Academic Press in 1989. In general there are good genetic, cytological, biochemical, and molecular genetic techniques available for *S. pombe*, although molecular genetics can be more difficult than with *S. cerevisiae*. We hope that this review of techniques will attract more groups to work on this very attractive and amenable organism.

Introduction to Biology of Fission Yeast

Schizosaccharomyces pombe grows as a cylinder around 3–4 μm in diameter and 7–15 μm in length. It has a typical eukaryotic cell cycle with discrete G_1, S, G_2, and M phases.[1] In normal minimal or complex media the generation time is between 2 and 4 hr; G_2 is about 0.7 of a cell cycle, and the remaining phases are each of about 0.1 of a cell cycle length. *Schizosaccharomyces pombe* is not very versatile in using different carbon sources, and it is difficult to vary the generation time substantially in batch culture. However, in chemostat cultures generation times can be increased to over 10 hr, and in these circumstances most of the expansion in the cell cycle occurs in G_1 phase.

Normally *S. pombe* cells are haploid and may be of two mating types known as h^+ and h^- (Fig. 1). Starvation induces the haploid cells of opposite mating types to mate in pairs, forming diploid zygotes which are heterozygous at the mating type locus (h^+/h^-). The zygotes then undergo meiosis to form four haploid spores, which germinate when nutrient conditions are improved to produce haploid clones.[2] Thus, normally the haploid phase predominates. However, h^+/h^- diploid strains can also be maintained, using procedures we describe later; thus, mutations can easily be isolated in a haploid strain and then tested for complementation and dominance in a diploid strain.

Isolates of *S. pombe* from the wild are homothallic *(h^{90})* strains; that is, they can switch their mating type between h^+ and h^- every other generation. This means that a single cell gives rise to a colony containing h^+ and h^- cells which then can mate with one another when nutritional conditions become limiting. Mutations and rearrangements at the mating-type locus give rise to h^+ and h^- strains which either cannot switch or switch rarely. These are called heterothallic strains. Heterothallic h^{+N} and h^{-U} strains do revert at low frequency (10^{-4}) to h^{90}; h^{-S} strains, however, are the result of a deletion and are therefore stable, and so it is convenient to use h^{-S} strains for experimental analysis.[3]

More than 270 *S. pombe* genes have been mapped genetically, and they define three linkage groups with a meiotic map length of 1200 centimorgans (cM).[4,5] Three chromosomes can be visualized microscopically.[6] Re-

---

[1] M. Mitchison, *in* "Methods in Cell Physiology" (D. M. Prescott, ed.), Vol. 4, p. 131. Academic Press, New York, 1970.
[2] U. Leupold, *in* "Methods in Cell Physiology" (D. M. Prescott, ed.), Vol. 4, p. 169. Academic Press, New York, 1970.
[3] H. Gutz, H. Heslot, U. Leupold, and N. Loprieno, *in* "Handbook of Genetics" (R. C. King, ed.), Vol. 1, p. 395, Plenum, New York, 1974.
[4] A. Gygax and P. Thuriuax, *Curr. Genet.* **8**, 85 (1984).

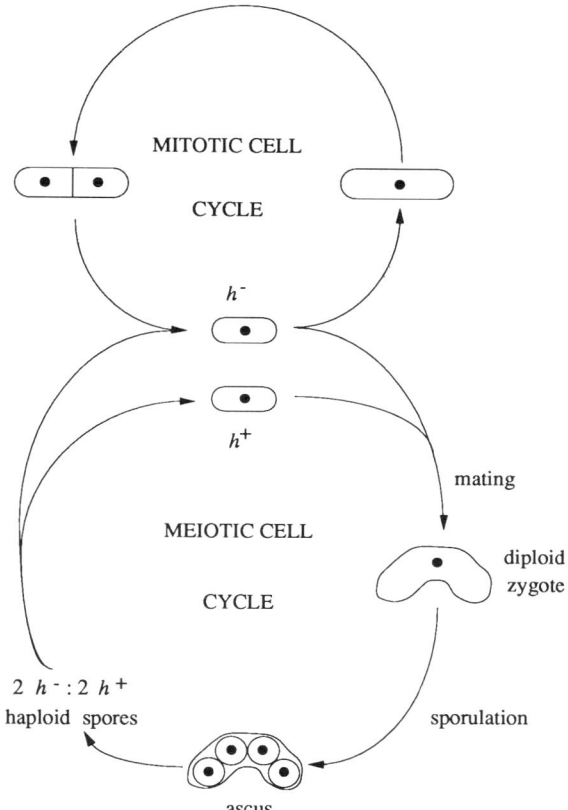

FIG. 1. *Schizosaccharomyces pombe* mitotic and meiotic cell cycles.

cently *S. pombe* chromosomes have been separated using pulsed-field gel electrophoresis (PFGE);[7] they have a size of 5.7, 4.6, and 3.5 Mb. There is also available a *Not*I macrorestriction map of the *S. pombe* genome with 14 detectable *Not*I sites.[8] Therefore, it is now possible to map a cloned gene within a particular region of any of the three chromosomes by using PFGE and Southern blotting.

---

[5] J. Kohli, *Curr. Genet.* **11**, 575 (1987).
[6] C. F. Robinow, *Genetics* **87**, 491 (1977).
[7] C. L. Smith, T. Matsumoto, O. Niwa, S. Klco, J.-B. Fan, M. Yanagida, and C. R. Cantor, *Nucleic Acids Res.* **15**, 4481 (1988).
[8] J.-B. Fan, Y. Chikashige, C. L. Smith, O. Niwa, M. Yanagida, and C. R. Cantor, *Nucleic Acids Res.* **17**, 2801 (1989).

Schizosaccharomyces pombe Strains

A standard collection of *S. pombe* mutant strains has been established by Alan Coddington in Norwich, England. A catalog is available and strains can be obtained, for a small fee, from Peter J. H. Jackman, National Collection of Yeast Cultures, AFRC Institute of Food Research, Colney Lane, Norwich NR4 7UA, UK. The nomenclature for *S. pombe* strains and genes differs from that of *S. cerevisiae* and follows lowercase italic symbols.[5]

All *S. pombe* strains have been derived from the haploid wild-type strains $972h^-$, $975h^+$ and $968h^{90}$ and are generally isogenic. They grow on media as described in Table I.[1,3,9] Long-term storage of *S. pombe* strains is on glycerol stocks at $-70°$. These are prepared as follows:

1. Grow up cells in 1 ml YES medium at $25°-32°$ for 2 days.
2. Mix with 1 ml of YES containing 30% glycerol in a cryotube. Snap-freeze on liquid nitrogen or dry ice/ethanol. The cells can then be stored at $-70°$. Strains kept in this way remain viable for several years at least. It is wise to make a duplicate each time and store it in a different freezer. For short-term storage cells can be kept as patches on YES slants or agar plates at 4° for up to 2 months. Strains do not store well on minimal medium or phloxin B-supplemented medium.

Reisolation of Fission Yeast Strains

For strains stored on glycerol at $-70°$, the following procedure is recommended:

1. With a sterile spatula scrape off a small amount of frozen glycerol stock and then transfer to a YES plate.
2. Incubate at $25°-32°$ for 1-4 days, depending on the strain.
3. When colonies are visible streak out to single colonies on a YEP (YES + phloxin B) plate and incubate at $25°-32°$ for 2-3 days.

Strains stored on slants or patches are streaked out onto YEP plates directly, and incubated at $25°-32°$ as appropriate. Before any genetic or molecular procedure is carried out the phenotype of the strain should be checked.

Testing the Phenotype of a Strain

Haploid/Diploid. It is important to check the ploidy because certain strains of *S. pombe* diploidize at a high frequency. Haploid cells divide at

---

[9] P. Nurse, *Nature (London)* **256**, 547 (1975).

approximately 12–15 μm in length and are 3–4 μm in width. Diploid cells are both longer (20–25 μm at division) and wider (4–5 μm); they are also less viable than haploid cells, and a diploid colony contains more dead cells (1–5%). Phloxin B is a stain that accumulates in dead cells, which become dark red. By growing a strain in YEP plates it is possible to screen for haploid colonies, which will be stained light pink whereas diploid colonies will be darker pink.[10] This can be confirmed by microscopic examination of the cells.

Mating Type. To test for the presence of homothallic h^{90}, the strain is streaked out to single colonies on YE and then replica plated to malt extract, incubated below 30° for 3 days to allow conjugation and sporulation to occur, and then held over a petri dish containing iodine crystals for about 1–5 min. h^{90} colonies will be stained black owing to the presence of starch in the spores. Often sectored colonies are seen. To check mating type, the strain should be crossed to h^+ and h^- tester strains (see later) and tested as above.

Temperature Sensitivity. Many temperature-sensitive mutants *(ts)* have been isolated in *S. pombe*. They can be checked by replica plating onto YEP (YES + phloxin B) and incubating at the restrictive temperature. Phloxin B will stain the dead cells and these can be examined under the light microscope for checking the ts phenotype.

Auxotrophy. The auxotrophic markers most commonly used in *S. pombe* are adenine, glutamic acid, histidine, leucine, lysine, and uracil, although others are available. To test for auxotrophy the strain is grown up to single colonies on YES and then replica plated to minimal medium with and without the appropriate supplement. The plates are incubated for 1–2 days and then examined for growth under these conditions.

After testing a strain in these various ways, it can be stored as a patch at 4° and generally used for 2–4 weeks without further testing.

Growing *Schizosaccharomyces pombe* Cells

Haploid strains of *S. pombe* grow with the generation times shown in Table II. As *S. pombe* cells enter stationary phase the cells generally accumulate in G_1 or G_2, depending on whether they are deprived of nitrogen or glucose, respectively,[11] and the cells become rounder and more refractile under phase microscopy. In supplemented yeast extract medium and minimal medium, glucose is usually limiting and cells accumulate in G_2.[11]

[10] J. Kohli, H. Hottinger, P. Munz, A. Strauss, and P. Thuriaux, *Genetics* **87**, 471 (1977).
[11] P. Nurse, P. Thuriaux, and K. Nasmyth, *Mol. Gen. Genet.* **146**, 167 (1976).

TABLE I
MEDIA FOR FISSION YEAST *Schizosaccharomyces pombe*[a]

Type of medium	Name	Use	Recipe	Concentration
Edinburgh minimal medium	EMM[b]	Vegetative growth	3 g/liter Potassium hydrogen phthalate	14.7 mM
			2.2 g/liter Na$_2$HPO$_4$	15.5 mM
			5 g/liter NH$_4$Cl	93.5 mM
			2% (w/v) Glucose	111 mM
			20 ml/liter Salts (50× stock)	
			1 ml/liter Vitamins (1000× stock)	
			0.1 ml/liter Minerals (10,000× stock)	
			Salts (50× stock)	
			52.5 g/liter MgCl$_2$ · 6H$_2$O	0.26 M
			0.735 mg/liter CaCl$_2$ · 2H$_2$O	4.99 mM
			50 g/liter KCl	0.67 M
			2 g/liter Na$_2$SO$_4$	14.1 mM
			Vitamins (1000× stock)	
			1 g/liter Pantothenic acid	4.20 mM
			10 g/liter Nicotinic acid	81.2 mM
			10 g/liter *myo*-inositol	55.5 mM
			10 mg/liter Biotin	40.8 μM
			Minerals (10,000× stock)	
			5 g/liter Boric acid	80.9 mM
			4 g/liter MnSO$_4$	23.7 mM
			4 g/liter ZnSO$_4$ · 7H$_2$O	13.9 mM
			2 g/liter FeCl$_2$ · 6H$_2$O	7.40 mM
			0.4 g/liter Molybdic acid	2.47 mM
			1 g/liter KI	6.02 mM
			0.4 g/liter CuSO$_4$ · 5H$_2$O	1.60 mM
			10 g/liter Citric acid	47.6 mM
			After autoclaving, a few drops of preservative (1:1:2, chlorobenzene/dichloroethane/chlorobutane) is added	

Medium	Abbreviation	Use	Composition
Minimal supplement	—	Vegetative growth	EMM plus 50–250 mg/liter supplements as required
Minimal low glucose	—	Yeast transformations	As EMM, but 0.5% (w/v) glucose instead of 2% (w/v)
Minimal sorbitol	EMMS	To grow up transformants	As EMM, but add 1.2 M sorbitol
Minimal glutamate	EMMG	Sporulating diploids in liquid	As EMM, but replace NH_4Cl with 1 g/liter sodium glutamate (5.91 mM)
Minimal phosphate-free	EMMP	[^{32}P]Phosphate labeling	As EMM, but remove Na_2-HPO_4 and replace potassium hydrogen phthalate with 2 g/liter sodium acetate trihydrate (14.6 mM) and adjust to pH 5.5
Yeast extract	YE[c]	Vegetative growth, inhibits conjugation and sporulation	0.5% (w/v) Oxoid yeast extract, 3.0% (w/v) glucose
Yeast extract + supplements	YES	Vegetative growth	YE plus 50–250 mg/liter adenine, histidine, leucine, uracil, and lysine hydrochloride
Yeast extract + phloxin B	YEP	Checking ploidy	YES plus 2.5 mg/liter phloxin B (Sigma P 4030), added when the medium has cooled below 60° from a 5 g/liter stock solution in sterile distilled water
Malt extract	ME[c]	Conjugation and sporulation	3% (w/v) Bacto-malt extract (DiaMalt AG or Difco) plus supplements added as for YES except lysine; adjust to pH 5.5 with NaOH

[a] Solid medium is made by adding 2% Difco Bacto-agar. All media are prepared in bulk. Media are sterilized by autoclaving at 10 psi for 20 min. At this pressure very little caramelization of glucose takes place. Media are stored in 500-ml bottles, and agar is remelted in a microwave oven before using.

[b] M. Mitchison, in "Methods in Cell Physiology" (D. M. Prescott, ed.), Vol. 4, p. 131. Academic Press, New York, 1970; P. Nurse, *Nature (London)* **256**, 547 (1975).

[c] H. Gutz, H. Heslot, U. Leupold, and N. Loprieno, in "Handbook of Genetics" (R. C. King, ed.), Vol. 1, p. 395. Plenum, New York, 1974.

TABLE II
GENERATION TIMES FOR FISSION YEAST
Schizosaccharomyces pombe

Medium	Temperature (°C)	Generation time
YE	25	3 hr
	29	2 hr 30 min
	32	2 hr 10 min
	35.5	2 hr
EMM	25	4 hr
	29	3 hr
	32	2 hr 30 min
	35.5	2 hr 20 min

For physiological experiments it is important that cultures be maintained in mid-exponential growth between 2×10^6 and 1×10^7 cells/ml. The optical density (OD) of a culture can be used to measure the concentration of cells. It is necessary to establish a ratio between OD_{595} and cell concentration. In our hands for wild-type strains, an OD_{595} of 0.1 is equivalent to 2×10^6 cells/ml, but this will vary between strains, from one spectrophotometer to another, and among different growth conditions.

To generate cultures and mid-exponential growth, use a fresh patch of a strain of checked phenotype to inoculate 10 ml YES (or minimal medium) and incubate for 1–2 days at the appropriate temperature until cells are in early stationary phase. At this point no shaking is required as *S. pombe* can grow in partially anaerobic conditions. This preculture can then be used to inoculate a larger culture, taking into consideration the generation times shown in Table II. A preculture normally contains $2-5 \times 10^7$ cells/ml, and one generation is necessary for cells to recover from stationary phase and reenter exponential growth. Typically a culture in 100 ml minimal medium grown at 25° overnight for 16 hr (i.e., 4 generations) to an OD_{595} of 0.25 (5×10^6 cells/ml) will require an inoculum of 1 ml from the preculture.

When growing liquid cultures for physiological experiments gentle shaking is advisable to maintain uniform growth conditions. Generally minimal medium is used, as the growth conditions can be more accurately defined. Phthalate is used as a buffer as it reduces clumping, which can be a problem with some strains.

The media for growing *S. pombe* strains are given in Table I. YES liquid or agar-containing solid medium is used for vegetative growth. Yeast extract from most sources when used in YES inhibits conjugation and sporulation. However *S. pombe* cells can grow, mate, and sporulate in the YEPD rich medium commonly used for growth of *S. cerevisiae*. For physiological experiments cells are generally cultured in EMM with the required supplements added. Crosses are carried out on ME, with each supplement added, or in YEPD.

Schizosaccharomyces pombe Classical Genetic Techniques

Genetic Crosses

Conjugation and sporulation cannot take place in *S. pombe* except under conditions of nutrient starvation. ME medium is generally used for genetic crosses. This medium also supports some vegetative growth; mixed strains will undergo several rounds of cell division before running out of nutrients. The cells will then be able to conjugate and sporulate. Alternatively, cells can be grown and sporulated in YEPD [1% yeast extract, 2% peptone, 2% glucose (all w/v)].

Several crosses can be carried out on a 9-cm ME plate. To cross two strains, a loopful of h^- and a loopful of h^+ are mixed together on a ME plate. A loopful of sterile distilled water is then used to thoroughly mix the cells on the agar plate to an area of about 1 cm^2. The cross is left to dry and is then incubated below 30°, as conjugation is severely reduced above this temperature. Fully formed four-spore asci can be seen after 2–3 days of incubation (see morphology of the vegetative cells, ascus, and spores in Fig. 1). When carrying out genetic crosses for recombination mapping, it is useful to have unlinked markers to ensure that recombination has occurred.

A cross between a homothallic and a heterothallic strain mostly generates asci of the homothallic parent. This is because yeast cells prefer to mate with sister cells which are of different mating type owing to homothallic switching. It is preferable to first select a diploid hybrid between these strains (see below) and then subject this to tetrad or random spore analysis.

Tetrad Analysis

A 2-day-old cross is usually used for tetrad analysis. At this stage the ascus wall has not yet started to break down. Asci are placed in a line about 2 mm apart on a YES plate using a micromanipulator (we use one from Zeiss, Jena, East Germany). The asci walls are then left to break down at 37° for about 3–5 hr, or at 20° overnight. Each ascus is micromanipulated to give a line of four isolated spores, separated by about 3–5 mm. The spores are incubated until colonies form at the appropriate temperature for the cross. Digestion of the ascus wall with enzymatic treatment is not practiced since the ascus easily falls apart, giving free spores. Also, unlike *S. cerevisiae* spores, *S. pombe* spores do not stick to each other. This means the technique of tetrad dissection is generally easier than with *S. cerevisiae*.

Random Spore Analysis

Using a 3-day-old cross, check for the presence of asci under the light microscope. Random spore analysis allows many more spores to be examined than in tetrad analysis, and in this way recombination mapping and strain construction can be carried out. However, it is important that all the classes of spores are viable when studying recombination frequencies.

One ml of sterile distilled water is inoculated with a loopful of the cross, 20 μl of a 1 in 10 dilution of Helicase *(Helix pomatia* juice, IBF Biotechnics No. 213473, Paris, France) is added, and the mixture is incubated overnight at 25°–29° or for at least 6 hr at 29°. Helicase is a crude snail enzyme that breaks down the ascus wall and kills vegetative cells. Five microliters of Glusulase (Biotechnology Systems, NEE-154, Boston, MA 02118) in 1 ml of sterile distilled water can be substituted for the Helicase. The number of spores per milliliter is counted using a hemacytometer. Between 200 and 1000 spores/plate can be plated out on YES–agar or selective medium. The plates are then incubated until colonies form.

Isolation of Diploid Strains

Diploid cells arise spontaneously in most *S. pombe* strains, probably as a result of endomitosis. This characteristic can be used to isolate homozygous diploids of any strain. The strain is streaked out to single colonies on YEP (YE + phloxin B) solid medium and incubated until colonies form. Diploids can be identified as clones that stain dark red with phloxin B and which contain large cells on microscopic examination. These diploids can undergo a diploid mitotic cycle and when starved of nutrients can conjugate with diploid cells of the opposite mating type to form a tetraploid zygote. This can be sporulated to generate four diploid spores, although aberrant segregation of chromosomes can occur during the tetraploid meiosis. Homothallic strains also generate diploid cells which can be induced to undergo meiosis. The cells can directly produce azygotic asci without mating and thus can be unambiguously identified.

Sporulating diploids can be isolated by crossing h^- and h^+ haploids with complementary growth requirements, for example, using strains with the markers *leu1-32-h^-* and *ura4-d18 h^+*. About 12 hr after crossing, when conjugation has occurred, a loopful of the cross is streaked out onto minimal medium which will only allow growth of a conjugated diploid. However, since mating is rapidly followed by meiosis and sporulation, many of the colonies growing on selective media will be prototrophic haploid segregants. To circumvent this problem, *ade6-M210* and *ade6-M216* mutations are commonly used. *ade6-M216* colonies are light pink and *ade6-M210* dark pink on plates of YE medium or EMM containing 10 μg/ml adenine because of accumulation of a red adenine precursor. On media containing adenine the red color is not observed. Diploid cells containing both mutations grow in the absence of adenine, and the colonies formed are white owing to intragenic complementation between the two *ade6* alleles. Because the alleles are tightly linked and there is infrequent gene conversion, spores generated by meiosis are unlikely to be adenine prototrophs and will not form colonies on the selective media.

Sporulating diploids are very unstable and will generate spores if they enter stationary phase from minimal medium. They can be maintained on yeast extract medium which inhibits sporulation. Alternatively, nonsporulating diploids can be derived by crossing-over at the mating-type locus, which leads to homozygosis at the mating-type locus. These arise fairly frequently and can be screened for by replica plating onto malt extract and looking for non-iodine-positive colonies which are not undergoing sporulation.

Sometimes it is required to cross a diploid and a haploid strain. Such a cross yields less than 10% spore viability, and most of the segregants are slow growing because of aneuploidy. Since *S. pombe* has only three linkage groups, normal haploid and diploid segregants can be obtained at a reasonable frequency from random spores; they are easily identified since they grow well.

Mutagenesis of Yeast Strains

Both ethylmethane sulfonate (EMS) and nitrosoguanidine can be used for mutagenesis; the former is safer to use, but the latter is a more effective mutagen.

Ethylmethane Sulfonate Mutagenesis[12]

1. Grow up 100 ml of cells in minimal medium to an OD_{595} of 0.2–0.5 ($4 \times 10^6 - 1 \times 10^7$ cells/ml).

2. Harvest the cells at 3000 rpm for 5 min in 50-ml plastic tubes and resuspend at 1×10^8 cells/ml in fresh medium.

3. Take 2 ml and transfer to a 50-ml plastic capped tube and add EMS (Sigma, St. Louis, MO) to a final concentration of 2%.

[12] J. Hayles, S. Aves, and P. Nurse, *EMBO J.* **5**, 3373 (1986).

4. Leave the cells at room temperature in the fume hood for 3 hr with gentle shaking.

5. Transfer 1 ml to an Eppendorf tube, harvest the cells, and wash 3 times with 1 ml sterile 0.9% NaCl.

6. Dilute the cells as required and plate out onto YES. The percentage of survivors should be approximately 50% (this percentage is strain dependent).

7. EMS is inactivated using an excess of 5% sodium thiosulfate.

Nitrosoguanidine Mutagenesis[13]

1. Grow up 100 ml of cells in minimal medium to an OD_{595} of 0.2–0.5 ($4 \times 10^6 - 1 \times 10^7$ cells/ml).

2. Take 10 ml and harvest cells at 3000 rpm for 5 min in 50-ml plastic tubes. Wash once with 10 ml of TM (50 mM Tris–maleate, pH 6).

3. Resuspend in TM at 1.4×10^8 cells/ml. Mix 700 μl of cells with 300 μl of 1 mg/ml nitrosoguanidine (NG, Sigma) in the same buffer. Incubate at 30° for 30, 60, and 90 min with occasional vortexing.

4. Remove 100 μl of cells and dilute with 900 μl of TM. Wash twice with 1 ml of TM and once with YES medium.

5. Cells are then resuspended in 1 ml of YES and incubated for 4 hr at 25° and plated into YES. The percentage of survivors should be approximately 40, 15, and 3% for 30, 60, and 90 min, respectively (these percentages are strain dependent).

6. NG is inactivated using bleach overnight.

After mutagenesis it is advisable to backcross the strains at least 3 times.

Schizosaccharomyces pombe Molecular Genetic Techniques

Schizosaccharomyces pombe molecular genetic techniques are based on procedures developed for *S. cerevisiae,* but there are differences in behavior between the two organisms. Modifications of standard *S. cerevisiae* procedures have been made to optimize their use in *S. pombe.* Most notably, homologous recombination between introduced DNA and the chromosome occurs at a lower level than in *S. cerevisiae,* and experiments involving gene integration and replacement are not always so straightforward to carry out.

Schizosaccharomyces pombe Plasmids

Schizosaccharomyces pombe plasmids consist of a bacterial origin of replication and selectable marker, a yeast selectable marker, and an equiva-

---

[13] T. Uemura and M. Yanagida, *EMBO J.* **3,** 1737 (1987).

lent to an autonomous replication sequence *(ars)* which is responsible for a high frequency of transformation.

Yeast Markers. Budding yeast markers used in *S. pombe* are the *LEU2* and *URA3* genes. Plasmids containing these markers complement the *S. pombe* mutations leu1⁻ and ura4⁻. The *URA3* gene is expressed very poorly in *S. pombe* and does not rescue the ura4⁻ mutation when it is present as a single copy. *Schizosaccharomyces pombe* markers commonly used are *ura4⁺* and *sup3-5*. The latter marker is an opal nonsense suppressing tRNA gene which suppresses *ade6-704*. This marker has a deleterious effect for the cell when present in several copies. On minimal medium supplemented with 10 μg/ml adenine or yeast extract medium *ade6-704* mutant colonies are red, but when suppressed by *sup3-5* they are white. If a *sup3-5*-containing plasmid is not integrated into the genome then instability leads to cells lacking *sup3-5* and hence to the formation of pink colonies. This contrasts with clones containing one copy of the integrated plasmid, which are white, and enables a rapid distinction to be made between integrated and nonintegrated clones.[14]

Autonomous Replication Sequences (ars). In contrast to *S. cerevisiae*, in *S. pombe* a bacterial plasmid such as pBR322 carrying a marker gene such as *LEU2* is able to replicate, often to high copy number. However, the transformation frequency obtained when using such plasmids is very low.[15] The addition of *S. pombe ars1⁺* sequences or the *S. cerevisiae* 2-μm origin leads to a high frequency of transformation and reduction in the copy number.[15] So it seems that in *S. pombe* a high frequency of transformation and effective replication capacity are to some extent independent phenomena.

Plasmid vectors based on 2 μm (pDB248, YEp13) are mitotically unstable, their copy number is low (5–10), they are much more prone to rearrangements (tandem duplications or deletions), and they are more difficult to recover from fission yeast than plasmids carrying *S. pombe ars1⁺*. Plasmids containing *ars1⁺* are also very unstable (with the exception of pFL20 and pMB332); their copy number is higher (15–80), and they tend to produce polymers with various numbers of repeat units. pFL20 and pMB332 yield rather stable transformants both mitotically and meiotically, owing to the presence of *stb* (stable) element. This element is not an *ars* sequence, nor it is a centromeric sequence. Plasmids containing this element still segregated asymmetrically 10 times more frequently during mitosis than *S. cerevisiae CEN* plasmids.[15]

Expression Vectors. Plasmids derived from the ones described above have been used to increase the expression of certain gene products. pSM1

---

[14] H. Hottinger, D. Pearson, F. Yamao, V. Gamulin, L. Cooley, T. Cooper, and D. Söll, *Mol. Gen. Genet.* **188**, 219 (1982).

[15] W.-D. Heyer, M. Sipiczki, and J. Kohli, *Mol. Cell. Biol.* **6**, 80 (1986).

and pSM2 are derivatives from pDB248, made by inserting the SV40 early promoter. Genes linked to this promoter are expressed at moderate levels. pEVP11 and pEVP12 contain the *S. pombe adh*$^+$ promoter inserted into YEp13; pEVP12 contains the *URA3* marker instead of *LEU2*. pART1 and pMB332 also have the *S. pombe adh*$^+$ promoter inserted into pIRT2 and pFL20, respectively. Genes linked to any of this element are expressed at high levels (5–20 times higher than those linked to the SV40 early promoter). Very recently, plasmids containing inducible promoters have been developed.[16,17] Table III lists the *S. pombe* plasmids most commonly used in our laboratories. Restriction maps and a full description of many of the plasmids can be found in Ref. 18.

Schizosaccharomyces pombe Transformation

Transformation is very straightforward, and because most laboratory *S. pombe* strains are isogenic the frequencies are generally high regardless of the strain used.

Protoplast Procedure[19]

1. Grow a 200-ml culture to an OD_{595} of 0.2–0.5 ($4 \times 10^6 - 1 \times 10^7$ cells/ml) in minimal medium containing 0.5% glucose and supplements.

2. Harvest cells, decant supernatant, and resuspend the pellet in 10 ml of

20 mM citrate/phosphate, pH 5.6 (2.82 g/liter Na_2HPO_4, 4.2 g/liter citric acid)

40 mM EDTA, pH 8.0

30 mM 2-mercaptoethanol (0.2%, v/v) (added after autoclaving)[20]

and transfer to a 50-ml plastic centrifuge tube.

3. Harvest cells and resuspend the contents of each tube in 5 ml of

50 mM citrate/phosphate, pH 5.6 (7.1 g/liter Na_2HPO_4, 11.5 g/liter citric acid)

1.2 M sorbitol

Adjust to pH 5.6 with 5 M NaOH and add

30 mM 2-mercaptoethanol (0.2%, v/v) (added after autoclaving)[20]

25 mg NovoZym 234 (Novo Industri A/S, Bagsvaerd, Denmark) (added after autoclaving)

[16] C. S. Hoffman and F. Winston, *Gene* **84,** 473 (1989).
[17] K. Maundrell, *J. Biol. Chem.* **265,** 10857 (1990).
[18] P. Russell, *in* "The Molecular Biology of the Fission Yeast" (A. Nasim, P. Young, and B. F. Johnson, eds.), p. 244. Academic Press, San Diego, California, 1989.
[19] D. Beach and P. Nurse, *Nature (London)* **290,** 140 (1981).
[20] Only necessary when the cell number of the culture is higher than the margin given above.

Incubate at 37° for 15–30 min until spheroplasts have formed.

4. Add 35 ml of
   10 mM Tris-HCl, pH 7.6
   1.2 M sorbitol

and divide between 2–4 tubes (there should be no more than 3×10^8 spheroplasts/tube). Spin gently at 2000 rpm for 5 min.

5. Wash twice more in 20 ml of the Tris–sorbitol solution, each time resuspending gently in 1 ml first. At the last resuspension take a sample and count the number of protoplasts with a hemacytometer.

6. Resuspend at $2–5 \times 10^8$ protoplasts/ml in
   10 mM Tris-HCl, pH 7.6
   10 mM CaCl$_2$
   1.2 M sorbitol

and combine the tubes.

7. Using 100 μl protoplast/transformation add 1–10 μg of transforming plasmid in up to $\frac{1}{10}$ total volume. Incubate at room temperature for 15 min.

8. Add 1 ml of
   10 mM Tris-HCl, pH 7.6
   10 mM CaCl$_2$
   20% polyethylene glycol (PEG) 4000

and incubate at room temperature for 15 min.

9. Spin at 2000 rpm for 5 min, drain well, and resuspend the protoplast in 0.2–0.5 ml of
   10 mM Tris-HCl, pH 7.6
   10 mM CaCl$_2$
   1.2 M sorbitol
   0.5 mg/ml yeast extract
   5 μg/ml supplements (Leu, Ura, Ade, His)

Incubate at 30° for 30–60 min.

10. Plate out 0.2-ml aliquots onto well-dried minimal sorbitol plates. Transformants appear in 2–5 days at 29°–32°.

Transformation frequency is about $1 \times 10^4 – 5 \times 10^4$ transformants/μg DNA. This transformation frequency can be increased to 7.5×10^5 transformants/μg DNA using lipofectin (BRL, Gaithersburg, MD).[21] For this purpose follow the protocol to Stage 7 and after the 15-min incubation of the protoplasts with DNA add 100 μl of 10 mM Tris-HCl, pH 7.6, 10 mM CaCl$_2$, 1.2 M sorbitol, 66 μg/ml lipofectin and incubate for a further 15 min at room temperature. Proceed to Stage 8 and follow the rest of the

---

[21] R. C. Allshire, *Proc. Natl. Acad. Sci. U.S.A.* **87**, 4043 (1990).

TABLE III
Schizosaccharomyces pombe Plasmids

Plasmid	Bacterial marker	Yeast marker	Promoter	Characteristics	Suggested uses	Reference
pDB262	Tet[a]	LEU2		2-μm based	Gene banks	Wright et al., 1986[b]
pWH4/5	Amp + Tet[a]	LEU2		2-μm based	Gene banks	Wright et al., 1986[b]
pDB248X	Amp + Tet	LEU2		2-μm based	All purposes	Durkacz et al., 1985[c]
pFL20	Amp + Tet	URA3		S. pombe ars1+, very stable, monomeric	Gene banks	Losson and Lacroute, 1983[d]
YEp13	Amp	LEU2		2-μm based	All purposes	Broach et al., 1979[e]
pIRT2	Amp	LEU2		S. pombe ars1+, f1 ORI, polylinker	All purposes	Booher and Beach, 1986[f]
pIRT2U	Amp	ura4+		S. pombe ars1+, f1 ORI, polylinker	All purposes	Carr et al., 1989[g]
pURA4	Amp + Tet	ura4+		S. pombe ars6+	All purposes	F. Lacroute, unpublished results[h]
pSTA12	Amp	sup3-5		pUC12 based, polylinker	Integration	A. Carr, unpublished results[i]
Constitutive promoters						
pSM1/2	Amp	LEU2	SV40 early promoter	Derived from pDB248, polylinker	Moderate level of expression	Jones et al., 1988[j]
pEVP11	Amp	LEU2	S. pombe adh+ promoter	Derived from YEp13, polylinker	High level of expression	Russell and Nurse, 1986[k]
pART1	Amp	LEU2	S. pombe adh+ promoter	Derived from pIRT2, polylinker	High level of expression	McLeod et al., 1987[l]

Plasmid	Marker	Promoter	Features	Expression	Reference
pMB332	Amp	URA3	S. pombe adh^+ promoter	Derived from pFL20, polylinker	High level of expression	Bröker, 1989[m]
Inducible promoters					
pCHY21	Amp	URA3	S. pombe fbp^+ promoter	Derived from pFL20, polylinker	Off in EMM + 8% glucose, on in EMM + 0.1% glucose + 3% glycerol	Hoffman and Winston, 1989[n]
pREP 1	Amp	LEU2	S. pombe nmI^+ promoter	Derived from pUC119, S. pombe $arsI^+$ autonomous replication	Off in EEM + thiamin, on in EMM	Maundrell, 1990[o]
pRIP 1/S	Amp	LEU2	S. pombe nmI^+ promoter	Derived from pUC119, S. pombe sup3-5 integration	Off in EEM + thiamin, on in EMM	Maundrell, 1990[o]

[a] Transcription from the Tet promoter is repressed by the λcI repressor. Cloning fragments into the λcI repressor gene inactivates its gene product and allows expression of the Tet gene which is fused to the λ P_R promoter.
[b] A. P. H. Wright, K. Maundrell, W.-D. Heyer, D. Beach, and P. Nurse, *Plasmid* **15**, 156 (1986).
[c] B. Durkackz, D. Beach, J. Hayles, and P. Nurse, *Mol. Gen. Genet.* **201**, 543 (1985).
[d] R. Losson and F. Lacroute, *Cell* (Cambridge, Mass.) **32**, 371 (1983).
[e] J. R. Broach, J. N. Strathern, and J. B. Hicks, *Gene* **8**, 121 (1979).
[f] R. Booher and D. Beach, *Mol. Cell Biol.* **6**, 3523 (1986).
[g] A. M. Carr, S. A. MacNeill, J. Hayles, and P. Nurse, *Mol. Gen. Genet.* **218**, 41 (1989).
[h] Françoise Lacroute, unpublished results (1987).
[i] Antony Carr, unpublished results (1989).
[j] R. H. Jones, S. Moreno, P. Nurse, and N. C. Jones, *Cell* (Cambridge, Mass.) **53**, 659 (1988).
[k] P. Russell and P. Nurse, *Cell* (Cambridge, Mass.) **45**, 145 (1986).
[l] M. McLeod, M. Stein, and D. Beach, *EMBO J.* **6**, 729 (1987).
[m] M. Bröker, *FEBS Lett.* **248**, 105 (1989).
[n] C. S. Hoffman and F. Winston, *Gene* **84**, 473 (1989).
[o] K. Maundrell, *J. Biol. Chem* **265**, 10857 (1990).

protocol. This method has also been used successfully for the transformation of minichromosomes greater than 500 kilobases.[21]

Protoplasts can be aliquotted, stored at $-70°$ in 10 mM Tris-HCl, pH 7.6, 10 mM CaCl$_2$, 1.2 M sorbitol (Stage 6) and used for at least 2 months.[22] The frequency of transformation is 1×10^3 transformants/μg DNA.

Lithium Chloride Procedure[23]

1. Grow a 50-ml culture to saturation in YEPD medium (1% yeast extract, 2% peptone, 2% glucose) with shaking at $25° - 35°$ for 24-48 hr.

2. Use 10 ml of this culture to inoculate 40 ml of fresh YEPD medium and incubate for 4-5 hr.

3. Harvest cells at 3000 rpm for 5 min. Wash once in sterile distilled water and resuspend in 0.6 ml of buffer I (20 mM Tris-HCl, pH 7.5, 2 mM EDTA, 0.2 M LiCl) to give a total volume of about 1.2 ml and a final concentration of 2×10^9 cells/ml.

4. Incubate at $30°$ for 1 hr with gentle shaking.

5. In an Eppendorf tube, mix 200 μl of competent yeast cells (4×10^8 cells) with 0.1-1 μg plasmid DNA and incubate at $30°$ for 30 min without shaking.

6. Add 700 μl of buffer II [40% (w/v) PEG 4000, 0.1 M LiCl in TE buffer, sterilized with 0.2-μm filters]. Mix by inverting the tube gently. Incubate at $30°$ for 30 min.

7. Heat-shock at $46°$ for 25 min.

8. Spread the cell suspension directly into YNB [0.67% yeast nitrogen base without amino acids, 2% glucose, 1.5% agar (all w/v)]. Colonies appear after 4-6 days at $30°$. The transformation frequency is 4×10^3 transformants/μg of plasmid DNA.

Lithium Acetate Procedure[24]

1. Grow a 150-ml culture in MB medium[25] to a density of $0.5-1 \times 10^7$ cells/ml (OD$_{600}$ 0.2-0.5).

---

[22] Juan Jimenez, personal communication (1990).
[23] M. Bröker, *BioTechniques* **5**, 516 (1987).
[24] K. Okazaky, N. Okazaky, and H. Okayama, submitted (1990).
[25] MB medium: 0.5 g/liter KH$_2$PO$_4$; 0.36 g/liter potassium acetate; 0.5 g/liter MgSO$_4 \cdot$7H$_2$O; 0.1 g/liter NaCl; 0.1 g/liter CaCl$_2 \cdot$2H$_2$O; 5 g/liter (NH$_4$)$_2$SO$_4$; 500 μg/liter H$_3$BO$_4$; 40 μg/liter CuSO$_4 \cdot$5H$_2$O; 100 μg/liter KI; 200 μg/liter FeCl$_3 \cdot$6H$_2$O; 400 μg/liter MnSO$_4 \cdot$H$_2$O; 200 μg/liter Na$_2$MoO$_4 \cdot$2H$_2$O; 400 μg/liter ZnSO$_4 \cdot$7H$_2$O; 5 g/liter glucose; 10 μg/liter biotin; 1 mg/liter calcium pantothenate; 10 mg/liter nicotinic acid; 10 mg/liter *myo*-inositol; and 150 mg/liter uracil (for ura4$^-$ strains) or 150 mg/liter leucine (for leu1$^-$ strains). Sterilize by filtering through a 0.45-μm pore size filter.

2. Harvest the cells at 3000 rpm for 5 min at room temperature.
3. Wash cells in 40 ml of water and spin them down as before.
4. Resuspend the cells at 1×10^9 cells/ml in 0.1 M lithium acetate (adjusted to pH 4.9 with acetic acid) and dispense 100-μl aliquots into Eppendorf tubes. Incubate at 30° (25° for *ts* mutants) for 60–120 min. Cells will sediment at this stage.
5. Add 1 μg of plasmid DNA in 15 μl TE (pH 7.5) to each tube and mix by gentle vortexing, completely resuspending cells sedimented during the incubation. Do not allow the tubes to cool down at this stage. Add 290 μl of 50% (w/v) PEG 4000 prewarmed at 30° (25° for *ts* mutants). Mix by gentle vortexing and incubate at 30° (25° for *ts* mutants) for 50 min.
6. Heat-shock at 43° for 15 min. Cool the tubes to room temperature for 10 min.
7. Centrifuge at 5000 rpm for 2 min in an Eppendorf centrifuge. Carefully remove the supernatant by aspiration.
8. Resuspend the cells in 1 ml of $\frac{1}{2}$ YEL–uracil (0.25% yeast extract, 1.5% glucose, and 30 μg/ml uracil) for ura4⁻ strains or $\frac{1}{2}$ YEL–leucine (0.25% yeast extract, 1.5% glucose, and 30 μg/ml leucine) for leu1⁻ strains, by pipetting up and down with a Pipetman P1000.
9. Transfer the suspension to a 50-ml flask and dilute with 9 ml of $\frac{1}{2}$ YEL–uracil or $\frac{1}{2}$ YEL–leucine. Incubate with shaking at 32° (25° for *ts* mutants) for 60 min or longer.
10. Plate aliquots of less than 0.3 ml onto MMA plates (MMA is the same as MB except that the amount of KH_2PO_4 and glucose are 1 and 10 g, respectively, potassium acetate and uracil or leucine are omitted, and 20 g of agar is added). If necessary, centrifuge the cells at this stage and resuspend in 1 ml of medium to spread more cells on a plate. This method gives the highest frequency of transformation described for *S. pombe*. The transformation frequency is between 3×10^5 and 2×10^6 transformants/μg of plasmid DNA.

Stability Test

The stability test is used to check the stability of a transformed plasmid. If the plasmid is replicating autonomously it will be lost in the absence of selection; on the other hand, if the plasmid has integrated or if there has been a reversion or gene conversion event, the phenotype is maintained after relaxing the selection. The procedure is as follows:

1. Take the transformant colony and streak out to single colonies on YES agar with no selection for about 3 days until colonies form.
2. Replica plate to selective medium (e.g., 35° on YES for a *ts* strain, 25° on minimal medium for the auxotrophic marker) and score for coin-

stability of the auxotrophic phenotypes. The *sup3-5/ade6-704* system described above is particularly useful for this purpose.

Integration of a Plasmid into the Genome

In *S. pombe* integration by homologous recombination is usually more frequent than nonhomologous recombination; however, for certain loci homologous recombination may only represent about 5–10% of the integration events. On average about 0.1% of the transformants obtained after transformation with an *ars* plasmid will have an integrated copy of the plasmid at the homologous locus. The frequency of integration can be enhanced up to 10-fold by a single cut of the plasmid in the region of interest to facilitate the recombination event. There can be problems concerning the selective markers used that may complicate the integration of a plasmid in *S. pombe*. The *S. cerevisiae URA3* gene on a multicopy plasmid complements *S. pombe* ura4⁻ mutations but is poorly expressed in *S. pombe*, and most of the integrated versions of *URA3* fail to complement ura4⁻ mutations. Therefore, *LEU2*, ura4⁺, or *sup3-5* markers should be used. For ura4⁺ plasmids the best strain to use is *ura4-D18*, which contains a complete deletion of the *S. pombe* ura4⁺ gene,[19] thus avoiding integration by homologous recombination at the *ura4* locus. To isolate an integrant, use the following procedure:

1. Transform a yeast strain with the plasmid of interest.
2. Isolate a transformant colony and grow up in 100 ml of YES medium (i.e., nonselective conditions) for about 20 generations (reinoculate 1 ml of this culture in 100 ml of fresh YES medium 2–3 times).
3. Plate out about 10^5 cells/plate onto selective medium and incubate until colonies form. The colonies should be stable owing to integration of the plasmid into the genome. This can be tested by replica plating to YES medium twice and then back to selective medium.
4. Confirm the integration by Southern blotting (cutting total DNA from the integrant strain with a restriction enzyme that cuts the vector once, but not the insert, should generate two fragments, whereas the wild-type will generate only one).

Recovering Plasmids from Schizosaccharomyces pombe[26]

Plasmid recovery from *S. pombe* is difficult as plasmids often seem to form multimers, and rearrangements (tandem duplication or deletions) are frequent.[15] This problem can be avoided by using the plasmid pFL20 or derivatives that contain the *stb* element and thus remain as monomers. To recover a plasmid from a transformant, proceed as follows:

[26] I. Hagan, J. Hayles, and P. Nurse, *J. Cell Sci.* **91**, 587 (1988).

1. Grow up 10 ml of cells under selective conditions to an OD_{595} of 1 (2×10^7 cells/ml).
2. Spin down the cells at 3000 rpm for 5 min.
3. Resuspend in 1.5 ml of
   50 mM citrate/phosphate, pH 5.6 (7.1 g/liter Na_2HPO_4, 11.5 g/liter citric acid)
   1.2 M sorbitol
Adjust to pH 5.6 with 5 M NaOH and add
   2 mg/ml Zymolyase-20T (Seikagaku Kogyo Co., Ltd., Tokyo, Japan) (add after autoclaving)
Transfer to an Eppendorf tube and incubate at 37° for 1 hr.
4. Pellet the cells in an Eppendorf centrifuge for 30 sec. Resuspend in 300 μl TE.
5. Add 35 μl of 10% sodium dodecyl sulfate (SDS), mix, and incubate at 65° for 5 min.
6. Add 100 μl of 5 M potassium acetate, mix, and leave on ice for 30 min.
7. Spin down at 4° for 10 min.
8. Add 50 μl of supernatant to 100 μl of NaI solution (Geneclean Kit, Stratech Scientific Ltd.) with 5 μl glassmilk (Geneclean Kit, Stratech Scientific Ltd.).
9. Incubate for 5 min at room temperature.
10. Spin for 5 sec (maximum) at room temperature, discard the supernatant, and wash the pellet 3 times with 400 μl of ice-cold NEW wash (Geneclean Kit, Stratech Scientific Ltd.).
11. Elute DNA twice with 10 μl of TE at 55° for 3 min each time.
12. Spin out the glassmilk and keep the supernatant.
13. Transform 5 μl of the supernatant into 100 μl of competent *E. coli* JA226 cells.

The use of the Geneclean kit improves the transformation frequency by at least 10-fold; also it is very important to use a *recBC E. coli* strain such as JA226 when using 2-μm or non-*ars*-containing plasmids. For *ars1*+-based plasmids *recA* strains like *E. coli* DH5 can be used.

Cloning Genes by Gap Repair[27-29]

This technique is designed to clone chromosomal mutant alleles of previously cloned genes.

---

[27] T. K. Orr-Weaver, J. W. Szostack, and R. J. Rothstein, this series, Vol. 101, p. 228.
[28] R. Rothstein, this volume [19].
[29] P. Russell and P. Nurse, *Cell (Cambridge, Mass.)* **45**, 145 (1986).

1. Construct a plasmid containing a selectable marker *(LEU2)* and the wild-type copy of the entire chromosomal region of interest.

2. Digest the plasmid with a restriction enzyme to completely remove the open reading frame (ORF). Purify the linear fragment containing the plasmid with the upstream and downstream flanking DNA sequences.

3. Transform 1 μg of this fragment into the strain containing the allele of interest. Identify transformants expressing the selectable marker. The gap in the plasmid is repaired using the mutant chromosomal sequences as a template.

4. Recover the plasmid from yeast as described before.

The efficiency of recovery of "repaired" plasmids versus "recircularized" plasmids appears to depend on the size of the flanking sequences (1 kb either side is recommended). On average 25% of the plasmids obtained are repaired. If overexpression of the mutant allele is deleterious to the cell, it may not be possible to recover the repaired plasmid. An alternative procedure involves integrating a *sup3-5*-containing plasmid adjacent to the mutant allele and then cutting out this plasmid with the mutant allele and recovering it in *E. coli*.[30]

Gene Disruption and Gene Replacement[28,31,32]

1. Make a disruption of the gene of interest by inserting the $ura4^+$ marker or the *LEU2* marker gene in the ORF. If feasible, delete as much of the ORF as possible and conserve at least 1 kb either side.

2. Purify the linear fragment.

3. Transform either a h^-/h^+ *ura4-D18/ura4-D18 ade6-M210/ade6-M216* or a h^-/h^+ *leu1-32/leu1-32 ade6-M210/ade6-M216* diploid strain, depending on the marker used, with 1, 5, 10, and 20 μg of the fragment. Select diploid transformants expressing the $ura4^+$ or *LEU2* gene.

The transformant diploid can be sporulated in minimal glutamate or malt extract and the spores plated out to see whether the gene deleted is essential or not. If the gene is not essential the deletion strain can be maintained as a haploid. A nonsporulating diploid h^+/h^+ or h^-/h^- generated by endomitosis can also be used to make the deletion. This can be mated to homozygous diploid of the opposite mating type to produce a tetraploid zygote which can sporulate and form four diploid spores. Some of these diploids will be heterozygous at the mating-type locus and have one chromosome with a deleted copy of the gene of interest. These can be

---

[30] A. M. Carr, S. A. MacNeill, J. Hayles, and P. Nurse, *Mol. Gen. Genet.* **218,** 41 (1989).
[31] R. J. Rothstein, this series, Vol. 101, p. 202.
[32] C. Grimm, J. Kohli, J. Murray, and K. Maundrell, *Mol. Gen. Genet.* **215,** 81 (1988).

selected for and then sporulated to generate haploid spores which can be plated out as above. This approach is useful when it is necessary to analyze the effect of the deletion in different genetic backgrounds.[29] Alternatively, h^{90} strains can be generated spontaneously from an h^+/h^+ diploid and identified by replica plating onto malt extract medium followed by iodine staining. Tetrad analysis can then be carried out using this strain.

Gene replacements in *S. pombe* can be carried out using the same approach. A gene replacement event can be selected using a diploid strain in which one copy of the gene is disrupted with the *ura4*$^+$ gene, because ura4$^-$ cells generated when the disrupted gene is replaced will be resistant to the drug 5-fluoroorotic acid (5-FOA).[32,33] Thus, replacement of the *ura4*$^+$-disrupted gene with a linear DNA fragment containing the *in vitro* altered gene will convert the cells to a 5-FOA-resistant phenotype.

Schizosaccharomyces pombe Biochemistry

We have included protocols for making DNA, RNA, and protein extracts and for labeling cells.

Preparing Schizosaccharomyces pombe Chromosomal DNA[34]

1. Grow 100 ml of cells in YE to an OD_{595} of 2–3 (early stationary phase) with shaking at 25°–35°.
2. Spin down at 3000 rpm for 5 min and resuspend in 5 ml of
    50 mM citrate/phosphate, pH 5.6 (7.1 g/liter Na_2HPO_4, 11.5 g/liter citric acid)
    40 mM EDTA, pH 8.0
    1.2 M sorbitol
3. Add 15 mg Zymolyase 20T (Seikagaku Kogyo) and incubate at 37° for 30–60 min.
4. Check digestion of cell walls using a phase-contrast microscope on a 10 µl sample to which 1 µl of 10% SDS has been added (the cells lose their characteristic refringence and become black).
5. Spin down at 3000 rpm for 5 min.
6. Resuspend in 15 ml of 5× TE (50 mM Tris-HCl, pH 7.5, 5 mM EDTA). Add 1.5 ml of 10% SDS and mix well. Recheck the lysis (if necessary, cells can be incubated at 65° for 5 min).
7. Add 5 ml of 5 M potassium acetate and keep on ice for 30 min. Centrifuge at 5000 rpm for 15 min. Pass the supernatant through a gauze and add 20 ml ice-cold 2-propanol and leave for 5 min at −20°.

[33] J. D. Boeke, F. Lacroute, and G. R. Fink, *Mol. Gen. Genet.* **197**, 345 (1984).
[34] B. Durkackz, D. Beach, J. Hayles, and P. Nurse, *Mol. Gen. Genet.* **201**, 543 (1985).

8. Centrifugate at 10,000 rpm for 10 min, drain well, and dry the pellet.

9. Resuspend in 3 ml of 5× TE and add RNase to a final concentration of 20 μg/ml and incubate for 2 hr at 37°.

10. Add 3 ml of phenol/chloroform (1 : 1, v/v), mix well, and transfer to a 15-ml Corex tube. Spin down at 10,000 rpm for 10 min.

11. Transfer the upper aqueous phase to another 15-ml Corex tube, add 0.3 ml of 3 M sodium acetate and 7.5 ml of ethanol, mix, and incubate on dry ice for 1 hr or at −20° for 4–5 hr. Precipitate the DNA by spinning at 10,000 rpm for 10 min. Wash the pellet with 5 ml of cold 70% ethanol and dry under reduced pressure.

12. Finally, resuspend the DNA in 0.2 ml of TE. Read the OD_{260} (1 unit = 50 μg DNA). One haploid cell contains a genome of 14,000 kb, which is equivalent to 1.53×10^{-14} g of DNA. This DNA can be used for restriction digestion and Southern blotting. For molecular cloning, such as library construction, the DNA should be further purified by CsCl centrifugation.

Preparing Schizosaccharomyces pombe Total RNA[35]

1. Grow up a 200-ml culture in minimal medium until the OD_{595} reaches 0.5 (1×10^7 cells/ml).

2. Pellet the cells at 3000 rpm for 5 min. Wash once in 10 ml of ice-cold water and transfer to a 15-ml plastic capped tube.

3. Spin down at 3000 rpm for 5 min and thoroughly drain the pellet.

4. Resuspend in 50 μl of ice-cold HE buffer (50 mM HEPES, pH 7.9, 5 mM EDTA, 100 mM NaCl) and add 2 ml of baked acid-washed glass beads. Vortex for 1–2 min (see below for cell breakage).

5. Wash quickly twice with 1 ml of HE buffer. Spin the extract for 30 sec in an Eppendorf centrifuge.

6. Quickly add the supernatant to a sterile plastic tube containing 1.5 ml of 200 mM HEPES, pH 7.9, 10 mM EDTA, 200 mM NaCl, 2% SDS, 200 μg/ml proteinase K (added fresh just before use) and incubate for 60 min at 37°.

7. Add 3 ml of phenol/chloroform (1 : 1, v/v), mix well, and transfer to a 15-ml Corex tube. Spin down at 10,000 rpm for 10 min.

8. Transfer the upper aqueous phase to another 15-ml Corex tube, add 0.3 ml of 3 M sodium acetate and 7.5 ml of ethanol, mix, and incubate on dry ice for 1 hr or at −20° for 4–5 hr. Precipitate the RNA by spinning at 10,000 rpm for 10 min. Wash the pellet with 5 ml of cold 70% ethanol and dry under reduced pressure.

[35] N. F. Kaüfer, V. Simanis, and P. Nurse, *Nature (London)* **318**, 78 (1985).

9. Finally, resuspend the RNA in 0.5 ml of sterile water. Read the OD_{260} (1 unit = 40 μg RNA). This RNA can be used for Northern blots, 5' primer extension, *in vitro* translation, etc. For poly(A)$^+$ RNA isolation use the protocol described by Clemens.[36]

Autoclave all the solutions and bake the glassware at 140° for 4 hr, including the glass beads, before use.

Preparing Schizosaccharomyces pombe Protein Extracts

Large-Scale Extracts. This procedure is used when it is necessary to make an extract from 1×10^{10} cells or more.

1. Harvest a culture in mid-log phase by centrifuging at 3000 rpm for 5 min at 4°.
2. Wash once with ice-cold buffer stop buffer (150 mM NaCl, 50 mM NaF, 10 mM EDTA, 1 mM NaN$_3$ pH 8).
3. Resuspend the cells at 1×10^{10} cells/ml in ice-cold HB buffer [25 mM MOPS, pH 7.2, 60 mM β-glycerophosphate, 15 mM *p*-nitrophenyl phosphate, 15 mM MgCl$_2$, 15 mM EGTA, 1 mM dithiothreitol (DTT), 0.1 mM sodium vanadate, 1% Triton X-100, 1 mM phenylmethylsulfonyl fluoride (PMSF), 20 μg/ml leupeptin, 40 μg/ml aprotinin). Break open the cells using one of the following procedures:

a. French pressure cell press at 18,000–20,000 psi. Concentrated extract (40–60 mg/ml) can be made by this procedure.

b. Add 2 volumes of acid-washed glass beads (0.5 mm diameter, Sigma, G-9268) and break for 2–3 min using a Braun MSK homogenizer. To check that the cells are broken, mix 1 μl of extract with 10 μl of buffer and observe the mixture under a phase-contrast microscope. More than 90% of the cells should lose their characteristic refringence and become black.

Small-Scale Extracts

1. Grow up a 50-ml culture to an OD_{595} of 0.25 (5×10^6 cells/ml) in minimal medium. The total cell number should be 2.5×10^8.
2. Harvest the cells at 3000 rpm for 5 min. Wash once in 5 ml of ice-cold stop buffer (150 mM NaCl, 50 mM NaF, 10 mM EDTA, 1 mM NaN$_3$, pH 8). Spin down at 5000 rpm for 5 min. Drain the pellet well. The extract can be made using one of the following protocols:

---

[36] M. J. Clemens, *in* "Transcription and Translation: A Practical Approach" (B. D. Hames and S. J. Higgins, eds.), p. 157. IRL Press, Oxford and Washington, D.C., 1984.

Native extract

3a. Resuspend in 20 μl of HB buffer (25 mM MOPS, pH 7.2, 60 mM β-glycerophosphate, 15 mM p-nitrophenyl phosphate, 15 mM MgCl$_2$, 15 mM EGTA, 1 mM DTT, 0.1 mM sodium vanadate, 1% Triton X-100, 1 mM PMSF, 20 μg/ml leupeptin, 40 μg/ml aprotinin). Add 1.5 ml of acid-washed glass beads (0.5 mm diameter, Sigma, G-9268) and vortex vigorously for 1 min.

4a. Wash the beads with 1 ml of the HB buffer and centrifuge the extract 15 min at 4° in an Eppendorf centrifuge. The protein concentration of such extracts is about 1.5 mg/ml.

Denatured extract

3b. Resuspend in 20 μl of RIPA buffer (10 mM sodium phosphate, pH 7, 1% Triton X-100, 0.1% SDS, 2 mM EDTA, 150 mM NaCl, 50 mM NaF, 0.1 mM sodium vanadate, 4 μg/ml leupeptin, 1 mM PMSF). Add 1.5 ml of acid-washed glass beads (0.5 mm diameter, Sigma, G-9268) and vortex for 1 min. Add 100 μl of 1% SDS and boil for 3 min.

4b. Wash the beads with 1 ml of RIPA buffer and centrifuge the extract for 15 min at 4° in an Eppendorf centrifuge. Alternatively, after breaking the cells open, add 1 ml of SDS-PAGE sample buffer to the extract and boil it directly in the same tube. Collect the extract and spin for 5 min in a Eppendorf centrifuge to remove the cell debris.

The buffers indicated in this section are the ones that we use for our own research purposes. They should be modified according to your needs.

[^{35}S]Methionine Labeling

1. Grow up a 50-ml culture in minimal medium to an OD$_{595}$ of 0.2 (4 × 10^6 cells/ml).

2. Take 10 ml and add 1 mCi of [^{35}S]methionine (Amersham, SJ1015). Incubate for 3-4 hr at 32° (1.5-2 generations).

3. Harvest the cells and break them open using the conditions described above.

Normally, from 1 ml extract we obtain an incorporation of 2 × 10^5 cpm/μl with a protein concentration of around 1 mg/ml.

Ortho[^{32}P]phosphate Labeling

1. Take an inoculum from a preculture in YES and inoculate 100 ml of phosphate-free minimal medium (MMP) plus 1 mM phosphate (added from a 0.5 M NaH$_2$PO$_4$ stock solution).

2. Filter the cells when they reach an OD$_{595}$ of 0.2-0.4 (4 × 10^6-8 × 10^6 cells/ml).

3. Resuspend them at 2×10^6 cells/ml in fresh phosphate-free minimal medium (MMP) containing 50–100 μM phosphate.

4. Take 5 ml of cells, add 1–2 mCi ortho[^{32}P]phosphoric acid (NEN, NEX054), and label for 3–4 hr. Add 10 ml of a culture containing 10^7 cells/ml in 1 mM azide.

5. Harvest the cells and break them open using the conditions described above.

If larger amounts of labeled proteins or RNA are desired, the ortho[^{32}P]phosphoric acid can be increased to 5 mCi. For immunoprecipitations of proteins, labeled RNA may contribute to background. To overcome this problem, perform a final wash of the protein A-Sepharose in 50 mM Tris-HCl, pH 8, to get rid of the SDS and then incubate in the same buffer containing 100 μg/ml RNase for 30 min at 4°. Spin down the protein A-Sepharose, resuspend in SDS–PAGE sample buffer, and boil for 3 min before loading onto the gel.

Schizosaccharomyces pombe Cell Biology

Staining Nuclei

Schizosaccharomyces pombe nuclei can be stained using DAPI (4′,6′-diamidino-2-phenylindole dihydrochloride, Sigma, D-1388).

1. Fix 900 μl of exponentially growing cells with 100 μl of 37% formaldehyde for 30 min at the same temperature that the cells are growing.

2. Spin the cells down 10 sec in an Eppendorf centrifuge.

3. Wash the cells once in 1 ml of PBS (10 mM sodium phosphate, pH 7.2, 150 mM NaCl, 1 mM NaN$_3$), once in 1 ml of PBS containing 1% Triton X-100, and resuspend them in 100 μl of PBS. At this point the cells can be stored at 4° and stained later (at least to 2 weeks).

4. Coat a coverslip with 1 mg/ml of poly(L-lysine) (Sigma, P-1399). Put a large drop on the coverslip and then immediately remove. Leave the coverslip to dry at room temperature or dry using a hair drier.

5. Apply cells to the coverslip as a monolayer. This can best be achieved by placing about 20 μl of the cell suspension on the coverslip, then removing most of the cells such that only a thin film of cells remains behind on the coverslip. Leave the coverslip to dry at room temperature.

6. Drop the coverslip with the cell monolayer facing down onto 2–3μl of DAPI mounting solution [1 μg/ml DAPI, 1 mg/ml *p*-phenylenediamine (Sigma, P-1519), 50% glycerol] on a microscope slide. Keep the mounting solution in the dark at $-20°$.

7. Seal the edges of the coverslip with nail polish and examine under

the fluorescence microscope. Slides can be stored in the dark at 4° for a few weeks.

Schizosaccharomyces pombe Immunofluorescence[37]

Cell Fixation. Different protocols of fixation are suitable for different antigens. In principle the best approach is to try all of them and then decide which one to use.

Methanol fixation

1. Filter 20 ml of early log-phase cells (OD_{595} 0.2, 4×10^6 cells/ml) onto a glass microfiber filter (Whatmann, 1822025).
2. Fill a centrifuge tube with methanol that has been precooled at $-20°$. Using forceps, immerse the filter in the methanol. After about 30 sec shake the filter gently. The cells should flake off.
3. Incubate for 10 min at $-20°$. Pellet the cells at 3000 rpm for 5 min. Remove as much methanol as possible.
4. Rehydrate cells by washing them for 5 min each in 75, 50, 25, and 0% methanol in PBS (10 mM sodium phosphate, pH 7.2, 150 mM NaCl).
5. Cells can be stored in PBS containing 1 mM NaN$_3$ at 4°.

Formaldehyde fixation. Prepare a fresh 17.5% formaldehyde solution (5×) using the following recipe: Weigh out 8.75 g of paraformaldehyde (Sigma, P-6148) and add PBS to 50 ml. Add 1 ml of 1 M NaOH and incubate at 65° for 15–20 min. Shake to solubilize and spin at 3000 rpm for 5 min to remove polymers. Take the clear supernatant and use it to fix the cells.

To fix the cells, proceed as follows:

1. Take a 20-ml sample of early log-phase cells (OD_{595} 0.2, 4×10^6 cells/ml) and add 5 ml of the 5× formaldehyde solution. Mix thoroughly and incubate for 30 min on a rotating wheel.
2. Spin down the cells at 3000 rpm for 5 min and wash 3 times with PBS. As before the cells can be kept at 4° in PBS containing 1 mM NaN$_3$.

Formaldehyde plus glutaraldehyde fixation. The protocol is exactly as for the formaldehyde fixation, but after the addition of the formaldehyde shake the cells for about 20 sec and add 100 µl of glutaraldehyde (BDH, Poole, UK). Wash the cells as before.

Digestion of Cell Wall

1. Pellet the cells and resuspend them in PBS containing 1.2 M sorbitol.

---

[37] I. M. Hagan and J. S. Hyams, *J. Cell Sci.* **89**, 343 (1988).

2. Add NovoZym 234 (Novo Industri A/S) and Zymolyase 20T (Seikagaku Kogyo) to 0.5 mg/ml each. Incubate at room temperature for 5–20 min. The time for the cell wall digestion is highly variable. Check to see if the digest is complete by mixing 10 μl of cells with 1 μl of 10% SDS. Cells with completely digested cell walls lose their refringence under the phase-contrast microscope. It is best not to wait until all the cells are lysable.

3. Immediately fill the tube with PBS containing 1% Triton X-100 and spin the cells down.

4. Wash the cells 3 times in PBS.

Cells fixed with glutaraldehyde have to be treated with sodium borohydride. To do this wash the cells 3 times with a freshly made 1 mg/ml sodium borohydride solution in PBS and then 3 times in PBS alone.

Antibody Staining

1. Resuspend the cells in 1 ml of PBAL [PBS containing 100 mM lysine-HCl and 1% fatty acid-free bovine serum albumin (BSA) (both from Sigma)]. Incubate for 30 min on a wheel at room temperature.

2. Pellet the cells and resuspend in 100 μl of a $\frac{1}{100}$ dilution of primary antibody in PBAL. Incubate at room temperature on a wheel from 1 hr to overnight, depending on the affinity of the antibody.

3. Wash the cells 3 times in PBAL. Leave the cells for 20 min on a wheel during the last wash.

4. Resuspend the cells in 100 μl of a $\frac{1}{100}$ dilution of secondary antibody. Incubate at room temperature on a wheel from 1 hr to overnight.

5. Wash as in Step 3.

6. Mount the cells for microscopy as described above for DAPI staining. Cells can be stored at 4° in the dark for at least several months.

Acknowledgments

We would like to thank Rachel Bartlett, Kevin Crawford, Susan Dorrington, Tamar Enoch, Ursula Fleig, Kathy Gould, Iain Hagan, Jacky Hayles, Juan Jimenez, Stuart MacNeill, Chris Norbury, and Shelly Sazer for providing protocols and for valuable comments on the manuscript, and Audrey Richards for preparation of the manuscript. Robin Allshire, Fred Winston, Kinsey Maundrell, Hiroto Okayama are also thanked for communicating unpublished results. S.M. is a fellow of the Consejo Superior de Investigaciones Científicas, C.S.I.C., Spain. This work has been funded by the Imperial Cancer Research Fund and the Medical Research Council (to P.N.) and the National Cancer Institute, U.S. Department of Health and Human Services, under Contract No. N01-C0-74101 with Bionetics Research, Inc. (to A.K.). The contents of this publication do not necessarily reflect the views or policies of the Department of Health and Human Services, nor does mention of trade names, commercial products, or organizations imply endorsements by the U.S. Government.

Section VII

Appendix

[57] Genetic and Physical Maps of *Saccharomyces cerevisiae*

By ROBERT K. MORTIMER, DAVID SCHILD, C. REBECCA CONTOPOULOU, and JONATHAN A. KANS

The genetic and physical maps of the chromosomes of *Saccharomyces cerevisiae* are based on the data presented in the genetic map of *Saccharomyces cerevisiae*, Edition 10.[1] The genetic maps are composed of solid vertical lines (representing map distances determined by tetrad analysis) or dashed lines (indicating linkages established by mitotic recombination analysis), with centromeres represented as circles and with the left arm above and the right arm below the centromere. Horizontal lines indicate the genetic position of a gene. In crowded regions, gene names often are not lined up with their corresponding horizontal lines. Vertical bars indicate a region within which a gene maps. Gene symbols are defined in Table I. Synonyms (alleles) (see Table II) are separated by commas. Genes listed below the chromosome have been assigned to that chromosome but not to a specific location. In this group of genes, capitalized gene symbols do not necessarily indicate dominance. Instead, they signify that the cloned gene was used to assign the gene to that chromosome. The physical maps are drawn as vertical helical lines. Gene symbols are in capital letters on the physical maps. The telomeres, when their positions relative to genes were known, are drawn as open arrows. Vertical bars next to the helical lines indicate the size of the restriction fragment on which a gene is located or region of uncertainty about the physical location of genes. Parentheses are no longer used to indicate uncertain gene order.

---

[1] R. K. Mortimer, D. Schild, C. R. Contopoulou, and J. A. Kans, *Yeast* **5**, 321–403. (1989). Reprinted with permission from Wiley, New York.

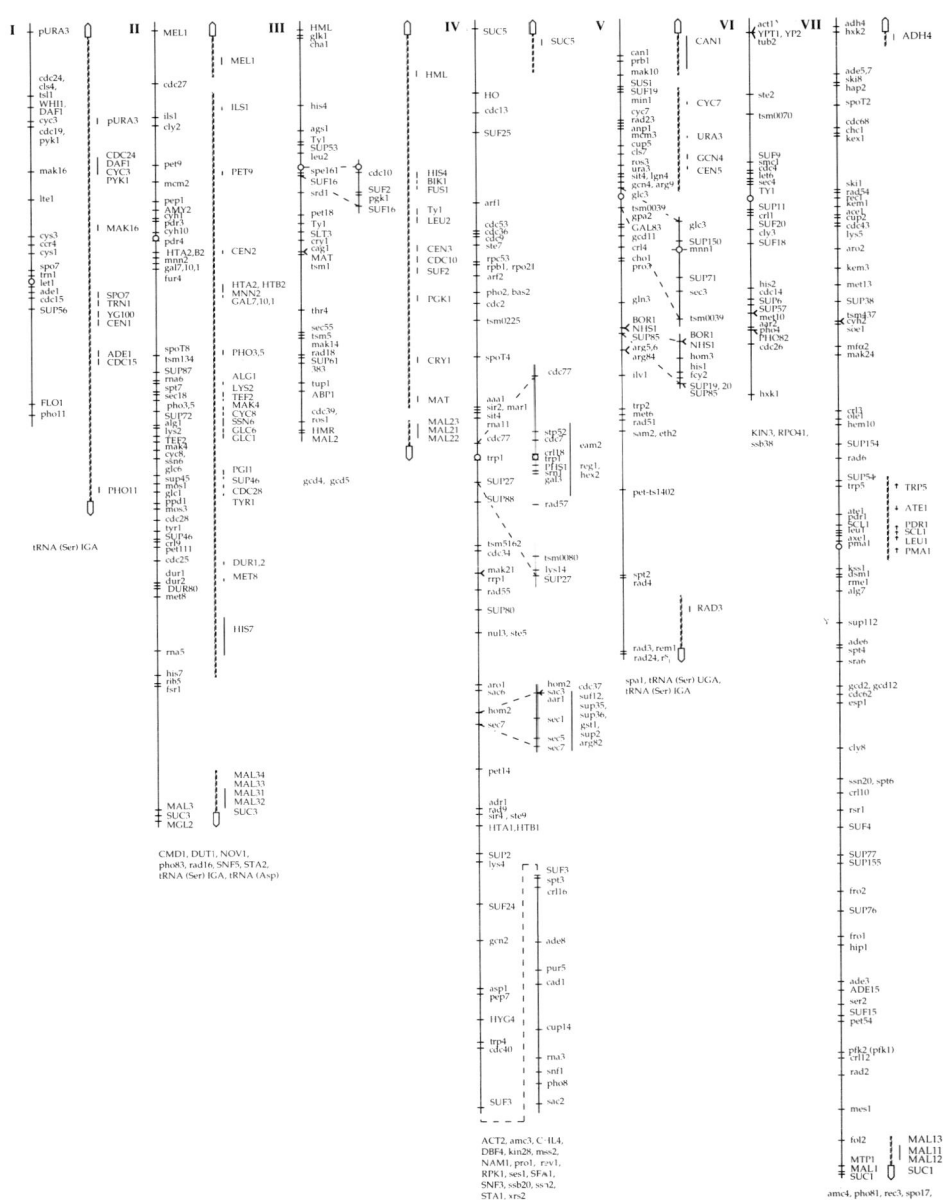

GENETIC AND PHYSICAL MAPS

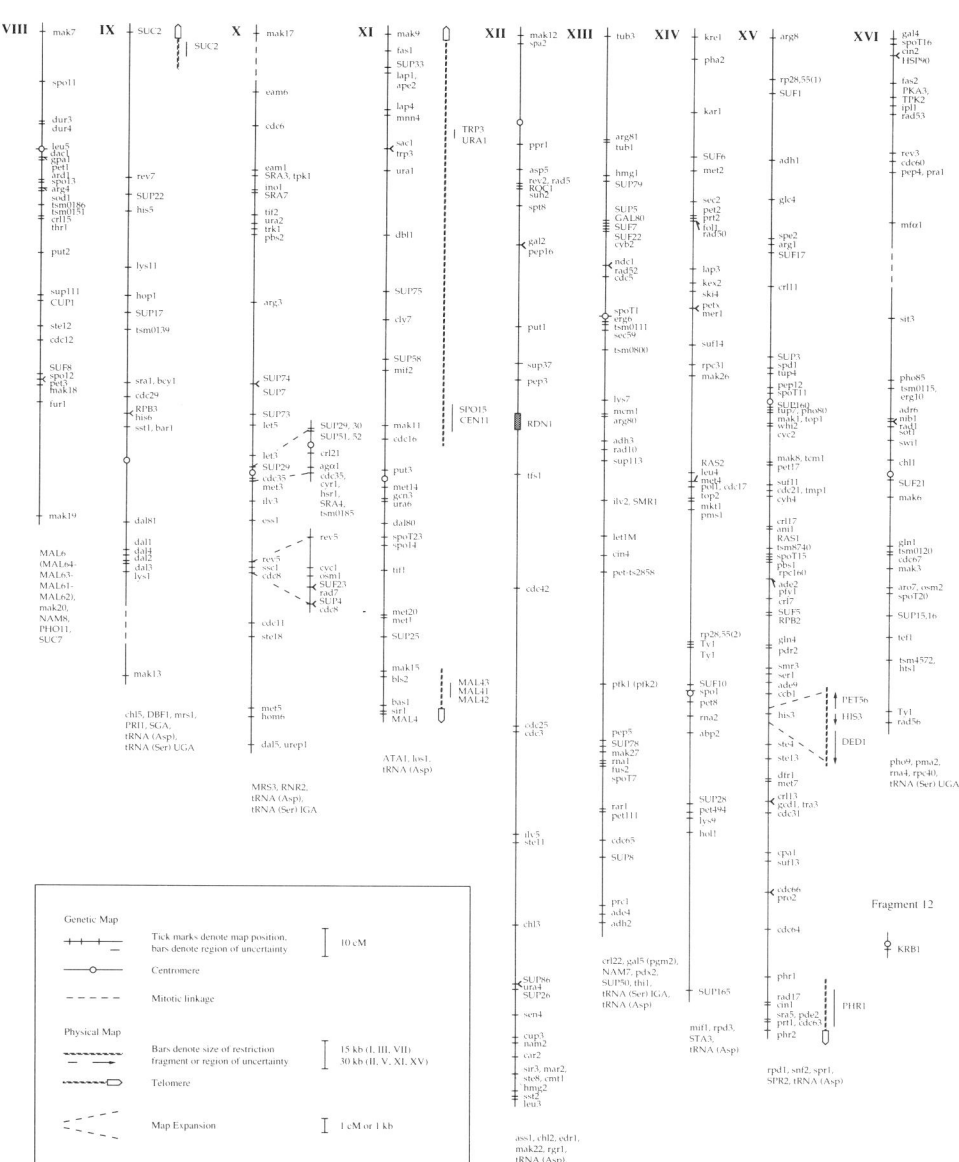

TABLE I
GLOSSARY OF MAPPED GENE SYMBOLS

Symbol	Definition[a]	Symbol	Definition
aaa	Amino terminal, amino acetyl transferase	cys	Cysteine requiring
aar	Amino acid analog resistance	dac	Division arrest control for mating pheromones
aas	Amino acid analog sensitive (see also gcn)	daf	Dominant α-factor resistance
abp	Actin binding protein	dal	Allantoin degradation deficient
ace	Activation of CUP1 expression	dbf	Dumbbell formation
act	Actin	dbl	Alcian blue dye binding deficient
ade	Adenine requiring	ded	Defines essential domain, lethal
adh	Alcohol dehydrogenase defective	dex	Dextran utilization
adr	Dehydrogenase regulation defective	dfr	Dihydrofolate reductase
agα	α-Cell specific sexual agglutination	dsm	Premeiotic DNA synthesis deficient
ags	Aminoglycoside sensitive	dur	Urea degradation deficient
alg	Asparagine-linked glycosylation deficient	eam	Endogenous ethanolamine biosynthesis
amc	Artificial minichromosome maintenance	edr	Enhanced δ recombination
amy	Antimycin resistance	erg	Ergosterol biosynthesis defective; may also be nystatin resistant
ani	Anisomycin resistance	esp	Extra spindle pole bodies
anp	ANP and osmotic sensitive	ess	Essential
ant	Antibiotic resistance	eth	Ethionine resistance
ape	Aminopeptidase	fas	Fatty acid synthetase deficient
ard	Arrest at start of cell cycle defective	fdp	Unable to grow on glucose, fructose, sucrose, or mannose
arf	ADP-ribosylation factor	flk	Flaky
arg	Arginine requiring	flo	Flocculation
aro	Aromatic amino acid requiring	fol	Folinic acid requiring
asp	Aspartic acid requiring	fro	Frothing
ass	Aspartyl-tRNA synthetase	fun	Function unknown
ata	Sporulation-specific gene characterized by ATA sequences	fur	Uracil permease
		fsr	Fluphenazine resistance
ate	Arginyl-tRNA-protein transferase deficient	fus	Fusion defective
axe	Axenomycin resistance	gal	Galactose nonutilizer
bar	a cells lack barrier effect on α-factor		

bas	Basal level control		gcd	General control of amino acid synthesis derepressed
bcy	Adenylate-cyclase and cAMP-dependent protein kinase deficient		gcn	General control of amino acid synthesis nonderepressible (see also aas)
bik	Nuclear fusion (bikaryon)		gln	Glucosamine accumulation
bls	Blasticidin-S resistance		glc	Glycogen storage
bor	Borrelidin resistance		glk	Glucokinase deficient
cad	Cadmium resistance		gln	Glutamine synthetase nonderepressible
cag	Constitutively agglutinable		gpα	G protein α homologous gene
can	Canavanine resistance		gst	G_1 to S transition
car	Catabolism of arginine defective		hap	Global regulator of respiratory genes
ccb	Cross-complementation of budding defect		hem	Heme synthesis deficient
ccr	Carbon catabolite repression		hex	Hexose metabolism regulation
cdc	Cell division cycle blocked at 36°		hip	Histidine specific permease
cen	Centromere		his	Histidine requiring
cha	Catabolism of hydroxy amino acids		hmg	HMG-CoA reductase
chc	Clathrin heavy chain gene		hml	Mating type cassette — left
chl	Chromosome loss		hmr	Mating type cassette — right
cho	Choline requiring		ho	Homothallic switching
cin	Chromosome instability		hol	Histidinol uptake proficient
cls	Calcium sensitive		hom	Homoserine requiring
cly	Cell lysis at 36°		hop	Homolog pairing
cmd	Calmodulin gene		hsp	Heat-shock protein
cmt	Control of mating type		hsr	Heat-shock resistance
cpa	Arginine requiring in presence of excess uracil		hta	Histone A genes
crl	Cycloheximide-resistant temperature-sensitive lethal		htb	Histone B genes
cry	Cryptopleurine resistance		hts	Histidinyl-tRNA synthetase
cup	Copper resistance		hxk	Hexokinase deficient
cyb	Cytochrome b_2 deficiency		hyg	Hygromycin resistance
cyc	Cytochrome c deficiency		ils	Isoleucyl-tRNA synthetase deficient; no growth at 36°
cyh	Cycloheximide resistance		ilv	Isoleucine-plus-valine requiring
cyr	Adenylate cyclase deficient		ino	Inositol deficient

(Continued)

TABLE I (Continued)

Symbol	Definition	Symbol	Definition
ipl	Increase in ploidy	pka	Protein kinase catalytic subunit
kar	Karyogamy defective	pma	Plasma membrane ATPase mutations
kem	kar enhancing mutation	pms	Postmeiotic segregation increased
kex	Killer expression defective	pol	DNA polymerase
kin	Protein kinase	ppd	Phosphoprotein phosphatase deficient
krb	Suppression of some mak mutations	ppr	Defective in pyrimidine biosynthetic pathway regulation
kre	Killer resistance	pra	Proteinase A deficient
kss	Protein kinases	prb	Proteinase B deficient
lap	Leucine aminopeptidase deficient	prc	Proteinase C deficient
let	Lethal	pri	DNA primase
leu	Leucine requiring	pro	Proline requiring
lgn	Sporulation-induced transcripts (see also sit)	prt	Protein synthesis defective at 36°
los	Loss of suppression and defective in tRNA processing	pur	Purine excretion
lte	Low temperature essential	put	Proline nonutilizer
lts	Low temperature sensitive	pyk	Pyruvate kinase deficient
lys	Lysine requiring	rad	Radiation (ultraviolet or ionizing) sensitive
mak	Maintenance of killer deficient	rar	Regulation of autonomous replication
mal	Maltose fermentation	ras	Homologous to RAS protooncogene
mar	Mating-type cassette expression	rdn	Ribosomal RNA structural genes
mat	Mating-type locus	rec	Recombination deficient
mcm	Minichromosome maintenance deficient	reg	Regulation of galactose pathway enzymes
mel	Melibiose fermentation	rev	Revertibility decreased
mer	Meiotic recombination	rgr	Resistant to glucose repression
mes	Methionyl-tRNA synthetase deficient; no growth at 36°	rib	Riboflavin biosynthesis
met	Methionine requiring	rme	Meiosis independent of mating-type heterozygosity
mfα	α-Mating factor	rna	RNA synthesis defective; unable to grow at 36°
mgl	α-Methylglucoside fermentation	rnr	Ribonucleotide reductase
mif	Mitotic frequency of chromosome transmission	roc	Roccal resistance

min	Methionine inhibited		ros	Relaxation of sterility
mkt	Maintenance of K₂ killer factor		rpo	RNA polymerase B
mmn	Mannan synthesis defective		rpb	RNA polymerase B
mos	Modifier of ochre suppressors		rpc	RNA polymerase C
mrs	Mitochondrial RNA splicing		rpd	Reduced potassium dependency
mss	Suppression of a mitochondrial RNA splice defect		rpk	Regulatory protein kinase
mtp	Melezitose fermentation		rrp	rRNA processing
nam	Nuclear suppressor of mitochondrial mutations		r⁵1	Radiation sensitive
ndc	Nuclear division cycle		rsr	Ras-related
nhs	Hydrogen sulfide production inhibitor		sac	Suppressor of actin mutations
nib	Nibbled colony phenotype due to 2-μm DNA		sam	S-Adenosylmethionine synthesis
nra	Neutral red accumulation		scl	Dominant suppression of *ts* lethality of *crl3*
nov	Novobiocin resistance		sec	Secretion deficient
nul	Nonmater		sen	Splicing endonuclease
ole	Oleic acid requiring		ser	Serine requiring
oli	Oligomycin resistance		ses	Seryl-tRNA synthetase
osm	Low osmotic pressure sensitive		sfa	Sensitive to formaldehyde
pbs	Polymyxin B resistance		sga	Suppression of growth arrest
pde	Phosphodiesterase (cAMP)		sir	Silent mating-type information regulation
pdr	Pleiotropic drug resistance		sit	Suppression of initiation of transcription
pdx	Pyridoxin requiring		sit	Sporulation-induced transcripts (see also lgn)
pep	Proteinase deficient		ski	Superkiller
pet	Petite; unable to grow on nonfermentable carbon sources		slt	Suppression at low temperature
pfk	Phosphofructokinase		smc	Stability of minichromosomes
pfy	Profilin of yeast		smr	Sulfometuron methyl resistance
pgi	Phosphoglucose isomerase deficient		snf	Deficient in derepression of many glucose-repressible genes
pgk	3-Phosphoglycerate kinase deficient		sod	Manganese-superoxide dismutase
pgm	Phosphoglucomutase deficient		soe	Suppression of *cdc8*
pha	Phenylalanine requiring		sot	Suppression of deoxythymidine monophosphate uptake
pho	Phosphatase deficient		spa	Spindle pole antigen
phr	Photoreactivation repair deficient		spd	Sporulation not repressed on rich medium
phs	Hydrogen sulfide production deficient		spe	Spermidine resistance

(Continued)

TABLE I (Continued)

Symbol	Definition	Symbol	Definition
spe	Stationary phase entry	thi	Thiamin requiring
spo	Sporulation deficient	thr	Threonine requiring
spr	Sporulation regulated genes	tif	Translation initiation factor
spt	Suppressors of Ty transcription	til	Thioisoleucine resistance
sra	Suppressors of the ras mutation	tmp	Thymidine monophosphate requiring
srd	Suppressor of *rrp1*	top	Topoisomerase deficient
srm	Suppressor of yeast *rna1-1*	tpk	Threonine/serine protein kinase
ssb	Single-strand binding protein	tra	Triazylalanine resistant
ssc	HSP70-related gene	trk	Transport of potassium
ssn	Suppressor of *snf1*	trn	Proline-tRNA gene
sst	Supersensitive to α-factor	trp	Tryptophan requiring
sta	Starch hydrolysis	tsl	Temperature sensitive lethal
ste	Sterile	tsm	Temperature sensitive lethal mutations
stp	Ste pseudorevertants	tub	Tubulin; MBC resistance
suc	Sucrose fermentation	tup	Deoxythymidine monophosphate uptake
suf	Suppression of frameshift mutation	tyr	Tyrosine requiring
suh	Suppression of *his2-1*	TY	Transposable element
sup	Suppression of nonsense mutation	umr	Ultraviolet mutability reduced
sus	Suppression of *ser1*	ura	Uracil requiring
swi	Homothallic switching deficient	urep	Ureidosuccinate permease
tcm	Tricodermin resistance	whi	Small cell size
tef	Translation elongation factor	YG100	Heat-shock gene
tel	Telomere	ypt	GTP-binding protein
tfs	Cdc25 suppressor	xrs	X-Ray sensitive

[a] Three gene symbols have two definitions: gcn, sit, and spe.

TABLE II
LIST OF MAPPED GENES[a]

Gene/synonym	Map position	Ref.[b]	Gene/synonym	Map position	Ref.[b]
aaa1	**4L**	F.-J. S. Lee, L.-W. Lin, and J. A. Smith, p.c.	*adh3*	13R	K. Kaneko and V. Williamson, p.c.
aar1	**4R**	McCusker and Haber (1988a)	*adh4*	**7L**	
aar2	**6R**	McCusker and Haber (1988a)	*adr1*	**4R**	
aas1	4R		*adr6*	**16L**	Taguchi and Young (1987)
gcn2			*agα1*	**10R**	Suzuki and Yanagishima (1986); J. Kurjan, p.c.
aas2	11R				
gcn3			*ags1*	**3L**	Ernst and Chan (1985)
aas3	**5L**	Olson *et al.* (1986); Thircos *et al.* (1984); Hinnebusch and Fink (1983)	*alg1*	2R	
arg9			*alg7*	7R	
gcn4			*amc3*	**4**	Larionov *et al.* (1988)
ABP1	**3R**	D. Drubin and D. Botstein, p.c.	*amc4*	7	Larionov *et al.* (1988)
abp2	**14R**	K. Wertman and D. Botstein, p.c.	*AMY1*	7L	Saunders and Rank (1981); Balzi *et al.* (1987)
ace1	**7L**	Thiele (1988)	*pdr1*		
act1	6L		*AMY2*	2L	Balzi *et al.* (1987)
ACT2	**4**	Schwob *et al.* (1988a,b)	*ani1*	**15R**	McCusker and Haber (1988b)
ade1	1R		*anp1*	5L	
ade2	15R		*ant1*	7L	Saunders and Rank (1981); Balzi *et al.* (1987)
ade3	7R		*pdr1*		
ade4	13R	Schild and Mortimer (1985)	*ape2*	**11L**	Hirsch *et al.* (1988)
ade5,7	7L		*lap1*		
ade6	7R		*ard1*	8R	Whiteway and Szostak (1989)
ade8	4R		*arf1*	**4L**	T. Stearns, R. Kahn, and D. Botstein, p.c.
ade9	15R		*arf2*	**4L**	T. Stearns, R. Kahn, and D. Botstein, p.c.
ADE15	7R		*arg1*	15L	
adh1	15L		*arg3*	10L	
adh2	13R		*arg4*	8R	

(*Continued*)

TABLE II (Continued)

Gene/synonym	Map position	Ref.[b]	Gene/synonym	Map position	Ref.[b]
arg5,6	5R		tsl1		Johnson et al. (1987)
arg8	15L		**cdc25**	**12R**	Portillo and Mazon (1986)
arg9	**5L**	Olson et al. (1986); Hinnebusch and Fink (1983); Thireos et al. (1984)	**cdc25**	**2R**	
aas3			cdc26	6R	
gcn4			**cdc27**	**2L**	J. Trueheart, p.c.
arg80	13R		cdc28	2R	
arg81	13L		cdc29	9L	
arg82	4R		cdc31	15R	Schild and Mortimer (1985); Baum et al. (1986)
arg84	5R				
aro1	4R		**cdc34**	**4R**	M. G. Goebl, p.c.
aro2	7L		cdc35	10R	Boutelet et al. (1985); Casperson et al. (1985)
aro7	16R	Ball et al. (1986)	cyr1		
osm2			**hsr1**		
asp1	4R		**SRA4**		
asp5	12R		tsm0185		
ass1	**12**	Kolman et al. (1988)	cdc36	4L	
ATA1	**11**	M. Breitenbach, p.c.	cdc37	4R	
ate1	7L	Balzi et al. (1988)	cdc39	3R	Jenness et al. (1987)
AXE1	7L	E. Balzi, p.c.	ros1		
bar1	9L		cdc40	4R	Kassir et al. (1985)
sst1			**cdc42**	**12R**	Johnson et al. (1987)
bas1	**11R**	Arndt et al. (1987)	cdc43	7L	Adams (1984)
bas2	**4L**	G. R. Fink, p.c.	**cdc53**	**4L**	M. G. Goebl, p.c.
pho2			cdc60	16L	Hanic-Joyce (1985)
bcy1	9L	Matsumoto et al. (1982); Toda et al. (1987); Cannon and Tatchell (1987)	cdc62	7R	Hanic-Joyce (1985)
sra1		Trueheart et al. (1987)	cdc63	15R	Hanic-Joyce (1985)
BIK1	**3L**		prt1		
bls2	11R	Ishiguro and Hayashi (1986)	cdc64	15R	Hanic-Joyce (1985)

Gene	Loc	Reference	Gene	Loc	Reference
BOR1	5R	G. Johnston, p.c.	*cdc65*	13R	G. Johnston, p.c.
BOR2	7L	Saunders and Rank (1982); Balzi *et al.* (1987)	*cdc66*	15R	
pdr1			*cdc67*	16R	J. Prendergast, L. Murray, and G. Johnston, p.c.
cad1	**4R**	A. Januska, p.c.			
cag1	**3R**	Doi and Yoshimura (1985)	*cdc68*	7L	A. Rowley and G. Johnston, p.c.
can1	5L		*cdc77*	4L	I. Villadsen, p.c.
car2	12R		*ndc2*		
ccb1	15R	A. Bender, p.c.	*CHA1*	3L	Bornaes *et al.* (1988)
ccr4	**1L**	C. Denis, p.c.	*CHC1*	7L	Payne *et al.* (1987); S. Lemmon, C. Freund, and E. Jones, p.c.
cdc2	4L				
cdc3	12R	Johnson *et al.* (1987)	*chl1*	16L	
cdc4	6L		*chl2*	12R	Kouprina *et al.* (1988)
cdc5	13L		*chl3*	12R	A. Tsouladze and V. Larionov, p.c.; Kouprina *et al.* (1988)
cdc6	10L				
cdc7	4L		*chl4*	4R	Kouprina *et al.* (1988)
cdc8	10R		*chl5*	9L	Kouprina *et al.* (1988)
cdc9	4L		*cho1*	5R	
cdc10	3R	Yeh *et al.* (1986)	*cin1*	15R	T. Stearns and D. Botstein, p.c.
cdc11	10R		*cin2*	16L	T. Stearns and D. Botstein, p.c.
cdc12	8R		*cin4*	13R	T. Stearns and D. Botstein, p.c.
cdc13	**4L**	B. Garvik and L. Hartwell, p.c.	*cls4*	1L	Ohya *et al.* (1986a)
cdc14	6R		*cdc24*		
cdc15	1R		*tsl1*		
cdc16	11L		*cls7*	5L	Ohya *et al.* (1986b)
cdc17	**14L**	Carson (1987)	*cly2*	2L	
pol1			*cly3*	6R	
cdc19	1L		*cly7*	11L	
pyk1			*cly8*	7R	
cdc21	15R		***cmd1***	**2**	Davis *et al.* (1986)
tmp1			*cmt1*	12R	Rine and Herskowitz (1987)
cdc24	1L	Ohya *et al.* (1986a)	*mar2*		
cls4			*sir3*		

(Continued)

837

TABLE II (Continued)

Gene/synonym	Map position	Ref.[b]	Gene/synonym	Map position	Ref.[b]
ste18	15R		**DBF4**	**4**	J. W. Chapman and L. H. Johnston, p.c.
cpa1	6R		dbl1	11L	
crl1	**7L**	McCusker and Haber (1988b)	**ded1**	**15R**	Struhl (1985)
crl3	**5R**	McCusker and Haber (1988b)	**dfr1**	**15R**	Barclay et al. (1988)
crl4	**15R**	McCusker and Haber (1988b)	dsm1	7R	M. Esposito, p.c.
crl7	**2R**	McCusker and Haber (1988b)	dur1	2R	Genbauffe and Cooper (1986)
crl9	**7R**	McCusker and Haber (1988b)	dur2	2R	Genbauffe and Cooper (1986)
crl10	**15L**	McCusker and Haber (1988b)	dur3	8L	
crl11	**7R**	McCusker and Haber (1988b)	dur4	8L	
crl12	**15R**	McCusker and Haber (1988b)	DUR80	2R	
crl13	**8R**	McCusker and Haber (1988b)	**DUT1**	**2**	Godsen et al. (1986)
crl15	**4R**	McCusker and Haber (1988b)	**eam1**	**10L**	K. Atkinson, p.c.
crl16	**15R**	McCusker and Haber (1988b)	**eam2**	**4**	K. Atkinson, p.c.
crl17	**4L**	McCusker and Haber (1988b)	**eam6**	**10L**	K. Atkinson, p.c.
crl18	**10R**	McCusker and Haber (1988b)	**edr1**	**12**	W. L. Arthur and R. Rohstein, p.c.
crl21	**13R**	McCusker and Haber (1988b)	**erg6**	**13R**	R. F. Gaber, B. K. Kennedy, D. Copple, and M. Bard, p.c.
crl22	**3R**		**erg10**	**16R**	Dequin et al. (1988)
cry1	8R		tsm0115		
CUP1	7L	Welch et al. (1989)	**esp1**	**7R**	P. Baum and B. Byers, p.c.
cup2	12R	Welch et al. (1989)	**ess1**	**10R**	Hanes (1988)
cup3	5L	Welch et al. (1989)	eth2	5R	Schild and Mortimer (1985)
cup5	4R	Welch et al. (1989)	sam2		
cup14	13L	T. Lodi, B. Guiard, and I. Ferrero, p.c.	fas1	11L	
cyb2	**10R**		**fas2**	**16L**	E. Schweizer, p.c.
cyc1	15R		**fcy2**	**5R**	E. Weber, p.c.
cyc2	1L		fdp1	2R	
cyc3	5L	Verdiere et al. (1988)	flk1	3R	
cyc7					

Gene	Location	Gene	Location	Reference
cyp3	2R	cyc9	1R	
cyc8		tup1	14L	
ssn6	3R	umr7	7R	
cyc9		FLO1	7R	
flk1	2L	fol1	7R	
tup1	7L	fol2	2R	Matsumoto et al. (1986)
umr7	7L	fro1	8R	Jenness et al. (1987)
cyh1		fro2	2R	Weber et al. (1986)
cyh2		**fsr1**	3L	Trueheart et al. (1987)
cyh3		**fur1**	13R	J. Trueheart, p.c.
pdr1	15R	**fur4**	2R	
cyh4	2L	**FUS1**	12R	
cyh10	10R	**fus2**	4R	
cyr1		gal1	16L	
cdc35		gal2	13R	G. McKnight, J. Hopper, and D. Oh, p.c.
hsr1		gal3		Fraenkel (1982)
SRA4		gal4	2R	
tsm0185		**gal5**	2R	
cys1	1L	pgm2	13L	
cys3	1L	gal7	5R	Schild and Mortimer (1985)
dac1	8R	gal10	15R	
gpa1	1L	gal80		
DAF1		GAL83	7R	
WHI1		gcd1		Niederberger et al. (1986)
dal1	9R	tra3		
dal2	9R	**gcd2**	3	Skvirsky et al. (1986)
dal3	9R	**gcd12**	3	Greenberg et al. (1986)
dal4		**gcd4**	5R	Harashima and Hinnebusch (1986)
dal5	10R	**gcd5**	7R	C. J. Paddon, p.c.
urep1		**gcd11**		
dal80	11R	**gcd12**		
dal81	9R	**gcd2**		
DBF1	9			J. W. Chapman and L. H. Johnston, p.c.

(Continued)

TABLE II (Continued)

Gene/synonym	Map position	Ref.[b]	Gene/synonym	Map position	Ref.[b]
gcn2	4R		ilv1	5R	
aas1			ilv2	13R	
gcn3	11R		SMR1		
aas2			ilv3	10R	
gcn4	**5L**	Olson et al. (1986); Hinnebusch and Fink (1983); Thireos et al. (1984)	ilv5	12R	
aas3			ino1	10L	
arg9			**ipl1**	**16L**	T. Stearns and D. Botstein, p.c.
glc1	2R		kar1	14L	
glc3	5L		**kem1**	**7L**	J. Kim and G. R. Fink, p.c.
glc4	15L		**kem3**	**7L**	J. Kim and G. R. Fink, p.c.
glc6	2R		kex1	7L	
glk1	3L		kex2	14L	
gln1	16R		**KIN3**	**6**	Jones and Rosamond (1988)
gln3	5R		**kin28**	**4**	Boulet et al. (1988)
gln4	**15R**	Ludmerer and Schimmel (1985)	KRB1	F12	
gpa1	**8R**	Miyajima et al. (1987)	**kre1**	**14L**	C. Boone and H. Bussey, p.c.
dac1			**kss1**	**7R**	W. Courchesne and J. Thorner, p.c.
gpa2	**5R**	Nakafuku et al. (1988)	lap1	11L	Hirsch et al. (1988)
gst1	**4R**	Kikuchi et al. (1988); Y. Kikuchi, p.c.	**ape2**		
suf12			lap3	14L	
sup2			lap4	11L	
sup35			let1	1R	
sup36			let1M	13R	
hap2	**7L**	Pinkham and Guarente (1985)	let3	10L	
hem10	7L		let5	10L	
hex2	**4R**	Niederacher and Entian (1987)	let6	6L	
reg1			leu1	7L	
hip1	7R		leu2	3L	

840

his1	5R		*leu3*	**12R**	Brisco *et al.* (1987)
his2	6R		*leu4*	14L	
his3	15R		**leu5**	**8C**	Drain and Schimmel (1986)
his8			*lgn4*	**5L**	Gottlin-Ninfa and Kaback (1986)
his4	3L		**sit4**		
his5	9L		**los1**	**11**	Hurt *et al.* (1987)
his6	9L		**lte1**	**1L**	D. Kaback, p.c.; Wickner *et al.* (1987)
his7	2R		*lts1*	7L	
hmg1	13L	Basson *et al.* (1987)	*lts3*	7L	
hmg2	12R	Basson *et al.* (1987)	*lts4*	4R	
HML	3L		*lts10*	9R	
HMR	3R		*lys1*	2R	
HO	4L		*lys2*	4R	
hol1	14R		*lys4*	7L	
hom2	4R		*lys5*	13R	
hom3	5R		*lys7*	14R	Borell *et al.* (1984)
hom6	10R	Schild and Mortimer (1985)	*lys9*	9L	
hop1	9L	Hollingsworth and Byers (1989)	**lys13**	**14R**	Borell *et al.* (1984)
HSP90	16L	T. Stearns and D. Botstein, p.c.	*lys11*		
hsr1	**10R**	H. Iida, p.c.	**lys13**		
cdc35			*lys9*		
cyr1			**lys14**	**4R**	C. R. Contopoulou, p.c.
SR44			*mak1*	15R	Thrash *et al.* (1985)
tsm0185			*top1*		
HTA2,B2	2R	Norris and Osley (1987)	*mak3*	16R	
HTA1,B1	4R	Norris and Osley (1987)	*mak4*	2R	
hts1	16R	G. Natsoulis, F. Hilger, and G. Fink, p.c.; M. Sandbaken and M. Culbertson, p.c.	*mak5*	2R	
tsm4572			*mak6*	16R	
hxk1	6R		*mak7*	8L	
hxk2	7L		*mak8*	15R	
HYG4	**4R**	McCusker (1987)	*tcm1*		
ils1	2L		*mak9*	11L	

(Continued)

TABLE II (Continued)

Gene/synonym	Map position	Ref.[b]	Gene/synonym	Map position	Ref.[b]
mak10	5L		met14	11R	
mak11	11L		met20	11R	
mak12	12L		**mfα1**	16L	Flessel et al. (1989)
mak13	9R		mfα2	7L	
mak14	3R		MGL2	2R	
mak15	11R		**mif1**	14	M. T. Brown and L. Hartwell, p.c.
mak16	1L		MIF2	11	M. T. Brown and L. Hartwell, p.c.
mak17	10L		min1	5L	
mak18	8R		**mkt1**	14L	Wickner (1987)
mak19	8R		mnn1	5C	
mak20	8		mnn2	2R	
mak21	4R		mnn4	11L	
mak22	12		**mos1**	2R	Gelugne and Bell (1988)
mak24	7L		**mos3**	2R	Gelugne and Bell (1988)
mak26	14L		**mrs1**	9R	Kreike et al. (1986); S. Lotz, p.c.
mak27	13R		MRS3	10	Schmidt et al. (1987)
MAL1	7R	Charron et al. (1989)	mss2	4L	Boulet al. (1988)
MAL11			MTP1	7R	Perkins and Needleman (1988)
MAL12			mut1	—	
MAL13			mut2	—	
MAL2	3R	Charron et al. (1989)	NAM1	4	Altamura et al. (1988)
MAL21			nam2	12R	
MAL22			NAM7	13	Altamura et al. (1988)
MAL23			NAM8	8	Altamura et al. (1988)
MAL3	2R	Charron et al. (1989)	**ndc1**	13L	Thomas and Botstein (1986)
MAL31			**ndc2**	4L	Villadsen (1988); I. Villadsen, p.c.
MAL32			cdc77		
MAL33			NHS1	5R	

842

Gene	Chr	Reference	Gene	Chr	Reference
MAL34	11R		*nib1*	16L	Pocklington and Orr (1986)
MAL4		Charron et al. (1989)	*NOV1*	2	R. Preston and E. Jones, p.c.
MAL41			*NRA2*	7L	
MAL42			**pdr1**		
MAL43			*nul3*	4R	
MAL6	8	Needleman et al. (1984); Cohen et al. (1985); Dubin (1987); Charron et al. (1989); Dubin et al. (1988)	*ste5*	7L	Saunders and Rank (1982); Balzi et al. (1987)
MAL61			*ole1*	7L	
MAL62			*oli1*		
MAL63			**pdr1**		
MAL64			*osm1*	10R	Ball et al. (1986)
mar1	4L		*osm2*	16R	
sir2			*aro7*		
mar2	12R		**pbs1**	15R	Boguslawski (1985)
cmt1			**pbs2**	10L	Boguslawski and Polazzi (1987)
ste8			**pde2**	15R	Sass et al. (1986); Wilson and Tatchell (1988)
sir3			**sra5**		
MAT	3R		**pdr1**	7L	Saunders and Rank (1982); Balzi et al. (1987)
mcm1	13R	Maine (1984)	*AMY1*		
mcm2	2L	Gibson (1989)	*ant1*		
mcm3	5L	Gibson (1989)	*BOR2*		
MEL1	2L	Hawthorne (1955); Vollrath et al. (1988)	*cyh3*		
mer1	14L	Engelbrecht and Roeder (1989)	*NRA2*		
mes1	7R		*oli1*		
met1	11R		*smr2*		
met2	14L		*til1*		
met3	10L		*pdr2*	15R	John Golin, p.c.
met4	14L		**pdr3**	2L	**Subik** et al. (1986)
met5	10R	Schild and Mortimer (1985)	*pdr4*	2C	R. Preston and E. Jones, p.c.
met6	5R	Schild and Mortimer (1985)	**pdx2**	13R	G. McKnight, J. Hopper, and D. Oh, p.c.
met7	15R		**pep1**	2L	R. Preston, L. Daniels, and E. Jones, p.c.
met8	2R		*pep3*	12R	
met10	6R		*pep4*	16L	Mechler et al. (1987)
met13	7L				

(Continued)

TABLE II (Continued)

Gene/synonym	Map position	Ref.[b]	Gene/synonym	Map position	Ref.[b]
pra1 / pep5	13R	C. Woolford, C. Dixon, P. Walters, and E. Jones, p.c.	**pra1**	16L	Mechler et al. (1987)
pep7	4R		pep4	5L	Moehle et al. (1987)
pep12	15L		prb1	13R	
pep16	12R		prc1	9	Lucchini et al. (1987)
pet1	8R		**PR11**	4	M. Brandriss, p.c.
pet2	14L		**pro1**	15R	Tomenchok and Brandriss (1987)
pet3	8R		pro2	5R	Tomenchok and Brandriss (1987)
pet8	14R		pro3	15R	
pet9	2L		prt1		
pet11	2R		cdc63	14L	
pet14	4R		prt2	4R	
pet17	15R		**prt3**	—	
pet18	3R		pur5	12R	Wang and Brandriss (1986)
pet54	7R	M. Contanzo, E. Seaver, and T. Fox, p.c.	**put1**	8R	
pet56	15R	Struhl (1985)	put2	11L	Brandriss (1987)
pet111	13R	Poutre and Fox (1987)	**put3**	1L	
pet494	14R		pyk1		
pet-ts1402	5R		cdc19	16L	
pet-ts2858	13R		rad1	7R	
petx	14L		rad2	5R	Sitney (1987); Montelone et al. (1988)
pfk1	13R	Lobo and Maitra (1983)	rad3		
pfk2			**rem1**	5R	Sitney (1987)
pfk2	7R	Heinisch and von Borstel (1988)	rad4	12R	
pfk1			rad5	7L	
			rev2	10R	
pfy1	15R	A. Adams, V. Oechsner, W. Bandlav, and D. Botstein, p.c.; Magdolen et al. (1988)	rad6	4R	
			rad7		
			rad9		

Gene	Location	Reference
pgi1	2R	
pgk1	3R	
pgm2	**13R**	Fraenkel (1982)
gal5	14L	
pha2	4L	G. R. Fink, p.c.
pho2		
bas2	2R	
pho3,5	6R	
pho4	4R	
pho8	16	Yoshida et al. (1988); K. Yoshida, p.c.
pho9	**1R**	D. Kaback, p.c.; de Jonge et al. (1988)
pho11,1	8	de Jonge et al. (1988)
pho11,2	15R	
pho80		
tup7	7	Yoshida et al. (1988)
pho81	6R	
PHO82	**2**	Yoshida et al. (1988); K. Yoshida, p.c.
pho83	16L	
pho85	15R	
phr1	15R	
phr2	4R	
PHS1	16L	A. Petitjean and K. Tatchell, p.c.
PK43	7R	McCusker et al. (1987); Ulaszewski et al. (1987)
TPK2	16L	
pma1	7R	Schlesser and Goffeau (1988)
pma2	16L	Williamson et al. (1985)
pms1	14L	Budd and Campbell (1987); Lucchini et al. (1988)
poll	14L	
cdc17		
ppd1	2R	Matsumoto et al. (1985)
ppr1	12R	
rad10	**13R**	Weiss and Friedberg (1985)
rad16	**2R**	J. Game, p.c.
rad17	**15R**	J. Game, p.c.
rad18	3L	
rad23	5L	
rad24	5R	Eckardt-Schupp et al. (1987); Sitney (1987)
r^s_1		
rad50	14L	
rad51	5R	
rad52	13L	
rad53	**16L**	K. Sitney, p.c.
rad54	7L	
rad55	4R	
rad56	16R	
rad57	4R	
r^s_1	5R	
rad24		
xrs2	**4R**	I. A. Zakharov, p.c.
rar1	**13R**	Kearsey and Edwards (1987)
RAS1	15R	
RAS2	14L	
RDN1	12R	
rec1	7L	Esposito et al. (1988)
rec3	7L	Esposito et al. (1988)
reg1	4	Matsumoto et al. (1983); Cannon and Tatchell (1987)
hex2		
rem1	5R	Montelone et al. (1988); Sitney (1987)
rad3		
rev1	4	F. Larimer, p.c.
rev2	12R	
rad5		
rev3	16L	C. Lawrence, p.c.

(Continued)

TABLE II *(Continued)*

Gene/synonym	Map position	Ref.[b]
rev5	10R	
rev7	9L	Lawrence et al. (1985)
rgr1	**12**	Sakai et al. (1988)
rib5	**2R**	de los Angeles Santos et al. (1988)
rme1	7R	
rna1	13R	
rna2	14R	
rna3	4R	S. Petersen-Bjørn and J. Friesen, p.c.
rna4	**16**	
rna5	2R	
rna6	2R	
tsm7269		
rna11	4L	
rn2	**10C**	Elledge and Davis (1987)
ROC1	12R	
ros1	**3R**	Jenness et al. (1987)
cdc39		
ros3	5L	Jenness et al. (1987)
rpb1	**4L**	M. L. Nonet, p.c.; Nonet et al. (1987)
rpo21		
RPB2	**15R**	C. Scafe, M. Nonet, and R. Young, p.c.
RPB3	**9L**	P. Kolodziej and R. Young, p.c.
rpc31	**14L**	Mosrin et al. (1988); P. Thuriaux, p.c.
rpc40	**16**	C. Mann and I. Treich, p.c.
rpc53	**4L**	C. Mann and I. Treich, p.c.
rpc160	**15R**	Gudenus et al. (1988)
rpd1	**15**	Vidal et al. (1988)
rpd3	**14**	Vidal et al. (1988)
mar1		
sir3	12R	Ivy et al. (1985); Rine and Herskowitz (1987)
mar2		
ste8		
cmt1	**4R**	Ivy et al. (1985)
sir4		
ste9		
sit3	**16L**	Arndt et al. (1989)
sit4	**4L**	Arndt et al. (1989)
sit4	**5L**	Gottlin-Ninfa and Kaback (1986)
lgn4		
ski1	7L	
ski4	14L	
ski8	**7L**	Sommer and Wickner (1987)
SLT3	3R	Inge-Vechtomov and Karpova (1984)
smc1	**6L**	Larionov et al. (1985); V. Larionov, p.c.
SMR1	13R	
ilv2		
smr2	7L	Saunders and Rank (1981); Balzi et al. (1987)
pdr1		
smr3	15R	
snf1	4R	
snf2	**15**	J. Celenza, L. Neigeborn, and M. Carlson, p.c.
SNF3	4	J. Celenza and M. Carlson, p.c.
SNF5	2	J. Celenza and M. Carlson, p.c.
sod1	**8R**	van Loon et al. (1986)
soe1	**7L**	J.-Y. Su and R. A. Sclafani, p.c.
sot1	16L	

Gene	Loc	Reference	Gene	Loc	Reference
RPK1	4	Schwob *et al.* (1988a,b)	*spa1*	5	Snyder and Davis (1988)
rpo21	4L	Himmelfarb *et al.* (1987)	*spa2*	12L	Snyder (1989)
rpb1			*spd1*	15L	
RPO41	6	Greenleaf *et al.* (1986)	*spe2*	15L	
rp28,55,1	15L	Papciak and Pearson (1987)	*spe161*	3R	Crouzet *et al.* (1988)
rp28,55,2	14L	Papciak and Pearson (1987)	*spo1*	14L	
rrp1	4R	Fabian and Hopper (1987)	*spo7*	1L	
rsr1	7R	A. Bender, p.c.	*spo11*	8L	
sac1	11L	Novick *et al.* (1989)	*spo12*	8R	
sac2	4R	Novick *et al.* (1989)	*spo13*	8R	Wang *et al.* (1987)
sac3	4R	Novick *et al.* (1989)	*spo14*	11R	
sac6	4R	Adams and Botstein (1989)	*spo15*	11C	Yeh *et al.* (1986)
sam2	5R		*spo17*	7L	Smith *et al.* (1988); Kennedy and Magee (1988)
eth2					
SCL1	7L	McCusker *et al.* (1987)	*spoT1*	13C	
sec1	4R		*spoT2*	7L	
sec2	14L		*spoT4*	4L	
sec3	5R		*spoT7*	13R	Tanaka and Tsuboi (1985)
sec4	6L		*spoT8*	2R	
sec5	4R		*spoT11*	15L	
sec7	4R		*spoT15*	15R	
sec18	2R		*spoT16*	16L	
sec55	3R		*spoT20*	16R	
sec59	13R		*spoT23*	11R	
sen1	12R	Winey and Culbertson (1988)	*spr1*	15	Primerano *et al.* (1988)
ser1	15R		*SPR2*	15	Primerano *et al.* (1988)
ser2	7R		*spt2*	5R	
ses1	4	Kolman *et al.* (1988)	*spt3*	4R	
SFA1	4	Mack and M. Brendel, p.c.	*spt4*	7R	J. Fassler and F. Winston, p.c.
SGA	9	Pretorius and Marmur (1988)	*spt6*	7R	Clark-Adams and Winston (1987)
sir1	11R	Ivy *et al.* (1985)	*ssn20*		
sir2	4L		*spt7*	2R	Winston *et al.* (1987)

(Continued)

TABLE II (Continued)

Gene/synonym	Map position	Ref.[b]	Gene/synonym	Map position	Ref.[b]
spt8	**12R**	Winston et al. (1987)	SUF4	7R	
sra1	**9L**	Cannon et al. (1986); Cannon and Tatchell (1987)	SUF5	15R	
bcy1			SUF6	14L	
SRA3	**10L**	Cannon et al. (1986)	SUF7	13L	
tpk1			SUF8	8R	
SRA4	**10R**	Cannon et al. (1986)	**SUF9**	**6L**	M. Winey and M. R. Culbertson, p.c.
cyr1			SUF10	14L	
cdc35			suf11	15R	
hsr1			suf12	4R	Kikuchi et al. (1988); Y. Kikuchi, p.c.; Ono et al. (1984)
tsm0185					
sra5	15R	Wilson and Tatchell (1988)	**gst1**		
pde2			**sup2**		
sra6	7R	Cannon et al. (1986)	sup35		
SRA7	**10L**	J. F. Cannon, p.c.	sup36		
srd1	3R	G. R. Fabian, S. Hess, and A. K. Hopper, p.c.	suf13	15R	
srn1	4R	L. S. Nolan, N. S. Atkinson, R. W. Durnst, and A. K. Hopper, p.c.	suf14	14L	
			SUF15	7R	
			SUF16	3R	
ssb20	4	Sugino et al. (1986)	SUF17	15L	
ssb38	6	Sugino et al. (1986)	SUF18	6R	
ssc1	**10R**	Craig et al. (1987)	SUF19	5L	
ssn2	4R		SUF20	6R	
ssn6	2R		SUF21	16R	
cyc8			SUF22	13L	
ssn20	**7R**	Neigeborn et al. (1987)	SUF23	10R	
spt6			SUF24	4R	
sst1	9L		SUF25	4L	
bar1			suh2	12R	
sst2	**12R**	Dietzel and Kurjan (1987)	SUP-1A	—	

STA1	4	Pretorius and Marmur (1988)	SUP2	4R	Ono et al. (1984); Kikuchi et al. (1988); Y. Kikuchi, p.c.
DEX2			**sup2**	4R	
MAL5			**gst1**		
STA2	2	Pretorius and Marmur (1988)	suf12		
DEX1			sup35		
STA3	14	Pretorius and Marmur (1988)	sup36		
DEX3			SUP3	15L	
ste2	6L	Jenness et al. (1987)	SUP4	10R	
ste4	15R		SUP5	13L	Schild and Mortimer (1985)
ste5	4R		SUP6	6L	
nul3			SUP7	10L	
ste7	4L		SUP8	13R	
ste8	12R		SUP11	6R	
cmt1			SUP15,16	16R	
mar2			SUP17	9L	
sir3			SUP19	5R	
ste9	4R		SUP20		
sir4			SUP19	5R	
ste11	12R	Chaleff and Tatchell (1985); Jenness et al. (1987); Johnson et al. (1987)	SUP20		
ste12	8R	Jenness et al. (1987)	SUP22	9L	
ste13	15R		SUP25	11R	
ste18	10R	Whiteway et al. (1988)	SUP26	12R	
stp52	4L	Katz et al. (1987)	SUP27	4R	
SUC1	7R		SUP28	14R	Ono et al. (1985)
SUC2	9L		SUP29	10C	
SUC3	2R		SUP30		
SUC5	4L		SUP29	10C	
SUC7	8	J. L. Celenza and H. Carlson, p.c.	SUP33	11L	Ono et al. (1985)
SUF1	15L		sup35	4R	Ono et al. (1984); Kikuchi et al. (1988); Y. Kikuchi, p.c.
SUF2	3R		**gst1**		
SUF3	4R		suf12		

(Continued)

849

TABLE II (Continued)

Gene/synonym	Map position	Ref.[b]	Gene/synonym	Map position	Ref.[b]
sup2			**thi1**	**13R**	G. McKnight, J. H. Hopper, and D. Oh, p.c.
sup36			thr1	8R	
sup36	4R	Kikuchi et al. (1988)	thr4	3R	
gst1			**tif1**	**11R**	P. Müller and P. Linder, p.c.
suf12			**tif2**	**10L**	P. Müller and P. Linder, p.c.
sup2			till	7L	Saunders and Rank (1982); Balzi et al. (1987)
sup35			**pdr1**		
SUP37	12R		tmp1	15R	
SUP38	7	All-Robyn et al. (1988)	cdc21		
SUP40	2	All-Robyn et al. (1988)	**top1**	**15R**	Thrash et al. (1985)
SUP44	7	All-Robyn et al. (1988)	**mak1**		
sup45	2R	Ono et al. (1984)	top2	14L	Voelkel-Meiman et al. (1986)
sup1			**tpk1**	**10L**	Cannon et al. (1986)
supQ			**SRA3**		
sup47			tra3	15R	
SUP46	2R		gcd1		
sup47	2R	Ono et al. (1984)	**trk1**	**10L**	Gaber et al. (1988)
sup1			**TRK2**	**16L**	A. Petitjean and K. Tatchell, p.c.
sup45			**PKA3**		
supQ			tm1	1R	
SUP50	**13R**	G. McKnight, J. H. Hopper, and D. Oh, p.c.	trp1	4R	Cummins et al. (1985)
SUP51	10C		trp2	5R	
SUP52			trp3	11L	
SUP52	10C		trp4	4R	
SUP51			trp5	7L	
SUP53	3L		tsl1	1L	Ohya et al. (1986a)
SUP54	7L		cdc24		
SUP56	1R		**cls4**		

Gene	Locus	Reference	Gene	Locus	Reference
SUP57	6R		tsm1	3R	
SUP58	11L		tsm5	3R	
SUP61	3R		tsm0039	5R	
SUP71	5R		tsm0070	6L	
SUP72	2R		tsm0080	4R	
SUP73	10L		tsm0111	13R	
SUP74	10L		tsm0115	16L	Dequin et al. (1988)
SUP75	11L		**erg10**		
SUP76	7R		tsm0119	7L	
SUP77	7R		tsm0120	16R	
SUP78	13R		tsm134	2R	
SUP79	13L		tsm0139	9L	
SUP80	4R		tsm0151	8R	
SUP85	5R		tsm0185	10R	Boutelet et al. (1985)
SUP86	12R		cdc35		
SUP87	2R		cyr1		
SUP88	4R		**hrs1**		
sup111	8R	Ono et al. (1986)	**SR44**		
sup112	7R	Ono et al. (1986)	tsm0186	8R	
sup113	13R	Ono et al. (1986)	tsm0225	4L	
SUP150	**5L**	Ono et al. (1988)	tsm437	7L	
SUP154	**7L**	Ono et al. (1988)	tsm0800	13R	
SUP155	7	Ono et al. (1988)	tsm4572	16R	
SUP160	**15R**	Ono et al. (1988)	hts1		
SUP165	**14R**	Ono et al. (1988)	tsm5162	4R	
SUS1	5L		tsm7269	2R	
swi1	16L		rna6		
tcm1	15R		tsm8740	15R	
mak8			**tub1**	**13L**	Schatz et al. (1986)
tef1	**16R**	Sandbaken and Culbertson (1988)	tub2	6L	
TEF2	2R	Schirmaier and Philippsen (1984)	**tub3**	**13L**	Schatz et al. (1986)
tfs1	**12R**	L. C. Robinson and K. Tatchell, p.c.	tup1	3R	

(Continued)

TABLE II (Continued)

Gene/ synonym	Map position	Ref.[b]	Gene/ synonym	Map position	Ref.[b]
cyc9			ura4	12R	Liljelund and Lacroute (1986)
flk1			ura6	11R	
umr7			urep1	10R	Turoscy and Cooper (1987)
tup4	15L		dal5		
tup7	15R		WHI1	1L	Nash et al. (1988)
pho80			DAF1		
tyr1	2R		whi2	15R	P. E. Sudbery, p.c.
umr7	3R		YG100	1L	Ingolia et al. (1982); M. Slater, p.c.
cyc9			YP2	6L	Segev and Botstein (1987); Gallwitz et al. (1983); Schmitt et al. (1986)
flk1			YPT1		
tup1			YPT1	6L	
			YP2		
ura1	11L		xrs2	4R	I. A. Zakharov, p.c.
ura2	10L		383	3R	C. Thrash-Bingham and W. L. Fangman, p.c.
ura3	5L				

[a] New genes and their positions are indicated in boldface type. There are three cases of the same gene name being used to describe two gene loci: cdc25 on chromosomes II and XII, sit4 on chromosomes IV and V, and pfk1 or pfk2 on chromosomes VII and XIII. There are also many cases of gene names with synonyms; for several of these genes, there is lack of agreement about which synonym to use.

[b] Only new references since the 1985 mapping review (Mortimer and Schild, 1985) have been included. (p.c., Personal communication.)

References for Table II

Adams, A. E. M. (1984). Cellular morphogenesis in the yeast *Saccharomyces cerevisiae*. Ph.D. Thesis, University of Michigan.

Adams, A. E. M., and Botstein, D. (1989). Dominant suppressors of yeast actin mutations that are reciprocally suppressed. *Genetics* **121**, 675.

All-Robyn, J. A., Kelley-Geraghty, D. C., and Leibman, S. W. (1988). Cloning of omnipotent suppressors in yeast. *Genome* **30**, Suppl. 1, 296.

Altamura, N., Ben Asher, E., Dujardin, G., Groudinsky, O., Kermorgant, M., and Slonimski, P. P. (1988). *Yeast* **4**, Special issue, S209.

Arndt, K. T., Styles, C., and Fink, G. R. (1987). Multiple global regulators control *HIS4* transcription in yeast. *Science* **237**, 874.

Arndt, K. T., Styles, C., and Fink, G. R. (1989). A suppressor of a *HIS4* transcriptional defect encodes a protein with homology to the catalytic subunit of protein phosphatases. *Cell (Cambridge, Mass.)* **56**, 527.

Ball, S. G., Wickner, R. B., Cottarel, G., Schaus, M., and Tirtiaux, C. (1986). Molecular cloning and characterization of *ARO7-OSM2*, a single yeast gene necessary for chorismate mutase activity and growth in hypertonic medium. *Mol. Gen. Genet.* **205**, 326.

Balzi, E., Chen, W., Ulaszewski, S., Capjeaux, E., and Goffeau, E. (1987). The multi drug resistance gene *PDR1* from *Saccharomyces cerevisiae*. *J. Biol. Chem.* **262**, 1687.

Balzi, E., Chen, W., and Goffeau, A. (1988). The arginyl-tRNA protein transferase gene *ATE1* of *Saccharomyces cerevisiae*. *Yeast* **4**, Special issue, S317.

Barclay, B. J., Huang, T., Nagel, M. G., Misener, V. L., Game, J. C., and Wahl, G. M. (1988). Mapping and sequencing of the dihydrofolate reductase gene *(DFR1)* of *Saccharomyces cerevisiae*. *Gene* **63**, 175.

Basson, M. E., Moore, R. L., O'Rear, J., and Rine, J. (1987). Identifying mutations in duplicated functions in *Saccharomyces cerevisiae*: Recessive mutations in *HMG-CoA* reductase genes. *Genetics* **117**, 645.

Baum, P., Furlong, C., and Byers, B. (1986). Yeast gene required for spindle pole body duplication: Homology of its product with Ca^{2+}-binding proteins. *Proc. Natl. Acad. Sci. U.S.A.* **83**, 5512.

Boguslawski, G. (1985). Effects of polymyxin B sulfate and polymyxin B nonapeptide on growth and permeability of the yeast *Saccharomyces cerevisiae*. *Mol. Gene. Genet.* **199**, 401.

Boguslawski, G., and Polazzi, J. O. (1987). Complete nucleotide sequence of a gene conferring polymyxin B resistance on yeast; similarity of the predicted polypeptide to protein kinases. *Proc. Natl. Acad. Sci. U.S.A.* **84**, 5848.

Borell, C. W., Urrestarazu, L. A., and Bhattacharjee, J. K. (1984). Two unlinked lysine genes (*LYS9* and *LYS14*) are required for the synthesis of saccharopine reductase in *Saccharomyces cerevisiae*. *J. Bacteriol.* **159**, 429.

Bornaes, C., Holmberg, S., and Petersen, J. G. L. (1988). The yeast *CHA1* gene encodes the catabolic L-serine (L-threonine) dehydratase. *Yeast* **4**, Special issue, S321.

Boulet, A., Simon M., and Faye, G. (1988). Isolation and characterization of the cell division cycle *CDC2* gene of *Saccharomyces cerevisiae*. *Yeast* **4**, Special issue, S119.

Boutelet, F., Petitjean, A., and Hilger, F. (1985). Yeast *cdc35* mutants are defective in adenylate cyclase and are allelic with *cyr1* mutants while *CAS1*, a new gene, is involved in the regulation of adenylate cyclase. *EMBO J.* **4**, 2635.

Brandriss, M. C. (1987). Evidence for positive regulation of the proline utilization pathway in *Saccharomyces cerevisiae*. *Genetics* **117**, 429.

Brisco, P. R. G., Cunningham, R. S., and Kohlaw, G. B. (1987). Cloning, disruption, and chromosomal mapping of yeast *LEU3*, a putative regulatory gene. *Genetics* **115**, 91.

Budd, M., and Campbell, J. L. (1987). Temperature-sensitive mutations in the yeast DNA polymerase I gene. *Proc. Natl. Acad. Sci. U.S.A.* **84**, 2838.

Cannon, J. F., and Tatchell, K. (1987). Characterization of *Saccharomyces cerevisiae* genes encoding subunits of cyclic AMP-dependent protein kinase. *Mol. Cell. Biol.* **7**, 2653.

Cannon, J. F., Gibbs, J. B., and Tatchell, K. (1986). Suppressors of the *ras2* mutation of *Saccharomyces cerevisiae*. *Genetics* **113**, 247.

Carson, M. J. (1987). CDC17, the structural gene for DNA polymerase I of yeast: Mitotic hyperrecombination and effects on telomere metabolism. Ph.D. Thesis, University of Washington, Seattle, Washington.

Casperson, G. F., Walker, N., and Bourne, H. R. (1985). Isolation of the gene encoding adenylate cyclase in *Saccharomyces cerevisiae*. *Proc. Natl. Acad. Sci. U.S.A.* **82**, 5060.

Chaleff, D. T., and Tatchell, K. (1985). Molecular cloning and characterization of the *STE7* and *STE11* genes of *Saccharomyces cerevisiae*. *Mol. Cell. Biol.* **5**, 1878.

Charron, M. J., Read, E., Haut, S. R., and Michels, C. A. (1989). Telomere-associated MAL loci of *Saccharomyces cerevisiae*. *Genetics* **122**, 307.

Chisholm, V. T., Lea, H. Z., Rai, R., and Cooper, T. G. (1987). Regulation of allantoate transport in wild-type and mutant strains of *Saccharomyces cerevisiae*. *J. Bacteriol.* **169**, 1684.

Clark-Adams, C. D., and Winston, F. (1987). The *SPT6* gene is essential for growth and is required for δ-mediated transcription in *Saccharomyces cerevisiae*. *Mol. Cell. Biol.* **7**, 679.

Cohen, J., Goldenthal, M. J., Chow, T., Buchferer, B., and Marmur, J. (1985). Organization of the MAL loci of *Saccharomyces*: Physical identification and functional characterization of three genes at the *MAL6* locus. *Mol. Gen. Genet.* **200**, 1.

Craig, E. A., Kramer, J., and Kosic-Smithers, J. (1987). SSC1, a member of the 70-kDa heat shock protein multigene family of *Saccharomyces cerevisiae*, is essential for growth. *Proc. Natl. Acad. Sci. U.S.A.* **84**, 4156.

Cross, F. R. (1988). *DAF1*, a mutant gene affecting size control, pheromone arrest, and cell cycle kinetics of *Saccharomyces cerevisiae*. *Mol. Cell. Biol.* **8**, 4675.

Crouzet, M., Bauer, F., and Aigle, M. (1988). Mutants of *Saccharomyces cerevisiae* impaired in stationary phase entry. *Yeast* **4**, Special issue, S39.

Cummins, C. M., Culbertson, M. R., and Knapp, G. (1985). Frameshift suppressor mutations outside the anticodon in yeast proline tRNAs containing an intervening sequence. *Mol. Cell. Biol.* **5**, 1760.

Davis, T. N., Urdea, M. S., Masiarz, F. R., and Thorner, J. (1986). Isolation of the yeast calmodulin gene: Calmodulin is an essential protein. *Cell (Cambridge, Mass.)* **47**, 423.

deJonge, P., Kaptein, A., Kaback, D. B., and Steensma, H. Y. (1988). Localization of the *Saccharomyces cerevisiae PHO11* gene near the ends of chromosomes I and VIII. *Yeast* **4**, Special issue, S79.

de los Angeles Santos, M., Iturriaga, E., and Eslava, A. (1988). Mapping of the *rib5* gene in *Saccharomyces cerevisiae* using UV light as an enhancer of *rad52*-mediated chromosome loss. *Curr. Genet.* **14**, 419.

Dequin, S., Gloeckler, R., Herbert, C. J., and Boutelet, F. (1988). Cloning, sequencing and analysis of the yeast *S. uvarum ERG10* gene encoding acetoacetyl CoA thiolase. *Curr. Genet.* **13**, 471.

Dietzel, C., and Kurjan, J. (1987). Pheromonal regulation and sequence of the *Saccharomyces cerevisiae SST2* gene: A model for desensitization to pheromone. *Mol. Cell. Biol.* **7**, 4169.

Doi, K. S., and Yoshimura, M. (1985). Alpha mating type-specific expression of mutations leading to constitutive agglutinability in *Saccharomyces cerevisiae*. *J. Bacteriol.* **161**, 596.

Drain, P., and Schimmel, P. (1986). Yeast *LEU5* is a *PET*-like gene that is not essential for leucine biosynthesis. *Mol. Gen. Genet.* **204**, 397.

Dubin, R. A. (1987). Molecular organization of the *MAL6* locus of *Saccharomyces carlsbergensis*. Ph.D. Thesis, City University of New York, New York, New York.

Dubin, R. A., Charron, M. J., Haut, S. R., Needleman, R. B., and Michels, C. A. (1988). Constitutive expression of the maltose fermentation enzymes in *Saccharomyces carlsbergensis* is dependent upon the mutational activation of a non-essential homolog of *MAL63*. *Mol. Cell. Biol.* **8**, 1027.

Eckardt-Schupp, F., Siede, W., and Game, J. C. (1987). The $RAD24$ ($=R_1^s$) gene product of *Saccharomyces cerevisiae* participates in two different pathways of DNA repair. *Genetics* **115**, 83.

Elledge, S. J., and Davis, R. W. (1987). Identification and isolation of the gene encoding the small subunit of ribonucleotide reductase from *Saccharomyces cerevisiae*: DNA damage-inducible gene required for mitotic viability. *Mol. Cell. Biol.* **7**, 2783.

Engebrecht, J., and Roeder, S. G. (1989). Yeast *mer1* mutants display reduced levels of meiotic recombination. *Genetics* **121**, 237.

Ernst, J. F., and Chan, R. K. (1985). Characterization of *Saccharomyces cerevisiae* mutants supersensitive to aminoglycoside antibiotics. *J. Bacteriol.* **163**, 8.

Esposito, M. S., Brown, J. T., and Rudin, N. (1988). The *REC1* gene of *Saccharomyces cerevisiae* is required for spontaneous mitotic gene conversion, intragenic recombination, intergenic recombination, genomic stability and sporulation. *In vivo* and *in vitro* properties of the temperature sensitive mutation *rec1-1*. *Yeast* **4**, Special issue, S308.

Fabian, G. R., and Hopper, A. K. (1987). *RRP1*, a *Saccharomyces cerevisiae* gene affecting rRNA processing and production of mature ribosomal subunits. *J. Bacteriol.* **169**, 1571.

Flessel, M. C., Brake, A. J., and Thorner, J. (1989). The *MFa1* gene of *Saccharomyces cerevisiae*: Genetic mapping and mutational analysis of promoter elements. *Genetics* **121**, 223.

Fraenkel, D. (1982). Carbohydrate metabolism. *in* "The Molecular Biology of the Yeast *Saccharomyces cerevisiae*: Metabolism and Gene Expression" (J. N. Strathern, E. W. Jones, and J. R. Broach, eds.). Cold Spring Harbor Laboratory, Cold Spring Harbor, New York.

Fujimura, H-A. (1989). The yeast G-protein homolog is involved in the mating pheromone signal transduction system. *Mol. Cell Biol.* **9**, 152.

Gaber, R. F., Styles, C. A., and Fink, G. R. (1988). *TRK1* encodes a plasma membrane protein required for high-affinity potassium transport in *Saccharomyces cerevisiae*. *Mol. Cell. Biol.* **8**, 2848.

Gallwitz, D., Donath, C., and Sander, C. (1983). A yeast gene encoding a protein homologous to the human *c-has/bas* proto-oncogene product. *Nature (London)* **306**, 704.

Gelugne, J.-P., and Bell, J. B. (1988). Modifiers of ochre suppressors in *Saccharomyces cerevisiae* that exhibit ochre suppressor-dependent amber suppression. *Curr. Genet.* **14**, 345.

Genbauffe, F. S., and Cooper, T. G. (1986). Induction and repression of the urea amidolyase gene in *Saccharomyces cerevisiae. Mol. Cell. Biol.* **6**, 3954.

Gibson, S. I. (1989). Analysis of *mcm3* in minichromosome maintenance mutant of yeast with a cell division cycle arrest phenotype. Ph.D. Thesis, Cornell University, Ithaca, New York.

Godsen, M. H., McIntosh, E. M., and Haynes, R. H. (1986). The isolation of dUTPase gene *(DUT1)* from *Saccharomyces cerevisiae. Yeast* **2**, Special issue, S121.

Gottlin-Ninfa, E., and Kaback, K. D. (1986). Isolation and functional analysis of sporulation-induced transcribed sequences from *Saccharomyces cerevisiae. Mol. Cell. Biol.* **6**, 2185.

Greenberg, M. L., Myers, P. L., Skvirsky, R. C., and Greer, H. (1986). New positive and negative regulators for general control of amino acid biosynthesis in *Saccharomyces cerevisiae. Mol. Cell. Biol.* **6**, 1820.

Greenleaf, A. L., Kelly, J. L., and Lehman, I. R. (1986). Yeast *RPO41* gene product is required for transcription and maintenance of the mitochondrial genome. *Proc. Natl. Acad. Sci. U.S.A.* **83**, 3391.

Gudenus, R., Mariotte, S., Moenne, A., Ruet, A., Memet, S., Buhler, J.-M., Sentenac, A., and Thuriaux, P. (1988). Conditional mutants of *RPC160*, the gene encoding the largest subunit of RNA polymerase C in *Saccharomyces cerevisiae. Genetics* **119**, 517.

Hanes, S. D. (1988). Ph.D. Thesis, Brown University, Providence, Rhode Island.

Hanic-Joyce, P. J. (1985). Mapping *cdc* mutations in the yeast *S. cerevisiae* by *rad52*-mediated chromosome loss. *Genetics* **110**, 591.

Harashima, S., and Hinnebusch, A. (1986). Multiple *GCD* genes required for repression of *GCN4*, a transcriptional activator of amino acid biosynthetic genes in *Saccharomyces cerevisiae. Mol. Cell. Biol.* **6**, 3990.

Hawthorne, D. C. (1955). Chromosome mapping in *Saccharomyces*. Ph.D. Thesis, University of Washington, Seattle, Washington.

Heinisch, J., and von Borstel, R. C. (1988). Comparison of the yeast phosphofructokinase sequences and mapping of the genes. *Yeast* **4**, Special issue, S335.

Himmelfarb, H. J., Simpson, E. M., and Friesen, J. D. (1987). Isolation and characterization of temperature sensitive RNA polymerase II mutants of *Saccharomyces cerevisiae. Mol. Cell. Biol.* **7**, 2155.

Hinnebusch, A. G., and Fink, G. R. (1983). Positive regulation in the general amino acid control of *Saccharomyces cerevisiae. Proc. Natl. Acad. Sci. U.S.A.* **80**, 5374.

Hirsch, H. H., Suarez Rendueles, P., Achstetter, T., and Wolf, D. H. (1988). Aminopeptidase yscII of yeast. Isolation of mutants and their biochemical and genetic analysis. *Eur. J. Biochem.* **173**, 589.

Hollingsworth, N. M., and Byers, B. (1989). *HOP1*: A yeast meiotic pairing gene. *Genetics* **121**, 445.

Hurt, D. J., Wang, S. S., Lin, Y.-H., and Hopper, A. K. (1987). Cloning and characterization of *LOS1*, a *Saccharomyces cerevisiae* gene that affects tRNA splicing. *Mol. Cell. Biol.* **7**, 1208.

Inge-Vechtomov, S. G., and Karpova, T. S. (1984). Dominant suppressors effective at low temperature (SLT) in *Saccharomyces cerevisiae. Genetica* **20**, 1620.

Ingolia, T. D., Slater, M. R., and Craig, E. A. (1982). *Saccharomyces cerevisiae* contains a complete antigene family related to the major heat shock-induced gene of *Drosophila*. *Mol. Cell. Biol.* **2**, 1388.

Ishiguro, J., and Hayashi, M. (1986). Genetic mapping of blasticidin resistant gene, *bls2*, in *Saccharomyces cerevisiae*. *Jpn. J. Genet.* **61**, 529.

Ivy, J. M., Hicks, J. B., and Klar, A. J. S. (1985). Map positions of yeast genes *sir1*, *sir3* and *sir4*. *Genetics* **111**, 735.

Jenness, D. D., Goldman, B. S., and Hartwell, L. H. (1987). *Saccharomyces cerevisiae* mutants unresponsive to a-factor pheromone: **a**-Factor binding and extragenic suppression. *Mol. Cell. Biol.* **7**, 1311.

Johnson, D. I., Jacobs, C. W., Pringle, J. R., Robinson, L. C., Carle, G. F., and Olson, M. V. (1987). Mapping of the *Saccharomyces cerevisiae CDC3*, *CDC25*, and *CDC42* genes to chromosome XII by chromosome blotting and tetrad analysis. *Yeast* **3**, 243.

Jones, D. G. L., and Rosamond, J. (1988). Identification of a gene encoding a novel protein kinase in yeast. *Yeast* **4**, Special issue, S45.

Kassir, Y., Kupiec, M., Shalom, A., and Simchen, G. (1985). Cloning and mapping of *CDC40*, a *Saccharomyces cerevisiae* gene with a role in DNA repair. *Genetics* **9**, 253.

Katz, M. E., Ferguson, J., and Reed, S. I. (1987). Temperature-sensitive lethal pseudorevertants of ste mutations in *Saccharomyces cerevisiae*. *Genetics* **115**, 627.

Kearsey, S. E., and Edwards, J. (1987). Mutations that increase the mitotic stability of minichromosomes in yeast: Characterization of *RAR1*. *Mol. Gen. Genet.* **210**, 509.

Kikuchi, Y., Shimatake, H., and Kikuchi, A. (1988). A yeast gene required for the G_1-to-S transition encodes a protein containing an A-kinase target site and GTPase domain. *EMBO J.* **7**, 1175.

Kolman, C. J., Snyder, M., and Soll, D. (1988). Genomic organization of tRNA and aminoacyl-tRNA synthetase genes for two amino acids in *Saccharomyces cerevisiae*. *Genomics* **3**, 201.

Kouprina, N. Y., Pashina, O. B., Nikolaishwili, N. T., Tsouladze, A. M., and Larionov, V. L. (1988). Genetic control of chromosome stability in the yeast *Saccharomyces cerevisiae*. *Yeast* **4**, Special issue, S87.

Kreike, J., Schulze, M., Pillar, T., Korte, A., and Rodel, G. (1986). Cloning of a nuclear gene *MRS1* involved in the excision of a single group I intron (b13) from the mitochondrial *COB* transcript in *S. cerevisiae*. *Curr. Genet.* **11**, 185.

Larionov, V., Karpova, T., Kouprina, N., and Gouravleva, G. (1985). A mutant of *S. cerevisiae* with impaired maintenance of centromere plasmids. *Curr. Genet.* **10**, 15.

Larionov, V. L., Kouprina, N. Y., Strunnikov, A. V., Vlassov, A. V., and Pirozhkov, V. A. (1988). Direct selection procedure for the isolation of yeast mutants with impaired segregation of artificial minichromosomes. *Yeast* **4**, Special issue, S89.

Lawrence, C. W., Das, G., and Christensen, R. B. (1985). *REV7*, a new gene concerned with UV mutagenesis in yeast. *Mol. Gen. Genet.* **200**, 80.

Liljelund, P., and Lacroute, F. (1986). Genetic characterization and isolation of the *Saccharomyces cerevisiae* gene coding for uridine monophospho-kinase. *Mol. Gen. Genet.* **205**, 74.

Lobo, Z., and Maitra, P. K. (1983). Phosphofructokinase mutants of yeast. *J. Biol. Chem.* **258**, 1444.

Lucchini, G., Francesconi, S., Foiani, M., Badaracco, G., and Plevani, P. (1987). Yeast DNA polymerase–DNA primase complex: Cloning of *PRI1*, a single essential gene related to DNA primase activity. *EMBO J.* **6**, 737.

Lucchini, G., Mazza, C., Scacheri, E., and Plevani, P. (1988). Genetic mapping of *S. cerevisiae* DNA polymerase I gene and characterization of a *pol1* temperature-sensitive mutant altered in the DNA primase–polymerase complex stability. *Mol. Gen. Genet.* **212**, 459.

Ludmerer, S. W., and Schimmel, P. (1985). Cloning of *GLN4*: An essential gene that encodes glutaminyl-tRNA synthetase in *Saccharomyces cerevisiae*. *J. Bacteriol.* **163**, 763.

McCusker, J. H. (1987). Pleiotropic drug resistance mutations in *Saccharomyces cerevisiae*. Ph.D. Thesis, Brandeis University, Waltham, Massachusetts.

McCusker, J. H., and Haber, J. E. (1988a). Mutations in *Saccharomyces cerevisiae* which confer resistance to several amino acid analogs. *J. Bacteriol.* (submitted).

McCusker, J. H., and Haber, J. E. (1988b). Cyclohexamide-resistant temperature-sensitive lethal mutations of *Saccharomyces cerevisiae*. *Genetics* **119**, 303.

McCusker, J. H., Perlin, D. S., and Haber, J. E. (1987). Pleiotropic plasma membrane ATPase mutations of *Saccharomyces cerevisiae*. *Mol. Cell. Biol.* **7**, 4082.

Magdolen, V., Oechsner, U., Muller, G., and Bandlow, W. (1988). The intron-containing gene for yeast profilin *(PFY)* encodes a vital function. *Mol. Cell. Biol.* **8**, 5108.

Maine, G. T. (1984). Ph.D. Thesis, Cornell University, Ithaca, New York.

Matsumoto, K., Uno, I., Oshima, Y., and Ishikawa, T. (1982). Isolation of characterization of yeast mutants deficient in adenylate cyclase and cAMP-dependent protein kinase. *Proc. Natl. Acad. Sci. U.S.A.* **79**, 2355.

Matsumoto, K., Yoshimatsu, T., and Oshima, Y. (1983). Recessive mutations conferring resistance to carbon catabolite repression of galactokinase synthesis in *Saccharomyces cerevisiae*. *J. Bacteriol.* **153**, 1405.

Matsumoto, K., Uno, I., Kato, K., and Ishikawa, T. (1985). Isolation and characterization of a phosphoprotein phosphatase-deficient mutant in yeast. *Yeast* **1**, 25.

Matsumoto, K., Uno, I., and Ishikawa, T. (1986). Fluphenazine-resistant *Saccharomyces cerevisiae* mutants defective in the cell division cycle. *J. Bacteriol.* **168**, 1352.

Mechler, B., Müller, H., and Wolf, D. H. (1987). Maturation of vacuolar (lysosomal) enzymes in yeast; proteinase yscA and proteinase yscB are catalysts of the processing and activation event of carboxypeptidase yscY. *EMBO J.* **6**, 2157.

Miyajima, I., Nakafuku, M., Nakayama, N., Brenner, C., Miyajima, A., Kaibuchi, K., Arai, K., Kaziro, Y., and Matsumoto, K. (1987). *GPA1*, a haploid-specific essential gene, encodes a yeast homolog of mammalian G protein which may be involved in mating factor signal transduction. *Cell (Cambridge, Mass.)* **50**, 1011.

Moehle, C. M., Aynardi, M. W., Kolodny, M. R., Park, F. J., and Jones, E. W. (1987). Protease B of *Saccharomyces cerevisiae*: Isolation and regulation of the *PRB1* structural gene. *Genetics* **115**, 255.

Montelone, B. A., Hoekstra, M. F., and Malone, R. E. (1988). Spontaneous mitotic recombination in yeast; the hyperrecombinational *rem1* mutations are alleles of the *RAD3* gene. *Genetics* **119**, 299.

Mortimer, R. K., and Schild, D. (1985). Genetic map of *Saccharomyces cerevisiae*, Edition 9. *Microbiol. Rev.* **49**, 181.

Mosrin, C., Moenne, A., Mariotte, S., Sentenac, A., and Thuriaux, P. (1988). Cloning and *in vitro* mutagenesis of three genes of RNA polymerase c(III) in *S. cerevisiae*. *Yeast* **4**, Special issue, S494.

Nakafuku, M., Obara, T., Kaibuchi, K., Miyajima, I., Miyajima, A., Itoh, H., Nakamura, S., Arai, K., Matsumoto, K., and Kaziro, Y. (1988). Isolation of second G protein homologous gene (*GPA2*) of yeast *Saccharomyces cerevisiae* and studies on its possible functions. *Proc. Natl. Acad. Sci. U.S.A.* **88**, 1374.

Nash, R., Tokiwa, G., Anand, S., Stojcic, C., Hazlett, M., Erickson, K., and Futcher, B. (1988). Cloning and partial characterization of the *WHI1* gene of *S. cerevisiae*. *Yeast* **4**, Special issue, S51.

Needleman, R. B., Kaback, D. B., Dubin, R. A., Perkins, E. L., Rosenberg, N. G., Sutherland, K. A., Forrest, D. B., and Michels, C. A. (1984). *MAL6* of *Saccharomyces*: A complex genetic locus containing three genes required for maltose fermentation. *Proc. Natl. Acad. Sci. U.S.A.* **81**, 2811.

Niederacher, D., and Entian, K.-D. (1987). Isolation and characterization of the regulatory *HEX2* gene necessary for glucose repression in yeast. *Mol. Gen. Genet.* **206**, 505.

Niederberger, P., Aebi, M., and Hütter, R. (1986). Identification and characterization of four new *GCD* genes in *Saccharomyces cerevisiae*. *Curr. Genet.* **10**, 657.

Neigeborn, L., Celenza, J. L., and Carlson, M. (1987). *SSN20* is an essential gene with mutant alleles that suppress defects in *SUC2* transcription in *Saccharomyces cerevisiae*. *Mol. Cell. Biol.* **7**, 672.

Nonet, M., Scafe, C., Sexton, J., and Young, R. (1987). Eucaryotic RNA polymerase conditional mutant that rapidly ceases mRNA synthesis. *Mol. Cell. Biol.* **7**, 1602.

Norris, D., and Osley, M. A. (1987). The two gene pairs encoding H2A and H2B play different roles in the *Saccharomyces ceresviae* life cycle. *Mol. Cell. Biol.* **7**, 3473.

Novick P., Osmond, B. C., and Botstein, D. (1989). Suppressors of yeast actin mutations. *Genetics* **121**, 659.

Ohya, Y. M., Miyamoto, S., Ohsumi, Y., and Anraku, Y. (1986a). Calcium-sensitive *cls4* mutant of *Saccharomyces cerevisiae* with a defect in bud formation. *J. Bacteriol.* **165**, 28.

Ohya, Y., Ohsumi, Y., and Anraku, Y. (1986b). Isolation and characterization of Ca^{2+}-sensitive mutants of *Saccharomyces cerevisiae*. *J. Gen. Microbiol.* **132**, 979.

Olson, M. V., Dutchik, J. E., Graham, M. Y., Brodeur, G. M., Helms, C., Frank, M., MacCollin, M., Sheinman, R., and Frank, T. (1986). Random-clone strategy for genomic restriction mapping in yeast. *Proc. Natl. Acad. Sci. U.S.A.* **83**, 7826.

Ono, B-i., Moriga, N., Ishihara, K., Ishiguro, J., Ishimo, Y., and Shinoda, S. (1984). Omnipotent suppressors effective in y$^+$ strains of *Saccharomyces cerevisiae*: Recessiveness and dominance. *Genetics* **107**, 219.

Ono, B-i., Ishino-Arao, Y., Shirai, T., Maeda, N., and Shinoda, S. (1985). Genetic mapping of leucine-inserting UAA suppressors in *Saccharomyces cerevisiae*. *Curr. Genet.* **9**, 197.

Ono, B-i., Ishino-Arao, Y., Tanaka, M., Awano, I., and Shinoda, S. (1986). Recessive nonsense suppressors in *Saccharomyces cerevisiae*: Action spectra, complementation groups and map positions. *Genetics* **114**, 363.

Ono, B-i., Fujimoto, R., Ohno, Y., Maeda, N., Tsuchiya, Y., Usui, T., and Ishino-Arao, Y. (1988). UGA-suppressors in *Saccharomyces cerevisiae*: Allelism, action spectra and map positions. *Genetics* **118**, 41.

Papciak, S. M., and Pearson, N. J. (1987). Genetic mapping of two pairs of linked ribosomal protein genes in *Saccharomyces cerevisiae*. *Curr. Genet.* **11**, 445.

Payne, G. S., Hasson, T. B., Hasson, M. S., and Schekman, R. (1987). Genetic and biochemical characterization of clathrin-deficient *Saccharomyces cerevisiae*. *Mol. Cell. Biol.* **7**, 3888.

Perkins, E. L., and Needleman, R. B. (1988). *MAL64ᶜ* is a global regulator of α-glucoside fermentation: Identification of a new gene involved in melezitose fermentation. *Curr. Genet.* **13**, 369.

Pinkham, J. L., and Guarente, L. (1985). Cloning and molecular analysis of the *HAP2* locus: A global regulator of respiratory genes in *Saccharomyces cerevisiae*. *Mol. Cell. Biol.* **5**, 3410.

Pocklington, M., and Orr, E. (1986). Novobiocin-resistant mutants in yeast. *Yeast* **2**, Special issue, S305.

Portillo, F., and Mazon, M. (1986). The *Saccharomyces cerevisiae* start mutant carrying the *cdc25* mutation is defective in activation of plasma membrane ATPase by glucose. *J. Bacteriol.* **168**, 1254.

Poutre, C. G., and Fox, T. D. (1987). *PET111*, a *Saccharomyces cerevisiae* nuclear gene required for translation of the mitochondrial mRNA encoding cytochrome *c* oxidase subunit II. *Genetics* **115**, 637.

Pretorius, I. S., and Marmur, J. (1988). Localization of yeast glucoamylase genes by PFGE and OFAGE. *Curr. Genet.* **14**, 9.

Primerano, D., Muthukumar, G., Suhng, S. H., and Magee, P. T. (1988). Molecular characterization of two sporulation regulated (SPR) genes, one of which is involved in spore development. *Yeast* **4**, Special issue, S54.

Rine, J., and Herskowitz, I. (1987). Four genes responsible for a position effect on expression from *HML* and *HMR* in *Saccharomyces cerevisiae*. *Genetics* **116**, 9.

Sakai, A., Shimizu, Y., and Hishinuma, F. (1988). Isolation and characterization of mutants which show an oversecretion phenotype in *Saccharomyces cerevisiae*. *Genetics* **119**, 499.

Sandbaken, M. G., and Culbertson, M. R. (1988). Mutations in elongation factor EF-1a affect the frequency of frameshifting and amino acid misincorporation in *Saccharomyces cerevisiae*. *Genetics* **120**, 923.

Sass, P., Field, J., Nikawa, J., Toda, T., and Wigler, M. (1986). Cloning and characterization of the high-affinity cAMP phosphodiesterase of *Saccharomyces cerevisiae*. *Proc. Natl. Acad. Sci. U.S.A.* **83**, 9303.

Saunders, G. W., and Rank, G. H. (1982). Allelism of pleiotropic drug resistance in *Saccharomyces cerevisiae*. *Can. J. Genet. Cytol.* **24**, 493.

Schatz, P. J., Solomon, F., and Botstein, D. (1986). Genetically essential and nonessential α-tubulin genes specify functionally interchangeable proteins. *Mol. Cell. Biol.* **6**, 3722.

Schild, D., and Mortimer, R. K. (1985). A mapping method for *Saccharomyces cerevisiae* using *rad52*-induced chromosome loss. *Genetics* **110**, 569.

Schirmaier, F., and Philippsen, P. (1984). Identification of two genes coding for the translation elongation factor EF-1a of *S. cerevisiae*. *EMBO J.* **3**, 3311.

Schlesser, A., and Goffeau, A. (1988). A second transport-ATPase gene in *Saccharomyces cerevisiae*. *Yeast* **4**, Special issue, S359.

Schmidt, C., Söllner, T., and Schweyen, R. J. (1987). Nuclear suppression of a mitochondrial RNA splice defect: Nucleotide sequence and disruption of the *MRS3* gene. *Mol. Gen. Genet.* **210**, 145.

Schmitt, H. D., Wagner, P., Pfaff, E., and Gallwitz, D. (1986). The *ras*-related *YPT1* gene product in yeast; A GTP-binding protein that might be involved in microtubule organization. *Cell (Cambridge, Mass.)* **47**, 401.

Schwob, E., Alt, G., Andres, S., Dirheimer, G., and Martin, R. P. (1988a). *ACT2*, a novel yeast split gene coding for an actin-like protein. *Yeast* **4**, Special issue, S108.

Schwob, E., Andres, S., Alt, G., Dirheimer, G., and Martin, R. P. (1988b). *RPK1*, a new protein kinase gene in *Saccharomyces cerevisiae*. *Yeast* **4**, Special issue, S57.

Segev, N., and Botstein, D. (1987). The *ras*-like yeast *YPT1* gene is itself essential for growth, sporulation, and starvation response. *Mol. Cell. Biol.* **7**, 2367.

Sitney, K. (1987). Genetic and molecular studies of the *RAD24* gene of *Saccharomyces*. Ph.D. Thesis, University of California, Berkeley, California.

Skvirsky, R. C., Greenberg, M. L., Myers, P. S., and Greer, H. (1986). A new negative control gene for amino acid biosynthesis in *Saccharomyces cerevisiae*. *Curr. Genet.* **10**, 495.

Smith, L. M., Robbins, L. G., Kennedy, A., and Magee, P. T. (1988). Identification and characterization of mutations affecting sporulation in *Saccharomyces cerevisiae*. *Genetics* **120**, 899.

Sommer, S., and Wickner, R. B. (1987). Gene disruption indicates that the only essential function of the *SKI8* chromosomal gene is to protect *Saccharomyces cerevisiae* from viral cytopathology. *Virology* **157**, 252.

Snyder, M. (1989). The SPA2 protein of yeast localizes to the sites of cell growth. *J. Cell Biol.* **108**, 1419.

Snyder, M., and Davis, R. W. (1988). *SPA1*: A gene important for chromosome segregation and other mitotic functions in *S. cerevisiae*. *Cell (Cambridge, Mass.)* **54**, 743.

Struhl, K. (1985). Nucleotide sequence and transcriptional mapping of the yeast *pet56-his3-ded1* gene region. *Nucleic Acids. Res.* **13**, 8587.

Subik, J., Ulaszewski, S., and Goffeau, A. (1986). Genetic mapping of nuclear mucidin resistance mutations in *Saccharomyces cerevisiae*. *Curr. Genet.* **10**, 665.

Sugino, A., Hamatake, R., Eberly, S., Sakai, A., Alexander, P., Desai, R., and Clark, A. (1986). Biochemical and genetical studies of DNA replication proteins in yeast. *Yeast* **2**, Special issue, S374.

Suzuki, K., and Yanagishima, N. (1986). Genetic characterization of an **a**-specific gene responsible for sexual agglutinability in *Saccharomyces cerevisiae*: Mapping and gene dose effect. *Curr. Genet.* **10**, 353.

Taguchi, A. K. W., and Young, E. T. (1987). Cloning and mapping of *ADR6*, a gene required for sporulation and expression of the alcohol dehydrogenase II isozyme from *Saccharomyces cerevisiae*. *Genetics* **116**, 531.

Tanaka, H., and Tsuboi, M. (1985). Cloning and mapping of the sporulation gene, *spoT7*, in *Saccharomyces cerevisiae*. *Mol. Gen. Genet.* **199**, 21.

Thiele, D. J. (1988). *ACE1* regulates expression of the *Saccharomyces cerevisiae* metallothionein gene. *Mol. Cell. Biol.* **8**, 2745.

Thireos, G., Driscoll, P. M., and Greer, H. (1984). 5'-Untranslated sequences are required for the translational control of a yeast regulatory gene. *Proc. Natl. Acad. Sci. U.S.A.* **81**, 5096.

Thomas, J. H., and Botstein, D. (1986). A gene required for the separation of chromosomes on the spindle apparatus in yeast. *Cell (Cambridge, Mass.)* **44**, 65.

Thrash, C., Bankier, A. T., Barrell, B. G., and Sternglanz, R. (1985). Cloning, characterization, and sequence of the yeast DNA topoisomerase I gene. *Proc. Natl. Acad. Sci. U.S.A.* **82**, 4374.

Toda, T., Cameron, S., Sass, P., Zoller, M., Scott, J. D., McMullen, B., Hurwitz, M., Krebs, E. G., and Wigler, M. (1987). Cloning and characterization of *BCY1*, a locus encoding a regulatory subunit of the cycle AMP-dependent protein kinase in *Saccharomyces cerevisiae*. *Mol. Cell. Biol.* **7**, 1371.

Tomenchok, D. M., and Brandriss, M. C. (1987). Gene–enzyme relationships in the proline biosynthetic pathway of *Saccharomyces cerevisiae*. *J. Bacteriol.* **169**, 5364.

Trueheart, J., Boeke, J. D., and Fink, G. R. (1987). Two genes required for cell fusion during yeast conjugation: Evidence for a pheromone-induced surface protein. *Mol. Cell. Biol.* **7**, 2316.

Turoscy, V., and Cooper, T. G. (1987). Ureidosuccinate is transported by the allantoate transport system in *Saccharomyces cerevisiae*. *J. Bacteriol.* **169**, 2598.

Ulaszewski, S., Balzi, E., and Goffeau A. (1987). Genetic and molecular mapping of the *pma1* mutation conferring vanadate resistance to the plasma membrane ATPase from *Saccharomyces cerevisiae*. *Mol. Gen. Genet.* **207**, 38.

van Loon, A. P. G. M., Pesoed-Hurt, B., and Schatz, G. (1986). A yeast mutant lacking mitochondrial manganese-superoxide dismutase is hypersensitive to oxygen. *Proc. Natl. Acad. Sci. U.S.A.* **83**, 3820.

Verdiere, J., Gaisne, M., Guiard, B., and Defranous, N. (1988). A single missense mutation in *CYP1* (*HAP1*) regulatory gene switches the expression of two structural genes encoding isocytochromes. *c. Yeast* **4**, Special issue, S425.

Vidal, M., Hilger, F., Burd, C. G., and Gaber, R. F. (1988). Mutations in *RPD1* and *RPD3* alter potassium transport in yeast. *Yeast* **4**, Special issue, S65.

Villadsen, I. S. (1988). *NDC2*, a gene that affects chromosome stability in yeast. *Yeast* **4**, Special issue, S98.

Voelkel-Meiman, K., DiNardo, S., and Sternglanz, R. (1986). Molecular cloning and genetic mapping of the DNA topoisomerase II gene of *Saccharomyces cerevisiae*. *Gene* **42**, 193.

Vollrath, D., Davis, R. W., Connelly, C., and Hieter, P. (1988). Physical mapping of large DNA by chromosome fragmentation. *Proc. Natl. Acad. Sci. U.S.A.* **85**, 6027.

Wang, H.-T., Frackman, S., Kowalisyn, J., Easton Esposito, R., and Elder, R. (1987). Developmental regulation of *SPO13*, a gene required for separation of homologous chromosomes at meiosis I. *Mol. Cell. Biol.* **7**, 1425.

Wang, S.-S., and Brandriss, M. C. (1986). Proline utilization in *Saccharomyces cerevisiae*: Analysis of the cloned *PUT1* gene. *Mol. Cell. Biol.* **6**, 2638.

Weber, E., Jund, R., and Chevallier, M-R. (1986). Chromosomal mapping of the uracil permease gene of *Saccharomyces cerevisiae*. *Curr. Genet.* **11**, 93.

Weiss, W. A., and Friedberg, E. C. (1985). Molecular cloning and characterization of the yeast *RAD10* gene and expression of *RAD10* protein in *E. coli*. *EMBO J.* **4**, 1575.

Welch, J. W., Fogel, S., Buchman, C., and Karin, M. (1989). The *CUP2* gene product regulates the expression of *CUP1* gene coding for yeast metallothionine. *EMBO J.* **8**, 255.

Whiteway, M., and Szostak, J. W. (1985). The *ARD1* gene of yeast functions in the switch between the mitotic and alternative developmental pathways. *Cell (Cambridge, Mass.)* **43,** 483.

Whiteway, M., Hougan, L., and Thomas, D. Y. (1988). Expression of *MFα1* in *MATα* cells supersensitive to **a**-factor leads to self-arrest. *Mol. Gen. Genet.* **214,** 85.

Wickner, R. B. (1987). *MKT1*, a non-essential *Saccharomyces cerevisiae* gene with a temperature-dependent effect on replication of M_2 double-stranded RNA. *J. Bacteriol.* **169,** 4941.

Wickner, R. B., Koh, T. J., Crowley, J. C., O'Neil, J., and Kaback, D. (1987). Molecular cloning of chromosome I DNA from *Saccharomyces cerevisiae*: Isolation of the *MAK16* gene and analysis of an adjacent gene essential for growth at low temperatures. *Yeast* **3,** 51.

Williamson, M. S., Game, J. C., and Fogel, S. (1985). Meiotic gene conversion mutants in *Saccharomyces cerevisiae*. I. Isolation and characterization of *pms1-1* and *pms1-2*. *Genetics* **110,** 609.

Wilson, R. B., and Tatchell, K. (1988). *SRA5* encodes the low-K_m cyclic AMP phosphodiesterase of *Saccharomyces cerevisiae*. *Mol. Cell. Biol.* **8,** 505–510.

Winey, M., and Culbertson, M. R. (1988). Mutations affecting the tRNA-splicing endonuclease activity of *Saccharomyces cerevisiae*. *Genetics* **118,** 609.

Winston, F., Dollard, C., Malone, E. A., Clare, J., Kapakos, J. G., Farabaugh, P., and Minehart, P. L. (1987). Three genes are required for trans-activation of Ty transcription in yeast. *Genetics* **115,** 649.

Yeh, E., Carbon, J., and Bloom, K. (1986). Tightly centromere-linked gene *(SPO15)* essential for meiosis in the yeast *Saccharomyces cerevisiae*. *Mol. Cell. Biol.* **6,** 158.

Yoshida, K., Ogawa, N., Okada, S., Hiraoka, E., and Oshima, Y. (1988). Regulatory circuit for phosphatase synthesis in *Saccharomyces cerevisiae*. *Yeast* **4,** Special issue, S374.

Author Index

Numbers in parentheses are footnote reference numbers and indicate that an author's work is referred to although the name is not cited in the text.

A

Abe, A., 380
Abe, M., 111
Abelson, J. N., 415
Abelson, J., 429, 568, 595(40), 597(40), 600(40), 608, 620, 719, 726, 727(8, 21)
Aboul-Enein, H. Y., 311
Abovich, N, 459, 461(20), 465
Abraham, J. A., 133
Abraham, J., 105, 294, 339
Achstetter, T., 429, 430, 431(4), 433(5), 435(2), 443(4), 450(4), 451(4), 857
Adams, 836
Adams, A. E. M., 477, 489, 565, 567, 570(3), 573(7), 587(3, 6, 7), 588(7), 598(7), 600(6, 7, 8, 26), 608, 645, 729, 730, 731(7), 732, 733, 734(10, 11), 735(5), 853
Adams, A. E., 718
Adams, A., 718, 844, 847
Adams, B. G., 384
Adams, E., 431
Adelberg, E. A., 8
Aebersold, R., 433
Aebi, M., 860
Aeschbach, R., 121
Agoston, D. V., 568, 599(34)
Ahlstrom-Jonasson, L., 137
Aibara, S., 448
Aigle, M., 12, 855
Ajam, N., 124, 125(56)
Akada, R., 105, 108(39), 110(39)
Akai, A., 152, 155
Al, E. J. M., 431, 446(23)
Alakhov, Y. B., 543
Alam, S. N., 311
Alani, E., 292, 303, 378
Alber, T., 492, 494(10)
Alberghina, L., 568, 599(33)
Alen, J. L., 736

Alexander, P., 863
All-Robyn, 850
All-Robyn, J. A., 12, 853
Allen, J. L., 574
Allen, R. D., 594
Allshire, R. C., 252, 809, 812(21)
Alouani, S., 496
Alt, G., 862
Altamura, 842
Altamura, N., 853
Altherton, J. P., 663, 667(10)
Amadò, R., 121
Ammer, H., 683, 693(5)
Ammerer, G., 375, 377, 378(19), 383(9), 387, 389, 438, 497, 675
Amos, W. B., 573
Anderegg, R. J., 89
Anderson, R. G. W., 697
Anderton, B. H., 578, 588(102)
Andreadis, A., 379
Andres, S., 862
Angelier, N., 582
Angerer, L. M., 857
Anraku, Y., 214, 215(39), 645, 861
Antebi, A., 492, 499(6), 502, 503(6)
Anziano, P. Q., 162, 163(47)
Arai, K., 78, 79, 86(19), 245, 859, 860
Arai, K.-I., 79
Aris, 622, 667(2)
Aris, J. P., 433, 568, 574(53), 600(48, 53), 665, 666(14), 719, 727(9), 736, 737
Armstrong, K. A., 196, 206(8)
Armstrong, K., 568, 600(51)
Arndt, 836, 846
Arndt, E., 467
Arndt, K. T., 251, 386, 853
Arnheim, N., 551
Arnold, E., 690
Arthur, W. L., 838
Asser, U., 517
Astell, C. R., 137, 854

Atcheson, C., 124
Atkinson, K., 838
Atkinson, N. S., 848
Aves, S., 805
Avgerinos, G. C., 491, 494(5), 496, 498(5)
Awano, I., 861
Axelrod, J. D., 551, 552(6)
Aynardi, M. W., 430, 435(19), 453(19), 859
Azpiroz, R., 569, 602(65)
Azubalis, D. A., 306, 307(11), 353
Azum, M., 723

B

Bacallao, R., 645
Baccari, E., 468
Bachrach, H. L., 479
Badaracco, G., 433, 858
Bai, Y., 433
Baier, G., 468
Baim, S. B., 362
Bajwa, W., 348
Baker, A., 602
Baker, B. S., 239
Baker, D., 520, 609
Baker, R. F., 479
Baldari, C., 206, 207(23), 376
Baldwin, A. S., 230, 234(5), 235(5)
Ball, 836, 843
Ball, S. G., 853
Ballesta, J. P. G., 465, 467, 468, 469(61), 472(61), 473(61), 477(61)
Ballou, C. E., 683, 688, 774
Balzi, 835, 836, 839, 843, 846, 850
Balzi, E., 836, 853, 854, 863
Bandlav, W., 844
Bandlow, W., 539, 729, 859
Banerjee, S. K., 860
Bankaitis, V. A., 644
Bankier, A. T., 863
Banks, A. R., 492, 497
Banta, L. M., 431, 644, 646(13)
Baranov, V. I., 543
Baranowski, H., 153
Barbone, F. P., 544
Barclay, B. J., 853
Bard, M., 838
Barlay, 838
Barnes, D. A., 308, 379, 702

Barnes, G., 142
Barnett, J. A., 181
Barr, P. J., 5, 491
Barratt, R., 853
Barrell, B. G., 863
Barry, K., 853
Bartel, B., 317, 508
Basson, 841
Basson, M. E., 242, 853
Basson, M., 249, 568, 595(54), 601(54), 603, 607(5), 609, 614(8), 665, 666(15)
Battey, J. F., 322, 328(17), 329(17)
Bauer, F., 855
Bauer, K., 746
Bauerle, R., 479
Baum, M. P., 252, 260(3)
Baum, P., 838, 853
Bautz, F. A., 729, 730(10)
Beach, D., 206, 207(26), 808, 811, 814(19), 817
Bechet, J., 363
Beck, A. K., 860
Becker, D. M., 92, 93(54), 196
Becker, J. M., 702
Beek, J. G., 465, 466(31)
Beggs, J. D., 281, 291(2), 319, 376
Beier, D. R., 387
Bel, 842
Belhumeur, P., 467
Bell, J. B., 856
Bell, L., 79
Beltrame, M., 467
Ben Aser, E., 853
Bendayan, M., 611, 626(10)
Bender, A., 79, 837, 847
Bender, W., 320
Benjamin, D. C., 234
Benkovic, S. J., 314
Bennett, C., 4, 12(4)
Benton, W., 320
Beran, K., 732
Berg, P. E., 160
Berk, A. J., 481
Berlin, V., 569, 782, 784
Bernard, M., 725
Bernard, S., 491
Bernardi, G., 150
Bernstein, M., 665, 669, 670(12, 30), 674(12), 688
Bernstine, E. G., 860

Berzofsky, J. A., 234
Betlach, M. C., 284, 481
Bettany, A. J. E., 418
Betz, H., 431
Betz, R., 78, 89, 774
Bhattacharjee, J. K., 853
Bianchi, M. E., 467
Bicknell, J. N., 181
Biemann, K., 683, 684(3), 689(3)
Bienz, M., 710
Bilanchone, V. W., 176
Birchmeier, C., 198, 199(15), 214(15)
Birky, C. W., Jr., 154
Birnboim, H. C., 327
Bishop, J. M., 510
Bisson, L. F., 690
Bitter, G. A., 391, 397(11), 493, 497
Bitter, G., 517
Bitter, W., 683
Björklund, B., 570
Bjornstad, K., 757
Black, D. A., 130
Blackburn, E. H., 282
Blair, L. C., 77, 80(1), 84, 104, 79(6)
Blair, L., 135, 250
Blinder, D., 246
Bliss, J. G., 183
Blobel, G., 433, 526, 568, 574(53), 600, 622, 665, 666(14), 667(2), 674, 675, 677(2), 681(2), 682(2), 719, 727(9), 735, 736, 737
Blodel, G., 568, 600(48)
Bloom, B., 233
Bloom, K., 773, 865
Blumer, K. J., 702
Blumer, K., 672
Bobek, L., 538
Böck, A., 465, 466(43), 467
Bock, M. R., 698
Bockus, B., 753
Boeke, J. D., 18, 30, 79, 138, 144, 175, 177(14), 179(14), 239, 241(1), 286, 292, 296(37), 303, 311(5), 318(5), 330, 342, 343, 346, 351(9), 353(2), 355(1), 356, 359(11), 361, 362, 509, 568, 694, 764, 776, 817, 863
Boeke, J., 75, 306, 307(10), 784, 787(12)
Boel, E., 497
Boerner, P., 161
Bogorad, L., 454

Boguslawski, 843
Boguslawski, G., 853
Böhni, P. C., 663, 666(7)
Bohni, P., 669, 670(31)
Boileau, G., 467
Bolivar, F., 231, 284, 481
Bollen, G. H. P. M., 454, 456(3), 457(3)
Bollrath, D., 74
Bolotin-Fukuhara, M., 151
Bond, J. F., 578
Bonnanfant-Jais, M.-L., 582
Bonner, W. M., 541
Booher, R., 811
Boone, C., 840
Borell, 841
Borell, C. W., 853
Borisy, G. G., 594
Bornaes, 837
Bornaes, C., 853
Borst, P., 152, 158, 159(8)
Bostian, K. A., 4, 196, 209(6), 379, 385, 392, 397(13)
Botstein, 842, 852
Botstein, D., 42, 100, 163, 174, 189, 206, 207(21), 214, 215(21, 36), 226, 240 248, 282, 293, 301, 303, 314, 319, 321, 322, 330, 379, 489, 497, 567, 568, 585(57), 586(57), 587(24, 28), 600(10, 15, 17, 24, 25, 26, 28), 601(57), 669, 670(39), 693, 694, 729, 758, 760(14), 835, 837, 840, 841, 844, 847, 862, 863
Boulet, 840, 842
Boulet, A., 853
Bourgeois, H. J., 721, 725(17), 728(17)
Bourne, H. R., 854
Bouteille, C. A., 721, 725(17), 728(17)
Boutelet, 836, 851
Boutelet, F., 7, 854, 855
Bouvier, S., 244
Bowers, B., 732
Boyd, A., 855
Boyde, A., 573
Boyer, H. W., 231, 284, 481
Bozzoni, I., 465, 466(24)
Bradford, M. M., 548
Bradley, G., 429, 431(6), 442(6), 443(6), 450(6)
Brake, A. J., 5, 491, 856
Brake, A., 250, 434, 688
Brakenhoff, G. J., 573

Brammer, W. J., 490
Branch, J. R., 379
Brand, A. H., 294, 339
Brand, R. C., 471
Brandl, C. J., 520, 533(8)
Brandli, A. W., 521
Brandriss, 844
Brandriss, M. C., 854, 863, 864
Brandriss, M.,
Branton, D., 578, 596(98), 598(98)
Braun, R. J., 100
Brearly, I., 73
Breeden, L., 294, 339
Breitbart, M., 153
Breitenbach, 836
Breitenbach, M., 13, 107, 110, 111, 115, 117(34, 35), 121, 122(34), 123, 124(35, 36), 125, 126(51)
Brendel, M., 847
Brennan, M., 189, 321
Brennan, V., 538
Brenner, C., 79, 245, 859
Brennwald, P., 406
Bretscher, A., 567, 600(27), 729
Brewer, B. J., 199, 253
Brill, J., 784
Brisco, 841
Brisco, P. R. G., 854
Briza, P., 115, 117(34, 35), 121, 122(34), 123, 124(35), 125(49, 51), 126(51)
Broach, J. A., 105
Broach, J. R., 7, 9(30), 12(6, 30), 42, 123, 124(55), 169, 175(3), 189, 196, 197, 200(9), 206(8), 207(9), 214(7), 240, 251, 294, 306, 319, 376, 383, 397, 568, 600(51), 736, 811
Broach, J., 105, 108
Brodeur, G. M., 861
Brodsky, G., 243, 283, 302
Broek, D., 92, 508
Broeze, R. J., 491, 494(5), 496, 498(5)
Bröeker, M., 811, 812
Brooks, I. M., 597, 598(145)
Brown, A. J. P., 418
Brown, D. D., 533
Brown, D. T., 594
Brown, J. T., 856
Brown, M. S., 697
Brown, M. T., 842
Brown, N., 356, 360(25)

Brown, R. J., 117
Brown, S. J., 467
Brown, S. S., 580, 584, 600
Brown, S. W., 491
Brown, W. R. A., 252
Brownstein, B. H., 252, 265(5), 266(5), 268(5)
Bruce, A. G., 409
Bruenn, J., 538
Bruhl, K.-H., 347
Brunner, J., 602
Brushi, C., 757
Brusick, D. J., 274
Buchferer, B., 855
Buchman, A. R., 433, 521, 533(9), 545
Buchman, C., 864
Bucking-Throm, E., 79, 86(22), 774
Buckley, C. M., 492, 499(6), 502(6), 503(6)
Budd, 845
Budd, M. E., 862
Budd, M., 303, 318(7), 854
Buhler, J., 303, 305, 318(6, 8)
Buhler, J.-M., 857
Bujard, H., 520
Bukhari, A., 231
Bulawa, C., 123
Buratowski, S., 374
Burd, C. G., 863
Burger, G., 539
Burger, M., 690
Burgers, P. M. J., 263
Burgett, S. G., 378
Burke, D. C., 374, 383(7), 389
Burke, D. T., 251, 252, 253(1), 254(1), 261(1), 263(1), 265(5), 266(5), 268(5)
Burke, D., 767
Burke, J. F., 327
Burke, J., 136, 137(14), 139(14), 145(14), 146(14)
Burke, R. L., 214, 215(37)
Burke, W., 117
Burkholder, A. C., 78, 245, 699, 702(15, 16), 703(15), 776
Burnette, W. N., 581
Burton, K., 98
Busby, S., 315
Busch, H., 720
Bush, H., 720
Bussey, H., 840
Butler, E. T., 523

Butow, R. A., 156, 157, 162, 163(47), 569, 602(65)
Button, L. L., 854
Byers, 841
Byers, B., 95, 113, 115(30), 117(30), 119(30), 120(30), 123, 126(53), 129, 565, 588(4), 603, 615, 718, 727(5), 774, 792(9), 838, 853, 854, 857

C

Cabib, E., 429, 433(12), 435, 732
Cafferkey, R. C., 241
Cahalan, M. D., 570
Cameron, S., 246, 863
Campbell, 845
Campbell, D., 42, 44(12), 854
Campbell, I., 5
Campbell, J. L., 303, 318(7), 433, 434, 568, 608, 719, 720(7), 723(7), 727(7), 854, 862
Cannon, 836, 845, 848, 850
Cannon, J. F., 848, 854
Cantor, C. R., 47, 174, 252, 797
Cantor, C., 58, 59(2), 60(2)
Cao, L., 292, 378
Capieaux, E., 854, 853
Capucci, L., 433
Caravallo, D., 491
Carbon, J., 199, 252, 260(3), 281, 282, 377, 378(20), 379, 397, 766, 865
Carbonetti, W., 331
Cardillo, T. W., 353
Carey, M., 545
Carle, G. F., 7, 9(31), 47, 49, 169, 175, 251, 252(1), 253(1), 254(1), 255, 261(1), 262, 263(1, 26), 854, 857
Carle, G., 58, 59, 60(7), 252, 760, 767
Carlemalm, E., 613
Carlson, H., 849
Carlson, M., 4, 90, 214, 215(36), 240, 319, 379, 431, 568, 599(32), 694, 846, 854, 860
Carpenter, A. T. C., 239
Carr, A. M., 811, 816
Carr, A., 811
Carr, S. A., 89
Carson, 837
Carter, B. L. A., 241

Casadaban, M. J., 339
Casperson, 836, 839
Casperson, G. F., 854
Castagnoli, L., 481
Celenca, J., 846
Celenza, J. L., 568, 599(32), 849, 854, 860
Celenza, J., 854
Ceriotti, G., 741
Cesareni, G., 206, 207(23, 24), 376, 481
Chaleff, 849
Chaleff, D. T., 345, 357(7), 361, 854
Chaleff, D., 246, 251(29)
Chamberlin, M., 523
Chambon, P., 394
Chan, 835
Chan, C. S. M., 177, 179(17)
Chan, C. S., 282
Chan, R. K., 84, 89(42), 90(42), 777, 856
Chan, Y.-L., 465, 466(44), 467, 468
Chanda, P. K., 537
Chang, L. M. S., 433, 434
Chang, M., 567, 600(14)
Chang, T.-H., 568, 600(40), 620, 719, 726(8), 727(8)
Chang, X., 578
Chang, Y.-H., 444, 450(68)
Chapman, J. W., 838, 839
Charnas, L., 274
Charnay, P., 231
Charron, 842, 843
Charron, M. J., 854, 855
Chasman, D. I., 521, 533(9), 545
Chatoo, B. B., 306, 307(11)
Chattoo, B. B., 353, 362
Chelstowska, A., 665, 666(24)
Chemla, Y., 403
Chen, C. Y., 397, 492, 494(11)
Chen, E. Y., 135, 140, 226, 293, 330
Chen, E., 698
Chen, I.-T., 467
Chen, K. K., 497
Chen, L. B., 587, 588(124)
Chen, L., 753
Chen, W., 494, 853, 854
Cheng, J.-F., 252
Cheng, S.-C., 429
Cherbet, G., 243
Cherenson, A., 510
Chevallier, M.-R., 864
Chiang, A., 568

Chikashige, Y., 797
Chirico, W. J., 675
Chisholm, 839
Chisholm, R. L., 581
Chisholm, V. T., 854
Chou, J., 339
Chow, T., 855
Chrebet, G., 283, 302
Christensen, R. B., 858, 860
Christianson, T., 160
Christie, G. E., 478
Christou, P., 163, 164(48)
Chu, F. K., 688
Chu, G., 47, 58, 59(4), 257, 259(22), 356
Chumley, F., 169, 313, 320, 357, 760
Church, G. M., 551, 552
Churchill, M. E. A., 533
Chvatchko, Y., 698, 699, 700, 707(6), 709(6)
Ciejek, E., 777
Cigan, A. M., 521, 523(17)
Ciriacy, M., 347
Cirillo, V. P., 667
Clare, J., 865
Clark, A. C., 244
Clark, A. J., 8
Clark, A., 863
Clark, D. J., 176
Clark, K. L., 83
Clark, M. W., 568, 595(40), 597(40), 600(40), 608, 620, 719, 720(7), 723(7), 726, 727(7, 8, 21)
Clark, W. G., 191, 192(11)
Clark-Adams, 847
Clark-Adams, C. D., 854
Clarke, L., 199, 252, 260(3), 282, 377, 378(20)
Claus, T. E., 253
Claverie, J. M., 467
Clay, F. J., 79, 86(20)
Clemens, M. J., 819
Clements, J. M., 493
Coen, D., 151
Cohen, 843
Cohen, J., 855
Cohen, L. H., 459, 460(17)
Cohen, S. N., 191, 192(11)
Coit, D. G., 491
Colby, D., 408
Coleman, K., 718
Colicelli, J., 198, 199(15), 214(15)

Colliex, C., 613
Collins, F. S., 254, 552
Conde, J., 103, 155, 784, 790, 791(26, 27)
Connelly, C., 7, 18(25), 49, 51, 58, 59, 66(5), 73(5, 8), 74, 75, 144, 169, 174(2), 283, 356, 757, 758(12, 13), 759(12), 762, 768, 864
Connolly, B., 143
Connolly, J. A., 579
Connolly, M., 510
Conquet, F., 468
Conrad, M. N., 736
Consul, S., 282, 301(10)
Contanzo, M., 844
Contopoloulou, C. R., 6
Contopoulou, C. R., 40, 42, 43(11), 241, 827, 841
Contopoulou, R., 40
Cook, P. R., 174
Cooke, H. J., 252
Cooke, H., 254
Cooley, L., 807
Cooper, 838, 852
Cooper, T. G., 699, 854, 856, 863
Cooper, T., 807
Copeland, C., 574, 601(87)
Copple, D., 838
Cornelissen, M. T. E., 459, 460(17)
Cornelissen, M., 481
Coruzzi, G., 153
Costantini, F., 283
Costanzo, M. C., 153, 157, 169, 174(4), 280
Cottarel, G., 757, 758(11), 759(11), 761, 762(11), 764(11), 765(11), 853
Coulson, A. R., 160, 550, 556
Coulson, A., 251
Courchesne, W., 840
Courtney, M., 491
Cox, B. S., 12, 140, 376
Cox, B., 73
Cox, E., 316
Crabb, J. W., 89
Craft, J., 568, 600(50), 718, 719(6), 727(6)
Craig, 848
Craig, E. A., 385, 675, 710, 711, 712, 715(5), 716, 855, 857
Craven, G. R., 471
Crawford, I. P., 478, 479
Crea, R., 231, 481
Croft, J., 753

Crosa, J. H., 481
Cross, 839
Cross, F. R., 855
Cross, F., 5, 77, 776
Cross, S. H., 252
Cross, S., 254
Crothers, D., 529, 532(27)
Crouch, R., 75
Crouzet, 847
Crouzet, M., 855
Crow, M. T., 598
Crowe, J., 669, 670(36)
Crowley, J. C., 864
Cryer, C. R., 169, 174(5)
Cryer, D. R., 151, 218, 359
Culbertson, 847, 848, 851
Culbertson, M. R., 42, 45(13), 153, 274, 276, 279(9), 280(9), 855, 862, 865
Culbertson, M., 841
Cumberledge, S., 766
Cummins, 850
Cummins, C. M., 855
Cunningham, R. S., 854
Cuny, M., 465
Curcio, M. J., 345, 356(4), 357(4)
Cusick, M. E., 600(40), 620, 719, 726(8), 727(8)
Cutler, D., 520

D

D'Ari, L., 93, 249, 568, 595(54), 601(54), 603, 607(5), 609, 614(8), 665, 666(15)
D'Urso, M., 252
Dackowski, W. R., 492
Dagert, M., 335
Dahlberg, J. E., 415
Dalbec, J. M., 431, 434(38)
Daniels, L. B., 438
Daniels, L., 843
Darnell, J. R., Jr., 427
Das, G., 18, 858
Daum, G., 602, 663, 666(7), 669, 670(35)
Davidow, L. S., 492, 499(6), 502(6), 503(6), 544
Davidson, N., 857
Davidson, R. S., 576
Davies, C., 339
Davies, J., 346, 348(12), 378, 379(24), 418

Davis, 837, 846, 847
Davis, A. R., 231
Davis, B. K., 239
Davis, F. C., Jr., 539
Davis, L. G., 322, 328(17), 329(17)
Davis, L. I., 600
Davis, L., 568, 600(16, 56)
Davis, R. W., 58, 59(4), 144, 174, 181, 189, 190, 198, 199, 230, 231(1), 232(1), 233, 234, 235, 237(2), 238(1, 2, 21), 240, 242(7), 252, 253, 254, 257, 259(22), 264(18), 270(18), 282, 283, 284(5), 286, 293, 294(6), 319, 321, 330, 356, 375, 378, 379, 380, 381(31), 382(12), 388, 389, 433, 434, 493, 494(14), 578, 590(16), 856, 863, 864
Davis, R., 47, 51, 58, 66(5), 68, 71(12), 73(5), 118, 119(47), 320, 509, 510(8), 756, 757(6)
Davis, T. N., 433, 855, 858
Davison, J., 481
Dawes, I. W., 107, 108, 111, 124, 125(10, 56)
Dawes, W., 110, 112(3), 124(3), 146
Dawson, D., 766
Day, C. R., 242
Day, P. R., 5,
De Crombrugghe, B., 315
de Deken, M., 363
de Fonbrune, P., 23
de Jonge, 845
de Jonge, P., 858
de los Angeles Santos, 846
de los Angeles Santos, M., 855
de Maro, C., 536
de Regt, V. C. H. F., 467, 469, 470(63, 88), 471(89), 472(61, 63, 89), 473(61, 63, 89), 477(61, 88, 89)
de Zamaroczy, M., 150
Dean-Johnson, M., 279
Deboben, A., 729, 730(10)
DeChiara, T. M., 383, 386, 389
Degnen, G., 316
deJonge, P., 74, 75, 855
Dellaporta, S. L., 320
DeMay, J., 604, 611
Demerec, M., 8
Denell, R. E., 467
Deng, Y., 578
Denis, C. L., 387, 389
Denis, C., 837

Dennell, R. E., 239
Dennis, E. S., 362
Dennis, P. P., 467, 468
Dequin, 838, 851
Dequin, S., 855
Deranous, N., 863
Derynck, R., 397, 492, 494(11)
Desai, R., 863
Deschamps, J., 353
Deshaies, R. J., 385, 669, 670(35), 675
Deshaies, R., 669, 670(31)
Deshies, R., 665, 669(13), 670(13), 672(13), 674(13)
Destree, A. T., 598
Deutsch, J., 151
Dewerchin, M., 343, 346(3)
Diaz, R., 698, 699(9)
Dibner, M. D., 322, 328(17), 329(17)
Dickinson, J. R., 110
Dickson, R. C., 348
DiDomenio, B., 124
Die, J. F., 362
Dietzel, 848
Dietzel, C., 79, 245, 777, 855
Dignard, D., 79
Dihanih, M., 544
DiNardo, S., 243, 864
Dirheimer, G., 862
Distel, B., 431, 446(23)
Dixit, A., 467
Dixon, C. K., 431, 453(26), 644
Dixon, C., 844
Dobberstein, B., 475, 520
Dobson, M. J., 12, 374, 375, 383(7, 10), 387(7), 389
Doctor, J. S., 42, 44(12)
Dodd, J., 305, 318(8)
Doel, S. M., 241
Doi, K. S., 855
Dollard, C., 865
Doly, J., 327
Domdey, H., 162
Donahue, P. N., 663, 667(10)
Donahue, T. F., 153, 276, 279(9), 280(9), 379, 521, 523(17)
Donath, C., 856
Donelson, J. E., 176, 397
Donis-Keller, H., 473
Donlan, J. W., 246
Doolittle, M. M., 42, 44(12)

Dorman, T. E., 492, 499(6), 502(6), 503(6)
Douglas, H. C., 181
Douglas, M. G., 240, 574, 736
Douglas, M., 156, 157(29)
Dowbenko, D., 231, 479
Dowding, A. J., 578, 588(102)
Dower, W. J., 183
Dowling, C. E., 551
Doxiadis, I., 539
Doy, Yoshimura, 837
Drain, 841
Drain, P., 855
Dresser, M. E., 127, 129, 600, 855
Dresser, M., 13
Driscoll, P. M., 863
Drubin, D. G., 248, 489, 565, 567, 570(3), 584, 585(25, 119), 586(25), 587(3, 25), 588(25), 592(25), 593(25), 597(25), 599(3), 600(25, 26), 645, 729, 730, 731(7), 732, 733, 734(11), 735(5)
Drubin, D., 835, 847
Drummond, R., 551
Drutsa, V., 314
Dubin, 843
Dubin, D. T., 538
Dubin, R. A., 855, 860
Dubois, E., 353
Dudl, R. J., 433
Duffus, J. H., 5, 735
Duhamel, R. C., 597
Dujardin, G., 853
Dujon, B., 7, 9(29), 12(29), 149, 150(1), 151, 154(1)
Dulic, V., 644, 646, 699, 708, 709(28)
Dumais, D. R., 491, 492(3), 494(3, 5), 496, 498(5)
Dumont, M. E., 38
Duncan, M. J., 491, 494(2), 495(2), 496(2), 497(2), 498(2), 499(2), 501(2)
Dunn, B., 66
Dunn, E. J., 408
Duntz, W., 774
Duntze, W., 78, 79, 86(22), 89, 774
Durán, A., 734
Durbin, K. J., 351, 361
Durin, D., 847
Durkackz, B., 811, 817
Durnst, R. W., 848
Dürr, M., 665, 666(18)
Dutcher, S., 757

Dutchik, J. E., 861
Dutchik, J., 50

E

East, I. J., 234
Easton Esposito, R., 858, 864
Eberly, S., 863
Eccleshall, R., 169, 174(5), 218, 359
Eckardt-Schupp, 845
Eckardt-Schupp, F., 855
Eckstein, F., 510
Edelman, I., 42, 45(13)
Edman, J. C., 231
Edwards, 845
Edwards, A. M., 544
Edwards, J., 858
Edwards, P. A. W., 597, 598(145)
Egan, K. M., 391, 493
Egerton, M., 646, 855
Ehlers, U., 604
Ehmann, C., 429, 430, 431(4), 433(5, 9), 441, 443(4), 449(17), 450(4), 451(4)
Ehrlich, S. D., 335
Eibel, H., 353
Eichinger, D. J., 30
Einerhand, S. W. C., 467, 469(61), 472(61), 473(61), 477(61)
El-Badry, H. M., 21
El-Baradi, T. T. A. L., 467, 468, 469, 470(63, 88), 471(89), 472(61, 63, 89), 473(61, 63, 85, 89), 477(61, 88, 89)
Elder, R., 115, 124, 864
Elguindi, I., 646
Elion, E. A., 427, 789
Elion, E., 784
Elledge, 846
Elledge, S. J., 388, 578, 856
Elledge, S., 235, 238(21), 293, 330, 509, 510(8)
Ellege, S., 234
Ellinger, A., 115, 117(34), 121(34), 122(34), 123, 125(51), 126(51)
Elliot, S., 492
Ellis, V., 503
Elorza, M. V., 734
Emeis, C. C., 125, 148
Emr, S. D., 240, 431, 601, 644, 646(13)
Emtage, J. S., 375, 383(10)

Emter, O., 429, 430(5), 433(5)
Eng, F., 854
Engebrecht, 843
Engebrecht, J. E., 147, 148(3)
Engebrecht, J., 856
Engelke, D. R., 551, 552, 554, 557(15), 558(15), 559, 561
Enger-Valk, B. E., 490
England, T. E., 409
Enquist, L. W., 316
Ensink, J. W., 433
Entian, 840
Entian, K.-D., 860
Ephrussi, A., 552
Erhart, E., 376
Erickson, A. H., 526
Erickson, K., 860
Erlich, H. A., 551
Erlich, H., 509
Ernst, 835
Ernst, J. F., 38, 365, 507, 856
Errada, P., 293, 330
Errede, B., 136, 137(14), 139(14), 145(14), 146(14), 246, 251(29), 339, 353
Eslava, A., 855
Esmon, B., 669, 670(40)
Esposito, 845
Esposito, M. S., 20, 97, 108, 111, 112, 145, 856
Esposito, M., 757, 838
Esposito, R. E., 13, 20, 97, 99, 101(10), 108, 110, 111, 112, 113, 114, 115, 117(1, 26, 33), 118, 119(32, 43), 120, 124, 142, 145, 421
Esposito, R., 40, 46(8)
Etcheverry, T., 408, 663, 664(8), 665, 666(8, 20), 667(8), 668(20), 669(8), 671(20), 672(20)
Eustice, D. C., 362
Evan, G., 510
Evans, C., 718
Evans, E. A., 674
Evans, R. M., 532, 533(29)

F

Fabian, 847
Fabian, G. R., 856
Fahrney, D., 433

Fakan, S., 726
Falco, S. C., 42, 189, 206, 321
Falcow, S., 481
Faloona, F., 551
Fan, J.-B., 797
Fangman, W. L., 199, 253, 322, 852, 856
Fangman, W., 74
Farabaugh, P. J., 345, 357(7), 379
Farabaugh, P., 865
Farr, A. L., 741
Farr, A., 445
Farrelly, F., 162
Fassler, J. S., 244
Fassler, J., 847
Fasullo, M. T., 283, 856
Fasullo, M., 118, 119(47)
Faulstich, H., 729, 730(10)
Favian, G. R., 848
Faye, G., 853
Fehrenbacher, G., 78, 774
Feinberg, A. P., 266
Feinberg, B., 536
Feldman, F., 152
Feldman, J., 105, 294
Feldman, R. I., 688
Feldman, R., 665, 669(12), 670(12)
Feldmann, H., 176, 413
Felluga, B., 128
Fenimore, C. M., 196, 209(6), 379
Fennel, D. F., 95
Fennell, D. J., 158, 159(35)
Ferguson, B., 74, 856
Ferguson, J., 79, 387, 389, 858
Ferrero, I., 838
Ferro, S., 669, 670(32), 671(32)
Ferro-Novick, S., 669, 670(33, 37), 684, 685(9), 693(9)
Fessl, F., 537
Festin, R., 570
Feuersanger, J. H., 42, 44(12)
Field, C., 669, 670(37, 38), 684, 685(9), 693(9), 698
Field, J., 92, 508, 862
Fields, S., 246
Fieschko, J. C., 493
Fiil, N. P., 497
Filipowicz, W., 537
Fincham, J. R. S., 5, 9(19)
Findeli, A., 491
Fink, 835, 836, 840

Fink, G. J., 840
Fink, G. R., 79, 84, 118, 119(46), 138, 144, 155, 160, 169, 182, 186(2), 206, 207(21), 215(21), 229, 230(49), 239, 241(1), 251, 264, 274, 281, 284, 285(3), 291(3), 292, 296(37), 303, 306, 307(10), 311, 313, 316, 318(5), 320, 330, 343, 345, 346, 351, 353, 357, 361, 362, 379, 386, 393, 394(19), 396(19), 401, 492, 499, 502(6), 503(6, 28), 509, 512, 536, 567, 568, 569, 595(13), 600(13, 56), 694, 760, 764, 776, 782, 783, 784(22), 787(12, 23), 789, 790, 791(22, 26), 792, 817, 836, 845, 853, 856, 857, 863
Fink, G., 73, 103, 319, 321, 337, 841
Finkelstein, D., 156, 157(29)
Finley, D., 459, 461(18), 508
Finnegan, J., 189
Finnsysesn, T., 347, 357(14), 360(14)
Firestone, R. A., 700
Firoozan, M., 197
Fischer, E. P., 431
Fisher, P. A., 581, 736
Fitsgerald-Hayes, M., 377, 252, 766
Fjellstedt, T. A., 306, 307(11), 353
Flanagan, P. M., 544, 546, 548(6)
Flavell, R. A., 158
Fleer, R., 40, 229
Flessel, 842
Flessel, M. C., 856
Floor, E., 247
Floy, K., 75
Flury, U., 152
Flynn, G. C., 698
Fogel, S., 19, 112, 854, 864
Foiani, M., 858
Folley, L. S., 169, 174(4), 280
Ford, S. K., 580, 587(109), 588(109), 589(109), 600
Fordham, M., 573
Forrest, D. B., 860
Forss, S., 231
Foss, M., 93
Foury, F., 155
Fowell, R. R., 13
Fowler, R., 316
Fox, 844
Fox, M. S., 301
Fox, T. D., 153, 155, 157, 160, 161, 169, 174(4), 280, 861

Fox, T., 844
Frackman, S., 115, 124, 864
Fradin, A., 413
Fraenkel, 839, 845
Fraenkel, D. G., 493, 690
Fraenkel, D., 856
Francesconi, S., 858
Frank, M., 47, 50, 262, 263(26), 861
Frank, R., 475
Frank, T., 50, 861
Franze, R., 231
Franzusoff, A., 544, 568, 574(58), 588(58), 596(58), 597(58), 598(58), 601(58), 667, 674
Freedman, L. P., 391, 393(10)
Freedman, R., 245
Freidel, K., 347
Freidfelder, D., 16
Freisen, J. D., 857
French, S., 479
Freund, C., 837
Frevert, J., 688
Frey, J., 430, 431(16), 443(16), 450(16), 451(16)
Fridovich-Keil, J. L., 578
Fried, H. M., 297, 306, 308, 421, 426, 465, 476
Fried, H., 465
Fried, M., 529, 532(27)
Friedberg, 845
Friedberg, E. C., 5, 40, 197, 229, 864
Friesen, J. D., 240, 465
Friesen, J., 846
Friis, J., 317
Fritsch, E. E., 313
Fritsch, E. F., 160, 161(42), 220, 260, 266(24), 286, 332, 334(7), 336(7), 359, 413, 483, 550
Fritz, H. J., 314
Froehlich, K. U., 226, 293
Froehlich, K.-U., 330
Fuccilo, D. A., 596
Fugit, D., 38
Fujimoto, R., 861
Fujimura, 839
Fujimura, H.-A., 856
Fujiwake, H., 594
Fukada, Y., 182, 337
Fukami, K., 182
Fukasawa, T., 214, 215(32), 244, 380

Fukuda, Y., 264, 291, 321, 365, 393, 394(18), 396(18), 503
Fukuhara, H., 152
Fukui, S., 736
Fuller, 596, 598(138)
Fuller, R. S., 434, 688
Fuller, R., 568, 574(58), 588(58), 596(58), 597(58), 598(58), 601(58), 665, 666(17)
Funkuda, Y., 311
Furlong, C., 853
Futcher, A. B., 12, 376
Futcher, B., 860

G

Gaber, 850
Gaber, R. F., 42, 45(13), 838, 856, 864
Gaber, R., 229
Gabrielesen, O. S., 177
Gaisne, M., 863
Galeotti, C. L., 376
Galibert, F., 231
Galiger, A. E., 728
Gallagher, M. J., 467
Gallwitz, D., 669, 670(34), 856, 862
Galwitz, 852
Game, J. C., 100, 853, 855, 864
Game, J. G., 117
Game, J., 845
Gamulin, V., 807
Garabedian, M. J., 397
Garcea, R. L., 568
Garcia, P. D., 521, 527(10), 544, 675, 680(1), 681(1), 682(1)
Gardiner, K., 48
Garfinkel, D. J., 18, 198, 215(17), 343, 345, 353(2), 356(4, 5), 357(4)
Garguilo, G., 467
Garner, M., 529, 532(26)
Garnjobst, 853
Garrett, J., 240
Garvik, B., 74, 247, 757, 837
Gas, N., 723
Gasior, E., 527, 536, 537
Gasser, S., 602
Gaudet, A., 766
Gayathri-Devi, K. R., 468
Geiduschek, E. P., 557
Gelfand, D. H., 551, 552

Gelfand, D., 509
Geller, B. L., 240
Gelugne, 842
Gelugne, J.-P., 856
Genbauffe, 838
Genbauffe, F. S., 856
George, J. P., 106
George-Nacimento, C., 453
Gerace, L., 601
Gerring, S. L., 7, 18(25), 49, 51(25), 64, 71(9), 72(9), 75, 169, 174(2), 762
Gerring, S., 757, 758(13)
Gervais, M., 231
Gesteland, R. F., 861
Gethman, R. C., 239
Ghiara, P., 376
Ghild, G. M., 538
Ghosh, B. K., 623
Gibbs, J. B., 854
Gibson, 843
Gibson, S. I., 856
Gietz, R. D., 206, 207(25)
Gifford, P., 380, 381(32), 382(12), 389
Giguere, V., 532, 533(29)
Gilbert, M., 568, 608, 719, 720(7), 723(7), 727(7)
Gilbert, W., 160, 162(24), 550, 551, 552
Gilmore, R. A., 365
Giloh, H., 576
Giniger, E., 380, 552
Giroux, C. N., 127, 129, 273, 600, 855
Glazer, A. N., 570
Gloeckler, R., 855
Godnay, A., 568, 599(34)
Godowski, P. J., 393, 397(15)
Godsen, 838
Godsen, M. H., 856
Goebels, M., 330, 342(5)
Goebl, M. G., 836
Goeddel, D. V., 231, 377, 378(19), 383, 492
Goegl, M. G., 836
Goetsch, L., 95, 123, 126(53), 129, 565, 588(4), 603, 615, 774, 792(9), 854
Goff, C. G., 499
Goff, S. A., 362
Goffeau, 845
Goffeau, A., 853, 854, 862, 863
Goffeau, E., 853
Gold, A. M., 433
Goldberg, A. L., 231
Goldenthal, M. J., 855
Goldman, B. S., 857
Goldman, R. D., 572, 581
Goldring, E. S., 151
Goldstein, J. L., 697
Goldstein, L. S. B., 582
Golin, J., 843
Golombek, J., 111
Gonzales, R., 751
Goodey, A. R., 241
Goodman, D., 93
Goodman, H. M., 231
Goodman, J. M., 663, 667(10)
Goodwin, D., 576
Gordon, J., 511, 512(15), 581
Gorenstein, C. G., 457, 459(6)
Gorman, J. A., 214, 215(33)
Goto, T., 100, 303
Gottlieb, S., 114, 117(33), 118, 119(43)
Gottlin-Ninfa, 841, 846
Gottlin-Ninfa, E., 856
Goud, B., 568, 601(59), 695, 696(31)
Gould-Somero, M., 239
Gounon, P., 582
Gouravleva, G., 858
Gourse, R. L., 468
Gown, A. M., 587, 588(125)
Graham, M. Y., 861
Graham, M., 50
Grandori, R., 568, 599(33)
Granot, D., 102, 106(29), 107(29), 108(29), 113
Grant, C. M., 197
Grant, F. J., 79
Grant, P. G., 101
Gravius, T. C., 499
Graziani, F., 467
Greaves, M., 517
Green, K. J., 581
Green, M. R., 520
Greenberg, 839
Greenberg, M. L., 856, 862
Greene, P. J., 284, 481
Greenhouse, J. J., 345
Greenleaf, 847
Greenleaf, A. L., 857
Greenwalt, D. E., 588, 596(126)
Greer, C., 729
Greer, H., 856, 862, 863
Greer, R. P., 860

Gregory, R. J., 473
Griffin, J., 492
Griffiths, E., 700
Griffiths, G., 625, 698
Grimm, C., 816, 817(32)
Grindley, N. D. F., 346
Grisafi, P. L., 789
Grisafi, P., 163, 314, 379, 497
Gritz, L., 378, 379(24)
Grivell, L. A., 12, 150
Grodeur, G., 50
Groner, Y., 426, 538
Gross, C. H., 489
Gross, S. R., 279
Grosskinsky, C., 233
Grossman, L. I., 151, 162
Grossman, L., 5
Groudinsky, O., 853
Groves, W. E., 539
Grubman, M. J., 479
Gruhl, H., 413
Grunstein, M., 320, 385, 567, 600(14)
Guarente, 840
Guarente, L. P., 388
Guarente, L., 5, 92, 93(54), 196, 365, 373, 374, 375, 379, 380, 381(3, 32), 382(12), 383, 388, 389, 433, 716, 861
Guba, R., 154
Gudenus, 846
Gudenus, R., 857
Guerra-Santos, L., 491
Guiard, B., 838, 863
Gull, K., 347, 357(14), 360(14)
Gunsalus, R. P., 535
Gurd, F. R. N., 234
Gustafsson, J. Å., 393
Guthrie, B. A., 567, 600(18)
Guthrie, C., 408, 409, 410(5), 413(5), 415
Gutz, H., 125, 148, 796, 798(3), 801
Guyer, M. S., 332, 334(11)
Gygax, A., 796

H

Haaijman, J. J., 572
Haarer, B. K., 489, 565, 567, 568, 570(3), 579(21, 29), 580(29), 581(29), 587(3), 599(3), 600, 645, 729, 730, 732, 733, 734, 735(5)
Haarer, B., 718
Haber, 835, 838
Haber, J. E., 106, 111, 137, 140, 141(27), 142(27), 143, 282, 293(7), 402, 859
Hadwiger, J. A., 250
Haeuptle, M. T., 475
Hagan, I. M., 822
Hagan, I., 813
Hagan, M. I., 608
Hagen, D. C., 78, 86, 89(45)
Hagen, D., 135
Hagendoorn, M. J. M., 465
Hagie, F. E., 377, 378(19), 397, 492, 494(11)
Hahn, S., 374, 379, 388, 433
Hahn, W., 230, 234(3)
Hahnenberger, K. M., 252, 260(3)
Haid, A., 154, 214, 215(31)
Hall, B. D., 113, 137, 255, 262(20), 377, 378(19), 387, 416, 554
Hall, J. C., 239
Hall, M. N., 313, 388, 476, 568
Hallewell, R. A., 231
Halvorson, H. O., 111, 125, 148
Halvorson, H., 430, 431(15)
Hamada, F., 383, 389
Hamatake, R., 863
Hamilton, B., 537
Hampsey, D. M., 365
Han, M., 385, 567, 600(14)
Hanahan, D., 335, 485
Hanes, 838
Hanes, S. D., 4, 857
Hanic-Joyce, 836
Hanic-Joyce, P. J., 857
Hannig, E. M., 538
Hannum, C., 234
Hansen, M. T., 497
Hansen, R. J., 431
Hansen, W. J., 665, 666(20), 668(20), 671(20), 672(20)
Hansen, W., 240, 242(7), 378, 521, 527(10), 544, 675, 680(1), 681(1), 682(1)
Harashima, 839
Harashima, S., 137, 857
Hard, K., 683
Hardeman, E., 240, 242(7), 378
Hardeman, K. J., 860
Harder, W., 665, 666(25), 667(25)
Hardie, I. D., 111, 124(10), 125(10), 146
Hardin, J., 568, 600(50), 718, 719(6), 727(6)

Hardy, R. R., 570, 598(72,74)
Hardy, R. W., 239
Hardy, S. J. S., 455
Harlow, E., 511
Harris, H., 440
Harrison, G. I., 183
Harrison, J. S., 5, 695
Hartig, A., 111, 125
Hartl, F.-U., 520
Hartley, J. L., 176, 397
Hartman, P. E., 8
Hartwell, L. H., 5, 77, 78, 79, 83, 86(22), 87(24), 100, 245, 645, 699, 702(15, 16), 703(15), 774, 776, 857
Hartwell, L., 74, 247, 678, 756, 757, 837, 842
Hasek, J., 567, 600(11,12)
Haselbeck, A., 688, 693
Hashimoto, H., 182, 214, 215(32), 244
Hasilik, A., 438
Haslik, A., 678
Hasson, M. S., 678, 861
Hasson, T. B., 678, 861
Hata, T., 433, 448
Hatakeyama, T., 467
Haugland, R. P., 570
Haut, S. R., 854, 855
Hawlitschek, G., 520
Hawthorne, 843
Hawthorne, C., 105
Hawthorne, D. C., 317, 857, 860
Hawthorne, D., 678
Hay, R., 602
Hayakawa, K., 570, 598(72)
Hayashi, 836
Hayashi, M., 857
Hayashi, R., 433, 448
Hayashibe, M., 732
Hayat, M. A., 588, 604, 612, 625(12), 626
Hayden, J. H., 594
Hayes, J., 805
Hayflick, J. S., 397, 492, 494(11)
Hayles, J., 811, 813, 816, 817
Haynes, R. H., 206, 215(28), 856
Haywood, L. J., 176
Hazlett, M., 860
Heberlein, U. A., 491
Heckman, J. E., 467
Hedin, L. O., 169, 174(4), 280
Heerikhuizen, H. v., 495

Heffron, F., 18, 133, 135, 140, 141(26), 142(26), 143, 226, 293, 330, 332, 334(10), 339
Hegemann, J., 75, 757, 758(11), 759(11), 761, 762(11), 764(11), 765(11), 768, 770(28)
Heinemann, J. A., 188, 189, 191(4), 193(4)
Heinisch, 844
Heinisch, J., 857
Heiter, P., 356
Hejmova, L., 690
Held, W., 538
Heller, G., 465, 466(43)
Helms, C., 50, 861
Hemmings, B. A., 438
Hemmings, B., 678
Hennessy, K., 584, 596(120), 597(120)
Henríquez, R., 737
Henry, S. A., 153, 276, 279, 280(9)
Henze, W. J., 698
Herbert, C. J., 855
Hereford, L. M., 79, 245
Hereford, L., 313, 388, 568
Hereford, M., 476
Herken, R., 604
Hermodson, M., 379
Hernandez-Verdun, D., 721, 725(17), 728(17)
Herrera, F., 537
Herrick, D., 416
Herrmann, B., 604
Herskowitz, 837, 846
Herskowitz, I., 5, 10, 77, 78, 80, 84, 86, 89(40), 90(40), 91(14), 92, 101, 102, 103(21), 104, 105, 106, 110(26), 114, 132, 133(3), 134(1, 2), 135, 136, 137(2, 4, 14), 138(1), 139(14), 145(3, 4, 14), 146(4, 14), 192, 214, 215(41), 229(41), 283, 313, 388, 497, 532, 568, 777, 862
Herzberg, M., 403
Herzenberg, L. A., 570, 598(72,74)
Heslot, H., 796, 798(3), 801
Hess, S., 848
Hessler, A., 239
Heyer, W.-D., 206, 207(26), 807, 811, 814(15)
Heym, G., 447
Heynecker, H. L., 284, 481
Heyneker, H. L., 231
Heyting, C., 152, 159(8)

Hibbs, A. R., 669, 670(36)
Hicke, L., 520
Hicks, J. B., 18, 20(51), 41, 53(10), 78, 81, 86, 87(35), 91(14), 103, 105, 132, 133, 134, 135, 145(3, 4), 146(4), 160, 182, 186(2), 189, 197, 200(9), 207(9), 264, 281, 284, 285(3), 291(3), 294, 306, 311, 319, 320, 339, 379, 393, 394(19), 396(19), 401, 499, 503(28), 512, 811, 857
Hicks, J., 5, 101, 103(21), 104(21), 105, 337
Hieter, P. A., 453
Hieter, P., 7, 18(25), 49, 51, 58, 59, 64, 66(5), 68, 71(9, 12), 72(9), 73(5, 8), 74, 75, 144, 169, 174(2), 206, 207(22), 214, 215(38), 283, 305, 308, 312, 316(9), 317, 509, 510(9), 756, 757, 758, 759(11, 12), 761, 762, 764(8, 11), 765(11), 768, 770(17, 28), 864
Higa, 326
Higgins, D. R., 196, 226(3)
Higo, K., 459, 460(9, 10), 462(9, 10), 463(9), 465, 466(26)
Higuchi, R. G., 551
Higuchi, R., 509
Hilger, F., 7, 841, 854, 863
Hill, A., 773
Hill, D. E., 362
Hill, J. E., 240
Hill, J., 284
Hilliker, S., 383, 386, 389
Himmelfarb, 847
Himmelfarb, H. J., 857
Hinnebusch, 835, 836, 839, 840
Hinnebusch, A. G., 857
Hinnebusch, A., 74, 857
Hinnen, A., 182, 186(2), 281, 285(3), 291(3), 337, 386
Hinze, H., 431
Hirano, M., 105, 108(39), 110(39)
Hiraoka, E., 865
Hiraoka, Y., 594
Hirose, S., 446
Hirose, T., 231
Hirsch, 835, 840
Hirsch, H. H., 644, 857
Hishinuma, F., 862
Hiti, A. L., 231
Hitzeman, R. A., 377, 378(19), 383, 397, 492

Hitzeman, R., 492, 494(11)
Hiusman, O., 226
Ho, K. H., 111, 112(9), 123(9)
Ho, Y.-S., 232
Hoekstra, M. F., 18, 140, 141(26), 142(26), 339, 860
Hoener, P. A., 552
Hofbauer, R., 537
Hoffman, C. S., 808, 811
Hoffman, C., 322
Hoffman, W., 306, 309
Hoffmann, W., 297
Hoffschulte, H., 433, 434(41)
Hoflack, B., 698
Hofschneider, P. H., 231
Hogness, D. S., 320
Hogness, D., 320
Holan, Z., 732
Holborow, E. J., 576
Holcomb, C. L., 665, 666(20), 668(20), 671(20), 672(20)
Holcomb, C., 663, 664(8), 666(8), 667(8), 669(8)
Holland, J. P., 387, 392, 493, 494(12)
Holland, M. J., 387, 392, 493, 494(12)
Hollenberg, C. P., 376
Hollenberg, S. M., 532, 533(29)
Hollingsworth, 841
Hollingsworth, N. M., 113, 115(30), 117(30), 119(30), 120(30), 857
Holloman, W. K., 856
Holly, J., 78, 706
Holm, C., 7, 100, 303, 322, 758, 760(14)
Holm, K., 406
Holmberg, S., 853
Holzer, H., 430, 431, 433, 434(41)
Honigberg, S. M., 120
Hope, I. A., 520, 521, 527(4), 528(3), 529(4), 532(3), 533(3, 4, 13), 534(14), 535(14)
Hopkins, N. H., 4
Hopper, 847
Hopper, A. K., 105, 113, 848, 856, 857
Hopper, J. E., 348
Hopper, J. H., 850
Hopper, J., 839
Horio, T., 594
Horn, G. T., 551
Horn, G., 509
Horowitz, H., 478
Horz, W., 386

Hotani, H., 594
Hottinger, H., 799, 807
Hougan, L., 79, 864
Houghten, R., 510
Howald, I., 644, 698, 699(6), 707(6), 709(6)
Howell, W. M., 130, 720
Hoyt, A., 317
Hoyt, M. A., 226, 293, 330
Hsiao, C., 379
Hsu, T. C., 720
Hsu, Y., 379
HsuChen, C.-C., 538
Huang, C.-J., 332, 334(10)
Huang, T., 853
Hubbell, H. R., 720
Huberman, J. A., 174, 736
Hudspeth, M. E. S., 162
Huffaker, T. C., 567, 600(15), 685, 690
Hughes, S. H., 345
Huibregtse, J. M., 551, 552, 557(15), 558(15), 559, 561
Huie, P., 723
Huisman, O., 293, 330
Hunt, 841
Hunt, E. C., 736
Hunt, T., 526
Hunter, C. P., 498, 644
Hunter, C., 438
Hurt, D. J., 857
Hurt, E. C., 568, 590(49, 52), 719, 727(10)
Hurwitz, M., 863
Hussain, I., 537, 540(17), 541(17)
Hütter, R., 860
Huynh, T. V., 233
Hyams, J. S., 608, 822
Hyman, R. W., 254, 264(18), 270(18)
Hynes, R. O., 598

Innis, M. A., 438, 452(54, 59), 552
Inoué, S., 594
Inoue, S., 750
Insley, M. Y., 78, 706
Irani, M., 315
Isenerg, I., 736
Ish-Horowicz, D., 327
Ishiguro, 836
Ishiguro, J., 857, 861
Ishihara, K., 861
Ishii, S.-I., 446
Ishikawa, K., 467
Ishikawa, T., 100, 107(18), 108(18), 182, 568, 859
Ishimo, Y., 861
Ishino-Arao, Y., 861
Ismail, S., 105
Itakura, K., 231
Ito, H., 182, 264, 291, 311, 321, 337, 365, 393, 394(18), 396(18), 446, 503
Itoh, H., 860
Itoh, T., 459, 460(9, 15), 462(9), 463(9), 465, 466(15, 26), 467, 596
Iturriaga, E., 855
Ivanyi, J., 233
Ivy, 846, 847
Ivy, J. M., 133, 857

J

Jackson, C., 5, 77, 776
Jackson, D. A., 174
Jackson, J. A., 118, 119(46)
Jackson, M., 365
Jackson, R. J., 526, 536
Jacobs, C. W., 567, 600(8), 857
Jacobs, C., 718
Jacobs, E., 343, 346(3)
Jacobs, P. A., 239
Jacobs-Lornea, M., 467
Jacobsen, H., 533
Jacobsen, K., 712
Jacobson, A., 416, 422
Jaeger, D., 75
Jaehning, J. A., 427, 736
Jäger, D., 177
Jagger, J., 279
Jahng, K.-Y., 79
James, G. T., 434

I

Ibrahimi, I., 520
Ide, G. H., 736
Iida, H., 841
Imanaka, T., 601, 602
Inchauspe, G., 723
Inge-Vechtomov, 846
Inge-Vechtomov, S. G., 857
Ingolia, 852
Ingolia, T. D., 378, 857

Jansen, A. E. M., 459, 460(12), 463(12)
Jansen, H. W., 314
Januska, A., 837
Jasin, M., 297, 306, 308, 465
Jayaram, M., 383
Jeffcoate, S. L., 433
Jelinek, W., 427
Jemtland, R., 177
Jenness, 836, 839, 846, 849
Jenness, D. D., 78, 245, 246, 699, 702(15, 16), 703(15), 707(17), 776, 857
Jensen, L., 114, 119(32)
Jensen, R. E., 10, 80, 102
Jensen, R., 132, 133(4), 135, 136, 137(4, 14), 139(14), 145(4, 14), 146(4, 14), 784
Jerome, J. F., 427, 736
Jett, J. H., 570
Jimenez, A., 346, 348(12), 418
Jimenez, J., 812
Jimerez, A., 101
Joachimiak, A. J., 535
Johnson, 836, 837, 849
Johnson, A. D., 532
Johnson, D. A., 597
Johnson, D. I., 857
Johnson, G. D., 576
Johnson, J. R., 6
Johnson, J. S., 489
Johnson, L. M., 433, 434, 644
Johnson, L., 753
Johnson, M. T., 274
Johnson, R. S., 683, 684(3), 689(3)
Johnson, T. K., 467
Johnston, G., 837
Johnston, J. R., 32
Johnston, L. H., 95, 838, 839
Johnston, M., 5, 190, 243, 375, 380, 381(28), 382(12), 389, 493, 494(14)
Johnston, S. A., 162, 163(47)
Jones, 840
Jones, D. G. L., 857
Jones, E. W., 4, 12(6), 393, 429, 430, 431, 433(8, 10), 435(1, 10, 19), 436, 437(8), 438, 439(54), 442, 443(67), 446(23), 447(10), 448(8), 452, 453, 489, 529, 565, 570(3), 587(3), 599(3), 644, 645, 646, 699, 730, 732, 733(5), 735(5), 859
Jones, E., 678, 837, 843, 844
Jones, M., 493
Jones, N. C., 396, 811
Jones, R. H., 396, 811
Jones, R., 467
Jong, A. Y. S., 433
Jong, A. Y.-S., 608, 719, 720(7), 723(7), 727(7)
Joyce, C. M., 346
Ju, Q., 235
Juan-Bidales, F., 468
Julius, D., 250, 512, 674, 684
Jund, R., 864
Juni, E., 447
Jurtz, C., 231

K

Kaback, 841, 846
Kaback, D. B., 855, 857, 858, 860
Kaback, D., 74, 841, 845, 864
Kaback, K. D., 856
Kahn, L., 617
Kahn, R., 835
Kaibuchi, K., 79, 245, 859
Kaiser, C. A., 497
Kaisho, Y., 506
Kaiuchi, K., 860
Kakkar, V. V., 503
Kalb, V. F., 242
Kalchhauser, H., 121, 123(49), 125(49)
Kalnins, V. I., 579
Kaltschmidt, E., 454, 457(1)
Kamerling, J. P., 683
Kane, P. M., 644, 645
Kane, P., 568, 601(60)
Kaneko, K., 835
Kans, J. A., 6, 40, 42, 43(11), 241, 827
Kapakos, J. G., 865
Kaptein, A., 74, 75, 855
Karin, M., 864
Karpova, 846
Karpova, T. S., 857
Karpova, T., 858
Karube, I., 182
Kasimos, J. M., 117
Kassavetis, G. A., 557
Kassir, 836
Kassir, Y., 83, 93(36), 95, 97(3), 100(3), 101, 102, 103(25), 104, 105, 106(22, 25, 29), 107(29), 108(29), 109(3), 110(29), 111, 113, 781, 858

Kaster, K. R., 378
Kataoka, T., 108
Kato, H., 182
Kato, K., 859
Katohda, S., 732
Katz, 849
Katz, M. E., 858
Katz, W. S., 567, 600(19)
Kaudewitz, F., 214, 215(31), 539
Kaufer, A. F., 306, 308
Kaüfer, N. F., 297, 465, 818
Kawaguchi, H., 105, 108(39), 110(39)
Kawanabe, Y., 774
Kawasaki, G., 492, 494(10)
Kay, L. M., 433
Kay, M. A., 467
Kayne, P., 385
Kazazian, J. J., Jr., *551
Kaziro, Y., 79, 245, 859, 860
Kazuo, N., 506
Kearsey, 845
Kearsey, S. E., 858
Keefer, D. A., 612
Keidler, S. H., 356, 360(25)
Keierleber, C., 536
Keil, R. L., 311
Kelleher, R. J., III, 544
Kellenberger, J., 613
Keller, R. A., 570
Keller, W., 231
Kelley, M., 510
Kelley, R. L., 535
Kelley-Geraghty, D. C., 853
Kelly, J. L., 857
Kelly, R. B., 584, 585(119)
Kelsay, K., 79, 86(20)
Kennedy, 847, 858
Kennedy, A., 863
Kennedy, B. K., 838
Kent, J., 756, 757(7)
Kepes, F., 669, 670(30), 684
Kermorgant, M., 853
Kidd, S., 510
Kif, J., 251
Kikuchi, 840, 848, 849, 850
Kikuchi, A., 858
Kikuchi, Y., 214, 215(32), 840, 848, 849, 858
Kilmartin, J. V., 477, 567, 573(7), 587(7), 588(7), 598(7), 600(7, 103), 608, 718, 729, 730(1)

Kim, C. H., 416, 425
Kim, H. B., 489, 567, 579(21), 584, 600
Kim, J., 840
Kim, K. E., 860
Kim, U., 385
Kim, U.-J., 567, 600(14)
Kimmerly, W. J., 241
Kimura, A., 182, 264, 291, 311, 321, 337, 365, 393, 394(18), 396(18), 503
Kimura, I., 467
Kimura, M., 467
King, R. M., 374, 383(7), 387(7), 389
Kingsman, A. J., 347, 357(14), 360(14), 374, 375, 383(7, 10), 387(7), 389
Kingsman, S. M., 347, 357(14), 360(14), 374, 375, 383(7, 10), 387(7), 389
Kinnaird, J. H., 108
Kirkman, C., 246
Kirsch, D. R., 242
Kirschner, M., 577, 595(93), 596(93), 598(93)
Klapholz, S., 40, 46(8), 99, 100(10), 101(10), 110, 112(1), 113, 114, 115(24, 26), 117(1, 26, 28), 118(1, 28, 32), 119(32), 120, 124(1), 254, 264(18), 270(18), 858
Klar, A. J. S., 103, 105, 133, 134, 135, 140, 142(24), 320, 339, 857
Klar, A. J., 294
Klar, A., 105, 430, 431(15), 781
Klco, S., 797
Kleckner, N., 226, 292, 293, 303, 330, 378
Kleid, D. G., 231, 479
Klein, H. L., 286, 357
Klein, T. M., 162
Klenow, H., 533
Klionsky, D. J., 431, 601, 644, 646(13)
Klis, F., 568, 599(35), 688
Klootwijk, J., 468, 471, 495
Knapp, G., 855
Kobayashi, T., 596
Koch, B. D., 385
Koch, D. B., 675
Koerner, T. J., 240
Koh, T. J., 864
Kohalmi, S. E., 273
Kohara, Y., 251
Kohlaw, G. B., 854
Kohlhaw, G., 379
Kohli, J., 797, 799, 807, 814(15), 816, 817(32)
Kohno, T., 499

Kohr, W. J., 492
Kohr, W. T., 383
Köhrer, K., 162
Koji, Y., 506
Kojima, H., 858
Kolb, J. M., 305, 318(8)
Kolman, 836, 847
Kolman, C. J., 858
Kolodkin, A. L., 140, 141(25), 142(24, 25)
Kolodny, M. R., 430, 435(19), 453(19), 859
Kolodziej, P., 508, 517, 846
Koltin, Y., 214, 215(33)
Kominami, E., 433, 434(41)
Kominami, K., 105, 108(39), 110(39)
Konopka, J. B., 5, 77
Konopka, J. H., 776
Köpke, A. K. E., 468
Koren, R., 4
Kornberg, R. D., 433, 521, 533(9), 544, 545, 546, 548(6)
Kornfeld, S., 698
Korsmeyer, S. J., 252, 265(5), 266(5), 268(5)
Korte, A., 858
Kosaka, K., 596
Koshland, D., 74, 317, 599, 600(148, 149), 734, 752, 756, 757(7), 758(8), 764
Kosic-Smithers, J., 855
Koski, R. A., 493
Kostriken, R., 133
Kotyk, A., 690
Kotylak, Z., 152
Kouprina, 837
Kouprina, N. Y., 858
Kouprina, N., 858
Koval, M., 599, 600(148), 734, 752
Kowalisyn, J., 115, 864
Kowallik, K. V., 347
Kozak, M., 374, 521, 523(16), 524(16)
Kozloff, E. N., 728
Kraakman, P., 459, 460(13), 463(13)
Kraig, E., 402
Kramer, B., 314
Kramer, J., 855
Kramer, R. A., 181, 383, 386, 389, 392, 397(13)
Kramer, W., 314
Krazewski, A., 231
Kreader, C. A., 467
Krebs, E. G., 863
Kreger-van-Rij, N. J. W., 123
Kreike, 842

Kreike, J., 858
Krieg, P. A., 520, 521, 522, 524(15), 675, 676(12)
Krieger, M., 433
Kruger, W., 214, 215(41), 229(41)
Kruiswijk, T., 457, 459(8)
Krupnick, D., 151
Kuang, W. J., 698
Kubisz, P., 582
Kumazaki, T., 467
Kunau, W. H., 665, 666(25), 667(25)
Kunes, S., 301, 693
Kung, H. H., 537
Kunisawa, T., 858
Kunkel, T. A., 313, 362
Kunkel, T., 510
Küntzel, H., 568, 599(34)
Kunz, B. A., 273
Kupiec, M., 858
Kupper, H., 231
Kuranda, M. J., 683, 684(6), 691(6), 694
Kurjan, 848
Kurjan, J., 78, 79, 86, 245, 497, 777, 855
Kurland, C. G., 455
Kuroiwa, T., 858
Kushner, P. J., 104
Kusumi, T., 137
Kuwana, Y., 467
Kuwano, Y., 467

L

Laas, W., 48
Labella, T., 252
Labieniec, L., 392
Lacaná, E., 568, 599(33)
Lacatena, R. M., 481
Lachkovics, E., 110
Lacks, S. A., 281
Lacroute, 852
Lacroute, F., 12, 144, 292, 296(37), 306, 307(10), 346, 362, 416, 811, 817, 858
Laemmli, U. K., 157, 486, 487(18), 511, 516(18), 672
Laemmli, V., 541
Lagosky, P. A., 206, 215(28)
Lai, M. H., 242
Lai, P.-H., 497
Laionov, V., 858
Lake, J., 617

Lalanne, J.-L., 465, 466(45)
Lam, K. B., 858
Lambie, E. J., 13, 111, 124(17), 125(17), 147, 568, 600(50), 718, 719(6), 727(6)
Lambris, J. D., 544
Landonio, L., 761
Lane, D., 511
Lang, B., 539
Langevin, G. L., 568
Langone, J. J., 578
Lanier, L. L., 570, 598(75)
Larimer, F., 845
Larionov, 835, 846
Larionov, V. L., 858
Larionov, V., 837, 846
Larkin, J. C., 459, 461(21), 466(21)
Laskey, R. A., 541
Last, R. L., 568
Lau, E. P., 492
Lausten, O., 32
Lavergne, J.-P., 468
Lawrence, 846
Lawrence, C. W., 153, 154(15), 783, 784(22), 791(22), 858, 860
Lawrence, C., 18, 845
Lawyer, F. C., 551
Laymon, R. A., 582
Laz, T., 493
Lazarow, P. B., 601, 602
Lazarow, P., 569, 602(62)
le Moullec, J.-M., 465, 466(45)
Lea, H. Z., 854
Leach, S. J., 234
Leatherwood, J., 545
LeBowitz, J. H., 230, 234(5), 235(5)
Lechner, K., 465, 466(43)
Lederberg, E. M., 18
Lederberg, J., 18, 188
LeDizet, M., 578
Lee, C.-H., 332, 334(10)
Lee, F.-J. S., 835
Lee, G. S. F., 273
Lee, J. C., 457
Lee, S., 757, 758(12), 759(12), 762(12)
Leer, R. J., 459, 460(11, 13, 17), 463(13), 465
Lehle, L., 682, 683(1)
Lehman, I. R., 857
Leibman, S. W., 853
Leibowitz, M. J., 537, 538, 540(17), 541(17), 544

Lemire, J. M., 385
Lemmon, S., 837
Lemoine, Y., 491
Lemontt, J. F., 492, 860
Lenard, J., 700
Lenney, J. F., 431, 433(36), 434(36, 38)
Lenney, J., 444
Lerner, R., 508, 510
Leser, G., 726
Lester, H. E., 279
Letnansky, K., 736
Leung, D. W., 383, 492
Leupold, U., 796, 798(3), 801, 857
Levine, H. L., 377, 378(19), 492
Levine, L. H., 383
LeVitre, J., 492, 499(6), 502(6), 503(6)
Levy, A., 465
Lewin, A. S., 154
Lewin, A., 160
Lewis, G., 510
Lewis, R. S., 570
Lewis, W. H. P., 440
Li, J., 311
Li, Y., 42, 383
Liang, C., 303
Liao, S.-M., 517
Liao, X., 406
Liebman, S. W., 12, 356, 360
Liebman, S., 197
Liebowitz, M. J., 92
Liljelund, 852
Liljelund, P., 858
Lillie, S. H., 581, 600
Lillie, S., 718
Lilly, M., 214, 215(34)
Lin, A., 467, 468
Lin, L.-W., 835
Lin, R.-J., 429
Lin, Y.-H., 857
Lindegren, C. C., 6, 28
Linder, P., 850
Lindquist, S., 710
Lindsley, D. L., 239
Link, A. J., 49
Linkens, M. N. K., 174
Linnane, A. W., 663, 666(6)
Linnemans, W. A. M., 573
Lipchitz, L. R., 860
Lipke, J. D., 79
Lipke, P. N., 78, 774
Lischwe, M. A., 720

Lisziewicz, J., 568, 599(34)
Litherland, S. A., 856
Little, R. D., 252, 265(5), 266(5), 268(5)
Little, S. H., 580
Littlewood, B. S., 276
Littlewood, R., 73
Liu, H., 567, 600(27), 729
Lobo, 844
Lobo, Z., 858
Lochmüller, H., 176
Locke, M., 723
Lodi, T., 838
Lodish, H. F., 521
Lohr, D., 736, 749(5)
Loison, G., 491, 496
Loken, L. L., 570, 598(75)
Loper, J. C., 242
Lopes, T. S., 495
Loprieno, N., 796, 798(3), 801
Lorincz, A., 322
Losson, R., 416, 811
Lott, J. B., 465, 466(46)
Lotz, S., 842
Loughney, K., 255, 262(20)
Louie, K. A., 468
Louise, A., 231
Lowe, D. A., 375, 383(10)
Lowry, O. H., 741
Lowry, O., 445
Lucchini, 844, 845
Lucchini, G., 858, 859
Lucero, M., 465, 466(45)
Lucioli, A., 465, 466(24)
Lucocq, J., 699
Ludmerer, S. W., 859
Ludmeyer, 840
Lue, N. F., 433, 521, 533(9), 544, 545, 546, 548(6)
Luft, J., 625
Lukens, H. B., 663, 666(6)
Lund, P. M., 12
Lundblad, V., 758
Lusnak, K., 854
Lustig, A. J., 620, 719, 726(8), 727(8)
Luzzatto, L., 255

M

Ma, C., 57, 859
Ma, H., 301, 693
McCabe, D. E., 163, 164(48)
McCaffrey, G., 78, 79, 86(20)
McCammon, M. T., 276
McCarty, G., 127, 129(64)
MacCollin, M., 50, 861
McCollum, D., 569, 602(63)
McConaughy, B. L., 198, 214(14)
McConnell, S., 569, 602(64)
McCormick, M., 768
McCready, S. M., 347, 357(14)
McCusker, 835, 838, 841, 845, 847
McCusker, J. H., 859
MacDonald, B., 508
McDonald, K., 607
McDowall, A., 568, 590(49), 719, 727(10), 736
McGill, C., 103, 108, 132, 133(4), 134(5), 145(4), 146(4)
McGill, N. I., 252
Machin, N., 140, 141(28), 142(28)
Machlum, E., 177
McIntosh, E. M., 856
McIntosh, J. R., 579, 582
Mack, 847
McKay, S. J., 252
Mackay, V. I., 101, 103(20)
MacKay, V. L., 78, 79, 83(7), 105, 246, 706
McKercher, P. D., 479
Mackie, G. A., 465, 466(46)
MacNeill, S. A., 811, 816
McKnight, G. L., 198, 214(14)
McKnight, G. S., 427
McKnight, G., 839, 850
McLaughlin, C. S., 12, 536, 537
McLeod, M., 811
McMullen, B., 863
McMullin, T. W., 155, 162(24), 169, 174(4), 280
McNally, J., 467, 468
McNamee, K. C., 576
McNeil, J. B., 240
Maeda, N., 861
Magdolen, 844
Magdolen, V., 729, 859
Magee, 847, 858
Magee, P. T., 110, 862, 863
Mager, W. H., 454, 456(3), 457(3), 459, 460(11, 12, 16, 17), 463(12), 465, 466(31)
Magni, G., 736
Mahadevan, S., 521, 533(13), 534(14), 535(14)

Mahler, H. R., 152
Maine, 843
Maine, G. T., 859
Maitra 844
Maitra, P. K., 858
Majors, J., 551, 552(6)
Mak, A., 409, 410(5), 413(5)
Makarow, M., 698, 699
Makayama, T., 467
Maleas, D., 757
Maler, B. A., 393, 397(15)
Maley, F., 683, 684(3, 4), 689(3, 4), 695
Malim, M. H., 347, 357(14), 360(14)
Malone, E. A., 865
Malone, R. E., 118, 120(26), 142, 860
Maloney, D., 408
Malva, C., 467
Mandel, 326
Maniatis, T., 160, 161(42), 220, 260, 266(24), 286, 313, 314, 332, 334(7), 336(7), 359, 413, 483, 520, 550
Mann, 846
Mann, C., 68, 71(12), 74, 75, 252, 286, 303, 318(6), 756, 757(6), 846, 859
Mann, K., 736
Manney, T. R., 78, 79, 83(7), 86(22), 101, 103(20), 431, 706, 774
Mans, R. J., 540
Manzi, A., 467
Mao, J., 492, 499(6), 502(6), 503(6)
Marak, J. J., 99
Marak, J., 603, 718
Marchuk, D., 254
Mares-Guia, M., 433, 434(45)
Margoliash, E., 234
Margolskee, J. P., 106
Margolskee, J., 103
Mariotte, S., 857, 860
Marmur, 847, 849
Marmur, J., 151, 169, 174(5), 218, 359, 855, 858, 861
Marquardt, O., 231
Marquet, M., 491, 496
Marriot, M., 667
Marrs, B. L., 191, 192(11)
Marshall-Carlson, L., 568, 599(32)
Marshallsay, B., 544
Martin, J. C., 570
Martin, K., 479
Martin, R. P., 862

Martin, T. E., 726
Martinez-Arias, A., 339
Martinuci, G. B., 128
Masiarz, F. R., 491, 855
Mason, T. L., 155, 161
Mason, T., 153
Mastrangelo, M. F., 198, 215(17), 345, 356(5)
Mateblowski, M., 665, 666(25), 667(25)
Mather, I. H., 588, 596(126)
Matheson, A. T., 468
Mathison, L., 42, 45(13), 274
Matile, P., 444, 645
Matsubara, K., 383, 389
Matsuda, Y., 431
Matsui, K. A., 459, 460(15)
Matsumoto, 836, 839, 845
Matsumoto, K., 79, 100, 107(18), 108(18), 245, 859
Matsumoto, S., 567, 594, 600(20)
Matsumoto, T., 797
Matsuoka, H., 182
Matsuomoto, K., 860
Matsuzaki, F., 567, 600(20)
Matsuzaki, Y., 244
Maundrell, K., 206, 207(26), 808, 811, 816, 817(32)
Maxam, A. M., 160
Maxam, A., 550
May, R., 735, 736(3)
Mayer, D., 729
Mayer, F., 604
Mayorga, L. S., 698, 699(9)
Mazetti, P., 468
Mazia, D., 128
Mazon, 836
Mazon, M., 861
Mazza, C., 859
Mechler, 843, 844
Mechler, B., 442, 644, 859
Mecke, D., 736
Medina, A., 387
Meechler, B., 438
Meek-Wagner, D. W., 322
Meeks-Wagner, D. W., 247
Mehnert, D., 306, 307(11)
Mehra, V., 235
Mehvert, D., 353
Mellman, I., 698

Mellor, J., 347, 357(14), 360(14), 375, 383(10)
Melnick, L. M., 491, 494(5), 496, 498(5)
Melnick, L., 853
Melton, D. A., 520, 521, 522, 524(15), 675, 676(12)
Memet, S., 857
Mendenhall, M. D., 250
Merryweather, J. P., 491
Mesiarz, F. R., 433
Messing, J., 284, 332, 334(8), 397, 481, 484(14), 485(11), 509
Mets, L. J., 454
Metzger, D., 394
Mevarech, M., 214, 215(33)
Meyer, D. I., 520, 622, 667(3), 669, 670(36)
Meyer, J., 444
Meyers, J., 153
Meyers, R. M., 314
Meyne, J., 252
Michael, A. F., 576
Michael, J. G., 234
Michael, T., 198, 199(15), 214(15)
Michaelis, S., 74, 86, 92
Michaels, C. A., 855
Michel, S., 457
Michels, C. A., 854, 860
Mierendorf, R., 234
Mieseld, R., 393, 397(15)
Miklos, G. L. G., 239
Miler, J. B., 598
Miller, A., 234
Miller, D. K., 700
Miller, J. F., 183
Miller, J. H., 192, 314, 332, 335(12)
Miller, J. J., 110, 111, 112(9), 123(8, 9)
Miller, K. G., 248, 567, 585(25), 586(25), 587(25), 588(25), 592(25), 593(25), 597(25), 600(25), 729
Miller, M. W., 732
Miller, O. L., 479
Miller, R., 379
Miller, S. M., 239
Milner, J. S., 214, 215(40)
Milstein, C., 578, 600(103)
Minehart, P. L., 865
Minton, A. P., 545
Miranda, M., 732
Mirsky, R., 578, 588(102)
Mis, J. R. A., 273

Misener, V. L., 853
Misra, L. M., 206, 207(27), 212(27), 568, 665, 666(16), 792
Mitchell, A. P., 102, 105, 107(38), 110(26, 38), 113, 114, 245, 429, 433(10), 435(10), 447(10), 452
Mitchison, M., 796, 798(1), 801
Mitchison, T., 594
Mitra, G., 465
Mitsui, K., 459, 465, 466(15)
Miura, T., 730
Miyajima, 840
Miyajima, A., 78, 79, 86(19), 245, 859, 860
Miyajima, I., 79, 859, 860
MiYakawa, I., 858
Miyakawa, T., 736
Miyamoto, S., 861
Miyanohara, A., 383, 389
Miyijama, I., 245
Moehle, C. M., 430, 431, 435(19), 438, 452(59), 453(19, 21, 26), 644, 859
Moenne, A., 857, 860
Moens, P. B., 603
Moerschell, R. P., 18, 362, 366(6), 368(6)
Moir, D. T., 491, 492, 494(2, 3, 5), 496, 497(2), 498, 499, 501(2), 502(6, 24), 503(6), 544
Moldave, K., 5, 527, 536, 537
Molenaar, C. M. T., 459, 460(11, 12, 16), 463(12), 465
Molenaar, I., 718
Monaghan, P., 597, 598(145)
Montegomery, D. L., 416
Montelone, 844, 845
Montelone, B. A., 117, 860
Moor, H., 645
Moore, D. M., 479
Moore, R. L., 853
Moore, S. A., 86
Mor, J.-R., 587
Mora, G., 455
Moreau, N., 582
Moreland, R. B., 476, 568
Moreno, S., 396, 811
Morgan, D. O., 479
Morgan, S. A., 426
Moriga, N., 861
Morikawa, H., 182
Morikawa, K., 594
Morimoto, R., 154

Morrison, A., 860
Morse, D. E., 479
Mortimer, 835, 836, 838, 839, 841, 843, 849, 852
Mortimer, K., 40
Mortimer, R. K., 6, 16, 32, 40, 42, 43(11), 53, 57, 58, 72(1), 96, 100, 138, 241, 784, 827, 860, 862
Mortimer, R., 678, 859
Moses, M. J., 127, 130
Moses, M., 127, 129(62, 64)
Mosrin, 846
Mosrin, C., 860
Mosteller, R. D., 479
Mott, J. E., 398
Mount, D. W., 231
Mourant, J. R., 183
Mousset, M., 363*
Mowshowitz, D. B., 38
Moyzis, R. K., 252
Mucke, E., 125
Mueckler, M., 521
Mulero, J. J., 169, 174(4), 280
Mulholland, J., 489, 568, 585(57), 586(57), 601(57), 669, 670(39)
Mullenbach, G. T., 491
Muller, F., 347
Müller, G., 729, 859
Muller, H., 438, 644, 859
Müller, P. P., 155
Müller, P., 850
Mulligan, R. C., 578
Mullis, K. B., 551
Mullis, K., 509
Munro, S., 508
Munz, P., 799
Murakami, H., 675
Murakami, K., 446
Murata, K., 182, 264, 291, 311, 321, 337, 365, 393, 394(18), 396(18), 503
Murata, S., 446
Murphy, R. F., 645, 646, 699
Murray, A. W., 253, 283
Murray, A., 766, 860
Murray, J. A. H., 206, 207(24), 376
Murray, J., 816, 817(32)
Murray, L., 837
Musters, W., 468
Musti, A. M., 392, 397(13)
Muthukumar, G., 862

Muto, A., 473
Myambo, K., 551
Myers, A. M., 240, 489
Myers, D. I., 675, 681(3), 682(3)
Myers, P. L., 856
Myers, P. S., 862

N

Nadakavukaren, K. K., 587, 588(124)
Nagel, M. B., 853
Naider, F., 702
Najarian, R., 214, 215(37)
Nakafuka, M., 245
Nakafuka, N., 245
Nakafuku, 840
Nakafuku, M., 79, 859
Nakai, S., 53
Nakajuku, M., 860
Nakamura, S., 860
Nakayama, N., 78, 79, 859
Nam, H. G., 421, 476
Nanninga, N., 476, 573
Nash, 852
Nash, R., 860
Nasmyth, K. A., 103, 132, 133, 137, 240, 282, 294, 339
Nasmyth, K. H., 105
Nasmyth, K., 80, 294, 297, 799
Nassal, M., 729
Natsoulis, G., 75, 303, 311(5), 318(5), 356, 509, 694, 764, 841
Naumovski, L., 197
Nayak, D. P., 231
Naysmyth, K. A., 339
Nazar, R. N., 465
Nebes, V. L., 430
Necas, O., 399
Needleman, 842, 843
Needleman, R. B., 152, 155, 855, 860, 861
Neff, N. F., 163
Negri, R., 557
Neigeborn, 848
Neigeborn, L., 431, 846, 860
Neighbors, B. W., 579
Nelson, H. C. M., 502
Netter, P., 151
Neukom, H., 121
Neupert, W., 520

Nevalainen, L. T., 699
Newlon, C. S., 105, 860
Newlon, C., 758
Newman, A. J., 429
Newman, A. P., 669, 670(33)
Newman, R., 517
Newmeyer, D., 853
Newton, C. H., 468
Nguyen, D. C., 570
Nguyen, L. H., 557
Nguyen-Juilleret, M., 491, 496
Nichols, B. P., 478
Nichols, D. L., 416
Nicklen, S., 160, 550
Nickoloff, J. A., 135, 140, 141(26), 142(26), 143
Nickoloff, J., 18
Nicolas, A., 289
Nicolet, C. M., 40, 229, 711
Niederacher, 840
Niederacher, D., 860
Niederberger, 839
Niederberger, P., 860
Nierhaus, K. H., 469, 471(89), 472(89), 473(89), 477(89)
Nieuwint, R. T. M., 459, 460(16)
Nikawa, J., 206, 207, 214, 215(35), 508, 862
Nikolaishwili, N. T., 858
Nikolics, K., 578, 596(98), 598(98)
Niman, H., 510
Niwa, O., 797
Noble, J. A., 438, 452(59)
Nogi, Y., 137, 214, 215(32), 244, 380
Nogueira Araujo, G. M. de C., 576
Nolan, L. S., 848
Noll, H., 261
Nomura, M., 305, 318(8), 468
Nonet, 846
Nonet, M. L., 846
Nonet, M., 419, 846, 860
Norris, 841
Norris, D., 244, 860
Norris, K., 497
Nosaki, C., 383
Novelli, G. D., 540
Novick, P. J., 568, 601(59), 695, 696(31), 698
Novick, P. N., 663, 666(9), 667(9)
Novick, P., 206, 207(21), 215(21), 303, 319, 321, 567, 568, 587(24, 28), 600(24, 28), 669, 670(32, 37), 671(32), 684, 685(9), 688, 693(9, 15), 729, 847
Nozaki, C., 389
Nozawa, H., 239
Nurse, P., 206, 207(26), 396, 798, 799, 801, 805, 808, 811, 813, 814(19), 815, 816, 817, 818

O

O'Hara, P. O., 79
O'Mally, K., 214, 215(34)
O'Neill, J., 864
O'Neill, K., 198, 199(15), 214(15)
O'Rear, J., 853
O'Sullivan, J., 242
Obara, T., 860
Obonai, K., 380
Ochoa, S., 536
Ochs, R. L., 720
Oechsner, V., 729, 844, 859
Oehm, A., 568, 608, 719, 720(7), 723(7), 727(7)
Ogata, K., 467
Ogawa, N., 865
Ogur, M., 306, 307(11), 353
Oh, D., 843, 850
Ohga, R., 623
Ohno, Y., 861
Ohsawa, T., 446
Ohsumi, Y., 214, 215(39), 645, 861
Ohtomo, N., 383, 389
Ohya, 837, 850
Ohya, Y. M., 861
Ohya, Y., 214, 215(39)
Oi, V. T., 570
Okada, S., 594, 865
Okayama, H., 812
Okazaky, K., 812
Okazaky, N., 812
Okret, C., 393, 397(15)
Olempska-Beer, Z., 110, 111(5)
Olesen, J. T., 388
Olesen, J., 433
Oliphant, A. R., 362
Olmsted, J. B., 572, 581, 617
Olson, 835, 836, 840
Olson, L. W., 603, 615
Olson, M. O. J., 720

Olson, M. V., 47, 49, 50(27), 51(27), 169, 175, 251, 252, 253(1), 254(1), 255, 257, 259(23), 261(1), 262, 263(1, 23, 26), 265(5), 266(5), 268(5), 554, 857, 861
Olson, M., 7, 9(31), 50, 58, 59, 60(7), 760, 767, 854
Ono, 848, 849, 850, 851
Ono, R.-i., 861
Ooi, T., 467
Orlean, P., 683, 685, 687, 690, 693(5), 694(11)
Orr, 843
Orr, E., 861
Orr-Weaver, T. K., 815
Orr-Weaver, T. L., 282, 286(9), 288(9), 289, 298
Orr-Weaver, T., 66, 320
Osawa, S., 457, 459, 460(10), 462(10), 465
Osbourne, B. I., 375
Oshima, 865
Oshima, T., 568
Oshima, Y., 132, 134(1), 137, 138(1), 386, 859
Osley, 841
Osley, M. A., 244, 860
Osmond, B. C., 568, 587(28), 600(28)
Osmond, B., 847
Osumi, M., 774
Otaka, E., 457, 459, 460(9, 10, 14, 15), 462(9, 10), 463(9), 465, 466(15, 26), 467
Otte, C. A., 84, 89(42), 90(42), 777
Overbye, K., 301
Overgaard-Hansen, K., 533
Ovodov, S. Y., 543
Owens, K., 623
Oyen, T. B., 177
Özkaynak, E., 459, 461(18)

P

Pabo, C. O., 533
Pachnis, V., 283
Paddon, C. J., 839
Paddon, C., 74
Page, G. S., 554
Pain, D., 675
Palaçian, E., 468
Paleologue, A., 468
Palmer, R., 599, 600(148, 149), 734, 752, 764

Palmiter, R. D., 427
Palzbell, T. G., 860
Pammer, M., 123
Panzeri, L., 761
Papciak, 847
Papciak, S. M., 861
Paquin, C. E., 349
Pardue, M., 66
Parent, S. A., 196, 209(6), 379
Parish, R. W., 665, 666(23)
Park, F. J., 430, 435(19), 438, 452(54), 453(19)
Park, H. O., 716
Park, M. R., 859
Parkash, S., 117
Parker, J. H., 365
Parker, M. L., 78, 706
Parker, R. R., 438, 453(57)
Parker, R., 416, 422
Parks, D. R., 570, 598(72, 74)
Parks, L. W., 276
Parry, D. M., 239
Parry, E. M., 140
Parry, E., 73
Pashina, O. B., 858
Patel, T., 375, 383(10)
Paterna, G. D., 467
Paterson, B. M., 392, 397(13), 536
Patshne, M., 716
Patterson, B., 415
Patterson, D., 48
Payne, 837
Payne, G. S., 579, 609, 678, 861
Payne, R. W., 181
Peacock, W. J., 362
Pearse, B. M. F., 697
Pearson, 847
Pearson, D., 807
Pearson, N. J., 861
Pelham, H. R. B., 393, 536, 710
Pelham, H., 508
Perbal, B., 485
Percival, K. J., 263
Percy, C., 234
Perkins, 842
Perkins, D. D., 55, 56(34)
Perkins, D., 853, 861
Perkins, E. L., 860, 861
Perkins, E., 4, 12(4)
Perlin, D. S., 859
Perlman, P. S., 154

Perlman, R. E., 345
Perrodin, G., 753
Perry, K., 78, 774
Perry, L. J., 383, 492
Perry, R. P., 467
Pesoed-Hurt, B., 863
Peterkosky, A., 623
Petersen, J. G. L., 853
Petersen-Bjorn, S., 846
Petes, T. D., 118, 119(44), 286, 357
Petes, T., 330, 342(5)
Petitjean, A., 845, 850, 854
Petrochilo, E., 151
Petropoulos, C. J., 345
Petterson, T., 479
Pevny, L., 283
Pfaff, E., 862
Pfeffer, S. R., 584, 585(119)
Pfeifer, K., 383, 384
Pflugfelder, M., 314
Philippsen, P., 7, 75, 177, 181, 182, 196, 218(1), 353, 757, 758(11), 759(11), 761, 762(11), 764(11), 765(11), 862
Philips, M. M., 140
Philipssen, 851
Phillips, H. S., 578, 596(98), 598(98)
Phillips, M. N., 101
Picard, D., 394, 395(21), 396(21)
Picologlou, S., 197
Pierce, M. K., 273
Pietras, D. F., 362, 853
Pietri, R., 623
Piggott, J. R., 241
Pillar, T., 858
Pillus, L., 567, 578, 596, 600(9)
Pinkham, 840
Pinkham, J. L., 388, 861
Pinkham, J., 379
Pinon, R., 95, 101
Pinto-Toro, J. A., 468
Piper, P. W., 554
Piperno, G., 578
Pirozhkov, V. A., 858
Pisetsky, D., 127, 129(64)
Planta, R. J., 454, 456(3), 457, 459, 460(11, 12, 13, 16, 17), 463(12, 13), 465, 466(31), 467, 468, 469, 470(63, 88), 471, 472(61, 63, 89), 473(61, 63, 85, 89), 477(61, 88, 89), 495
Platt, J. L., 576
Platt, T., 398, 478

Plesset, J., 536, 537(10)
Plevani, P., 433, 858, 859
Plizzi, C. M., 252, 260(3)
Pockling, 843
Pocklington, M., 861
Pohlig, G., 644, 665, 666(19)
Pokorný, V., 732
Polaina, J., 790, 791(27)
Polak, J. M., 604
Polazzi, 843
Polazzi, J. O., 853
Popolo, L., 568, 599(33)
Porta, G., 252
Porter, G., 406
Porter, R. R., 533
Portillo, 836
Portillo, F., 861
Potashkin, J. A., 736
Potter, S. S., 268
Potter, V. R., 735
Poutre, 844
Poutre, C. G., 861
Pouwels, P. H., 490
Powell, K. T., 183
Powers, S., 92, 108
Poyton, R. O., 246
Prager, E. M., 234
Prakash, L., 117
Pratt, P., 214, 215(34)
Prazmo, W., 153
Prendergast, J., 837
Prescott, D., 5
Preston, R. A., 489, 565, 569, 570(3), 587(3), 599(3), 601(61), 645, 646, 699, 730, 732, 733(5), 735(5), 843
Presutti, C., 465, 466(24)
Pretorius, 847, 849
Pretorius, I. S., 861
Preuss, D., 497
Price, B. R., 79
Price-Tillotson, B., 491, 494(5), 496, 498(5)
Primerano, 847
Primerano, D., 862
Pringle, J. R., 4, 12(6), 78, 79(6), 100, 429, 433(7), 489, 565, 567, 568, 570(3), 571, 579(21, 29), 580, 581(29), 584, 587, 588(109), 589(109), 599(3), 600, 608, 645, 646, 718, 729, 730, 731(7), 732, 733, 734, 735(5), 774, 857
Procter, D. P., 695
Pruss, R. M., 578, 588(102)

Ptashne, M., 380, 533, 545, 552
Puccierelli, G., 465
Puma, P., 491, 494(5), 496, 498(5)
Pure, G. A., 40, 197, 229
Purvis, I. J., 418
Putrament, A., 153
Puzicha, M., 669, 670(34)
Pypaert, M., 699

Q

Qian, S., 467
Qin, S., 536

R

Rabinowitz, M., 152, 154, 160
Radford, A., 5,
Radin, D. N., 19
Raff, M. C., 578, 588(102)
Rafti, F., 467
Ragsdale, C. W., 183
Rahe, B., 580, 587(109), 588(109), 589(109)
Rai, R., 854
Ramirez, C., 468
Ramsay, G., 510
Randall, R. J., 741
Randall, R., 445
Randolph, A., 453
Rank, 835, 837, 839, 843, 846, 850
Rank, G. H., 862
Rapport, E., 603
Raths, S. K., 702
Raths, S., 646
Ratzkin, B., 281
Raué, H. A., 467, 468, 469, 470(63, 88), 471(89), 472(61, 63, 89), 473(61, 63, 85, 89), 474, 476, 477(61, 88, 89), 495
Ray, A., 140, 141(25, 28), 142(25, 28)
Raymond, C. K., 644
Raymond, C., 568, 601(60)
Raymond, W., 226, 293, 330
Read, E., 854
Rebagliati, M. R., 520
Reboud, A.-M., 468
Reboud, J.-P., 468
Rechsteiner, M., 467
Redding, 596, 598(138)

Redding, K., 568, 574(58), 588(58), 596(58), 597(58), 598(58), 601(58)
Reddy, V. A., 683, 684(3), 689(3)
Redman, K. L., 467
Reed, S. I., 79, 250, 282, 568, 585(30), 858
Reese, T. S., 594
Reichlin, M., 234
Reid, G., 511
Reif, M. K., 155
Reineke, J., 672
Reith, A., 720, 721(16), 725(16)
Remacha, M., 465
Reneke, J. E., 702
Resnick, M. A., 4, 12(4), 110, 112(6), 117
Resnick, M., 100
Revzin, A., 529, 532(26)
Rexach, M., 520
Rhoads, D. D., 467
Ribas, J. C., 734
Rich, B. E., 467
Richardson, C. C., 551
Richardson, S. L., 568, 585(30)
Rickner, R., 40
Rico, H., 734
Riedel, N., 409, 410(5), 413(5)
Riehl-Bellon, N., 491
Riethman, H. C., 252
Riezman, H., 602, 608, 644, 646, 698, 699, 700(5), 702(5), 707(6), 708, 709(6), 709(28)
Rigaut, J. P., 720, 721(16), 725(16)
Riggs, A. D., 231
Riggs, D. L., 557
Riggs, M., 198, 199(15), 214(15)
Rine, 837, 846
Rine, J. D., 102, 106
Rine, J., 4, 93, 135, 142, 240, 241, 242, 249, 378, 565, 568, 588(5), 595(54), 596, 601(54), 603, 607(5), 609, 614(8), 665, 666(15), 853, 862
Ritch, T., 493
Ritzel, R. G., 273
Robash, M., 465
Robbins, L. G., 863
Robbins, P. W., 685, 690, 694
Roberts, C. J., 644, 665, 666(19)
Roberts, C., 568, 601(60)
Roberts, G. E., 536
Roberts, J. D., 362
Roberts, J. W., 4

Roberts, J., 510
Roberts, N. A., 374, 383(7, 10), 387(7), 389
Roberts, R. L., 732
Robertson, B. H., 479
Robertson, J., 214, 215(34)
Robinow, C. F., 99, 603, 797
Robinson, C. F., 718
Robinson, D. G., 604
Robinson, G. W., 197
Robinson, J. S., 644, 646(13)
Robinson, J., 718
Robinson, L. C., 107, 851, 857
Rockmill, B., 13, 111, 113, 124(17), 125(17), 147, 148(2)
Rodel, G., 858
Rodgers, L., 508
Rodriguez, R. L., 284, 481
Roeder, 843
Roeder, G. S., 13, 111, 113, 124(17), 125(17), 147, 148(2, 3), 345, 357(7), 548
Roeder, S. G., 21, 856
Rogers, D. T., 137, 385
Roghmann, M.-C., 356
Röhm, K. H., 430, 431(16), 443(16), 450(16), 451(16)
Rohrmann, G. F., 489
Roitsch, C., 491
Rolfe, M., 243, 283, 302
Rollin, C. F., 465
Roman, H. L., 96, 101(6)
Roman, H., 21, 53, 101, 140
Roncero, C., 734
Rosamond, 840
Rosamond, J., 857
Rosbash, M., 459, 461(19)
Rose, A. B., 196, 345, 376, 397
Rose, A. H., 5
Rose, A., 74
Rose, M. D., 79, 123, 124(55), 196, 197, 214(7), 206, 207(21, 27), 212(27), 215(21), 229, 230(12, 49), 316, 357, 359(29), 360(29), 567, 568, 595(13), 595(22), 600(13, 22, 51), 601(22), 665, 666(16), 792
Rose, M., 42, 59, 290, 309, 311(16), 319, 321, 379, 758, 770(17)
Rosebrough, N. J., 741
Rosebrough, N., 445
Rosenberg, M., 232
Rosenberg, N. G., 860

Rosenberg, S., 587, 588(124)
Rosenbluh, A., 214, 215(33)
Rosenfeld, M. G., 532, 533(29)
Rosser, D. S. E., 481
Roth, J. R., 174
Roth, R. M., 100, 117
Roth, R., 112
Rothblatt, 622, 665(1), 671(1)
Rothblatt, J. A., 520, 669, 670(35), 675, 681(3), 682(3)
Rothblatt, J., 544, 622, 667(3)
Rothblum, L. I., 720
Rothman, J. E., 698, 699(9), 710
Rothman, J. H., 498, 644, 645, 665, 666(19)
Rothman, J., 438
Rothstein, R. J., 20, 38, 282, 286(9), 288(9), 290(8), 298(11), 313, 377, 781, 815, 816, 856
Rothstein, R., 66, 115, 196, 227(4), 243, 283, 286, 289(28), 291(28), 293, 302, 760, 763(19), 768(19), 815, 816(28), 838
Roufa, D. J., 467
Rousseau, P., 125
Rout, M., 743
Rowley, A., 837
Roy, C., 465
Royan, S., 600
Rozijin, T. H., 718, 735
Rubin, G. M., 423
Ruby, S. W., 198
Rudin, N., 140, 141(27), 142(27), 856
Rudolph, H. R., 492, 499(6), 502(6), 503(6)
Rudolph, H., 386
Ruet, A., 857
Ruggiere, S., 736
Ruis, H., 537
Rupeš, I., 567, 600(12)
Rusconi, S., 393, 397(15)
Russel, M., 510
Russell, D. W., 136, 137, 139(14), 145(14), 146(14), 697
Russell, G., 576
Russell, P., 808, 811, 815, 817(29)
Rutgers, C. A., 474, 476(94)
Rutter, W. J., 231, 387
Ryabova, L. A., 543
Ryan, T., 523
Rytka, J., 665, 666(24)

S

Saari, G. C., 78, 79, 706
Saari, G., 438
Sadler, I., 214, 215(31), 568
Sadnik, I., 537
Sáenz-Robles, M. T., 465
Sáenz-Rogles, M., 468
Saheki, T., 430, 431
Saiki, R. K., 551
Saiki, R., 509
Sakai, 846
Sakai, A., 862, 863
Sakakibara, S., 446
Sakata, Y., 105, 108(39), 110(39)
Sale, W. J., 128
Salminen, A., 568, 601(59), 695, 696(31)
Salts, Y., 95, 101
Saltzgaber, J., 156
Salvato, K. A., 491, 494(5), 496, 498(5)
Sambrook, J., 160, 161(42), 220, 260, 266(24), 286, 313, 332, 334(7), 336(7), 359, 413, 483, 550
Sammak, P. J., 594
Sanchez, L., 101
Sánchez-Madrid, F., 468
Sandbaken, 841, 851
Sandbaken, M. G., 862
Sander, C., 856
Sanders, J. P. M., 152, 158, 159(8)
Sanders, N. J., 198, 215(17), 345, 356(4, 5), 357(4)
Sanders, S., 669, 670(35)
Sandler, L., 239
Sandmeyer, S. B., 176
Sando, N., 111, 858
Sands, S. M., 96, 101, 140
Sanford, J. C., 155, 162, 163(47)
Sanger, F., 160, 550, 556
Santa Anna-A., S., 92
Santiago, T. C., 418
Santos, M. J., 602
Sanz, R., 669, 670(36)
Sapperstein, S., 74
Sass, 843
Sass, P., 206, 207, 246, 862, 863
Sato, K., 596
Sauer, B., 341
Sauer, R. T., 244, 502, 533
Saunders, 835, 837, 839, 843, 846, 850

Saunders, C. A., 736
Saunders, G. W., 862
Savage, E. A., 273
Savarese, J. J., 125
Sawadogo, M., 548
Sayers, J., 510
Scacheri, E., 859
Scafe, C., 419, 860
Scarborough, G. A., 667
Scatchard, G., 707
Schaap, P. J., 465, 476
Schaber, M. D., 383, 386, 389
Schafe, C., 846
Schaffer, B., 73
Schaller, H., 231
Scharf, S., 509, 551
Schatten, G., 128
Schatz, 851
Schatz, G., 156, 214, 215(31), 511, 521, 602, 663, 665, 666(7, 22), 863
Schatz, P. J., 693, 862
Schatz, P., 589, 600(17)
Schaus, M., 853
Scheinman, R., 50
Schekman, R., 385, 399, 410, 512, 520, 544, 546, 579, 609, 622, 663, 664(8), 665, 666(8, 20), 667(4, 8), 668(20), 669, 670(12, 13, 30, 31, 32, 35, 37, 38), 671(1, 20), 672(13, 20), 674, 675, 678, 684, 685(9), 688, 693(9, 15), 698, 701, 729, 861
Schellenberg, M., 444
Schena, M., 391, 393(10), 394, 395(9, 21), 396(9, 21), 398(9)
Scherer, S., 189, 190, 282, 284(5), 286(6), 294(6), 319, 321, 379
Scherf, C., 7, 196, 218(1)
Schild, 835, 836, 838, 839, 841, 843, 849, 852
Schild, D., 6, 40, 42, 43(11), 58, 72(1), 96, 241, 827, 860, 862
Schimmang, T., 568, 590(49), 719, 727(10), 736
Schimmel, 840, 841
Schimmel, P., 379, 855, 859
Schirmaier, F., 182, 862
Schlesser, 845
Schlesser, A., 862
Schlessinger, D., 252, 265(5), 266(5), 268(5)
Schleyer, M., 520

Schloemer, R. H., 332, 334(10)
Schmid, G., 467
Schmidt, 842
Schmidt, B., 520
Schmidt, C., 862
Schmidt, W., 510
Schmitt, 852
Schmitt, H. D., 669, 670(34), 862
Schnapp, B. J., 594
Schneider, C., 517
Schneider, G., 602
Schneider, H., 520
Schneider, J. C., 365, 389
Schneider, W., 697
Schnell, R., 93
Schoppink, P. J., 459, 460(17)
Schroeder, R., 125
Schultes, N. P., 860
Schultz, L. D., 546
Schultz, L., 465
Schulz, J., 90
Schulze, E., 577, 595(93), 596(93), 598(93)
Schulze, M., 858
Schurmann, F.-W., 604
Schwartz, D. C., 47, 174
Schwartz, D., 58, 59(2), 60(2)
Schweingruber, A. M., 524
Schwenke, J., 429, 433(11)
Schweyen, R. J., 154, 862
Schwindinger, W. F., 297, 306, 308, 465
Schwob, 835, 847
Schwob, E., 862
Sclafani, R. A., 846
Scott, C. W., 663, 667(10)
Scott, J. D., 863
Scott, J. F., 241
Scott, J. H., 399, 410, 546
Scott, J., 663, 701
Scott, K. E., 28
Scully, M. F., 503
Seaver, E. C., 153
Seaver, E., 844
Sedat, J. W., 576
Seeburg, P. H., 397, 481, 492, 494(11), 578, 596(98), 598(98)
Seeliger, A., 730
Segall, J., 115, 117(34), 121(34), 122(34), 123(34)
Segawa, T., 380
Segev, 852

Segev, N., 100, 489, 568, 585(57), 586(57), 601(57), 669, 670(39), 862
Seifert, H. S., 18, 226, 293, 330
Sels, B. M., 539
Sena, E. P., 19
Sengstag, C., 155
Senior, A., 601
Sentandreu, R., 734
Sentenac, A., 303, 305, 318(6, 8), 554, 857, 860
Sercarz, E. E., 234
Server, P., 48
Sexton, J., 419, 860
Shafer, B. K., 198, 215(17), 345, 356(5)
Shakuto, S., 736
Shalom, A., 858
Shapira, S. K., 339
Shark, K., 162, 163(47)
Sharp, P. A., 230, 234(5), 235(5)
Sharp, P., 374
Sharrock, R. A., 468
Shatzman, A., 232
Shaw, E., 433, 434(45)
Sheetz, M. P., 594
Sheinman, R., 861
Shenk, T. E., 538
Shepard, C., 433
Shepherd, E. L., 587, 588(124)
Sherman, F., 7, 18, 20(51, 52), 21, 24, 38, 41, 42, 48(14), 53(10), 80, 81, 87(35), 160, 192, 193(12), 264, 282, 284, 301(10), 306, 307(11), 311, 353, 362, 363, 365, 366(6), 368(6), 375, 393, 394(19), 396(19), 401, 493, 499, 503(28), 512, 524, 783, 784, 791(22), 853, 861
Shero, J., 74, 75, 757, 758(11), 759(11), 761, 762(11), 764(11), 765(11), 768, 770(28)
Sherratt, D., 189
Shida, H., 623
Shida, M., 623
Shilo, B., 107
Shilo, V., 107
Shima, H., 105, 108(39), 110(39)
Shimada, H., 244
Shimatake, H., 858
Shimizu, Y., 862
Shimmin, L. C., 467, 468
Shimoda, C., 774
Shinoda, S., 861
Shio, H., 601, 602

Shirai, T., 861
Shires, T. K., 311
Shirmaier, 851
Shore, D., 80
Short Russell, S. R., 362
Shortle, D., 226, 282, 293(7), 303, 314
Shulman, R. W., 425, 426, 427(4)
Shuster, J. R., 453
Shwartzbauer, J., 471
Shweizer, E., 838
Sibum, C. P., 471
Sickerham, L. J., 671
Siddiqi, I., 140, 141(25), 142(25)
Siechertová, O., 732
Siede, W., 855
Signer, E., 192
Sikela, J., 230, 234(3)
Sikorski, R. S., 18, 206, 207(22), 283, 305, 308, 312, 316(9)
Sikorski, R., 75, 509, 510(9)
Sillevis-Smith, W. W., 718
Silver, P. A., 476, 568
Silverman, G. A., 252, 265(5), 266(5), 268(5)
Silverstein, S. C., 700
Simanis, V., 818
Simchen, G., 83, 93(36), 95, 97(3), 100, 101, 102, 103(25), 104, 105(22, 25), 106(22, 25, 29), 107, 108(29), 109(3), 110(29), 111, 113, 353, 781, 858
Simon, M., 853
Simons, K., 625, 698
Simpson, E. M., 857
Simpson, M., 594
Sinaud, V. I., 331
Singer, B., 646
Singer, J. D., 140, 141(26), 142(26), 143
Singer, R. H., 568
Singh, A., 363
Singh, H., 230, 234(5), 235(5)
Singh, K., 362
Sipiczki, M., 807, 814(15)
Sitney, 844
Sitney, K. C., 862
Sitney, K., 862
Skoneczny, M., 665, 666(24)
Skup, D., 467
Skvirsky, 839
Skvirsky, R. C., 856, 862
Slater, M. L., 732, 777, 791(17)
Slater, M. R., 716, 857

Slater, M., 852
Sleigh, M. J., 396
Sloat, B. F., 732
Slonimski, P. P., 150, 151, 152, 853
Slonimski, P., 753
Small, B., 479
Small, G. M., 601, 602
Smetana, K., 720
Smith, 847
Smith, A. R. W., 5
Smith, C. A., 188
Smith, C. L., 252, 797
Smith, D. E., 581
Smith, H. E., 105, 107(38), 110(38), 245
Smith, H., 113
Smith, J. A., 444, 450(68), 835
Smith, L. M., 863
Smith, M., 136, 137, 139(14), 145(14), 146(14), 314, 362
Smith, P. F., 612
Smith, P., 736
Smith, R. A., 491, 494(2), 495(2), 496(2), 497(2), 498(2), 499, 501(2)
Smith-Gill, S. J., 234
Smolinska, U., 154
Smythe, E., 699
Sninsky, J. J., 552
Snow, C. M., 601
Snow, R., 28, 276, 863
Snyder, 847
Snyder, M., 68, 71(12), 234, 235, 238(21), 293, 330, 433, 434, 567, 568, 574, 590(16), 596(23), 597(23), 600(16, 23, 50), 601(87), 718, 719(6), 727(6), 756, 757(6), 858, 863
So, M., 226, 293, 330, 331
Söll, D., 807, 858
Söllner, T., 862
Solomon, F., 567, 568, 578, 600(9, 17, 19), 601(46), 862
Solomon, M. J., 459, 461(18)
Som, T., 196, 206(8)
Sommer, 846
Sommers, S., 863
Song, J. M., 197
Sorger, P. K., 393
Sparling, P. F., 331
Spatola, E., 103, 132, 133(4), 134(5), 145(4), 146(4)
Spatrick, P., 699, 707(17)

Spector, D. L., 720
Spencer, D. M., 5
Spencer, F., 64, 71(9), 72(9), 74, 75, 214, 215(38), 757, 758(12, 13), 759(12), 762(12)
Spencer, J. F. T., 5
Spencer, M., 608
Spierer, P., 320
Spindler, K. R., 481
Spirin, A. S., 543
Spohn, W. H., 720
Sprague, G. F., Jr., 77, 78, 79, 83, 84, 86, 89(40, 45), 90(40), 102, 106, 132, 137(4), 145(4), 146(4), 188, 189, 191(4), 193(4), 777
Sprague, G., 250
Srb, A. M., 363
Sripati, C. E., 425, 426, 427(4), 465, 538
St. John, A. C., 231
St. John, T. P., 380, 381(31)
Staehelin, T., 511, 512(15), 581
Stahl, F. W., 140, 141(25, 28), 142(24, 25, 28)
Stahl, P. D., 698, 699(9)
Stanley, K., 231
Stark, H. C., 38
Stark, M. J., 214, 215(40)
Stearns, T., 489, 645, 730, 758, 760(14)
Steden, M., 78
Steele, G. D., 587, 588(124)
Steensma, H. Y., 855, 858
Steensma, H., 74, 75
Stein, M., 811
Steitz, J. A., 4, 467
Sternglanz, R., 41, 243, 294, 339, 863, 864
Sterns, T., 565, 570(3), 587(3), 599(3), 732, 733(5), 735(5), 835, 837, 840, 841
Steuer, E., 594
Stevens, B., 723
Stevens, T. H., 498, 568, 601(60), 644, 645, 665, 666(19)
Stevens, T., 438, 669, 670(40)
Stewart, G. G., 18
Stewart, J. W., 365, 524, 861
Stewart, M. J., 467
Stewart, S. E., 189, 206, 321
Stiles, J. I., 282, 301(10), 853
Stinchcomb, D. T., 189, 190, 199, 253, 282, 284(5), 319, 321, 379
Stiney, 845

Stiney, K., 845
Stockdale, F. E., 598
Stoffel, S., 509
Stoffer, S., 551
Stöffler, G., 467
Stöffler-Meilicke, M., 467
Stojcic, C., 860
Stollar, B. D., 538
Stollar, V., 538
Stone, D. E., 710, 715(5)
Storrie, B., 578
Stotz, A., 7, 196, 218(1), 761
Strathern, J. A., 339
Strathern, J. N., 4, 18, 86, 101, 103(21), 104(21), 105, 132, 133, 134, 135, 189, 196, 197, 198, 200(9), 207(9), 215(17), 226(3), 294, 306, 319, 320, 345, 356(5), 379, 811
Strathern, J., 105, 108, 135, 291
Strathern, N. J., 103
Strausberg, R. L., 154, 157
Strauss, A., 799
Strauss, F., 529, 532(28)
Streiblová, E., 567, 600(11, 12)
Strobel, O., 467
Strommaier, K., 231
Stroud, R. M., 433
Struhl, 838, 844
Struhl, K., 5, 189, 190, 199, 253, 282, 284(5), 297, 298(48), 319, 321, 347, 362, 379, 494, 520, 521, 527(4), 528(3), 529(4), 532(3), 533(3, 4, 13), 534(14), 535(14), 863
Strunnikov, A. V., 858
Stryer, L., 570
Stucka, R., 176
Stueber, D., 520
Sturtevant, J. M., 533
Styles, C. A., 84, 251, 343, 353, 361, 782, 856
Styles, C., 386, 853
Su, J.-Y., 846
Suarez Rendueles, P., 857
Subik, J., 863
Subramani, S., 569, 602(63)
Subramaniam, M. K., 600
Suda, K., 214, 215(31)
Sudbery, P. E., 852
Sugarman, E., 140, 141(27), 142(27)
Suggs, S., 492

Sugimoto, K., 546, 548(6)
Sugimoto, T., 596
Sugino, 848
Sugino, A., 206, 207(25), 863
Suhng, S. H., 862
Sulston, J., 251
Surosky, R. T., 421
Surosky, R., 770, 771(30)
Susek, R. E., 710
Susskind, M. M., 244
Sutcliffe, J. G., 252
Sutherland, D., 517
Sutherland, K. A., 860
Sutrave, P., 345
Sutter, T. R., 242
Suwa, S., 596
Suzuki, 835
Suzuki, K., 78, 459, 460(14), 863
Suzuki, Y., 380
Svobodová, J., 567, 600(11, 12)
Swain, W. F., 163, 164(48)
Swanson, J. A., 700
Sweeny, R., 7
Sweetser, D., 234, 235, 238(21)
Swerdlow, H., 408, 409, 410(5), 413(5)
Swimmer, C., 392
Switzer, R. L., 431
Synn, S., 860
Szaniszlo, P. J., 567, 600(8)
Szauter, P., 66
Szczesna, E., 537
Szent-Gyorgyi, C., 736
Szostack, J. W., 815
Szostak, J. W., 118, 119(45), 198, 245, 253, 282, 283, 286(9), 288(9), 289, 298, 300(10), 758, 864
Szostak, J., 66, 320, 766, 860

T

Tabak, H. F., 431, 446(23)
Tabor, S., 551
Tada, K., 111
Taguchi, 835
Taguchi, A. K. W., 863
Takada, K., 446
Takano, I., 137
Takeda, M., 240
Takenawa, T., 182

Talens, L. T., 732
Talian, J. C., 572, 581(80)
Tamiya, E., 182
Tamura, S., 467
Tanaka, 847
Tanaka, H., 863
Tanaka, M., 861
Tanaka, S., 467
Tanaka, T., 467, 596
Tanner, W., 568, 599(35), 667, 682, 683, 687, 688, 690, 693
Tarentino, A. L., 695
Tatchell, 836, 845, 848, 849
Tatchell, K., 103, 107, 132, 137, 240, 845, 850, 851, 854, 864
Tatum, E. L., 188
Tauton-Rigby, A., 499
Taylor, A., 774
Taylor, D. L., 570, 594
Taylor, D. P., 191, 192(11)
Taylor, G. R., 206, 215(28)
Teague, M. A., 246, 251(29)
Teem, J. L., 459, 461(19), 465
Teixido, J., 467, 469(61), 472(61), 473(61), 477(61)
Tekamp-Olson, P., 214, 215(37)
Thatchell, 843
Thiebbaut, F., 720, 721(16), 725(16)
Thiele, 835
Thiele, D. J., 538, 551, 552(5), 559, 863
Thieringer, R., 569, 602(62)
Thim, L., 497
Thireos, 835, 836, 840
Thireos, G., 863
Thomas, 842
Thomas, C. A., 268
Thomas, C. L., 473
Thomas, C. M., 188
Thomas, D. Y., 79, 539, 864
Thomas, D., 233
Thomas, J. H., 163, 206, 207(21), 215(21), 319, 321, 567, 600(10, 15), 863
Thomas, J., 784
Thomas, M., 181
Thomas, W., 356
Thompson, J. R., 459, 461(21), 466(21)
Thorner, J., 77, 78, 80(1), 245, 250, 308, 379, 433, 434, 512, 665, 666(17), 672, 674, 684, 688, 702, 774, 777, 784, 840, 855, 856, 858

Thorpe, R., 578, 588(102)
Thorsness, M., 242
Thorsness, P. E., 169, 174(4), 280
Thrash, 841, 850
Thrash, C., 863
Thrash-Bingham, C., 852
Thuriaux, P., 799, 846, 857, 860, 796
Tiano, H. F., 127
Timko, K., 538
Tiollais, P., 231
Tipper, D. J., 418
Tirtiaux, C., 853
Toda, 836
Toda, T., 246, 862, 863
Todd, P. E., 234
Toh-e, A., 383, 389
Tokiwa, G., 860
Tokuhisa, J. G., 362
Tokuyasu, K. T., 625
Tollervey, D., 408
Tomenchok, 844
Tomenchok, D. M., 863
Tonegawa, S., 552
Tonino, G. J. M., 735
Torchia, T. E., 348
Tötterman, T. H., 570
Towbin, H., 511, 512(15), 581
Townsend, M., 120
Toyn, J., 669, 670(36)
Traut, R. R., 457
Traver, C. N., 254, 264(18), 270(18)
Treich, I., 74, 75, 303, 318(6), 846, 859
Trembath, M. K., 153
Trevelyan, W. E., 695
Triech, 846
Trimble, R. B., 683, 684(3, 4), 689(3, 4), 695
Trischmann, H., 729
Tropschug, M., 520
Trueblood, C. E., 246
Truehart, J., 694
Trueheart, 836, 839
Trueheart, J., 79, 86(19), 138, 303, 311(5), 318(5), 509, 568, 764, 776, 784, 787(12), 788, 839, 863
Trumbly, R., 429, 431(6), 442(6), 443(6), 450(6)
Tschopp, J., 622, 666(4)
Tschumper, G., 379, 397
Tsouladze, A. M., 858
Tsouladze, A., 837

Tsuboi, 847
Tsuboi, M., 863
Tsuchiya, E., 736
Tsuchiya, Y., 861
Tsuji, K., 596
Tsunasawa, S., 362, 366(6), 368(6)
Tsurugi, K., 459, 465
Tuite, M. F., 12, 197, 347, 357(14), 360(14), 374, 383(7, 10), 387(7), 389, 536, 537(10)
Tullius, T. D., 533
Turner, B. G., 491, 494(5), 496, 498(5)
Turoscy, 852
Turoscy, V., 863
Twu, J.-S., 332, 334(10)
Tye, B. K., 177, 179(17), 282
Tye, B.-K., 770, 771(30)
Tzagoloff, A., 150, 152, 153, 155, 240

U

Uchida, E., 645
Ucker, D. S., 397
Udem, S. A., 426
Ueda, M., 231
Ueki, K., 111
Uemura, T., 594, 806
Ulane, R. E., 429, 433(12), 434
Ulaszewski, 845
Ulaszewski, S., 853, 863
Ullrich, A., 698
Underwood, M. R., 426
Unlenbeck, O. C., 409
Uno, I., 100, 107(18), 108(18), 182, 568, 859
Urdea, M. S., 433, 491, 855, 858
Urrestarazu, L. A., 853
Usui, T., 861

V

Vai, M., 568, 599(33)
Valdivieso, M. H., 734
Vale, R. D., 594
Valent, B., 361
Valenzuela, P., 5, 231, 387, 491
Vallen, E., 197, 230(12)
Valls, L. A., 498, 644
Valls, L., 438

Van Arsdell, J. N., 438, 452(54)
Van Arsdell, J., 398
Van Bommel, J. H., 459, 460(16)
van der Aar, P. C., 495
van der Voort, H. T. M., 573
van Loon, 846
van Loon, A. P. G. M., 863
van Raamsdonk-Duin, M. M. C., 459, 460(11, 13, 16, 17), 463(13), 465
van Slyke, B., 467
van Spronsen, E. A., 573
van Tuinen, E., 544, 608, 609
Van Vunakis, H., 578
van't Riet, J., 474, 476
Van, A. S., 245
VanCleemput, M., 478
Vandekerckhove, J., 729
vanLoon, A. P. G. M., 521
Varndell, I. M., 604
Varnum, S., 552
Varshavsky, A., 317, 459, 461(18), 508, 529, 532(28)
Vassarotti, A., 240
Veenhius, M., 665, 666(25), 667(25)
Veenstra, A. E., 495
Verbree, E. C., 467, 469(63), 470(63), 472(63), 473(63)
Verdiere, 838
Verdiere, J., 863
Vershon, A., 244
Vestweber, D., 602
Vezinhet, F., 108, 124, 125(56)
Vidal, 846
Vidal, M., 863
Vieira, J., 284, 397, 509
Vieria, J., 481, 484(14), 485(11)
Vierra, J., 332, 334(8)
Villadsen, 842
Villadsen, I. S., 864
Villadsen, I., 837, 842
Villela, M. D., 465
Vioque, A., 468
Vlak, J. M., 718
Vlassov, A. V., 858
Vliegenthart, J. F. G., 683
Voelkel-Meiman, 850
Voelkel-Meiman, K., 864
Voelkil, K., 243
Vogel, A. M., 587, 588(125)
Vogel, J. P., 206, 207(27), 212(27), 568

Vogel, J. R., 665, 666(16)
Vogel, K., 386
Vogel, J. R., 792
Vogelstein, B., 266
Volkert, F. C., 196, 206(8), 251
Vollrath, 843
Vollrath, D., 47, 51, 58, 59(4), 66(5), 73(5), 144, 257, 259(22), 283, 356, 864
von Borstel, 844
von Borstel, R. C., 273, 857
Voynow, P., 455
Vrana, K. E., 533
Vu, L., 305, 318(8)

W

Waddell, C. S., 113, 114, 117(28), 119(32), 120(28)
Waggoner, A. S., 570
Wagner, P., 862
Wagstaff, J. E., 113, 114, 115, 119(32)
Wagstaff, J., 118, 119(43), 120(26)
Wahl, G. M., 853
Wakem, L. P., 7, 18, 20(52), 42, 48(14)
Walker, E. S., 587, 588(124)
Walker, N., 854
Wallis, J. W., 243, 283, 302
Walsh, M., 753
Walter, P., 521, 527(10), 544, 675, 680(1), 681(1), 682(1)
Walters, P., 844
Walworth, N. C., 568, 601(59), 663, 666(9), 667(9), 695, 696(31)
Wang, 844, 847
Wang, H. T., 115
Wang, H.-T., 864
Wang, J. C., 303
Wang, J., 100
Wang, S. S., 857
Wang, S.-S., 864
Wang, Y.-L., 570, 594
Warner, J. A., 457, 459(6)
Warner, J. R., 306, 309, 416, 424, 425, 426, 427, 465, 538
Warner, J., 175, 235, 297, 536
Warnke, R., 597
Warren, G., 625, 699
Warren, R., 127, 129(64)
Wassenaar, G. M., 465, 466(31)

Waters, G., 675, 677(2), 681(2), 682(2)
Waters, M. G., 674, 675
Waterston, R., 251
Watson, J. D., 4
Watson, M. E., 241
Watzele, M., 568, 599(35), 683, 688, 693(5)
Weaver, J. C., 183
Webb, J. R., 675
Weber, 839
Weber, E., 838, 864
Weber, J., 427
Webster, T. D., 348
Wei, C.-M., 492
Wei, R., 379
Weidman, P. J., 698, 699(9)
Weiffenbach, B., 137
Weiner, A. M., 4
Weinstein, B., 568, 601(46)
Weisberg, R. A., 316
Weiser, U., 431, 440, 449(63)
Weisman, L. S., 645
Weiss, 845
Weiss, D. G., 594
Weiss, R. B., 554
Weiss, W. A., 864
Weiss-Brummer, B., 154
Welch, 838
Welch, J. W., 864
Welch, S. K., 78, 706
Well, A. M., 18
Werner, D., 403
Werner-Washburne, M., 385, 675, 710, 715(5)
Wertman, K., 835
Wheals, A. E., 16, 241
White, C. I., 143
White, J. C., 570
White, J. G., 573
White, J. H., 394
White, J., 12
White, N., 433
White, S., 375, 383(10)
White, T. J., 552
Whiteway, 849
Whiteway, M., 79, 245, 864
Wiame, J. M., 353, 363
Wickerham, L. J., 12, 14(42), 701
Wickner, 841, 842, 846
Wickner, R. B., 7, 9(32), 12, 92, 173, 853, 863, 864

Wickner, R., 864
Wickner, W., 567, 600(18), 645
Wiedemann, L. M., 467
Wieland, T., 729, 730
Wiemken, A., 444, 622, 645, 665, 666(5, 18), 667(5)
Wigler, M., 92, 108, 198, 199(15), 214(15), 246, 508, 862, 863
Wikström, A.-C., 393, 397(15)
Wilcox, C. A., 698
Wilke, C. M., 356, 360
Wilkins, B., 188
Wilkinson, L. E., 78, 79(6), 774
Willets, N., 188
Williams, R. C., Jr., 579
Williams, R. H., 433
Williams, R. S., 683, 684(3), 689(3)
Williamson, 845
Williamson, D. H., 95, 158, 159(35)
Williamson, M. S., 864
Williamson, V. W., 349
Williamson, V., 835
Willick, G. E., 465
Wills, N., 861
Willsky, G. R., 665, 666(21)
Wilson, 843, 848
Wilson, A. C., 234
Wilson, D. W., 251, 698
Wilson, I. A., 508
Wilson, I., 510
Wilson, R. B., 864
Winey, 847
Winey, M., 848, 865
Winge, Ö., 32
Winkleman, D., 617
Winkler, G., 121, 123, 125(49, 51), 126(51), 473
Winston, 847, 848
Winston, F., 169, 244, 290, 309, 311(16), 313, 316, 320, 322, 351, 353, 356, 357, 361, 506, 758, 760, 770(17), 808, 811, 847, 854, 865
Wintersberger, E., 111
Wintersberger, U., 736
Wintz, P., 751
Wise, J. A., 406, 408, 409, 410(5), 413(5)
Witsenboer, H. M. A., 459, 460(11)
Witte, C., 602
Wittekind, M., 305, 318(8), 536
Wittenberg, C., 250, 568, 585(30)

Wittmann, H. G., 454, 457(1)
Wittmann-Liebold, B., 465, 467, 468
Wojciechowicz, D., 78
Woldringh, C. L., 476
Wolf, D. H., 429, 430, 431(4), 433(5, 9), 435(2), 438, 440, 441, 442, 443(4), 449(17, 63), 450(4), 451(4), 644, 857, 859
Wolf, E. D., 162
Wong, C., 551
Wood, G. S., 597
Wood, J. S., 40, 247
Wood, J., 757
Woods, C. W., 242
Woody, S. T., 860
Wool, I. G., 467, 465, 466(44), 468
Woolford, C. A., 436, 438, 439(54), 452(54, 59)
Woolford, C., 844
Woolford, J. L., Jr., 459, 461(21), 466(21), 568
Woolford, J., 465
Worthington, M., 554
Woudt, L. P., 459, 460(12), 463(12), 465, 466(31)
Woychik, N., 517
Wreschner, D., 403
Wright, A. P. H., 811
Wright, A., 206, 207(26)
Wright, B., 578, 600(103)
Wright, J. F., 124, 125(56)
Wright, R., 249, 565, 568, 588(5), 595(54), 601(54), 603, 607(5), 609, 614(8), 665, 666(15)
Wu, A. M., 478
Wu, C., 383
Wu, L. C., 736
Wu, R., 5, 118, 119(45), 162, 282, 301(10)
Wulf, E., 729, 730(10)

X

Xu, H., 239, 241(1), 330, 346, 351(9), 359(11)

Y

Yaffe, M. P., 158, 161(34)
Yaffe, M., 569, 602(64)
Yaguchi, M., 465
Yahara, I., 567, 600(20)
Yamada, Y., 182
Yamamoto, K. R. Y., 397
Yamamoto, K. R., 391, 393, 394, 395, 396(9, 21), 397, 398(9)
Yamamoto, S., 214, 215(39)
Yamao, F., 807
Yamasaki, E., 499
Yamashiro, C. T., 644, 645
Yamashita, I., 105, 108(39), 110(39)
Yamashita, S., 214, 215(35)
Yanagashima, N., 774
Yanagida, M., 594, 797, 807
Yanagishima, 835
Yanagishima, N., 78, 863
Yang, C. H., 568, 600(50), 718, 719(6), 727(6)
Yanisch-Perron, C., 284, 397, 481, 484(14), 509
Yanish-Perron, C., 332, 334(8)
Yanofsky, C., 478, 479, 535
Yansura, D. G., 231
Yansura, D., 479
Yarrow, D., 181
Yeh, 837, 847
Yeh, E., 865
Yirinec, B. D., 700
Yocum, R. R., 380, 381(32), 382(12), 389
Yocum, R., 377
Yokosawa, H., 446
Yoshid, 845
Yoshida, K., 774, 845, 865
Yoshida, M., 105, 108(39), 110(39)
Yoshimatsu, T., 859
Yoshimura, M., 855
Youderian, P., 244
Young, 835
Young, A. T., 282, 301(10)
Young, D., 230, 235(4)
Young, E. T., 387, 389, 863
Young, R. A., 198, 230, 231(1), 232(1), 233, 234, 235, 237(2), 238(1, 2, 21)
Young, R., 419, 508, 517, 846, 860

Z

Zakain, V. A., 736
Zakharov, I. A., 845, 852

Zakian, V. A., 7, 241
Zakour, R. A., 362
Zakour, R., 510
Zamb, T. J., 100
Zaret, K. S., 306, 375
Zaslavsky, V. G., 231
Zassenhaus, H. P., 162
Zealy, G. R., 241
Zechel, K., 248
Zehner, Z., 392, 397(13)
Zeigel, R. F., 736
Zhang, J.-Y., 467
Zhou, P. B., 559
Zickle, D., 615
Zickler, D., 603

Ziegler, F. D., 683, 684(3, 4), 689(3, 4)
Zimmerman, F. K., 53
Zimmermann, R. A., 473
Zinn, K., 520
Zipser, D., 231
Zitomer, R. S., 416
Zoller, M. J., 136, 137(14), 139(14), 145(14), 146(14), 314, 362
Zoller, M., 137, 246, 863
Zollinger, M., 611, 626(10)
Zonghou, S., 155
Zubenko, G. S., 429, 433(10), 435(10), 436, 438, 442, 447(10), 452, 453(57)
Zubenko, G., 678
Zulch, G., 155

Subject Index

A

N-Acetylglucosamine transferase, structural gene for, 242
Acridine mustard ICR-170, as mutagen, 274
Acridine orange, 571
Actin
 staining, with fluorochrome-conjugated phalloidin, 729–731
 visualization of, 729
 with immunofluorescence methods, 600
Actin-associated proteins, 248
Actin-binding proteins, 248
ADH1 gene
 overexpression, in glucose, 387
 promoter, 383, 389
ADP/ATP carrier, *in vitro* transcription and translation of, for mitochondrial protein import assays, in isolated mitochondria, 639
a-factor. *See also* Mating factors
 response to, assay for, problem with, 88–89
Agglutinins, 78–79, 774
AgNOR (silver-binding protein), 720
Alcohol dehydrogenase isoenzyme III, for mitochondrial protein import assays, 632
 in vitro transcription and translation of, in isolated mitochondria, 639
Allele rescue, 298–301
Alleles
 dominant, 10
 nomenclature for, 8
 recessive, 10
 nomenclature for, 8
Allelic complementation, 19–20, 41
α-factor, 78. *See also* Mating factors
 degradation of, assay of, 709–710
 internalization, assay of, 707–709
 as marker for endocytosis in yeast, 699–700, 702–710
 production, rate-limiting step of, analysis using gene overproduction, 249–250
 response to, assay for, 87–88
 ^{35}S-labeled
  binding of, to Mata cells, 706–707
  preparation of, 702–706
 transit forms of, in evaluation of secretory pathway, 674
AMCA. *See* 7-Amino-4-methylcoumarin-3-acetic acid
American Type Culture Collection, 6
α-Aminoadipate, in counterselection techniques, 306, 309
 medium for, 310
7-Amino-4-methylcoumarin-3-acetic acid, 570–572
Aminopeptidase I, 429–430
Aminopeptidases
 agar overlay test
  Jones method, 443–444
  Trumbly and Bradley method, 442–443
 enzymatic assay of, 450–451
 genetic analysis, 442–444
Aminopeptidase yscCo, 429–430
Ammoniacal silver solution, 728
Anthranilate synthase hybrid proteins, production of, 478–481
Antibodies, for immunofluorescence methods, for yeast, 578–586
 affinity purification
  on columns, 584–586
  importance of, 579–581
  on nitrocellulose blots, 581–584
 primary
  sources of, 578–579
  type of, 578–579
  type of immunogen for, 579
 secondary, 586
Antibody probes, used in bacteriophage λgt11 vector–host system, 233–234
Antigenic determinants
 assembled topographic, 234
 segmental, 234
ARG4 marker, 379

ARS elements, 199, 282
Asci, 94
  digestion with snail juice, 32–34
  dissection of, 18, 32–34
  micromanipulation, 21
Ascospores
  isolation of, 53
  micromanipulation, 34–35
  pure and viable, preparation of, 125–126
  relocation and transfer of, 21
  RNA from, isolation of, 401–402
  separation of, 34–37
  sequential separation of cluster of, 5 mm apart on petri dishes, steps for, 35–37
Ascus. *See* Asci
ATPase
  F_1-
    α-subunit, *in vitro* transcription and translation of, for mitochondrial protein import assays, in isolated mitochondria, 639
    β-subunit
      in vitro transcription and translation of, for mitochondrial protein import assays, in isolated mitochondria, 639
      for mitochondrial protein import studies *in vivo*, 632
  H^+-
    azide-sensitive, assay, in assessment of purity of vacuole preparations, 658
    vacuolar, proton-pumping activity, assay of, 659
    vacuolar membrane-associated activity, assay, 655–656
    vanadate-sensitive, assay, in assessment of purity of vacuole preparations, 658

B

B23 (silver-binding protein), 720
Bacteriophage λ
  yeast gene bank, 50–51
  yeast inserts, 40
  hybridization tests, 50–51
Bacteriophage λgt11 vector–host system
  antibody probes used in, 233–234
  DNA library used in, 232–233
  and expression of foreign DNA in recombinants, 230–232
  gene isolation with, 230–238
  library screening
    procedure, 235–238
    by protein activity, 234–235
  phage and bacterial strains used for, 238
Barrier activity, 78
  assay for, 89–91
  visualized on petri plates, 89–90
Barrier confrontation assay, 91–92
Benzamidine, as protease inhibitor, 433–434
Biotin-avidin techniques, in immunofluorescence methods, 596–597
BUD5 gene, mutations, and mating-type switching, 138
Bud scars, Calcofluor staining, 717–718, 732–735

C

C23 (silver-binding protein), 720
Calcofluor, 571
Calcofluor staining, of bud scars, 717–718, 732–735
Canavanine selection technique, 306–307
  medium for, 310
  strains for, 309
Candida boidinii, peroxisomes from, 667
CAN1 gene, as counterselectable marker, 297, 305–306
5-(and 6)-Carboxy-2′,7′-dichlorofluorescein diacetate. *See* CDCFDA
6-Carboxyfluorescein diacetate. *See* CFDA
Carboxypeptidase S, 429–430
  coupled assay for, 439–442
  enzymatic assay of, 449–450
  genetic analysis, 439–442
Carboxypeptidase Y, 429–430
  activity
    APE overlay test for, 437–439
    well test for, 439
  enzymatic assay of, 448–449
  genetic analysis, 436–439
Carboxypeptidase Y activity filter assay, 659–660
CDCFDA, staining of yeast vacuoles, 646–648

CDC46 gene product, localization of, 596
CDC genes, 100
cdc mutants, cell division, rate-limiting step of, analysis using gene overproduction, 250
cDNA library, 232-233
  for complementing specific mutations in yeast, construction of, 212-214
  vectors for, 212-214
Cell cycle, yeast
  arrest, in mating reaction, 774-775
  assay for, 87
  G^1-S transition, rate-limiting step of, analysis using gene overproduction, 250
  of mating pair, 79
  reviews of, 4
Cell division, of yeast, reviews of, 4
Cell extracts, for *in vitro* protein synthesis, 536-537
Cell-free translation systems, 536
  yeast, 536-537
Cell fusion
  in mating reaction, 774-775
  assays of, 776-780
  mutants. *See also* Fusion mutants
  isolation of, 787-790
Cell structure, effect of overproduction of proteins on, 248-249
Cellufluor, staining of chitin-rich structures, 732
Cellular orientation, correlation with nuclear component orientation, 717-718
Cellular regulation, of yeast, reviews of, 4
Cell wall
  dissolution, in mating reaction, 774
  removal, for electron microscopy, 603, 605
  visualization, with immunofluorescence methods, 599
CEN mutations, mitotic analysis of, 758-766
  plasmids for, 758-762
  strains for, 758-762
Centraalbureau voor Schimmelcultures, 6
Centromere linkage, 41-42, 55
CFDA, staining of yeast vacuoles, 646-649
CHEF. *See* Contour-clamped homogeneous electric-field gel electrophoresis

Chitin, cell wall, staining of, with Calcofluor, 732-735
Chitinase, 684
  electrophoretic mobility, response to conditions affecting glycosylation, 686-688
  O-linked mannooligosaccharides on, analysis of, 690-696
  secreted, isolation of, by chitin binding, 696
Chromatin, structure, effect on gene expression, 244
Chromosomal DNA, *S. pombe*, preparation of, 817-818
Chromosomal footprinting, 552, 557-562
Chromosomal genes, yeast, nomenclature for, 8-11
Chromosomal mapping, 2-μm procedure, 42-43
Chromosome VII, 59
Chromosome IX, mutations on, mapping, 47
Chromosome XII, 7, 59
Chromosome XV, 59
Chromosome XVI, mapping of *CTF1* to, 64-65
Chromosome XVII, 7
Chromosome blotting, 59-66
  application of, 64-66
  principle of, 59-60
Chromosome exchange, in *spo13* meiosis, systems used to monitor, 118-119
Chromosome fragmentation, 51, 66-76
  advantages and disadvantages of, 72-73
  application of, 71-73
  events, 74-75
  in vivo, 58
  principle of, 66-67
  procedure, 68-72
  segmental aneuploidy produced by, 73-75
  vectors, 66-67
Chromosome fragments
  analysis of mitotic chromosome segregation in *S. cerevisiae* using, 756-766
  chromosome fragment strain construction for, 758-762
  manipulation of chromosome fragment sequences, 762-764
  principle of, 757-758

by quantitation of chromosome stability, 764–766
analysis of mutant centromere DNA sequences using, 764–766
centromere structure, changing, 763–764
endogenous sequences on long arm of, changing, 762
manipulation of chromosome fragment sequences, 762–764
markers on short arm of, changing, 762–763
meiotic segregation errors associated with, 771–772
nomenclature for, 76
Chromosome-length polymorphisms, 59
Chromosomes. *See also* Yeast artificial chromosomes
immunocytological analysis, 127
large, cleavage, using *HO* endonuclease, 143–144
structure, study of, using *HO* endonuclease, 143–144
transmission of, fidelity, and gene overexpression, 247
yeast, 6–7
genetic and physical maps of, 827–829
Chromosome segregation, in *S. cerevisiae*
analysis of, 749–773
during meiosis, using yeast artificial chromosomes, 765–773
during mitosis
by digital imaging microscopy of nuclear DNA movement, 750–756
using chromosome fragments, 756–766
meiotic
associated with chromosome fragments, 771–772
yeast artificial chromosomes as markers for, 771–773
Chromosome structure, of yeast, reviews of, 4
CHY2 gene, as counterselectable marker, 297, 305–307
Chymostatin, as protease inhibitor, 433
cir° tester strains, 42–47
Citrate synthase
in vitro transcription and translation of,

for mitochondrial protein import assays, in isolated mitochondria, 639
for mitochondrial protein import studies *in vivo*, 632
Cloned genes
chromosomal assignments of, 42
complementing activity, localization, 226–227
confirmation of, 226–228
expression in yeast, vectors for, 373–388
hybridization to chromosomes, 47–49
hybridization to fragments, 49–51
mapping, 38–40, 57–77
positional mapping of, 57–77
toxicity in *E. coli*, 229–230, 232
CLP. *See* Chromosome-length polymorphisms
Colony color sectoring assay, 68
Colony immunoblotting, to detect vacuolar antigens, 660–661
Complementation
gene cloning in yeast by, 195–230, 319
of recessive alleles, 196–197
Complementation analysis, 19–20
and confirmation of cloned genes, 226–228
Complementation assays
in cell and nuclear fusion mutants, 780–782
facilitation of, with mating-type switch using *HO* gene, 137–138
physiological, in analysis of heteroplasmic zygotes, 155
Complemention groups within gene, nomenclature for, 10
Concanavalin A, in analysis of glycosylation state of proteins, 689
Conditional mutations, in essential genes, pop-in/pop-out strategy for, 296–297
Conjugation
in bacteria, 187–188
between *E. coli* and *S. cerevisiae*, 188–195
by conjugation on permissive media, 190–191, 194
direct isolation of transconjugants on selective medium, 190–194
in *S. cerevisiae*, 774–775
Contour-clamped homogeneous electric-field gel electrophoresis, 47–49, 59–60

electrophoretic karyotypes obtained by, 60–61
procedure, 63
Counterselection
   for allele replacement, 297–298
   media for, 310
   methods of, comparative advantages and disadvantages of, 307
   removal of wild-type gene by, 304–305, 317
Cross-complementation, 197–198
Crossing-over, mitotic, 51–52
Cross-streaking, 19
CTF1 gene, mapping of, to chromosome XVI
   by chromosome blotting, 64–66
   by chromosome fragmentation, 71–73
cycl-31 mutants, transformation with synthetic oligonucleotides, 363–369
cycl-812 mutants, transformation with synthetic oligonucleotides, 363–369
Cycloheximide selection technique, 306
   medium for, 310
   strains for, 307–309
Cytochrome b_2
   in vitro transcription and translation of, for mitochondrial protein import assays, in isolated mitochondria, 639
   for mitochondrial protein import studies *in vivo*, 632
Cytochrome c_1, for mitochondrial protein import studies *in vivo*, 632
Cytochrome c peroxidase, for mitochondrial protein import studies *in vivo*, 632
Cytochrome oxidase
   subunit IV, *in vitro* transcription and translation of, for mitochondrial protein import assays, in isolated mitochondria, 639
   subunit V, *in vitro* transcription and translation of, for mitochondrial protein import assays, in isolated mitochondria, 639
Cytochrome P-450 lanosterol demethylase, structural gene for, 242
Cytoduction, in fusion mutants, 778–779
Cytoplasm, enzymatic markers for, assay, in assessment of purity of vacuole preparations, 657–658

Cytoplasmic translation extracts, preparation of, 677–680
Czechoslovak Collection of Yeasts, 6

D

DAPI, 571
   for nuclear staining in meiosis, 99
   staining
      in analysis of chromosome segregation in *S. cerevisiae*, 750–756
      of *S. pombe* nuclei, 821–822
Diaminobenzoic acid, colorimetric assay of DNA, 98
4′,6-Diamino-2-phenylindole. *See* DAPI
Digital imaging microscopy, of nuclear DNA movement during mitosis, 750–756
Dipeptidyl aminopeptidase B, vacuolar membrane-associated activity, assay, 655–657
Diphenylamine, assay of DNA, 98–99
Diploids
   a/a
      formation of, situations leading to, 101–102
      sporulation of, 101–103
   a/α, 77, 79, 132. *See also HO* gene
   α/α
      formation of, situations leading to, 101–102
      sporulation of, 101–103
   buds, 16
   composition of, 17
   formation, assay of, 80
   gene disruption in, 291–292
   homozygosity of heterozygous markers in, 51–52
   isolation, by pheromone tests, 83
   from nonmating mutants, 92–93
   size of, 16–17
   unsporulated, eliminating or reducing, from culture, 20
3,3′-Dipropylthiocarbocyanide iodide, in determination of mitochondrial membrane potential, 636
Direct immunofluorescence, 572
diS-C_3-(5). *See* 3,3′-Dipropylthiocarbocyanide iodide
DNA. *See also* Mitochondrial DNA

chromosomal, size of, 169
chromosomal-sized, from *S. cerevisiae*,
    47–49
  preparation of, in agarose, 60–62
  pulsed-field gel electrophoresis of, 58
colorimetric assay of, 98
complementing, determination of genetic
    locus of, 227–228
diphenylamine assay of, 98–99
encoding secretory proteins, *in vitro*
    transcription of, 675–677
  with bacteriophage SP6 RNA polymerase, 675–677
  with bacteriophage T7 RNA polymerase, 675, 677
fractionation, by cesium chloride
    gradients, 174–175
genomic
  direct sequence analysis, 550–552
    chain-termination method, 550–551
    methods, 550–551
    protocol for, 552–557
    using modified T7 DNA polymerase,
        551–556
    using *Taq* DNA polymerase, 551,
        556–557
    footprinting, 552, 557–562
  partially digested, size fractionation of,
      220–221
  high-molecular-weight
    human, isolation of, 255–257
    partial digestion of, 257–259
    preparation of, 218–219
    size fractionation, in YAC cloning
      consequences of, before and after ligation, 268–270
      importance of, 267–269
      on sucrose density gradient, 261–263
in vivo labeling, 423
integrative transformation of yeast,
    281–301
isolation, 169–175
  by cesium chloride gradients, 174
  small-scale procedure, 174
  from yeast strains containing plasmids,
      322–323
premeiotic synthesis of, 97–99
preparation of, as agarose gel wafers, 49
radioactive labeling of, 97–98
repair, studies of, using double-strand
    breaks, 142–143
repeated sequences, restriction maps of,
    175–179
restriction spectra, 169
  correlation between bands in and
    repeated DNA sequences, 175–179
  transmission to yeast, by conjugation
    between yeast and bacteria, 188–195
yeast, 169–182
  in yeast nucleus, immunoelectron
    microscopy of, 622–624
DNA-binding proteins, analysis of, reverse
    band-shift assay for, 529–532
DNA fragments, liberation, with *HO*
    endonuclease, 144
DNA sequence analysis
  of mitochondrial DNA mutations,
    160–161
  shuttle mutagenesis elements as mobile
    priming sites in, 341–342
Domains, within gene, nomenclature for, 10
Dominance tests, in cell and nuclear fusion
    mutants, 780–782
Double-label immunofluorescence, for
    yeast, 597–598
  nucleolar labeling by, 719
Double-strand breaks
  delivery of, 140–143, 282
  use of
    rationale for, 140
    for studies of DNA repair, 142–143
    for studies of recombination, 140–142,
      282
Double-stranded (ds) RNA viruses, 7–8
Doubling time, of normal laboratory
    haploid strains, 15–16
DPAP A, assay, in assessment of purity of
    vacuole preparations, 658–659

E

EGTA, as protease inhibitor, 434
Electron microscopy, 565. *See also*
    Immunogold labeling
  nucleolar staining for, 724–728
  thin-section, preparation of yeast cells for,
    602–608
    chemicals for, 604
    enzymes for, 604
    materials for, 604
    methods, 604–608

modified procedures
  for cell types with walls stable to
    digestion, 606-607
  for sporulating cells, 607
  for stationary-phase cells, 606-607
  osmium tetroxide-ferrocyanide
    fixation for visualization of membrane structures, 607-608
  reagents for standard procedures,
    604-605
  standard procedures for fixing and
    embedding vegetatively grown
    cells, 605-606
of yeast
  difficulties of, 602-603
  principles of, 603
Electrophoretic karyotype. *See also specific strains*
  for *S. cerevisiae*, development of, 59-60
Electroporation
  mutagenized inserts returned to yeast by,
    337
  one-step gene disruption with, 291
  transformation of yeast by, 182-187
    plating, 185-186
    preparation of electrocompetent cells,
      183-184
    procedure, 184-185
EMS. *See* Ethylmethane sulfonate
Endocytosis
  definition of, 697
  fluid-phase, 697
  functions of, 697
  receptor-mediated, 697-698
  in yeast
    α-factor as specific marker for,
      699-700, 702-710
    assays, 697-710
    Lucifer Yellow as nonspecific marker
      for, 698-702
    markers for, 698
Endoglycosidase H, in analysis of glycosylation state of proteins, 683-688
Endoplasmic reticulum
  enrichment of, 665-667
  enzymatic markers for, assay, in
    assessment of purity of vacuole
    preparations, 658
  immunological markers, 666
  isolation of, 662-663
  marker enzyme analysis, 668-669

visualization, with immunofluorescence
  methods, 601
Endoproteinase A, 429
Endoproteinase B, 429
Enzymes
  formed by allelic complementation, 19
  structural genes for, isolation of, 242-243
Epitope-tagged protein
  initial examination of, 511-512
  preparative immunochromatography of,
    517
Epitope tagging, 387-388
  advantages of, 508
  applications of, 508
  epitope addition
    methods, 509-510
    principles, 509
  principle of, 508
  protein surveillance using, 508
    buffers for, 512-513
    immunoprecipitation procedure,
      515-516
    in vivo labeling of yeast proteins,
      513-514
    materials, 512
    media, 512
    preparation of cell extracts, 514-515
    troubleshooting, 516
    useful antibodies for, 510-511
Epoxy resins
  alternative etching methods with, 626
  embedding with, 613-614
  removal of, from ultrathin sections,
    625-626
Escherichia coli
  calcium shocked
    preparation of, 325-326
    transformation of, 325-326, 328
  cloned genes toxic in, 229-230, 232
  expression of foreign DNA in, 230-232
  lon mutants, 231
  strains, used in shuttle mutagenesis, 332,
    334
  transformation, using DNA from yeast
    strains containing plasmids, 323-326
Ether test, for screening of colonies for
  spore viability and/or sporulation, 146-147
Ethidium bromide treatment, for induction
  of rho° and rho^- strains, 151

Ethylenediaminetetraacetate, as protease inhibitor, 433
Ethylmethane sulfonate mutagenesis, 273
  method, 277-278
  of *S. pombe*, 805-806
  safety considerations with, 276-277

F

Field-inversion gel electrophoresis, 47-49
FIGE. *See* Field-inversion gel electrophoresis
Fission yeast. *See Schizosaccharomyces pombe*
FITC. *See* Fluorescein isothiocyanate
FITC-dextran, accumulation of, in yeast, 699
Fluorescein isothiocyanate, 570-572
Fluorescence microscopy, vacuole visualization with, 645-653
Fluorescence microscopy techniques. *See also* Immunofluorescence methods
5-Fluoroorotic acid
  in counterselection techniques, 305
  medium for, 310
  in pop-in/pop-out gene replacement, 296-297
Fluorophores, 570-572
Flying promoter transposon, 340, 342
Footprinting, 552, 557-562
Formaldehyde
  fixation of yeast cells with, 586-587, 611-612
  stocks, deterioration of, 588
  stock solution, 611
  formula for, 625
Formaldehyde developer, for nucleolar staining, 728
Forward genetics, 302
Frameshift mutations, 274
Fuchsin, basic, decolorized, 728
Fuchsin-sulfurous acid, 728
FUS1 gene, 79
FUS genes, mutations. *See* Fusion mutants
FUS1 induction, assay of, 86-87
Fusion-gene product overexpression, for immunofluorescence methods, 594-595
Fusion mutants
  assays for, 776-780
  with bilateral block in cell or nuclear fusion, isolation of, 782-787

complementation and dominance tests in, 780-782
cytoduction in, 778-779
cytological analysis
  by nuclear staining, 777
  for septum formation, 778
defective in cellular fusion, isolation of, 787-790
defective in nuclear fusion, isolation of, 790-792
isogenic *MAT*a and *MAT*α strains, 780
signal reception and transduction assay in, 776-777
zygotes, clonal analysis of, by micromanipulation, 778-780
Fusion proteins, in immunofluorescence methods, 579
fus1 mutants. *See* Fusion mutants
fus2 mutants. *See* Fusion mutants

G

Galactose regulon, study of, using gene overexpression, 243-244
GAL promoter, 389
  in method for measurement of mRNA decay, 416-417, 421
  regulated expression with, 379-385
  used to curtail protein expression, 384-385
GAL1 promoter, 493
Gene cloning, 3. *See also* Cloned genes
  approaches to, 198-199
  by complementation, 195-230
  cloning of dominant alleles, 197
  complementation of recessive alleles, 196-197
  and high-copy cross-suppression, 197-198, 230
  for isolation of regulated promoters, 198
  strategies for, 196-199
  in yeast, to isolate specific genes from other organisms, 198
  trouble-shooting guide, 228-230
Gene clusters, nomenclature for, 10
Gene disruption, 282-283, 303
  in diploid cell, 291-292
  by internal fragment disruption, 293-294
  media for, 283-284
  one-step, 290-293
  in *S. pombe*, 816-817

transposon-induced, 292-293
  yeast markers for, 284-285
  yeast strains for, 283-284
Gene dosage
  alteration
    for analysis of biological pathways, 245-247
    for analysis of rate-limiting steps in biological pathways, 249-250
    and macromolecular fidelity, 247-248
    for study of cell structure, 248-249
    for study of macromolecular interactions, 243-245
  selection of recombinant clones through effects of, 241-243
Gene expression
  constitutive, vectors for, 389-394
  inducible, vectors for, 389-391, 394-397
  regulated, with GAL promoter, 379-385
    and overproduction of toxic protein, 381-384
  regulation, by inorganic phosphate, 385-386
  translational regulation of, 536
Gene mapping, 18, 38-57
  hybridization to chromosomes, 47-49
  hybridization to fragments, 49-51
  2-μm method, 40, 42-47
  positional, 38
    using mutations, 57-58
  procedures, 41-57
  single-gene segregation and centromere linkage, 41-42
  strategy, 38-41
    current, 40
    far future, 40-41
    near future, 40
Gene overexpression, 239-251
  future perspectives, 250-251
  in glucose, 387
  for immunofluorescence methods, 594-595
  methods of achieving, 239-241
  uses of, 241-250
Gene replacement, 3-4
  direct, use of counterselectable marker to facilitate, 297-298
  one-step, 195, 290-293
    mating-type switching by, 137
  pop-in/pop-out, 294-297
  in *S. pombe*, 816-817

techniques, 293-298
Genes, yeast, 7. *See also* Cloned genes; Mapped genes; Resistance genes; Sensitivity genes
  isolation of, with λgt11 system, 230-238
  meiosis-specific, 99-100
  mitotic cell cycle, 100
Gene symbols, glossary of, 830-834
Gene targeting, 282-283, 303
  media for, 283-284
  methods, 285-290
  yeast strains for, 283-284
Genetic cassettes, 132
Genetic engineering, of yeast, 3-4
Genetic loci, nomenclature for, 8
Genetic map, of *S. cerevisiae*, 827-829
Genetic mapping, 66
  nomenclature for, 11
Genetic nomenclature
  committee for, 11
  of yeast, 8-12
Genetics, of yeast, reviews of, 4
Genome, yeast, 6-8
  of diploid cell, 9
  physical map of, 49-50
Genomic banks, plasmid-borne. *See also* Plasmid banks
  acquisition of, 214-224
    by mail, 214-216
  de novo construction of, 216-224
    initial considerations, 216-218
    ligation step, 217
    using dephosphorylated vector DNA, 217-218, 221-224
    using partially filled-in overhangs, 218, 221, 224
  incomplete, 229
Genomic DNA library, 232-233
Giemsa stain, for nuclear staining in meiosis, 99
Glass fibers
  source, 30
  use of, for constructing microneedles, 29-31
Glucan cell wall, degradation, 171-172
Glucose, gene overexpression in, 387
Glucose-6-phosphate dehydrogenase, assay, in assessment of purity of vacuole preparations, 657-658
Glusulase, 749
  cell wall removal with, 605

source, 32, 604
N-Glycanase, in analysis of glycosylation state of proteins, 688
Glycoprotein localization, 662
Glycoproteins
  carbohydrate portion, radiolabeling, general considerations, 690, 694–696
  electrophoretic mobility, response to conditions affecting glycosylation, 685–688
  N-linked carbohydrate, analysis of, 689
  N-linked chains on, estimation of number of, 688–689
  O-linked mannooligosaccharides on, analysis of, 690–696
    autoradiography, 696
    Endo-H treatment, 695
    immunoprecipitation procedure, 694–695
    [^{3}H]mannose labeling, 695–696
    materials for, 693
    methods, 690–696
    radiolabeling procedure, 690, 694–696
    reagents and media for, 694
    strains used for, 693–694
  yeast, analysis of, 682–697
Glycosylation state of proteins, analysis of, 683–688
Golden retriever plasmids, 320
Golgi apparatus
  enrichment of, 666–667
  enzymatic markers for, assay, in assessment of purity of vacuole preparations, 658–659
  immunological markers, 666
  marker enzyme analysis, 668–669
Golgi complex, visualization, with immunofluorescence methods, 601
GPA1 gene, 79–80
GPD-promoter based vectors, 390–391
G protein, in mating reaction, 79
  and phenotypes caused by increased gene dosage, 246–247
Growth, of yeast, 12–17
  for biochemical analysis of proteases, 444
  for isolation of nuclei, 738–739
  for mRNA decay measurement experiments, 418
  temperature for, 171

Growth inhibitors, in gene isolation using multicopy plasmid libraries, 241–243

H

Halo test
  for detection of killer activity, 84–85
  for formation of pheromones, 84–86
Hansenula polymorpha, peroxisomes from, 667
HAP1 induction, from *GAL* UAS, 381–384
Haploid cells
  buds, 16
  composition of, 17
  phenotypes, 77
  size of, 16–17
Haploidization
  commitment to, 97
  monitoring, 96
Haploid meiosis, 117
  in identification of Rec$^-$ mutants, 118–119
  in isolation of spore-wall mutants, 121–123
  spo13, 113–115
  sytems used to monitor exchange during, in isolation of Rec mutants, 119–120
Haploid spores, of strain that carries *HO* allele, mutagenesis of, 108–109
Heat-shock protein genes, induction of, assay of, 711–717
Heat-shock proteins, 710
Heat-shock (stress) response, induction
  heat-shock protein–*lacZ* fusions as reporters for, 714–716
  HSP genes used as probes for, 712
  protein synthesis in, assay of, 712–714
  protocol for, 710–711
  and RNA levels, 711–712
  and thermotolerance, assay of, 714
  transcriptional fusions as reporters for, 716
  translational fusions as reporters for, 716
Heat-shock transcription factor, immunoelectron microscopy mapping of, 620
Hepatitis B surface antigen, expression in yeast, using *ADH1* and *PGK* transcription vectors, 387
Hgm^R marker, 378–379

HIS3 marker, 379
HIS4 marker, 379
Histone proteins, balance of, in chromosome assembly, 247
HML, 102, 132
  alleles, identification of, 104
  mutations, identification of, 104-106
HM loci, expression of, events causing, 105-106
HMR, 102, 132
  alleles, identification of, 104
  mutations, identification of, 104-106
Hoechst 33258, 571
HO endonuclease, 133-134
  cleavage of large chromosomes with, 143-144
  cleavage site, 133-135
  in delivery of double-strand breaks, 140-143
  DNA fragment liberation with, in detection of homology between DNA segment and YAC plasmid, 144
  in study of chromosome structure, 143-144
HO gene
  cloned, inactive, to convert homothallic strains to heterothallism, 144-146
  inactivation, 135-136
  induction of homothallic interconversion with, 102, 104, 112
  practical use of, for mating-type switching, 132-146
HO/HO strains, 782-787
Human u-PA, screening yeast colonies for, 502-506
Hybridization
  in chromosome blotting procedure, 63-64
  gel preparation for, 63
Hybridization experiments, 198
  with mitochondrial RNA, anomalous results with, 162
  in restriction spectra analysis, using *EcoRI* restriction spectra, 181-182
Hybridoma cell culture
  materials, 511
  methods, 511
Hydroxylamine, mutagenesis protocol using, 315-316
Hydroxymethylglutaryl-CoA reductase
  localization of, 595

overproduction of, effects on cell structure, 249
structural gene for, 242

I

IME1 gene, 102, 105, 245
  mutations, identification of, 106-107
IME2 gene, 245
IME21 gene, 108
Immunoblotting, for detection of yeast membrane and lumenal space components, 668-669
Immunoelectron microscopy, 604. *See also* Immunogold labeling
  cell fixation
    choice of fixatives, 611-612
    procedure, 611-612, 614-616
  cell growth for, 614
  chemicals for, 610
  colloidal gold secondary antibody conjugates for, 610
  data collection, 620-624
  embedding procedure, 614, 616
    choice of resins for, 612-614
  embedding resins for, 609
  equipment for, 610
  fixation of antibodies, 619
  fixatives for, 609
  grids for, 616-617
  grid staining in, vanadium stain for, 626
  heavy metal staining for, 610, 619-620
  immunostaining procedure, 618
  immunostaining solutions and antibody, preparation of, 617-618
  materials for, 609-610
  nucleolar staining for, 725
  on-section labeling technique, problems with, 624-625
  primary antibody staining, 618-619
  proteins for, 610
  secondary anti-IgG antibody-colloidal gold conjugates, 619
  sectioning procedure, 616-617
  using anti-DNA antibody, 622-624
  using anti-DNA plus dsDNA, 622-624
  using human autoimmune serum (#238), 620-624
  of yeast, planning, 609

Immunofluorescence methods
 for *S. pombe*, 822–823
 for yeast, 565–602, 648–650
  actin system visualization in, 600
  adsorption of affinity-purified
   antibodies to fixed yeast cells, 651
  antibodies for, 578–586
  antibody staining of fixed cells,
   652–653
  antigens immunolocalized with,
   648–650
  applications of, 567–569
  cell wall visualization in, 599
  cytoskeletal elements visualized in, 600
  detection of cell surface components,
   599–600
  detection of nonabundant antigens, 593
  detection of specific structures,
   599–602
  double-label immunofluorescence,
   597–598
  endoplasmic reticulum visualized in,
   601
  fixation for, 586–589, 651–652
   choice of fixative, 587–588
   duration of, 588–589
   rapid, importance of, 587
   resuspension in phosphate-buffered
    formaldehyde, 588
   standard protocol for, 586–587
  flexible, empirical approach in, 569
  fluorophores for, 570–572
  Golgi complex visualized in, 601
  growth conditions for, 569
  membrane-bounded organelles
   visualized in, 600–602
  microscopy and photomicroscopy,
   572–578
   bleaching problem in, 576–577
   field diaphragm, 574–575
   film for, 577
   filters, 574
   immobilization of cells, 576
   lamps, 574
   microscopes, 572–574
    confocal scanning, 573–574
    conventional, 572–573
   number of photographs in, 577–578
   objectives, 575–576
   slides for, 576
  microtubules visualized in, 600

  mitochondria visualized in, 602
  mounting medium for, 590
  multiple-label immunofluorescence,
   598
  neck filaments visualized in, 600
  nucleus visualized in, 600–601
  overexpression of gene products and
   fusion-gene products, 594–595
  permeabilization protocol for, 590–591
  peroxisomes visualized in, 601–602
  plasma membrane visualization in,
   599–600
  preparation of slides, 591
  procedures, 589–598
  reagents and materials for, suppliers of,
   566
  sandwiching methods, 595–598
  secretory vesicles visualized with, 601
  solutions for, 589–590, 650–651
  spindle-pole body visualization in, 600
  staining of cells with antibodies,
   591–593
  strains for, 569
  utility of, 567–569
  vacuoles visualized with, 601, 646,
   649–653
  value of positive control, 593
  video enhancement of signals, 594
Immunogold labeling. *See also* Immunoelectron microscopy
 on-section labeling technique, problems
  with, 624–625
 of ultrathin sections, 608–626
  data collection, 620–624
  procedure, 617–620
Immunoprecipitation, of proteins, in
 evaluation of secretory pathway, 672–
 674
Indicator media, 13, 16
Indirect immunofluorescence, 572, 648–650
Inorganic phosphate, regulation of gene
 expression, 385–387
Inositol starvation, method, for mutant
 enrichment, 279–281
Inositol starvation medium, 280–281
Interferon
 expression in yeast, using *ADH1* and
  PGK transcription vectors, 387
 phosphate- and temperature-regulated
  expression of, by *PHO5* UAS,
  386–387

Introns, 7
Invertase, 684
  cytoplasmic form, 685–686
  electrophoretic mobility, response to conditions affecting glycosylation, 685–686
  ER form, 685–686
  N-linked chains on, estimation of number of, 688–689
  secreted form, 685–686
Ion channels, of vacuolar membrane, 659
Iron-sulfur protein, *in vitro* transcription and translation of, for mitochondrial protein import assays, in isolated mitochondria, 639

K

Karmellae, 249
KAR1 product, localization of, 595
KEX2 gene product, localization of, 596
Kex2 protease, assay, in assessment of purity of vacuole preparations, 658
Killer strains, designations for, 11
Kilmartin monoclonal antitubulin antibody YOL1/34, 593
KIL-o mutants, 7–8
Kluyveromyces lactis, restriction spectra of, 180–181

L

Labeling techniques
  in vivo, for yeast proteins, 513–514
  in yeast, 423–428
Lectins, in analysis of glycosylation state of proteins, 689
LEU2, DNA fragment, yeast transformation, 281–282
LEU2 marker, 378–379
Leupeptin, as protease inhibitor, 433
Light microscopy. *See also* Immunofluorescence methods
  advantages of, 565–567
  nucleolar stains for, 720–724
Lithium acetate transformation, 182, 186, 240
  one-step gene disruption with, 291
  of *S. pombe*, 812–813
Lithium chloride transformation, of *S. pombe*, 812

Lithium salt-mediated transformation, mutagenized inserts returned to yeast by, 337
Locus numbers, 8
LR Gold, embedding with, 612–613
LR White, embedding with, 612–613
Lucifer Yellow
  cell-associated
    quantification of, 701–702
    visualization of, 700–701
  as marker for endocytosis in yeast, 698–702
LYS2 gene
  as counterselectable marker, 305–306, 309
  Ty mutagenesis, 353–355
LYS5 gene, Ty mutagenesis, 353–355
LYS2 marker, 379
Lyticase, 749
  source, 604

M

Macromolecular assemblies, fidelity of, and gene overexpression, 247–248
α-Mannosidase, vacuolar membrane-associated activity, assay, 655, 657
Mannosyltransferase I, assay, in assessment of purity of vacuole preparations, 658
Map distances, determination, 283
  in centrimorgans, 55–57
Mapped genes, list of, 835–852
Mapping. *See* Chromosomal mapping; Gene mapping; Meiotic mapping; Mitotic mapping
Marker rescue, 150
MAT, 80
  alleles, identification of, 103
  mutations at, identification of, 103–104
  signal transduction pathway, 104–107
*MAT*a, 80, 132
  nomenclature for, 10
  tester strains, in mating-type tests, 19, 96
*MAT*α, 80, 132
  nomenclature for, 10
  tester strains, in mating-type tests, 19, 96
Mating, 18–19
  of nonmating mutants, 91–93
  overview of, 77–80
  signal transduction pathway in, 78–80
Mating defect, bilateral, testing for, 138

Mating factors. See also **a**-factor; α-factor
  production, assays for, 137
  secreted, 77–78. See also Pheromones
Mating reaction, yeast, 77
  assay of, 77–93
    by complementation of nutritional
      requirements, 80–83
    qualitative, 80–83
    quantitative, 80, 83–84
  cell fusion in, 774–775. See also Cell
      fusion
  fusion pathway, 774–776
    mutants, assays of, 776–780
  nuclear fusion in, 774–775. See also
      Nuclear fusion
  signal transduction in, 774–776
  signal transduction pathway in, study of,
      245–246
Mating type **a**, 77–80
Mating type α, 77–80
Mating-type locus. See also MAT
  mating-type switching from α to **a** by
      changing genetic cassettes at,
      132–134
  mutant alleles of, nomenclature for, 10
  wild-type alleles of, 10
  nomenclature for, 10
Mating types, testing of, by replica plating,
    19
Mating-type switching, 132–146. See also
    HO gene
  for construction of isogenic strains
      differing only in mating-type locus,
      136–137
  in construction of polyploid strains,
      138–140
  for facilitation of complementation tests,
      137–138
  by one-step gene replacement, 137
  rationale for, 136
  in testing for bilateral mating defect, 138
Media. See also Indicator media; specific
    medium; Sporulation media
  complex, 13
  drug-containing, 284, 310
  inorganic phosphate depleted, 423
  for petri plates, preparation, 13–14
  SMII secretion, 505
  standard laboratory, ingredients of, 12–16
  synthetic, 12
    complete (SC), 12, 15

growth rates in, 15–16
minimal (SD), 12, 14
preparation, 15
used in counterselection, 310
Meiosis, 77, 94, 110–111
  analysis of chromosome segregation in,
      765–773
  commitment to, 96
  monitoring, 96–99
  mutations and genes regulating entry
      into, identification of, 104–110
  nuclear staining in, 99
  readiness for, 96
  single division, in spo13 mutants,
      113–115
  in identification of Rec⁻ mutants,
      117–118
  starvation signal transduction pathway,
      107–108
  study of, using mutations, 99–101
Meiotic analysis, 41
Meiotic mapping, 39–41, 53–58
Meiotic nuclei
  spreading, in visualization of synaptone-
      mal complexes, 127–131
  staining of, 99
Meiotic progeny, procedure for obtaining,
    20
Mendelian inheritance, in yeast, 8–9
Mercurials, as protease inhibitors, 433
Messenger RNA. See also Protein synthesis,
    in vitro
  chimeric, decay rates, in S. cerevisiae,
      measurement of, 422–423
  decay rates, in S. cerevisiae, measurement
      of, 415–423
    based on transcriptional inhibition,
      415–417
    methods for, 415–416
      comparison of, 416–417
      procedure, 421–422
    in vivo labeling, 415–417
  preparation of, by in vitro transcription,
      524–525
  transcription, inhibition of, 415–416
    with GAL promoters, 416–417, 421
    with 1,10-phenanthroline, 418–419
    with temperature-sensitive RNA
      polymerase II mutants, 419–420
    with thiolutin, 418–419
Metabolism, of yeast, reviews of, 4

Methionine, as source of labeled methyl groups in RNA, 426-427
[^{35}S]Methionine, metabolic labeling of proteins with, 671-672
N-Methyl-N'-nitro-N-nitrosoguanidine mutagenesis, 273
  method, 277-278
  safety considerations with, 276
Micromanipulation
  clonal analysis of zygotes by, 778-780
  diploids isolated by, 19
  for isolation of cells, 37
  separation of ascospores from asci by, 21
Micromanipulators
  C. H. Stoelting Co., 22
  Carl Zeiss Instruments, 22
  de Fonbrune, 22-23, 27-28
  inexpensive, designed for yeast genetic studies, 24-27
  Jena, 23
  joystick, 22-24
  Lawrence Precision Machine, 22, 24-25, 28
  Leitz manual, 22
  Mark I, 22
  mechanical stage I, 22
  microneedles, 28-32
  microscope requirements with, 28
  and microscope unit, 22, 24, 26
  Narishige USA Inc., 22
  Rainin Instrument Co., 22, 24, 26, 28
  Research Instruments Limited, 22
  Singer Instrument Co. Ltd., 22
  Singer MKIII, 22, 24, 27-28
  Singer MSM System, 22, 24, 27-28
  single control lever, 22
  Technical Products International, 22
  Wild Leitz, 22
  for yeast studies, 21-28
    design, 21-22
    operation, 21-22
Microneedles, 28-32
  construction of, 28-29
    from glass fibers, 29-31
  for isolation of cells, 37
  transfer of ascospores to, technique for, 34-35
Microscopes, 28
Microtubules, visualization, with immunofluorescence methods, 600
MIF1, 247

MIF2, 247
Mitochondria
  analysis of, in intact cells, 628-631
  assembly, research on, 627
  enrichment of, 666-667
  enzymatic markers for, assay, in assessment of purity of vacuole preparations, 658
  immunological markers, 666
  isolated
    functional assays, 634-643
    membrane potential, determination of, 636
    respiratory coupling index, determination of, 635-636
    respiratory rate, determination, 635
  isolation of, 631-634, 662-663
    differential centrifugation, 633-634
    for protein import assays, 638
  marker enzyme analysis, 668-669
  metabolism, research on, 627
  oxidative phosphorylation, research on, 627
  protein synthesis, analysis of, in isolated mitochondria, 636
  research on, 627-628
  visualization, with immunofluorescence methods, 602
Mitochondrial DNA, 169
  cloning in bacterial vectors, 159-160
  in construction of strains, 154-156
  in crosses, 154
  detection of, 159
  isolation of, 158-159
    small-scale, 159
  mutations
    in construction of strains, 154-156
    DNA sequence analysis of, 160-161
  restriction analysis, 159-160
  and restriction mapping, 179
  rho$^+$, replacement of wild-type by mutant gene in, 164-165
  yeast, 7
Mitochondrial function, analysis of, in intact cells, 628-629
Mitochondrial genes
  expression of, analysis by *in vivo* labeling of translation products, 156-158
  genetic and restriction maps, 149
  research on, 627
  size, 149

Mitochondrial mutants
  designations for, 11
  isolation of, 150–156
  large, that block all mitochondrial protein synthesis (cytoplasmic petite, rho^-), 150. See also entries under rho
    induction, 150–151
  limited lesions affecting single functions (mit^-), 150
    isolation of, 152–154
  research on, 627
Mitochondrial outer membrane, protein import assay, in isolated mitochondria, 642–643
Mitochondrial protein import
  analysis of, in intact cells, 629–632
  in isolated mitochondria
    analysis of, 637–643
    in vitro transcription and translation of precursor proteins for, 637–639
  research on, 627–628
Mitochondrial proteins, extraction, 631
Mitochondrial RNA
  hybridization, 162
    anomalous results with, 162
  preparation of, 161–162
Mitochondrial translation products, in vivo labeling of, 156–158
Mitosis, 77
  analysis of chromosome segregation in, 750–766
  nuclear DNA movement during, digital imaging microscopy of, 750–756
Mitotic gene conversion, 51–53
Mitotic mapping, 39–40, 51
mob genes, 188–189
Molecular biology, of yeast, reviews of, 4
Monoclonal antibodies, for protein surveillance, 510–511
mtDNA. See Mitochondrial DNA
Multiple-label immunofluorescence, for yeast, 598
Mutagenesis, 18. See also Transposon mutagenesis
  classical techniques, 273–281
    choice of mutagen, 273–274
    choice of strain with, 275–276
    dose of mutagen, 274
    growth conditions after mutagen treatment, 274–275
    methods, 277–281

  in vitro, in plasmid reshuffling, 314–316
  selective, of essential genes, 303
  UV-induced, 109
Mutagens
  choice of, 273–274
  dose of, 274
  safety considerations with, 276–277
Mutants
  defective, selection for, 362–363
  enrichment procedures for, 276
    inositol starvation method, 279–281
  identification of, after plasmid reshuffling, 317
  nonmating, mating of, 91–93
  novel, recovery systems for, 111–123
  protease-deficient, 434
  supersecreting, 492
Mutations
  chromosome assignment, 57–59
  conditional, in essential genes, pop-in/pop-out strategy for, 296–297
  genetic mapping of, 57–58
  plasmid-borne, mapping, 301
  recessive, mapping, 47
  in study of meiosis, 99–101
  unknown, characterizing, 41

N

NADPH-cytochrome-c reductase, assay, in assessment of purity of vacuole preparations, 658
$NaHSO_3$, as protease inhibitor, 434
National Collection of Yeast Cultures, 6
Neck filaments, visualization, with immunofluorescence methods, 600
Neurospora crassa, nuclei, isolation of, 736, 745–746
Nitrosoguanidine mutagenesis, of S. pombe, 806
Non-Mendelian determinants
  nomenclature for, 11–12
  of yeast, 12
Non-Mendelian inheritance, in yeast, 8–9
Northern blot analysis, of low molecular weight RNA, 414–415
Nuclear fusion
  in mating reaction, 774–775
  assays of, 776–780
  mutants. See also Fusion mutants
    isolation of, 790–792

Nuclear staining, in meiosis, 99
Nuclei
  enrichment of, 665-667
  immunological markers, 666
  marker enzyme analysis, 668-669
Nucleolar stains, 717-728
  for light microscopy, 720-724
  materials for, 720
Nucleolin, 720
Nucleolus
  antibody mapping, 719-720
  attachment to nuclear envelope, 718
  bismuth staining of, 723-724
  immunocytological analysis, 127
  as marker for yeast cell orientation, 718
  orientation of, in cell, 718-719, 723
  silver-staining procedure
    for electron microscopy, 725-726
    for light microscopy, 721-723
  stained with silver staining procedure, 722-723
Nucleus, yeast
  fractionation of, 746-749
  isolated, evaluation of, 741-744
  isolation of, 662-663, 735-749
    and amount of yeast in culture, 747
    from different yeast strains, 743, 745, 749
    media for, 737
    methods, 735-736
    procedure for, 739-740
    protease inhibitor cocktails for, 738
    reagents for, 737-738
    results, 741-743
    solutions for, 737
  staining, in assay for fusion mutants, 777
  structural arrangement in, 717-719
  visualization of
    in Ficoll suspension, 741-743
    with immunofluorescence methods, 600-601
    ribonucleoprotein staining of ultrathin sections for, 726-728
Null mutations, 283

O

OFAGE. *See* Orthogonal-field alternation gel electrophoresis
Oligonucleotides, synthetic
  synthesis of, 364-365
  transformation with, 3-4, 362-369
Organelles, yeast
  enrichment of, 665-668
  fractionation, 662-669
  marker enzyme analysis, 668-669
  proteins localized to, biogenesis of, 662
Organic precursors, *in vivo* labeling, 424-428
oriT site, 188-189
Orthogonal-field alternation gel electrophoresis, 47-49, 59
  electrophoretic karyotypes obtained by, 60
  procedure, 62-63
Ortho[^{32}P]phosphate labeling
  of RNA, from *S. cerevisiae*, 408-409
  in *Schizosaccharomyces pombe*, 820-821
Osmium tetroxide-ferrocyanide solution,
  postfixation in, 603, 611
  procedure, for visualization of membrane structures, 607-608
Oxalyticase, 749

P

Paraformaldehyde, stock solution, formula for, 625
pATH plasmids
  availability of, 490
  as cloning vectors, 483-485
  nomenclature, 490
  pATH1, 481-482
  for production of hybrid proteins, 478-490
    advantages of, 478-479
    alternative applications of, 489-490
    cells transformed with, induced expression of hybrid proteins by tryptophan starvation, 486-487
    hybrid proteins produced from
      analysis of, by denaturing electrophoresis, 487-488
      antisera against, 489
      extraction of, 487
      purification of hybrid polypeptides from, for antigenic use, 488
    materials and methods for, 485-490
    media for, 486
    multiple cloning sites, 478-479
      nucleotide sequences of, 481-483
      position of restriction enzyme cleavage

sites in reading frame, 481-482, 484
potential pitfalls with, 488-489
solutions for, 485-486
strains used with, 485
use of, 479-485
PEP4 mutants, 442
Pepstatin A, as protease inhibitor, 433
Peroxisomes
enrichment of, 666-667
immunological markers, 666
isolation of, 663
marker enzyme analysis, 668-669
visualization, with immunofluorescence methods, 601-602
PFGGE. *See* Pulsed-field gradient gel electrophoresis
PGK gene
overexpression, in glucose, 387
promoter, 383, 389
PGK promoter, 492
Phage library screens, 198
procedure, 235-238
using protein activity, 234-235
Phalloidin, fluorochrome-conjugated, actin staining with, 729-731
1,10-Phenanthroline, inhibition of messenger RNA transcription with, 418-419
Phenotypes, nomenclature for, 8-10
Pheromones, 77-78
formation of, assay for, 83-86
response to
assay for, 86-89
visualized on petri plates, 89
PHO5 gene
promoter, 382-383, 389
regulation, 385-386
UAS, 389
regulation of interferon expression, 386-387
Phosphoproteins, labeling, in yeast, 424-428
Phycobiliproteins, 570-571
B-Phycoerythrin, 571
C-Phycoerythrin, 571
Physical maps, 58
from chromosome fragmentation data, 75-76
of *S. cerevisiae*, 827-829

Pichia pastoris, peroxisomes from, 667
Plasma membrane
enrichment of, 666-667
enzymatic markers for, assay, in assessment of purity of vacuole preparations, 658
immunological markers, 666
marker enzyme analysis, 668-669
visualization, with immunofluorescence methods, 599-600
Plasmid, 3
ARS, 199
stability, 199
for changing markers on short arm of chromosome fragments, 762-763
conjugative
functions encoded by, 188
mobilizable, 188
self-transmissible, 188
copy number, regulation of, 240-241
DNA, introduction into yeast
by conjugation between bacteria and yeast, 188-194
replica plating technique in, 194
gap repair, 320
cloning genes by, 815-816
site mapping by, 299-301
insert-containing, recovery of, 206-207
integration in yeast
by homologous recombination, 285-290
targeted, 288
frequency of, variables affecting, 288-290
JBD207, 240
2-μm circle, 7, 169, 206, 239-240
for mitotic analysis of *CEN* mutations, 758-762
pAAH5, 383
pADNS, restriction map and genome organization of, 213
pAH9, 383
pAH10, 383
pAH21, 383
pAM82, 382
pATH, 478-490
pBM150, 382
pCGS732, 495-496
pCGS740, 495-496
pDPT51

in bacteria × yeast crosses, 191–193
conjugation function, 189
selectable markers, 189
yeast replication determinant, 189
pECG2, 335, 339
　physical and genetic map of, 334
pEMBLY, inserts in, 207
pEMBLYe24, properties of, 201
pFL1, previously constructed clone banks on, 215
pG-1, 390–391
　in constitutive gene expression, 391–394
　construction of, 397
　restriction sites in, 391–392
pG-2, 390–391
　in constitutive gene expression, 391–394
　construction of, 398
　restriction sites in, 391–392
pG-3, 390–391
　in constitutive gene expression, 391–394
　construction of, 398
　restriction sites in, 391–392
pGAL-HO, 135
　in construction of polyploid strains, 138–140
　in delivery of double-strand breaks, 140–143
　structure of, 139
pGTy, 343–346
pGTy2–917
　marker genes, 344
　properties of, 345–346
　restriction map, 344
pGTy1-H3
　marker genes, 344
　properties of, 344
　restriction map, 344
pHSS4, 335, 339
pHSS6, 335, 339
　physical and genetic map of, 334
pHSS8, 335, 339
　physical and genetic map of, 334
pHSS9, 335, 339
pHSS11, 335, 339
pHSS12, 335, 339
pHSS13, 335, 339
pHSS19, 335, 339

pHSS20, 335, 339
pHSS21, 335, 339
pHV1
　properties of, 200
　restriction map and genomic organization of, 208
pKP151, 382
pLGSD5, 374, 382, 384
pLGSD5-ATG, 382
pMA56, 383
pMA91, 383
pMA230, 383
pMFH18R-*Bst*X1, 335, 339
pMFH18R-*Bst*X1, physical and genetic map of, 334
pMR366
　previously constructed clone banks on, 215
　properties of, 205, 207–209
　restriction map and genome organization of, 211
pRS
　inserts in, 207
　for plasmid shuffling, restriction maps of, 308–309
pRS313, properties of, 203
pRS314, properties of, 204
pRS315, properties of, 204
pRS316
　properties of, 205
　restriction map of, 308–309
pRS317, restriction map of, 308–309
pRS318, restriction map of, 308–309
pRS319, restriction map of, 308–309
pSB32
　previously constructed clone banks on, 215
　properties of, 203
　restriction map and genome organization of, 211
pSB283, 780
pSEY8, 240
p(*spo13*)16, 115–116
p(*SPO13*)16, 115–117
pTV3, restriction map and genomic organization of, 209
pUC-HIS3, genomic organization, 285
pUC-LEU2, genomic organization, 285
pUC-URA3, genomic organization, 285

p2UG
　diagram of, 395
　in inducible gene expression, 394–397
　levels of expression from, 396–397
　restriction sites in, 392, 395
pUN, 388
pUV2, restriction map and genomic organization of, 208
pWB215, 717
pWH5
　inserts in, 207
　properties of, 202–203
pYAC3, 254
pYAC4
　for cloning of large fragments of human DNA, 254
　restriction map, 253
　sequence modules in, 252–254
　yeast selectable markers, 254
pYAC55, 254
pYACneo, 254
pYAC4-Neo, 254
pYAC-RC, 254
pYE4, 383
pYEp51, 382
pYEp52, 382
pZJHSE2–137, 716
recovery of
　into *E. coli*, 319–329
　　bacterial strains, 321
　　bacterial transformations, 323–326
　　materials, 321
　　methods, 321–327
　　plasmids for, 321
　　yeast DNA preparations, 322–323
　　yeast strain for, 321
　　yeast transformations, 321
　from *S. pombe*, 814–815
and recovery of mutant chromosomal alleles, 282
S. pombe, 806–808, 810–811
　autonomous replication sequences, 807
　expression vectors, 807–808
　yeast markers, 807
transfer, by conjugation, 187–188
　between yeast and *E. coli*, 188–195
as vector for gene cloning by complementation, 199–214
YCp (yeast centromeric), 199
　advanced vector technology, 206–209

for construction of yeast genomic banks, 210
restriction map and genome organization of, 210–211
stability, 199
YCp50, 375
　conjugation function, 189
　frequency of transmission of, in bacteria × yeast crosses, 190–191
　inserts in, 207
　previously constructed clone banks on, 215
　properties of, 203, 321
　recovery of, into *E. coli*, 321–329
　restriction map and genome organization of, 210
　selectable markers, 189
　yeast replication determinant, 189
YCpG11, previously constructed clone banks on, 215
YCp50-*HO*, 136
yeast. See also specific plasmid
　types of, 199
　yeast selectable marker, 199
YEp (yeast episomal), 199
　advanced vector technology, 206–209
　for construction of yeast genomic banks, 210
　mitotic stability, 206
　partitioning activity induced by, 199–206
　restriction map and genomic organization of, 207–209
YEp13
　conjugation function, 189
　inserts in, 207
　previously constructed clone banks on, 215
　properties of, 200
　restriction map and genomic organization of, 207
　selectable markers, 189
　transmission of, in bacteria × yeast crosses, 192–195
　　frequency of, 190–191
　yeast replication determinant, 189
YEp24, 46–47, 240
　conjugation function, 189
　frequency of transmission of, in bacteria × yeast crosses, 190–191

previously constructed clone banks on, 215
properties of, 200, 321
recovery of, into *E. coli*, 321-329
selectable markers, 189
yeast replication determinant, 189
YEp51, restriction map and genome organization of, 212
YEp52, restriction map and genome organization of, 213
YEp61, 382
YEp62, 374, 382
YEp351, 240
YEp352, 240
YEp13-*HO*, 137
YEpL3, properties of, 200
YEplac, inserts in, 207
YEplac112, properties of, 201
YEplac181, properties of, 202
YEplac195, properties of, 202
YEp1PT, 383
YIp5, 284
conjugation function, 189
frequency of transmission of, in bacteria × yeast crosses, 190-191
genomic organization, 285
properties of, 321
selectable markers, 189
yeast replication determinant, 189
YIp5 derivative, recovery of, into *E. coli*, 321-329
Plasmid banks
acquisition of, 214-224
incomplete, 229
use of, 224-226
recipient strains, 224-225
transformant selection, 225-226
Plasmid reshuffling
expected results, 318
identification of mutant alleles of interest, 317
Plasmid shuffling
E. coli strains for, 312-313
in vitro mutagenesis, 314-316
in vivo mutagenesis, 316
materials for, 307-313
methods, 313-317
preparation of host strain, 307
recipient yeast host strain, construction of, 313

removal of episomal wild-type gene by counterselection, 317
steps involved in, 303-305
yeast-*E. coli* shuttle vectors for, 312
yeast growth media for, 311
yeast strains for, 307-311
Ploidy, determination of, 139-140
PMSF, as protease inhibitor, 434
Poly/Bed-812, 613-614
embedding yeast cells in, 616
Polyploid strains, construction of, 138-140
Porin, 29-kDa, mitochondrial import assay, in isolated mitochondria, 643
Prochymosin, calf, screening yeast colonies for, 498-502
Promoters, regulatable, 373, 379-388
Protease A
enzymatic assays of, 444-447
indirect test for, 442
Protease B, 429
activity, HPA overlay test for, 435-436
cytosolic inhibitor of, 429
enzymatic assay of, 447-448
genetic analysis, 435-436
pH optimum, 429
Protease-deficient mutants, 434
cautions to users of, 453
most useful, design of, 451-453
Protease inhibitor cocktails, 432-434
for isolation of yeast nuclei, 738
Proteases, 428-453
activation of, 431-433
artifacts from, in biochemical studies, 429
biochemical analysis, 444-453
enzymatic assays of, 444
genetic analyses, 435-444
requirements of tests in, 435
in mating, 78
vacuolar, 429-430. *See also specific protease*
levels
effects of genotype on, 430-431
effects of growth stage on, 430-431
effects of medium composition on, 430-431
Protein
carbohydrate moieties, analysis, 683-684
functional and structural domains, mapping, using reverse biochemistry, 533-534

glycosylation, analysis of, 683–685
in vivo labeling, 423
N-glycosylation, in yeast, 682–683
O-mannosylation
  analysis of, 684–685
  in yeast, 682–683
  subunit structure, analysis, using reverse biochemistry, 534–535
Proteinase A, 430
Proteinase B, 430
Protein production, by secretion from yeast, 491–507
  advantages of, 492
  methods, 492–507
  mutant yeast host strains, 492
  promoters, 492–494
  secretion signals, 497
  vectors, 494–497
  yeast colony screens, 497–506
Protein synthesis, in vitro, 521–523, 536–550. See also Reverse biochemistry; RNA polymerase II, transcription, in vitro
  advantages of, 544
  applications of, 544
  cells for, 538
  chemicals for, 538
  enzymes for, 538
  materials for, 537–538
  media for, 538
  methods, 539–541
  micrococcal nuclease titration, 539–540
  preparation of translation extract, 539
  results, 541–544
  RNA for, 538
  strains for, 537–538
  translation reaction, 540–541
Protein tagging, 387–388
Protein translation assays, 681–682
Protein translocation
  across microsomal membranes, in vitro, 675–682
  assays
    microsomal membrane fractions for, 677, 680–681
    preparation of yeast cell extracts for, 677–681
    protocol for, 681–682
    reagents for, 675–681
    yeast cytoplasmic translation extracts for, 677–680

cotranslational, 675
posttranslational, 675
Protein transport, analysis of, by radiolabeling of wild-type and sec mutant cells, 669–674
Protoplast transformation, of S. pombe, 808–812
Prototrophic diploid colonies, selection of, 19
Pulsed-field gel electrophoresis, 47
  of chromosome-sized DNA, 58
  principle of, 59
Pulsed-field gradient gel electrophoresis, 47–49
PUT2 protein, in vitro transcription and translation of, for mitochondrial protein import assays, in isolated mitochondria, 639

Q

Quinacrine, staining of yeast vacuoles, 646–649

R

RAD genes, 100
Random spores, 20–21
Rec mutants, isolation of, 119–120
Rec⁻ mutants
  analysis, use of spo13 in, 117–119
  identification, use of spo13 in, 117–119
  isolation of, 120
Recombinant DNA procedures
  for determining chromosomal positions of genes, 38
  for yeast, review of, 4
Recombination
  commitment to, 97
  homologous, 320
    integration of plasmid DNA by, 285–290
  of transforming DNA, 282, 303
  monitoring, 96
  studies of, using double-strand breaks, 140–142
  in yeast mitochondria, 154
Regulatory proteins, titration, by increasing copy number of regulatory sites, 244–245
Replica plating, 18

in introduction of plasmid DNA into yeast, 194
Resistance genes, nomenclature for, 10
Respiratory defects, mutations causing, 150
Restriction enzyme digestion, 284
Retriever vectors, 320
Retrotransposons, 7
Reverse biochemistry, 520–535
 advantages of, 520–521
 applications, 529–535
 DNA template for, 523–524
 limitations of, 521
 methods, 523–529
 preparation of mRNA by *in vitro* transcription, 524–525
 preparation of protein by *in vitro* transcription, 525–529
 principle of, 521–523
Reverse genetics, 302
RGE. *See* Rotary gel electrophoresis
rho^- clones, identification of, 151–152
rho^0 mutants, 7
rho^0 strains, 151
 conversion to stable synthetic rho^- strains, by mitochondrial transformation, 162–165
 induction, ethidium bromide treatment for, 151
rho^+ strains, 150
 mutations affecting single genes (mit^-), isolation of, 152–154
rho^- strains, 151
 induction, ethidium bromide treatment for, 151
 synthetic, 162
  construction, by mitochondrial transformation with plasmids, using high-velocity microprojectile bombardment, 162–165
  identification of, 164
Ribonucleoprotein staining, of ultrathin sections, for visualization of yeast nucleus, 726–728
Ribosomal core particles, preparation of, 469–470
Ribosomal core proteins, preparation of, 469–470
Ribosomal DNA, yeast, restriction spectra of, 175–182
Ribosomal protein fragments, yeast, *in vitro* synthesis of, 474–475

Ribosomal protein–rRNA interactions, 468
 analysis of, 470–476
 identification of ribosomal protein binding site on rRNA, by RNase protection, 472–474
 identification of ribosomal protein in, on nitrocellulose filter assay, 470–472
Ribosomal proteins
 primary binding, 468
 yeast, 453–477
  analysis of, strain specificity, 477
  characterization of, 453–454
  functional analysis of, 467–477
  homologies with ribosomal proteins of other organisms, 465–466
  in vitro synthesis of, 474–475
  in vitro synthesized, rRNA binding assay for, 475–476
  nomenclature, 457–459
  nuclear localization signals of, identification of, 476–477
  preparation of, strain specificity, 477
  separation of, 453–454
  sequences of, 457–464
  two-dimensional gel electrophoresis, 453–454
   in minigel system, 455–457
Ribosomal RNA, yeast, 7
Ribosomal subunits
 isolation of, 454–455
 reconstitution, 467–468
 stripping of, 469
Ribosomes
 isolation of, 454–455
 protein extraction, 455
Ribsomal RNA. *See also* Ribosomal protein–rRNA interactions
RME1, 102, 105, 114, 245
RNA. *See also* Messenger RNA; Mitochondrial RNA; Ribosomal RNA; tRNA
 from ascospores, isolation of, 401–402
 high molecular weight
  preparation of, 398–405
   precautions against breakdown by endogenous nucleases, 398
  prepared, evaluation of, 404–405
  sources of, 398
  TBE-agarose gel electrophoresis, 404–405
 in vivo labeling, 423

in vivo ^{32}P-labeled, preparation of, 408–409, 423–424
low molecular weight
 analysis of, 406
 functions ascribed to, 405
 gel electrophoresis, 413–414
 Northern blot analysis, 414–415
 preparation of, 405–415
  precautions against breakdown by endogenous nucleases, 405
  size fractionation on DEAE resins, 412–413
nuclear, isolation of, 409–411
poly(A)$^+$
 isolation of, 402–403
 prepared, evaluation of quality of, 405
precursors, labeling, 424–427
small cytoplasmic, isolation of, 411–412
small nuclear, gel electrophoresis, 413–414
total
 extraction of, 406–408
 of *S. pombe*, preparation of, 818–819
 from vegetative yeast cells, isolation of, 399–401
RNA11 antigen, immunoelectron microscopy mapping of, 620
RNA polymerase II
 epitope tagging, 517–519
 in vivo labeling of, 513–514
 transcription, *in vitro*, 545–550
  assay for initiation by RNA polymerase II, 548–550
  nuclear extracts for, 545–546
   preparation of, 546
  procedure, 547–548
Rotary gel electrophoresis, 48–49
Run-on transcription, labeling with UTP by, 427–428

S

Saccharomyces bayanus, 5
Saccharomyces carlsbergensis, 5
 ribosomal core particles, preparation of, 469–470
Saccharomyces cerevisiae. See also Yeast
 140, restriction spectra of, 180–181
Saccharomyces chevalieri, 5
Saccharomyces chodati, 5

Saccharomyces dairensis, restriction spectra of, 180–181
Saccharomyces diastaticus, 5, 137
Saccharomyces douglasii, restriction spectra of, 180–181
Saccharomyces exiguus, restriction spectra of, 180–181
Saccharomyces kluyveri, restriction spectra of, 180–181
Saccharomyces servazii, restriction spectra of, 180–181
Saccharomyces telluris, restriction spectra of, 180–181
Saccharomyces unisporus, restriction spectra of, 180–181
Saccharomyces uvarum, restriction spectra of, 180–181
SAD duplication, 105
Schizosaccharomyces pombe, 5
 antibody staining, 823
 biochemistry, 817–821
 biology of, 796–797
 cell biology, 821–822
 cell cycle, 796–797
 cell fixation, 822
 cell wall, digestion of, 822–823
 centromeric regions, cloning, 251–252
 chromosomal DNA, preparation of, 817–818
 chromosome mapping, 796–797
 chromosomes, 796–797
 classical genetic techniques for, 803–806
 diploid strains, 796–797
  analysis of, 804–805
 gene cloning, by gap repair, 815–816
 gene disruption, 816–817
 gene mapping, 796–797
 generation times, 796, 799, 802
 gene replacement, 816–817
 genetic crosses, 803
 genome size, 795
 growth, 799–803
 haploid spores, 796–797
 heterothallic strains, 796
 homothallic strains, 796–797
 immunofluorescence, 822–823
 mating types, 796–797
 mating type switching, 796
 media for, 798, 800–803
 meiosis, 796–797
 [^{35}S]methionine labeling in, 820

molecular genetic analysis of, 795-823
molecular genetic techniques, 806
mutagenesis, 805-806
nuclei
 isolation of, 736, 743-746
 staining, 821-822
 ortho[^{32}P]phosphate labeling in, 820-821
plasmids, 806-808, 810-811
 integration into genome, 814
 recovery of, 814-815
 transformed, stability test, 813-814
protein extracts
 denatured, preparation of, 820
 native, preparation of, 820
 preparation of, 819-820
  large-scale, 819
  small-scale, 819
proteins, homologies with *S. cerevisiae* proteins, 795
random spore analysis, 804
RNA
 preparation of, 406
 total, preparation of, 818-819
strains, 798-799
 auxotrophy, testing for, 799
 haploid/diploid, testing for, 798-799
 mating type, testing for, 799
 reisolation of, 798
 storage, 799
 temperature sensitivity, testing for, 799
 testing phenotypes of, 798-799
tetrad analysis, 803-804
transformation, 808-813
zygotes, 796
SEC genes, 669-671
sec mutants, analysis of secretory pathway in, 669-674
Secretory pathway, yeast. *See also* Protein production, by secretion from yeast
 polypeptide transit through, analysis of, 662-674
Secretory vesicles
 enrichment of, 666-668
 immunological markers, 666
 isolation of, 662-663
 marker enzyme analysis, 668-669
 visualization, with immunofluorescence methods, 601
Segregation genes, identification and analysis of, use of *spo13* system in, 120
Sensitivity genes, nomenclature for, 10

Septum formation, in fusion mutants, 778
Shmoos, 79, 774
 formation, assay for, 86
Shuttle mutagenesis, 329-342
 applicability of, 342
 epitope insertion element, 337-338, 340-341
 procedure, 332-333
 transposons, 337-341
 utility of, 330
 vectors used for, 335
Shuttle vectors. *See also* Vectors, yeast; Yeast-*E. coli* shuttle vectors
 definition of, 319
 in transcription of cloned genes, 388
Signal transduction pathways, study of, alteration of gene dosage for, 245-246
Silver-binding proteins, 720
Silver staining procedure, nucleolar-specific
 for electron microscopy, 725-726
 for whole-mount yeast cells, for light microscopy, 721-723
Sindbis virus mRNA translation, *in vitro*, 541-544
Single-gene segregation, 41-42
SIR genes, 102, 106
Small ribonucleoproteins, preparation of, 409-411
Sm antigen, immunoelectron microscopy mapping of, 620
SMII secretion medium, 505
Snail juice
 digestion of ascus wall with, 20, 32-34
 source, 32
$^{35}SO_4^{2-}$, metabolic labeling of proteins with, 671-672
Southern transfer, gel preparation for, 63
Soybean inhibitor, as protease inhibitor, 434
SPA1 gene product, localization of, 596
SPA2 gene product, localization of, 596
Spheroplasts
 formation, 663-664, 678-679, 736
  for purification of vacuoles, 654
 lysis, 664-665, 679, 736
  for purification of vacuoles, 654
 preparation of
  for isolation of mitochondria, 631-633
  for isolation of nuclei, 739-740, 748-749
 preparation of extracts for *in vitro* protein synthesis from, 679-680

preparation of microsomes active in protein translocation from, 680–681
Spheroplast transformation, 182, 186–187
  mutagenized inserts returned to yeast by, 337
  one-step gene disruption with, 291
  with YAC cloning, 263–264
Spindle-pole body
  immunocytological analysis, 127
  visualization, with immunofluorescence methods, 600
spo gene symbol, 112
Spore enrichment, 146–149
Spores
  dissection, in yeast strains used for analysis of chromosome segregation in meiosis, 773
  hydrophobic isolation of, 147–149
  purification of, 124–126
Spore wall, purification of, 124–127
Spore wall genes, cloning, 123–124
Spore wall mutants, isolation of, 121–123
Sporulated culture, microscopic field of, 33
Sporulation, 77, 94, 110–111
  of **a/a** and α/α diploids, 101–103
  forced, of heteroplasmic zygotes, 155
  on plates, 95
  synchronous, in liquid media, 95
  temperature for, 95
Sporulation genes, cloning, 123–124
Sporulation media, 13, 17, 94–95
Sporulation-specific mutants or genes
  interaction between, 109–110
  isolation, 108–109, 111–123
  screening, 108–109
spo13 system, 113–115
  in identification and analysis of Rec⁻ mutants, 117–119
  in identification and analysis of segregation genes, 120
  introduction into laboratory strains, 115–117
Spotters, 19
spt3 mutants, and Ty mutagenesis, 360–361
SSB-1 protein, 723
Starvation signal transduction pathway, in meiosis, 107–108
STE3 gene, mutation, and mating efficiency, 78
STE4 gene, 79–80
STE18 gene, 79–80

STE genes, study of, 245–246
Strains, yeast, 3, 5–6
  A364A, for mutagenesis, 275
  AB972, 59
  for analysis of chromosome segregation in meiosis, 767–768
    spore dissection, 773
  chromosome fragment-bearing, for analysis of chromosome segregation in mitosis, 758–766
  chromosome fragment-containing YPH49-derived, 76–77
  containing artificial chromosomes
    for analysis of chromosome segregation in meiosis, 767–768
    construction of, 768–770
    genotypes of, 768–770
    meiotic chromosome segregation, quantitative analysis, 770–771
  containing endogenous 2-μm plasmids, designation of, 42
  D273–10B, 6
  for determining centromere linkage, 43–46
  EJ101, genotype of, 429
  for genetic and biochemical studies, 6
  HO/HO, 782–787
  isogenic
    *MAT*a and *MAT*α, containing fusion mutations, 780
    of opposite mating type, 275
    construction of, 136–137
  lacking 2-μm plasmids, designation of, 42
  mapping, 45–46, 48
    centromere-marked, 41–42, 44
    multiply marked, 43–44
  for meiotic analysis, 43–46
  PB12–1C, 117
  preservation, 18
  S288C, 6
    for mutagenesis, 275
    restriction spectra of DNA from, 177–181
  sources, 6
  wild-type, 6
  X2180–1A, 275
  X2180–1B, 275
  YPH49, genotype of, 60
  YPH80
    construction of, 60
    electrophoretic karyotype of, 60–61, 64

genotype of, 60
YPH149, 59–60
  chromosome lengths, 69
  electrophoretic karyotype of, 61, 64, 69
  genotype of, 60
  homology to pBR322 on two chromosome fragments, 64
Suc d'Helix pomatia, source, 32
SUP11 gene, 67–68
SUP4-o gene, as counterselectable marker, 297
Suppressors
  dominant, nomenclature for, 10
  frameshift, nomenclature describing, 11
  metabolic, nomenclature describing, 11
  recessive, nomenclature for, 10
Synaptonemal complexes, visualization of, by nuclear spreading, 127–131
  coating of slides with poly (L-lysine), 128
  coating of slides with polystyrene, 128
  electron microscopic analysis, 130–131
  light microscopic analysis, 129–130
  preparation of fixative, 128
  preparation of microscope slides, 128
  procedure, 129

T

TAFE. *See* Transverse alternating-field electrophoresis
TATA boxes, 374
TDH3 promoter, 493
Temperature-sensitive mutants, 112. *See also ts* mutants
Temperature-sensitive RNA polymerase II mutants, inhibition of messenger RNA transcription with, 419–420
Tetrad analysis, 53, 58
Tetrad Dissection System, 24, 26
Tetrads, 94
  different types of, origin of, 54–55
  PD/NPD ratios, parameters for deducing statistically significant deviations from 1:1 for, 55
Tetramethylrhodamine isothiocyanate, 570–572
Texas Red, 570–571
Thin-section electron microscopy, preparation of yeast cells for, 602–608
Thiolutin, inhibition of messenger RNA transcription with, 418–419

Thymidine kinase, heterologous, in yeast, 240–241
TPI1 promoter, 492
tra genes, 188–189
Transcription. *See also* Run-on transcription
Transcription cassettes
  for epitope tagging, 373
  origin of replication in, 375–378
  for protein fusion to an amino-terminal signal sequence, to induce secretion, 373
  for protein tagging, 373
  with regulatable promoters, 373
  requirements for, 373–379
  selectable markers used in, 378–379
  sites required for gene expression, 373–375
  types of, 373
Transformation
  DNA
    with circular molecules, 286–288
    with linear molecules, 288
    utility of, 282–283
  with synthetic oligonucleotides, 3–4, 362–369
    advantages of, 363–364
    cost of, 368
    disadvantages of, 364
    frequencies
      and cotransformation, 366
      factors affecting, 366–368
      and oligonucleotide concentrations, 366–367
      and oligonucleotide length, 366–368
      and oligonucleotide mismatches, 368
      strand dependency, 368
    multiple mutants constructed with, 369
    procedure for, 365–366
    for specific alterations, 369
    transformants, selection of, 362–363
  of yeast. *See* Electroporation; Lithium acetate transformation; Spheroplast transformation
Translation, biochemistry of, 536
Transplacon transposons, 330
  utility of, 330
Transposon insertion mutagenesis, 319
Transposon mutagenesis, 329–342. *See also* Shuttle mutagenesis; Ty mutagenesis
  Tn3-based system
    advantages and disadvantages of, 330–331

applicability of, 331
bacterial strains and media, 332, 334
materials and methods, 332-337
protocol for, 332-337
Tn*10*-based system, advantages and disadvantages of, 330-331
with yeast selectable marker genes, 292-293
Transposon tagging, 198
Transverse alternating-field electrophoresis, 47-49
TRITC. *See* Tetramethylrhodamine isothiocyanate
tRNA, gel electrophoresis, 413-414
tRNA ligase
immunoelectron microscopy mapping of, 620
nuclear localization of, 597
trpE fusion genes, construction of, 477-490
TRP1 marker, 379
ts mutants, 100. *See also* Temperature-sensitive mutants
multicopy plasmid transformants, enzymatic screening, for gene isolation, 243
Tunicamycin, in analysis of glycosylation state of proteins, 683-688
TUN^R marker, 378-379
Ty elements
marked, 343-347
transposition efficiency of, 351
marker genes for, 344, 346
Ty-induced mutations
chromosomal manipulations of, 356-357
cloning, 357-360
Ty insertions, marked, as portable genetic markers, 357
Ty mutagenesis, 342-361
assaying Ty transposition, 348-351
examples of, 353-361
experimental approach, 351-352
genetic characterization of mutants, 352-353
induced, with marked Ty elements, 343-353
rationale for, 343
strain requirements, 347-348
tagging essential genes in, 361
target genes, 355-356
Ty retrotransposons, 330

Ty transposition vectors, for gene overexpression in yeast, 241

U

Ultrathin sections
immunogold labeling, 608-626
removal of epoxy resins from, 625-626
ribonucleoprotein staining of, for visualization of yeast nucleus, 726-728
sectioning procedure, 616-617
Ultraviolet light mutagenesis, 273-274
method, 278-279
safety considerations with, 276
Upstream activator sequences, 373-374, 389
URA3 cassette, 292
URA3 gene, 67
as counterselectable marker, 305, 307, 309-310
URA3-hisG marker, 378-379
URA3 marker, 378-379

V

Vacuolar membrane, ion channels, 659
Vacuoles
enrichment of, 666-667
enzyme marker assays, 655-657
flotation, 655
immunological markers, 666
isolation of, 662-663
marker enzyme analysis, 668-669
methods for studying, 644-662
morphology of, 646, 652
protein sorting to, 644-645
assessment of fidelity of, 659-661
purification of, 653-659
cell growth for, 654
solutions for, 654
spheroplast formation for, 654
spheroplast lysis for, 654
purity, assessment of
enzyme marker assays for, 655-657
nonvacuolar enzyme marker assays for, 657-659
visualization of
by fluorescence microscopy, 645-653

with immunofluorescence methods, 601, 646, 649–653
vital staining of, 645–649
Vanadium stain, for grid staining, in immunoelectron microscopy, 626
Vectors, yeast, 284–285. *See also* Plasmid; Transcription cassettes; Yeast artificial chromosomes
–ATG, 374
centromeric, 377, 388
for constitutive gene expression, 389–394
construction of, 397–398
for construction of cDNA library for complementing specific mutations in yeast, 212–214
for construction of yeast genomic banks, selection of, 209–214
with dominant selectable markers for yeast transformation, 209
for expression of cloned genes, 373–388
for inducible gene expression, 389–391, 394–397
construction of, 398
integration into chromosome, 377–378
2-μm, 375–376
with very high copy origins, 376–377
2-μm circle, 389
plasmid, 199–214
with promoter cassettes, 382–383
selectable markers on, 210–212, 389
that propagate at high copy levels, 209, 230, 239–240
Tn3-free, 332, 334, 338–339
Vegetative cells
micromanipulation, 37
preferential killing of, 20–21
relocation and transfer of, 21
walls, purification of, 126–127
Vital staining, of yeast cells, 645–649
VPS genes, 644, 659

W

Wild-type genes, nomenclature for, 8

Y

YAC. *See* Yeast artificial chromosomes
Yeast
getting started with, 3–21

information on, 4–5
mammalian genes introduced into, 4
virtues of, for biological studies, 3–4
Yeast artificial chromosomes
analysis of chromosome segregation in *S. cerevisiae* using, 765–773
for cloning of DNA, 283
for cloning of large fragments of human DNA
colony-screening protocol for, 265–267
DNA preparation, 255–259
ligation procedure, 259–261, 267
storage of library, 265
sucrose density gradient fractionation, 261–263
transformant selection, 264, 267, 270
transformation, 263–264, 267
using large ligation mixtures LM-I and LM-II, 267–269
construction, 251
marker-changed, testing, 773
use of, 251
Yeast cells, vital staining of, 645–649
Yeast–*E. coli* shuttle vectors, 319
for plasmid shuffling, 312
Yeast Genetics Stock Center, 6, 76–77, 451
YEPD plates, formula for, 435
YEPG plates, formula for, 435
YPAD medium, 13
YPDG medium, 13
YPD medium, 12–13, 15
YPG medium, 13
Y/Z site, 133–135

Z

Zygotes
clonal analysis of, by micromanipulation, 778–780
heteroplasmic, 154–155
analysis of gene expression in, 157–158
forced sporulation of, 155
isolation, 81–83
from mating mixture, 19
micromanipulation, 21, 37
relocation and transfer of, 21
Zymolyase, 749
protease activity, 429
source, 32

DATE DUE

Demco, Inc. 38-293